RECIPROCALS OF BASIC FUNCTIONS

18. $\displaystyle\int \frac{1}{1 \pm \sin u}\, du = \tan u \mp \sec u + C$

19. $\displaystyle\int \frac{1}{1 \pm \cos u}\, du = -\cot u \pm \csc u + C$

20. $\displaystyle\int \frac{1}{1 \pm \tan u}\, du = \tfrac{1}{2}(u \pm \ln|\cos u \pm \sin u|) + C$

21. $\displaystyle\int \frac{1}{\sin u \cos u}\, du = \ln|\tan u| + C$

22. $\displaystyle\int \frac{1}{1 \pm \cot u}\, du = \tfrac{1}{2}(u \mp \ln|\sin u \pm \cos u|) + C$

23. $\displaystyle\int \frac{1}{1 \pm \sec u}\, du = u + \cot u \mp \csc u + C$

24. $\displaystyle\int \frac{1}{1 \pm \csc u}\, du = u - \tan u \pm \sec u + C$

25. $\displaystyle\int \frac{1}{1 \pm e^u}\, du = u - \ln(1 \pm e^u) + C$

POWERS OF TRIGONOMETRIC FUNCTIONS

26. $\displaystyle\int \sin^2 u\, du = \tfrac{1}{2}u - \tfrac{1}{4}\sin 2u + C$

27. $\displaystyle\int \cos^2 u\, du = \tfrac{1}{2}u + \tfrac{1}{4}\sin 2u + C$

28. $\displaystyle\int \tan^2 u\, du = \tan u - u + C$

29. $\displaystyle\int \sin^n u\, du = -\frac{1}{n}\sin^{n-1} u \cos u + \frac{n-1}{n}\int \sin^{n-2} u\, du$

30. $\displaystyle\int \cos^n u\, du = \frac{1}{n}\cos^{n-1} u \sin u + \frac{n-1}{n}\int \cos^{n-2} u\, du$

31. $\displaystyle\int \tan^n u\, du = \frac{1}{n-1}\tan^{n-1} u - \int \tan^{n-2} u\, du$

32. $\displaystyle\int \cot^2 u\, du = -\cot u - u + C$

33. $\displaystyle\int \sec^2 u\, du = \tan u + C$

34. $\displaystyle\int \csc^2 u\, du = -\cot u + C$

35. $\displaystyle\int \cot^n u\, du = -\frac{1}{n-1}\cot^{n-1} u - \int \cot^{n-2} u\, du$

36. $\displaystyle\int \sec^n u\, du = \frac{1}{n-1}\sec^{n-2} u \tan u + \frac{n-2}{n-1}\int \sec^{n-2} u\, du$

37. $\displaystyle\int \csc^n u\, du = -\frac{1}{n-1}\csc^{n-2} u \cot u + \frac{n-2}{n-1}\int \csc^{n-2} u\, du$

PRODUCTS OF TRIGONOMETRIC FUNCTIONS

38. $\displaystyle\int \sin mu \sin nu\, du = -\frac{\sin(m+n)u}{2(m+n)} + \frac{\sin(m-n)u}{2(m-n)} + C$

39. $\displaystyle\int \cos mu \cos nu\, du = \frac{\sin(m+n)u}{2(m+n)} + \frac{\sin(m-n)u}{2(m-n)} + C$

40. $\displaystyle\int \sin mu \cos nu\, du = -\frac{\cos(m+n)u}{2(m+n)} - \frac{\cos(m-n)u}{2(m-n)} + C$

41. $\displaystyle\int \sin^m u \cos^n u\, du = -\frac{\sin^{m-1} u \cos^{n+1} u}{m+n} + \frac{m-1}{m+n}\int \sin^{m-2} u \cos^n u\, du$

$$= \frac{\sin^{m+1} u \cos^{n-1} u}{m+n} + \frac{n-1}{m+n}\int \sin^m u \cos^{n-2} u\, du$$

PRODUCTS OF TRIGONOMETRIC AND EXPONENTIAL FUNCTIONS

42. $\displaystyle\int e^{au} \sin bu\, du = \frac{e^{au}}{a^2 + b^2}(a \sin bu - b \cos bu) + C$

43. $\displaystyle\int e^{au} \cos bu\, du = \frac{e^{au}}{a^2 + b^2}(a \cos bu + b \sin bu) + C$

POWERS OF u MULTIPLYING OR DIVIDING BASIC FUNCTIONS

44. $\displaystyle\int u \sin u\, du = \sin u - u \cos u + C$

45. $\displaystyle\int u \cos u\, du = \cos u + u \sin u + C$

46. $\displaystyle\int u^2 \sin u\, du = 2u \sin u + (2 - u^2)\cos u + C$

47. $\displaystyle\int u^2 \cos u\, du = 2u \cos u + (u^2 - 2)\sin u + C$

48. $\displaystyle\int u^n \sin u\, du = -u^n \cos u + n\int u^{n-1}\cos u\, du$

49. $\displaystyle\int u^n \cos u\, du = u^n \sin u - n\int u^{n-1}\sin u\, du$

50. $\displaystyle\int u^n \ln u\, du = \frac{u^{n+1}}{(n+1)^2}[(n+1)\ln u - 1] + C$

51. $\displaystyle\int u e^u\, du = e^u(u - 1) + C$

52. $\displaystyle\int u^n e^u\, du = u^n e^u - n\int u^{n-1} e^u\, du$

53. $\displaystyle\int u^n a^u\, du = \frac{u^n a^u}{\ln a} - \frac{n}{\ln a}\int u^{n-1} a^u\, du + C$

54. $\displaystyle\int \frac{e^u\, du}{u^n} = -\frac{e^u}{(n-1)u^{n-1}} + \frac{1}{n-1}\int \frac{e^u\, du}{u^{n-1}}$

55. $\displaystyle\int \frac{a^u\, du}{u^n} = -\frac{a^u}{(n-1)u^{n-1}} + \frac{\ln a}{n-1}\int \frac{a^u\, du}{u^{n-1}}$

56. $\displaystyle\int \frac{du}{u \ln u} = \ln|\ln u| + C$

POLYNOMIALS MULTIPLYING BASIC FUNCTIONS

57. $\displaystyle\int p(u)e^{au}\, du = \frac{1}{a}p(u)e^{au} - \frac{1}{a^2}p'(u)e^{au} + \frac{1}{a^3}p''(u)e^{au} - \cdots$ [signs alternate: $+ - + - \cdots$]

58. $\displaystyle\int p(u)\sin au\, du = -\frac{1}{a}p(u)\cos au + \frac{1}{a^2}p'(u)\sin au + \frac{1}{a^3}p''(u)\cos au - \cdots$ [signs alternate in pairs after first term: $+ + - - + + - - \cdots$]

59. $\displaystyle\int p(u)\cos au\, du = \frac{1}{a}p(u)\sin au + \frac{1}{a^2}p'(u)\cos au - \frac{1}{a^3}p''(u)\sin au - \cdots$ [signs alternate in pairs: $+ + - - + + - - \cdots$]

CALCULUS

BRIEF EDITION

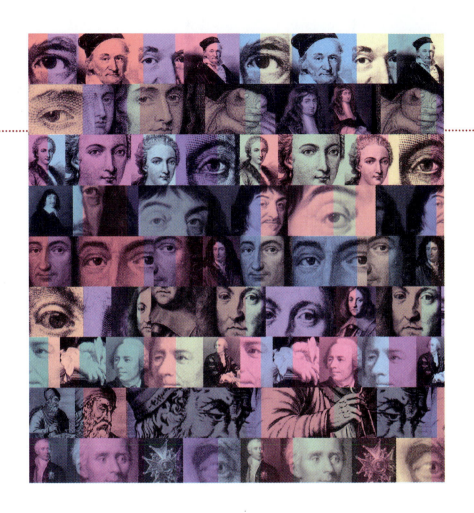

CALCULUS

BRIEF EDITION

SEVENTH EDITION

HOWARD ANTON
Drexel University

IRL BIVENS
Davidson College

STEPHEN DAVIS
Davidson College

JOHN WILEY & SONS, INC.

BP45

4/17/05

Senior Acquisitions Editor: Kimberly Murphy
Freelance Developmental Editor: Anne Scanlan-Rohrer
Marketing Manager: Julie Z. Lindstrom
Senior Production Editor: Ken Santor
Senior Designer: Harold Nolan
Cover Design: Norm Christensen
Photo Editor: Hilary Newman
Illustration Editor: Sigmund Malinowski
Illustration Studio: Techsetters, Inc.
Typesetting: Techsetters, Inc.

This book was set in Times Roman by Techsetters, Inc., and printed and bound by Von Hoffmann Press. The cover was printed by Von Hoffmann Press.

This book is printed on acid-free paper. ∞

The paper in this book was manufactured by a mill whose forest management programs include sustained yield harvesting of its timberlands. Sustained yield harvesting principles ensure that the numbers of trees cut each year does not exceed the amount of new growth.

ISBN 0-471-38158-6 (acid-free paper)

Printed in the United States of America

10 9 8 7 6 5 4

ABOUT
HOWARD ANTON

Howard Anton obtained his B.A. from Lehigh University, his M.A. from the University of Illinois, and his Ph.D. from the Polytechnic University of Brooklyn, all in mathematics. In the early 1960s he worked for Burroughs Corporation and Avco Corporation at Cape Canaveral, Florida, where he was involved with the manned space program. In 1968 he joined the Mathematics Department at Drexel University, where he taught full time until 1983. Since that time he has been an adjunct professor at Drexel and has devoted the majority of his time to textbook writing and activities for mathematical associations. Dr. Anton was president of the EPADEL Section of the Mathematical Association of America (MAA), served on the board of Governors of that organization, and guided the creation of the Student Chapters of the MAA. He has published numerous research papers in functional analysis, approximation theory, and topology, as well as pedagogical papers. He is best known for his textbooks in mathematics, which are among the most widely used in the world. There are currently more than one hundred versions of his books, including translations into Spanish, Arabic, Portuguese, Italian, Indonesian, French, Japanese, Chinese, Hebrew, and German. For relaxation, Dr. Anton enjoys traveling and photography.

ABOUT
IRL BIVENS

Irl C. Bivens, recipient of the George Polya Award and the Merten M. Hasse Prize for Expository Writing in Mathematics, received his A.B. from Pfeiffer College and his Ph.D. from the University of North Carolina at Chapel Hill, both in mathematics. Since 1982, he has taught at Davidson College, where he currently holds the position of professor of mathematics. A typical academic year sees him teaching courses in calculus, topology, and geometry. Dr. Bivens also enjoys mathematical history, and his annual History of Mathematics seminar is a perennial favorite with Davidson mathematics majors. He has published numerous articles on undergraduate mathematics, as well as research papers in his specialty, differential geometry. He is currently a member of the editorial board of the new MAA Problem Book series, a reviewer for *Mathematical Reviews*, associate editor of the *College Mathematics Journal*, and a coeditor of the Problems and Solutions section of the *College Mathematics Journal*. When he is not pursuing mathematics, Professor Bivens can be found perfecting his juggling technique or catching an action flick with his son Robert. He and Howard Anton met in 1987 on a mathematical lecture tour of the People's Republic of China.

ABOUT
STEPHEN DAVIS

Stephen L. Davis received his B.A. from Lindenwood College and his Ph.D. from Rutgers University in mathematics. Having previously taught at Rutgers University and Ohio State University, Dr. Davis came to Davidson College in 1981, where he is currently a professor of mathematics and chair of the Department of Mathematics. He regularly teaches calculus, linear algebra, abstract algebra, and computer science. A sabbatical in 1995–1996 took him to Swarthmore College as a visiting associate professor. Professor Davis has published numerous articles on calculus reform and testing, as well as research papers on finite group theory, his specialty. Professor Davis is currently secretary-treasurer of the Southeastern Section of the Mathematics Association of America, a faculty consultant for the Educational Testing Service Advanced Placement Calculus Test, a board member of the North Carolina Association of Advanced Placement Mathematics Teachers, and is actively involved in nurturing mathematically talented high school students through leadership in the Charlotte Mathematics Club. He was formerly North Carolina state director for the MAA. For relaxation, he plays basketball, juggles, and travels. Professor Davis and his wife Elisabeth have three children, Laura, Anne, and James, all former calculus students.

To
My Wife Pat
My Children: Brian, David, and Lauren

In Memory of
My Mother Shirley
My Father Benjamin
My Esteemed Colleague Albert Herr
My Benefactor Stephen Girard (1750–1831)

HA

To
My Son Robert

IB

To
My Wife Elisabeth
My Children: Laura, Anne, and James

SD

PREFACE

The primary goal of this edition is to foster **conceptual understanding** and an appreciation of the **applicability** of the subject matter. Some of the significant features of this edition are as follows:

Multiple Versions   For greater flexibility, there are now two versions of this text — *late transcendental* and *early transcendental*. The late transcendental version covers logarithmic, exponential, and inverse trigonometric functions *after* all of the basic material on differentiation and integration has been developed; in the early transcendental version, logarithmic, exponential, and inverse trigonometric functions are discussed earlier. The late transcendental version is organized along the lines of the fifth edition and the early transcendental version along the lines of the sixth edition. Both versions of this text are available in two volumes, a brief edition that covers the single variable material, and a multivariable edition that covers the multivariable material.

Technology   This edition provides many examples and exercises for instructors who want to use graphing calculators, computer algebra systems, or other programs. However, these are implemented in a way that allows the text to be used in courses where technology is used extensively, moderately, or not at all. To provide a sound foundation for the technology material, we have included a section entitled Graphing Functions on Calculators and Computers; Computer Algebra Systems (Section 1.3). The text is accompanied by a CD containing the program *Graphing Advantage Plus*, which is a Windows program that can be used for graphing, numerical integration, finding roots and intersections, and curve fitting by least squares.

Horizon Modules   Selected chapters end with modules called *Expanding the Calculus Horizon*. As the name implies, these modules are intended to take the student a step beyond the traditional calculus text. The modules, all of which are optional, can be assigned either as individual or group projects and can be used by instructors to tailor the calculus course to meet their specific needs and teaching philosophies. For example, there are modules that touch on iteration and dynamical systems, equations of motion, application of integration to railroad design, collision of comets with Earth, and hurricane modeling.

Mathematical Modeling   Mathematical modeling plays a prominent role in this edition. A new section on mathematical modeling has been added early in the text (Section 1.7). In Sections 9.3 and 9.4 we discuss mathematical modeling with differential equations, and in Section 10.10 we discuss mathematical modeling with Taylor series.

Applicability of Calculus   One of the primary goals of this edition is to link calculus to the real world and the student's own experience. This theme starts with the Introduction and is carried through in the examples, exercises, and modules. Applications given in the exercises have been chosen to provide the student a sense of how calculus can be applied.

Early Differential Equations   Basic ideas about differential equations, initial-value problems, direction fields, and integral curves are introduced concurrently with integration and then revisited in more detail in Chapter 9. We have also added new material on second-order differential equations.

For the Reader   At various points in the exposition the student is assigned a brief task. Some of these are appropriate for all readers, and others are appropriate only for readers who have a graphing utility or a CAS. The tasks for all readers are designed to immerse students more deeply into the text by asking them to think about an idea and reach some conclusion; the tasks for students using technology are designed to familiarize them with the procedures for using that technology by asking them to read their documentation and perform some text-related computation. Some instructors may want to make these tasks part of their assignments.

Logarithmic and Exponential Functions   Logarithmic and exponential functions are introduced in Section 4.2 from the exponent point of view and then revisited from the integral point of view in Section 6.9. The organization has been designed so that instructors who want to deemphasize the integral definition can do so without compromising the integrity of the material.

Early Parametric Option   In keeping with the current trend of discussing parametric equations early, parametric curves are introduced in Section 1.8 and then revisited in Chapter 11, where calculus-related matters are discussed. Instructors who prefer the traditional late discussion of parametric equations will have no problem deferring the material in Section 1.8 until the discussion of analytic geometry in Chapter 11.

More Variety in Exercises   The exercise sets have been revised and expanded to include more variety and better pairings between odd and even exercises. There are many more exercises on mathematical modeling and the use of technology. As in earlier editions, a large number of exercises focus on *conceptual understanding*. Exercises that require technology are marked with icons for easy identification. Supplementary exercises at the ends of the chapters draw on all of the concepts developed in the chapter.

Analysis of Functions   The traditional "curve sketching" is part of the *Analysis of Functions* (Sections 5.1–5.3). The approach has been updated to focus on the interplay between calculus and technology, with the goal of locating, describing, and analyzing all of the significant features of a graph.

Principles of Integral Evaluation   The traditional "Techniques of Integration" is now entitled "Principles of Integral Evaluation" to reflect its more modern approach to the material. The chapter emphasizes general methods and the role of technology rather than specific tricks for evaluating complicated or obscure integrals.

Appendix on Polynomial Equations   Since many calculus students are weak in solving polynomial equations, we have included an appendix (Appendix F) that reviews the Factor Theorem, the Remainder Theorem, and procedures for finding rational roots.

Rule of Four   The "rule of four" refers to presenting concepts from the verbal, algebraic, visual, and numerical points of view. In keeping with current pedagogical philosophy, we used this approach whenever appropriate.

Internet   This text is supplemented by an Internet site

http://www.wiley.com/college/anton

OTHER FEATURES

Flexibility   This edition has a built-in flexibility that is designed to serve a broad spectrum of calculus philosophies—from traditional to reform. Technology can be emphasized or not, and the order of many topics can be permuted freely to accommodate the instructor's specific needs.

Trigonometry Review   Deficiencies in trigonometry plague many students, so we have included a substantial trigonometry review in Appendix E.

Historical Notes   The biographies and historical notes have been a hallmark of this text from its first edition and have been maintained. All of the biographical materials have been distilled from standard sources with the goal of capturing the personalities of the great mathematicians and bringing them to life for the students.

Graded Exercise Sets   Section Exercise Sets are "graded" to begin with routine problems and progress gradually toward problems of greater difficulty. However, the Supplementary Exercises are not graded by level of difficulty, so as not to give the student a predisposition about the level of effort required.

Rigor   The challenge of writing a good calculus book is to strike the right balance between rigor and clarity. Our goal is to present precise mathematics to the fullest extent possible for the freshman audience. Where clarity and rigor conflict, we choose clarity; however, we believe it to be important that the student understand the difference between a careful proof and an informal argument, so we have tried to make it clear to the reader when the arguments being presented are informal or motivational. Theory involving ϵ-δ arguments appear in separate sections so that they can be covered or not, as preferred by the instructor.

Mathematical Level   This text is written at a mathematical level that will prepare students for a wide variety of careers that require a sound mathematics background, including engineering, the various sciences, and business.

Computer Graphics   This edition makes extensive use of modern computer graphics to clarify concepts and to develop the student's ability to visualize mathematical objects, particularly those in 3-space. For those students who are working with graphing technology, there are many exercises that are designed to develop the student's ability to generate and analyze mathematical curves and surfaces.

NEW FEATURES IN THE SEVENTH EDITION

▶ There are now both early transcendental and late transcendental versions of this text.

▶ About one-third of the exercises are new. They provide more exploratory and open-ended problems, more exercises involving tabular data, more exercises that use technology, and better pairing of odd and even exercises.

▶ There is more emphasis on mathematical modeling. There is a new early introduction to the topic (Section 1.7) and greater emphasis on modeling with differential equations.

▶ Section 9.4 on second-order differential equations is new.

▶ Chapter 10 on infinite series has been reorganized to allow instructors to get to the heart of the material more quickly.

▶ Section 11.5 on rotation of coordinate axes, which had been relegated to the Internet in the sixth edition, has been restored to the text.

A WELCOME TO MY NEW COAUTHORS

It is hard to believe that a quarter-century has passed since I first began writing this book in a small office tucked away in a corner of my bedroom. What I naively believed would be a two-year writing project took almost eight years. I have been told by many that this text established a new standard of clarity in mathematical exposition and changed the way in which calculus books are written — of that I am proud. However, after a lifetime working as a lone author, the inexorable passage of time has made it clear to me that the moment for coauthors has arrived — and so I welcome my new writing colleagues, Irl Bivens and Stephen Davis, both of whom have unique talents and years of teaching experience in calculus. They have been a pleasure to work with, and I thank them for their dedication and contributions to this new edition.

Howard Anton

SUPPLEMENTS FOR THE STUDENT

Student Resource Manual   0471-441732

The *Student Resource Manual* provides students with detailed solutions to odd-numbered exercises from the text, as well as including sample tests for each section and chapter of the text. Available for purchase at www.wiley.com/college

Student Resource and Survival CD   0471-441813

Available for both Windows and Macintosh users, the *Student Resource and Survival CD* contains detailed solutions to odd-numbered exercises from the text and sample tests for each section and chapter of the text. In addition, this CD features precalculus review material and a brief introduction to those aspects of linear algebra which are of immediate concern for the calculus student. Available for purchase at www.wiley.com/college

Graphing Advantage Plus™

This Windows program can be used to solve most of the technology exercises in this text that are labeled with the icon 〰, thereby eliminating the need to purchase a graphing calculator or other program for that purpose. The program also allows the user to export graphs to word processing documents for reports, and to display various kinds of course materials and Internet links in a convenient menu that can be customized. If a CAS is to be used in the course, then the user can create a customized menu that accesses the CAS with a quick click. The *Calculus Resource CD* packaged with this text contains a free time-limited version that will be active for 16 weeks (about a semester). Instructions for obtaining an unrestricted version are included on the CD.

Electronic Calculus Tutorial

This electronic calculus tutorial developed by Intelligent Environments is a software package that solves and documents calculus problems in real time. This intelligent software package steps students through problems and provides customized feedback from the text, allowing students to identify and learn from their mistakes more efficiently. Instructors can use the tutorial to preview problems and explore functions graphically and analytically. For more information, visit www.wiley.com/college/anton or speak with your Wiley representative.

SUPPLEMENTS FOR THE INSTRUCTOR

SUPPLEMENTS FOR THE INSTRUCTOR CAN BE OBTAINED BY SENDING A REQUEST ON YOUR INSTITUTIONAL LETTERHEAD TO MATHEMATICS MARKETING MANAGER, JOHN WILEY & SONS, INC., 605 THIRD AVENUE, NEW YORK, NY 10158-0012, OR BY CONTACTING YOUR LOCAL WILEY REPRESENTATIVE.

Complete Solutions Manual

Detailed solutions to all exercises in the text.

Test Bank

Contains a variety of questions and solutions for every section in the text.

Instructor's Resource CD-ROM

Contains the complete *Solutions Manual* and *Test Bank*, as well as precalculus review material and a brief introduction to linear algebra.

Graphing Advantage Plus

A version of *Graphing Advantage Plus* (described above) without time restrictions. The program can be used for classroom presentations, for creating overhead transparencies with colorized graphs, and for exporting graphs to word processing documents for creating examinations.

OTHER RESOURCES FOR THE INSTRUCTOR

eGrade

An on-line assessment system that contains a large bank of skill-building problems and solutions. Instructors can now automate the process of assigning, delivering, grading, and routing all kinds of homework, quizzes, and tests while providing students with immediate scoring and feedback on their work. Wiley *eGrade* "does the math"... and much more. For more information, visit www.wiley.com/college/egrade

Electronic Calculus Tutorial

This electronic calculus tutorial developed by Intelligent Environments is a software package that solves and documents calculus problems in real time. Instructors can use the tutorial to preview problems and explore functions graphically and analytically. For more information, visit www.wiley.com/college/anton or speak with your Wiley representative.

The Faculty Resource Network

The *Faculty Resource Network* is a peer-to-peer network of academic faculty dedicated to the effective use of technology in the classroom. This group can help you apply innovative classroom techniques, implement specific software packages, and tailor the technology experience to the specific needs of each individual class. Ask your Wiley representative for more details.

ACKNOWLEDGMENTS

It has been our good fortune to have the advice and guidance of many talented people whose knowledge and skills have enhanced this book in many ways. For their valuable help we thank:

Reviewers and Contributors to Earlier Editions

Edith Ainsworth, *University of Alabama*

Loren Argabright, *Drexel University*

David Armacost, *Amherst College*

Dan Arndt, *University of Texas at Dallas*

Ajay Arora, *McMaster University*

John Bailey, *Clark State Community College*

Robert C. Banash, *St. Ambrose University*

William H. Barker, *Bowdoin College*

George R. Barnes, *University of Louisville*

Scott E. Barnett, *Wayne State University*

Larry Bates, *University of Calgary*

John P. Beckwith, *Michigan Technological University*

Joan E. Bell, *Northeastern Oklahoma State University*

Harry N. Bixler, *Bernard M. Baruch College, CUNY*

Marilyn Blockus, *San Jose State University*

Ray Boersma, *Front Range Community College*

Barbara Bohannon, *Hofstra University*

David Bolen, *Virginia Military Institute*

Daniel Bonar, *Denison University*

George W. Booth, *Brooklyn College*

Phyllis Boutilier, *Michigan Technological University*

Linda Bridge, *Long Beach City College*

Mark Bridger, *Northeastern University*

Judith Broadwin, *Jericho High School*

John Brothers, *Indiana University*

Stephen L. Brown, *Olivet Nazarene University*

Virginia Buchanan, *Hiram College*

Robert C. Bucker, *Western Kentucky University*

Robert Bumcrot, *Hofstra University*

Christopher Butler, *Case Western Reserve University*

Carlos E. Caballero, *Winthrop University*

James Caristi, *Valparaiso University*

Stan R. Chadick, *Northwestern State University*

Hongwei Chen, *Christopher Newport University*

Chris Christensen, *Northern Kentucky University*

Robert D. Cismowski, *San Bernardino Valley College*

Patricia Clark, *Rochester Institute of Technology*

Hannah Clavner, *Drexel University*

Ted Clinkenbeard, *Des Moines Area Community College*

David Clydesdale, *Sauk Valley Community College*

David Cohen, *University of California, Los Angeles*

Michael Cohen, *Hofstra University*

Pasquale Condo, *University of Lowell*

Robert Conley, *Precision Visuals*

Mary Ann Connors, *U.S. Military Academy at West Point*

Cecil J. Coone, *State Technical Institute at Memphis*

Norman Cornish, *University of Detroit*

Terrance Cremeans, *Oakland Community College*

Lawrence Cusick, *California State University–Fresno*

Michael Dagg, *Numerical Solutions, Inc.*

Art Davis, *San Jose State University*

A. L. Deal, *Virginia Military Institute*

Charles Denlinger, *Millersville University*

William H. Dent, *Maryville College*

Blaise DeSesa, *Allentown College of St. Francis de Sales*

Dennis DeTurck, *University of Pennsylvania*

Jacqueline Dewar, *Loyola Marymount University*

Preston Dinkins, *Southern University*

Gloria S. Dion, *Educational Testing Service*

Irving Drooyan, *Los Angeles Pierce College*

Tom Drouet, *East Los Angeles College*

Clyde Dubbs, *New Mexico Institute of Mining and Technology*

Della Duncan, *California State University–Fresno*

Ken Dunn, *Dalhousie University*

Sheldon Dyck, *Waterloo Maple Software*

Hugh B. Easler, *College of William and Mary*

Scott Eckert, *Cuyamaca College*

Joseph M. Egar, *Cleveland State University*

Judith Elkins, *Sweet Briar College*

Brett Elliott, *Southeastern Oklahoma State University*

William D. Emerson, *Metropolitan State College*

Garret J. Etgen, *University of Houston*

Benny Evans, *Oklahoma State University*

Philip Farmer, *Diablo Valley College*

Victor Feser, *University of Maryland*

Iris Brann Fetta, *Clemson University*

James H. Fife, *Educational Testing Service*

Sally E. Fischbeck, *Rochester Institute of Technology*

Dorothy M. Fitzgerald, *Golden West College*

Barbara Flajnik, *Virginia Military Institute*

Daniel Flath, *University of South Alabama*

Ernesto Franco, *California State University–Fresno*

Nicholas E. Frangos, *Hofstra University*

Katherine Franklin, *Los Angeles Pierce College*

Marc Frantz, *Indiana University–Purdue University at Indianapolis*

Michael Frantz, *University of La Verne*

Susan L. Friedman, *Bernard M. Baruch College, CUNY*

William R. Fuller, *Purdue University*

Beverly Fusfield

Daniel B. Gallup, *Pasadena City College*

Susan Gerstein

Mahmood Ghamsary, *Long Beach City College*

G. S. Gill, *Brigham Young University*

Michael Gilpin, *Michigan Technological University*

Kaplana Godbole, *Michigan Technological Institute*

S. B. Gokhale, *Western Illinois University*

Morton Goldberg, *Broome Community College*

Mardechai Goodman, *Rosary College*

Sid Graham, *Michigan Technological University*

Raymond Greenwell, *Hofstra University*

Dixie Griffin, Jr., *Louisiana Tech University*

Gary Grimes, *Mt. Hood Community College*

David Gross, *University of Connecticut*

Jane Grossman, *University of Lowell*

Michael Grossman, *University of Lowell*

Dennis Hadah, *Saddleback Community College*

Diane Hagglund, *Waterloo Maple Software*

Douglas W. Hall, *Michigan State University*

Nancy A. Harrington, *University of Lowell*

Kent Harris, *Western Illinois University*

J. Derrick Head, *University of Minnesota–Morris*

Jim Hefferon, *St. Michael College*

Albert Herr, *Drexel University*

Peter Herron, *Suffolk County Community College*

Warland R. Hersey, *North Shore Community College*

Konrad J. Heuvers, *Michigan Technological University*

Dean Hickerson

Robert Higgins, *Quantics Corporation*

Rebecca Hill, *Rochester Institute of Technology*

Tommie Ann Hill-Natter, *Prairie View A&M University*

Holly Hirst, *Appalachian State University*

Edwin Hoefer, *Rochester Institute of Technology*

Louis F. Hoelzle, *Bucks County Community College*

Robert Homolka, *Kansas State University–Salina*

Henry Horton, *University of West Florida*

Hugh E. Huntley, *University of Michigan*

Fatenah Issa, *Loyola University of Chicago*

Emmett Johnson, *Grambling State University*

Jerry Johnson, *University of Nevada–Reno*

John M. Johnson, *George Fox College*

Wells R. Johnson, *Bowdoin College*

Herbert Kasube, *Bradley University*

Phil Kavanagh, *Mesa State College*

Maureen Kelley, *Northern Essex Community College*

Dan Kemp, *South Dakota State University*

Harvey B. Keynes, *University of Minnesota*

Lynn Kiaer, *Rose-Hulman Institute of Technology*

Cecilia Knoll, *Florida Institute of Technology*

Holly A. Kresch, *Diablo Valley College*

Richard Krikorian, *Westchester Community College*

Paul Kumpel, *SUNY, Stony Brook*

Fat C. Lam, *Gallaudet University*

Leo Lampone, *Quantics Corporation*

James F. Lanahan, *University of Detroit–Mercy*

Bruce Landman, *University of North Carolina at Greensboro*

Kuen Hung Lee, *Los Angeles Trade–Technology College*

Marshall J. Leitman, *Case Western Reserve University*

Benjamin Levy, *Lexington H.S., Lexington, Mass.*

Darryl A. Linde, *Northeastern Oklahoma State University*

Phil Locke, *University of Maine, Orono*

Leland E. Long, *Muscatine Community College*

John Lucas, *University of Wisconsin–Oshkosh*

Stanley M. Lukawecki, *Clemson University*

Nicholas Macri, *Temple University*

Michael Magill, *Purdue University*

Melvin J. Maron, *University of Louisville*

Mauricio Marroquin, *Los Angeles Valley College*

Thomas W. Mason, *Florida A&M University*

Majid Masso, *Brookdale Community College*

Larry Matthews, *Concordia College*

Thomas McElligott, *University of Lowell*

Phillip McGill, *Illinois Central College*

Judith McKinney, *California State Polytechnic University, Pomona*

Joseph Meier, *Millersville University*

Robert Meitz, *Arizona State University*

Laurie Haskell Messina, *University of Oklahoma*

Aileen Michaels, *Hofstra University*

Janet S. Milton, *Radford University*

Robert Mitchell, *Rowan College of New Jersey*

Marilyn Molloy, *Our Lady of the Lake University*

Ron Moore, *Ryerson Polytechnical Institute*

Barbara Moses, *Bowling Green State University*

David Nash, *VP Research, Autofacts, Inc.*

Kylene Norman, *Clark State Community College*

Roxie Novak, *Radford University*

Richard Nowakowski, *Dalhousie University*

Stanley Ocken, *City College–CUNY*

Ralph Okojie, *Elizabeth City State University*

Donald Passman, *University of Wisconsin*

David Patterson, *West Texas A&M*

Walter M. Patterson, *Lander University*

Steven E. Pav, *Alfred University*

Edward Peifer, *Ulster County Community College*

Gary L. Peterson, *James Madison University*

Robert Phillips, *University of South Carolina at Aiken*

Mark A. Pinsky, *Northeastern University*

Catherine H. Pirri, *Northern Essex Community College*

Father Bernard Portz, *Creighton University*

Irwin Pressman, *Carleton University*

Douglas Quinney, *University of Keele*

David Randall, *Oakland Community College*

Richard Remzowski, *Broome Community College*

Guanshen Ren, *College of Saint Scholastica*

William H. Richardson, *Wichita State University*

John Rickert, *Rose-Hulman Institute of Technology*

David Robbins, *Trinity College*

Lila F. Roberts, *Georgia Southern University*

David Rollins, *University of Central Florida*

Naomi Rose, *Mercer County Community College*

Sharon Ross, *DeKalb College*

David Ryeburn, *Simon Fraser University*

David Sandell, *U.S. Coast Guard Academy*

Avinash Sathaye, *University of Kentucky*

Ned W. Schillow, *Lehigh County Community College*

Dennis Schneider, *Knox College*

Dan Seth, *Morehead State University*

George Shapiro, *Brooklyn College*

Parashu R. Sharma, *Grambling State University*

Michael D. Shaw, *Florida Institute of Technology*

Donald R. Sherbert, *University of Illinois*

Howard Sherwood, *University of Central Florida*

Mary Margaret Shoaf-Grubbs, *College of New Rochelle*

Bhagat Singh, *University of Wisconsin Centers*

Martha Sklar, *Los Angeles City College*

Henry Smith, *Southeastern Louisiana University*

John L. Smith, *Rancho Santiago Community College*

Wolfe Snow, *Brooklyn College*

Ian Spatz, *Brooklyn College*

Jean Springer, *Mount Royal College*

Norton Starr, *Amherst College*

Mark Stevenson, *Oakland Community College*

Gary S. Stoudt, *University of Indiana of Pennsylvania*

John A. Suvak, *Memorial University of Newfoundland*

P. Narayana Swamy, *Southern Illinois University*

Richard B. Thompson, *The University of Arizona*

Skip Thompson, *Radford University*

Josef S. Torok, *Rochester Institute of Technology*

William F. Trench, *Trinity University*

Walter W. Turner, *Western Michigan University*

Thomas Vanden Eynden, *Thomas More College*

Paul Vesce, *University of Missouri–Kansas City*

Richard C. Vile, *Eastern Michigan University*

David Voss, *Western Illinois University*

Ronald Wagoner, *California State University–Fresno*

Shirley Wakin, *University of New Haven*

James E. Ward, *Bowdoin College*

James Warner, *Precision Visuals*

Peter Waterman, *Northern Illinois University*

Evelyn Weinstock, *Glassboro State College*

Bruce R. Wenner, *University of Missouri–Kansas City*

Candice A. Weston, *University of Lowell*

Bruce F. White, *Lander University*

Neil Wigley, *University of Windsor*

Gary L. Wood, *Azusa Pacific University*
Yihren Wu, *Hofstra University*

Richard Yuskaitis, *Precision Visuals*
Michael Zeidler, *Milwaukee Area Technical College*

Michael L. Zwilling, *Mount Union College*

Development Team for the Seventh Edition

The following people critiqued and reviewed various parts of the manuscript and suggested many of the ideas that found their way into this new edition:

Mary Lane Baggett, *University of Mississippi*
Kbenesh Blayneh, *Florida A&M University*
Christopher Butler, *Case Western Reserve University*
Cheryl Cantwell, *Seminole Community College*
Judith Carter, *North Shore Community College*
Fielden Cox, *Centennial College*
Gary Crown, *Wichita State University*
Debbie A. Desrochers, *Napa Valley College*
Bob Grant, *Mesa Community College*
Karl Havlak, *Angelo State University*
Joe Howe, *St. Charles County Community College*
Shirley Huffman, *Southwest Missouri State University*
Gary S. Itzkowitz, *Rowan University*
Kenneth Kalmanson, *Montclair State University*
David Keller, *Kirkwood Community College*
Vesna Kilibarda, *Indiana University Northwest*
Cecilia Knoll, *Florida Institute of Technology*

John Kubicek, *Southwest Missouri State University*
Theodore Lai, *Hudson County Community College*
Jeuel LaTorre, *Clemson University*
Phoebe Lutz, *Delta College*
Ernest Manfred, *U.S. Coast Guard Academy*
Doug Nelson, *Central Oregon Community College*
Lawrence J. Newberry, *Glendale College*
Judith Palagallo, *The University of Akron*
Lefkios Petevis, *Kirkwood Community College*
Thomas W. Polaski, *Winthrop University*
B. David Redman, Jr., *Delta College*
George W. Schultz, *St. Petersburg Junior College*
Richard B. Shad, *Florida Community College–Jacksonville*
Ann Sitomer, *Portland Community College*
Jeanne Smith, *Saddleback Community College*
Rajalakshmi Sriram, *Okaloosa-Walton Community College*
Ted Wilcox, *Rochester Institute of Technology*

The following people read the seventh edition at various stages for mathematical and pedagogical accuracy and/or assisted with the critically important job of preparing answers to exercises:

Blaise DeSesa, *Drexel University*
Bradley E. Garner, *Boise State University*
Carrie Garner
Rob Gilchrist, *U.S. Air Force Academy*
Dean Hickerson
Phoebe Lutz, *Delta College*
Eric Murphy, *U.S. Air Force Academy*

Ann Ostberg
Irwin Pressman, *Carleton University*
B. David Redman, Jr., *Delta College*
Irmgard Redman, *Delta College*
David Ryeburn, *Simon Fraser University*
Neil Wigley, *University of Windsor*

The following people created materials for tests and other supplements:

John L. Orr, *University of Nebraska, Lincoln*
Henry Smith, *Southeastern Louisiana University*

Neil Wigley, *University of Windsor*

We gratefully acknowledge permission to adapt various exercises, examples, and text materials from the following publications:

Physics, 3rd ed., Cutnell and Johnson, John Wiley & Sons, Inc., 1995.
Fundamentals of Physics, 4th ed., Halliday, Resnick, and Walker, John Wiley & Sons, Inc., 1993.
Applications of Calculus, Philip Straffin, Ed., MAA Notes Number 29, The Mathematical Association of America, 1993.
Calculus Problems for a New Century, Robert Fraga, Ed., MAA Notes Number 28, Vol. 2, The Mathematical Association of America, 1993.
Engineering Mechanics, Meriam and Kraige, 3rd ed., Vol. 2, John Wiley & Sons, Inc., 1992.

Special Contributions

A special debt of gratitude to:

Kimberly Murphy, editor, for her dedication and hard work in guiding this new edition through a hectic transition period.

Anne Scanlan-Rohrer, developmental editor, whose thoughtful advice and captivating spirit contributed to the quality of this text in ways that are too numerous to mention.

Mary Beth Bohman, assistant editor for supplements, for skillfully and successfully coordinating a very complicated set of supplements.

Lucille Buonocore, associate production manager, whose work behind the scenes is recognized and appreciated.

Kenneth Santor, senior production editor, whose keen eye for detail resolved a myriad of daily production issues.

Ray Paquin, new media editor, whose knowledge of the Internet and electronic media has guided this book into the twenty-first century.

Harry Nolan and Norm Christensen, cover designers, for the tasteful and beautifully executed cover designs.

Julie Lindstrom, marketing manager, for her enthusiasm and hard work in conveying the philosophy of this text to the marketplace.

Lilian Brady, copy editor, for her usual unerring eye for aesthetics and attention to detail.

Hilary Newman, photo editor, for successfully unearthing our most obscure photographic requests.

David Ryeburn and Dean Hickerson, technical advisors, for helping us to keep the mathematics sound and the exposition clear.

Stacy French, editorial assistant, for her excellent work in keeping all of the critical communications flowing.

Carol Sawyer, Rena Lam, John Rogosich, and the rest of the Techsetters staff, who helped us produce these beautiful books on a very tight schedule.

Barbara Holland, our former editor, who put together the new writing team and launched this new edition with enthusiasm and dedication.

CONTENTS

COMPLETE EDITION

FOR THE STUDENT

Calculus is a compilation of ideas that provides a way of viewing and analyzing the physical world. As with all mathematics courses, calculus involves equations and formulas. However, if you successfully learn to use all of the formulas and solve all of the problems in this text but don't master the underlying ideas, you will have missed the most important part of calculus. Keep in mind that every single problem in this text has already been solved by somebody, so your ability to solve those problems gives you nothing unique. However, if you master the ideas of calculus, then you will have the tools to go beyond what other people have done, limited only by your own talents and creativity.

Before starting your studies, you may find it helpful to leaf through this text to get a general feeling for its different parts.

▶ At the beginning of each chapter you will find a page that gives an overview of the chapter, and at the beginning of each section you will find an introduction that gives an overview of that section. To help you locate specific information, sections are divided into topics described by headings in the margin.

▶ Each section ends with a set of exercises. The answers to most odd-numbered exercises appear in the back of the book. Worked-out solutions to the odd-numbered exercises are given in the *Student Resource Manual* and on a CD, which are available as supplements to the text.

▶ Some of the exercises are tagged with icons to indicate that some kind of technology is required for their solution. If your calculus course does not incorporate the use of technology, then your instructor will probably not assign these. Those exercises tagged with the icon ⌇ require graphing technology, which might be either a graphing calculator or a computer program that produces graphs from equations. Those exercises tagged with the icon **c** require a computer algebra system (called a CAS), which is a program that can perform symbolic as well as numerical calculations. The most common CAS programs are *Mathematica*, *Maple*, and *Derive*. Some of the newer calculators incorporate CAS capabilities.

▶ Each chapter ends with a set of supplementary exercises, many of which involve a combination of ideas from various sections within the chapter.

▶ Near the end of the text you will find seven appendices. Appendices A–F review some precalculus material, including trigonometry, and Appendix G contains some proofs that may or may not be part of your course.

▶ There is also reference material on the endpapers that are inside the front and back covers of the text.

▶ Illustrations in the exposition are referenced using a triple-number system. For example, Figure 1.6.3 is the third figure in Section 1.6, and Figure 7.2.5 is the fifth figure in Section 7.2. The same numbering system is used for theorems and definitions. Illustrations in the exercises are identified by the exercise number with which they are associated. For example, in a particular exercise set, Figure Ex-7 would be associated with Exercise 7.

▶ The ideas in this text were created by real people with interesting personalities and backgrounds. Pictures and signatures of many of these people appear on the opening pages of the chapters, and biographical sketches of various mathematicians appear throughout the text as footnotes.

▶ At various places in the text you will see elements labeled "For the Reader," which are designed to reinforce ideas in the text. Some of these ask you to think about an idea, some ask you to perform a computation, and (for students using technology) some ask you to read your reference manual and then use the technology to perform a computation or to generate a graph.

As you read through this book, you will find some ideas that you understand immediately, others that you don't understand until you have read them several times, and others that you do not understand, even after numerous readings. Don't become discouraged—some calculus ideas take time to "percolate," and you may well find that the idea suddenly becomes clear later when you least expect it.

If you find that your answer to an exercise does not match that in the back of the book, do not presume immediately that your answer is incorrect—there may be more than one way to express the answer. For example, if your answer is $\sqrt{3}/3$ and the text answer is $1/\sqrt{3}$, then both are correct, since your answer can be obtained by rationalizing the text answer. In general, if your answer does not match that in the text, then your best first step is to look for an algebraic manipulation or a trigonometric identity that relates the two answers. In cases where the answer is a decimal approximation, your answer may differ from that in the text because of different choices in the number of decimal places used in the computations.

Some exercises require a verbal answer. Express those answers in complete, correctly punctuated, logical sentences—not fragmented phrases and formulas.

Introduction

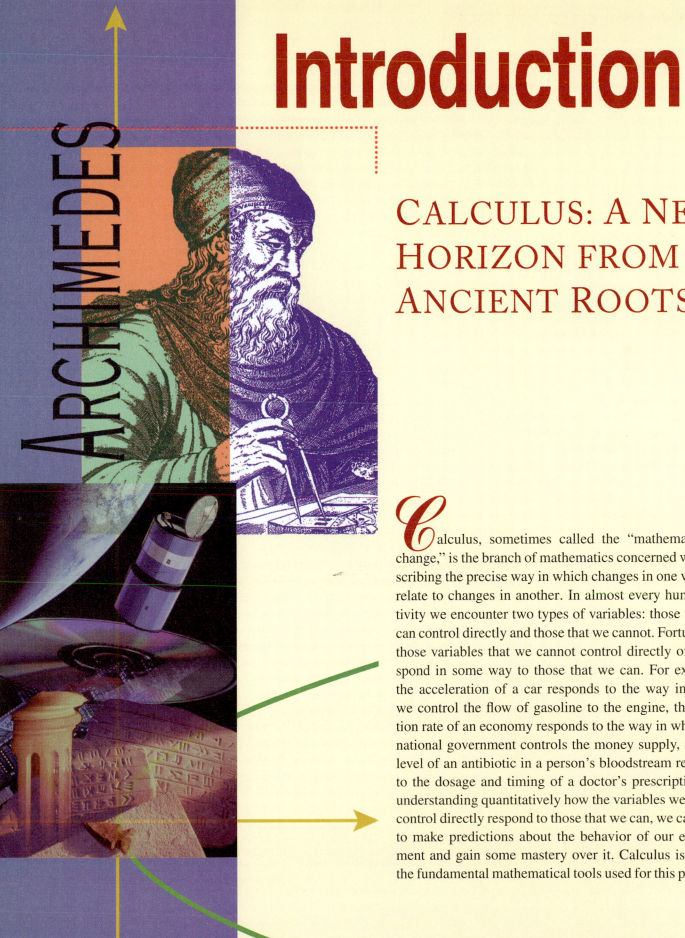

ARCHIMEDES

CALCULUS: A NEW HORIZON FROM ANCIENT ROOTS

*C*alculus, sometimes called the "mathematics of change," is the branch of mathematics concerned with describing the precise way in which changes in one variable relate to changes in another. In almost every human activity we encounter two types of variables: those that we can control directly and those that we cannot. Fortunately, those variables that we cannot control directly often respond in some way to those that we can. For example, the acceleration of a car responds to the way in which we control the flow of gasoline to the engine, the inflation rate of an economy responds to the way in which the national government controls the money supply, and the level of an antibiotic in a person's bloodstream responds to the dosage and timing of a doctor's prescription. By understanding quantitatively how the variables we cannot control directly respond to those that we can, we can hope to make predictions about the behavior of our environment and gain some mastery over it. Calculus is one of the fundamental mathematical tools used for this purpose.

Calculus has an enormous, but often unnoticed, impact on our daily lives. To provide some sense of how calculus affects us, we have selected a few of its applications to fields of contemporary research. All of these applications involve other branches of science and mathematics, but they all use calculus in some essential way. The first three applications are based on a new and exciting area of mathematics called the theory of *wavelets*. Wavelets make it possible to capture and store mathematical representations of images and signals using much less data than previously possible. As a result, the current research literature is exploding with new applications of wavelets to such diverse fields as astronomy, acoustics, nuclear engineering, image processing, neurophysiology, music, medicine, speech synthesization, earthquake prediction, and pure mathematics, to name only a few.

FBI Fingerprint Compression — The U.S. Federal Bureau of Investigation began collecting fingerprints and handprints in 1924 and now has more than 30 million such prints in its files, all of which are being digitized for storage on computer. It takes about 0.6 megabyte of storage space to record a fingerprint and 6 megabytes to record a pair of handprints, so that digitizing the current FBI archive would result in about 200×10^{12} bytes of data to be stored, which is the capacity of roughly 138 million floppy disks. At today's prices for computer equipment, storage media, and labor, this would cost roughly 200 million dollars. To reduce this cost, the FBI's Criminal Justice Information Service Division began working with the National Institute of Standards, the Los Alamos National Laboratory, and several other groups to devise compression methods for reducing the storage space. These methods, which are based on wavelets, are proving to be highly successful. Figure 1 is a good example—the image on the left is an original thumbprint and the one on the right is a mathematical reconstruction from a 26:1 data compression.

Music — Researchers with the Numerical Algorithms Research Group at Yale University have investigated the application of wavelets to sound synthesis (musical and voice). To approximate the sound of a musical instrument or voice, samples are taken and decomposed mathematically into numbers called *wavelet packet coefficients*. These coefficients can be stored on a computer and later the sound can be reconstructed (synthesized) from the computer data. This area of research makes it possible to reproduce complex sounds from a small amount of data and to transmit those data electronically in a highly compressed form. This research may eventually speed up the transmission of sound over the Internet, for example.

Removing Noise from Data — In fields ranging from planetary science to molecular spectroscopy, scientists are faced with the problem of recovering a true signal from incomplete or noisy data. For example, weak signals from deep space probes are often so overwhelmed with background noise that the signal itself is barely detectable, yet the signal must be used to produce a photograph or provide other information. Researchers at Stanford University and elsewhere have been working for several years on using wavelet methods to filter out such noise. For example, Figure 2 shows a signal from a medical imaging signal that has been cleaned up (de-noised) using wavelets.

Original   Reconstruction

Figure 1

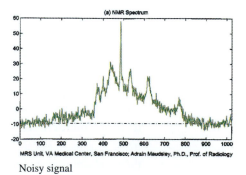 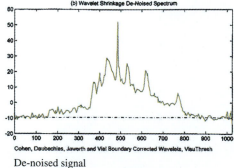

Noisy signal   De-noised signal

Figure 2

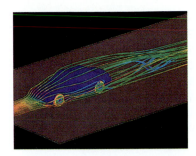

Airflow past a Saturn SL2

Figure 3

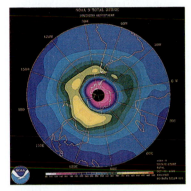

Ozone hole in the Southern Hemisphere

Figure 4

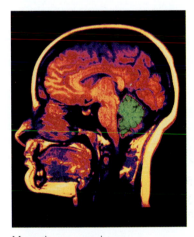

Magnetic resonance image

Figure 5

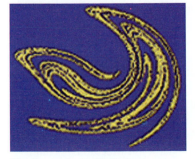

Chaotic ventricular fibrillation

Figure 6

Airflow Past an Automobile — Problems involving fluid flow (air, water, and blood, for example) are a major focus of scientific research. The Army High Performance Computing Research Center (AHPCRC) sponsors numerous unclassified research projects that involve teams of researchers from various science and engineering disciplines. One such project deals with airflow past an automobile (they use a General Motors Saturn SL2). The problem is quite complex since it takes into account the body contours, the wheels, the recessed headlights, and the spoiler. Figure 3 shows a simulation of airflow past an automobile that was produced using state-of-the-art mathematical methods and a Cray T3D supercomputer.

Weather Prediction — Modern meteorology is a marriage between mathematics and physics. Today's meteorologists are concerned with much more than predicting daily weather changes—their research delves into such areas as global warming, holes in the ozone layer (Figure 4), and weather patterns on other planets. In 1904 the Norwegian meteorologist Vilhelm Bjerknes (1862–1951) proposed that the state of the atmosphere at any future time can be determined by measuring appropriate variables at a single instant of time and then solving certain hydrodynamic equations. Although Bjerknes' idea is true in principle, it is difficult to apply because of uncertainties in measured variables, the enormous amounts of data to be processed, and technical complications involved with solving the equations. However, new mathematical discoveries have dramatically improved meteorological predictions and spawned enormous economic benefits. For example, it costs about 50 million dollars to prepare for a hurricane over 300 nautical miles of coastline, even if the hurricane does not hit the area. On the other hand, if the hurricane hits without adequate preparation, then the added costs can mount to billions of dollars (let alone the loss of life). Thus, each new mathematical breakthrough that produces more accurate hurricane prediction translates into enormous economic savings and preservation of human life.

Medical Imaging and DNA Structure — Advances in *nuclear magnetic resonance* (NMR) have made it possible to determine the structure of biological macromolecules, study DNA replication, and determine how proteins act as enzymes and antibodies. Related advances in *magnetic resonance imaging* (MRI) have made it possible to view internal human tissue without invasive surgery and to provide real-time images during surgical procedures (Figure 5). High-quality NMR and MRI would not be possible without mathematical discoveries that have occurred within the last decade.

Controlling Chaotic Behavior in the Human Heart — Chaos theory, which is one of the most exciting new branches of mathematics, is concerned with identifying regularities in phenomena that on the surface seem random and unpredictable (Figure 6). Today's research literature abounds with applications of chaos theory to almost every imaginable branch of science. Researchers at the Applied Chaos Laboratory at Georgia Tech University collaborated with physicians at the Emory University Medical Center in applying chaos theory to control the chaotic behavior of heart tissue that is undergoing ventricular fibrillation (cardiac arrest). The research, though experimental, is already showing promising results.

The World Model of the Future — In anticipation of the 1992 United Nations Earth Summit, researchers at the Institute for Economic Analysis (IEA) at New York University were commissioned by a number of world leaders with the daunting task of creating a model that would predict the economic and environmental future of the world. They started with the World Model and World Database developed by Nobel laureate Wassily Leontief and his colleagues at Harvard in the 1970s, but they expanded on the model by incorporating such environmental factors as the cost of controlling pollutant emissions (from mining, energy creation, and automobiles, for example). They also accounted for the effect of population growth rates on the added demand for energy and other natural resources. Models such as this require a team effort by government, academic, and industrial experts in a variety of fields and play an important role in guiding the decisions of governmental agencies.

Deep Space Exploration — Alexander Wolszczan of Penn State University may go down in history as the first scientist to identify a planetary system beyond our own. While searching the radio sky, Professor Wolszczan discovered a new pulsar, PSR1257+12, that seemed to wobble as it traveled through space. As a result of an extensive mathematical analysis, many scientists are now convinced that the wobble is caused by two or three planets orbiting PSR1257+12. Although scientists have been able to detect pulsars for some time by searching for faint periodic radio signals from outer space, it is only recently that the mathematical techniques have been developed to analyze the data in a way that stands up to scientific scrutiny. Wolszczan predicts that the planets orbiting PSR1257+12 are barren and inhospitable because of stellar winds, but his methods open the possibility of discovering new planetary systems that may sustain intelligent life.

THE ROOTS OF CALCULUS

Today's exciting applications of calculus have roots that can be traced to the work of the Greek mathematician Archimedes, but the actual discovery of the fundamental principles of calculus was made independently by Isaac Newton (English) and Gottfried Leibniz (German) in the late seventeenth century. The work of Newton and Leibniz was motivated by four major classes of scientific and mathematical problems of the time:

- Find the tangent line to a general curve at a given point.
- Find the area of a general region, the length of a general curve, and the volume of a general solid.
- Find the maximum or minimum value of a quantity—for example, the maximum and minimum distances of a planet from the Sun, or the maximum range attainable for a projectile by varying its angle of fire.
- Given a formula for the distance traveled by a body in any specified amount of time, find the velocity and acceleration of the body at any instant. Conversely, given a formula that specifies the acceleration of velocity at any instant, find the distance traveled by the body in a specified period of time.

THE DISCOVERY OF CALCULUS

Newton and Leibniz found a fundamental relationship between the problem of finding a tangent line to a curve and the problem of determining the area of a region. Their realization of this connection is considered to be the "discovery of calculus."

Although Newton saw how these two problems are related 10 years before Leibniz, Leibniz published his work 20 years before Newton. This situation led to a stormy debate over which of the two was the rightful discoverer of calculus. The debate engulfed Europe for half a century, with the scientists of the European continent supporting Leibniz and those from England supporting Newton. The conflict was extremely unfortunate because Newton's inferior notation badly hampered scientific development in England, and the Continent in turn lost the benefit of Newton's discoveries in astronomy and physics for nearly 50 years. In spite of it all, Newton and Leibniz were sincere admirers of each other's work.

ISAAC NEWTON (1642–1727)

Newton was born in the village of Woolsthorpe, England. His father died before he was born and his mother raised him on the family farm. As a youth he showed little evidence of his later brilliance, except for an unusual talent with mechanical devices—he apparently built a working water clock and a toy flour mill powered by a mouse. In 1661 he entered Trinity College in Cambridge with a deficiency in geometry. Fortunately, Newton caught the eye of Isaac Barrow, a gifted mathematician and teacher. Under Barrow's guidance Newton immersed himself in mathematics and science, but he graduated without any special distinction. Because the Plague was spreading rapidly through London, Newton returned to his home in Woolsthorpe and stayed there during the years of 1665 and 1666. In those two momentous years the entire framework of modern science was miraculously created in Newton's mind—he discovered calculus, recognized the underlying principles of planetary motion and gravity, and determined that "white" sunlight was composed of all colors, red to violet. For whatever reasons he kept his discoveries to himself. In 1667 he returned to Cambridge to obtain his Master's degree and upon graduation became a teacher at Trinity. Then in 1669 Newton succeeded his teacher, Isaac Barrow, to the Lucasian chair of mathematics at Trinity, one of the most honored chairs of mathematics in the world. Thereafter, brilliant discoveries flowed from Newton steadily. He formulated the law of gravitation and used it to explain the motion of the Moon, the planets, and the tides; he formulated basic theories of light, thermodynamics, and hydrodynamics; and he devised and constructed the first modern reflecting telescope.

Throughout his life Newton was hesitant to publish his major discoveries, revealing them only to a select circle of friends, perhaps because of a fear of criticism or controversy. In 1687, only after intense coaxing by the astronomer, Edmond Halley (Halley's comet), did Newton publish his masterpiece, *Philosophiae Naturalis Principia Mathematica* (The Mathematical Principles of Natural Philosophy). This work is generally considered to be the most important and influential scientific book ever written. In it Newton explained the workings of the solar system and formulated the basic laws of motion, which to this day are fundamental in engineering and physics. However, not even the pleas of his friends could convince Newton to publish his discovery of calculus. Only after Leibniz published his results did Newton relent and publish his own work on calculus.

After 25 years as a professor, Newton suffered depression and a nervous breakdown. He gave up research in 1695 to accept a position as warden and later master of the London Mint. During the 25 years that he worked at the mint, he did virtually no scientific or mathematical work. He was knighted in 1705 and on his death was buried in Westminster Abbey with all the honors his country could bestow. It is interesting to note that Newton was a learned theologian who viewed the primary value of his work to be its support of the existence of God. Throughout his life he worked passionately to date biblical events by relating them to astronomical phenomena. He was so consumed with this passion that he spent years searching the Book of Daniel for clues to the end of the world and the geography of hell.

Newton described his brilliant accomplishments as follows: "I seem to have been only like a boy playing on the seashore and diverting myself in now and then finding a smoother pebble or prettier shell than ordinary, whilst the great ocean of truth lay all undiscovered before me."

GOTTFRIED WILHELM LEIBNIZ (1646–1716)

This gifted genius was one of the last people to have mastered most major fields of knowledge—an impossible accomplishment in our own era of specialization. He was an expert in law, religion, philosophy, literature, politics, geology, metaphysics, alchemy, history, and mathematics.

Leibniz was born in Leipzig, Germany. His father, a professor of moral philosophy at the University of Leipzig, died when Leibniz was six years old. The precocious boy then gained access to his father's library and began reading voraciously on a wide range of subjects, a habit that he maintained throughout his life. At age 15 he entered the University of Leipzig as a law student and by the age of 20 received a doctorate from the University of Altdorf. Subsequently, Leibniz followed a career in law and international politics, serving as counsel to kings and princes.

During his numerous foreign missions, Leibniz came in contact with outstanding mathematicians and scientists who stimulated his interest in mathematics—most notably, the physicist Christian Huygens. In mathematics Leibniz was self-taught, learning the subject by reading papers and journals. As a result of this fragmented mathematical education, Leibniz often rediscovered the results of others, and this helped to fuel the debate over the discovery of calculus.

Leibniz never married. He was moderate in his habits, quick-tempered, but easily appeased, and charitable in his judgment of other people's work. In spite of his great achievements, Leibniz never received the honors showered on Newton, and he spent his final years as a lonely embittered man. At his funeral there was one mourner, his secretary. An eyewitness stated, "He was buried more like a robber than what he really was—an ornament of his country."

1

FUNCTIONS

René Descartes

*O*ne of the important themes in calculus is the analysis of relationships between physical or mathematical quantities. Such relationships can be described in terms of graphs, formulas, numerical data, or words. In this chapter we will develop the concept of a *function*, which is the basic idea that underlies almost all mathematical and physical relationships, regardless of the form in which they are expressed. We will study properties of some of the most basic functions that occur in calculus, and we will examine some familiar ideas involving lines, polynomials, and trigonometric functions from viewpoints that may be new. We will also discuss ideas relating to the use of graphing utilities such as graphing calculators and graphing software for computers. Before you start reading, you may want to scan through the appendices, since they contain various kinds of precalculus material that may be helpful if you need to review some of those ideas.

1.1 FUNCTIONS AND THE ANALYSIS OF GRAPHICAL INFORMATION

In this section we will define and develop the concept of a function. Functions are used by mathematicians and scientists to describe the relationships between variable quantities and hence play a central role in calculus and its applications.

SCATTER PLOTS AND TABULAR DATA

Many scientific laws are discovered by collecting, organizing, and analyzing experimental data. Since graphs play a major role in studying data, we will begin by discussing the kinds of information that a graph can convey.

To start, we will focus on paired data. For example, Table 1.1.1 shows the top qualifying speed by year in the Indianapolis 500 auto race from 1980 to 1999. This table pairs up each year t between 1980 and 1999 with the top qualifying speed S for that year. These paired data can be represented graphically in a number of ways:

Table 1.1.1

INDIANAPOLIS 500 QUALIFYING SPEEDS

YEAR t	SPEED S (mi/h)
1980	192.256
1981	200.546
1982	207.004
1983	207.395
1984	210.029
1985	212.583
1986	216.828
1987	215.390
1988	219.198
1989	223.885
1990	225.301
1991	224.113
1992	232.482
1993	223.967
1994	228.011
1995	231.604
1996	233.100
1997	218.263
1998	223.503
1999	225.179

- One possibility is to plot the paired data points in a rectangular tS-coordinate system (t horizontal and S vertical), in which case we obtain a **scatter plot** of S versus t (Figure 1.1.1a).

- A second possibility is to enhance the scatter plot visually by joining successive points with straight-line segments, in which case we obtain a **line graph** (Figure 1.1.1b).

- A third possibility is to represent the paired data by a **bar graph** (Figure 1.1.1c).

All three graphical representations reveal an upward trend in the data, as one would expect with improvements in automotive technology.

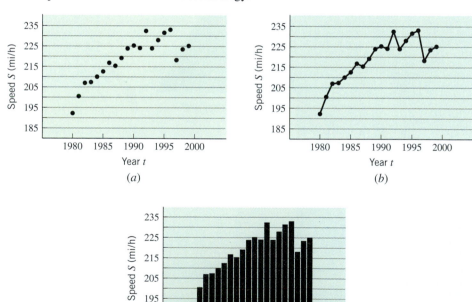

Figure 1.1.1

EXTRACTING INFORMATION FROM GRAPHS

One of the first books to use graphs for representing numerical data was *The Commercial and Political Atlas*, published in 1786 by the Scottish political economist William Playfair (1759–1823). Figure 1.1.2a shows an engraving from that work that compares exports and imports by England to Denmark and Norway (combined). In spite of its antiquity, the engraving is modern in spirit and provides a wealth of information. You should be able to extract the following information from Playfair's graphs:

- In the year 1700 imports were valued at about 70,000 pounds and exports at about 35,000 pounds.

- During the period from 1700 to about 1754 imports exceeded exports (a trade deficit for England).

- In the year 1754 the imports and exports were equal (a trade balance in today's economic terminology).

- From 1754 to 1780 exports exceeded imports (a trade surplus for England). The greatest surplus occurred in 1780, at which time exports exceeded imports by about 95,000 pounds.

- During the period from 1700 to 1725 imports were rising. They peaked in 1725, and then slowly fell until about 1760, at which time they bottomed out and began to rise again slowly until 1780.

- During the period from 1760 to 1780 exports and imports were both rising, but exports were rising more rapidly than imports, resulting in an ever-widening trade surplus for England.

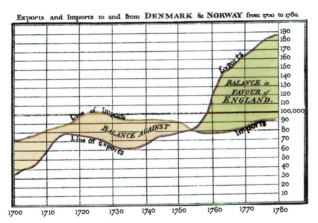

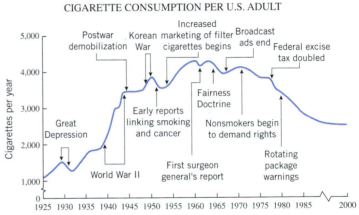

Playfair's Graph of 1786: The horizontal scale is in years from 1700 to 1780 and the vertical scale is in units of 1,000 pounds sterling from 0 to 200.

(a)

Figure 1.1.2

Source: U.S. Department of Health and Human Services.

(b)

Figure 1.1.2*b* is a more contemporary graph; it describes the per capita consumption of cigarettes in the United States between 1925 and 1995.

FOR THE READER.   Use the graph in Figure 1.1.2*b* to provide reasonable answers to the following questions:

- When did the maximum annual cigarette consumption per adult occur and how many were consumed?

- What factors are likely to cause sharp decreases in cigarette consumption?

- What factors are likely to cause sharp increases in cigarette consumption?

- What were the long- and short-term effects of the first surgeon general's report on the health risks of smoking?

GRAPHS OF EQUATIONS

Graphs can be used to describe mathematical equations as well as physical data. For example, consider the equation

$$y = x\sqrt{9 - x^2} \tag{1}$$

For each value of x in the interval $-3 \le x \le 3$, this equation produces a corresponding real value of y, which is obtained by substituting the value of x into the right side of the equation. Some typical values are shown in Table 1.1.2.

Table 1.1.2

x	-3	-2	-1	0	1	2	3
y	0	$-2\sqrt{5} \approx -4.47214$	$-2\sqrt{2} \approx -2.82843$	0	$2\sqrt{2} \approx 2.82843$	$2\sqrt{5} \approx 4.47214$	0

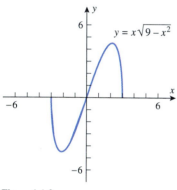

Figure 1.1.3

The set of *all* points in the xy-plane whose coordinates satisfy an equation in x and y is called the **graph** of that equation in the xy-plane. Figure 1.1.3 shows the graph of Equation (1) in the xy-plane. Notice that the graph extends only over the interval $[-3, 3]$. This is because values of x outside of this interval produce complex values of y, and in these cases the ordered pairs (x, y) do not correspond to points in the xy-plane. For example, if $x = 8$, then the corresponding value of y is $y = 8\sqrt{-55} = 8\sqrt{55}\,i$, and the ordered pair $(8, 8\sqrt{55}\,i)$ is not a point in the xy-plane.

Example 1 Figure 1.1.4 shows the graph of an unspecified equation that was used to obtain the values that appear in the shaded parts of the accompanying tables. Examine the graph and confirm that the values in the tables are reasonable approximations. ◀

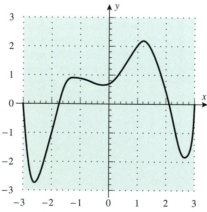

x	y
-3	0
-2	-1
-1	0.9
0	0.7
1	2
2	0.4
3	0

x	y
None	-3
$-2.8, -2.3$	-2
$-2.9, -2, 2.4, 2.9$	-1
$-3, -1.7, 2.1, 3$	0
$0.3, 1.8$	1
$1, 1.4$	2
None	3

Figure 1.1.4

FUNCTIONS

Tables, graphs, and equations provide three methods for describing how one quantity depends on another—numerical, visual, and algebraic. The fundamental importance of this idea was recognized by Leibniz in 1673 when he coined the term *function* to describe the dependence of one quantity on another. The following examples illustrate how this term is used:

- The area A of a circle depends on its radius r by the equation $A = \pi r^2$, so we say that A is a function of r.

- The velocity v of a ball falling freely in the Earth's gravitational field increases with time t until it hits the ground, so we say that v *is a function of* t.

- In a bacteria culture, the number n of bacteria present after 1 hour of growth depends on the number n_0 of bacteria present initially, so we say that n *is a function of* n_0.

This idea is captured in the following definition.

1.1.1 DEFINITION.   If a variable y depends on a variable x in such a way that each value of x determines exactly one value of y, then we say that **y is a function of x**.

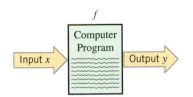

f

Figure 1.1.5

In the mid-eighteenth century the Swiss mathematician Leonhard Euler[*] (pronounced "oiler") conceived the idea of denoting functions by letters of the alphabet, thereby making it possible to describe functions without stating specific formulas, graphs, or tables. To understand Euler's idea, think of a function as a computer program that takes an *input x*, operates on it in some way, and produces exactly one *output y*. The computer program is an object in its own right, so we can give it a name, say f. Thus, the function f (the computer program) associates a unique output y with each input x (Figure 1.1.5). This suggests the following definition.

1.1.2 DEFINITION.   A **function** f is a rule that associates a unique output with each input. If the input is denoted by x, then the output is denoted by $f(x)$ (read "f of x").

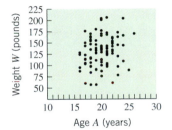

Figure 1.1.6

REMARK.   In this definition the term *unique* means "exactly one." Thus, a function cannot assign two different outputs to the same input. For example, Figure 1.1.6 shows a scatter plot of weight versus age for a random sample of 100 college students. This scatter plot does not describe the weight W as a function of the age A because there are some values of A with more than one corresponding value of W. This is to be expected, since two people with the same age need not have the same weight. In contrast, Table 1.1.1 describes S as a function of t because there is only one top qualifying speed in a given year; similarly, Equation (1) describes y as a function of x because each input x in the interval $-3 \le x \le 3$ produces exactly one output $y = x\sqrt{9 - x^2}$.

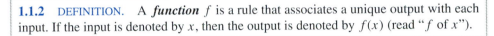

[*]LEONHARD EULER (1707–1783). Euler was probably the most prolific mathematician who ever lived. It has been said that "Euler wrote mathematics as effortlessly as most men breathe." He was born in Basel, Switzerland, and was the son of a Protestant minister who had himself studied mathematics. Euler's genius developed early. He attended the University of Basel, where by age 16 he obtained both a Bachelor of Arts degree and a Master's degree in philosophy. While at Basel, Euler had the good fortune to be tutored one day a week in mathematics by a distinguished mathematician, Johann Bernoulli. At the urging of his father, Euler then began to study theology. The lure of mathematics was too great, however, and by age 18 Euler had begun to do mathematical research. Nevertheless, the influence of his father and his theological studies remained, and throughout his life Euler was a deeply religious, unaffected person. At various times Euler taught at St. Petersburg Academy of Sciences (in Russia), the University of Basel, and the Berlin Academy of Sciences. Euler's energy and capacity for work were virtually boundless. His collected works form more than 100 quarto-sized volumes and it is believed that much of his work has been lost. What is particularly astonishing is that Euler was blind for the last 17 years of his life, and this was one of his most productive periods! Euler's flawless memory was phenomenal. Early in his life he memorized the entire *Aeneid* by Virgil and at age 70 could not only recite the entire work, but could also state the first and last sentence on each page of the book from which he memorized the work. His ability to solve problems in his head was beyond belief. He worked out in his head major problems of lunar motion that baffled Isaac Newton and once did a complicated calculation in his head to settle an argument between two students whose computations differed in the fiftieth decimal place.

Following the development of calculus by Leibniz and Newton, results in mathematics developed rapidly in a disorganized way. Euler's genius gave coherence to the mathematical landscape. He was the first mathematician to bring the full power of calculus to bear on problems from physics. He made major contributions to virtually every branch of mathematics as well as to the theory of optics, planetary motion, electricity, magnetism, and general mechanics.

WAYS TO DESCRIBE FUNCTIONS

Four common methods for representing functions are:

- Numerically by tables
- Geometrically by graphs
- Algebraically by formulas
- Verbally

The method of representation often depends on how the function arises. For example:

- Table 1.1.1 is a numerical representation of S as a function of t. This is the natural way in which data of this type are recorded.
- Figure 1.1.7 shows a record of the amount of deflection D of a seismograph needle during an earthquake. The variable D is a function of the time t that has elapsed since the shock wave left the earthquake's epicenter. In this case the function originates as a graph.
- Some of the most familiar examples of functions arise as formulas; for example, the formula $C = 2\pi r$ expresses the circumference C of a circle as a function of its radius r.
- Sometimes functions are described in words. For example, Isaac Newton's Law of Universal Gravitation is often stated as follows: The gravitational force of attraction between two bodies in the Universe is directly proportional to the product of their masses and inversely proportional to the square of the distance between them. This is the verbal description of the formula

$$F = G\frac{m_1 m_2}{r^2} \tag{2}$$

in which F is the force of attraction, m_1 and m_2 are the masses, r is the distance between them, and G is a constant.

Later in the text we will also show how to use methods of calculus to define new kinds of functions.

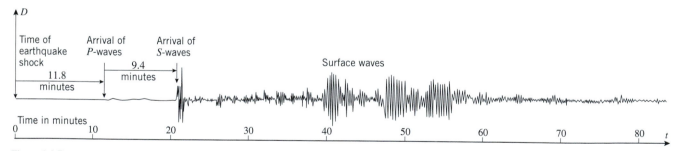

Figure 1.1.7

Sometimes it is desirable to convert one representation of a function into another. For example, in Figure 1.1.1 we converted the numerical relationship between S and t into a graphical relationship, and in writing Formula (2) we converted the verbal representation of the Law of Universal Gravitation into an algebraic relationship.

The problem of converting numerical representations of functions into algebraic formulas often requires special techniques known as ***curve fitting***. For example, Table 1.1.3 gives the U.S. population at 10-year intervals from 1790 to 1850. This table is a numerical representation of the function $P = f(t)$ that relates the U.S. population P to the year t. If we plot P versus t, we obtain the scatter plot in Figure 1.1.8a, and if we use curve-fitting methods

Table 1.1.3

U.S. POPULATION

YEAR t	POPULATION P (millions)
1790	3.9
1800	5.3
1810	7.2
1820	9.6
1830	12
1840	17
1850	23

Source: The World Almanac.

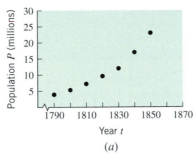

(a)

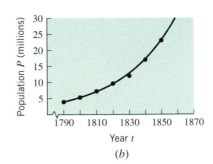

(b)

Figure 1.1.8

that will be discussed later, we can obtain the approximation

$$P \approx 3.94(1.03)^{t-1790}$$

Figure 1.1.8b shows the graph of this equation imposed on the scatter plot.

DISCRETE VERSUS CONTINUOUS DATA

Engineers and physicists distinguish between ***continuous data*** and ***discrete data***. Continuous data have values that vary *continuously* over an interval, whereas discrete data have values that make *discrete* jumps. For example, for the seismic data in Figure 1.1.7 both the time and intensity vary continuously, whereas in Table 1.1.3 and Figure 1.1.8a both the year and population make discrete jumps. As a rule, continuous data lead to graphs that are continuous, unbroken curves, whereas discrete data lead to scatter plots consisting of isolated points. Sometimes, as in Figure 1.1.8b, it is desirable to approximate a scatter plot by a continuous curve. This is useful for making conjectures about the values of the quantities between the recorded data points.

GRAPHS AS PROBLEM-SOLVING TOOLS

Sometimes a function is buried in the statement of a problem, and it is up to the problem solver to uncover it and use it in an appropriate way to solve the problem. Here is an example that illustrates the power of graphical representations of functions as a problem-solving tool.

Example 2 Figure 1.1.9a shows an offshore oil well located at a point W that is 5 km from the closest point A on a straight shoreline. Oil is to be piped from W to a shore point B that is 8 km from A. It costs \$1,000,000/km to lay pipe under water and \$500,000/km over land. In your role as project manager you receive three proposals for piping the oil from W to B. Proposal 1 claims that it is cheapest to pipe directly from W to B, since the shortest distance between two points is a straight line. Proposal 2 claims that it is cheapest to pipe directly to point A and then along the shoreline to B, thereby using the least amount of expensive underwater pipe. Proposal 3 claims that it is cheapest to compromise by piping under water

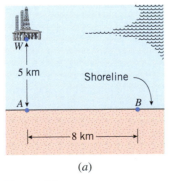

(a)

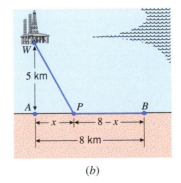

(b)

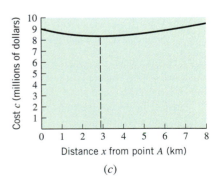

(c)

Figure 1.1.9

to some well-chosen point between A and B, and then piping along the shoreline to B. Which proposal is correct?

Solution. Let P be any point between A and B (Figure 1.1.9b), and let

$x = $ distance (in kilometers) between A and P

$c = $ cost (in millions of dollars) for the entire pipeline

Proposal 1 claims that $x = 8$ results in the least cost, Proposal 2 claims that it is $x = 0$, and Proposal 3 claims it is some value of x between 0 and 8. From Figure 1.1.9b the length of pipe along the shore is

$$8 - x \qquad (3)$$

and from the Theorem of Pythagoras, the length of pipe under water is

$$\sqrt{x^2 + 25} \qquad (4)$$

Thus, from (3) and (4) the total cost c (in millions of dollars) for the pipeline is

$$c = 1\left(\sqrt{x^2 + 25}\right) + 0.5(8 - x) = \sqrt{x^2 + 25} + 0.5(8 - x) \qquad (5)$$

where $0 \le x \le 8$. The graph of Equation (5), shown in Figure 1.1.9c, makes it clear that Proposal 3 is correct—the most cost-effective strategy is to pipe to a point a little less than 3 km from point A. ◀

EXERCISE SET 1.1   ∿ Graphing Utility

1. Use the cigarette consumption graph in Figure 1.1.2b to answer the following questions, making reasonable approximations where needed.
   (a) When did the annual cigarette consumption reach 3000 per adult for the first time?
   (b) When did the annual cigarette consumption per adult reach its peak, and what was the peak value?
   (c) Can you tell from the graph how many cigarettes were consumed in a given year? If not, what additional information would you need to make that determination?
   (d) What factors are likely to cause a sharp increase in annual cigarette consumption per adult?
   (e) What factors are likely to cause a sharp decline in annual cigarette consumption per adult?

2. The accompanying graph shows the median income in U.S. households (adjusted for inflation) between 1975 and 1995. Use the graph to answer the following questions, making reasonable approximations where needed.
   (a) When did the median income reach its maximum value, and what was the median income when that occurred?
   (b) When did the median income reach its minimum value, and what was the median income when that occurred?
   (c) The median income was declining during the 4-year period between 1989 and 1993. Was it declining more

rapidly during the first 2 years or the second 2 years of that period? Explain your reasoning.

MEDIAN U.S. HOUSEHOLD INCOME IN
THOUSANDS OF CONSTANT 1995 DOLLARS

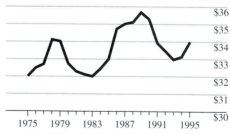

Source: Census Bureau, March 1996
[1996 measures 1995 income].

Figure Ex-2

3. Use the accompanying graph to answer the following questions, making reasonable approximations where needed.
   (a) For what values of x is $y = 1$?
   (b) For what values of x is $y = 3$?
   (c) For what values of y is $x = 3$?
   (d) For what values of x is $y \le 0$?
   (e) What are the maximum and minimum values of y and for what values of x do they occur?

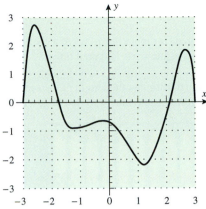

Figure Ex-3

4. Use the accompanying table to answer the questions posed in Exercise 3.

x	–2	–1	0	2	3	4	5	6
y	5	1	–2	7	–1	1	0	9

Table Ex-4

5. Use the equation $y = x^2 - 6x + 8$ to answer the following questions.
   (a) For what values of x is $y = 0$?
   (b) For what values of x is $y = -10$?
   (c) For what values of x is $y \geq 0$?
   (d) Does y have a minimum value? A maximum value? If so, find them.

6. Use the equation $y = 1 + \sqrt{x}$ to answer the following questions.
   (a) For what values of x is $y = 4$?
   (b) For what values of x is $y = 0$?
   (c) For what values of x is $y \geq 6$?
   (d) Does y have a minimum value? A maximum value? If so, find them.

7. (a) If you had a device that could record the Earth's population continuously, would you expect the graph of population versus time to be a continuous (unbroken) curve? Explain what might cause breaks in the curve.
   (b) Suppose that a hospital patient receives an injection of an antibiotic every 8 hours and that between injections the concentration C of the antibiotic in the bloodstream decreases as the antibiotic is absorbed by the tissues. What might the graph of C versus the elapsed time t look like?

8. (a) If you had a device that could record the temperature of a room continuously over a 24-hour period, would you expect the graph of temperature versus time to be a continuous (unbroken) curve? Explain your reasoning.
   (b) If you had a computer that could track the number of boxes of cereal on the shelf of a market continuously over a 1-week period, would you expect the graph of the number of boxes on the shelf versus time to be a continuous (unbroken) curve? Explain your reasoning.

9. A construction company wants to build a rectangular enclosure with an area of 1000 square feet by fencing in three sides and using its office building as the fourth side. Your objective as supervising engineer is to design the enclosure so that it uses the least amount of fencing. Proceed as follows.
   (a) Let x and y be the dimensions of the enclosure, where x is measured parallel to the building, and let L be the length of fencing required for those dimensions. Since the area must be 1000 square feet, we must have $xy = 1000$. Find a formula for L in terms of x and y, and then express L in terms of x alone by using the area equation.
   (b) Are there any restrictions on the value of x? Explain.
   (c) Make a graph of L versus x over a reasonable interval, and use the graph to estimate the value of x that results in the smallest value of L.
   (d) Estimate the smallest value of L.

10. A manufacturer constructs open boxes from sheets of cardboard that are 6 inches square by cutting small squares from the corners and folding up the sides (as shown in the accompanying figure). The Research and Development Department asks you to determine the size of the square that produces a box of greatest volume. Proceed as follows.
   (a) Let x be the length of a side of the square to be cut, and let V be the volume of the resulting box. Show that $V = x(6 - 2x)^2$.
   (b) Are there any restrictions on the value of x? Explain.
   (c) Make a graph of V versus x over an appropriate interval, and use the graph to estimate the value of x that results in the largest volume.
   (d) Estimate the largest volume.

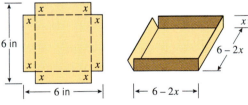

Figure Ex-10

11. A soup company wants to manufacture a can in the shape of a right circular cylinder that will hold 500 cm³ of liquid. The material for the top and bottom costs 0.02 cent/cm², and the material for the sides costs 0.01 cent/cm².
   (a) Use the method of Exercises 9 and 10 to estimate the radius r and height h of the can that costs the least to manufacture. [*Suggestion:* Express the cost C in terms of r.]
   (b) Suppose that the tops and bottoms of radius r are punched out from square sheets with sides of length $2r$ and the scraps are waste. If you allow for the cost of

the waste, would you expect the can of least cost to be taller or shorter than the one in part (a)? Explain.

(c) Estimate the radius, height, and cost of the can in part (b), and determine whether your conjecture was correct.

12. The designer of a sports facility wants to put a quarter-mile (1320 ft) running track around a football field, oriented as in the accompanying figure. The football field is 360 ft long (including the end zones) and 160 ft wide. The track consists of two straightaways and two semicircles, with the straightaways extending at least the length of the football field.

(a) Show that it is possible to construct a quarter-mile track around the football field. [*Suggestion:* Find the shortest track that can be constructed around the field.]

(b) Let L be the length of a straightaway (in feet), and let x be the distance (in feet) between a sideline of the football field and a straightaway. Make a graph of L versus x.

(c) Use the graph to estimate the value of x that produces the shortest straightaways, and then find this value of x exactly.

(d) Use the graph to estimate the length of the longest possible straightaways, and then find that length exactly.

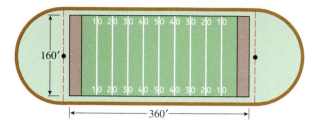

Figure Ex-12

1.2 PROPERTIES OF FUNCTIONS

In this section we will explore properties of functions in more detail. We will assume that you are familiar with the standard notation for intervals and the basic properties of absolute value. Reviews of these topics are provided in Appendices A and B.

INDEPENDENT AND DEPENDENT VARIABLES

Recall from the last section that a function f is a rule that associates a unique output $f(x)$ with each input x. This output is sometimes called the *value* of f at x or the *image* of x under f. Sometimes we will want to denote the output by a single letter, say y, and write

$$y = f(x)$$

This equation expresses y as a function of x; the variable x is called the *independent variable* (or *argument*) of f, and the variable y is called the *dependent variable* of f. This terminology is intended to suggest that x is free to vary, but that once x has a specific value a corresponding value of y is determined. For now we will only consider functions in which the independent and dependent variables are real numbers, in which case we say that f is a *real-valued function of a real variable*. Later, we will consider other kinds of functions as well.

Table 1.2.1 can be viewed as a numerical representation of a function of f. For this function we have

Table 1.2.1

x	0	1	2	3
y	3	4	-1	6

$f(0) = 3$ f associates $y = 3$ with $x = 0$.

$f(1) = 4$ f associates $y = 4$ with $x = 1$.

$f(2) = -1$ f associates $y = -1$ with $x = 2$.

$f(3) = 6$ f associates $y = 6$ with $x = 3$.

To illustrate how functions can be defined by equations, consider

$$y = 3x^2 - 4x + 2 \tag{1}$$

This equation has the form $y = f(x)$, where

$$f(x) = 3x^2 - 4x + 2 \tag{2}$$

The outputs of f (the y-values) are obtained by substituting numerical values for x in this formula. For example,

$$f(0) = 3(0)^2 - 4(0) + 2 = 2 \qquad \text{f associates $y = 2$ with $x = 0$.}$$

$$f(-1.7) = 3(-1.7)^2 - 4(-1.7) + 2 = 17.47 \qquad \text{f associates $y = 17.47$ with $x = -1.7$.}$$

$$f(\sqrt{2}) = 3(\sqrt{2})^2 - 4\sqrt{2} + 2 = 8 - 4\sqrt{2} \qquad \text{f associates $y = 8 - 4\sqrt{2}$ with $x = \sqrt{2}$.}$$

> REMARK.   Although f, x, and y are the most common notations for functions and variables, any letters can be used. For example, to indicate that the area A of a circle is a function of the radius r, it would be more natural to write $A = f(r)$ [where $f(r) = \pi r^2$]. Similarly, to indicate that the circumference C of a circle is a function of the radius r, we might write $C = g(r)$ [where $g(r) = 2\pi r$]. The area function and the circumference function are different, which is why we denoted them by different letters, f and g.

DOMAIN AND RANGE

If $y = f(x)$, then the set of all possible inputs (x-values) is called the ***domain*** of f, and the set of outputs (y-values) that result when x varies over the domain is called the ***range*** of f. For example, consider the equations

$$y = x^2 \quad \text{and} \quad y = x^2, \quad x \geq 2$$

In the first equation there is no restriction on x, so we may assume that any real value of x is an allowable input. Thus, the equation defines a function $f(x) = x^2$ with domain $-\infty < x < +\infty$. In the second equation, the inequality $x \geq 2$ restricts the allowable inputs to be greater than or equal to 2, so the equation defines a function $g(x) = x^2$, $x \geq 2$ with domain $2 \leq x < +\infty$.

As x varies over the domain of the function $f(x) = x^2$, the values of $y = x^2$ vary over the interval $0 \leq y < +\infty$, so this is the range of f. By comparison, as x varies over the domain of the function $g(x) = x^2$, $x \geq 2$, the values of $y = x^2$ vary over the interval $4 \leq y < +\infty$, so this is the range of g.

It is important to understand here that even though $f(x) = x^2$ and $g(x) = x^2$, $x \geq 2$ involve the same formula, we regard them to be different functions because they have different domains. In short, *to fully describe a function you must not only specify the rule that relates the inputs and outputs, but you must also specify the domain, that is, the set of allowable inputs.*

GRAPHS OF FUNCTIONS

If f is a real-valued function of a real variable, then the ***graph*** of f in the xy-plane is defined to be the graph of the equation $y = f(x)$. For example, the graph of the function $f(x) = x$ is the graph of the equation $y = x$, shown in Figure 1.2.1. That figure also shows the graphs

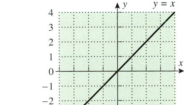
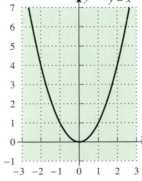
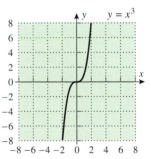
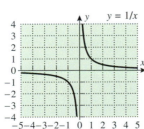
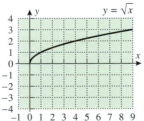
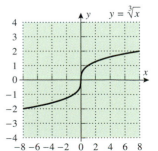

Figure 1.2.1

of some other basic functions that may already be familiar to you. Later in this chapter we will discuss techniques for graphing functions using graphing calculators and computers.

Graphs can provide useful visual information about a function. For example, because the graph of a function f in the xy-plane consists of all points whose coordinates satisfy the equation $y = f(x)$, the points on the graph of f are of the form $(x, f(x))$; hence each y-coordinate is the value of f at the x-coordinate (Figure 1.2.2a). Pictures of the domain and range of f can be obtained by projecting the graph of f onto the coordinate axes (Figure 1.2.2b). The values of x for which $f(x) = 0$ are the x-coordinates of the points where the graph of f intersects the x-axis (Figure 1.2.2c); these values of x are called the **zeros** of f, the **roots** of $f(x) = 0$, or the **x-intercepts** of $y = f(x)$.

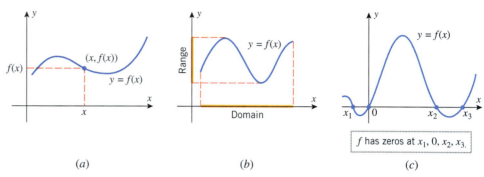

(a)   (b)   (c)

Figure 1.2.2

THE VERTICAL LINE TEST

Not every curve in the xy-plane is the graph of a function. For example, consider the curve in Figure 1.2.3, which is cut at two distinct points, (a, b) and (a, c), by a vertical line. This curve cannot be the graph of $y = f(x)$ for any function f; otherwise, we would have

$$f(a) = b \quad \text{and} \quad f(a) = c$$

which is impossible, since f cannot assign two different values to a. Thus, there is no function f whose graph is the given curve. This illustrates the following general result, which we will call the **vertical line test**.

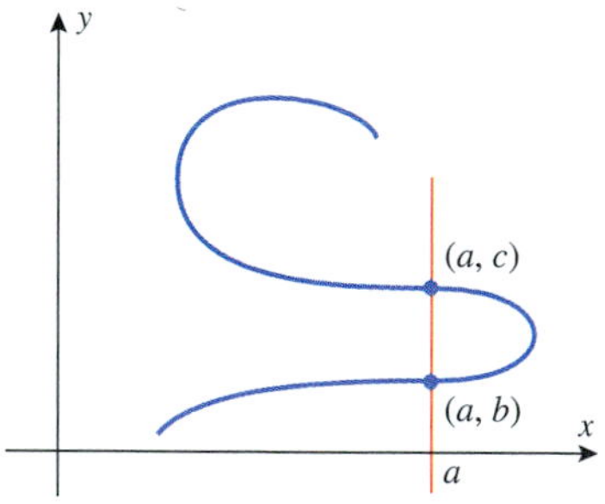

> **1.2.1** THE VERTICAL LINE TEST. *A curve in the xy-plane is the graph of some function f if and only if no vertical line intersects the curve more than once.*

Figure 1.2.3

Example 1 The graph of the equation

$$x^2 + y^2 = 25 \tag{3}$$

is a circle of radius 5, centered at the origin (see Appendix D for a review of circles), and hence there are vertical lines that cut the graph more than once. This can also be seen algebraically by solving (3) for y in terms of x:

$$y = \pm\sqrt{25 - x^2}$$

This equation does not define y as a function of x because the right side is "multiple valued" in the sense that values of x in the interval $(-5, 5)$ produce two corresponding values of y. For example, if $x = 4$, then $y = \pm 3$, and hence $(4, 3)$ and $(4, -3)$ are two points on the circle that lie on the same vertical line (Figure 1.2.4a). However, we can regard the circle as the union of two semicircles:

$$y = \sqrt{25 - x^2} \quad \text{and} \quad y = -\sqrt{25 - x^2}$$

(Figure 1.2.4b), each of which defines y as a function of x. ◀

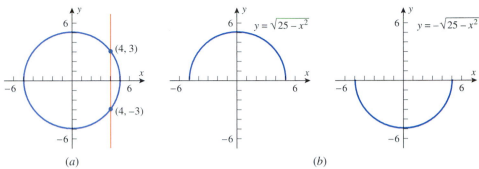

Figure 1.2.4

THE ABSOLUTE VALUE FUNCTION

Recall that the **absolute value** or **magnitude** of a real number x is defined by

$$|x| = \begin{cases} x, & x \geq 0 \\ -x, & x < 0 \end{cases}$$

The effect of taking the absolute value of a number is to strip away the minus sign if the number is negative and to leave the number unchanged if it is nonnegative. Thus,

$$|5| = 5, \quad \left|-\tfrac{4}{7}\right| = \tfrac{4}{7}, \quad |0| = 0$$

A more detailed discussion of the properties of absolute value is given in Appendix B. However, for convenience we provide the following summary of its algebraic properties.

1.2.2   PROPERTIES OF ABSOLUTE VALUE.   *If a and b are real numbers, then*

(a) $|-a| = |a|$ A number and its negative have the same absolute value.

(b) $|ab| = |a|\,|b|$ The absolute value of a product is the product of the absolute values.

(c) $|a/b| = |a|/|b|, \, b \neq 0$ The absolute value of a ratio is the ratio of the absolute values.

(d) $|a + b| \leq |a| + |b|$ The *triangle inequality*

REMARK.   Symbols such as $+x$ and $-x$ are deceptive, since it is tempting to conclude that $+x$ is positive and $-x$ is negative. However, this need not be so, since x itself can be positive or negative. For example, if x is negative, say $x = -3$, then $-x = 3$ is positive and $+x = -3$ is negative.

The graph of the function $f(x) = |x|$ can be obtained by graphing the two parts of the equation

$$y = \begin{cases} x, & x \geq 0 \\ -x, & x < 0 \end{cases}$$

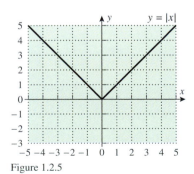

Figure 1.2.5

separately. For $x \geq 0$, the graph of $y = x$ is a ray of slope 1 with its endpoint at the origin, and for $x < 0$, the graph of $y = -x$ is a ray of slope -1 with its endpoint at the origin. Combining the two parts produces the V-shaped graph in Figure 1.2.5.

Absolute values have important relationships to square roots. To see why this is so, recall from algebra that every positive real number x has two square roots, one positive and one negative. By definition, the symbol $\sqrt{x}$ denotes the *positive* square root of x. To denote the negative square root you must write $-\sqrt{x}$. For example, the positive square root of 9 is $\sqrt{9} = 3$, and the negative square root is $-\sqrt{9} = -3$. (Do not make the mistake of writing $\sqrt{9} = \pm 3$.)

Care must be exercised in simplifying expressions of the form $\sqrt{x^2}$, since it is *not* always true that $\sqrt{x^2} = x$. This equation is correct if x is nonnegative, but it is false for negative x. For example, if $x = -4$, then

$$\sqrt{x^2} = \sqrt{(-4)^2} = \sqrt{16} = 4 \neq x$$

A statement that is correct for all real values of x is

$$\sqrt{x^2} = |x|$$

> **FOR THE READER.**   Verify this relationship by using a graphing utility to show that the equations $y = \sqrt{x^2}$ and $y = |x|$ have the same graph.

FUNCTIONS DEFINED PIECEWISE

The absolute value function $f(x) = |x|$ is an example of a function that is defined *piecewise* in the sense that the formula for f changes, depending on the value of x.

Example 2   Sketch the graph of the function defined piecewise by the formula

$$f(x) = \begin{cases} 0, & x \le -1 \\ \sqrt{1 - x^2}, & -1 < x < 1 \\ x, & x \ge 1 \end{cases}$$

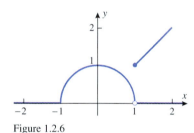

Figure 1.2.6

Solution.   The formula for f changes at the points $x = -1$ and $x = 1$. (We call these the *breakpoints* for the formula.) A good procedure for graphing functions defined piecewise is to graph the function separately over the open intervals determined by the breakpoints, and then graph f at the breakpoints themselves. For the function f in this example the graph is the horizontal ray $y = 0$ on the interval $(-\infty, -1)$, it is the semicircle $y = \sqrt{1 - x^2}$ on the interval $(-1, 1)$, and it is the ray $y = x$ on the interval $(1, +\infty)$. The formula for f specifies that the equation $y = 0$ applies at the breakpoint -1 [so $y = f(-1) = 0$], and it specifies that the equation $y = x$ applies at the breakpoint 1 [so $y = f(1) = 1$]. The graph of f is shown in Figure 1.2.6.   ◄

> **REMARK.**   In Figure 1.2.6 the solid dot and open circle at the breakpoint $x = 1$ serve to emphasize that the point on the graph lies on the ray and not the semicircle. There is no ambiguity at the breakpoint $x = -1$ because the two parts of the graph join together continuously there.

Example 3   Increasing the speed at which air moves over a person's skin increases the rate of moisture evaporation and makes the person feel cooler. (This is why we fan ourselves in hot weather.) The *windchill index* is the temperature at a wind speed of 4 mi/h that would produce the same sensation on exposed skin as the current temperature and wind speed combination. An empirical formula (i.e., a formula based on experimental data) for the windchill index W at $32°$F for a wind speed of v mi/h is

$$W = \begin{cases} 32, & 0 \le v \le 4 \\ 91.4 + 59.4(0.0203v - 0.304\sqrt{v} - 0.474), & 4 < v < 45 \\ -3.8, & v \ge 45 \end{cases}$$

A computer-generated graph of $W(v)$ is shown in Figure 1.2.7.   ◄

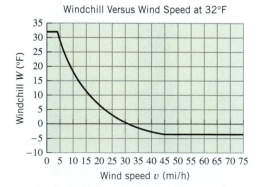

Figure 1.2.7

THE NATURAL DOMAIN

Sometimes, restrictions on the allowable values of an independent variable result from a mathematical formula that defines the function. For example, if $f(x) = 1/x$, then $x = 0$ must be excluded from the domain to avoid division by zero, and if $f(x) = \sqrt{x}$, then negative values of x must be excluded from the domain, since we are only considering real-valued functions of a real variable for now. We make the following definition.

1.2.3 DEFINITION. If a real-valued function of a real variable is defined by a formula, and if no domain is stated explicitly, then it is to be understood that the domain consists of all real numbers for which the formula yields a real value. This is called the ***natural domain*** of the function.

Example 4 Find the natural domain of

(a) $f(x) = x^3$ (b) $f(x) = 1/[(x-1)(x-3)]$

(c) $f(x) = \tan x$ (d) $f(x) = \sqrt{x^2 - 5x + 6}$

Solution (a). The function f has real values for all real x, so its natural domain is the interval $(-\infty, +\infty)$.

Solution (b). The function f has real values for all real x, except $x = 1$ and $x = 3$, where divisions by zero occur. Thus, the natural domain is

$$\{x : x \neq 1 \text{ and } x \neq 3\} = (-\infty, 1) \cup (1, 3) \cup (3, +\infty)$$

Solution (c). Since $f(x) = \tan x = \sin x / \cos x$, the function f has real values except where $\cos x = 0$, and this occurs when x is an odd integer multiple of $\pi/2$. Thus, the natural domain consists of all real numbers except

$$x = \pm\frac{\pi}{2}, \pm\frac{3\pi}{2}, \pm\frac{5\pi}{2}, \ldots$$

Solution (d). The function f has real values, except when the expression inside the radical is negative. Thus the natural domain consists of all real numbers x such that

$$x^2 - 5x + 6 = (x-3)(x-2) \geq 0$$

This inequality is satisfied if $x \leq 2$ or $x \geq 3$ (verify), so the natural domain of f is

$$(-\infty, 2] \cup [3, +\infty) \qquad \blacktriangleleft$$

REMARK. In some problems we will want to limit the domain of a function by imposing specific restrictions. For example, by writing

$$f(x) = x^2, \quad x \geq 0$$

we can limit the domain of f to the positive x-axis (Figure 1.2.8).

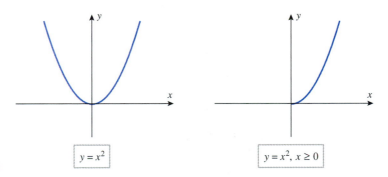

Figure 1.2.8

Algebraic expressions are frequently simplified by canceling common factors in the numerator and denominator. However, care must be exercised when simplifying formulas for functions in this way, since this process can alter the domain.

Example 5   The natural domain of the function

$$f(x) = \frac{x^2 - 4}{x - 2}$$

consists of all real x except $x = 2$. However, if we factor the numerator and then cancel the common factor in the numerator and denominator, we obtain

$$f(x) = \frac{(x - 2)(x + 2)}{x - 2} = x + 2$$

which *is* defined at $x = 2$ [since $f(2) = 4$ for the altered function f]. Thus, the algebraic simplification has altered the domain of the function. Geometrically, the graph of $y = x + 2$ is a line of slope 1 and y-intercept 2, whereas the graph of $y = (x^2 - 4)/(x - 2)$ is the same line, but with a hole in it at $x = 2$, since y is undefined there (Figure 1.2.9). Thus, the geometric effect of the algebraic cancellation is to eliminate the hole in the original graph. In some situations such minor alterations in the domain are irrelevant to the problem under consideration and can be ignored. However, if we wanted to preserve the domain in this example, then we would express the simplified form of the function as

$$f(x) = x + 2, \quad x \neq 2 \qquad \blacktriangleleft$$

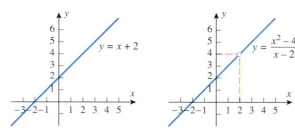

Figure 1.2.9

Example 6   Find the domain and range of

(a) $f(x) = 2 + \sqrt{x - 1}$ (b) $f(x) = (x + 1)/(x - 1)$

Solution (a).   Since no domain is stated explicitly, the domain of f is the natural domain $[1, +\infty)$. As x varies over the interval $[1, +\infty)$, the value of $\sqrt{x - 1}$ varies over the interval $[0, +\infty)$, so the value of $f(x) = 2 + \sqrt{x - 1}$ varies over the interval $[2, +\infty)$, which is the range of f. The domain and range are shown graphically in Figure 1.2.10*a*.

Solution (b).   The given function f is defined for all real x, except $x = 1$, so the natural domain of f is

$$\{x : x \neq 1\} = (-\infty, 1) \cup (1, +\infty)$$

To determine the range it will be convenient to introduce a dependent variable

$$y = \frac{x + 1}{x - 1} \qquad (4)$$

Although the set of possible y-values is not immediately evident from this equation, the graph of (4), which is shown in Figure 1.2.10*b*, suggests that the range of f consists of all

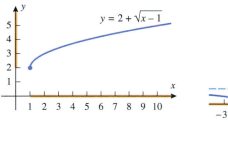
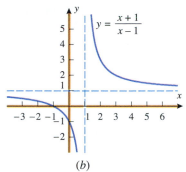

(a)   (b)

Figure 1.2.10

y, except $y = 1$. To see that this is so, we solve (4) for x in terms of y:

$$(x - 1)y = x + 1$$
$$xy - y = x + 1$$
$$xy - x = y + 1$$
$$x(y - 1) = y + 1$$
$$x = \frac{y + 1}{y - 1}$$

It is now evident from the right side of this equation that $y = 1$ is not in the range; otherwise we would have a division by zero. No other values of y are excluded by this equation, so the range of the function f is $\{y : y \neq 1\} = (-\infty, 1) \cup (1, +\infty)$, which agrees with the result obtained graphically. ◀

DOMAIN AND RANGE IN APPLIED PROBLEMS

In applications, physical considerations often impose restrictions on the domain and range of a function.

Example 7   An open box is to be made from a 16-inch by 30-inch piece of cardboard by cutting out squares of equal size from the four corners and bending up the sides (Figure 1.2.11a).

(a)   Let V be the volume of the box that results when the squares have sides of length x. Find a formula for V as a function of x.

(b)   Find the domain of V.

(c)   Use the graph of V given in Figure 1.2.11c to estimate the range of V.

(d)   Describe in words what the graph tells you about the volume.

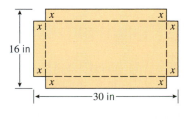
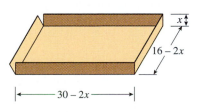
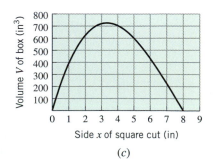

(a)   (b)   (c)

Figure 1.2.11

Solution (a). As shown in Figure 1.2.11*b*, the resulting box has dimensions $16 - 2x$ by $30 - 2x$ by x, so the volume $V(x)$ is given by

$$V(x) = (16 - 2x)(30 - 2x)x = 480x - 92x^2 + 4x^3$$

Solution (b). The domain is the set of x-values and the range is the set of V-values. Because x is a length, it must be nonnegative, and because we cannot cut out squares whose sides are more than 8 in long (why?), the x-values in the domain must satisfy

$$0 \le x \le 8$$

Solution (c). From the graph of V versus x in Figure 1.2.11*c* we estimate that the V-values in the range satisfy

$$0 \le V \le 725$$

Note that this is an approximation. Later we will show how to find the range exactly.

Solution (d). The graph tells us that the box of maximum volume occurs for a value of x that is between 3 and 4 and that the maximum volume is approximately 725 in^3. Moreover, the volume decreases toward zero as x gets closer to 0 or 8. ◀

In applications involving time, formulas for functions are often expressed in terms of a variable t whose starting value is taken to be $t = 0$.

Example 8   At 8:05 A.M. a car is clocked at 100 ft/s by a radar detector that is positioned at the edge of a straight highway. Assuming that the car maintains a constant speed between 8:05 A.M. and 8:06 A.M., find a function $D(t)$ that expresses the distance traveled by the car during that time interval as a function of the time t.

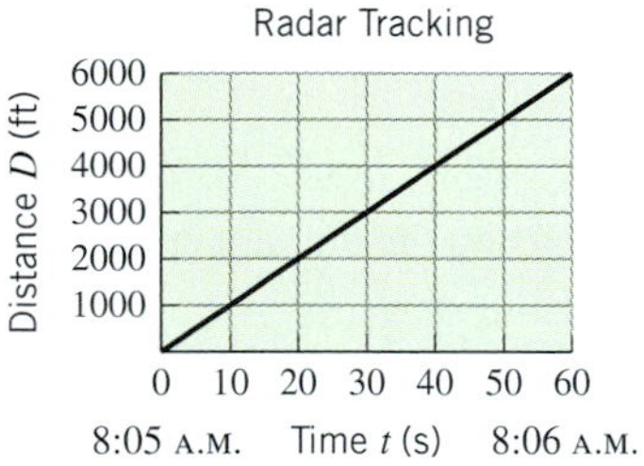

Solution. It would be clumsy to use clock time for the variable t, so let us agree to measure the elapsed time in seconds, starting with $t = 0$ at 8:05 A.M. and ending with $t = 60$ at 8:06 A.M. At each instant, the distance traveled (in ft) is equal to the speed of the car (in ft/s) multiplied by the elapsed time (in s). Thus,

$$D(t) = 100t, \quad 0 \le t \le 60$$

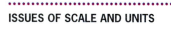

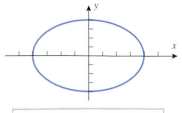

Figure 1.2.12

The graph of D versus t is shown in Figure 1.2.12. ◀

ISSUES OF SCALE AND UNITS

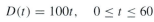

In geometric problems where you want to preserve the "true" shape of a graph, you must use units of equal length on both axes. For example, if you graph a circle in a coordinate system in which 1 unit in the y-direction is smaller than 1 unit in the x-direction, then the circle will be squashed vertically into an elliptical shape (Figure 1.2.13). You must also use units of equal length when you want to apply the distance formula

$$d = \sqrt{(x_2 - x_1)^2 + (y_2 - y_1)^2}$$

to calculate the distance between two points (x_1, y_1) and (x_2, y_2) in the xy-plane.

However, sometimes it is inconvenient or impossible to display a graph using units of equal length. For example, consider the equation

$$y = x^2$$

The circle is squashed because 1 unit on the y-axis has a smaller length than 1 unit on the x-axis.

Figure 1.2.13

If we want to show the portion of the graph over the interval $-3 \le x \le 3$, then there is no problem using units of equal length, since y only varies from 0 to 9 over that interval. However, if we want to show the portion of the graph over the interval $-10 \le x \le 10$, then there is a problem keeping the units equal in length, since the value of y varies between 0 and 100. In this case the only reasonable way to show all of the graph that occurs over the interval $-10 \le x \le 10$ is to compress the unit of length along the y-axis, as illustrated in Figure 1.2.14.

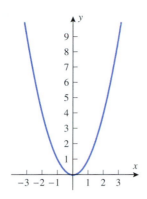

Figure 1.2.14

> • REMARK.   In applications where the variables on the two axes have unrelated units (say,
> centimeters on the y-axis and seconds on the x-axis), then nothing is gained by requiring
> the units to have equal lengths; choose the lengths to make the graph as clear as possible.

EXERCISE SET 1.2   ∿ Graphing Utility

1. Find $f(0)$, $f(2)$, $f(-2)$, $f(3)$, $f(\sqrt{2}\,)$, and $f(3t)$.

   (a) $f(x) = 3x^2 - 2$ (b) $f(x) = \begin{cases} \dfrac{1}{x}, & x > 3 \\ 2x, & x \le 3 \end{cases}$

2. Find $g(3)$, $g(-1)$, $g(\pi)$, $g(-1.1)$, and $g(t^2 - 1)$.

   (a) $g(x) = \dfrac{x+1}{x-1}$ (b) $g(x) = \begin{cases} \sqrt{x+1}, & x \ge 1 \\ 3, & x < 1 \end{cases}$

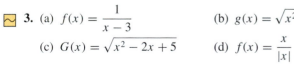

In Exercises 3–6, find the natural domain of the function algebraically, and confirm that your result is consistent with
the graph produced by your graphing utility. [*Note:* Set your
graphing utility to the radian mode when graphing trigonometric functions.]

∿ 3. (a) $f(x) = \dfrac{1}{x-3}$ (b) $g(x) = \sqrt{x^2 - 3}$

   (c) $G(x) = \sqrt{x^2 - 2x + 5}$ (d) $f(x) = \dfrac{x}{|x|}$

   (e) $h(x) = \dfrac{1}{1 - \sin x}$

∿ 4. (a) $f(x) = \dfrac{1}{5x + 7}$ (b) $h(x) = \sqrt{x - 3x^2}$

   (c) $G(x) = \sqrt{\dfrac{x^2 - 4}{x - 4}}$ (d) $f(x) = \dfrac{x^2 - 1}{x + 1}$

   (e) $h(x) = \dfrac{3}{2 - \cos x}$

∿ 5. (a) $f(x) = \sqrt{3 - x}$ (b) $g(x) = \sqrt{4 - x^2}$
   (c) $h(x) = 3 + \sqrt{x}$ (d) $G(x) = x^3 + 2$
   (e) $H(x) = 3 \sin x$

∿ 6. (a) $f(x) = \sqrt{3x - 2}$ (b) $g(x) = \sqrt{9 - 4x^2}$
   (c) $h(x) = \dfrac{1}{3 + \sqrt{x}}$ (d) $G(x) = \dfrac{3}{x}$
   (e) $H(x) = \sin^2 \sqrt{x}$

7. In each part of the accompanying figure, determine whether
   the graph defines y as a function of x.

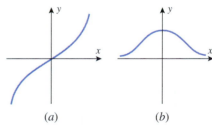

(a)   (b)

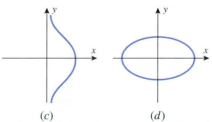

(c)   (d)

Figure Ex-7

8. Express the length L of a chord of a circle with radius 10 cm as a function of the central angle θ (see the accompanying figure).

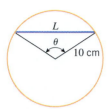

Figure Ex-8

9. As shown in the accompanying figure, a pendulum of constant length L makes an angle θ with its vertical position. Express the height h as a function of the angle θ.

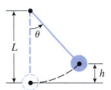

Figure Ex-9

10. A cup of hot coffee sits on a table. You pour in some cool milk and let it sit for an hour. Sketch a rough graph of the temperature of the coffee as a function of time.

11. A boat is bobbing up and down on some gentle waves. Suddenly it gets hit by a large wave and sinks. Sketch a rough graph of the height of the boat above the ocean floor as a function of time.

12. Make a rough sketch of your weight as a function of time from birth to the present.

In Exercises 13 and 14, express the function in piecewise form without using absolute values. [*Suggestion:* It may help to generate the graph of the function.]

 13. (a) $f(x) = |x| + 3x + 1$ (b) $g(x) = |x| + |x - 1|$
 14. (a) $f(x) = 3 + |2x - 5|$ (b) $g(x) = 3|x - 2| - |x + 1|$

15. As shown in the accompanying figure, an open box is to be constructed from a rectangular sheet of metal, 8 inches by 15 inches, by cutting out squares with sides of length x from each corner and bending up the sides.

(a) Express the volume V as a function of x.

(b) Find the natural domain and the range of the function, ignoring any physical restrictions on the values of the variables.

(c) Modify the domain and range appropriately to account for the physical restrictions on the values of V and x.

(d) In words, describe how the volume V of the box varies with x, and discuss how one might construct boxes of maximum volume and minimum volume.

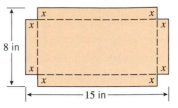

Figure Ex-15

16. As shown in the accompanying figure, a camera is mounted at a point 3000 ft from the base of a rocket launching pad. The shuttle rises vertically when launched, and the camera's elevation angle is continually adjusted to follow the bottom of the rocket.

(a) Choose letters to represent the height of the rocket and the elevation angle of the camera, and express the height as a function of the elevation angle.

(b) Find the natural domain and the range of the function, ignoring any physical restrictions on the values of the variables.

(c) Modify the domain and range appropriately to account for the physical restrictions on the values of the variables.

(d) Generate the graph of height versus the elevation on a graphing utility, and use it to estimate the height of the rocket when the elevation angle is $\pi/4 \approx 0.7854$ radian. Compare this estimate to the exact height. [*Suggestion:* If you are using a graphing calculator, the trace and zoom features will be helpful here.]

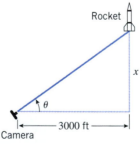

Figure Ex-16

In Exercises 17 and 18: (i) Explain why the function f has one or more holes in its graph, and state the x-values at which those holes occur. (ii) Find a function g whose graph is identical to that of f, but without the holes.

17. $f(x) = \dfrac{(x + 2)(x^2 - 1)}{(x + 2)(x - 1)}$ **18.** $f(x) = \dfrac{x + \sqrt{x}}{\sqrt{x}}$

19. For a given outside temperature T and wind speed v, the windchill index (WCI) is the equivalent temperature that exposed skin would feel with a wind speed of 4 mi/h. An empirical formula for the WCI (based on experience and observation) is

$$\text{WCI} = \begin{cases} T, & 0 \le v \le 4 \\ 91.4 + (91.4 - T)(0.0203v - 0.304\sqrt{v} - 0.474), & 4 < v < 45 \\ 1.6T - 55, & v \ge 45 \end{cases}$$

where T is the air temperature in $°F$, v is the wind speed in mi/h, and WCI is the equivalent temperature in $°F$. Find the WCI to the nearest degree if the air temperature is $25°F$ and

(a) $v = 3$ mi/h   (b) $v = 15$ mi/h
(c) $v = 46$ mi/h.
[Adapted from UMAP Module 658, *Windchill*, W. Bosch and L. Cobb, COMAP, Arlington, MA.]

In Exercises 20–22, use the formula for the windchill index described in Exercise 19.

20. Find the air temperature to the nearest degree if the WCI is reported as $-60°F$ with a wind speed of 48 mi/h.

21. Find the air temperature to the nearest degree if the WCI is reported as $-10°F$ with a wind speed of 8 mi/h.

22. Find the wind speed to the nearest mile per hour if the WCI is reported as $-15°F$ with an air temperature of $20°F$.

23. At 9:23 A.M. a lunar lander that is 1000 ft above the Moon's surface begins a vertical descent, touching down at 10:13 A.M. Assuming that the lander maintains a constant speed, find a function $D(t)$ that expresses the altitude of the lander above the Moon's surface as a function of t.

1.3 GRAPHING FUNCTIONS ON CALCULATORS AND COMPUTERS; COMPUTER ALGEBRA SYSTEMS

In this section we will discuss issues that relate to generating graphs of equations and functions with graphing utilities (graphing calculators and computers). Because graphing utilities vary widely, it is difficult to make general statements about them. Therefore, at various places in this section we will ask you to refer to the documentation for your own graphing utility for specific details about the way it operates.

GRAPHING CALCULATORS AND COMPUTER ALGEBRA SYSTEMS

The development of new technology has significantly changed how and where mathematicians, engineers, and scientists perform their work, as well as their approach to problem solving. Not only have portable computers and handheld calculators with graphing capabilities become standard tools in the scientific community, but there have been major new innovations in computer software. Among the most significant of these innovations are programs called *Computer Algebra Systems* (abbreviated CAS), the most common being *Mathematica*, *Maple*, and *Derive*.[*] Computer algebra systems not only have powerful graphing capabilities, but, as their name suggests, they can perform many of the symbolic computations that occur in algebra, calculus, and branches of higher mathematics. For example, it is a trivial task for a CAS to perform the factorization

$$x^6 + 23x^5 + 147x^4 - 139x^3 - 3464x^2 - 2112x + 23040 = (x + 5)(x - 3)^2(x + 8)^3$$

or the exact numerical computation

$$\left(\frac{63456}{3177295} - \frac{43907}{22854377} \right)^3 = \frac{225191245716420829125932023012 2866923}{38289595581936920444956594536 9203764688375}$$

Technology has also made it possible to generate graphs of equations and functions in seconds that in the past might have taken hours to produce. Graphing technology includes handheld graphing calculators, computer algebra systems, and software designed for that purpose. Figure 1.3.1 shows the graphs of the function $f(x) = x^4 - x^3 - 2x^2$ produced with various graphing utilities; the first two were generated with the CAS programs, *Mathematica* and *Maple*, and the third with a graphing calculator. Graphing calculators produce coarser graphs than most computer programs but have the advantage of being compact and portable.

---

[*] *Mathematica* is a product of Wolfram Research, Inc.; *Maple* is a product of Waterloo Maple Software, Inc.; and *Derive* is a product of Soft Warehouse, Inc.

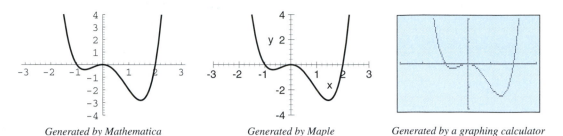

Generated by Mathematica   Generated by Maple   Generated by a graphing calculator

Figure 1.3.1

• •
VIEWING WINDOWS

Graphing utilities can only show a portion of the xy-plane in the viewing screen, so the first step in graphing an equation is to determine which rectangular portion of the xy-plane you want to display. This region is called the ***viewing window*** (or ***viewing rectangle***). For example, in Figure 1.3.1 the viewing window extends over the interval $[-3, 3]$ in the x-direction and over the interval $[-4, 4]$ in the y-direction, so we say that the viewing window is $[-3, 3] \times [-4, 4]$ (read "$[-3, 3]$ by $[-4, 4]$"). In general, if the viewing window is $[a, b] \times [c, d]$, then the window extends between $x = a$ and $x = b$ in the x-direction and between $y = c$ and $y = d$ in the y-direction. We will call $[a, b]$ the ***x-interval*** for the window and $[c, d]$ the ***y-interval*** for the window (Figure 1.3.2).

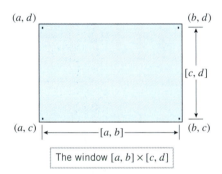

The window $[a, b] \times [c, d]$

Figure 1.3.2

Different graphing utilities designate viewing windows in different ways. For example, the first two graphs in Figure 1.3.1 were produced by the commands

```
Plot[x^4 - x^3 -2*x^2, {x, -3, 3}, PlotRange->{-4, 4}]
```
(*Mathematica*)
```
plot( x^4 - x^3 -2*x^2, x = -3..3, y = -4..4);
```
(*Maple*)

and the last graph was produced on a graphing calculator by pressing the GRAPH button after setting the following values for the variables that determine the x-interval and y-interval:

$$x\text{Min} = -3, \quad x\text{Max} = 3, \quad y\text{Min} = -4, \quad y\text{Max} = 4$$

FOR THE READER.   Use your own graphing utility to generate the graph of the function $f(x) = x^4 - x^3 - 2x^2$ in the window $[-3, 3] \times [-4, 4]$.

• •
TICK MARKS AND GRID LINES

To help locate points in a viewing window visually, graphing utilities provide methods for drawing ***tick marks*** (also called ***scale marks***) on the coordinate axes or at other locations in the viewing window. With computer programs such as *Mathematica* and *Maple*, there are specific commands for designating the spacing between tick marks, but if the user does not specify the spacing, then the programs make certain *default* choices. For example, in the first two parts of Figure 1.3.1, the tick marks shown were the default choices.

On some graphing calculators the spacing between tick marks is determined by two **scale variables** (also called **scale factors**), which we will denote by

$$x\text{Scl} \quad \text{and} \quad y\text{Scl}$$

(The notation varies among calculators.) These variables specify the spacing between the tick marks in the x- and y-directions, respectively. For example, in the third part of Figure 1.3.1 the window and tick marks were designated by the settings

$$x\text{Min} = -3 \qquad x\text{Max} = 3$$
$$y\text{Min} = -4 \qquad y\text{Max} = 4$$
$$x\text{Scl} = 1 \qquad y\text{Scl} = 1$$

Most graphing utilities allow for variations in the design and positioning of tick marks. For example, Figure 1.3.3 shows two variations of the graphs in Figure 1.3.1; the first was generated on a computer using an option for placing the ticks and numbers on the edges of a box, and the second was generated on a graphing calculator using an option for drawing grid lines to simulate graph paper.

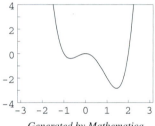 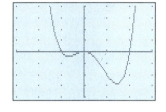

Generated by Mathematica   *Generated by a graphing calculator*

Figure 1.3.3

Example 1 Figure 1.3.4a shows the window $[-5, 5] \times [-5, 5]$ with the tick marks spaced 0.5 unit apart in the x-direction and 10 units apart in the y-direction. Note that no tick marks are actually visible in the y-direction because the tick mark at the origin is covered by the x-axis, and all other tick marks in the y-direction fall outside of the viewing window. ◄

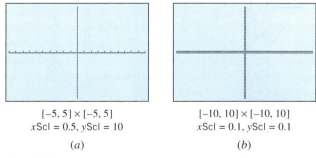

$[-5, 5] \times [-5, 5]$ $[-10, 10] \times [-10, 10]$
$x\text{Scl} = 0.5, y\text{Scl} = 10$ $x\text{Scl} = 0.1, y\text{Scl} = 0.1$

(a) (b)

Figure 1.3.4

Example 2 Figure 1.3.4b shows the window $[-10, 10] \times [-10, 10]$ with the tick marks spaced 0.1 unit apart in the x- and y-directions. In this case the tick marks are so close together that they create the effect of thick lines on the coordinate axes. When this occurs you will usually want to increase the scale factors to reduce the number of tick marks and make them legible. ◄

FOR THE READER. Graphing calculators provide a way of clearing all settings and returning them to *default values*. For example, on one calculator the default window is $[-10, 10] \times [-10, 10]$ and the default scale factors are $x\text{Scl} = 1$ and $y\text{Scl} = 1$. Check your documentation to determine the default values for your calculator and how to reset the calculator to its default configuration. If you are using a computer program, check your documentation to determine the commands for specifying the spacing between tick marks.

CHOOSING A VIEWING WINDOW

When the graph of a function extends indefinitely in some direction, no single viewing window can show the entire graph. In such cases the choice of the viewing window can drastically affect one's perception of how the graph looks. For example, Figure 1.3.5 shows a computer-generated graph of $y = 9 - x^2$, and Figure 1.3.6 shows four views of this graph generated on a calculator.

- In part (*a*) the graph falls completely outside of the window, so the window is blank (except for the ticks and axes).

- In part (*b*) the graph is broken into two pieces because it passes in and out of the window.

- In part (*c*) the graph appears to be a straight line because we have zoomed in on such a small segment of the curve.

- In part (*d*) we have a more complete picture of the graph shape because the window encompasses all of the important points, namely the high point on the graph and the intersections with the x-axis.

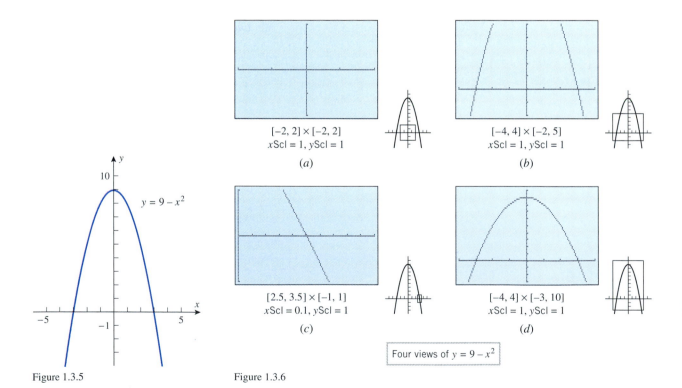

$[-2, 2] \times [-2, 2]$
$x\text{Scl} = 1, y\text{Scl} = 1$

(*a*)

$[-4, 4] \times [-2, 5]$
$x\text{Scl} = 1, y\text{Scl} = 1$

(*b*)

$[2.5, 3.5] \times [-1, 1]$
$x\text{Scl} = 0.1, y\text{Scl} = 1$

(*c*)

$[-4, 4] \times [-3, 10]$
$x\text{Scl} = 1, y\text{Scl} = 1$

(*d*)

Four views of $y = 9 - x^2$

$y = 9 - x^2$

Figure 1.3.5

Figure 1.3.6

For a function whose graph does not extend indefinitely in either the x- or y-directions, the domain and range of the function can be used to obtain a viewing window that contains the entire graph.

Example 3   Use the domain and range of the function $f(x) = \sqrt{12 - 3x^2}$ to determine a viewing window that contains the entire graph.

Solution.   The natural domain of f is $[-2, 2]$ and the range is $[0, \sqrt{12}]$ (verify), so the entire graph will be contained in the viewing window $[-2, 2] \times [0, \sqrt{12}]$. For clarity, it is desirable to use a slightly larger window to avoid having the graph too close to the edges of the screen. For example, taking the viewing window to be $[-3, 3] \times [-1, 4]$ yields the graph in Figure 1.3.7.   ◀

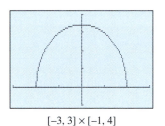

$[-3, 3] \times [-1, 4]$
xScl $= 1$, yScl $= 1$

Figure 1.3.7

If the graph of f extends indefinitely in either the x- or y-direction, then it will not be possible to show the entire graph in any one viewing window. In such cases one tries to choose the window to show all of the important features for the problem at hand. (Of course, what is important in one problem may not be important in another, so the choice of the viewing window will often depend on the objectives in the problem.)

Example 4   Graph the equation $y = x^3 - 12x^2 + 18$ in the following windows and discuss the advantages and disadvantages of each window.

(a) $[-10, 10] \times [-10, 10]$ with xScl $= 1$, yScl $= 1$
(b) $[-20, 20] \times [-20, 20]$ with xScl $= 1$, yScl $= 1$
(c) $[-20, 20] \times [-300, 20]$ with xScl $= 1$, yScl $= 20$
(d) $[-5, 15] \times [-300, 20]$ with xScl $= 1$, yScl $= 20$
(e) $[1, 2] \times [-1, 1]$ with xScl $= 0.1$, yScl $= 0.1$

Solution (a).   The window in Figure 1.3.8a has chopped off the portion of the graph that intersects the y-axis, and it shows only two of three possible real roots for the given cubic polynomial. To remedy these problems we need to widen the window in both the x- and y-directions.

Solution (b).   The window in Figure 1.3.8b shows the intersection of the graph with the y-axis and the three real roots, but it has chopped off the portion of the graph between the two positive roots. Moreover, the ticks in the y-direction are nearly illegible because they are so close together. We need to extend the window in the negative y-direction and increase yScl. We do not know how far to extend the window, so some experimentation will be required to obtain what we want.

Solution (c).   The window in Figure 1.3.8c shows all of the main features of the graph. However, we have some wasted space in the x-direction. We can improve the picture by shortening the window in the x-direction appropriately.

Solution (d).   The window in Figure 1.3.8d shows all of the main features of the graph without a lot of wasted space. However, the window does not provide a clear view of the roots. To get a closer view of the roots we must forget about showing all of the main features of the graph and choose windows that zoom in on the roots themselves.

Solution (e).   The window in Figure 1.3.8e displays very little of the graph, but it clearly shows that the root in the interval $[1, 2]$ is approximately 1.3.   ◀

FOR THE READER.   Sometimes you will want to determine the viewing window by choosing the x-interval for the window and allowing the graphing utility to determine a y-interval that encompasses the maximum and minimum values of the function over the x-interval. Most graphing utilities provide some method for doing this, so check your documentation to determine how to use this feature. Allowing the graphing utility to determine the y-interval of the window takes some of the guesswork out of problems like that in part (b) of the preceding example.

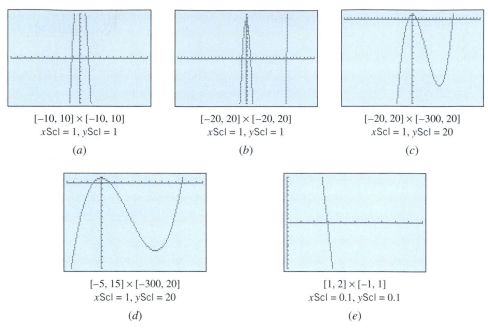

$$[-10, 10] \times [-10, 10]$$
$$x\text{Scl} = 1, y\text{Scl} = 1$$
$$(a)$$

$$[-20, 20] \times [-20, 20]$$
$$x\text{Scl} = 1, y\text{Scl} = 1$$
$$(b)$$

$$[-20, 20] \times [-300, 20]$$
$$x\text{Scl} = 1, y\text{Scl} = 20$$
$$(c)$$

$$[-5, 15] \times [-300, 20]$$
$$x\text{Scl} = 1, y\text{Scl} = 20$$
$$(d)$$

$$[1, 2] \times [-1, 1]$$
$$x\text{Scl} = 0.1, y\text{Scl} = 0.1$$
$$(e)$$

Figure 1.3.8

ZOOMING

The process of enlarging or reducing the size of a viewing window is called *zooming*. If you reduce the size of the window, you see less of the graph as a whole, but more detail of the part shown; this is called *zooming in*. In contrast, if you enlarge the size of the window, you see more of the graph as a whole, but less detail of the part shown; this is called *zooming out*. Most graphing calculators provide menu items for zooming in or zooming out by fixed factors. For example, on one calculator the amount of enlargement or reduction is controlled by setting values for two *zoom factors*, xFact and yFact. If

$$x\text{Fact} = 10 \quad \text{and} \quad y\text{Fact} = 5$$

then each time a zoom command is executed the viewing window is enlarged or reduced by a factor of 10 in the x-direction and a factor of 5 in the y-direction. With computer programs such as *Mathematica* and *Maple*, zooming is controlled by adjusting the x-interval and y-interval directly; however, there are ways to automate this by programming.

FOR THE READER. If you are using a graphing calculator, read your documentation to determine how to use the zooming feature.

COMPRESSION

Enlarging the viewing window for a graph has the geometric effect of compressing the graph, since more of the graph is packed into the calculator screen. If the compression is sufficiently great, then some of the detail in the graph may be lost. Thus, the choice of the viewing window frequently depends on whether you want to see more of the graph or more of the detail. Figure 1.3.9 shows two views of the equation

$$y = x^5(x - 2)$$

In part (a) of the figure the y-interval is very large, resulting in a vertical compression that obscures the detail in the vicinity of the x-axis. In part (b) the y-interval is smaller, and consequently we see more of the detail in the vicinity of the x-axis but less of the graph in the y-direction.

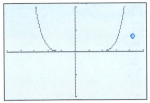

$[-5, 5] \times [-1000, 1000]$
xScl $= 1$, yScl $= 500$

(a)

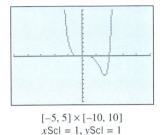

$[-5, 5] \times [-10, 10]$
xScl $= 1$, yScl $= 1$

(b)

Figure 1.3.9

Example 5 Describe the graph of the function $f(x) = x + 0.01 \sin 50\pi x$; then graph the function in the following windows and explain why the graphs do or do not differ from your description.

(a) $[-10, 10] \times [-10, 10]$ (b) $[-1, 1] \times [-1, 1]$

(c) $[-0.1, 0.1] \times [-0.1, 0.1]$ (d) $[-0.01, 0.01] \times [-0.01, 0.01]$

Solution. The formula for f is the sum of the function x (whose graph is a straight line) and the function $0.01 \sin 50\pi x$ (whose graph is a sinusoidal curve with an amplitude of 0.01 and a period of $2\pi/50\pi = 0.04$). Intuitively, this suggests that the graph of f will follow the general path of the line $y = x$ but will have small bumps resulting from the contributions of the sinusoidal oscillations.

To generate the four graphs, we first set the calculator to the radian mode.[*] Because the windows in successive parts of this example are decreasing in size by a factor of 10, it will be convenient to use the zoom in feature of the calculator with the zoom factors set to 10 in the x- and y-directions. In Figure 1.3.10a the graph appears to be a straight line because compression has hidden the small sinusoidal oscillations. (Keep in mind that the amplitude of the sinusoidal portion of the function is only 0.01.) In part (b) the oscillations have begun to appear since the y-interval has been reduced, and in part (c) the oscillations have become very clear because the vertical scale is more in keeping with the amplitude of the oscillations. In part (d) the graph appears to be a line segment because we have zoomed in on such a small portion of the curve. ◀

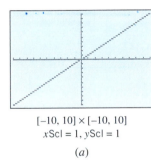

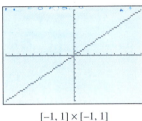

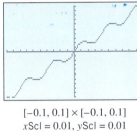

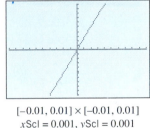

$[-10, 10] \times [-10, 10]$
xScl $= 1$, yScl $= 1$

(a)

$[-1, 1] \times [-1, 1]$
xScl $= 0.1$, yScl $= 0.1$

(b)

$[-0.1, 0.1] \times [-0.1, 0.1]$
xScl $= 0.01$, yScl $= 0.01$

(c)

$[-0.01, 0.01] \times [-0.01, 0.01]$
xScl $= 0.001$, yScl $= 0.001$

(d)

Figure 1.3.10

ASPECT RATIO DISTORTION

Figure 1.3.11a shows a circle of radius 5 and two perpendicular lines graphed in the window $[-10, 10] \times [-10, 10]$ with xScl $= 1$ and yScl $= 1$. However, the circle is distorted and the lines do not appear perpendicular because the calculator has not used the same length for 1 unit on the x-axis and 1 unit on the y-axis. (Compare the spacing between the ticks on the axes.) This is called ***aspect ratio distortion***. Many calculators provide a menu item for automatically correcting the distortion by adjusting the viewing window appropriately. For example, one calculator makes this correction to the viewing window $[-10, 10] \times [-10, 10]$ by changing it to

$$[-16.9970674487, 16.9970674487] \times [-10, 10]$$

(Figure 1.3.11b). With computer programs such as *Mathematica* and *Maple*, aspect ratio distortion is controlled with adjustments to the physical dimensions of the viewing window on the computer screen, rather than altering the x- and y-intervals of the viewing window.

FOR THE READER. Read the documentation for your graphing utility to determine how to control aspect ratio distortion.

---

[*]In this text we follow the convention that angles are measured in radians unless degree measure is specified.

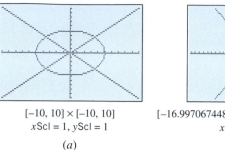 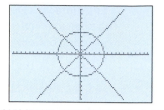

$$[-10, 10] \times [-10, 10]$$
$$x\text{Scl} = 1, y\text{Scl} = 1$$

$$[-16.9970674487, 16.9970674487] \times [-10, 10]$$
$$x\text{Scl} = 1, y\text{Scl} = 1$$

(a) (b)

Figure 1.3.11

PIXELS AND RESOLUTION

Sometimes graphing utilities produce unexpected results. For example, Figure 1.3.12 shows the graph of $y = \cos(10\pi x)$, which was generated on a graphing calculator in four different windows. (Your own calculator may produce different results.) The first graph has the correct shape, but the remaining three do not. To explain what is happening here we need to understand more precisely how graphing utilities generate graphs.

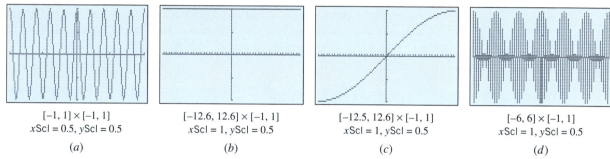

$$[-1, 1] \times [-1, 1]$$
$$x\text{Scl} = 0.5, y\text{Scl} = 0.5$$

$$[-12.6, 12.6] \times [-1, 1]$$
$$x\text{Scl} = 1, y\text{Scl} = 0.5$$

$$[-12.5, 12.6] \times [-1, 1]$$
$$x\text{Scl} = 1, y\text{Scl} = 0.5$$

$$[-6, 6] \times [-1, 1]$$
$$x\text{Scl} = 1, y\text{Scl} = 0.5$$

(a) (b) (c) (d)

Figure 1.3.12

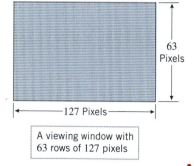

63 Pixels

127 Pixels

A viewing window with 63 rows of 127 pixels

Figure 1.3.13

An image on a graphing utility screen is made by activating a collection of rectangular dots called **pixels**. For black-and-white displays each pixel has two possible states—an activated (or dark) state and a deactivated (or light) state. A display's pixels are organized into a rectangular grid; for example, one calculator screen's pixels are arrayed in 63 rows, each containing 127 pixels (Figure 1.3.13). Since graphical elements are produced by activating pixels, a screen with more pixels can show images in finer detail. For example, a computer screen may have a **resolution** of 1600×1200 (1600 pixels per row, arrayed in 1200 rows), so the computer screen is capable of displaying much smoother graphs than the calculator.

FOR THE READER.   If you are using a graphing calculator, check the documentation to determine its resolution.

SAMPLING ERROR

The procedure that a graphing utility follows to generate a graph is similar to the procedure for plotting points by hand. When a viewing window is selected and an equation is entered, the graphing utility determines the x-coordinates of certain pixels on the x-axis and computes the corresponding points (x, y) on the graph. It then activates the pixels whose coordinates most closely match those of the calculated points and uses some built-in algorithm to activate additional intermediate pixels to create the curve shape. The point to keep in mind here is that *changing the window changes the points plotted by the graphing utility*. Thus, it is possible that a particular window will produce a false impression about the graph shape because significant characteristics of the graph occur *between* the plotted pixels. This is called **sampling error**. This is exactly what occurred in Figure 1.3.12 when we graphed $y = \cos(10\pi x)$. In part (b) of the figure the plotted pixels happened to fall at the peaks of the cosine curve, giving the false impression that the graph is a horizontal line at $y = 1$. In part (c) the plotted pixels fell at successively higher points along the graph, and in part (d) the plotted pixels fell in a strange way that created yet another misleading impression of the graph shape.

REMARK. Figure 1.3.12 suggests that for trigonometric graphs with rapid oscillations, restricting the x-interval to a few periods is likely to produce a more accurate representation about the graph shape.

FALSE GAPS

Sometimes graphs that are continuous appear to have gaps when they are generated on a calculator. These *false gaps* typically occur where the graph rises so rapidly that vertical space is opened up between successive pixels.

Example 6 Figure 1.3.14 shows the graph of the semicircle $y = \sqrt{9 - x^2}$ in two viewing windows. Although this semicircle has x-intercepts at the points $x = \pm 3$, part (a) of the figure shows false gaps at those points because there are no pixels with x-coordinates ± 3 in the window selected. In part (b) no gaps occur because there are pixels with x-coordinates $x = \pm 3$ in the window being used. ◄

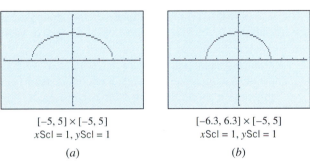

$[-5, 5] \times [-5, 5]$
$x\text{Scl} = 1, y\text{Scl} = 1$

(a)

$[-6.3, 6.3] \times [-5, 5]$
$x\text{Scl} = 1, y\text{Scl} = 1$

(b)

Figure 1.3.14

FALSE LINE SEGMENTS

In addition to creating false gaps in continuous graphs, calculators can err in the opposite direction by placing *false line segments* in the gaps of discontinuous curves.

Example 7 Figure 1.3.15a shows the graph of $y = 1/(x - 1)$ in the default window on a calculator. Although the graph appears to contain vertical line segments near $x = 1$, they should not be there. There is actually a gap in the curve at $x = 1$, since a division by zero occurs at that point (Figure 1.3.15b). ◄

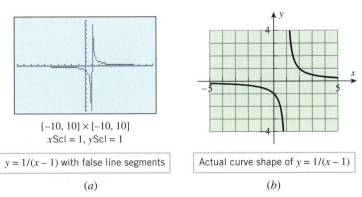

$[-10, 10] \times [-10, 10]$
$x\text{Scl} = 1, y\text{Scl} = 1$

$y = 1/(x - 1)$ with false line segments

(a)

Actual curve shape of $y = 1/(x - 1)$

(b)

Figure 1.3.15

....
ERRORS OF OMISSION

Most graphing utilities use logarithms to evaluate functions with fractional exponents such as $f(x) = x^{2/3} = \sqrt[3]{x^2}$. However, because logarithms are only defined for positive numbers, many (but not all) graphing utilities will omit portions of the graphs of functions with fractional exponents. For example, one calculator graphs $y = x^{2/3}$ as in Figure 1.3.16a, whereas the actual graph is as in Figure 1.3.16b. (See the discussion preceding Exercise 29 for a way of circumventing this problem.)

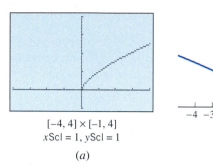

$[-4, 4] \times [-1, 4]$
xScl = 1, yScl = 1

(a)   (b)

Figure 1.3.16

FOR THE READER.   Determine whether your graphing utility produces the graph of the equation $y = x^{2/3}$ for both positive and negative values of x.

....
WHAT IS THE TRUE SHAPE OF A GRAPH?

Although graphing utilities are powerful tools for generating graphs quickly, they can produce misleading graphs as a result of compression, sampling error, false gaps, and false line segments. In short, *graphing utilities can suggest graph shapes, but they cannot establish them with certainty.* Thus, the more you know about the functions you are graphing, the easier it will be to choose good viewing windows, and the better you will be able to judge the reasonableness of the results produced by your graphing utility.

....
MORE INFORMATION ON GRAPHING AND CALCULATING UTILITIES

The main source of information about your graphing utility is its own documentation, and from time to time we will suggest that you refer to that documentation to learn some particular technique.

EXERCISE SET 1.3   ∿ Graphing Utility
....

1. Use a graphing utility to generate the graph of the function $f(x) = x^4 - x^2$ in the given viewing windows, and specify the window that you think gives the best view of the graph.
   (a) $[-50, 50] \times [-50, 50]$ (b) $[-5, 5] \times [-5, 5]$
   (c) $[-2, 2] \times [-2, 2]$ (d) $[-2, 2] \times [-1, 1]$
   (e) $[-1.5, 1.5] \times [-0.5, 0.5]$

2. Use a graphing utility to generate the graph of the function $f(x) = x^5 - x^3$ in the given viewing windows, and specify the window that you think gives the best view of the graph.
   (a) $[-50, 50] \times [-50, 50]$ (b) $[-5, 5] \times [-5, 5]$
   (c) $[-2, 2] \times [-2, 2]$ (d) $[-2, 2] \times [-1, 1]$
   (e) $[-1.5, 1.5] \times [-0.5, 0.5]$

3. Use a graphing utility to generate the graph of the function $f(x) = x^2 + 12$ in the given viewing windows, and specify

the window that you think gives the best view of the graph.
   (a) $[-1, 1] \times [13, 15]$ (b) $[-2, 2] \times [11, 15]$
   (c) $[-4, 4] \times [10, 28]$ (d) A window of your choice

4. Use a graphing utility to generate the graph of the function $f(x) = -12 - x^2$ in the given viewing windows, and specify the window that you think gives the best view of the graph.
   (a) $[-1, 1] \times [-15, -13]$ (b) $[-2, 2] \times [-15, -11]$
   (c) $[-4, 4] \times [-28, -10]$ (d) A window of your choice

In Exercises 5 and 6, use the domain and range of f to determine a viewing window that contains the entire graph, and generate the graph in that window.

5. $f(x) = \sqrt{16 - 2x^2}$ **6.** $f(x) = \sqrt{3 - 2x - x^2}$

7. Graph the function $f(x) = x^3 - 15x^2 - 3x + 45$ using the stated windows and tick spacing, and discuss the advantages and disadvantages of each window.
  - (a) $[-10, 10] \times [-10, 10]$ with $x\,\mathrm{Scl} = 1$ and $y\,\mathrm{Scl} = 1$
  - (b) $[-20, 20] \times [-20, 20]$ with $x\,\mathrm{Scl} = 1$ and $y\,\mathrm{Scl} = 1$
  - (c) $[-5, 20] \times [-500, 50]$ with $x\,\mathrm{Scl} = 5$ and $y\,\mathrm{Scl} = 50$
  - (d) $[-2, -1] \times [-1, 1]$ with $x\,\mathrm{Scl} = 0.1$ and $y\,\mathrm{Scl} = 0.1$
  - (e) $[9, 11] \times [-486, -484]$
    with $x\,\mathrm{Scl} = 0.1$ and $y\,\mathrm{Scl} = 0.1$

8. Graph the function $f(x) = -x^3 - 12x^2 + 4x + 48$ using the stated windows and tick spacing, and discuss the advantages and disadvantages of each window.
  - (a) $[-10, 10] \times [-10, 10]$ with $x\,\mathrm{Scl} = 1$ and $y\,\mathrm{Scl} = 1$
  - (b) $[-20, 20] \times [-20, 20]$ with $x\,\mathrm{Scl} = 1$ and $y\,\mathrm{Scl} = 1$
  - (c) $[-16, 4] \times [-250, 50]$ with $x\,\mathrm{Scl} = 2$ and $y\,\mathrm{Scl} = 25$
  - (d) $[-3, -1] \times [-1, 1]$ with $x\,\mathrm{Scl} = 0.1$ and $y\,\mathrm{Scl} = 0.1$
  - (e) $[-9, -7] \times [-241, -239]$
    with $x\,\mathrm{Scl} = 0.1$ and $y\,\mathrm{Scl} = 0.1$

In Exercises 9–16, generate the graph of f in a viewing window that you think is appropriate.

9. $f(x) = x^2 - 9x - 36$ **10.** $f(x) = \dfrac{x + 7}{x - 9}$

11. $f(x) = 2\cos 80x$ **12.** $f(x) = 12\sin(x/80)$

13. $f(x) = 300 - 10x^2 + 0.01x^3$

14. $f(x) = x(30 - 2x)(25 - 2x)$

15. $f(x) = x^2 + \dfrac{1}{x}$ **16.** $f(x) = \sqrt{11x - 18}$

In Exercises 17 and 18, generate the graph of f and determine whether your graphs contain false line segments. Sketch the actual graph and see if you can make the false line segments disappear by changing the viewing window.

17. $f(x) = \dfrac{x}{x^2 - 1}$ **18.** $f(x) = \dfrac{x^2}{4 - x^2}$

19. The graph of the equation $x^2 + y^2 = 16$ is a circle of radius 4 centered at the origin.
  - (a) Find a function whose graph is the upper semicircle and graph it.
  - (b) Find a function whose graph is the lower semicircle and graph it.
  - (c) Graph the upper and lower semicircles together. If the combined graphs do not appear circular, see if you can adjust the viewing window to eliminate the aspect ratio distortion.
  - (d) Graph the portion of the circle in the first quadrant.
  - (e) Is there a function whose graph is the right half of the circle? Explain.

20. In each part, graph the equation by solving for y in terms of x and graphing the resulting functions together.
  - (a) $x^2/4 + y^2/9 = 1$ (b) $y^2 - x^2 = 1$

21. Read the documentation for your graphing utility to determine how to graph functions involving absolute values, and graph the given equation.
  - (a) $y = |x|$ (b) $y = |x - 1|$
  - (c) $y = |x| - 1$ (d) $y = |\sin x|$
  - (e) $y = \sin|x|$ (f) $y = |x| - |x + 1|$

22. Based on your knowledge of the absolute value function, sketch the graph of $f(x) = |x|/x$. Check your result using a graphing utility.

23. Make a conjecture about the relationship between the graph of $y = f(x)$ and the graph of $y = |f(x)|$; check your conjecture with some specific functions.

24. Make a conjecture about the relationship between the graph of $y = f(x)$ and the graph of $y = f(|x|)$; check your conjecture with some specific functions.

25.
  - (a) Based on your knowledge of the absolute value function, sketch the graph of $y = |x - a|$, where a is a constant. Check your result using a graphing utility and some specific values of a.
  - (b) Sketch the graph of $y = |x - 1| + |x - 2|$; check your result with a graphing utility.

26. How are the graphs of $y = |x|$ and $y = \sqrt{x^2}$ related? Check your answer with a graphing utility.

Most graphing utilities provide some way of graphing functions that are defined piecewise; read the documentation for your graphing utility to find out how to do this. However, if your goal is just to find the general shape of the graph, you can graph each portion of the function separately and combine the pieces with a hand-drawn sketch. Use this method in Exercises 27 and 28.

27. Draw the graph of
$$f(x) = \begin{cases} \sqrt[3]{x - 2}, & x \le 2 \\ x^3 - 2x - 4, & x > 2 \end{cases}$$

28. Draw the graph of
$$f(x) = \begin{cases} x^3 - x^2, & x \le 1 \\ \dfrac{1}{1 - x}, & 1 < x < 4 \\ x^2 \cos\sqrt{x}, & 4 \le x \end{cases}$$

We noted in the text that for functions involving fractional exponents (or radicals), graphing utilities sometimes omit portions of the graph. If $f(x) = x^{p/q}$, where p/q is a positive fraction in *lowest terms*, then you can circumvent this problem as follows:
- If p is even and q is odd, then graph $g(x) = |x|^{p/q}$ instead of $f(x)$.
- If p is odd and q is odd, then graph $g(x) = (|x|/x)|x|^{p/q}$ instead of $f(x)$.

We will explain why this works in the exercises of the next section.

29. (a) Generate the graphs of $f(x) = x^{2/5}$ and $g(x) = |x|^{2/5}$, and determine whether your graphing utility missed part of the graph of f.

(b) Generate the graphs of the functions $f(x) = x^{1/5}$ and $g(x) = (|x|/x)|x|^{1/5}$, and then determine whether your graphing utility missed part of the graph of f.

(c) Generate a graph of the function $f(x) = (x-1)^{4/5}$ that shows all of its important features.

(d) Generate a graph of the function $f(x) = (x+1)^{3/4}$ that shows all of its important features.

30. The graphs of $y = (x^2 - 4)^{2/3}$ and $y = [(x^2 - 4)^2]^{1/3}$ should be the same. Does your graphing utility produce the same graph for both equations? If not, what do you think is happening?

31. In each part, graph the function for various values of c, and write a paragraph or two that describes how changes in c affect the graph in each case.

(a) $y = cx^2$

(b) $y = x^2 + cx$

(c) $y = x^2 + x + c$

32. The graph of an equation of the form $y^2 = x(x-a)(x-b)$ (where $0 < a < b$) is called a **bipartite cubic**. The accompanying figure shows a typical graph of this type.

(a) Graph the bipartite cubic $y^2 = x(x-1)(x-2)$ by solving for y in terms of x and graphing the two resulting functions.

(b) Find the x-intercepts of the bipartite cubic
$$y^2 = x(x-a)(x-b)$$
and make a conjecture about how changes in the values of a and b would affect the graph. Test your conjecture by graphing the bipartite cubic for various values of a and b.

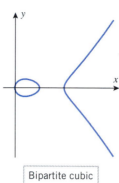

Bipartite cubic

Figure Ex-32

33. Based on your knowledge of the graphs of $y = x$ and $y = \sin x$, make a sketch of the graph of $y = x \sin x$. Check your conclusion using a graphing utility.

34. What do you think the graph of $y = \sin(1/x)$ looks like? Test your conclusion using a graphing utility. [*Suggestion:* Examine the graph on a succession of smaller and smaller intervals centered at $x = 0$.]

1.4 NEW FUNCTIONS FROM OLD

Just as numbers can be added, subtracted, multiplied, and divided to produce other numbers, so functions can be added, subtracted, multiplied, and divided to produce other functions. In this section we will discuss these operations and some others that have no analogs in ordinary arithmetic.

ARITHMETIC OPERATIONS ON FUNCTIONS

Two functions, f and g, can be added, subtracted, multiplied, and divided in a natural way to form new functions $f + g$, $f - g$, fg, and f/g. For example, $f + g$ is defined by the formula

$$(f + g)(x) = f(x) + g(x) \tag{1}$$

which states that for each input the value of $f + g$ is obtained by adding the values of f and g. For example, if

$$f(x) = x \quad \text{and} \quad g(x) = x^2$$

then

$$(f + g)(x) = f(x) + g(x) = x + x^2$$

Equation (1) provides a formula for $f + g$ but does not say anything about the domain of $f + g$. However, for the right side of this equation to be defined, x must lie in the domain of f *and* in the domain of g, so we define the domain of $f + g$ to be the intersection of those two domains. More generally, we make the following definition:

1.4.1   DEFINITION.   Given functions f and g, we define

$(f + g)(x) = f(x) + g(x)$

$(f - g)(x) = f(x) - g(x)$

$(fg)(x)\quad = f(x)g(x)$

$(f/g)(x)\quad = f(x)/g(x)$

For the functions $f + g$, $f - g$, and fg we define the domain to be the intersection of the domains of f and g, and for the function f/g we define the domain to be the intersection of the domains of f and g but with the points where $g(x) = 0$ excluded (to avoid division by zero).

REMARK.   If f is a constant function, say $f(x) = c$ for all x, then the product of f and g is cg, so multiplying a function by a constant is a special case of multiplying two functions.

Example 1   Let

$$f(x) = 1 + \sqrt{x - 2} \quad \text{and} \quad g(x) = x - 3$$

Find $(f + g)(x)$, $(f - g)(x)$, $(fg)(x)$, $(f/g)(x)$, and $(7f)(x)$; state the domains of $f + g$, $f - g$, fg, f/g, and $7f$.

Solution.   First, we will find formulas for the functions and then the domains. The formulas are

$$(f + g)(x) = f(x) + g(x) = (1 + \sqrt{x - 2}) + (x - 3) = x - 2 + \sqrt{x - 2} \tag{2}$$

$$(f - g)(x) = f(x) - g(x) = (1 + \sqrt{x - 2}) - (x - 3) = 4 - x + \sqrt{x - 2} \tag{3}$$

$$(fg)(x) = f(x)g(x) = (1 + \sqrt{x - 2})(x - 3) \tag{4}$$

$$(f/g)(x) = f(x)/g(x) = \frac{1 + \sqrt{x - 2}}{x - 3} \tag{5}$$

$$(7f)(x) = 7f(x) = 7 + 7\sqrt{x - 2} \tag{6}$$

In all five cases the natural domain determined by the formula is the same as the domain specified in Definition 1.4.1, so there is no need to state the domain explicitly in any of these cases. For example, the domain of f is $[2, +\infty)$, the domain of g is $(-\infty, +\infty)$, and the natural domain for $f(x) + g(x)$ determined by Formula (2) is $[2, +\infty)$, which is precisely the intersection of the domains of f and g.   ◀

REMARK.   There are situations in which the natural domain associated with the formula resulting from an operation on two functions is not the correct domain for the new function. For example, if $f(x) = \sqrt{x}$ and $g(x) = \sqrt{x}$, then according to Definition 1.4.1 the domain of fg should be $[0, +\infty) \cap [0, +\infty) = [0, +\infty)$. However, $(fg)(x) = \sqrt{x}\sqrt{x} = x$, which has a natural domain of $(-\infty, +\infty)$. Thus, to be precise in describing the formula for fg, we must write $(fg)(x) = x$, $x \geq 0$.

STRETCHES AND COMPRESSIONS

Multiplying a function f by a *positive* constant c has the geometric effect of stretching or compressing the graph of $y = f(x)$ in the y-direction. For example, examine the graphs of $y = f(x)$, $y = 2f(x)$, and $y = \frac{1}{2}f(x)$ shown in Figure 1.4.1a. Multiplying by 2 doubles each y-coordinate, thereby stretching the graph, and multiplying by $\frac{1}{2}$ cuts each y-coordinate in half, thereby compressing the graph. In general, if $c > 0$, then the graph of $y = cf(x)$ can be obtained from the graph of $y = f(x)$ by compressing the graph of $y = f(x)$ vertically by a factor of $1/c$ if $0 < c < 1$, or stretching it by a factor of c if $c > 1$.

Analogously, multiplying x by a ***positive*** constant c has the geometric effect of stretching or compressing the graph of $y = f(x)$ in the x-direction. For example, examine the graphs

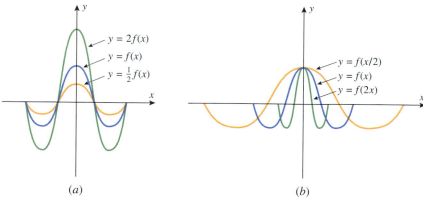

Figure 1.4.1

of $y = f(x)$, $y = f(2x)$, and $y = f(x/2)$ shown in Figure 1.4.1*b*. Multiplying x by 2 compresses the graph by a factor of 2 and multiplying x by $\frac{1}{2}$ stretches the graph by a factor of 2. [This is a little confusing, but think of it this way: The value of $2x$ changes twice as fast as the value of x, so a point moving along the x-axis will only have to move half as far from the origin for $y = f(2x)$ to have the same value as $y = f(x)$.] In general, if $c > 0$, then the graph of $y = f(cx)$ can be obtained from the graph of $y = f(x)$ by stretching the graph of $y = f(x)$ horizontally by a factor of $1/c$ if $0 < c < 1$, or compressing it by a factor of c if $c > 1$.

SUMS OF FUNCTIONS

Car Sales in Millions

Adding two functions can be accomplished geometrically by adding the corresponding y-coordinates of their graphs. For example, Figure 1.4.2 shows line graphs of yearly new car sales $N(t)$ and used car sales $U(t)$ in the United States between 1985 and 1995. The sum of these functions, $T(t) = N(t) + U(t)$, represents the yearly total car sales for that period. As illustrated in the figure, the graph of $T(t)$ can be obtained by adding the values of $N(t)$ and $U(t)$ together at each time t and plotting the resulting value.

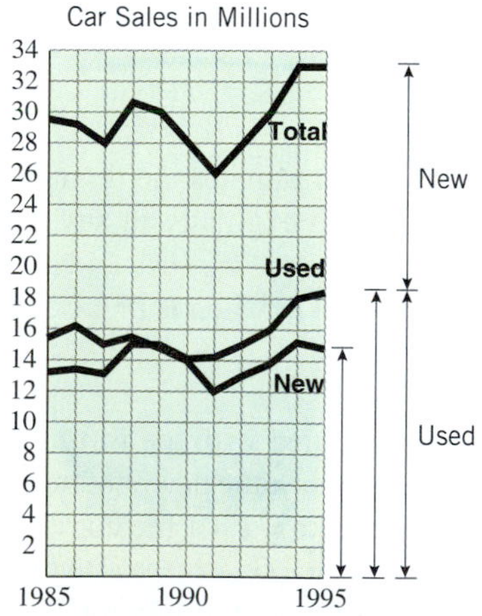

Example 2   Referring to Figure 1.2.1 for the graphs of $y = \sqrt{x}$ and $y = 1/x$, make a sketch that shows the general shape of the graph of $y = \sqrt{x} + 1/x$ for $x \geq 0$.

Solution.   To add the corresponding y-values of $y = \sqrt{x}$ and $y = 1/x$ graphically, just imagine them to be "stacked" on top of one another. This yields the sketch in Figure 1.4.3. ◀

Source: NADA.

Figure 1.4.2

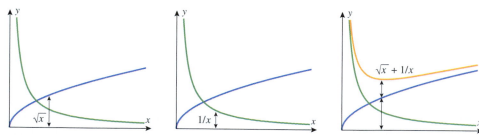

Figure 1.4.3

COMPOSITION OF FUNCTIONS

We now consider an operation on functions, called *composition*, which has no direct analog in ordinary arithmetic. Informally stated, the operation of composition is performed by substituting some function for the independent variable of another function. For example, suppose that

$$f(x) = x^2 \quad \text{and} \quad g(x) = x + 1$$

If we substitute $g(x)$ for x in the formula for f, we obtain a new function

$$f(g(x)) = (g(x))^2 = (x+1)^2$$

which we denote by $f \circ g$. Thus,

$$(f \circ g)(x) = f(g(x)) = (g(x))^2 = (x+1)^2$$

In general, we make the following definition.

1.4.2   DEFINITION.   Given functions f and g, the **composition** of f with g, denoted by $f \circ g$, is the function defined by

$$(f \circ g)(x) = f(g(x))$$

The domain of $f \circ g$ is defined to consist of all x in the domain of g for which $g(x)$ is in the domain of f.

REMARK.   Although the domain of $f \circ g$ may seem complicated at first glance, it makes sense intuitively: To compute $f(g(x))$ one needs x in the domain of g to compute $g(x)$, then one needs $g(x)$ in the domain of f to compute $f(g(x))$.

COMPOSITIONS VIEWED AS COMPUTER PROGRAMS

In Section 1.1 we noted that a function f can be viewed as a computer program that takes an input x, operates on it, and produces an output $f(x)$. From this viewpoint composition can be viewed as two programs, g and f, operating in succession: An input x is fed first to a program g, which produces the output $g(x)$; then this output is fed as input to a program f, which produces the output $f(g(x))$ (Figure 1.4.4). However, rather than have two separate programs operating in succession, we could create a **single** program that takes the input x and directly produces the output $f(g(x))$. This program is the composition $f \circ g$ since $(f \circ g)(x) = f(g(x))$.

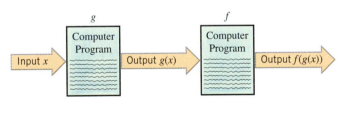

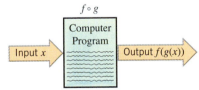

Figure 1.4.4

Example 3   Let $f(x) = x^2 + 3$ and $g(x) = \sqrt{x}$. Find

(a) $(f \circ g)(x)$ (b) $(g \circ f)(x)$

Solution (a).   The formula for $f(g(x))$ is

$$f(g(x)) = [g(x)]^2 + 3 = (\sqrt{x})^2 + 3 = x + 3$$

Since the domain of g is $[0, +\infty)$ and the domain of f is $(-\infty, +\infty)$, the domain of $f \circ g$ consists of all x in $[0, +\infty)$ such that $g(x) = \sqrt{x}$ lies in $(-\infty, +\infty)$; thus, the domain of $f \circ g$ is $[0, +\infty)$. Therefore,

$$(f \circ g)(x) = x + 3, \quad x \geq 0$$

Solution (b). The formula for $g(f(x))$ is

$$g(f(x)) = \sqrt{f(x)} = \sqrt{x^2 + 3}$$

Since the domain of f is $(-\infty, +\infty)$ and the domain of g is $[0, +\infty)$, the domain of $g \circ f$ consists of all x in $(-\infty, +\infty)$ such that $f(x) = x^2 + 3$ lies in $[0, +\infty)$. Thus, the domain of $g \circ f$ is $(-\infty, +\infty)$. Therefore,

$$(g \circ f)(x) = \sqrt{x^2 + 3}$$

There is no need to indicate that the domain is $(-\infty, +\infty)$, since this is the natural domain of $\sqrt{x^2 + 3}$. ◀

REMARK.   Note that the functions $f \circ g$ and $g \circ f$ in the preceding example are not the same. Thus, the order in which functions are composed can (and usually will) make a difference in the end result.

Compositions can also be defined for three or more functions; for example, $(f \circ g \circ h)(x)$ is computed as

$$(f \circ g \circ h)(x) = f(g(h(x)))$$

In other words, first find $h(x)$, then find $g(h(x))$, and then find $f(g(h(x)))$.

Example 4   Find $(f \circ g \circ h)(x)$ if

$$f(x) = \sqrt{x}, \quad g(x) = 1/x, \quad h(x) = x^3$$

Solution.

$$(f \circ g \circ h)(x) = f(g(h(x))) = f(g(x^3)) = f(1/x^3) = \sqrt{1/x^3} = 1/x^{3/2} \quad ◀$$

EXPRESSING A FUNCTION AS A COMPOSITION

Many problems in mathematics are attacked by "decomposing" functions into compositions of simpler functions. For example, consider the function h given by

$$h(x) = (x + 1)^2$$

To evaluate $h(x)$ for a given value of x, we would first compute $x + 1$ and then square the result. These two operations are performed by the functions

$$g(x) = x + 1 \quad \text{and} \quad f(x) = x^2$$

We can express h in terms of f and g by writing

$$h(x) = (x + 1)^2 = [g(x)]^2 = f(g(x))$$

so we have succeeded in expressing h as the composition $h = f \circ g$.

The thought process in this example suggests a general procedure for decomposing a function h into a composition $h = f \circ g$:

- Think about how you would evaluate $h(x)$ for a specific value of x, trying to break the evaluation into two steps performed in succession.
- The first operation in the evaluation will determine a function g and the second a function f.
- The formula for h can then be written as $h(x) = f(g(x))$.

For descriptive purposes, we will refer to g as the "inside function" and f as the "outside function" in the expression $f(g(x))$. The inside function performs the first operation and the outside function performs the second.

Example 5   Express $h(x) = (x - 4)^5$ as a composition of two functions.

Solution. To evaluate $h(x)$ for a given value of x we would first compute $x - 4$ and then raise the result to the fifth power. Therefore, the inside function (first operation) is

$$g(x) = x - 4$$

and the outside function (second operation) is

$$f(x) = x^5$$

so $h(x) = f(g(x))$. As a check,

$$f(g(x)) = [g(x)]^5 = (x - 4)^5 = h(x) \qquad \blacktriangleleft$$

Example 6   Express $\sin(x^3)$ as a composition of two functions.

Solution. To evaluate $\sin(x^3)$, we would first compute x^3 and then take the sine, so $g(x) = x^3$ is the inside function and $f(x) = \sin x$ the outside function. Therefore,

$$\sin(x^3) = f(g(x)) \qquad \boxed{g(x) = x^3 \text{ and } f(x) = \sin x} \qquad \blacktriangleleft$$

Example 7   Table 1.4.1 gives some more examples of decomposing functions into compositions.

Table 1.4.1

FUNCTION	$g(x)$ INSIDE	$f(x)$ OUTSIDE	COMPOSITION
$(x^2 + 1)^{10}$	$x^2 + 1$	x^{10}	$(x^2 + 1)^{10} = f(g(x))$
$\sin^3 x$	$\sin x$	x^3	$\sin^3 x = f(g(x))$
$\tan(x^5)$	x^5	$\tan x$	$\tan(x^5) = f(g(x))$
$\sqrt{4 - 3x}$	$4 - 3x$	$\sqrt{x}$	$\sqrt{4 - 3x} = f(g(x))$
$8 + \sqrt{x}$	$\sqrt{x}$	$8 + x$	$8 + \sqrt{x} = f(g(x))$
$\dfrac{1}{x + 1}$	$x + 1$	$\dfrac{1}{x}$	$\dfrac{1}{x + 1} = f(g(x))$

$\blacktriangleleft$

REMARK.   It should be noted that there is always more than one way to express a function as a composition. For example, here are two ways to express $(x^2 + 1)^{10}$ as a composition that differ from that in Table 1.4.1:

$$(x^2 + 1)^{10} = \left[(x^2 + 1)^2\right]^5 = f(g(x)) \qquad \boxed{g(x) = (x^2 + 1)^2 \text{ and } f(x) = x^5}$$

$$(x^2 + 1)^{10} = \left[(x^2 + 1)^3\right]^{10/3} = f(g(x)) \qquad \boxed{g(x) = (x^2 + 1)^3 \text{ and } f(x) = x^{10/3}}$$

SYMMETRY

Figure 1.4.5 shows the graphs of three curves that have certain obvious symmetries. The graph in part (*a*) is **symmetric about the x-axis** in the sense that for each point (x, y) on the graph the point $(x, -y)$ is also on the graph; the graph in part (*b*) is **symmetric about the y-axis** in the sense that for each point (x, y) on the graph the point $(-x, y)$ is also on the graph; and the graph in part (*c*) is **symmetric about the origin** in the sense that for each point (x, y) on the graph the point $(-x, -y)$ is also on the graph. Geometrically, symmetry about the origin occurs if rotating the graph $180°$ about the origin leaves the graph unchanged.

Symmetries can often be detected from the equation of a curve. For example, the graph of

$$y = x^3 \tag{7}$$

must be symmetric about the origin because for any point (x, y) whose coordinates satisfy

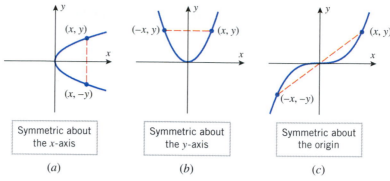

Figure 1.4.5

(7), the coordinates of the point $(-x, -y)$ also satisfy (7), since substituting these coordinates in (7) yields

$$-y = (-x)^3$$

which simplifies to (7). This suggests the following symmetry tests (Figure 1.4.6).

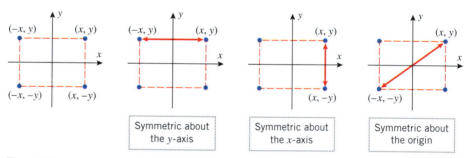

Figure 1.4.6

1.4.3 THEOREM (*Symmetry Tests*).

(a) *A plane curve is symmetric about the y-axis if and only if replacing x by $-x$ in its equation produces an equivalent equation.*

(b) *A plane curve is symmetric about the x-axis if and only if replacing y by $-y$ in its equation produces an equivalent equation.*

(c) *A plane curve is symmetric about the origin if and only if replacing both x by $-x$ and y by $-y$ in its equation produces an equivalent equation.*

EVEN AND ODD FUNCTIONS

For the graph of a function f to be symmetric about the y-axis, the equations $y = f(x)$ and $y = f(-x)$ must be equivalent; for this to happen we must have

$$f(x) = f(-x)$$

A function with this property is called an ***even function***. Some examples are x^2, x^4, x^6, and $\cos x$. Similarly, for the graph of a function f to be symmetric about the origin, the equations $y = f(x)$ and $-y = f(-x)$ must be equivalent; for this to happen we must have

$$f(x) = -f(-x)$$

A function with this property is called an ***odd function***. Some examples are x, x^3, x^5, and $\sin x$.

FOR THE READER.   Explain why the graph of a nonzero function cannot by symmetric about the x-axis.

TRANSLATIONS

Once you know the graph of an equation $y = f(x)$, there are some techniques that can be used to help visualize the graphs of the equations

$$y = f(x) + c, \quad y = f(x) - c, \quad y = f(x + c), \quad y = f(x - c)$$

where c is any positive constant.

If a positive constant is added to or subtracted from $f(x)$, the geometric effect is to translate the graph of $y = f(x)$ parallel to the y-axis; addition translates the graph in the positive direction and subtraction translates it in the negative direction. This is illustrated in Table 1.4.2. Similarly, if a positive constant is added to or subtracted from the independent variable x, the geometric effect is to translate the graph of the function parallel to the x-axis; subtraction translates the graph in the positive direction, and addition translates it in the negative direction. This is also illustrated in Table 1.4.2.

Table 1.4.2

OPERATION ON $y = f(x)$	Add a positive constant c to $f(x)$	Subtract a positive constant c from $f(x)$	Add a positive constant c to x	Subtract a positive constant c from x
NEW EQUATION	$y = f(x) + c$	$y = f(x) - c$	$y = f(x + c)$	$y = f(x - c)$
GEOMETRIC EFFECT	Translates the graph of $y = f(x)$ up c units	Translates the graph of $y = f(x)$ down c units	Translates the graph of $y = f(x)$ left c units	Translates the graph of $y = f(x)$ right c units
EXAMPLE				

Before proceeding to the following examples, it will be helpful to review the graphs in Figures 1.2.1 and 1.2.5.

Example 8   Sketch the graph of

  (a) $y = \sqrt{x - 3}$ (b) $y = \sqrt{x + 3}$

Solution. The graph of the equation $y = \sqrt{x - 3}$ can be obtained by translating the graph of $y = \sqrt{x}$ right 3 units, and the graph of $y = \sqrt{x + 3}$ by translating the graph of $y = \sqrt{x}$ left 3 units (Figure 1.4.7). ◄

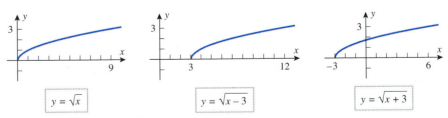

Figure 1.4.7

Example 9   Sketch the graph of $y = |x - 3| + 2$.

Solution. The graph can be obtained by two translations: first translate the graph of $y = |x|$ right 3 units to obtain the graph of $y = |x - 3|$, then translate this graph up 2 units to obtain the graph of $y = |x - 3| + 2$ (Figure 1.4.8). ◄

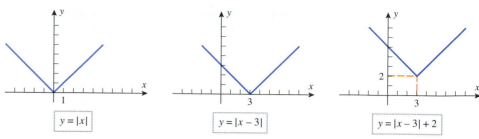

Figure 1.4.8

REMARK.   The graph in the preceding example could also have been obtained by performing the translations in the opposite order: first translating the graph of $y = |x|$ up 2 units to obtain the graph of $y = |x| + 2$, then translating this graph right 3 units to obtain the graph of $y = |x - 3| + 2$.

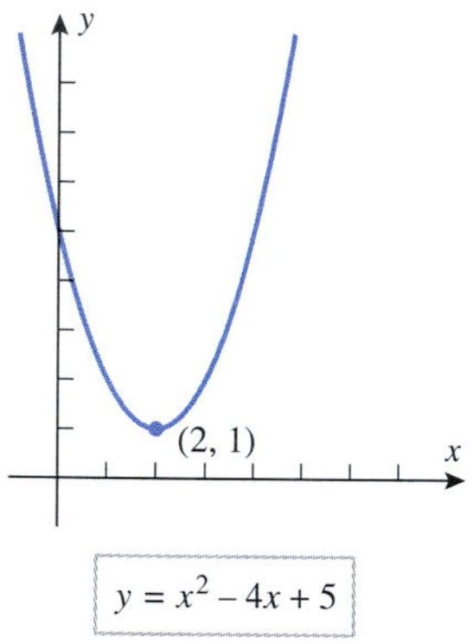

Example 10   Sketch the graph of $y = x^2 - 4x + 5$.

Solution.   Completing the square on the first two terms yields

$$y = (x^2 - 4x + 4) - 4 + 5 = (x - 2)^2 + 1$$

(see Appendix D for a review of this technique). In this form we see that the graph can be obtained by translating the graph of $y = x^2$ right 2 units because of the $x - 2$, and up 1 unit because of the $+1$ (Figure 1.4.9).   ◄

Example 11   By completing the square, an equation of the form $y = ax^2 + bx + c$ with $a \neq 0$ can be expressed as

$$y = a(x - h)^2 + k \tag{8}$$

Sketch the graph of this equation.

Solution.   We can build up Equation (8) in three steps from the equation $y = x^2$. First, we can multiply by a to obtain $y = ax^2$. If $a > 0$, this operation has the geometric effect of stretching or compressing the graph of $y = x^2$; and if $a < 0$, it has the geometric effect of reflecting the graph about the x-axis, in addition to stretching or compressing it. Since stretching or compressing does not alter the general parabolic shape of the original curve, the graph of $y = ax^2$ looks roughly like one of those in Figure 1.4.10a. Next, we can subtract h from x to obtain the equation $y = a(x - h)^2$, and then we can add k to obtain $y = a(x - h)^2 + k$. Subtracting h causes a horizontal translation (right or left, depending on the sign of h), and adding k causes a vertical translation (up or down, depending on the sign of k). Thus, the graph of (8) looks roughly like one of those in Figure 1.4.10b, which are shown with $h > 0$ and $k > 0$ for simplicity.   ◄

(figure at left, margin)

$y = x^2 - 4x + 5$

(2, 1)

Figure 1.4.9

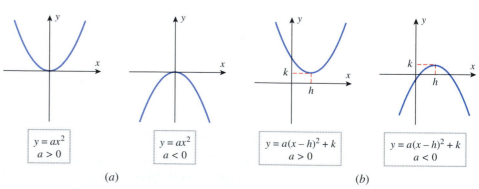

(a)

(b)

Figure 1.4.10

REFLECTIONS

The graph of $y = f(-x)$ is the reflection of the graph of $y = f(x)$ about the y-axis, and the graph of $y = -f(x)$ [or equivalently, $-y = f(x)$] is the reflection of the graph of $y = f(x)$ about the x-axis. Thus, if you know what the graph of $y = f(x)$ looks like, you can obtain the graphs of $y = f(-x)$ and $y = -f(x)$ by making appropriate reflections. This is illustrated in Table 1.4.3.

Table 1.4.3

OPERATION ON $y = f(x)$	Replace x by $-x$	Multiply $f(x)$ by -1
NEW EQUATION	$y = f(-x)$	$y = -f(x)$
GEOMETRIC EFFECT	Reflects the graph of $y = f(x)$ about the y-axis	Reflects the graph of $y = f(x)$ about the x-axis
EXAMPLE		

FOR THE READER. Describe the geometric effect of multiplying a function f by a *negative* constant in terms of reflection and stretching or compressing. What is the geometric effect of multiplying the independent variable of a function f by a *negative* constant?

Example 12 Sketch the graph of $y = \sqrt[3]{2 - x}$.

Solution. The graph can be obtained by a reflection and a translation: first reflect the graph of $y = \sqrt[3]{x}$ about the y-axis to obtain the graph of $y = \sqrt[3]{-x}$, then translate this graph right 2 units to obtain the graph of the equation $y = \sqrt[3]{-(x - 2)} = \sqrt[3]{2 - x}$ (Figure 1.4.11). ◀

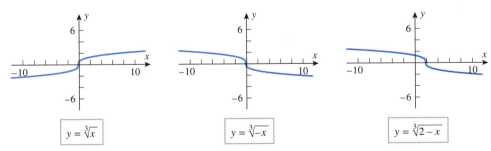

Figure 1.4.11

Example 13 Sketch the graph of $y = 4 - |x - 2|$.

Solution. The graph can be obtained by a reflection and two translations: first translate the graph of $y = |x|$ right 2 units to obtain the graph of $y = |x - 2|$; then reflect this graph about the x-axis to obtain the graph of $y = -|x - 2|$; and then translate this graph up 4 units to obtain the graph of the equation $y = -|x - 2| + 4 = 4 - |x - 2|$ (Figure 1.4.12). ◀

Figure 1.4.12

EXERCISE SET 1.4 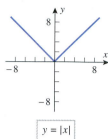 Graphing Utility

1. The graph of a function f is shown in the accompanying figure. Sketch the graphs of the following equations.
   (a) $y = f(x) - 1$ (b) $y = f(x - 1)$
   (c) $y = \frac{1}{2}f(x)$ (d) $y = f\left(-\frac{1}{2}x\right)$

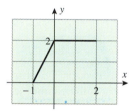

Figure Ex-1

2. Use the graph in Exercise 1 to sketch the graphs of the following equations.
   (a) $y = -f(-x)$ (b) $y = f(2 - x)$
   (c) $y = 1 - f(2 - x)$ (d) $y = \frac{1}{2}f(2x)$

3. The graph of a function f is shown in the accompanying figure. Sketch the graphs of the following equations.
   (a) $y = f(x + 1)$ (b) $y = f(2x)$
   (c) $y = |f(x)|$ (d) $y = 1 - |f(x)|$

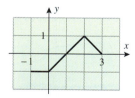

Figure Ex-3

4. Use the graph in Exercise 3 to sketch the graph of the equation $y = f(|x|)$.

In Exercises 5–12, sketch the graph of the equation by translating, reflecting, compressing, and stretching the graph of $y = x^2$ appropriately, and then use a graphing utility to confirm that your sketch is correct.

5. $y = 1 + (x - 2)^2$ 6. $y = 2 - (x + 1)^2$
7. $y = -2(x + 1)^2 - 3$ 8. $y = \frac{1}{2}(x - 3)^2 + 2$
9. $y = x^2 + 6x$ 10. $y = x^2 + 6x - 10$

11. $y = 1 + 2x - x^2$ 12. $y = \frac{1}{2}(x^2 - 2x + 3)$

In Exercises 13–16, sketch the graph of the equation by translating, reflecting, compressing, and stretching the graph of $y = \sqrt{x}$ appropriately, and then use a graphing utility to confirm that your sketch is correct.

13. $y = 3 - \sqrt{x + 1}$ 14. $y = 1 + \sqrt{x - 4}$
15. $y = \frac{1}{2}\sqrt{x} + 1$ 16. $y = -\sqrt{3x}$

In Exercises 17–20, sketch the graph of the equation by translating, reflecting, compressing, and stretching the graph of $y = 1/x$ appropriately, and then use a graphing utility to confirm that your sketch is correct.

17. $y = \dfrac{1}{x - 3}$ 18. $y = \dfrac{1}{1 - x}$
19. $y = 2 - \dfrac{1}{x + 1}$ 20. $y = \dfrac{x - 1}{x}$

In Exercises 21–24, sketch the graph of the equation by translating, reflecting, compressing, and stretching the graph of $y = |x|$ appropriately, and then use a graphing utility to confirm that your sketch is correct.

21. $y = |x + 2| - 2$ 22. $y = 1 - |x - 3|$
23. $y = |2x - 1| + 1$ 24. $y = \sqrt{x^2 - 4x + 4}$

In Exercises 25–28, sketch the graph of the equation by translating, reflecting, compressing, and stretching the graph of $y = \sqrt[3]{x}$ appropriately, and then use a graphing utility to confirm that your sketch is correct.

25. $y = 1 - 2\sqrt[3]{x}$ 26. $y = \sqrt[3]{x - 2} - 3$
27. $y = 2 + \sqrt[3]{x + 1}$ 28. $y + \sqrt[3]{x - 2} = 0$

29. (a) Sketch the graph of $y = x + |x|$ by adding the corresponding y-coordinates on the graphs of $y = x$ and $y = |x|$.

(b) Express the equation $y = x + |x|$ in piecewise form with no absolute values, and confirm that the graph you obtained in part (a) is consistent with this equation.

~ **30.** Sketch the graph of $y = x + (1/x)$ by adding corresponding y-coordinates on the graphs of $y = x$ and $y = 1/x$. Use a graphing utility to confirm that your sketch is correct.

In Exercises 31–34, find formulas for $f + g$, $f - g$, fg, and f/g, and state the domains of the functions.

31. $f(x) = 2x$, $g(x) = x^2 + 1$

32. $f(x) = 3x - 2$, $g(x) = |x|$

33. $f(x) = 2\sqrt{x} - 1$, $g(x) = \sqrt{x - 1}$

34. $f(x) = \dfrac{x}{1 + x^2}$, $g(x) = \dfrac{1}{x}$

35. Let $f(x) = \sqrt{x}$ and $g(x) = x^3 + 1$. Find

  (a) $f(g(2))$ (b) $g(f(4))$

  (c) $f(f(16))$ (d) $g(g(0))$.

36. Let $g(x) = \pi - x^2$ and $h(x) = \cos x$. Find

  (a) $g(h(0))$ (b) $h(g(\sqrt{\pi/2}\,))$

  (c) $g(g(1))$ (d) $h(h(\pi/2))$.

37. Let $f(x) = x^2 + 1$. Find

  (a) $f(t^2)$ (b) $f(t + 2)$ (c) $f(x + 2)$

  (d) $f\left(\dfrac{1}{x}\right)$ (e) $f(x + h)$ (f) $f(-x)$

  (g) $f(\sqrt{x}\,)$ (h) $f(3x)$.

38. Let $g(x) = \sqrt{x}$. Find

  (a) $g(5s + 2)$ (b) $g(\sqrt{x} + 2)$ (c) $3g(5x)$

  (d) $\dfrac{1}{g(x)}$ (e) $g(g(x))$ (f) $(g(x))^2 - g(x^2)$

  (g) $g(1/\sqrt{x}\,)$ (h) $g((x - 1)^2)$.

In Exercises 39–44, find formulas for $f \circ g$ and $g \circ f$, and state the domains of the functions.

39. $f(x) = 2x + 1$, $g(x) = x^2 - x$

40. $f(x) = 2 - x^2$, $g(x) = x^3$

41. $f(x) = x^2$, $g(x) = \sqrt{1 - x}$

42. $f(x) = \sqrt{x - 3}$, $g(x) = \sqrt{x^2 + 3}$

43. $f(x) = \dfrac{1 + x}{1 - x}$, $g(x) = \dfrac{x}{1 - x}$

44. $f(x) = \dfrac{x}{1 + x^2}$, $g(x) = \dfrac{1}{x}$

In Exercises 45 and 46, find a formula for $f \circ g \circ h$.

45. $f(x) = x^2 + 1$, $g(x) = \dfrac{1}{x}$, $h(x) = x^3$

46. $f(x) = \dfrac{1}{1 + x}$, $g(x) = \sqrt[3]{x}$, $h(x) = \dfrac{1}{x^3}$

In Exercises 47–50, express f as a composition of two functions; that is, find g and h such that $f = g \circ h$. [*Note:* Each exercise has more than one solution.]

47. (a) $f(x) = \sqrt{x + 2}$ (b) $f(x) = |x^2 - 3x + 5|$

48. (a) $f(x) = x^2 + 1$ (b) $f(x) = \dfrac{1}{x - 3}$

49. (a) $f(x) = \sin^2 x$ (b) $f(x) = \dfrac{3}{5 + \cos x}$

50. (a) $f(x) = 3\sin(x^2)$ (b) $f(x) = 3\sin^2 x + 4\sin x$

In Exercises 51 and 52, express F as a composition of three functions; that is, find f, g, and h such that $F = f \circ g \circ h$. [*Note:* Each exercise has more than one solution.]

51. (a) $F(x) = \left(1 + \sin(x^2)\right)^3$ (b) $F(x) = \sqrt{1 - \sqrt[3]{x}}$

52. (a) $F(x) = \dfrac{1}{1 - x^2}$ (b) $F(x) = |5 + 2x|$

53. Use the data in the accompanying table to make a scatter plot of $y = f(g(x))$.

x	-3	-2	-1	0	1	2	3
$f(x)$	-4	-3	-2	-1	0	1	2
$g(x)$	-1	0	1	2	3	-2	-3

Table Ex-53

54. Find the domain of $g \circ f$ for the functions f and g in Exercise 53.

55. Sketch the graph of $y = f(g(x))$ for the functions graphed in the accompanying figure.

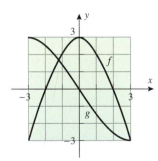

Figure Ex-55

56. Sketch the graph of $y = g(f(x))$ for the functions graphed in Exercise 55.

57. Use the graphs of f and g in Exercise 55 to estimate the solutions of the equations $f(g(x)) = 0$ and $g(f(x)) = 0$.

58. Use the table in Exercise 53 to solve the equations $f(g(x)) = 0$ and $g(f(x)) = 0$.

In Exercises 59–62, find

$$\frac{f(w) - f(x)}{w - x} \quad \text{and} \quad \frac{f(x + h) - f(x)}{h}$$

Simplify as much as possible.

59. $f(x) = 3x^2 - 5$ **60.** $f(x) = x^2 + 6x$

61. $f(x) = 1/x$ **62.** $f(x) = 1/x^2$

63. In each part of the accompanying figure determine whether the graph is symmetric about the x-axis, the y-axis, the origin, or none of the preceding.

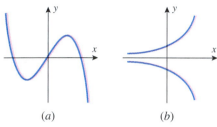

(a)   (b)

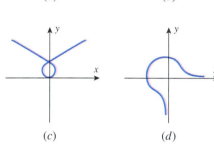

(c)   (d)

Figure Ex-63

64. The accompanying figure shows a portion of a graph. Complete the graph so that the entire graph is symmetric about
(a) the x-axis   (b) the y-axis   (c) the origin.

Figure Ex-64

65. Complete the accompanying table so that the graph of $y = f(x)$ (which is a scatter plot) is symmetric about
(a) the y-axis   (b) the origin.

x	-3	-2	-1	0	1	2	3
$f(x)$	1		-1	0		-5	

Table Ex-65

66. The accompanying figure shows a portion of the graph of a function f. Complete the graph assuming that
(a) f is an even function   (b) f is an odd function.

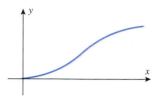

Figure Ex-66

67. Classify the functions graphed in the accompanying figure as even, odd, or neither.

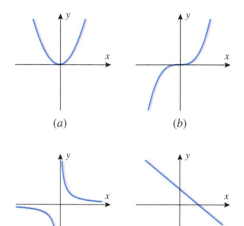

(a)   (b)

(c)   (d)

Figure Ex-67

68. Classify the functions whose values are given in the following table as even, odd, or neither.

x	-3	-2	-1	0	1	2	3
$f(x)$	5	3	2	3	1	-3	5
$g(x)$	4	1	-2	0	2	-1	-4
$h(x)$	2	-5	8	-2	8	-5	2

69. In each part, classify the function as even, odd, or neither.
(a) $f(x) = x^2$ (b) $f(x) = x^3$
(c) $f(x) = |x|$ (d) $f(x) = x + 1$
(e) $f(x) = \dfrac{x^5 - x}{1 + x^2}$ (f) $f(x) = 2$

In Exercises 70 and 71, use Theorem 1.4.3 to determine whether the graph has symmetries about the x-axis, the y-axis, or the origin.

70. (a) $x = 5y^2 + 9$ (b) $x^2 - 2y^2 = 3$
(c) $xy = 5$

71. (a) $x^4 = 2y^3 + y$ (b) $y = \dfrac{x}{3 + x^2}$
(c) $y^2 = |x| - 5$

In Exercises 72 and 73: (i) Use a graphing utility to graph the equation in the first quadrant. [*Note:* To do this you will have to solve the equation for y in terms of x.] (ii) Use symmetry to make a hand-drawn sketch of the entire graph. (iii) Confirm your work by generating the graph of the equation in the remaining three quadrants.

 72. $9x^2 + 4y^2 = 36$ **73.** $4x^2 + 16y^2 = 16$

 74. The graph of the equation $x^{2/3} + y^{2/3} = 1$, which is shown in the accompanying figure, is called a ***four-cusped hypo-cycloid***.

(a) Use Theorem 1.4.3 to confirm that this graph is symmetric about the x-axis, the y-axis, and the origin.

(b) Find a function f whose graph in the first quadrant coincides with the four-cusped hypocycloid, and use a graphing utility to confirm your work.

(c) Repeat part (b) for the remaining three quadrants.

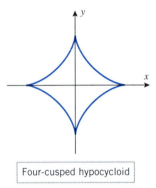

Four-cusped hypocycloid

Figure Ex-74

75. The equation $y = |f(x)|$ can be written as

$$y = \begin{cases} f(x), & f(x) \geq 0 \\ -f(x), & f(x) < 0 \end{cases}$$

which shows that the graph of $y = |f(x)|$ can be obtained from the graph of $y = f(x)$ by retaining the portion that lies on or above the x-axis and reflecting about the x-axis the portion that lies below the x-axis. Use this method to obtain the graph of $y = |2x - 3|$ from the graph of $y = 2x - 3$.

In Exercises 76 and 77, use the method described in Exercise 75.

76. Sketch the graph of $y = |1 - x^2|$.

77. Sketch the graph of

(a) $f(x) = |\cos x|$ (b) $f(x) = \cos x + |\cos x|$.

78. The ***greatest integer function***, $[x]$, is defined to be the greatest integer that is less than or equal to x. For example, $[2.7] = 2$, $[-2.3] = -3$, and $[4] = 4$. Sketch the graph of

(a) $f(x) = [x]$ (b) $f(x) = [x^2]$

(c) $f(x) = [x]^2$ (d) $f(x) = [\sin x]$.

79. Is it ever true that $f \circ g = g \circ f$ if f and g are nonconstant functions? If not, prove it; if so, give some examples for which it is true.

80. In the discussion preceding Exercise 29 of Section 1.3, we gave a procedure for generating a complete graph of $f(x) = x^{p/q}$ in which we suggested graphing the function $g(x) = |x|^{p/q}$ instead of $f(x)$ when p is even and q is odd and graphing $g(x) = (|x|/x)|x|^{p/q}$ if p is odd and q is odd. Show that in both cases $f(x) = g(x)$ if $x > 0$ or $x < 0$. [*Hint:* Show that $f(x)$ is an even function if p is even and q is odd and is an odd function if p is odd and q is odd.]

1.5 LINES

This section includes a quick review of precalculus material on lines. Readers who want to review this material in more depth are referred to Appendix C.

EQUATIONS OF LINES

An equation that is expressible in the form

$$Ax + By + C = 0 \tag{1}$$

where A and B are not both zero, is called a ***first-degree equation*** or a ***linear equation*** in x and y. It is shown in precalculus that every first-degree equation in x and y has a straight line as its graph and, conversely, every straight line can be represented by a first-degree equation in x and y. For this reason (1) is sometimes called the ***general equation*** of a line. Recall that equations of lines may be written in several different forms:

$$y = mx + b \qquad \boxed{\text{Slope-intercept form}} \tag{2}$$

$$y - y_1 = m(x - x_1) \qquad \boxed{\text{Point-slope form}} \tag{3}$$

$$\frac{x}{a} + \frac{y}{b} = 1 \qquad \boxed{\text{Double-intercept form}} \tag{4}$$

In these equations m is the slope of the line, a is the x-intercept, b is the y-intercept, and (x_1, y_1) is any point on the line (Figure 1.5.1). Keep in mind that these equations do not apply to vertical lines. For vertical lines the slope is *undefined*, or stated informally, a vertical line has infinite slope. Vertical and horizontal lines have particularly simple equations:

$x = a$ — The vertical line with x-intercept a (5)

$y = b$ — The horizontal line with y-intercept b (6)

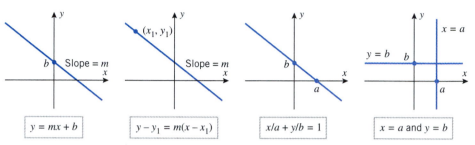

Figure 1.5.1

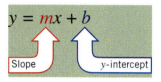

Figure 1.5.2

Equation (2) is especially useful because the slope and the y-intercept of the line can be determined by inspection: the slope is the coefficient of x, and the y-intercept is the constant term (Figure 1.5.2). This equation expresses y as a function of x, the function being $f(x) = mx + b$. A function of this form is called a ***linear function*** of x.

INTERPRETATIONS OF SLOPE

The slope m of a nonvertical line $y = mx + b$ has two important interpretations (which are related but different in viewpoint):

- m is a measure of the *steepness* of the line.
- m is the rate of change of y with respect to x.

The steepness interpretation has an analog in surveying. Surveyors measure the grade or slope of a hill as the ratio of its rise over its run (Figure 1.5.3*a*). The same idea applies to lines. Consider a particle that moves left to right along a nonvertical line from a point $P_1(x_1, y_1)$ to a point $P_2(x_2, y_2)$. In the course of its travel the point moves $y_2 - y_1$ units vertically as it travels $x_2 - x_1$ units horizontally (Figure 1.5.3*b*). The vertical change, which is denoted by $\Delta y = y_2 - y_1$, is called the ***rise***, and the horizontal change, which is denoted by $\Delta x = x_2 - x_1$, is called the ***run***. The ratio of the rise over the run is always equal to the

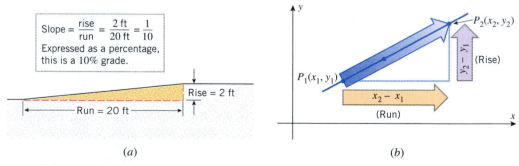

(*a*)   (*b*)

Figure 1.5.3

slope, regardless of where the points P_1 and P_2 are located on the line; that is,

$$m = \frac{\Delta y}{\Delta x} = \frac{y_2 - y_1}{x_2 - x_1} \tag{7}$$

REMARK. The symbols Δx and Δy should not be interpreted as products; rather, Δx should be viewed as a single entity representing the *change* in the value of x, and Δy as a single entity representing the *change* in the value of y. In general, if v is any variable whose value changes from an initial value of v_1 to a final value of v_2, then we call $\Delta v = v_2 - v_1$ (final value minus initial value) an ***increment*** in v. Increments can be positive or negative, depending on whether the final value is larger or smaller than the initial value.

ANGLE OF INCLINATION

The slope of a nonvertical line L is related to the angle that L makes with the positive x-axis. If ϕ is the smallest positive angle measured counterclockwise from the x-axis to L, then the slope of the line can be expressed as

$$m = \tan \phi \tag{8}$$

(Figure 1.5.4a). The angle ϕ, which is called the ***angle of inclination*** of the line, satisfies $0° \le \phi < 180°$ in degree measure (or, equivalently, $0 \le \phi < \pi$ in radian measure). If ϕ is an acute angle, then $m = \tan \phi$ is positive and the line slopes up to the right, and if ϕ is an obtuse angle, then $m = \tan \phi$ is negative and the line slopes down to the right. For example, a line whose angle of inclination is $45°$ has slope $m = \tan 45° = 1$, and a line whose angle of inclination is $135°$ has a slope of $m = \tan 135° = -1$ (Figure 1.5.4b). Figure 1.5.5 shows a convenient way of using the line $x = 1$ as a "ruler" for visualizing the relationship between lines of various slopes.

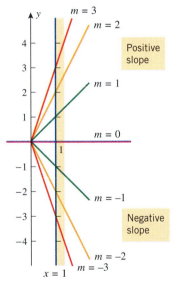

Figure 1.5.5

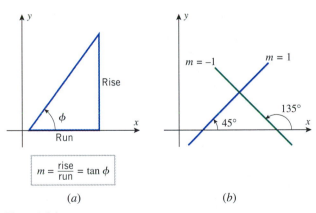

Figure 1.5.4

SLOPES OF LINES IN APPLIED PROBLEMS

In applied problems, changing the units of measurement can change the slope of a line, so it is essential to include the units when calculating the slope. The following example illustrates this.

Example 1 Suppose that a uniform rod of length 40 cm (= 0.4 m) is thermally insulated around the lateral surface and that the exposed ends of the rod are held at constant temperatures of 25°C and 5°C, respectively (Figure 1.5.6a). It is shown in physics that under appropriate conditions the graph of the temperature T versus the distance x from the left-hand end of the rod will be a straight line. Parts (b) and (c) of Figure 1.5.6 show two such graphs: one in which x is measured in centimeters and one in which it is measured in

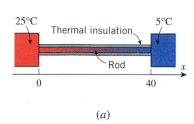

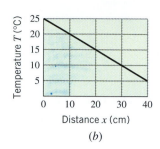

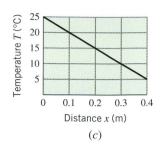

(a)   (b)   (c)

Figure 1.5.6

meters. The slopes in the two cases are

$$m = \frac{5 - 25}{40 - 0} = \frac{-20}{40} = -0.5 \qquad (9)$$

$$m = \frac{5 - 25}{0.4 - 0} = \frac{-20}{0.4} = -50 \qquad (10)$$

The slope in (9) implies that the temperature *decreases* at a rate of $0.5°C$ per centimeter of distance from the left end of the rod, and the slope in (10) implies that the temperature decreases at a rate of $50°C$ per meter of distance from the left end of the rod. The two statements are equivalent physically, even though the slopes differ. ◀

Example 2   Find the slope-intercept form of the equation of the temperature distribution in the preceding example if the temperature T is measured in degrees Celsius ($°C$) and the distance x is measured in (a) centimeters and (b) meters.

Solution (a). The slope is $m = -0.5$ and the intercept on the T-axis is 25, so

$$T = -0.5x + 25, \quad 0 \le x \le 40$$

where the restriction on x is required because the rod is 40 cm in length. The graph of this equation with the restriction is a line segment rather than a line.

Solution (b). The slope is $m = -50$, the intercept on the T-axis is 25, and the restriction on x is $0 \le x \le 0.4$. Thus, the equation is

$$T = -50x + 25, \quad 0 \le x \le 0.4 \qquad ◀$$

SLOPES AS RATE OF CHANGE

If y is a linear function of x, say $y = mx + b$, then it follows from (7) that

$$\Delta y = m \Delta x$$

Thus, a 1-unit increase in x ($\Delta x = 1$) produces an m-unit change in y ($\Delta y = m$). Moreover, this is true at every point on the line (Figure 1.5.7), so we say that y changes at a *constant rate* with respect to x, and we call m the ***rate of change of y with respect to x***. This idea can be summarized as follows.

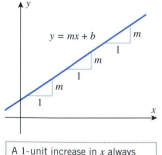

A 1-unit increase in x always produces an m-unit change in y.

Figure 1.5.7

1.5.1   CONSTANT RATE OF CHANGE.   *If a variable y is related to a variable x in such a way that the rate of change of y with respect to x is constant, say m, then y is a linear function of x of the form*

$$y = mx + b$$

Conversely, if y is a linear function of x whose graph has slope m, then the rate of change of y with respect to x is constant and equal to m.

It follows from this that linear functions are appropriate whenever experimentation or theory suggests that the rate of change of y with respect to x is constant.

UNIFORM RECTILINEAR MOTION

One of the important themes in calculus is the study of motion. To describe the motion of an object completely, one must specify its *speed* (how fast it is going) and the direction in which it is moving. The speed and the direction of motion together comprise what is called the *velocity* of the object. For example, knowing that the speed of an aircraft is 500 mi/h tells us how fast it is going, but not which way it is moving. In contrast, knowing that the velocity of the aircraft is 500 mi/h *due south* pins down the speed and the direction of motion.

Later, we will study the motion of particles that move along curves in two- or three-dimensional space, but for now we will focus on motion along a line; this is called *rectilinear motion*. In general rectilinear motion, a particle can move back and forth along the line (as with a piston moving up and down in a cylinder); however, for now we will only consider the simple case in which the particle moves in just *one direction* along a line (as with a car traveling on a straight road).

For simplicity, we will assume that the motion is along a coordinate line, such as an x-axis or y-axis, and that the particle is moving in the positive direction. In general discussions we will usually name the coordinate line the s-axis to avoid being specific. A graphical description of rectilinear motion along an s-axis can be obtained by making a plot of the s-coordinate of the particle versus the elapsed time t. This is called the *position versus time curve* for the particle. Figure 1.5.8a shows a typical position versus time curve for a car moving in the positive direction along an s-axis.

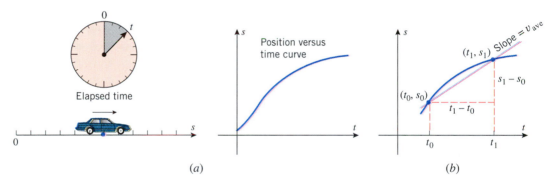

Figure 1.5.8

FOR THE READER.   How can you tell from the position versus time curve in Figure 1.5.8a that the car does not reverse direction?

Because we are assuming that the particle is moving in the positive direction of the s-axis, there is no ambiguity about the direction of motion, and hence the terms "speed" and "velocity" can be used interchangeably. However, later, when we consider general rectilinear motion or motion along a curved path, it will be necessary to distinguish between these terms, since the direction of motion may vary.

For a particle in rectilinear motion along a coordinate axis, we define the *average velocity* v_{ave} of the particle during the time interval from t_0 to t_1 to be

$$v_{ave} = \frac{s_1 - s_0}{t_1 - t_0} = \frac{\Delta s}{\Delta t} \tag{11}$$

where s_0 and s_1 are the s-coordinates of the particle at times t_0 and t_1, respectively. Geometrically, this is the slope of the secant line connecting the points (t_0, s_0) and (t_1, s_1) on the position versus time curve (Figure 1.5.8b). The quantity $\Delta s = s_1 - s_0$ is called the *displacement* or *change in position* of the particle during the time interval from t_0 to t_1. With this terminology, Formula (11) states that for a particle in rectilinear motion *the average velocity over a time interval is the displacement during the time interval divided by the length of the time interval.* For example, if a car moving in one direction along a straight road travels 75 miles in 3 hours, then its average velocity is $75/3 = 25$ mi/h.

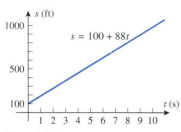

Figure 1.5.9

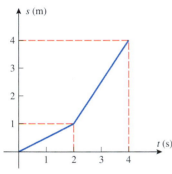

Position versus time curve for a particle with coordinate s_0 at time $t = 0$ and moving with constant velocity v

Figure 1.5.10

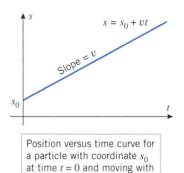

Figure 1.5.11

....

CONSTANT ACCELERATION

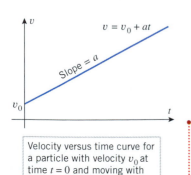

Velocity versus time curve for a particle with velocity v_0 at time $t = 0$ and moving with constant acceleration a

Figure 1.5.12

In the special case where the average velocity of a particle in rectilinear motion is the same over every time interval, the particle is said to have **constant velocity** or **uniform rectilinear motion**. If the average velocity over every time interval is v, then we will refer to v as the **velocity** of the particle (dropping the adjective "average").

For a particle with uniform rectilinear motion the displacement over *any* time interval is given by the formula

$$\text{displacement} = \text{velocity} \times \text{elapsed time} \tag{12}$$

Example 3   Suppose that a car moves with a constant velocity of 88 ft/s in the positive direction of an s-axis. Given that the s-coordinate of the car at time $t = 0$ is $s = 100$, find an equation for s as a function of t, and graph the position versus time curve.

Solution.   It follows from (12) that in a period of t seconds, the car will move $88t$ feet from its starting point, so its coordinate s at time t will be

$$s = 100 + 88t$$

The graph of this equation is the line in Figure 1.5.9.   ◀

It is not accidental that the position versus time curve turned out to be a line in the last example; this will always be the case for uniform rectilinear motion. To see why this is so, suppose that a particle moves with constant velocity v in the positive direction along an s-axis, starting at the point s_0 at time $t = 0$. It follows from (12) that in t units of time the particle will move vt units from its starting point s_0, so its coordinate s at time t will be

$$s = s_0 + vt$$

which is a line with s-intercept s_0 and slope v (Figure 1.5.10). It follows from this equation and 1.5.1 that we can view the velocity v as the rate of change of s with respect to t, that is, the rate of change of position with respect to time.

Example 4   Figure 1.5.11 shows the position versus time curve for a particle moving along an s-axis. Describe the motion of the particle in words.

Solution.   At time $t = 0$ the particle is at the origin. From time $t = 0$ to $t = 2$ the slope of the line segment is $\frac{1}{2}$, so the particle is moving with a constant velocity of $\frac{1}{2} = 0.5$ m/s. At time $t = 2$ the particle is at the point $s = 1$ (i.e., 1 meter from the origin). From time $t = 2$ to $t = 4$ the slope of the line segment is $\frac{3}{2}$, so the particle is moving with a constant velocity of $\frac{3}{2} = 1.5$ m/s. At time $t = 4$ it is at the point $s = 4$.   ◀

In everyday language we say that an object is "accelerating" if it is speeding up and "decelerating" if it is slowing down. Mathematically, the **acceleration** of a particle in rectilinear motion is defined to be the *rate of change of velocity with respect to time*, where the acceleration is positive if the velocity is increasing and negative if it is decreasing. Thus, for a particle that moves in the positive direction of an s-axis, negative acceleration means the particle is "decelerating" in everyday language. Acceleration, like velocity, can be variable or constant. For example, by pressing the gas pedal of a car toward the floor smoothly, the driver can make the car's velocity increase at a constant rate (a constant acceleration); however, if the driver suddenly slams the pedal to the floor, the car will lurch forward, reflecting a nonconstant acceleration. Later in the text we will study acceleration in more depth, but for now we will only consider the case in which acceleration is constant.

REMARK.   The units of acceleration are units of velocity divided by units of time. For example, if the velocity of a particle is increasing at a rate of 3 feet per second each second, then its acceleration is 3 ft/s/s (velocity in ft/s divided by time in s); this is usually written as 3 ft/s² (read "3 feet per second per second" or "3 feet per second squared"). Similarly, if the velocity of a particle is decreasing at a rate of 3 feet per second each second, then it has an acceleration of -3 ft/s².

Graphical information about the acceleration of a particle can be obtained from the graph of velocity versus time; this is called the **velocity versus time curve**. In the case where the particle has constant acceleration, the velocity versus time curve will be linear, and its slope, which is the rate of change of velocity with time, will be the acceleration (Figure 1.5.12).

Example 5   Suppose that a car moves in the positive direction of an s-axis in such a way that its velocity v increases at a constant rate of 2 ft/s^2.

(a)   Assuming that the velocity of the car is 88 ft/s at time $t = 0$, find an equation for v as a function of t.

(b)   Make a graph of velocity versus time, and mark the point on the graph at which the car attains a velocity of 100 ft/s.

Solution (a).   Since the rate of change of v with respect to t is 2 ft/s^2, and since $v = 88$ ft/s if $t = 0$, the equation for velocity as a function of time is

$$v = 88 + 2t \tag{13}$$

Solution (b).   To find the time it takes for the car to reach a velocity of 100 ft/s, we substitute $v = 100$ in (13) and solve for t. This yields $t = 6$. The graph of (13) and the point at which the velocity reaches 100 ft/s is shown in Figure 1.5.13.   ◄

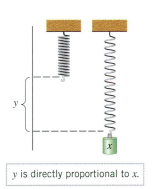

Velocity versus time curve for a particle with a velocity of 88 ft/s at time $t = 0$ and moving with a constant acceleration of 2 ft/s^2

Figure 1.5.13

DIRECT PROPORTION

Recall that a variable y is said to be **directly proportional** to a variable x if there is a positive constant k, called the **constant of proportionality**, such that

$$y = kx \tag{14}$$

The graph of this equation is a line through the origin whose slope k is the constant of proportionality. Thus, linear functions are appropriate in physical problems where one variable is directly proportional to another.

Hooke's law[*] in physics provides a nice example of direct proportion. It follows from this law that if a weight of x units is suspended from a spring, then the spring will be stretched by an amount y that is directly proportional to x, that is, $y = kx$ (Figure 1.5.14). The constant k depends on the stiffness of the spring: the stiffer the spring, the smaller the value of k (why?).

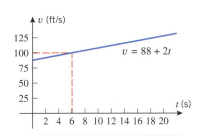

y is directly proportional to x.

Figure 1.5.14

Example 6   Figure 1.5.15 shows an old-fashioned spring scale that is calibrated in pounds.

(a)   Given that the pound scale marks are 0.5 in apart, find an equation that expresses the length y that the spring is stretched (in inches) in terms of the suspended weight x (in pounds).

(b)   Graph the equation obtained in part (a).

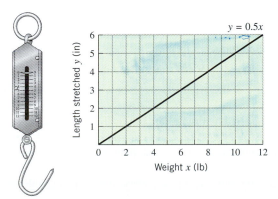

Figure 1.5.15

---

[*]Hooke's law, named for the English physicist Robert Hooke (1635–1703), applies only for small displacements that do not stretch the spring to the point of permanently distorting it.

Solution (a). It follows from Hooke's law that y is related to x by an equation of the form $y = kx$. To find k we rewrite this equation as $k = y/x$ and use the fact that a weight of $x = 1$ lb stretches the spring $y = 0.5$ in. Thus,

$$k = \frac{y}{x} = \frac{0.5}{1} = 0.5 \qquad \text{and hence} \qquad y = 0.5x$$

Solution (b). The graph of the equation $y = 0.5x$ is shown in Figure 1.5.15.   ◄

LINEAR DATA

One method for determining whether n points

$$(x_1, y_1), (x_2, y_2), \dots, (x_n, y_n)$$

lie on a line is to compare the slopes of the line segments joining successive points. The points lie on a line if and only if those slopes are equal (Figure 1.5.16).

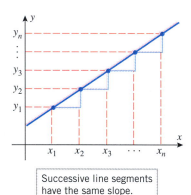

Successive line segments have the same slope.

Figure 1.5.16

Table 1.5.1

x	y
1.5	0.3
2.5	1.1
3.5	1.9
5.5	3.5
9.5	6.7

Example 7   Consider the data in Table 1.5.1.

(a)   Explain why a linear function is appropriate for the data in the table.

(b)   Find a linear equation that relates x and y, and graph the equation and the data together.

Solution (a). The five data points lie on a line, since each 1-unit increase in x produces a corresponding 0.8-unit increase in y. Thus, the slope of the line segment joining any two successive data points is

$$m = \frac{\Delta y}{\Delta x} = \frac{0.8}{1} = 0.8$$

Solution (b). A linear equation relating x and y can be obtained from the point-slope form of the line using the slope $m = 0.8$ calculated in part (a) and any one of the five data points. If we use the first data point, $(1.5, 0.3)$, we obtain

$$y - 0.3 = 0.8(x - 1.5)$$

or in slope-intercept form,

$$y = 0.8x - 0.9$$

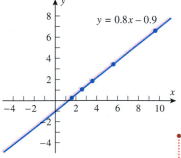

Figure 1.5.17

The graph of this equation together with the given data are shown in Figure 1.5.17.   ◄

REMARK.   Sometimes, data points that should theoretically lie on a line do not because of experimental error and other factors. In such cases curve-fitting techniques are used to find a line that most closely fits the data. Such techniques will be discussed later in the text.

OTHER APPLICATIONS OF LINEAR FUNCTIONS

Linear functions arise in a variety of practical problems. Here is a typical example.

Example 8   A university parking lot charges $3.00 per day but offers a $40.00 monthly sticker with which the student pays only $0.25 per day.

(a) Find equations for the cost C of parking for x days per month under both payment methods, and graph the equations for $0 \leq x \leq 30$. (Treat C as a continuous function of x, even though x only assumes integer values.)

(b) Find the value of x for which the graphs intersect, and discuss the significance of this value.

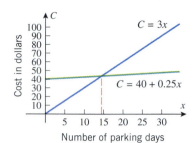

Figure 1.5.18

Solution (a). The cost in dollars of parking for x days at $3.00 per day is $C = 3x$, and the cost for the $40.00 sticker plus x days at $0.25 per day is $C = 40 + 0.25x$ (Figure 1.5.18).

Solution (b). The graphs intersect at the point where

$$3x = 40 + 0.25x$$

which is $x = 40/2.75 \approx 14.5$. This value of x is not an option for the student, since x must be an integer. However, it is the dividing point at which the monthly sticker method becomes less expensive than the daily payment method; that is, for $x \geq 15$ it is cheaper to buy the monthly sticker and for $x \leq 14$ it is cheaper to pay the daily rate. ◀

EXERCISE SET 1.5   Graphing Utility

Exercises 1–26 involve the basic properties of lines and slope. In some of these exercises you will need to use slopes to determine whether two lines are parallel or perpendicular. If you have forgotten how to do this, review Appendix C.

1. (a) Find the slopes of the sides of the triangle with vertices $(0, 3)$, $(2, 0)$, and $\left(6, \frac{8}{3}\right)$.

   (b) Is this a right triangle? Explain.

2. (a) Find the slopes of the sides of the quadrilateral with vertices $(-3, -1)$, $(5, -1)$, $(7, 3)$, and $(-1, 3)$.

   (b) Is this a parallelogram? Explain.

3. List the lines in the accompanying figure in the order of increasing slope.

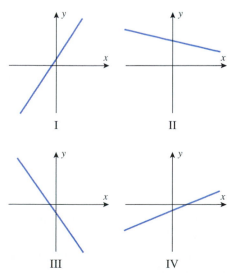

Figure Ex-3

4. List the lines in the accompanying figure in the order of increasing slope.

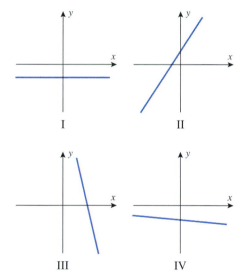

Figure Ex-4

5. Use slopes to determine whether the given points lie on the same line.

   (a) $(1, 1)$, $(-2, -5)$, and $(0, -1)$

   (b) $(-2, 4)$, $(0, 2)$, and $(1, 5)$

6. A particle, initially at $(7, 5)$, moves along a line of slope $m = -2$ to a new position (x, y).

   (a) Find y if $x = 9$.   (b) Find x if $y = 12$.

7. A particle, initially at $(1, 2)$, moves along a line of slope $m = 3$ to a new position (x, y).

   (a) Find y if $x = 5$.   (b) Find x if $y = -2$.

8. Find x and y if the line through $(0, 0)$ and (x, y) has slope $\frac{1}{2}$, and the line through (x, y) and $(7, 5)$ has slope 2.

9. Find x if the slope of the line through $(1, 2)$ and $(x, 0)$ is the negative of the slope of the line through $(4, 5)$ and $(x, 0)$.

In Exercises 10 and 11, find the angle of inclination of the line with slope m to the nearest degree. Use a calculating utility, where needed.

10. (a) $m = \frac{1}{2}$ (b) $m = -1$
(c) $m = 2$ (d) $m = -57$

11. (a) $m = -\frac{1}{2}$ (b) $m = 1$
(c) $m = -2$ (d) $m = 57$

In Exercises 12 and 13, find the angle of inclination of the line to the nearest degree. Use a calculating utility, where needed.

12. (a) $3y = 2 - \sqrt{3}x$ (b) $y - 4x + 7 = 0$

13. (a) $y = \sqrt{3}x + 2$ (b) $y + 2x + 5 = 0$

14. Find equations for the x- and y-axes.

In Exercises 15–22, find the slope-intercept form of the equation of the line satisfying the stated conditions, and check your answer using a graphing utility.

15. Slope $= -2$, y-intercept $= 4$

16. $m = 5$, $b = -3$

17. The line is parallel to $y = 4x - 2$ and y-intercept $= 7$.

18. The line is parallel to $3x + 2y = 5$ and passes through $(-1, 2)$.

19. The line is perpendicular to $y = 5x + 9$ and y-intercept $= 6$.

20. The line is perpendicular to $x - 4y = 7$ and passes through $(3, -4)$.

21. The line passes through $(2, 4)$ and $(1, -7)$.

22. The line passes through $(-3, 6)$ and $(-2, 1)$.

23. In each part, classify the lines as parallel, perpendicular, or neither.
(a) $y = 4x - 7$ and $y = 4x + 9$
(b) $y = 2x - 3$ and $y = 7 - \frac{1}{2}x$
(c) $5x - 3y + 6 = 0$ and $10x - 6y + 7 = 0$
(d) $Ax + By + C = 0$ and $Bx - Ay + D = 0$
(e) $y - 2 = 4(x - 3)$ and $y - 7 = \frac{1}{4}(x - 3)$

24. In each part, classify the lines as parallel, perpendicular, or neither.
(a) $y = -5x + 1$ and $y = 3 - 5x$
(b) $y - 1 = 2(x - 3)$ and $y - 4 = -\frac{1}{2}(x + 7)$
(c) $4x + 5y + 7 = 0$ and $5x - 4y + 9 = 0$
(d) $Ax + By + C = 0$ and $Ax + By + D = 0$
(e) $y = \frac{1}{2}x$ and $x = \frac{1}{2}y$

In Exercises 25 and 26, use the graph to find the equation of the line in slope-intercept form, and then check your result by using a graphing utility to graph the equation.

25.

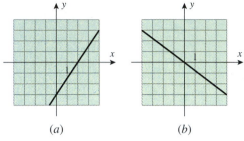

(a) (b)

Figure Ex-25

26.

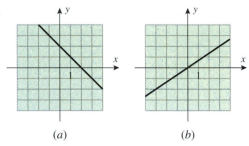

(a) (b)

Figure Ex-26

27. The accompanying figure shows the position versus time curve for a particle moving along an x-axis.
(a) What is the velocity of the particle?
(b) What is the x-coordinate of the particle at time $t = 0$?
(c) What is the x-coordinate of the particle at time $t = 2$?
(d) At what time does the particle have an x-coordinate of $x = 4$?

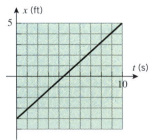

Figure Ex-27

28. A particle moving along an x-axis with constant velocity is at the point $x = 1$ when $t = 2$ and is at the point $x = 5$ when $t = 4$.
(a) Find the velocity of the particle if x is in meters and t is in seconds.
(b) Find an equation that expresses x as a function of t.
(c) What is the coordinate of the particle at time $t = 0$?

29. A particle moving along an x-axis with constant acceleration has velocity $v = 3$ ft/s at time $t = 1$ and velocity $v = -1$ ft/s at time $t = 4$.
    (a) Find the acceleration of the particle.
    (b) Find an equation that expresses v as a function of t.
    (c) What is the velocity of the particle at time $t = 0$?

30. The accompanying figure shows the velocity versus time curve for a particle moving along the x-axis.
    (a) What is the acceleration of the particle?
    (b) What is the velocity of the particle at time $t = 0$?
    (c) What is the velocity of the particle at time $t = 2$?
    (d) At what time does the particle have a velocity of $v = 3$ ft/s?

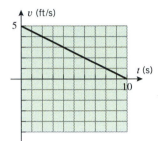

Figure Ex-30

31. The accompanying figure shows the position versus time curve for a particle moving along an x-axis.
    (a) Describe the motion of the particle in words.
    (b) Find the average velocity of the particle from $t = 0$ to $t = 10$.
    (c) Find the average speed of the particle from $t = 0$ to $t = 10$.

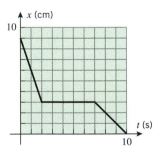

Figure Ex-31

32. The accompanying figure shows the velocity versus time curve for a particle moving along an x-axis. Describe the motion of the particle in words.

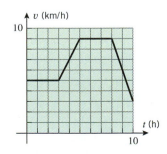

Figure Ex-32

33. A locomotive travels on a straight track at a constant speed of 40 mi/h, then reverses direction and returns to its starting point, traveling at a constant speed of 60 mi/h.
    (a) What is the average velocity for the round-trip?
    (b) What is the average speed for the round-trip?
    (c) What is the total distance traveled by the train if the total trip took 5 h?

34. A ball is tossed straight up at time $t = 0$ with an initial velocity of 64 ft/s. We will show later using basic principles of physics that the velocity of the ball as a function of time is $v = 64 - 32t$.
    (a) What direction is the ball traveling 3 s after it is released? Explain your reasoning.
    (b) At what time does the ball reach its maximum height above the ground? Explain your reasoning.
    (c) What can you say about the acceleration of the ball?

35. A car is stopped at a toll booth on a straight highway. Starting at time $t = 0$ it accelerates at a constant rate of 10 ft/s^2 for 10 s. It then travels at a constant speed of 100 ft/s for 90 s. At that time it begins to decelerate at a constant rate of 5 ft/s^2 for 20 s, at which point in time it reaches a full stop at a traffic light.
    (a) Sketch the velocity versus time curve.
    (b) Express v as a piecewise function of t.

36. Make a reasonable sketch of a position versus time curve for a particle that moves in the positive x-direction with positive constant acceleration.

37. A spring with a natural length of 15 in stretches to a length of 20 in when a 45-lb object is suspended from it.
    (a) Use Hooke's law to find an equation that expresses the amount y by which the spring is stretched (in inches) in terms of the suspended weight x (in pounds).
    (b) Graph the equation obtained in part (a).
    (c) Find the length of the spring when a 100-lb object is suspended from it.
    (d) What is the largest weight that can be suspended from the spring if the spring cannot be stretched to more than twice its natural length?

38. The spring in a heavy-duty shock absorber has a natural length of 3 ft and is compressed 0.2 ft by a load of 1 ton. An additional load of 5 tons compresses the spring an additional 1 ft.
    (a) Assuming that Hooke's law applies to compression as well as extension, find an equation that expresses the length y that the spring is compressed from its natural length (in feet) in terms of the load x (in tons).
    (b) Graph the equation obtained in part (a).
    (c) Find the amount that the spring is compressed from its natural length by a load of 3 tons.
    (d) Find the maximum load that can be applied if safety regulations prohibit compressing the spring to less than half its natural length.

In Exercises 39 and 40, confirm that a linear function is appropriate for the relationship between x and y. Find a linear equation relating x and y, and verify that the data points lie on the graph of your equation.

39.

x	0	1	2	4	6
y	2	3.2	4.4	6.8	9.2

40.

x	−1	0	2	5	8
y	12.6	10.5	6.3	0	−6.3

41. There are two common systems for measuring temperature, Celsius and Fahrenheit. Water freezes at $0°$ Celsius ($0°$C) and $32°$ Fahrenheit ($32°$F); it boils at $100°$C and $212°$F.

(a) Assuming that the Celsius temperature T_C and the Fahrenheit temperature T_F are related by a linear equation, find the equation.

(b) What is the slope of the line relating T_F and T_C if T_F is plotted on the horizontal axis?

(c) At what temperature is the Fahrenheit reading equal to the Celsius reading?

(d) Normal body temperature is $98.6°$F. What is it in $°$C?

42. Thermometers are calibrated using the so-called "triple point" of water, which is 273.16 K on the Kelvin scale and $0.01°$C on the Celsius scale. A one-degree difference on the Celsius scale is the same as a one-degree difference on the Kelvin scale, so there is a linear relationship between the temperature T_C in degrees Celsius and the temperature T_K in kelvins.

(a) Find an equation that relates T_C and T_K.

(b) Absolute zero (0 K on the Kelvin scale) is the temperature below which a body's temperature cannot be lowered. Express absolute zero in $°$C.

43. To the extent that water can be assumed to be incompressible, the pressure p in a body of water varies linearly with the distance h below the surface.

(a) Given that the pressure is 1 atmosphere (1 atm) at the surface and 5.9 atm at a depth of 50 m, find an equation that relates pressure to depth.

(b) At what depth is the pressure twice that at the surface?

44. A resistance thermometer is a device that determines temperature by measuring the resistance of a fine wire whose resistance varies with temperature. Suppose that the resistance R in ohms (Ω) varies linearly with the temperature T in $°$C and that $R = 123.4$ Ω when $T = 20°$C and that $R = 133.9$ Ω when $T = 45°$C.

(a) Find an equation for R in terms of T.

(b) If R is measured experimentally as 128.6 Ω, what is the temperature?

45. Suppose that the mass of a spherical mothball decreases with time, due to evaporation, at a rate that is proportional to its surface area. Assuming that it always retains the shape of a sphere, it can be shown that the radius r of the sphere decreases linearly with the time t.

(a) If, at a certain instant, the radius is 0.80 mm and 4 days later it is 0.75 mm, find an equation for r (in millimeters) in terms of the elapsed time t (in days).

(b) How long will it take for the mothball to completely evaporate?

46. The accompanying figure shows three masses suspended from a spring: a mass of 11 g, a mass of 24 g, and an unknown mass of W g.

(a) What will the pointer indicate on the scale if no mass is suspended?

(b) Find W.

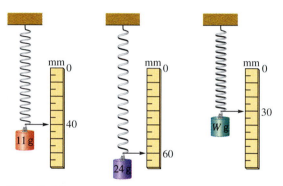

Figure Ex-46

47. The price for a round-trip bus ride from a university to center city is $2.00, but it is possible to purchase a monthly commuter pass for $25.00 with which each round-trip ride costs an additional $0.25.

(a) Find equations for the cost C of making x round-trips per month under both payment plans, and graph the equations for $0 \le x \le 30$ (treating C as a continuous function of x, even though x assumes only integer values).

(b) How many round-trips per month would a student have to make for the commuter pass to be worthwhile?

48. A student must decide between buying one of two used cars: car A for $4000 or car B for $5500. Car A gets 20 miles per gallon of gas, and car B gets 30 miles per gallon. The student estimates that gas will run $1.25 per gallon. Both cars are in excellent condition, so the student feels that repair costs should be negligible for the foreseeable future. How many miles would the student have to drive before car B becomes the better buy?

1.6 FAMILIES OF FUNCTIONS

Functions are often grouped into families according to the form of their defining formulas or other common characteristics. In this section we will discuss some of the most basic families of functions.

This section includes quick reviews of precalculus material on polynomials and trigonometry. Readers who want to review this material in more depth are referred to Appendices E and F. Instructors who want to spend some additional time on precalculus review can divide this section into two parts, covering the trigonometry material in a second lecture.

FAMILIES OF LINES

A function f whose values are all the same is called a **constant function**. For example, the formula $f(x) = c$ defines the constant function whose value is c for all x. The graph of the constant function $f(x) = c$ is the horizontal line $y = c$ (Figure 1.6.1a). If we vary c, then we obtain a set or **family** of horizontal lines (Figure 1.6.1b).

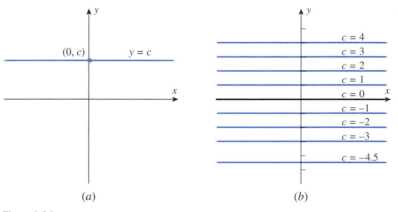

(a)   (b)

Figure 1.6.1

REMARK. The expression $f(x) = c$ can be confusing because it can be interpreted either as an equation that is satisfied for certain x (as in $x^2 = c$) or as an identity that is satisfied for all x; it is the latter interpretation that defines a constant function. Thus, when you see an expression of the form $f(x) = c$, you will have to determine from its context whether it is intended as an equation or a constant function.

The quantities m and b in the equation $y = mx + b$ can be viewed as unspecified constants whose values may change from one application to another; such changeable constants are called **parameters**.

If we keep b fixed and vary the parameter m in the equation $y = mx + b$, then we obtain a family of lines whose members all have y-intercept b (Figure 1.6.2a); and if we keep m fixed and vary the parameter b, then we obtain a family of parallel lines whose members all have slope m (Figure 1.6.2b).

Example 1

(a) Find an equation for the family of lines with slope $\frac{1}{2}$.

(b) Find the member of the family in part (a) that passes through the point $(4, 1)$.

(c) Find an equation for the family of lines whose members are perpendicular to the lines in part (a).

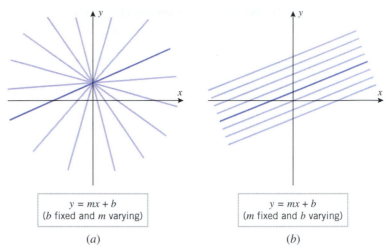

Figure 1.6.2

Solution (a). The lines of slope $\frac{1}{2}$ are of the form

$$y = \tfrac{1}{2}x + b \qquad (1)$$

where the parameter b can have any real value.

Solution (b). To find the line in the family that passes through the point $(4, 1)$, we must find the value of b for which the coordinates $x = 4$ and $y = 1$ satisfy (1). Substituting these coordinates into (1) and solving for b yields $b = -1$, and hence the equation of the line is

$$y = \tfrac{1}{2}x - 1 \qquad (2)$$

(Figure 1.6.3*a*).

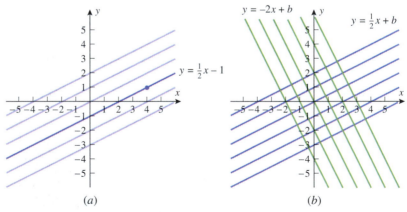

Figure 1.6.3

Solution (c). Since the slopes of perpendicular lines are negative reciprocals, it follows that the lines perpendicular to those in part (a) have slope -2 and hence are of the form

$$y = -2x + b$$

Some typical lines in families (1) and (2) are graphed in Figure 1.6.3*b*. ◀

THE FAMILY $y = x^n$

A function of the form $f(x) = x^p$, where p is constant, is called a ***power function***. Consider the case where p is a positive integer, say $p = n$. The graphs of the curves $y = x^n$ for $n = 1, 2, 3, 4,$ and 5 are shown in Figure 1.6.4. The first graph is the line $y = x$ with slope 1 that passes through the origin, and the second is a parabola that opens up and has its vertex at the origin (see Appendix D).

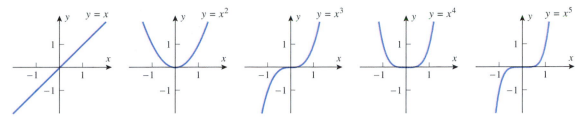

Figure 1.6.4

For $n > 2$ the shape of the graph of $y = x^n$ depends on whether n is even or odd (Figure 1.6.5). For even values of n the graphs have the same general shape as the parabola $y = x^2$ (though they are not actually parabolas if $n > 2$), and for odd values of n greater than 1 they have the same general shape as $y = x^3$. The graphs in the family $y = x^n$ share a number of important characteristics:

- For even values of n the functions $f(x) = x^n$ are even, and their graphs are symmetric about the y-axis; for odd values of n the functions $f(x) = x^n$ are odd, and their graphs are symmetric about the origin.

- For all values of n the graphs pass through the origin and the point $(1, 1)$. For even values of n the graphs pass through $(-1, 1)$, and for odd values of n they pass through $(-1, -1)$.

- Increasing n causes the graph to become flatter over the interval $-1 < x < 1$ and steeper over the intervals $x > 1$ and $x < -1$.

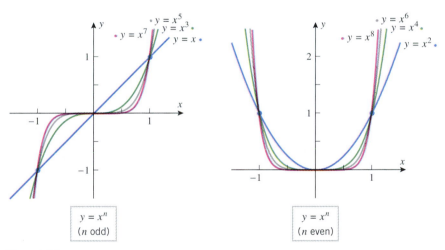

Figure 1.6.5

REMARK.   The last characteristic can be explained numerically by considering the effect of raising a real number x to successively higher powers. If x is a fraction, that is, $-1 < x < 1$, then the absolute value of x^n *decreases* as n increases (try raising $\frac{1}{2}$ or $-\frac{1}{2}$ to higher and higher powers, for example). This explains why successive graphs in Figure 1.6.5 become flatter over the interval $-1 < x < 1$. On the other hand, if $x > 1$ or $x < -1$, then the absolute value of x^n *increases* as n increases (try raising 2 or -2 to higher and higher powers). This explains why successive graphs become steeper if $x > 1$ or $x < -1$.

THE FAMILY $y = x^{-n}$

If p is a negative integer, say $p = -n$, then the power functions $f(x) = x^p$ have the form $f(x) = x^{-n} = 1/x^n$. Figure 1.6.6a shows the graphs of $y = 1/x$ and $y = 1/x^2$, and Figure 1.6.6b shows how these graphs relate to other members of the family. The graph of $y = 1/x$ is called an ***equilateral hyperbola*** (for reasons to be discussed later).

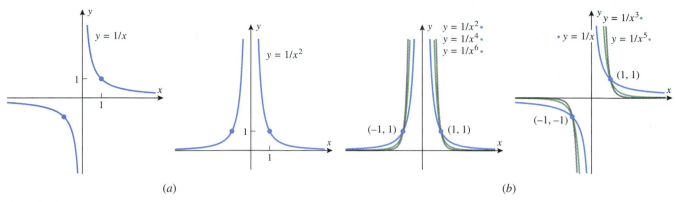

Figure 1.6.6

For odd values of n the graphs have the same general shape as $y = 1/x$, and for even values of n they have the same general shape as $y = 1/x^2$. The graphs in the family $y = 1/x^n$ share a number of important characteristics:

- For even values of n the functions $f(x) = 1/x^n$ are even, and their graphs are symmetric about the y-axis; for odd values of n the functions $f(x) = x^n$ are odd, and their graphs are symmetric about the origin.

- For all values of n the graphs pass through the point $(1, 1)$ and have a break (called a **discontinuity**) at $x = 0$. This is caused by the division by zero that occurs when $x = 0$. For even values of n the graphs pass through $(-1, 1)$, and for odd values of n they pass through $(-1, -1)$.

- Increasing n causes the graph to become steeper over the intervals $-1 < x < 0$ and $0 < x < 1$, and flatter over the intervals $x > 1$ and $x < -1$.

REMARK.   The last characteristic can be explained numerically by considering the effect of raising the reciprocal of a number x to successively higher powers. If x is a nonzero fraction, then it lies in the interval $-1 < x < 1$, and its reciprocal satisfies $1/x > 1$ or $1/x < -1$. Thus, as n increases the absolute value of $1/x^n$ also increases. This explains why successive graphs in Figure 1.6.6 become successively steeper over the interval $-1 < x < 1$. On the other hand, if $x > 1$ or $x < -1$, then $-1 < 1/x < 1$. Thus, as n increases the absolute value of $1/x^n$ *decreases*. This explains why successive graphs in Figure 1.6.6 get successively flatter if $x > 1$ or $x < -1$.

THE FAMILY $y = x^{1/n}$

If $p = 1/n$, where n is a positive integer, then the power functions $f(x) = x^p$ have the form $f(x) = x^{1/n} = \sqrt[n]{x}$. In particular, if $n = 2$, then $f(x) = \sqrt{x}$, and if $n = 3$, then $f(x) = \sqrt[3]{x}$. The graphs of these functions are shown in parts (a) and (b) of Figure 1.6.7.

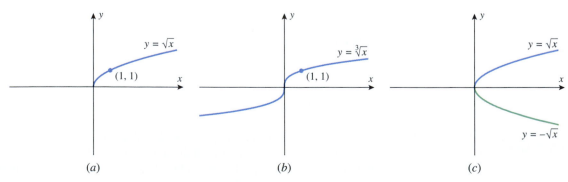

Figure 1.6.7

Observe that the graph of $y = \sqrt[3]{x}$ extends over the entire x-axis because $f(x) = \sqrt[3]{x}$ is defined for all real values of x (every real number has a cube root); in contrast, the graph of $y = \sqrt{x}$ only extends over the nonnegative x-axis (negative numbers have imaginary square roots). Observe also that the graph of $y = \sqrt{x}$ is the upper half of the parabola $x = y^2$ (Figure 1.6.7c).

For even values of n the graphs of $y = \sqrt[n]{x}$ have the same general shape as $y = \sqrt{x}$, and for odd values of n they have the same general shape as $y = \sqrt[3]{x}$.

FOR THE READER.   Sketch the graphs of $y = \sqrt[n]{x}$ for $n = 2, 4, 6$ on one set of axes and for $n = 3, 5, 7$ on another set. Use a graphing utility to check your work.

POWER FUNCTIONS WITH FRACTIONAL AND IRRATIONAL EXPONENTS

Power functions can also have fractional or irrational exponents. For example,

$$f(x) = x^{2/3}, \quad f(x) = \sqrt[5]{x^3}, \quad f(x) = x^{-7/8}, \quad \text{and} \quad f(x) = x^{\sqrt{2}} \tag{3}$$

are all power functions of this type; we will discuss power functions of these forms in later sections.

FOR THE READER.   The graph of $f(x) = x^{2/3}$ is given in Figure 1.3.16b. Read the note preceding Exercise 29 of Section 1.3, and use a graphing utility to generate graphs of $f(x) = \sqrt[5]{x}$ and $f(x) = x^{-7/8}$ that show all of their significant features.

INVERSE PROPORTIONS

Recall that a variable y is said to be *inversely proportional to a variable x* if there is a positive constant k, called the *constant of proportionality*, such that

$$y = \frac{k}{x} \tag{4}$$

Since k is assumed to be positive, the graph of this equation has the same basic shape as $y = 1/x$ but is compressed or stretched in the y-direction.

Observe that in Formula (4) doubling x decreases y by a factor of $1/2$, tripling x decreases y by a factor of $1/3$, and, more generally, increasing x by a factor of r decreases y by a factor of $1/r$.

Functions involving inverse proportion arise in various laws of physics. For example, *Boyle's law* in physics states that at a constant temperature the pressure P exerted by a fixed quantity of an ideal gas is inversely proportional to the volume V occupied by the gas, that is,

$$P = \frac{k}{V}$$

Figure 1.6.8 shows a temperature control unit that uses Boyle's law.

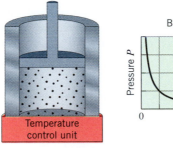

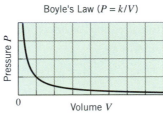

Boyle's Law ($P = k/V$)

Pressure P

Volume V

Temperature control unit

As the volume of the gas changes, the temperature control unit adds or removes heat to maintain a constant temperature.

Figure 1.6.8

If y is inversely proportional to x, then it follows from (4) that the product of y and x is constant, since $yx = k$. This provides a useful way of identifying inverse proportionality in experimental data.

Example 2   Table 1.6.1 shows some experimental data.

(a)   Explain why the data suggest that y is inversely proportional to x.

(b)   Express y as a function of x.

(c)   Graph your function and the data together for $x \geq 0$.

$y = \dfrac{5}{x}$

Figure 1.6.9

Table 1.6.1

EXPERIMENTAL DATA

x	0.8	1	2.5	4	6.25	10
y	6.25	5	2	1.25	0.8	0.5

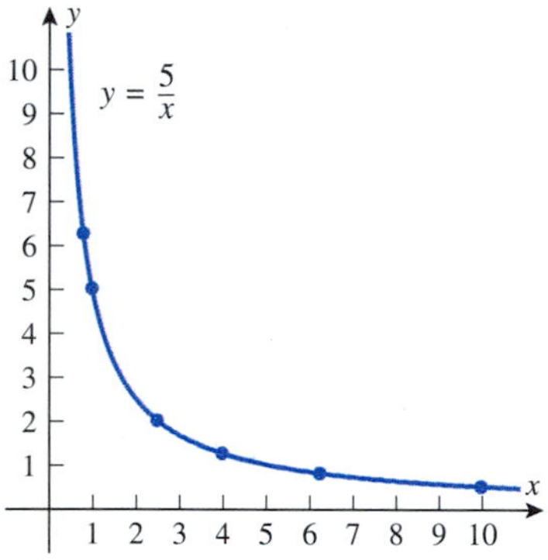

Solution.   For every data point we have $xy = 5$, so y is inversely proportional to x and $y = 5/x$. The graph of this equation with the data points is shown in Figure 1.6.9.   ◀

A QUICK REVIEW OF POLYNOMIALS

A detailed review of polynomials is given in Appendix F, but for convenience we will review some of the terminology here.

A **polynomial in x** is a function that is expressible as a sum of finitely many terms of the form cx^n, where c is a constant and n is a nonnegative integer. Some examples of polynomials are

$$2x + 1, \quad 3x^2 + 5x - \sqrt{2}, \quad x^3, \quad 4\,(= 4x^0), \quad 5x^7 - x^4 + 3$$

The function $\left(x^2 - 4\right)^3$ is also a polynomial because it can be expanded by the binomial formula (see the inside front cover) and expressed as a sum of terms of the form cx^n:

$$\left(x^2 - 4\right)^3 = \left(x^2\right)^3 - 3\left(x^2\right)^2(4) + 3(x^2)(4^2) - (4^3) = x^6 - 12x^4 + 48x^2 - 64 \quad (5)$$

A general polynomial can be written in either of the following forms, depending on whether one wants the powers of x in ascending or descending order:

$$c_0 + c_1 x + c_2 x^2 + \cdots + c_n x^n$$
$$c_n x^n + c_{n-1} x^{n-1} + \cdots + c_1 x + c_0$$

The constants $c_0, c_1, \ldots, c_n$ are called the **coefficients** of the polynomial. When a polynomial is expressed in one of these forms, the highest power of x that occurs with a nonzero coefficient is called the **degree** of the polynomial. Nonzero constant polynomials are considered to have degree 0, since we can write $c = cx^0$. Polynomials of degree 1, 2, 3, 4, and 5 are described as **linear**, **quadratic**, **cubic**, **quartic**, and **quintic**, respectively. For example,

$$3 + 5x \qquad \boxed{\text{Has degree 1 (linear)}}$$

$$x^2 - 3x + 1 \qquad \boxed{\text{Has degree 2 (quadratic)}}$$

$$2x^3 - 7 \qquad \boxed{\text{Has degree 3 (cubic)}}$$

$$8x^4 - 9x^3 + 5x - 3 \qquad \boxed{\text{Has degree 4 (quartic)}}$$

$$\sqrt{3} + x^3 + x^5 \qquad \boxed{\text{Has degree 5 (quintic)}}$$

$$\left(x^2 - 4\right)^3 \qquad \boxed{\text{Has degree 6 [see (5)]}}$$

The natural domain of a polynomial in x is $(-\infty, +\infty)$, since the only operations involved are multiplication and addition; the range depends on the particular polynomial. We already know that the graphs of polynomials of degree 0 and 1 are lines and that the graphs of polynomials of degree 2 are parabolas. Figure 1.6.10 shows the graphs of some typical polynomials of higher degree. Later, we will discuss polynomial graphs in detail, but for now it suffices to observe that graphs of polynomials are very well behaved in the sense that they have no discontinuities or sharp corners. As illustrated in Figure 1.6.10, the graphs of polynomials wander up and down for awhile in a roller-coaster fashion, but eventually that behavior stops and the graphs steadily rise or fall indefinitely as one travels along the curve in either the positive or negative direction. We will see later that the number of peaks and valleys is less than the degree of the polynomial.

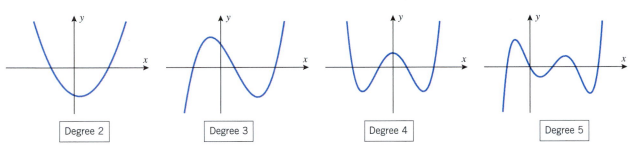

Figure 1.6.10

RATIONAL FUNCTIONS

A function that can be expressed as a ratio of two polynomials is called a *rational function*. If $P(x)$ and $Q(x)$ are polynomials, then the domain of the rational function

$$f(x) = \frac{P(x)}{Q(x)}$$

consists of all values of x such that $Q(x) \neq 0$. For example, the domain of the rational function

$$f(x) = \frac{x^2 + 2x}{x^2 - 1}$$

consists of all values of x, except $x = 1$ and $x = -1$. Its graph is shown in Figure 1.6.11 along with the graphs of two other typical rational functions.

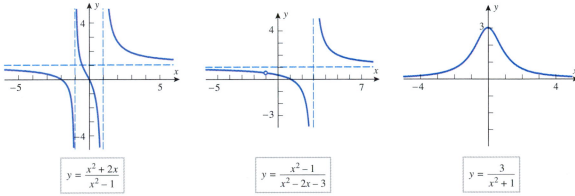

$$y = \frac{x^2 + 2x}{x^2 - 1} \qquad y = \frac{x^2 - 1}{x^2 - 2x - 3} \qquad y = \frac{3}{x^2 + 1}$$

Figure 1.6.11

The graphs of rational functions with nonconstant denominators differ from the graphs of polynomials in some essential ways:

- Unlike polynomials whose graphs are continuous (unbroken) curves, the graphs of rational functions have discontinuities at the points where the denominator is zero.

- Unlike polynomials, rational functions may have numbers at which they are not defined. Near such points, many (but not all) rational functions have graphs that approximate a vertical line, called a ***vertical asymptote***. These are represented by the dashed vertical lines in Figure 1.6.11.

- Unlike the graphs of nonconstant polynomials, which eventually rise or fall indefinitely, the graphs of many (but not all) rational functions eventually get closer and closer to some horizontal line, called a ***horizontal asymptote***, as one travels along the curve in either the positive or negative direction. The horizontal asymptotes are represented by the dashed horizontal lines in the first two parts of Figure 1.6.11; in the third part of the figure the x-axis is a horizontal asymptote.

ALGEBRAIC FUNCTIONS

Functions that can be constructed from polynomials by applying finitely many algebraic operations (addition, subtraction, division, and root extraction) are called ***algebraic functions***. Some examples are

$$f(x) = \sqrt{x^2 - 4}, \quad f(x) = 3\sqrt[3]{x}(2 + x), \quad f(x) = x^{2/3}(x + 2)^2$$

As illustrated in Figure 1.6.12, the graphs of algebraic functions vary widely, so it is difficult to make general statements about them. Later in this text we will develop general calculus methods for analyzing such functions.

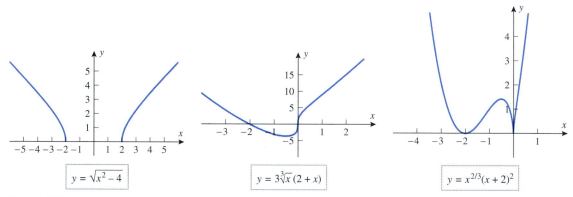

$$y = \sqrt{x^2 - 4} \qquad y = 3\sqrt[3]{x}\,(2 + x) \qquad y = x^{2/3}(x + 2)^2$$

Figure 1.6.12

A QUICK REVIEW OF TRIGONOMETRIC FUNCTIONS

A detailed review of trigonometric functions is given in Appendix E, but for convenience we will summarize some of the main ideas here.

It is often convenient to think of the trigonometric functions in terms of circles rather than triangles. For this purpose, consider a point that moves either clockwise or counterclockwise along the ***unit circle*** $u^2 + v^2 = 1$ in the uv-plane, starting at $(1, 0)$ and stopping at a point P (Figure 1.6.13a). Let x denote the ***signed*** arc length traveled by the moving point, taking x to be positive for counterclockwise motion and negative for clockwise motion. (We allow for the possibility that the point may traverse the circle more than once.) When convenient, the variable x can also be interpreted as the angle in radians that is swept out by the radial line from the origin to P, with the usual convention that angles are positive if generated by counterclockwise rotations and negative if generated by clockwise rotations. We can *define* $\cos x$ to be the u-coordinate of P and $\sin x$ to be the v-coordinate of P (Figure 1.6.13b).

The remaining trigonometric functions can be defined in terms of the functions $\sin x$ and $\cos x$:

$$\tan x = \frac{\sin x}{\cos x} \qquad \cot x = \frac{\cos x}{\sin x}$$

$$\sec x = \frac{1}{\cos x} \qquad \csc x = \frac{1}{\sin x}$$

The graphs of the six trigonometric functions in Figure 1.6.14 should already be familiar

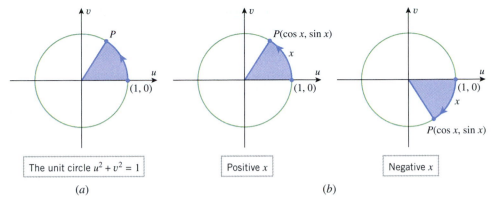

Figure 1.6.13

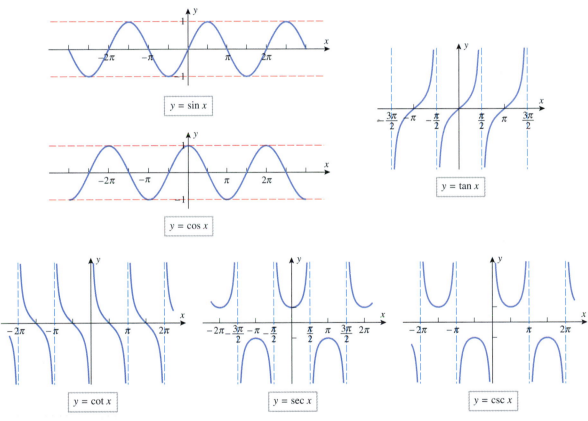

Figure 1.6.14

to you, but try generating them using a graphing utility, making sure to use radian measure for x.

⋮ REMARK.   In this text we will always assume that the independent variable in a trigonometric function is in radians unless specifically stated otherwise.

PROPERTIES OF sin x, cos x, AND tan x

Many of the basic properties of $\sin x$ and $\cos x$ can be deduced from the circle definitions of these functions. For example:

•   As the point $P(\cos x, \sin x)$ moves around the unit circle, its coordinates vary between -1 and 1, and hence

$$-1 \le \sin x \le 1 \quad \text{and} \quad -1 \le \cos x \le 1$$

- If x increases or decreases by 2π radians, then the point $P(\cos x, \sin x)$ makes one complete revolution around the unit circle, and the coordinates return to their starting values. Thus, $\sin x$ and $\cos x$ have period 2π; that is,

$$\sin(x \pm 2\pi) = \sin x$$
$$\cos(x \pm 2\pi) = \cos x$$

- As $P(\cos x, \sin x)$ moves around the unit circle, $\sin x$ is zero when P is on the horizontal axis (which occurs when x is an integer multiple of π), and $\cos x$ is zero when P is on the vertical axis (which occurs when x is an odd multiple of $\pi/2$). Thus,

$$\sin x = 0 \quad \text{if and only if} \quad x = 0, \pm\pi, \pm2\pi, \pm3\pi, \dots$$
$$\cos x = 0 \quad \text{if and only if} \quad x = \pm\pi/2, \pm3\pi/2, \pm5\pi/2, \dots$$

- As $P(\cos x, \sin x)$ moves around the unit circle $u^2 + v^2 = 1$, its coordinates satisfy this equation for all x, which produces the fundamental trigonometric identity

$$\cos^2 x + \sin^2 x = 1$$

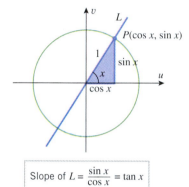

Slope of $L = \dfrac{\sin x}{\cos x} = \tan x$

Figure 1.6.15

Observe that the graph of $y = \tan x$ has vertical asymptotes at the points $x = \pm\pi/2$, $\pm3\pi/2, \pm5\pi/2, \dots$. This is to be expected since $\tan x = \sin x/\cos x$, and these are the values of x at which $\cos x$ is zero. What is less obvious, however, is the fact that $\tan x$ repeats every π radians (i.e., has period π), even though $\sin x$ and $\cos x$ have period 2π. This can be explained by interpreting

$$\tan x = \frac{\sin x}{\cos x}$$

as the slope of the line L that passes through the origin and the point $P(\cos x, \sin x)$ on the unit circle in the uv-plane (Figure 1.6.15). Each time x increases or decreases by π radians, the point P traverses half the circumference, and the line L rotates π radians, so its starting and ending slope are the same.

RADIANS AS A DIMENSIONLESS UNIT

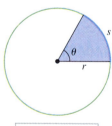

If θ is in radians, then $s = r\theta$.

Figure 1.6.16

The choice of radian measure as opposed to degree measure depends on the nature of the problem being considered; degree measure is usually chosen in engineering problems involving measurements of angles, and radian measure is usually chosen when the function properties of $\sin x$, $\cos x$, $\tan x$, ... are the primary focus. Radian measure is also usually chosen in problems involving arc lengths on circles because of the basic result in trigonometry which states that the arc length s of a sector with radius r and a central angle of θ (radians) is given by

$$s = r\theta \tag{6}$$

(Figure 1.6.16).

In applications involving angles, radians require special treatment to ensure that quantities are assigned proper units. To see why this is so, let us rewrite (6) as

$$\theta = \frac{s}{r}$$

The left side of this equation is in radians, and the right side is the ratio of two lengths, say meters/meters or feet/feet. However, because these units of length cancel, the right side of this equation is actually *dimensionless* (has no units). Thus, to ensure consistency between the two sides of the equation, we would have to omit the units of radians on the left side to make it dimensionless as well. In practical terms this means that units of radians can be used in intermediate computations, when convenient, but they need to be omitted in the end result to ensure consistency of units. This is confusing, to say the least, but the following example should clarify the idea.

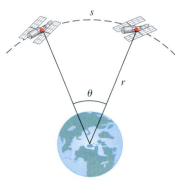

Figure 1.6.17

Example 3   Suppose that two satellites circle the equator in an orbit whose radius is $r = 4.23 \times 10^7$ m (Figure 1.6.17). Find the arc length s that separates the satellites if they have an angular separation of $\theta = 2.00°$.

Solution. To apply Formula (6), we must convert the angular separation to radians:

$$2.00° = \frac{\pi}{180}(2.00) \approx 0.0349 \text{ rad}$$

Thus, from (6)

$$s = r\theta \approx (4.23 \times 10^7 \text{ m})(0.0349 \text{ rad}) \approx 1.48 \times 10^6 \text{ m}$$

In this computation the product $r\theta$ produces units of meters × radians, but if we treat radians as dimensionless, we have meters × radians = meters, which correctly produces units of meters (m) for the arc length s. ◀

THE FAMILIES $y = A$ sin Bx AND $y = A$ cos Bx

Many important applications lead to trigonometric functions of the form

$$f(x) = A\sin(Bx - C) \quad \text{and} \quad g(x) = A\cos(Bx - C) \tag{7}$$

where A, B, and C are nonzero constants. The graphs of such functions can be obtained by stretching, compressing, translating, and reflecting the graphs of $y = \sin x$ and $y = \cos x$ appropriately. To see why this is so, let us start with the case where $C = 0$ and consider how the graphs of the equations

$$y = A\sin Bx \quad \text{and} \quad y = A\cos Bx$$

relate to the graphs of $y = \sin x$ and $y = \cos x$. If A and B are positive, then the effect of the constant A is to stretch or compress the graphs of $y = \sin x$ and $y = \cos x$ vertically and the effect of the constant B is to compress or stretch the graphs of $\sin x$ and $\cos x$ horizontally. For example, the graph of $y = 2\sin 4x$ can be obtained by stretching the graph of $y = \sin x$ vertically by a factor of 2 and compressing it horizontally by a factor of 4. (Recall from Section 1.4 that the multiplier of x *stretches* when it is less than 1 and *compresses* when it is greater than 1.) Thus, as shown in Figure 1.6.18, the graph of $y = 2\sin 4x$ varies between -2 and 2, and repeats every $2\pi/4 = \pi/2$ units.

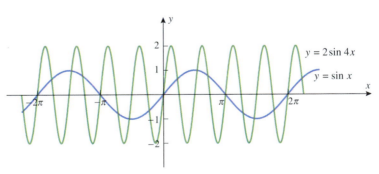

Figure 1.6.18

In general, if A and B are positive numbers, then the graphs of

$$y = A\sin Bx \quad \text{and} \quad y = A\cos Bx$$

oscillate between $-A$ and A and repeat every $2\pi/B$ units, so we say that these functions have ***amplitude*** A and ***period*** $2\pi/B$. In addition, we define the ***frequency*** of these functions to be the reciprocal of the period, that is, the frequency is $B/2\pi$. If A or B is negative, then these constants cause reflections of the graphs about the axes as well as compressing or stretching them; and in this case the amplitude, period, and frequency are given by

$$\text{amplitude} = |A|, \qquad \text{period} = \frac{2\pi}{|B|}, \qquad \text{frequency} = \frac{|B|}{2\pi}$$

Example 4   Make sketches of the following graphs that show the period and amplitude.

(a) $y = 3 \sin 2\pi x$ (b) $y = -3 \cos 0.5x$ (c) $y = 1 + \sin x$

Solution (a).   The equation is of the form $y = A \sin Bx$ with $A = 3$ and $B = 2\pi$, so the graph has the shape of a sine function, but with amplitude $A = 3$ and period $2\pi/B = 2\pi/2\pi = 1$ (Figure 1.6.19a).

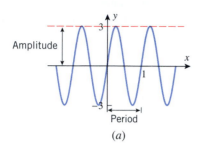 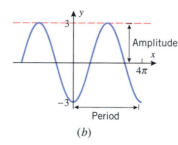 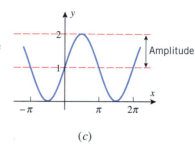

Figure 1.6.19   (a)   (b)   (c)

Solution (b).   The equation is of the form $y = A \cos Bx$ with $A = -3$ and $B = 0.5$, so the graph has the shape of a cosine curve that has been reflected about the x-axis (because $A = -3$ is negative), but with amplitude $|A| = 3$ and period $2\pi/B = 2\pi/0.5 = 4\pi$ (Figure 1.6.19b).

Solution (c).   The graph has the shape of a sine curve that has been translated up 1 unit (Figure 1.6.19c). ◀

THE FAMILIES $y = A \sin(Bx - C)$ AND $y = A \cos(Bx - C)$

To investigate the graphs of the more general families

$$y = A \sin(Bx - C) \quad \text{and} \quad y = A \cos(Bx - C)$$

it will be helpful to rewrite these equations as

$$y = A \sin\left[B\left(x - \frac{C}{B}\right)\right] \quad \text{and} \quad y = A \cos\left[B\left(x - \frac{C}{B}\right)\right]$$

In this form we see that the graphs of these equations can be obtained by translating the graphs of $y = A \sin Bx$ and $y = A \cos Bx$ to the left or right, depending on the sign of C/B. For example, if $C/B > 0$, then the graph of

$$y = A \sin[B(x - C/B)] = A \sin(Bx - C)$$

can be obtained by translating the graph of $y = A \sin Bx$ to the right by C/B units (Figure 1.6.20). If $C/B < 0$, the graph of $y = A \sin(Bx - C)$ is obtained by translating the graph of $y = A \sin Bx$ to the left by $|C/B|$ units.

Example 5   Find the amplitude and period of

$$y = 3 \cos\left(2x + \frac{\pi}{2}\right)$$

and determine how the graph of $y = 3 \cos 2x$ should be translated to produce the graph of this equation. Confirm your results by graphing the equation on a calculator or computer.

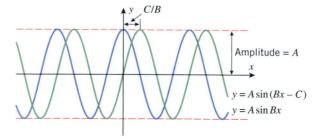

Figure 1.6.20

Solution. The equation can be rewritten as

$$y = 3\cos\left[2x - \left(-\frac{\pi}{2}\right)\right] = 3\cos\left[2\left(x - \left(-\frac{\pi}{4}\right)\right)\right]$$

which is of the form

$$y = A\cos\left[B\left(x - \frac{C}{B}\right)\right]$$

with $A = 3$, $B = 2$, and $C/B = -\pi/4$. Thus, the amplitude is $A = 3$, the period is $2\pi/B = \pi$, and the graph is obtained by translating the graph of $y = 3\cos 2x$ left by $|C/B| = \pi/4$ units (Figure 1.6.21). ◄

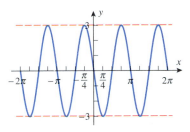

Figure 1.6.21

OTHER FAMILIES

In addition to the functions mentioned in this section, there are exponential and logarithmic functions, which we will study later, and various special functions that arise in physics and engineering. There are also many kinds of functions that have no names; indeed, one of the important themes of calculus is to provide methods for analyzing new types of functions.

EXERCISE SET 1.6 ∿ Graphing Utility

1. (a) Find an equation for the family of lines whose members have slope $m = 3$.
   (b) Find an equation for the member of the family that passes through $(-1, 3)$.
   (c) Sketch some members of the family, and label them with their equations. Include the line in part (b).

2. Find an equation for the family of lines whose members are perpendicular to those in Exercise 1.

3. (a) Find an equation for the family of lines with y-intercept $b = 2$.
   (b) Find an equation for the member of the family whose angle of inclination is $135°$.
   (c) Sketch some members of the family, and label them with their equations. Include the line in part (b).

4. Find an equation for
   (a) the family of lines that pass through the origin
   (b) the family of lines with x-intercept $a = 1$
   (c) the family of lines that pass through the point $(1, -2)$
   (d) the family of lines parallel to $2x + 4y = 1$.

In Exercises 5 and 6, state a geometric property common to all lines in the family, and sketch five of the lines.

5. (a) The family $y = -x + b$
   (b) The family $y = mx - 1$
   (c) The family $y = m(x + 4) + 2$
   (d) The family $x - ky = 1$

6. (a) The family $y = b$
   (b) The family $Ax + 2y + 1 = 0$
   (c) The family $2x + By + 1 = 0$
   (d) The family $y - 1 = m(x + 1)$

7. Find an equation for the family of lines tangent to the circle with center at the origin and radius 3.

8. Find an equation for the family of lines that pass through the intersection of $5x - 3y + 11 = 0$ and $2x - 9y + 7 = 0$.

9. The U.S. Internal Revenue Service uses a 10-year linear depreciation schedule to determine the value of various business items. This means that an item is assumed to have a value of zero at the end of the tenth year and that at intermediate times the value is a linear function of the elapsed time. Sketch some typical depreciation lines, and explain the practical significance of the y-intercepts.

10. Find all lines through $(6, -1)$ for which the product of the x- and y-intercepts is 3.

11. In each part, match the equation with one of the accompanying graphs.
   (a) $y = \sqrt[5]{x}$
   (b) $y = 2x^5$
   (c) $y = -1/x^8$
   (d) $y = \sqrt{x^2 - 1}$
   (e) $y = \sqrt[4]{x - 2}$
   (f) $y = -\sqrt[5]{x^2}$

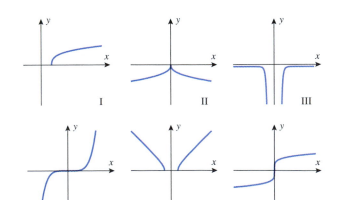

Figure Ex-11

12. The accompanying table gives approximate values of three functions: one of the form kx^2, one of the form kx^{-3}, and one of the form $kx^{3/2}$. Identify which is which, and estimate k in each case.

x	0.25	0.37	2.1	4.0	5.8	6.2	7.9	9.3
$f(x)$	640	197	1.08	0.156	0.0513	0.0420	0.0203	0.0124
$g(x)$	0.0312	0.0684	2.20	8.00	16.8	19.2	31.2	43.2
$h(x)$	0.250	0.450	6.09	16.0	27.9	30.9	44.4	56.7

Table Ex-12

In Exercises 13 and 14, sketch the graph of the equation for $n = 1, 3,$ and 5 in one coordinate system and for $n = 2, 4,$ and 6 in another coordinate system. Check your work with a graphing utility.

13. (a) $y = -x^n$ (b) $y = 2x^{-n}$ (c) $y = (x - 1)^{1/n}$

14. (a) $y = 2x^n$ (b) $y = -x^{-n}$
(c) $y = -3(x + 2)^{1/n}$

15. (a) Sketch the graph of $y = ax^2$ for $a = \pm 1, \pm 2,$ and ± 3 in a single coordinate system.
(b) Sketch the graph of $y = x^2 + b$ for $b = \pm 1, \pm 2,$ and ± 3 in a single coordinate system.
(c) Sketch some typical members of the family of curves $y = ax^2 + b$.

16. (a) Sketch the graph of $y = a\sqrt{x}$ for $a = \pm 1, \pm 2,$ and ± 3 in a single coordinate system.
(b) Sketch the graph of $y = \sqrt{x} + b$ for $b = \pm 1, \pm 2,$ and ± 3 in a single coordinate system.
(c) Sketch some typical members of the family of curves $y = a\sqrt{x} + b$.

In Exercises 17–20, sketch the graph of the equation by making appropriate transformations to the graph of a basic power function. Check your work with a graphing utility.

17. (a) $y = 2(x + 1)^2$ (b) $y = -3(x - 2)^3$
(c) $y = \dfrac{-3}{(x + 1)^2}$ (d) $y = \dfrac{1}{(x - 3)^5}$

18. (a) $y = 1 - \sqrt{x + 2}$ (b) $y = 1 - \sqrt[3]{x + 2}$
(c) $y = \dfrac{5}{(1 - x)^3}$ (d) $y = \dfrac{2}{(4 + x)^4}$

19. (a) $y = \sqrt[3]{x + 1}$ (b) $y = 1 - \sqrt{x - 2}$
(c) $y = (x - 1)^5 + 2$ (d) $y = \dfrac{x + 1}{x}$

20. (a) $y = 1 + \dfrac{1}{x - 2}$ (b) $y = \dfrac{1}{1 + 2x + x^2}$
(c) $y = -\dfrac{2}{x^7}$ (d) $y = x^2 + 2x$

21. Sketch the graph of $y = x^2 + 2x$ by completing the square and making appropriate transformations to the graph of $y = x^2$.

22. (a) Use the graph of $y = \sqrt{x}$ to help sketch the graph of $y = \sqrt{|x|}$.
(b) Use the graph of $y = \sqrt[3]{x}$ to help sketch the graph of $y = \sqrt[3]{|x|}$.

23. As discussed in this section, Boyle's law states that at a constant temperature the pressure P exerted by a gas is related to the volume V by the equation $P = k/V$.
(a) Find the appropriate units for the constant k if pressure (which is force per unit area) is in newtons per square meter (N/m^2) and volume is in cubic meters (m^3).
(b) Find k if the gas exerts a pressure of 20,000 N/m^2 when the volume is 1 liter (0.001 m^3).
(c) Make a table that shows the pressures for volumes of 0.25, 0.5, 1.0, 1.5, and 2.0 liters.
(d) Make a graph of P versus V.

24. A manufacturer of cardboard drink containers wants to construct a closed rectangular container that has a square base and will hold $\frac{1}{10}$ liter (100 cm^3). Estimate the dimension of the container that will require the least amount of material for its manufacture.

A variable y is said to be *inversely proportional to the square of a variable* x if y is related to x by an equation of the form $y = k/x^2$, where k is a nonzero constant, called the *constant of proportionality*. This terminology is used in Exercises 25 and 26.

25. According to *Coulomb's law*, the force F of attraction between positive and negative point charges is inversely proportional to the square of the distance x between them.
(a) Assuming that the force of attraction between two point charges is 0.0005 newton when the distance between them is 0.3 meter, find the constant of proportionality (with proper units).
(b) Find the force of attraction between the point charges when they are 3 meters apart.
(c) Make a graph of force versus distance for the two charges.
(d) What happens to the force as the particles get closer and closer together? What happens as they get farther and farther apart?

26. It follows from Newton's Law of Universal Gravitation that the weight W of an object (relative to the Earth) is inversely proportional to the square of the distance x between the object and the center of the Earth, that is, $W = C/x^2$.
(a) Assuming that a weather satellite weighs 2000 pounds on the surface of the Earth and that the Earth is a sphere of radius 4000 miles, find the constant C.

(b) Find the weight of the satellite when it is 1000 miles above the surface of the Earth.

(c) Make a graph of the satellite's weight versus its distance from the center of the Earth.

(d) Is there any distance from the center of the Earth at which the weight of the satellite is zero? Explain your reasoning.

27. In each part, match the equation with one of the accompanying graphs, and give the equations for the horizontal and vertical asymptotes.

(a) $y = \dfrac{x^2}{x^2 - x - 2}$

(b) $y = \dfrac{x - 1}{x^2 - x - 6}$

(c) $y = \dfrac{2x^4}{x^4 + 1}$

(d) $y = \dfrac{4}{(x + 2)^2}$

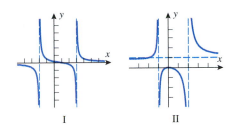

I   II

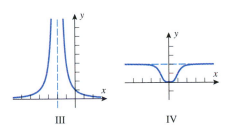

III   IV   Figure Ex-27

28. Find an equation of the form $y = k/(x^2 + bx + c)$ whose graph is a reasonable match to that in the accompanying figure. Check your work with a graphing utility.

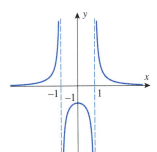

Figure Ex-28

In Exercises 29 and 30, draw a radial line from the origin with the given angle, and determine whether the six trigonometric functions are positive, negative, or undefined for that angle.

29. (a) $\dfrac{\pi}{3}$ (b) $-\dfrac{\pi}{2}$ (c) $\dfrac{2\pi}{3}$

(d) -1 (e) $\dfrac{5\pi}{4}$ (f) $\dfrac{11\pi}{6}$

30. (a) $\dfrac{3\pi}{2}$ (b) $-\dfrac{5\pi}{4}$ (c) π

(d) $\dfrac{5\pi}{2}$ (e) 4 (f) $-\dfrac{33\pi}{7}$

In Exercises 31 and 32, use a calculating utility set to the radian mode to confirm the approximations $\sin(\pi/5) \approx 0.588$ and $\cos(\pi/8) \approx 0.924$, and then use these values to approximate the given expressions by hand calculation. Check your answers using the trigonometric function operations of your calculating utility.

31. (a) $\sin \dfrac{4\pi}{5}$ (b) $\cos\left(-\dfrac{\pi}{8}\right)$ (c) $\sin \dfrac{11\pi}{5}$

(d) $\cos \dfrac{7\pi}{8}$ (e) $\cos^2 \dfrac{\pi}{5}$ (f) $\sin^2 \dfrac{2\pi}{5}$

32. (a) $\sin \dfrac{16\pi}{5}$ (b) $\cos\left(-\dfrac{17\pi}{8}\right)$ (c) $\sin \dfrac{41\pi}{5}$

(d) $\sin^2\left(-\dfrac{\pi}{16}\right)$ (e) $\cos \dfrac{27\pi}{8}$ (f) $\tan^2 \dfrac{\pi}{8}$

33. Assuming that $\sin \alpha = a$, $\cos \beta = b$, and $\tan \gamma = c$, express the stated quantities in terms of a, b, and c.

(a) $\sin(-\alpha)$ (b) $\cos(-\beta)$ (c) $\tan(-\gamma)$

(d) $\sin\left(\dfrac{\pi}{2} - \alpha\right)$ (e) $\cos(\pi - \beta)$ (f) $\sin(\alpha + \pi)$

(g) $\sin(2\beta)$ (h) $\cos(2\beta)$ (i) $\sec(\beta + 2\pi)$

(j) $\csc(\alpha + \pi)$ (k) $\cot(\gamma + 5\pi)$ (l) $\sin^2\left(\dfrac{\beta}{2}\right)$

34. A ship travels from a point near Hawaii at $20°$ N latitude directly north to a point near Alaska at $56°$ N latitude.

(a) Assuming the Earth to be a sphere of radius 4000 mi, find the actual distance traveled by the ship.

(b) What fraction of the Earth's circumference did the ship travel?

35. The Moon completes one revolution around the Earth in approximately 27.3 days. Assuming that the Moon's orbit is a circle with a radius of 0.38×10^9 m from the center of the Earth, find the arc length traveled by the Moon in 1 day.

36. A spoked wheel with a diameter of 3 ft rolls along a flat road without slipping. How far along the road does the wheel roll if the spokes turn through $225°$?

37. As illustrated in the accompanying figure (next page), suppose that you hold one quarter flat against a table while you rotate a second quarter around it without slippage. Through what angle will the second quarter have turned about its own center when it returns to its original location?

Figure Ex-37

38. Suppose that you begin cutting wedge-shaped pieces from a pie so that the arc length along the outer crust of each piece is equal to the radius. What fraction of the pie will remain after all pieces that can be cut in this way are eaten?

In Exercises 39 and 40, find an equation that has the form $y = D + A \sin Bx$ or $y = D + A \cos Bx$ for each graph.

39.

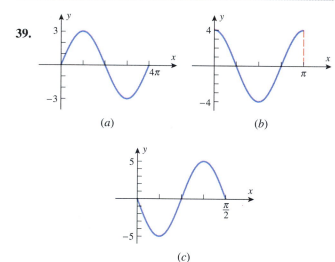

(a)

(b)

(c)

Figure Ex-39

40.

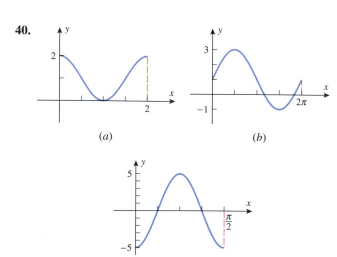

(a)

(b)

(c)

Figure Ex-40

41. In each part, find an equation for the graph that has the form $y = y_0 + A \sin(Bx - C)$.

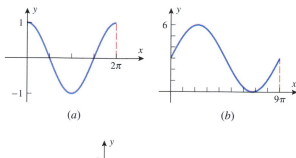

(a)

(b)

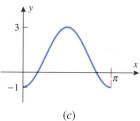

(c)

Figure Ex-41

42. In the United States, a standard electrical outlet supplies sinusoidal electrical current with a maximum voltage of $V = 120\sqrt{2}$ volts (V) at a frequency of 60 hertz (Hz). Write an equation that expresses V as a function of the time t, assuming that $V = 0$ if $t = 0$. [*Note:* 1 Hz = 1 cycle per second.]

In Exercises 43 and 44, find the amplitude, period, and phase shift, and sketch at least two periods of the graph by hand. Check your work with a graphing utility.

43. (a) $y = 3 \sin 4x$ (b) $y = -2 \cos \pi x$

(c) $y = 2 + \cos\left(\dfrac{x}{2}\right)$

44. (a) $y = -1 - 4 \sin 2x$ (b) $y = \frac{1}{2} \cos(3x - \pi)$

(c) $y = -4 \sin\left(\dfrac{x}{3} + 2\pi\right)$

45. Equations of the form

$$x = A_1 \sin \omega t + A_2 \cos \omega t$$

arise in the study of vibrations and other periodic motion.

(a) Use the trigonometric identity for $\sin(\alpha + \beta)$ to show that this equation can be expressed in the form

$$x = A \sin(\omega t + \theta)$$

(b) State formulas that express A and θ in terms of the constants A_1, A_2, and ω.

(c) Express the equation

$$x = 5\sqrt{3} \sin 2\pi t + \tfrac{5}{2} \cos 2\pi t$$

in the form $x = A \sin(\omega t + \theta)$, and use a graphing utility to confirm that both equations have the same graph.

46. Determine the number of solutions of $x = 2 \sin x$, and use a graphing or calculating utility to estimate them.

1.7 MATHEMATICAL MODELS

In this section we will introduce some simple mathematical models that are based on linear, quadratic, and trigonometric functions. The method of "least squares" will be discussed briefly within the context of linear and quadratic regression.

MATHEMATICAL MODELS

A **mathematical model** of a physical law or phenomenon is a description of that law or phenomenon in the language of mathematics. The modern scientific view about mathematical models was first expressed in the seventeenth century by the Italian mathematician and scientist Galileo Galilei (1564–1642) who wrote, "The book of nature is written in mathematics." This seemingly innocuous statement represented a major departure from the approach of the ancient Greek scientists and philosophers, who generally described scientific laws in words. For example, the Greek philosopher Aristotle (384 B.C.–322 B.C.) would have described the motion of a falling body by saying that such bodies "seek their natural position," whereas today we would use functions, equations, or other mathematical means to describe the motion precisely.

Mathematical models allow us to use mathematical methods to deduce results about the physical world that are not evident or have never been observed. For example, the possibility of placing a satellite in orbit around the Earth was deduced mathematically from Isaac Newton's model of mechanics nearly 200 years before the launching of *Sputnik*, and Albert Einstein (1879–1955) gave a relativistic model of mechanics in 1915 that explained a precession (position shift) in the perihelion of the planet Mercury that was not confirmed by physical measurement until 1967.

One of the most important steps in creating a mathematical model of a physical phenomenon is deciding which factors to consider and which to ignore—the more factors one takes into account the more complicated the formulas and equations of the model tend to become, so there is always a balance to be struck between keeping a model mathematically simple and considering enough factors to make the model useful. For example, if a meteorologist were trying to model the relationship between the speed of a raindrop when it hits the ground and the height of the cloud in which it was formed, then he or she would certainly have to take air resistance into account, but with equal certainty he or she would ignore the infinitesimal effect that Pluto's gravitational pull has on the raindrop. The danger is that in trying to keep a mathematical model from becoming too complicated one might oversimplify to the point where the results it produces do not agree with reality. We are reminded of this by Einstein's admonition: "Everything must be as simple as possible, but not simpler." A good mathematical model is one that produces results that are consistent with the physical world. If a time comes when the mathematical results produced by the model do not agree with real-world observations, then the model must be abandoned or modified in favor of a new model that does. This is the nature of the **scientific method**—old models constantly being replaced by new models that more accurately describe the real world.

FUNCTIONS AS MODELS

In this section we will consider some simple models that involve only two variables. In our general discussion we will refer to these variables as x and y, but in specific examples other letters will be more appropriate. We will assume that the data for the phenomenon being modeled consist of a collection of ordered pairs of measurements

$$(x_1, y_1), \ (x_2, y_2), \ (x_3, y_3), \ \ldots, (x_n, y_n)$$

that relate corresponding values of the variables x and y. We distinguish between two types of phenomena—**deterministic phenomena** in which each value of x determines one value of y and **probabilistic phenomena** in which the value of y associated with a specific x is not uniquely determined, but rather depends on probabilities in some way. For example, if y is the amount that a spring is stretched by a force x, then for a given spring the value of y is uniquely determined by the value of x, so this is a deterministic phenomenon. In contrast, if y is the weight of a person whose height is x, then the value of y is not uniquely

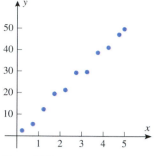

Figure 1.7.1

· ·

LINEAR FUNCTIONS AS MODELS

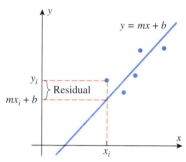

Figure 1.7.2

determined by x, but there is a ***correlation*** between weight and height that makes it more likely for a taller person to weigh more, so this is a probabilistic phenomenon.

In a deterministic model the variable y is a function of x, so the goal is to use the data in some reasonable way to find a formula $y = f(x)$ for this function. For example, Figure 1.7.1 strongly suggests that the relationship between x and y is linear, so in absence of additional information it would be natural to look for a linear function $y = mx + b$ as a model. In a probabilistic model the variable y need not be a function of x, so the goal is to find an equation $y = f(x)$ that specifies the *average value* of y that can be expected to occur for a given x. A more precise explanation of what is meant by "average value" and "expected to occur" requires ideas from probability and statistics, so we will depend on your intuition here.

Let us suppose that we have decided to model a particular phenomenon with a (yet to be determined) linear function $y = mx + b$. Ideally, we would then like to choose the parameters m and b such that the line $y = mx + b$ passes through all our data points. In practice, this may be impossible, either because of errors in our measurements or because there is not a strictly linear relationship between the variables x and y. We are then faced with the problem of finding the line $y = mx + b$ that "best fits" our set of data points. The key to determining this line is based on the following idea: For any proposed linear function $y = mx + b$, we draw a vertical connector from each data point (x_i, y_i) to the point $(x_i, mx_i + b)$ on the line and consider the differences $y_i - (mx_i + b)$. These differences, which are called ***residuals***, may be viewed as "errors" that result when the line is used to model the data (Figure 1.7.2). Data points above the line have positive errors, those below the line have negative errors, and those on the line have no error. One of the most common procedures is to look for a line such that *the sum of the squares* of the residuals is as small as possible. This line, known as a ***least squares line*** or ***linear regression line***, is one choice for a line that "best fits" a given set of data. Most graphing calculators, spreadsheets, and CAS programs provide methods for finding regression lines. We will assume that you have access to some such method in this section.

It is possible to compute a regression line, even in cases where the data have no apparent linear pattern. Thus, it is important to have some quantitative method of determining whether a linear model is appropriate for the data. The most common measure of linearity in data is called the ***correlation coefficient***. Following convention, we denote the correlation coefficient by the letter r. Although a detailed discussion of correlation coefficients is beyond the scope of this text, here are some of the basic facts:

- The values of r are in the interval $-1 \leq r \leq 1$, where r has the same sign as the slope of the regression line.

- If r is equal to 1 or -1, then the data points all lie on a line, so a linear model is a perfect fit for the data.

- If $r = 0$, then the data points exhibit no linear tendency, so a linear model is inappropriate for the data.

The closer r is to 1 or -1, the more tightly the data points hug the regression line and the more appropriate the regression line is as a model; the closer r is to 0, the more scattered the points and the less appropriate the regression line is as a model (Figure 1.7.3).

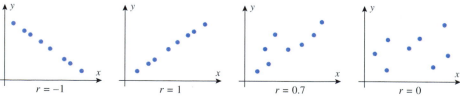

Figure 1.7.3

Roughly stated, the value of r^2 is a measure of the percentage of data points that fall in a "tight linear band." Thus $r = 0.5$ means that 25% of the points fall in a tight linear band, and $r = 0.9$ means that 81% of the points fall in a tight linear band. (A precise explanation of what is meant by a "tight linear band" requires ideas from statistics.)

Example 1   Table 1.7.1 gives a set of data points relating the pressure p in atmospheres (atm) and the temperature T (in °C) of a fixed quantity of carbon dioxide in a closed cylinder. The associated scatter plot in Figure 1.7.4a suggests that there is a linear relationship between the pressure and the temperature.

Table 1.7.1

TEMPERATURE T (°C)	PRESSURE p (atm)
0	2.54
50	3.06
100	3.46
150	4.00
200	4.41

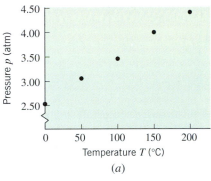

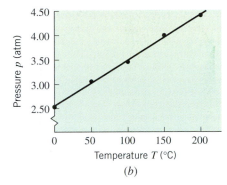

Figure 1.7.4

(a) Use your calculating utility to find the least squares line for the data. If your utility can produce the correlation coefficient, then find it.

(b) Use the model obtained in part (a) to predict the pressure when the temperature is 250°C.

(c) Use the model obtained in part (a) to predict a temperature at which the pressure of the gas will be zero.

Solution (a). The least squares line is given by $p = 0.00936T + 2.558$ (Figure 1.7.4b) with correlation coefficient $r = 0.998979$.

Solution (b). If $T = 250$, then $p = (0.00936)(250) + 2.558 = 4.898$ (atm).

Solution (c). Solving the equation $0 = p = 0.00936T + 2.558$ yields $T \approx -273.291$°C. ◀

It is not always convenient (or necessary) to obtain the least squares line for a linear phenomenon in order to create a model. In some cases, more elementary methods suffice. Here is an example.

Example 2   Figure 1.7.5a shows a graph of temperature versus altitude that was transmitted by the *Magellan* spacecraft when it entered the atmosphere of Venus in October 1991. The graph strongly suggests that there is a linear relationship between temperature and altitude for altitudes between 35 km and 60 km.

(a) Use the graph transmitted by the *Magellan* spacecraft to find a linear model of temperature versus altitude in the Venusian atmosphere that is valid for altitudes between 35 km and 60 km.

(b) Use the model obtained in part (a) to estimate the temperature at the surface of Venus, and discuss the assumptions you are making in obtaining the estimate.

Solution (a). Let T be the temperature in kelvins and h the altitude in kilometers. We will first estimate the slope m of the linear portion of the graph, then estimate the coordinates of a data point (h_1, T_1) on that portion of the graph, and then use the point-slope form of a line

$$T - T_1 = m(h - h_1) \tag{1}$$

The graph nearly passes through the point (60, 250), so we will take $h_1 \approx 60$ and $T_1 \approx 250$. In Figure 1.7.5*b* we have sketched a line that closely approximates the linear portion of the data. Using the intersections of that line with the edges of the grid box, we estimate the slope to be

$$m \approx \frac{100 - 490}{78 - 30} = -\frac{390}{48} = -8.125 \text{ K/km}$$

Substituting our estimates of h_1, T_1, and m into (1) yields the equation

$$T - 250 = -8.125(h - 60)$$

or equivalently,

$$T = -8.125h + 737.5 \tag{2}$$

Solution (b). The *Magellan* spacecraft stopped transmitting data at an altitude of approximately 35 km, so we cannot be certain that the linear model applies at lower altitudes. However, if we *assume* that the model is valid at all lower altitudes, then we can approximate the temperature at the surface of Venus by setting $h = 0$ in (2). We obtain $T \approx 737.5$ K. ◄

REMARK. The method of the preceding example is crude, at best, since it relies on extracting rough estimates of numerical data from a graph. Nevertheless, the final result is quite good, since the most recent information from NASA places the surface temperature of Venus at about 740 K (hot enough to melt lead).

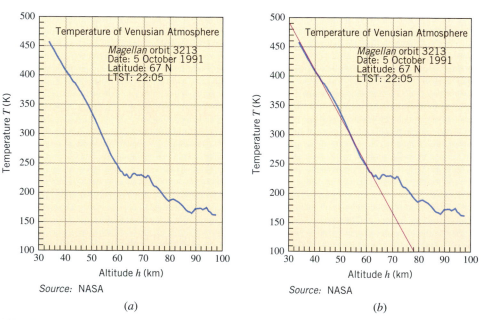

Figure 1.7.5

QUADRATIC AND TRIGONOMETRIC FUNCTIONS AS MODELS

Although models based on linear functions $y = mx + b$ are simple, the relationship between the variables x and y associated with a particular phenomenon may be nonlinear, in which case replacing the function $y = mx + b$ by the quadratic function $y = ax^2 + bx + c$ may provide a better model. Most calculators, spreadsheets, and CAS programs will perform a least squares quadratic regression in a manner that is similar to linear regression.

Table 1.7.2

TIME t (s)	HEIGHT h (cm)
0.008333	98.4
0.025	96.9
0.04167	95.1
0.05833	92.9
0.075	90.8
0.09167	88.1
0.10833	85.3
0.125	82.1
0.14167	78.6
0.15833	74.9

Example 3   A student in a physics lab is studying the equations of motion of a falling body. She collects the data displayed in Table 1.7.2, which gives the height of the object at a number of times over a 0.15-s period of time. She knows that if air resistance is negligible and if the acceleration of the object due to gravity is assumed to be constant, then the height h of the object should be a quadratic function of time t. A scatter plot of the data is provided in Figure 1.7.6a, which suggests a portion of an inverted parabola.

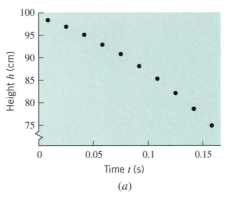 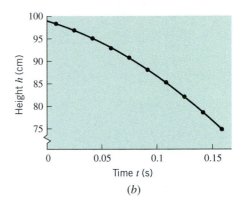

(a) (b)

Figure 1.7.6

(a)   Determine the quadratic regression curve for the data in Table 1.7.2.

(b)   According to the model obtained in part (a), when will the object strike the ground?

Solution (a).   Using the quadratic regression routine on a calculator, we find that the quadratic curve that best fits the data in Table 1.7.2 has equation

$$h = 99.02 - 73.21t - 499.13t^2$$

Figure 1.7.6b shows the data points and the graph of this quadratic function on the same set of axes. It appears that we have excellent agreement between our curve and the data.

Solution (b).   Solving the equation $0 = h = 99.02 - 73.21t - 499.13t^2$, we find that the object will strike the ground at $t \approx 0.38$ s.   ◀

The trigonometric functions $y = A \sin(Bx - C)$ and $y = A \cos(Bx - C)$ are particularly useful for modeling periodic phenomena.

Example 4   Figure 1.7.7a shows a table and scatter plot of temperature data recorded over a 24-hour period in the city of Philadelphia.[*] Find a function that models the data, and graph your function and data together.

Solution.   The pattern of the data suggests that the relationship between the temperature T and the time t can be modeled by a sinusoidal function that has been translated both horizontally and vertically, so we will look for an equation of the form

$$T = D + A \sin[Bt - C] = D + A \sin\left[B\left(t - \frac{C}{B}\right)\right] \tag{3}$$

Since the highest temperature is $95°$F and the lowest temperature is $75°$F, we take $2A = 20$ or $A = 10$. The midpoint between the high and low is $85°$F, so we have a vertical shift of $D = 85$. The period seems to be about 24, so $2\pi/B = 24$ or $B = \pi/12$. The horizontal shift appears to be about 10 (verify), so $C/B = 10$. Substituting these values in (3) yields

---

[*]This example is based on the article "Everybody Talks About It!—Weather Investigations," by Gloria S. Dion and Iris Brann Fetta, *The Mathematics Teacher*, Vol. 89, No. 2, February 1996, pp. 160–165.

PHILADELPHIA TEMPERATURES
FROM 1:00 A.M. TO 12:00 MIDNIGHT ON 27 AUGUST 1993
(t = HOURS AFTER MIDNIGHT AND T = DEGREES FAHRENHEIT)

	A.M.		P.M.	
	t	T	t	T
1:00	1	78°	13	91°
2:00	2	77°	14	93°
3:00	3	77°	15	94°
4:00	4	76°	16	95°
5:00	5	76°	17	93°
6:00	6	75°	18	92°
7:00	7	75°	19	89°
8:00	8	77°	20	86°
9:00	9	79°	21	84°
10:00	10	83°	22	83°
11:00	11	87°	23	81°
12:00	12	90°	24	79°

Source: Philadelphia Inquirer, 28 August 1993.

Figure 1.7.7

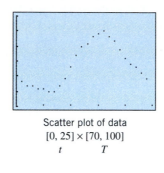

Scatter plot of data
$[0, 25] \times [70, 100]$
t T

(a)

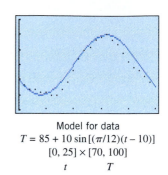

Model for data
$T = 85 + 10 \sin[(\pi/12)(t - 10)]$
$[0, 25] \times [70, 100]$
t T

(b)

the equation

$$T = 85 + 10 \sin\left[\frac{\pi}{12}(t - 10)\right]$$

(Figure 1.7.7*b*).   ◄

Note that in Example 4 we did not use a regression routine to fit the curve to the data. Some calculators may not be equipped to compute regression for trigonometric functions. In this case, we can use the calculator's graphing capability to see that a proposed model gives a *reasonable* fit to the data points, though it may not be the *best* fit.

FOR THE READER.   Using regression, a best fit to the data in Example 4 is

$$y = 84.2037 + 9.5964 \sin(0.2849t - 2.9300)$$

How does the graph of this best-fit curve compare with that found in Example 4?

EXERCISE SET 1.7   ⌇ Graphing Utility

1. One of the lines in the accompanying figure is the regression line. Which one is it?

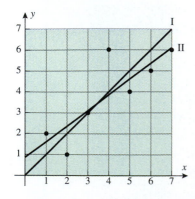

Figure Ex-1

2. In each part, find a model (linear, quadratic, or trigonometric) that reasonably describes the scatter plot in the accompanying figure.

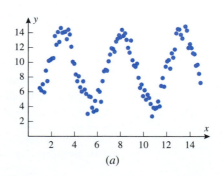

(a)

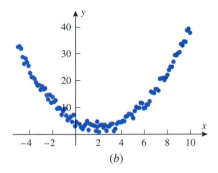

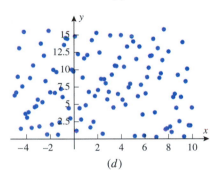

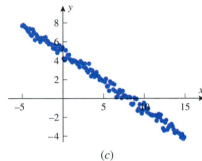

Figure Ex-2

5. A 20-liter container holds 100 g of N_2. The pressure p of this gas is measured at various temperatures T (see the accompanying table).
   (a) Find the least squares line for this collection of data points. If your calculating utility can produce the correlation coefficient, then find it.
   (b) Use the model obtained in part (a) to predict the pressure of the gas at a temperature of $-50°C$.
   (c) Use the model obtained in part (a) to predict a temperature at which the pressure of the gas will be zero.

TEMPERATURE T (°C)	PRESSURE p (atm)
0	3.99
25	4.34
50	4.70
75	5.08
100	5.45

Table Ex-5

6. A 40-liter container holds 20 g of H_2. The pressure p of this gas is measured at various temperatures T (see the accompanying table).
   (a) Find the least squares line for this collection of data points. If your calculating utility can produce the correlation coefficient, then find it.
   (b) Use the model obtained in part (a) to predict a temperature at which the pressure of the gas will be zero.
   (c) At approximately what temperature of the gas will a $10°C$ increase in temperature result in a 5% increase in pressure?

TEMPERATURE T (°C)	PRESSURE p (atm)
0	5.55
30	6.13
60	6.75
90	7.35
120	7.98

Table Ex-6

3. Table 1.1.1 provides data for the top qualifying speeds at the Indianapolis 500 from 1980 to 1999. Find the least squares line for these data. What is the correlation coefficient? Sketch the least squares line on a scatter plot of the data points.

4. A 25-liter container holds 150 g of O_2. The pressure p of the gas is measured at various temperatures T (see the accompanying table).
   (a) Determine the least squares line for the data given in the table.
   (b) Use the model obtained in part (a) to estimate the pressure of the gas at a temperature of $-50°C$.

TEMPERATURE T (°C)	PRESSURE p (atm)
0	4.18
50	4.96
100	5.74
150	6.49
200	7.26

Table Ex-4

7. The *resistivity* of a metal is a measure of the extent to which a wire made from the metal will resist the flow of electrical current. (The actual *resistance* of the wire will depend on both the resistivity of the metal and the dimensions of the wire.) A common unit for resistivity is the ohm-meter $(\Omega \cdot m)$. Experiments show that lowering the temperature of a metal also lowers its resistivity. The accompanying table gives the resistivity of copper at various temperatures.
   (a) Find the least squares line for this collection of data points.
   (b) Using the model obtained in part (a), at what temperature will copper have a resistivity of zero?

TEMPERATURE (°C)	RESISTIVITY ($10^{-8}\,\Omega \cdot m$)
−100	0.82
−50	1.19
0	1.54
50	1.91
100	2.27
150	2.63

Table Ex-7

8. The accompanying table gives the resistivity of tungsten at various temperatures.
   (a) Find the least squares line for this collection of data points.
   (b) Using the model obtained in part (a), at what temperature will tungsten have a resistivity of zero?

TEMPERATURE (°C)	RESISTIVITY ($10^{-8}\,\Omega \cdot m$)
−100	2.43
−50	3.61
0	4.78
50	5.96
100	7.16
150	8.32

Table Ex-8

9. The accompanying table gives the number of inches that a spring is stretched by various attached weights.
   (a) Use linear regression to express the amount of stretch of the spring as a function of the weight attached.
   (b) Use the model obtained in part (a) to determine the weight required to stretch the spring 8 in.

WEIGHT (lb)	STRETCH (in)
0	0
2	0.99
4	2.01
6	2.99
8	4.00
10	5.03
12	6.01

Table Ex-9

10. The accompanying table gives the number of inches that a spring is stretched by various attached weights.
   (a) Use linear regression to express the amount of stretch of the spring as a function of the weight attached.
   (b) Suppose that the spring has been stretched a certain amount by a weight and that adding another 5 lb to the weight doubles the stretch of the spring. Use the model obtained in part (a) to determine the original amount that the spring was stretched.

11. The accompanying table provides the heights and rebounds per minute for players on the 1998–1999 Davidson Col-

lege women's basketball team who played more than 100 minutes during the season.
   (a) Find the least squares line for these data. If your calculating utility can produce the correlation coefficient, then find it.
   (b) Sketch the least squares line on a scatter plot of the data points.
   (c) Is the least squares line a good model for these data? Explain.

WEIGHT (lb)	STRETCH (in)
0	0
1	0.73
2	1.50
3	2.24
4	3.02
5	3.77

Table Ex-10

HEIGHT	REBOUNDS PER MINUTE
5′11″	0.25
6′2″	0.176
5′6″	0.141
5′11″	0.162
6′1″	0.167
5′8″	0.091
5′11″	0.278
6′3″	0.167
6′0″	0.214

Table Ex-11

12. The accompanying table provides the heights and weights for players on the 1999–2000 Davidson College men's basketball team.
   (a) Find the least squares line for these data. If your calculating utility can produce the correlation coefficient, then find it.
   (b) Sketch the least squares line on a scatter plot of the data points.
   (c) Use this model to predict the weight of the team's new 7-ft recruit.

HEIGHT	WEIGHT (lb)
6′0″	165
6′0″	180
6′4″	195
6′3″	185
6′7″	210
6′4″	190
6′3″	190
6′9″	240
7′2″	280
5′10″	175
6′7″	215
6′7″	235
6′8″	225

Table Ex-12

13. (**The Age of the Universe**) In the early 1900s the astronomer Edwin P. Hubble (1889–1953) noted an unexpected relationship between the radial velocity of a galaxy and its distance d from any reference point (Earth, for example). That relationship, now known as **Hubble's law**, states that the galaxies are receding with a velocity v that is directly proportional to the distance d. This is usually expressed as $v = Hd$, where

H (the constant of proportionality) is called **Hubble's constant**. When applying this formula it is usual to express v in kilometers per second (km/s) and d in millions of light-years (Mly), in which case H has units of km/s/Mly. The accompanying figure shows an original plot and trend line of the velocity-distance relationship obtained by Hubble and a collaborator Milton L. Humason (1891–1972).

(a) Use the trend line in the figure to estimate Hubble's constant.

(b) An estimate of the age of the universe can be obtained by assuming that the galaxies move with constant velocity v, in which case v and d are related by $d = vt$. Assuming that the Universe began with a "big bang" that initiated its expansion, show that the Universe is roughly 1.5×10^{10} years old. [Use the conversion 1 Mly $\approx 9.048 \times 10^{18}$ km and take $H = 20$ km/s/Mly, which is in keeping with current estimates that place H between 15 and 27 km/s/Mly. (Note that the current estimates are significantly less than that resulting from Hubble's data.)]

(c) In a more realistic model of the Universe, the velocity v would decrease with time. What effect would that have on your estimate in part (b)?

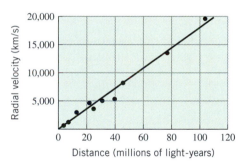

Figure Ex-13

14. A professor wishes to use midterm grades as a predictor of final grades in a small seminar that he teaches once a year. The midterm grades and final grades for last year's seminar are listed in the accompanying table.

(a) Find the linear regression model that expresses the final grade in terms of the midterm grade.

(b) Suppose that a student in this year's seminar earned a midterm grade of 88. Use the model obtained in part (a) to predict the student's final grade in the seminar.

MIDTERM GRADE	FINAL GRADE
78	78
94	91
78	76
84	82
95	92
96	93
77	75

Table Ex-14

15. A student is studying the equations of motion for an object moving along a number line with constant acceleration. The accompanying table gives the position in meters of the object at various times.

(a) Use quadratic regression to model the position of the object as a function of time.

(b) Based on the model obtained in part (a), what will be the position of the object after 2 s?

TIME (s)	POSITION (m)
0.2537	0.045
0.4064	0.09
0.5981	0.165
0.75	0.24
0.8781	0.315
1.032	0.42
1.1846	0.54
1.3208	0.66

Table Ex-15

16. Table 1.1.3 gives data for the U.S. population at 10-year intervals from 1790 to 1850. Use quadratic regression to model the U.S. population as a function of time since 1790. What does your model predict as the population of the United States in the year 2000? How accurate is this prediction?

17. The accompanying table gives the minutes of daylight predicted for Davidson, North Carolina, in 10-day increments during the year 2000. Find a function that models the data in this table, and graph your function on a scatter plot of the data.

DAY	DAYLIGHT (min)	DAY	DAYLIGHT (min)
10	716	190	986
20	727	200	975
30	744	210	961
40	762	220	944
50	783	230	926
60	804	240	905
70	826	250	883
80	848	260	861
90	872	270	839
100	894	280	817
110	915	290	795
120	935	300	774
130	954	310	755
140	969	320	738
150	982	330	723
160	990	340	712
170	993	350	706
180	992	360	706

Table Ex-17

18. The accompanying table gives the fraction of the Moon that is illuminated (as seen from Earth) at midnight (Eastern

Standard Time) in 2-day intervals for the first 60 days of 1999. Find a function that models the data in this table, and graph your function on a scatter plot of the data.

DAY	ILLUMINATION	DAY	ILLUMINATION
2	1	32	1
4	0.94	34	0.93
6	0.81	36	0.79
8	0.63	38	0.62
10	0.44	40	0.43
12	0.26	42	0.25
14	0.12	44	0.10
16	0.02	46	0.01
18	0	48	0.01
20	0.07	50	0.11
22	0.22	52	0.29
24	0.43	54	0.51
26	0.66	56	0.73
28	0.85	58	0.90
30	0.97	60	0.99

Table Ex-18

19. The accompanying table provides data about the relationship between distance d traveled in meters and elapsed time t in seconds for an object dropped near the Earth's surface. Plot time versus distance and make a guess at a "square-root function" that provides a reasonable model for t in terms of d. Use a graphing utility to confirm the reasonableness of your guess.

d (meters)	0	2.5	5	10	15	20	25
t (seconds)	0	0.7	1.0	1.4	1.7	2	2.3

Table Ex-19

20. (a) The accompanying table below provides data on five moons of the planet Saturn. In this table r is the *orbital radius* (the average distance between the moon and Saturn) and t is the time in days required for the moon to complete one orbit around Saturn. For each data pair calculate $tr^{-3/2}$, and use your results to find a reasonable model for r as a function of t.

(b) Use the model obtained in part (a) to estimate the orbital radius of the moon Enceladus, given that its orbit time is $t \approx 1.370$ days.

(c) Use the model obtained in part (a) to estimate the orbit time of the moon Tethys, given that its orbital radius is $r \approx 2.9467 \times 10^5$ km.

MOON	RADIUS (100,000 km)	ORBIT TIME (days)
1980S28	1.3767	0.602
1980S27	1.3935	0.613
1980S26	1.4170	0.629
1980S3	1.5142	0.694
1980S1	1.5147	0.695

Table Ex-20

1.8 PARAMETRIC EQUATIONS

Thus far, our study of graphs has focused on graphs of functions. However, because such graphs must pass the vertical line test, this limitation precludes curves with self-intersections or even such basic curves as circles. In this section we will study an alternative method for describing curves algebraically that is not subject to the severe restriction of the vertical line test.

This material is placed here to provide an early parametric option. However, it can be deferred until Chapter 11, if preferred.

PARAMETRIC EQUATIONS

Suppose that a particle moves along a curve C in the xy-plane in such a way that its x- and y-coordinates, as functions of time, are

$$x = f(t), \quad y = g(t)$$

We call these the **parametric equations** of motion for the particle and refer to C as the **trajectory** of the particle or the **graph** of the equations (Figure 1.8.1). The variable t is called the **parameter** for the equations.

Example 1   Sketch the trajectory over the time interval $0 \le t \le 10$ of the particle whose parametric equations of motion are

$$x = t - 3\sin t, \quad y = 4 - 3\cos t \tag{1}$$

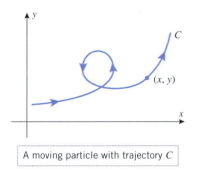

A moving particle with trajectory C

Figure 1.8.1

Solution. One way to sketch the trajectory is to choose a representative succession of times, plot the (x, y) coordinates of points on the trajectory at those times, and connect the points with a smooth curve. The trajectory in Figure 1.8.2 was obtained in this way from Table 1.8.1 in which the approximate coordinates of the particle are given at time increments of 1 unit. Observe that there is no t-axis in the picture; the values of t appear only as labels on the plotted points, and even these are usually omitted unless it is important to emphasize the location of the particle at specific times. ◄

Table 1.8.1

t	x	y
0	0.0	1.0
1	−1.5	2.4
2	−0.7	5.2
3	2.6	7.0
4	6.3	6.0
5	7.9	3.1
6	6.8	1.1
7	5.0	1.7
8	5.0	4.4
9	7.8	6.7
10	11.6	6.5

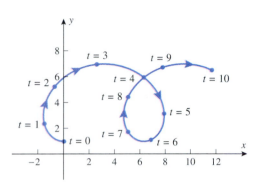

Figure 1.8.2

FOR THE READER.   Read the documentation for your graphing utility to learn how to graph parametric equations, and then generate the trajectory in Example 1. Explore the behavior of the particle beyond time $t = 10$.

Although parametric equations commonly arise in problems of motion with time as the parameter, they arise in other contexts as well. Thus, unless the problem dictates that the parameter t in the equations

$$x = f(t), \quad y = g(t)$$

represents time, it should be viewed simply as an independent variable that varies over some interval of real numbers. (In fact, there is no need to use the letter t for the parameter; any letter not reserved for another purpose can be used.) If no restrictions on the parameter are stated explicitly or implied by the equations, then it is understood that it varies from $-\infty$ to $+\infty$. To indicate that a parameter t is restricted to an interval $[a, b]$, we will write

$$x = f(t), \quad y = g(t) \quad (a \le t \le b)$$

Example 2 Find the graph of the parametric equations

$$x = \cos t, \quad y = \sin t \quad (0 \le t \le 2\pi) \tag{2}$$

Solution. One way to find the graph is to eliminate the parameter t by noting that

$$x^2 + y^2 = \sin^2 t + \cos^2 t = 1$$

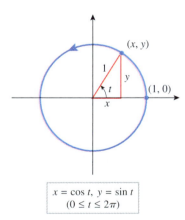

$x = \cos t, \ y = \sin t$
$(0 \le t \le 2\pi)$

Figure 1.8.3

Thus, the graph is contained in the unit circle $x^2 + y^2 = 1$. Geometrically, t can be interpreted as the angle swept out by the radial line from the origin to the point $(x, y) = (\cos t, \sin t)$ on the unit circle (Figure 1.8.3). As t increases from 0 to 2π, the point traces the circle counterclockwise, starting at $(1, 0)$ when $t = 0$ and completing one full revolution when $t = 2\pi$. One can obtain different portions of the circle by varying the interval over which the parameter varies. For example,

$$x = \cos t, \quad y = \sin t \quad (0 \le t \le \pi) \tag{3}$$

represents just the upper semicircle in Figure 1.8.3. ◄

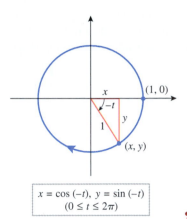

$$x = \cos(-t),\ y = \sin(-t)$$
$$(0 \le t \le 2\pi)$$

Figure 1.8.4

The direction in which the graph of a pair of parametric equations is traced as the parameter increases is called the ***direction of increasing parameter*** or sometimes the ***orientation*** imposed on the curve by the equations. Thus, we make a distinction between a ***curve***, which is a set of points, and a ***parametric curve***, which is a curve with an orientation imposed on it by a set of parametric equations. For example, we saw in Example 2 that the circle represented parametrically by (2) is traced counterclockwise as t increases and hence has *counterclockwise orientation*. As shown in Figures 1.8.2 and 1.8.3, the orientation of a parametric curve can be indicated by arrowheads.

To obtain parametric equations for the unit circle with *clockwise orientation*, we can replace t by $-t$ in (2), and use the identities $\cos(-t) = \cos t$ and $\sin(-t) = -\sin t$. This yields

$$x = \cos t, \quad y = -\sin t \qquad (0 \le t \le 2\pi)$$

Here, the circle is traced clockwise by a point that starts at $(1, 0)$ when $t = 0$ and completes one full revolution when $t = 2\pi$ (Figure 1.8.4).

FOR THE READER.   When parametric equations are graphed using a calculator, the orientation can often be determined by watching the direction in which the graph is traced on the screen. However, many computers graph so fast that it is often hard to discern the orientation. See if you can use your graphing utility to confirm that (3) has a counterclockwise orientation.

Example 3   Graph the parametric curve

$$x = 2t - 3, \quad y = 6t - 7$$

by eliminating the parameter, and indicate the orientation on the graph.

Solution.   To eliminate the parameter we will solve the first equation for t as a function of x, and then substitute this expression for t into the second equation:

$$t = \left(\tfrac{1}{2}\right)(x + 3)$$

$$y = 6\left(\tfrac{1}{2}\right)(x + 3) - 7$$

$$y = 3x + 2$$

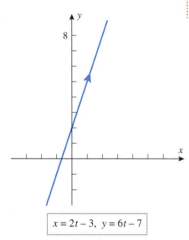

$$x = 2t - 3,\ y = 6t - 7$$

Figure 1.8.5

Thus, the graph is a line of slope 3 and y-intercept 2. To find the orientation we must look to the original equations; the direction of increasing t can be deduced by observing that x increases as t increases *or* by observing that y increases as t increases. Either piece of information tells us that the line is traced left to right as shown in Figure 1.8.5.   ◀

REMARK.   Not all parametric equations produce curves with definite orientations; if the equations are badly behaved, then the point tracing the curve may leap around sporadically or move back and forth, failing to determine a definite direction. For example, if

$$x = \sin t, \quad y = \sin^2 t$$

then the point (x, y) moves along the parabola $y = x^2$. However, the value of x varies periodically between -1 and 1, so the point (x, y) moves periodically back and forth along the parabola between the points $(-1, 1)$ and $(1, 1)$ (as shown in Figure 1.8.6). Later in the text we will discuss restrictions that eliminate such erratic behavior, but for now we will just avoid such complications.

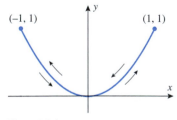

Figure 1.8.6

An equation $y = f(x)$ can be expressed in parametric form by introducing the parameter $t = x$; this yields the parametric equations $x = t$, $y = f(t)$. For example, the portion of the curve $y = \cos x$ over the interval $[-2\pi, 2\pi]$ can be expressed parametrically as

$$x = t, \quad y = \cos t \qquad (-2\pi \le t \le 2\pi)$$

(Figure 1.8.7).

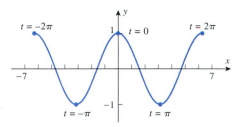

Figure 1.8.7

**GENERATING PARAMETRIC
CURVES WITH GRAPHING UTILITIES**

Many graphing utilities allow you to graph equations of the form $y = f(x)$ but not equations of the form $x = g(y)$. Sometimes you will be able to rewrite $x = g(y)$ in the form $y = f(x)$; however, if this is inconvenient or impossible, then you can graph $x = g(y)$ by introducing a parameter $t = y$ and expressing the equation in the parametric form $x = g(t)$, $y = t$. (You may have to experiment with various intervals for t to produce a complete graph.)

Example 4   Use a graphing utility to graph the equation $x = 3y^5 - 5y^3 + 1$.

Solution. If we let $t = y$ be the parameter, then the equation can be written in parametric form as

$$x = 3t^5 - 5t^3 + 1, \quad y = t$$

Figure 1.8.8 shows the graph of these equations for $-1.5 \le t \le 1.5$.   ◀

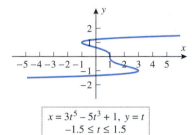

$x = 3t^5 - 5t^3 + 1, \; y = t$
$-1.5 \le t \le 1.5$

Figure 1.8.8

Some parametric curves are so complex that it is virtually impossible to visualize them without using some kind of graphing utility. Figure 1.8.9 shows three such curves.

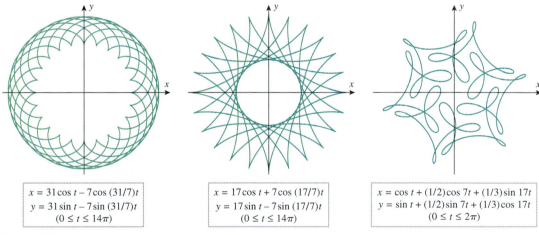

$x = 31\cos t - 7\cos (31/7)t$
$y = 31\sin t - 7\sin (31/7)t$
$(0 \le t \le 14\pi)$

$x = 17\cos t + 7\cos (17/7)t$
$y = 17\sin t - 7\sin (17/7)t$
$(0 \le t \le 14\pi)$

$x = \cos t + (1/2)\cos 7t + (1/3)\sin 17t$
$y = \sin t + (1/2)\sin 7t + (1/3)\cos 17t$
$(0 \le t \le 2\pi)$

Figure 1.8.9

FOR THE READER.   Without spending too much time, try your hand at generating some parametric curves with a graphing utility that you think are interesting or beautiful.

TRANSLATION

If a parametric curve C is given by the equations $x = f(t)$, $y = g(t)$, then adding a constant to $f(t)$ translates the curve C in the x-direction, and adding a constant to $g(t)$ translates it in the y-direction. Thus, a circle of radius r, centered at (x_0, y_0) can be represented parametrically as

$$x = x_0 + r\cos t, \quad y = y_0 + r\sin t \qquad (0 \le t \le 2\pi) \tag{4}$$

(Figure 1.8.10). If desired, we can eliminate the parameter from these equations by noting

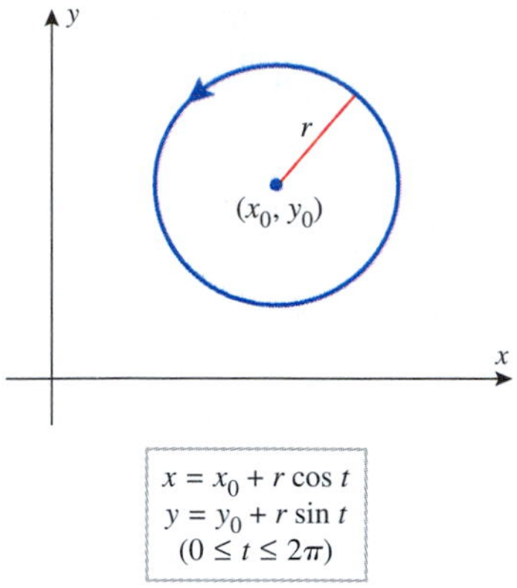

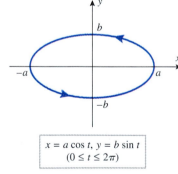

$$x = x_0 + r \cos t$$
$$y = y_0 + r \sin t$$
$$(0 \le t \le 2\pi)$$

Figure 1.8.10

that

$$(x - x_0)^2 + (y - y_0)^2 = (r \cos t)^2 + (r \sin t)^2 = r^2$$

Thus, we have obtained the familiar equation in rectangular coordinates for a circle of radius r, centered at (x_0, y_0):

$$(x - x_0)^2 + (y - y_0)^2 = r^2 \tag{5}$$

FOR THE READER. Use the parametric capability of your graphing utility to generate a circle of radius 5 that is centered at $(3, -2)$.

SCALING

If a parametric curve C is given by the equations $x = f(t)$, $y = g(t)$, then multiplying $f(t)$ by a constant stretches or compresses C in the x-direction, and multiplying $g(t)$ by a constant stretches or compresses C in the y-direction. For example, we would expect the parametric equations

$$x = 3 \cos t, \quad y = 2 \sin t \quad (0 \le t \le 2\pi)$$

to represent an ellipse, centered at the origin, since the graph of these equations results from stretching the unit circle

$$x = \cos t, \quad y = \sin t \quad (0 \le t \le 2\pi)$$

by a factor of 3 in the x-direction and a factor of 2 in the y-direction. In general, if a and b are positive constants, then the parametric equations

$$x = a \cos t, \quad y = b \sin t \quad (0 \le t \le 2\pi) \tag{6}$$

represent an ellipse, centered at the origin, and extending between $-a$ and a on the x-axis and between $-b$ and b on the y-axis (Figure 1.8.11). The numbers a and b are called the *semiaxes* of the ellipse. If desired, we can eliminate the parameter t in (6) and rewrite the equations in rectangular coordinates as

$$\frac{x^2}{a^2} + \frac{y^2}{b^2} = 1 \tag{7}$$

$x = a \cos t, \ y = b \sin t$
$(0 \le t \le 2\pi)$

Figure 1.8.11

FOR THE READER. Use the parametric capability of your graphing utility to generate an ellipse that is centered at the origin and that extends between -4 and 4 in the x-direction and between -3 and 3 in the y-direction. Generate an ellipse with the same dimensions, but translated so that its center is at $(2, 3)$.

LISSAJOUS CURVES

In the mid-1850s the French physicist Jules Antoine Lissajous (1822–1880) became interested in parametric equations of the form

$$x = \sin at, \quad y = \sin bt \tag{8}$$

in the course of studying vibrations that combine two perpendicular sinusoidal motions.

The first equation in (8) describes a sinusoidal oscillation in the x-direction with frequency $a/2\pi$, and the second describes a sinusoidal oscillation in the y-direction with frequency $b/2\pi$. If a/b is a rational number, then the combined effect of the oscillations is a periodic motion along a path called a ***Lissajous curve***. Figure 1.8.12 shows some typical Lissajous curves.

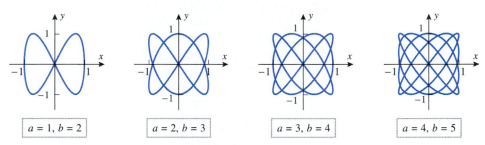

$a = 1, b = 2$ $a = 2, b = 3$ $a = 3, b = 4$ $a = 4, b = 5$

Figure 1.8.12

FOR THE READER.   Generate some Lissajous curves on your graphing utility, and also see if you can figure out when each of the curves in Figure 1.8.12 begins to repeat.

CYCLOIDS

If a wheel rolls in a straight line along a flat road, then a point on the rim of the wheel will trace a curve called a ***cycloid*** (Figure 1.8.13). This curve has a fascinating history, which we will discuss shortly; but first we will show how to obtain parametric equations for it. For this purpose, let us assume that the wheel has radius a and rolls along the positive x-axis of a rectangular coordinate system. Let $P(x, y)$ be the point on the rim that traces the cycloid, and assume that P is initially at the origin. We will take as our parameter the angle θ that is swept out by the radial line to P as the wheel rolls (Figure 1.8.13). It is standard here to regard θ to be positive, even though it is generated by a clockwise rotation.

The motion of P is a combination of the movement of the wheel's center parallel to the x-axis and the rotation of P around the center. As the radial line sweeps out an angle θ, the point P traverses an arc of length $a\theta$, and the wheel moves a distance $a\theta$ along the x-axis (why?). Thus, as suggested by Figure 1.8.14, the center moves to the point $(a\theta, a)$, and the coordinates of $P(x, y)$ are

$$x = a\theta - a\sin\theta, \quad y = a - a\cos\theta \tag{9}$$

These are the equations of the cycloid in terms of the parameter θ.

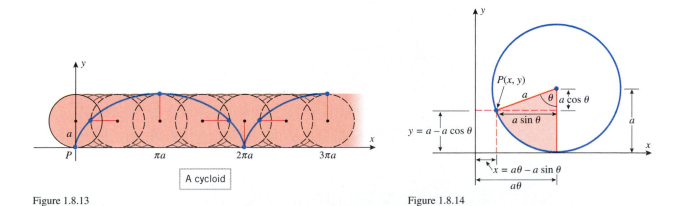

A cycloid

Figure 1.8.13   Figure 1.8.14

FOR THE READER.   Use your graphing utility to generate two "arches" of the cycloid produced by a point on the rim of a wheel of radius 1.

THE ROLE OF THE CYCLOID IN MATHEMATICS HISTORY

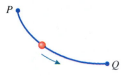

P

Q

Figure 1.8.15

The cycloid is of interest because it provides the solution to two famous mathematical problems—the ***brachistochrone problem*** (from Greek words meaning "shortest time") and the ***tautochrone problem*** (from Greek words meaning "equal time"). The brachistochrone problem is to determine the shape of a wire along which a bead might slide from a point P to another point Q, not directly below, in the *shortest time*. The tautochrone problem is to find the shape of a wire from P to Q such that two beads started at any points on the wire between P and Q reach Q in the same amount of time. The solution to both problems turns out to be an inverted cycloid (Figure 1.8.15).

In June of 1696, Johann Bernoulli[*] posed the brachistochrone problem in the form of a challenge to other mathematicians. At first, one might conjecture that the wire should form a straight line, since that shape results in the shortest distance from P to Q. However, the inverted cycloid allows the bead to fall more rapidly at first, building up sufficient initial speed to reach Q in the shortest time, even though it travels a longer distance. The problem was solved by Newton and Leibniz as well as by Johann Bernoulli and his older brother

---

[*]BERNOULLI. An amazing Swiss family that included several generations of outstanding mathematicians and scientists. Nikolaus Bernoulli (1623–1708), a druggist, fled from Antwerp to escape religious persecution and ultimately settled in Basel, Switzerland. There he had three sons, Jakob I (also called Jacques or James), Nikolaus, and Johann I (also called Jean or John). The Roman numerals are used to distinguish family members with identical names (see the family tree below). Following Newton and Leibniz, the Bernoulli brothers, Jakob I and Johann I, are considered by some to be the two most important founders of calculus. Jakob I was self-taught in mathematics. His father wanted him to study for the ministry, but he turned to mathematics and in 1686 became a professor at the University of Basel. When he started working in mathematics, he knew nothing of Newton's and Leibniz' work. He eventually became familiar with Newton's results, but because so little of Leibniz' work was published, Jakob duplicated many of Leibniz' results.

Jakob's younger brother Johann I was urged to enter into business by his father. Instead, he turned to medicine and studied mathematics under the guidance of his older brother. He eventually became a mathematics professor at Groningen in Holland, and then, when Jakob died in 1705, Johann succeeded him as mathematics professor at Basel. Throughout their lives, Jakob I and Johann I had a mutual passion for criticizing each other's work, which frequently erupted into ugly confrontations. Leibniz tried to mediate the disputes, but Jakob, who resented Leibniz' superior intellect, accused him of siding with Johann, and thus Leibniz became entangled in the arguments. The brothers often worked on common problems that they posed as challenges to one another. Johann, interested in gaining fame, often used unscrupulous means to make himself appear the originator of his brother's results; Jakob occasionally retaliated. Thus, it is often difficult to determine who deserves credit for many results. However, both men made major contributions to the development of calculus. In addition to his work on calculus, Jakob helped establish fundamental principles in probability, including the Law of Large Numbers, which is a cornerstone of modern probability theory.

Among the other members of the Bernoulli family, Daniel, son of Johann I, is the most famous. He was a professor of mathematics at St. Petersburg Academy in Russia and subsequently a professor of anatomy and then physics at Basel. He did work in calculus and probability, but is best known for his work in physics. A basic law of fluid flow, called Bernoulli's principle, is named in his honor. He won the annual prize of the French Academy 10 times for work on vibrating strings, tides of the sea, and kinetic theory of gases.

Johann II succeeded his father as professor of mathematics at Basel. His research was on the theory of heat and sound. Nikolaus I was a mathematician and law scholar who worked on probability and series. On the recommendation of Leibniz, he was appointed professor of mathematics at Padua and then went to Basel as a professor of logic and then law. Nikolaus II was professor of jurisprudence in Switzerland and then professor of mathematics at St. Petersburg Academy. Johann III was a professor of mathematics and astronomy in Berlin and Jakob II succeeded his uncle Daniel as professor of mathematics at St. Petersburg Academy in Russia. Truly an incredible family!

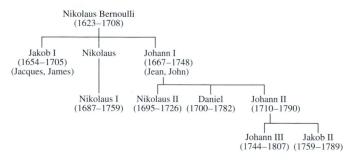

Jakob; it was formulated and solved *incorrectly* years earlier by Galileo, who thought the answer was a circular arc.

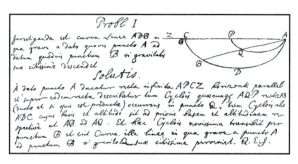

Newton's solution of the brachistochrone problem in his own handwriting

EXERCISE SET 1.8   ∿ Graphing Utility

1. (a) By eliminating the parameter, sketch the trajectory over the time interval $0 \le t \le 5$ of the particle whose parametric equations of motion are

    $$x = t - 1, \quad y = t + 1$$

   (b) Indicate the direction of motion on your sketch.

   (c) Make a table of x- and y-coordinates of the particle at times $t = 0, 1, 2, 3, 4, 5$.

   (d) Mark the position of the particle on the curve at the times in part (c), and label those positions with the values of t.

2. (a) By eliminating the parameter, sketch the trajectory over the time interval $0 \le t \le 1$ of the particle whose parametric equations of motion are

    $$x = \cos(\pi t), \quad y = \sin(\pi t)$$

   (b) Indicate the direction of motion on your sketch.

   (c) Make a table of x- and y-coordinates of the particle at times $t = 0, 0.25, 0.5, 0.75, 1$.

   (d) Mark the position of the particle on the curve at the times in part (c), and label those positions with the values of t.

In Exercises 3–12, sketch the curve by eliminating the parameter, and indicate the direction of increasing t.

3. $x = 3t - 4, \ y = 6t + 2$

4. $x = t - 3, \ y = 3t - 7 \quad (0 \le t \le 3)$

5. $x = 2\cos t, \ y = 5\sin t \quad (0 \le t \le 2\pi)$

6. $x = \sqrt{t}, \ y = 2t + 4$

7. $x = 3 + 2\cos t, \ y = 2 + 4\sin t \quad (0 \le t \le 2\pi)$

8. $x = \sec t, \ y = \tan t \quad (\pi \le t < 3\pi/2)$

9. $x = \cos 2t, \ y = \sin t \quad (-\pi/2 \le t \le \pi/2)$

10. $x = 4t + 3, \ y = 16t^2 - 9$

11. $x = 2\sin^2 t, \ y = 3\cos^2 t$

12. $x = \sec^2 t, \ y = \tan^2 t$

In Exercises 13–18, find parametric equations for the curve, and check your work by generating the curve with a graphing utility.

∿ 13. A circle of radius 5, centered at the origin, oriented clockwise.

∿ 14. The portion of the circle $x^2 + y^2 = 1$ that lies in the third quadrant, oriented counterclockwise.

∿ 15. A vertical line intersecting the x-axis at $x = 2$, oriented upward.

∿ 16. The ellipse $\frac{x^2}{4} + \frac{y^2}{9} = 1$, oriented counterclockwise.

∿ 17. The portion of the parabola $x = y^2$ joining $(1, -1)$ and $(1, 1)$, oriented down to up.

∿ 18. The circle of radius 4, centered at $(1, -3)$, oriented counterclockwise.

19. In each part, match the parametric equation with one of the curves labeled (I)–(VI), and explain your reasoning.

   (a) $x = \sqrt{t}, \ y = \sin 3t$ (b) $x = 2\cos t, \ y = 3\sin t$

   (c) $x = t\cos t, \ y = t\sin t$ (d) $x = \dfrac{3t}{1 + t^3}, \ y = \dfrac{3t^2}{1 + t^3}$

   (e) $x = \dfrac{t^3}{1 + t^2}, \ y = \dfrac{2t^2}{1 + t^2}$ (f) $x = 2\cos t, \ y = \sin 2t$

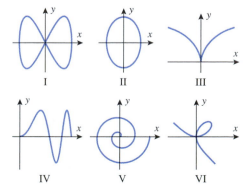

Figure Ex-19

20. Use a graphing utility to generate the curves in Exercise 19, and in each case identify the orientation.

21. (a) Use a graphing utility to generate the trajectory of a particle whose equations of motion over the time interval $0 \le t \le 5$ are

$$x = 6t - \tfrac{1}{2}t^3, \quad y = 1 + \tfrac{1}{2}t^2$$

(b) Make a table of x- and y-coordinates of the particle at times $t = 0, 1, 2, 3, 4, 5$.

(c) At what times is the particle on the y-axis?

(d) During what time interval is $y < 5$?

(e) At what time is the x-coordinate of the particle maximum?

22. (a) Use a graphing utility to generate the trajectory of a paper airplane whose equations of motion for $t \ge 0$ are

$$x = t - 2 \sin t, \quad y = 3 - 2 \cos t$$

(b) Assuming that the plane flies in a room in which the floor is at $y = 0$, explain why the plane will not crash into the floor. [For simplicity, ignore the physical size of the plane by treating it as a particle.]

(c) How high must the ceiling be to ensure that the plane does not touch or crash into it?

In Exercises 23 and 24, graph the equation using a graphing utility.

23. (a) $x = y^2 + 2y + 1$
   (b) $x = \sin y, \; -2\pi \le y \le 2\pi$

24. (a) $x = y + 2y^3 - y^5$
   (b) $x = \tan y, \; -\pi/2 < y < \pi/2$

25. (a) By eliminating the parameter, show that the equations

$$x = x_0 + (x_1 - x_0)t, \quad y = y_0 + (y_1 - y_0)t$$

represent the line passing through the points (x_0, y_0) and (x_1, y_1).

(b) Show that if $0 \le t \le 1$, then the equations in part (a) represent the line segment joining (x_0, y_0) and (x_1, y_1), oriented in the direction from (x_0, y_0) to (x_1, y_1).

(c) Use the result in part (b) to find parametric equations for the line segment joining the points $(1, -2)$ and $(2, 4)$, oriented in the direction from $(1, -2)$ to $(2, 4)$.

(d) Use the result in part (b) to find parametric equations for the line segment in part (c), but oriented in the direction from $(2, 4)$ to $(1, -2)$.

26. Use the result in Exercise 25 to find

(a) parametric equations for the line segment joining the points $(-3, -4)$ and $(-5, 1)$, oriented from $(-3, -4)$ to $(-5, 1)$

(b) parametric equations for the line segment traced from $(0, b)$ to $(a, 0)$, oriented from $(0, b)$ to $(a, 0)$.

27. (a) Suppose that the line segment from the point $P(x_0, y_0)$ to $Q(x_1, y_1)$ is represented parametrically by

$$x = x_0 + (x_1 - x_0)t,$$
$$y = y_0 + (y_1 - y_0)t \qquad (0 \le t \le 1)$$

and that $R(x, y)$ is the point on the line segment corresponding to a specified value of t (see the accompanying figure). Show that $t = r/q$, where r is the distance from P to R and q is the distance from P to Q.

(b) What value of t produces the midpoint between points P and Q?

(c) What value of t produces the point that is three-fourths of the way from P to Q?

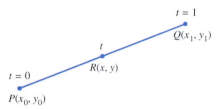

Figure Ex-27

28. Find parametric equations for the line segment joining $P(2, -1)$ and $Q(3, 1)$, and use the result in Exercise 27 to find

(a) the midpoint between P and Q

(b) the point that is one-fourth of the way from P to Q

(c) the point that is three-fourths of the way from P to Q.

29. Explain why the parametric curve

$$x = t^2, \quad y = t^4 \qquad (-1 \le t \le 1)$$

does not have a definite orientation.

30. (a) In parts (a) and (b) of Exercise 25 we obtained parametric equations for a line segment in which the parameter varied from $t = 0$ to $t = 1$. Sometimes it is desirable to have parametric equations for a line segment in which the parameter varies over some other interval, say $t_0 \le t \le t_1$. Use the ideas in Exercise 25 to show that the line segment joining the points (x_0, y_0) and (x_1, y_1) can be represented parametrically as

$$x = x_0 + (x_1 - x_0)\frac{t - t_0}{t_1 - t_0},$$
$$\qquad\qquad\qquad\qquad (t_0 \le t \le t_1)$$
$$y = y_0 + (y_1 - y_0)\frac{t - t_0}{t_1 - t_0}$$

(b) Which way is the line segment oriented?

(c) Find parametric equations for the line segment traced from $(3, -1)$ to $(1, 4)$ as t varies from 1 to 2, and check your result with a graphing utility.

31. (a) By eliminating the parameter, show that if a and c are not both zero, then the graph of the parametric equations

$$x = at + b, \quad y = ct + d \qquad (t_0 \le t \le t_1)$$

is a line segment.

(b) Sketch the parametric curve

$$x = 2t - 1, \quad y = t + 1 \qquad (1 \leq t \leq 2)$$

and indicate its orientation.

32. (a) What can you say about the line in Exercise 31 if a or c (but not both) is zero?

(b) What do the equations represent if a and c are both zero?

33. Parametric curves can be defined piecewise by using different formulas for different values of the parameter. Sketch the curve that is represented piecewise by the parametric equations

$$x = 2t, \quad y = 4t^2 \qquad \left(0 \leq t \leq \tfrac{1}{2}\right)$$

$$x = 2 - 2t, \quad y = 2t \qquad \left(\tfrac{1}{2} \leq t \leq 1\right)$$

34. Find parametric equations for the rectangle in the accompanying figure, assuming that the rectangle is traced counterclockwise as t varies from 0 to 1, starting at $\left(\tfrac{1}{2}, \tfrac{1}{2}\right)$ when $t = 0$. [*Hint:* Represent the rectangle piecewise, letting t vary from 0 to $\tfrac{1}{4}$ for the first edge, from $\tfrac{1}{4}$ to $\tfrac{1}{2}$ for the second edge, and so forth.]

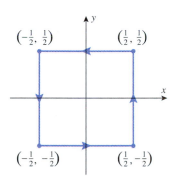

$\left(-\tfrac{1}{2}, \tfrac{1}{2}\right)$ $\left(\tfrac{1}{2}, \tfrac{1}{2}\right)$

$\left(-\tfrac{1}{2}, -\tfrac{1}{2}\right)$ $\left(\tfrac{1}{2}, -\tfrac{1}{2}\right)$

Figure Ex-34

35. (a) Find parametric equations for the ellipse that is centered at the origin and has intercepts $(4, 0)$, $(-4, 0)$, $(0, 3)$, and $(0, -3)$.

(b) Find parametric equations for the ellipse that results by translating the ellipse in part (a) so that its center is at $(-1, 2)$.

(c) Confirm your results in parts (a) and (b) using a graphing utility.

36. We will show later in the text that if a projectile is fired from ground level with an initial speed of v_0 meters per second at an angle α with the horizontal, and if air resistance is neglected, then its position after t seconds, relative to the coordinate system in the accompanying figure is

$$x = (v_0 \cos \alpha)t, \quad y = (v_0 \sin \alpha)t - \tfrac{1}{2}gt^2$$

where $g \approx 9.8 \text{ m/s}^2$.

(a) By eliminating the parameter, show that the trajectory is a parabola.

(b) Sketch the trajectory if $\alpha = 30°$ and $v_0 = 1000 \text{ m/s}$.

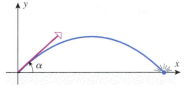

Figure Ex-36

37. A shell is fired from a cannon at an angle of $\alpha = 45°$ with an initial speed of $v_0 = 800 \text{ m/s}$.

(a) Find parametric equations for the shell's trajectory relative to the coordinate system in Figure Ex-36.

(b) How high does the shell rise?

(c) How far does the shell travel horizontally?

38. A robot arm, designed to buff flat surfaces on an automobile, consists of two attached rods, one that moves back and forth horizontally, and a second, with the buffing pad at the end, that moves up and down (see the accompanying figure).

(a) Suppose that the horizontal arm of the robot moves so that the x-coordinate of the buffer's center at time t is $x = 25 \sin \pi t$ and the vertical arm moves so that the y-coordinate of the buffer's center at time t is $y = 12.5 \sin \pi t$. Graph the trajectory of the center of the buffing pad.

(b) Suppose that the x- and y-coordinates in part (a) are $x = 25 \sin \pi at$ and $y = 12.5 \sin \pi bt$, where the constants a and b can be controlled by programming the robot arm. Graph the trajectory of the center of the pad if $a = 4$ and $b = 5$.

(c) Investigate the trajectories that result in part (b) for various choices of a and b.

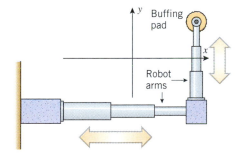

Figure Ex-38

39. Describe the family of curves described by the parametric equations

$$x = a \cos t + h, \quad y = b \sin t + k \qquad (0 \leq t \leq 2\pi)$$

if

(a) h and k are fixed but a and b can vary

(b) a and b are fixed but h and k can vary

(c) $a = 1$ and $b = 1$, but h and k vary so that $h = k + 1$.

40. A *hypocycloid* is a curve traced by a point P on the circumference of a circle that rolls inside a larger fixed circle. Suppose that the fixed circle has radius a, the rolling circle has radius b, and the fixed circle is centered at the origin. Let ϕ be the angle shown in the following figure, and assume that the point P is at $(a, 0)$ when $\phi = 0$. Show that the hypocycloid generated is given by the parametric equations

$$x = (a - b) \cos \phi + b \cos \left(\frac{a - b}{b} \phi \right)$$

$$y = (a - b) \sin \phi - b \sin \left(\frac{a - b}{b} \phi \right)$$

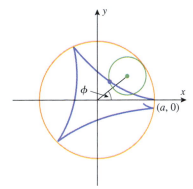

Figure Ex-40

41. If $b = \frac{1}{4} a$ in Exercise 40, then the resulting curve is called a four-cusped hypocycloid.
   (a) Sketch this curve.
   (b) Show that the curve is given by the parametric equations
      $x = a \cos^3 \phi, \quad y = a \sin^3 \phi.$
   (c) Show that the curve is given by the equation

   $$x^{2/3} + y^{2/3} = a^{2/3}$$

   in rectangular coordinates.

42. (a) Use a graphing utility to study how the curves in the family

   $$x = 2a \cos^2 t, \quad y = 2a \cos t \sin t \quad (-2\pi < t < 2\pi)$$

   change as a varies from 0 to 5.
   (b) Confirm your conclusion algebraically.
   (c) Write a brief paragraph that describes your findings.

SUPPLEMENTARY EXERCISES

 Graphing Utility   CAS

1. Referring to the cigarette consumption graph in Figure 1.1.2b, during what 5-year period was the annual cigarette consumption per adult increasing most rapidly on average? Explain your reasoning.

2. Use the graphs of the functions f and g in the accompanying figure to solve the following problems.
   (a) Find the values of $f(-2)$ and $g(3)$.
   (b) For what values of x is $f(x) = g(x)$?
   (c) For what values of x is $f(x) < 2$?
   (d) What are the domain and range of f?
   (e) What are the domain and range of g?
   (f) Find the zeros of f and g.

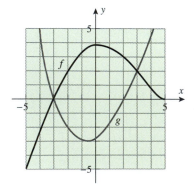

Figure Ex-2

3. A glass filled with water that has a temperature of $40°$F is placed in a room in which the temperature is a constant $70°$F. Sketch a rough graph that reasonably describes the temperature of the water in the glass as a function of the elapsed time.

4. A student begins driving toward school but 5 minutes into the trip remembers that he forgot his homework. He drives home hurriedly, retrieves his notes, and then drives at great speed toward school, hitting a tree 5 minutes after leaving home. Sketch a rough graph that reasonably describes the student's distance from home as a function of the elapsed time.

5. A rectangular storage container with an open top and a square base has a volume of 8 cubic meters. Material for the base costs \$5 per square meter, and material for the sides \$2 per square meter. Express the total cost of the materials as a function of the length of a side of the base.

6. You want to paint the top of a circular table. Find a formula that expresses the amount of paint required as a function of the radius, and discuss all of the assumptions you have made in finding the formula.

7. Sketch the graph of the function

$$f(x) = \begin{cases} -1, & x \le -5 \\ \sqrt{25 - x^2}, & -5 < x < 5 \\ x - 5, & x \ge 5 \end{cases}$$

8. A ball of radius 3 inches is coated uniformly with plastic. Express the volume of the plastic as a function of its thickness.

9. A box with a closed top is to be made from a 6-ft by 10-ft piece of cardboard by cutting out four squares of equal size (see the accompanying figure), folding along the dashed lines, and tucking the two extra flaps inside.
   (a) Find a formula that expresses the volume of the box as a function of the length of the sides of the cut-out squares.

(b) Find an inequality that specifies the domain of the function in part (a).

(c) Estimate the dimensions of the box of largest volume.

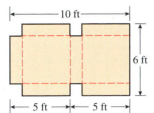

Figure Ex-9

10. Let $f(x) = -x^2$ and $g(x) = 1/\sqrt{x}$. Find the natural domains of $f \circ g$ and $g \circ f$.

11. Given that $f(x) = x^2 + 1$ and $g(x) = 3x + 2$, find all values of x such that $f(g(x)) = g(f(x))$.

12. Let $f(x) = (2x - 1)/(x + 1)$ and $g(x) = 1/(x - 1)$.
(a) Find $f(g(x))$.
(b) Is the natural domain of the function $f(g(x))$ obtained in part (a) the same as the domain of $f \circ g$? Explain.

13. Find $f(g(h(x)))$, given that

$$f(x) = \frac{x}{x - 1}, \quad g(x) = \frac{1}{x}, \quad h(x) = x^2 - 1$$

14. Given that $f(x) = 2x + 1$ and $h(x) = 2x^2 + 4x + 1$, find a function g such that $f(g(x)) = (x)$.

15. Complete the following table.

x	-4	-3	-2	-1	0	1	2	3	4
$f(x)$	0	-1	2	1	3	-2	-3	4	-4
$g(x)$	3	2	1	-3	-1	-4	4	-2	0
$(f \circ g)(x)$									
$(g \circ f)(x)$									

16. (a) Write an equation for the graph that is obtained by reflecting the graph of $y = |x - 1|$ about the y-axis, then stretching that graph vertically by a factor of 2, then translating that graph down 3 units, and then reflecting that graph about the x-axis.
(b) Sketch the original graph and the final graph.

17. In each part, classify the function as even, odd, or neither.
(a) $x^2 \sin x$ (b) $\sin^2 x$
(c) $x + x^2$ (d) $\sin x \tan x$

18. (a) Find exact values for all x-intercepts of

$$y = \cos x - \sin 2x$$

in the interval $-2\pi \le x \le 2\pi$.
(b) Find the coordinates of all intersections of the graphs of $y = \cos x$ and $y = \sin 2x$ if $-2\pi \le x \le 2\pi$, and use a graphing utility to check your answer.

19. (a) A surveyor measures the angle of elevation α of a tower from a point A due south of the tower and also measures the angle of elevation β from a point B that is d feet due east of the point A (see the accompanying figure). Show that the height h of the tower in feet is given by

$$h = \frac{d \tan \alpha \tan \beta}{\sqrt{\tan^2 \alpha - \tan^2 \beta}}$$

(b) Use a calculating utility to approximate the height of the tower to the nearest tenth of a foot if $\alpha = 17°$, $\beta = 12°$, and $d = 1000$ ft.

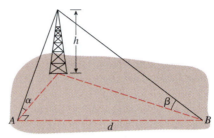

Figure Ex-19

20. Suppose that the expected low temperature in Anchorage, Alaska (in °F), is modeled by the equation

$$T = 50 \sin \frac{2\pi}{365} (t - 101) + 25$$

where t is in days and $t = 0$ corresponds to January 1.
(a) Sketch the graph of T versus t for $0 \le t \le 365$.
(b) Use the model to predict when the coldest day of the year will occur.
(c) Based on this model, how many days during the year would you expect the temperature to be below $0°$F?

21. The accompanying figure shows the graphs of the equations $y = 1 + 2 \sin x$ and $y = 2 \sin(x/2) + 2 \cos(x/2)$ for $-2\pi \le x \le 2\pi$. Without the aid of a calculator, label each curve by its equation, and find the coordinates of the points A, B, C, and D. Explain your reasoning.

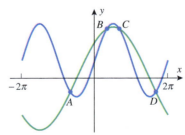

Figure Ex-21

22. The accompanying figure shows a model for the tide variation in an inlet to San Francisco Bay during a 24-hour period. Find an equation of the form $y = y_0 + y_1 \sin(at + b)$ for the model, assuming that $t = 0$ corresponds to midnight.

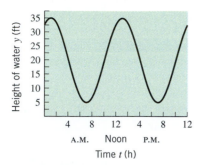

Figure Ex-22

23. In each part describe the family of curves.
    (a) $(x - a)^2 + (y - a^2)^2 = 1$
    (b) $y = a + (x - 2a)^2$

 24. (a) Suppose that the equations $x = f(t)$, $y = g(t)$ describe a curve C as t increases from 0 to 1. Find parametric equations that describe the same curve C but traced in the opposite direction as t increases from 0 to 1.
    (b) Check your work using the parametric graphing feature of a graphing utility by generating the line segment between $(1, 2)$ and $(4, 0)$ in both possible directions as t increases from 0 to 1.

25. Sketch the graph of the equation $x^2 - 4y^2 = 0$.

26. Find an equation for a parabola that passes through the points $(2, 0)$, $(8, 18)$, and $(-8, 18)$.

27. Sketch the curve described by the parametric equations

$$x = t \cos(2\pi t), \quad y = t \sin(2\pi t)$$

and check your result with a graphing utility.

28. The electrical resistance R in ohms (Ω) for a pure metal wire is related to its temperature T in $^\circ$C by the formula

$$R = R_0(1 + kT)$$

in which R_0 and k are positive constants.
    (a) Make a hand-drawn sketch of the graph of R versus T, and explain the geometric significance of R_0 and k for your graph.
    (b) In theory, the resistance R of a pure metal wire drops to zero when the temperature reaches absolute zero ($T = -273^\circ$C). What information does this give you about k?
    (c) A tungsten bulb filament has a resistance of 1.1 Ω at a temperature of 20°C. What information does this give you about R_0 for the filament?
    (d) At what temperature will a tungsten filament have a resistance of 1.5 Ω?

Most of the following exercises require access to graphing and calculating utilities. When you are asked to *find* an answer or to *solve* an equation, you may find either an exact result or a numerical approximation, depending on the particular technology you are using and your own imagination.

29. Find the distance between the point $P(1, 2)$ and an arbitrary point $(x, \sqrt{x})$ on the curve $y = \sqrt{x}$. Graph this distance versus x, and use the graph to find the x-coordinate of the point on the curve that is closest to the point P.

30. Find the distance between the point $P(1, 0)$ and an arbitrary point $(x, 1/x)$ on the curve $y = 1/x$, where $x > 0$. Graph this distance versus x, and use the graph to find the x-coordinate of the point on the curve that is closest to the point P.

In Exercises 31 and 32, use **Archimedes' principle**: *A body wholly or partially immersed in a fluid is buoyed up by a force equal to the weight of the fluid that it displaces.*

31. A hollow metal sphere of diameter 5 feet weighs 108 pounds and floats partially submerged in seawater. Assuming that seawater weighs 63.9 pounds per cubic foot, how far below the surface is the bottom of the sphere? [*Hint:* If a sphere of radius r is submerged to a depth h, then the volume V of the submerged portion is given by the formula $V = \pi h^2(r - h/3)$.]

32. Suppose that a hollow metal sphere of diameter 5 feet and weight w pounds floats in seawater. (See Exercise 31.)
    (a) Graph w versus h for $0 \le h \le 5$.

    (b) Find the weight of the sphere if exactly half of the sphere is submerged.

33. A breeding group of 20 bighorn sheep is released in a protected area in Colorado. It is expected that with careful management the number of sheep, N, after t years will be given by the formula

$$N = \frac{220}{1 + 10(0.83^t)}$$

and that the sheep population will be able to maintain itself without further supervision once the population reaches a size of 80.
    (a) Graph N versus t.

    (b) How many years must the state of Colorado maintain a program to care for the sheep?

    (c) How many bighorn sheep can the environment in the protected area support? [*Hint:* Examine the graph of N versus t for large values of t.]

In Exercises 34 and 35, use the following empirical formula for the windchill index (WCI) [see Example 3 of Section 1.2]:

$$\text{WCI} = \begin{cases} T, & 0 \le v \le 4 \\ 91.4 + (91.4 - T)(0.0203v - 0.304\sqrt{v} - 0.474), & 4 < v < 45 \\ 1.6T - 55, & v \ge 45 \end{cases}$$

where T is the air temperature in $^\circ$F, v is the wind speed in mi/h, and WCI is the equivalent temperature in $^\circ$F.

34. (a) Graph T versus v over the interval $4 \le v \le 45$ for WCI = 0.

  (b) Use your graph to estimate the values of T for WCI = 0 corresponding to $v = 10, 20, 30$, to the nearest degree.

35. (a) Graph WCI versus v over the interval $0 \le v \le 50$ for $T = 20$.

  (b) Use your graph to estimate the values of the WCI corresponding to $v = 10, 20, 30, 40$, to the nearest degree.

  (c) Use your graph to estimate the values of v corresponding to WCI $= -20, -10, 0, 10$, to the nearest mile per hour.

36. Find the domain and range of the function

$$f(x) = \frac{\sin x}{x^4 + x^3 + 5}$$

37. Find the domain and range of the function

$$f(x) = x^2 - \sqrt{1 + x - x^4}$$

38. An oven is preheated and then remains at a constant temperature. A potato is placed in the oven to bake. Suppose that the temperature T (in °F) of the potato t minutes later is given by $T = 400 - 325(0.97^t)$. The potato will be considered done when its temperature is anywhere between 260°F and 280°F.

  (a) During what interval of time would the potato be considered done?

  (b) How long does it take for the temperature of the potato to reach 95% of the oven temperature?

39. Suppose that a package of medical supplies is dropped from a helicopter straight down by parachute into a remote area. The velocity v (in feet per second) of the package t seconds after it is released is given by $v = 24.61(1 - 0.273^t)$.

  (a) Graph v versus t.

  (b) Show that the graph has a horizontal asymptote $v = c$.

  (c) The constant c is called the **terminal velocity**. Explain what the terminal velocity means in practical terms.

  (d) Can the package actually reach its terminal velocity? Explain.

  (e) How long does it take for the package to reach 98% of its terminal velocity?

40. An ancient Babylonian tablet known as Plimpton 322 contains a sequence of numbers that appear to be the squares of secants of various angles ranging from about 45° to 31°. The secants of these angles are listed in the accompanying table.

  (a) Using linear regression, find a function that (approximately) expresses these secants in terms of their position number within the table.

  (b) Do you see any connection between your linear function from part (a) and the fact that the base for the Babylonian number system was 60?

ENTRY NUMBER	FUNCTION VALUE
1	1.4083
2	1.3961
3	1.3852
4	1.3734
5	1.3472
6	1.3361
7	1.3110
8	1.3010
9	1.2817
10	1.2594
11	1.2500
12	1.2204
13	1.2042
14	1.1959
15	1.1778

Table Ex-40

41. An important problem addressed by calculus is that of finding a good linear approximation to the function $f(x)$ near a particular x-value. One possible approach (not the best) is to sample values of the function near the specified x-value, find the least squares line for this sample, and translate the least squares line so that it passes through the point on the graph of $y = f(x)$ corresponding to the given x-value. Let $f(x) = x^2 \sin x$.

  (a) Make a table of $(x, f(x))$ values for $x = 1.9, 1.92, 1.94, \ldots, 2.1$.

  (b) Find a least squares line for the data in part (a).

  (c) Find the equation of the line passing through the point $(2, f(2))$ and parallel to the least squares line.

  (d) Using a graphing utility with a graphing window containing $(2, f(2))$, graph $y = f(x)$ and the line you found in part (c). How do the graphs compare as you zoom closer to the point $(2, f(2))$?

  [*Note:* The best linear approximation to $y = x^2 \sin x$ near $x = 2$ is given by $y \approx 1.9726x - 0.308015$. Later, we will see how to use the tools of calculus to find this answer.]

42. An extension of the linear approximation problem is finding a good polynomial approximation to the function $f(x)$ near a particular x-value. One possible approach (not the best) is to sample values of the function near the specified x-value, apply polynomial regression to this sample, and translate the regression curve so that it passes through the point on the graph of $y = f(x)$ corresponding to the given x-value. Let $f(x) = \cos x$.

  (a) Make a table of $(x, f(x))$ values for $x = -0.1, -0.08, -0.06, \ldots, 0.1$.

  (b) Use quadratic regression to model the data in part (a) with a quadratic polynomial.

  (c) Translate your quadratic modeling function from part (b) to obtain a quadratic function that passes through the point $(0, f(0))$.

(d) Using a graphing utility with a graphing window containing $(0, f(0))$, graph $y = f(x)$ and the polynomial you found in part (c). How do the graphs compare as you zoom closer to the point $(0, f(0))$?

[*Note:* The best quadratic approximation to $y = \cos x$ near $x = 0$ is given by $y \approx -0.5x^2 + 1$. In Chapter 10, we will see how to use the tools of calculus to find this answer.]

43. The accompanying table gives the water level (in meters above the mean low-water mark) at a Cape Hatteras, North Carolina, fishing pier, recorded in 2-hour increments starting from midnight, July 1, 1999. Why should we expect that a trigonometric function should fit these data? Find a function that models the data, and graph your function on a scatter plot.

HOUR	WATER LEVEL (m)	HOUR	WATER LEVEL (m)
0	0.526	36	0.534
2	0.157	38	0.192
4	0.161	40	0.141
6	0.486	42	0.426
8	0.779	44	0.849
10	0.740	46	1.032
12	0.412	48	0.765
14	0.141	50	0.281
16	0.260	52	0.042
18	0.633	54	0.157
20	1.015	56	0.587
22	1.021	58	0.777
24	0.670	60	0.620
26	0.231	62	0.241
28	0.128	64	0.045
30	0.345	66	0.195
32	0.697	68	0.613
34	0.821	70	0.945

Table Ex-43

EXPANDING THE CALCULUS HORIZON

Iteration and Dynamical Systems

*W*hat do the four figures below have in common? The answer is that all of them are of interest in contemporary research and all involve a mathematical process called **iteration**. In this module we will introduce this concept and touch on some of the fascinating ideas to which it leads.

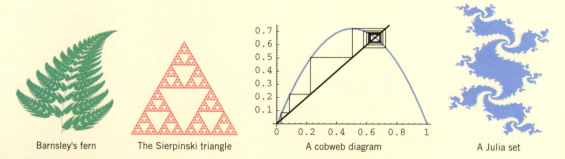

Barnsley's fern   The Sierpinski triangle   A cobweb diagram   A Julia set

Iterative Processes

Recall that in the notation $y = f(x)$, the variable x is called an **input** of the function f, and the variable y is called the corresponding **output**. Suppose that we start with some input, say $x = c$, and each time we compute an output we feed it back into f as an input. This generates the following sequence of numbers:

$$f(c), \quad f(f(c)), \quad f(f(f(c))), \quad f(f(f(f(c)))), \ldots$$

This is called an **iterated function sequence** for f (from the Latin word *iteratus*, meaning "to repeat"). The number c is called the **seed value** for the sequence, the terms in the sequence are called **iterates**, and each time f is applied we say that we have performed an **iteration**. Iterated function sequences arise in a wide variety of physical processes that are collectively called **dynamical systems**.

···········
Exercise 1   Let $f(x) = x^2$.

(a) Calculate the first 10 iterates in the iterated function sequence for f, starting with seed values of $c = 0.5, 1$, and 2. In each case make a conjecture about the ***long-term behavior*** of the iterates, that is, the behavior of the iterates as more and more iterations are performed.

(b) Try your own seed values, and make a conjecture about the effect of a seed value on the long-term behavior of the iterates.

Recursion Formulas

The proliferation of parentheses in an iterated function sequence can become confusing, so for simplicity let us introduce the following notation for the successive iterates

$$y_0 = c, \quad y_1 = f(c), \quad y_2 = f(f(c)), \quad y_3 = f(f(f(c))), \quad y_4 = f(f(f(f(c)))), \dots$$

or expressed more simply,

$$y_0 = c, \quad y_1 = f(y_0), \quad y_2 = f(y_1), \quad y_3 = f(y_2), \quad y_4 = f(y_3), \dots$$

Thus, successive terms in the sequence are related by the formulas

$$y_0 = c, \quad y_{n+1} = f(y_n) \quad (n = 0, 1, 2, 3, \dots)$$

These two formulas, taken together, comprise what is called a *recursion formula* for the iterated function sequence. In general, a ***recursion formula*** is any formula or set of formulas that provides a method for generating the terms of a sequence from the preceding terms and a seed value. For example, the recursion formula for the iterated function sequence of $f(x) = x^2$ with seed value c is

$$y_0 = c, \quad y_{n+1} = y_n^2$$

As another example, the recursion formula

$$y_0 = 1, \quad y_{n+1} = \frac{1}{2}\left(y_n + \frac{p}{y_n}\right) \tag{1}$$

produces an iterated function sequence whose iterates can be used to approximate $\sqrt{p}$ to any degree of accuracy.

···········
Exercise 2   Use (1) to approximate $\sqrt{5}$ by generating successive iterates on a calculator until you encounter two successive iterates that are the same. Compare this approximation of $\sqrt{5}$ to that produced directly by your calculator.

···········
Exercise 3

(a) Find iterates y_1 up to y_6 of the sequence that is generated by the recursion formula

$$y_0 = 1, \quad y_{n+1} = \tfrac{1}{2}y_n$$

(b) By examining the terms generated in part (a), find a formula that expresses y_n as a function of n.

···········
Exercise 4   Suppose that you deposit \$1000 in a bank at 5% interest per year and allow it to accumulate value without making withdrawals.

(a) If y_n denotes the value of the account at the end of the nth year, how could you find the value of y_{n+1} if you knew the value of y_n?

(b) Starting with $y_0 = 1000$ (dollars), use the result in part (a) to calculate y_1, y_2, y_3, y_4, and y_5.

(c) Find a recursion formula for the sequence of yearly account values assuming that $y_0 = 1000$.

(d) Find a formula that expresses y_n as a function of n, and use that formula to calculate the value of the account at the end of the 15th year.

Exploring Iterated Function Sequences

Iterated function sequences for a function f can be explored in various ways. Here are three possibilities:

- Choose a specific seed value, and investigate the long-term behavior of the iterates (as in Exercise 1).

- Let the seed value be a variable x (in which case the iterates become functions of x), and investigate what happens to the graphs of the iterates as more and more iterations are performed.

- Choose a specific iterate, say the 10th, and investigate how the value of this iterate varies with different seed values.

Exercise 5   Let $f(x) = \sqrt{x}$.

(a) Find formulas for the first five iterates in the iterated function sequence for f, taking the seed value to be x.

(b) Graph the iterates in part (a) in the same coordinate system, and make a conjecture about the behavior of the graphs as more and more iterations are performed.

Continued Fractions and Fibonacci Sequences

If $f(x) = 1/x$, and the seed value is x, then the iterated function sequence for f flip-flops between x and $1/x$:

$$y_1 = \frac{1}{x}, \quad y_2 = \frac{1}{1/x} = x, \quad y_3 = \frac{1}{x}, \quad y_4 = \frac{1}{1/x} = x, \dots$$

However, if $f(x) = 1/(x+1)$, then the iterated function sequence becomes a sequence of fractions that, if continued indefinitely, is an example of a *continued fraction*:

$$\frac{1}{1+x}, \quad \cfrac{1}{1+\cfrac{1}{1+x}}, \quad \cfrac{1}{1+\cfrac{1}{1+\cfrac{1}{1+x}}}, \quad \cfrac{1}{1+\cfrac{1}{1+\cfrac{1}{1+\cfrac{1}{1+x}}}}, \dots$$

Exercise 6   Let $f(x) = 1/(x + 1)$ and $c = 1$.

(a) Find *exact values* for the first 10 terms in the iterated function sequence for f; that is, express each term as a fraction p/q with no common factors in the numerator and denominator.

(b) Write down the numerators from part (a) in sequence, and see if you can discover how each term after the first two is related to its predecessors. The sequence of numerators is called a *Fibonacci sequence* [in honor of its medieval discoverer Leonardo ("Fibonacci") da Pisa]. Do some research on Fibonacci and his sequence, and write a paper on the subject.

(c) Use the pattern you discovered in part (b) to write down the exact values of the second 10 terms in the iterated function sequence.

(d) Find a recursion formula that will generate all the terms in the Fibonacci sequence after the first two.

(e) It can be proved that the terms in the iterated function sequence for f get closer and closer to one of the two solutions of the equation $q = 1/(1+q)$. Which solution is it? This solution is a number known as the *golden ratio*. Do some research on the golden ratio, and write a paper on the subject.

Applications to Ecology

There are numerous models for predicting the growth and decline of populations (flowers, plants, people, animals, etc.). One way to model populations is to give a recursion formula that describes how the number of individuals in each generation relates to the number of individuals in the preceding generation. One of the simplest such models, called the ***exponential model***, assumes that the number of individuals in each generation is a fixed percentage of the number of individuals in the preceding generation. Thus, if there are c individuals initially and if the number of individuals in any generation is r times the number of individuals in the preceding generation, then the growth through successive generations is given by the recursion formula

$$y_0 = c, \quad y_{n+1} = r y_n \qquad (n = 0, 1, 2, 3, \ldots)$$

Exercise 7   Suppose that a population with an exponential growth model has c individuals initially.

(a) Express the iterates $y_1, y_2, y_3,$ and y_4 in terms of c and r.

(b) Find a formula for y_{n+1} in terms of c and r.

(c) Describe the eventual fate of the population if $r = 1$, $r < 1$, and $r > 1$.

There is a more sophisticated model of population growth, called the ***logistic model***, that takes environmental constraints into account. In this model, it is assumed that there is some maximum population that can be supported by the environment, and the population is expressed as a fraction of the maximum. Thus, in each generation the population is represented as a number in the interval $0 \leq y_n \leq 1$. When y_n is near 0 the population has lots of room to grow, but when y_n is near 1 the population is close to the maximum and the environmental factors tend to inhibit further growth. Models of this type are given by recursion formulas of the form

$$y_0 = c, \quad y_{n+1} = k y_n (1 - y_n) \qquad (2)$$

in which k is a positive constant that depends on the ecological conditions.

Figure 1 illustrates a graphical method for tracking the growth of a population described by (2). That figure, which is called a ***cobweb diagram***, shows graphs of the line $y = x$ and the curve $y = kx(1 - x)$.

Exercise 8   Explain why the values $y_1, y_2,$ and y_3 are the populations for the first three generations of the logistic growth model given by (2).

Exercise 9   The cobweb diagram in Figure 2 tracks the growth of a population with a logistical growth model given by the recursion formula

$$y_0 = 0.1, \quad y_{n+1} = 2.9 y_n (1 - y_n)$$

(a) Find the populations $y_1, y_2, \ldots, y_5$ of the first five generations.

(b) What happens to the population over the long term?

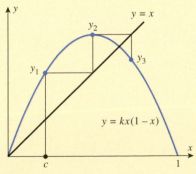

Figure 1

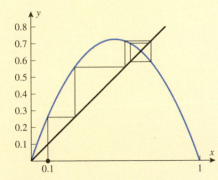

Figure 2

Chaos and Fractals

Observe that (2) is a recursion formula for the iterated function sequence of $f(x) = kx(1 - x)$. Iterated function sequences of this form are called *iterated quadratic systems*. These are important not only in modeling populations but also in the study of *chaos* and *fractals*—two important fields of contemporary research.

Module by: *C. Lynn Kiaer, Rose-Hulman Institute of Technology*
Howard Anton, Drexel University

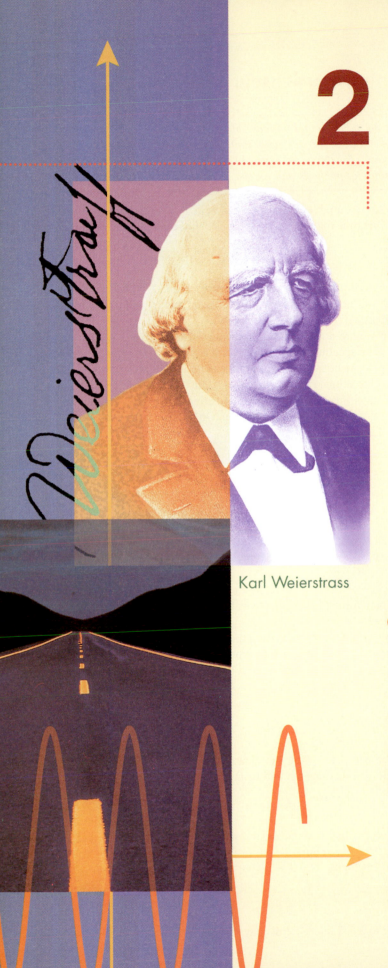

Karl Weierstrass

2

LIMITS AND CONTINUITY

The problem of defining and calculating instantaneous rates such as speed and acceleration attracted almost all the mathematicians of the seventeenth century.

—Morris Kline

The development of calculus in the seventeenth century by Newton and Leibniz provided scientists with their first real understanding of what is meant by an "instantaneous rate of change" such as velocity and acceleration. Once the idea was understood conceptually, efficient computational methods followed, and science took a quantum leap forward. The fundamental building block on which rates of change rest is the concept of a "limit," an idea that is so important that all other calculus concepts are now based on it.

In this chapter we will develop the concept of a limit in stages, proceeding from an informal, intuitive notion to a precise mathematical definition. We will also develop theorems and procedures for calculating limits, and we will conclude the chapter by using the limits to study "continuous" curves.

2.1 LIMITS (AN INTUITIVE APPROACH)

The concept of a limit is the fundamental building block on which all other calculus concepts are based. In this section we will study limits informally, with the goal of developing an "intuitive feel" for the basic ideas. In the following three sections we will focus on the computational methods and precise definitions.

INSTANTANEOUS VELOCITY AND THE SLOPE OF A CURVE

Recall from Formula (11) of Section 1.5 that if a particle moves along an s-axis, then the average velocity v_{ave} over the time interval from t_0 to t_1 is defined as

$$v_{ave} = \frac{\Delta s}{\Delta t} = \frac{s_1 - s_0}{t_1 - t_0} \tag{1}$$

where s_0 and s_1 are the s-coordinates of the particle at times t_0 and t_1, respectively. Geometrically, v_{ave} is the slope of the line joining the points (t_0, s_0) and (t_1, s_1) on the position versus time curve for the particle (Figure 2.1.1).

Suppose, however, that we are not interested in average velocity over a time interval, but rather the velocity v_{inst} at a specific instant in time. It is not a simple matter of applying Formula (1), since the displacement and the elapsed time in an instant are both zero. However, intuition suggests that the velocity at an instant $t = t_0$ can be approximated by finding the position of the particle at a time t_1 just before, or just after, time t_0 and computing the average velocity over the brief time interval between the two moments. That is,

$$v_{inst} \approx v_{ave} = \frac{s_1 - s_0}{t_1 - t_0} \tag{2}$$

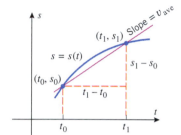

Figure 2.1.1

provided $\Delta t = t_1 - t_0$ is small. Moreover, if we are able to make very precise measurements, the closer t_1 is to t_0, the better v_{ave} approximates v_{inst}. That is, as we sample at times t_1, closer and closer to t_0, v_{ave} approaches a *limiting value* that we understand to be v_{inst}.

Example 1 Suppose that a ball is thrown vertically upward and the height in feet of the ball t seconds after its release is modeled by the function

$$s(t) = -16t^2 + 29t + 6, \qquad 0 \le t \le 2$$

What is a reasonable estimate for the instantaneous velocity of the ball at time $t = 0.5$ s?

Solution. At any time $0 \le t \le 2$ we may envision the height $s(t)$ of the ball as a position on a (vertical) s-axis, where $s = 0$ corresponds to ground level (Figure 2.1.2). The height of the ball at time $t = 0.5$ s is $s(0.5) = 16.5$ ft, and the height of the ball 0.01 s later is $s(0.51) = 16.6284$ ft. Therefore, the average velocity of the ball over the time interval from $t = 0.5$ to $t = 0.51$ is

$$v_{ave} = \frac{16.6284 - 16.5}{0.51 - 0.5} = \frac{0.1284}{0.01} = 12.84 \text{ ft/s}$$

Similarly, the height of the ball 0.49 s after its release is $s(0.49) = 16.3684$ ft, and the average velocity of the ball over the time interval from $t = 0.49$ to $t = 0.5$ is

$$v_{ave} = \frac{16.3684 - 16.5}{0.49 - 0.5} = \frac{-0.1316}{-0.01} = 13.16 \text{ ft/s}$$

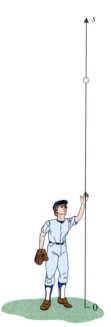

Figure 2.1.2

Consequently, we would expect the instantaneous velocity of the ball at time $t = 0.5$ to be between 12.84 ft/s and 13.16 ft/s. To improve our estimate of this instantaneous velocity, we can compute the average velocity

$$v_{ave}(t_1) = \frac{s(t_1) - 16.5}{t_1 - 0.5} = \frac{-16t_1^2 + 29t_1 + 6 - 16.5}{t_1 - 0.5} = \frac{-16t_1^2 + 29t_1 - 10.5}{t_1 - 0.5}$$

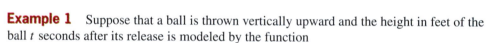

for values of t_1 even closer to 0.5. Table 2.1.1 displays the results of several such computa-

Table 2.1.1

TIME t_1 (s)	$v_{ave}(t_1) = \dfrac{-16t_1^2 + 29t_1 - 10.5}{t_1 - 0.5}$ (ft/s)
0.5010	12.9840
0.5005	12.9920
0.5001	12.9984
0.4999	13.0016
0.4995	13.0080
0.4990	13.0160

tions. It appears from these computations that a reasonable estimate for the instantaneous velocity of the ball at time $t = 0.5$ s is 13 ft/s.   ◀

FOR THE READER.   The domain of the height function $s(t) = -16t^2 + 29t + 6$ in Example 1 is the closed interval $[0, 2]$. Why do we not consider values of t less than 0 or greater than 2 for this function? In Table 2.1.1, why is there not a value of $v_{ave}(t_1)$ for $t_1 = 0.5$?

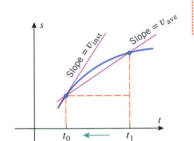

Figure 2.1.3

We can interpret v_{inst} geometrically from the interpretation of v_{ave} as the slope of the line joining the points (t_0, s_0) and (t_1, s_1) on the position versus time curve for the particle. When $\Delta t = t_1 - t_0$ is small, the points (t_0, s_0) and (t_1, s_1) are very close to each other on the curve. As the sampling point (t_1, s_1) is selected closer to our anchoring point (t_0, s_0), the slope v_{ave} more nearly approximates what we might reasonably call the *slope* of the position curve at time $t = t_0$. Thus, v_{inst} can be viewed as the slope of the position curve at time $t = t_0$ (Figure 2.1.3). We will explore this connection more fully in Section 3.1.

LIMITS

In Example 1 it appeared that choosing values of t_1 close to (but not equal to) 0.5 resulted in values of $v_{ave}(t_1)$ that were close to 13. One way of describing this behavior is to say that the *limiting value* of $v_{ave}(t_1)$ as t_1 approaches 0.5 is 13 or, equivalently, that 13 is the *limit* of $v_{ave}(t_1)$ as t_1 approaches 0.5. More generally, we will see that the concept of the limit of a function provides a foundation for the tools of calculus. Thus, it is appropriate to start a study of calculus by focusing on the limit concept itself.

The most basic use of limits is to describe how a function behaves as the independent variable approaches a given value. For example, let us examine the behavior of the function

$$f(x) = x^2 - x + 1$$

for x-values closer and closer to 2. It is evident from the graph and table in Figure 2.1.4 that the values of $f(x)$ get closer and closer to 3 as values of x are selected closer and closer to 2 on either the left or the right side of 2. We describe this by saying that the "limit of $x^2 - x + 1$ is 3 as x approaches 2 from either side," and we write

$$\lim_{x \to 2} (x^2 - x + 1) = 3 \tag{3}$$

Observe that in our investigation of $\lim_{x \to 2} (x^2 - x + 1)$ we are only concerned with the values of $f(x)$ *near* $x = 2$ and not the value of $f(x)$ *at* $x = 2$.

This leads us to the following general idea.

2.1.1   LIMITS (AN INFORMAL VIEW).   If the values of $f(x)$ can be made as close as we like to L by taking values of x sufficiently close to a (but not equal to a), then we write

$$\lim_{x \to a} f(x) = L \tag{4}$$

which is read "the limit of $f(x)$ as x approaches a is L."

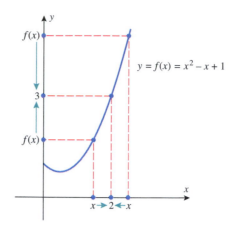

x	1.0	1.5	1.9	1.95	1.99	1.995	1.999	2	2.001	2.005	2.01	2.05	2.1	2.5	3.0
$f(x)$	1.000000	1.750000	2.710000	2.852500	2.970100	2.985025	2.997001		3.003001	3.015025	3.030100	3.152500	3.310000	4.750000	7.000000

Left side ⟶   ⟵ Right side

Figure 2.1.4

Equation (4) is also commonly written as

$$f(x) \to L \quad \text{as} \quad x \to a$$

With this notation we can express (3) as

$$x^2 - x + 1 \to 3 \quad \text{as} \quad x \to 2$$

In order to investigate $\lim_{x \to a} f(x)$, we ask ourselves the question, "If x is close to, but different from, a, is there a particular number to which $f(x)$ is close?" This question presumes that the function f is defined "everywhere near a," in other words, that f is defined at all points x in some open interval containing a, except possibly at $x = a$. The value of f at a, if it exists at all, is not relevant to the determination of $\lim_{x \to a} f(x)$. Many important applications of the limit concept involve contexts in which the domain of the function excludes a. Indeed, our discussion of instantaneous velocity concluded that v_{inst} could be interpreted as a limit of the average velocities, even though the average velocity at an instant is not defined.

The process of determining a limit generally involves a *discovery* phase, followed by a *verification* phase. The discovery phase begins with sampled x-values, and ends with a conjecture for the limit. Figure 2.1.4 illustrates the discovery phase for the problem of finding the value of $\lim_{x \to 2} (x^2 - x + 1)$. We sampled values for x near 2 and found that the corresponding values of $f(x)$ were close to 3. Indeed, values of x nearer to 2 produced values of $f(x)$ closer to 3. Our conjecture that $\lim_{x \to 2} (x^2 - x + 1) = 3$ concluded the discovery phase for this limit. However, a complete treatment of any limit also involves a verification phase in which it is shown that the conjectured limit is actually correct. For example, consider our conjecture that $\lim_{x \to 2} (x^2 - x + 1) = 3$. We can only sample a relatively few values of x near 2, even by using a graphing utility. We cannot sample *all* values of x near 2, for no matter how close to 2 we take an x-value, there are infinitely many values of x nearer yet to 2. To verify that $\lim_{x \to 2} (x^2 - x + 1)$ is indeed 3, we need to resort to an analysis that can overcome this dilemma. This analysis will require a more mathematically precise definition of limit and is the focus of Section 2.4. In this section, we concentrate on the discovery phase for limit problems.

Example 2   Make a conjecture about the value of the limit

$$\lim_{x \to 0} \frac{x}{\sqrt{x+1} - 1} \tag{5}$$

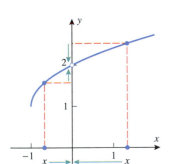

Figure 2.1.5

Solution. Observe that the function

$$f(x) = \frac{x}{\sqrt{x+1}-1}$$

is not defined at $x = 0$. However, f is defined for $x > -1$, $x \neq 0$, so the domain of f contains values of x "everywhere near 0." Table 2.1.2 shows samples of x-values approaching 0 from the left side and from the right side. In both cases the values of $f(x)$, calculated to six decimal places, appear to get closer and closer to 2, and hence we conjecture that

$$\lim_{x \to 0} \frac{x}{\sqrt{x+1}-1} = 2 \tag{6}$$

A graphing utility could be used to produce Figure 2.1.5, providing more evidence in support of our conjecture. In the next section we will see that the graph of $f(x)$ is identical to that of $y = \sqrt{x+1}+1$, except for a hole at $(0, 2)$. ◄

Table 2.1.2

x	−0.01	−0.001	−0.0001	−0.00001	0	0.00001	0.0001	0.001	0.01
$f(x)$	1.994987	1.999500	1.999950	1.999995		2.000005	2.000050	2.000500	2.004988

Left side ⟶ ⟵ Right side

FOR THE READER. Using a graphing utility, find a window about $x = 0$ in which all values of $f(x)$ are within 0.5 of $y = 2$. Find a window in which all values of $f(x)$ are within 0.1 of $y = 2$.

Example 3 Make a conjecture about the value of the limit

$$\lim_{x \to 0} \frac{\sin x}{x} \tag{7}$$

Solution. The function $f(x) = (\sin x)/x$ is not defined at $x = 0$, but, as discussed previously, this has no bearing on the limit. With the help of a calculating utility set in radian mode, we obtain the table in Figure 2.1.6. The data in the table suggest that

$$\lim_{x \to 0} \frac{\sin x}{x} = 1 \tag{8}$$

The result is consistent with the graph of $f(x) = (\sin x)/x$ shown in the figure. Later in this chapter we will give a geometric argument to prove that our conjecture is correct. ◄

x (RADIANS)	$y = \dfrac{\sin x}{x}$
±1.0	0.84147
±0.9	0.87036
±0.8	0.89670
±0.7	0.92031
±0.6	0.94107
±0.5	0.95885
±0.4	0.97355
±0.3	0.98507
±0.2	0.99335
±0.1	0.99833
±0.01	0.99998

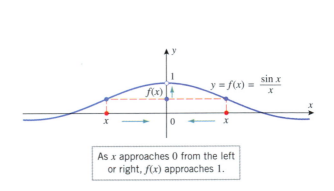

As x approaches 0 from the left or right, $f(x)$ approaches 1.

Figure 2.1.6

FOR THE READER. Use a calculating utility to sample x-values closer to 0 than those in the table in Figure 2.1.6. Does the limit change if x is in degrees?

SAMPLING PITFALLS

Although numerical and graphical evidence is helpful for guessing at limits, we can be misled by an insufficient or poorly selected sample. For example, the table in Figure 2.1.7 shows values of $f(x) = \sin(\pi/x)$ at selected values of x on both sides of 0. The data *incorrectly* suggest that

$$\cdot\text{W R O N G} \cdot \text{W R O}$$
$$\text{N} \lim_{x \to 0} \sin\left(\frac{\pi}{x}\right) = 0 \text{W}$$
$$\text{R O N G} \cdot \text{W R O N G} \cdot$$

The fact that this is incorrect is evidenced by the graph of f shown in the figure. This graph indicates that as $x \to 0$, the values of f oscillate between -1 and 1 with increasing rapidity, and hence do not approach a limit. The data are deceiving because the table consists only of sample values of x that are x-intercepts for $f(x)$.

x (RADIANS)	$\dfrac{\pi}{x}$	$f(x) = \sin\left(\dfrac{\pi}{x}\right)$
$x = \pm 1$	$\pm\pi$	$\sin(\pm\pi) = 0$
$x = \pm 0.1$	$\pm 10\pi$	$\sin(\pm 10\pi) = 0$
$x = \pm 0.01$	$\pm 100\pi$	$\sin(\pm 100\pi) = 0$
$x = \pm 0.001$	$\pm 1000\pi$	$\sin(\pm 1000\pi) = 0$
$x = \pm 0.0001$	$\pm 10,000\pi$	$\sin(\pm 10,000\pi) = 0$
$\vdots$	$\vdots$	$\vdots$

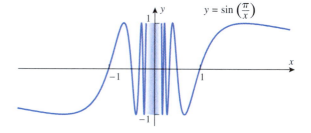

Figure 2.1.7

Numerical evidence can lead to incorrect conclusions about limits because of roundoff error or because the sample of values used is not extensive enough to give a good indication of the behavior of the function. Thus, when a limit is conjectured from a table of values, it is important to look for corroborating evidence to support the conjecture.

ONE-SIDED LIMITS

The limit in (4) is commonly called a *two-sided limit* because it requires the values of $f(x)$ to get closer and closer to L as values of x are taken from *either* side of $x = a$. However, some functions exhibit different behaviors on the two sides of an x-value a, in which case it is necessary to distinguish whether values of x near a are on the left side or on the right side of a for purposes of investigating limiting behavior. For example, consider the function

$$f(x) = \frac{|x|}{x} = \begin{cases} 1, & x > 0 \\ -1, & x < 0 \end{cases}$$

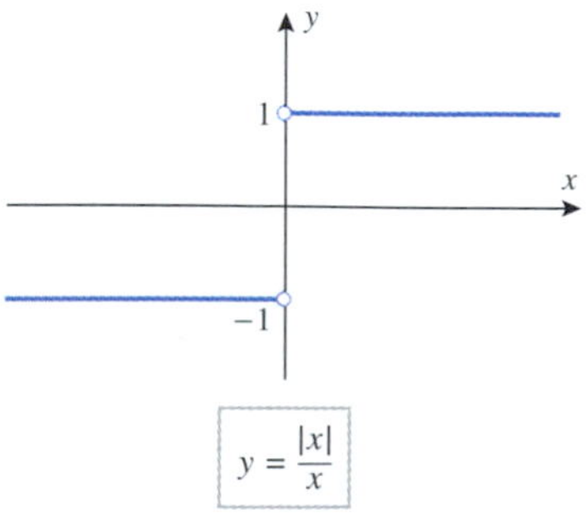

(Figure 2.1.8). Note that x-values approaching 0 and to the *right* of 0 produce $f(x)$ values that approach 1 (in fact, they are exactly 1 for all such values of x). On the other hand, x-values approaching 0 and to the *left* of 0 produce $f(x)$ values that approach -1. We describe these two statements by saying that "the limit of $f(x) = |x|/x$ is 1 as x approaches 0 from the right" and that "the limit of $f(x) = |x|/x$ is -1 as x approaches 0 from the left." We denote these limits by writing

$$\lim_{x \to 0^+} \frac{|x|}{x} = 1 \quad \text{and} \quad \lim_{x \to 0^-} \frac{|x|}{x} = -1 \tag{9--10}$$

With this notation, the superscript "$+$" indicates a limit from the right and the superscript "$-$" indicates a limit from the left.

This leads to the following general idea.

$y = \dfrac{|x|}{x}$

Figure 2.1.8

2.1.2 ONE-SIDED LIMITS (AN INFORMAL VIEW). If the values of $f(x)$ can be made as close as we like to L by taking values of x sufficiently close to a (but greater than a), then we write

$$\lim_{x \to a^+} f(x) = L \tag{11}$$

which is read "the limit of $f(x)$ as x approaches a from the right is L." Similarly, if the values of $f(x)$ can be made as close as we like to L by taking values of x sufficiently close to a (but less than a), then we write

$$\lim_{x \to a^-} f(x) = L \tag{12}$$

which is read "the limit of $f(x)$ as x approaches a from the left is L."

Equations (11) and (12), which are called ***one-sided limits***, are also commonly written as

$$f(x) \to L \text{ as } x \to a^+ \quad \text{and} \quad f(x) \to L \text{ as } x \to a^-$$

respectively. With this notation (9) and (10) can be expressed as

$$\frac{|x|}{x} \to 1 \text{ as } x \to 0^+ \quad \text{and} \quad \frac{|x|}{x} \to -1 \text{ as } x \to 0^-$$

THE RELATIONSHIP BETWEEN ONE-SIDED LIMITS AND TWO-SIDED LIMITS

In general, there is no guarantee that a function will have a limit at a specified location. If the values of $f(x)$ do not get closer and closer to some *single* number L as $x \to a$, then we say that the limit of $f(x)$ as x approaches a ***does not exist*** (and similarly for one-sided limits). For example, the two-sided limit $\lim_{x \to 0} |x|/x$ does not exist because the values of $f(x)$ do not approach a single number as $x \to 0$; the values approach -1 from the left and 1 from the right.

In general, the following condition must be satisfied for the two-sided limit of a function to exist.

2.1.3 THE RELATIONSHIP BETWEEN ONE-SIDED AND TWO-SIDED LIMITS. The two-sided limit of a function $f(x)$ exists at a if and only if both of the one-sided limits exist at a and have the same value; that is,

$$\lim_{x \to a} f(x) = L \quad \text{if and only if} \quad \lim_{x \to a^-} f(x) = L = \lim_{x \to a^+} f(x)$$

REMARK.   Sometimes, one or both of the one-sided limits may fail to exist (which, in turn, implies that the two-sided limit does not exist). For example, we saw earlier that the one-sided limits of $f(x) = \sin(\pi/x)$ do not exist as x approaches 0 because the function keeps oscillating between -1 and 1, failing to settle on a single value. This implies that the two-sided limit does not exist as x approaches 0.

Example 4 For the functions in Figure 2.1.9, find the one-sided and two-sided limits at $x = a$ if they exist.

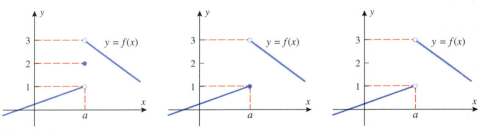

Figure 2.1.9

Solution. The functions in all three figures have the same one-sided limits as $x \to a$, since the functions are identical, except at $x = a$. These limits are

$$\lim_{x \to a^+} f(x) = 3 \quad \text{and} \quad \lim_{x \to a^-} f(x) = 1$$

In all three cases the two-sided limit does not exist as $x \to a$ because the one-sided limits are not equal. ◀

Example 5 For the functions in Figure 2.1.10, find the one-sided and two-sided limits at $x = a$ if they exist.

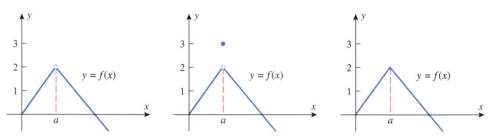

Figure 2.1.10

Solution. As in the preceding example, the value of f at $x = a$ has no bearing on the limits as $x \to a$, so that in all three cases we have

$$\lim_{x \to a^+} f(x) = 2 \quad \text{and} \quad \lim_{x \to a^-} f(x) = 2$$

Since the one-sided limits are equal, the two-sided limit exists and

$$\lim_{x \to a} f(x) = 2$$ ◀

INFINITE LIMITS AND VERTICAL ASYMPTOTES

Sometimes one-sided or two-sided limits will fail to exist because the values of the function increase or decrease indefinitely. For example, consider the behavior of $f(x) = 1/x$ for values of x near 0. It is evident from the table and graph in Figure 2.1.11 that as x-values are taken closer and closer to 0 from the right, the values of $f(x) = 1/x$ are positive and increase indefinitely; and as x-values are taken closer and closer to 0 from the left, the values of $f(x) = 1/x$ are negative and decrease indefinitely. We describe these limiting behaviors

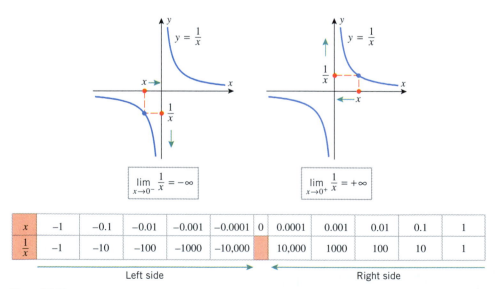

x	-1	-0.1	-0.01	-0.001	-0.0001	0	0.0001	0.001	0.01	0.1	1
$\frac{1}{x}$	-1	-10	-100	-1000	$-10,000$		10,000	1000	100	10	1

Left side → ← Right side

Figure 2.1.11

by writing

$$\lim_{x \to 0^+} \frac{1}{x} = +\infty \quad \text{and} \quad \lim_{x \to 0^-} \frac{1}{x} = -\infty$$

More generally:

2.1.4   INFINITE LIMITS (AN INFORMAL VIEW).   If the values of $f(x)$ increase indefinitely as x approaches a from the right or left, then we write

$$\lim_{x \to a^+} f(x) = +\infty \quad \text{or} \quad \lim_{x \to a^-} f(x) = +\infty$$

as appropriate, and we say that $f(x)$ ***increases without bound***, or $f(x)$ ***approaches*** $+\infty$, as $x \to a^+$ or as $x \to a^-$. Similarly, if the values of $f(x)$ decrease indefinitely as x approaches a from the right or left, then we write

$$\lim_{x \to a^+} f(x) = -\infty \quad \text{or} \quad \lim_{x \to a^-} f(x) = -\infty$$

as appropriate, and say that $f(x)$ ***decreases without bound***, or $f(x)$ ***approaches*** $-\infty$, as $x \to a^+$ or as $x \to a^-$. Moreover, if both one-sided limits are $+\infty$, then we write

$$\lim_{x \to a} f(x) = +\infty$$

and if both one-sided limits are $-\infty$, then we write

$$\lim_{x \to a} f(x) = -\infty$$

REMARK.   It should be emphasized that the symbols $+\infty$ and $-\infty$ are *not* real numbers. The phrase "$f(x)$ approaches $+\infty$" is akin to saying that "$f(x)$ approaches the unapproachable"; it is a colloquialism for "$f(x)$ increases without bound." The symbols $+\infty$ and $-\infty$ are used here to encapsulate a particular way in which limits fail to exist. To say, for example, that $f(x) \to +\infty$ as $x \to a^+$ is to indicate that $\lim_{x \to a^+} f(x)$ *does not exist* and that it fails to exist because the values of $f(x)$ increase without bound as x approaches a from the right. Furthermore, since $+\infty$ and $-\infty$ are not numbers, it is inappropriate to manipulate these symbols using rules of algebra. For example, it is *not* correct to equate $(+\infty) - (+\infty)$ with 0.

Example 6   For the functions in Figure 2.1.12, describe the limits at $x = a$ in appropriate limit notation.

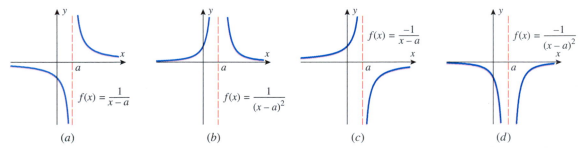

| (a) | (b) | (c) | (d) |

Figure 2.1.12

Solution (a).   In Figure 2.1.12a, the function increases indefinitely as x approaches a from the right and decreases indefinitely as x approaches a from the left. Thus,

$$\lim_{x \to a^+} \frac{1}{x - a} = +\infty \quad \text{and} \quad \lim_{x \to a^-} \frac{1}{x - a} = -\infty$$

Solution (b). In Figure 2.1.12b, the function increases indefinitely as x approaches a from both the left and right. Thus,

$$\lim_{x \to a} \frac{1}{(x-a)^2} = \lim_{x \to a^+} \frac{1}{(x-a)^2} = \lim_{x \to a^-} \frac{1}{(x-a)^2} = +\infty$$

Solution (c). In Figure 2.1.12c, the function decreases indefinitely as x approaches a from the right and increases indefinitely as x approaches a from the left. Thus,

$$\lim_{x \to a^+} \frac{-1}{x-a} = -\infty \quad \text{and} \quad \lim_{x \to a^-} \frac{-1}{x-a} = +\infty$$

Solution (d). In Figure 2.1.12d, the function decreases indefinitely as x approaches a from both the left and right. Thus,

$$\lim_{x \to a} \frac{-1}{(x-a)^2} = \lim_{x \to a^+} \frac{-1}{(x-a)^2} = \lim_{x \to a^-} \frac{-1}{(x-a)^2} = -\infty \qquad \blacktriangleleft$$

Geometrically, if $f(x) \to +\infty$ as $x \to a^-$ or as $x \to a^+$, then the graph of $y = f(x)$ rises without bound and squeezes closer to the vertical line $x = a$ on the indicated side of $x = a$. If $f(x) \to -\infty$ as $x \to a^-$ or as $x \to a^+$, then the graph of $y = f(x)$ falls without bound and squeezes closer to the vertical line $x = a$ on the indicated side of $x = a$. In these cases, we call the line $x = a$ a *vertical asymptote*. ("Asymptote" comes from the Greek *asymptotos*, meaning "nonintersecting." We will soon see that taking "asymptote" to be synonymous with "nonintersecting" is a bit misleading.)

2.1.5 DEFINITION.   A line $x = a$ is called a ***vertical asymptote*** of the graph of a function f if $f(x) \to +\infty$ or $f(x) \to -\infty$ as x approaches a from the left or right.

Example 7   The four functions graphed in Figure 2.1.12 all have a vertical asymptote at $x = a$, which is indicated by the dashed vertical lines in the figure. $\blacktriangleleft$

LIMITS AT INFINITY AND HORIZONTAL ASYMPTOTES

Thus far, we have used limits to describe the behavior of $f(x)$ as x approaches a. However, sometimes we will not be concerned with the behavior of $f(x)$ near a specific x-value, but rather with how the values of $f(x)$ behave as x increases without bound or decreases without bound. This is sometimes called the ***end behavior*** of the function because it describes how the function behaves for values of x that are far from the origin. For example, it is evident from the table and graph in Figure 2.1.13 that as x increases without bound, the values of

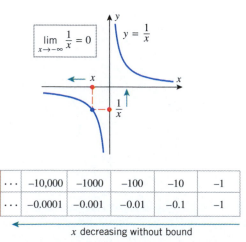

x	$\cdots$	$-10{,}000$	-1000	-100	-10	-1
$f(x)$	$\cdots$	-0.0001	-0.001	-0.01	-0.1	-1

x decreasing without bound

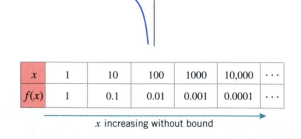

x	1	10	100	1000	$10{,}000$	$\cdots$
$f(x)$	1	0.1	0.01	0.001	0.0001	$\cdots$

x increasing without bound

Figure 2.1.13

$f(x) = 1/x$ are positive, but get closer and closer to 0; and as x decreases without bound, the values of $f(x) = 1/x$ are negative, and also get closer and closer to 0. We indicate these limiting behaviors by writing

$$\lim_{x \to +\infty} \frac{1}{x} = 0 \quad \text{and} \quad \lim_{x \to -\infty} \frac{1}{x} = 0$$

More generally:

2.1.6 **LIMITS AT INFINITY (AN INFORMAL VIEW).** If the values of $f(x)$ eventually get closer and closer to a number L as x increases without bound, then we write

$$\lim_{x \to +\infty} f(x) = L \quad \text{or} \quad f(x) \to L \text{ as } x \to +\infty \tag{13}$$

Similarly, if the values of $f(x)$ eventually get closer and closer to a number L as x decreases without bound, then we write

$$\lim_{x \to -\infty} f(x) = L \quad \text{or} \quad f(x) \to L \text{ as } x \to -\infty \tag{14}$$

Geometrically, if $f(x) \to L$ as $x \to +\infty$, then the graph of $y = f(x)$ eventually gets closer and closer to the line $y = L$ as the graph is traversed in the positive direction (Figure 2.1.14a); and if $f(x) \to L$ as $x \to -\infty$, then the graph of $y = f(x)$ eventually gets closer and closer to the line $y = L$ as the graph is traversed in the negative x-direction (Figure 2.1.14b). In either case we call the line $y = L$ a *horizontal asymptote* of the graph of f. For example, the function in Figure 2.1.13 has $y = 0$ as a horizontal asymptote.

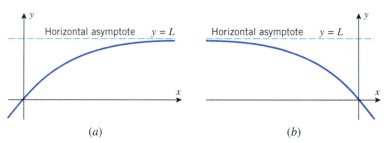

Figure 2.1.14

2.1.7 **DEFINITION.** A line $y = L$ is called a ***horizontal asymptote*** of the graph of a function f if

$$\lim_{x \to +\infty} f(x) = L \quad \text{or} \quad \lim_{x \to -\infty} f(x) = L$$

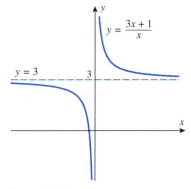

Figure 2.1.15

Sometimes the existence of a horizontal asymptote of a function f will be readily apparent from the formula for f. For example, it is evident that the function

$$f(x) = \frac{3x + 1}{x} = 3 + \frac{1}{x}$$

has a horizontal asymptote at $y = 3$ (Figure 2.1.15), since the value of $1/x$ approaches 0 as $x \to +\infty$ or as $x \to -\infty$. For more complicated functions, algebraic manipulations or special techniques that we will study in the next section may have to be applied to confirm the existence of horizontal asymptotes.

HOW LIMITS AT INFINITY CAN FAIL TO EXIST

Limits at infinity can fail to exist for various reasons. One possibility is that the values of $f(x)$ may increase or decrease without bound as $x \to +\infty$ or as $x \to -\infty$. For example, the values of $f(x) = x^3$ increase without bound as $x \to +\infty$ and decrease without bound as

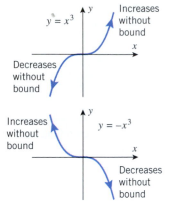

Increases without bound

$y = x^3$

Decreases without bound

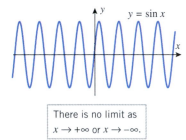

Increases without bound

$y = -x^3$

Decreases without bound

Figure 2.1.16

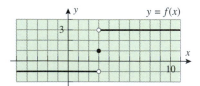

$y = \sin x$

There is no limit as $x \to +\infty$ or $x \to -\infty$.

Figure 2.1.17

$x \to -\infty$; and for $f(x) = -x^3$ the values decrease without bound as $x \to +\infty$ and increase without bound as $x \to -\infty$ (Figure 2.1.16). We denote this by writing

$$\lim_{x \to +\infty} x^3 = +\infty, \quad \lim_{x \to -\infty} x^3 = -\infty, \quad \lim_{x \to +\infty} (-x^3) = -\infty, \quad \lim_{x \to -\infty} (-x^3) = +\infty$$

More generally:

2.1.8   INFINITE LIMITS AT INFINITY (AN INFORMAL VIEW).   If the values of $f(x)$ increase without bound as $x \to +\infty$ or as $x \to -\infty$, then we write

$$\lim_{x \to +\infty} f(x) = +\infty \quad \text{or} \quad \lim_{x \to -\infty} f(x) = +\infty$$

as appropriate; and if the values of $f(x)$ decrease without bound as $x \to +\infty$ or as $x \to -\infty$, then we write

$$\lim_{x \to +\infty} f(x) = -\infty \quad \text{or} \quad \lim_{x \to -\infty} f(x) = -\infty$$

as appropriate.

Limits at infinity can also fail to exist because the graph of the function oscillates indefinitely in such a way that the values of the function do not approach a fixed number and do not increase or decrease without bound; the trigonometric functions $\sin x$ and $\cos x$ have this property, for example (Figure 2.1.17). In such cases we say that the limit *fails to exist because of oscillation.*

EXERCISE SET 2.1   ~ Graphing Utility   c CAS

. . . .

> In each of Exercises 1–14, assume that all of the significant features of the indicated function, including its end behavior, can be correctly surmised from the graph shown in the figure.

1. For the function f graphed in the accompanying figure, find
 (a) $\lim\limits_{x \to 3^-} f(x)$ (b) $\lim\limits_{x \to 3^+} f(x)$ (c) $\lim\limits_{x \to 3} f(x)$
 (d) $f(3)$ (e) $\lim\limits_{x \to -\infty} f(x)$ (f) $\lim\limits_{x \to +\infty} f(x)$.

$y = f(x)$

Figure Ex-1

2. For the function f graphed in the accompanying figure, find
 (a) $\lim\limits_{x \to 2^-} f(x)$ (b) $\lim\limits_{x \to 2^+} f(x)$ (c) $\lim\limits_{x \to 2} f(x)$
 (d) $f(2)$ (e) $\lim\limits_{x \to -\infty} f(x)$ (f) $\lim\limits_{x \to +\infty} f(x)$.

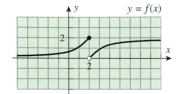

$y = f(x)$

Figure Ex-2

3. For the function g graphed in the accompanying figure, find
 (a) $\lim\limits_{x \to 4^-} g(x)$ (b) $\lim\limits_{x \to 4^+} g(x)$ (c) $\lim\limits_{x \to 4} g(x)$
 (d) $g(4)$ (e) $\lim\limits_{x \to -\infty} g(x)$ (f) $\lim\limits_{x \to +\infty} g(x)$.

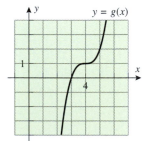

$y = g(x)$

Figure Ex-3

4. For the function g graphed in the accompanying figure, find
 (a) $\lim\limits_{x \to 0^-} g(x)$ (b) $\lim\limits_{x \to 0^+} g(x)$ (c) $\lim\limits_{x \to 0} g(x)$
 (d) $g(0)$ (e) $\lim\limits_{x \to -\infty} g(x)$ (f) $\lim\limits_{x \to +\infty} g(x)$.

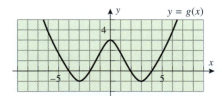

$y = g(x)$

Figure Ex-4

5. For the function F graphed in the accompanying figure, find
(a) $\lim\limits_{x \to -2^-} F(x)$
(b) $\lim\limits_{x \to -2^+} F(x)$
(c) $\lim\limits_{x \to -2} F(x)$
(d) $F(-2)$
(e) $\lim\limits_{x \to -\infty} F(x)$
(f) $\lim\limits_{x \to +\infty} F(x)$.

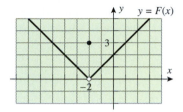

Figure Ex-5

6. For the function F graphed in the accompanying figure, find
(a) $\lim\limits_{x \to 3^-} F(x)$
(b) $\lim\limits_{x \to 3^+} F(x)$
(c) $\lim\limits_{x \to 3} F(x)$
(d) $F(3)$
(e) $\lim\limits_{x \to -\infty} F(x)$
(f) $\lim\limits_{x \to +\infty} F(x)$.

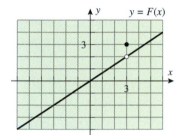

Figure Ex-6

7. For the function ϕ graphed in the accompanying figure, find
(a) $\lim\limits_{x \to -2^-} \phi(x)$
(b) $\lim\limits_{x \to -2^+} \phi(x)$
(c) $\lim\limits_{x \to -2} \phi(x)$
(d) $\phi(-2)$
(e) $\lim\limits_{x \to -\infty} \phi(x)$
(f) $\lim\limits_{x \to +\infty} \phi(x)$.

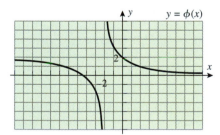

Figure Ex-7

8. For the function ϕ graphed in the accompanying figure, find
(a) $\lim\limits_{x \to 4^-} \phi(x)$
(b) $\lim\limits_{x \to 4^+} \phi(x)$
(c) $\lim\limits_{x \to 4} \phi(x)$
(d) $\phi(4)$
(e) $\lim\limits_{x \to -\infty} \phi(x)$
(f) $\lim\limits_{x \to +\infty} \phi(x)$.

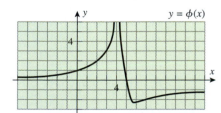

Figure Ex-8

9. For the function f graphed in the accompanying figure, find
(a) $\lim\limits_{x \to 3^-} f(x)$
(b) $\lim\limits_{x \to 3^+} f(x)$
(c) $\lim\limits_{x \to 3} f(x)$
(d) $f(3)$
(e) $\lim\limits_{x \to -\infty} f(x)$
(f) $\lim\limits_{x \to +\infty} f(x)$.

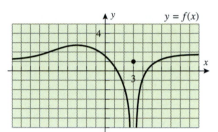

Figure Ex-9

10. For the function f graphed in the accompanying figure, find
(a) $\lim\limits_{x \to 0^-} f(x)$
(b) $\lim\limits_{x \to 0^+} f(x)$
(c) $\lim\limits_{x \to 0} f(x)$
(d) $f(0)$
(e) $\lim\limits_{x \to -\infty} f(x)$
(f) $\lim\limits_{x \to +\infty} f(x)$.

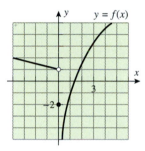

Figure Ex-10

11. For the function G graphed in the accompanying figure, find
(a) $\lim\limits_{x \to 0^-} G(x)$
(b) $\lim\limits_{x \to 0^+} G(x)$
(c) $\lim\limits_{x \to 0} G(x)$
(d) $G(0)$
(e) $\lim\limits_{x \to -\infty} G(x)$
(f) $\lim\limits_{x \to +\infty} G(x)$.

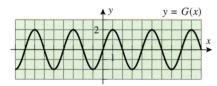

Figure Ex-11

12. For the function G graphed in the accompanying figure, find
(a) $\lim\limits_{x \to 0^-} G(x)$
(b) $\lim\limits_{x \to 0^+} G(x)$
(c) $\lim\limits_{x \to 0} G(x)$
(d) $G(0)$
(e) $\lim\limits_{x \to -\infty} G(x)$
(f) $\lim\limits_{x \to +\infty} G(x)$.

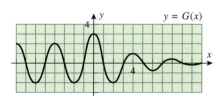

Figure Ex-12

13. Consider the function g graphed in the accompanying figure. For what values of x_0 does $\lim\limits_{x \to x_0} g(x)$ exist?

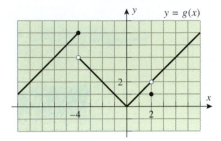

Figure Ex-13

14. Consider the function f graphed in the accompanying figure. For what values of x_0 does $\lim\limits_{x \to x_0} f(x)$ exist?

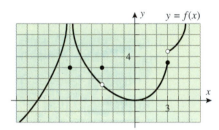

Figure Ex-14

In Exercises 15–18, sketch a possible graph for a function f with the specified properties. (Many different solutions are possible.)

15. (i) $f(0) = 2$ and $f(2) = 1$

(ii) $\lim\limits_{x \to 1^-} f(x) = +\infty$ and $\lim\limits_{x \to 1^+} f(x) = -\infty$

(iii) $\lim\limits_{x \to +\infty} f(x) = 0$ and $\lim\limits_{x \to -\infty} f(x) = +\infty$

16. (i) $f(0) = f(2) = 1$

(ii) $\lim\limits_{x \to 2^-} f(x) = +\infty$ and $\lim\limits_{x \to 2^+} f(x) = 0$

(iii) $\lim\limits_{x \to -1^-} f(x) = -\infty$ and $\lim\limits_{x \to -1^+} f(x) = +\infty$

(iv) $\lim\limits_{x \to +\infty} f(x) = 2$ and $\lim\limits_{x \to -\infty} f(x) = +\infty$

17. (i) $f(x) = 0$ if x is an integer and $f(x) \neq 0$ if x is not an integer

(ii) $\lim\limits_{x \to +\infty} f(x) = 0$ and $\lim\limits_{x \to -\infty} f(x) = 0$

18. (i) $f(x) = 1$ if x is a positive integer and $f(x) \neq 1$ if $x > 0$ is not a positive integer

(ii) $f(x) = -1$ if x is a negative integer and $f(x) \neq -1$ if $x < 0$ is not a negative integer

(iii) $\lim\limits_{x \to +\infty} f(x) = 1$ and $\lim\limits_{x \to -\infty} f(x) = -1$

In Exercises 19–22: (i) Make a guess at the limit (if it exists) by evaluating the function at the specified x-values. (ii) Confirm your conclusions about the limit by graphing the function over an appropriate interval. (iii) If you have a CAS, then use it to find the limit. [*Note:* For the trigonometric functions, be sure to set your calculating and graphing utilities to the radian mode.]

[c] 19. (a) $\lim\limits_{x \to 1} \dfrac{x - 1}{x^3 - 1}$; $x = 2, 1.5, 1.1, 1.01, 1.001, 0, 0.5, 0.9,$
$\qquad 0.99, 0.999$

(b) $\lim\limits_{x \to 1^+} \dfrac{x + 1}{x^3 - 1}$; $x = 2, 1.5, 1.1, 1.01, 1.001, 1.0001$

(c) $\lim\limits_{x \to 1^-} \dfrac{x + 1}{x^3 - 1}$; $x = 0, 0.5, 0.9, 0.99, 0.999, 0.9999$

[c] 20. (a) $\lim\limits_{x \to 0} \dfrac{\sqrt{x + 1} - 1}{x}$; $x = \pm 0.25, \pm 0.1, \pm 0.001,$
$\qquad \pm 0.0001$

(b) $\lim\limits_{x \to 0^+} \dfrac{\sqrt{x + 1} + 1}{x}$; $x = 0.25, 0.1, 0.001, 0.0001$

(c) $\lim\limits_{x \to 0^-} \dfrac{\sqrt{x + 1} + 1}{x}$; $x = -0.25, -0.1, -0.001,$
$\qquad -0.0001$

[c] 21. (a) $\lim\limits_{x \to 0} \dfrac{\sin 3x}{x}$; $x = \pm 0.25, \pm 0.1, \pm 0.001, \pm 0.0001$

(b) $\lim\limits_{x \to -1} \dfrac{\cos x}{x + 1}$; $x = 0, -0.5, -0.9, -0.99, -0.999,$
$\qquad -1.5, -1.1, -1.01, -1.001$

[c] 22. (a) $\lim\limits_{x \to -1} \dfrac{\tan(x + 1)}{x + 1}$; $x = 0, -0.5, -0.9, -0.99, -0.999,$
$\qquad -1.5, -1.1, -1.01, -1.001$

(b) $\lim\limits_{x \to 0} \dfrac{\sin(5x)}{\sin(2x)}$; $x = \pm 0.25, \pm 0.1, \pm 0.001, \pm 0.0001$

23. Consider the motion of the ball described in Example 1. By interpreting instantaneous velocity as a limit of average velocity, make a conjecture for the value of the instantaneous velocity of the ball 0.25 s after its release.

24. Consider the motion of the ball described in Example 1. By interpreting instantaneous velocity as a limit of average velocity, make a conjecture for the value of the instantaneous velocity of the ball 0.75 s after its release.

In Exercises 25 and 26: (i) Approximate the y-coordinates of all horizontal asymptotes of $y = f(x)$ by evaluating f at the x-values $\pm 10, \pm 100, \pm 1000, \pm 100{,}000$, and $\pm 100{,}000{,}000$. (ii) Confirm your conclusions by graphing $y = f(x)$ over an appropriate interval. (iii) If you have a CAS, then use it to find the horizontal asymptotes.

[c] 25. (a) $f(x) = \dfrac{2x + 3}{x + 4}$

(b) $f(x) = \left(1 + \dfrac{3}{x}\right)^x$

(c) $f(x) = \dfrac{x^2 + 1}{x + 1}$

c **26.** (a) $f(x) = \dfrac{x^2 - 1}{5x^2 + 1}$ (b) $f(x) = \left(2 + \dfrac{1}{x}\right)^x$

(c) $f(x) = \dfrac{\sin x}{x}$

27. Assume that a particle is accelerated by a constant force. The two curves $v = n(t)$ and $v = e(t)$ in the accompanying figure provide velocity versus time curves for the particle as predicted by classical physics and by the special theory of relativity, respectively. The parameter c represents the speed of light. Using the language of limits, describe the differences in the long-term predictions of the two theories.

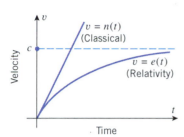

Figure Ex-27

28. Let $T = f(t)$ denote the temperature of a baked potato t minutes after it has been removed from a hot oven. The accompanying figure shows the temperature versus time curve for the potato, where r is the temperature of the room.
(a) What is the physical significance of $\displaystyle\lim_{t \to 0^+} f(t)$?
(b) What is the physical significance of $\displaystyle\lim_{t \to +\infty} f(t)$?

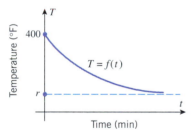

Figure Ex-28

In Exercises 29 and 30: (i) Conjecture a limit from numerical evidence. (ii) Use the substitution $t = 1/x$ to express the limit as an equivalent limit in which $t \to 0^+$ or $t \to 0^-$, as appropriate. (iii) Use a graphing utility to make a conjecture about your limit in (ii).

29. (a) $\displaystyle\lim_{x \to +\infty} x \sin\left(\dfrac{1}{x}\right)$ (b) $\displaystyle\lim_{x \to +\infty} \dfrac{1 - x}{1 + x}$

(c) $\displaystyle\lim_{x \to -\infty} \left(1 + \dfrac{2}{x}\right)^x$

30. (a) $\displaystyle\lim_{x \to +\infty} \dfrac{\cos(\pi/x)}{\pi/x}$ (b) $\displaystyle\lim_{x \to +\infty} \dfrac{x}{1 + x}$

(c) $\displaystyle\lim_{x \to -\infty} (1 - 2x)^{1/x}$

31. Suppose that $f(x)$ denotes a function such that
$$\lim_{t \to 0} f(1/t) = L$$
What can be said about
$$\lim_{x \to +\infty} f(x) \quad \text{and} \quad \lim_{x \to -\infty} f(x)?$$

32. (a) Do any of the trigonometric functions, $\sin x$, $\cos x$, $\tan x$, $\cot x$, $\sec x$, $\csc x$, have horizontal asymptotes?
(b) Do any of them have vertical asymptotes? Where?

33. (a) Let
$$f(x) = \left(1 + x^2\right)^{1.1/x^2}$$
Graph f in the window $[-1, 1] \times [2.5, 3.5]$ and use the calculator's trace feature to make a conjecture about the limit of $f(x)$ as $x \to 0$.
(b) Graph f in the window $[-0.001, 0.001] \times [2.5, 3.5]$ and use the calculator's trace feature to make a conjecture about the limit of $f(x)$ as $x \to 0$.
(c) Graph f in the window
$$[-0.000001, 0.000001] \times [2.5, 3.5]$$
and use the calculator's trace feature to make a conjecture about the limit of $f(x)$ as $x \to 0$.
(d) Later we will be able to show that
$$\lim_{x \to 0} \left(1 + x^2\right)^{1.1/x^2} \approx 3.00416602$$
What flaw do your graphs reveal about using numerical evidence (as revealed by the graphs you obtained) to make conjectures about limits?

Roundoff error is one source of inaccuracy in calculator and computer computations. Another source of error, called *catastrophic subtraction*, occurs when two nearly equal numbers are subtracted, and the result is used as part of another calculation. For example, by hand calculation we have
$$(0.123456789012345 - 0.123456789012344) \times 10^{15} = 1$$
However, a calculator that can only store 14 decimal digits produces a value of 0 for this computation, since the numbers being subtracted are identical in the first 14 digits. Catastrophic subtraction can sometimes be avoided by rearranging formulas algebraically, but your best defense is to be aware that it can occur. Watch out for it in the next exercise.

c **34.** (a) Let
$$f(x) = \dfrac{x - \sin x}{x^3}$$
Make a conjecture about the limit of f as $x \to 0^+$ by evaluating $f(x)$ at $x = 0.1, 0.01, 0.001, 0.0001$.
(b) Evaluate the function $f(x)$ at $x = 0.000001$, $0.0000001, 0.00000001, 0.000000001, 0.0000000001$, and make another conjecture.
(c) What flaw does this reveal about using numerical evidence to make conjectures about limits?
(d) If you have a CAS, use it to show that the exact value of the limit is $\frac{1}{6}$.

35. (a) The accompanying figure shows two different views of the graph of the function in Exercise 34, as generated by *Mathematica*. What is happening?

(b) Use your graphing utility to generate the graphs, and see whether the same problem occurs.

(c) Would you expect a similar problem to occur in the vicinity of $x = 0$ for the function

$$f(x) = \frac{1 - \cos x}{x}?$$

See if it does.

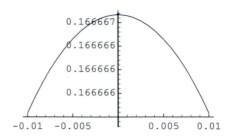

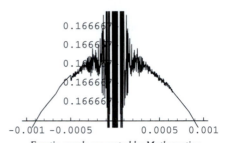

Erratic graph generated by Mathematica

Figure Ex-35

2.2 COMPUTING LIMITS

In this section we will discuss algebraic techniques for computing limits of many functions. We base these results on the informal development of the limit concept discussed in the preceding section. A more formal derivation of these results is possible after Section 2.4.

SOME BASIC LIMITS

Our strategy for finding limits algebraically has two parts:

• First we will obtain the limits of some simple functions.

• Then we will develop a repertoire of theorems that will enable us to use the limits of those simple functions as building blocks for finding limits of more complicated functions.

We start with the cases of a constant function $f(x) = k$, the identity function $f(x) = x$, and the reciprocal function $f(x) = 1/x$.

2.2.1   THEOREM.   *Let a and k be real numbers.*

$$\lim_{x \to a} k = k \qquad\qquad \lim_{x \to a} x = a$$

$$\lim_{x \to 0^-} \frac{1}{x} = -\infty \qquad\qquad \lim_{x \to 0^+} \frac{1}{x} = +\infty$$

The four limits in Theorem 2.2.1 should be evident from inspection of the function graphs shown in Figure 2.2.1.

In the case of the constant function $f(x) = k$, the values of $f(x)$ do not change as x varies, so the limit of $f(x)$ is k, regardless of at which number a the limit is taken. For example,

$$\lim_{x \to -25} 3 = 3, \qquad \lim_{x \to 0} 3 = 3, \qquad \lim_{x \to \pi} 3 = 3$$

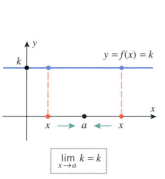

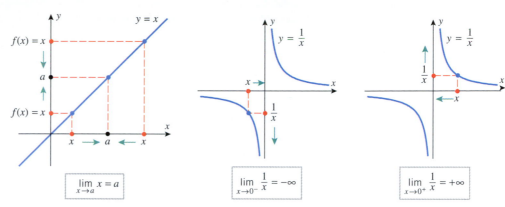

$$\lim_{x \to a} k = k$$

$$\lim_{x \to a} x = a$$

$$\lim_{x \to 0^-} \frac{1}{x} = -\infty$$

$$\lim_{x \to 0^+} \frac{1}{x} = +\infty$$

Figure 2.2.1

Since the identity function $f(x) = x$ just echoes its input, it is clear that $f(x) = x \to a$ as $x \to a$. In terms of our informal definition of limits (2.1.1), if we decide just how close to a we would like the value of $f(x) = x$ to be, we need only restrict its input x to be just as close to a.

The one-sided limits of the reciprocal function $f(x) = 1/x$ about 0 should conform with your experience with fractions: making the denominator closer to zero increases the magnitude of the fraction (i.e., increases its absolute value). This is illustrated in Table 2.2.1.

Table 2.2.1

	VALUES					CONCLUSION
x	-1	-0.1	-0.01	-0.001	-0.0001 $\cdots$	As $x \to 0^-$ the value of $1/x$
$1/x$	-1	-10	-100	-1000	$-10,000$ $\cdots$	decreases without bound.
x	1	0.1	0.01	0.001	0.0001 $\cdots$	As $x \to 0^+$ the value of $1/x$
$1/x$	1	10	100	1000	$10,000$ $\cdots$	increases without bound.

The following theorem, parts of which are proved in Appendix G, will be our basic tool for finding limits algebraically.

2.2.2   THEOREM.   *Let a be a real number, and suppose that*

$$\lim_{x \to a} f(x) = L_1 \quad and \quad \lim_{x \to a} g(x) = L_2$$

That is, the limits exist and have values L_1 and L_2, respectively. Then:

(a) $\displaystyle \lim_{x \to a} [f(x) + g(x)] = \lim_{x \to a} f(x) + \lim_{x \to a} g(x) = L_1 + L_2$

(b) $\displaystyle \lim_{x \to a} [f(x) - g(x)] = \lim_{x \to a} f(x) - \lim_{x \to a} g(x) = L_1 - L_2$

(c) $\displaystyle \lim_{x \to a} [f(x)g(x)] = \left(\lim_{x \to a} f(x) \right) \left(\lim_{x \to a} g(x) \right) = L_1 L_2$

(d) $\displaystyle \lim_{x \to a} \frac{f(x)}{g(x)} = \frac{\lim_{x \to a} f(x)}{\lim_{x \to a} g(x)} = \frac{L_1}{L_2}, \quad provided \ L_2 \neq 0$

(e) $\displaystyle \lim_{x \to a} \sqrt[n]{f(x)} = \sqrt[n]{\lim_{x \to a} f(x)} = \sqrt[n]{L_1}, \quad provided \ L_1 > 0 \ if \ n \ is \ even.$

Moreover, these statements are also true for the one-sided limits as $x \to a^-$ or as $x \to a^+$.

A casual restatement of this theorem is as follows:

> (a)   *The limit of a sum is the sum of the limits.*
>
> (b)   *The limit of a difference is the difference of the limits.*
>
> (c)   *The limit of a product is the product of the limits.*
>
> (d)   *The limit of a quotient is the quotient of the limits, provided the limit of the denominator is not zero.*
>
> (e)   *The limit of an nth root is the nth root of the limit.*

REMARK.   Although results (*a*) and (*c*) in Theorem 2.2.2 are stated for two functions, they hold for any finite number of functions. For example, if the limits of $f(x)$, $g(x)$, and $h(x)$ exist as $x \to a$, then the limit of their sum and the limit of their product also exist as $x \to a$ and are given by the formulas

$$\lim_{x \to a} [f(x) + g(x) + h(x)] = \lim_{x \to a} f(x) + \lim_{x \to a} g(x) + \lim_{x \to a} h(x)$$

$$\lim_{x \to a} [f(x)g(x)h(x)] = \left(\lim_{x \to a} f(x) \right) \left(\lim_{x \to a} g(x) \right) \left(\lim_{x \to a} h(x) \right)$$

In particular, if $f(x) = g(x) = h(x)$, then this yields

$$\lim_{x \to a} [f(x)]^3 = \left(\lim_{x \to a} f(x) \right)^3$$

More generally, if n is a positive integer, then the limit of the *n*th power of a function is the *n*th power of the function's limit. Thus,

$$\lim_{x \to a} x^n = \left(\lim_{x \to a} x \right)^n = a^n \tag{1}$$

For example,

$$\lim_{x \to 3} x^4 = 3^4 = 81$$

Another useful result follows from part (*c*) of Theorem 2.2.2 in the special case when one of the factors is a constant k:

$$\lim_{x \to a} (k \cdot f(x)) = \left(\lim_{x \to a} k \right) \cdot \left(\lim_{x \to a} f(x) \right) = k \cdot \left(\lim_{x \to a} f(x) \right) \tag{2}$$

and similarly for $\lim_{x \to a}$ replaced by a one-sided limit, $\lim_{x \to a^+}$ or $\lim_{x \to a^-}$. Rephrased, this last statement says:

> *A constant factor can be moved through a limit symbol.*

LIMITS OF POLYNOMIALS AND RATIONAL FUNCTIONS AS $x \to a$

Example 1   Find $\lim_{x \to 5} (x^2 - 4x + 3)$ and justify each step.

Solution.   First note that $\lim_{x \to 5} x^2 = 5^2 = 25$ by Equation (1). Also, from Equation (2), $\lim_{x \to 5} 4x = 4(\lim_{x \to 5} x) = 4(5) = 20$. Since $\lim_{x \to 5} 3 = 3$ by Theorem 2.2.1, we may appeal to Theorem 2.2.2(*a*) and (*b*) to write

$$\lim_{x \to 5} (x^2 - 4x + 3) = \lim_{x \to 5} x^2 - \lim_{x \to 5} 4x + \lim_{x \to 5} 3 = 25 - 20 + 3 = 8$$

However, for conciseness, it is common to reverse the order of this argument and simply write

$$\lim_{x \to 5} (x^2 - 4x + 3) = \lim_{x \to 5} x^2 - \lim_{x \to 5} 4x + \lim_{x \to 5} 3 \qquad \boxed{\text{Theorem 2.2.2}(a), (b)}$$

$$= \left(\lim_{x \to 5} x \right)^2 - 4 \lim_{x \to 5} x + \lim_{x \to 5} 3 \qquad \boxed{\text{Equations (1), (2)}}$$

$$= 5^2 - 4(5) + 3 \qquad \boxed{\text{Theorem 2.2.1}}$$

$$= 8 \qquad\qquad\qquad\qquad\qquad \blacktriangleleft$$

REMARK. In our presentation of limit arguments, we will adopt the convention of providing just a concise, reverse argument, bearing in mind that the validity of each equality may be conditional upon the successful resolution of the remaining limits.

Our next result will show that the limit of a polynomial $p(x)$ at $x = a$ is the same as the value of the polynomial at $x = a$. This greatly simplifies the computation of limits of polynomials by allowing us to simply evaluate the polynomial.

2.2.3 THEOREM. *For any polynomial*

$$p(x) = c_0 + c_1 x + \cdots + c_n x^n$$

and any real number a,

$$\lim_{x \to a} p(x) = c_0 + c_1 a + \cdots + c_n a^n = p(a)$$

Proof.

$$\lim_{x \to a} p(x) = \lim_{x \to a} \left(c_0 + c_1 x + \cdots + c_n x^n \right)$$

$$= \lim_{x \to a} c_0 + \lim_{x \to a} c_1 x + \cdots + \lim_{x \to a} c_n x^n$$

$$= \lim_{x \to a} c_0 + c_1 \lim_{x \to a} x + \cdots + c_n \lim_{x \to a} x^n$$

$$= c_0 + c_1 a + \cdots + c_n a^n = p(a) \qquad\qquad\qquad \blacksquare$$

Recall that a rational function is a ratio of two polynomials. Theorem 2.2.3 and Theorem 2.2.2(d) can often be used in combination to compute limits of rational functions. We illustrate this in the following example and prove a more general result in Theorem 2.2.4.

Example 2 Find $\displaystyle\lim_{x \to 2} \frac{5x^3 + 4}{x - 3}$.

Solution.

$$\lim_{x \to 2} \frac{5x^3 + 4}{x - 3} = \frac{\displaystyle\lim_{x \to 2} (5x^3 + 4)}{\displaystyle\lim_{x \to 2} (x - 3)} \qquad \boxed{\text{Theorem 2.2.2}(d)}$$

$$= \frac{5 \cdot 2^3 + 4}{2 - 3} = -44 \qquad \boxed{\text{Theorem 2.2.3}} \qquad \blacktriangleleft$$

Here is a general theorem on limits of rational functions.

2.2.4 THEOREM. *Consider the rational function*

$$f(x) = \frac{p(x)}{q(x)}$$

where $p(x)$ *and* $q(x)$ *are polynomials. For any real number* a,

(a) *if* $q(a) \neq 0$, *then* $\displaystyle\lim_{x \to a} f(x) = f(a)$.

(b) *if* $q(a) = 0$ *but* $p(a) \neq 0$, *then* $\displaystyle\lim_{x \to a} f(x)$ *does not exist.*

Proof. If $q(a) \neq 0$, then

$$\lim_{x \to a} f(x) = \lim_{x \to a} \frac{p(x)}{q(x)}$$

$$= \frac{\displaystyle\lim_{x \to a} p(x)}{\displaystyle\lim_{x \to a} q(x)} \qquad \boxed{\text{Theorem 2.2.2}(d)}$$

$$= \frac{p(a)}{q(a)} = f(a) \qquad \boxed{\text{Theorem 2.2.3}}$$

If $q(a) = 0$ and $p(a) \neq 0$, then we again appeal to your experience with fractions. For values of x sufficiently near a, the value of $p(x)$ will be near $p(a)$ and not zero. Thus, since $0 = q(a) = \lim_{x \to a} q(x)$, as values of x approach a, the magnitude (absolute value) of the fraction $p(x)/q(x)$ will increase without bound, so $\lim_{x \to a} f(x)$ does not exist. ∎

As an illustration of part (*b*) of Theorem 2.2.4, consider

$$\lim_{x \to 3} \frac{5x^3 + 4}{x - 3}$$

Note that $\lim_{x \to 3}(5x^3 + 4) = 5 \cdot 3^3 + 4 = 139$ and $\lim_{x \to 3}(x - 3) = 3 - 3 = 0$. It is evident from Table 2.2.2 that

$$\lim_{x \to 3} \frac{5x^3 + 4}{x - 3}$$

does not exist.

Table 2.2.2

	VALUES				CONCLUSION
x	2.99	2.999	2.9999	. . .	The value of $\dfrac{5x^3+4}{x-3}$ decreases
$\dfrac{5x^3+4}{x-3}$	−13,765.45	−138,865.04	−1,389,865.00	. . .	without bound as $x \to 3^-$.
x	3.01	3.001	3.0001	. . .	The value of $\dfrac{5x^3+4}{x-3}$ increases
$\dfrac{5x^3+4}{x-3}$	14,035.45	139,135.05	1,390,135.00	. . .	without bound as $x \to 3^+$.

In Theorem 2.2.4(*b*), where the limit of the denominator is zero but the limit of the numerator is not zero, the response "does not exist" can be elaborated upon in one of the following three ways.

- The limit may be $-\infty$.
- The limit may be $+\infty$.
- The limit may be $-\infty$ from one side and $+\infty$ from the other.

Figure 2.2.2 illustrates these three possibilities graphically for rational functions of the form $1/(x - a)$, $1/(x - a)^2$, and $-1/(x - a)^2$.

Example 3   Find

(a) $\displaystyle\lim_{x \to 4^-} \frac{2 - x}{(x - 4)(x + 2)}$ (b) $\displaystyle\lim_{x \to 4^+} \frac{2 - x}{(x - 4)(x + 2)}$ (c) $\displaystyle\lim_{x \to 4} \frac{2 - x}{(x - 4)(x + 2)}$

Solution.   With $n(x) = 2 - x$ and $d(x) = (x - 4)(x + 2)$, we see that $n(4) = -2$ and $d(4) = 0$. By Theorem 2.2.4(*b*), each of the limits does not exist. To be more specific, we

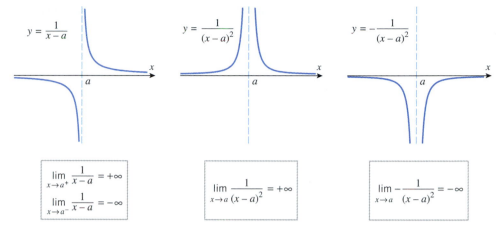

Figure 2.2.2

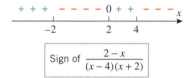

Figure 2.2.3

analyze the sign of the ratio $n(x)/d(x)$ near $x = 4$. The sign of the ratio, which is given in Figure 2.2.3, is determined by the signs of $2 - x$, $x - 4$, and $x + 2$. (The method of test values, discussed in Appendix A, provides a simple way of finding the sign of the ratio here.) It follows from this figure that as x approaches 4 from the left, the ratio is always positive; and as x approaches 4 from the right, the ratio is always negative. Thus,

$$\lim_{x \to 4^-} \frac{2 - x}{(x - 4)(x + 2)} = +\infty \quad \text{and} \quad \lim_{x \to 4^+} \frac{2 - x}{(x - 4)(x + 2)} = -\infty$$

Because the one-sided limits have opposite signs, all we can say about the two-sided limit is that it does not exist. ◀

INDETERMINATE FORMS OF TYPE 0/0

The missing case in Theorem 2.2.4 is when *both* the numerator and the denominator of a rational function $f(x) = n(x)/d(x)$ have a zero at $x = a$. In this case, $n(x)$ and $d(x)$ will each have a factor of $x - a$, and canceling this factor *may* result in a rational function to which Theorem 2.2.4 applies.

Example 4 Find $\lim\limits_{x \to 2} \dfrac{x^2 - 4}{x - 2}$.

Solution. Since 2 is a zero of both the numerator and the denominator, they share a common factor of $x - 2$. The limit can be obtained as follows:

$$\lim_{x \to 2} \frac{x^2 - 4}{x - 2} = \lim_{x \to 2} \frac{(x - 2)(x + 2)}{x - 2} = \lim_{x \to 2} (x + 2) = 4 \qquad ◀$$

REMARK. Although correct, the second equality in the preceding computation needs some justification, since canceling the factor $x - 2$ alters the function by expanding its domain. However, as discussed in Example 5 of Section 1.2, the two functions are identical, except at $x = 2$ (Figure 1.2.9). From our discussions in the last section, we know that this difference has no effect on the limit as x approaches 2.

Example 5 Find

(a) $\lim\limits_{x \to 3} \dfrac{x^2 - 6x + 9}{x - 3}$ (b) $\lim\limits_{x \to -4} \dfrac{2x + 8}{x^2 + x - 12}$ (c) $\lim\limits_{x \to 5} \dfrac{x^2 - 3x - 10}{x^2 - 10x + 25}$

Solution (a). The numerator and the denominator both have a zero at $x = 3$, so there is a common factor of $x - 3$. Then

$$\lim_{x \to 3} \frac{x^2 - 6x + 9}{x - 3} = \lim_{x \to 3} \frac{(x - 3)^2}{x - 3} = \lim_{x \to 3} (x - 3) = 0$$

Solution (b). The numerator and the denominator both have a zero at $x = -4$, so there is a common factor of $x - (-4) = x + 4$. Then

$$\lim_{x \to -4} \frac{2x + 8}{x^2 + x - 12} = \lim_{x \to -4} \frac{2(x + 4)}{(x + 4)(x - 3)} = \lim_{x \to -4} \frac{2}{x - 3} = -\frac{2}{7}$$

Solution (c). The numerator and the denominator both have a zero at $x = 5$, so there is a common factor of $x - 5$. Then

$$\lim_{x \to 5} \frac{x^2 - 3x - 10}{x^2 - 10x + 25} = \lim_{x \to 5} \frac{(x - 5)(x + 2)}{(x - 5)(x - 5)} = \lim_{x \to 5} \frac{x + 2}{x - 5}$$

However,

$$\lim_{x \to 5} (x + 2) = 7 \neq 0 \quad \text{and} \quad \lim_{x \to 5} (x - 5) = 0$$

By Theorem 2.2.4(*b*),

$$\lim_{x \to 5} \frac{x^2 - 3x - 10}{x^2 - 10x + 25} = \lim_{x \to 5} \frac{x + 2}{x - 5}$$

does not exist.   ◀

The case of a limit of a quotient,

$$\lim_{x \to a} \frac{f(x)}{g(x)}$$

where $\lim_{x \to a} f(x) = 0$ and $\lim_{x \to a} g(x) = 0$, is called an ***indeterminate form of type 0/0***. Note that the limits in Examples 4 and 5 produced a variety of answers. The word "indeterminate" here refers to the fact that the limiting behavior of the quotient cannot be determined without further study. The expression "0/0" is just a mnemonic device to describe the circumstance of a limit of a quotient in which both the numerator and denominator approach 0.

LIMITS INVOLVING RADICALS

Example 6   Find $\displaystyle\lim_{x \to 0} \frac{x}{\sqrt{x + 1} - 1}$.

Solution.   Recall that in Example 2 of Section 2.1 we conjectured this limit to be 2. Note that this limit expression is an indeterminate form of type 0/0, so Theorem 2.2.2(*d*) does not apply. One strategy for resolving this limit is to first rationalize the denominator of the function. This yields

$$\frac{x}{\sqrt{x + 1} - 1} = \frac{x(\sqrt{x + 1} + 1)}{(x + 1) - 1} = \sqrt{x + 1} + 1, \quad x \neq 0$$

Therefore,

$$\lim_{x \to 0} \frac{x}{\sqrt{x + 1} - 1} = \lim_{x \to 0} (\sqrt{x + 1} + 1) = 2 \qquad ◀$$

LIMITS OF PIECEWISE-DEFINED FUNCTIONS

For functions that are defined piecewise, a two-sided limit at an x-value where the formula changes is best obtained by first finding the one-sided limits at that number.

Example 7   Let

$$f(x) = \begin{cases} 1/(x + 2), & x < -2 \\ x^2 - 5, & -2 < x \le 3 \\ \sqrt{x + 13}, & x > 3 \end{cases}$$

Find

(a) $\displaystyle\lim_{x \to -2} f(x)$ (b) $\displaystyle\lim_{x \to 0} f(x)$ (c) $\displaystyle\lim_{x \to 3} f(x)$

Solution (a). As x approaches -2 from the left, the formula for f is

$$f(x) = \frac{1}{x+2}$$

so that

$$\lim_{x \to 2^-} f(x) = \lim_{x \to 2^-} \frac{1}{x+2} = -\infty$$

As x approaches -2 from the right, the formula for f is

$$f(x) = x^2 - 5$$

so that

$$\lim_{x \to -2^+} f(x) = \lim_{x \to 2^+} (x^2 - 5) = (-2)^2 - 5 = -1$$

Thus, $\lim_{x \to -2} f(x)$ does not exist.

Solution (b). As x approaches 0 from either the left or the right, the formula for f is

$$f(x) = x^2 - 5$$

Thus,

$$\lim_{x \to 0} f(x) = \lim_{x \to 0} (x^2 - 5) = 0^2 - 5 = -5$$

Solution (c). As x approaches 3 from the left, the formula for f is

$$f(x) = x^2 - 5$$

so that

$$\lim_{x \to 3^-} f(x) = \lim_{x \to 3^-} (x^2 - 5) = 3^2 - 5 = 4$$

As x approaches 3 from the right, the formula for f is

$$f(x) = \sqrt{x + 13}$$

so that

$$\lim_{x \to 3^+} f(x) = \lim_{x \to 3^+} \sqrt{x + 13} = \sqrt{\lim_{x \to 3^+} (x + 13)} = \sqrt{3 + 13} = 4$$

Since the one-sided limits are equal, we have

$$\lim_{x \to 3} f(x) = 4$$

We note that the limit calculations in parts (a), (b), and (c) are consistent with the graph of f shown in Figure 2.2.4. ◀

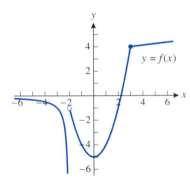

Figure 2.2.4

EXERCISE SET 2.2

1. In each part, find the limit by inspection.

   (a) $\lim_{x \to 8} 7$ (b) $\lim_{x \to 0^+} \pi$

   (c) $\lim_{x \to -2} 3x$ (d) $\lim_{y \to 3^+} 12y$

2. In each part, find the stated limit of $f(x) = x/|x|$ by inspection.

   (a) $\lim_{x \to 5} f(x)$ (b) $\lim_{x \to -5} f(x)$

   (c) $\lim_{x \to 0^+} f(x)$ (d) $\lim_{x \to 0^-} f(x)$

3. Given that

   $$\lim_{x \to a} f(x) = 2, \quad \lim_{x \to a} g(x) = -4, \quad \lim_{x \to a} h(x) = 0$$

   find the limits that exist. If the limit does not exist, explain why.

   (a) $\lim_{x \to a} [f(x) + 2g(x)]$ (b) $\lim_{x \to a} [h(x) - 3g(x) + 1]$

   (c) $\lim_{x \to a} [f(x)g(x)]$ (d) $\lim_{x \to a} [g(x)]^2$

   (e) $\lim_{x \to a} \sqrt[3]{6 + f(x)}$ (f) $\lim_{x \to a} \frac{2}{g(x)}$

   (g) $\lim_{x \to a} \frac{3f(x) - 8g(x)}{h(x)}$ (h) $\lim_{x \to a} \frac{7g(x)}{2f(x) + g(x)}$

4. Use the graphs of f and g in the accompanying figure to find the limits that exist. If the limit does not exist, explain why.

(a) $\lim\limits_{x \to 2} [f(x) + g(x)]$

(b) $\lim\limits_{x \to 0} [f(x) + g(x)]$

(c) $\lim\limits_{x \to 0^+} [f(x) + g(x)]$

(d) $\lim\limits_{x \to 0^-} [f(x) + g(x)]$

(e) $\lim\limits_{x \to 2} \dfrac{f(x)}{1 + g(x)}$

(f) $\lim\limits_{x \to 2} \dfrac{1 + g(x)}{f(x)}$

(g) $\lim\limits_{x \to 0^+} \sqrt{f(x)}$

(h) $\lim\limits_{x \to 0^-} \sqrt{f(x)}$

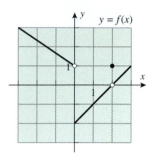

$y = f(x)$

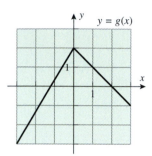

$y = g(x)$

Figure Ex-4

In Exercises 5–30, find the limits.

5. $\lim\limits_{y \to 2^-} \dfrac{(y - 1)(y - 2)}{y + 1}$

6. $\lim\limits_{x \to 3} \dfrac{x^2 - 2x}{x + 1}$

7. $\lim\limits_{x \to 4} \dfrac{x^2 - 16}{x - 4}$

8. $\lim\limits_{x \to 0} \dfrac{6x - 9}{x^3 - 12x + 3}$

9. $\lim\limits_{x \to 1^+} \dfrac{x^4 - 1}{x - 1}$

10. $\lim\limits_{t \to -2} \dfrac{t^3 + 8}{t + 2}$

11. $\lim\limits_{x \to -1} \dfrac{x^2 + 6x + 5}{x^2 - 3x - 4}$

12. $\lim\limits_{x \to 2} \dfrac{x^2 - 4x + 4}{x^2 + x - 6}$

13. $\lim\limits_{t \to 2} \dfrac{t^3 + 3t^2 - 12t + 4}{t^3 - 4t}$

14. $\lim\limits_{t \to 1} \dfrac{t^3 + t^2 - 5t + 3}{t^3 - 3t + 2}$

15. $\lim\limits_{x \to 3^+} \dfrac{x}{x - 3}$

16. $\lim\limits_{x \to 3^-} \dfrac{x}{x - 3}$

17. $\lim\limits_{x \to 3} \dfrac{x}{x - 3}$

18. $\lim\limits_{x \to 2^+} \dfrac{x}{x^2 - 4}$

19. $\lim\limits_{x \to 2^-} \dfrac{x}{x^2 - 4}$

20. $\lim\limits_{x \to 2} \dfrac{x}{x^2 - 4}$

21. $\lim\limits_{y \to 6^+} \dfrac{y + 6}{y^2 - 36}$

22. $\lim\limits_{y \to 6^-} \dfrac{y + 6}{y^2 - 36}$

23. $\lim\limits_{y \to 6} \dfrac{y + 6}{y^2 - 36}$

24. $\lim\limits_{x \to 4^+} \dfrac{3 - x}{x^2 - 2x - 8}$

25. $\lim\limits_{x \to 4^-} \dfrac{3 - x}{x^2 - 2x - 8}$

26. $\lim\limits_{x \to 4} \dfrac{3 - x}{x^2 - 2x - 8}$

27. $\lim\limits_{x \to 2^+} \dfrac{1}{|2 - x|}$

28. $\lim\limits_{x \to 3^-} \dfrac{1}{|x - 3|}$

29. $\lim\limits_{x \to 9} \dfrac{x - 9}{\sqrt{x} - 3}$

30. $\lim\limits_{y \to 4} \dfrac{4 - y}{2 - \sqrt{y}}$

31. Verify the limit in Example 1 of Section 2.1. That is, find
$$\lim\limits_{t_1 \to 0.5} \dfrac{-16t_1^2 + 29t_1 - 10.5}{t_1 - 0.5}$$

32. Let $s(t) = -16t^2 + 29t + 6$. Find
$$\lim\limits_{t \to 1.5} \dfrac{s(t) - s(1.5)}{t - 1.5}$$

33. Let
$$f(x) = \begin{cases} x - 1, & x \le 3 \\ 3x - 7, & x > 3 \end{cases}$$
Find
(a) $\lim\limits_{x \to 3^-} f(x)$
(b) $\lim\limits_{x \to 3^+} f(x)$
(c) $\lim\limits_{x \to 3} f(x)$.

34. Let
$$g(t) = \begin{cases} t^2, & t \ge 0 \\ t - 2, & t < 0 \end{cases}$$
Find
(a) $\lim\limits_{t \to 0^-} g(t)$
(b) $\lim\limits_{t \to 0^+} g(t)$
(c) $\lim\limits_{t \to 0} g(t)$.

35. Let $f(x) = \dfrac{x^3 - 1}{x - 1}$.
(a) Find $\lim\limits_{x \to 1} f(x)$.
(b) Sketch the graph of $y = f(x)$.

36. Let
$$f(x) = \begin{cases} \dfrac{x^2 - 9}{x + 3}, & x \ne -3 \\ k, & x = -3 \end{cases}$$
(a) Find k so that $f(-3) = \lim\limits_{x \to -3} f(x)$.
(b) With k assigned the value $\lim\limits_{x \to -3} f(x)$, show that $f(x)$ can be expressed as a polynomial.

37. (a) Explain why the following calculation is incorrect.
$$\lim\limits_{x \to 0^+} \left(\dfrac{1}{x} - \dfrac{1}{x^2} \right) = \lim\limits_{x \to 0^+} \dfrac{1}{x} - \lim\limits_{x \to 0^+} \dfrac{1}{x^2}$$
$$= +\infty - (+\infty) = 0$$
(b) Show that $\lim\limits_{x \to 0^+} \left(\dfrac{1}{x} - \dfrac{1}{x^2} \right) = -\infty$.

38. Find $\lim\limits_{x \to 0^-} \left(\dfrac{1}{x} + \dfrac{1}{x^2} \right)$.

In Exercises 39 and 40, first rationalize the numerator, then find the limit.

39. $\lim\limits_{x \to 0} \dfrac{\sqrt{x + 4} - 2}{x}$

40. $\lim\limits_{x \to 0} \dfrac{\sqrt{x^2 + 4} - 2}{x}$

41. Let $p(x)$ and $q(x)$ be polynomials, and suppose $q(x_0) = 0$. Discuss the behavior of the graph of $y = p(x)/q(x)$ in the vicinity of $x = x_0$. Give examples to support your conclusions.

2.3 COMPUTING LIMITS: END BEHAVIOR

In this section we will discuss algebraic techniques for computing limits at $\pm\infty$ for many functions. We base these results on the informal development of the limit concept discussed in Section 2.1. A more formal development of these results is possible after Section 2.4.

SOME BASIC LIMITS

The behavior of a function toward the extremes of its domain is sometimes called its **end behavior**. Here we will use limits to investigate the end behavior of a function as $x \to -\infty$ or as $x \to +\infty$. As in the last section, we will begin by obtaining limits of some simple functions and then use these as building blocks for finding limits of more complicated functions.

2.3.1 THEOREM. *Let k be a real number.*

$$\lim_{x \to -\infty} k = k \qquad \lim_{x \to +\infty} k = k$$

$$\lim_{x \to -\infty} x = -\infty \qquad \lim_{x \to +\infty} x = +\infty$$

$$\lim_{x \to -\infty} \frac{1}{x} = 0 \qquad \lim_{x \to +\infty} \frac{1}{x} = 0$$

The six limits in Theorem 2.3.1 should be evident from inspection of the function graphs in Figure 2.3.1.

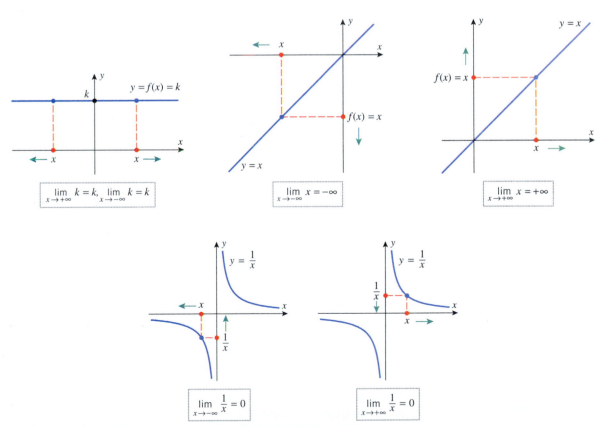

Figure 2.3.1

The limits of the reciprocal function $f(x) = 1/x$ should make sense to you intuitively, based on your experience with fractions: increasing the magnitude of x makes its reciprocal closer to zero. This is illustrated in Table 2.3.1.

Table 2.3.1

			VALUES				CONCLUSION
x	-1	-10	-100	-1000	$-10,000$	$\cdots$	As $x \to -\infty$ the value of $1/x$
$1/x$	-1	-0.1	-0.01	-0.001	-0.0001	$\cdots$	increases toward zero.
x	1	10	100	1000	$10,000$	$\cdots$	As $x \to +\infty$ the value of $1/x$
$1/x$	1	0.1	0.01	0.001	0.0001	$\cdots$	decreases toward zero.

The following theorem mirrors Theorem 2.2.2 as our tool for finding limits at $\pm\infty$ algebraically. (The proof is similar to that of the portions of Theorem 2.2.2 that are proved in Appendix G.)

2.3.2 THEOREM. *Suppose that*

$$\lim_{x \to +\infty} f(x) = L_1 \quad and \quad \lim_{x \to +\infty} g(x) = L_2$$

That is, the limits exist and have values L_1 and L_2, respectively. Then:

(a) $\displaystyle \lim_{x \to +\infty} [f(x) + g(x)] = \lim_{x \to +\infty} f(x) + \lim_{x \to +\infty} g(x) = L_1 + L_2$

(b) $\displaystyle \lim_{x \to +\infty} [f(x) - g(x)] = \lim_{x \to +\infty} f(x) - \lim_{x \to +\infty} g(x) = L_1 - L_2$

(c) $\displaystyle \lim_{x \to +\infty} [f(x)g(x)] = \left(\lim_{x \to +\infty} f(x) \right) \left(\lim_{x \to +\infty} g(x) \right) = L_1 L_2$

(d) $\displaystyle \lim_{x \to +\infty} \frac{f(x)}{g(x)} = \frac{\lim_{x \to +\infty} f(x)}{\lim_{x \to +\infty} g(x)} = \frac{L_1}{L_2}, \quad provided\ L_2 \neq 0$

(e) $\displaystyle \lim_{x \to +\infty} \sqrt[n]{f(x)} = \sqrt[n]{\lim_{x \to +\infty} f(x)} = \sqrt[n]{L_1}, \quad provided\ L_1 > 0\ if\ n\ is\ even.$

Moreover, these statements are also true for limits as $x \to -\infty$.

REMARK. As in the remark following Theorem 2.2.2, results (a) and (c) can be extended to sums or products of any finite number of functions. In particular, for any positive integer n,

$$\lim_{x \to +\infty} (f(x))^n = \left(\lim_{x \to +\infty} f(x) \right)^n \qquad \lim_{x \to -\infty} (f(x))^n = \left(\lim_{x \to -\infty} f(x) \right)^n$$

Also, since $\lim_{x \to +\infty}(1/x) = 0$, if n is a positive integer, then

$$\lim_{x \to +\infty} \frac{1}{x^n} = \left(\lim_{x \to +\infty} \frac{1}{x} \right)^n = 0 \qquad \lim_{x \to -\infty} \frac{1}{x^n} = \left(\lim_{x \to -\infty} \frac{1}{x} \right)^n = 0 \qquad (1)$$

For example,

$$\lim_{x \to +\infty} \frac{1}{x^4} = 0 \quad and \quad \lim_{x \to -\infty} \frac{1}{x^4} = 0$$

Another useful result follows from part (c) of Theorem 2.3.2 in the special case where one of the factors is a constant k:

$$\lim_{x \to +\infty} (k \cdot f(x)) = \left(\lim_{x \to +\infty} k \right) \cdot \left(\lim_{x \to +\infty} f(x) \right) = k \cdot \left(\lim_{x \to +\infty} f(x) \right) \qquad (2)$$

and similarly, for $\lim_{x \to +\infty}$ replaced by $\lim_{x \to -\infty}$. Rephrased, this last statement says:

A constant factor can be moved through a limit symbol.

LIMITS OF x^n AS $x \to \pm\infty$

In Figure 2.3.2 we have graphed the polynomials of the form x^n for $n = 1, 2, 3$, and 4. Below each figure we have indicated the limits as $x \to +\infty$ and as $x \to -\infty$. The results in the figure are special cases of the following general results:

$$\lim_{x \to +\infty} x^n = +\infty, \quad n = 1, 2, 3, \dots \tag{3}$$

$$\lim_{x \to -\infty} x^n = \begin{cases} -\infty, & n = 1, 3, 5, \dots \\ +\infty, & n = 2, 4, 6, \dots \end{cases} \tag{4}$$

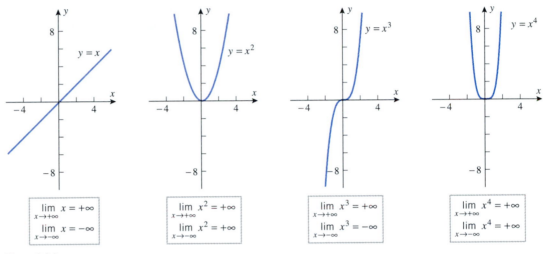

Figure 2.3.2

Multiplying x^n by a positive real number does not affect limits (3) and (4), but multiplying by a negative real number reverses the sign.

Example 1

$$\lim_{x \to +\infty} 2x^5 = +\infty, \qquad \lim_{x \to -\infty} 2x^5 = -\infty$$

$$\lim_{x \to +\infty} -7x^6 = -\infty, \qquad \lim_{x \to -\infty} -7x^6 = -\infty \qquad \blacktriangleleft$$

LIMITS OF POLYNOMIALS AS $x \to \pm\infty$

There is a useful principle about polynomials which, expressed informally, states that:

The end behavior of a polynomial matches the end behavior of its highest degree term.

More precisely, if $c_n \neq 0$ then

$$\lim_{x \to -\infty} \left(c_0 + c_1 x + \dots + c_n x^n \right) = \lim_{x \to -\infty} c_n x^n \tag{5}$$

$$\lim_{x \to +\infty} \left(c_0 + c_1 x + \dots + c_n x^n \right) = \lim_{x \to +\infty} c_n x^n \tag{6}$$

We can motivate these results by factoring out the highest power of x from the polynomial and examining the limit of the factored expression. Thus,

$$c_0 + c_1 x + \cdots + c_n x^n = x^n \left(\frac{c_0}{x^n} + \frac{c_1}{x^{n-1}} + \cdots + c_n \right)$$

As $x \to -\infty$ or $x \to +\infty$, it follows from (1) that all of the terms with positive powers of x in the denominator approach 0, so (5) and (6) are certainly plausible.

Example 2

$$\lim_{x \to -\infty} (7x^5 - 4x^3 + 2x - 9) = \lim_{x \to -\infty} 7x^5 = -\infty$$

$$\lim_{x \to -\infty} (-4x^8 + 17x^3 - 5x + 1) = \lim_{x \to -\infty} -4x^8 = -\infty \qquad \blacktriangleleft$$

LIMITS OF RATIONAL FUNCTIONS AS $x \to \pm\infty$

A useful technique for determining the end behavior of a rational function $f(x) = n(x)/d(x)$ is to factor and cancel the highest power of x that occurs in the denominator $d(x)$ from both $n(x)$ and $d(x)$. The denominator of the resulting fraction then has a (nonzero) limit equal to the leading coefficient of $d(x)$, so the limit of the resulting fraction can be quickly determined using (1), (5), and (6). The following examples illustrate this technique.

Example 3 Find $\displaystyle\lim_{x \to +\infty} \frac{3x + 5}{6x - 8}$.

Solution. Divide the numerator and denominator by the highest power of x that occurs in the denominator; that is, $x^1 = x$. We obtain

$$\lim_{x \to +\infty} \frac{3x + 5}{6x - 8} = \lim_{x \to +\infty} \frac{x(3 + 5/x)}{x(6 - 8/x)} = \lim_{x \to +\infty} \frac{3 + 5/x}{6 - 8/x} = \frac{\displaystyle\lim_{x \to +\infty} (3 + 5/x)}{\displaystyle\lim_{x \to +\infty} (6 - 8/x)}$$

$$= \frac{\displaystyle\lim_{x \to +\infty} 3 + \lim_{x \to +\infty} 5/x}{\displaystyle\lim_{x \to +\infty} 6 - \lim_{x \to +\infty} 8/x} = \frac{3 + 5 \displaystyle\lim_{x \to +\infty} 1/x}{6 - 8 \displaystyle\lim_{x \to +\infty} 1/x}$$

$$= \frac{3 + (5 \cdot 0)}{6 - (8 \cdot 0)} = \frac{1}{2} \qquad \blacktriangleleft$$

Example 4 Find

(a) $\displaystyle\lim_{x \to -\infty} \frac{4x^2 - x}{2x^3 - 5}$ (b) $\displaystyle\lim_{x \to -\infty} \frac{5x^3 - 2x^2 + 1}{3x + 5}$

Solution (a). Divide the numerator and denominator by the highest power of x that occurs in the denominator, namely x^3. We obtain

$$\lim_{x \to -\infty} \frac{4x^2 - x}{2x^3 - 5} = \lim_{x \to -\infty} \frac{x^3(4/x - 1/x^2)}{x^3(2 - 5/x^3)} = \lim_{x \to -\infty} \frac{4/x - 1/x^2}{2 - 5/x^3}$$

$$= \frac{\displaystyle\lim_{x \to -\infty} (4/x - 1/x^2)}{\displaystyle\lim_{x \to -\infty} (2 - 5/x^3)} = \frac{(4 \cdot 0) - 0}{2 - (5 \cdot 0)} = \frac{0}{2} = 0$$

Solution (b). Divide the numerator and denominator by x to obtain

$$\lim_{x \to -\infty} \frac{5x^3 - 2x^2 + 1}{3x + 5} = \lim_{x \to -\infty} \frac{5x^2 - 2x + 1/x}{3 + 5/x} = +\infty$$

where the final step is justified by the fact that

$$5x^2 - 2x \to +\infty, \quad \frac{1}{x} \to 0, \quad \text{and} \quad 3 + \frac{5}{x} \to 3$$

as $x \to -\infty$. $\qquad \blacktriangleleft$

LIMITS INVOLVING RADICALS

Example 5   Find $\lim\limits_{x\to+\infty}\sqrt[3]{\dfrac{3x+5}{6x-8}}$.

Solution.

$$\lim_{x\to+\infty}\sqrt[3]{\frac{3x+5}{6x-8}}=\sqrt[3]{\lim_{x\to+\infty}\frac{3x+5}{6x-8}}\qquad\boxed{\text{Theorem 2.3.2(e)}}$$

$$=\sqrt[3]{\frac{1}{2}}\qquad\boxed{\text{Example 3}}\qquad\blacktriangleleft$$

Example 6   Find

(a) $\lim\limits_{x\to+\infty}\dfrac{\sqrt{x^2+2}}{3x-6}$ (b) $\lim\limits_{x\to-\infty}\dfrac{\sqrt{x^2+2}}{3x-6}$

In both parts it would be helpful to manipulate the function so that the powers of x are transformed to powers of $1/x$. This can be achieved in both cases by dividing the numerator and denominator by $|x|$ and using the fact that $\sqrt{x^2}=|x|$.

Solution (a).   As $x\to+\infty$, the values of x under consideration are positive, so we can replace $|x|$ by x where helpful. We obtain

$$\lim_{x\to+\infty}\frac{\sqrt{x^2+2}}{3x-6}=\lim_{x\to+\infty}\frac{\sqrt{x^2+2}/|x|}{(3x-6)/|x|}=\lim_{x\to+\infty}\frac{\sqrt{x^2+2}/\sqrt{x^2}}{(3x-6)/x}$$

$$=\lim_{x\to+\infty}\frac{\sqrt{1+2/x^2}}{3-6/x}=\frac{\lim\limits_{x\to+\infty}\sqrt{1+2/x^2}}{\lim\limits_{x\to+\infty}(3-6/x)}$$

$$=\frac{\sqrt{\lim\limits_{x\to+\infty}(1+2/x^2)}}{\lim\limits_{x\to+\infty}(3-6/x)}=\frac{\sqrt{\left(\lim\limits_{x\to+\infty}1\right)+\left(2\lim\limits_{x\to+\infty}1/x^2\right)}}{\left(\lim\limits_{x\to+\infty}3\right)-\left(6\lim\limits_{x\to+\infty}1/x\right)}$$

$$=\frac{\sqrt{1+(2\cdot0)}}{3-(6\cdot0)}=\frac{1}{3}$$

Solution (b).   As $x\to-\infty$, the values of x under consideration are negative, so we can replace $|x|$ by $-x$ where helpful. We obtain

$$\lim_{x\to-\infty}\frac{\sqrt{x^2+2}}{3x-6}=\lim_{x\to-\infty}\frac{\sqrt{x^2+2}/|x|}{(3x-6)/|x|}=\lim_{x\to-\infty}\frac{\sqrt{x^2+2}/\sqrt{x^2}}{(3x-6)/(-x)}$$

$$=\lim_{x\to-\infty}\frac{\sqrt{1+2/x^2}}{-3+6/x}=-\frac{1}{3}\qquad\blacktriangleleft$$

FOR THE READER.   Use a graphing utility to explore the end behavior of

$$f(x)=\frac{\sqrt{x^2+2}}{3x-6}$$

Your investigation should support the results of Example 6.

Example 7   Find

(a) $\lim\limits_{x\to+\infty}(\sqrt{x^6+5}-x^3)$ (b) $\lim\limits_{x\to+\infty}(\sqrt{x^6+5x^3}-x^3)$

Solution.   Graphs of the functions $f(x)=\sqrt{x^6+5}-x^3$ and $g(x)=\sqrt{x^6+5x^3}-x^3$ for $x\geq0$ are shown in Figure 2.3.3. From the graphs we might conjecture that the limits are 0 and 2.5, respectively. To confirm this, we treat each function as a fraction with denominator

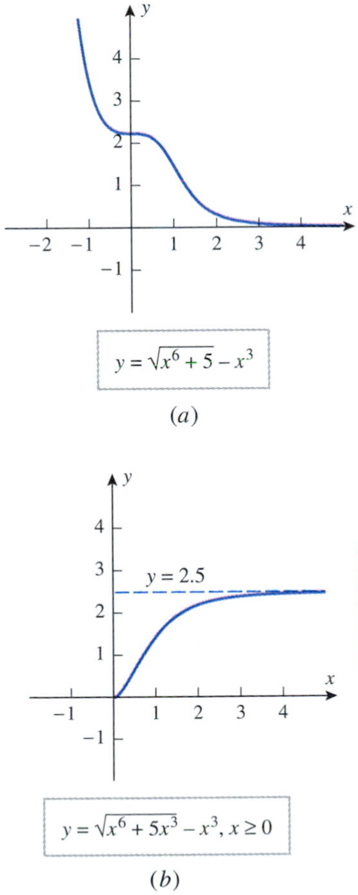

$y=\sqrt{x^6+5}-x^3$

(a)

$y=\sqrt{x^6+5x^3}-x^3,\ x\geq0$

(b)

Figure 2.3.3

1 and rationalize the numerator.

$$\lim_{x \to +\infty} (\sqrt{x^6 + 5} - x^3) = \lim_{x \to +\infty} (\sqrt{x^6 + 5} - x^3) \left(\frac{\sqrt{x^6 + 5} + x^3}{\sqrt{x^6 + 5} + x^3} \right)$$

$$= \lim_{x \to +\infty} \frac{(x^6 + 5) - x^6}{\sqrt{x^6 + 5} + x^3} = \lim_{x \to +\infty} \frac{5}{\sqrt{x^6 + 5} + x^3}$$

$$= \lim_{x \to +\infty} \frac{5/x^3}{\sqrt{1 + 5/x^6} + 1} \qquad \boxed{\sqrt{x^6} = x^3 \text{ for } x > 0}$$

$$= \frac{0}{\sqrt{1 + 0} + 1} = 0$$

$$\lim_{x \to +\infty} (\sqrt{x^6 + 5x^3} - x^3) = \lim_{x \to +\infty} (\sqrt{x^6 + 5x^3} - x^3) \left(\frac{\sqrt{x^6 + 5x^3} + x^3}{\sqrt{x^6 + 5x^3} + x^3} \right)$$

$$= \lim_{x \to +\infty} \frac{(x^6 + 5x^3) - x^6}{\sqrt{x^6 + 5x^3} + x^3} = \lim_{x \to +\infty} \frac{5x^3}{\sqrt{x^6 + 5x^3} + x^3}$$

$$= \lim_{x \to +\infty} \frac{5}{\sqrt{1 + 5/x^3} + 1} \qquad \boxed{\sqrt{x^6} = x^3 \text{ for } x > 0}$$

$$= \frac{5}{\sqrt{1 + 0} + 1} = \frac{5}{2}$$

◀

REMARK. Example 7 illustrates an *indeterminate form of type* $\infty - \infty$. Exercises 31–34 explore more examples of this type.

EXERCISE SET 2.3 ⌐ Graphing Utility

1. In each part, find the limit by inspection.
   (a) $\lim_{x \to -\infty} (-3)$
   (b) $\lim_{h \to +\infty} (-2h)$

2. In each part, find the stated limit of $f(x) = x/|x|$ by inspection.
   (a) $\lim_{x \to +\infty} f(x)$
   (b) $\lim_{x \to -\infty} f(x)$

3. Given that
$$\lim_{x \to +\infty} f(x) = 3, \quad \lim_{x \to +\infty} g(x) = -5, \quad \lim_{x \to +\infty} h(x) = 0$$
find the limits that exist. If the limit does not exist, explain why.
   (a) $\lim_{x \to +\infty} [f(x) + 3g(x)]$
   (b) $\lim_{x \to +\infty} [h(x) - 4g(x) + 1]$
   (c) $\lim_{x \to +\infty} [f(x)g(x)]$
   (d) $\lim_{x \to +\infty} [g(x)]^2$
   (e) $\lim_{x \to +\infty} \sqrt[3]{5 + f(x)}$
   (f) $\lim_{x \to +\infty} \frac{3}{g(x)}$
   (g) $\lim_{x \to +\infty} \frac{3h(x) + 4}{x^2}$
   (h) $\lim_{x \to +\infty} \frac{6f(x)}{5f(x) + 3g(x)}$

4. Given that
$$\lim_{x \to -\infty} f(x) = 7, \quad \lim_{x \to -\infty} g(x) = -6$$
find the limits that exist. If the limit does not exist, explain why.

   (a) $\lim_{x \to -\infty} [2f(x) - g(x)]$
   (b) $\lim_{x \to -\infty} [6f(x) + 7g(x)]$
   (c) $\lim_{x \to -\infty} [x^2 + g(x)]$
   (d) $\lim_{x \to -\infty} [x^2 g(x)]$
   (e) $\lim_{x \to -\infty} \sqrt[3]{f(x)g(x)}$
   (f) $\lim_{x \to -\infty} \frac{g(x)}{f(x)}$
   (g) $\lim_{x \to -\infty} \left[f(x) + \frac{g(x)}{x} \right]$
   (h) $\lim_{x \to -\infty} \frac{xf(x)}{(2x + 3)g(x)}$

In Exercises 5–28, find the limits.

5. $\lim_{x \to -\infty} (3 - x)$
6. $\lim_{x \to -\infty} \left(5 - \frac{1}{x} \right)$
7. $\lim_{x \to +\infty} (1 + 2x - 3x^5)$
8. $\lim_{x \to +\infty} (2x^3 - 100x + 5)$
9. $\lim_{x \to +\infty} \sqrt{x}$
10. $\lim_{x \to -\infty} \sqrt{5 - x}$
11. $\lim_{x \to +\infty} \frac{3x + 1}{2x - 5}$
12. $\lim_{x \to +\infty} \frac{5x^2 - 4x}{2x^2 + 3}$
13. $\lim_{y \to -\infty} \frac{3}{y + 4}$
14. $\lim_{x \to +\infty} \frac{1}{x - 12}$
15. $\lim_{x \to -\infty} \frac{x - 2}{x^2 + 2x + 1}$
16. $\lim_{x \to +\infty} \frac{5x^2 + 7}{3x^2 - x}$
17. $\lim_{x \to +\infty} \sqrt[3]{\frac{2 + 3x - 5x^2}{1 + 8x^2}}$
18. $\lim_{s \to +\infty} \sqrt[3]{\frac{3s^7 - 4s^5}{2s^7 + 1}}$

19. $\lim\limits_{x \to -\infty} \dfrac{\sqrt{5x^2 - 2}}{x + 3}$

20. $\lim\limits_{x \to +\infty} \dfrac{\sqrt{5x^2 - 2}}{x + 3}$

21. $\lim\limits_{y \to -\infty} \dfrac{2 - y}{\sqrt{7 + 6y^2}}$

22. $\lim\limits_{y \to +\infty} \dfrac{2 - y}{\sqrt{7 + 6y^2}}$

23. $\lim\limits_{x \to -\infty} \dfrac{\sqrt{3x^4 + x}}{x^2 - 8}$

24. $\lim\limits_{x \to +\infty} \dfrac{\sqrt{3x^4 + x}}{x^2 - 8}$

25. $\lim\limits_{x \to +\infty} \dfrac{7 - 6x^5}{x + 3}$

26. $\lim\limits_{t \to -\infty} \dfrac{5 - 2t^3}{t^2 + 1}$

27. $\lim\limits_{t \to +\infty} \dfrac{6 - t^3}{7t^3 + 3}$

28. $\lim\limits_{x \to -\infty} \dfrac{x + 4x^3}{1 - x^2 + 7x^3}$

29. Let
$$f(x) = \begin{cases} 2x^2 + 5, & x < 0 \\ \dfrac{3 - 5x^3}{1 + 4x + x^3}, & x \geq 0 \end{cases}$$

Find

(a) $\lim\limits_{x \to -\infty} f(x)$

(b) $\lim\limits_{x \to +\infty} f(x)$.

30. Let
$$g(t) = \begin{cases} \dfrac{2 + 3t}{5t^2 + 6}, & t < 1{,}000{,}000 \\ \dfrac{\sqrt{36t^2 - 100}}{5 - t}, & t > 1{,}000{,}000 \end{cases}$$

Find

(a) $\lim\limits_{t \to -\infty} g(t)$

(b) $\lim\limits_{t \to +\infty} g(t)$.

In Exercises 31–34, find the limits.

31. $\lim\limits_{x \to +\infty} (\sqrt{x^2 + 3} - x)$

32. $\lim\limits_{x \to +\infty} (\sqrt{x^2 - 3x} - x)$

33. $\lim\limits_{x \to +\infty} (\sqrt{x^2 + ax} - x)$

34. $\lim\limits_{x \to +\infty} (\sqrt{x^2 + ax} - \sqrt{x^2 + bx})$

35. Discuss the limits of $p(x) = (1 - x)^n$ as $x \to +\infty$ and $x \to -\infty$ for positive integer values of n.

36. Let $p(x) = (1 - x)^n$ and $q(x) = (1 - x)^m$. Discuss the limits of $p(x)/q(x)$ as $x \to +\infty$ and $x \to -\infty$ for positive integer values of m and n.

37. Let $p(x)$ be a polynomial of degree n. Discuss the limits of $p(x)/x^m$ as $x \to +\infty$ and $x \to -\infty$ for positive integer values of m.

38. In each part, find examples of polynomials $p(x)$ and $q(x)$ that satisfy the stated condition and such that $p(x) \to +\infty$ and $q(x) \to +\infty$ as $x \to +\infty$.

(a) $\lim\limits_{x \to +\infty} \dfrac{p(x)}{q(x)} = 1$

(b) $\lim\limits_{x \to +\infty} \dfrac{p(x)}{q(x)} = 0$

(c) $\lim\limits_{x \to +\infty} \dfrac{p(x)}{q(x)} = +\infty$

(d) $\lim\limits_{x \to +\infty} [p(x) - q(x)] = 3$

39. Assuming that m and n are positive integers, find
$$\lim\limits_{x \to -\infty} \dfrac{2 + 3x^n}{1 - x^m}$$

[*Hint:* Your answer will depend on whether $m < n$, $m = n$, or $m > n$.]

40. Find
$$\lim\limits_{x \to +\infty} \dfrac{c_0 + c_1 x + \cdots + c_n x^n}{d_0 + d_1 x + \cdots + d_m x^m}$$

where $c_n \neq 0$ and $d_m \neq 0$. [*Hint:* Your answer will depend on whether $m < n$, $m = n$, or $m > n$.]

The notion of an asymptote can be extended to include curves as well as lines. Specifically, we say that $f(x)$ *is asymptotic to* $g(x)$ *as* $x \to +\infty$ if
$$\lim\limits_{x \to +\infty} [f(x) - g(x)] = 0$$

and that $f(x)$ *is asymptotic to* $g(x)$ *as* $x \to -\infty$ if
$$\lim\limits_{x \to -\infty} [f(x) - g(x)] = 0$$

Informally stated, if $f(x)$ is asymptotic to $g(x)$ as $x \to +\infty$, then the graph of $y = f(x)$ gets closer and closer to the graph of $y = g(x)$ as $x \to +\infty$, and if $f(x)$ is asymptotic to $g(x)$ as $x \to -\infty$, then the graph of $y = f(x)$ gets closer and closer to the graph of $y = g(x)$ as $x \to -\infty$. For example, if
$$f(x) = x^2 + \dfrac{2}{x - 1} \quad \text{and} \quad g(x) = x^2$$

then $f(x)$ is asymptotic to $g(x)$ as $x \to +\infty$ and as $x \to -\infty$ since
$$\lim\limits_{x \to +\infty} [f(x) - g(x)] = \lim\limits_{x \to +\infty} \dfrac{2}{x - 1} = 0$$
$$\lim\limits_{x \to -\infty} [f(x) - g(x)] = \lim\limits_{x \to -\infty} \dfrac{2}{x - 1} = 0$$

This asymptotic behavior is illustrated in the following figure, which also shows the vertical asymptote of $f(x)$ at $x = 1$.

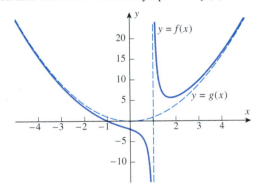

In Exercises 41–46, determine a simpler function $g(x)$ to which $f(x)$ is asymptotic as $x \to +\infty$ or $x \to -\infty$. Use a graphing utility to generate the graphs of $y = f(x)$ and $y = g(x)$ and identify all vertical asymptotes.

41. $f(x) = \dfrac{x^2 - 2}{x - 2}$

42. $f(x) = \dfrac{x^3 - x + 3}{x}$

43. $f(x) = \dfrac{-x^3 + 3x^2 + x - 1}{x - 3}$

44. $f(x) = \dfrac{x^5 - x^3 + 3}{x^2 - 1}$

45. $f(x) = \sin x + \dfrac{1}{x - 1}$

46. $f(x) = \sqrt{\dfrac{x^3 - x^2 + 2}{x - 1}}$

2.4 LIMITS (DISCUSSED MORE RIGOROUSLY)

Thus far, our discussion of limits has been based on our intuitive feeling of what it means for the values of a function to get closer and closer to a limiting value. However, this level of informality can only take us so far, so our goal in this section is to define limits precisely. From a purely mathematical point of view these definitions are needed to establish limits with certainty and to prove theorems about them. However, they will also provide us with a deeper understanding of the limit concept, making it possible for us to visualize some of the more subtle properties of functions.

MOTIVATION FOR THE DEFINITION OF A LIMIT

In Sections 2.1 to 2.3 our emphasis was on the discovery of values of limits, either through the sampling of selected x-values or through the application of limit theorems. In the preceding sections we interpreted $\lim_{x \to a} f(x) = L$ to mean that the values of $f(x)$ can be made as close as we like to L by selecting x-values sufficiently close to a (but not equal to a). Although this informal definition is sufficient for many purposes, we need a more precise definition to verify that a conjectured limit is actually correct, or to prove the limit theorems in Sections 2.2 and 2.3. One of our goals in this section is to give the informal phrases "as close as we like to L" and "sufficiently close to a" a precise mathematical interpretation. This will enable us to replace the informal definition of limit given in Definition 2.1.1 with a more fully developed version that may be used in proofs.

To start, consider the function f graphed in Figure 2.4.1a for which $f(x) \to L$ as $x \to a$. We have intentionally placed a hole in the graph at $x = a$ to emphasize that the function f need not be defined at $x = a$ to have a limit there. Also, to simplify the discussion, we have chosen a function that is increasing on an open interval containing a.

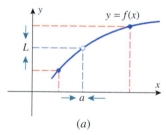 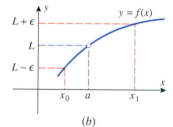 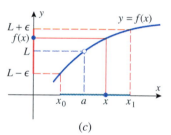

(a)   (b)   (c)

Figure 2.4.1

To motivate an appropriate definition for a two-sided limit, suppose that we choose *any* positive number, say ϵ, and draw horizontal lines from $L + \epsilon$ and $L - \epsilon$ on the y-axis to the curve $y = f(x)$ and then draw vertical lines from those points on the curve to the x-axis. As shown in Figure 2.4.1b, let x_0 and x_1 be points where the vertical lines intersect the x-axis.

Next, imagine that x gets closer and closer to a (from either side). Eventually, x will lie inside the interval (x_0, x_1), which is marked in green in Figure 2.4.1c; and when this happens, the value of $f(x)$ will fall between $L - \epsilon$ and $L + \epsilon$, marked in red in the figure. Thus, we conclude:

> If $f(x) \to L$ as $x \to a$, then for any positive number ϵ, we can find an open interval (x_0, x_1) on the x-axis that contains a and has the property that for each x in that interval (except possibly for $x = a$), the value of $f(x)$ is between $L - \epsilon$ and $L + \epsilon$.

FOR THE READER.   Consider the limit, $\lim_{x \to 0} (\sin x)/x$, conjectured to be 1 in Example 3 of Section 2.1. Draw a figure similar to Figure 2.4.1 that illustrates the preceding analysis for this limit.

What is important about this result is that it holds no matter how small we make ϵ. However, making ϵ smaller and smaller forces $f(x)$ *closer and closer* to L—which is precisely the concept we were trying to capture mathematically.

Observe that in Figure 2.4.1c the interval (x_0, x_1) extends farther on the right side of a than on the left side. However, for many purposes it is preferable to have an interval that extends the same distance on both sides of a. For this purpose, let us choose any positive number δ that is smaller than both $x_1 - a$ and $a - x_0$, and consider the interval $(a - \delta, a + \delta)$. This interval extends the same distance δ on both sides of a and lies inside of the interval (x_0, x_1) (Figure 2.4.2). Moreover, the condition $L - \epsilon < f(x) < L + \epsilon$ holds for every x in this interval (except possibly $x = a$), since this condition holds on the larger interval (x_0, x_1). This is illustrated by graphing f in the window $[a - \delta, a + \delta] \times [L - \epsilon, L + \epsilon]$ and observing that the graph "exits" the window at the sides, not at the top or bottom (except possibly at $x = a$).

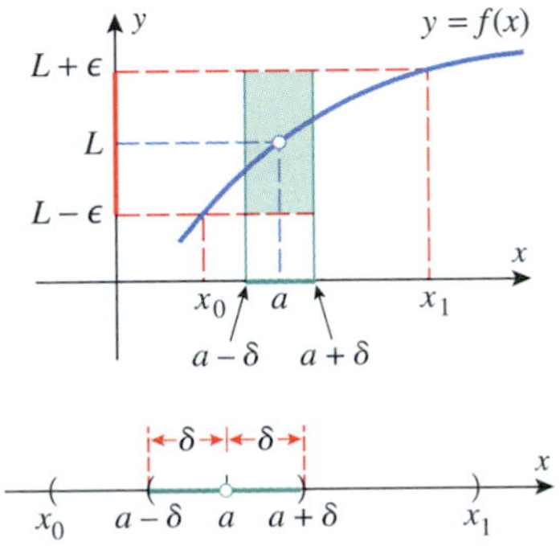

Example 1 Let $f(x) = \frac{1}{2}x + \frac{1}{4}\sin(\pi x/2)$. It can be shown that $\lim_{x \to 1} f(x) = L = 0.75$. Let $\epsilon = 0.05$.

(a) Use a graphing utility to find an open interval (x_0, x_1) containing $a = 1$ such that for each x in this interval, $f(x)$ is between $L - \epsilon = 0.75 - \epsilon = 0.75 - 0.05 = 0.70$ and $L + \epsilon = 0.75 + \epsilon = 0.75 + 0.05 = 0.80$.

(b) Find a value of δ such that $f(x)$ is between 0.70 and 0.80 for every x in the interval $(1 - \delta, 1 + \delta)$.

Solution (a). Figure 2.4.3 displays the graph of f. With a graphing utility, we discover that (to five decimal places) the points $(0.90769, 0.70122)$ and $(1.09231, 0.79353)$ are on the graph of f. Suppose that we take $x_0 = 0.908$ and $x_1 = 1.09$. Since the graph of f rises from left to right, we see that for $x_0 = 0.908 < x < 1.090 = x_1$, we have $0.90769 < x < 1.09231$ and therefore $0.7 < 0.70122 < f(x) < 0.79353 < 0.8$.

Solution (b). Since $x_1 - a = 1.09 - 1 = 0.09$ and $a - x_0 = 1 - 0.908 = 0.902$, any value or δ that is less than 0.09 will be acceptable. For example, for $\delta = 0.08$, if x belongs to the interval $(1 - \delta, 1 + \delta) = (0.92, 1.08)$, then $f(x)$ will lie between 0.70 and 0.80. ◄

Note that the condition $L - \epsilon < f(x) < L + \epsilon$ can be expressed as

$$|f(x) - L| < \epsilon$$

and the condition that x lies in the interval $(a - \delta, a + \delta)$, but $x \neq a$, can be expressed as

$$0 < |x - a| < \delta$$

Thus, we can summarize this discussion in the following definition.

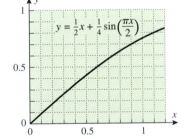

Figure 2.4.2

Figure 2.4.3

2.4.1 LIMIT DEFINITION. Let $f(x)$ be defined for all x in some open interval containing the number a, with the possible exception that $f(x)$ need not be defined at a. We will write

$$\lim_{x \to a} f(x) = L$$

if given any number $\epsilon > 0$ we can find a number $\delta > 0$ such that

$$|f(x) - L| < \epsilon \quad \text{if} \quad 0 < |x - a| < \delta$$

REMARK. With this definition we have made the transition from informal to formal in the definition of a two-sided limit. The phrase "as close as we like to L" has been given quantitative meaning by the number $\epsilon > 0$, and the phrase "sufficiently close to a" has been made precise by the number $\delta > 0$. Commonly known as the "ϵ-δ definition" of a limit, Definition 2.4.1 was developed primarily by the German mathematician Karl Weierstrass[*] in the nineteenth century.

The definitions for one-sided limits are similar to Definition 2.4.1. For example, in the definition of $\lim_{x \to a^+} f(x)$ we assume that $f(x)$ is defined for all x in an interval of the form (a, b) and replace the condition $0 < |x - a| < \delta$ by the condition $a < x < a + \delta$. Comparable changes are made in the definition of $\lim_{x \to a^-} f(x)$.

In the preceding sections we illustrated various numerical and graphical methods for *guessing* at limits. Now that we have a precise definition to work with, we can actually confirm the validity of those guesses with mathematical proof. Here is a typical example of such a proof.

Example 2 Use Definition 2.4.1 to prove that $\lim_{x \to 2} (3x - 5) = 1$.

Solution. We must show that given any positive number ϵ, we can find a positive number δ such that

$$| \underbrace{(3x - 5)}_{f(x)} - \underbrace{1}_{L} | < \epsilon \quad \text{if} \quad 0 < |x - \underbrace{2}_{a}| < \delta \tag{1}$$

There are two things to do. First, we must *discover* a value of δ for which this statement holds, and then we must *prove* that the statement holds for that δ. For the discovery part we begin by simplifying (1) and writing it as

$$|3x - 6| < \epsilon \quad \text{if} \quad 0 < |x - 2| < \delta$$

Next, we will rewrite this statement in a form that will facilitate the discovery of an appropriate δ:

$$3|x - 2| < \epsilon \quad \text{if} \quad 0 < |x - 2| < \delta$$
$$|x - 2| < \epsilon/3 \quad \text{if} \quad 0 < |x - 2| < \delta \tag{2}$$

It should be self-evident that this last statement holds if $\delta = \epsilon/3$, which completes the discovery portion of our work. Now we need to prove that (1) holds for this choice of δ. However, statement (1) is equivalent to (2), and (2) holds with $\delta = \epsilon/3$, so (1) also holds with $\delta = \epsilon/3$. This proves that $\lim_{x \to 2} (3x - 5) = 1$. ◄

---

[*]KARL WEIERSTRASS (1815–1897). Weierstrass, the son of a customs officer, was born in Ostenfelde, Germany. As a youth Weierstrass showed outstanding skills in languages and mathematics. However, at the urging of his dominant father, Weierstrass entered the law and commerce program at the University of Bonn. To the chagrin of his family, the rugged and congenial young man concentrated instead on fencing and beer drinking. Four years later he returned home without a degree. In 1839 Weierstrass entered the Academy of Münster to study for a career in secondary education, and he met and studied under an excellent mathematician named Christof Gudermann. Gudermann's ideas greatly influenced the work of Weierstrass. After receiving his teaching certificate, Weierstrass spent the next 15 years in secondary education teaching German, geography, and mathematics. In addition, he taught handwriting to small children. During this period much of Weierstrass's mathematical work was ignored because he was a secondary schoolteacher and not a college professor. Then, in 1854, he published a paper of major importance that created a sensation in the mathematics world and catapulted him to international fame overnight. He was immediately given an honorary Doctorate at the University of Königsberg and began a new career in college teaching at the University of Berlin in 1856. In 1859 the strain of his mathematical research caused a temporary nervous breakdown and led to spells of dizziness that plagued him for the rest of his life. Weierstrass was a brilliant teacher and his classes overflowed with multitudes of auditors. In spite of his fame, he never lost his early beer-drinking congeniality and was always in the company of students, both ordinary and brilliant. Weierstrass was acknowledged as the leading mathematical analyst in the world. He and his students opened the door to the modern school of mathematical analysis.

REMARK. This example illustrates the general form of a limit proof: We *assume* that we are given a positive number ϵ, and we try to *prove* that we can find a positive number δ such that

$$|f(x) - L| < \epsilon \quad \text{if} \quad 0 < |x - a| < \delta \tag{3}$$

This is done by first discovering δ, and then proving that the discovered δ works. Since the argument has to be general enough to work for all positive values of ϵ, the quantity δ has to be expressed as a function of ϵ. In Example 2 we found the function $\delta = \epsilon/3$ by some simple algebra; however, most limit proofs require a little more algebraic and logical ingenuity. Thus, if you find our ensuing discussion of "ϵ-δ" proofs challenging, do not become discouraged; the concepts and techniques are intrinsically difficult. In fact, a precise understanding of limits evaded the finest mathematical minds for more than 150 years after the basic concepts of calculus were discovered.

Example 3 Prove that $\displaystyle\lim_{x \to 0^+} \sqrt{x} = 0$.

Solution. Note that the domain of $\sqrt{x}$ is $0 \le x$, so it is valid to discuss the limit as $x \to 0^+$. We must show that given $\epsilon > 0$, there exists a $\delta > 0$ such that

$$|\sqrt{x} - 0| < \epsilon \quad \text{if} \quad 0 < x < 0 + \delta$$

or more simply,

$$\sqrt{x} < \epsilon \quad \text{if} \quad 0 < x < \delta \tag{4}$$

But, by squaring both sides of the inequality $\sqrt{x} < \epsilon$, we can rewrite (4) as

$$x < \epsilon^2 \quad \text{if} \quad 0 < x < \delta \tag{5}$$

It should be self-evident that (5) is true if $\delta = \epsilon^2$; and since (5) is a reformulation of (4), we have shown that (4) holds with $\delta = \epsilon^2$. This proves that $\lim_{x \to 0^+} \sqrt{x} = 0$. ◀

REMARK. In this example the limit from the left and the two-sided limit do not exist at $x = 0$ because the domain of $\sqrt{x}$ includes no numbers to the left of 0.

THE VALUE OF δ IS NOT UNIQUE

In preparation for our next example, we note that the value of δ in Definition 2.4.1 is not unique; once we have found a value of δ that fulfills the requirements of the definition, then any *smaller* positive number δ_1 will also fulfill those requirements. That is, if it is true that

$$|f(x) - L| < \epsilon \quad \text{if} \quad 0 < |x - a| < \delta$$

then it will also be true that

$$|f(x) - L| < \epsilon \quad \text{if} \quad 0 < |x - a| < \delta_1$$

This is because $\{x : 0 < |x - a| < \delta_1\}$ is a subset of $\{x : 0 < |x - a| < \delta\}$ (Figure 2.4.4), and hence if $|f(x) - L| < \epsilon$ is satisfied for all x in the larger set, then it will automatically be satisfied for all x in the subset. Thus, in Example 2, where we used $\delta = \epsilon/3$, we could have used any smaller value of δ such as $\delta = \epsilon/4$, $\delta = \epsilon/5$, or $\delta = \epsilon/6$.

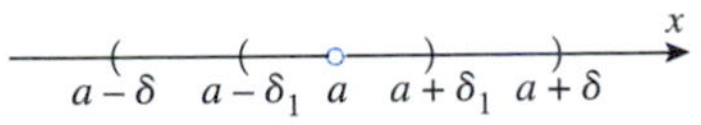

Figure 2.4.4

Example 4 Prove that $\displaystyle\lim_{x \to 3} x^2 = 9$.

Solution. We must show that given any positive number ϵ, we can find a positive number δ such that

$$|x^2 - 9| < \epsilon \quad \text{if} \quad 0 < |x - 3| < \delta \tag{6}$$

Because $|x - 3|$ occurs on the right side of this "if statement," it will be helpful to factor the left side to introduce a factor of $|x - 3|$. This yields the following alternative form of (6)

$$|x + 3||x - 3| < \epsilon \quad \text{if} \quad 0 < |x - 3| < \delta \tag{7}$$

Using the triangle inequality, we see that

$$|x + 3| = |(x - 3) + 6| \leq |x - 3| + 6$$

Therefore, if $0 < |x - 3| < \delta$, then

$$|x + 3||x - 3| \leq (|x - 3| + 6)|x - 3| < (\delta + 6)\delta$$

It follows that (7) will be satisfied for any positive value of δ such that $(\delta + 6)\delta \leq \epsilon$. Let us agree to restrict our attention to positive values of δ such that $\delta \leq 1$. (This is justified because of our earlier observation that once a value of δ is found, then any smaller positive value of δ can be used.) With this restriction, $(\delta + 6)\delta \leq 7\delta$, so that (7) will be satisfied as long as it is also the case that $7\delta \leq \epsilon$. We can achieve this by taking δ to be the minimum of the numbers $\epsilon/7$ and 1, which is sometimes written as $\delta = \min(\epsilon/7, 1)$. This proves that $\lim_{x \to 3} x^2 = 9$. ◀

REMARK. You may have wondered how we knew to make the restriction $\delta \leq 1$ (as opposed to $\delta \leq \frac{1}{2}$ or $\delta \leq 5$, for example). Actually, it does not matter; any restriction of the form $\delta \leq c$ would work equally well.

LIMITS AS $x \to \pm\infty$

In Section 2.1 we discussed the limits

$$\lim_{x \to +\infty} f(x) = L \quad \text{and} \quad \lim_{x \to -\infty} f(x) = L$$

from an intuitive viewpoint. We interpreted the first statement to mean that the values of $f(x)$ eventually get closer and closer to L as x increases indefinitely, and we interpreted the second statement to mean that the values of $f(x)$ eventually get closer and closer to L as x decreases indefinitely. These ideas are captured more precisely in the following definitions and are illustrated in Figure 2.4.5.

2.4.2 DEFINITION. Let $f(x)$ be defined for all x in some infinite open interval extending in the positive x-direction. We will write

$$\lim_{x \to +\infty} f(x) = L$$

if given any number $\epsilon > 0$, there corresponds a positive number N such that

$$|f(x) - L| < \epsilon \quad \text{if} \quad x > N$$

2.4.3 DEFINITION. Let $f(x)$ be defined for all x in some infinite open interval extending in the negative x-direction. We will write

$$\lim_{x \to -\infty} f(x) = L$$

if given any number $\epsilon > 0$, there corresponds a negative number N such that

$$|f(x) - L| < \epsilon \quad \text{if} \quad x < N$$

To see how these definitions relate to our informal concepts of these limits, suppose that $f(x) \to L$ as $x \to +\infty$, and for a given ϵ let N be the positive number described in Definition 2.4.2. If x is allowed to increase indefinitely, then eventually x will lie in the interval $(N, +\infty)$, which is marked in green in Figure 2.4.5a; when this happens, the value of $f(x)$ will fall between $L - \epsilon$ and $L + \epsilon$, marked in red in the figure. Since this is true for all positive values of ϵ (no matter how small), we can force the values of $f(x)$ as close as we like to L by making N sufficiently large. This agrees with our informal concept of this limit. Similarly, Figure 2.4.5b illustrates Definition 2.4.3.

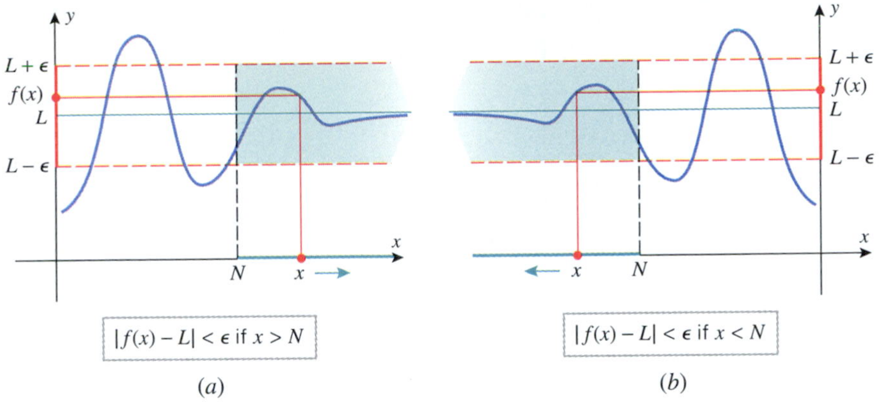

$$|f(x) - L| < \epsilon \text{ if } x > N$$

(a)

$$|f(x) - L| < \epsilon \text{ if } x < N$$

(b)

Figure 2.4.5

Example 5 Prove that $\lim\limits_{x \to +\infty} \dfrac{1}{x} = 0$.

Solution. Applying Definition 2.4.2 with $f(x) = 1/x$ and $L = 0$, we must show that given $\epsilon > 0$, we can find a number $N > 0$ such that

$$\left| \frac{1}{x} - 0 \right| < \epsilon \quad \text{if} \quad x > N \tag{8}$$

Because $x \to +\infty$ we can assume that $x > 0$. Thus, we can eliminate the absolute values in this statement and rewrite it as

$$\frac{1}{x} < \epsilon \quad \text{if} \quad x > N$$

or, on taking reciprocals,

$$x > \frac{1}{\epsilon} \quad \text{if} \quad x > N \tag{9}$$

It is self-evident that $N = 1/\epsilon$ satisfies this requirement, and since (9) and (8) are equivalent for $x > 0$, the proof is complete. ◀

INFINITE LIMITS

In Section 2.1 we discussed limits of the following type from an intuitive viewpoint:

$$\lim_{x \to a} f(x) = +\infty, \qquad \lim_{x \to a} f(x) = -\infty \tag{10}$$

$$\lim_{x \to a^+} f(x) = +\infty, \qquad \lim_{x \to a^+} f(x) = -\infty \tag{11}$$

$$\lim_{x \to a^-} f(x) = +\infty, \qquad \lim_{x \to a^-} f(x) = -\infty \tag{12}$$

Recall that each of these expressions describes a particular way in which the limit fails to exist. The $+\infty$ indicates that the limit fails to exist because $f(x)$ increases without bound, and the $-\infty$ indicates that the limit fails to exist because $f(x)$ decreases without bound. These ideas are captured more precisely in the following definitions and are illustrated in Figure 2.4.6.

2.4.4 DEFINITION. Let $f(x)$ be defined for all x in some open interval containing a, except that $f(x)$ need not be defined at a. We will write

$$\lim_{x \to a} f(x) = +\infty$$

if given any positive number M, we can find a number $\delta > 0$ such that $f(x)$ satisfies

$$f(x) > M \quad \text{if} \quad 0 < |x - a| < \delta$$

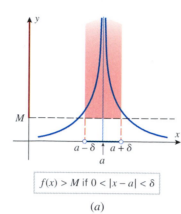
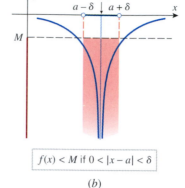

$$\boxed{f(x) > M \text{ if } 0 < |x-a| < \delta}$$

$$\boxed{f(x) < M \text{ if } 0 < |x-a| < \delta}$$

(a)   (b)

Figure 2.4.6

2.4.5   DEFINITION.   Let $f(x)$ be defined for all x in some open interval containing a, except that $f(x)$ need not be defined at a. We will write

$$\lim_{x \to a} f(x) = -\infty$$

if given any negative number M, we can find a number $\delta > 0$ such that $f(x)$ satisfies

$$f(x) < M \quad \text{if} \quad 0 < |x-a| < \delta$$

To see how these definitions relate to our informal concepts of these limits, suppose that $f(x) \to +\infty$ as $x \to a$, and for a given M let δ be the corresponding positive number described in Definition 2.4.4. Next, imagine that x gets closer and closer to a (from either side). Eventually, x will lie in the interval $(a - \delta, a + \delta)$, which is marked in green in Figure 2.4.6a; when this happens the value of $f(x)$ will be greater than M, marked in red in the figure. Since this is true for any positive value of M (no matter how large), we can force the values of $f(x)$ to be as large as we like by making x sufficiently close to a. This agrees with our informal concept of this limit. Similarly, Figure 2.4.6b illustrates Definition 2.4.5.

REMARK.   The definitions for the one-sided limits are similar. For example, in the definition of $\lim_{x \to a^-} f(x) = +\infty$ we assume that $f(x)$ is defined for all x in some interval of the form (c, a) and replace the condition $0 < |x - a| < \delta$ by the condition $a - \delta < x < a$.

Example 6   Prove that $\lim_{x \to 0} \dfrac{1}{x^2} = +\infty$.

Solution.   Applying Definition 2.4.4 with $f(x) = 1/x^2$ and $a = 0$, we must show that given a number $M > 0$, we can find a number $\delta > 0$ such that

$$\frac{1}{x^2} > M \quad \text{if} \quad 0 < |x - 0| < \delta \tag{13}$$

or, on taking reciprocals and simplifying,

$$x^2 < \frac{1}{M} \quad \text{if} \quad 0 < |x| < \delta \tag{14}$$

But $x^2 < 1/M$ if $|x| < 1/\sqrt{M}$, so that $\delta = 1/\sqrt{M}$ satisfies (14). Since (13) is equivalent to (14), the proof is complete. ◄

FOR THE READER.   How would you define

$$\lim_{x \to +\infty} f(x) = +\infty, \qquad \lim_{x \to +\infty} f(x) = -\infty$$

$$\lim_{x \to -\infty} f(x) = +\infty, \qquad \lim_{x \to -\infty} f(x) = -\infty? \tag{15}$$

EXERCISE SET 2.4 ◠ Graphing Utility

1. (a) Find the largest open interval, centered at the origin on the x-axis, such that for each x in the interval the value of the function $f(x) = x + 2$ is within 0.1 unit of the number $f(0) = 2$.
   (b) Find the largest open interval, centered at $x = 3$, such that for each x in the interval the value of the function $f(x) = 4x - 5$ is within 0.01 unit of the number $f(3) = 7$.
   (c) Find the largest open interval, centered at $x = 4$, such that for each x in the interval the value of the function $f(x) = x^2$ is within 0.001 unit of the number $f(4) = 16$.

2. In each part, find the largest open interval, centered at $x = 0$, such that for each x in the interval the value of $f(x) = 2x + 3$ is within ϵ units of the number $f(0) = 3$.

   (a) $\epsilon = 0.1$ (b) $\epsilon = 0.01$

   (c) $\epsilon = 0.0012$

3. (a) Find the values of x_1 and x_2 in the accompanying figure.
   (b) Find a positive number δ such that $|\sqrt{x} - 2| < 0.05$ if $0 < |x - 4| < \delta$.

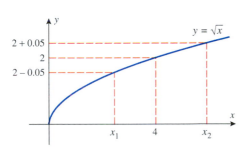

Not drawn to scale

Figure Ex-3

4. (a) Find the values of x_1 and x_2 in the accompanying figure.
   (b) Find a positive number δ such that $|(1/x) - 1| < 0.1$ if $0 < |x - 1| < \delta$.

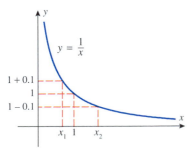

Not drawn to scale

Figure Ex-4

◠ 5. Generate the graph of $f(x) = x^3 - 4x + 5$ with a graphing utility, and use the graph to find a number δ such that $|f(x) - 2| < 0.05$ if $0 < |x - 1| < \delta$. [*Hint*: Show

that the inequality $|f(x) - 2| < 0.05$ can be rewritten as $1.95 < x^3 - 4x + 5 < 2.05$, and estimate the values of x for which $x^3 - 4x + 5 = 1.95$ and $x^3 - 4x + 5 = 2.05$.]

◠ 6. Use the method of Exercise 5 to find a number δ such that $|\sqrt{5x + 1} - 4| < 0.5$ if $0 < |x - 3| < \delta$.

◠ 7. Let $f(x) = x + \sqrt{x}$ with $L = \lim_{x \to 1} f(x)$ and let $\epsilon = 0.2$. Use a graphing utility and its trace feature to find a positive number δ such that $|f(x) - L| < \epsilon$ if $0 < |x - 1| < \delta$.

◠ 8. Let $f(x) = (\sin 2x)/x$ and use a graphing utility to conjecture the value of $L = \lim_{x \to 0} f(x)$. Then let $\epsilon = 0.1$ and use the graphing utility and its trace feature to find a positive number δ such that $|f(x) - L| < \epsilon$ if $0 < |x| < \delta$.

In Exercises 9–18, a positive number ϵ and the limit L of a function f at a are given. Find a number δ such that $|f(x) - L| < \epsilon$ if $0 < |x - a| < \delta$.

9. $\lim_{x \to 4} 2x = 8$; $\epsilon = 0.1$ 10. $\lim_{x \to -2} \frac{1}{2}x = -1$; $\epsilon = 0.1$

11. $\lim_{x \to -1} (7x + 5) = -2$; $\epsilon = 0.01$

12. $\lim_{x \to 3} (5x - 2) = 13$; $\epsilon = 0.01$

13. $\lim_{x \to 2} \frac{x^2 - 4}{x - 2} = 4$; $\epsilon = 0.05$

14. $\lim_{x \to -1} \frac{x^2 - 1}{x + 1} = -2$; $\epsilon = 0.05$

15. $\lim_{x \to 4} x^2 = 16$; $\epsilon = 0.001$ 16. $\lim_{x \to 9} \sqrt{x} = 3$; $\epsilon = 0.001$

17. $\lim_{x \to 5} \frac{1}{x} = \frac{1}{5}$; $\epsilon = 0.05$ 18. $\lim_{x \to 0} |x| = 0$; $\epsilon = 0.05$

In Exercises 19–32, use Definition 2.4.1 to prove that the stated limit is correct.

19. $\lim_{x \to 5} 3x = 15$ 20. $\lim_{x \to 3} (4x - 5) = 7$

21. $\lim_{x \to 2} (2x - 7) = -3$ 22. $\lim_{x \to -1} (2 - 3x) = 5$

23. $\lim_{x \to 0} \frac{x^2 + x}{x} = 1$ 24. $\lim_{x \to -3} \frac{x^2 - 9}{x + 3} = -6$

25. $\lim_{x \to 1} 2x^2 = 2$ 26. $\lim_{x \to 3} (x^2 - 5) = 4$

27. $\lim_{x \to 1/3} \frac{1}{x} = 3$ 28. $\lim_{x \to -2} \frac{1}{x + 1} = -1$

29. $\lim_{x \to 4} \sqrt{x} = 2$ 30. $\lim_{x \to 6} \sqrt{x + 3} = 3$

31. $\lim_{x \to 1} f(x) = 3$, where $f(x) = \begin{cases} x + 2, & x \neq 1 \\ 10, & x = 1 \end{cases}$

32. $\lim_{x \to 2} (x^2 + 3x - 1) = 9$

33. (a) Find the smallest positive number N such that for each x in the interval $(N, +\infty)$, the value of the function $f(x) = 1/x^2$ is within 0.1 unit of $L = 0$.

(b) Find the smallest positive number N such that for each x in the interval $(N, +\infty)$, the value of $f(x) = x/(x+1)$ is within 0.01 unit of $L = 1$.

(c) Find the largest negative number N such that for each x in the interval $(-\infty, N)$, the value of the function $f(x) = 1/x^3$ is within 0.001 unit of $L = 0$.

(d) Find the largest negative number N such that for each x in the interval $(-\infty, N)$, the value of the function $f(x) = x/(x + 1)$ is within 0.01 unit of $L = 1$.

34. In each part, find the smallest positive value of N such that for each x in the interval $(N, +\infty)$, the function $f(x) = 1/x^3$ is within ϵ units of the number $L = 0$.
   (a) $\epsilon = 0.1$ (b) $\epsilon = 0.01$ (c) $\epsilon = 0.001$

35. (a) Find the values of x_1 and x_2 in the accompanying figure.

(b) Find a positive number N such that

$$\left| \frac{x^2}{1 + x^2} - 1 \right| < \epsilon$$

for $x > N$.

(c) Find a negative number N such that

$$\left| \frac{x^2}{1 + x^2} - 1 \right| < \epsilon$$

for $x < N$.

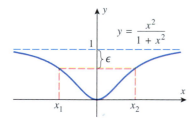

$$y = \frac{x^2}{1 + x^2}$$

Not drawn to scale

Figure Ex-35

36. (a) Find the values of x_1 and x_2 in the accompanying figure.

(b) Find a positive number N such that

$$\left| \frac{1}{\sqrt[3]{x}} - 0 \right| = \left| \frac{1}{\sqrt[3]{x}} \right| < \epsilon$$

for $x > N$.

(c) Find a negative number N such that

$$\left| \frac{1}{\sqrt[3]{x}} - 0 \right| = \left| \frac{1}{\sqrt[3]{x}} \right| < \epsilon$$

for $x < N$.

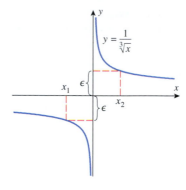

$$y = \frac{1}{\sqrt[3]{x}}$$

Figure Ex-36

In Exercises 37–40, a positive number ϵ and the limit L of a function f at $+\infty$ are given. Find a positive number N such that $|f(x) - L| < \epsilon$ if $x > N$.

37. $\lim\limits_{x \to +\infty} \dfrac{1}{x^2} = 0$; $\epsilon = 0.01$

38. $\lim\limits_{x \to +\infty} \dfrac{1}{x + 2} = 0$; $\epsilon = 0.005$

39. $\lim\limits_{x \to +\infty} \dfrac{x}{x + 1} = 1$; $\epsilon = 0.001$

40. $\lim\limits_{x \to +\infty} \dfrac{4x - 1}{2x + 5} = 2$; $\epsilon = 0.1$

In Exercises 41–44, a positive number ϵ and the limit L of a function f at $-\infty$ are given. Find a negative number N such that $|f(x) - L| < \epsilon$ if $x < N$.

41. $\lim\limits_{x \to -\infty} \dfrac{1}{x + 2} = 0$; $\epsilon = 0.005$

42. $\lim\limits_{x \to -\infty} \dfrac{1}{x^2} = 0$; $\epsilon = 0.01$

43. $\lim\limits_{x \to -\infty} \dfrac{4x - 1}{2x + 5} = 2$; $\epsilon = 0.1$

44. $\lim\limits_{x \to -\infty} \dfrac{x}{x + 1} = 1$; $\epsilon = 0.001$

In Exercises 45–52, use Definition 2.4.2 or 2.4.3 to prove that the stated limit is correct.

45. $\lim\limits_{x \to +\infty} \dfrac{1}{x^2} = 0$ **46.** $\lim\limits_{x \to -\infty} \dfrac{1}{x} = 0$

47. $\lim\limits_{x \to -\infty} \dfrac{1}{x + 2} = 0$ **48.** $\lim\limits_{x \to +\infty} \dfrac{1}{x + 2} = 0$

49. $\lim\limits_{x \to +\infty} \dfrac{x}{x + 1} = 1$ **50.** $\lim\limits_{x \to -\infty} \dfrac{x}{x + 1} = 1$

51. $\lim\limits_{x \to -\infty} \dfrac{4x - 1}{2x + 5} = 2$ **52.** $\lim\limits_{x \to +\infty} \dfrac{4x - 1}{2x + 5} = 2$

53. (a) Find the largest open interval, centered at the origin on the x-axis, such that for each x in the interval, other than the center, the values of $f(x) = 1/x^2$ are greater than 100.

(b) Find the largest open interval, centered at $x = 1$, such that for each x in the interval, other than the center, the values of the function
$$f(x) = 1/|x - 1|$$
are greater than 1000.

(c) Find the largest open interval, centered at $x = 3$, such that for each x in the interval, other than the center, the values of the function
$$f(x) = -1/(x - 3)^2$$
are less than -1000.

(d) Find the largest open interval, centered at the origin on the x-axis, such that for each x in the interval, other than the center, the values of $f(x) = -1/x^4$ are less than $-10,000$.

54. In each part, find the largest open interval, centered at $x = 1$, such that for each x in the interval the value of $f(x) = 1/(x - 1)^2$ is greater than M.
(a) $M = 10$ (b) $M = 1000$ (c) $M = 100,000$

In Exercises 55–60, use Definition 2.4.4 or 2.4.5 to prove that the stated limit is correct.

55. $\lim\limits_{x \to 3} \dfrac{1}{(x - 3)^2} = +\infty$ **56.** $\lim\limits_{x \to 3} \dfrac{-1}{(x - 3)^2} = -\infty$

57. $\lim\limits_{x \to 0} \dfrac{1}{|x|} = +\infty$ **58.** $\lim\limits_{x \to 1} \dfrac{1}{|x - 1|} = +\infty$

59. $\lim\limits_{x \to 0} \left(-\dfrac{1}{x^4}\right) = -\infty$ **60.** $\lim\limits_{x \to 0} \dfrac{1}{x^4} = +\infty$

In Exercises 61–66, use the remark following Definition 2.4.1 to prove that the stated limit is correct.

61. $\lim\limits_{x \to 2^+} (x + 1) = 3$ **62.** $\lim\limits_{x \to 1^-} (3x + 2) = 5$

63. $\lim\limits_{x \to 4^+} \sqrt{x - 4} = 0$ **64.** $\lim\limits_{x \to 0^-} \sqrt{-x} = 0$

65. $\lim\limits_{x \to 2^+} f(x) = 2$, where $f(x) = \begin{cases} x, & x > 2 \\ 3x, & x \le 2 \end{cases}$

66. $\lim\limits_{x \to 2^-} f(x) = 6$, where $f(x) = \begin{cases} x, & x > 2 \\ 3x, & x \le 2 \end{cases}$

In Exercises 67 and 68, use the remark following Definitions 2.4.4 and 2.4.5 to prove that the stated limit is correct.

67. (a) $\lim\limits_{x \to 1^+} \dfrac{1}{1 - x} = -\infty$ (b) $\lim\limits_{x \to 1^-} \dfrac{1}{1 - x} = +\infty$

68. (a) $\lim\limits_{x \to 0^+} \dfrac{1}{x} = +\infty$ (b) $\lim\limits_{x \to 0^-} \dfrac{1}{x} = -\infty$

For Exercises 69 and 70, write out definitions of the four limits in (15), and use your definitions to prove that the stated limits are correct.

69. (a) $\lim\limits_{x \to +\infty} (x + 1) = +\infty$ (b) $\lim\limits_{x \to -\infty} (x + 1) = -\infty$

70. (a) $\lim\limits_{x \to +\infty} (x^2 - 3) = +\infty$ (b) $\lim\limits_{x \to -\infty} (x^3 + 5) = -\infty$

71. Prove the result in Example 4 under the assumption that $\delta \le 2$ rather than $\delta \le 1$.

72. (a) In Definition 2.4.1 there is a condition requiring that $f(x)$ be defined for all x in some open interval containing a, except possibly at a itself. What is the purpose of this requirement?
(b) Why is $\lim\limits_{x \to 0} \sqrt{x} = 0$ an incorrect statement?
(c) Is $\lim\limits_{x \to 0.01} \sqrt{x} = 0.1$ a correct statement?

2.5 CONTINUITY

A moving object cannot vanish at some point and reappear someplace else to continue its motion. Thus, we perceive the path of a moving object as an unbroken curve, without gaps, breaks, or holes. In this section, we translate "unbroken curve" into a precise mathematical formulation called continuity, and develop some fundamental properties of continuous curves.

DEFINITION OF CONTINUITY

Recall from Theorem 2.2.3 that if $p(x)$ is a polynomial and c is a real number, then $\lim_{x \to c} p(x) = p(c)$ (see Figure 2.5.1). Together with Theorem 2.2.2, we are able to calculate limits of a variety of combinations of functions by evaluating the combination.

That is, we saw many examples of functions $f(x)$ such that $\lim_{x \to c} f(x) = f(c)$ if $f(x)$ is defined on an interval containing a number c. In this case, function values $f(x)$ can be guaranteed to be near $f(c)$ for any x-value selected close enough to c. (See Exercise 53 for a precise formulation of this statement.)

On the other hand, we have also seen functions for which this nice property is not true. For example, the function

$$f(x) = \begin{cases} \sin(\pi/x), & x \neq 0 \\ 0, & x = 0 \end{cases}$$

does not satisfy $\lim_{x \to 0} f(x) = f(0)$, since $\lim_{x \to 0} f(x)$ fails to exist (Figure 2.5.2).

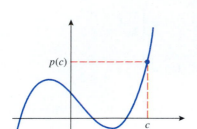

$$\lim_{x \to c} p(x) = p(c)$$

Figure 2.5.1

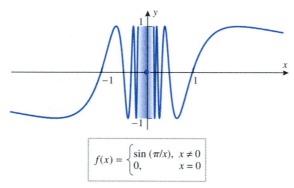

$$f(x) = \begin{cases} \sin(\pi/x), & x \neq 0 \\ 0, & x = 0 \end{cases}$$

Figure 2.5.2

The term *continuous* is used to describe the useful circumstance where the calculation of a limit can be accomplished by mere evaluation of the function.

2.5.1   DEFINITION.   A function f is said to be ***continuous at $x = c$*** provided the following conditions are satisfied:

1. $f(c)$ is defined.

2. $\lim_{x \to c} f(x)$ exists.

3. $\lim_{x \to c} f(x) = f(c)$.

If one or more of the conditions of this definition fails to hold, then we will say that f has a ***discontinuity at $x = c$***. Each function drawn in Figure 2.5.3 illustrates a discontinuity at $x = c$. In Figure 2.5.3a, the function is not defined at c, violating the first condition of Definition 2.5.1. In Figures 2.5.3b and 2.5.3c, $\lim_{x \to c} f(x)$ does not exist, violating the second condition of Definition 2.5.1. In Figure 2.5.3d, the function is defined at c and $\lim_{x \to c} f(x)$ exists, but these two values are not equal, violating the third condition of Definition 2.5.1.

From such graphs we can develop an intuitive, geometric feel for where a function is continuous and where it is discontinuous. Observe that continuity at c may fail due to a "break" in the graph of the function, either due to a hole or to a jump as in Figure 2.5.3, or perhaps due to a wild oscillation as in Figure 2.5.2. Although the intuitive interpretation of "f is continuous at c" as "the graph of f is unbroken at c" lacks precision, it is a useful guide in most circumstances.

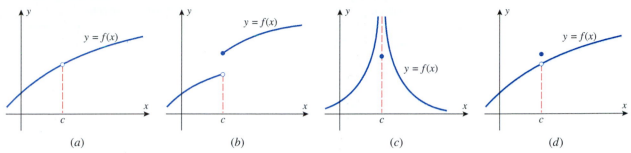

Figure 2.5.3

REMARK. Note that the third condition of Definition 2.5.1 really implies the first two conditions, since it is understood in the statement $\lim_{x \to c} f(x) = f(c)$ that the limit on the left exists, the expression $f(c)$ on the right is defined and has a finite value, and that quantities on the two sides are equal. Thus, when we want to establish continuity of a function at a point, our usual procedure will be to establish the validity of the third condition only.

Example 1 Determine whether the following functions are continuous at $x = 2$.

$$f(x) = \frac{x^2 - 4}{x - 2}, \qquad g(x) = \begin{cases} \dfrac{x^2 - 4}{x - 2}, & x \neq 2 \\ 3, & x = 2, \end{cases} \qquad h(x) = \begin{cases} \dfrac{x^2 - 4}{x - 2}, & x \neq 2 \\ 4, & x = 2 \end{cases}$$

Solution. In each case we must determine whether the limit of the function as $x \to 2$ is the same as the value of the function at $x = 2$. In all three cases the functions are identical, except at $x = 2$, and hence all three have the same limit at $x = 2$, namely

$$\lim_{x \to 2} f(x) = \lim_{x \to 2} g(x) = \lim_{x \to 2} h(x) = \lim_{x \to 2} \frac{x^2 - 4}{x - 2} = \lim_{x \to 2} (x + 2) = 4$$

The function f is undefined at $x = 2$, and hence is not continuous at $x = 2$ (Figure 2.5.4a). The function g is defined at $x = 2$, but its value there is $g(2) = 3$, which is not the same as the limit as x approaches 2; hence, g is also not continuous at $x = 2$ (Figure 2.5.4b). The value of the function h at $x = 2$ is $h(2) = 4$, which is the same as the limit as x approaches 2; hence, h is continuous at $x = 2$ (Figure 2.5.4c). (Note that the function h could have been written more simply as $h(x) = x + 2$, but we wrote it in piecewise form to emphasize its relationship to f and g.) ◄

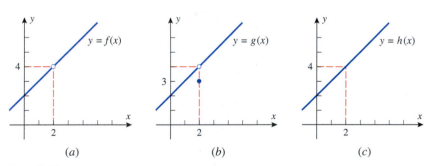

Figure 2.5.4

CONTINUITY IN APPLICATIONS

In applications, discontinuities often signal the occurrence of important physical phenomena. For example, Figure 2.5.5a is a graph of voltage versus time for an underground cable that is accidentally cut by a work crew at time $t = t_0$ (the voltage drops to zero when the line

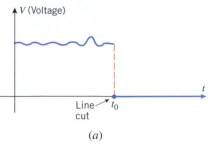

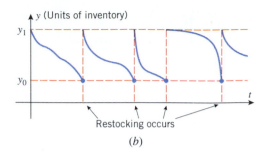

(a)   *(b)*

Figure 2.5.5

is cut). Figure 2.5.5*b* shows the graph of inventory versus time for a company that restocks its warehouse to y_1 units when the inventory falls to y_0 units. The discontinuities occur at those times when restocking occurs.

Given the possible physical significance of discontinuities, it is important to be able to identify discontinuities for specific functions, and to be able to make general statements about the continuity properties of entire families of functions. This is our next goal.

CONTINUITY ON AN INTERVAL AND CONTINUITY OF POLYNOMIALS

If a function f is continuous at each number in an open interval (a, b), then we say that f is *continuous on (a, b)*. This definition applies to infinite open intervals of the form $(a, +\infty)$, $(-\infty, b)$, and $(-\infty, +\infty)$. In the case where f is continuous on $(-\infty, +\infty)$, we will say that f is *continuous everywhere*.

The general procedure for showing that a function is continuous everywhere is to show that it is continuous at an *arbitrary* real number. For example, we showed in Theorem 2.2.3 that if $p(x)$ is a polynomial and a is any real number, then

$$\lim_{x \to a} p(x) = p(a)$$

Thus, we have the following result.

2.5.2   THEOREM.   *Polynomials are continuous everywhere.*

Example 2   Show that $|x|$ is continuous everywhere (Figure 1.2.5).

Solution.   We can write $|x|$ as

$$|x| = \begin{cases} x & \text{if } x > 0 \\ 0 & \text{if } x = 0 \\ -x & \text{if } x < 0 \end{cases}$$

so $|x|$ is the same as the polynomial x on the interval $(0, +\infty)$ and is the same as the polynomial $-x$ on the interval $(-\infty, 0)$. But polynomials are continuous everywhere, so $x = 0$ is the only possible discontinuity for $|x|$. Since $|0| = 0$, to prove the continuity at $x = 0$ we must show that

$$\lim_{x \to 0} |x| = 0 \tag{1}$$

Because the formula for $|x|$ changes at 0, it will be helpful to consider the one-sided limits at 0 rather than the two-sided limit. We obtain

$$\lim_{x \to 0^+} |x| = \lim_{x \to 0^+} x = 0 \quad \text{and} \quad \lim_{x \to 0^-} |x| = \lim_{x \to 0^-} (-x) = 0$$

Thus, (1) holds and $|x|$ is continuous at $x = 0$.   ◄

**SOME PROPERTIES OF
CONTINUOUS FUNCTIONS**

The following theorem, which is a consequence of Theorem 2.2.2, will enable us to reach conclusions about the continuity of functions that are obtained by adding, subtracting, multiplying, and dividing continuous functions.

2.5.3 THEOREM. *If the functions f and g are continuous at c, then*

(a) $f + g$ *is continuous at* c.
(b) $f - g$ *is continuous at* c.
(c) fg *is continuous at* c.
(d) f/g *is continuous at* c *if* $g(c) \neq 0$ *and has a discontinuity at* c *if* $g(c) = 0$.

We will prove part (d). The remaining proofs are similar and will be omitted.

Proof.   First, consider the case where $g(c) = 0$. In this case $f(c)/g(c)$ is undefined, so the function f/g has a discontinuity at c.

Next, consider the case where $g(c) \neq 0$. To prove that f/g is continuous at c, we must show that

$$\lim_{x \to c} \frac{f(x)}{g(x)} = \frac{f(c)}{g(c)} \tag{2}$$

Since f and g are continuous at c,

$$\lim_{x \to c} f(x) = f(c) \quad \text{and} \quad \lim_{x \to c} g(x) = g(c)$$

Thus, by Theorem 2.2.2(d)

$$\lim_{x \to c} \frac{f(x)}{g(x)} = \frac{\lim\limits_{x \to c} f(x)}{\lim\limits_{x \to c} g(x)} = \frac{f(c)}{g(c)}$$

which proves (2).

**CONTINUITY OF RATIONAL
FUNCTIONS**

Since polynomials are continuous everywhere, and since rational functions are ratios of polynomials, part (d) of Theorem 2.5.3 yields the following result.

2.5.4 THEOREM.   *A rational function is continuous at every number where the denominator is nonzero.*

Example 3   For what values of x is there a hole or a gap in the graph of

$$y = \frac{x^2 - 9}{x^2 - 5x + 6}?$$

Solution.   The function being graphed is a rational function, and hence is continuous at every number where the denominator is nonzero. Solving the equation

$$x^2 - 5x + 6 = 0$$

yields discontinuities at $x = 2$ and at $x = 3$.   ◄

FOR THE READER.   If you use a graphing utility to generate the graph of the equation in this example, then there is a good chance that you will see the discontinuity at $x = 2$ but not at $x = 3$. Try it, and explain what you think is happening.

CONTINUITY OF COMPOSITIONS

The following theorem, whose proof is given in Appendix G, will be useful for calculating limits of icompositions of functions.

2.5.5 THEOREM. *If* $\lim_{x \to c} g(x) = L$ *and if the function* f *is continuous at* L, *then* $\lim_{x \to c} f(g(x)) = f(L)$. *That is,*

$$\lim_{x \to c} f(g(x)) = f\left(\lim_{x \to c} g(x) \right)$$

This equality remains valid if $\lim_{x \to c}$ *is replaced everywhere by one of* $\lim_{x \to c^{+}}$, $\lim_{x \to c^{-}}$, $\lim_{x \to +\infty}$, *or* $\lim_{x \to -\infty}$.

In words, this theorem states:

> *A limit symbol can be moved through a function sign provided the limit of the expression inside the function sign exists and the function is continuous at this limit.*

Example 4 We know from Example 2 that the function $|x|$ is continuous everywhere; thus, it follows that if $\lim_{x \to a} g(x)$ exists, then

$$\lim_{x \to a} |g(x)| = \left| \lim_{x \to a} g(x) \right| \tag{3}$$

That is, a limit symbol can be moved through an absolute value sign, provided the limit of the expression inside the absolute value signs exists. For example,

$$\lim_{x \to 3} |5 - x^2| = \left| \lim_{x \to 3} (5 - x^2) \right| = |-4| = 4 \qquad \blacktriangleleft$$

The following theorem is concerned with the continuity of compositions of functions; the first part deals with continuity at a specific number, and the second part with continuity everywhere.

2.5.6 THEOREM.
(a) If the function g is continuous at c, and the function f is continuous at $g(c)$, then the composition $f \circ g$ is continuous at c.
(b) If the function g is continuous everywhere and the function f is continuous everywhere, then the composition $f \circ g$ is continuous everywhere.

Proof. We will prove part (a) only; the proof of part (b) can be obtained by applying part (a) at an arbitrary number c. To prove that $f \circ g$ is continuous at c, we must show that the value of $f \circ g$ and the value of its limit are the same at $x = c$. But this is so, since we can write

$$\lim_{x \to c} (f \circ g)(x) = \lim_{x \to c} f(g(x)) = f(\lim_{x \to c} g(x)) = f(g(c)) = (f \circ g)(c) \qquad \blacksquare$$

Theorem 2.5.5 ⎵   ⎵ g is continuous at c.

We know from Example 2 that the function $|x|$ is continuous everywhere. Thus, if $g(x)$ is continuous at c, then by part (a) of Theorem 2.5.6, the function $|g(x)|$ must also be continuous at c; and, more generally, if $g(x)$ is continuous everywhere, then so is $|g(x)|$. Stated informally:

> *The absolute value of a continuous function is continuous.*

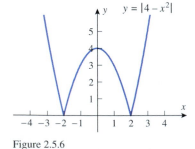

Figure 2.5.6

For example, the polynomial $g(x) = 4 - x^2$ is continuous everywhere, so we can conclude that the function $|4 - x^2|$ is also continuous everywhere (Figure 2.5.6).

FOR THE READER. Can the absolute value of a function that is not continuous be continuous? Justify your answer.

**CONTINUITY FROM THE LEFT
AND RIGHT**

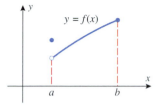

Figure 2.5.7

Because Definition 2.5.1 involves a two-sided limit, that definition does not generally apply at the endpoints of a closed interval $[a, b]$ or at the endpoint of an interval of the form $[a, b)$, $(a, b]$, $(-\infty, b]$, or $[a, +\infty)$. To remedy this problem, we will agree that a function is continuous at an endpoint of an interval if its value at the endpoint is equal to the appropriate one-sided limit at that endpoint. For example, the function graphed in Figure 2.5.7 is continuous at the right endpoint of the interval $[a, b]$ because

$$\lim_{x \to b^-} f(x) = f(b)$$

but it is not continuous at the left endpoint because

$$\lim_{x \to a^+} f(x) \neq f(a)$$

In general, we will say a function f is *continuous from the left* at c if

$$\lim_{x \to c^-} f(x) = f(c)$$

and is *continuous from the right* at c if

$$\lim_{x \to c^+} f(x) = f(c)$$

Using this terminology we define continuity on a closed interval as follows.

2.5.7 DEFINITION. A function f is said to be *continuous on a closed interval $[a, b]$* if the following conditions are satisfied:
1. f is continuous on (a, b).
2. f is continuous from the right at a.
3. f is continuous from the left at b.

FOR THE READER. We leave it for you to modify this definition appropriately so that it applies to intervals of the form $[a, +\infty)$, $(-\infty, b]$, $(a, b]$, and $[a, b)$.

Example 5 What can you say about the continuity of the function $f(x) = \sqrt{9 - x^2}$?

Solution. Because the natural domain of this function is the closed interval $[-3, 3]$, we will need to investigate the continuity of f on the open interval $(-3, 3)$ and at the two endpoints. If c is any number in the interval $(-3, 3)$, then it follows from Theorem 2.2.2(e) that

$$\lim_{x \to c} f(x) = \lim_{x \to c} \sqrt{9 - x^2} = \sqrt{\lim_{x \to c} (9 - x^2)} = \sqrt{9 - c^2} = f(c)$$

which proves f is continuous at each number in the interval $(-3, 3)$. The function f is also continuous at the endpoints since

$$\lim_{x \to 3^-} f(x) = \lim_{x \to 3^-} \sqrt{9 - x^2} = \sqrt{\lim_{x \to 3^-} (9 - x^2)} = 0 = f(3)$$

$$\lim_{x \to -3^+} f(x) = \lim_{x \to -3^+} \sqrt{9 - x^2} = \sqrt{\lim_{x \to -3^+} (9 - x^2)} = 0 = f(-3)$$

Thus, f is continuous on the closed interval $[-3, 3]$. ◀

**THE INTERMEDIATE-VALUE
THEOREM**

Figure 2.5.8 shows the graph of a function that is continuous on the closed interval $[a, b]$. The figure suggests that if we draw any horizontal line $y = k$, where k is between $f(a)$ and $f(b)$, then that line will cross the curve $y = f(x)$ at least once over the interval $[a, b]$. Stated in numerical terms, if f is continuous on $[a, b]$, then the function f must take on every value k between $f(a)$ and $f(b)$ at least once as x varies from a to b. For example, the polynomial $p(x) = x^5 - x + 3$ has a value of 3 at $x = 1$ and a value of 33 at $x = 2$. Thus, it follows from the continuity of p that the equation $x^5 - x + 3 = k$ has at least one

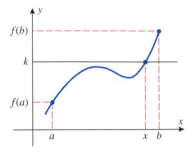

Figure 2.5.8

. . . .

APPROXIMATING ROOTS USING THE INTERMEDIATE-VALUE THEOREM

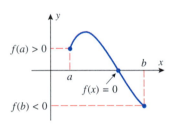

Figure 2.5.9

solution in the interval $[1, 2]$ for every value of k between 3 and 33. This idea is stated more precisely in the following theorem.

2.5.8   THEOREM (*Intermediate-Value Theorem*).   *If f is continuous on a closed interval $[a, b]$ and k is any number between $f(a)$ and $f(b)$, inclusive, then there is at least one number x in the interval $[a, b]$ such that $f(x) = k$.*

Although this theorem is intuitively obvious, its proof depends on a mathematically precise development of the real number system, which is beyond the scope of this text.

A variety of problems can be reduced to solving an equation $f(x) = 0$ for its roots. Sometimes it is possible to solve for the roots exactly using algebra, but often this is not possible and one must settle for decimal approximations of the roots. One procedure for approximating roots is based on the following consequence of the Intermediate-Value Theorem.

2.5.9   THEOREM.   *If f is continuous on $[a, b]$, and if $f(a)$ and $f(b)$ are nonzero and have opposite signs, then there is at least one solution of the equation $f(x) = 0$ in the interval (a, b).*

This result, which is illustrated in Figure 2.5.9, can be proved as follows.

Proof.   Since $f(a)$ and $f(b)$ have opposite signs, 0 is between $f(a)$ and $f(b)$. Thus, by the Intermediate-Value Theorem there is at least one number x in the interval $[a, b]$ such that $f(x) = 0$. However, $f(a)$ and $f(b)$ are nonzero, so x must lie in the interval (a, b), which completes the proof.   ∎

Before we illustrate how this theorem can be used to approximate roots, it will be helpful to discuss some standard terminology for describing errors in approximations. If x is an approximation to a quantity x_0, then we call

$$\epsilon = |x - x_0|$$

the ***absolute error*** or (less precisely) the ***error*** in the approximation. The terminology in Table 2.5.1 is used to describe the size of such errors:

Table 2.5.1

ERROR	DESCRIPTION
$|x - x_0| \leq 0.1$	x approximates x_0 with an error of at most 0.1.
$|x - x_0| \leq 0.01$	x approximates x_0 with an error of at most 0.01.
$|x - x_0| \leq 0.001$	x approximates x_0 with an error of at most 0.001.
$|x - x_0| \leq 0.0001$	x approximates x_0 with an error of at most 0.0001.
$|x - x_0| \leq 0.5$	x approximates x_0 to the nearest integer.
$|x - x_0| \leq 0.05$	x approximates x_0 to 1 decimal place (i.e., to the nearest tenth).
$|x - x_0| \leq 0.005$	x approximates x_0 to 2 decimal places (i.e., to the nearest hundredth).
$|x - x_0| \leq 0.0005$	x approximates x_0 to 3 decimal places (i.e., to the nearest thousandth).

Example 6   The equation

$$x^3 - x - 1 = 0$$

cannot be solved algebraically very easily because the left side has no simple factors. However, if we graph $p(x) = x^3 - x - 1$ with a graphing utility (Figure 2.5.10), then we are led to conjecture that there is one real root and that this root lies inside the interval $[1, 2]$.

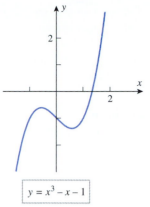

$y = x^3 - x - 1$

Figure 2.5.10

The existence of a root in this interval is also confirmed by Theorem 2.5.9, since $p(1) = -1$ and $p(2) = 5$ have opposite signs. Approximate this root to two decimal-place accuracy.

Solution. Our objective is to approximate the unknown root x_0 with an error of at most 0.005. It follows that if we can find an interval of length 0.01 that contains the root, then the midpoint of that interval will approximate the root with an error of at most $0.01/2 = 0.005$, which will achieve the desired accuracy.

We know that the root x_0 lies in the interval $[1, 2]$. However, this interval has length 1, which is too large. We can pinpoint the location of the root more precisely by dividing the interval $[1, 2]$ into 10 equal parts and evaluating p at the points of subdivision using a calculating utility (Table 2.5.2). In this table $p(1.3)$ and $p(1.4)$ have opposite signs, so we know that the root lies in the interval $[1.3, 1.4]$. This interval has length 0.1, which is still too large, so we repeat the process by dividing the interval $[1.3, 1.4]$ into 10 parts and evaluating p at the points of subdivision; this yields Table 2.5.3, which tells us that the root is inside the interval $[1.32, 1.33]$ (Figure 2.5.11). Since this interval has length 0.01, its midpoint 1.325 will approximate the root with an error of at most 0.005. Thus, $x_0 \approx 1.325$ to two decimal-place accuracy. ◀

Table 2.5.2

x	1	1.1	1.2	1.3	1.4	1.5	1.6	1.7	1.8	1.9	2
$f(x)$	−1	−0.77	−0.47	−0.10	0.34	0.88	1.50	2.21	3.03	3.96	5

Table 2.5.3

x	1.3	1.31	1.32	1.33	1.34	1.35	1.36	1.37	1.38	1.39	1.4
$f(x)$	−0.103	−0.062	−0.020	0.023	0.066	0.110	0.155	0.201	0.248	0.296	0.344

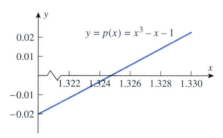

Figure 2.5.11

APPROXIMATING ROOTS BY ZOOMING WITH A GRAPHING UTILITY

The method illustrated in Example 6 can also be implemented with a graphing utility as follows.

Step 1. Figure 2.5.12*a* shows the graph of f in the window $[-5, 5] \times [-5, 5]$ with xScl $= 1$ and yScl $= 1$. That graph places the root between $x = 1$ and $x = 2$.

Step 2. Since we know that the root lies between $x = 1$ and $x = 2$, we will zoom in by regraphing f over an x-interval that extends between these values and in which xScl $= 0.1$. The y-interval and yScl are not critical, as long as the y-interval extends above and below the x-axis. Figure 2.5.12*b* shows the graph of f in the window $[1, 2] \times [-1, 1]$ with xScl $= 0.1$ and yScl $= 0.1$. That graph places the root between $x = 1.3$ and $x = 1.4$.

Step 3.   Since we know that the root lies between $x = 1.3$ and $x = 1.4$, we will zoom in again by regraphing f over an x-interval that extends between these values and in which $x\,\text{Scl} = 0.01$. Figure 2.5.12c shows the graph of the function f in the window $[1.3, 1.4] \times [-0.1, 0.1]$ with $x\,\text{Scl} = 0.01$ and $y\,\text{Scl} = 0.01$. That graph places the root between $x = 1.32$ and $x = 1.33$.

Step 4.   Since the interval in Step 3 has length 0.01, its midpoint 1.325 approximates the root with an error of at most 0.005, so $x_0 \approx 1.325$ to two decimal-place accuracy.

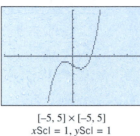
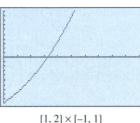
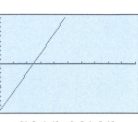

$[-5, 5] \times [-5, 5]$ $[1, 2] \times [-1, 1]$ $[1.3, 1.4] \times [-0.1, 0.1]$
$x\text{Scl} = 1, y\text{Scl} = 1$ $x\text{Scl} = 0.1, y\text{Scl} = 0.1$ $x\text{Scl} = 0.01, y\text{Scl} = 0.01$

Figure 2.5.12   (*a*)   (*b*)   (*c*)

• REMARK.   To say that x approximates x_0 to n decimal places does *not* mean that the first n decimal places of x and x_0 will be the same when the numbers are rounded to n decimal places. For example, $x = 1.084$ approximates $x_0 = 1.087$ to two decimal places because $|x - x_0| = 0.003 \, (<0.005)$. However, if we round these values to two decimal places, then we obtain $x \approx 1.08$ and $x_0 \approx 1.09$. Thus, if you approximate a number to n decimal places, then you should display that approximation to at least $n + 1$ decimal places to preserve the accuracy.

• FOR THE READER.   Use a graphing or calculating utility to show that the root x_0 in Example 6 can be approximated as $x_0 \approx 1.3245$ to three decimal-place accuracy.

EXERCISE SET 2.5   Graphing Utility

In Exercises 1–4, let f be the function whose graph is shown. On which of the following intervals, if any, is f continuous?
(a) $[1, 3]$ (b) $(1, 3)$ (c) $[1, 2]$
(d) $(1, 2)$ (e) $[2, 3]$ (f) $(2, 3)$
For each interval on which f is not continuous, indicate which conditions for the continuity of f do not hold.

3.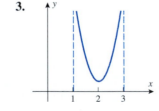

4.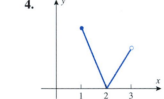

1.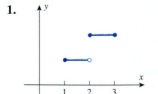

2.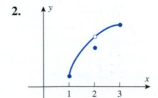

In Exercises 5 and 6, find all values of c such that the specified function has a discontinuity at $x = c$. For each such value of c, determine which conditions of Definition 2.5.1 fail to be satisfied.

5. (a) The function f in Exercise 1 of Section 2.1.
   (b) The function F in Exercise 5 of Section 2.1.
   (c) The function f in Exercise 9 of Section 2.1.

6. (a) The function f in Exercise 2 of Section 2.1.
   (b) The function F in Exercise 6 of Section 2.1.
   (c) The function f in Exercise 10 of Section 2.1.

7. Suppose that f and g are continuous functions such that
$f(2) = 1$ and $\lim\limits_{x \to 2} [f(x) + 4g(x)] = 13$. Find
   (a) $g(2)$ (b) $\lim\limits_{x \to 2} g(x)$.

8. Suppose that f and g are continuous functions such that
$\lim\limits_{x \to 3} g(x) = 5$ and $f(3) = -2$. Find $\lim\limits_{x \to 3} [f(x)/g(x)]$.

9. In each part sketch the graph of a function f that satisfies
the stated conditions.
   (a) f is continuous everywhere except at $x = 3$, at which
   point it is continuous from the right.
   (b) f has a two-sided limit at $x = 3$, but it is not continuous
   at $x = 3$.
   (c) f is not continuous at $x = 3$, but if its value at $x = 3$
   is changed from $f(3) = 1$ to $f(3) = 0$, it becomes
   continuous at $x = 3$.
   (d) f is continuous on the interval $[0, 3)$ and is defined on
   the closed interval $[0, 3]$; but f is not continuous on the
   interval $[0, 3]$.

10. Find formulas for some functions that are continuous on the
intervals $(-\infty, 0)$ and $(0, +\infty)$, but are not continuous on the
interval $(-\infty, +\infty)$.

11. A student parking lot at a university charges \$2.00 for the
first half hour (or any part) and \$1.00 for each subsequent
half hour (or any part) up to a daily maximum of \$10.00.
   (a) Sketch a graph of cost as a function of the time parked.
   (b) Discuss the significance of the discontinuities in the
   graph to a student who parks there.

12. In each part determine whether the function is continuous
or not, and explain your reasoning.
   (a) The Earth's population as a function of time.
   (b) Your exact height as a function of time.
   (c) The cost of a taxi ride in your city as a function of the
   distance traveled.
   (d) The volume of a melting ice cube as a function of time.

In Exercises 13–24, find the values of x (if any) at which f
is not continuous.

13. $f(x) = x^3 - 2x + 3$ **14.** $f(x) = (x - 5)^{17}$

15. $f(x) = \dfrac{x}{x^2 + 1}$ **16.** $f(x) = \dfrac{x}{x^2 - 1}$

17. $f(x) = \dfrac{x - 4}{x^2 - 16}$ **18.** $f(x) = \dfrac{3x + 1}{x^2 + 7x - 2}$

19. $f(x) = \dfrac{x}{|x| - 3}$ **20.** $f(x) = \dfrac{5}{x} + \dfrac{2x}{x + 4}$

21. $f(x) = |x^3 - 2x^2|$ **22.** $f(x) = \dfrac{x + 3}{|x^2 + 3x|}$

23. $f(x) = \begin{cases} 2x + 3, & x \le 4 \\ 7 + \dfrac{16}{x}, & x > 4 \end{cases}$

24. $f(x) = \begin{cases} \dfrac{3}{x - 1}, & x \ne 1 \\ 3, & x = 1 \end{cases}$

25. Find a value for the constant k, if possible, that will make
the function continuous everywhere.
   (a) $f(x) = \begin{cases} 7x - 2, & x \le 1 \\ kx^2, & x > 1 \end{cases}$

   (b) $f(x) = \begin{cases} kx^2, & x \le 2 \\ 2x + k, & x > 2 \end{cases}$

26. On which of the following intervals is
$$f(x) = \dfrac{1}{\sqrt{x - 2}}$$
continuous?
   (a) $[2, +\infty)$ (b) $(-\infty, +\infty)$ (c) $(2, +\infty)$ (d) $[1, 2)$

A function f is said to have a **_removable discontinuity_** at
$x = c$ if $\lim_{x \to c} f(x)$ exists but f is not continuous at $x = c$,
either because f is not defined at c or because the definition
for $f(c)$ differs from the value of the limit. This terminology
will be needed in Exercises 27–30.

27. (a) Sketch the graph of a function with a removable discontinuity at $x = c$ for which $f(c)$ is undefined.
   (b) Sketch the graph of a function with a removable discontinuity at $x = c$ for which $f(c)$ is defined.

28. (a) The terminology *removable discontinuity* is appropriate because a removable discontinuity of a function f
   at $x = c$ can be "removed" by redefining the value of
   f appropriately at $x = c$. What value for $f(c)$ removes
   the discontinuity?
   (b) Show that the following functions have removable discontinuities at $x = 1$, and sketch their graphs.
   $$f(x) = \dfrac{x^2 - 1}{x - 1} \quad \text{and} \quad g(x) = \begin{cases} 1, & x > 1 \\ 0, & x = 1 \\ 1, & x < 1 \end{cases}$$
   (c) What values should be assigned to $f(1)$ and $g(1)$ to
   remove the discontinuities?

In Exercises 29 and 30, find the values of x (if any) at which
f is not continuous, and determine whether each such value
is a removable discontinuity.

29. (a) $f(x) = \dfrac{|x|}{x}$ (b) $f(x) = \dfrac{x^2 + 3x}{x + 3}$

   (c) $f(x) = \dfrac{x - 2}{|x| - 2}$

30. (a) $f(x) = \dfrac{x^2 - 4}{x^3 - 8}$

(b) $f(x) = \begin{cases} 2x - 3, & x \le 2 \\ x^2, & x > 2 \end{cases}$

(c) $f(x) = \begin{cases} 3x^2 + 5, & x \ne 1 \\ 6, & x = 1 \end{cases}$

31. (a) Use a graphing utility to generate the graph of the function $f(x) = (x + 3)/(2x^2 + 5x - 3)$, and then use the graph to make a conjecture about the number and locations of all discontinuities.

(b) Check your conjecture by factoring the denominator.

32. (a) Use a graphing utility to generate the graph of the function $f(x) = x/(x^3 - x + 2)$, and then use the graph to make a conjecture about the number and locations of all discontinuities.

(b) Use the Intermediate-Value Theorem to approximate the location of all discontinuities to two decimal places.

33. Prove that $f(x) = x^{3/5}$ is continuous everywhere, carefully justifying each step.

34. Prove that $f(x) = 1/\sqrt{x^4 + 7x^2 + 1}$ is continuous everywhere, carefully justifying each step.

35. Let f and g be discontinuous at c. Give examples to show that

(a) $f + g$ can be continuous or discontinuous at c

(b) fg can be continuous or discontinuous at c.

36. Prove Theorem 2.5.4.

37. Prove:

(a) part (*a*) of Theorem 2.5.3

(b) part (*b*) of Theorem 2.5.3

(c) part (*c*) of Theorem 2.5.3.

38. Prove: If f and g are continuous on $[a, b]$, and $f(a) > g(a)$, $f(b) < g(b)$, then there is at least one solution of the equation $f(x) = g(x)$ in (a, b). [*Hint:* Consider $f(x) - g(x)$.]

39. Give an example of a function f that is defined on a closed interval, and whose values at the endpoints have opposite signs, but for which the equation $f(x) = 0$ has no solution in the interval.

40. Use the Intermediate-Value Theorem to show that there is a square with a diagonal length that is between r and $2r$ and an area that is half the area of a circle of radius r.

41. Use the Intermediate-Value Theorem to show that there is a right circular cylinder of height h and radius less than r whose volume is equal to that of a right circular cone of height h and radius r.

In Exercises 42 and 43, show that the equation has at least one solution in the given interval.

---

42. $x^3 - 4x + 1 = 0$; $[1, 2]$ **43.** $x^3 + x^2 - 2x = 1$; $[-1, 1]$

44. Prove: If $p(x)$ is a polynomial of odd degree, then the equation $p(x) = 0$ has at least one real solution.

45. The accompanying figure shows the graph of $y = x^4 + x - 1$. Use the method of Example 6 to approximate the x-intercepts with an error of at most 0.05.

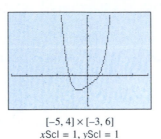

$[-5, 4] \times [-3, 6]$
xScl $= 1$, yScl $= 1$

Figure Ex-45

46. Use a graphing utility to solve the problem in Exercise 45 by zooming.

47. The accompanying figure shows the graph of $y = 5 - x - x^4$. Use the method of Example 6 to approximate the roots of the equation $5 - x - x^4 = 0$ to two decimal-place accuracy.

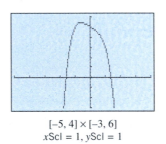

$[-5, 4] \times [-3, 6]$
xScl $= 1$, yScl $= 1$

Figure Ex-47

48. Use a graphing utility to solve the problem in Exercise 47 by zooming.

49. Use the fact that $\sqrt{5}$ is a solution of $x^2 - 5 = 0$ to approximate $\sqrt{5}$ with an error of at most 0.005.

50. Prove that if a and b are positive, then the equation

$$\frac{a}{x - 1} + \frac{b}{x - 3} = 0$$

has at least one solution in the interval $(1, 3)$.

51. A sphere of unknown radius x consists of a spherical core and a coating that is 1 cm thick (see the accompanying figure). Given that the volume of the coating and the volume of the core are the same, approximate the radius of the sphere to three decimal-place accuracy.

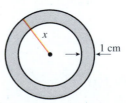

Figure Ex-51

52. A monk begins walking up a mountain road at 12:00 noon and reaches the top at 12:00 midnight. He meditates and rests until 12:00 noon the next day, at which time he begins walking down the same road, reaching the bottom at 12:00 midnight. Show that there is at least one point on the road that he reaches at the same time of day on the way up as on the way down.

53. Let f be defined at c. Prove that f is continuous at c if, given $\epsilon > 0$, there exists a $\delta > 0$ such that $|f(x) - f(c)| < \epsilon$ if $|x - c| < \delta$.

2.6 LIMITS AND CONTINUITY OF TRIGONOMETRIC FUNCTIONS

In this section we will investigate the continuity properties of the trigonometric functions, and we will discuss some important limits involving these functions.

CONTINUITY OF TRIGONOMETRIC FUNCTIONS

Before we begin, recall that in the expressions $\sin x$, $\cos x$, $\tan x$, $\cot x$, $\sec x$, and $\csc x$ it is understood that x is in radian measure.

In trigonometry, the graphs of $\sin x$ and $\cos x$ are drawn as continuous curves (Figure 2.6.1). To actually prove that these functions are continuous everywhere, we must show that the following equalities hold for every real number c:

$$\lim_{x \to c} \sin x = \sin c \quad \text{and} \quad \lim_{x \to c} \cos x = \cos c \qquad (1\text{--}2)$$

Although we will not formally prove these results, we can make them plausible by considering the behavior of the point $P(\cos x, \sin x)$ as it moves around the unit circle. For this purpose, view c as a fixed angle in radian measure, and let $Q(\cos c, \sin c)$ be the corresponding point on the unit circle. As $x \to c$ (i.e., as the angle x approaches the angle c), the point P moves along the circle toward Q, and this implies that the coordinates of P approach the corresponding coordinates of Q; that is, $\cos x \to \cos c$, and $\sin x \to \sin c$ (Figure 2.6.2).

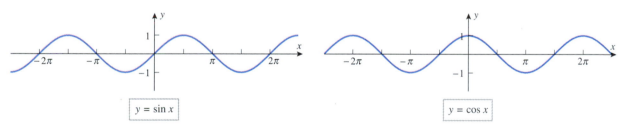

$y = \sin x$ $\qquad$ $y = \cos x$

Figure 2.6.1

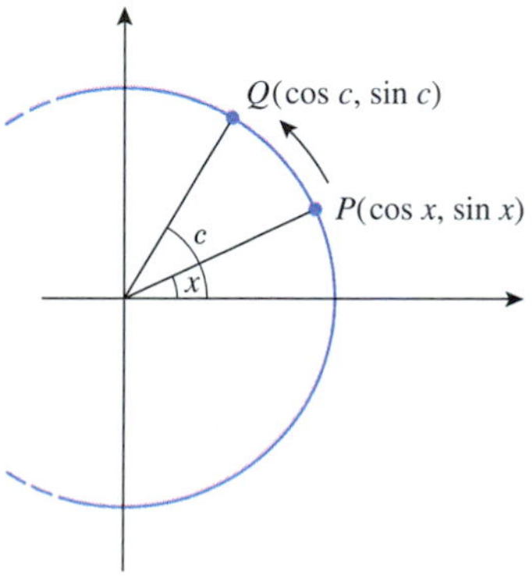

Formulas (1) and (2) can be used to find limits of the remaining trigonometric functions by expressing them in terms of $\sin x$ and $\cos x$; for example, if $\cos c \neq 0$, then

$$\lim_{x \to c} \tan x = \lim_{x \to c} \frac{\sin x}{\cos x} = \frac{\sin c}{\cos c} = \tan c$$

Thus, we are led to the following theorem.

Figure 2.6.2

> **2.6.1 THEOREM.** *If c is any number in the natural domain of the stated trigonometric function, then*
>
> $$\lim_{x \to c} \sin x = \sin c \qquad \lim_{x \to c} \cos x = \cos c \qquad \lim_{x \to c} \tan x = \tan c$$
>
> $$\lim_{x \to c} \csc x = \csc c \qquad \lim_{x \to c} \sec x = \sec c \qquad \lim_{x \to c} \cot x = \cot c$$

It follows from this theorem, for example, that $\sin x$ and $\cos x$ are continuous everywhere and that $\tan x$ is continuous, except at the points where it is undefined.

Example 1 Find the limit

$$\lim_{x \to 1} \cos\left(\frac{x^2 - 1}{x - 1}\right)$$

Solution. Recall from the last section that since the cosine function is continuous everywhere,

$$\lim_{x \to 1} \cos(g(x)) = \cos\left(\lim_{x \to 1} g(x)\right)$$

provided $\lim_{x \to 1} g(x)$ exists. Thus,

$$\lim_{x \to 1} \cos\left(\frac{x^2 - 1}{x - 1}\right) = \lim_{x \to 1} \cos(x + 1) = \cos\left(\lim_{x \to 1} (x + 1)\right) = \cos 2 \quad \blacktriangleleft$$

OBTAINING LIMITS BY SQUEEZING

In Section 2.1 we used the numerical evidence in the table in Figure 2.1.6 to *conjecture* that

$$\lim_{x \to 0} \frac{\sin x}{x} = 1 \tag{3}$$

However, it is not a simple matter to establish this limit with certainty. The difficulty is that the numerator and denominator both approach zero as $x \to 0$. As discussed in Section 2.2, such limits are called indeterminate forms of type $0/0$. Sometimes indeterminate forms of this type can be established by manipulating the ratio algebraically, but in this case no simple algebraic manipulation will work, so we must look for other methods.

The problem with indeterminate forms of type $0/0$ is that there are two conflicting influences at work: as the numerator approaches 0 it drives the magnitude of the ratio toward 0, and as the denominator approaches 0 it drives the magnitude of the ratio toward $\pm\infty$ (depending on the sign of the expression). The limiting behavior of the ratio is determined by the precise way in which these influences offset each other. Later in this text we will discuss general methods for attacking indeterminate forms, but for the limit in (3) we can use a method called *squeezing*.

In the method of squeezing one proves that a function f has a limit L at a number c by trapping the function between two other functions, g and h, whose limits at c are known to be L (Figure 2.6.3). This is the idea behind the following theorem, which we state without proof.

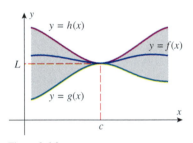

Figure 2.6.3

2.6.2 THEOREM (*The Squeezing Theorem*). *Let f, g, and h be functions satisfying*

$$g(x) \le f(x) \le h(x)$$

for all x in some open interval containing the number c, with the possible exception that the inequalities need not hold at c. If g and h have the same limit as x approaches c, say

$$\lim_{x \to c} g(x) = \lim_{x \to c} h(x) = L$$

then f also has this limit as x approaches c, that is,

$$\lim_{x \to c} f(x) = L$$

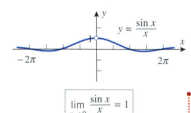

FOR THE READER. The Squeezing Theorem also holds for one-sided limits and limits at $+\infty$ or $-\infty$. How do you think the hypotheses of the theorem would change in those cases?

The usefulness of the Squeezing Theorem will be evident in our proof of the following theorem (Figure 2.6.4).

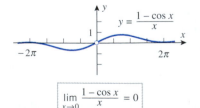

Figure 2.6.4

2.6.3 THEOREM.

(a) $\displaystyle\lim_{x \to 0} \frac{\sin x}{x} = 1$ (b) $\displaystyle\lim_{x \to 0} \frac{1 - \cos x}{x} = 0$

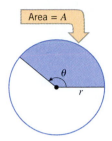

Figure 2.6.5

However, before giving the proof, it will be helpful to review the formula for the area A of a sector with radius r and a central angle of θ radians (Figure 2.6.5). The area of the sector can be derived by setting up the following proportion to the area of the entire circle:

$$\frac{A}{\pi r^2} = \frac{\theta}{2\pi} \qquad \left[\frac{\text{area of the sector}}{\text{area of the circle}} = \frac{\text{central angle of the sector}}{\text{central angle of the circle}} \right]$$

From this we obtain the formula

$$A = \tfrac{1}{2} r^2 \theta \tag{4}$$

Now we are ready for the proof of Theorem 2.6.3.

Proof (a). In this proof we will interpret x as an angle in radian measure, and we will assume to start that $0 < x < \pi/2$. It follows from Formula (4) that the area of a sector of radius 1 and central angle x is $x/2$. Moreover, it is suggested by Figure 2.6.6 that the area of this sector lies between the areas of two triangles, one with area $(\tan x)/2$ and one with area $(\sin x)/2$. Thus,

$$\frac{\tan x}{2} \geq \frac{x}{2} \geq \frac{\sin x}{2}$$

Multiplying through by $2/(\sin x)$ yields

$$\frac{1}{\cos x} \geq \frac{x}{\sin x} \geq 1$$

and then taking reciprocals and reversing the inequalities yields

$$\cos x \leq \frac{\sin x}{x} \leq 1 \tag{5}$$

Moreover, these inequalities also hold for $-\pi/2 < x < 0$, since replacing x by $-x$ in (5) and using the identities $\sin(-x) = -\sin x$ and $\cos(-x) = \cos x$ leaves the inequalities unchanged (verify). Finally, since the functions $\cos x$ and 1 both have limits of 1 as $x \to 0$, it follows from the Squeezing Theorem that $(\sin x)/x$ also has a limit of 1 as $x \to 0$.

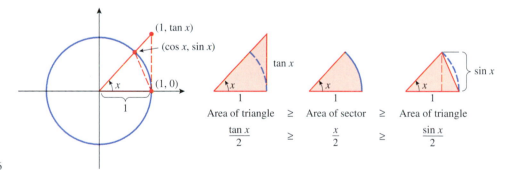

Figure 2.6.6

Proof (b). For this proof we will use the limit in part (a), the continuity of the sine function, and the trigonometric identity $\sin^2 x = 1 - \cos^2 x$. We obtain

$$\lim_{x \to 0} \frac{1 - \cos x}{x} = \lim_{x \to 0} \left[\frac{1 - \cos x}{x} \cdot \frac{1 + \cos x}{1 + \cos x} \right] = \lim_{x \to 0} \frac{\sin^2 x}{(1 + \cos x)x}$$

$$= \left(\lim_{x \to 0} \frac{\sin x}{x} \right) \left(\lim_{x \to 0} \frac{\sin x}{1 + \cos x} \right) = (1) \left(\frac{0}{1 + 1} \right) = 0 \qquad \blacksquare$$

Example 2 Find

(a) $\displaystyle\lim_{x \to 0} \frac{\tan x}{x}$ (b) $\displaystyle\lim_{\theta \to 0} \frac{\sin 2\theta}{\theta}$ (c) $\displaystyle\lim_{x \to 0} \frac{\sin 3x}{\sin 5x}$

Solution (a).

$$\lim_{x \to 0} \frac{\tan x}{x} = \lim_{x \to 0} \left(\frac{\sin x}{x} \cdot \frac{1}{\cos x} \right) = (1)(1) = 1$$

Solution (b). The trick is to multiply and divide by 2, which will make the denominator the same as the argument of the sine function [just as in Theorem 2.6.3(a)]:

$$\lim_{\theta \to 0} \frac{\sin 2\theta}{\theta} = \lim_{\theta \to 0} 2 \cdot \frac{\sin 2\theta}{2\theta} = 2 \lim_{\theta \to 0} \frac{\sin 2\theta}{2\theta}$$

Now make the substitution $x = 2\theta$, and use the fact that $x \to 0$ as $\theta \to 0$. This yields

$$\lim_{\theta \to 0} \frac{\sin 2\theta}{\theta} = 2 \lim_{\theta \to 0} \frac{\sin 2\theta}{2\theta} = 2 \lim_{x \to 0} \frac{\sin x}{x} = 2(1) = 2$$

Solution (c).

$$\lim_{x \to 0} \frac{\sin 3x}{\sin 5x} = \lim_{x \to 0} \frac{\dfrac{\sin 3x}{x}}{\dfrac{\sin 5x}{x}} = \lim_{x \to 0} \frac{3 \cdot \dfrac{\sin 3x}{3x}}{5 \cdot \dfrac{\sin 5x}{5x}} = \frac{3 \cdot 1}{5 \cdot 1} = \frac{3}{5}$$ ◀

• FOR THE READER.   Use a graphing utility to confirm the limits in the last example graphically, and if you have a CAS, then use it to obtain the limits.

Example 3   Make conjectures about the limits

(a) $\displaystyle \lim_{x \to 0} \sin \left(\frac{1}{x} \right)$ (b) $\displaystyle \lim_{x \to 0} x \sin \left(\frac{1}{x} \right)$

and confirm your conclusions by generating the graphs of the functions near $x = 0$ using a graphing utility.

Solution (a). Since $1/x \to +\infty$ as $x \to 0^+$, we can view $\sin(1/x)$ as the sine of an angle that increases indefinitely as $x \to 0^+$. As this angle increases, the function $\sin(1/x)$ keeps oscillating between -1 and 1 without approaching a limit. Similarly, there is no limit from the left since $1/x \to -\infty$ as $x \to 0^-$. These conclusions are consistent with the graph of $y = \sin(1/x)$ shown in Figure 2.6.7a. Observe that the oscillations become more and more rapid as x approaches 0 because $1/x$ increases (or decreases) more and more rapidly as x approaches 0.

Solution (b). If $x > 0$, $-x \le x \sin(1/x) \le x$, and if $x < 0$, $x \le x \sin(1/x) \le -x$. Thus, for $x \ne 0$, $-|x| \le x \sin(1/x) \le |x|$. Since both $|x| \to 0$ and $-|x| \to 0$ as $x \to 0$, the Squeezing Theorem applies and we can conclude that $x \sin(1/x) \to 0$ as $x \to 0$. This is illustrated in Figure 2.6.7b. ◀

• REMARK.   It follows from part (b) of this example that the function

$$f(x) = \begin{cases} x \sin(1/x), & x \ne 0 \\ 0, & x = 0 \end{cases}$$

is continuous at $x = 0$, since the value of the function and the value of the limit are the same at 0. This shows that the behavior of a function can be very complex in the vicinity of an x-value c, even though the function is continuous at c.

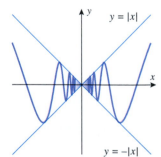

$y = \sin \left(\frac{1}{x} \right)$

(a)

$y = |x|$

$y = -|x|$

$y = x \sin \left(\frac{1}{x} \right)$

(b)

Figure 2.6.7

EXERCISE SET 2.6   ~ Graphing Utility

In Exercises 1–10, find the discontinuities, if any.

1. $f(x) = \sin(x^2 - 2)$

2. $f(x) = \cos\left(\dfrac{x}{x - \pi}\right)$

3. $f(x) = \cot x$

4. $f(x) = \sec x$

5. $f(x) = \csc x$

6. $f(x) = \dfrac{1}{1 + \sin^2 x}$

7. $f(x) = |\cos x|$

8. $f(x) = \sqrt{2 + \tan^2 x}$

9. $f(x) = \dfrac{1}{1 - 2\sin x}$

10. $f(x) = \dfrac{3}{5 + 2\cos x}$

11. Use Theorem 2.5.6 to show that the following functions are continuous everywhere by expressing them as compositions of simpler functions that are known to be continuous.
(a) $\sin(x^3 + 7x + 1)$ (b) $|\sin x|$
(c) $\cos^3(x + 1)$ (d) $\sqrt{3 + \sin 2x}$
(e) $\sin(\sin x)$ (f) $\cos^5 x - 2\cos^3 x + 1$

12. (a) Prove that if $g(x)$ is continuous everywhere, then so are $\sin(g(x))$, $\cos(g(x))$, $g(\sin(x))$, and $g(\cos(x))$.
(b) Illustrate the result in part (a) with some of your own choices for g.

Find the limits in Exercises 13–35.

13. $\lim\limits_{x \to +\infty} \cos\left(\dfrac{1}{x}\right)$

14. $\lim\limits_{x \to +\infty} \sin\left(\dfrac{2}{x}\right)$

15. $\lim\limits_{x \to +\infty} \sin\left(\dfrac{\pi x}{2 - 3x}\right)$

16. $\lim\limits_{h \to 0} \dfrac{\sin h}{2h}$

17. $\lim\limits_{\theta \to 0} \dfrac{\sin 3\theta}{\theta}$

18. $\lim\limits_{\theta \to 0^+} \dfrac{\sin \theta}{\theta^2}$

19. $\lim\limits_{x \to 0^-} \dfrac{\sin x}{|x|}$

20. $\lim\limits_{x \to 0} \dfrac{\sin^2 x}{3x^2}$

21. $\lim\limits_{x \to 0^+} \dfrac{\sin x}{5\sqrt{x}}$

22. $\lim\limits_{x \to 0} \dfrac{\sin 6x}{\sin 8x}$

23. $\lim\limits_{x \to 0} \dfrac{\tan 7x}{\sin 3x}$

24. $\lim\limits_{\theta \to 0} \dfrac{\sin^2 \theta}{\theta}$

25. $\lim\limits_{h \to 0} \dfrac{h}{\tan h}$

26. $\lim\limits_{h \to 0} \dfrac{\sin h}{1 - \cos h}$

27. $\lim\limits_{\theta \to 0} \dfrac{\theta^2}{1 - \cos \theta}$

28. $\lim\limits_{x \to 0} \dfrac{x}{\cos\left(\frac{1}{2}\pi - x\right)}$

29. $\lim\limits_{\theta \to 0} \dfrac{\theta}{\cos \theta}$

30. $\lim\limits_{t \to 0} \dfrac{t^2}{1 - \cos^2 t}$

31. $\lim\limits_{h \to 0} \dfrac{1 - \cos 5h}{\cos 7h - 1}$

32. $\lim\limits_{x \to 0^+} \sin\left(\dfrac{1}{x}\right)$

33. $\lim\limits_{x \to 0^+} \cos\left(\dfrac{1}{x}\right)$

34. $\lim\limits_{x \to 0} \dfrac{x^2 - 3\sin x}{x}$

35. $\lim\limits_{x \to 0} \dfrac{2x + \sin x}{x}$

In Exercises 36–39: (i) Construct a table to estimate the limit by evaluating the function near the limiting value. (ii) Find the exact value of the limit.

36. $\lim\limits_{x \to 5} \dfrac{\sin(x - 5)}{x^2 - 25}$

37. $\lim\limits_{x \to 2} \dfrac{\sin(2x - 4)}{x^2 - 4}$

38. $\lim\limits_{x \to -2} \dfrac{\sin(x^2 + 3x + 2)}{x + 2}$

39. $\lim\limits_{x \to -1} \dfrac{\sin(x^2 + 3x + 2)}{x^3 + 1}$

40. Find a value for the constant k that makes

$$f(x) = \begin{cases} \dfrac{\sin 3x}{x}, & x \neq 0 \\ k, & x = 0 \end{cases}$$

continuous at $x = 0$.

41. Find a nonzero value for the constant k that makes

$$f(x) = \begin{cases} \dfrac{\tan kx}{x}, & x < 0 \\ 3x + 2k^2, & x \geq 0 \end{cases}$$

continuous at $x = 0$.

42. Is

$$f(x) = \begin{cases} \dfrac{\sin x}{|x|}, & x \neq 0 \\ 1, & x = 0 \end{cases}$$

continuous at $x = 0$?

43. In each part, find the limit by making the indicated substitution.
(a) $\lim\limits_{x \to +\infty} x \sin \dfrac{1}{x}$; $t = \dfrac{1}{x}$
(b) $\lim\limits_{x \to -\infty} x \left(1 - \cos \dfrac{1}{x}\right)$; $t = \dfrac{1}{x}$
(c) $\lim\limits_{x \to \pi} \dfrac{\pi - x}{\sin x}$; $t = \pi - x$

44. Find $\lim\limits_{x \to 2} \dfrac{\cos(\pi/x)}{x - 2}$. $\left[\textit{Hint: Let } t = \dfrac{\pi}{2} - \dfrac{\pi}{x}.\right]$

45. Find $\lim\limits_{x \to 1} \dfrac{\sin(\pi x)}{x - 1}$.

46. Find $\lim\limits_{x \to \pi/4} \dfrac{\tan x - 1}{x - \pi/4}$.

~ **47.** Use the Squeezing Theorem to show that

$$\lim\limits_{x \to 0} x \cos \dfrac{50\pi}{x} = 0$$

and illustrate the principle involved by using a graphing utility to graph $y = |x|$, $y = -|x|$, and $y = x\cos(50\pi/x)$ on the same screen in the window $[-1, 1] \times [-1, 1]$.

~ **48.** Use the Squeezing Theorem to show that

$$\lim\limits_{x \to 0} x^2 \sin\left(\dfrac{50\pi}{\sqrt[3]{x}}\right) = 0$$

and illustrate the principle involved by using a graphing utility to graph $y = x^2$, $y = -x^2$, and $y = x^2 \sin(50\pi/\sqrt[3]{x})$ on the same screen in the window $[-0.5, 0.5] \times [-0.25, 0.25]$.

49. Sketch the graphs of $y = 1 - x^2$, $y = \cos x$, and $y = f(x)$, where f is a function that satisfies the inequalities

$$1 - x^2 \le f(x) \le \cos x$$

for all x in the interval $(-\pi/2, \pi/2)$. What can you say about the limit of $f(x)$ as $x \to 0$? Explain your reasoning.

50. Sketch the graphs of $y = 1/x$, $y = -1/x$, and $y = f(x)$, where f is a function that satisfies the inequalities

$$-\frac{1}{x} \le f(x) \le \frac{1}{x}$$

for all x in the interval $[1, +\infty)$. What can you say about the limit of $f(x)$ as $x \to +\infty$? Explain your reasoning.

51. Find formulas for functions g and h such that $g(x) \to 0$ and $h(x) \to 0$ as $x \to +\infty$ and such that

$$g(x) \le \frac{\sin x}{x} \le h(x)$$

for positive values of x. What can you say about the limit

$$\lim_{x \to +\infty} \frac{\sin x}{x}?$$

Explain your reasoning.

52. Draw pictures analogous to Figure 2.6.3 that illustrate the Squeezing Theorem for limits of the forms $\lim_{x \to +\infty} f(x)$ and $\lim_{x \to -\infty} f(x)$.

Recall that unless stated otherwise the variable x in trigonometric functions such as $\sin x$ and $\cos x$ is assumed to be in radian measure. The limits in Theorem 2.6.3 are based on that assumption. Exercises 53 and 54 explore what happens to those limits if degree measure is used for x.

---

53. (a) Show that if x is in degrees, then

$$\lim_{x \to 0} \frac{\sin x}{x} = \frac{\pi}{180}$$

(b) Confirm that the limit in part (a) is consistent with the results produced by your calculating utility by setting the utility to degree measure and calculating $(\sin x)/x$ for some values of x that get closer and closer to 0.

54. What is the limit of $(1 - \cos x)/x$ as $x \to 0$ if x is in degrees?

55. It follows from part (a) of Theorem 2.6.3 that if θ is small (near zero) and measured in radians, then one should expect the approximation

$$\sin \theta \approx \theta$$

to be good.
(a) Find $\sin 10°$ using a calculating utility.
(b) Estimate $\sin 10°$ using the approximation above.

56. (a) Use the approximation of $\sin \theta$ that is given in Exercise 55 together with the identity $\cos 2\alpha = 1 - 2 \sin^2 \alpha$ with $\alpha = \theta/2$ to show that if θ is small (near zero)

and measured in radians, then one should expect the approximation

$$\cos \theta \approx 1 - \tfrac{1}{2}\theta^2$$

to be good.
(b) Find $\cos 10°$ using a calculating utility.
(c) Estimate $\cos 10°$ using the approximation above.

57. It follows from part (a) of Example 2 that if θ is small (near zero) and measured in radians, then one should expect the approximation

$$\tan \theta \approx \theta$$

to be good.
(a) Find $\tan 5°$ using a calculating utility.
(b) Find $\tan 5°$ using the approximation above.

58. Referring to the accompanying figure, suppose that the angle of elevation of the top of a building, as measured from a point L feet from its base, is found to be α degrees.
(a) Use the relationship $h = L \tan \alpha$ to calculate the height of a building for which $L = 500$ ft and $\alpha = 6°$.
(b) Show that if L is large compared to the building height h, then one should expect good results in approximating h by $h \approx \pi L\alpha/180$.
(c) Use the result in part (b) to approximate the building height h in part (a).

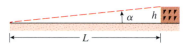

Figure Ex-58

59. (a) Use the Intermediate-Value Theorem to show that the equation $x = \cos x$ has at least one solution in the interval $[0, \pi/2]$.
(b) Show graphically that there is exactly one solution in the interval.
(c) Approximate the solution to three decimal places.

60. (a) Use the Intermediate-Value Theorem to show that the equation $x + \sin x = 1$ has at least one solution in the interval $[0, \pi/6]$.
(b) Show graphically that there is exactly one solution in the interval.
(c) Approximate the solution to three decimal places.

61. In the study of falling objects near the surface of the Earth, the **_acceleration g due to gravity_** is commonly taken to be 9.8 m/s^2 or 32 ft/s^2. However, the elliptical shape of the Earth and other factors cause variations in this constant that are latitude dependent. The following formula, known as the Geodetic Reference Formula of 1967, is commonly used to predict the value of g at a latitude of ϕ degrees (either north or south of the equator):

$$g = 9.7803185(1.0 + 0.005278895 \sin^2 \phi \\ - 0.000023462 \sin^4 \phi) \text{ m/s}^2$$

(a) Observe that g is an even function of ϕ. What does this suggest about the shape of the Earth, as modeled by the Geodetic Reference Formula?

(b) Show that $g = 9.8$ m/s^2 somewhere between latitudes of 38° and 39°.

62. Let

$$f(x) = \begin{cases} 1 & \text{if } x \text{ is a rational number} \\ 0 & \text{if } x \text{ is an irrational number} \end{cases}$$

(a) Make a conjecture about the limit of $f(x)$ as $x \to 0$.

(b) Make a conjecture about the limit of $xf(x)$ as $x \to 0$.

(c) Prove your conjectures.

SUPPLEMENTARY EXERCISES

 Graphing Utility [C] CAS

1. For the function f graphed in the accompanying figure, find the limit if it exists.

(a) $\lim_{x \to 1} f(x)$ (b) $\lim_{x \to 2} f(x)$ (c) $\lim_{x \to 3} f(x)$

(d) $\lim_{x \to 4} f(x)$ (e) $\lim_{x \to +\infty} f(x)$ (f) $\lim_{x \to -\infty} f(x)$

(g) $\lim_{x \to 3^+} f(x)$ (h) $\lim_{x \to 3^-} f(x)$ (i) $\lim_{x \to 0} f(x)$

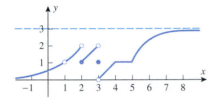

Figure Ex-1

2. (a) Find a formula for a rational function that has a vertical asymptote at $x = 1$ and a horizontal asymptote at $y = 2$.

(b) Check your work by using a graphing utility to graph the function.

3. (a) Write a paragraph or two that describes how the limit of a function can fail to exist at $x = a$. Accompany your description with some specific examples.

(b) Write a paragraph or two that describes how the limit of a function can fail to exist as $x \to +\infty$ or $x \to -\infty$. Also, accompany your description with some specific examples.

(c) Write a paragraph or two that describes how a function can fail to be continuous at $x = a$. Accompany your description with some specific examples.

4. Show that the conclusion of the Intermediate-Value Theorem may be false if f is not continuous on the interval $[a, b]$.

5. In each part, evaluate the function for the stated values of x, and make a conjecture about the value of the limit. Confirm your conjecture by finding the limit algebraically.

(a) $f(x) = \dfrac{x-2}{x^2-4}$; $\lim_{x \to 2^+} f(x)$; $x = 2.5, 2.1, 2.01,$
  $2.001, 2.0001, 2.00001$

(b) $f(x) = \dfrac{\tan 4x}{x}$; $\lim_{x \to 0} f(x)$; $x = \pm 1.0, \pm 0.1, \pm 0.01,$
  $\pm 0.001, \pm 0.0001, \pm 0.00001$

6. In each part, find the horizontal asymptotes, if any.

(a) $y = \dfrac{2x-7}{x^2-4x}$ (b) $y = \dfrac{x^3 - x^2 + 10}{3x^2 - 4x}$

(c) $y = \dfrac{2x^2 - 6}{x^2 + 5x}$

7. (a) Approximate the value for the limit

$$\lim_{x \to 0} \frac{3^x - 2^x}{x}$$

to three decimal places by constructing an appropriate table of values.

(b) Confirm your approximation using graphical evidence.

8. According to Ohm's law, when a voltage of V volts is applied across a resistor with a resistance of R ohms, a current of $I = V/R$ amperes flows through the resistor.

(a) How much current flows if a voltage of 3.0 volts is applied across a resistance of 7.5 ohms?

(b) If the resistance varies by ± 0.1 ohm, and the voltage remains constant at 3.0 volts, what is the resulting range of values for the current?

(c) If temperature variations cause the resistance to vary by $\pm \delta$ from its value of 7.5 ohms, and the voltage remains constant at 3.0 volts, what is the resulting range of values for the current?

(d) If the current is not allowed to vary by more than $\epsilon = \pm 0.001$ ampere at a voltage of 3.0 volts, what variation of $\pm \delta$ from the value of 7.5 ohms is allowable?

(e) Certain alloys become **superconductors** as their temperature approaches absolute zero ($-273°$C), meaning that their resistance approaches zero. If the voltage remains constant, what happens to the current in a superconductor as $R \to 0^+$?

9. Suppose that f is continuous on the interval $[0, 1]$ and that $0 \leq f(x) \leq 1$ for all x in this interval.

(a) Sketch the graph of $y = x$ together with a possible graph for f over the interval $[0, 1]$.

(b) Use the Intermediate-Value Theorem to help prove that there is at least one number c in the interval $[0, 1]$ such that $f(c) = c$.

10. Use algebraic methods to find

(a) $\lim_{\theta \to 0} \tan\left(\dfrac{1 - \cos\theta}{\theta}\right)$ (b) $\lim_{t \to 1} \dfrac{t-1}{\sqrt{t}-1}$

(c) $\displaystyle\lim_{x \to +\infty} \frac{(2x-1)^5}{(3x^2+2x-7)(x^3-9x)}$

(d) $\displaystyle\lim_{\theta \to 0} \cos\left(\frac{\sin(\theta+\pi)}{2\theta}\right).$

11. Suppose that f is continuous on the interval $[0, 1]$, that $f(0) = 2$, and that f has no zeros in the interval. Prove that $f(x) > 0$ for all x in $[0, 1]$.

12. Suppose that

$$f(x) = \begin{cases} -x^4 + 3, & x \leq 2 \\ x^2 + 9, & x > 2 \end{cases}$$

Is f continuous everywhere? Justify your conclusion.

13. Show that the equation $x^4 + 5x^3 + 5x - 1 = 0$ has at least two real solutions in the interval $[-6, 2]$.

14. Use the Intermediate-Value Theorem to approximate $\sqrt{11}$ to three decimal places, and check your answer by finding the root directly with a calculating utility.

15. Suppose that f is continuous at x_0 and that $f(x_0) > 0$. Give either an ϵ-δ proof or a convincing verbal argument to show that there must be an open interval containing x_0 on which $f(x) > 0$.

16. Sketch the graph of $f(x) = |x^2 - 4|/(x^2 - 4)$.

17. In each part, approximate the discontinuities of f to three decimal places.

(a) $f(x) = \dfrac{x+1}{x^2+2x-5}$

(b) $f(x) = \dfrac{x+3}{|2\sin x - x|}$

18. In Example 3 of Section 2.6 we used the Squeezing Theorem to prove that

$$\lim_{x \to 0} x \sin\left(\frac{1}{x}\right) = 0$$

Why couldn't we have obtained the same result by writing

$$\lim_{x \to 0} x \sin\left(\frac{1}{x}\right) = \lim_{x \to 0} x \cdot \lim_{x \to 0} \sin\left(\frac{1}{x}\right)$$

$$= 0 \cdot \lim_{x \to 0} \sin\left(\frac{1}{x}\right) = 0?$$

In Exercises 19 and 20, find $\displaystyle\lim_{x \to a} f(x)$, if it exists, where a is replaced by 0, 5^+, -5^-, -5, 5, $-\infty$, $+\infty$.

19. (a) $f(x) = \sqrt{5-x}$ (b) $f(x) = (x^2-25)/(x-5)$

20. (a) $f(x) = (x+5)/(x^2-25)$

(b) $f(x) = \begin{cases} (x-5)/|x-5|, & x \neq 5 \\ 0, & x = 5 \end{cases}$

In Exercises 21–28, find the indicated limit, if it exists.

21. $\displaystyle\lim_{x \to 0} \frac{\tan ax}{\sin bx}$ $(a \neq 0, b \neq 0)$

22. $\displaystyle\lim_{x \to 0} \frac{\sin 3x}{\tan 3x}$

23. $\displaystyle\lim_{\theta \to 0} \frac{\sin 2\theta}{\theta^2}$

24. $\displaystyle\lim_{x \to 0} \frac{x \sin x}{1 - \cos x}$

25. $\displaystyle\lim_{x \to 0^+} \frac{\sin x}{\sqrt{x}}$

26. $\displaystyle\lim_{x \to 0} \frac{\sin^2(kx)}{x^2}$, $k \neq 0$

27. $\displaystyle\lim_{x \to 0} \frac{3x - \sin(kx)}{x}$, $k \neq 0$

28. $\displaystyle\lim_{x \to +\infty} \frac{2x + x \sin 3x}{5x^2 - 2x + 1}$

29. One dictionary describes a continuous function as "one whose value at each point is closely approached by its values at neighboring points."

(a) How would you explain the meaning of the terms "neighboring points" and "closely approached" to a nonmathematician?

(b) Write a paragraph that explains why the dictionary definition is consistent with Definition 2.5.1.

30. (a) Show by rationalizing the numerator that

$$\lim_{x \to 0} \frac{\sqrt{x^2+4}-2}{x^2} = \frac{1}{4}$$

(b) Evaluate $f(x)$ for

$$x = \pm 1.0, \pm 0.1, \pm 0.01, \pm 0.001, \pm 0.0001, \pm 0.00001$$

and explain why the values are not getting closer and closer to the limit.

(c) The accompanying figure shows the graph of f generated with a graphing utility and zooming in on the origin. Explain what is happening.

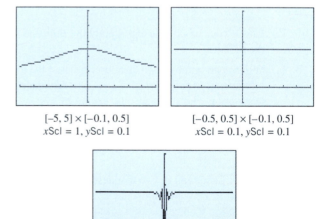

$[-5, 5] \times [-0.1, 0.5]$
$x\text{Scl} = 1, y\text{Scl} = 0.1$

$[-0.5, 0.5] \times [-0.1, 0.5]$
$x\text{Scl} = 0.1, y\text{Scl} = 0.1$

$[-5 \times 10^{-6}, 5 \times 10^{-6}] \times [-0.1, 0.5]$
$x\text{Scl} = 10^{-6}, y\text{Scl} = 0.1$

Figure Ex-30

In Exercises 31–36, approximate the limit of the function by looking at its graph and calculating values for some appropriate choices of x. Compare your answer with the value produced by a CAS.

31. $\displaystyle\lim_{x \to 0} (1+x)^{1/x}$

32. $\displaystyle\lim_{x \to 3} \frac{2^x - 8}{x - 3}$

c 33. $\lim_{x \to 1} \dfrac{\sin x - \sin 1}{x - 1}$ **c 34.** $\lim_{x \to 0^+} x^{-2}(1.001)^{-1/x}$

c 35. $\lim_{x \to +\infty} \left(\sqrt{x + \sqrt{x}} - \sqrt{x} \right)$

c 36. $\lim_{x \to +\infty} (3^x + 5^x)^{1/x}$

37. The limit

$$\lim_{x \to 0} \frac{\sin x}{x} = 1$$

ensures that there is a number δ such that

$$\left| \frac{\sin x}{x} - 1 \right| < 0.001$$

if $0 < |x| < \delta$. Estimate the largest such δ.

38. If \$1000 is invested in an account that pays 7% interest compounded n times each year, then in 10 years there will be $1000(1 + 0.07/n)^{10n}$ dollars in the account. How much money will be in the account in 10 years if the interest is compounded quarterly ($n = 4$)? Monthly ($n = 12$)? Daily ($n = 365$)? Estimate the amount of money that will be in the account in 10 years if the interest is compounded *continuously*, that is, as $n \to +\infty$?

39. There are various numerical methods other than the method discussed in Section 2.5 to obtain approximate solutions of equations of the form $f(x) = 0$. One such method requires that the equation be expressed in the form $x = g(x)$, so that a solution $x = c$ can be interpreted as the value of x where the line $y = x$ intersects the curve $y = g(x)$, as shown in the accompanying figure. If x_1 is an initial estimate of c and the graph of $y = g(x)$ is not too steep in the vicinity of c, then a better approximation can be obtained from $x_2 = g(x_1)$ (see the figure). An even better approximation is obtained from $x_3 = g(x_2)$, and so forth. The formula $x_{n+1} = g(x_n)$ for $n = 1, 2, 3, \ldots$ generates successive approximations $x_2, x_3, x_4, \ldots$ that get closer and closer to c.

(a) The equation $x^3 - x - 1 = 0$ has only one real solution. Show that this equation can be written as

$$x = g(x) = \sqrt[3]{x + 1}$$

(b) Graph $y = x$ and $y = g(x)$ in the same coordinate system for $-1 \le x \le 3$.

(c) Starting with an arbitrary estimate x_1, make a sketch that shows the location of the successive iterates

$$x_2 = g(x_1), \quad x_3 = g(x_2), \ldots$$

(d) Use $x_1 = 1$ and calculate $x_2, x_3, \ldots$, continuing until you obtain two consecutive values that differ by less than 10^{-4}. Experiment with other starting values such as $x_1 = 2$ or $x_1 = 1.5$.

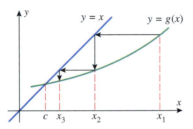
Figure Ex-39

40. The method described in Exercise 39 will not always work.

(a) The equation $x^3 - x - 1 = 0$ can be expressed as $x = g(x) = x^3 - 1$. Graph $y = x$ and $y = g(x)$ in the same coordinate system. Starting with an arbitrary estimate x_1, make a sketch illustrating the locations of the successive iterates $x_2 = g(x_1)$, $x_3 = g(x_2)$, $\ldots$.

(b) Use $x_1 = 1$ and calculate the successive iterates x_n for $n = 2, 3, 4, 5, 6$.

In Exercises 41 and 42, use the method of Exercise 39 to approximate the roots of the equation.

41. $x^5 - x - 2 = 0$ **42.** $x - \cos x = 0$

3

Sir Isaac Newton

THE DERIVATIVE

*M*any real-world phenomena involve chang-ing quantities—the speed of a rocket, the inflation of cur-rency, the number of bacteria in a culture, the shock inten-sity of an earthquake, the voltage of an electrical signal, and so forth. In this chapter we will develop the concept of a *derivative*, which is the mathematical tool that is used to study rates at which quantities change. In Section 3.1 we will interpret both average and instantaneous velocity geometrically, and we will define the slope of a curve at a point. In Sections 3.2 to 3.6 we will provide a precise defi-nition of the derivative and we will develop mathematical tools for calculating derivatives efficiently. In Section 3.7 we will show how these methods of differentiation can be applied to problems involving rates of change.

One of the important themes of calculus is that many nonlinear functions can be closely approximated by linear functions. In Section 3.8 we will show how derivatives can be used to generate such approximations.

3.1 SLOPES AND RATES OF CHANGE

In this section we will explore the connection between velocity at an instant, the slope of a curve at a point, and rate of change. Our work here is intended to be informal and introductory, and all of the ideas that we develop will be revisited in more detail in later sections.

VELOCITY AND SLOPES

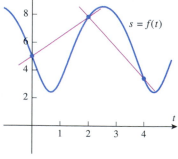

In Section 2.1 we interpreted the *instantaneous velocity* of a particle moving along an s-axis as a limit of average velocities. We begin our introduction to the derivative with another visit to the topic of velocity.

For purposes of illustration, consider a bell ringer practicing for her part in a change-ringing group at an English bell tower. The ringer controls a rope, pulling periodically to ring the bell. We will concentrate on the position of the *sally* (the handgrip on the rope), measured in feet above the floor of the ringing room. Imagine the s-axis as the line of travel of the sally. Figure 3.1.1a shows a sequence of "snapshots" of one such scenario, taken at times $t = 0, 1, 2, 3,$ and 4 s.

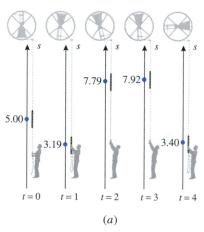

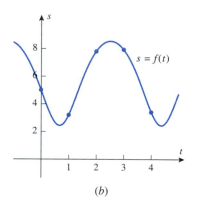

Figure 3.1.1

We may be able to record the height of the sally at various times, as in Table 3.1.1, or even model the motion of the sally by a function, as depicted in the graph in Figure 3.1.1b. The velocity of the sally measures the rate of ascent of the sally in its motion during the ringing of the bell. For example, during the first 2 s ($t = 0$ to $t = 2$), the displacement of the sally is $f(2) - f(0) = 7.79 - 5.00 = 2.79$ ft, so the average velocity of the sally during these 2 s is

$$v_{\text{ave}} = \frac{7.79 - 5.00}{2 - 0} \approx 1.39 \text{ ft/s}$$

The average velocity during the next 2 s ($t = 2$ to $t = 4$) is

$$v_{\text{ave}} = \frac{3.40 - 7.79}{4 - 2} \approx -2.19 \text{ ft/s}$$

Note that the displacement of the sally is negative during this latter time interval, since its position at time $t = 4$ is below that at time $t = 2$. Thus, the average velocity is also negative.

Table 3.1.1

t (seconds)	0.0	0.5	1.0	1.5	2.0	2.5	3.0	3.5	4.0
$s = f(t)$ (ft)	5.00	2.66	3.19	5.78	7.79	8.52	7.92	6.02	3.40

We can see from the graph of $s = f(t)$ (Figure 3.1.2) that these average velocities are equal to the slopes of the lines through the points (0, 5.00) and (2, 7.79), and through

Figure 3.1.2

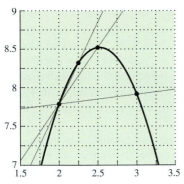

Figure 3.1.3

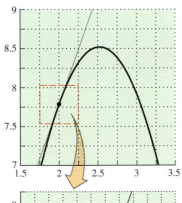

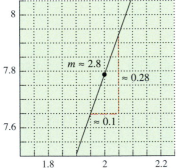

Figure 3.1.4

SLOPE OF A CURVE

(2, 7.79) and (4, 3.40). Thus, average velocity can be interpreted as a geometric property of the graph of the position function.

3.1.1 GEOMETRIC INTERPRETATION OF AVERAGE VELOCITY. *If an object moves along an s-axis, and if the position versus time curve is* $s = f(t)$, *then the average velocity of the object between times* t_0 *and* t_1,

$$v_{\text{ave}} = \frac{f(t_1) - f(t_0)}{t_1 - t_0}$$

is represented geometrically by the slope of the line joining the points $(t_0, f(t_0))$ *and* $(t_1, f(t_1))$.

Now, from the graph of $s = f(t)$ in Figure 3.1.1*b* we can see that the sally is rising more quickly during the time interval $1.5 \leq t \leq 2$ than during the interval $2 \leq t \leq 2.5$. This is numerically revealed using the data in Table 3.1.1 to obtain average velocities of 4.02 ft/s and 1.46 ft/s, respectively, for these two time intervals. But what of the velocity, v_{inst}, of the sally at the instant our clock strikes $t = 2$ s? How should v_{inst} be defined? Does it have a geometric interpretation as well? We argued in Section 2.1 that the "instantaneous velocity" at a particular moment in time should be the limiting value of average velocities. This suggests that we define the instantaneous velocity of the sally at time $t = 2$ by

$$v_{\text{inst}} = \lim_{t_1 \to 2} \frac{f(t_1) - f(2)}{t_1 - 2}$$

It follows that we can estimate v_{inst} at $t = 2$ by calculating average velocities over ever smaller intervals anchored at 2. That is, we would expect that the fractions

$$\frac{f(2.2) - f(2)}{2.2 - 2}, \qquad \frac{f(2.1) - f(2)}{2.1 - 2}, \qquad \frac{f(2.01) - f(2)}{2.01 - 2}$$

would, in turn, each yield a better estimate for v_{inst}. Since Table 3.1.1 is lacking for such refined data, consider the portion of the graph of $s = f(t)$ near $t = 2$ shown in Figure 3.1.3.

The ratios that produce average velocities on an interval $2 \leq t \leq t_1$ are slopes of lines through the points $(2, f(2))$ and $(t_1, f(t_1))$. Figure 3.1.3 shows such lines for $t_1 = 3, 2.5,$ and 2.25. We can infer the limiting value of these slopes as t_1 approaches 2 by magnifying a portion of the graph of f near the point $(2, f(2))$. This is illustrated in Figure 3.1.4, from which it appears that the limiting value is about 2.8. Thus, subject to our crude measuring devices, the instantaneous velocity at time $t = 2$ is given by $v_{\text{inst}} \approx 2.8$ ft/s.

The preceding discussion of average and instantaneous velocities could be cast as an investigation of slopes related to the position curve. The slope of a general function curve at a point can be translated into useful information in many applications, so a consideration of the notion of the *slope of a curve* is warranted.

Consider the function $y = f(x)$ whose graph is shown in Figure 3.1.5. We focus on the point $P(x_0, f(x_0))$. One has an intuitive notion that the "steepness" of the curve varies at different points. For example, view the graph of $y = f(x)$ in Figure 3.1.5 as the cross

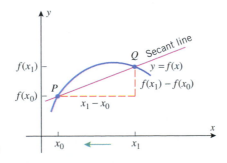

Figure 3.1.5

section of a hill and imagine a hiker walking the hill from left to right. The hiker will find the trek fairly arduous at point P, but the climb gets easier as she approaches the summit. Rather than rely on comparative notions of "less steep" or "more steep," we seek a numeric value to attach to each point on the curve that will describe "how steep" the curve is at that point. For straight lines, steepness is the same at every point, and the measure used to describe steepness is the slope of the line. (Note that slope not only describes "how steep" a line is, but also whether the line rises or falls.) Our goal is to define slope for our curve $y = f(x)$, even though $f(x)$ is not linear.

Since we know how to calculate the slope of a line through two points, let us consider a line joining point P with another point $Q(x_1, f(x_1))$ on the curve. By analogy with secants to circles, a line determined by two points on a curve is called a **secant line** to the curve. The slope of the secant line PQ is given by

$$m_{\text{sec}} = \frac{f(x_1) - f(x_0)}{x_1 - x_0} \tag{1}$$

As the sampling point $Q(x_1, f(x_1))$ is chosen closer to P, that is, as x_1 is selected closer to x_0, the slopes m_{sec} more nearly approximate what we might reasonably call the "slope" of the curve $y = f(x)$ at the point P. Thus, from (1), the slope of the curve $y = f(x)$ at $P(x_0, f(x_0))$ should be defined by

$$m_{\text{curve}} = \lim_{x_1 \to x_0} \frac{f(x_1) - f(x_0)}{x_1 - x_0} \tag{2}$$

Example 1 Consider the function $f(x) = 6x - x^2$ and the point $P(2, f(2)) = (2, 8)$.

(a) Find the slopes of secant lines to the graph of $y = f(x)$ determined by P and points on the graph at $x = 3$ and $x = 1.5$.

(b) Find the slope of the graph of $y = f(x)$ at the point P.

Solution (a). The secant line to the graph of f through P and $Q(3, f(3)) = (3, 9)$ has slope

$$m_{\text{sec}} = \frac{9 - 8}{3 - 2} = 1$$

The secant line to the graph of f through P and $Q(1.5, f(1.5)) = (1.5, 6.75)$ has slope

$$m_{\text{sec}} = \frac{6.75 - 8}{1.5 - 2} = 2.5$$

Solution (b). The slope of the graph of f at the point P is

$$m_{\text{curve}} = \lim_{x_1 \to 2} \frac{f(x_1) - f(2)}{x_1 - 2} = \lim_{x_1 \to 2} \frac{6x_1 - x_1^2 - 8}{x_1 - 2}$$

$$= \lim_{x_1 \to 2} \frac{(4 - x_1)(x_1 - 2)}{x_1 - 2} = \lim_{x_1 \to 2} (4 - x_1) = 4 - 2 = 2 \quad \blacktriangleleft$$

Recall our discussion of instantaneous velocity as a limit of average velocities, in which average velocities corresponded to slopes of secant lines on the position curve. We now have an interpretation of such a limit of slopes of secant lines as the slope of the position curve at the instant in question. This provides a geometric interpretation of instantaneous velocity as the slope of the graph of the position curve.

3.1.2 GEOMETRIC INTERPRETATION OF INSTANTANEOUS VELOCITY. *If a particle moves along an s-axis, and if the position versus time curve is $s = f(t)$, then the instantaneous velocity of the particle at time t_0,*

$$v_{\text{inst}} = \lim_{t_1 \to t_0} \frac{f(t_1) - f(t_0)}{t_1 - t_0}$$

is represented geometrically by the slope of the curve at the point $(t_0, f(t_0))$.

Velocity or slope can be viewed as *rate of change*—the rate of change of position with respect to time, or the rate of change of a function's value with respect to its input. Rates of change occur in many applications. For example:

- A microbiologist might be interested in the rate at which the number of bacteria in a colony changes with time.

- An engineer might be interested in the rate at which the length of a metal rod changes with temperature.

- An economist might be interested in the rate at which production cost changes with the quantity of a product that is manufactured.

- A medical researcher might be interested in the rate at which the radius of an artery changes with the concentration of alcohol in the bloodstream.

In general, if x and y are quantities related by an equation $y = f(x)$, we can consider the rate at which y changes with x. As with velocity, we distinguish between an average rate of change, represented by the slope of a secant line to the graph of $y = f(x)$, and an instantaneous rate of change, represented by the slope of the curve at a point.

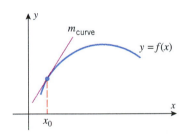

m_{sec} is the average rate of change of y with respect to x over the interval $[x_0, x_1]$.

(*a*)

3.1.3 DEFINITION.   If $y = f(x)$, then the ***average rate of change of y with respect to x over the interval*** $[x_0, x_1]$ is

$$r_{\text{ave}} = \frac{f(x_1) - f(x_0)}{x_1 - x_0} \qquad (3)$$

Geometrically, the average rate of change of y with respect to x over the interval $[x_0, x_1]$ is the slope of the secant line to the graph of $y = f(x)$ through the points $(x_0, f(x_0))$ and $(x_1, f(x_1))$:

$$r_{\text{ave}} = m_{\text{sec}}$$

(see Figure 3.1.6*a*).

m_{curve} is the instantaneous rate of change of y with respect to x when $x = x_0$.

(*b*)

Figure 3.1.6

3.1.4 DEFINITION.   If $y = f(x)$, then the ***instantaneous rate of change of y with respect to x when x = x_0*** is

$$r_{\text{inst}} = \lim_{x_1 \to x_0} \frac{f(x_1) - f(x_0)}{x_1 - x_0} \qquad (4)$$

Geometrically, the instantaneous rate of change of y with respect to x when $x = x_0$ is the slope of the graph of $y = f(x)$ at the point $(x_0, f(x_0))$:

$$r_{\text{inst}} = m_{\text{curve}}$$

(see Figure 3.1.6*b*).

Example 2   Let $y = x^2 + 1$.

(a)   Find the average rate of change of y with respect to x over the interval $[3, 5]$.

(b)   Find the instantaneous rate of change of y with respect to x when $x = -4$.

(c)   Find the instantaneous rate of change of y with respect to x at the general point corresponding to $x = x_0$.

Solution (a).   We apply Formula (3) with $f(x) = x^2 + 1$, $x_0 = 3$, and $x_1 = 5$. This yields

$$r_{\text{ave}} = \frac{f(x_1) - f(x_0)}{x_1 - x_0} = \frac{f(5) - f(3)}{5 - 3} = \frac{26 - 10}{2} = 8$$

Thus, on the average, y increases 8 units per unit increase in x over the interval $[3, 5]$.

Solution (b). We apply Formula (4) with $f(x) = x^2 + 1$ and $x_0 = -4$. This yields

$$r_{inst} = \lim_{x_1 \to x_0} \frac{f(x_1) - f(x_0)}{x_1 - x_0} = \lim_{x_1 \to -4} \frac{f(x_1) - f(-4)}{x_1 - (-4)} = \lim_{x_1 \to -4} \frac{(x_1^2 + 1) - 17}{x_1 + 4}$$

$$= \lim_{x_1 \to -4} \frac{x_1^2 - 16}{x_1 + 4} = \lim_{x_1 \to -4} \frac{(x_1 + 4)(x_1 - 4)}{x_1 + 4} = \lim_{x_1 \to -4} (x_1 - 4) = -8$$

Thus, for a small change in x from $x = -4$, the value of y will change approximately eight times as much *in the opposite direction*. That is, because the instantaneous rate of change is negative, the value of y *decreases* as values of x move through $x = -4$ from left to right.

Solution (c). We proceed as in part (b):

$$r_{inst} = \lim_{x_1 \to x_0} \frac{f(x_1) - f(x_0)}{x_1 - x_0} = \lim_{x_1 \to x_0} \frac{(x_1^2 + 1) - (x_0^2 + 1)}{x_1 - x_0} = \lim_{x_1 \to x_0} \frac{x_1^2 - x_0^2}{x_1 - x_0}$$

$$= \lim_{x_1 \to x_0} \frac{(x_1 + x_0)(x_1 - x_0)}{x_1 - x_0} = \lim_{x_1 \to x_0} (x_1 + x_0) = 2x_0$$

Thus, the instantaneous rate of change of y with respect to x at $x = x_0$ is $2x_0$. Observe that the result in part (b) can be obtained from this more general result by setting $x_0 = -4$. ◀

RATES OF CHANGE IN APPLICATIONS

In applied problems, average and instantaneous rates of change must be accompanied by appropriate units. In general, the units for a rate of change of y with respect to x are obtained by "dividing" the units of y by the units of x and then simplifying according to the standard rules of algebra. Here are some examples:

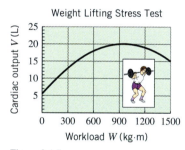

- If y is in degrees Fahrenheit (°F) and x is in inches (in), then a rate of change of y with respect to x has units of degrees Fahrenheit per inch (°F/in).

- If y is in feet per second (ft/s) and x is in seconds (s), then a rate of change of y with respect to x has units of feet per second per second (ft/s/s), which would usually be written as ft/s^2.

- If y is in newton-meters (N·m) and x is in meters (m), then a rate of change of y with respect to x has units of newtons (N), since N·m/m = N.

- If y is in foot-pounds (ft·lb) and x is in hours (h), then a rate of change of y with respect to x has units of foot-pounds per hour (ft·lb/h).

Figure 3.1.7

Example 3 The limiting factor in athletic endurance is cardiac output, that is, the volume of blood that the heart can pump per unit of time during an athletic competition. Figure 3.1.7 shows a stress-test graph of cardiac output V in liters (L) of blood versus workload W in kilogram-meters (kg·m) for 1 minute of weight lifting. This graph illustrates the known medical fact that cardiac output increases with the workload, but after reaching a peak value begins to decrease.

(a) Use the secant line shown in Figure 3.1.8a to estimate the average rate of change of cardiac output with respect to workload as the workload increases from 300 to 1200 kg·m.

(b) Use the line segment shown in Figure 3.1.8b to estimate the instantaneous rate of change of cardiac output with respect to workload at the point where the workload is 300 kg·m.

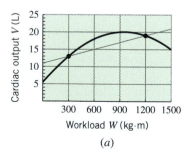

Solution (a). Using the estimated points (300, 13) and (1200, 19), the slope of the secant line indicated in Figure 3.1.8a is

$$m_{sec} \approx \frac{19 - 13}{1200 - 300} \approx 0.0067 \; \frac{\text{L}}{\text{kg·m}}$$

Since $r_{ave} = m_{sec}$, the average rate of change of cardiac output with respect to workload

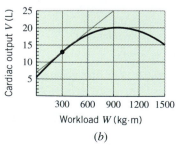

Figure 3.1.8

over the interval is approximately 0.0067 L/kg·m. This means that on the average a 1-unit increase in workload produced a 0.0067-L increase in cardiac output over the interval.

Solution (b). We estimate the slope of the cardiac output curve at $W = 300$ by sketching a line that appears to meet the curve at $W = 300$ with slope equal to that of the curve (Figure 3.1.8b). Estimating points $(0, 7)$ and $(900, 25)$ on this line, we obtain

$$r_{inst} \approx \frac{25 - 7}{900 - 0} = 0.02 \frac{\text{L}}{\text{kg·m}}$$

◀

EXERCISE SET 3.1

1. The accompanying figure shows the position versus time curve for an elevator that moves upward a distance of 60 m and then discharges its passengers.
   (a) Estimate the instantaneous velocity of the elevator at $t = 10$ s.
   (b) Sketch a velocity versus time curve for the motion of the elevator for $0 \leq t \leq 20$.

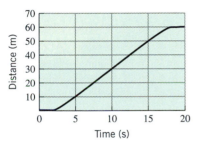

Figure Ex-1

2. The accompanying figure shows the position versus time curve for a certain particle moving along a straight line. Estimate each of the following from the graph:
   (a) the average velocity over the interval $0 \leq t \leq 3$
   (b) the values of t at which the instantaneous velocity is zero
   (c) the values of t at which the instantaneous velocity is either a maximum or a minimum
   (d) the instantaneous velocity when $t = 3$ s.

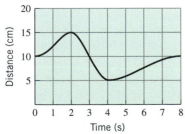

Figure Ex-2

3. The accompanying figure shows the position versus time curve for a certain particle moving on a straight line.
   (a) Is the particle moving faster at time t_0 or time t_2? Explain.
   (b) The portion of the curve near the origin is horizontal. What does this tell us about the initial velocity of the particle?
   (c) Is the particle speeding up or slowing down in the interval $[t_0, t_1]$? Explain.
   (d) Is the particle speeding up or slowing down in the interval $[t_1, t_2]$? Explain.

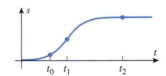

Figure Ex-3

4. An automobile, initially at rest, begins to move along a straight track. The velocity increases steadily until suddenly the driver sees a concrete barrier in the road and applies the brakes sharply at time t_0. The car decelerates rapidly, but it is too late—the car crashes into the barrier at time t_1 and instantaneously comes to rest. Sketch a position versus time curve that might represent the motion of the car.

5. If a particle moves at constant velocity, what can you say about its position versus time curve?

6. The accompanying figure shows the position versus time curves of four different particles moving on a straight line. For each particle, determine whether its instantaneous velocity is increasing or decreasing with time.

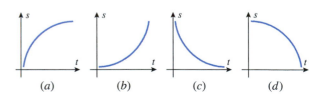

Figure Ex-6

In Exercises 7–10, a function $y = f(x)$ and values of x_0 and x_1 are given.
(a) Find the average rate of change of y with respect to x over the interval $[x_0, x_1]$.
(b) Find the instantaneous rate of change of y with respect to x at the given value of x_0.
(c) Find the instantaneous rate of change of y with respect to x at a general x-value x_0.
(d) Sketch the graph of $y = f(x)$ together with the secant line whose slope is given by the result in part (a), and indicate graphically the slope of the curve given by the result in part (b).

7. $y = \frac{1}{2}x^2$; $x_0 = 3$, $x_1 = 4$

8. $y = x^3$; $x_0 = 1$, $x_1 = 2$

9. $y = 1/x$; $x_0 = 2$, $x_1 = 3$

10. $y = 1/x^2$; $x_0 = 1$, $x_1 = 2$

In Exercises 11–14, a function $y = f(x)$ and an x-value x_0 are given.
(a) Find the slope of the graph of f at a general x-value x_0.
(b) Find the slope of the graph of f at the x-value specified by the given x_0.

11. $f(x) = x^2 + 1$; $x_0 = 2$

12. $f(x) = x^2 + 3x + 2$; $x_0 = 2$

13. $f(x) = \sqrt{x}$; $x_0 = 1$

14. $f(x) = 1/\sqrt{x}$; $x_0 = 4$

15. Suppose that the outside temperature versus time curve over a 24-hour period is as shown in the accompanying figure.
(a) Estimate the maximum temperature and the time at which it occurs.
(b) The temperature rise is fairly linear from 8 A.M. to 2 P.M. Estimate the rate at which the temperature is increasing during this time period.
(c) Estimate the time at which the temperature is decreasing most rapidly. Estimate the instantaneous rate of change of temperature with respect to time at this instant.

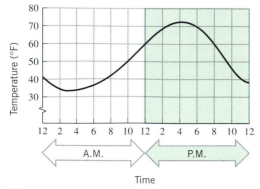

Figure Ex-15

16. The accompanying figure shows the graph of the pressure p in atmospheres (atm) versus the volume V in liters (L) of 1 mole of an ideal gas at a constant temperature of 300 K (kelvins). Use the line segments shown in the figure to estimate the rate of change of pressure with respect to volume at the points where $V = 10$ L and $V = 25$ L.

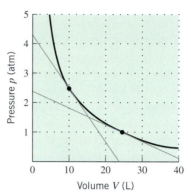

Figure Ex-16

17. The accompanying figure shows the graph of the height h in centimeters versus the age t in years of an individual from birth to age 20.
(a) When is the growth rate greatest?
(b) Estimate the growth rate at age 5.
(c) At approximately what age between 10 and 20 is the growth rate greatest? Estimate the growth rate at this age.
(d) Draw a rough graph of the growth rate versus age.

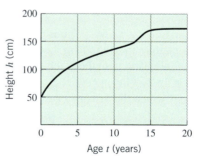

Figure Ex-17

In Exercises 18–21, use geometric interpretations 3.1.1 and 3.1.2 to find the average and instantaneous velocity.

18. A rock is dropped from a height of 576 ft and falls toward Earth in a straight line. In t seconds the rock drops a distance of $s = 16t^2$ ft.
(a) How many seconds after release does the rock hit the ground?

(b) What is the average velocity of the rock during the time it is falling?

(c) What is the average velocity of the rock for the first 3 s?

(d) What is the instantaneous velocity of the rock when it hits the ground?

19. During the first 40 s of a rocket flight, the rocket is propelled straight up so that in t seconds it reaches a height of $s = 5t^3$ ft.

(a) How high does the rocket travel in 40 s?

(b) What is the average velocity of the rocket during the first 40 s?

(c) What is the average velocity of the rocket during the first 135 ft of its flight?

(d) What is the instantaneous velocity of the rocket at the end of 40 s?

20. A particle moves on a line away from its initial position so that after t hours it is $s = 3t^2 + t$ miles from its initial position.

(a) Find the average velocity of the particle over the interval $[1, 3]$.

(b) Find the instantaneous velocity at $t = 1$.

21. A particle moves in the positive direction along a straight line so that after t minutes its distance is $s = 6t^4$ feet from the origin.

(a) Find the average velocity of the particle over the interval $[2, 4]$.

(b) Find the instantaneous velocity at $t = 2$.

3.2 THE DERIVATIVE

In this section we will introduce the concept of a "derivative," the primary mathematical tool that is used to calculate rates of change and slopes of curves.

SLOPE OF A CURVE AND TANGENT LINES

In the preceding section we argued that the slope of the graph of $y = f(x)$ at $x = x_0$ should be given by

$$m_{\text{curve}} = \lim_{x_1 \to x_0} \frac{f(x_1) - f(x_0)}{x_1 - x_0} \tag{1}$$

The ratio

$$\frac{f(x_1) - f(x_0)}{x_1 - x_0}$$

is called a **difference quotient**. As we saw in the last section, the difference quotient can also be interpreted as the average rate of change of $f(x)$ over the interval $[x_0, x_1]$, and its limit as $x_1 \to x_0$ is the instantaneous rate of change of $f(x)$ at $x = x_0$.

The geometric problem of finding the slope of a curve, and the somewhat paradoxical notions of instantaneous velocity and instantaneous rate of change, are all resolved by a limit of a difference quotient. The fact that problems in such disparate areas are unified by this expression is celebrated in the definition of the *derivative* of a function at a value in its domain.

3.2.1 DEFINITION. Suppose that x_0 is a number in the domain of a function f. If

$$\lim_{x_1 \to x_0} \frac{f(x_1) - f(x_0)}{x_1 - x_0}$$

exists, then the value of this limit is called the **derivative of f at x = x_0** and is denoted by $f'(x_0)$. That is,

$$f'(x_0) = \lim_{x_1 \to x_0} \frac{f(x_1) - f(x_0)}{x_1 - x_0} \tag{2}$$

(see Figure 3.2.1). If the limit of the difference quotient exists, $f'(x_0)$ is the **slope of the graph of f at the point P $(x_0, f(x_0))$** (or at $x = x_0$). If this limit does not exist, then the slope of the graph of f is **undefined** at P (or at $x = x_0$).

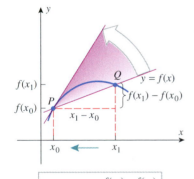

$$f'(x_0) = \lim_{x_1 \to x_0} \frac{f(x_1) - f(x_0)}{x_1 - x_0}$$

Figure 3.2.1

Now that we have defined the derivative of a function, we can begin to answer a question that fueled much of the early development of calculus. Mathematicians of the seventeenth century were perplexed by the problem of defining a *tangent line* to a general curve. Of course, in the case of a circle the definition was apparent: a line is tangent to a circle if it meets the circle at a single point. But, it was also clear that this simple definition would not suffice in many cases. For example, the y-axis intersects the parabola $y = x^2$ at a single point but does not appear to be "tangent" to the curve (Figure 3.2.2a). On the other hand, the line $y = 1$ does seem to be tangent to the graph of $y = \sin x$, even though it intersects this graph infinitely often (Figure 3.2.2b).

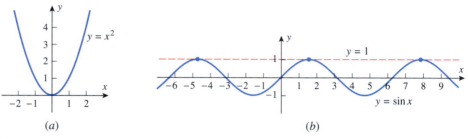

(a)   (b)

Figure 3.2.2

By the end of the first half of the seventeenth century, mathematicians such as Descartes and Fermat had developed a variety of procedures for *constructing* tangent lines. However, a general definition of a tangent line to a curve was still missing. Roughly speaking, a line should be tangent to the graph of a function $y = f(x)$ at a point $(x_0, f(x_0))$ provided the line has the same *direction* as the graph at the point. Since the direction of a line is determined by its slope, we would expect a line to be tangent to the graph at $(x_0, f(x_0))$ if the slope of the line is equal to the slope of the graph of f at x_0. Thus, we can now use the derivative to define the tangent line to a curve when the curve is the graph of a function $y = f(x)$. (Later we will extend this definition to more general curves.)

3.2.2 DEFINITION. Suppose that x_0 is a number in the domain of a function f. If

$$f'(x_0) = \lim_{x_1 \to x_0} \frac{f(x_1) - f(x_0)}{x_1 - x_0}$$

exists, then we define the ***tangent line to the graph of f at the point $P(x_0, f(x_0))$*** to be the line whose equation is

$$y - f(x_0) = f'(x_0)(x - x_0) \tag{3}$$

We also call this the ***tangent line to the graph of f at $x = x_0$***.

WARNING. Tangent lines to graphs do not have all of the properties of tangent lines to circles. For example, a tangent line to a circle intersects the circle only at the point of tangency, whereas a tangent line to a general graph may intersect the graph at points other than the point of tangency (Figure 3.2.3).

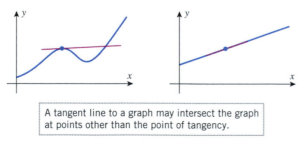

A tangent line to a graph may intersect the graph at points other than the point of tangency.

Figure 3.2.3

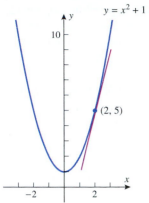

$y = x^2 + 1$

Figure 3.2.4

Example 1 Find the slope of the graph of $y = x^2 + 1$ at the point $(2, 5)$, and use it to find the equation of the tangent line to $y = x^2 + 1$ at $x = 2$ (Figure 3.2.4).

Solution. From (2), the slope of the graph of $y = x^2 + 1$ at the point $(2, 5)$ is given by

$$f'(2) = \lim_{x_1 \to 2} \frac{f(x_1) - f(2)}{x_1 - 2} = \lim_{x_1 \to 2} \frac{(x_1^2 + 1) - 5}{x_1 - 2} = \lim_{x_1 \to 2} \frac{x_1^2 - 4}{x_1 - 2}$$

$$= \lim_{x_1 \to 2} \frac{(x_1 - 2)(x_1 + 2)}{x_1 - 2} = \lim_{x_1 \to 2} (x_1 + 2) = 4$$

The tangent line is the line through the point $(2, 5)$ with slope 4,

$$y - 5 = 4(x - 2)$$

which we may also write in slope-intercept form as $y = 4x - 3$. ◀

SLOPE OF A CURVE BY ZOOMING

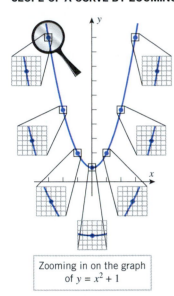

Zooming in on the graph of $y = x^2 + 1$

Figure 3.2.5

The slope of a curve at a point can be estimated by zooming on a graphing utility. The idea is to zoom in on the point until the surrounding portion of the curve appears to be a straight line (Figure 3.2.5). The utility's trace operation can then be used to estimate the slope. Figure 3.2.6 illustrates this procedure for the tangent line in Example 1. The first part of the figure shows the graph of $y = x^2 + 1$ in the window

$$[-6.3, 6.3] \times [0, 6.2]$$

and the second part shows the graph after we have zoomed in on the point $(2, 5)$ by a factor of 10. The trace operation produces the points

$$(2.05, 5.2025) \quad \text{and} \quad (1.95, 4.8025)$$

on the curve, so the slope of the tangent line can be approximated as

$$f'(2) \approx \frac{5.2025 - 4.8025}{2.05 - 1.95} = \frac{0.4}{0.1} = 4.0$$

which happens to agree exactly with the result in Example 1. It is important to understand, however, that the exact agreement in this case is accidental; in general, this method will not produce exact results because of roundoff errors in the computations, and also because the magnified portion of the curve may have a slight curvature, even though it appears straight on the screen.

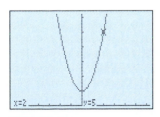 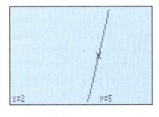 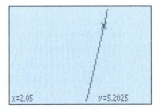 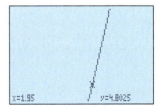

Figure 3.2.6

THE DERIVATIVE

In general, the slope of a curve $y = f(x)$ will depend on the point $(x, f(x))$ at which the slope is computed. That is, the slope is itself a function of x. To illustrate this, let us use (2) to compute $f'(x_0)$ at a general x-value x_0 for the curve $y = x^2 + 1$. The computations are similar to those in Example 1.

$$f'(x_0) = \lim_{x_1 \to x_0} \frac{f(x_1) - f(x_0)}{x_1 - x_0} = \lim_{x_1 \to x_0} \frac{(x_1^2 + 1) - (x_0^2 + 1)}{x_1 - x_0} = \lim_{x_1 \to x_0} \frac{x_1^2 - x_0^2}{x_1 - x_0}$$

$$= \lim_{x_1 \to x_0} \frac{(x_1 - x_0)(x_1 + x_0)}{x_1 - x_0} = \lim_{x_1 \to x_0} (x_1 + x_0) = 2x_0 \tag{4}$$

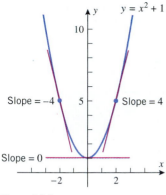

Figure 3.2.7

Now we can use the general formula $f'(x_0) = 2x_0$ to compute the slope of the tangent line at any point along the curve $y = x^2 + 1$ simply by substituting the appropriate value for $x = x_0$. For example, if $x_0 = 2$, $2x_0 = 4$, so $f'(2) = 4$, agreeing with the result in Example 1. Similarly, if $x_0 = 0$, then $2x_0 = 0$, so $f'(0) = 0$; and if $x_0 = -2$, then $2x_0 = -4$, so $f'(-2) = -4$ (Figure 3.2.7).

To generalize this idea, replacing x_0 by x in (2), the slope of the graph of $y = f(x)$ at a general point $(x, f(x))$ is given by

$$f'(x) = \lim_{x_1 \to x} \frac{f(x_1) - f(x)}{x_1 - x} \tag{5}$$

The fact that this describes a "slope-producing function" is so important that there is a common terminology associated with it. [To simplify notation, we use w in the place of x_1 in (5).]

3.2.3   DEFINITION.   The function f' defined by the formula

$$f'(x) = \lim_{w \to x} \frac{f(w) - f(x)}{w - x} \tag{6}$$

is called the ***derivative of f with respect to x***. The domain of f' consists of all x in the domain of f for which the limit exists.

REMARK.   Despite the presence of the symbol w in the definition, Formula (6) defines the function f' as a function of the single variable x. To calculate the value of $f'(x)$ at a particular input value x, we fix the value of x and let $w \to x$ in (6). The answer to this limit no longer involves the symbol w; w "disappears" at the step in which the limit is evaluated.

REMARK.   This is our first encounter with what was alluded to in Section 1.1 as a function that is the result of a "continuing process of incremental refinement." That is, the derivative function f' is *derived* from the function f via a limit. The use of a limiting process to define a new object is a fundamental tool in calculus and will be employed again in later chapters.

Recalling from the last section that the slope of the graph of $y = f(x)$ can be interpreted as the instantaneous rate of change of y with respect to x, it follows that the derivative of a function f can be interpreted in several ways:

Interpretations of the Derivative.   The derivative f' of a function f can be interpreted as a function whose value at x is the slope of the graph of $y = f(x)$ at x, or, alternatively, it can be interpreted as a function whose value at x is the instantaneous rate of change of y with respect to x at x. In particular, when $y = f(t)$ describes the position at time t of an object moving along a straight line, then $f'(t)$ describes the (instantaneous) velocity of the object at time t.

Example 2

(a)   Find the derivative with respect to x of $f(x) = x^3 - x$.

(b)   Graph f and f' together, and discuss the relationship between the two graphs.

Solution (a).   Later in this chapter we will develop efficient methods for finding derivatives, but for now we will find the derivative directly from Formula (6) in the definition of f'. The

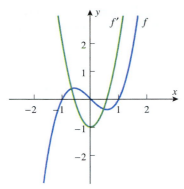

Figure 3.2.8

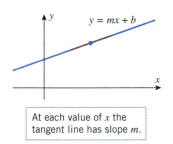

At each value of x the tangent line has slope m.

Figure 3.2.9

computations are as follows:

$$f'(x) = \lim_{w \to x} \frac{f(w) - f(x)}{w - x} = \lim_{w \to x} \frac{(w^3 - w) - (x^3 - x)}{w - x} = \lim_{w \to x} \frac{(w^3 - x^3) - (w - x)}{w - x}$$

$$= \lim_{w \to x} \frac{(w - x)[(w^2 + wx + x^2) - 1]}{w - x} = \lim_{w \to x} (w^2 + wx + x^2 - 1)$$

$$= x^2 + x^2 + x^2 - 1 = 3x^2 - 1$$

Solution (b). Since $f'(x)$ can be interpreted as the slope of the graph of $y = f(x)$ at x, the derivative $f'(x)$ is positive where the graph of f has positive slope, it is negative where the graph of f has negative slope, and it is zero where the graph of f is horizontal. We leave it for the reader to verify that this is consistent with the graphs of $f(x) = x^3 - x$ and $f'(x) = 3x^2 - 1$ shown in Figure 3.2.8. ◄

Example 3 At each value of x, the tangent line to a line $y = mx + b$ coincides with the line itself (Figure 3.2.9), and hence all tangent lines have slope m. This suggests geometrically that if $f(x) = mx + b$, then $f'(x) = m$ for all x. This is confirmed by the following computations:

$$f'(x) = \lim_{w \to x} \frac{f(w) - f(x)}{w - x} = \lim_{w \to x} \frac{(mw + b) - (mx + b)}{w - x} = \lim_{w \to x} \frac{mw - mx}{w - x}$$

$$= \lim_{w \to x} \frac{m(w - x)}{w - x} = \lim_{w \to x} m = m$$ ◄

Example 4

(a) Find the derivative with respect to x of $f(x) = \sqrt{x}$.

(b) Find the slope of the curve $y = \sqrt{x}$ at $x = 9$.

(c) Find the limits of $f'(x)$ as $x \to 0^+$ and as $x \to +\infty$, and explain what those limits say about the graph of f.

Solution (a). From Definition 3.2.3,

$$f'(x) = \lim_{w \to x} \frac{f(w) - f(x)}{w - x} = \lim_{w \to x} \frac{\sqrt{w} - \sqrt{x}}{w - x} = \lim_{w \to x} \frac{\sqrt{w} - \sqrt{x}}{w - x} \cdot \frac{\sqrt{w} + \sqrt{x}}{\sqrt{w} + \sqrt{x}}$$

$$= \lim_{w \to x} \frac{w - x}{(w - x)(\sqrt{w} + \sqrt{x})} = \lim_{w \to x} \frac{1}{\sqrt{w} + \sqrt{x}} = \frac{1}{\sqrt{x} + \sqrt{x}} = \frac{1}{2\sqrt{x}}$$

Solution (b). The slope of the curve $y = \sqrt{x}$ at $x = 9$ is $f'(9)$. From part (a), this slope is $f'(9) = 1/(2\sqrt{9}) = \frac{1}{6}$.

Solution (c). The graphs of $f(x) = \sqrt{x}$ and $f'(x) = 1/(2\sqrt{x})$ are shown in Figure 3.2.10. Observe that $f'(x) > 0$ if $x > 0$, which means that all tangent lines to the graph of $y = \sqrt{x}$ have positive slope at all points in this interval. Since

$$\lim_{x \to 0^+} \frac{1}{2\sqrt{x}} = +\infty \quad \text{and} \quad \lim_{x \to +\infty} \frac{1}{2\sqrt{x}} = 0$$

the graph becomes more and more vertical as $x \to 0^+$ and more and more horizontal as $x \to +\infty$. ◄

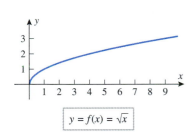

$y = f(x) = \sqrt{x}$

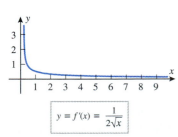

$y = f'(x) = \dfrac{1}{2\sqrt{x}}$

Figure 3.2.10

FOR THE READER. Use a graphing utility to estimate the slope of the curve $y = \sqrt{x}$ at $x = 9$ by zooming, and compare your result to the exact value obtained in the last example. If you have a CAS, read the documentation to determine how it can be used to find derivatives, and then use it to confirm the derivative obtained in Example 4(a).

Example 5 Consider the situation in Example 1 of Section 2.1 where a ball is thrown vertically upward so that the height (in feet) of the ball above the ground t seconds after its release is modeled by the function

$$s(t) = -16t^2 + 29t + 6, \quad 0 \le t \le 2$$

(a) Use the derivative of $s(t)$ at $t = 0.5$ to determine the instantaneous velocity of the ball at time $t = 0.5$ s.

(b) Find the velocity function $v(t) = s'(t)$ for $0 < t < 2$. What is the velocity of the ball just before impacting the ground at time $t = 2$ s?

Solution (a). When $t = 0.5$ s, the height of the ball is $s(0.5) = 16.5$ ft. The ball's instantaneous velocity at time $t = 0.5$ is given by the derivative of s at $t = 0.5$, that is, $s'(0.5)$. From Definition 3.2.1,

$$s'(0.5) = \lim_{w \to 0.5} \frac{s(w) - s(0.5)}{w - 0.5} = \lim_{w \to 0.5} \frac{(-16w^2 + 29w + 6) - 16.5}{w - 0.5}$$

$$= \lim_{w \to 0.5} \frac{-16w^2 + 29w - 10.5}{w - 0.5} \cdot \frac{2}{2} = \lim_{w \to 0.5} \frac{-32w^2 + 58w - 21}{2w - 1}$$

$$= \lim_{w \to 0.5} \frac{(2w - 1)(-16w + 21)}{2w - 1} = \lim_{w \to 0.5} (-16w + 21) = -8 + 21 = 13$$

Thus, the velocity of the ball at time $t = 0.5$ s is $s'(0.5) = 13$ ft/s, which agrees with our estimate from numerical evidence in Example 1 of Section 2.1.

Solution (b). From Definition 3.2.3,

$$v(t) = s'(t) = \lim_{w \to t} \frac{s(w) - s(t)}{w - t} = \lim_{w \to t} \frac{(-16w^2 + 29w + 6) - (-16t^2 + 29t + 6)}{w - t}$$

$$= \lim_{w \to t} \frac{-16(w^2 - t^2) + 29(w - t) + (6 - 6)}{w - t}$$

$$= \lim_{w \to t} \frac{-16(w - t)(w + t) + 29(w - t)}{w - t} = \lim_{w \to t} \frac{(w - t)[-16(w + t) + 29]}{w - t}$$

$$= \lim_{w \to t} [-16(w + t) + 29] = -16(t + t) + 29 = -32t + 29$$

Thus, for $0 < t < 2$, the velocity of the ball is given by $v(t) = s'(t) = -32t + 29$. As $t \to 2^-$, $s'(t) = -32t + 29 \to -64 + 29 = -35$ ft/s. That is, the ball is falling at a speed approaching 35 ft/s when its impact with the ground is imminent. ◀

DIFFERENTIABILITY

Observe that a function f must be defined at $x = x_0$ in order for the difference quotient

$$\frac{f(w) - f(x_0)}{w - x_0}$$

to make sense, since this quotient references a value for $f(x_0)$. Since a value for $f(x_0)$ is required before the limit of this quotient can be considered, values in the domain of the derivative function f' must also be in the domain of f.

For a number x_0 in the domain of a function f, we say that f is **differentiable at x_0**, or that **the derivative of f exists at x_0**, if

$$\lim_{w \to x_0} \frac{f(w) - f(x_0)}{w - x_0}$$

exists. Thus, the domain of f' consists of all values of x at which f is differentiable. If x_0 is not in the domain of f or if the limit does not exist, then we say that f is **not differentiable at x_0**, or that **the derivative of f does not exist at x_0**. If f is differentiable at every value of x in an open interval (a, b), then we say that f is **differentiable on (a, b)**. This definition also applies to infinite open intervals of the form $(a, +\infty)$, $(-\infty, b)$, and $(-\infty, +\infty)$. In the case where f is differentiable on $(-\infty, +\infty)$ we will say that f is **differentiable everywhere**.

Geometrically, if f is differentiable at a value x_0 for x, then the graph of f has a tangent line at x_0. If f is defined at x_0 but is not differentiable at x_0, then either the graph of f has no well-defined tangent line at x_0 or it has a vertical tangent line at x_0. Informally, the most commonly encountered circumstances of nondifferentiability occur where the graph of f has

- a corner,
- a vertical tangent line, or
- a discontinuity.

Figure 3.2.11 illustrates each of these situations.

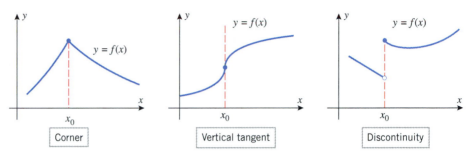

Figure 3.2.11

It makes sense intuitively that a function is not differentiable where its graph has a corner, since there is no reasonable way to define the graph's slope at a corner. For example, Figure 3.2.12a shows a typical corner point $P(x_0, f(x_0))$ on the graph of a function f. At this point, the slopes of secant lines joining P and nearby points Q have different limiting values, depending on whether Q is to the left or to the right of P. Hence, the slopes of the secant lines do not have a two-sided limit.

A *vertical tangent line* occurs at a place on a continuous curve where the slopes of secant lines approach $+\infty$ or approach $-\infty$ (Figure 3.2.12b). Since an infinite limit is a special way of saying that a limit does not exist, a function f is not differentiable at a point of vertical tangency.

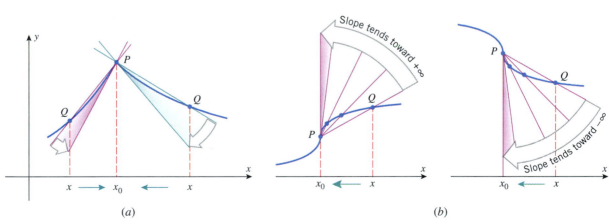

Figure 3.2.12

We will explore the relationship between differentiability and continuity later in this section. It should be noted that there are other, less common, circumstances under which a function may fail to be differentiable. See Exercise 45 for one such example.

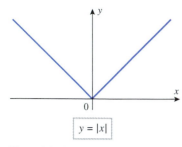

Figure 3.2.13

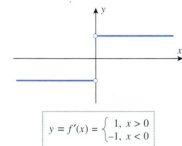

Figure 3.2.14

Example 6   The graph of $y = |x|$ in Figure 3.2.13 has a corner at $x = 0$, which implies that $f(x) = |x|$ is not differentiable at $x = 0$.

(a)   Prove that $f(x) = |x|$ is not differentiable at $x = 0$ by showing that the limit in Definition 3.2.3 does not exist at $x = 0$.

(b)   Find a formula for $f'(x)$.

Solution (a).   From Formula (6) with $x = 0$, the value of $f'(0)$, if it were to exist, would be given by

$$f'(0) = \lim_{w \to 0} \frac{f(w) - f(0)}{w - 0} = \lim_{w \to 0} \frac{|w| - |0|}{w} = \lim_{w \to 0} \frac{|w|}{w}$$

But

$$\frac{|w|}{w} = \begin{cases} 1, & w > 0 \\ -1, & w < 0 \end{cases}$$

so that

$$\lim_{w \to 0^-} \frac{|w|}{w} = -1 \quad \text{and} \quad \lim_{w \to 0^+} \frac{|w|}{w} = 1$$

Thus,

$$f'(0) = \lim_{w \to 0} \frac{|w|}{w}$$

does not exist because the one-sided limits are not equal.

Solution (b).   A formula for the derivative of $f(x) = |x|$ can be obtained by writing $|x|$ in piecewise form and treating the cases $x > 0$ and $x < 0$ separately. If $x > 0$, then $f(x) = x$ and $f'(x) = 1$; if $x < 0$, then $f(x) = -x$ and $f'(x) = -1$. Thus,

$$f'(x) = \begin{cases} 1, & x > 0 \\ -1, & x < 0 \end{cases}$$

The graph of f' is shown in Figure 3.2.14. Observe that f' is not continuous at $x = 0$, so this example shows that a function that is continuous everywhere may have a derivative that fails to be continuous everywhere.   ◀

DIFFERENTIABILITY AND CONTINUITY

It makes sense intuitively that a function f cannot be differentiable where it has a "jump" discontinuity, since the value of the function changes precipitously at the "jump." The following theorem shows that a function f must be continuous at a value x_0 in order for it to be differentiable there (or stated another way, a function f cannot be differentiable where it is not continuous).

3.2.4   **THEOREM.**   *If f is differentiable at $x = x_0$, then f must also be continuous at x_0.*

Proof.   We are given that f is differentiable at x_0, so it follows from (6) that $f'(x_0)$ exists and is given by

$$f'(x_0) = \lim_{w \to x_0} \frac{f(w) - f(x_0)}{w - x_0} \tag{7}$$

To show that f is continuous at x_0, we must show that

$$\lim_{w \to x_0} f(w) = f(x_0)$$

or equivalently,

$$\lim_{w \to x_0} [f(w) - f(x_0)] = 0$$

However, this can be proved using (7) as follows:

$$\lim_{w \to x_0} [f(w) - f(x_0)] = \lim_{w \to x_0} \left[\frac{f(w) - f(x_0)}{w - x_0} \cdot (w - x_0) \right]$$

$$= \lim_{w \to x_0} \left[\frac{f(w) - f(x_0)}{w - x_0} \right] \cdot \lim_{w \to x_0} (w - x_0)$$

$$= f'(x_0) \cdot 0 = 0$$

REMARK.   The converse to Theorem 3.2.4 is false. That is, *a function may be continuous at an input value, but not differentiable there.* For example, the function $f(x) = |x|$ is continuous at $x = 0$ but not differentiable at $x = 0$ (see Example 6). In fact, any function whose graph has a corner and is continuous at the location of the corner will be continuous but not differentiable at the corner.

The relationship between continuity and differentiability was of great historical significance in the development of calculus. In the early nineteenth century mathematicians believed that if a continuous function had many points of nondifferentiability, these points, like the tips of a sawblade, would have to be separated from each other and joined by smooth curve segments (Figure 3.2.15). This misconception was shattered by a series of discoveries beginning in 1834. In that year a Bohemian priest, philosopher, and mathematician named Bernhard Bolzano[*] discovered a procedure for constructing a continuous function that is not differentiable at any point. Later, in 1860, the great German mathematician, Karl Weierstrass produced the first formula for such a function. The graphs of such functions are impossible to draw; it is as if the corners are so numerous that any segment of the curve, when suitably enlarged, reveals more corners. The discovery of these pathological functions was important in that it made mathematicians distrustful of their geometric intuition and more reliant on precise mathematical proof. However, these functions remained only mathematical curiosities until the early 1980s, when applications of them began to emerge. During recent decades, such functions have started to play a fundamental role in the study of geometric objects called *fractals*. Fractals have revealed an order to natural phenomena that were previously dismissed as random and chaotic.

Figure 3.2.15

DERIVATIVE NOTATION

The process of finding a derivative is called *differentiation*. You can think of differentiation as an operation on functions that associates a function f' with a function f. When the

[*] BERNHARD BOLZANO (1781–1848). Bolzano, the son of an art dealer, was born in Prague, Bohemia (Czech Republic). He was educated at the University of Prague, and eventually won enough mathematical fame to be recommended for a mathematics chair there. However, Bolzano became an ordained Roman Catholic priest, and in 1805 he was appointed to a chair of Philosophy at the University of Prague. Bolzano was a man of great human compassion; he spoke out for educational reform, he voiced the right of individual conscience over government demands, and he lectured on the absurdity of war and militarism. His views so disenchanted Emperor Franz I of Austria that the emperor pressed the Archbishop of Prague to have Bolzano recant his statements. Bolzano refused and was then forced to retire in 1824 on a small pension. Bolzano's main contribution to mathematics was philosophical. His work helped convince mathematicians that sound mathematics must ultimately rest on rigorous proof rather than intuition. In addition to his work in mathematics, Bolzano investigated problems concerning space, force, and wave propagation.

independent variable is x, the differentiation operation is often denoted by

$$\frac{d}{dx}[f(x)]$$

which is read "*the derivative of $f(x)$ with respect to x.*" Thus,

$$\frac{d}{dx}[f(x)] = f'(x) \tag{8}$$

For example, with this notation the derivatives obtained in Examples 2, 3, and 4 can be expressed as

$$\frac{d}{dx}[x^3 - x] = 3x^2 - 1, \quad \frac{d}{dx}[mx + b] = m, \quad \frac{d}{dx}[\sqrt{x}] = \frac{1}{2\sqrt{x}} \tag{9}$$

To denote the value of the derivative at a specific value $x = x_0$ with the notation in (8), we would write

$$\frac{d}{dx}[f(x)]\bigg|_{x=x_0} = f'(x_0) \tag{10}$$

For example, from (9)

$$\frac{d}{dx}[x^3 - x]\bigg|_{x=1} = 3(1^2) - 1 = 2, \quad \frac{d}{dx}[mx + b]\bigg|_{x=5} = m, \quad \frac{d}{dx}[\sqrt{x}]\bigg|_{x=9} = \frac{1}{2\sqrt{9}} = \frac{1}{6}$$

Notations (8) and (10) are convenient when no dependent variable is involved. However, if there is a dependent variable, say $y = f(x)$, then (8) and (10) can be written as

$$\frac{d}{dx}[y] = f'(x) \quad \text{and} \quad \frac{d}{dx}[y]\bigg|_{x=x_0} = f'(x_0)$$

It is common to omit the brackets on the left side and write these expressions as

$$\frac{dy}{dx} = f'(x) \qquad \text{and} \qquad \frac{dy}{dx}\bigg|_{x=x_0} = f'(x_0)$$

where dy/dx is read as "the derivative of y with respect to x." For example, if $y = \sqrt{x}$, then

$$\frac{dy}{dx} = \frac{1}{2\sqrt{x}}, \quad \frac{dy}{dx}\bigg|_{x=x_0} = \frac{1}{2\sqrt{x_0}}, \quad \frac{dy}{dx}\bigg|_{x=9} = \frac{1}{2\sqrt{9}} = \frac{1}{6}$$

REMARK.   Later, the symbols dy and dx will be defined separately. However, for the time being, dy/dx should not be regarded as a ratio; rather, it should be considered as a single symbol denoting the derivative.

When letters other than x and y are used for the independent and dependent variables, then the various notations for the derivative must be adjusted accordingly. For example, if $y = f(u)$, then the derivative with respect to u would be written as

$$\frac{d}{du}[f(u)] = f'(u) \quad \text{and} \quad \frac{dy}{du} = f'(u)$$

In particular, if $y = \sqrt{u}$, then

$$\frac{dy}{du} = \frac{1}{2\sqrt{u}}, \quad \frac{dy}{du}\bigg|_{u=u_0} = \frac{1}{2\sqrt{u_0}}, \quad \frac{dy}{du}\bigg|_{u=9} = \frac{1}{2\sqrt{9}} = \frac{1}{6}$$

OTHER NOTATIONS

Some writers denote the derivative as $D_x[f(x)] = f'(x)$, but we will not use this notation in this text. In problems where the name of the independent variable is clear from the context, there are some other possible notations for the derivative. For example, if $y = f(x)$, but it is clear from the problem that the independent variable is x, then the derivative with respect to x might be denoted by y' or f'.

Often, you will see Definition 3.2.3 expressed using h or Δx for the difference $w - x$. With $h = w - x$, then $w = x + h$ and $w \to x$ is equivalent to $h \to 0$. Thus, Formula (6) has the form

$$f'(x) = \lim_{h \to 0} \frac{f(x + h) - f(x)}{h} \tag{11}$$

Or, using Δx instead of h for $w - x$, Formula (6) has the form

$$f'(x) = \lim_{\Delta x \to 0} \frac{f(x + \Delta x) - f(x)}{\Delta x} \tag{12}$$

If $y = f(x)$, then it is also common to let

$$\Delta y = f(w) - f(x) = f(x + \Delta x) - f(x)$$

in which case

$$\frac{dy}{dx} = \lim_{\Delta x \to 0} \frac{\Delta y}{\Delta x} = \lim_{\Delta x \to 0} \frac{f(x + \Delta x) - f(x)}{\Delta x} \tag{13}$$

The geometric interpretations of Δx and Δy are shown in Figure 3.2.16.

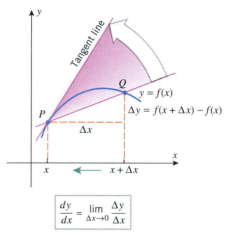

$$\frac{dy}{dx} = \lim_{\Delta x \to 0} \frac{\Delta y}{\Delta x}$$

Figure 3.2.16

DERIVATIVES AT THE ENDPOINTS OF AN INTERVAL

If a function f is defined on a closed interval $[a, b]$ and is not defined outside that interval, then the derivative $f'(x)$ is not defined at the endpoints because

$$f'(x) = \lim_{w \to x} \frac{f(w) - f(x)}{w - x}$$

is a two-sided limit and only a one-sided limit makes sense at an endpoint. To deal with this situation, we define *derivatives from the left and right*. These are denoted by f'_- and f'_+, respectively, and are defined by

$$f'_-(x) = \lim_{w \to x^-} \frac{f(w) - f(x)}{w - x} \quad \text{and} \quad f'_+(x) = \lim_{w \to x^+} \frac{f(w) - f(x)}{w - x}$$

At points where $f'_+(x)$ exists we say that the function f is *differentiable from the right*, and at points where $f'_-(x)$ exists we say that the function f is *differentiable from the left*. Geometrically, $f'_+(x)$ is the limit of the slopes of the secant lines approaching x from the right, and $f'_-(x)$ is the limit of the slopes of the secant lines approaching x from the left (Figure 3.2.17).

It can be proved that a function f is continuous from the left at those points where it is differentiable from the left, and f is continuous from the right at those points where it is differentiable from the right.

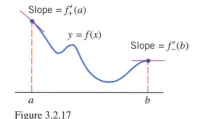

Figure 3.2.17

We say a function f is ***differentiable on an interval*** of the form $[a, b]$, $[a, +\infty)$, $(-\infty, b]$, $[a, b)$, or $(a, b]$ if f is differentiable at all numbers inside the interval, and it is differentiable at the endpoint(s) from the left or right, as appropriate.derivatives from the left and right

EXERCISE SET 3.2   ~ Graphing Utility

1. Use the graph of $y = f(x)$ in the accompanying figure to estimate the value of $f'(1)$, $f'(3)$, $f'(5)$, and $f'(6)$.

2. For the function graphed in the accompanying figure, arrange the numbers 0, $f'(-3)$, $f'(0)$, $f'(2)$, and $f'(4)$ in increasing order.

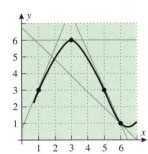

Figure Ex-1

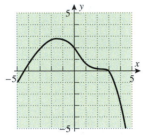

Figure Ex-2

3. (a) If you are given an equation for the tangent line at the point $(a, f(a))$ on a curve $y = f(x)$, how would you go about finding $f'(a)$?
   (b) Given that the tangent line to the graph of $y = f(x)$ at the point $(2, 5)$ has the equation $y = 3x - 1$, find $f'(2)$.
   (c) For the function $y = f(x)$ in part (b), what is the instantaneous rate of change of y with respect to x at $x = 2$?

4. Given that the tangent line to $y = f(x)$ at the point $(-1, 3)$ passes through the point $(0, 4)$, find $f'(-1)$.

5. Sketch the graph of a function f for which $f(0) = 1$, $f'(0) = 0$, $f'(x) > 0$ if $x < 0$, and $f'(x) < 0$ if $x > 0$.

6. Sketch the graph of a function f for which $f(0) = 0$, $f'(0) = 0$, and $f'(x) > 0$ if $x < 0$ or $x > 0$.

7. Given that $f(3) = -1$ and $f'(3) = 5$, find an equation for the tangent line to the graph of $y = f(x)$ at $x = 3$.

8. Given that $f(-2) = 3$ and $f'(-2) = -4$, find an equation for the tangent line to the graph of $y = f(x)$ at $x = -2$.

In Exercises 9–14, use Definition 3.2.3 to find $f'(x)$, and then find the equation of the tangent line to $y = f(x)$ at $x = a$.

9. $f(x) = 3x^2$; $a = 3$
10. $f(x) = x^4$; $a = -2$
11. $f(x) = x^3$; $a = 0$
12. $f(x) = 2x^3 + 1$; $a = -1$
13. $f(x) = \sqrt{x + 1}$; $a = 8$
14. $f(x) = \sqrt{2x + 1}$; $a = 4$

In Exercises 15–20, use Formula (13) to find dy/dx.

15. $y = \dfrac{1}{x}$

16. $y = \dfrac{1}{x + 1}$

17. $y = ax^2 + b$
   (a, b constants)

18. $y = x^2 - x$

19. $y = \dfrac{1}{\sqrt{x}}$

20. $y = \dfrac{1}{x^2}$

In Exercises 21 and 22, use Definition 3.2.3 (with appropriate change in notation) to obtain the derivative requested.

21. Find $f'(t)$ if $f(t) = 4t^2 + t$.

22. Find dV/dr if $V = \frac{4}{3}\pi r^3$.

23. Match the graphs of the functions shown in (a)–(f) with the graphs of their derivatives in (A)–(F).

(a)
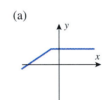

(b)
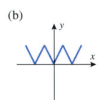

(c)
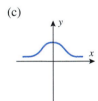

(d)

(e)
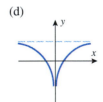

(f)
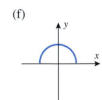

(A)
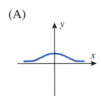

(B)
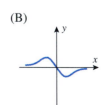

(C)
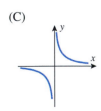

(D)
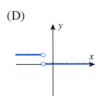

(E)

(F)
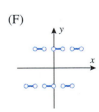

24. Find a function f such that $f'(x) = 1$ for all x, and give an informal argument to justify your answer.

In Exercises 25 and 26, sketch the graph of the derivative of the function whose graph is shown.

25. (a)   (b)   (c)

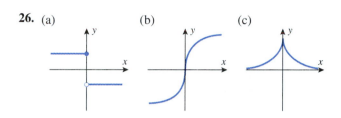

26. (a)   (b)   (c)

In Exercises 27 and 28, the limit represents $f'(a)$ for some function f and some number a. Find $f(x)$ and a in each case.

27. (a) $\displaystyle \lim_{x_1 \to 3} \frac{x_1^2 - 9}{x_1 - 3}$ (b) $\displaystyle \lim_{\Delta x \to 0} \frac{\sqrt{1 + \Delta x} - 1}{\Delta x}$

28. (a) $\displaystyle \lim_{x \to 1} \frac{x^7 - 1}{x - 1}$ (b) $\displaystyle \lim_{h \to 0} \frac{\cos(\pi + h) + 1}{h}$

29. Find $dy/dx|_{x=1}$, given that $y = 4x^2 + 1$.

30. Find $dy/dx|_{x=-2}$, given that $y = (5/x) + 1$.

31. Find an equation for the line that is tangent to the curve $y = x^3 - 2x + 1$ at the point $(0, 1)$, and use a graphing utility to graph the curve and its tangent line on the same screen.

32. Use a graphing utility to graph the following on the same screen: the curve $y = x^2/4$, the tangent line to this curve at $x = 1$, and the secant line joining the points $(0, 0)$ and $(2, 1)$ on this curve.

33. Let $f(x) = 2^x$. Estimate $f'(1)$ by
(a) using a graphing utility to zoom in at an appropriate point until the graph looks like a straight line, and then estimating the slope
(b) using a calculating utility to estimate the limit in Definition 3.2.3 by making a table of values for a succession of values of w approaching 1.

34. Let $f(x) = \sin x$. Estimate $f'(\pi/4)$ by
(a) using a graphing utility to zoom in at an appropriate point until the graph looks like a straight line, and then estimating the slope
(b) using a calculating utility to estimate the limit in Definition 3.2.3 by making a table of values for a succession of values of w approaching $\pi/4$.

35. Suppose that the cost of drilling x feet for an oil well is $C = f(x)$ dollars.
(a) What are the units of $f'(x)$?

(b) In practical terms, what does $f'(x)$ mean in this case?
(c) What can you say about the sign of $f'(x)$?
(d) Estimate the cost of drilling an additional foot, starting at a depth of 300 ft, given that $f'(300) = 1000$.

36. A paint manufacturing company estimates that it can sell $g = f(p)$ gallons of paint at a price of p dollars.
(a) What are the units of dg/dp?
(b) In practical terms, what does dg/dp mean in this case?
(c) What can you say about the sign of dg/dp?
(d) Given that $dg/dp|_{p=10} = -100$, what can you say about the effect of increasing the price from \$10 per gallon to \$11 per gallon?

37. It is a fact that when a flexible rope is wrapped around a rough cylinder, a small force of magnitude F_0 at one end can resist a large force of magnitude F at the other end. The size of F depends on the angle θ through which the rope is wrapped around the cylinder (see the accompanying figure). That figure shows the graph of F (in pounds) versus θ (in radians), where F is the magnitude of the force that can be resisted by a force with magnitude $F_0 = 10$ lb for a certain rope and cylinder.
(a) Estimate the values of F and $dF/d\theta$ when the angle $\theta = 10$ radians.
(b) It can be shown that the force F satisfies the equation $dF/d\theta = \mu F$, where the constant μ is called the *coefficient of friction*. Use the results in part (a) to estimate the value of μ.

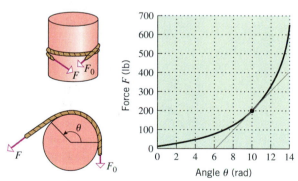

Figure Ex-37

38. According to the U. S. Bureau of the Census, the estimated and projected midyear world population, N, in billions for the years 1950, 1975, 2000, 2025, and 2050 was 2.555, 4.088, 6.080, 7.841, and 9.104, respectively. Although the increase in population is not a continuous function of the time t, we can apply the ideas in this section if we are willing to approximate the graph of N versus t by a continuous curve, as shown in the accompanying figure.
(a) Use the tangent line at $t = 2000$ shown in the figure to approximate the value of dN/dt there. Interpret your result as a rate of change.
(b) The instantaneous *growth rate* is defined as

$$\frac{dN/dt}{N}$$

Use your answer to part (a) to approximate the instantaneous growth rate at the start of the year 2000. Express the result as a percentage and include the proper units.

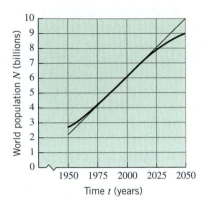

Figure Ex-38

39. According to *Newton's Law of Cooling*, the rate of change of an object's temperature is proportional to the difference between the temperature of the object and that of the surrounding medium. The accompanying figure shows the graph of the temperature T (in degrees Fahrenheit) versus time t (in minutes) for a cup of coffee, initially with a temperature of $200°\,$F, that is allowed to cool in a room with a constant temperature of $75°\,$F.

(a) Estimate T and dT/dt when $t = 10$ min.

(b) Newton's Law of Cooling can be expressed as

$$\frac{dT}{dt} = k(T - T_0)$$

where k is the constant of proportionality and T_0 is the temperature (assumed constant) of the surrounding medium. Use the results in part (a) to estimate the value of k.

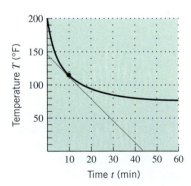

Figure Ex-39

40. Write a paragraph that explains what it means for a function to be differentiable. Include some examples of functions that are not differentiable, and explain the relationship between differentiability and continuity.

41. Show that $f(x) = \sqrt[3]{x}$ is continuous at $x = 0$ but not differentiable at $x = 0$. Sketch the graph of f.

42. Show that $f(x) = \sqrt[3]{(x - 2)^2}$ is continuous at $x = 2$ but not differentiable at $x = 2$. Sketch the graph of f.

43. Show that

$$f(x) = \begin{cases} x^2 + 1, & x \le 1 \\ 2x, & x > 1 \end{cases}$$

is continuous and differentiable at $x = 1$. Sketch the graph of f.

44. Show that

$$f(x) = \begin{cases} x^2 + 2, & x \le 1 \\ x + 2, & x > 1 \end{cases}$$

is continuous but not differentiable at $x = 1$. Sketch the graph of f.

45. Show that

$$f(x) = \begin{cases} x \sin(1/x), & x \ne 0 \\ 0, & x = 0 \end{cases}$$

is continuous but not differentiable at $x = 0$. Sketch the graph of f near $x = 0$. (See Figure 2.6.7b and the remark following Example 3 in Section 2.6.)

46. Show that

$$f(x) = \begin{cases} x^2 \sin(1/x), & x \ne 0 \\ 0, & x = 0 \end{cases}$$

is continuous and differentiable at $x = 0$. Sketch the graph of f near $x = 0$.

47. Suppose that a function f is differentiable at $x = 1$ and

$$\lim_{h \to 0} \frac{f(1 + h)}{h} = 5$$

Find $f(1)$ and $f'(1)$.

48. Suppose that f is a differentiable function with the property that

$$f(x + y) = f(x) + f(y) + 5xy \quad \text{and} \quad \lim_{h \to 0} \frac{f(h)}{h} = 3$$

Find $f(0)$ and $f'(x)$.

49. Suppose that f has the property $f(x + y) = f(x)f(y)$ for all values of x and y and that $f(0) = f'(0) = 1$. Show that f is differentiable and $f'(x) = f(x)$. [*Hint:* Start by expressing $f'(x)$ as a limit.]

3.3 TECHNIQUES OF DIFFERENTIATION

In the last section we defined the derivative of a function f as a limit, and we used that limit to calculate a few simple derivatives. In this section we will develop some important theorems that will enable us to calculate derivatives more efficiently.

DERIVATIVE OF A CONSTANT

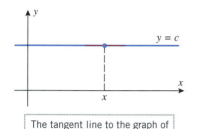

The tangent line to the graph of $f(x) = c$ has slope 0 for all x.

Figure 3.3.1

The graph of a constant function $f(x) = c$ is the horizontal line $y = c$, and hence the tangent line to this graph has slope 0 at every value of x (Figure 3.3.1). Thus, we should expect the derivative of a constant function to be 0 for all x.

3.3.1 THEOREM. *The derivative of a constant function is 0; that is, if c is any real number, then*

$$\frac{d}{dx}[c] = 0$$

Proof. Let $f(x) = c$. Then from the definition of the derivative,

$$\frac{d}{dx}[c] = f'(x) = \lim_{w \to x} \frac{f(w) - f(x)}{w - x} = \lim_{w \to x} \frac{c - c}{w - x} = \lim_{w \to x} 0 = 0$$ ∎

Example 1 If $f(x) = 5$ for all x, then $f'(x) = 0$ for all x; that is,

$$\frac{d}{dx}[5] = 0$$ ◀

For our next derivative rule, we will need the algebraic identity

$$w^n - x^n = (w - x)(w^{n-1} + w^{n-2}x + w^{n-3}x^2 + \cdots + wx^{n-2} + x^{n-1})$$

which is valid for any positive integer n. This identity may be verified by expanding the right-hand side of the equation and noting the cancellation of terms. For example, with $n = 4$ we have

$$
\begin{aligned}
(w - x)(w^3 + w^2x + wx^2 + x^3) \\
= w^4 + (w^3x - xw^3) + (w^2x^2 - xw^2x) + (wx^3 - xwx^2) - x^4 \\
= w^4 + 0 + 0 + 0 - x^4 = w^4 - x^4
\end{aligned}
$$

3.3.2 THEOREM (*The Power Rule*). *If n is a positive integer, then*

$$\frac{d}{dx}[x^n] = nx^{n-1}$$

Proof. Let $f(x) = x^n$. Then from the definition of the derivative we obtain

$$\frac{d}{dx}[x^n] = f'(x) = \lim_{w \to x} \frac{f(w) - f(x)}{w - x} = \lim_{w \to x} \frac{w^n - x^n}{w - x}$$

$$= \lim_{w \to x} \frac{(w - x)(w^{n-1} + w^{n-2}x + w^{n-3}x^2 + \cdots + wx^{n-2} + x^{n-1})}{w - x}$$

$$= \lim_{w \to x} w^{n-1} + w^{n-2}x + w^{n-3}x^2 + \cdots + wx^{n-2} + x^{n-1}$$

$$= \underbrace{x^{n-1} + x^{n-1} + \cdots + x^{n-1}}_{\text{n terms in all}}$$

$$= nx^{n-1}$$ ∎

In words, *the derivative of x raised to a positive integer power is the product of the integer exponent and x raised to the next lower integer power.*

Example 2

$$\frac{d}{dx}[x^5] = 5x^4, \quad \frac{d}{dx}[x] = 1 \cdot x^0 = 1, \quad \frac{d}{dx}[x^{12}] = 12x^{11}$$

◀

DERIVATIVE OF A CONSTANT TIMES A FUNCTION

3.3.3 THEOREM. *If f is differentiable at x and c is any real number, then cf is also differentiable at x and*

$$\frac{d}{dx}[cf(x)] = c\frac{d}{dx}[f(x)]$$

Proof.

$$\frac{d}{dx}[cf(x)] = \lim_{w \to x}\frac{cf(w) - cf(x)}{w - x} = \lim_{w \to x} c\left[\frac{f(w) - f(x)}{w - x}\right]$$

$$= c\lim_{w \to x}\frac{f(w) - f(x)}{w - x} = c\frac{d}{dx}[f(x)]$$

A constant factor can be moved through a limit sign.

In function notation, Theorem 3.3.3 states

$$(cf)' = cf'$$

In words, *a constant factor can be moved through a derivative sign.*

Example 3

$$\frac{d}{dx}[4x^8] = 4\frac{d}{dx}[x^8] = 4[8x^7] = 32x^7$$

$$\frac{d}{dx}[-x^{12}] = (-1)\frac{d}{dx}[x^{12}] = -12x^{11}$$

$$\frac{d}{dx}\left[\frac{x}{\pi}\right] = \frac{1}{\pi}\frac{d}{dx}[x] = \frac{1}{\pi}$$

◀

DERIVATIVES OF SUMS AND DIFFERENCES

3.3.4 THEOREM. *If f and g are differentiable at x, then so are f + g and f − g and*

$$\frac{d}{dx}[f(x) + g(x)] = \frac{d}{dx}[f(x)] + \frac{d}{dx}[g(x)]$$

$$\frac{d}{dx}[f(x) - g(x)] = \frac{d}{dx}[f(x)] - \frac{d}{dx}[g(x)]$$

Proof.

$$\frac{d}{dx}[f(x) + g(x)] = \lim_{w \to x}\frac{[f(w) + g(w)] - [f(x) + g(x)]}{w - x}$$

$$= \lim_{w \to x}\frac{[f(w) - f(x)] + [g(w) - g(x)]}{w - x}$$

$$= \lim_{w \to x} \frac{f(w) - f(x)}{w - x} + \lim_{w \to x} \frac{g(w) - g(x)}{w - x}$$

The limit of a sum is the sum of the limits.

$$= \frac{d}{dx}[f(x)] + \frac{d}{dx}[g(x)]$$

The proof for $f - g$ is similar. ∎

In function notation, Theorem 3.3.4 states

$$(f + g)' = f' + g' \qquad (f - g)' = f' - g'$$

In words, *the derivative of a sum equals the sum of the derivatives*, and *the derivative of a difference equals the difference of the derivatives.*

Example 4

$$\frac{d}{dx}[x^4 + x^2] = \frac{d}{dx}[x^4] + \frac{d}{dx}[x^2] = 4x^3 + 2x$$

$$\frac{d}{dx}[6x^{11} - 9] = \frac{d}{dx}[6x^{11}] - \frac{d}{dx}[9] = 66x^{10} - 0 = 66x^{10} \qquad \blacktriangleleft$$

Although Theorem 3.3.4 was stated for sums and differences of two terms, it can be extended to any mixture of finitely many sums and differences of differentiable functions. For example,

$$\frac{d}{dx}[3x^8 - 2x^5 + 6x + 1] = \frac{d}{dx}[3x^8] - \frac{d}{dx}[2x^5] + \frac{d}{dx}[6x] + \frac{d}{dx}[1]$$

$$= 24x^7 - 10x^4 + 6$$

DERIVATIVE OF A PRODUCT

3.3.5 THEOREM (*The Product Rule*). *If f and g are differentiable at x, then so is the product $f \cdot g$, and*

$$\frac{d}{dx}[f(x)g(x)] = f(x)\frac{d}{dx}[g(x)] + g(x)\frac{d}{dx}[f(x)]$$

Proof. The earlier proofs in this section were straightforward applications of the definition of the derivative. However, this proof requires a trick—adding and subtracting the quantity $f(w)g(x)$ to the numerator in the derivative definition as follows:

$$\frac{d}{dx}[f(x)g(x)] = \lim_{w \to x} \frac{f(w) \cdot g(w) - f(x) \cdot g(x)}{w - x}$$

$$= \lim_{w \to x} \frac{f(w)g(w) - f(w)g(x) + f(w)g(x) - f(x)g(x)}{w - x}$$

$$= \lim_{w \to x} \left[f(w) \cdot \frac{g(w) - g(x)}{w - x} + g(x) \cdot \frac{f(w) - f(x)}{w - x} \right]$$

$$= \lim_{w \to x} f(w) \cdot \lim_{w \to x} \frac{g(w) - g(x)}{w - x} + \lim_{w \to x} g(x) \cdot \lim_{w \to x} \frac{f(w) - f(x)}{w - x}$$

$$= [\lim_{w \to x} f(w)]\frac{d}{dx}[g(x)] + [\lim_{w \to x} g(x)]\frac{d}{dx}[f(x)]$$

$$= f(x)\frac{d}{dx}[g(x)] + g(x)\frac{d}{dx}[f(x)]$$

[*Note:* In the last step $f(w) \to f(x)$ as $w \to x$ because f is continuous at x by Theorem 3.2.4, and $g(x) \to g(x)$ as $w \to x$ because $g(x)$ does not involve w and hence remains constant.] ∎

The product rule can be written in function notation as

$$(f \cdot g)' = f \cdot g' + g \cdot f'$$

In words, *the derivative of a product of two functions is the first function times the derivative of the second plus the second function times the derivative of the first.*

WARNING. Note that in general $(f \cdot g)' \neq f' \cdot g'$; that is, the derivative of a product is *not* generally the product of the derivatives!

Example 5 Find dy/dx if $y = (4x^2 - 1)(7x^3 + x)$.

Solution. There are two methods that can be used to find dy/dx. We can either use the product rule or we can multiply out the factors in y and then differentiate. We will give both methods.

Method I. (*Using the Product Rule*)

$$\frac{dy}{dx} = \frac{d}{dx}[(4x^2 - 1)(7x^3 + x)]$$

$$= (4x^2 - 1)\frac{d}{dx}[7x^3 + x] + (7x^3 + x)\frac{d}{dx}[4x^2 - 1]$$

$$= (4x^2 - 1)(21x^2 + 1) + (7x^3 + x)(8x) = 140x^4 - 9x^2 - 1$$

Method II. (*Multiplying First*)

$$y = (4x^2 - 1)(7x^3 + x) = 28x^5 - 3x^3 - x$$

Thus,

$$\frac{dy}{dx} = \frac{d}{dx}[28x^5 - 3x^3 - x] = 140x^4 - 9x^2 - 1$$

which agrees with the result obtained using the product rule. ◀

DERIVATIVE OF A QUOTIENT

3.3.6 THEOREM (*The Quotient Rule*). *If f and g are differentiable at x and $g(x) \neq 0$, then f/g is differentiable at x and*

$$\frac{d}{dx}\left[\frac{f(x)}{g(x)}\right] = \frac{g(x)\dfrac{d}{dx}[f(x)] - f(x)\dfrac{d}{dx}[g(x)]}{[g(x)]^2}$$

Proof.

$$\frac{d}{dx}\left[\frac{f(x)}{g(x)}\right] = \lim_{w \to x} \frac{\dfrac{f(w)}{g(w)} - \dfrac{f(x)}{g(x)}}{w - x} = \lim_{w \to x} \frac{f(w) \cdot g(x) - f(x) \cdot g(w)}{(w - x) \cdot g(x) \cdot g(w)}$$

Adding and subtracting $f(x) \cdot g(x)$ in the numerator yields

$$\frac{d}{dx}\left[\frac{f(x)}{g(x)}\right] = \lim_{w \to x} \frac{f(w) \cdot g(x) - f(x) \cdot g(x) - f(x) \cdot g(w) + f(x) \cdot g(x)}{(w - x) \cdot g(x) \cdot g(w)}$$

$$= \lim_{w \to x} \frac{\left[g(x) \cdot \dfrac{f(w) - f(x)}{w - x}\right] - \left[f(x) \cdot \dfrac{g(w) - g(x)}{w - x}\right]}{g(x) \cdot g(w)}$$

$$= \frac{\displaystyle\lim_{w \to x} g(x) \cdot \lim_{w \to x} \frac{f(w) - f(x)}{w - x} - \lim_{w \to x} f(x) \cdot \lim_{w \to x} \frac{g(w) - g(x)}{w - x}}{\displaystyle\lim_{w \to x} g(x) \cdot \lim_{w \to x} g(w)}$$

$$= \frac{[\lim_{w \to x} g(x)] \cdot \frac{d}{dx}[f(x)] - [\lim_{w \to x} f(x)] \cdot \frac{d}{dx}[g(x)]}{\lim_{w \to x} g(x) \cdot \lim_{w \to x} g(w)}$$

$$= \frac{g(x)\frac{d}{dx}[f(x)] - f(x)\frac{d}{dx}[g(x)]}{[g(x)]^2}$$

[See the note at the end of the proof of Theorem 3.3.5 for an explanation of the last step.] ■

The quotient rule can be written in function notation as

$$\left(\frac{f}{g}\right)' = \frac{g \cdot f' - f \cdot g'}{g^2}$$

In words, *the derivative of a quotient of two functions is the denominator times the derivative of the numerator minus the numerator times the derivative of the denominator, all divided by the denominator squared.*

WARNING.   Note that in general $(f/g)' \neq f'/g'$; that is, the derivative of a quotient is *not* generally the quotient of the derivatives.

Example 6   Let $f(x) = \dfrac{x^2 - 1}{x^4 + 1}$.

(a)   Graph $y = f(x)$, and use your graph to make rough estimates of the locations of all horizontal tangent lines.

(b)   By differentiating, find the exact locations of the horizontal tangent lines.

Solution (a). In Figure 3.3.2 we have shown the graph of the equation $y = f(x)$ in the window $[-2.5, 2.5] \times [-1, 1]$. This graph suggests that horizontal tangent lines occur at $x = 0$, $x \approx 1.5$, and $x \approx -1.5$.

Solution (b). To find the exact locations of the horizontal tangent lines, we must find the points where $dy/dx = 0$ (why?). We start by finding dy/dx:

$$\frac{dy}{dx} = \frac{d}{dx}\left[\frac{x^2 - 1}{x^4 + 1}\right] = \frac{(x^4 + 1)\frac{d}{dx}[x^2 - 1] - (x^2 - 1)\frac{d}{dx}[x^4 + 1]}{(x^4 + 1)^2}$$

$$= \frac{(x^4 + 1)(2x) - (x^2 - 1)(4x^3)}{(x^4 + 1)^2} \qquad \boxed{\begin{array}{l}\text{The differentiation is complete.}\\ \text{The rest is simplification.}\end{array}}$$

$$= \frac{-2x^5 + 4x^3 + 2x}{(x^4 + 1)^2} = -\frac{2x(x^4 - 2x^2 - 1)}{(x^4 + 1)^2}$$

Now we will set $dy/dx = 0$ and solve for x. We obtain

$$-\frac{2x(x^4 - 2x^2 - 1)}{(x^4 + 1)^2} = 0$$

The solutions of this equation are the values of x for which the numerator is 0:

$$2x(x^4 - 2x^2 - 1) = 0$$

The first factor yields the solution $x = 0$. Other solutions can be found by solving the

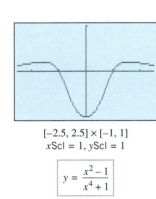

$[-2.5, 2.5] \times [-1, 1]$
$x\text{Scl} = 1, y\text{Scl} = 1$

$$y = \frac{x^2 - 1}{x^4 + 1}$$

Figure 3.3.2

equation

$$x^4 - 2x^2 - 1 = 0$$

This can be treated as a quadratic equation in x^2 and solved by the quadratic formula. This yields

$$x^2 = \frac{2 \pm \sqrt{8}}{2} = 1 \pm \sqrt{2}$$

The minus sign yields imaginary values of x, which we ignore since they are not relevant to the problem. The plus sign yields the solutions

$$x = \pm\sqrt{1 + \sqrt{2}}$$

In summary, horizontal tangent lines occur at

$$x = 0, \quad x = \sqrt{1 + \sqrt{2}} \approx 1.55, \quad \text{and} \quad x = -\sqrt{1 + \sqrt{2}} \approx -1.55$$

which is consistent with the rough estimates that we obtained graphically in part (a). ◄

THE POWER RULE FOR INTEGER EXPONENTS

In Theorem 3.3.2 we established the formula

$$\frac{d}{dx}[x^n] = nx^{n-1}$$

for *positive* integer values of n. Eventually, we will show that this formula applies if n is any real number. As our first step in this direction we will show that it applies for *all integer* values of n.

3.3.7 THEOREM. *If n is any integer, then*

$$\frac{d}{dx}[x^n] = nx^{n-1} \tag{1}$$

Proof. The result has already been established in the case where $n > 0$. If $n < 0$, then let $m = -n$ so that

$$f(x) = x^{-m} = \frac{1}{x^m}$$

From Theorem 3.3.6,

$$f'(x) = \frac{d}{dx}\left[\frac{1}{x^m}\right] = \frac{x^m \dfrac{d}{dx}[1] - 1 \dfrac{d}{dx}[x^m]}{(x^m)^2} = -\frac{\dfrac{d}{dx}[x^m]}{(x^m)^2}$$

Since $n < 0$, it follows that $m > 0$, so x^m can be differentiated using Theorem 3.3.2. Thus,

$$f'(x) = -\frac{mx^{m-1}}{x^{2m}} = -mx^{m-1-2m} = -mx^{-m-1} = nx^{n-1}$$

which proves (1). In the case $n = 0$ Formula (1) reduces to

$$\frac{d}{dx}[1] = 0 \cdot x^{-1} = 0$$

which is correct by Theorem 3.3.1. ■

Example 7

$$\frac{d}{dx}[x^{-9}] = -9x^{-9-1} = -9x^{-10}$$

$$\frac{d}{dx}\left[\frac{1}{x}\right] = \frac{d}{dx}[x^{-1}] = (-1)x^{-1-1} = -x^{-2} = -\frac{1}{x^2}$$

◄

In Example 4 of Section 3.2 we showed that

$$\frac{d}{dx}[\sqrt{x}] = \frac{1}{2\sqrt{x}}$$
(2)

which shows that Formula (1) also works with $n = \frac{1}{2}$, since

$$\frac{d}{dx}[x^{1/2}] = \frac{1}{2x^{1/2}} = \frac{1}{2}x^{-1/2}$$

HIGHER DERIVATIVES

If the derivative f' of a function f is itself differentiable, then the derivative of f' is denoted by f'' and is called the **second derivative** of f. As long as we have differentiability, we can continue the process of differentiating derivatives to obtain third, fourth, fifth, and even higher derivatives of f. The successive derivatives of f are denoted by

$$f', \quad f'' = (f')', \quad f''' = (f'')', \quad f^{(4)} = (f''')', \quad f^{(5)} = (f^{(4)})', \dots$$

These are called the first derivative, the second derivative, the third derivative, and so forth. Beyond the third derivative, it is too clumsy to continue using primes, so we switch from primes to integers in parentheses to denote the **order** of the derivative. In this notation it is easy to denote a derivative of arbitrary order by writing

$$f^{(n)} \qquad \boxed{\text{The } n\text{th derivative of } f}$$

The significance of the derivatives of order 2 and higher will be discussed later.

Example 8   If $f(x) = 3x^4 - 2x^3 + x^2 - 4x + 2$, then

$$f'(x) \quad = 12x^3 - 6x^2 + 2x - 4$$
$$f''(x) \quad = 36x^2 - 12x + 2$$
$$f'''(x) \quad = 72x - 12$$
$$f^{(4)}(x) = 72$$
$$f^{(5)}(x) = 0$$
$$\vdots$$
$$f^{(n)}(x) = 0 \quad (n \geq 5)$$
◀

Successive derivatives can also be denoted as follows:

$$f'(x) \quad = \frac{d}{dx}[f(x)]$$

$$f''(x) \quad = \frac{d}{dx}\left[\frac{d}{dx}[f(x)]\right] = \frac{d^2}{dx^2}[f(x)]$$

$$f'''(x) \quad = \frac{d}{dx}\left[\frac{d^2}{dx^2}[f(x)]\right] = \frac{d^3}{dx^3}[f(x)]$$

$$\vdots \qquad\qquad\qquad \vdots$$

In general, we write

$$f^{(n)}(x) = \frac{d^n}{dx^n}[f(x)]$$

which is read "the nth derivative of f with respect to x."

When a dependent variable is involved, say $y = f(x)$, then successive derivatives can be denoted by writing

$$\frac{dy}{dx}, \quad \frac{d^2y}{dx^2}, \quad \frac{d^3y}{dx^3}, \quad \frac{d^4y}{dx^4}, \dots, \frac{d^ny}{dx^n}, \dots$$

or more briefly,

$$y', \quad y'', \quad y''', \quad y^{(4)}, \dots, y^{(n)}, \dots$$

EXERCISE SET 3.3   Graphing Utility

In Exercises 1–12, find dy/dx.

1. $y = 4x^7$

2. $y = -3x^{12}$

3. $y = 3x^8 + 2x + 1$

4. $y = \frac{1}{2}(x^4 + 7)$

5. $y = \pi^3$

6. $y = \sqrt{2}x + (1/\sqrt{2})$

7. $y = -\frac{1}{3}(x^7 + 2x - 9)$

8. $y = \dfrac{x^2 + 1}{5}$

9. $y = ax^3 + bx^2 + cx + d$ (a, b, c, d constant)

10. $y = \dfrac{1}{a}\left(x^2 + \dfrac{1}{b}x + c\right)$ (a, b, c constant)

11. $y = -3x^{-8} + 2\sqrt{x}$

12. $y = 7x^{-6} - 5\sqrt{x}$

In Exercises 13–20, find $f'(x)$.

13. $f(x) = x^{-3} + \dfrac{1}{x^7}$

14. $f(x) = \sqrt{x} + \dfrac{1}{x}$

15. $f(x) = (3x^2 + 6)\left(2x - \frac{1}{4}\right)$

16. $f(x) = (2 - x - 3x^3)(7 + x^5)$

17. $f(x) = (x^3 + 7x^2 - 8)(2x^{-3} + x^{-4})$

18. $f(x) = \left(\dfrac{1}{x} + \dfrac{1}{x^2}\right)(3x^3 + 27)$

19. $f(x) = (3x^2 + 1)^2$

20. $f(x) = (x^5 + 2x)^2$

In Exercises 21 and 22, find $y'(1)$.

21. $y = \dfrac{1}{5x - 3}$

22. $y = \dfrac{3}{\sqrt{x} + 2}$

In Exercises 23 and 24, find dx/dt.

23. $x = \dfrac{3t}{2t + 1}$

24. $x = \dfrac{t^2 + 1}{3t}$

In Exercises 25–28, find $dy/dx|_{x=1}$.

25. $y = \dfrac{2x - 1}{x + 3}$

26. $y = \dfrac{4x + 1}{x^2 - 5}$

27. $y = \left(\dfrac{3x + 2}{x}\right)(x^{-5} + 1)$

28. $y = (2x^7 - x^2)\left(\dfrac{x - 1}{x + 1}\right)$

In Exercises 29 and 30, approximate $f'(1)$ by considering difference quotients

$$\dfrac{f(x_1) - f(1)}{x_1 - 1}$$

for values of x_1 near 1, and then find the exact value of $f'(1)$ by differentiating.

29. $f(x) = x^3 - 3x + 1$

30. $f(x) = x\sqrt{x}$

In Exercises 31 and 32, use a graphing utility to estimate the value of $f'(1)$ by zooming in on the graph of f, and then compare your estimate to the exact value obtained by differentiating.

31. $f(x) = \dfrac{x}{x^2 + 1}$

32. $f(x) = \dfrac{x^2 - 1}{x^2 + 1}$

In Exercises 33–36, find the indicated derivative.

33. $\dfrac{d}{dt}[16t^2]$

34. $\dfrac{dC}{dr}$, where $C = 2\pi r$

35. $V'(r)$, where $V = \pi r^3$

36. $\dfrac{d}{d\alpha}[2\alpha^{-1} + \alpha]$

37. A spherical balloon is being inflated.
  (a) Find a general formula for the instantaneous rate of change of the volume V with respect to the radius r, given that $V = \frac{4}{3}\pi r^3$.
  (b) Find the rate of change of V with respect to r at the instant when the radius is $r = 5$.

38. Find $\dfrac{d}{d\lambda}\left[\dfrac{\lambda\lambda_0 + \lambda^6}{2 - \lambda_0}\right]$ (λ_0 is constant).

39. Find $g'(4)$ given that $f(4) = 3$ and $f'(4) = -5$.
  (a) $g(x) = \sqrt{x}\,f(x)$
  (b) $g(x) = \dfrac{f(x)}{x}$

40. Find $g'(3)$ given that $f(3) = -2$ and $f'(3) = 4$.
  (a) $g(x) = 3x^2 - 5f(x)$
  (b) $g(x) = \dfrac{2x + 1}{f(x)}$

41. Find $F'(2)$ given that $f(2) = -1$, $f'(2) = 4$, $g(2) = 1$, and $g'(2) = -5$.
  (a) $F(x) = 5f(x) + 2g(x)$ (b) $F(x) = f(x) - 3g(x)$
  (c) $F(x) = f(x)g(x)$ (d) $F(x) = f(x)/g(x)$

42. Find $F'(\pi)$ given that $f(\pi) = 10$, $f'(\pi) = -1$, $g(\pi) = -3$, and $g'(\pi) = 2$.
  (a) $F(x) = 6f(x) - 5g(x)$ (b) $F(x) = x(f(x) + g(x))$
  (c) $F(x) = 2f(x)g(x)$ (d) $F(x) = \dfrac{f(x)}{4 + g(x)}$

43. Find an equation of the tangent line to the graph of $y = f(x)$ at $x = -3$ if $f(-3) = 2$ and $f'(-3) = 5$.

44. Find an equation for the line that is tangent to the curve $y = (1 - x)/(1 + x)$ at $x = 2$.

In Exercises 45 and 46, find d^2y/dx^2.

45. (a) $y = 7x^3 - 5x^2 + x$ (b) $y = 12x^2 - 2x + 3$
  (c) $y = \dfrac{x + 1}{x}$ (d) $y = (5x^2 - 3)(7x^3 + x)$

46. (a) $y = 4x^7 - 5x^3 + 2x$ (b) $y = 3x + 2$
  (c) $y = \dfrac{3x - 2}{5x}$ (d) $y = (x^3 - 5)(2x + 3)$

In Exercises 47 and 48, find y'''.

47. (a) $y = x^{-5} + x^5$ (b) $y = 1/x$
(c) $y = ax^3 + bx + c$ (a, b, c constant)

48. (a) $y = 5x^2 - 4x + 7$ (b) $y = 3x^{-2} + 4x^{-1} + x$
(c) $y = ax^4 + bx^2 + c$ (a, b, c constant)

49. Find
(a) $f'''(2)$, where $f(x) = 3x^2 - 2$
(b) $\dfrac{d^2 y}{dx^2}\Big|_{x=1}$, where $y = 6x^5 - 4x^2$
(c) $\dfrac{d^4}{dx^4}[x^{-3}]\Big|_{x=1}$

50. Find
(a) $y'''(0)$, where $y = 4x^4 + 2x^3 + 3$
(b) $\dfrac{d^4 y}{dx^4}\Big|_{x=1}$, where $y = \dfrac{6}{x^4}$.

51. Show that $y = x^3 + 3x + 1$ satisfies $y''' + xy'' - 2y' = 0$.

52. Show that if $x \neq 0$, then $y = 1/x$ satisfies the equation $x^3 y'' + x^2 y' - xy = 0$.

53. Find a general formula for $F''(x)$ if $F(x) = xf(x)$ and f and f' are differentiable at x.

54. Suppose that the function f is differentiable everywhere and $F(x) = xf(x)$.
(a) Express $F'''(x)$ in terms of x and derivatives of f.
(b) For $n \geq 2$, conjecture a formula for $F^n(x)$.

In Exercises 55 and 56, use a graphing utility to make rough estimates of the locations of all horizontal tangent lines, and then find their exact locations by differentiating.

55. $y = \frac{1}{3}x^3 - \frac{3}{2}x^2 + 2x$ **56.** $y = \dfrac{x}{x^2 + 9}$

57. Find a function $y = ax^2 + bx + c$ whose graph has an x-intercept of 1, a y-intercept of -2, and a tangent line with a slope of -1 at the y-intercept.

58. Find k if the curve $y = x^2 + k$ is tangent to the line $y = 2x$.

59. Find the x-coordinate of the point on the graph of $y = x^2$ where the tangent line is parallel to the secant line that cuts the curve at $x = -1$ and $x = 2$.

60. Find the x-coordinate of the point on the graph of $y = \sqrt{x}$ where the tangent line is parallel to the secant line that cuts the curve at $x = 1$ and $x = 4$.

61. Find the coordinates of all points on the graph of $y = 1 - x^2$ at which the tangent line passes through the point $(2, 0)$.

62. Show that any two tangent lines to the parabola $y = ax^2$, $a \neq 0$, intersect at a point that is on the vertical line halfway between the points of tangency.

63. Suppose that L is the tangent line at $x = x_0$ to the graph of the cubic equation $y = ax^3 + bx$. Find the x-coordinate of the point where L intersects the graph a second time.

64. Show that the segment of the tangent line to the graph of $y = 1/x$ that is cut off by the coordinate axes is bisected by the point of tangency.

65. Show that the triangle that is formed by any tangent line to the graph of $y = 1/x$, $x > 0$, and the coordinate axes has an area of 2 square units.

66. Find conditions on a, b, c, and d so that the graph of the polynomial $f(x) = ax^3 + bx^2 + cx + d$ has
(a) exactly two horizontal tangents
(b) exactly one horizontal tangent
(c) no horizontal tangents.

67. Newton's Law of Universal Gravitation states that the magnitude F of the force exerted by a point with mass M on a point with mass m is
$$F = \dfrac{GmM}{r^2}$$
where G is a constant and r is the distance between the bodies. Assuming that the points are moving, find a formula for the instantaneous rate of change of F with respect to r.

68. In the temperature range between $0°C$ and $700°C$ the resistance R [in ohms (Ω)] of a certain platinum resistance thermometer is given by
$$R = 10 + 0.04124T - 1.779 \times 10^{-5}T^2$$
where T is the temperature in degrees Celsius. Where in the interval from $0°C$ to $700°C$ is the resistance of the thermometer most sensitive and least sensitive to temperature changes? [*Hint:* Consider the size of dR/dT in the interval $0 \leq T \leq 700$.]

In Exercises 69 and 70, use a graphing utility to make rough estimates of the intervals on which $f'(x) > 0$, and then find those intervals exactly by differentiating.

69. $f(x) = x - \dfrac{1}{x}$ **70.** $f(x) = \dfrac{5x}{x^2 + 4}$

71. Apply the product rule (3.3.5) twice to show that if f, g, and h are differentiable functions, then $f \cdot g \cdot h$ is differentiable, and
$$(f \cdot g \cdot h)' = f' \cdot g \cdot h + f \cdot g' \cdot h + f \cdot g \cdot h'$$

72. Based on the result in Exercise 71, make a conjecture about a formula for differentiating a product of n functions.

73. Use the formula in Exercise 71 to find
(a) $\dfrac{d}{dx}\left[(2x + 1)\left(1 + \dfrac{1}{x}\right)(x^{-3} + 7)\right]$
(b) $\dfrac{d}{dx}[(x^7 + 2x - 3)^3]$.

74. Use the formula you obtained in Exercise 72 to find
(a) $\dfrac{d}{dx}[x^{-5}(x^2 + 2x)(4 - 3x)(2x^9 + 1)]$
(b) $\dfrac{d}{dx}[(x^2 + 1)^{50}]$.

200 The Derivative

In Exercises 75–78, you are asked to determine whether a piecewise-defined function f is differentiable at a value $x = x_0$, where f is defined by different formulas on different sides of x_0. You may use the following result, which is a consequence of the Mean-Value Theorem (discussed in Section 4.8). **Theorem.** *Let f be continuous at x_0 and suppose that $\lim_{x \to x_0} f'(x)$ exists. Then f is differentiable at x_0, and $f'(x_0) = \lim_{x \to x_0} f'(x)$.*

75. Show that
$$f(x) = \begin{cases} x^2 + x + 1, & x \le 1 \\ 3x, & x > 1 \end{cases}$$
is continuous at $x = 1$. Determine whether f is differentiable at $x = 1$. If so, find the value of the derivative there. Sketch the graph of f.

76. Let
$$f(x) = \begin{cases} x^2 - 16x, & x < 9 \\ 12\sqrt{x}, & x \ge 9 \end{cases}$$
Is f continuous at $x = 9$? Determine whether f is differentiable at $x = 9$. If so, find the value of the derivative there.

77. Let
$$f(x) = \begin{cases} x^2, & x \le 1 \\ \sqrt{x}, & x > 1 \end{cases}$$
Determine whether f is differentiable at $x = 1$. If so, find the value of the derivative there.

78. Let
$$f(x) = \begin{cases} x^3 + \frac{1}{16}, & x < \frac{1}{2} \\ \frac{3}{4}x^2, & x \ge \frac{1}{2} \end{cases}$$

Determine whether f is differentiable at $x = \frac{1}{2}$. If so, find the value of the derivative there.

79. Find all points where f fails to be differentiable. Justify your answer.
(a) $f(x) = |3x - 2|$ (b) $f(x) = |x^2 - 4|$

80. In each part compute f', f'', f''' and then state the formula for $f^{(n)}$.
(a) $f(x) = 1/x$ (b) $f(x) = 1/x^2$
[*Hint:* The expression $(-1)^n$ has a value of 1 if n is even and -1 if n is odd. Use this expression in your answer.]

81. (a) Prove:
$$\frac{d^2}{dx^2}[cf(x)] = c\frac{d^2}{dx^2}[f(x)]$$
$$\frac{d^2}{dx^2}[f(x) + g(x)] = \frac{d^2}{dx^2}[f(x)] + \frac{d^2}{dx^2}[g(x)]$$
(b) Do the results in part (a) generalize to nth derivatives? Justify your answer.

82. Prove:
$$(f \cdot g)'' = f'' \cdot g + 2f' \cdot g' + f \cdot g''$$

83. (a) Find $f^{(n)}(x)$ if $f(x) = x^n$.
(b) Find $f^{(n)}(x)$ if $f(x) = x^k$ and $n > k$, where k is a positive integer.
(c) Find $f^{(n)}(x)$ if
$$f(x) = a_0 + a_1 x + a_2 x^2 + \cdots + a_n x^n$$

84. Let $f(x) = x^8 - 2x + 3$; find
$$\lim_{w \to 2} \frac{f'(w) - f'(2)}{w - 2}$$

85. (a) Prove: If $f''(x)$ exists for each x in (a, b), then both f and f' are continuous on (a, b).
(b) What can be said about the continuity of f and its derivatives if $f^{(n)}(x)$ exists for each x in (a, b)?

3.4 DERIVATIVES OF TRIGONOMETRIC FUNCTIONS

The main objective of this section is to obtain formulas for the derivatives of trigonometric functions.

DERIVATIVES OF THE TRIGONOMETRIC FUNCTIONS

For the purpose of finding derivatives of the trigonometric functions $\sin x$, $\cos x$, $\tan x$, $\cot x$, $\sec x$, and $\csc x$, we will assume that x is measured in radians. We will also need the following limits, which were stated in Theorem 2.6.3 (with x rather than h as the variable):

$$\lim_{h \to 0} \frac{\sin h}{h} = 1 \quad \text{and} \quad \lim_{h \to 0} \frac{1 - \cos h}{h} = 0$$

We begin with the problem of differentiating $\sin x$. Using the alternative form

$$f'(x) = \lim_{h \to 0} \frac{f(x + h) - f(x)}{h}$$

for the definition of a derivative (Formula (11) of Section 3.2), we have

$$\frac{d}{dx}[\sin x] = \lim_{h \to 0} \frac{\sin(x+h) - \sin x}{h}$$

$$= \lim_{h \to 0} \frac{\sin x \cos h + \cos x \sin h - \sin x}{h} \qquad \boxed{\text{By the addition formula for sine}}$$

$$= \lim_{h \to 0} \left[\sin x \left(\frac{\cos h - 1}{h} \right) + \cos x \left(\frac{\sin h}{h} \right) \right]$$

$$= \lim_{h \to 0} \left[\cos x \left(\frac{\sin h}{h} \right) - \sin x \left(\frac{1 - \cos h}{h} \right) \right]$$

Since $\sin x$ and $\cos x$ do not involve h, they remain constant as $h \to 0$; thus,

$$\lim_{h \to 0} (\sin x) = \sin x \quad \text{and} \quad \lim_{h \to 0} (\cos x) = \cos x$$

Consequently,

$$\frac{d}{dx}[\sin x] = \cos x \cdot \lim_{h \to 0} \left(\frac{\sin h}{h} \right) - \sin x \cdot \lim_{h \to 0} \left(\frac{1 - \cos h}{h} \right)$$

$$= \cos x \cdot (1) - \sin x \cdot (0) = \cos x$$

Thus, we have shown that

$$\frac{d}{dx}[\sin x] = \cos x \qquad (1)$$

The derivative of $\cos x$ can be obtained similarly, resulting in the formula

$$\frac{d}{dx}[\cos x] = -\sin x \qquad (2)$$

The derivatives of the remaining trigonometric functions are

$$\frac{d}{dx}[\tan x] = \sec^2 x \qquad \frac{d}{dx}[\sec x] = \sec x \tan x \qquad (3\text{–}4)$$

$$\frac{d}{dx}[\cot x] = -\csc^2 x \qquad \frac{d}{dx}[\csc x] = -\csc x \cot x \qquad (5\text{–}6)$$

These can all be obtained from (1) and (2) using the relationships

$$\tan x = \frac{\sin x}{\cos x}, \qquad \cot x = \frac{\cos x}{\sin x}, \qquad \sec x = \frac{1}{\cos x}, \qquad \csc x = \frac{1}{\sin x}$$

For example,

$$\frac{d}{dx}[\tan x] = \frac{d}{dx}\left[\frac{\sin x}{\cos x} \right] = \frac{\cos x \cdot \frac{d}{dx}[\sin x] - \sin x \cdot \frac{d}{dx}[\cos x]}{\cos^2 x}$$

$$= \frac{\cos x \cdot \cos x - \sin x \cdot (-\sin x)}{\cos^2 x} = \frac{\cos^2 x + \sin^2 x}{\cos^2 x} = \frac{1}{\cos^2 x} = \sec^2 x$$

REMARK.   The derivative formulas for the trigonometric functions should be memorized. An easy way of doing this is discussed in Exercise 42. Moreover, we emphasize again that in all of the derivative formulas for the trigonometric functions, x is measured in radians.

Example 1   Find $f'(x)$ if $f(x) = x^2 \tan x$.

Solution. Using the product rule and Formula (3), we obtain

$$f'(x) = x^2 \cdot \frac{d}{dx}[\tan x] + \tan x \cdot \frac{d}{dx}[x^2] = x^2 \sec^2 x + 2x \tan x$$ ◄

Example 2   Find dy/dx if $y = \dfrac{\sin x}{1 + \cos x}$.

Solution. Using the quotient rule together with Formulas (1) and (2) we obtain

$$\frac{dy}{dx} = \frac{(1 + \cos x) \cdot \dfrac{d}{dx}[\sin x] - \sin x \cdot \dfrac{d}{dx}[1 + \cos x]}{(1 + \cos x)^2}$$

$$= \frac{(1 + \cos x)(\cos x) - (\sin x)(-\sin x)}{(1 + \cos x)^2}$$

$$= \frac{\cos x + \cos^2 x + \sin^2 x}{(1 + \cos x)^2} = \frac{\cos x + 1}{(1 + \cos x)^2} = \frac{1}{1 + \cos x}$$ ◄

Example 3   Find $y''(\pi/4)$ if $y(x) = \sec x$.

Solution.

$$y'(x) = \sec x \tan x$$

$$y''(x) = \sec x \cdot \frac{d}{dx}[\tan x] + \tan x \cdot \frac{d}{dx}[\sec x]$$

$$= \sec x \cdot \sec^2 x + \tan x \cdot \sec x \tan x$$

$$= \sec^3 x + \sec x \tan^2 x$$

Thus,

$$y''(\pi/4) = \sec^3(\pi/4) + \sec(\pi/4) \tan^2(\pi/4)$$

$$= (\sqrt{2})^3 + (\sqrt{2})(1)^2 = 3\sqrt{2}$$ ◄

Example 4   On a sunny day, a 50-ft flagpole casts a shadow that changes with the angle of elevation of the Sun. Let s be the length of the shadow and θ the angle of elevation of the Sun (Figure 3.4.1). Find the rate at which the length of the shadow is changing with respect to θ when $\theta = 45°$. Express your answer in units of feet/degree.

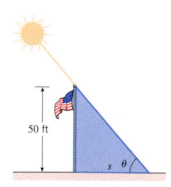

50 ft

Figure 3.4.1

Solution. The variables s and θ are related by $\tan \theta = 50/s$, or equivalently,

$$s = 50 \cot \theta \qquad (7)$$

If θ is measured in radians, then Formula (5) is applicable, which yields

$$\frac{ds}{d\theta} = -50 \csc^2 \theta$$

which is the rate of change of shadow length with respect to the elevation angle θ in units of feet/radian. When $\theta = 45°$ (or equivalently, $\theta = \pi/4$ radians), we obtain

$$\left. \frac{ds}{d\theta} \right|_{\theta = \pi/4} = -50 \csc^2(\pi/4) = -100 \text{ feet/radian}$$

Converting radians (rad) to degrees (deg) yields

$$-100 \frac{\text{ft}}{\text{rad}} \cdot \frac{\pi}{180} \frac{\text{rad}}{\text{deg}} = -\frac{5}{9} \pi \frac{\text{ft}}{\text{deg}} \approx -1.75 \text{ ft/deg}$$

Thus, when $\theta = 45°$, the shadow length is decreasing (because of the minus sign) at an approximate rate of 1.75 ft/deg increase in the angle of elevation. ◄

EXERCISE SET 3.4 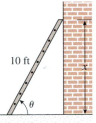 Graphing Utility

In Exercises 1–18, find $f'(x)$.

1. $f(x) = 2\cos x - 3\sin x$ **2.** $f(x) = \sin x \cos x$

3. $f(x) = \dfrac{\sin x}{x}$ **4.** $f(x) = x^2 \cos x$

5. $f(x) = x^3 \sin x - 5\cos x$ **6.** $f(x) = \dfrac{\cos x}{x \sin x}$

7. $f(x) = \sec x - \sqrt{2}\tan x$ **8.** $f(x) = (x^2 + 1)\sec x$

9. $f(x) = \sec x \tan x$ **10.** $f(x) = \dfrac{\sec x}{1 + \tan x}$

11. $f(x) = \csc x \cot x$

12. $f(x) = x - 4\csc x + 2\cot x$

13. $f(x) = \dfrac{\cot x}{1 + \csc x}$ **14.** $f(x) = \dfrac{\csc x}{\tan x}$

15. $f(x) = \sin^2 x + \cos^2 x$ **16.** $f(x) = \dfrac{1}{\cot x}$

17. $f(x) = \dfrac{\sin x \sec x}{1 + x \tan x}$

18. $f(x) = \dfrac{(x^2 + 1)\cot x}{3 - \cos x \csc x}$

In Exercises 19–24, find d^2y/dx^2.

19. $y = x \cos x$ **20.** $y = \csc x$

21. $y = x \sin x - 3\cos x$ **22.** $y = x^2 \cos x + 4\sin x$

23. $y = \sin x \cos x$ **24.** $y = \tan x$

25. Find the equation of the line tangent to the graph of $\tan x$ at
(a) $x = 0$ (b) $x = \pi/4$ (c) $x = -\pi/4$.

26. Find the equation of the line tangent to the graph of $\sin x$ at
(a) $x = 0$ (b) $x = \pi$ (c) $x = \pi/4$.

27. (a) Show that $y = x \sin x$ is a solution to $y'' + y = 2\cos x$.
(b) Show that $y = x \sin x$ is a solution of the equation $y^{(4)} + y'' = -2\cos x$.

28. (a) Show that $y = \cos x$ and $y = \sin x$ are solutions of the equation $y'' + y = 0$.
(b) Show that $y = A \sin x + B \cos x$ is a solution of the equation $y'' + y = 0$ for all constants A and B.

29. Find all values in the interval $[-2\pi, 2\pi]$ at which the graph of f has a horizontal tangent line.
(a) $f(x) = \sin x$ (b) $f(x) = x + \cos x$
(c) $f(x) = \tan x$ (d) $f(x) = \sec x$

 30. (a) Use a graphing utility to make rough estimates of the values in the interval $[0, 2\pi]$ at which the graph of $y = \sin x \cos x$ has a horizontal tangent line.
(b) Find the exact locations of the points where the graph has a horizontal tangent line.

31. A 10-ft ladder leans against a wall at an angle θ with the horizontal, as shown in the accompanying figure. The top of the ladder is x feet above the ground. If the bottom of the ladder is pushed toward the wall, find the rate at which

x changes with respect to θ when $\theta = 60°$. Express the answer in units of feet/degree.

32. An airplane is flying on a horizontal path at a height of 3800 ft, as shown in the accompanying figure. At what rate is the distance s between the airplane and the fixed point P changing with respect to θ when $\theta = 30°$? Express the answer in units of feet/degree.

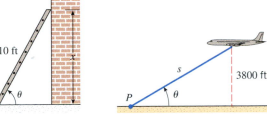

Figure Ex-31   Figure Ex-32

33. A searchlight is trained on the side of a tall building. As the light rotates, the spot it illuminates moves up and down the side of the building. That is, the distance D between ground level and the illuminated spot on the side of the building is a function of the angle θ formed by the light beam and the horizontal (see the accompanying figure). If the searchlight is located 50 m from the building, find the rate at which D is changing with respect to θ when $\theta = 45°$. Express your answer in units of meters/degree.

34. An Earth-observing satellite can see only a portion of the Earth's surface. The satellite has horizon sensors that can detect the angle θ shown in the accompanying figure. Let r be the radius of the Earth (assumed spherical) and h the distance of the satellite from the Earth's surface.
(a) Show that $h = r(\csc\theta - 1)$.
(b) Using $r = 6378$ km, find the rate at which h is changing with respect to θ when $\theta = 30°$. Express the answer in units of kilometers/degree. [Adapted from *Space Mathematics*, NASA, 1985.]

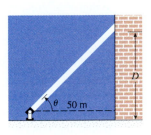   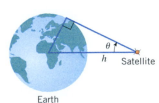

Figure Ex-33   Figure Ex-34

In Exercises 35 and 36, make a conjecture about the derivative by calculating the first few derivatives and observing the resulting pattern.

35. (a) $\dfrac{d^{87}}{dx^{87}}[\sin x]$ (b) $\dfrac{d^{100}}{dx^{100}}[\cos x]$

36. $\dfrac{d^{17}}{dx^{17}}[x\sin x]$

37. In each part, determine where f is differentiable.
  (a) $f(x)=\sin x$ (b) $f(x)=\cos x$
  (c) $f(x)=\tan x$ (d) $f(x)=\cot x$
  (e) $f(x)=\sec x$ (f) $f(x)=\csc x$
  (g) $f(x)=\dfrac{1}{1+\cos x}$ (h) $f(x)=\dfrac{1}{\sin x\cos x}$
  (i) $f(x)=\dfrac{\cos x}{2-\sin x}$

38. (a) Derive Formula (2) using the definition of a derivative.
  (b) Use Formulas (1) and (2) to obtain (5).
  (c) Use Formula (2) to obtain (4).
  (d) Use Formula (1) to obtain (6).

39. Let $f(x)=\cos x$. Find all positive integers n for which $f^{(n)}(x)=\sin x$.

40. (a) Show that $\displaystyle\lim_{h\to 0}\frac{\tan h}{h}=1$.
  (b) Use the result in part (a) to help derive the formula for the derivative of $\tan x$ directly from the definition of a derivative.

41. Without using any trigonometric identities, find
$$\lim_{x\to 0}\frac{\tan(x+y)-\tan y}{x}$$
[*Hint:* Relate the given limit to the definition of the derivative of an appropriate function of y.]

42. Let us agree to call the functions $\cos x$, $\cot x$, and $\csc x$ the *cofunctions* of $\sin x$, $\tan x$, and $\sec x$, respectively. Convince yourself that the derivative of any cofunction can be obtained from the derivative of the corresponding function by introducing a minus sign and replacing each function in the derivative by its cofunction. Memorize the derivatives of $\sin x$, $\tan x$, and $\sec x$ and then use the above observation to deduce the derivatives of the cofunctions.

43. The derivative formulas for $\sin x$, $\cos x$, $\tan x$, $\cot x$, $\sec x$, and $\csc x$ were obtained under the assumption that x is measured in radians. This exercise shows that different (more complicated) formulas result if x is measured in degrees. Prove that if h and x are degree measures, then
  (a) $\displaystyle\lim_{h\to 0}\frac{\cos h-1}{h}=0$ (b) $\displaystyle\lim_{h\to 0}\frac{\sin h}{h}=\frac{\pi}{180}$
  (c) $\dfrac{d}{dx}[\sin x]=\dfrac{\pi}{180}\cos x$.

3.5 THE CHAIN RULE

In this section we will derive a formula that expresses the derivative of a composition $f\circ g$ in terms of the derivatives of f and g. This formula will enable us to differentiate complicated functions using known derivatives of simpler functions.

DERIVATIVES OF COMPOSITIONS

3.5.1 PROBLEM. *If we know the derivatives of f and g, how can we use this information to find the derivative of the composition $f\circ g$?*

The key to solving this problem is to introduce dependent variables
$$y=(f\circ g)(x)=f(g(x))\quad\text{and}\quad u=g(x)$$
so that $y=f(u)$. We are interested in using the known derivatives
$$\frac{dy}{du}=f'(u)\quad\text{and}\quad\frac{du}{dx}=g'(x)$$
to find the unknown derivative
$$\frac{dy}{dx}=\frac{d}{dx}[f(g(x))]$$
Stated another way, we are interested in using the known rates of change dy/du and du/dx to find the unknown rate of change dy/dx. But intuition suggests that rates of change multiply. For example, if y changes at 4 times the rate of change of u and u changes at 2 times the rate of change of x, then y changes at $4\times 2=8$ times the rate of change of x. This suggests that
$$\frac{dy}{dx}=\frac{dy}{du}\cdot\frac{du}{dx}$$
These ideas are formalized in the following theorem.

3.5.2 THEOREM (*The Chain Rule*). *If g is differentiable at x and f is differentiable at g(x), then the composition f ∘ g is differentiable at x. Moreover,*

$$(f \circ g)'(x) = f'(g(x))g'(x)$$

Alternatively, if

$$y = f(g(x)) \quad and \quad u = g(x)$$

then y = f(u) and

$$\frac{dy}{dx} = \frac{dy}{du} \cdot \frac{du}{dx} \tag{1}$$

The proof of this result is given in Appendix G.

Example 1   Find $h'(x)$ if $h(x) = 4\cos(x^3)$.

Solution.   We first find functions f and g such that $f \circ g = h$. Observe that if $g(x) = x^3$ and $f(u) = 4\cos u$, then

$$(f \circ g)(x) = f(g(x)) = 4\cos(g(x)) = 4\cos(x^3) = h(x)$$

Also,

$$f'(u) = -4\sin u \quad and \quad g'(x) = 3x^2$$

Using the chain rule,

$$h'(x) = f'(g(x))\,g'(x) = (-4\sin g(x))(3x^2) = -12x^2\sin(x^3)$$

Alternatively, set $y = h(x)$ and let $u = x^3$. Then $y = 4\cos u$. By the form of the chain rule in Formula (1),

$$h'(x) = \frac{dy}{dx} = \frac{dy}{du} \cdot \frac{du}{dx} = \frac{d}{du}[4\cos u] \cdot \frac{d}{dx}[x^3]$$
$$= (-4\sin u) \cdot (3x^2) = (-4\sin(x^3)) \cdot (3x^2) = -12x^2\sin(x^3) \quad \blacktriangleleft$$

Formula (1) is easy to remember because the left side is exactly what results if we "cancel" the du's on the right side. This "canceling" device provides a good way to remember the chain rule when variables other than x, y, and u are used.

Example 2   Find dw/dt if $w = \tan x$ and $x = 4t^3 + t$.

Solution.   In this case the chain rule takes the form

$$\frac{dw}{dt} = \frac{dw}{dx} \cdot \frac{dx}{dt} = \frac{d}{dx}[\tan x] \cdot \frac{d}{dt}[4t^3 + t]$$
$$= (\sec^2 x)(12t^2 + 1) = (12t^2 + 1)\sec^2(4t^3 + t) \quad \blacktriangleleft$$

AN ALTERNATIVE APPROACH TO USING THE CHAIN RULE

Although Formula (1) is useful, it is sometimes unwieldy because it involves so many variables. As you become more comfortable with the chain rule, you may want to dispense with actually writing out all these variables. To accomplish this, it is helpful to note that since $(f \circ g)(x) = f(g(x))$, the chain rule may be written in the form

$$\frac{d}{dx}[f(g(x))] = (f \circ g)'(x) = f'(g(x))g'(x)$$

If we call $g(x)$ the "inside function" and f the "outside function," then this equation states that:

The derivative of $f(g(x))$ is the derivative of the outside function evaluated at the inside function times the derivative of the inside function.

That is,

$$\frac{d}{dx}[f(g(x))] = \underbrace{f'(g(x))}_{\substack{\text{Derivative of} \\ \text{the outside} \\ \text{evaluated at} \\ \text{the inside}}} \cdot \underbrace{g'(x)}_{\substack{\text{Derivative} \\ \text{of the inside}}} \tag{2}$$

For example,

$$\frac{d}{dx}[\cos(x^2 + 9)] = \underbrace{-\sin(x^2 + 9)}_{\substack{\text{Derivative of the} \\ \text{outside evaluated} \\ \text{at the inside}}} \cdot \underbrace{2x}_{\substack{\text{Derivative} \\ \text{of the inside}}}$$

$$\frac{d}{dx}[\tan^2 x] = \frac{d}{dx}\left[(\tan x)^2\right] = \underbrace{(2\tan x)}_{\substack{\text{Derivative of} \\ \text{the outside} \\ \text{evaluated at} \\ \text{the inside}}} \cdot \underbrace{(\sec^2 x)}_{\substack{\text{Derivative} \\ \text{of the inside}}} = 2\tan x \sec^2 x$$

Substituting $u = g(x)$ into (2) yields the following alternative form:

$$\frac{d}{dx}[f(u)] = f'(u)\frac{du}{dx} \tag{3}$$

For example, to differentiate the function

$$f(x) = \left(x^2 - x + 1\right)^{23} \tag{4}$$

we can let $u = x^2 - x + 1$ and then apply (3) to obtain

$$\frac{d}{dx}\left[(x^2 - x + 1)^{23}\right] = \frac{d}{dx}[u^{23}] = \underbrace{23u^{22}}_{f'(u)}\frac{du}{dx}$$

$$= 23\left(x^2 - x + 1\right)^{22}\frac{d}{dx}[x^2 - x + 1]$$

$$= 23\left(x^2 - x + 1\right)^{22} \cdot (2x - 1)$$

More generally, if u were any other differentiable function of x, the pattern of computations would be virtually the same. For example, if $u = \cos x$, then

$$\frac{d}{dx}[\cos^{23} x] = \frac{d}{dx}[u^{23}] = 23u^{22}\frac{du}{dx} = 23\cos^{22} x \frac{d}{dx}[\cos x]$$

$$= 23\cos^{22} x \cdot (-\sin x) = -23\sin x \cos^{22} x$$

In both of the preceding computations, the chain rule took the form

$$\frac{d}{dx}[u^{23}] = 23u^{22}\frac{du}{dx} \tag{5}$$

This formula is a generalization of the more basic formula

$$\frac{d}{dx}[x^{23}] = 23x^{22} \tag{6}$$

In fact, in the special case where $u = x$, Formula (5) reduces to (6) since

$$\frac{d}{dx}[u^{23}] = 23u^{22}\frac{du}{dx} = 23x^{22}\frac{d[x]}{dx} = 23x^{22}$$

Table 3.5.1 contains a list of *generalized derivative formulas* that are consequences of (3).

Table 3.5.1

GENERALIZED DERIVATIVE FORMULAS

$$\frac{d}{dx}[u^n] = nu^{n-1}\frac{du}{dx} \quad (n \text{ an integer}) \qquad \frac{d}{dx}[\sqrt{u}] = \frac{1}{2\sqrt{u}}\frac{du}{dx}$$

$$\frac{d}{dx}[\sin u] = \cos u \frac{du}{dx} \qquad\qquad \frac{d}{dx}[\cos u] = -\sin u \frac{du}{dx}$$

$$\frac{d}{dx}[\tan u] = \sec^2 u \frac{du}{dx} \qquad\qquad \frac{d}{dx}[\cot u] = -\csc^2 u \frac{du}{dx}$$

$$\frac{d}{dx}[\sec u] = \sec u \tan u \frac{du}{dx} \qquad\qquad \frac{d}{dx}[\csc u] = -\csc u \cot u \frac{du}{dx}$$

Example 3 Find

(a) $\dfrac{d}{dx}[\sin(2x)]$ (b) $\dfrac{d}{dx}[\tan(x^2+1)]$ (c) $\dfrac{d}{dx}[\sqrt{x^3+\csc x}]$

(d) $\dfrac{d}{dx}\left[(1+x^5\cot x)^{-8}\right]$ (e) $\dfrac{d}{dx}\left[\dfrac{1}{x^3+2x-3}\right]$

Solution (a). Taking $u = 2x$ in the generalized derivative formula for $\sin u$ yields

$$\frac{d}{dx}[\sin(2x)] = \frac{d}{dx}[\sin u] = \cos u \frac{du}{dx} = \cos 2x \cdot \frac{d}{dx}[2x] = \cos 2x \cdot 2 = 2\cos 2x$$

Solution (b). Taking $u = x^2 + 1$ in the generalized derivative formula for $\tan u$ yields

$$\frac{d}{dx}[\tan(x^2+1)] = \frac{d}{dx}[\tan u] = \sec^2 u \frac{du}{dx}$$

$$= \sec^2(x^2+1) \cdot \frac{d}{dx}[x^2+1] = \sec^2(x^2+1) \cdot 2x$$

$$= 2x\sec^2(x^2+1)$$

Solution (c). Taking $u = x^3 + \csc x$ in the generalized derivative formula for $\sqrt{u}$ yields

$$\frac{d}{dx}[\sqrt{x^3+\csc x}] = \frac{d}{dx}[\sqrt{u}] = \frac{1}{2\sqrt{u}}\frac{du}{dx} = \frac{1}{2\sqrt{x^3+\csc x}} \cdot \frac{d}{dx}[x^3+\csc x]$$

$$= \frac{1}{2\sqrt{x^3+\csc x}} \cdot (3x^2 - \csc x \cot x) = \frac{3x^2 - \csc x \cot x}{2\sqrt{x^3+\csc x}}$$

Solution (d). Taking $u = 1 + x^5\cot x$ in the generalized derivative formula for u^{-8} yields

$$\frac{d}{dx}\left[(1+x^5\cot x)^{-8}\right] = \frac{d}{dx}[u^{-8}] = -8u^{-9}\frac{du}{dx}$$

$$= -8\left(1+x^5\cot x\right)^{-9} \cdot \frac{d}{dx}[1+x^5\cot x]$$

$$= -8\left(1+x^5\cot x\right)^{-9} \cdot (x^5(-\csc^2 x) + 5x^4\cot x)$$

$$= (8x^5\csc^2 x - 40x^4\cot x)\left(1+x^5\cot x\right)^{-9}$$

Solution (e). Taking $u = x^3 + 2x - 3$ in the generalized derivative formula for u^{-1} yields

$$\frac{d}{dx}\left[\frac{1}{x^3+2x-3}\right] = \frac{d}{dx}[(x^3+2x-3)^{-1}] = \frac{d}{dx}[u^{-1}]$$

$$= -u^{-2}\frac{du}{dx} = -(x^3+2x-3)^{-2}\frac{d}{dx}[x^3+2x-3]$$

$$= -(x^3+2x-3)^{-2}(3x^2+2) = -\frac{3x^2+2}{(x^3+2x-3)^2} \qquad \blacktriangleleft$$

Sometimes you will have to make adjustments in notation or apply the chain rule more than once to calculate a derivative.

Example 4 Find

(a) $\dfrac{d}{dx}[\sin(\sqrt{1+\cos x})]$ (b) $\dfrac{d\mu}{dt}$ if $\mu = \sec\sqrt{\omega t}$ (ω constant)

Solution (a). Taking $u = \sqrt{1+\cos x}$ in the generalized derivative formula for $\sin u$ yields

$$\frac{d}{dx}[\sin(\sqrt{1+\cos x})] = \frac{d}{dx}[\sin u] = \cos u\,\frac{du}{dx}$$

$$= \cos(\sqrt{1+\cos x}) \cdot \frac{d}{dx}[\sqrt{1+\cos x}]$$

> We use the generalized derivative formula for $\sqrt{u}$ with $u = 1+\cos x$.

$$= \cos(\sqrt{1+\cos x}) \cdot \frac{-\sin x}{2\sqrt{1+\cos x}}$$

$$= -\frac{\sin x\cos(\sqrt{1+\cos x})}{2\sqrt{1+\cos x}}$$

Solution (b).

$$\frac{d\mu}{dt} = \frac{d}{dt}[\sec\sqrt{\omega t}] = \sec\sqrt{\omega t}\,\tan\sqrt{\omega t}\,\frac{d}{dt}[\sqrt{\omega t}]$$

> We used the generalized derivative formula for $\sec u$ with $u = \sqrt{\omega t}$.

$$= \sec\sqrt{\omega t}\,\tan\sqrt{\omega t}\,\frac{\omega}{2\sqrt{\omega t}}$$

> We used the generalized derivative formula for $\sqrt{u}$ with $u = \omega t$.

◀

DIFFERENTIATING USING COMPUTER ALGEBRA SYSTEMS

Although the chain rule makes it possible to differentiate extremely complicated functions, the computations can be time-consuming to execute by hand. For complicated derivatives engineers and scientists often use computer algebra systems such as *Mathematica*, *Maple*, and *Derive*. For example, although we have all of the mathematical tools to perform the differentiation

$$\frac{d}{dx}\left[\frac{(x^2+1)^{10}\sin^3(\sqrt{x})}{\sqrt{1+\csc x}}\right] \tag{7}$$

by hand, the computations are sufficiently tedious that it would be more efficient to use a computer algebra system.

• FOR THE READER.   If you have a CAS, use it to perform the differentiation in (7).

EXERCISE SET 3.5   ~ Graphing Utility   C CAS

1. Given that $f'(0) = 2$, $g(0) = 0$, and $g'(0) = 3$, find $(f \circ g)'(0)$.

2. Given that $f'(9) = 5$, $g(2) = 9$, and $g'(2) = -3$, find $(f \circ g)'(2)$.

3. Let $f(x) = x^5$ and $g(x) = 2x - 3$.
   (a) Find $(f \circ g)(x)$ and $(f \circ g)'(x)$.
   (b) Find $(g \circ f)(x)$ and $(g \circ f)'(x)$.

4. Let $f(x) = 5\sqrt{x}$ and $g(x) = 4 + \cos x$.
   (a) Find $(f \circ g)(x)$ and $(f \circ g)'(x)$.
   (b) Find $(g \circ f)(x)$ and $(g \circ f)'(x)$.

5. Given the following table of values, find the indicated derivatives in parts (a) and (b).

x	$f(x)$	$f'(x)$	$g(x)$	$g'(x)$
3	5	-2	5	7
5	3	-1	12	4

   (a) $F'(3)$, where $F(x) = f(g(x))$
   (b) $G'(3)$, where $G(x) = g(f(x))$

6. Given the following table of values, find the indicated derivatives in parts (a) and (b).

x	$f(x)$	$f'(x)$	$g(x)$	$g'(x)$
-1	2	3	2	-3
2	0	4	1	-5

(a) $F'(-1)$, where $F(x) = f(g(x))$
(b) $G'(-1)$, where $G(x) = g(f(x))$

In Exercises 7–26, find $f'(x)$.

7. $f(x) = (x^3 + 2x)^{37}$

8. $f(x) = (3x^2 + 2x - 1)^6$

9. $f(x) = \left(x^3 - \dfrac{7}{x}\right)^{-2}$

10. $f(x) = \dfrac{1}{(x^5 - x + 1)^9}$

11. $f(x) = \dfrac{4}{(3x^2 - 2x + 1)^3}$

12. $f(x) = \sqrt{x^3 - 2x + 5}$

13. $f(x) = \sqrt{4 + \sqrt{3x}}$

14. $f(x) = \sin^3 x$

15. $f(x) = \sin(x^3)$

16. $f(x) = \cos^2(3\sqrt{x})$

17. $f(x) = 4\cos^5 x$

18. $f(x) = \csc(x^3)$

19. $f(x) = \sin\left(\dfrac{1}{x^2}\right)$

20. $f(x) = \tan^4(x^3)$

21. $f(x) = 2\sec^2(x^7)$

22. $f(x) = \cos^3\left(\dfrac{x}{x+1}\right)$

23. $f(x) = \sqrt{\cos(5x)}$

24. $f(x) = \sqrt{3x - \sin^2(4x)}$

25. $f(x) = [x + \csc(x^3 + 3)]^{-3}$

26. $f(x) = [x^4 - \sec(4x^2 - 2)]^{-4}$

In Exercises 27–40, find dy/dx.

27. $y = x^3 \sin^2(5x)$

28. $y = \sqrt{x} \tan^3(\sqrt{x})$

29. $y = x^5 \sec(1/x)$

30. $y = \dfrac{\sin x}{\sec(3x + 1)}$

31. $y = \cos(\cos x)$

32. $y = \sin(\tan 3x)$

33. $y = \cos^3(\sin 2x)$

34. $y = \dfrac{1 + \csc(x^2)}{1 - \cot(x^2)}$

35. $y = (5x + 8)^{13} (x^3 + 7x)^{12}$

36. $y = (2x - 5)^2 (x^2 + 4)^3$

37. $y = \left(\dfrac{x - 5}{2x + 1}\right)^3$

38. $y = \left(\dfrac{1 + x^2}{1 - x^2}\right)^{17}$

39. $y = \dfrac{(2x + 3)^3}{(4x^2 - 1)^8}$

40. $y = [1 + \sin^3(x^5)]^{12}$

In Exercises 41 and 42, use a CAS to find dy/dx.

C **41.** $y = [x \sin 2x + \tan^4(x^7)]^5$

C **42.** $y = \tan^4\left(2 + \dfrac{(7 - x)\sqrt{3x^2 + 5}}{x^3 + \sin x}\right)$

In Exercises 43–50, find an equation for the tangent line to the graph at the specified value of x.

43. $y = x \cos 3x$, $x = \pi$

44. $y = \sin(1 + x^3)$, $x = -3$

45. $y = \sec^3\left(\dfrac{\pi}{2} - x\right)$, $x = -\dfrac{\pi}{2}$

46. $y = \left(x - \dfrac{1}{x}\right)^3$, $x = 2$

47. $y = \tan(4x^2)$, $x = \sqrt{\pi}$

48. $y = 3\cot^4 x$, $x = \dfrac{\pi}{4}$

49. $y = x^2\sqrt{5 - x^2}$, $x = 1$

50. $y = \dfrac{x}{\sqrt{1 - x^2}}$, $x = 0$

In Exercises 51–54, find d^2y/dx^2.

51. $y = x \cos(5x) - \sin^2 x$

52. $y = \sin(3x^2)$

53. $y = \dfrac{1 + x}{1 - x}$

54. $y = x \tan\left(\dfrac{1}{x}\right)$

In Exercises 55–58, find the indicated derivative.

55. $y = \cot^3(\pi - \theta)$; find $\dfrac{dy}{d\theta}$.

56. $\lambda = \left(\dfrac{au + b}{cu + d}\right)^6$; find $\dfrac{d\lambda}{du}$ $(a, b, c, d$ constants$)$.

57. $\dfrac{d}{d\omega}[a \cos^2 \pi\omega + b \sin^2 \pi\omega]$ $(a, b$ constants$)$.

58. $x = \csc^2\left(\dfrac{\pi}{3} - y\right)$; find $\dfrac{dx}{dy}$.

59. (a) Use a graphing utility to obtain the graph of the function $f(x) = x\sqrt{4 - x^2}$.
(b) Use the graph in part (a) to make a rough sketch of the graph of f'.
(c) Find $f'(x)$, and then check your work in part (b) by using the graphing utility to obtain the graph of f'.
(d) Find the equation of the tangent line to the graph of f at $x = 1$, and graph f and the tangent line together.

60. (a) Use a graphing utility to obtain the graph of the function $f(x) = \sin x^2 \cos x$ over the interval $[-\pi/2, \pi/2]$.
(b) Use the graph in part (a) to make a rough sketch of the graph of f' over the interval.
(c) Find $f'(x)$, and then check your work in part (b) by using the graphing utility to obtain the graph of f' over the interval.
(d) Find the equation of the tangent line to the graph of f at $x = 1$, and graph f and the tangent line together over the interval.

61. If an object suspended from a spring is displaced vertically from its equilibrium position by a small amount and re-leased, and if the air resistance and the mass of the spring are ignored, then the resulting oscillation of the object is called **simple harmonic motion**. Under appropriate conditions the displacement y from equilibrium in terms of time t is given by

$$y = A \cos \omega t$$

where A is the initial displacement at time $t = 0$, and ω is a constant that depends on the mass of the object and the stiffness of the spring (see the accompanying figure). The constant $|A|$ is called the **amplitude** of the motion and ω the **angular frequency**.

(a) Show that

$$\frac{d^2 y}{dt^2} = -\omega^2 y$$

(b) The **period** T is the time required to make one complete oscillation. Show that $T = 2\pi/\omega$.

(c) The **frequency** f of the vibration is the number of oscillations per unit time. Find f in terms of the period T.

(d) Find the amplitude, period, and frequency of an object that is executing simple harmonic motion given by $y = 0.6 \cos 15t$, where t is in seconds and y is in centimeters.

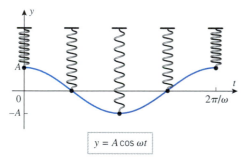

$$y = A \cos \omega t$$

Figure Ex-61

62. Find the value of the constant A so that $y = A \sin 3t$ satisfies the equation

$$\frac{d^2 y}{dt^2} + 2y = 4 \sin 3t$$

63. The accompanying figure shows the graph of atmospheric pressure p (lb/in^2) versus the altitude h (mi) above sea level.

(a) From the graph and the tangent line at $h = 2$ shown on the graph, estimate the values of p and dp/dh at an altitude of 2 mi.

(b) If the altitude of a space vehicle is increasing at the rate of 0.3 mi/s at the instant when it is 2 mi above sea level, how fast is the pressure changing with time at this instant?

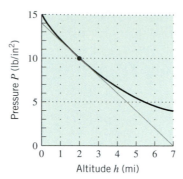

Figure Ex-63

64. The force F (in pounds) acting at an angle θ with the horizontal that is needed to drag a crate weighing W pounds along a horizontal surface at a constant velocity is given by

$$F = \frac{\mu W}{\cos \theta + \mu \sin \theta}$$

where μ is a constant called the **coefficient of sliding friction** between the crate and the surface (see the accompanying figure). Suppose that the crate weighs 150 lb and that $\mu = 0.3$.

(a) Find $dF/d\theta$ when $\theta = 30°$. Express the answer in units of pounds/degree.

(b) Find dF/dt when $\theta = 30°$ if θ is decreasing at the rate of $0.5°$/s at this instant.

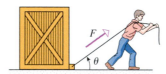

Figure Ex-64

65. Recall that

$$\frac{d}{dx}(|x|) = \begin{cases} 1, & x > 0 \\ -1, & x < 0 \end{cases}$$

Use this result and the chain rule to find

$$\frac{d}{dx}(|\sin x|)$$

for nonzero x in the interval $(-\pi, \pi)$.

66. Use the derivative formula for $\sin x$ and the identity

$$\cos x = \sin\left(\frac{\pi}{2} - x\right)$$

to obtain the derivative formula for $\cos x$.

67. Let

$$f(x) = \begin{cases} x \sin \dfrac{1}{x}, & x \neq 0 \\ 0, & x = 0 \end{cases}$$

(a) Show that f is continuous at $x = 0$.

(b) Use Definition 3.2.1 to show that $f'(0)$ does not exist.

(c) Find $f'(x)$ for $x \neq 0$.

(d) Determine whether $\lim_{x \to 0} f'(x)$ exists.

68. Let

$$f(x) = \begin{cases} x^2 \sin \dfrac{1}{x}, & x \neq 0 \\ 0, & x = 0 \end{cases}$$

(a) Show that f is continuous at $x = 0$.

(b) Use Definition 3.2.1 to find $f'(0)$.

(c) Find $f'(x)$ for $x \neq 0$.

(d) Show that f' is not continuous at $x = 0$.

69. Given the following table of values, find the indicated derivatives in parts (a) and (b).

x	$f(x)$	$f'(x)$
2	1	7
8	5	-3

(a) $g'(2)$, where $g(x) = [f(x)]^3$

(b) $h'(2)$, where $h(x) = f(x^3)$

70. Given that $f'(x) = \sqrt{3x+4}$ and $g(x) = x^2 - 1$, find $F'(x)$ if $F(x) = f(g(x))$.

71. Given that $f'(x) = \dfrac{x}{x^2+1}$ and $g(x) = \sqrt{3x-1}$, find $F'(x)$ if $F(x) = f(g(x))$.

72. Find $f'(x^2)$ if $\dfrac{d}{dx}[f(x^2)] = x^2$.

73. Find $\dfrac{d}{dx}[f(x)]$ if $\dfrac{d}{dx}[f(3x)] = 6x$.

74. Recall that a function f is **even** if $f(-x) = f(x)$ and **odd** if $f(-x) = -f(x)$, for all x in the domain of f. Assuming that f is differentiable, prove:

(a) f' is odd if f is even

(b) f' is even if f is odd.

75. Draw some pictures to illustrate the results in Exercise 74, and write a paragraph that gives an informal explanation of why the results are true.

76. Let $y = f_1(u)$, $u = f_2(v)$, $v = f_3(w)$, and $w = f_4(x)$. Express dy/dx in terms of dy/du, dw/dx, du/dv, and dv/dw.

77. Find a formula for

$$\frac{d}{dx}[f(g(h(x)))]$$

3.6 IMPLICIT DIFFERENTIATION

In earlier sections we were concerned with differentiating functions that were given by equations of the form $y = f(x)$. In this section we will consider methods for differentiating functions for which it is inconvenient or impossible to express them in this form.

FUNCTIONS DEFINED EXPLICITLY AND IMPLICITLY

An equation of the form $y = f(x)$ is said to define y **explicitly** as a function of x because the variable y appears alone on one side of the equation. However, sometimes functions are defined by equations in which y is not alone on one side; for example, the equation

$$yx + y + 1 = x \tag{1}$$

is not of the form $y = f(x)$. However, this equation still defines y as a function of x since it can be rewritten as

$$y = \frac{x-1}{x+1}$$

Thus, we say that (1) defines y **implicitly** as a function of x, the function being

$$f(x) = \frac{x-1}{x+1}$$

An equation in x and y can implicitly define more than one function of x; for example, if we solve the equation

$$x^2 + y^2 = 1 \tag{2}$$

for y in terms of x, we obtain $y = \pm\sqrt{1-x^2}$, so we have found two functions that are defined implicitly by (2), namely

$$f_1(x) = \sqrt{1 - x^2} \quad \text{and} \quad f_2(x) = -\sqrt{1 - x^2} \tag{3}$$

The graphs of these functions are the upper and lower semicircles of the circle $x^2 + y^2 = 1$ (Figure 3.6.1).

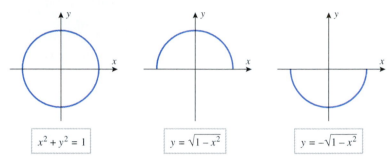

$$\boxed{x^2 + y^2 = 1} \qquad \boxed{y = \sqrt{1 - x^2}} \qquad \boxed{y = -\sqrt{1 - x^2}}$$

Figure 3.6.1

Observe that the complete circle $x^2 + y^2 = 1$ does not pass the vertical line test, and hence is not itself the graph of a function of x. However, the upper and lower semicircles (which are only portions of the entire circle) do pass the vertical line test, and hence are graphs of functions. In general, if we have an equation in x and y, then any portion of its graph that passes the vertical line test can be viewed as the graph of a function defined by the equation. Thus, we make the following definition.

> **3.6.1** DEFINITION. We will say that a given equation in x and y defines the function f *implicitly* if the graph of $y = f(x)$ coincides with a portion of the graph of the equation.

Thus, for example, the equation $x^2 + y^2 = 1$ defines the functions $f_1(x) = \sqrt{1 - x^2}$ and $f_2(x) = -\sqrt{1 - x^2}$ implicitly, since the graphs of these functions are contained in the circle $x^2 + y^2 = 1$.

Sometimes it may be difficult or impossible to solve an equation in x and y for y in terms of x. For example, with persistence the equation

$$x^3 + y^3 = 3xy \tag{4}$$

can be solved for y in terms of x, but the algebra is tedious and the resulting formulas are complicated. On the other hand, the equation

$$\sin(xy) = y$$

cannot be solved for y in terms of x by any elementary method. Thus, even though an equation in x and y may define one or more functions of x, it may not be practical or possible to find explicit formulas for those functions.

GRAPHS OF EQUATIONS IN x AND y

When an equation in x and y cannot be solved for y in terms of x (or x in terms of y), it may be difficult or time-consuming to obtain even a rough sketch of the graph, so the graphing of such equations is usually best left for graphing utilities. In particular, the CAS programs *Mathematica* and *Maple* both have "implicit plotting" capabilities for graphing such equations. For example, Figure 3.6.2 shows the graph of Equation (4), which is called the ***Folium of Descartes***.

FOR THE READER. Figure 3.6.3 shows the graphs of two functions (in blue) that are defined implicitly by (4). Sketch some more.

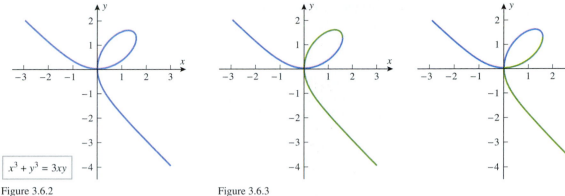

Figure 3.6.2

$x^3 + y^3 = 3xy$

Figure 3.6.3

IMPLICIT DIFFERENTIATION

In general, it is not necessary to solve an equation for y in terms of x in order to differentiate the functions defined implicitly by the equation. To illustrate this, let us consider the simple equation

$$xy = 1 \tag{5}$$

One way to find dy/dx is to rewrite this equation as

$$y = \frac{1}{x} \tag{6}$$

from which it follows that

$$\frac{dy}{dx} = -\frac{1}{x^2} \tag{7}$$

However, there is another way to obtain this derivative. We can differentiate both sides of (5) *before* solving for y in terms of x, treating y as a (temporarily unspecified) differentiable function of x. With this approach we obtain

$$\frac{d}{dx}[xy] = \frac{d}{dx}[1]$$

$$x\frac{d}{dx}[y] + y\frac{d}{dx}[x] = 0$$

$$x\frac{dy}{dx} + y = 0$$

$$\frac{dy}{dx} = -\frac{y}{x}$$

If we now substitute (6) into the last expression, we obtain

$$\frac{dy}{dx} = -\frac{1}{x^2}$$

which agrees with (7). This method of obtaining derivatives is called ***implicit differentiation***.

Example 1   Use implicit differentiation to find dy/dx if $5y^2 + \sin y = x^2$.

$$\frac{d}{dx}[5y^2 + \sin y] = \frac{d}{dx}[x^2]$$

$$5\frac{d}{dx}[y^2] + \frac{d}{dx}[\sin y] = 2x$$

$$5\left(2y\frac{dy}{dx}\right) + (\cos y)\frac{dy}{dx} = 2x \qquad \begin{array}{l} \text{The chain rule was} \\ \text{used here because} \\ y \text{ is a function of } x. \end{array}$$

$$10y\frac{dy}{dx} + (\cos y)\frac{dy}{dx} = 2x$$

Solving for dy/dx we obtain

$$\frac{dy}{dx} = \frac{2x}{10y + \cos y} \tag{8}$$

Note that this formula involves both x and y. In order to obtain a formula for dy/dx that involves x alone, we would have to solve the original equation for y in terms of x and then substitute in (8). However, it is impossible to do this, so we are forced to leave the formula for dy/dx in terms of x and y. ◀

Example 2 Use implicit differentiation to find d^2y/dx^2 if $4x^2 - 2y^2 = 9$.

Solution. Differentiating both sides of $4x^2 - 2y^2 = 9$ implicitly yields

$$8x - 4y\frac{dy}{dx} = 0$$

from which we obtain

$$\frac{dy}{dx} = \frac{2x}{y} \tag{9}$$

Differentiating both sides of (9) implicitly yields

$$\frac{d^2y}{dx^2} = \frac{(y)(2) - (2x)(dy/dx)}{y^2} \tag{10}$$

Substituting (9) into (10) and simplifying using the original equation, we obtain

$$\frac{d^2y}{dx^2} = \frac{2y - 2x(2x/y)}{y^2} = \frac{2y^2 - 4x^2}{y^3} = -\frac{9}{y^3} \qquad ◀$$

In Examples 1 and 2, the resulting formulas for dy/dx involved both x and y. Although it is usually more desirable to have the formula for dy/dx expressed in terms of x alone, having the formula in terms of x and y is not an impediment to finding slopes and equations of tangent lines provided the x- and y-coordinates of the point of tangency are known. This is illustrated in the following example.

Example 3 Find the slopes of the curve $y^2 - x + 1 = 0$ at the points $(2, -1)$ and $(2, 1)$.

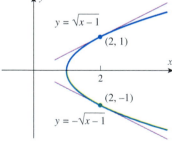

Figure 3.6.4

Solution. We could proceed by solving the equation for y in terms of x, and then evaluating the derivative of $y = \sqrt{x-1}$ at $(2, 1)$ and the derivative of $y = -\sqrt{x-1}$ at $(2, -1)$ (Figure 3.6.4). However, implicit differentiation is more efficient since it gives the slopes of *both* functions. Differentiating implicitly yields

$$\frac{d}{dx}[y^2 - x + 1] = \frac{d}{dx}[0]$$

$$\frac{d}{dx}[y^2] - \frac{d}{dx}[x] + \frac{d}{dx}[1] = \frac{d}{dx}[0]$$

$$2y\frac{dy}{dx} - 1 = 0$$

$$\frac{dy}{dx} = \frac{1}{2y}$$

At $(2, -1)$ we have $y = -1$, and at $(2, 1)$ we have $y = 1$, so the slopes of the curve at those points are

$$\frac{dy}{dx}\bigg|_{\substack{x=2 \\ y=-1}} = -\frac{1}{2} \quad \text{and} \quad \frac{dy}{dx}\bigg|_{\substack{x=2 \\ y=1}} = \frac{1}{2} \qquad ◀$$

Example 4

(a) Use implicit differentiation to find dy/dx for the Folium of Descartes $x^3 + y^3 = 3xy$.

(b) Find an equation for the tangent line to the Folium of Descartes at the point $\left(\frac{3}{2}, \frac{3}{2}\right)$.

(c) At what point(s) in the first quadrant is the tangent line to the Folium of Descartes horizontal?

Solution (a). Differentiating both sides of the given equation implicitly yields

$$\frac{d}{dx}[x^3 + y^3] = \frac{d}{dx}[3xy]$$

$$3x^2 + 3y^2 \frac{dy}{dx} = 3x \frac{dy}{dx} + 3y$$

$$x^2 + y^2 \frac{dy}{dx} = x \frac{dy}{dx} + y$$

$$(y^2 - x) \frac{dy}{dx} = y - x^2$$

$$\frac{dy}{dx} = \frac{y - x^2}{y^2 - x} \tag{11}$$

Solution (b). At the point $\left(\frac{3}{2}, \frac{3}{2}\right)$, we have $x = \frac{3}{2}$ and $y = \frac{3}{2}$, so from (11) the slope $m_{\tan}$ of the tangent line at this point is

$$m_{\tan} = \frac{dy}{dx}\bigg|_{\substack{x=3/2 \\ y=3/2}} = \frac{(3/2) - (3/2)^2}{(3/2)^2 - (3/2)} = -1$$

Thus, the equation of the tangent line at the point $\left(\frac{3}{2}, \frac{3}{2}\right)$ is

$$y - \tfrac{3}{2} = -1\left(x - \tfrac{3}{2}\right) \quad \text{or} \quad x + y = 3$$

which is consistent with Figure 3.6.5.

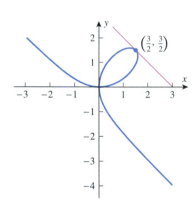

Figure 3.6.5

Solution (c). The tangent line is horizontal at the points where $dy/dx = 0$, and from (11) this occurs only where $y - x^2 = 0$ or

$$y = x^2 \tag{12}$$

Substituting this expression for y in the equation $x^3 + y^3 = 3xy$ for the curve yields

$$x^3 + \left(x^2\right)^3 = 3x^3$$

$$x^6 - 2x^3 = 0$$

$$x^3(x^3 - 2) = 0$$

whose solutions are $x = 0$ and $x = 2^{1/3}$. From (12), the solutions $x = 0$ and $x = 2^{1/3}$ yield the points $(0, 0)$ and $(2^{1/3}, 2^{2/3}) \approx (1.26, 1.59)$, respectively. Of these two, only $(2^{1/3}, 2^{2/3})$ is in the first quadrant. Substituting $x = 2^{1/3}$, $y = 2^{2/3}$ into (11) yields

$$\frac{dy}{dx}\bigg|_{\substack{x=2^{1/3} \\ y=2^{2/3}}} = \frac{0}{2^{4/3} - 2^{2/3}} = 0$$

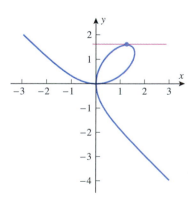

Figure 3.6.6

We conclude that $(2^{1/3}, 2^{2/3})$ is the only point on the Folium of Descartes in the first quadrant at which the tangent line is horizontal (Figure 3.6.6). ◀

> **REMARK.** Note that (11) gives an undefined expression for dy/dx at $(0,0)$. However, using more advanced techniques it can be shown that the x-axis is tangent to a portion of the Folium of Descartes at the origin.

DIFFERENTIABILITY OF FUNCTIONS DEFINED IMPLICITLY

When differentiating implicitly, it is assumed that y represents a differentiable function of x. If this is not so, then the resulting calculations may be nonsense. For example, if we differentiate the equation

$$x^2 + y^2 + 1 = 0 \tag{13}$$

we obtain

$$2x + 2y\frac{dy}{dx} = 0 \quad \text{or} \quad \frac{dy}{dx} = -\frac{x}{y}$$

However, this derivative is meaningless because (13) does not define a function of x. (The left side of the equation is greater than zero.)

In general, differentiability of implicitly defined functions can be difficult to determine analytically. For example, the first function in Figure 3.6.3 appears to have zero derivative at the origin, whereas the second function in that figure is not differentiable at the origin. However, from Example 4(a) we note that the formula derived for the implicit derivative cannot be evaluated at the origin. This results from the ambiguity created by the curve crossing itself at the origin. We leave a more careful discussion of differentiability for implicitly defined functions for an advanced course in analysis.

DERIVATIVES OF RATIONAL POWERS OF x

In Theorem 3.3.7 and the discussion immediately following it, we showed that the formula

$$\frac{d}{dx}[x^n] = nx^{n-1} \tag{14}$$

holds for integer values of n and for $n = \frac{1}{2}$. We will now use implicit differentiation to show that this formula holds for any rational exponent. More precisely, we will show that if r is a rational number, then

$$\frac{d}{dx}[x^r] = rx^{r-1} \tag{15}$$

wherever x^r and x^{r-1} are defined. For now, we will assume without proof that x^r is differentiable; the justification for this will be considered later.

Let $y = x^r$. Since r is a rational number, it can be expressed as a ratio of integers $r = m/n$. Thus, $y = x^r = x^{m/n}$ can be written as

$$y^n = x^m \quad \text{so that} \quad \frac{d}{dx}[y^n] = \frac{d}{dx}[x^m]$$

By differentiating implicitly with respect to x and using (14), we obtain

$$ny^{n-1}\frac{dy}{dx} = mx^{m-1} \tag{16}$$

But

$$y^{n-1} = \left[x^{m/n}\right]^{n-1} = x^{m-(m/n)}$$

Thus, (16) can be written as

$$nx^{m-(m/n)}\frac{dy}{dx} = mx^{m-1}$$

so that

$$\frac{dy}{dx} = \frac{m}{n}x^{(m/n)-1} = rx^{r-1}$$

which establishes (15).

Example 5 From (15)

$$\frac{d}{dx}[x^{4/5}] = \frac{4}{5}x^{(4/5)-1} = \frac{4}{5}x^{-1/5}$$

$$\frac{d}{dx}[x^{-7/8}] = -\frac{7}{8}x^{(-7/8)-1} = -\frac{7}{8}x^{-15/8}$$

$$\frac{d}{dx}[\sqrt[3]{x}] = \frac{d}{dx}[x^{1/3}] = \frac{1}{3}x^{-2/3} = \frac{1}{3\sqrt[3]{x^2}}$$ ◀

If u is a differentiable function of x, and r is a rational number, then the chain rule yields the following generalization of (15):

$$\frac{d}{dx}[u^r] = ru^{r-1} \cdot \frac{du}{dx} \tag{17}$$

Example 6

$$\frac{d}{dx}\left[x^2 - x + 2\right]^{3/4} = \frac{3}{4}\left(x^2 - x + 2\right)^{-1/4} \cdot \frac{d}{dx}[x^2 - x + 2]$$

$$= \frac{3}{4}\left(x^2 - x + 2\right)^{-1/4}(2x - 1)$$

$$\frac{d}{dx}[(\sec \pi x)^{-4/5}] = -\frac{4}{5}(\sec \pi x)^{-9/5} \cdot \frac{d}{dx}[\sec \pi x]$$

$$= -\frac{4}{5}(\sec \pi x)^{-9/5} \cdot \sec \pi x \tan \pi x \cdot \pi$$

$$= -\frac{4\pi}{5}(\sec \pi x)^{-4/5} \tan \pi x$$ ◀

EXERCISE SET 3.6 C CAS

In Exercises 1–8, find dy/dx.

1. $y = \sqrt[3]{2x - 5}$

2. $y = \sqrt[3]{2 + \tan(x^2)}$

3. $y = \left(\dfrac{x-1}{x+2}\right)^{3/2}$

4. $y = \sqrt{\dfrac{x^2+1}{x^2-5}}$

5. $y = x^3 \left(5x^2 + 1\right)^{-2/3}$

6. $y = \dfrac{(3 - 2x)^{4/3}}{x^2}$

7. $y = [\sin(3/x)]^{5/2}$

8. $y = \left[\cos(x^3)\right]^{-1/2}$

In Exercises 9 and 10: (a) Find dy/dx by differentiating implicitly. (b) Solve the equation for y as a function of x, and find dy/dx from that equation. (c) Confirm that the two results are consistent by expressing the derivative in part (a) as a function of x alone.

9. $x^3 + xy - 2x = 1$

10. $\sqrt{y} - \sin x = 2$

In Exercises 11–20, find dy/dx by implicit differentiation.

11. $x^2 + y^2 = 100$

12. $x^3 - y^3 = 6xy$

13. $x^2y + 3xy^3 - x = 3$

14. $x^3y^2 - 5x^2y + x = 1$

15. $\dfrac{1}{y} + \dfrac{1}{x} = 1$

16. $x^2 = \dfrac{x+y}{x-y}$

17. $\sin(x^2y^2) = x$

18. $x^2 = \dfrac{\cot y}{1 + \csc y}$

19. $\tan^3(xy^2 + y) = x$

20. $\dfrac{xy^3}{1 + \sec y} = 1 + y^4$

In Exercises 21–26, find d^2y/dx^2 by implicit differentiation.

21. $3x^2 - 4y^2 = 7$

22. $x^3 + y^3 = 1$

23. $x^3y^3 - 4 = 0$

24. $2xy - y^2 = 3$

25. $y + \sin y = x$

26. $x \cos y = y$

In Exercises 27 and 28, find the slope of the tangent line to the curve at the given points in two ways: first by solving for y in terms of x and differentiating and then by implicit differentiation.

27. $x^2 + y^2 = 1$; $(1/\sqrt{2}, 1/\sqrt{2})$, $(1/\sqrt{2}, -1/\sqrt{2})$

28. $y^2 - x + 1 = 0$; $(10, 3)$, $(10, -3)$

In Exercises 29–32, use implicit differentiation to find the slope of the tangent line to the curve at the specified point, and check that your answer is consistent with the accompanying graph.

29. $x^4 + y^4 = 16$; $(1, \sqrt[4]{15})$ [*Lamé's special quartic*]

30. $y^3 + yx^2 + x^2 - 3y^2 = 0$; $(0, 3)$ [*trisectrix*]

31. $2\left(x^2 + y^2\right)^2 = 25(x^2 - y^2)$; $(3, 1)$ [*lemniscate*]

32. $x^{2/3} + y^{2/3} = 4$; $(-1, 3\sqrt{3})$ [*four-cusped hypocycloid*]

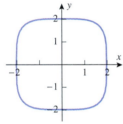

Figure Ex-29

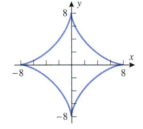

Figure Ex-30

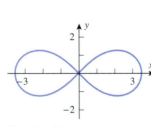

Figure Ex-31

Figure Ex-32

c **33.** If you have a CAS, read the documentation on "implicit plotting," and then generate the four curves in Exercises 29–32.

c **34.** Curves with equations of the form $y^2 = x(x - a)(x - b)$, where $a < b$ are called *bipartite cubics*.
   (a) Use the implicit plotting capability of a CAS to graph the bipartite cubic $y^2 = x(x - 1)(x - 2)$.
   (b) At what points does the curve in part (a) have a horizontal tangent line?
   (c) Solve the equation in part (a) for y in terms of x, and use the result to explain why the graph consists of two separate parts (i.e., is *bipartite*).
   (d) Graph the equation in part (a) without using the implicit plotting capability of the CAS.

c **35.** (a) Use the implicit plotting capability of a CAS to graph the rotated ellipse $x^2 - xy + y^2 = 4$.
   (b) Use the graph to estimate the x-coordinates of all horizontal tangent lines.
   (c) Find the exact values for the x-coordinates in part (b).

In Exercises 36–39, use implicit differentiation to find the specified derivative.

36. $\sqrt{u} + \sqrt{v} = 5$; du/dv **37.** $a^4 - t^4 = 6a^2t$; da/dt

38. $y = \sin x$; dx/dy.

39. $a^2\omega^2 + b^2\lambda^2 = 1$ $(a, b$ constants); $d\omega/d\lambda$

40. At what point(s) is the tangent line to the curve $y^2 = 2x^3$ perpendicular to the line $4x - 3y + 1 = 0$?

41. Find the values of a and b for the curve $x^2y + ay^2 = b$ if the point $(1, 1)$ is on its graph and the tangent line at $(1, 1)$ has the equation $4x + 3y = 7$.

42. Find the coordinates of the point in the first quadrant at which the tangent line to the curve $x^3 - xy + y^3 = 0$ is parallel to the x-axis.

43. Find equations for two lines through the origin that are tangent to the curve $x^2 - 4x + y^2 + 3 = 0$.

44. Use implicit differentiation to show that the equation of the tangent line to the curve $y^2 = kx$ at (x_0, y_0) is

$$y_0 y = \tfrac{1}{2}k(x + x_0)$$

45. Find dy/dx if

$$2y^3 t + t^3 y = 1 \quad \text{and} \quad \frac{dt}{dx} = \frac{1}{\cos t}$$

In Exercises 46 and 47, find dy/dt in terms of x, y, and dx/dt, assuming that x and y are differentiable functions of the variable t. [*Hint:* Differentiate both sides of the given equation with respect to t.]

46. $x^3y^2 + y = 3$ **47.** $xy^2 = \sin 3x$

48. (a) Show that $f(x) = x^{4/3}$ is differentiable at 0, but not twice differentiable at 0.
   (b) Show that $f(x) = x^{7/3}$ is twice differentiable at 0, but not three times differentiable at 0.
   (c) Find an exponent k such that $f(x) = x^k$ is $(n-1)$ times differentiable at 0, but not n times differentiable at 0.

In Exercises 49 and 50, find all rational values of r such that $y = x^r$ satisfies the given equation.

49. $3x^2y'' + 4xy' - 2y = 0$ **50.** $16x^2y'' + 24xy' + y = 0$

Two curves are said to be ***orthogonal*** if their tangent lines are perpendicular at each point of intersection, and two families of curves are said to be ***orthogonal trajectories*** of one another if each member of one family is orthogonal to each member of the other family. This terminology is used in Exercises 51 and 52.

51. The accompanying figure shows some typical members of the families of circles $x^2 + (y - c)^2 = c^2$ (black curves) and $(x - k)^2 + y^2 = k^2$ (gray curves). Show that these families are orthogonal trajectories of one another. [*Hint:* For the tangent lines to be perpendicular at a point of intersection, the slopes of those tangent lines must be negative reciprocals of one another.]

52. The accompanying figure shows some typical members of the families of hyperbolas $xy = c$ (black curves) and $x^2 - y^2 = k$ (gray curves), where $c \neq 0$ and $k \neq 0$. Use the hint in Exercise 51 to show that these families are orthogonal trajectories of one another.

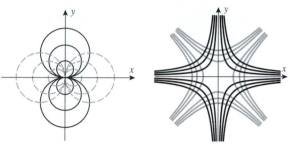

Figure Ex-51   Figure Ex-52

3.7 RELATED RATES

In this section we will study related rates problems. In such problems one tries to find the rate at which some quantity is changing by relating the quantity to other quantities whose rates of change are known.

DIFFERENTIATING EQUATIONS TO RELATE RATES

Figure 3.7.1 shows a liquid draining through a conical filter. As the liquid drains, its volume V, height h, and radius r are functions of the elapsed time t, and at each instant these variables are related by the equation

$$V = \frac{\pi}{3}r^2 h$$

If we differentiate both sides of this equation with respect to t, then we obtain

$$\frac{dV}{dt} = \frac{\pi}{3}\left[r^2\frac{dh}{dt} + h\left(2r\frac{dr}{dt}\right)\right] = \frac{\pi}{3}\left(r^2\frac{dh}{dt} + 2rh\frac{dr}{dt}\right)$$

Thus, if at a given instant we have values for r, h, and two of the three rates in this equation, then we can solve for the value of the third rate at this instant. In this section we present some specific examples that use this basic idea.

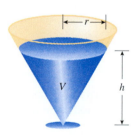

Figure 3.7.1

Example 1   Assume that oil spilled from a ruptured tanker spreads in a circular pattern whose radius increases at a constant rate of 2 ft/s. How fast is the area of the spill increasing when the radius of the spill is 60 ft?

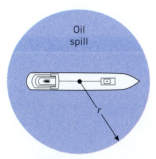

Figure 3.7.2

Solution. Let

t = number of seconds elapsed from the time of the spill

r = radius of the spill in feet after t seconds

A = area of the spill in square feet after t seconds

(Figure 3.7.2). We know the rate at which the radius is increasing, and we want to find the rate at which the area is increasing at the instant when $r = 60$; that is, we want to find

$$\frac{dA}{dt}\bigg|_{r=60} \quad \text{given that} \quad \frac{dr}{dt} = 2 \text{ ft/s}$$

From the formula for the area of a circle we obtain

$$A = \pi r^2 \tag{1}$$

Because A and r are functions of t, we can differentiate both sides of (1) with respect to t to obtain

$$\frac{dA}{dt} = 2\pi r \frac{dr}{dt}$$

Thus, when $r = 60$ the area of the spill is increasing at the rate of

$$\frac{dA}{dt}\bigg|_{r=60} = 2\pi(60)(2) = 240\pi \text{ ft}^2/\text{s}$$

or approximately 754 ft²/s. ◀

With only minor variations, the method used in Example 1 can be used to solve a variety of related rates problems. The method consists of five steps:

A Strategy for Solving Related Rates Problems

Step 1. Identify the rates of change that are known and the rate of change that is to be found. Interpret each rate as a derivative of a variable with respect to time, and provide a description of each variable involved.

Step 2. Find an equation relating those quantities whose rates are identified in Step 1. In a geometric problem, this is aided by drawing an appropriately labeled figure that illustrates a relationship involving these quantities.

Step 3. Obtain an equation involving the rates in Step 1 by differentiating both sides of the equation in Step 2 with respect to the time variable.

Step 4. Evaluate the equation found in Step 3 using the known values for the quantities and their rates of change at the moment in question.

Step 5. Solve for the value of the remaining rate of change at this moment.

WARNING. Do not substitute prematurely; that is, always perform the differentiation in Step 3 *before* performing the substitution in Step 4.

Example 2 A baseball diamond is a square whose sides are 90 ft long (Figure 3.7.3). Suppose that a player running from second base to third base has a speed of 30 ft/s at the instant when he is 20 ft from third base. At what rate is the player's distance from home plate changing at that instant?

Solution. The rate we wish to find is the rate of change of the distance from the player to home plate. We are given the speed of the player as he moves along the base path from second to third base, which tells us both the speed with which he is moving away from

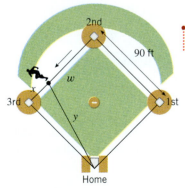

Figure 3.7.3

second base and the speed with which he is approaching third base. Let

t = number of seconds since the player left second base
w = distance in feet from the player to second base
x = distance in feet from the player to third base
y = distance in feet from the player to home plate

Thus, we want to find

$$\left.\frac{dy}{dt}\right|_{x=20} \quad \text{given that} \quad \left.\frac{dw}{dt}\right|_{x=20} = 30 \text{ ft/s} \quad \text{and} \quad \left.\frac{dx}{dt}\right|_{x=20} = -30 \text{ ft/s}$$

[Note that $(dy/dx)_{x=20}$ is negative because x is decreasing with respect to t.]
From the Theorem of Pythagoras,

$$x^2 + 90^2 = y^2 \tag{2}$$

Differentiating both sides of this equation with respect to t yields

$$2x\frac{dx}{dt} = 2y\frac{dy}{dt} \quad \text{or} \quad x\frac{dx}{dt} = y\frac{dy}{dt} \tag{3}$$

To evaluate (3) at the instant when $x = 20$ we need a value for y at this instant. Substituting $x = 20$ into (2) yields

$$400 + 8100 = (y|_{x=20})^2 \quad \text{or} \quad y|_{x=20} = \sqrt{8500} = 10\sqrt{85}$$

Then, evaluating (3) when $x = 20$ yields

$$20 \cdot (-30) = 10\sqrt{85} \cdot \left.\frac{dy}{dt}\right|_{x=20} \quad \text{or} \quad \left.\frac{dy}{dt}\right|_{x=20} = \frac{-600}{10\sqrt{85}} = -\frac{60}{\sqrt{85}} \approx -6.51 \text{ ft/s}$$

The negative sign in the answer tells us that y is decreasing, which makes sense in the physical situation of the problem (Figure 3.7.3).   ◀

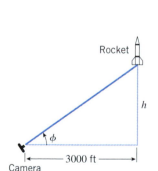

Rocket

Elevation
angle

3000 ft

Camera   Launching
pad

Figure 3.7.4

FOR THE READER.   In our solution for Example 2 we chose to relate x and y. An alternative approach would be to relate w and y. Solve the problem using this alternative approach.

Example 3   In Figure 3.7.4 we have shown a camera mounted at a point 3000 ft from the base of a rocket launching pad. If the rocket is rising vertically at 880 ft/s when it is 4000 ft above the launching pad, how fast must the camera elevation angle change at that instant to keep the camera aimed at the rocket?

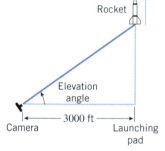

Rocket

h

ϕ

3000 ft

Camera

Figure 3.7.5

Solution.   Let

t = number of seconds elapsed from the time of launch
ϕ = camera elevation angle in radians after t seconds
h = height of the rocket in feet after t seconds

(Figure 3.7.5). At each instant the rate at which the camera elevation angle must change is $d\phi/dt$, and the rate at which the rocket is rising is dh/dt. We want to find

$$\left.\frac{d\phi}{dt}\right|_{h=4000} \quad \text{given that} \quad \left.\frac{dh}{dt}\right|_{h=4000} = 880 \text{ ft/s}$$

From Figure 3.7.5 we see that

$$\tan \phi = \frac{h}{3000} \tag{4}$$

Because ϕ and h are functions of t, we can differentiate both sides of (4) with respect to t to obtain

$$(\sec^2 \phi)\frac{d\phi}{dt} = \frac{1}{3000}\frac{dh}{dt} \tag{5}$$

When $h = 4000$, it follows that

$$(\sec \phi)_{h=4000} = \frac{5000}{3000} = \frac{5}{3}$$

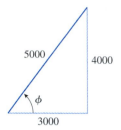

Figure 3.7.6

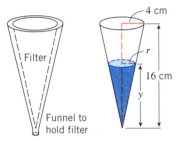

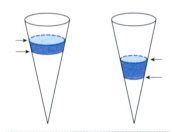

The same volume has drained, but the change in height is greater near the bottom than near the top.

Figure 3.7.8

Figure 3.7.7

(Figure 3.7.6), so that from (5)

$$\left(\frac{5}{3}\right)^2 \frac{d\phi}{dt}\bigg|_{h=4000} = \frac{1}{3000} \cdot 880 = \frac{22}{75}$$

$$\frac{d\phi}{dt}\bigg|_{h=4000} = \frac{22}{75} \cdot \frac{9}{25} = \frac{66}{625} \approx 0.11 \text{ rad/s} \approx 6.05 \text{ deg/s} \quad \blacktriangleleft$$

Example 4 Suppose that liquid is to be cleared of sediment by allowing it to drain through a conical filter that is 16 cm high and has a radius of 4 cm at the top (Figure 3.7.7). Suppose also that the liquid flows out of the cone at a constant rate of 2 cm^3/min.

(a) Do you think that the depth of the liquid will decrease at a constant rate? Give a verbal argument that justifies your conclusion.

(b) Find a formula that expresses the rate at which the depth of the liquid is changing in terms of the depth, and use that formula to determine whether your conclusion in part (a) is correct.

(c) At what rate is the depth of the liquid changing at the instant when the liquid in the cone is 8 cm deep?

Solution (a). For the volume of liquid to decrease by a *fixed amount*, it requires a greater decrease in depth when the cone is close to empty than when it is almost full (Figure 3.7.8). This suggests that for the volume to decrease at a constant rate, the depth must decrease at an increasing rate.

Solution (b). Let

$t = $ time elapsed from the initial observation (min)
$V = $ volume of liquid in the cone at time t (cm^3)
$y = $ depth of the liquid in the cone at time t (cm)
$r = $ radius of the liquid surface at time t (cm)

(Figure 3.7.7). At each instant the rate at which the volume of liquid is changing is dV/dt, and the rate at which the depth is changing is dy/dt. We want to express dy/dt in terms of y given that dV/dt has a constant value of $dV/dt = -2$. (We must use a minus sign here because V *decreases* as t increases.)

From the formula for the volume of a cone, the volume V, the radius r, and the depth y are related by

$$V = \tfrac{1}{3}\pi r^2 y \tag{6}$$

If we differentiate both sides of (6) with respect to t, the right side will involve the quantity dr/dt. Since we have no direct information about dr/dt, it is desirable to eliminate r from (6) before differentiating. This can be done using similar triangles. From Figure 3.7.7 we see that

$$\frac{r}{y} = \frac{4}{16} \quad \text{or} \quad r = \frac{1}{4}y$$

Substituting this expression in (6) gives

$$V = \frac{\pi}{48} y^3 \tag{7}$$

Differentiating both sides of (7) with respect to t we obtain

$$\frac{dV}{dt} = \frac{\pi}{48}\left(3y^2 \frac{dy}{dt}\right)$$

or

$$\frac{dy}{dt} = \frac{16}{\pi y^2}\frac{dV}{dt} = \frac{16}{\pi y^2}(-2) = -\frac{32}{\pi y^2} \tag{8}$$

which expresses dy/dt in terms of y. The minus sign tells us that y is decreasing with time, and

$$\left|\frac{dy}{dt}\right| = \frac{32}{\pi y^2}$$

tells us how fast y is decreasing. From this formula we see that $|dy/dt|$ increases as y decreases, which confirms our conjecture in part (a) that the depth of the liquid decreases more quickly as the liquid drains through the filter.

Solution (c). The rate at which the depth is changing when the depth is 8 cm can be obtained from (8) with $y = 8$:

$$\left.\frac{dy}{dt}\right|_{y=8} = -\frac{32}{\pi(8^2)} = -\frac{1}{2\pi} \approx -0.16 \text{ cm/min}$$ ◀

EXERCISE SET 3.7

In Exercises 1–4, both x and y denote functions of t that are related by the given equation. Use this equation and the given derivative information to find the specified derivative.

1. Equation: $y = 3x + 5$.
  (a) Given that $dx/dt = 2$, find dy/dt when $x = 1$.
  (b) Given that $dy/dt = -1$, find dx/dt when $x = 0$.

2. Equation: $x + 4y = 3$.
  (a) Given that $dx/dt = 1$, find dy/dt when $x = 2$.
  (b) Given that $dy/dt = 4$, find dx/dt when $x = 3$.

3. Equation: $x^2 + y^2 = 1$.
  (a) Given that $dx/dt = 1$, find dy/dt when

$$(x, y) = \left(\frac{1}{2}, \frac{\sqrt{3}}{2}\right)$$

  (b) Given that $dy/dt = -2$, find dx/dt when

$$(x, y) = \left(\frac{\sqrt{2}}{2}, \frac{\sqrt{2}}{2}\right)$$

4. Equation: $x^2 + y^2 = 2x$.
  (a) Given that $dx/dt = -2$, find dy/dt when $(x, y) = (1, 1)$.
  (b) Given that $dy/dt = 3$, find dx/dt when

$$(x, y) = \left(\frac{2 + \sqrt{2}}{2}, \frac{\sqrt{2}}{2}\right)$$

5. Let A be the area of a square whose sides have length x, and assume that x varies with the time t.
  (a) Draw a picture of the square with the labels A and x placed appropriately.
  (b) Write an equation that relates A and x.
  (c) Use the equation in part (b) to find an equation that relates dA/dt and dx/dt.
  (d) At a certain instant the sides are 3 ft long and increasing at a rate of 2 ft/min. How fast is the area increasing at that instant?

6. Let A be the area of a circle of radius r, and assume that r increases with the time t.
  (a) Draw a picture of the circle with the labels A and r placed appropriately.
  (b) Write an equation that relates A and r.
  (c) Use the equation in part (b) to find an equation that relates dA/dt and dr/dt.
  (d) At a certain instant the radius is 5 cm and increasing at the rate of 2 cm/s. How fast is the area increasing at that instant?

7. Let V be the volume of a cylinder having height h and radius r, and assume that h and r vary with time.
  (a) How are dV/dt, dh/dt, and dr/dt related?
  (b) At a certain instant, the height is 6 in and increasing at 1 in/s, while the radius is 10 in and decreasing at 1 in/s. How fast is the volume changing at that instant? Is the volume increasing or decreasing at that instant?

8. Let l be the length of a diagonal of a rectangle whose sides have lengths x and y, and assume that x and y vary with time.
  (a) How are dl/dt, dx/dt, and dy/dt related?
  (b) If x increases at a constant rate of $\frac{1}{2}$ ft/s and y decreases at a constant rate of $\frac{1}{4}$ ft/s, how fast is the size of the diagonal changing when $x = 3$ ft and $y = 4$ ft? Is the diagonal increasing or decreasing at that instant?

9. Let θ (in radians) be an acute angle in a right triangle, and let x and y, respectively, be the lengths of the sides adjacent to and opposite θ. Suppose also that x and y vary with time.
  (a) How are $d\theta/dt$, dx/dt, and dy/dt related?
  (b) At a certain instant, $x = 2$ units and is increasing at 1 unit/s, while $y = 2$ units and is decreasing at $\frac{1}{4}$ unit/s. How fast is θ changing at that instant? Is θ increasing or decreasing at that instant?

10. Suppose that $z = x^3y^2$, where both x and y are changing with time. At a certain instant when $x = 1$ and $y = 2$, x is decreasing at the rate of 2 units/s, and y is increasing at the rate of 3 units/s. How fast is z changing at this instant? Is z increasing or decreasing?

11. The minute hand of a certain clock is 4 in long. Starting from the moment when the hand is pointing straight up, how fast is the area of the sector that is swept out by the hand increasing at any instant during the next revolution of the hand?

12. A stone dropped into a still pond sends out a circular ripple whose radius increases at a constant rate of 3 ft/s. How rapidly is the area enclosed by the ripple increasing at the end of 10 s?

13. Oil spilled from a ruptured tanker spreads in a circle whose area increases at a constant rate of 6 mi^2/h. How fast is the radius of the spill increasing when the area is 9 mi^2?

14. A spherical balloon is inflated so that its volume is increasing at the rate of 3 ft^3/min. How fast is the diameter of the balloon increasing when the radius is 1 ft?

15. A spherical balloon is to be deflated so that its radius decreases at a constant rate of 15 cm/min. At what rate must air be removed when the radius is 9 cm?

16. A 17-ft ladder is leaning against a wall. If the bottom of the ladder is pulled along the ground away from the wall at a constant rate of 5 ft/s, how fast will the top of the ladder be moving down the wall when it is 8 ft above the ground?

17. A 13-ft ladder is leaning against a wall. If the top of the ladder slips down the wall at a rate of 2 ft/s, how fast will the foot be moving away from the wall when the top is 5 ft above the ground?

18. A 10-ft plank is leaning against a wall. If at a certain instant the bottom of the plank is 2 ft from the wall and is being pushed toward the wall at the rate of 6 in/s, how fast is the acute angle that the plank makes with the ground increasing?

19. A softball diamond is a square whose sides are 60 ft long. Suppose that a player running from first to second base has a speed of 25 ft/s at the instant when she is 10 ft from second base. At what rate is the player's distance from home plate changing at that instant?

20. A rocket, rising vertically, is tracked by a radar station that is on the ground 5 mi from the launchpad. How fast is the rocket rising when it is 4 mi high and its distance from the radar station is increasing at a rate of 2000 mi/h?

21. For the camera and rocket shown in Figure 3.7.4, at what rate is the camera-to-rocket distance changing when the rocket is 4000 ft up and rising vertically at 880 ft/s?

22. For the camera and rocket shown in Figure 3.7.4, at what rate is the rocket rising when the elevation angle is $\pi/4$ radians and increasing at a rate of 0.2 radian/s?

23. A satellite is in an elliptical orbit around the Earth. Its distance r (in miles) from the center of the Earth is given by

$$r = \frac{4995}{1 + 0.12\cos\theta}$$

where θ is the angle measured from the point on the orbit nearest the Earth's surface (see the accompanying figure).

(a) Find the altitude of the satellite at *perigee* (the point nearest the surface of the Earth) and at *apogee* (the point farthest from the surface of the Earth). Use 3960 mi as the radius of the Earth.

(b) At the instant when θ is 120°, the angle θ is increasing at the rate of 2.7°/min. Find the altitude of the satellite and the rate at which the altitude is changing at this instant. Express the rate in units of mi/min.

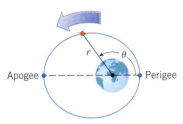

Figure Ex-23

24. An aircraft is flying horizontally at a constant height of 4000 ft above a fixed observation point (see the accompanying figure). At a certain instant the angle of elevation θ is 30° and decreasing, and the speed of the aircraft is 300 mi/h.

(a) How fast is θ decreasing at this instant? Express the result in units of degrees/s.

(b) How fast is the distance between the aircraft and the observation point changing at this instant? Express the result in units of ft/s. Use 1 mi = 5280 ft.

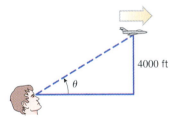

Figure Ex-24

25. A conical water tank with vertex down has a radius of 10 ft at the top and is 24 ft high. If water flows into the tank at a rate of 20 ft^3/min, how fast is the depth of the water increasing when the water is 16 ft deep?

26. Grain pouring from a chute at the rate of 8 ft^3/min forms a conical pile whose altitude is always twice its radius. How fast is the altitude of the pile increasing at the instant when the pile is 6 ft high?

27. Sand pouring from a chute forms a conical pile whose height is always equal to the diameter. If the height increases at a constant rate of 5 ft/min, at what rate is sand pouring from the chute when the pile is 10 ft high?

28. Wheat is poured through a chute at the rate of 10 ft^3/min, and falls in a conical pile whose bottom radius is always half the altitude. How fast will the circumference of the base be increasing when the pile is 8 ft high?

29. An aircraft is climbing at a 30° angle to the horizontal. How fast is the aircraft gaining altitude if its speed is 500 mi/h?

30. A boat is pulled into a dock by means of a rope attached to a pulley on the dock (see the accompanying figure). The rope is attached to the bow of the boat at a point 10 ft below the pulley. If the rope is pulled through the pulley at a rate of 20 ft/min, at what rate will the boat be approaching the dock when 125 ft of rope is out?

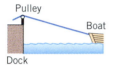

Figure Ex-30

31. For the boat in Exercise 30, how fast must the rope be pulled if we want the boat to approach the dock at a rate of 12 ft/min at the instant when 125 ft of rope is out?

32. A man 6 ft tall is walking at the rate of 3 ft/s toward a streetlight 18 ft high (see the accompanying figure).
(a) At what rate is his shadow length changing?
(b) How fast is the tip of his shadow moving?

Figure Ex-32

33. A beacon that makes one revolution every 10 s is located on a ship anchored 4 kilometers from a straight shoreline. How fast is the beam moving along the shoreline when it makes an angle of 45° with the shore?

34. An aircraft is flying at a constant altitude with a constant speed of 600 mi/h. An antiaircraft missile is fired on a straight line perpendicular to the flight path of the aircraft so that it will hit the aircraft at a point P (see the accompanying figure). At the instant the aircraft is 2 mi from the impact point P the missile is 4 mi from P and flying at 1200 mi/h. At that instant, how rapidly is the distance between missile and aircraft decreasing?

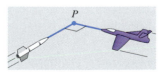

Figure Ex-34

35. Solve Exercise 34 under the assumption that the angle between the flight paths is 120° instead of the assumption that the paths are perpendicular. [*Hint:* Use the law of cosines.]

36. A police helicopter is flying due north at 100 mi/h and at a constant altitude of $\frac{1}{2}$ mi. Below, a car is traveling west on a highway at 75 mi/h. At the moment the helicopter crosses over the highway the car is 2 mi east of the helicopter.
(a) How fast is the distance between the car and helicopter changing at the moment the helicopter crosses the highway?
(b) Is the distance between the car and helicopter increasing or decreasing at that moment?

37. A particle is moving along the curve whose equation is

$$\frac{xy^3}{1+y^2} = \frac{8}{5}$$

Assume that the x-coordinate is increasing at the rate of 6 units/s when the particle is at the point $(1, 2)$.
(a) At what rate is the y-coordinate of the point changing at that instant?
(b) Is the particle rising or falling at that instant?

38. A point P is moving along the curve whose equation is $y = \sqrt{x^3 + 17}$. When P is at $(2, 5)$, y is increasing at the rate of 2 units/s. How fast is x changing?

39. A point P is moving along the line whose equation is $y = 2x$. How fast is the distance between P and the point $(3, 0)$ changing at the instant when P is at $(3, 6)$ if x is decreasing at the rate of 2 units/s at that instant?

40. A point P is moving along the curve whose equation is $y = \sqrt{x}$. Suppose that x is increasing at the rate of 4 units/s when $x = 3$.
(a) How fast is the distance between P and the point $(2, 0)$ changing at this instant?
(b) How fast is the angle of inclination of the line segment from P to $(2, 0)$ changing at this instant?

41. A particle is moving along the curve $y = x/(x^2 + 1)$. Find all values of x at which the rate of change of x with respect to time is three times that of y. [Assume that dx/dt is never zero.]

42. A particle is moving along the curve $16x^2 + 9y^2 = 144$. Find all points (x, y) at which the rates of change of x and y with respect to time are equal. [Assume that dx/dt and dy/dt are never both zero at the same point.]

43. The *thin lens equation* in physics is

$$\frac{1}{s} + \frac{1}{S} = \frac{1}{f}$$

where s is the object distance from the lens, S is the image distance from the lens, and f is the focal length of the lens. Suppose that a certain lens has a focal length of 6 cm and that an object is moving toward the lens at the rate of 2 cm/s. How fast is the image distance changing at the instant when the object is 10 cm from the lens? Is the image moving away from the lens or toward the lens?

44. Water is stored in a cone-shaped reservoir (vertex down). Assuming the water evaporates at a rate proportional to the surface area exposed to the air, show that the depth of the water will decrease at a constant rate that does not depend on the dimensions of the reservoir.

45. A meteor enters the Earth's atmosphere and burns up at a rate that, at each instant, is proportional to its surface area. Assuming that the meteor is always spherical, show that the radius decreases at a constant rate.

46. On a certain clock the minute hand is 4 in long and the hour hand is 3 in long. How fast is the distance between the tips of the hands changing at 9 o'clock?

47. Coffee is poured at a uniform rate of 20 cm^3/s into a cup whose inside is shaped like a truncated cone (see the accompanying figure). If the upper and lower radii of the cup are 4 cm and 2 cm and the height of the cup is 6 cm, how fast will the coffee level be rising when the coffee is halfway up? [*Hint:* Extend the cup downward to form a cone.]

Figure Ex-47

3.8 LOCAL LINEAR APPROXIMATION; DIFFERENTIALS

In this section we will show how derivatives can be used to approximate nonlinear functions by simpler linear functions. We will also define the differentials dy and dx and use them to interpret the derivative dy/dx as a ratio of differentials.

LOCAL LINEAR APPROXIMATION

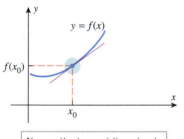

Near x_0 the tangent line closely approximates the curve.

Figure 3.8.1

In the solution of certain problems, it can be useful (and sometimes even necessary) to approximate a nonlinear function by a linear function. For example, the equations that describe the motion of a swinging pendulum may be greatly simplified by using the fact that if x is close to 0, then $\sin x \approx x$. The existence of such linear approximations provides us with a geometric interpretation of differentiability. We saw in Section 3.2 that if a function f is differentiable at a number x_0, then the tangent line to the graph of f through the point $P = (x_0, f(x_0))$ will very closely approximate the graph of f for values of x near x_0 (Figure 3.8.1). This linear approximation may be described informally in terms of the behavior of the graph of f under magnification: if f is differentiable at x_0, then stronger and stronger magnifications at P eventually make the curve segment containing P look more and more like a nonvertical line segment, that line being the tangent line to the graph of f at P. For this reason, a function that is differentiable at x_0 is said to be *locally linear* at the point $P(x_0, f(x_0))$ (Figure 3.8.2*a*). By contrast, the graph of a function that is not differentiable at x_0 due to a corner at the point $P(x_0, f(x_0))$ cannot be magnified to resemble a straight line segment at that point (Figure 3.8.2*b*).

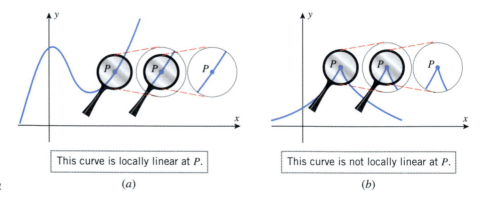

This curve is locally linear at *P*.

This curve is not locally linear at *P*.

Figure 3.8.2   (*a*)   (*b*)

To capture this intuitive idea analytically, assume that a function f is differentiable at x_0 and recall that the equation of the tangent line to the graph of the function f through $P = (x_0, f(x_0))$ is $y = f(x_0) + f'(x_0)(x - x_0)$. Since this line closely approximates the graph of f for values of x near x_0, it follows that

$$f(x) \approx f(x_0) + f'(x_0)(x - x_0) \tag{1}$$

provided x is close to x_0. We call (1) the ***local linear approximation of f at x_0***. Furthermore, it can be shown that (1) is actually the *best* linear approximation of f near x_0 in the sense that any other linear function will fail to give as good an approximation to f for values of x very close to x_0. An alternative version of this formula can be obtained by letting $\Delta x = x - x_0$, in which case (1) can be expressed as

$$f(x_0 + \Delta x) \approx f(x_0) + f'(x_0)\, \Delta x \tag{2}$$

Example 1

(a)   Find the local linear approximation of $f(x) = \sqrt{x}$ at $x_0 = 1$.

(b)   Use the local linear approximation obtained in part (a) to approximate $\sqrt{1.1}$, and compare your approximation to the result produced directly by a calculating utility.

Solution (a). Since $f'(x) = 1/(2\sqrt{x})$, it follows from (1) that the local linear approximation of $\sqrt{x}$ at $x_0 = 1$ is

$$\sqrt{x} \approx \sqrt{1} + \frac{1}{2\sqrt{1}}(x - 1) = 1 + \frac{1}{2}(x - 1) = \frac{1}{2}(x + 1)$$

In other words, if x is close to 1, then we expect $\sqrt{x}$ to be about $\frac{1}{2}(x + 1)$. Figure 3.8.3 shows both the graph of $f(x) = \sqrt{x}$ and the local linear approximation $y = \frac{1}{2}(x + 1)$.

Solution (b). Applying the local linear approximation from part (a) yields

$$\sqrt{1.1} \approx \tfrac{1}{2}(1.1 + 1) = 1.05$$

Since the tangent line $y = \frac{1}{2}(x + 1)$ in Figure 3.8.3 lies above the graph of $f(x) = \sqrt{x}$, we would expect this approximation to be slightly too large. This expectation is confirmed by the calculator approximation $\sqrt{1.1} \approx 1.04881$.   ◀

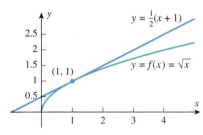

Figure 3.8.3

Example 2

(a)   Show that if x is close to 0, then $\sin x \approx x$.

(b)   Use the approximation from part (a) to approximate $\sin 2°$, and compare your approximation to the result produced directly by your calculating utility.

Solution (a). Since we are interested in approximating $\sin x$ for values of x close to 0, we compute the local linear approximation of $f(x) = \sin x$ at $x_0 = 0$. With $f(x) = \sin x$, $f'(x) = \cos x$, and $x_0 = 0$, the approximation in (1) becomes

$$\sin x \approx \sin 0 + (\cos 0)(x - 0) = 0 + 1(x) = x$$

Figure 3.8.4 shows both the graph of $f(x) = \sin x$ and the local linear approximation $y = x$.

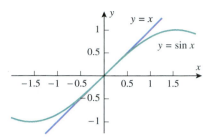

Figure 3.8.4

Solution (b). In the approximation $\sin x \approx x$, the variable x is in radian measure, so we must first convert $2°$ to radians before we can apply this approximation. Since

$$2° = 2\left(\frac{\pi}{180}\right) = \frac{\pi}{90} \approx 0.0349066 \text{ radian}$$

it follows that $\sin 2° \approx 0.0349066$. Comparing the two graphs in Figure 3.8.4, we would expect this approximation to be slightly too large. The calculator approximation $\sin 2° \approx 0.0348995$ shows that this is indeed the case. ◀

REMARK.   Part (b) in both Example 1 and Example 2 is meant to be illustrative only. We are not suggesting that you replace individual calculator computations with the local linear approximation. Local linear approximations are significant because they allow us to model a complicated function by a simple one. This idea will be pursued in greater detail later in the text.

ERROR IN LOCAL LINEAR APPROXIMATIONS

As a general rule, the accuracy of the local linear approximation to $f(x)$ at x_0 will deteriorate as x gets progressively farther from x_0. To illustrate this for the approximation $\sin x \approx x$ in Example 2, let us graph the function

$$E(x) = |\sin x - x|$$

which is the absolute value of the error in the approximation (Figure 3.8.5).

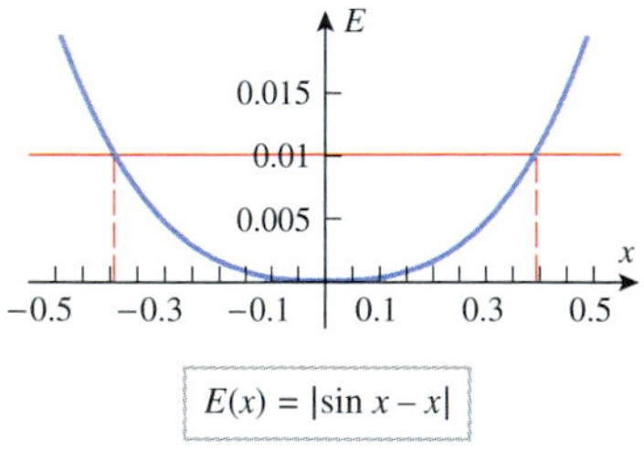

In Figure 3.8.5, the graph shows how the absolute error in the local linear approximation of $\sin x$ increases as x moves progressively farther from 0 in either the positive or negative direction. The graph also tells us that for values of x between the two vertical lines, the absolute error does not exceed 0.01. Thus, for example, we could use the local linear approximation $\sin x \approx x$ for all values of x in the interval $-0.35 < x < 0.35$ (radians) with confidence that the approximation is within 0.01 of the exact value.

Figure 3.8.5

$$E(x) = |\sin x - x|$$

DIFFERENTIALS

Newton and Leibniz independently developed different notations for the derivative. This created a notational divide between Britain and the European continent that lasted for more than 50 years. The ***Leibniz notation*** dy/dx eventually prevailed for its superior utility. For example, we have already mentioned that the Leibniz notation makes the chain rule

$$\frac{dy}{dx} = \frac{dy}{du} \cdot \frac{du}{dx}$$

easy to remember.

Up to now we have been interpreting dy/dx as a single entity representing the derivative of y with respect to x, but we have not attached any meaning to the individual symbols "dy" and "dx." Early in the development of calculus, these symbols represented "infinitely small changes" in the variables y and x and the derivative dy/dx was thought to be a ratio of these infinitely small changes. However, the precise meaning of an "infinitely small change" in a variable turned out to be logically elusive and eventually such arguments were replaced by an analysis that was based on the more modern concept of a limit.

Our next objective is to define the symbols dy and dx so that dy/dx can actually be treated as a ratio. We begin by defining the symbol "dx" to be a *variable* that can assume any real number as its value. The variable dx is called the **differential of x**. If we are given a function $y = f(x)$ that is differentiable at $x = x_0$, then we define the **differential of f at x_0** to be the function of dx given by the formula

$$dy = f'(x_0)\,dx \tag{3}$$

where the symbol "dy" is simply the dependent variable of this function. The variable dy is called the differential of y and we note that it is proportional to dx with constant of proportionality $f'(x_0)$. If $dx \neq 0$, then we can divide both sides of (3) by dx to obtain

$$\frac{dy}{dx} = f'(x_0)$$

Thus, we have achieved our goal of defining dy and dx so that their ratio is a derivative. It is customary to omit the subscript on x and simply write the differential dy as

$$dy = f'(x)\,dx \tag{4}$$

where it is understood that x is regarded as fixed at some value.

Because $f'(x)$ is equal to the slope of the tangent line to the graph of f at the point $(x, f(x))$, the differentials dy and dx can be viewed as a corresponding rise and run of this tangent line (Figure 3.8.6).

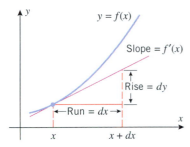

Figure 3.8.6

Example 3 Given the function $y = x^2$, geometrically interpret the relationship between the differentials dx and dy when $x = 3$.

Solution. Since $dy/dx = 2x$, we have $dy = 2x\,dx = 6\,dx$ when $x = 3$. This tells us that if we travel along the tangent line to the curve $y = x^2$ at the point $(3, 9)$, then any change of dx units in the horizonal direction produces a change of $dy = 6\,dx$ units in the vertical direction. ◀

Recall that given a function $y = f(x)$, we defined $\Delta y = f(x + \Delta x) - f(x)$ to denote the signed change in y from its value at some initial number x to its value at a new number $x + \Delta x$. It is important to understand the distinction between the increment Δy and the differential dy. To see the difference, let us assign the independent variables dx and Δx the same value, so $dx = \Delta x$. Then Δy represents the change in y that occurs when we start at x and travel *along the curve $y = f(x)$* until we have moved $\Delta x\ (= dx)$ units in the x-direction, and dy represents the change in y that occurs if we start at x and travel *along the tangent line* until we have moved $dx\ (= \Delta x)$ units in the x-direction (Figure 3.8.7).

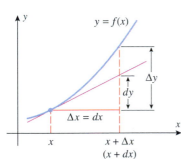

Figure 3.8.7

Example 4 Let $y = \sqrt{x}$. Find dy and Δy at $x = 4$ with $dx = \Delta x = 3$. Then make a sketch of $y = \sqrt{x}$, showing dy and Δy in the picture.

Solution. With $f(x) = \sqrt{x}$ we obtain

$$\Delta y = f(x + \Delta x) - f(x) = \sqrt{x + \Delta x} - \sqrt{x} = \sqrt{7} - \sqrt{4} \approx 0.65$$

If $y = \sqrt{x}$, then

$$\frac{dy}{dx} = \frac{1}{2\sqrt{x}}, \quad \text{so} \quad dy = \frac{1}{2\sqrt{x}}\,dx = \frac{1}{2\sqrt{4}}(3) = \frac{3}{4} = 0.75$$

Figure 3.8.8 shows the curve $y = \sqrt{x}$ together with dy and Δy. ◀

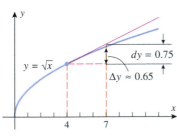

Figure 3.8.8

Although Δy and dy are generally different, the differential dy will nonetheless be a good approximation for Δy provided $dx = \Delta x$ is close to 0. To see this, recall from Section

3.2 that

$$f'(x) = \lim_{\Delta x \to 0} \frac{\Delta y}{\Delta x}$$

It follows that if Δx is close to 0, then we will have $f'(x) \approx \Delta y / \Delta x$ or, equivalently,

$$\Delta y \approx f'(x) \Delta x = f'(x)\,dx = dy \tag{5}$$

As the reader might guess by comparing Figure 3.8.1 with Figure 3.8.7, the approximation $\Delta y \approx dy$ is simply a restatement of the local linear approximation of a function.

FOR THE READER. Obtain the approximation $\Delta y \approx dy$ directly from the local linear approximation (2) by renaming some parameters and using some algebra.

ERROR PROPAGATION IN APPLICATIONS

In applications, small errors invariably occur in measured quantities. When these quantities are used in computations, those errors are propagated in turn to the computed quantities. For example, suppose that in an application the variables x and y are related by a function $y = f(x)$. If x_a is the actual value of x, but it is measured to be x_0, then we define the difference $dx = x_0 - x_a$ to be the *error* in the measurement of x. Note that if the error is positive, the measured value is larger than the actual value, and if the error is negative, the measured value is smaller than the actual value. Since y is determined from x by the function $y = f(x)$, the true value of y is $f(x_a)$ and the value of y computed from the measured value of x is $f(x_0)$. The *propagated error* in the computed value of y is then defined to be $f(x_0) - f(x_a)$. Note that if the propagated error is positive, the calculated value of y will be too large, and if this error is negative, the calculated value of y will be too small. If f is differentiable at the measured value x_0, and if the error in the measurement of x is close to 0, then the local linear approximation (1) (with x replaced by x_a) becomes

$$f(x_a) \approx f(x_0) + f'(x_0)(x_a - x_0) = f(x_0) - f'(x_0)(x_0 - x_a) = f(x_0) - f'(x_0)\,dx$$

We can now use this approximation in our formula for the propagated error to obtain

$$f(x_0) - f(x_a) \approx f(x_0) - (f(x_0) - f'(x_0)\,dx) = f'(x_0)\,dx$$

In other words, the propagated error may be approximated by

$$f(x_0) - f(x_a) \approx dy \tag{6}$$

where $dy = f'(x_0)\,dx$ is the value of the differential of f at x_0 when $dx = x_0 - x_a$ is the error in the measurement of x.

Unfortunately, this approximation cannot be used directly in applied problems because the measurement error $dx = x_0 - x_a$ will in general be unknown. (Keep in mind that the only value of x that is available to the researcher is the measured value x_0.) However, although the exact value of the error in measuring x will generally be unknown, it is often possible to determine upper and lower bounds for this error. Upper and lower bounds for the propagated error can then be approximated by using the differential $dy = f'(x_0)\,dx$.

Example 5 Suppose that the side of a square is measured with a ruler to be 10 inches with a measurement error of at most $\pm\frac{1}{32}$ of an inch.

(a) Use a differential to estimate the error in the computed area of the square.

(b) Compare the estimate from part (a) with the actual possible error computed using a calculating utility.

Solution (a). The side of a square x and the area of the square y are related by the equation $y = x^2$. Since $dy = 2x\,dx$, if we set $x = 10$, then $dy = 20\,dx$. To say that the measurement error is at most $\pm\frac{1}{32}$ of an inch means that the measurement error $dx = x_0 - x_a$ satisfies the inequalities $-\frac{1}{32} \leq dx \leq \frac{1}{32}$. Multiplying each term by 20 yields the equivalent inequalities

$$20\left(-\tfrac{1}{32}\right) \leq dy \leq 20\left(\tfrac{1}{32}\right) \quad \text{or} \quad -\tfrac{5}{8} \leq dy \leq \tfrac{5}{8}$$

Since we are using the differential dy to approximate the propagated error, we estimate this propagated error to be between $-\frac{5}{8}$ and $\frac{5}{8}$ of a square inch. In other words, we estimate the propagated error to be at most $\pm\frac{5}{8}$ of a square inch.

Solution (b). The area of the square is computed to be 100 square inches, but the actual area could be as much as

$$\left(10 + \tfrac{1}{32}\right)^2 = 100 + \tfrac{5}{8} + \tfrac{1}{1024} \quad \text{or as little as} \quad \left(10 - \tfrac{1}{32}\right)^2 = 100 - \tfrac{5}{8} + \tfrac{1}{1024}$$

The propagated error is therefore between $-\frac{5}{8} + \frac{1}{1024}$ and $\frac{5}{8} + \frac{1}{1024}$. Therefore, the upper and lower bounds for the propagated error that we found in part (a) differ from the actual upper and lower bounds by $\frac{1}{1024}$ of a square inch.   ◀

FOR THE READER.   Examine a ruler and explain why a measurement error of at most $\frac{1}{32}$ of an inch is reasonable.

The ratio of the error in some measured or calculated quantity to the true value of the quantity is called the ***relative error*** of the measurement or calculation. When expressed as a percentage, the relative error is called the ***percentage error***. For example, suppose that the side of a square is measured to be 10 inches, but the actual length of the side is 9.98 inches. The relative error in this measurement is then $0.02/9.98 \approx 0.002004008$ or about 0.2004008%. However, as a practical matter the relative error cannot be computed exactly, since both the error and the true value of the quantity are usually unknown. To approximate the relative error in the measurement or computation of some quantity q, we use the ratio dq/q, where q is the measured or calculated value of the quantity. If q is a measured quantity, the numerator dq of this ratio denotes a measurement error, and if q is a computed quantity, dq is an estimate of the propagated error given by (6).

Example 6   The radius of a sphere is measured with a percentage error within $\pm0.04\%$. Estimate the percentage error in the calculated volume of the sphere.

Solution.   The volume V of a sphere is $V = \frac{4}{3}\pi r^3$, so $dV = 4\pi r^2 \, dr$. It then follows from the formulas for V and dV that

$$\frac{dV}{V} = \frac{4\pi r^2 \, dr}{\frac{4}{3}\pi r^3} = 3\frac{dr}{r}$$

If dr denotes the error in measurement of the radius of the sphere, then the relative error in this measurement is estimated by the ratio dr/r, where r is the measured value of the radius. Our assumption that the percentage error in this measurement is within $\pm0.04\%$ then becomes $-0.0004 \leq dr/r \leq 0.0004$. Multiplying each term by 3 yields the equivalent inequalities

$$-0.0012 = 3(-0.0004) \leq dV/V \leq 3(0.0004) = 0.0012$$

Since we are using dV/V to approximate the relative error in the calculated volume of the sphere, we estimate this percentage error to be within $\pm0.12\%$.   ◀

MORE NOTATION; DIFFERENTIAL FORMULAS

The symbol df is another common notation for the differential of a function $y = f(x)$. For example, if $f(x) = \sin x$, then we can write $df = \cos x \, dx$. We can also view the symbol "d" as an *operator* that acts on a function to produce the corresponding differential.

For example, $d[x^2] = 2x\,dx$, $d[\sin x] = \cos x\,dx$, and so on. All of the general rules of differentiation then have corresponding differential versions:

DERIVATIVE FORMULA	DIFFERENTIAL FORMULA
$\dfrac{d}{dx}[c] = 0$	$d[c] = 0$
$\dfrac{d}{dx}[cf] = c\dfrac{df}{dx}$	$d[cf] = c\,df$
$\dfrac{d}{dx}[f + g] = \dfrac{df}{dx} + \dfrac{dg}{dx}$	$d[f + g] = df + dg$
$\dfrac{d}{dx}[fg] = f\dfrac{dg}{dx} + g\dfrac{df}{dx}$	$d[fg] = f\,dg + g\,df$
$\dfrac{d}{dx}\left[\dfrac{f}{g}\right] = \dfrac{g\dfrac{df}{dx} - f\dfrac{dg}{dx}}{g^2}$	$d\left[\dfrac{f}{g}\right] = \dfrac{g\,df - f\,dg}{g^2}$

For example,

$$d[x^2 \sin x] = (x^2 \cos x + 2x \sin x)\,dx$$
$$= x^2(\cos x\,dx) + (2x\,dx)\sin x$$
$$= x^2 d[\sin x] + (\sin x)d[x^2]$$

illustrates the differential version of the product rule.

EXERCISE SET 3.8   ⌐ Graphing Utility

1. (a) Use Formula (1) to obtain the local linear approximation of x^3 at $x_0 = 1$.
   (b) Use Formula (2) to rewrite the approximation obtained in part (a) in terms of Δx.
   (c) Use the result obtained in part (a) to approximate $(1.02)^3$, and confirm that the formula obtained in part (b) produces the same result.

2. (a) Use Formula (1) to obtain the local linear approximation of $1/x$ at $x_0 = 2$.
   (b) Use Formula (2) to rewrite the approximation obtained in part (a) in terms of Δx.
   (c) Use the result obtained in part (a) to approximate $1/2.05$, and confirm that the formula obtained in part (b) produces the same result.

3. (a) Find the local linear approximation of $f(x) = \sqrt{1+x}$ at $x_0 = 0$, and use it to approximate $\sqrt{0.9}$ and $\sqrt{1.1}$.
   (b) Graph f and its tangent line at x_0 together, and use the graphs to illustrate the relationship between the exact values and the approximations of $\sqrt{0.9}$ and $\sqrt{1.1}$.

4. (a) Find the local linear approximation of $f(x) = 1/\sqrt{x}$ at $x_0 = 4$, and use it to approximate $1/\sqrt{3.9}$ and $1/\sqrt{4.1}$.
   (b) Graph f and its tangent line at x_0 together, and use the graphs to illustrate the relationship between the exact values and the approximations of $1/\sqrt{3.9}$ and $1/\sqrt{4.1}$.

In Exercises 5–8, confirm that the stated formula is the local linear approximation at $x_0 = 0$.

5. $(1 + x)^{15} \approx 1 + 15x$

6. $\dfrac{1}{\sqrt{1 - x}} \approx 1 + \tfrac{1}{2}x$

7. $\tan x \approx x$

8. $\dfrac{1}{1 + x} \approx 1 - x$

In Exercises 9–12, confirm that the stated formula is the local linear approximation of f at $x_0 = 1$, where $\Delta x = x - 1$.

9. $f(x) = x^4$; $(1 + \Delta x)^4 \approx 1 + 4\Delta x$

10. $f(x) = \sqrt{x}$; $\sqrt{1 + \Delta x} \approx 1 + \tfrac{1}{2}\Delta x$

11. $f(x) = \dfrac{1}{2 + x}$; $\dfrac{1}{3 + \Delta x} \approx \dfrac{1}{3} - \dfrac{1}{9}\Delta x$

12. $f(x) = (4 + x)^3$; $(5 + \Delta x)^3 \approx 125 + 75\Delta x$

In Exercises 13–16, confirm that the formula is the local linear approximation at $x_0 = 0$, and use a graphing utility to estimate an interval of x-values on which the error is at most ± 0.1.

13. $\sqrt{x + 3} \approx \sqrt{3} + \dfrac{1}{2\sqrt{3}}x$

14. $\dfrac{1}{\sqrt{9 - x}} \approx \dfrac{1}{3} + \dfrac{1}{54}x$

15. $\tan 2x \approx 2x$

16. $\dfrac{1}{(1 + 2x)^5} \approx 1 - 10x$

17. (a) Use the local linear approximation of $\sin x$ at $x_0 = 0$ obtained in Example 2 to approximate $\sin 1°$, and compare the approximation to the result produced directly by your calculating device.

(b) How would you choose x_0 to approximate $\sin 44°$?

(c) Approximate $\sin 44°$; compare the approximation to the result produced directly by your calculating device.

18. (a) Use the local linear approximation of $\tan x$ at $x_0 = 0$ to approximate $\tan 2°$, and compare the approximation to the result produced directly by your calculating device.

(b) How would you choose x_0 to approximate $\tan 61°$?

(c) Approximate $\tan 61°$; compare the approximation to the result produced directly by your calculating device.

In Exercises 19–27, use an appropriate local linear approximation to estimate the value of the given quantity.

19. $(3.02)^4$ **20.** $(1.97)^3$ **21.** $\sqrt{65}$

22. $\sqrt{24}$ **23.** $\sqrt{80.9}$ **24.** $\sqrt{36.03}$

25. $\sin 0.1$ **26.** $\tan 0.2$ **27.** $\cos 31°$

28. The approximation $(1 + x)^k \approx 1 + kx$ is commonly used by engineers for quick calculations.

(a) Derive this result, and use it to make a rough estimate of $(1.001)^{37}$.

(b) Compare your estimate to that produced directly by your calculating device.

(c) Show that this formula produces a very bad estimate of $(1.1)^{37}$, and explain why.

29. (a) Let $y = x^2$. Find dy and Δy at $x = 2$ with $dx = \Delta x = 1$.

(b) Sketch the graph of $y = x^2$, showing dy and Δy in the picture.

30. (a) Let $y = x^3$. Find dy and Δy at $x = 1$ with $dx = \Delta x = 1$.

(b) Sketch the graph of $y = x^3$, showing dy and Δy in the picture.

31. (a) Let $y = 1/x$. Find dy and Δy at $x = 1$ with $dx = \Delta x = -0.5$.

(b) Sketch the graph of $y = 1/x$, showing dy and Δy in the picture.

32. (a) Let $y = \sqrt{x}$. Find dy and Δy at $x = 9$ with $dx = \Delta x = -1$.

(b) Sketch the graph of $y = \sqrt{x}$, showing dy and Δy in the picture.

In Exercises 33–36, find formulas for dy and Δy.

33. $y = x^3$ **34.** $y = 8x - 4$

35. $y = x^2 - 2x + 1$ **36.** $y = \sin x$

In Exercises 37–40, find the differential dy.

37. (a) $y = 4x^3 - 7x^2$ (b) $y = x \cos x$

38. (a) $y = 1/x$ (b) $y = 5 \tan x$

39. (a) $y = x\sqrt{1 - x}$ (b) $y = (1 + x)^{-17}$

40. (a) $y = \dfrac{1}{x^3 - 1}$ (b) $y = \dfrac{1 - x^3}{2 - x}$

In Exercises 41–44, use dy to approximate Δy when x changes as indicated.

41. $y = \sqrt{3x - 2}$; from $x = 2$ to $x = 2.03$

42. $y = \sqrt{x^2 + 8}$; from $x = 1$ to $x = 0.97$

43. $y = \dfrac{x}{x^2 + 1}$; from $x = 2$ to $x = 1.96$

44. $y = x\sqrt{8x + 1}$; from $x = 3$ to $x = 3.05$

45. The side of a square is measured to be 10 ft, with a possible error of ±0.1 ft.

(a) Use differentials to estimate the error in the calculated area.

(b) Estimate the percentage errors in the side and the area.

46. The side of a cube is measured to be 25 cm, with a possible error of ±1 cm.

(a) Use differentials to estimate the error in the calculated volume.

(b) Estimate the percentage errors in the side and volume.

47. The hypotenuse of a right triangle is known to be 10 in exactly, and one of the acute angles is measured to be $30°$, with a possible error of $\pm1°$.

(a) Use differentials to estimate the errors in the sides opposite and adjacent to the measured angle.

(b) Estimate the percentage errors in the sides.

48. One side of a right triangle is known to be 25 cm exactly. The angle opposite to this side is measured to be $60°$, with a possible error of $\pm0.5°$.

(a) Use differentials to estimate the errors in the adjacent side and the hypotenuse.

(b) Estimate the percentage errors in the adjacent side and hypotenuse.

49. The electrical resistance R of a certain wire is given by $R = k/r^2$, where k is a constant and r is the radius of the wire. Assuming that the radius r has a possible error of $\pm5\%$, use differentials to estimate the percentage error in R. (Assume k is exact.)

50. A 12-foot ladder leaning against a wall makes an angle θ with the floor. If the top of the ladder is h feet up the wall, express h in terms of θ and then use dh to estimate the change in h if θ changes from $60°$ to $59°$.

51. The area of a right triangle with a hypotenuse of H is calculated using the formula $A = \frac{1}{4}H^2 \sin 2\theta$, where θ is one of the acute angles. Use differentials to approximate the error in calculating A if $H = 4$ cm (exactly) and θ is measured to be $30°$, with a possible error of $\pm15'$.

52. The side of a square is measured with a possible percentage error of $\pm1\%$. Use differentials to estimate the percentage error in the area.

53. The side of a cube is measured with a possible percentage error of ±2%. Use differentials to estimate the percentage error in the volume.

54. The volume of a sphere is to be computed from a measured value of its radius. Estimate the maximum permissible percentage error in the measurement if the percentage error in the volume must be kept within ±3%. ($V = \frac{4}{3}\pi r^3$ is the volume of a sphere of radius r.)

55. The area of a circle is to be computed from a measured value of its diameter. Estimate the maximum permissible percentage error in the measurement if the percentage error in the area must be kept within ±1%.

56. A steel cube with 1-in sides is coated with 0.01 in of copper. Use differentials to estimate the volume of copper in the coating. [*Hint:* Let ΔV be the change in the volume of the cube.]

57. A metal rod 15 cm long and 5 cm in diameter is to be covered (except for the ends) with insulation that is 0.001 cm thick. Use differentials to estimate the volume of insulation. [*Hint:* Let ΔV be the change in volume of the rod.]

58. The time required for one complete oscillation of a pendulum is called its *period*. If L is the length of the pendulum, then the period is given by $P = 2\pi\sqrt{L/g}$, where g is a con-

stant called *the acceleration due to gravity*. Use differentials to show that the percentage error in P is approximately half the percentage error in L.

59. If the temperature T of a metal rod of length L is changed by an amount ΔT, then the length will change by the amount $\Delta L = \alpha L \, \Delta T$, where α is called the **coefficient of linear expansion**. For moderate changes in temperature α is taken as constant.

    (a) Suppose that a rod 40 cm long at 20°C is found to be 40.006 cm long when the temperature is raised to 30°C. Find α.

    (b) If an aluminum pole is 180 cm long at 15°C, how long is the pole if the temperature is raised to 40°C? [Take $\alpha = 2.3 \times 10^{-5}/°C$.]

60. If the temperature T of a solid or liquid of volume V is changed by an amount ΔT, then the volume will change by the amount $\Delta V = \beta V \, \Delta T$, where β is called the **coefficient of volume expansion**. For moderate changes in temperature β is taken as constant. Suppose that a tank truck loads 4000 gallons of ethyl alcohol at a temperature of 35°C and delivers its load sometime later at a temperature of 15°C. Using $\beta = 7.5 \times 10^{-4}/°C$ for ethyl alcohol, find the number of gallons delivered.

SUPPLEMENTARY EXERCISES

 Graphing Utility   [C] CAS

1. State the definition of a derivative, and give two interpretations of it.

2. Explain the difference between average and instantaneous rate of change, and discuss how they are calculated.

3. Given that $y = f(x)$, explain the difference between dy and Δy. Draw a picture that illustrates the relationship between these quantities.

4. Use the definition of a derivative to find dy/dx, and check your answer by calculating the derivative using appropriate derivative formulas.
    (a) $y = \sqrt{9 - 4x}$ (b) $y = \dfrac{x}{x + 1}$

In Exercises 5–8, find the values of x at which the curve $y = f(x)$ has a horizontal tangent line.

---

5. $f(x) = (2x + 7)^6(x - 2)^5$ **6.** $f(x) = \dfrac{(x - 3)^4}{x^2 + 2x}$

7. $f(x) = \sqrt{3x + 1}(x - 1)^2$ **8.** $f(x) = \left(\dfrac{3x + 1}{x^2}\right)^3$

9. The accompanying figure shows the graph of $y = f'(x)$ for an unspecified function f.
    (a) For what values of x does the curve $y = f(x)$ have a horizontal tangent line?

    (b) Over what intervals does the curve $y = f(x)$ have tangent lines with positive slope?

    (c) Over what intervals does the curve $y = f(x)$ have tangent lines with negative slope?

    (d) Given that $g(x) = f(x) \sin x$, and $f(0) = -1$, find $g''(0)$.

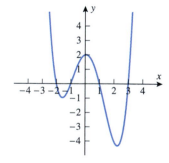

Figure Ex-9

10. In each part, evaluate the expression given that $f(1) = 1$, $g(1) = -2$, $f'(1) = 3$, and $g'(1) = -1$.

    (a) $\dfrac{d}{dx}[f(x)g(x)]\Big|_{x=1}$ (b) $\dfrac{d}{dx}\left[\dfrac{f(x)}{g(x)}\right]\Big|_{x=1}$

    (c) $\dfrac{d}{dx}[\sqrt{f(x)}]\Big|_{x=1}$ (d) $\dfrac{d}{dx}[f(1)g'(1)]$

11. Find the equations of all lines through the origin that are tangent to the curve $y = x^3 - 9x^2 - 16x$.

12. Find all values of x for which the tangent line to $y = 2x^3 - x^2$ is perpendicular to the line $x + 4y = 10$.

13. Find all values of x for which the line that is tangent to $y = 3x - \tan x$ is parallel to the line $y - x = 2$.

14. Suppose that $f(x) = \begin{cases} x^2 - 1, & x \leq 1 \\ k(x - 1), & x > 1. \end{cases}$

For what values of k is f
(a) continuous?   (b) differentiable?

15. Let $f(x) = x^2$. Show that for any distinct values of a and b, the slope of the tangent line to $y = f(x)$ at $x = \frac{1}{2}(a+b)$ is equal to the slope of the secant line through the points (a, a^2) and (b, b^2). Draw a picture to illustrate this result.

16. A car is traveling on a straight road that is 120 mi long. For the first 100 mi the car travels at an average velocity of 50 mi/h. Show that no matter how fast the car travels for the final 20 mi it cannot bring the average velocity up to 60 mi/h for the entire trip.

17. In each part, use the given information to find Δx, Δy, and dy.
(a) $y = 1/(x - 1)$; x decreases from 2 to 1.5.
(b) $y = \tan x$; x increases from $-\pi/4$ to 0.
(c) $y = \sqrt{25 - x^2}$; x increases from 0 to 3.

18. Use the formula $V = l^3$ for the volume of a cube of side l to find
(a) the average rate at which the volume of a cube changes with l as l increases from $l = 2$ to $l = 4$
(b) the instantaneous rate at which the volume of a cube changes with l when $l = 5$.

19. The amount of water in a tank t minutes after it has started to drain is given by $W = 100(t - 15)^2$ gal.
(a) At what rate is the water running out at the end of 5 min?
(b) What is the average rate at which the water flows out during the first 5 min?

20. Use an appropriate local linear approximation to estimate the value of $\cot 46°$, and compare your answer to the value obtained with a calculating device.

21. The base of the Great Pyramid at Giza is a square that is 230 m on each side.
(a) As illustrated in the accompanying figure, suppose that an archaeologist standing at the center of a side measures the angle of elevation of the apex to be $\phi = 51°$ with an error of $\pm 0.5°$. What can the archaeologist reasonably say about the height of the pyramid?
(b) Use differentials to estimate the allowable error in the elevation angle that will ensure an error in the height is at most ± 5 m.

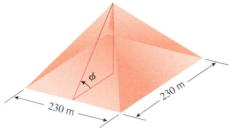
Figure Ex-21

22. The period T of a clock pendulum (i.e., the time required for one back-and-forth movement) is given in terms of its length L by $T = 2\pi\sqrt{L/g}$, where g is the gravitational constant.
(a) Assuming that the length of a clock pendulum can vary (say, due to temperature changes), find the rate of change of the period T with respect to the length L.
(b) If L is in meters (m) and T is in seconds (s), what are the units for the rate of change in part (a)?
(c) If a pendulum clock is running slow, should the length of the pendulum be increased or decreased to correct the problem?
(d) The constant g generally decreases with altitude. If you move a pendulum clock from sea level to a higher elevation, will it run faster or slower?
(e) Assuming the length of the pendulum to be constant, find the rate of change of the period T with respect to g.
(f) Assuming that T is in seconds (s) and g is in meters per second per second (m/s^2), find the units for the rate of change in part (e).

In Exercises 23 and 24, zoom in on the graph of f on an interval containing $x = x_0$ until the graph looks like a straight line. Estimate the slope of this line and then check your answer by finding the exact value of $f'(x_0)$.

23. (a) $f(x) = x^2 - 1$, $x_0 = 1.8$
(b) $f(x) = \dfrac{x^2}{x - 2}$, $x_0 = 3.5$

24. (a) $f(x) = x^3 - x^2 + 1$, $x_0 = 2.3$
(b) $f(x) = \dfrac{x}{x^2 + 1}$, $x_0 = -0.5$

In Exercises 25 and 26, approximate $f'(2)$ by considering the difference quotients

$$\frac{f(x_1) - f(2)}{x_1 - 2}$$

for values of x_1 near 2. If you have a CAS, see if it can find the exact value of the limit of these difference quotients as $x_1 \to 2$.

C **25.** $f(x) = 2^x$ **C** **26.** $f(x) = x^{\sin x}$

27. At time $t = 0$ a car moves into the passing lane to pass a slow-moving truck. The average velocity of the car from $t = 1$ to $t = 1 + h$ is

$$v_{ave} = \frac{3(h+1)^{2.5} + 580h - 3}{10h}$$

Estimate the instantaneous velocity of the car at $t = 1$, where time is in seconds and distance is in feet.

28. A sky diver jumps from an airplane. Suppose that the distance she falls during the first t seconds before her parachute opens is $s(t) = 986((0.835)^t - 1) + 176t$, where s is in feet and $t \geq 1$. Graph s versus t for $1 \leq t \leq 20$, and use your graph to estimate the instantaneous velocity at $t = 15$.

29. Approximate the values of x at which the tangent line to the graph of $y = x^3 - \sin x$ is horizontal.

30. Use a graphing utility to graph the function

$$f(x) = |x^4 - x - 1| - x$$

and find the values of x where the derivative of this function does not exist.

31. Use a CAS to find the derivative of f from the definition

$$f'(x) = \lim_{w \to x} \frac{f(w) - f(x)}{w - x}.$$

and check the result by finding the derivative by hand.

(a) $f(x) = x^5$

(b) $f(x) = 1/x$

(c) $f(x) = 1/\sqrt{x}$

(d) $f(x) = \dfrac{2x + 1}{x - 1}$

(e) $f(x) = \sqrt{3x^2 + 5}$

(f) $f(x) = \sin 3x$

In Exercises 32–37: (a) Use a CAS to find $f'(x)$ via Definition 3.2.3; (b) use the CAS to find $f''(x)$.

32. $f(x) = x^2 \sin x$

33. $f(x) = \sqrt{x} + \cos^2 x$

34. $f(x) = \dfrac{2x^2 - x + 5}{3x + 2}$

35. $f(x) = \dfrac{\tan x}{1 + x^2}$

36. $f(x) = \dfrac{1}{x} \sin \sqrt{x}$

37. $f(x) = \dfrac{\sqrt{x^4 - 3x + 2}}{x(2 - \cos x)}$

In Exercises 38 and 39, find the equation of the tangent line at the specified point.

38. $x^{2/3} - y^{2/3} - y = 1$; $(1, -1)$

39. $\sin xy = y$; $(\pi/2, 1)$

40. The hypotenuse of a right triangle is growing at a constant rate of a centimeters per second and one leg is decreasing at a constant rate of b centimeters per second. How fast is the acute angle between the hypotenuse and the other leg changing at the instant when both legs are 1 cm?

EXPANDING THE CALCULUS HORIZON

Robotics

*R*obin designs and sells room dividers to defray college expenses. She is soon overwhelmed with orders and decides to build a robot to spray paint her dividers. As in most engineering projects, Robin begins with a simplified model that she will eventually refine to be more realistic. However, Robin quickly discovers that robotics (the design and control of robots) involves a considerable amount of mathematics, some of which we will discuss in this module.

The Design Plan

Robin's plan is to develop a two-dimensional version of the robot arm in Figure 1. As shown in Figure 2, Robin's robot arm will consist of two links of fixed length, each of which will rotate independently about a pivot point. A paint sprayer will be attached to the end of the second link, and a computer will vary the angles θ_1 and θ_2, thereby allowing the robot to paint a region of the xy-plane.

The Mathematical Analysis

To analyze the motion of the robot arm, Robin denotes the coordinates of the paint sprayer by (x, y), as in Figure 3, and she derives the following equations that express x and y in terms of the angles θ_1 and θ_2 and the lengths l_1 and l_2 of the links:

$$x = l_1 \cos\theta_1 + l_2 \cos(\theta_1 + \theta_2)$$
$$y = l_1 \sin\theta_1 + l_2 \sin(\theta_1 + \theta_2) \tag{1}$$

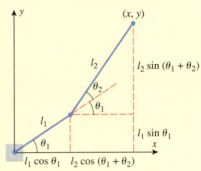

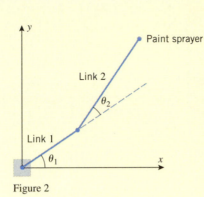

Figure 1   Figure 2   Figure 3

Exercise 1   Use Figure 3 to confirm the equations in (1).

In the language of robotics, θ_1 and θ_2 are called the **control angles**, the point (x, y) is called the **end effector**, and the equations in (1) are called the **forward kinematic equations** (from the Greek word *kinema*, meaning "motion").

Exercise 2   What is the region of the plane that can be reached by the end effector if:
(a) $l_1 = l_2$, (b) $l_1 > l_2$, and (c) $l_1 < l_2$?

Exercise 3   What are the coordinates of the end effector if $l_1 = 2$, $l_2 = 3$, $\theta_1 = \pi/4$, and $\theta_2 = \pi/6$?

Simulating Paint Patterns

Robin recognizes that if θ_1 and θ_2 are regarded as functions of time, then the forward kinematic equations can be expressed as

$$x = l_1 \cos\theta_1(t) + l_2 \cos(\theta_1(t) + \theta_2(t))$$

$$y = l_1 \sin\theta_1(t) + l_2 \sin(\theta_1(t) + \theta_2(t))$$

which are parametric equations for the curve traced by the end effector. For example, if the arms extend horizontally along the positive x-axis at time $t = 0$, and if links 1 and 2 rotate at the constant rates of ω_1 and ω_2 radians per second (rad/s), respectively, then

$$\theta_1(t) = \omega_1 t \quad \text{and} \quad \theta_2(t) = \omega_2 t$$

and the parametric equations of motion for the end effector become

$$x = l_1 \cos\omega_1 t + l_2 \cos(\omega_1 t + \omega_2 t)$$

$$y = l_1 \sin\omega_1 t + l_2 \sin(\omega_1 t + \omega_2 t)$$

Exercise 4   Show that if $l_1 = l_2 = 1$, and if $\omega_1 = 2$ rad/s and $\omega_2 = 3$ rad/s, then the parametric equations of motion are

$$x = \cos 2t + \cos 5t$$

$$y = \sin 2t + \sin 5t$$

Use a graphing utility to show that the curve traced by the end effector over the time interval $0 \le t \le 2\pi$ is as shown in Figure 4. This would be the painting pattern of Robin's paint sprayer.

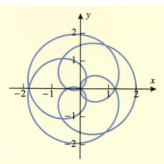

Figure 4

· · · · · · · · · · ·

Exercise 5   Use a graphing utility to explore how the rotation rates of the links affect the spray patterns of a robot arm for which $l_1 = l_2 = 1$.

· · · · · · · · · · ·

Exercise 6   Suppose that $l_1 = l_2 = 1$, and a malfunction in the robot arm causes the second link to lock at $\theta_2 = 0$, while the first link rotates at a constant rate of 1 rad/s. Make a conjecture about the path of the end effector, and confirm your conjecture by finding parametric equations for its motion.

Controlling the Position of the End Effector

Robin's plan is to make the robot paint the dividers in vertical strips, sweeping from the bottom up. After a strip is painted, she will have the arm return to the bottom of the divider and then move horizontally to position itself for the next upward sweep. Since the sections of her dividers will be 3 ft wide by 5 ft high, Robin decides on a robot with two 3-ft links whose base is positioned near the lower left corner of a divider section, as in Figure 5*a*. Since the fully extended links span a radius of 6 ft, she feels that this arrangement will work.

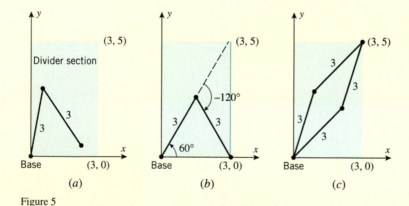

Figure 5

Robin starts with the problem of painting the far right edge from $(3, 0)$ to $(3, 5)$. With the help of some basic geometry (Figure 5*b*), she determines that the end effector can be placed at the point $(3, 0)$ by taking the control angles to be $\theta_1 = \pi/3 \ (= 60°)$ and $\theta_2 = -2\pi/3 \ (= -120°)$ (verify). However, the problem of finding the control angles that correspond to the point $(3, 5)$ is more complicated, so she starts by substituting the link lengths $l_1 = l_2 = 3$ into the forward kinematic equations in (1) to obtain

$$x = 3\cos\theta_1 + 3\cos(\theta_1 + \theta_2)$$
$$y = 3\sin\theta_1 + 3\sin(\theta_1 + \theta_2)$$

(2)

Thus, to put the end effector at the point $(3, 5)$, the control angles must satisfy the equations

$$\cos\theta_1 + \cos(\theta_1 + \theta_2) = 1$$
$$3\sin\theta_1 + 3\sin(\theta_1 + \theta_2) = 5$$

(3)

Solving these equations for θ_1 and θ_2 challenges Robin's algebra and trigonometry skills, but she manages to do it using the procedure in the following exercise.

··········
Exercise 7

(a) Use the equations in (3) and the identity

$$\sin^2(\theta_1 + \theta_2) + \cos^2(\theta_1 + \theta_2) = 1$$

to show that

$$15\sin\theta_1 + 9\cos\theta_1 = 17$$

(b) Solve the last equation for $\sin\theta_1$ in terms of $\cos\theta_1$ and substitute in the identity

$$\sin^2\theta_1 + \cos^2\theta_1 = 1$$

to obtain

$$153\cos^2\theta_1 - 153\cos\theta_1 + 32 = 0$$

(c) Treat this as a quadratic equation in $\cos\theta_1$, and use the quadratic formula to obtain

$$\cos\theta_1 = \frac{1}{2} \pm \frac{5\sqrt{17}}{102}$$

(d) Use the arccosine (inverse cosine) operation of a calculating utility to solve the equations in part (c) to obtain

$$\theta_1 \approx 0.792436 \text{ rad} \approx 45.4032° \quad \text{and} \quad \theta_1 \approx 1.26832 \text{ rad} \approx 72.6693°$$

(e) Substitute each of these angles into the first equation in (3), and solve for the corresponding values of θ_2.

At first, Robin was surprised that the solutions for θ_1 and θ_2 were not unique, but her sketch in Figure 5c quickly made it clear that there will ordinarily be two ways of positioning the links to put the end effector at a specified point.

Controlling the Motion of the End Effector

Now that Robin has figured out how to place the end effector at the points $(3, 0)$ and $(3, 5)$, she turns to the problem of making the robot paint the vertical line segment between those points. She recognizes that not only must she make the end effector move on a vertical line, but she must control its velocity—if the end effector moves too quickly, the paint will be too thin, and if it moves too slowly, the paint will be too thick.

After some experimentation, she decides that the end effector should have a constant velocity of 1 ft/s. Thus, Robin's mathematical problem is to determine the rotation rates $d\theta_1/dt$ and $d\theta_2/dt$ (in rad/s) that will make $dx/dt = 0$ and $dy/dt = 1$. The first condition will ensure that the end effector moves vertically (no horizontal velocity), and the second condition will ensure that it moves upward at 1 ft/s.

To find formulas for dx/dt and dy/dt, Robin uses the chain rule to differentiate the forward kinematic equations in (2) and obtains

$$\frac{dx}{dt} = -3\sin\theta_1 \frac{d\theta_1}{dt} - [3\sin(\theta_1 + \theta_2)]\left(\frac{d\theta_1}{dt} + \frac{d\theta_2}{dt}\right)$$

$$\frac{dy}{dt} = 3\cos\theta_1 \frac{d\theta_1}{dt} + [3\cos(\theta_1 + \theta_2)]\left(\frac{d\theta_1}{dt} + \frac{d\theta_2}{dt}\right)$$

She uses the forward kinematic equations again to simplify these formulas and she then substitutes $dx/dt = 0$ and $dy/dt = 1$ to obtain

$$-y\frac{d\theta_1}{dt} - 3\sin(\theta_1 + \theta_2)\frac{d\theta_2}{dt} = 0$$

$$x\frac{d\theta_1}{dt} + 3\cos(\theta_1 + \theta_2)\frac{d\theta_2}{dt} = 1$$

(4)

· · · · · · · · · · ·

Exercise 8   Confirm Robin's computations.

The equations in (4) will be used in the following way: At a given time t, the robot will report the control angles θ_1 and θ_2 of its links to the computer, the computer will use the forward kinematic equations in (2) to calculate the x- and y-coordinates of the end effector, and then the values of θ_1, θ_2, x, and y will be substituted into (4) to produce two equations in the two unknowns $d\theta_1/dt$ and $d\theta_2/dt$. The computer will solve these equations to determine the required rotation rates for the links.

· · · · · · · · · · ·

Exercise 9   In each part, use the given information to sketch the position of the links, and then calculate the rotation rates for the links in rad/s that will make the end effector of Robin's robot move upward with a velocity of 1 ft/s from that position.

(a) $\theta_1 = \pi/3$, $\theta_2 = -2\pi/3$ (b) $\theta_1 = \pi/2$, $\theta_2 = -\pi/2$

Module by Mary Ann Connors, USMA, West Point, and Howard Anton, Drexel University, and based on the article "Moving a Planar Robot Arm" by Walter Meyer, MAA Notes Number 29, The Mathematical Association of America, 1993.

Leonhard Euler

4

EXPONENTIAL, LOGARITHMIC, AND INVERSE TRIGONOMETRIC FUNCTIONS

*I*n this chapter we will expand our collection of "elementary" functions to include the exponential, logarithmic, and inverse trigonometric functions. The heart of the chapter is Section 4.1 on inverse functions, in which we develop fundamental ideas that link a function and its inverse numerically, algebraically, and graphically. Our focus will be on those aspects of inverse functions that relate to calculus. In particular, we will see that there is an important connection between the derivative of a function and the derivative of its inverse. This connection will allow us to develop a number of derivative formulas for exponential, logarithmic, and inverse trigonometric functions. With the aid of these formulas, we will discuss a powerful tool for evaluating limits known as L'Hôpital's rule.

4.1 INVERSE FUNCTIONS

*In everyday language the term "inversion" conveys the idea of a reversal. For example, in meteorology a temperature inversion is a reversal in the usual temperature proper-ties of air layers; in music, a melodic inversion reverses an ascending interval to the corresponding descending interval; and in grammar an inversion is a reversal of the normal order of words. In mathematics the term **inverse** is used to describe functions that are reverses of one another in the sense that each undoes the effect of the other. The purpose of this section is to discuss this fundamental mathematical idea.*

INVERSE FUNCTIONS

The idea of solving an equation $y = f(x)$ for x as a function of y, say $x = g(y)$, is one of the most important ideas in mathematics. Sometimes, solving an equation is a simple process; for example, using basic algebra the equation

$$y = x^3 + 1 \qquad \boxed{y = f(x)}$$

can be solved for x as a function of y:

$$x = \sqrt[3]{y - 1} \qquad \boxed{x = g(y)}$$

The first equation is better for computing y if x is known, and the second is better for computing x if y is known (Figure 4.1.1).

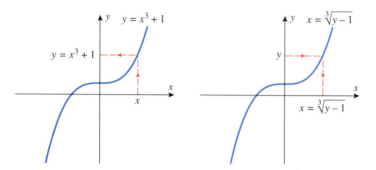

Figure 4.1.1

Our primary interest in this section is to identify relationships that may exist between the functions f and g when an equation $y = f(x)$ is expressed as $x = g(y)$, or conversely. For example, consider the functions $f(x) = x^3 + 1$ and $g(y) = \sqrt[3]{y - 1}$ discussed above. When these functions are composed in either order they cancel out the effect of one another in the sense that

$$g(f(x)) = \sqrt[3]{f(x) - 1} = \sqrt[3]{(x^3 + 1) - 1} = x$$

$$f(g(y)) = [g(y)]^3 + 1 = (\sqrt[3]{y - 1})^3 + 1 = y \tag{1}$$

The first of these equations states that each output of the composition $g(f(x))$ is the same as the input, and the second states that each output of the composition $f(g(y))$ is the same as the input. Pairs of functions with these two properties are so important that there is some terminology for them.

4.1.1   DEFINITION.   If the functions f and g satisfy the two conditions

$$g(f(x)) = x \text{ for every } x \text{ in the domain of } f$$

$$f(g(y)) = y \text{ for every } y \text{ in the domain of } g$$

then we say that f and g are **inverses**. Moreover, we call **f an inverse function for g** and **g an inverse function for f**.

•   REMARK.   For brevity we will often refer to an inverse function for f as an **inverse of f**.

Example 1 It follows from (1) that $f(x) = x^3 + 1$ and $g(y) = \sqrt[3]{y-1}$ are inverses. ◄

It can be shown that a function cannot have two different inverse functions. Thus, if a function f has an inverse function, then the inverse is unique, and we are entitled to talk about *the* inverse of f. The inverse of a function f is commonly denoted by f^{-1} (read "f inverse"). Thus, instead of using g in Example 1, the inverse of $f(x) = x^3 + 1$ could have been expressed as $f^{-1}(y) = \sqrt[3]{y-1}$.

⋮ WARNING. The symbol f^{-1} should always be interpreted as the inverse of f and *never* as the reciprocal $1/f$.

It is important to understand that a function is determined by the relationship that it establishes between its inputs and outputs and not by the letter used for the independent variable. Thus, even though the formulas $f(x) = 3x$ and $f(y) = 3y$ use different independent variables, they define the *same* function f, since the two formulas have the same "form" and hence assign the same value to each input; for example, in either notation $f(2) = 6$. As we progress through this text, there will be certain occasions on which we will want the independent variables for f and f^{-1} to be the same, and other occasions on which we will want them to be different. Thus, in Example 1 we could have expressed the inverse of $f(x) = x^3 + 1$ as $f^{-1}(x) = \sqrt[3]{x-1}$ had we wanted f and f^{-1} to have the same independent variable.

If we use the notation f^{-1} (rather than g) in Definition 4.1.1, and if we use x as the independent variable in the formulas for both f and f^{-1}, then the defining equations relating these functions are

$$f^{-1}(f(x)) = x \quad \text{for every } x \text{ in the domain of } f$$
$$f(f^{-1}(x)) = x \quad \text{for every } x \text{ in the domain of } f^{-1} \tag{2}$$

Example 2 Confirm each of the following.

(a) The inverse of $f(x) = 2x$ is $f^{-1}(x) = \frac{1}{2}x$.

(b) The inverse of $f(x) = x^3$ is $f^{-1}(x) = x^{1/3}$.

Solution (a).

$$f^{-1}(f(x)) = f^{-1}(2x) = \frac{1}{2}(2x) = x$$
$$f(f^{-1}(x)) = f\left(\frac{1}{2}x\right) = 2\left(\frac{1}{2}x\right) = x$$

Solution (b).

$$f^{-1}(f(x)) = f^{-1}(x^3) = \left(x^3\right)^{1/3} = x$$
$$f(f^{-1}(x)) = f(x^{1/3}) = \left(x^{1/3}\right)^3 = x \qquad ◄$$

⋮ REMARK. The results in Example 2 should make sense to you intuitively, since the operations of multiplying by 2 and multiplying by $\frac{1}{2}$ in either order cancel the effect of one another, as do the operations of cubing and taking a cube root.

DOMAIN AND RANGE OF INVERSE FUNCTIONS

The equations in (2) imply certain relationships between the domains and ranges of f and f^{-1}. For example, in the first equation the quantity $f(x)$ is an input of f^{-1}, so points in the range of f lie in the domain of f^{-1}; and in the second equation the quantity $f^{-1}(x)$ is an input of f, so points in the range of f^{-1} lie in the domain of f. All of this suggests the following relationships, which we state without formal proof:

$$\text{domain of } f^{-1} = \text{range of } f$$
$$\text{range of } f^{-1} = \text{domain of } f \tag{3}$$

At the beginning of this section we solved the equation $y = f(x) = x^3 + 1$ for x as a function of y to obtain $x = g(y) = \sqrt[3]{y-1}$, and we observed in Example 1 that g is the inverse of f. This was not accidental—whenever an equation $y = f(x)$ is solved for x as a function of y, say $x = g(y)$, then f and g will be inverses. We can see why this is so by making two substitutions:

- Substitute $y = f(x)$ into $x = g(y)$. This yields $x = g(f(x))$, which is the first equation in Definition 4.1.1.
- Substitute $x = g(y)$ into $y = f(x)$. This yields $y = f(g(y))$, which is the second equation in Definition 4.1.1.

Since f and g satisfy the two conditions in Definition 4.1.1, we conclude that they are inverses. In summary:

If an equation $y = f(x)$ can be solved for x as a function of y, then f has an inverse function and the resulting equation is $x = f^{-1}(y)$.

A METHOD FOR FINDING INVERSES

Example 3 Find the inverse of $f(x) = \sqrt{3x-2}$.

Solution. From the discussion above we can find a formula for $f^{-1}(y)$ by solving the equation

$$y = \sqrt{3x-2}$$

for x as a function of y. The computations are

$$y^2 = 3x - 2$$
$$x = \tfrac{1}{3}(y^2 + 2)$$

from which it follows that

$$f^{-1}(y) = \tfrac{1}{3}(y^2 + 2)$$

At this point we have successfully produced a formula for f^{-1}; however, we are not quite done, since there is no guarantee that the natural domain associated with this formula is the correct domain for f^{-1}. To determine whether this is so, we will examine the range of $y = f(x) = \sqrt{3x-2}$. The range consists of all y in the interval $[0, +\infty)$, so from (3) this interval is also the domain of $f^{-1}(y)$; thus, the inverse of f is given by the formula

$$f^{-1}(y) = \tfrac{1}{3}(y^2 + 2), \quad y \geq 0 \qquad \blacktriangleleft$$

REMARK. When a formula for f^{-1} is obtained by solving the equation $y = f(x)$ for x as a function of y, the resulting formula has y as the independent variable. If it is preferable to have x as the independent variable for f^{-1}, then there are two ways to proceed: you can solve $y = f(x)$ for x as a function of y, and then replace y by x in the *final* formula for f^{-1}, or you can interchange x and y in the *original* equation and solve the equation $x = f(y)$ for y in terms of x, in which case the final equation will be $y = f^{-1}(x)$. In Example 3, either of these procedures will produce $f^{-1}(x) = \tfrac{1}{3}(x^2 + 2), x \geq 0$.

Solving $y = f(x)$ for x as a function of y not only provides a method for finding the inverse of a function f, but it also provides an interpretation of what the values of f^{-1} represent. This tells us that for a given y, the quantity $f^{-1}(y)$ is that number x with the property that $f(x) = y$. For example, if $f^{-1}(1) = 4$, then you know that $f(4) = 1$; and similarly, if $f(3) = 7$, then you know that $f^{-1}(7) = 3$.

EXISTENCE OF INVERSE FUNCTIONS

Not every function has an inverse function. In general, in order for a function f to have an inverse function it must assign distinct outputs to distinct inputs. To see why this is so, consider the function $f(x) = x^2$. Since $f(2) = f(-2) = 4$, the function f assigns the same output to two distinct inputs. If f were to have an inverse function, then the equation $f(2) = 4$ would imply that $f^{-1}(4) = 2$, and the equation $f(-2) = 4$ would imply that $f^{-1}(4) = -2$. This is obviously impossible, since f^{-1} cannot be a function and have two different values for $f^{-1}(4)$. Thus, $f(x) = x^2$ has no inverse. Another way to see that $f(x) = x^2$ has no inverse is to attempt to find the inverse by solving the equation $y = x^2$ for x in terms of y. We run into trouble immediately because the resulting equation, $x = \pm\sqrt{y}$, does not express x as a *single* function of y.

Functions that assign distinct outputs to distinct inputs are sufficiently important that there is a name for them—they are said to be **one-to-one** or **invertible**. Stated algebraically, a function f is one-to-one if $f(x_1) \neq f(x_2)$ whenever $x_1 \neq x_2$; and stated geometrically, a function f is one-to-one if the graph of $y = f(x)$ is cut at most once by any horizontal line (Figure 4.1.2).

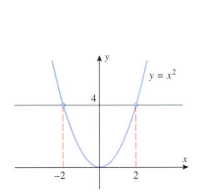

Figure 4.1.3

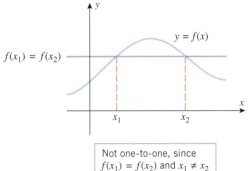

Not one-to-one, since
$f(x_1) = f(x_2)$ and $x_1 \neq x_2$

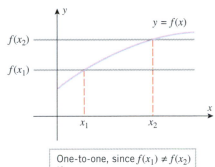

One-to-one, since $f(x_1) \neq f(x_2)$
if $x_1 \neq x_2$

Figure 4.1.2

One can prove that a function f has an inverse function if and only if it is one-to-one, and this provides us with the following geometric test for determining whether a function has an inverse function.

4.1.2   THEOREM (*The Horizontal Line Test*).   *A function f has an inverse function if and only if its graph is cut at most once by any horizontal line.*

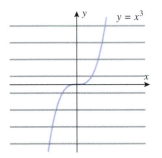

Figure 4.1.4

Example 4   We observed above that the function $f(x) = x^2$ does not have an inverse function. This is confirmed by the horizontal line test, since the graph of $y = x^2$ is cut more than once by certain horizontal lines (Figure 4.1.3).   ◀

Example 5   We saw in Example 2(b) that the function $f(x) = x^3$ has an inverse [namely, $f^{-1}(x) = x^{1/3}$]. The existence of an inverse is confirmed by the horizontal line test, since the graph of $y = x^3$ is cut at most once by any horizontal line (Figure 4.1.4).   ◀

Example 6   Explain why the function f that is graphed in Figure 4.1.5 has an inverse function, and find $f^{-1}(3)$.

Solution.   The function f has an inverse function since its graph passes the horizontal line test. To evaluate $f^{-1}(3)$, we view $f^{-1}(3)$ as that number x for which $f(x) = 3$. From the graph we see that $f(2) = 3$, so $f^{-1}(3) = 2$.   ◀

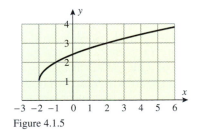

Figure 4.1.5

GRAPHS OF INVERSE FUNCTIONS

Our next objective is to explore the relationship between the graphs of f and f^{-1}. For this purpose, it will be desirable to use x as the independent variable for both functions, which means that we will be comparing the graphs of $y = f(x)$ and $y = f^{-1}(x)$.

If (a, b) is a point on the graph $y = f(x)$, then $b = f(a)$. This is equivalent to the statement that $a = f^{-1}(b)$, which means that (b, a) is a point on the graph of $y = f^{-1}(x)$. In short, reversing the coordinates of a point on the graph of f produces a point on the graph of f^{-1}. Similarly, reversing the coordinates of a point on the graph of f^{-1} produces a point on the graph of f (verify). However, the geometric effect of reversing the coordinates of a point is to reflect that point about the line $y = x$ (Figure 4.1.6), and hence the graphs of $y = f(x)$ and $y = f^{-1}(x)$ are reflections of one another about this line (Figure 4.1.7). In summary, we have the following result.

4.1.3 THEOREM. *If f has an inverse function f^{-1}, then the graphs of $y = f(x)$ and $y = f^{-1}(x)$ are reflections of one another about the line $y = x$; that is, each is the mirror image of the other with respect to that line.*

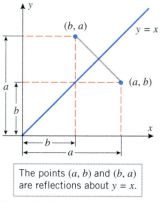

The points (a, b) and (b, a) are reflections about $y = x$.

Figure 4.1.6

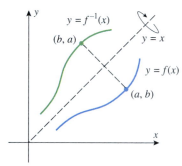

Figure 4.1.7

Example 7 Figure 4.1.8 shows the graphs of the inverse functions discussed in Examples 2 and 3. ◀

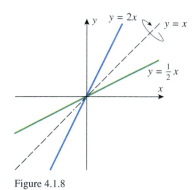
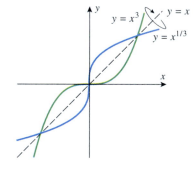
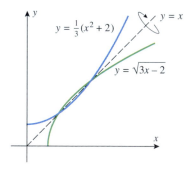

Figure 4.1.8

INCREASING OR DECREASING FUNCTIONS ARE INVERTIBLE

If the graph of a function f is always increasing or always decreasing over the domain of f, then a horizontal line will cut the graph of f in at most one point (Figure 4.1.9), so f must have an inverse function. One way to tell whether the graph of a function is increasing or decreasing over an interval is by examining the slope of its graph. We will prove in the next chapter that f must be increasing on any interval on which $f'(x) > 0$ (since the graph has positive slope) and that f must be decreasing on any interval on which $f'(x) < 0$ (since the graph has negative slope). These intuitive observations suggest the following theorem, which we state without formal proof.

4.1.4 THEOREM. *If the domain of a function f is an interval on which $f'(x) > 0$ or on which $f'(x) < 0$, then f has an inverse function.*

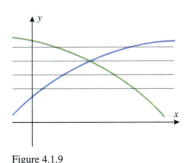

Figure 4.1.9

Example 8 The graph of $f(x) = x^5 + x + 1$ is always increasing on $(-\infty, +\infty)$ since

$$f'(x) = 5x^4 + 1 > 0$$

for all x. However, there is no easy way to solve the equation $y = x^5 + x + 1$ for x in terms of y (try it), so even though we know that f has an inverse function f^{-1}, we cannot produce a formula for $f^{-1}(x)$. ◀

REMARK.   What is important to understand here is that our inability to find an explicit formula for the inverse function does not negate the existence of the inverse. In this case the inverse function $x = f^{-1}(y)$ is implicitly defined by the equation $y = x^5 + x + 1$, so we can use implicit differentiation (Section 3.6) to investigate properties of the inverse function determined by its derivative.

RESTRICTING DOMAINS FOR INVERTIBILITY

Frequently, the domain of a function that is not one-to-one can be partitioned into intervals so that the "piece" of the function defined on each interval in the partition is one-to-one. Thus, the function may be viewed as piecewise defined in terms of one-to-one functions. For example, the function $f(x) = x^2$ is not one-to-one on its natural domain, $-\infty < x < +\infty$, but consider

$$f(x) = \begin{cases} x^2, & x < 0 \\ x^2, & x \geq 0 \end{cases}$$

(Figure 4.1.10). The "piece" of $f(x)$ given by

$$g(x) = x^2, \quad x \geq 0$$

is increasing, and so is one-to-one, on its specified domain. Thus, g has an inverse function g^{-1}. Solving

$$y = x^2, \quad x \geq 0$$

for x yields $x = \sqrt{y}$, so $g^{-1}(y) = \sqrt{y}$. Similarly, if

$$h(x) = x^2, \quad x \leq 0$$

then h has an inverse function, $h^{-1}(y) = -\sqrt{y}$. Geometrically, the graphs of $g(x) = x^2$, $x \geq 0$ and $g^{-1}(x) = \sqrt{x}$ are reflections of one another about the line $y = x$, as are the graphs of $h(x) = x^2, x \leq 0$ and $h^{-1}(x) = -\sqrt{x}$ (Figure 4.1.11).

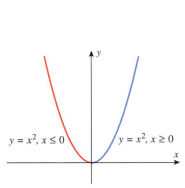

Figure 4.1.10

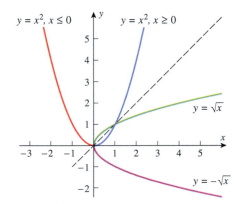

Figure 4.1.11

The functions $g(x)$ and $h(x)$ in the last paragraph are called **restrictions** of the function $f(x)$ because each is obtained from $f(x)$ merely by placing a restriction on its domain. In particular, we say that $g(x)$ is the restriction of $f(x)$ to the interval $[0, +\infty)$ and that $h(x)$ is the restriction of $f(x)$ to the interval $(-\infty, 0]$.

CONTINUITY OF INVERSE FUNCTIONS

Since the graphs of a one-to-one function f and its inverse function f^{-1} are reflections of one another about the line $y = x$, it is intuitively clear that if the graph of f has no breaks, then neither will the graph of f^{-1}. This suggests the following result, which we state without proof.

> **4.1.5 THEOREM.** *Suppose that f is a function with domain D and range R. If D is an interval and f is continuous and one-to-one on D, then R is an interval and the inverse of f is continuous on R.*

For example, the function $f(x) = x^5 + x + 1$ in Example 8 has domain and range $(-\infty, +\infty)$, and f is continuous and one-to-one on $(-\infty, +\infty)$. Thus, we can conclude that f^{-1} is continuous on $(-\infty, +\infty)$, despite our inability to find a formula for $f^{-1}(x)$.

DIFFERENTIABILITY OF INVERSE FUNCTIONS

Suppose that f is a function whose domain D is an open interval and that f is continuous and one-to-one on D. Informally, the places where f fails to be differentiable occur where the graph of f has a corner or a vertical tangent line. Similarly, f^{-1} will be differentiable on its domain except where the graph of f^{-1} has a corner or a vertical tangent line. Note that a corner in the graph of f will reflect about the line $y = x$ to a corner in the graph of f^{-1}, and vice versa. However, since a vertical line is the reflection of a horizontal line about the graph of $y = x$ (Figure 4.1.12), a point of vertical tangency on the graph of f^{-1} will correspond to a point of horizontal tangency on the graph of f. Thus, f^{-1} will fail to be differentiable at a point $(x, f^{-1}(x))$ on its graph if $f'(f^{-1}(x)) = 0$.

Now suppose that f is differentiable at a point (a, b) and that $f'(a) \neq 0$. Then

$$y - b = f'(a)(x - a)$$

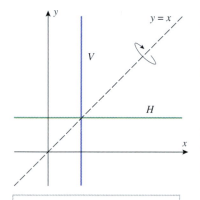

The vertical line V reflects into the horizontal line H and conversely.

Figure 4.1.12

is the equation of the tangent line to the graph of f at (a, b). The reflection of this line about the graph of $y = x$ should carry it to a tangent line L to the graph of $y = f^{-1}(x)$ at the point (b, a). The equation of L is

$$x - b = f'(a)(y - a) \quad \text{or} \quad y - a = \frac{1}{f'(a)}(x - b)$$

which tells us that the slope of the curve $y = f^{-1}(x)$ at (b, a) and the slope of the curve $y = f(x)$ at (a, b) are reciprocals (Figure 4.1.13). Using $a = f^{-1}(b)$ we obtain

$$(f^{-1})'(b) = \frac{1}{f'(f^{-1}(b))}$$

In summary, we have the following result.

> **4.1.6 THEOREM (Differentiability of Inverse Functions).** *Suppose that f is a function whose domain D is an open interval, and let R be the range of f. If f is differentiable and one-to-one on D, then f^{-1} is differentiable at any value x in R for which $f'(f^{-1}(x)) \neq 0$. Furthermore, if x is in R with $f'(f^{-1}(x)) \neq 0$, then*
>
> $$\frac{d}{dx}[f^{-1}(x)] = \frac{1}{f'(f^{-1}(x))} \tag{4}$$

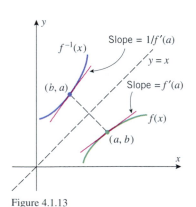

Figure 4.1.13

As an immediate consequence of Theorems 4.1.4 and 4.1.5 we have the following result.

> **4.1.7 COROLLARY.** *If the domain of a function f is an interval on which $f'(x) > 0$ or on which $f'(x) < 0$, then f has an inverse function f^{-1} and $f^{-1}(x)$ is differentiable at any value x in the range of f. The derivative of f^{-1} is given by Formula (4).*

REMARK.   A careful proof of Theorem 4.1.6 would involve the definition of the derivative of $f^{-1}(x)$. As a sketch of this argument, consider the special case where f is differentiable and increasing on D and set $g(x) = f^{-1}(x)$. Then g is increasing and continuous on R. Now, $g(x)$ is differentiable at those values of x for which

$$\lim_{w \to x} \frac{g(w) - g(x)}{w - x}$$

exists. For w and x in R with $w \neq x$, set $r = g(w)$ and $s = g(x)$, so $f(r) = w$, $f(s) = x$, and $r \neq s$. Then

$$\frac{g(w) - g(x)}{w - x} = \frac{r - s}{f(r) - f(s)} = \frac{1}{\left[\dfrac{f(r) - f(s)}{r - s} \right]}$$

Using the facts that f and g are continuous and increasing on their domains and that f and g are inverse functions, we can argue that $w \to x$ if and only if $r \to s$. Thus,

$$\lim_{w \to x} \frac{g(w) - g(x)}{w - x}$$

exists provided

$$\lim_{r \to s} \frac{f(r) - f(s)}{r - s}$$

exists and is not zero. That is, $y = f^{-1}(x)$ is differentiable at x if f is differentiable at y and $f'(y) \neq 0$.

Formula (4) can be expressed in a less forbidding form by setting

$$y = f^{-1}(x) \quad \text{so that} \quad x = f(y)$$

Thus,

$$\frac{dy}{dx} = (f^{-1})'(x) \quad \text{and} \quad \frac{dx}{dy} = f'(y) = f'(f^{-1}(x))$$

Substituting these expressions into Formula (4) yields the following alternative version of that formula:

$$\frac{dy}{dx} = \frac{1}{dx/dy} \tag{5}$$

If an explicit formula can be obtained for the inverse of a function, then the differentiability of the inverse function can generally be deduced from that formula. However, if no explicit formula for the inverse can be obtained, then Theorem 4.1.6 is the primary tool for establishing differentiability of the inverse function. Once the differentiability has been established, a derivative function for the inverse function can be obtained either by implicit differentiation or by using Formula (4) or (5).

Example 9   We saw in Example 8 that the function $f(x) = x^5 + x + 1$ is invertible.

(a)   Show that f^{-1} is differentiable on the interval $(-\infty, +\infty)$.

(b)   Find a formula for the derivative of f^{-1} using Formula (5).

(c)   Find a formula for the derivative of f^{-1} using implicit differentiation.

Solution (a).   Both the range and domain of f are $(-\infty, +\infty)$. Since

$$f'(x) = 5x^4 + 1 > 0$$

for all x, f^{-1} is differentiable at every x in its domain, $(-\infty, +\infty)$.

Solution (b).   If we let $y = f^{-1}(x)$, then

$$x = f(y) = y^5 + y + 1 \tag{6}$$

from which it follows that $dx/dy = 5y^4 + 1$. Then, from Formula (5),

$$\frac{dy}{dx} = \frac{1}{dx/dy} = \frac{1}{5y^4 + 1} \tag{7}$$

Since we were unable to solve (6) for y in terms of x, we must leave (7) in terms of y.

Solution (c). Differentiating (6) implicitly with respect to x yields

$$\frac{d}{dx}[x] = \frac{d}{dx}[y^5 + y + 1]$$

$$1 = (5y^4 + 1)\frac{dy}{dx}$$

$$\frac{dy}{dx} = \frac{1}{5y^4 + 1}$$

which agrees with (7). ◀

GRAPHING INVERSE FUNCTIONS WITH GRAPHING UTILITIES

Most graphing utilities cannot graph inverse functions directly. However, there is a way of graphing inverse functions by expressing the graphs parametrically. To see how this can be done, suppose that we are interested in graphing the inverse of a one-to-one function f. We observed in Section 1.8 that the equation $y = f(x)$ can be expressed parametrically as

$$x = t, \quad y = f(t) \tag{8}$$

Moreover, we know that the graph of f^{-1} can be obtained by interchanging x and y, since this reflects the graph of f about the line $y = x$. Thus, from (8) the graph of f^{-1} can be represented parametrically as

$$x = f(t), \quad y = t \tag{9}$$

For example, Figure 4.1.14 shows the graph of $f(x) = x^5 + x + 1$ and its inverse generated with a graphing utility. The graph of f was generated from the parametric equations

$$x = t, \quad y = t^5 + t + 1$$

and the graph of f^{-1} was generated from the parametric equations

$$x = t^5 + t + 1, \quad y = t$$

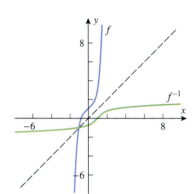

Figure 4.1.14

EXERCISE SET 4.1   ∿ Graphing Utility

1. In (a)–(d), determine whether f and g are inverse functions.
(a) $f(x) = 4x, \ g(x) = \frac{1}{4}x$
(b) $f(x) = 3x + 1, \ g(x) = 3x - 1$
(c) $f(x) = \sqrt[3]{x - 2}, \ g(x) = x^3 + 2$
(d) $f(x) = x^4, \ g(x) = \sqrt[4]{x}$

∿ **2.** Check your answers to Exercise 1 with a graphing utility by determining whether the graphs of f and g are reflections of one another about the line $y = x$.

3. In each part, determine whether the function f defined by the table is one-to-one.

(a)
x	1	2	3	4	5	6
$f(x)$	−2	−1	0	1	2	3

(b)
x	1	2	3	4	5	6
$f(x)$	4	−7	6	−3	1	4

4. In each part, determine whether the function f is one-to-one, and justify your answer.
(a) $f(t)$ is the number of people in line at a movie theater at time t.
(b) $f(x)$ is your weight on your xth birthday.
(c) $f(v)$ is the weight of v cubic inches of lead.

5. In each part, use the horizontal line test to determine whether the function f is one-to-one.
(a) $f(x) = 3x + 2$ (b) $f(x) = \sqrt{x - 1}$
(c) $f(x) = |x|$ (d) $f(x) = x^3$
(e) $f(x) = x^2 - 2x + 2$ (f) $f(x) = \sin x$

∿ **6.** In each part, generate the graph of the function f with a graphing utility, and determine whether f is one-to-one.
(a) $f(x) = x^3 - 3x + 2$ (b) $f(x) = x^3 - 3x^2 + 3x - 1$

7. In each part, determine whether f is one-to-one.
(a) $f(x) = \tan x$
(b) $f(x) = \tan x, \quad -\pi < x < \pi, x \neq \pm\pi/2$
(c) $f(x) = \tan x, \quad -\pi/2 < x < \pi/2$

8. In each part, determine whether f is one-to-one.
(a) $f(x) = \cos x$
(b) $f(x) = \cos x, \quad -\pi/2 \le x \le \pi/2$
(c) $f(x) = \cos x, \quad 0 \le x \le \pi$

9. (a) The accompanying figure shows the graph of a function f over its domain $-8 \le x \le 8$. Explain why f has an inverse, and use the graph to find $f^{-1}(2)$, $f^{-1}(-1)$, and $f^{-1}(0)$.
(b) Find the domain and range of f^{-1}.
(c) Sketch the graph of f^{-1}.

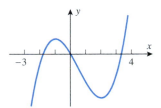

Figure Ex-9

10. (a) Explain why the function f graphed in the accompanying figure has no inverse function on its domain $-3 \le x \le 4$.
(b) Subdivide the domain into three adjacent intervals on each of which the function f has an inverse.

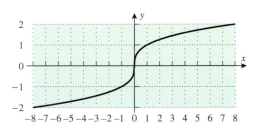

Figure Ex-10

In Exercises 11 and 12, determine whether the function f is one-to-one by examining the sign of $f'(x)$.

11. (a) $f(x) = x^2 + 8x + 1$
(b) $f(x) = 2x^5 + x^3 + 3x + 2$
(c) $f(x) = 2x + \sin x$

12. (a) $f(x) = x^3 + 3x^2 - 8$
(b) $f(x) = x^5 + 8x^3 + 2x - 1$
(c) $f(x) = \dfrac{x}{x+1}$

In Exercises 13–23, find a formula for $f^{-1}(x)$.

13. $f(x) = x^5$

14. $f(x) = 6x$

15. $f(x) = 7x - 6$

16. $f(x) = \dfrac{x+1}{x-1}$

17. $f(x) = 3x^3 - 5$

18. $f(x) = \sqrt[5]{4x+2}$

19. $f(x) = \sqrt[3]{2x-1}$

20. $f(x) = 5/(x^2+1), \quad x \ge 0$

21. $f(x) = 3/x^2, \quad x < 0$

22. $f(x) = \begin{cases} 2x, & x \le 0 \\ x^2, & x > 0 \end{cases}$

23. $f(x) = \begin{cases} 5/2 - x, & x < 2 \\ 1/x, & x \ge 2 \end{cases}$

24. Find a formula for $p^{-1}(x)$, given that
$$p(x) = x^3 - 3x^2 + 3x - 1$$

In Exercises 25–29, find a formula for $f^{-1}(x)$, and state the domain of f^{-1}.

25. $f(x) = (x+2)^4, \quad x \ge 0$

26. $f(x) = \sqrt{x+3}$

27. $f(x) = -\sqrt{3-2x}$

28. $f(x) = 3x^2 + 5x - 2, \quad x \ge 0$

29. $f(x) = x - 5x^2, \quad x \ge 1$

30. The formula $F = \frac{9}{5}C + 32$, where $C \ge -273.15$ expresses the Fahrenheit temperature F as a function of the Celsius temperature C.
(a) Find a formula for the inverse function.
(b) In words, what does the inverse function tell you?
(c) Find the domain and range of the inverse function.

31. (a) One meter is about 6.214×10^{-4} miles. Find a formula $y = f(x)$ that expresses a length x in meters as a function of the same length y in miles.
(b) Find a formula for the inverse of f.
(c) In practical terms, what does the formula $x = f^{-1}(y)$ tell you?

32. Suppose that f is a one-to-one, continuous function such that $\lim_{x \to 3} f(x) = 7$. Find $\lim_{x \to 7} f^{-1}(x)$, and justify your reasoning.

33. Let $f(x) = x^2, x > 1$, and $g(x) = \sqrt{x}$.
(a) Show that $f(g(x)) = x, x > 1$, and $g(f(x)) = x, x > 1$.
(b) Show that f and g are *not* inverses by showing that the graphs of $y = f(x)$ and $y = g(x)$ are not reflections of one another about $y = x$.
(c) Do parts (a) and (b) contradict one another? Explain.

34. Let $f(x) = ax^2 + bx + c, a > 0$. Find f^{-1} if the domain of f is restricted to
(a) $x \ge -b/(2a)$
(b) $x \le -b/(2a)$.

35. (a) Show that $f(x) = (3 - x)/(1 - x)$ is its own inverse.
(b) What does the result in part (a) tell you about the graph of f?

36. Suppose that a line of nonzero slope m intersects the x-axis at $(x_0, 0)$. Find an equation for the reflection of this line about $y = x$.

37. (a) Show that $f(x) = x^3 - 3x^2 + 2x$ is not one-to-one on $(-\infty, +\infty)$.
(b) Find the largest value of k such that f is one-to-one on the interval $(-k, k)$.

38. (a) Show that the function $f(x) = x^4 - 2x^3$ is not one-to-one on $(-\infty, +\infty)$.
(b) Find the smallest value of k such that f is one-to-one on the interval $[k, +\infty)$.

39. Let $f(x) = 2x^3 + 5x + 3$. Find x if $f^{-1}(x) = 1$.

40. Let $f(x) = \dfrac{x^3}{x^2 + 1}$. Find x if $f^{-1}(x) = 2$.

In Exercises 41–44, use a graphing utility and parametric equations to display the graphs of f and f^{-1} on the same screen.

41. $f(x) = x^3 + 0.2x - 1$, $-1 \le x \le 2$

42. $f(x) = \sqrt{x^2 + 2} + x$, $-5 \le x \le 5$

43. $f(x) = \cos(\cos 0.5x)$, $0 \le x \le 3$

44. $f(x) = x + \sin x$, $0 \le x \le 6$

In Exercises 45–48, find the derivative of f^{-1} by using Formula (5), and check your result by differentiating implicitly.

45. $f(x) = 5x^3 + x - 7$ **46.** $f(x) = 1/x^2,\ x > 0$

47. $f(x) = 2x^5 + x^3 + 1$

48. $f(x) = 5x - \sin 2x$, $-\dfrac{\pi}{4} < x < \dfrac{\pi}{4}$

49. Prove that if $a^2 + bc \ne 0$, then the graph of

$$f(x) = \frac{ax + b}{cx - a}$$

is symmetric about the line $y = x$.

50. (a) Prove: If f and g are one-to-one, then so is the composition $f \circ g$.

(b) Prove: If f and g are one-to-one, then

$$(f \circ g)^{-1} = g^{-1} \circ f^{-1}$$

51. Sketch the graph of a function that is one-to-one on $(-\infty, +\infty)$, yet not increasing on $(-\infty, +\infty)$ and not decreasing on $(-\infty, +\infty)$.

52. Prove: A one-to-one function f cannot have two different inverse functions.

53. Let $F(x) = f(2g(x))$ where $f(x) = x^4 + x^3 + 1$ for $0 \le x \le 2$, and $g(x) = f^{-1}(x)$. Find $F(3)$.

4.2 EXPONENTIAL AND LOGARITHMIC FUNCTIONS

When logarithms were introduced in the seventeenth century as a computational tool, they provided scientists of that period computing power that was previously unimaginable. Although computers and calculators have largely replaced logarithms for numerical calculations, the logarithmic functions and their relatives have wide-ranging applications in mathematics and science. Some of these will be introduced in this section.

IRRATIONAL EXPONENTS

In algebra, integer and rational powers of a number b are defined by

$$b^n = b \times b \times \cdots \times b \quad (n \text{ factors}), \qquad b^{-n} = \frac{1}{b^n}, \qquad b^0 = 1,$$

$$b^{p/q} = \sqrt[q]{b^p} = (\sqrt[q]{b})^p, \qquad b^{-p/q} = \frac{1}{b^{p/q}}$$

If b is negative, then some of the fractional powers of b will have imaginary values; for example, $(-2)^{1/2} = \sqrt{-2}$. To avoid this complication we will assume throughout this section that $b \ge 0$, even if it is not stated explicitly.

Observe that the preceding definitions do not include *irrational* powers of b such as

$$2^\pi, \quad 3^{\sqrt{2}}, \quad \text{and} \quad \pi^{-\sqrt{7}}$$

There are various methods for defining irrational powers. One approach is to define irrational powers of b as limits of rational powers of b. For example, to define 2^π we can start with the decimal representation of π, namely,

$$3.1415926\ldots$$

From this decimal we can form a sequence of rational numbers that gets closer and closer to π, namely,

$$3.1, \quad 3.14, \quad 3.141, \quad 3.1415, \quad 3.14159$$

and from these we can form a sequence of *rational* powers of 2:

$$2^{3.1}, \quad 2^{3.14}, \quad 2^{3.141}, \quad 2^{3.1415}, \quad 2^{3.14159}$$

Table 4.2.1

x	2^x
3	8.000000
3.1	8.574188
3.14	8.815241
3.141	8.821353
3.1415	8.824411
3.14159	8.824962
3.141592	8.824974

Since the exponents of the terms in this sequence approach a limit of π, it seems plausible that the terms themselves approach a limit, and it would seem reasonable to *define* 2^π to be this limit. Table 4.2.1 provides numerical evidence that the sequence does, in fact, have a limit and that to four decimal places the value of this limit is $2^\pi \approx 8.8250$. More generally, for any irrational exponent p and positive number b, we can define b^p as the limit of the rational powers of b created from the decimal expansion of p.

FOR THE READER.   Confirm the approximation $2^\pi \approx 8.8250$ by computing 2^π directly using your calculating utility.

Although our definition of b^p for irrational p certainly seems reasonable, there is a lot of tedious mathematical detail required to make the definition precise. We will not be concerned with such matters here and will accept without proof that the following familiar laws hold for all real exponents:

$$b^p b^q = b^{p+q}, \quad \frac{b^p}{b^q} = b^{p-q}, \quad (b^p)^q = b^{pq}$$

THE FAMILY OF EXPONENTIAL FUNCTIONS

A function of the form $f(x) = b^x$, where $b > 0$ and $b \neq 1$, is called an *exponential function with base b*. Some examples are

$$f(x) = 2^x, \quad f(x) = \left(\tfrac{1}{2}\right)^x, \quad f(x) = \pi^x$$

Note that an exponential function has a constant base and variable exponent. Thus, functions such as $f(x) = x^2$ and $f(x) = x^\pi$ would not be classified as exponential functions, since they have a variable base and a constant exponent.

It can be shown that exponential functions are continuous and have one of the basic two shapes shown in Figure 4.2.1a, depending on whether $0 < b < 1$ or $b > 1$. Figure 4.2.1b shows the graphs of some specific exponential functions.

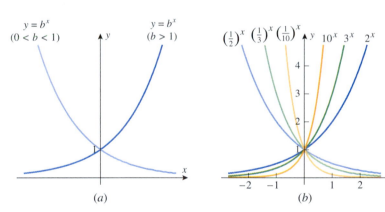

Figure 4.2.1

REMARK.   If $b = 1$, then the function b^x is constant, since $b^x = 1^x = 1$. This case is of no interest to us here, so we have excluded it from the family of exponential functions.

FOR THE READER.   Use your graphing utility to confirm that the graphs of $y = \left(\tfrac{1}{2}\right)^x$ and $y = 2^x$ agree with Figure 4.2.1b, and explain why the two graphs are reflections of one another about the y-axis.

Since it is not our objective in this section to develop the properties of exponential functions in rigorous mathematical detail, we will simply observe without proof that the following properties of exponential functions are consistent with the graphs shown in Figure 4.2.1.

> **4.2.1   THEOREM.**   *If $b > 0$ and $b \neq 1$, then:*
>
> (a)   *The function $f(x) = b^x$ is defined for all real values of x, so its natural domain is $(-\infty, +\infty)$.*
>
> (b)   *The function $f(x) = b^x$ is continuous on the interval $(-\infty, +\infty)$, and its range is $(0, +\infty)$.*

LOGARITHMS

Recall from algebra that a logarithm is an exponent. More precisely, if $b > 0$ and $b \neq 1$, then for positive values of x the **logarithm to the base b of x** is denoted by

$$\log_b x$$

and is defined to be that exponent to which b must be raised to produce x. For example,

$$\log_{10} 100 = 2, \quad \log_{10}(1/1000) = -3, \quad \log_2 16 = 4, \quad \log_b 1 = 0, \quad \log_b b = 1$$

| $10^2 = 100$ | $10^{-3} = 1/1000$ | $2^4 = 16$ | $b^0 = 1$ | $b^1 = b$ |

Historically, the first logarithms ever studied were the logarithms with base 10, called **common logarithms**. For such logarithms it is usual to suppress explicit reference to the base and write $\log x$ rather than $\log_{10} x$. More recently, logarithms with base 2 have played a role in computer science, since they arise naturally in the binary number system. However, the most widely used logarithms in applications are the **natural logarithms**, which have an irrational base denoted by the letter e in honor of the Swiss mathematician Leonhard Euler (biography on p. 11), who first suggested its application to logarithms in an unpublished paper written in 1728. This constant, whose value to six decimal places is

$$e \approx 2.718282 \tag{1}$$

arises as the horizontal asymptote of the graph of the equation

$$y = \left(1 + \frac{1}{x}\right)^x \tag{2}$$

(Figure 4.2.2).

THE VALUES OF $(1 + 1/x)^x$
APPROACH e AS $x \to +\infty$

x	$1 + \dfrac{1}{x}$	$\left(1 + \dfrac{1}{x}\right)^x$
1	2	≈ 2.000000
10	1.1	2.593742
100	1.01	2.704814
1000	1.001	2.716924
10,000	1.0001	2.718146
100,000	1.00001	2.718268
1,000,000	1.000001	2.718280

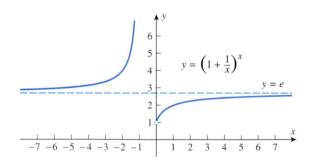

Figure 4.2.2

The fact that $y = e$ is a horizontal asymptote of (2) as $x \to +\infty$ and as $x \to -\infty$ is expressed by the limits

$$e = \lim_{x \to +\infty} \left(1 + \frac{1}{x}\right)^x \qquad \text{and} \qquad e = \lim_{x \to -\infty} \left(1 + \frac{1}{x}\right)^x \tag{3-4}$$

Later, we will show that these limits can be derived from the limit

$$e = \lim_{x \to 0} (1 + x)^{1/x} \tag{5}$$

which is sometimes taken as the definition of the number e.

It is standard to denote the natural logarithm of x by $\ln x$ (read "ell en of x"), rather than $\log_e x$. Thus, $\ln x$ can be viewed as that power to which e must be raised to produce x. For example,

$$\ln 1 = 0, \qquad \ln e = 1, \qquad \ln 1/e = -1, \qquad \ln(e^2) = 2$$

| Since $e^0 = 1$ | Since $e^1 = e$ | Since $e^{-1} = 1/e$ | Since $e^2 = e^2$ |

In general, the statements

$$y = \ln x \quad \text{and} \quad x = e^y$$

are equivalent.

The exponential function $f(x) = e^x$ is called the ***natural exponential function***. To simplify typography, this function is sometimes written as $\exp x$. Thus, for example, you might see the relationship $e^{x_1+x_2} = e^{x_1} e^{x_2}$ expressed as

$$\exp(x_1 + x_2) = \exp(x_1) \exp(x_2)$$

This notation is also used by graphing and calculating utilities, and it is typical to access the function e^x with some variation of the command EXP.

FOR THE READER. Most scientific calculating utilities provide some way of evaluating common logarithms, natural logarithms, and powers of e. Check your documentation to see how this is done, and then confirm the approximation $e \approx 2.718282$ and the values that appear in the table in Figure 4.2.2.

LOGARITHMIC FUNCTIONS

Figure 4.2.1a suggests that if $b > 0$ and $b \neq 1$, then the graph of $y = b^x$ passes the horizontal line test, and this implies that the function $f(x) = b^x$ has an inverse function. To find a formula for this inverse (with x as the independent variable), we can solve the equation $x = b^y$ for y as a function of x. This can be done by taking the logarithm to the base b of both sides of this equation. This yields

$$\log_b x = \log_b(b^y) \tag{6}$$

However, if we think of $\log_b(b^y)$ as that exponent to which b must be raised to produce b^y, then it becomes evident that $\log_b(b^y) = y$. Thus, (6) can be rewritten as

$$y = \log_b x$$

from which we conclude that the inverse of $f(x) = b^x$ is $f^{-1}(x) = \log_b x$. This implies that the graphs of $y = b^x$ and $y = \log_b x$ are reflections of one another about the line $y = x$ (Figure 4.2.3). We call $\log_b x$ the ***logarithmic function with base b***.

Recall from Section 4.1 that a one-to-one function f and its inverse satisfy the equations

$$f^{-1}(f(x)) = x \quad \text{for every } x \text{ in the domain of } f$$

$$f(f^{-1}(x)) = x \quad \text{for every } x \text{ in the domain of } f^{-1}$$

In particular, if we take $f(x) = b^x$ and $f^{-1}(x) = \log_b x$, and if we keep in mind that the domain of f^{-1} is the same as the range of f, then we obtain

$$\begin{aligned} \log_b(b^x) &= x \quad \text{for all real values of } x \\ b^{\log_b x} &= x \quad \text{for } x > 0 \end{aligned} \tag{7}$$

In the special case where $b = e$, these equations become

$$\begin{aligned} \ln(e^x) &= x \quad \text{for all real values of } x \\ e^{\ln x} &= x \quad \text{for } x > 0 \end{aligned} \tag{8}$$

In words, the equations in (7) tell us that the functions b^x and $\log_b x$ cancel out the effect of one another when composed in either order; for example,

$$\log 10^x = x, \quad 10^{\log x} = x, \quad \ln e^x = x, \quad e^{\ln x} = x, \quad \ln e^5 = 5, \quad e^{\ln \pi} = \pi$$

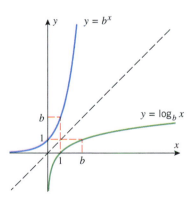

Figure 4.2.3

REMARK.   Figure 4.2.4 shows computer-generated tables and graphs of $y = e^x$ and $y = \ln x$. The values of $y = e^x$ and $y = \ln x$ have been rounded to the second decimal place in the tables. This explains why the column under $y = e^x$ in the second table is not identical to the column under x in the first table.

The inverse relationship between b^x and $\log_b x$ allows us to translate properties of exponential functions into properties of logarithmic functions, and vice versa.

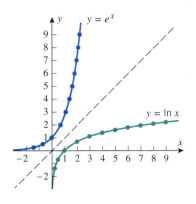

x	$y = \ln x$	x	$y = e^x$
0.25	−1.39	−1.39	0.25
0.50	−0.69	−0.69	0.50
1	0	0	1.00
2	0.69	0.69	1.99
3	1.10	1.10	3.00
4	1.39	1.39	4.01
5	1.61	1.61	5.00
6	1.79	1.79	5.99
7	1.95	1.95	7.03
8	2.08	2.08	8.00
9	2.20	2.20	9.03

Figure 4.2.4

4.2.2   THEOREM (*Comparison of Exponential and Logarithmic Functions*).   *If $b > 0$ and $b \neq 1$, then*:

$b^0 = 1$ $\qquad\qquad\qquad\qquad$ $\log_b 1 = 0$

$b^1 = b$ $\qquad\qquad\qquad\qquad$ $\log_b b = 1$

range $b^x = (0, +\infty)$ $\qquad\qquad$ domain $\log_b x = (0, +\infty)$

domain $b^x = (-\infty, +\infty)$ $\qquad$ range $\log_b x = (-\infty, +\infty)$

$y = b^x$ *is continuous on* $(-\infty, +\infty)$ $\qquad$ $y = \log_b x$ *is continuous on* $(0, +\infty)$

You should recall the following algebraic properties of logarithms from your earlier studies.

4.2.3   THEOREM (*Algebraic Properties of Logarithms*).   *If $b > 0$, $b \neq 1$, $a > 0$, $c > 0$, and r is any real number, then*:

$\log_b(ac) = \log_b a + \log_b c$ $\qquad$ Product property

$\log_b(a/c) = \log_b a - \log_b c$ $\qquad$ Quotient property

$\log_b(a^r) = r \log_b a$ $\qquad$ Power property

$\log_b(1/c) = -\log_b c$ $\qquad$ Reciprocal property

These properties are often used to expand a single logarithm into sums, differences, and multiples of other logarithms and, conversely, to condense sums, differences, and multiples of logarithms into a single logarithm. For example,

$$\log \frac{xy^5}{\sqrt{z}} = \log xy^5 - \log \sqrt{z} = \log x + \log y^5 - \log z^{1/2} = \log x + 5 \log y - \tfrac{1}{2} \log z$$

$$5 \log 2 + \log 3 - \log 8 = \log 32 + \log 3 - \log 8 = \log \frac{32 \cdot 3}{8} = \log 12$$

$$\tfrac{1}{3} \ln x - \ln(x^2 - 1) + 2 \ln(x + 3) = \ln x^{1/3} - \ln(x^2 - 1) + \ln(x + 3)^2 = \ln \frac{\sqrt[3]{x}(x+3)^2}{x^2 - 1}$$

REMARK.   Expressions of the form $\log_b(u + v)$ and $\log_b(u - v)$ have no useful simplifications in terms of $\log_b u$ and $\log_b v$. In particular,

$$\log_b(u + v) \neq \log_b u + \log_b v$$

$$\log_b(u - v) \neq \log_b u - \log_b v$$

SOLVING EQUATIONS INVOLVING EXPONENTIALS AND LOGARITHMS

The equation $y = e^x$ can be solved for x in terms of y as $x = \ln y$, provided (of course) that y is in the domain of the natural logarithm function and x is in the domain of the natural exponential function; that is, $y > 0$ and x is any real number. Thus,

$y = e^x$ is equivalent to $x = \ln y$ if $y > 0$ and x is any real number

More generally, if $b > 0$ and $b \neq 1$, then

$y = b^x$ is equivalent to $x = \log_b y$ if $y > 0$ and x is any real number

Equations of the form $\log_b x = k$ can be solved by converting them to the exponential form $x = b^k$, and equations of the form $b^x = k$ can be solved by taking a logarithm of both sides (usually log or ln).

Example 1 Find x such that

(a) $\log x = \sqrt{2}$ (b) $\ln(x + 1) = 5$ (c) $5^x = 7$

Solution (a). Converting the equation to exponential form yields

$$x = 10^{\sqrt{2}} \approx 25.95$$

Solution (b). Converting the equation to exponential form yields

$$x + 1 = e^5 \quad \text{or} \quad x = e^5 - 1 \approx 147.41$$

Solution (c). Taking the natural logarithm of both sides and using the power property of logarithms yields

$$x \ln 5 = \ln 7 \quad \text{or} \quad x = \frac{\ln 7}{\ln 5} \approx 1.21 \qquad \blacktriangleleft$$

Example 2 A satellite that requires 7 watts of power to operate at full capacity is equipped with a radioisotope power supply whose power output P in watts is given by the equation

$$P = 75e^{-t/125}$$

where t is the time in days that the supply is used. How long can the satellite operate at full capacity?

Solution. The power P will fall to 7 watts when

$$7 = 75e^{-t/125}$$

The solution for t is as follows:

$$7/75 = e^{-t/125}$$
$$\ln(7/75) = \ln(e^{-t/125})$$
$$\ln(7/75) = -t/125$$
$$t = -125 \ln(7/75) \approx 296.4$$

so the satellite can operate at full capacity for about 296 days. $\blacktriangleleft$

Here is a more complicated example.

Example 3 Solve $\dfrac{e^x - e^{-x}}{2} = 1$ for x.

Solution. Multiplying both sides of the given equation by 2 yields

$$e^x - e^{-x} = 2$$

or equivalently,

$$e^x - \frac{1}{e^x} = 2$$

Multiplying through by e^x yields

$$e^{2x} - 1 = 2e^x \quad \text{or} \quad e^{2x} - 2e^x - 1 = 0$$

This is really a quadratic equation in disguise, as can be seen by rewriting it in the form

$$\left(e^x\right)^2 - 2e^x - 1 = 0$$

and letting $u = e^x$ to obtain

$$u^2 - 2u - 1 = 0$$

Solving for u by the quadratic formula yields

$$u = \frac{2 \pm \sqrt{4+4}}{2} = \frac{2 \pm \sqrt{8}}{2} = 1 \pm \sqrt{2}$$

or, since $u = e^x$,

$$e^x = 1 \pm \sqrt{2}$$

But e^x cannot be negative, so we discard the negative value $1 - \sqrt{2}$; thus,

$$e^x = 1 + \sqrt{2}$$
$$\ln e^x = \ln(1 + \sqrt{2})$$
$$x = \ln(1 + \sqrt{2}) \approx 0.881 \qquad \blacktriangleleft$$

CHANGE OF BASE FORMULA FOR LOGARITHMS

Scientific calculators generally provide keys for evaluating common logarithms and natural logarithms but have no keys for evaluating logarithms with other bases. However, this is not a serious deficiency because it is possible to express a logarithm with any base in terms of logarithms with any other base (see Exercise 40). For example, the following formula expresses a logarithm with base b in terms of natural logarithms:

$$\log_b x = \frac{\ln x}{\ln b} \qquad (9)$$

We can derive this result by letting $y = \log_b x$, from which it follows that $b^y = x$. Taking the natural logarithm of both sides of this equation we obtain $y \ln b = \ln x$, from which (9) follows.

Example 4 Use a calculating utility to evaluate $\log_2 5$ by expressing this logarithm in terms of natural logarithms.

Solution. From (9) we obtain

$$\log_2 5 = \frac{\ln 5}{\ln 2} \approx 2.321928 \qquad \blacktriangleleft$$

LOGARITHMIC SCALES IN SCIENCE AND ENGINEERING

Logarithms are used in science and engineering to deal with quantities whose units vary over an excessively wide range of values. For example, the "loudness" of a sound can be measured by its *intensity I* (in watts per square meter), which is related to the energy transmitted by the sound wave—the greater the intensity, the greater the transmitted energy, and the louder the sound is perceived by the human ear. However, intensity units are unwieldy because they vary over an enormous range. For example, a sound at the threshold of human hearing has an intensity of about 10^{-12} W/m^2, a close whisper has an intensity that is about 100 times the hearing threshold, and a jet engine at 50 meters has an intensity that is about $1,000,000,000,000 = 10^{12}$ times the hearing threshold. To see how logarithms can be used

Table 4.2.2

β (dB)	I/I_0
0	$10^0 = 1$
10	$10^1 = 10$
20	$10^2 = 100$
30	$10^3 = 1,000$
40	$10^4 = 10,000$
50	$10^5 = 100,000$
$\vdots$	$\vdots$
120	$10^{12} = 1,000,000,000,000$

to reduce this wide spread, observe that if

$$y = \log x$$

then increasing x by a *factor* of 10 *adds* 1 unit to y since

$$\log 10x = \log 10 + \log x = 1 + y$$

Physicists and engineers take advantage of this property by measuring loudness in terms of the **sound level** β, which is defined by

$$\beta = 10 \log(I/I_0)$$

where $I_0 = 10^{-12}$ W/m^2 is a reference intensity close to the threshold of human hearing. The units of β are **decibels** (dB), named in honor of the telephone inventor Alexander Graham Bell. With this scale of measurement, *multiplying* the intensity I by a factor of 10 *adds* 10 dB to the sound level β (verify). This results in a more tractable scale than intensity for measuring sound loudness (Table 4.2.2). Some other familiar logarithmic scales are the **Richter scale** used to measure earthquake intensity and the **pH scale** used to measure acidity in chemistry, both of which are discussed in the exercises.

Example 5   In 1976 the rock group The Who set the record for the loudest concert: 120 dB. By comparison, a jackhammer positioned at the same spot as The Who would have produced a sound level of 92 dB. What is the ratio of the sound intensity of The Who to the sound intensity of a jackhammer?

Solution.   Let I_1 and $\beta_1(= 120$ dB$)$ denote the intensity and sound level of The Who, and let I_2 and β_2 $(= 92$ dB$)$ denote the intensity and sound level of the jackhammer. Then

$$I_1/I_2 = (I_1/I_0)/(I_2/I_0)$$
$$\log(I_1/I_2) = \log(I_1/I_0) - \log(I_2/I_0)$$
$$10\log(I_1/I_2) = 10\log(I_1/I_0) - 10\log(I_2/I_0)$$
$$10\log(I_1/I_2) = \beta_1 - \beta_2 = 120 - 92 = 28$$
$$\log(I_1/I_2) = 2.8$$

Thus, $I_1/I_2 = 10^{2.8} \approx 631$, which tells us that the sound intensity of The Who was 630 times greater than a jackhammer!   ◀

Peter Townsend of The Who sustained permanent hearing reduction due to the high decibel level of his band's music.

EXPONENTIAL AND LOGARITHMIC GROWTH

Table 4.2.3

x	e^x	$\ln x$
1	2.72	0.00
2	7.39	0.69
3	20.09	1.10
4	54.60	1.39
5	148.41	1.61
6	403.43	1.79
7	1096.63	1.95
8	2980.96	2.08
9	8103.08	2.20
10	22026.47	2.30
100	2.69×10^{43}	4.61
1000	1.97×10^{434}	6.91

The growth patterns of e^x and $\ln x$ illustrated by Table 4.2.3 are worth noting. Both functions increase as x increases, but they increase in dramatically different ways—e^x increases extremely rapidly and $\ln x$ increases extremely slowly. For example, at $x = 10$ the value of e^x is over 22,000, but at $x = 1000$ the value of $\ln x$ has not even reached 7.

The table strongly suggests that $e^x \to +\infty$ as $x \to +\infty$. However, the growth of $\ln x$ is so slow that its limiting behavior as $x \to +\infty$ is not clear from the table. In spite of its slow growth, it is still true that $\ln x \to +\infty$ as $x \to +\infty$. To see that this is so, choose any positive number M (as large as you like). The value of $\ln x$ will reach M when $x = e^M$, since

$$\ln x = \ln(e^M) = M$$

Since $\ln x$ increases as x increases, we can conclude that $\ln x > M$ for $x > e^M$; hence, $\ln x \to +\infty$ as $x \to +\infty$ since the values of $\ln x$ eventually exceed any positive number M (Figure 4.2.5).

In summary,

$$\lim_{x \to +\infty} e^x = +\infty \qquad \lim_{x \to +\infty} \ln x = +\infty \qquad (10\text{--}11)$$

The following limits, which are consistent with Figure 4.2.5, can be deduced numerically by constructing appropriate tables of values (verify):

$$\lim_{x \to -\infty} e^x = 0 \qquad \lim_{x \to 0^+} \ln x = -\infty \qquad (12\text{–}13)$$

The following limits can be deduced numerically, but they can be seen more readily by noting that the graph of $y = e^{-x}$ is the reflection about the y-axis of the graph of $y = e^x$ (Figure 4.2.6):

$$\lim_{x \to +\infty} e^{-x} = 0 \qquad \lim_{x \to -\infty} e^{-x} = +\infty \qquad (14\text{–}15)$$

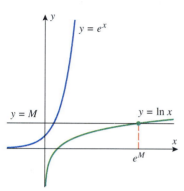

Figure 4.2.5

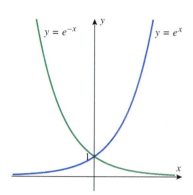

Figure 4.2.6

EXERCISE SET 4.2 ～ Graphing Utility

In Exercises 1 and 2, simplify the expression without using a calculating utility.

1. (a) $-8^{2/3}$ (b) $(-8)^{2/3}$ (c) $8^{-2/3}$

2. (a) 2^{-4} (b) $4^{1.5}$ (c) $9^{-0.5}$

In Exercises 3 and 4, use a calculating utility to approximate the expression. Round your answer to four decimal places.

3. (a) $2^{1.57}$ (b) $5^{-2.1}$

4. (a) $\sqrt[5]{24}$ (b) $\sqrt[8]{0.6}$

In Exercises 5 and 6, find the exact value of the expression without using a calculating utility.

5. (a) $\log_2 16$ (b) $\log_2 \left(\frac{1}{32}\right)$
   (c) $\log_4 4$ (d) $\log_9 3$

6. (a) $\log_{10}(0.001)$ (b) $\log_{10}(10^4)$
   (c) $\ln(e^3)$ (d) $\ln(\sqrt{e})$

In Exercises 7 and 8, use a calculating utility to approximate the expression. Round your answer to four decimal places.

7. (a) $\log 23.2$ (b) $\ln 0.74$

8. (a) $\log 0.3$ (b) $\ln \pi$

In Exercises 9 and 10, use the logarithm properties in Theorem 4.2.3 to rewrite the expression in terms of r, s, and t, where $r = \ln a$, $s = \ln b$, and $t = \ln c$.

9. (a) $\ln a^2 \sqrt{bc}$ (b) $\ln \dfrac{b}{a^3 c}$

10. (a) $\ln \dfrac{\sqrt[3]{c}}{ab}$ (b) $\ln \sqrt{\dfrac{ab^3}{c^2}}$

In Exercises 11 and 12, expand the logarithm in terms of sums, differences, and multiples of simpler logarithms.

11. (a) $\log(10x\sqrt{x-3})$ (b) $\ln \dfrac{x^2 \sin^3 x}{\sqrt{x^2+1}}$

12. (a) $\log \dfrac{\sqrt[3]{x+2}}{\cos 5x}$ (b) $\ln \sqrt{\dfrac{x^2+1}{x^3+5}}$

In Exercises 13–15, rewrite the expression as a single logarithm.

13. $4 \log 2 - \log 3 + \log 16$

14. $\frac{1}{2}\log x - 3\log(\sin 2x) + 2$

15. $2\ln(x + 1) + \frac{1}{3}\ln x - \ln(\cos x)$

In Exercises 16–25, solve for x without using a calculating utility.

16. $\log_{10}(1 + x) = 3$ **17.** $\log_{10}(\sqrt{x}) = -1$

18. $\ln(x^2) = 4$ **19.** $\ln(1/x) = -2$

20. $\log_3(3^x) = 7$ **21.** $\log_5(5^{2x}) = 8$

22. $\log_{10} x^2 + \log_{10} x = 30$

23. $\log_{10} x^{3/2} - \log_{10} \sqrt{x} = 5$

24. $\ln 4x - 3\ln(x^2) = \ln 2$

25. $\ln(1/x) + \ln(2x^3) = \ln 3$

In Exercises 26–31, solve for x without using a calculating utility. Use the natural logarithm anywhere that logarithms are needed.

26. $3^x = 2$ **27.** $5^{-2x} = 3$

28. $3e^{-2x} = 5$ **29.** $2e^{3x} = 7$

30. $e^x - 2xe^x = 0$ **31.** $xe^{-x} + 2e^{-x} = 0$

In Exercises 32 and 33, rewrite the given equation as a quadratic equation in u, where $u = e^x$; then solve for x.

32. $e^{2x} - e^x = 6$ **33.** $e^{-2x} - 3e^{-x} = -2$

In Exercises 34–36, sketch the graph of the equation without using a graphing utility.

34. (a) $y = 1 + \ln(x - 2)$ (b) $y = 3 + e^{x-2}$

35. (a) $y = \left(\frac{1}{2}\right)^{x-1} - 1$ (b) $y = \ln|x|$

36. (a) $y = 1 - e^{-x+1}$ (b) $y = 3\ln\sqrt[3]{x - 1}$

37. Use a calculating utility and the change of base formula (9) to find the values of $\log_2 7.35$ and $\log_5 0.6$, rounded to four decimal places.

In Exercises 38 and 39, graph the functions on the same screen of a graphing utility. [Use the change of base formula (9), where needed.]

38. $y = \ln x$, $y = e^x$, $\log x$, 10^x

39. $y = \log_2 x$, $\ln x$, $\log_5 x$, $\log x$

40. (a) Derive the general change of base formula
$$\log_b x = \frac{\log_a x}{\log_a b}$$
(b) Use the result in part (a) to find the exact value of $(\log_2 81)(\log_3 32)$ without using a calculating utility.

41. Use a graphing utility to estimate the two points of intersection of the graphs of $y = x^{0.2}$ and $y = \ln x$.

42. The United States public debt D, in billions of dollars, has been modeled as $D = 0.051517(1.1306727)^x$, where x is the number of years since 1900. Based on this model, when did the debt first reach one trillion dollars?

43. (a) Is the curve in the accompanying figure the graph of an exponential function? Explain your reasoning.
(b) Find the equation of an exponential function that passes through the point $(4, 2)$.
(c) Find the equation of an exponential function that passes through the point $\left(2, \frac{1}{4}\right)$.
(d) Use a graphing utility to generate the graph of an exponential function that passes through the point $(2, 5)$.

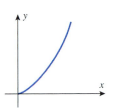

Figure Ex-43

44. (a) Make a conjecture about the general shape of the graph of $y = \log(\log x)$, and sketch the graph of this equation and $y = \log x$ in the same coordinate system.
(b) Check your work in part (a) with a graphing utility.

45. Find the fallacy in the following "proof" that $\frac{1}{8} > \frac{1}{4}$. Multiply both sides of the inequality $3 > 2$ by $\log \frac{1}{2}$ to get
$$3\log \frac{1}{2} > 2\log \frac{1}{2}$$
$$\log \left(\frac{1}{2}\right)^3 > \log \left(\frac{1}{2}\right)^2$$
$$\log \frac{1}{8} > \log \frac{1}{4}$$
$$\frac{1}{8} > \frac{1}{4}$$

46. Prove the four algebraic properties of logarithms in Theorem 4.2.3.

47. If equipment in the satellite of Example 2 requires 15 watts to operate correctly, what is the operational lifetime of the power supply?

48. The equation $Q = 12e^{-0.055t}$ gives the mass Q in grams of radioactive potassium-42 that will remain from some initial quantity after t hours of radioactive decay.
(a) How many grams were there initially?
(b) How many grams remain after 4 hours?
(c) How long will it take to reduce the amount of radioactive potassium-42 to half of the initial amount?

49. The acidity of a substance is measured by its pH value, which is defined by the formula
$$pH = -\log[H^+]$$
where the symbol $[H^+]$ denotes the concentration of hydrogen ions measured in moles per liter. Distilled water has a pH of 7; a substance is called *acidic* if it has pH < 7 and

basic if it has pH > 7. Find the pH of each of the following substances and state whether it is acidic or basic.

SUBSTANCE		$[H^+]$
(a)	Arterial blood	3.9×10^{-8} mol/L
(b)	Tomatoes	6.3×10^{-5} mol/L
(c)	Milk	4.0×10^{-7} mol/L
(d)	Coffee	1.2×10^{-6} mol/L

50. Use the definition of pH in Exercise 49 to find $[H^+]$ in a solution having a pH equal to
   (a) 2.44 (b) 8.06

51. The perceived loudness β of a sound in decibels (dB) is related to its intensity I in watts/square meter (W/m²) by the equation

$$\beta = 10 \log(I/I_0)$$

where $I_0 = 10^{-12}$ W/m². Damage to the average ear occurs at 90 dB or greater. Find the decibel level of each of the following sounds and state whether it will cause ear damage.

SOUND		I
(a)	Jet aircraft (from 500 ft)	1.0×10^2 W/m²
(b)	Amplified rock music	1.0 W/m²
(c)	Garbage disposal	1.0×10^{-4} W/m²
(d)	TV (mid volume from 10 ft)	3.2×10^{-5} W/m²

In Exercises 52–54, use the definition of the decibel level of a sound (see Exercise 51).

52. If one sound is three times as intense as another, how much greater is its decibel level?

53. According to one source, the noise inside a moving automobile is about 70 dB, whereas an electric blender generates 93 dB. Find the ratio of the intensity of the noise of the blender to that of the automobile.

54. Suppose that the decibel level of an echo is $\frac{2}{3}$ the decibel level of the original sound. If each echo results in another echo, how many echoes will be heard from a 120-dB sound given that the average human ear can hear a sound as low as 10 dB?

55. On the *Richter scale*, the magnitude M of an earthquake is related to the released energy E in joules (J) by the equation

$$\log E = 4.4 + 1.5M$$

   (a) Find the energy E of the 1906 San Francisco earthquake that registered $M = 8.2$ on the Richter scale.
   (b) If the released energy of one earthquake is 10 times that of another, how much greater is its magnitude on the Richter scale?

56. Suppose that the magnitudes of two earthquakes differ by 1 on the Richter scale. Find the ratio of the released energy of the larger earthquake to that of the smaller earthquake. [*Note:* See Exercise 55 for terminology.]

In Exercises 57 and 58, use Formula (3) or (5), as appropriate, to find the limit.

57. Find $\lim\limits_{x \to 0} (1 - 2x)^{1/x}$. [*Hint:* Let $t = -2x$.]

58. Find $\lim\limits_{x \to +\infty} (1 + 3/x)^x$. [*Hint:* Let $t = 3/x$.]

4.3 DERIVATIVES OF LOGARITHMIC AND EXPONENTIAL FUNCTIONS

In this section we will obtain derivative formulas for logarithmic and exponential functions, and we will discuss the general relationship between the derivative of a one-to-one function and its inverse function.

DERIVATIVES OF LOGARITHMIC FUNCTIONS

The natural logarithm plays a special role in calculus that can be motivated by differentiating $\log_b x$, where b is an arbitrary base. For this purpose, recall that $\log_b x$ is continuous for $x > 0$. We will also need the limit

$$\lim_{v \to 0} (1 + v)^{1/v} = e$$

that was given in Formula (5) of Section 4.2 (with x rather than v as the variable).

Using the definition of a derivative, we obtain

$$\frac{d}{dx}[\log_b x] = \lim_{w \to x} \frac{\log_b w - \log_b x}{w - x}$$

$$= \lim_{w \to x} \left[\frac{1}{w - x} \log_b \left(\frac{w}{x} \right) \right]$$

The quotient property of logarithms in Theorem 4.2.3

$$= \lim_{w \to x} \left[\frac{1}{w - x} \log_b \left(\frac{x + (w - x)}{x} \right) \right]$$

$$= \lim_{w \to x} \left[\frac{1}{w - x} \log_b \left(1 + \frac{w - x}{x} \right) \right]$$

$$= \lim_{w \to x} \left[\frac{1}{x} \frac{x}{w - x} \log_b \left(1 + \frac{w - x}{x} \right) \right]$$

$$= \lim_{v \to 0} \left[\frac{1}{x} \frac{1}{v} \log_b (1 + v) \right]$$

Let $v = (w - x)/x$ and note that $v \to 0$ if and only if $w \to x$.

$$= \frac{1}{x} \lim_{v \to 0} \left[\frac{1}{v} \log_b (1 + v) \right]$$

x is fixed for this limit computation, so $1/x$ can be moved through the limit sign.

$$= \frac{1}{x} \lim_{v \to 0} \left[\log_b (1 + v)^{1/v} \right]$$

The power property of logarithms in Theorem 4.2.3

$$= \frac{1}{x} \log_b \left[\lim_{v \to 0} (1 + v)^{1/v} \right]$$

$\log_b x$ is continuous on $(0, +\infty)$, so we can move the limit through the function symbol.

$$= \frac{1}{x} \log_b e$$

Formula (5) of Section 4.2

Thus,

$$\frac{d}{dx}[\log_b x] = \frac{1}{x} \log_b e, \quad x > 0$$

But from Formula (9) of Section 4.2 we have $\log_b e = 1/\ln b$, so we can rewrite this derivative formula as

$$\frac{d}{dx}[\log_b x] = \frac{1}{x \ln b}, \quad x > 0 \tag{1}$$

In the special case where $b = e$, we have $\ln e = 1$, so this formula becomes

$$\frac{d}{dx}[\ln x] = \frac{1}{x}, \quad x > 0 \tag{2}$$

Thus, among all possible bases, the base $b = e$ produces the simplest formula for the derivative of $\log_b x$. This is one of the reasons why the natural logarithm function is preferred over other logarithms in calculus.

Example 1

(a) Figure 4.3.1 shows the graph of $y = \ln x$ and its tangent lines at the points $x = \frac{1}{2}, 1,$ 3, and 5. Find the slopes of those tangent lines.

(b) Does the graph of $y = \ln x$ have any horizontal tangent lines? Use the derivative of $\ln x$ to justify your answer.

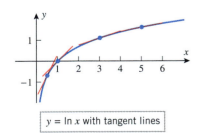

y = ln x with tangent lines

Figure 4.3.1

Solution (a). From (2), the slopes of the tangent lines at the points $x = \frac{1}{2}, 1, 3,$ and 5 are $1/x = 2, 1, \frac{1}{3},$ and $\frac{1}{5}$, which is consistent with Figure 4.3.1.

Solution (b). From the graph of $y = \ln x$, it does not appear that there are any horizontal tangent lines. This is confirmed by the fact that $dy/dx = 1/x$ is not equal to zero for any real value of x. ◀

If u is a differentiable function of x, and if $u(x) > 0$, then applying the chain rule to (1) and (2) produces the following generalized derivative formulas:

$$\frac{d}{dx}[\log_b u] = \frac{1}{u \ln b} \cdot \frac{du}{dx} \qquad \text{and} \qquad \frac{d}{dx}[\ln u] = \frac{1}{u} \cdot \frac{du}{dx} \qquad (3\text{--}4)$$

Example 2   Find $\dfrac{d}{dx}[\ln(x^2 + 1)]$.

Solution.   From (4) with $u = x^2 + 1$,

$$\frac{d}{dx}[\ln(x^2 + 1)] = \frac{1}{x^2 + 1} \cdot \frac{d}{dx}[x^2 + 1] = \frac{1}{x^2 + 1} \cdot 2x = \frac{2x}{x^2 + 1} \qquad ◀$$

When possible, the properties of logarithms in Theorem 4.2.3 should be used to convert products, quotients, and exponents into sums, differences, and constant multiples *before* differentiating a function involving logarithms.

Example 3

$$\frac{d}{dx}\left[\ln\left(\frac{x^2 \sin x}{\sqrt{1+x}}\right)\right] = \frac{d}{dx}\left[2 \ln x + \ln(\sin x) - \frac{1}{2}\ln(1+x)\right]$$

$$= \frac{2}{x} + \frac{\cos x}{\sin x} - \frac{1}{2(1+x)}$$

$$= \frac{2}{x} + \cot x - \frac{1}{2+2x} \qquad ◀$$

Example 4   Find $\dfrac{d}{dx}[\ln|x|]$.

Solution.   The function $\ln|x|$ is defined for all x, except $x = 0$; we will consider the cases $x > 0$ and $x < 0$ separately.
   If $x > 0$, then $|x| = x$, so

$$\frac{d}{dx}[\ln|x|] = \frac{d}{dx}[\ln x] = \frac{1}{x}$$

If $x < 0$, then $|x| = -x$, so from (4) we have

$$\frac{d}{dx}[\ln|x|] = \frac{d}{dx}[\ln(-x)] = \frac{1}{(-x)} \cdot \frac{d}{dx}[-x] = \frac{1}{x}$$

Since the same formula results in both cases, we have shown that

$$\frac{d}{dx}[\ln|x|] = \frac{1}{x} \quad \text{if } x \neq 0 \qquad (5)$$

Example 5   From (5) and the chain rule,

$$\frac{d}{dx}[\ln|\sin x|] = \frac{1}{\sin x} \cdot \frac{d}{dx}[\sin x] = \frac{\cos x}{\sin x} = \cot x \qquad ◀$$

LOGARITHMIC DIFFERENTIATION

We now consider a technique called *logarithmic differentiation* that is useful for differentiating functions that are composed of products, quotients, and powers.

Example 6 The derivative of

$$y = \frac{x^2 \sqrt[3]{7x - 14}}{\left(1 + x^2\right)^4} \tag{6}$$

is messy to calculate directly. However, if we first take the natural logarithm of both sides and then use its properties, we can write

$$\ln y = 2 \ln x + \tfrac{1}{3} \ln(7x - 14) - 4 \ln(1 + x^2)$$

Differentiating both sides with respect to x yields

$$\frac{1}{y} \frac{dy}{dx} = \frac{2}{x} + \frac{7/3}{7x - 14} - \frac{8x}{1 + x^2} \tag{7}$$

Thus, on solving for dy/dx and using (6) we obtain

$$\frac{dy}{dx} = \frac{x^2 \sqrt[3]{7x - 14}}{\left(1 + x^2\right)^4} \left[\frac{2}{x} + \frac{1}{3x - 6} - \frac{8x}{1 + x^2} \right] \tag{8}$$

REMARK. Since $\ln y$ is defined only for $y > 0$, logarithmic differentiation of $y = f(x)$ is valid only on intervals where $f(x)$ is positive. Thus, the derivative obtained in the preceding example is valid on the interval $(2, +\infty)$, since the given function is positive for $x > 2$. However, the formula is actually valid on the interval $(-\infty, 2)$ as well. This can be seen by taking absolute values before proceeding with the logarithmic differentiation and noting that $\ln |y|$ is defined for all y except $y = 0$. If we do this and simplify using properties of logarithms and absolute values, we obtain

$$\ln |y| = 2 \ln |x| + \tfrac{1}{3} \ln |7x - 14| - 4 \ln |1 + x^2|$$

Differentiating both sides with respect to x yields (7), and hence results in (8).

In general, if the derivative of $y = f(x)$ is to be obtained by logarithmic differentiation, then the same formula for dy/dx will result regardless of whether one first takes absolute values or not. Thus, a derivative formula obtained by logarithmic differentiation will be valid except perhaps at points where $f(x)$ is zero. The formula may, in fact, be valid at those points as well, but it is not guaranteed.

DERIVATIVES OF IRRATIONAL POWERS OF x

We know from Formula (15) of Section 3.6 that the differentiation formula

$$\frac{d}{dx}[x^r] = rx^{r-1} \tag{9}$$

holds for rational values of r. We will now use logarithmic differentiation to show that this formula holds if r is *any* real number (rational or irrational). In our computations we will assume that x^r is a differentiable function and that the familiar laws of exponents hold for real exponents.

Let $y = x^r$, where r is a real number. The derivative dy/dx can be obtained by logarithmic differentiation as follows:

$$\ln |y| = \ln |x^r| = r \ln |x|$$

$$\frac{d}{dx}[\ln |y|] = \frac{d}{dx}[r \ln |x|]$$

$$\frac{1}{y} \frac{dy}{dx} = \frac{r}{x}$$

$$\frac{dy}{dx} = \frac{r}{x} y = \frac{r}{x} x^r = rx^{r-1}$$

This establishes (9) for real values of r. Thus, for example,

$$\frac{d}{dx}[x^\pi] = \pi x^{\pi - 1} \quad \text{and} \quad \frac{d}{dx}\left[x^{\sqrt{2}}\right] = \sqrt{2} x^{\sqrt{2} - 1} \tag{10}$$

DERIVATIVES OF EXPONENTIAL FUNCTIONS

By (1) we know that

$$\frac{d}{dx}[\log_b x]$$

is a nonzero function, so Theorem 4.1.6 establishes that the inverse function for $\log_b x$ is differentiable on $(-\infty, +\infty)$.

To obtain a derivative formula for the exponential function with base b, we rewrite $y = b^x$ as

$$x = \log_b y$$

and differentiate implicitly using (3) to obtain

$$1 = \frac{1}{y \ln b} \cdot \frac{dy}{dx}$$

Solving for dy/dx and replacing y by b^x we have

$$\frac{dy}{dx} = y \ln b = b^x \ln b$$

Thus, we have shown that

$$\frac{d}{dx}[b^x] = b^x \ln b \tag{11}$$

In the special case where $b = e$ we have $\ln e = 1$, so that (11) becomes

$$\frac{d}{dx}[e^x] = e^x \tag{12}$$

Moreover, if u is a differentiable function of x, then it follows from (11) and (12) that

$$\frac{d}{dx}[b^u] = b^u \ln b \cdot \frac{du}{dx} \quad \text{and} \quad \frac{d}{dx}[e^u] = e^u \cdot \frac{du}{dx} \tag{13--14}$$

REMARK. It is important to distinguish between differentiating the exponential function b^x (variable exponent and constant base) and the power function x^b (variable base and constant exponent). For example, compare the derivative of x^π in (10) to the following derivative of π^x, which is obtained from (11):

$$\frac{d}{dx}[\pi^x] = \pi^x \ln \pi$$

Example 7 The following computations use Formulas (13) and (14).

$$\frac{d}{dx}[2^{\sin x}] = (2^{\sin x})(\ln 2) \cdot \frac{d}{dx}[\sin x] = (2^{\sin x})(\ln 2)(\cos x)$$

$$\frac{d}{dx}[e^{-2x}] = e^{-2x} \cdot \frac{d}{dx}[-2x] = -2e^{-2x}$$

$$\frac{d}{dx}\left[e^{x^3}\right] = e^{x^3} \cdot \frac{d}{dx}[x^3] = 3x^2 e^{x^3}$$

$$\frac{d}{dx}[e^{\cos x}] = e^{\cos x} \cdot \frac{d}{dx}[\cos x] = -(\sin x)e^{\cos x} \qquad \blacktriangleleft$$

The rules

$$\frac{d}{dx}(u^n) = n \cdot u^{n-1} \frac{du}{dx} \quad \text{if } n \text{ is a real number}$$

$$\frac{d}{dx}(b^u) = b^u \ln b \cdot \frac{du}{dx} \quad \text{if } b > 0, b \neq 1$$

deal with derivatives of exponential expressions in which either the base or the exponent of the expression is a number. The following example illustrates the application of logarithmic differentiation for finding dy/dx when y is an expression of the form $y = u^v$ where *both* u and v are nonconstant functions of x.

Example 8 Use logarithmic differentiation to find $\dfrac{d}{dx}[(x^2 + 1)^{\sin x}]$.

Solution. Setting $y = (x^2 + 1)^{\sin x}$ we have

$$\ln y = \ln[(x^2 + 1)^{\sin x}] = (\sin x)\ln(x^2 + 1)$$

Then

$$\frac{d}{dx}(\ln y) = \frac{1}{y} \cdot \frac{dy}{dx}$$

$$= \frac{d}{dx}[(\sin x)\ln(x^2 + 1)] = (\sin x)\frac{1}{x^2 + 1}(2x) + (\cos x)\ln(x^2 + 1)$$

Thus,

$$\frac{dy}{dx} = y\left[\frac{2x \sin x}{x^2 + 1} + (\cos x)\ln(x^2 + 1)\right]$$

$$= (x^2 + 1)^{\sin x}\left[\frac{2x \sin x}{x^2 + 1} + (\cos x)\ln(x^2 + 1)\right]$$ ◀

EXERCISE SET 4.3   ⬚ Graphing Utility

In Exercises 1–30, find dy/dx.

1. $y = \ln 2x$

2. $y = \ln(x^3)$

3. $y = (\ln x)^2$

4. $y = \ln(\sin x)$

5. $y = \ln|\tan x|$

6. $y = \ln(2 + \sqrt{x})$

7. $y = \ln\left(\dfrac{x}{1 + x^2}\right)$

8. $y = \ln(\ln x)$

9. $y = \ln|x^3 - 7x^2 - 3|$

10. $y = x^3 \ln x$

11. $y = \sqrt{\ln x}$

12. $y = \sqrt{1 + \ln^2 x}$

13. $y = \cos(\ln x)$

14. $y = \sin^2(\ln x)$

15. $y = x^3 \log_2(3 - 2x)$

16. $y = x\left[\log_2(x^2 - 2x)\right]^3$

17. $y = \dfrac{x^2}{1 + \log x}$

18. $y = \dfrac{\log x}{1 + \log x}$

19. $y = e^{7x}$

20. $y = e^{-5x^2}$

21. $y = x^3 e^x$

22. $y = e^{1/x}$

23. $y = \dfrac{e^x - e^{-x}}{e^x + e^{-x}}$

24. $y = \sin(e^x)$

25. $y = e^{x \tan x}$

26. $y = \dfrac{e^x}{\ln x}$

27. $y = e^{(x - e^{3x})}$

28. $y = \exp(\sqrt{1 + 5x^3})$

29. $y = \ln(1 - xe^{-x})$

30. $y = \ln(\cos e^x)$

In Exercises 31 and 32, find dy/dx by implicit differentiation.

31. $y + \ln xy = 1$

32. $y = \ln(x \tan y)$

In Exercises 33 and 34, use the method of Example 3 to help perform the indicated differentiation.

33. $\dfrac{d}{dx}\left[\ln\dfrac{\cos x}{\sqrt{4 - 3x^2}}\right]$

34. $\dfrac{d}{dx}\left[\ln\sqrt{\dfrac{x - 1}{x + 1}}\right]$

In Exercises 35–38, find dy/dx using the method of logarithmic differentiation.

35. $y = x\sqrt[3]{1 + x^2}$

36. $y = \sqrt[5]{\dfrac{x - 1}{x + 1}}$

37. $y = \dfrac{(x^2 - 8)^{1/3}\sqrt{x^3 + 1}}{x^6 - 7x + 5}$

38. $y = \dfrac{\sin x \cos x \tan^3 x}{\sqrt{x}}$

In Exercises 39–42, find $f'(x)$ by Formula (13) and then by logarithmic differentiation.

39. $f(x) = 2^x$

40. $f(x) = 3^{-x}$

41. $f(x) = \pi^{\sin x}$

42. $f(x) = \pi^{x \tan x}$

In Exercises 43–46, find dy/dx using the method of logarithmic differentiation.

43. $y = (x^3 - 2x)^{\ln x}$

44. $y = x^{\sin x}$

45. $y = (\ln x)^{\tan x}$

46. $y = (x^2 + 3)^{\ln x}$

47. Find $f'(x)$ if $f(x) = x^e$.

48. (a) Explain why Formula (11) cannot be used to find $(d/dx)[x^x]$.

(b) Find this derivative by logarithmic differentiation.

49. Find

(a) $\dfrac{d}{dx}[\log_x e]$

(b) $\dfrac{d}{dx}[\log_x 2]$.

50. Find

(a) $\dfrac{d}{dx}[\log_{(1/x)} e]$

(b) $\dfrac{d}{dx}[\log_{(\ln x)} e]$.

51. Let $f(x) = e^{kx}$ and $g(x) = e^{-kx}$. Find

(a) $f^{(n)}(x)$

(b) $g^{(n)}(x)$.

52. Find dy/dt if $y = e^{-\lambda t}(A \sin \omega t + B \cos \omega t)$, where A, B, λ, and ω are constants.

53. Find $f'(x)$ if

$$f(x) = \frac{1}{\sqrt{2\pi}\sigma} \exp\left[-\frac{1}{2}\left(\frac{x-\mu}{\sigma}\right)^2\right]$$

where μ and σ are constants and $\sigma \neq 0$.

54. Show that for any constants A and k, the function $y = Ae^{kt}$ satisfies the equation $dy/dt = ky$.

55. Show that for any constants A and B, the function

$$y = Ae^{2x} + Be^{-4x}$$

satisfies the equation

$$y'' + 2y' - 8y = 0$$

56. Show that

(a) $y = xe^{-x}$ satisfies the equation $xy' = (1 - x)y$

(b) $y = xe^{-x^2/2}$ satisfies the equation $xy' = (1 - x^2)y$.

In Exercises 57 and 58, find the limit by interpreting the expression as an appropriate derivative.

57. (a) $\displaystyle\lim_{w \to 1} \frac{\ln w}{w - 1}$

(b) $\displaystyle\lim_{w \to 0} \frac{10^w - 1}{w}$

58. (a) $\displaystyle\lim_{\Delta x \to 0} \frac{\ln(e^2 + \Delta x) - 2}{\Delta x}$

(b) $\displaystyle\lim_{w \to 1} \frac{2^w - 2}{w - 1}$

4.4 INVERSE TRIGONOMETRIC FUNCTIONS AND THEIR DERIVATIVES

A common problem in trigonometry is to find an angle whose trigonometric functions are known. As you may recall, problems of this type involve the computation of "arc functions" such as $\arcsin x$, $\arccos x$, $\arctan x$, and so forth. In this section we will consider this idea from the viewpoint of inverse functions, with the goal of developing derivative formulas for the inverse trigonometric functions.

INVERSE TRIGONOMETRIC FUNCTIONS

None of the six basic trigonometric functions is one-to-one because they all repeat periodically and hence do not pass the horizontal line test. Thus, to define inverse trigonometric functions we must first restrict the domains of the trigonometric functions to make them one-to-one. The top part of Figure 4.4.1 shows how these restrictions are made for $\sin x$, $\cos x$, $\tan x$, and $\sec x$. (Inverses of $\cot x$ and $\csc x$ are of lesser importance and will be left for the exercises.) The inverses of these restricted functions are denoted by

$$\sin^{-1} x, \quad \cos^{-1} x, \quad \tan^{-1} x, \quad \sec^{-1} x$$

(or alternatively by $\arcsin x$, $\arccos x$, $\arctan x$, $\text{arcsec } x$) and are defined as follows:

4.4.1 DEFINITION. The ***inverse sine function***, denoted by $\sin^{-1}$, is defined to be the inverse of the restricted sine function

$$\sin x, \quad -\pi/2 \leq x \leq \pi/2$$

4.4.2 DEFINITION. The ***inverse cosine function***, denoted by $\cos^{-1}$, is defined to be the inverse of the restricted cosine function

$$\cos x, \quad 0 \leq x \leq \pi$$

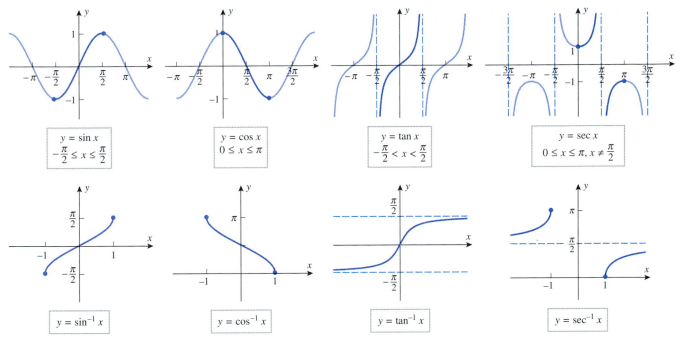

$$y = \sin x$$
$$-\frac{\pi}{2} \le x \le \frac{\pi}{2}$$

$$y = \cos x$$
$$0 \le x \le \pi$$

$$y = \tan x$$
$$-\frac{\pi}{2} < x < \frac{\pi}{2}$$

$$y = \sec x$$
$$0 \le x \le \pi, x \ne \frac{\pi}{2}$$

$$y = \sin^{-1} x$$

$$y = \cos^{-1} x$$

$$y = \tan^{-1} x$$

$$y = \sec^{-1} x$$

Figure 4.4.1

4.4.3   DEFINITION.   The *inverse tangent function*, denoted by $\tan^{-1}$, is defined to be the inverse of the restricted tangent function

$$\tan x, \quad -\pi/2 < x < \pi/2$$

4.4.4   DEFINITION.*   The *inverse secant function*, denoted by $\sec^{-1}$, is defined to be the inverse of the restricted secant function

$$\sec x, \quad 0 \le x \le \pi \text{ with } x \ne \pi/2$$

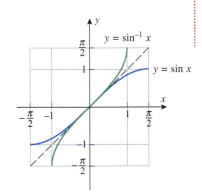

Figure 4.4.2

REMARK.   The notations $\sin^{-1} x$, $\cos^{-1} x, \ldots$ are reserved exclusively for the inverse trigonometric functions and are not used for reciprocals of the trigonometric functions. For example, to denote the reciprocal $1/\sin x$ in exponent form, we would write $(\sin x)^{-1}$ and *never* $\sin^{-1} x$.

The graphs of the inverse trigonometric functions, which are shown in the bottom part of Figure 4.4.1, are obtained by reflecting the graphs in the top part of the figure about the line $y = x$. If you have trouble visualizing these relationships, then look at Figure 4.4.2 for a more detailed illustration for the inverse sine. It may also help to keep in mind that reflection about $y = x$ converts vertical lines to horizontal lines, and vice versa, and that x-intercepts reflect into y-intercepts, and vice versa.

Table 4.4.1 summarizes the basic properties of the inverse sine, cosine, tangent, and secant functions. You should confirm that the domains and ranges listed in this table are consistent with the graphs in the bottom part of Figure 4.4.1.

---

*There is no universal agreement on the definition of $\sec^{-1} x$, and some mathematicians prefer to restrict the domain of $\sec x$ so that $0 \le x < \pi/2$ or $\pi \le x < 3\pi/2$, which was the definition used in some earlier editions of this text. Each definition has advantages and disadvantages, but we have changed to the current definition to conform with the conventions used by the CAS programs *Mathematica*, *Maple*, and *Derive*.

Table 4.4.1

FUNCTION	DOMAIN	RANGE	BASIC RELATIONSHIPS
$\sin^{-1}$	$[-1, 1]$	$[-\pi/2, \pi/2]$	$\sin^{-1}(\sin x) = x$ if $-\pi/2 \le x \le \pi/2$ $\sin(\sin^{-1} x) = x$ if $-1 \le x \le 1$
$\cos^{-1}$	$[-1, 1]$	$[0, \pi]$	$\cos^{-1}(\cos x) = x$ if $0 \le x \le \pi$ $\cos(\cos^{-1} x) = x$ if $-1 \le x \le 1$
$\tan^{-1}$	$(-\infty, +\infty)$	$(-\pi/2, \pi/2)$	$\tan^{-1}(\tan x) = x$ if $-\pi/2 < x < \pi/2$ $\tan(\tan^{-1} x) = x$ if $-\infty < x < +\infty$
$\sec^{-1}$	$(-\infty, -1] \cup [1, +\infty)$	$[0, \pi/2) \cup (\pi/2, \pi]$	$\sec^{-1}(\sec x) = x$ if $0 \le x \le \pi, x \ne \pi/2$ $\sec(\sec^{-1} x) = x$ if $|x| \ge 1$

EVALUATING INVERSE TRIGONOMETRIC FUNCTIONS

A common problem in trigonometry is to find an angle whose sine is known. For example, you might want to find an angle x in radian measure such that

$$\sin x = \tfrac{1}{2} \tag{1}$$

and, more generally, for a given value of y in the interval $-1 \le y \le 1$ you might want to solve the equation

$$\sin x = y \tag{2}$$

Because $\sin x$ repeats periodically, such equations have infinitely many solutions for x; however, if we solve this equation as

$$x = \sin^{-1} y$$

then we isolate the specific solution that lies in the interval $[-\pi/2, \pi/2]$, since this is the range of the inverse sine. For example, Figure 4.4.3 shows four solutions of Equation (1), namely, $-11\pi/6$, $-7\pi/6$, $\pi/6$, and $5\pi/6$. Of these, $\pi/6$ is the solution in the interval $[-\pi/2, \pi/2]$, so

$$\sin^{-1}\left(\tfrac{1}{2}\right) = \pi/6 \tag{3}$$

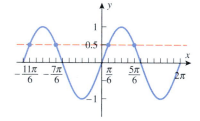

Figure 4.4.3

FOR THE READER.   Refer to the documentation for your calculating utility to determine how to calculate inverse sines, inverse cosines, and inverse tangents; and then confirm Equation (3) numerically by showing that

$$\sin^{-1}(0.5) \approx 0.523598775598\ldots \approx \pi/6$$

In general, if we view $x = \sin^{-1} y$ as an angle in radian measure whose sine is y, then the restriction $-\pi/2 \le \theta \le \pi/2$ imposes the geometric requirement that the angle x terminate in either the first or fourth quadrant or on an axis adjacent to those quadrants.

Example 1   Find exact values of

   (a) $\sin^{-1}(1/\sqrt{2})$ (b) $\sin^{-1}(-1)$

by inspection, and confirm your results numerically using a calculating utility.

Solution (a). Because $\sin^{-1}(1/\sqrt{2}) > 0$, we can view $x = \sin^{-1}(1/\sqrt{2})$ as that angle in the first quadrant such that $\sin \theta = 1/\sqrt{2}$. Thus, $\sin^{-1}(1/\sqrt{2}) = \pi/4$. You can confirm this with your calculating utility by showing that $\sin^{-1}(1/\sqrt{2}) \approx 0.785 \approx \pi/4$.

Solution (b). Because $\sin^{-1}(-1) < 0$, we can view $x = \sin^{-1}(-1)$ as an angle in the fourth quadrant (or an adjacent axis) such that $\sin x = -1$. Thus, $\sin^{-1}(-1) = -\pi/2$. You can confirm this with your calculating utility by showing that $\sin^{-1}(-1) \approx -1.57 \approx -\pi/2$.

◄

FOR THE READER.   If $x = \cos^{-1} y$ is viewed as an angle in radian measure whose cosine is y, in what possible quadrants can x lie? Answer the same question for $x = \tan^{-1} y$ and $x = \sec^{-1} y$.

FOR THE READER.   Most calculators do not provide a direct method for calculating inverse secants. In such situations the identity

$$\sec^{-1} x = \cos^{-1}(1/x) \tag{4}$$

is useful (Exercise 16). Use this formula to show that

$$\sec^{-1}(2.25) \approx 1.11 \quad \text{and} \quad \sec^{-1}(-2.25) \approx 2.03$$

If you have a calculating utility (such as a CAS) that can find $\sec^{-1} x$ directly, use it to check these values.

IDENTITIES FOR INVERSE TRIGONOMETRIC FUNCTIONS

If we interpret $\sin^{-1} x$ as an angle in radian measure whose sine is x, and if that angle is *nonnegative*, then we can represent $\sin^{-1} x$ geometrically as an angle in a right triangle in which the hypotenuse has length 1 and the side opposite to the angle $\sin^{-1} x$ has length x (Figure 4.4.4a). By the Theorem of Pythagoras the side adjacent to the angle $\sin^{-1} x$ has length $\sqrt{1 - x^2}$. Moreover, the third angle in Figure 4.4.4a is $\cos^{-1} x$, since the cosine of that angle is x (Figure 4.4.4b). This triangle motivates a number of useful identities involving inverse trigonometric functions that are valid for $-1 \le x \le 1$; for example,

$$\sin^{-1} x + \cos^{-1} x = \frac{\pi}{2} \tag{5}$$

$$\cos(\sin^{-1} x) = \sqrt{1 - x^2} \tag{6}$$

$$\sin(\cos^{-1} x) = \sqrt{1 - x^2} \tag{7}$$

$$\tan(\sin^{-1} x) = \frac{x}{\sqrt{1 - x^2}} \tag{8}$$

In a similar manner, $\tan^{-1} x$ and $\sec^{-1} x$ can be represented as angles in the right triangles shown in Figures 4.4.4c and 4.4.4d (verify). Those triangles reveal additional useful identities; for example,

$$\sec(\tan^{-1} x) = \sqrt{1 + x^2} \tag{9}$$

$$\sin(\sec^{-1} x) = \frac{\sqrt{x^2 - 1}}{x} \quad (x \ge 1) \tag{10a}$$

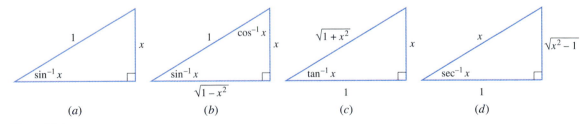

(a)   (b)   (c)   (d)

Figure 4.4.4

REMARK.   We leave it as an exercise to use (4) and (7) to obtain the following identity that is valid for $x \ge 1$ and $x \le -1$ (Exercise 46):

$$\sin(\sec^{-1} x) = \frac{\sqrt{x^2 - 1}}{|x|} \quad (|x| \ge 1) \tag{10b}$$

REMARK.   There is nothing to be gained by memorizing these identities; what is important to understand is the *method* that was used to obtain them.

Referring to Figure 4.4.1, observe that the inverse sine and inverse tangent are odd functions; that is,

$$\sin^{-1}(-x) = -\sin^{-1}(x) \quad \text{and} \quad \tan^{-1}(-x) = -\tan^{-1}(x) \tag{11–12}$$

Example 2 Figure 4.4.5 shows a computer-generated graph of $y = \sin^{-1}(\sin x)$. One might think that this graph should be the line $y = x$, since $\sin^{-1}(\sin x) = x$. Why isn't it?

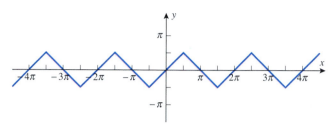

Figure 4.4.5

Solution. The relationship $\sin^{-1}(\sin x) = x$ is valid on the interval $-\pi/2 \le x \le \pi/2$, so we can say with certainty that the graphs of $y = \sin^{-1}(\sin x)$ and $y = x$ coincide on this interval (which is confirmed by Figure 4.4.5). However, outside of this interval the relationship $\sin^{-1}(\sin x) = x$ does not hold. For example, if the quantity x lies in the interval $\pi/2 \le x \le 3\pi/2$, then the quantity $x - \pi$ lies in the interval $-\pi/2 \le x \le \pi/2$, so

$$\sin^{-1}[\sin(x - \pi)] = x - \pi$$

Thus, by using the identity $\sin(x - \pi) = -\sin x$ and the fact that $\sin^{-1}$ is an odd function, we can express $\sin^{-1}(\sin x)$ as

$$\sin^{-1}(\sin x) = \sin^{-1}[-\sin(x - \pi)] = -\sin^{-1}[\sin(x - \pi)] = -(x - \pi)$$

This shows that on the interval $\pi/2 \le x \le 3\pi/2$ the graph of $y = \sin^{-1}(\sin x)$ coincides with the line $y = -(x - \pi)$, which has slope -1 and an x-intercept at $x = \pi$. This agrees with Figure 4.4.5. ◀

DERIVATIVES OF THE INVERSE TRIGONOMETRIC FUNCTIONS

Recall that if f is a one-to-one function whose derivative is known, then there are two basic ways to obtain a derivative formula for $f^{-1}(x)$—we can rewrite the equation $y = f^{-1}(x)$ as $x = f(y)$, and differentiate implicitly, or we can apply Formula (4) or (5) of Section 4.1. Here we will use implicit differentiation to obtain the derivative formula for $y = \sin^{-1} x$. Rewriting this equation as $x = \sin y$ and differentiating implicitly, we obtain

$$\frac{d}{dx}[x] = \frac{d}{dx}[\sin y]$$

$$1 = \cos y \cdot \frac{dy}{dx}$$

$$\frac{dy}{dx} = \frac{1}{\cos y} = \frac{1}{\cos(\sin^{-1} x)}$$

At this point we have succeeded in obtaining the derivative; however, this derivative formula can be simplified by applying Formula (6), which is derived from the triangle in Figure 4.4.6. This yields

$$\frac{dy}{dx} = \frac{1}{\sqrt{1 - x^2}}$$

Thus, we have shown that

$$\frac{d}{dx}[\sin^{-1} x] = \frac{1}{\sqrt{1 - x^2}} \quad (-1 < x < 1) \tag{13}$$

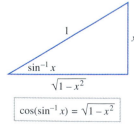

$$\cos(\sin^{-1} x) = \sqrt{1 - x^2}$$

Figure 4.4.6

If u is a differentiable function of x, then (13) and the chain rule produce the following generalized derivative formula:

$$\frac{d}{dx}[\sin^{-1} u] = \frac{1}{\sqrt{1 - u^2}} \frac{du}{dx} \qquad (-1 < u < 1) \tag{14}$$

The method used to obtain this formula can also be used to obtain generalized derivative formulas for the other inverse trigonometric functions. These formulas are

$$\frac{d}{dx}[\cos^{-1} u] = \frac{-1}{\sqrt{1 - u^2}} \frac{du}{dx} \qquad (-1 < u < 1) \tag{15}$$

$$\frac{d}{dx}[\tan^{-1} u] = \frac{1}{1 + u^2} \frac{du}{dx} \qquad (-\infty < u < +\infty) \tag{16}$$

$$\frac{d}{dx}[\sec^{-1} u] = \frac{1}{|u|\sqrt{u^2 - 1}} \frac{du}{dx} \qquad (1 < |u|) \tag{17}$$

DIFFERENTIABILITY OF THE INVERSE TRIGONOMETRIC FUNCTIONS

In the derivation of (13) we *assumed* that $\sin^{-1} x$ is differentiable. However, we can establish the differentiability with the help of Theorem 4.1.6. Since $f(x) = \sin x$ and $f'(x) = \cos x$, it follows from that theorem that the function $f^{-1}(x) = \sin^{-1} x$ will be differentiable at any value of x where $\cos(\sin^{-1} x) \neq 0$ or from (6) where $\sqrt{1 - x^2} \neq 0$. Thus, $\sin^{-1} x$ is differentiable on the interval $(-1, 1)$. The differentiability of the remaining inverse trigonometric functions can be deduced similarly.

REMARK.   Observe that $\sin^{-1} x$ is only differentiable on the interval $(-1, 1)$, even though its domain is $[-1, 1]$. However, it can be seen geometrically that $\sin^{-1}$ cannot be differentiable at $x = \pm 1$. Just observe that the graph of $y = \sin x$ has horizontal tangent lines at $(\pi/2, 1)$ and $(-\pi/2, -1)$ and that these become points of vertical tangency for $y = \sin^{-1} x$ when reflected around the line $y = x$.

Example 3   Find dy/dx if

(a) $y = \sin^{-1}(x^3)$ (b) $y = \sec^{-1}(e^x)$

Solution (a). From (14)

$$\frac{dy}{dx} = \frac{1}{\sqrt{1 - (x^3)^2}}(3x^2) = \frac{3x^2}{\sqrt{1 - x^6}}$$

Solution (b). From (17)

$$\frac{dy}{dx} = \frac{1}{e^x \sqrt{(e^x)^2 - 1}}(e^x) = \frac{1}{\sqrt{e^{2x} - 1}} \qquad \blacktriangleleft$$

EXERCISE SET 4.4   ~ Graphing Utility

1. Find the exact value of
   (a) $\sin^{-1}(-1)$ (b) $\cos^{-1}(-1)$
   (c) $\tan^{-1}(-1)$ (d) $\sec^{-1}(1)$.

2. Find the exact value of
   (a) $\sin^{-1}\left(\frac{1}{2}\sqrt{3}\right)$ (b) $\cos^{-1}\left(\frac{1}{2}\right)$
   (c) $\tan^{-1}(1)$ (d) $\sec^{-1}(-2)$.

3. Given that $\theta = \sin^{-1}\left(-\frac{1}{2}\sqrt{3}\right)$, find the exact values of $\cos\theta$, $\tan\theta$, $\cot\theta$, $\sec\theta$, and $\csc\theta$.

4. Given that $\theta = \cos^{-1}\left(\frac{1}{2}\right)$, find the exact values of $\sin\theta$, $\tan\theta$, $\cot\theta$, $\sec\theta$, and $\csc\theta$.

5. Given that $\theta = \tan^{-1}\left(\frac{4}{3}\right)$, find the exact values of $\sin\theta$, $\cos\theta$, $\cot\theta$, $\sec\theta$, and $\csc\theta$.

6. Given that $\theta = \sec^{-1} 2.6$, find the exact values of $\sin\theta$, $\cos\theta$, $\tan\theta$, $\cot\theta$, and $\csc\theta$.

7. Find the exact value of
   (a) $\sin^{-1}(\sin \pi/7)$ (b) $\sin^{-1}(\sin \pi)$
   (c) $\sin^{-1}(\sin 5\pi/7)$ (d) $\sin^{-1}(\sin 630)$.

8. Find the exact value of
   (a) $\cos^{-1}(\cos \pi/7)$ (b) $\cos^{-1}(\cos \pi)$
   (c) $\cos^{-1}(\cos 12\pi/7)$ (d) $\cos^{-1}(\cos 200)$.

9. For which values of x is it true that
   (a) $\cos^{-1}(\cos x) = x$ (b) $\cos(\cos^{-1} x) = x$
   (c) $\tan^{-1}(\tan x) = x$ (d) $\tan(\tan^{-1} x) = x$

In Exercises 10 and 11, find the exact value of the given quantity.

10. $\sec\left[\sin^{-1}\left(-\frac{3}{4}\right)\right]$ **11.** $\sin\left[2\cos^{-1}\left(\frac{3}{5}\right)\right]$

In Exercises 12 and 13, complete the identities using the triangle method (Figure 4.4.4).

12. (a) $\sin(\cos^{-1} x) = ?$ (b) $\tan(\cos^{-1} x) = ?$
   (c) $\csc(\tan^{-1} x) = ?$ (d) $\sin(\tan^{-1} x) = ?$

13. (a) $\cos(\tan^{-1} x) = ?$ (b) $\tan(\cos^{-1} x) = ?$
   (c) $\sin(\sec^{-1} x) = ?$ (d) $\cot(\sec^{-1} x) = ?$

 14. (a) Use a calculating utility set to radian measure to make tables of values of $y = \sin^{-1} x$ and $y = \cos^{-1} x$ for $x = -1, -0.8, -0.6, \ldots, 0, 0.2, \ldots, 1$. Round your answers to two decimal places.
   (b) Plot the points obtained in part (a), and use the points to sketch the graphs of $y = \sin^{-1} x$ and $y = \cos^{-1} x$. Confirm that your sketches agree with those in Figure 4.4.1.
   (c) Use your graphing utility to graph $y = \sin^{-1} x$ and $y = \cos^{-1} x$; confirm that the graphs agree with those in Figure 4.4.1.

The function $\cot^{-1} x$ is defined to be the inverse of the restricted cotangent function
$$\cot x, \quad 0 < x < \pi$$
and the function $\csc^{-1} x$ is defined to be the inverse of the restricted cosecant function
$$\csc x, \quad -\pi/2 < x < \pi/2, \quad x \neq 0$$
Use these definitions in Exercises 15 and 16 and in all subsequent exercises that involve these functions.

15. (a) Sketch the graphs of $\cot^{-1} x$ and $\csc^{-1} x$.
   (b) Find the domain and range of $\cot^{-1} x$ and $\csc^{-1} x$.

16. Show that
   (a) $\cot^{-1} x = \begin{cases} \tan^{-1}(1/x), & \text{if } x > 0 \\ \pi + \tan^{-1}(1/x), & \text{if } x < 0 \end{cases}$

   (b) $\sec^{-1} x = \cos^{-1}\dfrac{1}{x}$,   if $|x| \geq 1$

   (c) $\csc^{-1} x = \sin^{-1}\dfrac{1}{x}$,   if $|x| \geq 1$.

17. Most scientific calculators have keys for the values of only $\sin^{-1} x$, $\cos^{-1} x$, and $\tan^{-1} x$. The formulas in Exercise 16 show how a calculator can be used to obtain values of $\cot^{-1} x$, $\sec^{-1} x$, and $\csc^{-1} x$ for positive values of x. Use these formulas and a calculator to find numerical values for each of the following inverse trigonometric functions. Express your answers in degrees, rounded to the nearest tenth of a degree.
   (a) $\cot^{-1} 0.7$ (b) $\sec^{-1} 1.2$ (c) $\csc^{-1} 2.3$

18. (a) Use Theorem 4.1.6 to prove that
$$\left.\frac{d}{dx}[\cot^{-1} x]\right|_{x=0} = -1$$
   (b) Use part (a) above, part (a) of Exercise 16, and the chain rule to show that
$$\frac{d}{dx}[\cot^{-1} x] = -\frac{1}{1+x^2}$$
   for $-\infty < x < +\infty$.
   (c) Conclude from part (b) that
$$\frac{d}{dx}[\cot^{-1} u] = -\frac{1}{1+u^2}\frac{du}{dx}$$
   for $-\infty < u < +\infty$.

19. (a) Use part (c) of Exercise 16 and the chain rule to show that
$$\frac{d}{dx}[\csc^{-1} x] = -\frac{1}{|x|\sqrt{x^2-1}}$$
   for $1 < |x|$.
   (b) Conclude from part (a) that
$$\frac{d}{dx}[\csc^{-1} u] = -\frac{1}{|u|\sqrt{u^2-1}}\frac{du}{dx}$$
   for $1 < |u|$.

In Exercises 20–22, use a calculating utility to approximate the solution of each equation. Where radians are used, express your answer to four decimal places, and where degrees are used, express it to the nearest tenth of a degree. [*Note:* In each part, the solution is not in the range of the relevant inverse trigonometric function.]

20. (a) $\sin x = 0.37$, $\pi/2 < x < \pi$
   (b) $\sin\theta = -0.61$, $180° < \theta < 270°$

21. (a) $\cos x = -0.85$, $\pi < x < 3\pi/2$
   (b) $\cos\theta = 0.23$, $-90° < \theta < 0°$

22. (a) $\tan x = 3.16$, $-\pi < x < -\pi/2$
   (b) $\tan\theta = -0.45$, $90° < \theta < 180°$

In Exercises 23–30, find dy/dx.

23. (a) $y = \sin^{-1}\left(\frac{1}{3}x\right)$ (b) $y = \cos^{-1}(2x+1)$

24. (a) $y = \tan^{-1}(x^2)$ (b) $y = \cot^{-1}(\sqrt{x})$

25. (a) $y = \sec^{-1}(x^7)$ (b) $y = \csc^{-1}(e^x)$

26. (a) $y = (\tan x)^{-1}$ (b) $y = \dfrac{1}{\tan^{-1} x}$

27. (a) $y = \sin^{-1}(1/x)$ (b) $y = \cos^{-1}(\cos x)$

28. (a) $y = \ln(\cos^{-1} x)$ (b) $y = \sqrt{\cot^{-1} x}$

29. (a) $y = e^x \sec^{-1} x$ (b) $y = x^2 \left(\sin^{-1} x\right)^3$

30. (a) $y = \sin^{-1} x + \cos^{-1} x$ (b) $y = \sec^{-1} x + \csc^{-1} x$

> In Exercises 31 and 32, find dy/dx by implicit differentiation.

31. $x^3 + x \tan^{-1} y = e^y$

32. $\sin^{-1}(xy) = \cos^{-1}(x - y)$

 33. In each part, sketch the graph and check your work with a graphing utility.

(a) $y = \sin^{-1} 2x$ (b) $y = \tan^{-1} \frac{1}{2}x$

34. (a) Use a calculating utility to evaluate $\sin^{-1}(\sin^{-1} 0.25)$ and $\sin^{-1}(\sin^{-1} 0.9)$, and explain what you think is happening in the second calculation.

(b) For what values of x in the interval $-1 \le x \le 1$ will your calculating utility produce a real value for the function $\sin^{-1}(\sin^{-1} x)$?

35. An Earth-observing satellite has horizon sensors that can measure the angle θ shown in the accompanying figure. Let R be the radius of the Earth (assumed spherical) and h the distance between the satellite and the Earth's surface.

(a) Show that $\sin \theta = \dfrac{R}{R + h}$.

(b) Find θ, to the nearest degree, for a satellite that is 10,000 km from the Earth's surface (use $R = 6378$ km).

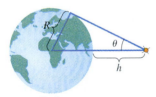

Earth

Figure Ex-35

36. The number of hours of daylight on a given day at a given point on the Earth's surface depends on the latitude λ of the point, the angle γ through which the Earth has moved in its orbital plane during the time period from the vernal equinox (March 21), and the angle of inclination ϕ of the Earth's axis of rotation measured from ecliptic north ($\phi \approx 23.45°$). The number of hours of daylight h can be approximated by the formula

$$h = \begin{cases} 24, & D \ge 1 \\ 12 + \frac{2}{15} \sin^{-1} D, & |D| < 1 \\ 0, & D \le -1 \end{cases}$$

where

$$D = \frac{\sin\phi \, \sin\gamma \, \tan\lambda}{\sqrt{1 - \sin^2\phi \, \sin^2\gamma}}$$

and $\sin^{-1} D$ is in degree measure. Given that Fairbanks, Alaska, is located at a latitude of $\lambda = 65°$ N and also that

$\gamma = 90°$ on June 20 and $\gamma = 270°$ on December 20, approximate

(a) the maximum number of daylight hours at Fairbanks to one decimal place

(b) the minimum number of daylight hours at Fairbanks to one decimal place.

[*Note:* This problem was adapted from *TEAM, A Path to Applied Mathematics*, The Mathematical Association of America, Washington, D.C., 1985.]

37. A soccer player kicks a ball with an initial speed of 14 m/s at an angle θ with the horizontal (see the accompanying figure). The ball lands 18 m down the field. If air resistance is neglected, then the ball will have a parabolic trajectory and the horizontal range R will be given by

$$R = \frac{v^2}{g} \sin 2\theta$$

where v is the initial speed of the ball and g is the acceleration due to gravity. Using $g = 9.8$ m/s^2, approximate two values of θ, to the nearest degree, at which the ball could have been kicked. Which angle results in the shorter time of flight? Why?

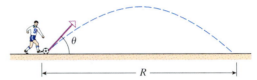

Figure Ex-37

38. The *law of cosines* states that

$$c^2 = a^2 + b^2 - 2ab \cos \theta$$

where a, b, and c are the lengths of the sides of a triangle and θ is the angle formed by sides a and b. Find θ, to the nearest degree, for the triangle with $a = 2$, $b = 3$, and $c = 4$.

39. An airplane is flying at a constant height of 3000 ft above water at a speed of 400 ft/s. The pilot is to release a survival package so that it lands in the water at a sighted point P. If air resistance is neglected, then the package will follow a parabolic trajectory whose equation relative to the coordinate system in the accompanying figure is

$$y = 3000 - \frac{g}{2v^2}x^2$$

where g is the acceleration due to gravity and v is the speed of the airplane. Using $g = 32$ ft/s^2, find the "line of sight" angle θ, to the nearest degree, that will result in the package hitting the target point.

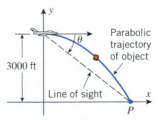

Figure Ex-39

40. (a) A camera is positioned x feet from the base of a missile launching pad (see the accompanying figure). If a missile of length a feet is launched vertically, show that when the base of the missile is b feet above the camera lens, the angle θ subtended at the lens by the missile is

$$\theta = \cot^{-1}\frac{x}{a+b} - \cot^{-1}\frac{x}{b}$$

(b) How far from the launching pad should the camera be positioned to maximize the angle θ subtended at the lens by the missile?

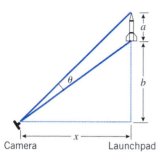

Camera   Launchpad

Figure Ex-40

41. Prove:
  (a) $\sin^{-1}(-x) = -\sin^{-1}x$
  (b) $\tan^{-1}(-x) = -\tan^{-1}x$.

42. Prove:
  (a) $\cos^{-1}(-x) = \pi - \cos^{-1}x$
  (b) $\sec^{-1}(-x) = \pi - \sec^{-1}x$

43. Prove:
  (a) $\sin^{-1}x = \tan^{-1}\dfrac{x}{\sqrt{1-x^2}}$ $(|x| < 1)$
  (b) $\cos^{-1}x = \dfrac{\pi}{2} - \tan^{-1}\dfrac{x}{\sqrt{1-x^2}}$ $(|x| < 1)$

44. Prove:

$$\tan^{-1}x + \tan^{-1}y = \tan^{-1}\left(\frac{x+y}{1-xy}\right)$$

provided $-\pi/2 < \tan^{-1}x + \tan^{-1}y < \pi/2$. [*Hint:* Use an identity for $\tan(\alpha + \beta)$.]

45. Use the result in Exercise 44 to show that
  (a) $\tan^{-1}\frac{1}{2} + \tan^{-1}\frac{1}{3} = \pi/4$
  (b) $2\tan^{-1}\frac{1}{3} + \tan^{-1}\frac{1}{7} = \pi/4$.

46. Use identities (4) and (7) to obtain identity (10b).

4.5 L'HÔPITAL'S RULE; INDETERMINATE FORMS

In this section we will discuss a general method for using derivatives to find limits. This method will enable us to establish limits with certainty that earlier in the text we were only able to conjecture using numerical or graphical evidence. The method that we will discuss in this section is an extremely powerful tool that is used internally by many computer programs to calculate limits of various types.

INDETERMINATE FORMS OF TYPE 0/0

In earlier sections we discussed limits that can be determined by inspection or by some appropriate algebraic manipulation. Two special exceptions to this were the limits in Theorem 2.6.3,

$$\lim_{x\to 0}\frac{\sin x}{x} = 1 \quad \text{and} \quad \lim_{x\to 0}\frac{1-\cos x}{x} = 0 \tag{1–2}$$

Equation (1) was shown using the Squeezing Theorem (2.6.2) and some careful manipulation of inequalities, and Equation (2) then followed using the identity $\sin^2 x + \cos^2 x = 1$. These in turn were used in Section 3.4 to derive the derivatives of the sine and cosine functions. Equations (1) and (2) are actually special cases of these derivatives, as can be seen by

$$\lim_{x\to 0}\frac{\sin x}{x} = \lim_{x\to 0}\frac{\sin x - \sin 0}{x-0} = \frac{d}{dx}(\sin x)\Big|_{x=0} = \cos 0 = 1$$

and

$$\lim_{x\to 0}\frac{1-\cos x}{x} = \lim_{x\to 0}-\frac{\cos x - 1}{x} = -\left(\lim_{x\to 0}\frac{\cos x - \cos 0}{x-0}\right)$$

$$= -\left(\frac{d}{dx}(\cos x)\Big|_{x=0}\right) = \sin 0 = 0$$

What makes the limits in (1) and (2) bothersome is the fact that the numerator and denominator both approach 0 as $x \to 0$. Such limits are called ***indeterminate forms of*** *type* **0/0.** As illustrated above, the definition of a derivative provides an important class of examples of indeterminate forms of type 0/0. Our goal here is to develop a general method, based on the derivative, for evaluating indeterminate forms.

L'HÔPITAL'S RULE

Consider the limit

$$\lim_{x \to 0} \frac{e^{2x} - 1}{\sin x} \tag{3}$$

Unlike (1) and (2), the limit in (3) is not easily seen as the evaluation of the derivative of a function at $x = 0$. However, (3) can be expressed as the ratio of two derivatives:

$$\lim_{x \to 0} \frac{e^{2x} - 1}{\sin x} = \lim_{x \to 0} \frac{(e^{2x} - e^{2(0)})/(x - 0)}{(\sin x - \sin 0)/(x - 0)} = \frac{\dfrac{d}{dx}(e^{2x})\Big|_{x=0}}{\dfrac{d}{dx}(\sin x)\Big|_{x=0}} = \frac{2e^0}{\cos 0} = 2 \tag{4}$$

The method of (4) can be stated more generally. Suppose that f and g are differentiable functions at $x = a$ and that

$$\lim_{x \to a} \frac{f(x)}{g(x)} \tag{5}$$

is an indeterminate form of type 0/0, that is,

$$\lim_{x \to a} f(x) = 0 \quad \text{and} \quad \lim_{x \to a} g(x) = 0 \tag{6}$$

In particular, the differentiability of f and g at $x = a$ implies that f and g are continuous at $x = a$, and hence from (6)

$$f(a) = \lim_{x \to a} f(x) = 0 \quad \text{and} \quad g(a) = \lim_{x \to a} g(x) = 0$$

Furthermore, since f and g are differentiable at $x = a$,

$$\lim_{x \to a} \frac{f(x)}{x - a} = \lim_{x \to a} \frac{f(x) - f(a)}{x - a} = f'(a)$$

and

$$\lim_{x \to a} \frac{g(x)}{x - a} = \lim_{x \to a} \frac{g(x) - g(a)}{x - a} = g'(a)$$

If $g'(a) \neq 0$ then the indeterminate form in (5) can be evaluated as the ratio of derivative values:

$$\lim_{x \to a} \frac{f(x)}{g(x)} = \lim_{x \to a} \frac{f(x)/(x - a)}{g(x)/(x - a)} = \frac{\displaystyle\lim_{x \to a} \frac{f(x) - f(a)}{x - a}}{\displaystyle\lim_{x \to a} \frac{g(x) - g(a)}{x - a}} = \frac{f'(a)}{g'(a)} \tag{7}$$

If $f'(x)$ and $g'(x)$ are continuous at $x = a$, the result in (7) is a special case of ***L'Hôpital's***[*] ***rule***, which converts an indeterminate form of type 0/0 into a new limit involving derivatives. Moreover, L'Hôpital's rule is also true for limits at $-\infty$ and at $+\infty$. We state the following result without proof.

---

[*] GUILLAUME FRANCOIS ANTOINE DE L'HÔPITAL (1661–1704). French mathematician. L'Hôpital, born to parents of the French high nobility, held the title of Marquis de Sainte-Mesme Comte d'Autrement. He showed mathematical talent quite early and at age 15 solved a difficult problem about cycloids posed by Pascal. As a young man he served briefly as a cavalry officer, but resigned because of nearsightedness. In his own time he gained fame as the author of the first textbook ever published on differential calculus, *L'Analyse des Infiniment Petits pour l'Intelligence des Lignes Courbes* (1696). L'Hôpital's rule appeared for the first time in that book. Actually, L'Hôpital's rule and most of the material in the calculus text were due to John Bernoulli, who was L'Hôpital's teacher. L'Hôpital dropped his plans for a book on integral calculus when Leibniz informed him that he intended to write such a text. L'Hôpital was apparently generous and personable, and his many contacts with major mathematicians provided the vehicle for disseminating major discoveries in calculus throughout Europe.

4.5.1 THEOREM (*L'Hôpital's Rule for Form 0/0*). *Suppose that f and g are differentiable functions on an open interval containing $x = a$, except possibly at $x = a$, and that*

$$\lim_{x \to a} f(x) = 0 \quad and \quad \lim_{x \to a} g(x) = 0$$

If $\lim_{x \to a} [f'(x)/g'(x)]$ has a finite limit, or if this limit is $+\infty$ or $-\infty$, then

$$\lim_{x \to a} \frac{f(x)}{g(x)} = \lim_{x \to a} \frac{f'(x)}{g'(x)}$$

Moreover, this statement is also true in the case of a limit as $x \to a^-$, $x \to a^+$, $x \to -\infty$, or as $x \to +\infty$.

REMARK. Note that in L'Hôpital's rule the numerator and denominator are differentiated separately, which is not the same as differentiating $f(x)/g(x)$.

In the following examples we will apply L'Hôpital's rule using the following three-step process:

Step 1. Check that the limit of $f(x)/g(x)$ is an indeterminate form. If it is not, then L'Hôpital's rule cannot be used.

Step 2. Differentiate f and g separately.

Step 3. Find the limit of $f'(x)/g'(x)$. If this limit is finite, $+\infty$, or $-\infty$, then it is equal to the limit of $f(x)/g(x)$.

Example 1 In each part confirm that the limit is an indeterminate form of type $0/0$, and evaluate it using L'Hôpital's rule.

(a) $\displaystyle\lim_{x \to 2} \frac{x^2 - 4}{x - 2}$ 　(b) $\displaystyle\lim_{x \to 0} \frac{\sin 2x}{x}$ 　(c) $\displaystyle\lim_{x \to \pi/2} \frac{1 - \sin x}{\cos x}$ 　(d) $\displaystyle\lim_{x \to 0} \frac{e^x - 1}{x^3}$

(e) $\displaystyle\lim_{x \to 0^-} \frac{\tan x}{x^2}$ 　(f) $\displaystyle\lim_{x \to 0} \frac{1 - \cos x}{x^2}$ 　(g) $\displaystyle\lim_{x \to +\infty} \frac{x^{-4/3}}{\sin(1/x)}$

Solution (a). The numerator and denominator have a limit of 0, so the limit is an indeterminate form of type $0/0$. Applying L'Hôpital's rule yields

$$\lim_{x \to 2} \frac{x^2 - 4}{x - 2} = \lim_{x \to 2} \frac{\dfrac{d}{dx}[x^2 - 4]}{\dfrac{d}{dx}[x - 2]} = \lim_{x \to 2} \frac{2x}{1} = 4$$

This limit can also be recognized as the derivative of $y = x^2$ at $x = 2$,

$$\lim_{x \to 2} \frac{x^2 - 4}{x - 2} = \frac{d}{dx}(x^2)\bigg|_{x=2} = 2 \cdot 2 = 4$$

Finally, observe that this limit could have been obtained by factoring

$$\lim_{x \to 2} \frac{x^2 - 4}{x - 2} = \lim_{x \to 2} \frac{(x - 2)(x + 2)}{x - 2} = \lim_{x \to 2} (x + 2) = 4$$

Solution (b). The numerator and denominator have a limit of 0, so the limit is an indeterminate form of type $0/0$. Applying L'Hôpital's rule yields

$$\lim_{x \to 0} \frac{\sin 2x}{x} = \lim_{x \to 0} \frac{\dfrac{d}{dx}[\sin 2x]}{\dfrac{d}{dx}[x]} = \lim_{x \to 0} \frac{2 \cos 2x}{1} = 2$$

Observe that this result agrees with that obtained by substitution in Example 2(b) of Section 2.6.

Solution (c). The numerator and denominator have a limit of 0, so the limit is an indeterminate form of type 0/0. Applying L'Hôpital's rule yields

$$\lim_{x \to \pi/2} \frac{1 - \sin x}{\cos x} = \lim_{x \to \pi/2} \frac{\dfrac{d}{dx}[1 - \sin x]}{\dfrac{d}{dx}[\cos x]} = \lim_{x \to \pi/2} \frac{-\cos x}{-\sin x} = \frac{0}{-1} = 0$$

Solution (d). The numerator and denominator have a limit of 0, so the limit is an indeterminate form of type 0/0. Applying L'Hôpital's rule yields

$$\lim_{x \to 0} \frac{e^x - 1}{x^3} = \lim_{x \to 0} \frac{\dfrac{d}{dx}[e^x - 1]}{\dfrac{d}{dx}[x^3]} = \lim_{x \to 0} \frac{e^x}{3x^2} = +\infty$$

Solution (e). The numerator and denominator have a limit of 0, so the limit is an indeterminate form of type 0/0. Applying L'Hôpital's rule yields

$$\lim_{x \to 0^-} \frac{\tan x}{x^2} = \lim_{x \to 0^-} \frac{\sec^2 x}{2x} = -\infty$$

Solution (f). The numerator and denominator have a limit of 0, so the limit is an indeterminate form of type 0/0. Applying L'Hôpital's rule yields

$$\lim_{x \to 0} \frac{1 - \cos x}{x^2} = \lim_{x \to 0} \frac{\sin x}{2x}$$

Since the new limit is another indeterminate form of type 0/0, we apply L'Hôpital's rule again:

$$\lim_{x \to 0} \frac{1 - \cos x}{x^2} = \lim_{x \to 0} \frac{\sin x}{2x} = \lim_{x \to 0} \frac{\cos x}{2} = \frac{1}{2}$$

Solution (g). The numerator and denominator have a limit of 0, so the limit is an indeterminate form of type 0/0. Applying L'Hôpital's rule yields

$$\lim_{x \to +\infty} \frac{x^{-4/3}}{\sin(1/x)} = \lim_{x \to +\infty} \frac{-\frac{4}{3}x^{-7/3}}{(-1/x^2)\cos(1/x)} = \lim_{x \to +\infty} \frac{\frac{4}{3}x^{-1/3}}{\cos(1/x)} = \frac{0}{1} = 0 \quad \blacktriangleleft$$

WARNING. Applying L'Hôpital's rule to limits that are not indeterminate forms can lead to incorrect results. For example, in the limit

$$\lim_{x \to 0} \frac{x + 6}{x + 2} = \frac{6}{2} = 3$$

the numerator approaches 6 and the denominator approaches 2, so the limit is not an indeterminate form of type 0/0. However, if we ignore this and blindly apply L'Hôpital's rule, we reach the following *erroneous* conclusion:

$$\lim_{x \to 0} \frac{x + 6}{x + 2} = \lim_{x \to 0} \frac{\dfrac{d}{dx}[x + 6]}{\dfrac{d}{dx}[x + 2]} = \lim_{x \to 0} \frac{1}{1} = 1$$

· WRONG · WRONG · WRONG · WRONG · WRONG · WRONG · WRONG · WRON G · WRONG · WRONG · WRONG · WRO NG · WRONG · WRONG · WR ONG · WRONG · WRONG · W RONG · WRONG · WRONG · WRONG ·

INDETERMINATE FORMS OF TYPE ∞/∞

When we want to indicate that the limit (or the one-sided limits) of a function are $+\infty$ or $-\infty$ without being specific about the sign, we will say that the limit is ∞. For example,

$$\lim_{x \to a^+} f(x) = \infty \quad \text{means} \quad \lim_{x \to a^+} f(x) = +\infty \quad \text{or} \quad \lim_{x \to a^+} f(x) = -\infty$$

$$\lim_{x \to +\infty} f(x) = \infty \quad \text{means} \quad \lim_{x \to +\infty} f(x) = +\infty \quad \text{or} \quad \lim_{x \to +\infty} f(x) = -\infty$$

$$\lim_{x \to a} f(x) = \infty \quad \text{means} \quad \lim_{x \to a^+} f(x) = \pm\infty \quad \text{and} \quad \lim_{x \to a^-} f(x) = \pm\infty$$

The limit of a ratio, $f(x)/g(x)$, in which the numerator has limit ∞ and the denominator has limit ∞ is called an ***indeterminate form of type*** ∞/∞. The following version of L'Hôpital's rule, which we state without proof, can often be used to evaluate limits of this type.

4.5.2   THEOREM (*L'Hôpital's Rule for Form* ∞/∞).   *Suppose that f and g are differentiable functions on an open interval containing $x = a$, except possibly at $x = a$, and that*

$$\lim_{x \to a} f(x) = \infty \quad and \quad \lim_{x \to a} g(x) = \infty$$

If $\lim_{x \to a} [f'(x)/g'(x)]$ *has a finite limit, or if this limit is* $+\infty$ *or* $-\infty$, *then*

$$\lim_{x \to a} \frac{f(x)}{g(x)} = \lim_{x \to a} \frac{f'(x)}{g'(x)}$$

Moreover, this statement is also true in the case of a limit as $x \to a^-$, $x \to a^+$, $x \to -\infty$, *or as* $x \to +\infty$.

Example 2   In each part confirm that the limit is an indeterminate form of type ∞/∞ and apply L'Hôpital's rule.

(a) $\displaystyle \lim_{x \to +\infty} \frac{x}{e^x}$ (b) $\displaystyle \lim_{x \to 0^+} \frac{\ln x}{\csc x}$

Solution (a).   The numerator and denominator both have a limit of $+\infty$, so we have an indeterminate form of type ∞/∞. Applying L'Hôpital's rule yields

$$\lim_{x \to +\infty} \frac{x}{e^x} = \lim_{x \to +\infty} \frac{1}{e^x} = 0$$

Solution (b).   The numerator has a limit of $-\infty$ and the denominator has a limit of $+\infty$, so we have an indeterminate form of type ∞/∞. Applying L'Hôpital's rule yields

$$\lim_{x \to 0^+} \frac{\ln x}{\csc x} = \lim_{x \to 0^+} \frac{1/x}{-\csc x \cot x} \tag{8}$$

This last limit is again an indeterminate form of type ∞/∞. Moreover, any additional applications of L'Hôpital's rule will yield powers of $1/x$ in the numerator and expressions involving $\csc x$ and $\cot x$ in the denominator; thus, repeated application of L'Hôpital's rule simply produces new indeterminate forms. We must try something else. The last limit in (8) can be rewritten as

$$\lim_{x \to 0^+} \left(-\frac{\sin x}{x} \tan x \right) = -\lim_{x \to 0^+} \frac{\sin x}{x} \cdot \lim_{x \to 0^+} \tan x = -(1)(0) = 0$$

Thus,

$$\lim_{x \to 0^+} \frac{\ln x}{\csc x} = 0 \qquad \blacktriangleleft$$

ANALYZING THE GROWTH OF EXPONENTIAL FUNCTIONS USING L'HÔPITAL'S RULE

If n is any positive integer, then $x^n \to +\infty$ as $x \to +\infty$. Such integer powers of x are sometimes used as "measuring sticks" to describe how rapidly other functions grow. For example, we know that $e^x \to +\infty$ as $x \to +\infty$ and that the growth of e^x is very rapid (Table 4.2.3); however, the growth of x^n is also rapid when n is a high power, so it is reasonable to ask whether high powers of x grow more or less rapidly than e^x. One way to investigate this is to examine the behavior of the ratio x^n/e^x as $x \to +\infty$. For example, Figure 4.5.1a shows the graph of $y = x^5/e^x$. This graph suggests that $x^5/e^x \to 0$ as $x \to +\infty$, and this implies that the growth of the function e^x is sufficiently rapid that its values eventually overtake those of x^5 and force the ratio toward zero. Stated informally, "e^x eventually grows more rapidly than x^5." The same conclusion could have been reached

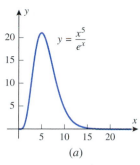
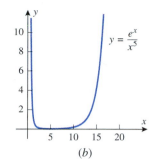

Figure 4.5.1

by putting e^x on top and examining the behavior of e^x/x^5 as $x \to +\infty$ (Figure 4.5.1*b*). In this case the values of e^x eventually overtake those of x^5 and force the ratio toward $+\infty$. More generally, we can use L'Hôpital's rule to show that e^x *eventually grows more rapidly than any positive integer power of x*, that is,

$$\lim_{x \to +\infty} \frac{x^n}{e^x} = 0 \qquad \text{and} \qquad \lim_{x \to +\infty} \frac{e^x}{x^n} = +\infty \tag{9--10}$$

Both limits are indeterminate forms of type ∞/∞ that can be evaluated using L'Hôpital's rule. For example, to establish (9), we will need to apply L'Hôpital's rule n times. For this purpose, observe that successive differentiations of x^n reduce the exponent by 1 each time, thus producing a constant for the nth derivative. For example, the successive derivatives of x^3 are $3x^2, 6x$, and 6. In general, the nth derivative of x^n is the constant $n(n-1)(n-2)\cdots 1 = n!$ (verify).[*] Thus, applying L'Hôpital's rule n times to (9) yields

$$\lim_{x \to +\infty} \frac{x^n}{e^x} = \lim_{x \to +\infty} \frac{n!}{e^x} = 0$$

Limit (10) can be established similarly.

INDETERMINATE FORMS OF TYPE $0 \cdot \infty$

Thus far we have discussed indeterminate forms of type $0/0$ and ∞/∞. However, these are not the only possibilities; in general, the limit of an expression that has one of the forms

$$\frac{f(x)}{g(x)}, \quad f(x) \cdot g(x), \quad f(x)^{g(x)}, \quad f(x) - g(x), \quad f(x) + g(x)$$

is called an *indeterminate form* if the limits of $f(x)$ and $g(x)$ individually exert conflicting influences on the limit of the entire expression. For example, the limit

$$\lim_{x \to 0^+} x \ln x$$

is an ***indeterminate form of type $0 \cdot \infty$*** because the limit of the first factor is 0, the limit of the second factor is $-\infty$, and these two limits exert conflicting influences on the product. On the other hand, the limit

$$\lim_{x \to +\infty} [\sqrt{x}(1 - x^2)]$$

is not an indeterminate form because the first factor has a limit of $+\infty$, the second factor has a limit of $-\infty$, and these influences work together to produce a limit of $-\infty$ for the product.

> **WARNING.**   It is tempting to argue that an indeterminate form of type $0 \cdot \infty$ has value 0 since "zero times anything is zero." However, this is fallacious since $0 \cdot \infty$ is not a product of numbers, but rather a statement about limits. For example, the following limits are of the form $0 \cdot \infty$:
>
> $$\lim_{x \to 0^+} x \cdot \frac{1}{x} = 1, \qquad \lim_{x \to 0^+} x^2 \cdot \frac{1}{x} = 0, \qquad \lim_{x \to 0^+} \sqrt{x} \cdot \frac{1}{x} = +\infty$$

---

[*] Recall that for $n \geq 1$ the expression $n!$, read ***n-factorial***, denotes the product of the first n positive integers.

Indeterminate forms of type $0 \cdot \infty$ can sometimes be evaluated by rewriting the product as a ratio, and then applying L'Hôpital's rule for indeterminate forms of type $0/0$ or ∞/∞.

Example 3   Evaluate

(a) $\lim\limits_{x \to 0^+} x \ln x$ (b) $\lim\limits_{x \to \pi/4} (1 - \tan x) \sec 2x$

Solution (a).   The factor x has a limit of 0 and the factor $\ln x$ has a limit of $-\infty$, so the stated problem is an indeterminate form of type $0 \cdot \infty$. There are two possible approaches: we can rewrite the limit as

$$\lim_{x \to 0^+} \frac{\ln x}{1/x} \quad \text{or} \quad \lim_{x \to 0^+} \frac{x}{1/\ln x}$$

the first being an indeterminate form of type ∞/∞ and the second an indeterminate form of type $0/0$. However, the first form is the preferred initial choice because the derivative of $1/x$ is less complicated than the derivative of $1/\ln x$. That choice yields

$$\lim_{x \to 0^+} x \ln x = \lim_{x \to 0^+} \frac{\ln x}{1/x} = \lim_{x \to 0^+} \frac{1/x}{-1/x^2} = \lim_{x \to 0^+} (-x) = 0$$

Solution (b).   The stated problem is an indeterminate form of type $0 \cdot \infty$. We will convert it to an indeterminate form of type $0/0$:

$$\lim_{x \to \pi/4} (1 - \tan x) \sec 2x = \lim_{x \to \pi/4} \frac{1 - \tan x}{1/\sec 2x} = \lim_{x \to \pi/4} \frac{1 - \tan x}{\cos 2x}$$

$$= \lim_{x \to \pi/4} \frac{-\sec^2 x}{-2 \sin 2x} = \frac{-2}{-2} = 1 \quad \blacktriangleleft$$

··········

INDETERMINATE FORMS OF TYPE $\infty - \infty$

A limit problem that leads to one of the expressions

$$(+\infty) - (+\infty), \quad (-\infty) - (-\infty),$$

$$(+\infty) + (-\infty), \quad (-\infty) + (+\infty)$$

is called an ***indeterminate form of type $\infty - \infty$***. Such limits are indeterminate because the two terms exert conflicting influences on the expression: one pushes it in the positive direction and the other pushes it in the negative direction. However, limit problems that lead to one of the expressions

$$(+\infty) + (+\infty), \quad (+\infty) - (-\infty),$$

$$(-\infty) + (-\infty), \quad (-\infty) - (+\infty)$$

are not indeterminate, since the two terms work together (those on the top produce a limit of $+\infty$ and those on the bottom produce a limit of $-\infty$).

Indeterminate forms of type $\infty - \infty$ can sometimes be evaluated by combining the terms and manipulating the result to produce an indeterminate form of type $0/0$ or ∞/∞.

Example 4   Evaluate $\lim\limits_{x \to 0^+} \left(\dfrac{1}{x} - \dfrac{1}{\sin x} \right)$.

Solution.   Both terms have a limit of $+\infty$, so the stated problem is an indeterminate form of type $\infty - \infty$. Combining the two terms yields

$$\lim_{x \to 0^+} \left(\frac{1}{x} - \frac{1}{\sin x} \right) = \lim_{x \to 0^+} \left(\frac{\sin x - x}{x \sin x} \right)$$

which is an indeterminate form of type $0/0$. Applying L'Hôpital's rule twice yields

$$\lim_{x \to 0^+} \left(\frac{\sin x - x}{x \sin x} \right) = \lim_{x \to 0^+} \frac{\cos x - 1}{\sin x + x \cos x}$$

$$= \lim_{x \to 0^+} \frac{-\sin x}{\cos x + \cos x - x \sin x} = \frac{0}{2} = 0 \quad \blacktriangleleft$$

INDETERMINATE FORMS OF TYPE 0^0, ∞^0, 1^∞

Limits of the form

$$\lim f(x)^{g(x)}$$

give rise to ***indeterminate forms of the types*** 0^0, ∞^0, ***and*** 1^∞. (The interpretations of these symbols should be clear.) For example, the limit

$$\lim_{x \to 0^+} (1 + x)^{1/x}$$

whose value we know to be e [see Formula (5) of Section 4.2] is an indeterminate form of type 1^∞. It is indeterminate because the expressions $1 + x$ and $1/x$ exert two conflicting influences: the first approaches 1, which drives the expression toward 1, and the second approaches $+\infty$, which drives the expression toward $+\infty$.

Indeterminate forms of types 0^0, ∞^0, and 1^∞ can sometimes be evaluated by first introducing a dependent variable

$$y = f(x)^{g(x)}$$

and then calculating the limit of $\ln y$ by expressing it as

$$\lim \ln y = \lim [\ln(f(x)^{g(x)})] = \lim [g(x) \ln f(x)]$$

Once the limit of $\ln y$ is known, the limit of $y = f(x)^{g(x)}$ itself can generally be obtained by a method that we will illustrate in the next example.

Example 5 Show that $\displaystyle\lim_{x \to 0} (1 + x)^{1/x} = e$.

Solution. As discussed above, we begin by introducing a dependent variable

$$y = (1 + x)^{1/x}$$

and taking the natural logarithm of both sides:

$$\ln y = \ln(1 + x)^{1/x} = \frac{1}{x} \ln(1 + x) = \frac{\ln(1 + x)}{x}$$

Thus,

$$\lim_{x \to 0} \ln y = \lim_{x \to 0} \frac{\ln(1 + x)}{x}$$

which is an indeterminate form of type 0/0, so by L'Hôpital's rule

$$\lim_{x \to 0} \ln y = \lim_{x \to 0} \frac{\ln(1 + x)}{x} = \lim_{x \to 0} \frac{1/(1 + x)}{1} = 1$$

Since we have shown that $\ln y \to 1$ as $x \to 0$, the continuity of the exponential function implies that $e^{\ln y} \to e^1$ as $x \to 0$, and this implies that $y \to e$ as $x \to 0$. Thus,

$$\lim_{x \to 0} (1 + x)^{1/x} = e \qquad \blacktriangleleft$$

EXERCISE SET 4.5 Graphing Utility CAS

In Exercises 1 and 2, evaluate the given limit without using L'Hôpital's rule, and then check that your answer is correct using L'Hôpital's rule.

1. (a) $\displaystyle\lim_{x \to 2} \frac{x^2 - 4}{x^2 + 2x - 8}$ (b) $\displaystyle\lim_{x \to +\infty} \frac{2x - 5}{3x + 7}$

2. (a) $\displaystyle\lim_{x \to 0} \frac{\sin x}{\tan x}$ (b) $\displaystyle\lim_{x \to 1} \frac{x^2 - 1}{x^3 - 1}$

In Exercises 3–36, find the limit.

3. $\displaystyle\lim_{x \to 1} \frac{\ln x}{x - 1}$

4. $\displaystyle\lim_{x \to 0} \frac{\sin 2x}{\sin 5x}$

5. $\displaystyle\lim_{x \to 0} \frac{e^x - 1}{\sin x}$

6. $\displaystyle\lim_{x \to 3} \frac{x - 3}{3x^2 - 13x + 12}$

7. $\displaystyle\lim_{\theta \to 0} \frac{\tan \theta}{\theta}$

8. $\displaystyle\lim_{t \to 0} \frac{te^t}{1 - e^t}$

9. $\displaystyle\lim_{x \to \pi^+} \frac{\sin x}{x - \pi}$

10. $\displaystyle\lim_{x \to 0^+} \frac{\sin x}{x^2}$

11. $\displaystyle\lim_{x \to +\infty} \frac{\ln x}{x}$

12. $\displaystyle\lim_{x \to +\infty} \frac{e^{3x}}{x^2}$

13. $\lim_{x \to 0^+} \dfrac{\cot x}{\ln x}$

14. $\lim_{x \to 0^+} \dfrac{1 - \ln x}{e^{1/x}}$

15. $\lim_{x \to +\infty} \dfrac{x^{100}}{e^x}$

16. $\lim_{x \to 0^+} \dfrac{\ln(\sin x)}{\ln(\tan x)}$

17. $\lim_{x \to 0} \dfrac{\sin^{-1} 2x}{x}$

18. $\lim_{x \to 0} \dfrac{x - \tan^{-1} x}{x^3}$

19. $\lim_{x \to +\infty} x e^{-x}$

20. $\lim_{x \to \pi^-} (x - \pi) \tan \tfrac{1}{2} x$

21. $\lim_{x \to +\infty} x \sin \dfrac{\pi}{x}$

22. $\lim_{x \to 0^+} \tan x \ln x$

23. $\lim_{x \to \pi/2^-} \sec 3x \cos 5x$

24. $\lim_{x \to \pi} (x - \pi) \cot x$

25. $\lim_{x \to +\infty} (1 - 3/x)^x$

26. $\lim_{x \to 0} (1 + 2x)^{-3/x}$

27. $\lim_{x \to 0} (e^x + x)^{1/x}$

28. $\lim_{x \to +\infty} (1 + a/x)^{bx}$

29. $\lim_{x \to 1} (2 - x)^{\tan[(\pi/2)x]}$

30. $\lim_{x \to +\infty} [\cos(2/x)]^{x^2}$

31. $\lim_{x \to 0} (\csc x - 1/x)$

32. $\lim_{x \to 0} \left(\dfrac{1}{x^2} - \dfrac{\cos 3x}{x^2} \right)$

33. $\lim_{x \to +\infty} (\sqrt{x^2 + x} - x)$

34. $\lim_{x \to 0} \left(\dfrac{1}{x} - \dfrac{1}{e^x - 1} \right)$

35. $\lim_{x \to +\infty} [x - \ln(x^2 + 1)]$

36. $\lim_{x \to +\infty} [\ln x - \ln(1 + x)]$

C 37. Use a CAS to check the answers you obtained in Exercises 31–36.

38. Show that for any positive integer n

(a) $\lim_{x \to +\infty} \dfrac{\ln x}{x^n} = 0$

(b) $\lim_{x \to +\infty} \dfrac{x^n}{\ln x} = +\infty$

39. (a) Find the error in the following calculation:

$$\lim_{x \to 1} \frac{x^3 - x^2 + x - 1}{x^3 - x^2} = \lim_{x \to 1} \frac{3x^2 - 2x + 1}{3x^2 - 2x}$$

$$= \lim_{x \to 1} \frac{6x - 2}{6x - 2} = 1$$

(b) Find the correct answer.

40. Find $\lim_{x \to 1} \dfrac{x^4 - 4x^3 + 6x^2 - 4x + 1}{x^4 - 3x^3 + 3x^2 - x}$.

In Exercises 41–44, make a conjecture about the limit by graphing the function involved with a graphing utility; then check your conjecture using L'Hôpital's rule.

~ 41. $\lim_{x \to +\infty} \dfrac{\ln(\ln x)}{\sqrt{x}}$

~ 42. $\lim_{x \to 0^+} x^x$

~ 43. $\lim_{x \to 0^+} (\sin x)^{3/\ln x}$

~ 44. $\lim_{x \to (\pi/2)^-} \dfrac{4 \tan x}{1 + \sec x}$

In Exercises 45–48, make a conjecture about the equations of horizontal asymptotes, if any, by graphing the equation with a graphing utility; then check your answer using L'Hôpital's rule.

~ 45. $y = \ln x - e^x$

~ 46. $y = x - \ln(1 + 2e^x)$

~ 47. $y = (\ln x)^{1/x}$

~ 48. $y = \left(\dfrac{x + 1}{x + 2} \right)^x$

49. Limits of the type

$$0/\infty, \quad \infty/0, \quad 0^\infty, \quad \infty \cdot \infty, \quad +\infty + (+\infty),$$

$$+\infty - (-\infty), \quad -\infty + (-\infty), \quad -\infty - (+\infty)$$

are *not* indeterminate forms. Find the following limits by inspection.

(a) $\lim_{x \to 0^+} \dfrac{x}{\ln x}$

(b) $\lim_{x \to +\infty} \dfrac{x^3}{e^{-x}}$

(c) $\lim_{x \to (\pi/2)^-} (\cos x)^{\tan x}$

(d) $\lim_{x \to 0^+} (\ln x) \cot x$

(e) $\lim_{x \to 0^+} \left(\dfrac{1}{x} - \ln x \right)$

(f) $\lim_{x \to -\infty} (x + x^3)$

50. There is a myth that circulates among beginning calculus students which states that all indeterminate forms of types 0^0, ∞^0, and 1^∞ have value 1 because "anything to the zero power is 1" and "1 to any power is 1." The fallacy is that 0^0, ∞^0, and 1^∞ are not powers of numbers, but rather descriptions of limits. The following examples, which were suggested by Prof. Jack Staib of Drexel University, show that such indeterminate forms can have any positive real value:

(a) $\lim_{x \to 0^+} \left[x^{(\ln a)/(1 + \ln x)} \right] = a$ (form 0^0)

(b) $\lim_{x \to +\infty} \left[x^{(\ln a)/(1 + \ln x)} \right] = a$ (form ∞^0)

(c) $\lim_{x \to 0} \left[(x + 1)^{(\ln a)/x} \right] = a$ (form 1^∞).

Verify these results.

In Exercises 51–54, verify that L'Hôpital's rule is of no help in finding the limit, then find the limit, if it exists, by some other method.

51. $\lim_{x \to +\infty} \dfrac{x + \sin 2x}{x}$

52. $\lim_{x \to +\infty} \dfrac{2x - \sin x}{3x + \sin x}$

53. $\lim_{x \to +\infty} \dfrac{x(2 + \sin 2x)}{x + 1}$

54. $\lim_{x \to +\infty} \dfrac{x(2 + \sin x)}{x^2 + 1}$

55. The accompanying schematic diagram represents an electrical circuit consisting of an electromotive force that produces a voltage V, a resistor with resistance R, and an inductor with inductance L. It is shown in electrical circuit theory that if the voltage is first applied at time $t = 0$, then the current I flowing through the circuit at time t is given by

$$I = \frac{V}{R} (1 - e^{-Rt/L})$$

What is the effect on the current at a fixed time t if the resistance approaches 0 (i.e., $R \to 0^+$)?

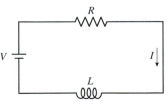

Figure Ex-55

56. (a) Show that $\lim\limits_{x \to \pi/2} (\pi/2 - x)\tan x = 1$.

(b) Show that

$$\lim_{x \to \pi/2} \left(\frac{1}{\pi/2 - x} - \tan x \right) = 0$$

(c) It follows from part (b) that the approximation

$$\tan x \approx \frac{1}{\pi/2 - x}$$

should be good for values of x near $\pi/2$. Use a calculator to find $\tan x$ and $1/(\pi/2 - x)$ for $x = 1.57$; compare the results.

c 57. (a) Use a CAS to show that if k is a positive constant, then

$$\lim_{x \to +\infty} x(k^{1/x} - 1) = \ln k$$

(b) Confirm this result using L'Hôpital's rule. [*Hint:* Express the limit in terms of $t = 1/x$.]

(c) If n is a positive integer, then it follows from part (a) with $x = n$ that the approximation

$$n(\sqrt[n]{k} - 1) \approx \ln k$$

should be good when n is large. Use this result and the square root key on a calculator to approximate the values of $\ln 0.3$ and $\ln 2$ with $n = 1024$, then compare

the values obtained with values of the logarithms generated directly from the calculator. [*Hint:* The nth roots for which n is a power of 2 can be obtained as successive square roots.]

~ 58. Let $f(x) = x^2 \sin(1/x)$.

(a) Are the limits $\lim\limits_{x \to 0^+} f(x)$ and $\lim\limits_{x \to 0^-} f(x)$ indeterminate forms?

(b) Use a graphing utility to generate the graph of f, and use the graph to make conjectures about the limits in part (a).

(c) Use the Squeezing Theorem (2.6.2) to confirm that your conjectures in part (b) are correct.

59. Find all values of k and ℓ such that

$$\lim_{x \to 0} \frac{k + \cos \ell x}{x^2} = -4$$

60. (a) Explain why L'Hôpital's rule does not apply to the problem

$$\lim_{x \to 0} \frac{x^2 \sin(1/x)}{\sin x}$$

(b) Find the limit.

61. Find $\lim\limits_{x \to 0^+} \dfrac{x \sin(1/x)}{\sin x}$ if it exists.

SUPPLEMENTARY EXERCISES

~ Graphing Utility

1. (a) State conditions under which two functions, f and g, will be inverses, and give several examples of such functions.

(b) In words, what is the relationship between the graphs of $y = f(x)$ and $y = g(x)$ when f and g are inverse functions?

(c) What is the relationship between the domains and ranges of inverse functions f and g?

(d) What condition must be satisfied for a function f to have an inverse? Give some examples of functions that do not have inverses.

(e) If f and g are inverse functions and f is continuous, must g be continuous? Give a reasonable informal argument to support your answer.

(f) If f and g are inverse functions and f is differentiable, must g be differentiable? Give a reasonable informal argument to support your answer.

2. (a) State the restrictions on the domains of $\sin x$, $\cos x$, $\tan x$, and $\sec x$ that are imposed to make those functions one-to-one in the definitions of $\sin^{-1} x$, $\cos^{-1} x$, $\tan^{-1} x$, and $\sec^{-1} x$.

(b) Sketch the graphs of the restricted trigonometric functions in part (a) and their inverses.

3. In each part, find $f^{-1}(x)$ if the inverse exists.

(a) $f(x) = 8x^3 - 1$ **(b)** $f(x) = x^2 - 2x + 1$

(c) $f(x) = (e^x)^2 + 1$ **(d)** $f(x) = (x + 2)/(x - 1)$

4. Let $f(x) = (ax + b)/(cx + d)$. What conditions on a, b, c, d guarantee that f^{-1} exists? Find $f^{-1}(x)$.

5. Express the following function as a rational function of x:

$$3 \ln \left(e^{2x}(e^x)^3 \right) + 2 \exp(\ln 1)$$

6. In each part, find the exact numerical value of the given expression.

(a) $\cos[\cos^{-1}(4/5) + \sin^{-1}(5/13)]$

(b) $\sin[\sin^{-1}(4/5) + \cos^{-1}(5/13)]$

7. In each part, find $(f^{-1})'(x)$ using Formula (4) of Section 4.1, and check your answer by differentiating f^{-1} directly.

(a) $f(x) = 3/(x + 1)$ **(b)** $f(x) = \sqrt{e^x}$

8. Suppose that $y = Ce^{kt}$, where C and k are constants, and let $Y = \ln y$. Show that the graph of Y versus t is a line, and state its slope and Y-intercept.

~ 9. (a) Sketch the curves $y = \pm e^{-x/2}$ and $y = e^{-x/2} \sin 2x$ for $-\pi/2 \le x \le 3\pi/2$ in the same coordinate system, and check your work using a graphing utility.

(b) Find all x-intercepts of the curve $y = e^{-x/2} \sin 2x$ in the stated interval, and find the x-coordinates of all points where this curve intersects the curves $y = \pm e^{-x/2}$.

10. In each part, sketch the graph, and check your work with a graphing utility.
   (a) $f(x) = 3 \sin^{-1}(x/2)$
   (b) $f(x) = \cos^{-1} x - \pi/2$
   (c) $f(x) = 2 \tan^{-1}(-3x)$
   (d) $f(x) = \cos^{-1} x + \sin^{-1} x$

11. (a) Show that the graphs of $y = \ln x$ and $y = x^{0.2}$ intersect.
   (b) Approximate the solution(s) of the equation $\ln x = x^{0.2}$ to three decimal places.

12. (a) Show that for $x > 0$ and $k \neq 0$ the equations
   $$x^k = e^x \quad \text{and} \quad \frac{\ln x}{x} = \frac{1}{k}$$
   have the same solutions.
   (b) Use the graph of $y = (\ln x)/x$ to determine the values of k for which the equation $x^k = e^x$ has two distinct positive solutions.
   (c) Estimate the positive solution(s) of $x^8 = e^x$.

13. Show that the rate of change of $y = 5000e^{1.07x}$ is proportional to y.

14. Show that the rate of change of $y = 3^{2x}5^{7x}$ is proportional to y.

15. In each part, use any appropriate method to find dy/dx.
   (a) $y = e^{\ln(x^3+1)}$
   (b) $y = \dfrac{a}{1 + be^{-x}}$
   (c) $y = \ln\left(\dfrac{\sqrt{x}\sqrt[3]{x+1}}{\sin x \sec x}\right)$
   (d) $y = (1 + x)^{1/x}$
   (e) $y = x^{(e^x)}$
   (f) $y = \ln\left(\dfrac{1 + e^x + e^{2x}}{1 - e^{3x}}\right)$

16. Show that the function $y = e^{ax} \sin bx$ satisfies
   $$y'' - 2ay' + (a^2 + b^2)y = 0$$
   for any real constants a and b.

17. Show that the function $y = \tan^{-1} x$ satisfies
   $$y'' = -2 \sin y \cos^3 y$$

18. (a) Suppose that the graph of $y = \log x$ is drawn with equal scales of 1 inch per unit in both the x- and y-directions. If a bug wants to walk along the graph until it reaches a height of 5 ft above the x-axis, how many miles to the right of the origin will it have to travel?
   (b) Suppose that the graph of $y = 10^x$ is drawn with equal scales of 1 inch per unit in both the x- and y-directions. If a bug wants to walk along the graph until it reaches a height of 100 mi above the x-axis, how many feet to the right of the origin will it have to travel?

19. Find the value of b so that the line $y = x$ is tangent to the graph of $y = \log_b x$. Confirm your result by graphing both $y = x$ and $y = \log_b x$ in the same coordinate system.

20. In each part, find the value of k for which the graphs of $y = f(x)$ and $y = \ln x$ share a common tangent line at their point of intersection. Confirm your result by graphing $y = f(x)$ and $y = \ln x$ in the same coordinate system.
   (a) $f(x) = \sqrt{x} + k$
   (b) $f(x) = k\sqrt{x}$

21. A particle is moving along the curve $y = x \ln x$. Find all values of x at which the rate of change of y with respect to time is three times that of x. [Assume that dx/dt is never zero.]

22. Find a point on the graph of $y = e^{3x}$ at which the tangent line passes through the origin.

23. The equilibrium constant k of a balanced chemical reaction changes with the absolute temperature T according to the law
   $$k = k_0 \exp\left(-\frac{q(T - T_0)}{2T_0 T}\right)$$
   where k_0, q, and T_0 are constants. Find the rate of change of k with respect to T.

24. Recall from Section 4.2 that the loudness β of a sound in decibels (dB) is given by $\beta = 10 \log(I/I_0)$, where I is the intensity of the sound in watts per square meter (W/m^2) and I_0 is a constant that is approximately the intensity of a sound at the threshold of human hearing. Find the rate of change of β with respect to I at the point where
   (a) $I/I_0 = 10$
   (b) $I/I_0 = 100$
   (c) $I/I_0 = 1000$

25. Suppose that the population of deer on an island is modeled by the equation
   $$P(t) = \frac{95}{5 - 4e^{-t/4}}$$
   where $P(t)$ is the number of deer t weeks after an initial observation at time $t = 0$.
   (a) Use a graphing utility to graph the function $P(t)$.
   (b) In words, explain what happens to the population over time. Check your conclusion by finding $\lim\limits_{t \to +\infty} P(t)$.
   (c) In words, what happens to the *rate* of population growth over time? Check your conclusion by graphing $P'(t)$.

26. Suppose that the population of oxygen-dependent bacteria in a pond is modeled by the equation
   $$P(t) = \frac{60}{5 + 7e^{-t}}$$
   where $P(t)$ is the population (in billions) t days after an initial observation at time $t = 0$.
   (a) Use a graphing utility to graph the function $P(t)$.
   (b) In words, explain what happens to the population over time. Check your conclusion by finding $\lim\limits_{t \to +\infty} P(t)$.
   (c) In words, what happens to the *rate* of population growth over time? Check your conclusion by graphing $P'(t)$.

27. (a) Make a conjecture about the shape of the graph of $y = \frac{1}{2}x - \ln x$, and draw a rough sketch.

(b) Check your conjecture by graphing the equation over the interval $0 < x < 5$ with a graphing utility.

(c) Show that the slopes of the tangent lines to the curve at $x = 1$ and $x = e$ have opposite signs.

(d) What does part (c) imply about the existence of a horizontal tangent line to the curve? Explain your reasoning.

(e) Find the exact x-coordinates of all horizontal tangent lines to the curve.

28. Suppose that $\lim f(x) = \pm\infty$ and $\lim g(x) = \pm\infty$. In each of the four possible cases, state whether $\lim [f(x) - g(x)]$ is an indeterminate form, and give a reasonable informal argument to support your answer.

29. (a) Under what conditions will a limit of the form

$$\lim_{x \to a} [f(x)/g(x)]$$

be an indeterminate form?

(b) If $\lim_{x \to a} g(x) = 0$, must $\lim_{x \to a} [f(x)/g(x)]$ be an indeterminate form? Give some examples to support your answer.

30. In each part, find the limit.

(a) $\lim_{x \to +\infty} (e^x - x^2)$

(b) $\lim_{x \to 1} \sqrt{\dfrac{\ln x}{x^4 - 1}}$

(c) $\lim_{x \to 0} \dfrac{a^x - 1}{x}, \quad a > 0$

5

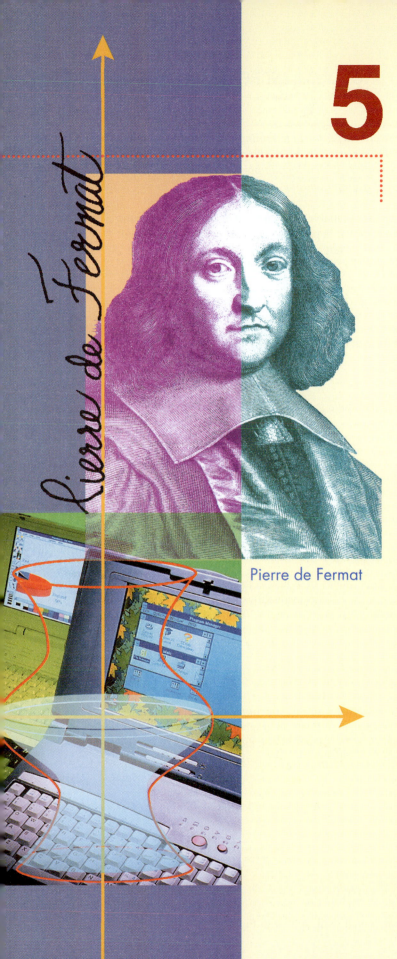

Pierre de Fermat

THE DERIVATIVE IN GRAPHING AND APPLICATIONS

*I*n this chapter we will study various applications of the derivative. For example, we will use methods of calculus to analyze functions and their graphs. In the process, we will show how calculus and graphing utilities, working together, can provide most of the important information about the behavior of functions. Another important application of the derivative will be in the solution of *optimization problems*. For example, if time is the main consideration in a problem, we might be interested in finding the quickest way to perform a task, and if cost is the main consideration, we might be interested in finding the least expensive way to perform a task. Mathematically, optimization problems can be reduced to finding the largest or smallest value of a function on some interval, and determining where the largest or smallest value occurs. Using the derivative, we will develop the mathematical tools necessary for solving such problems. We will also use the derivative to study the motion of a particle moving along a line, and we will show how the derivative can help us to approximate solutions of equations.

5.1 ANALYSIS OF FUNCTIONS I: INCREASE, DECREASE, AND CONCAVITY

Although graphing utilities are useful for determining the general shape of a graph, many problems require more precision than graphing utilities are capable of producing. The purpose of this section is to develop mathematical tools that can be used to determine the exact shape of a graph and the precise locations of its key features.

INCREASING AND DECREASING FUNCTIONS

Suppose that a function f is differentiable at x_0 and that $f'(x_0) > 0$. Since the slope of the graph of f at the point $P(x_0, f(x_0))$ is positive, we would expect that a point $Q(x, f(x))$ on the graph of f that is just to the left of P would be *lower* than P, and we would expect that Q would be *higher* than P if Q is just to the right of P. Analytically, to see why this is the case, recall that

$$f'(x_0) = \lim_{x \to x_0} \frac{f(x) - f(x_0)}{x - x_0}$$

(Definition 3.2.1 with x_1 replaced by x). Since $0 < f'(x_0)$, it follows that

$$0 < \frac{f(x) - f(x_0)}{x - x_0}$$

for values of x very close to (but not equal to) x_0. However, for the difference quotient

$$\frac{f(x) - f(x_0)}{x - x_0}$$

to be positive, its numerator $f(x) - f(x_0)$ and its denominator $x - x_0$ must have the same sign. Therefore, for values of x very close to x_0, we must have

$$f(x) - f(x_0) < 0 \quad \text{when} \quad x - x_0 < 0$$

and

$$0 < f(x) - f(x_0) \quad \text{when} \quad 0 < x - x_0$$

Equivalently, $f(x) < f(x_0)$ for values of x just to the left of x_0, and $f(x_0) < f(x)$ for values of x just to the right of x_0. These inequalities confirm our expectation about the relative positions of P and Q. Similarly, if $f'(x_0) < 0$, then $f(x) > f(x_0)$ for values of x just to the left of x_0, and $f(x_0) > f(x)$ for values of x just to the right of x_0. Geometrically, this means that our point Q would be *higher* than P if Q is just to the left of P, and that Q would be *lower* than P if Q is just to the right of P.

Our next goal is to relate the sign of the derivative of a function f and the relative positions of points on the graph of f over an entire interval. The terms *increasing*, *decreasing*, and *constant* are used to describe the behavior of a function over an interval as we travel left to right along its graph. For example, the function graphed in Figure 5.1.1 can be described as increasing on the interval $(-\infty, 0]$, decreasing on the interval $[0, 2]$, increasing again on the interval $[2, 4]$, and constant on the interval $[4, +\infty)$.

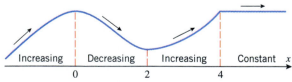

Figure 5.1.1

The following definition, which is illustrated in Figure 5.1.2, expresses these intuitive ideas precisely.

5.1.1   DEFINITION.   Let f be defined on an interval, and let x_1 and x_2 denote numbers in that interval.

(a) f is ***increasing*** on the interval if $f(x_1) < f(x_2)$ whenever $x_1 < x_2$.

(b) f is ***decreasing*** on the interval if $f(x_1) > f(x_2)$ whenever $x_1 < x_2$.

(c) f is ***constant*** on the interval if $f(x_1) = f(x_2)$ for all x_1 and x_2.

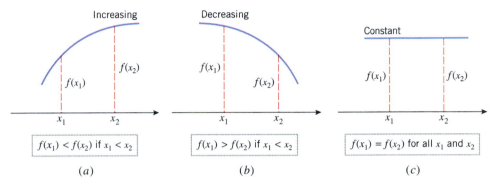

Figure 5.1.2

Figure 5.1.3 suggests that a differentiable function f is increasing on any interval where its graph has positive slope, is decreasing on any interval where its graph has negative slope, and is constant on any interval where its graph has zero slope. This intuitive observation suggests the following important theorem that will be proved in Section 5.8.

5.1.2   THEOREM.   *Let f be a function that is continuous on a closed interval $[a, b]$ and differentiable on the open interval (a, b).*

(a)   *If $f'(x) > 0$ for every value of x in (a, b), then f is increasing on $[a, b]$.*

(b)   *If $f'(x) < 0$ for every value of x in (a, b), then f is decreasing on $[a, b]$.*

(c)   *If $f'(x) = 0$ for every value of x in (a, b), then f is constant on $[a, b]$.*

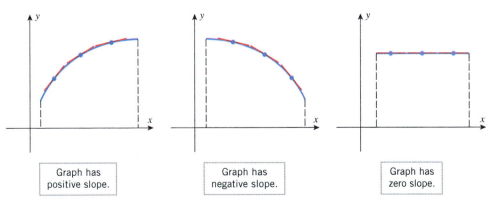

Figure 5.1.3

REMARK.   Observe that in Theorem 5.1.2 it is only necessary to examine the derivative of f on the open interval (a, b) to determine whether f is increasing, decreasing, or constant on the closed interval $[a, b]$. Moreover, although this theorem was stated for a closed interval $[a, b]$, it is applicable to any interval I on which f is continuous and inside of which f is differentiable. For example, if f is continuous on $[a, +\infty)$ and $f'(x) > 0$ for each x in the interval $(a, +\infty)$, then f is increasing on $[a, +\infty)$; and if $f'(x) < 0$ on $(-\infty, +\infty)$, then f is decreasing on $(-\infty, +\infty)$ [the continuity on $(-\infty, +\infty)$ follows from the differentiability].

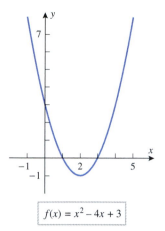

$$f(x) = x^2 - 4x + 3$$

Figure 5.1.4

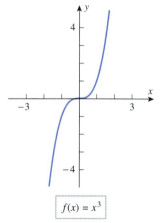

$$f(x) = x^3$$

Figure 5.1.5

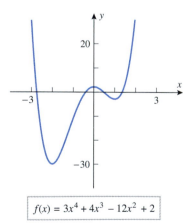

$$f(x) = 3x^4 + 4x^3 - 12x^2 + 2$$

Figure 5.1.6

Example 1   Find the intervals on which the following functions are increasing and the intervals on which they are decreasing.

(a) $f(x) = x^2 - 4x + 3$ (b) $f(x) = x^3$

Solution (a). The graph of f in Figure 5.1.4 suggests that f is decreasing for $x \leq 2$ and increasing for $x \geq 2$. To confirm this, we differentiate f to obtain

$$f'(x) = 2x - 4 = 2(x - 2)$$

It follows that

$$f'(x) < 0 \quad \text{if} \quad -\infty < x < 2$$
$$f'(x) > 0 \quad \text{if} \quad 2 < x < +\infty$$

Since f is continuous at $x = 2$, it follows from Theorem 5.1.2 and the subsequent remark that

f is decreasing on $(-\infty, 2]$

f is increasing on $[2, +\infty)$

These conclusions are consistent with the graph of f in Figure 5.1.4.

Solution (b). The graph of f in Figure 5.1.5 suggests that f is increasing over the entire x-axis. To confirm this, we differentiate f to obtain $f'(x) = 3x^2$. Thus,

$$f'(x) > 0 \quad \text{if} \quad -\infty < x < 0$$
$$f'(x) > 0 \quad \text{if} \quad 0 < x < +\infty$$

Since f is continuous at $x = 0$,

f is increasing on $(-\infty, 0]$

f is increasing on $[0, +\infty)$

Hence f is increasing over the entire interval $(-\infty, +\infty)$, which is consistent with the graph in Figure 5.1.5 (see Exercise 53).   ◄

Example 2

(a)   Use the graph of $f(x) = 3x^4 + 4x^3 - 12x^2 + 2$ in Figure 5.1.6 to make a conjecture about the intervals on which f is increasing or decreasing.

(b)   Use Theorem 5.1.2 to determine whether your conjecture is correct.

Solution (a). The graph suggests that f is decreasing if $x \leq -2$, increasing if $-2 \leq x \leq 0$, decreasing if $0 \leq x \leq 1$, and increasing if $x \geq 1$.

Solution (b). Differentiating f we obtain

$$f'(x) = 12x^3 + 12x^2 - 24x = 12x(x^2 + x - 2) = 12x(x + 2)(x - 1)$$

The sign analysis of f' in Table 5.1.1 can be obtained using the method of test values discussed in Appendix A. The conclusions in that table confirm the conjecture in part (a).   ◄

Table 5.1.1

INTERVAL	$(12x)(x+2)(x-1)$	$f'(x)$	CONCLUSION
$x < -2$	$(-)(-)(-)$	$-$	f is decreasing on $(-\infty, -2]$
$-2 < x < 0$	$(-)(+)(-)$	$+$	f is increasing on $[-2, 0]$
$0 < x < 1$	$(+)(+)(-)$	$-$	f is decreasing on $[0, 1]$
$1 < x$	$(+)(+)(+)$	$+$	f is increasing on $[1, +\infty)$

CONCAVITY

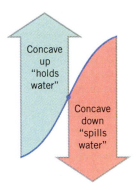

Figure 5.1.7

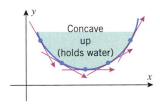

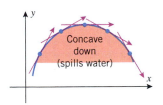

Figure 5.1.8

Although the sign of the derivative of f reveals where the graph of f is increasing or decreasing, it does not reveal the direction of *curvature*. For example, on both sides of the point in Figure 5.1.7 the graph is increasing, but on the left side it has an upward curvature ("holds water") and on the right side it has a downward curvature ("spills water"). On intervals where the graph of f has upward curvature we say that f is *concave up*, and on intervals where the graph has downward curvature we say that f is *concave down*.

For differentiable functions, the direction of curvature can be characterized in terms of the tangent lines in two ways: As suggested by Figure 5.1.8, the graph of a function f has upward curvature on intervals where the graph lies above its tangent lines, and it has downward curvature on intervals where it lies below its tangent lines. Alternatively, the graph has upward curvature on intervals where the tangent lines have increasing slopes and downward curvature on intervals where they have decreasing slopes. We will use this latter characterization as our formal definition.

> **5.1.3** DEFINITION. If f is differentiable on an open interval I, then f is said to be *concave up* on I if f' is increasing on I, and f is said to be *concave down* on I if f' is decreasing on I.

To apply this definition we need some way to determine the intervals on which f' is increasing or decreasing. One way to do this is to apply Theorem 5.1.2 (and the remark that follows it) to the function f'. It follows from that theorem and remark that f' will be increasing where its derivative f'' is positive and will be decreasing where its derivative f'' is negative. This is the idea behind the following theorem.

> **5.1.4** THEOREM. *Let f be twice differentiable on an open interval I.*
> *(a) If $f''(x) > 0$ on I, then f is concave up on I.*
> *(b) If $f''(x) < 0$ on I, then f is concave down on I.*

Example 3 Find open intervals on which the following functions are concave up and open intervals on which they are concave down.

(a) $f(x) = x^2 - 4x + 3$ (b) $f(x) = x^3$ (c) $f(x) = x^3 - 3x^2 + 1$

Solution (a). Calculating the first two derivatives we obtain

$$f'(x) = 2x - 4 \quad \text{and} \quad f''(x) = 2$$

Since $f''(x) > 0$ for all x, the function f is concave up on $(-\infty, +\infty)$. This is consistent with Figure 5.1.4.

Solution (b). Calculating the first two derivatives we obtain

$$f'(x) = 3x^2 \quad \text{and} \quad f''(x) = 6x$$

Since $f''(x) < 0$ if $x < 0$ and $f''(x) > 0$ if $x > 0$, the function f is concave down on $(-\infty, 0)$ and concave up on $(0, +\infty)$. This is consistent with Figure 5.1.5.

Solution (c). Calculating the first two derivatives we obtain

$$f'(x) = 3x^2 - 6x \quad \text{and} \quad f''(x) = 6x - 6 = 6(x - 1)$$

Since $f''(x) > 0$ if $x > 1$ and $f''(x) < 0$ if $x < 1$, we conclude that

f is concave up on $(1, +\infty)$

f is concave down on $(-\infty, 1)$

which is consistent with the graph in Figure 5.1.9. ◀

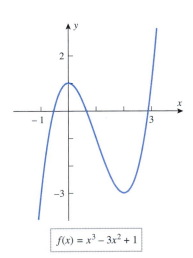

$f(x) = x^3 - 3x^2 + 1$

Figure 5.1.9

INFLECTION POINTS

Points where a graph changes from concave up to concave down, or vice versa, are of special interest, so there is some terminology associated with them.

5.1.5 DEFINITION. If f is continuous on an open interval containing a value x_0, and if f changes the direction of its concavity at the point $(x_0, f(x_0))$, then we say that f has an **inflection point at x_0**, and we call the point $(x_0, f(x_0))$ on the graph of f an **inflection point** of f (Figure 5.1.10).

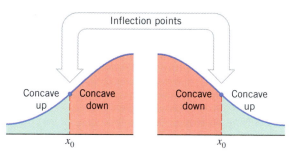

Figure 5.1.10

For example, the function $f(x) = x^3$ has an inflection point at $x = 0$ (Figure 5.1.5), the function $f(x) = x^3 - 3x^2 + 1$ has an inflection point at $x = 1$ (Figure 5.1.9), and the function $f(x) = x^2 - 4x + 3$ has no inflection points (Figure 5.1.4).

Example 4 Use the graph in Figure 5.1.6 to make rough estimates of the locations of the inflection points of $f(x) = 3x^4 + 4x^3 - 12x^2 + 2$, and check your estimates by finding the exact locations of the inflection points.

Solution. The graph changes from concave up to concave down somewhere between -2 and -1, say roughly at $x = -1.25$; and the graph changes from concave down to concave up somewhere between 0 and 1, say roughly at $x = 0.5$. To find the exact locations of the inflection points, we start by calculating the second derivative of f:

$$f'(x) = 12x^3 + 12x^2 - 24x$$
$$f''(x) = 36x^2 + 24x - 24 = 12(3x^2 + 2x - 2)$$

We could analyze the sign of f'' by factoring this function and applying the method of test values (as in Table 5.1.1). However, here is another approach. The graph of f'' is a parabola that opens up, and the quadratic formula shows that the equation $f''(x) = 0$ has the roots

$$x = \frac{-1 - \sqrt{7}}{3} \approx -1.22 \quad \text{and} \quad x = \frac{-1 + \sqrt{7}}{3} \approx 0.55 \tag{1}$$

(verify). Thus, from the rough graph of f'' in Figure 5.1.11 we obtain the sign analysis of f'' in Table 5.1.2; this implies that f has inflection points at the values in (1). ◄

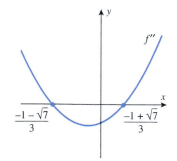

Figure 5.1.11

Table 5.1.2

INTERVAL	SIGN OF f''	CONCLUSION
$x < \dfrac{-1 - \sqrt{7}}{3}$	+	f is concave up
$\dfrac{-1 - \sqrt{7}}{3} < x < \dfrac{-1 + \sqrt{7}}{3}$	–	f is concave down
$\dfrac{-1 + \sqrt{7}}{3} < x$	+	f is concave up

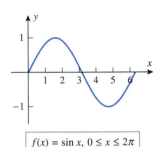

$f(x) = xe^{-x}$

(a)

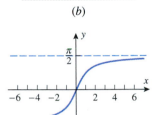

$f(x) = \sin x, \; 0 \le x \le 2\pi$

(b)

$f(x) = \tan^{-1} x$

(c)

Figure 5.1.12

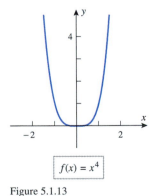

$f(x) = x^4$

Figure 5.1.13

INFLECTION POINTS IN
APPLICATIONS

Example 5 Find the inflection points of the following functions, and confirm that your results are consistent with the graphs of the functions.

(a) $f(x) = xe^{-x}$ (b) $f(x) = \sin x, \quad 0 \le x \le 2\pi$ (c) $f(x) = \tan^{-1} x$

Solution (a). Calculating the first two derivatives of f we obtain

$$f'(x) = (1 - x)e^{-x}, \quad f''(x) = (x - 2)e^{-x}$$

(verify). Keeping in mind that e^{-x} is always positive, it follows that the sign of f'' is determined by the factor $x - 2$. Thus, $f''(x) < 0$ if $x < 2$, and $f''(x) > 0$ if $x > 2$, which implies that the graph is concave down for $x < 2$ and concave up for $x > 2$. Thus, there is an inflection point at $x = 2$ (Figure 5.1.12a).

Solution (b). Calculating the first two derivatives of f we obtain

$$f'(x) = \cos x, \quad f''(x) = -\sin x$$

Thus, $f''(x) < 0$ if $0 < x < \pi$, and $f''(x) > 0$ if $\pi < x < 2\pi$, which implies that the graph is concave down for $0 < x < \pi$ and concave up for $\pi < x < 2\pi$. Thus, there is an inflection point at $x = \pi \approx 3.14$ (Figure 5.1.12b).

Solution (c). Calculating the first two derivatives of f we obtain

$$f'(x) = \frac{1}{1 + x^2}, \quad f''(x) = -\frac{2x}{\left(1 + x^2\right)^2}$$

(verify). Thus, $f''(x) > 0$ if $x < 0$, and $f''(x) < 0$ if $x > 0$, which implies that the graph is concave up for $x < 0$ and concave down for $x > 0$. Thus, there is an inflection point at $x = 0$ (Figure 5.1.12c). ◄

FOR THE READER. If you have a CAS, devise a method for using it to find exact values for the inflection points of a function f, and use your method to find the inflection points of $f(x) = x/(x^2 + 1)$. Verify that your results are consistent with the graph of f.

In the preceding examples the inflection points of f occurred where $f''(x) = 0$. However, inflection points do not always occur where $f''(x) = 0$. Here is a specific example.

Example 6 Find the inflection points, if any, of $f(x) = x^4$.

Solution. Calculating the first two derivatives of f we obtain

$$f'(x) = 4x^3, \quad f''(x) = 12x^2$$

Here $f''(x) > 0$ for $x < 0$ and for $x > 0$, which implies that f is concave up for $x < 0$ and for $x > 0$. [In fact, f is concave up on $(-\infty, +\infty)$.] Thus, there are no inflection points; and in particular, there is no inflection point at $x = 0$, even though $f''(0) = 0$ (Figure 5.1.13). ◄

FOR THE READER. An inflection point may occur at a point of nondifferentiability. Verify that this is the case for $f(x) = x^{1/3}$ at $x = 0$.

Up to now we have viewed the inflection points of a curve $y = f(x)$ as those points where the curve changes the direction of its concavity. However, inflection points also mark the points on the curve where the slopes of the tangent lines change from increasing to decreasing, or vice versa (Figure 5.1.14); stated another way:

Inflection points mark the places on the curve $y = f(x)$ where the rate of change of y with respect to x changes from increasing to decreasing, or vice versa.

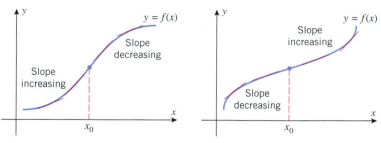

Figure 5.1.14

Note that we are dealing with a rather subtle concept here—a change of a rate of change. However, the following physical example should help to clarify the idea: Suppose that water is added to the flask in Figure 5.1.15 in such a way that the volume increases at a constant rate, and let us examine the rate at which the water level y rises with the time t. Initially, the level y will rise at a slow rate because of the wide base. However, as the diameter of the flask narrows, the rate at which the level y rises will increase until the level is at the narrow point in the neck. From that point on the rate at which the level rises will decrease as the diameter gets wider and wider. Thus, the narrow point in the neck is the point at which the rate of change of y with respect to t changes from increasing to decreasing.

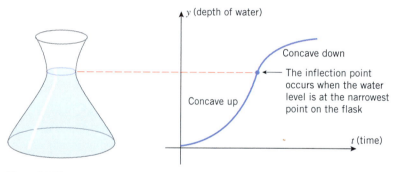

Figure 5.1.15

LOGISTIC CURVES

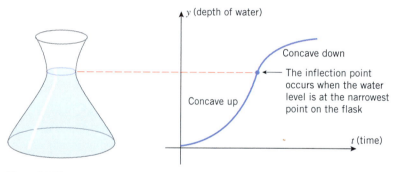

Logistic growth curve

Figure 5.1.16

When a population grows in an environment in which space or food is limited, the graph of population versus time is typically an S-shaped curve of the form shown in Figure 5.1.16. The scenario described by this curve is a population that grows slowly at first and then more and more rapidly as the number of individuals producing offspring increases. However, at a certain point in time (where the inflection point occurs) the environmental factors begin to show their effect, and the growth rate begins a steady decline. Over an extended period of time the population approaches a limiting value that represents the upper limit on the number of individuals that the available space or food can sustain. Population growth curves of this type are called *logistic growth curves*.

Example 7   We will show in a later chapter that logistic growth curves arise from equations of the form

$$y = \frac{L}{1 + Ae^{-kt}} \tag{2}$$

where y is the population at time t $(t \geq 0)$ and A, k, and L are positive constants. Show that Figure 5.1.17 correctly describes the graph of this equation.

Solution.   We leave it for you to confirm that at time $t = 0$ the value of y is

$$y = \frac{L}{1 + A}$$

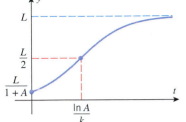

Figure 5.1.17

and that for $t \geq 0$ the population y satisfies

$$\frac{L}{1 + A} \leq y < L$$

This is consistent with the graph in Figure 5.1.17. The horizontal asymptote at $y = L$ is confirmed by the limit

$$\lim_{t \to +\infty} \frac{L}{1 + Ae^{-kt}} = \frac{L}{1 + 0} = L$$

Physically, L represents the upper limit on the size of the population.

To investigate intervals of increase or decrease, concavity, and inflection points, we need the first and second derivatives of y with respect to t. We leave it for you to confirm that

$$\frac{dy}{dt} = \frac{k}{L} y (L - y) \tag{3}$$

$$\frac{d^2 y}{dt^2} = \frac{k^2}{L^2} y (L - y)(L - 2y) \tag{4}$$

Since $k > 0$, $y > 0$, and $L - y > 0$, it follows from (3) that $dy/dt > 0$ for all t. Thus, y is always increasing, which is consistent with Figure 5.1.17.

Since $y > 0$ and $L - y > 0$, it follows from (4) that

$$\frac{d^2 y}{dt^2} > 0 \quad \text{if} \quad L - 2y > 0$$

$$\frac{d^2 y}{dt^2} < 0 \quad \text{if} \quad L - 2y < 0$$

Thus, the graph of y versus t is concave up if $y < L/2$, concave down if $y > L/2$, and has an inflection point where $y = L/2$, all of which is consistent with Figure 5.1.17.

Finally, we leave it as an exercise for you to confirm that the inflection point occurs at time

$$t = \frac{1}{k} \ln A = \frac{\ln A}{k} \tag{5}$$

by solving the equation

$$\frac{L}{2} = \frac{L}{1 + Ae^{-kt}}$$

for t. ◀

EXERCISE SET 5.1   ⌇ Graphing Utility   [c] CAS

1. In each part, sketch the graph of a function f with the stated properties, and discuss the signs of f' and f''.
   (a) The function f is concave up and increasing on the interval $(-\infty, +\infty)$.
   (b) The function f is concave down and increasing on the interval $(-\infty, +\infty)$.
   (c) The function f is concave up and decreasing on the interval $(-\infty, +\infty)$.
   (d) The function f is concave down and decreasing on the interval $(-\infty, +\infty)$.

2. In each part, sketch the graph of a function f with the stated properties.
   (a) f is increasing on $(-\infty, +\infty)$, has an inflection point at the origin, and is concave up on $(0, +\infty)$.
   (b) f is increasing on $(-\infty, +\infty)$, has an inflection point at the origin, and is concave down on $(0, +\infty)$.
   (c) f is decreasing on $(-\infty, +\infty)$, has an inflection point at the origin, and is concave up on $(0, +\infty)$.
   (d) f is decreasing on $(-\infty, +\infty)$, has an inflection point at the origin, and is concave down on $(0, +\infty)$.

3. Use the graph of the equation $y = f(x)$ in the accompanying figure to find the signs of dy/dx and d^2y/dx^2 at the points A, B, and C.

4. Use the graph of the equation $y = f'(x)$ in the accompanying figure to find the signs of dy/dx and d^2y/dx^2 at the points A, B, and C.

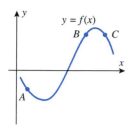

Figure Ex-3

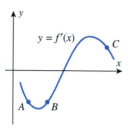

Figure Ex-4

5. Use the graph of $y = f''(x)$ in the accompanying figure to determine the x-coordinates of all inflection points of f. Explain your reasoning.

6. Use the graph of $y = f'(x)$ in the accompanying figure to replace the question mark with $<$, $=$, or $>$, as appropriate. Explain your reasoning.
   (a) $f(0)$? $f(1)$ (b) $f(1)$? $f(2)$ (c) $f'(0)$? 0
   (d) $f'(1)$? 0 (e) $f''(0)$? 0 (f) $f''(2)$? 0

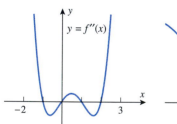

Figure Ex-5

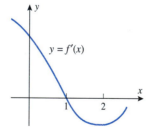

Figure Ex-6

7. In each part, use the graph of $y = f(x)$ in the accompanying figure to find the requested information.
   (a) Find the intervals on which f is increasing.
   (b) Find the intervals on which f is decreasing.
   (c) Find the open intervals on which f is concave up.
   (d) Find the open intervals on which f is concave down.
   (e) Find all values of x at which f has an inflection point.

Figure Ex-7

8. Use the graph in Exercise 7 to make a table that shows the signs of f' and f'' over the intervals $(1, 2)$, $(2, 3)$, $(3, 4)$, $(4, 5)$, $(5, 6)$, and $(6, 7)$.

In Exercises 9 and 10, a sign chart is presented for the first and second derivatives of a function f. Assuming that f is continuous everywhere, find: (a) the intervals on which f is increasing, (b) the intervals on which f is decreasing, (c) the open intervals on which f is concave up, (d) the open intervals on which f is concave down, and (e) the x-coordinates of all inflection points.

9.

INTERVAL	SIGN OF $f'(x)$	SIGN OF $f''(x)$
$x < 1$	$-$	$+$
$1 < x < 2$	$+$	$+$
$2 < x < 3$	$+$	$-$
$3 < x < 4$	$-$	$-$
$4 < x$	$-$	$+$

10.

INTERVAL	SIGN OF $f'(x)$	SIGN OF $f''(x)$
$x < 1$	$+$	$+$
$1 < x < 3$	$+$	$-$
$3 < x$	$+$	$+$

In Exercises 11–26, find: (a) the intervals on which f is increasing, (b) the intervals on which f is decreasing, (c) the open intervals on which f is concave up, (d) the open intervals on which f is concave down, and (e) the x-coordinates of all inflection points.

11. $f(x) = x^2 - 5x + 6$ **12.** $f(x) = 4 - 3x - x^2$
13. $f(x) = (x + 2)^3$ **14.** $f(x) = 5 + 12x - x^3$
15. $f(x) = 3x^4 - 4x^3$ **16.** $f(x) = x^4 - 8x^2 + 16$
17. $f(x) = \dfrac{x^2}{x^2 + 2}$ **18.** $f(x) = \dfrac{x}{x^2 + 2}$
19. $f(x) = \sqrt[3]{x + 2}$ **20.** $f(x) = x^{2/3}$
21. $f(x) = x^{1/3}(x + 4)$ **22.** $f(x) = x^{4/3} - x^{1/3}$
23. $f(x) = e^{-x^2/2}$ **24.** $f(x) = xe^{x^2}$
25. $f(x) = \ln(1 + x^2)$ **26.** $f(x) = x^2 \ln x$

In Exercises 27–32, analyze the trigonometric function f over the specified interval, stating where f is increasing, decreasing, concave up, and concave down, and stating the x-coordinates of all inflection points. Confirm that your results are consistent with the graph of f generated with a graphing utility.

27. $f(x) = \cos x$; $[0, 2\pi]$ **28.** $f(x) = \sin^2 2x$; $[0, \pi]$
29. $f(x) = \tan x$; $(-\pi/2, \pi/2)$
30. $f(x) = 2x + \cot x$; $(0, \pi)$
31. $f(x) = \sin x \cos x$; $[0, \pi]$

32. $f(x) = \cos^2 x - 2\sin x;\ [0, 2\pi]$

33. In each part sketch a continuous curve $y = f(x)$ with the stated properties.
  (a) $f(2) = 4,\ f'(2) = 0,\ f''(x) > 0$ for all x
  (b) $f(2) = 4,\ f'(2) = 0,\ f''(x) < 0$ for $x < 2,\ f''(x) > 0$ for $x > 2$
  (c) $f(2) = 4,\ f''(x) < 0$ for $x \neq 2$ and $\lim\limits_{x \to 2^+} f'(x) = +\infty,$
  $\lim\limits_{x \to 2^-} f'(x) = -\infty$

34. In each part sketch a continuous curve $y = f(x)$ with the stated properties.
  (a) $f(2) = 4,\ f'(2) = 0,\ f''(x) < 0$ for all x
  (b) $f(2) = 4,\ f'(2) = 0,\ f''(x) > 0$ for $x < 2,\ f''(x) < 0$ for $x > 2$
  (c) $f(2) = 4,\ f''(x) > 0$ for $x \neq 2$ and $\lim\limits_{x \to 2^+} f'(x) = -\infty,$
  $\lim\limits_{x \to 2^-} f'(x) = +\infty$

35. In each part, assume that a is a constant and find the inflection points, if any.
  (a) $f(x) = (x - a)^3$ (b) $f(x) = (x - a)^4$

36. Given that a is a constant and n is a positive integer, what can you say about the existence of inflection points of the function $f(x) = (x - a)^n$? Justify your answer.

If f is increasing on an interval $[0, b)$, then it follows from Definition 5.1.1 that $f(0) < f(x)$ for each x in the interval. Use this result in Exercises 37–42.

---

37. Show that $\sqrt[3]{1 + x} < 1 + \frac{1}{3}x$ if $x > 0$, and confirm the inequality with a graphing utility. [*Hint:* Show that the function $f(x) = 1 + \frac{1}{3}x - \sqrt[3]{1 + x}$ is increasing on $[0, +\infty)$.]

38. Show that $x < \tan x$ if $0 < x < \pi/2$, and confirm the inequality with a graphing utility. [*Hint:* Show that the function $f(x) = \tan x - x$ is increasing on $[0, \pi/2)$.]

39. Use a graphing utility to make a conjecture about the relative sizes of x and $\sin x$ for $x \geq 0$, and prove your conjecture.

40. Use a graphing utility to make a conjecture about the relative sizes of $1 - x^2/2$ and $\cos x$ for $x \geq 0$, and prove your conjecture. [*Hint:* Use the result of Exercise 39.]

41. (a) Show that $\ln(x + 1) \leq x$ if $x \geq 0$.
  (b) Show that $\ln(x + 1) \geq x - \frac{1}{2}x^2$ if $x \geq 0$.
  (c) Confirm the inequalities in parts (a) and (b) with a graphing utility.

42. (a) Show that $e^x \geq 1 + x$ if $x \geq 0$.
  (b) Show that $e^x \geq 1 + x + \frac{1}{2}x^2$ if $x \geq 0$.
  (c) Confirm the inequalities in parts (a) and (b) with a graphing utility.

In Exercises 43 and 44, use a graphing utility to generate the graphs of f' and f'' over the stated interval; then use those graphs to estimate the x-coordinates of the inflection points of f, the intervals on which f is concave up or down, and the intervals on which f is increasing or decreasing. Check your estimates by graphing f.

---

43. $f(x) = x^4 - 24x^2 + 12x,\ -5 \leq x \leq 5$

44. $f(x) = \dfrac{1}{1 + x^2},\ -5 \leq x \leq 5$

In Exercises 45 and 46, use a CAS to find f'' and to approximate the x-coordinates of the inflection points to six decimal places. Confirm that your answer is consistent with the graph of f.

---

45. $f(x) = \dfrac{10x - 3}{3x^2 - 5x + 8}$ **46.** $f(x) = \dfrac{x^3 - 8x + 7}{\sqrt{x^2 + 1}}$

47. Use Definition 5.1.1 to prove that $f(x) = x^2$ is increasing on $[0, +\infty)$.

48. Use Definition 5.1.1 to prove that $f(x) = 1/x$ is decreasing on $(0, +\infty)$.

49. In each part, determine whether the statement is true or false. If it is false, find functions for which the statement fails to hold.
  (a) If f and g are increasing on an interval, then so is $f + g$.
  (b) If f and g are increasing on an interval, then so is $f \cdot g$.

50. In each part, find functions f and g that are increasing on $(-\infty, +\infty)$ and for which $f - g$ has the stated property.
  (a) $f - g$ is decreasing on $(-\infty, +\infty)$.
  (b) $f - g$ is constant on $(-\infty, +\infty)$.
  (c) $f - g$ is increasing on $(-\infty, +\infty)$.

51. (a) Prove that a general cubic polynomial
  $$f(x) = ax^3 + bx^2 + cx + d \quad (a \neq 0)$$
  has exactly one inflection point.
  (b) Prove that if a cubic polynomial has three x-intercepts, then the inflection point occurs at the average value of the intercepts.
  (c) Use the result in part (b) to find the inflection point of the cubic polynomial $f(x) = x^3 - 3x^2 + 2x$, and check your result by using f'' to determine where f is concave up and concave down.

52. From Exercise 51, the polynomial $f(x) = x^3 + bx^2 + 1$ has one inflection point. Use a graphing utility to reach a conclusion about the effect of the constant b on the location of the inflection point. Use f'' to explain what you have observed graphically.

53. Use Definition 5.1.1 to prove:
  (a) If f is increasing on the intervals $(a, c]$ and $[c, b)$, then f is increasing on (a, b).
  (b) If f is decreasing on the intervals $(a, c]$ and $[c, b)$, then f is decreasing on (a, b).

54. Use part (a) of Exercise 53 to show that $f(x) = x + \sin x$ is increasing on the interval $(-\infty, +\infty)$.

55. Use part (b) of Exercise 53 to show that $f(x) = \cos x - x$ is decreasing on the interval $(-\infty, +\infty)$.

56. Let $y = 1/(1 + x^2)$. Find the values of x for which y is increasing most rapidly or decreasing most rapidly.

In Exercises 57 and 58, suppose that water is flowing at a constant rate into the container shown. Make a rough sketch of the graph of the water level y versus the time t. Make sure that your sketch conveys where the graph is concave up and concave down, and label the y-coordinates of the inflection points.

57.

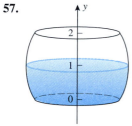

58.

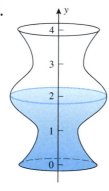

59. Suppose that a population y grows according to the logistic model given by Formula (2).
(a) At what rate is y increasing at time $t = 0$?
(b) In words, describe how the rate of growth of y varies with time.
(c) At what time is the population growing most rapidly?

 60. Suppose that the number of individuals at time t in a certain wildlife population is given by

$$N(t) = \frac{340}{1 + 9(0.77)^t}, \quad t \geq 0$$

where t is in years. Use a graphing utility to estimate the time at which the size of the population is increasing most rapidly.

 61. Suppose that the spread of a flu virus on a college campus is modeled by the function

$$y(t) = \frac{1000}{1 + 999e^{-0.9t}}$$

where $y(t)$ is the number of infected students at time t (in days, starting with $t = 0$). Use a graphing utility to estimate the day on which the virus is spreading most rapidly.

62. Assuming that A, k, and L are positive constants, verify that the graph of $y = L/(1 + Ae^{-kt})$ has an inflection point at $(\frac{1}{k} \ln A, \frac{1}{2}L)$.

63. Suppose that $g(x)$ is a function that is defined and differentiable for all real numbers x and that $g(x)$ has the following properties:
(i) $g(0) = 2$ and $g'(0) = -\frac{2}{3}$.
(ii) $g(4) = 3$ and $g'(4) = 3$.
(iii) $g(x)$ is concave up for $x < 4$ and concave down for $x > 4$.
(iv) $g(x) \geq -10$ for all x.
(a) How many zeros does g have?
(b) How many zeros does g' have?
(c) Exactly one of the following limits is possible:

$$\lim_{x \to +\infty} g'(x) = -5, \quad \lim_{x \to +\infty} g'(x) = 0, \quad \lim_{x \to +\infty} g'(x) = 5$$

Identify which of these results is possible and draw a rough sketch of the graph of such a function $g(x)$. Explain why the other two results are impossible.

5.2 ANALYSIS OF FUNCTIONS II: RELATIVE EXTREMA; FIRST AND SECOND DERIVATIVE TESTS

In this section we will discuss methods for finding the high and low points on the graph of a function. The ideas we develop here will have important applications.

RELATIVE MAXIMA AND MINIMA

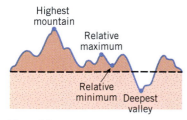

Figure 5.2.1

If we imagine the graph of a function f to be a two-dimensional mountain range with hills and valleys, then the tops of the hills are called *relative maxima*, and the bottoms of the valleys are called *relative minima* (Figure 5.2.1).

The relative maxima are the high points in their *immediate vicinity*, and the relative minima are the low points. Note that a relative maximum need not be the highest point in the entire mountain range, and a relative minimum need not be the lowest point—they are just high and low points *relative* to the nearby terrain. These ideas are captured in the following definition.

5.2.1   DEFINITION.   A function f is said to have a *relative maximum* at x_0 if there is an open interval containing x_0 on which $f(x_0)$ is the largest value, that is, $f(x_0) \geq f(x)$ for all x in the interval. Similarly, f is said to have a *relative minimum* at x_0 if there is an open interval containing x_0 on which $f(x_0)$ is the smallest value, that is, $f(x_0) \leq f(x)$ for all x in the interval. If f has either a relative maximum or a relative minimum at x_0, then f is said to have a *relative extremum* at x_0.

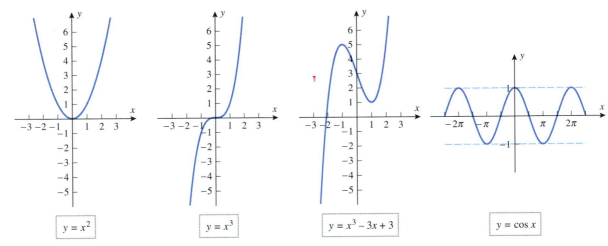

Figure 5.2.2

Example 1 Locate the relative extrema of the four functions graphed in Figure 5.2.2.

Solution.

(a) The function $f(x) = x^2$ has a relative minimum at $x = 0$ but no relative maxima.

(b) The function $f(x) = x^3$ has no relative extrema.

(c) The function $f(x) = x^3 - 3x + 3$ has a relative maximum at $x = -1$ and a relative minimum at $x = 1$.

(d) The function $f(x) = \cos x$ has relative maxima at all even multiples of π and relative minima at all odd multiples of π. ◄

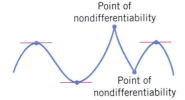

Point of nondifferentiability

Point of nondifferentiability

Figure 5.2.3

Points at which relative extrema occur can be viewed as the transition points that separate the regions where a graph is increasing from those where it is decreasing. As suggested by Figure 5.2.3, the relative extrema of a continuous function f occur at points where the graph of f either has a horizontal tangent line or is not differentiable. This is the content of the following theorem.

> **5.2.2 THEOREM.** *Suppose that f is a function defined on an open interval containing the number x_0. If f has a relative extremum at $x = x_0$, then either $f'(x_0) = 0$ or f is not differentiable at x_0.*

Proof. Assume that f has a relative extreme value at x_0. There are two possibilities— either f is differentiable at x_0 or it is not. If it is not, then we are done. If f is differentiable at x_0, then we must show that $f'(x_0) = 0$. It cannot be the case that $f'(x_0) > 0$, for then f would not have a relative extreme value at x_0. (See the discussion at the beginning of Section 5.1.) For the same reason, it cannot be the case that $f'(x_0) < 0$. We conclude that if f has a relative extreme value at x_0 and if f is differentiable at x_0, then $f'(x_0) = 0$. ∎

CRITICAL NUMBERS

Values in the domain of f at which either $f'(x) = 0$ or f is not differentiable are called *critical numbers* of f. Thus, Theorem 5.2.2 can be rephrased as follows:

> *If a function is defined on an open interval, its relative extrema on the interval, if any, occur at critical numbers.*

Sometimes we will want to distinguish critical numbers at which $f'(x) = 0$ from those at which f is not differentiable. We will call a point on the graph of f at which $f'(x) = 0$ a *stationary point* of f.

It is important not to read too much into Theorem 5.2.2—the theorem asserts that the set of critical numbers is a complete set of *candidates* for locations of relative extrema, but it does not say that a critical number must yield a relative extremum. That is, there may be critical numbers at which a relative extremum does not occur. For example, for the eight critical numbers shown in Figure 5.2.4, relative extrema occur at each x_0 marked in the top row, but not at any x_0 marked in the bottom row.

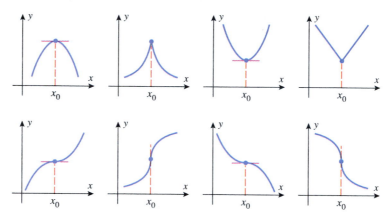

Figure 5.2.4

FIRST DERIVATIVE TEST

To develop an effective method for finding relative extrema of a function f, we need some criteria that will enable us to distinguish between the critical numbers where relative extrema occur and those where they do not. One such criterion can be motivated by examining the sign of the first derivative of f on each side of the eight critical numbers in Figure 5.2.4:

- At the two relative maxima in the top row, f' is positive to the left of x_0 and negative to the right.

- At the two relative minima in the top row, f' is negative to the left of x_0 and positive to the right.

- At the first two critical numbers in the bottom row, f' is positive on both sides of x_0.

- At the last two critical numbers in the bottom row, f' is negative on both sides of x_0.

These observations suggest that a function f will have relative extrema at those critical numbers where f' changes sign. Moreover, if the sign changes from positive to negative, then a relative maximum occurs; and if the sign changes from negative to positive, then a relative minimum occurs. This is the content of the following theorem.

5.2.3 THEOREM (*First Derivative Test*). *Suppose f is continuous at a critical number x_0.*

(a) *If $f'(x) > 0$ on an open interval extending left from x_0 and $f'(x) < 0$ on an open interval extending right from x_0, then f has a relative maximum at x_0.*

(b) *If $f'(x) < 0$ on an open interval extending left from x_0 and $f'(x) > 0$ on an open interval extending right from x_0, then f has a relative minimum at x_0.*

(c) *If $f'(x)$ has the same sign [either $f'(x) > 0$ or $f'(x) < 0$] on an open interval extending left from x_0 and on an open interval extending right from x_0, then f does not have a relative extremum at x_0.*

We will prove part (*a*) and leave parts (*b*) and (*c*) as exercises.

Proof (a). We are assuming that $f'(x) > 0$ on the interval (a, x_0) and that $f'(x) < 0$ on the interval (x_0, b), and we want to show that

$$f(x_0) \geq f(x)$$

for all x in the interval (a, b). However, the two hypotheses, together with Theorem 5.1.2 (and its following remark), imply that f is increasing on the interval $(a, x_0]$ and decreasing on the interval $[x_0, b)$. Thus, $f(x_0) \geq f(x)$ for all x in (a, b) with equality only at x_0.

Example 2

(a) Locate the relative maxima and minima of $f(x) = 3x^{5/3} - 15x^{2/3}$.

(b) Confirm that the results in part (a) agree with the graph of f.

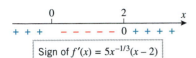

Sign of $f'(x) = 5x^{-1/3}(x-2)$

Figure 5.2.5

Solution (a). The function f is defined and continuous for all real values of x, and its derivative is

$$f'(x) = 5x^{2/3} - 10x^{-1/3} = 5x^{-1/3}(x-2) = \frac{5(x-2)}{x^{1/3}}$$

Since $f'(x)$ does not exist if $x = 0$, and since $f'(x) = 0$ if $x = 2$, there are critical numbers at $x = 0$ and $x = 2$. To apply the first derivative test, we examine the sign of $f'(x)$ on intervals extending to the left and right of the critical numbers (Figure 5.2.5). Since the sign of the derivative changes from positive to negative at $x = 0$, there is a relative maximum there, and since it changes from negative to positive at $x = 2$, there is a relative minimum there.

Solution (b). The result in part (a) agrees with the graph of f shown in Figure 5.2.6. ◄

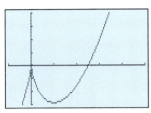

$[-2, 10] \times [-15, 20]$
$x\text{Scl} = 2, y\text{Scl} = 5$

$f(x) = 3x^{5/3} - 15x^{2/3}$

Figure 5.2.6

FOR THE READER. As discussed in the subsection of Section 1.3 entitled Errors of Omission, many graphing utilities omit portions of the graphs of functions with fractional exponents and must be "tricked" into producing complete graphs; and indeed, for the function in the last example both a calculator and a CAS failed to produce the portion of the graph over the negative x-axis. To generate the graph in Figure 5.2.6, apply the techniques discussed in Exercise 29 of Section 1.3 to each term in the formula for f. Use a graphing utility to generate this graph.

Example 3
Locate the relative extrema of $f(x) = x^3 - 3x^2 + 3x - 1$, if any.

Solution. Since f is differentiable everywhere, the only possible critical numbers correspond to stationary points. Differentiating f yields

$$f'(x) = 3x^2 - 6x + 3 = 3(x-1)^2$$

Solving $f'(x) = 0$ yields only $x = 1$. However, $3(x-1)^2 \geq 0$ for all x, so $f'(x)$ does not change sign at $x = 1$; consequently, f does not have a relative extremum at $x = 1$. Thus, f has no relative extrema (Figure 5.2.7). ◄

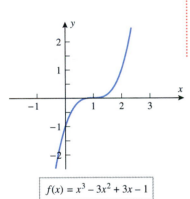

$f(x) = x^3 - 3x^2 + 3x - 1$

Figure 5.2.7

FOR THE READER. How many relative extrema can a polynomial of degree n have? Explain your reasoning.

SECOND DERIVATIVE TEST

There is another test for relative extrema that is often easier to apply than the first derivative test. It is based on the geometric observation that a function f has a relative maximum at a stationary point if the graph of f is concave down on an open interval containing the point, and it has a relative minimum if it is concave up (Figure 5.2.8).

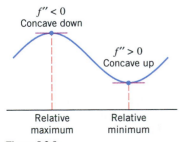

Figure 5.2.8

5.2.4 THEOREM (Second Derivative Test). *Suppose that f is twice differentiable at x_0.*

(a) *If $f'(x_0) = 0$ and $f''(x_0) > 0$, then f has a relative minimum at x_0.*

(b) *If $f'(x_0) = 0$ and $f''(x_0) < 0$, then f has a relative maximum at x_0.*

(c) *If $f'(x_0) = 0$ and $f''(x_0) = 0$, then the test is inconclusive; that is, f may have a relative maximum, a relative minimum, or neither at x_0.*

We will prove parts (a) and (c) and leave part (b) as an exercise.

Proof (a). We are assuming that $f'(x_0) = 0$ and $f''(x_0) > 0$, and we want to show that f has a relative minimum at x_0. It follows from our discussion at the beginning of Section 5.1 (with the function f replaced by f') that if $f''(x_0) > 0$, then $f'(x) < f'(x_0) = 0$ for x just to the left of x_0, and $f'(x) > f'(x_0) = 0$ for x just to the right of x_0. But then f has a relative minimum at x_0 by the first derivative test.

Proof (c). Consider the functions $f(x) = x^3$, $f(x) = x^4$, and $f(x) = -x^4$. It is easy to check that in all three cases $f'(0) = 0$ and $f''(0) = 0$; but from Figure 1.6.4, $f(x) = x^4$ has a relative minimum at $x = 0$, $f(x) = -x^4$ has a relative maximum at $x = 0$ (why?), and $f(x) = x^3$ has neither a relative maximum nor a relative minimum at $x = 0$. ■

Example 4   Locate the relative maxima and minima of $f(x) = x^4 - 2x^2$, and confirm that your results are consistent with the graph of f.

Solution.
$$f'(x) = 4x^3 - 4x = 4x(x-1)(x+1)$$
$$f''(x) = 12x^2 - 4$$

Solving $f'(x) = 0$ yields stationary points at $x = 0$, $x = 1$, and $x = -1$. Evaluating f'' at these points yields
$$f''(0) = -4 < 0$$
$$f''(1) = 8 > 0$$
$$f''(-1) = 8 > 0$$

so there is a relative maximum at $x = 0$ and relative minima at $x = 1$ and at $x = -1$ (Figure 5.2.9). ◀

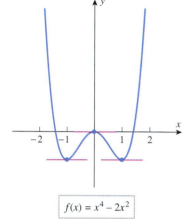

$f(x) = x^4 - 2x^2$

Figure 5.2.9

Example 5   Use a graphing utility to generate a graph of $f(x) = (\ln x)/x$ and identify the exact locations of all relative extrema.

Solution.   Note that the domain of $f(x)$ is $(0, +\infty)$. Figure 5.2.10 shows a graph of $f(x) = (\ln x)/x$ obtained from a graphing utility. This figure suggests that the graph has one relative maximum.

$$f'(x) = \frac{x\left(\frac{1}{x}\right) - (\ln x)(1)}{x^2} = \frac{1 - \ln x}{x^2}$$

$$f''(x) = \frac{x^2\left(-\frac{1}{x}\right) - (1 - \ln x)(2x)}{x^4} = \frac{2x\ln x - 3x}{x^4} = \frac{2\ln x - 3}{x^3}$$

Solving $f'(x) = 0$ yields the stationary point $x = e$. Now,
$$f''(e) = \frac{2-3}{e^3} = -\frac{1}{e^3} < 0$$

so there is a relative maximum at $x = e$. ◀

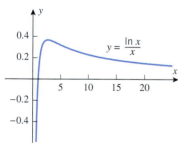

$y = \dfrac{\ln x}{x}$

Figure 5.2.10

MORE ON THE SIGNIFICANCE OF INFLECTION POINTS

In Section 5.1 we observed that the inflection points of a curve $y = f(x)$ mark the points where the slopes of the tangent lines change from increasing to decreasing, or vice versa. Thus, in the case where f is differentiable, $f'(x)$ will have a relative maximum or relative minimum at any inflection point of f (Figure 5.2.11); stated another way:

For a differentiable function $y = f(x)$, the rate of change of y with respect to x will have a relative extremum at any inflection point of f. That is, an inflection point identifies a place on the graph of $y = f(x)$ where the graph is steepest or where the graph is least steep in the vicinity of the point.

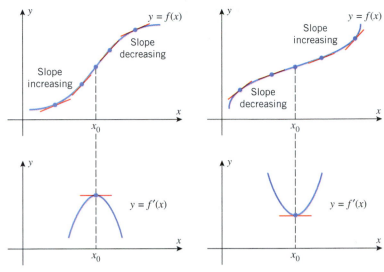

Figure 5.2.11

As an illustration of this principle, consider the flask shown in Figure 5.1.15. We observed in Section 5.1 that if water is poured into the flask so that the volume increases at a constant rate, then the graph of y versus t has an inflection point when y is at the narrow point in the neck. However, this is also the place where the water level is rising most rapidly.

EXERCISE SET 5.2   ~ Graphing Utility   [C] CAS

1. In each part, sketch the graph of a continuous function f with the stated properties.
   (a) f is concave up on the interval $(-\infty, +\infty)$ and has exactly one relative extremum.
   (b) f is concave up on the interval $(-\infty, +\infty)$ and has no relative extrema.
   (c) The function f has exactly two relative extrema on the interval $(-\infty, +\infty)$, and $f(x) \to +\infty$ as $x \to +\infty$.
   (d) The function f has exactly two relative extrema on the interval $(-\infty, +\infty)$, and $f(x) \to -\infty$ as $x \to +\infty$.

2. In each part, sketch the graph of a continuous function f with the stated properties.
   (a) f has exactly one relative extremum on $(-\infty, +\infty)$, and $f(x) \to 0$ as $x \to +\infty$ and as $x \to -\infty$.
   (b) f has exactly two relative extrema on $(-\infty, +\infty)$, and $f(x) \to 0$ as $x \to +\infty$ and as $x \to -\infty$.
   (c) f has exactly one inflection point and one relative extremum on $(-\infty, +\infty)$.
   (d) f has infinitely many relative extrema, and $f(x) \to 0$ as $x \to +\infty$ and as $x \to -\infty$.

3. (a) Use both the first and second derivative tests to show that $f(x) = 3x^2 - 6x + 1$ has a relative minimum at $x = 1$.

   (b) Use both the first and second derivative tests to show that $f(x) = x^3 - 3x + 3$ has a relative minimum at $x = 1$ and a relative maximum at $x = -1$.

4. (a) Use both the first and second derivative tests to show that $f(x) = \sin^2 x$ has a relative minimum at $x = 0$.
   (b) Use both the first and second derivative tests to show that $g(x) = \tan^2 x$ has a relative minimum at $x = 0$.
   (c) Give an informal verbal argument to explain without calculus why the functions in parts (a) and (b) have relative minima at $x = 0$.

5. (a) Show that both of the functions $f(x) = (x - 1)^4$ and $g(x) = x^3 - 3x^2 + 3x - 2$ have stationary points at $x = 1$.
   (b) What does the second derivative test tell you about the nature of these stationary points?
   (c) What does the first derivative test tell you about the nature of these stationary points?

6. (a) Show that $f(x) = 1 - x^5$ and $g(x) = 3x^4 - 8x^3$ both have stationary points at $x = 0$.
   (b) What does the second derivative test tell you about the nature of these stationary points?
   (c) What does the first derivative test tell you about the nature of these stationary points?

In Exercises 7–12, locate the critical numbers and identify which critical numbers correspond to stationary points.

7. (a) $f(x) = x^3 + 3x^2 - 9x + 1$
   (b) $f(x) = x^4 - 6x^2 - 3$

8. (a) $f(x) = 2x^3 - 6x + 7$ (b) $f(x) = 3x^4 - 4x^3$

9. (a) $f(x) = \dfrac{x}{x^2 + 2}$ (b) $f(x) = x^{2/3}$

10. (a) $f(x) = \dfrac{x^2 - 3}{x^2 + 1}$ (b) $f(x) = \sqrt[3]{x + 2}$

11. (a) $f(x) = x^{1/3}(x + 4)$ (b) $f(x) = \cos 3x$

12. (a) $f(x) = x^{4/3} - 6x^{1/3}$ (b) $f(x) = |\sin x|$

In Exercises 13–16, use the graph of f' shown in the figure to estimate all values of x at which f has (a) relative minima, (b) relative maxima, and (c) inflection points.

13.

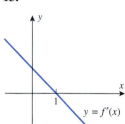

14.

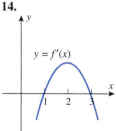

$y = f'(x)$

15.

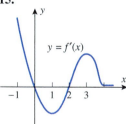

$y = f'(x)$

16.

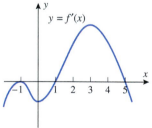

$y = f'(x)$

In Exercises 17–20, use the given derivative to find all critical numbers of f, and at each critical number determine whether a relative maximum, relative minimum, or neither occurs.

17. (a) $f'(x) = x^3(x^2 - 5)$ (b) $f'(x) = \dfrac{x^2 - 1}{x^2 + 1}$

18. (a) $f'(x) = x^2(2x + 1)(x - 1)$
   (b) $f'(x) = \dfrac{9 - 4x^2}{\sqrt[3]{x + 1}}$

19. (a) $f'(x) = xe^{-x}$ (b) $f'(x) = (e^x - 2)(e^x + 3)$

20. (a) $f'(x) = \ln\left(\dfrac{2}{1 + x^2}\right)$
   (b) $f'(x) = (1 - x)\ln x, \quad x > 0$

In Exercises 21–24, find the relative extrema using both the first and second derivative tests.

21. $f(x) = 1 - 4x - x^2$ **22.** $f(x) = 2x^3 - 9x^2 + 12x$

23. $f(x) = \sin^2 x, \quad 0 < x < 2\pi$

24. $f(x) = \frac{1}{2}x - \sin x, \quad 0 < x < 2\pi$

In Exercises 25–38, use any method to find the relative extrema of the function f.

25. $f(x) = x^3 + 5x - 2$ **26.** $f(x) = x^4 - 2x^2 + 7$

27. $f(x) = x(x - 1)^2$ **28.** $f(x) = x^4 + 2x^3$

29. $f(x) = 2x^2 - x^4$ **30.** $f(x) = (2x - 1)^5$

31. $f(x) = x^{4/5}$ **32.** $f(x) = 2x + x^{2/3}$

33. $f(x) = \dfrac{x^2}{x^2 + 1}$ **34.** $f(x) = \dfrac{x}{x + 2}$

35. $f(x) = \ln(1 + x^2)$ **36.** $f(x) = x^2 e^x$

37. $f(x) = |x^2 - 4|$ **38.** $f(x) = \begin{cases} 9 - x, & x \le 3 \\ x^2 - 3, & x > 3 \end{cases}$

In Exercises 39–42, find the relative extrema in the interval $0 < x < 2\pi$, and confirm that your results are consistent with the graph of f generated by a graphing utility.

39. $f(x) = |\sin 2x|$ **40.** $f(x) = \sqrt{3}x + 2\sin x$

41. $f(x) = \cos^2 x$ **42.** $f(x) = \dfrac{\sin x}{2 - \cos x}$

In Exercises 43–46, use a graphing utility to make a conjecture about the relative extrema of f, and then check your conjecture using either the first or second derivative test.

43. $f(x) = x \ln x$ **44.** $f(x) = \dfrac{2}{e^x + e^{-x}}$

45. $f(x) = x^2 e^{-2x}$ **46.** $f(x) = 10 \ln x - x$

In Exercises 47 and 48, use a graphing utility to generate the graphs of f' and f'' over the stated interval, and then use those graphs to estimate the x-coordinates of the relative extrema of f. Check that your estimates are consistent with the graph of f.

47. $f(x) = x^4 - 24x^2 + 12x, \quad -5 \le x \le 5$

48. $f(x) = \sin\frac{1}{2}x \cos x, \quad -\pi/2 \le x \le \pi/2$

In Exercises 49–52, use a CAS to graph f' and f'', and then use those graphs to estimate the x-coordinates of the relative extrema of f. Check that your estimates are consistent with the graph of f.

49. $f(x) = \dfrac{10x - 3}{3x^2 - 5x + 8}$ **50.** $f(x) = \dfrac{x^3 - 8x + 7}{\sqrt{x^2 + 1}}$

c **51.** $f(x) = \dfrac{x^3 - x^2}{x^2 + 1}$ **c** **52.** $f(x) = \sqrt{x^4 - \sin^2 x + 1}$

53. In each part, find k so that f has a relative extremum at the point where $x = 3$.

   (a) $f(x) = x^2 + \dfrac{k}{x}$ (b) $f(x) = \dfrac{x}{x^2 + k}$

c **54.** (a) Use a CAS to graph the function

$$f(x) = \frac{x^4 + 1}{x^2 + 1}$$

   and use the graph to estimate the x-coordinates of the relative extrema.

   (b) Find the exact x-coordinates by using the CAS to solve the equation $f'(x) = 0$.

55. The two graphs in the accompanying figure depict a function $r(x)$ and its derivative $r'(x)$.

   (a) Approximate the coordinates of each inflection point on the graph of $y = r(x)$.

   (b) Suppose that $f(x)$ is a function that is continuous everywhere and whose *derivative* satisfies

$$f'(x) = (x^2 - 4) \cdot r(x)$$

   (i) What are the critical numbers for $f(x)$? At each critical number, identify whether $f(x)$ has a (relative) maximum, minimum, or neither a maximum or minimum.

   (ii) Approximate $f''(1)$.

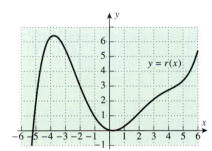

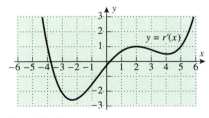

Figure Ex-55

56. With $r(x)$ as provided in Exercise 55, let $g(x)$ be a function that is continuous everywhere such that $g'(x) = x - r(x)$. For which values of x does $g(x)$ have an inflection point?

57. Find values of a, b, c, and d so that the function

$$f(x) = ax^3 + bx^2 + cx + d$$

has a relative minimum at $(0, 0)$ and a relative maximum at $(1, 1)$.

~ **58.** Functions of the form

$$f(x) = cx^n e^{-x}, \quad x > 0$$

where n is a positive integer and $c = 1/n!$, arise in the statistical study of traffic flow.

   (a) Use a graphing utility to generate the graph of f for $n = 2$, 3, 4, and 5, and make a conjecture about the number and locations of the relative extrema of f.

   (b) Confirm your conjecture using the first derivative test.

~ **59.** Functions of the form

$$f(x) = \frac{1}{\sqrt{2\pi}} e^{-x^2/2}$$

arise in a wide variety of statistical problems.

   (a) Use the first derivative test to show that f has a relative maximum at $x = 0$, and confirm this by using a graphing utility to graph f.

   (b) Sketch the graph of

$$f(x) = \frac{1}{\sqrt{2\pi}} e^{-(x-\mu)^2/2}$$

   where μ is a constant, and label the coordinates of the relative extrema.

60. Let h and g have relative maxima at x_0. Prove or disprove:

   (a) $h + g$ has a relative maximum at x_0

   (b) $h - g$ has a relative maximum at x_0.

61. Sketch some curves that show that the three parts of the first derivative test (Theorem 5.2.3) can be false without the assumption that f is continuous at x_0.

5.3 ANALYSIS OF FUNCTIONS III: APPLYING TECHNOLOGY AND THE TOOLS OF CALCULUS

In this section we will discuss how to use technology and the tools of calculus that we developed in the last two sections to analyze various types of graphs that occur in applications.

PROPERTIES OF GRAPHS

In many problems, the properties of interest in the graph of a function are:

- symmetries
- x-intercepts
- relative extrema
- intervals of increase and decrease
- asymptotes

- periodicity
- y-intercepts
- inflection points
- concavity
- behavior as $x \to +\infty$ or as $x \to -\infty$

Some of these properties may not be relevant in certain cases; for example, asymptotes are characteristic of rational functions but not of polynomials, and periodicity is characteristic of trigonometric functions but not of polynomial or rational functions. Thus, when analyzing the graph of a function f, it helps to know something about the general properties of the family to which it belongs.

In a given problem you will usually have a definite objective for your analysis. For example, you may be interested in finding a graph that highlights all of the important characteristics of f; or you may be interested in something specific, say the exact locations of the relative extrema or the behavior of the graph as $x \to +\infty$ or as $x \to -\infty$. However, regardless of your objectives, you will usually find it helpful to begin your analysis by generating a graph with a graphing utility. As discussed in Section 1.3, some of the function's important characteristics may be obscured by compression or resolution problems. However, with this graph as a starting point, you can often use calculus to complete the analysis and resolve any ambiguities.

A PROCEDURE FOR ANALYZING GRAPHS

There are no hard and fast rules that are guaranteed to produce all of the information you may need about the graph of a function f, but here is one possible way of organizing the analysis of a function (the order of the steps can be varied).

Step 1. Use a graphing utility to generate the graph of f in some reasonable window, taking advantage of any general knowledge you have about the function to help in choosing the window.

Step 2. See if the graph suggests the existence of symmetries, periodicity, or domain restrictions. If so, try to confirm those properties analytically.

Step 3. Find the intercepts, if needed.

Step 4. Investigate the behavior of the graph as $x \to +\infty$ and as $x \to -\infty$, and identify all horizontal and vertical asymptotes, if any.

Step 5. Calculate $f'(x)$ and $f''(x)$, and use these derivatives to determine the critical numbers, the intervals on which f is increasing or decreasing, the intervals on which f is concave up and concave down, and the inflection points.

Step 6. If you have discovered that some of the significant features did not fall within the graphing window in Step 1, then try adjusting the window to include them. However, it is possible that compression or resolution problems may prevent you from showing all of the features of interest in a single window, in which case you may need to use different windows to focus on different features. In some cases you may even find that a hand-drawn sketch labeled with the location of the significant features is clearer or more informative than a graph generated with a graphing utility.

ANALYSIS OF POLYNOMIALS

Polynomials are among the simplest functions to graph and analyze. Their significant features are symmetry, intercepts, relative extrema, inflection points, and the behavior as $x \to +\infty$ and as $x \to -\infty$. Figure 5.3.1 shows the graphs of four typical polynomials in x.

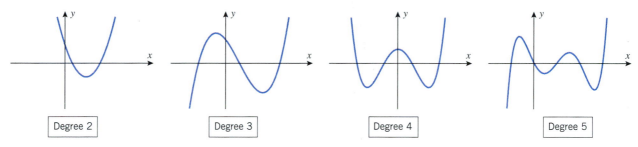

Degree 2   Degree 3   Degree 4   Degree 5

Figure 5.3.1

The graphs in Figure 5.3.1 have properties that are common to all polynomials:

- The natural domain of a polynomial in x is the entire x-axis, since the only operations involved in its formula are additions, subtractions, and multiplications; the range depends on the particular polynomial.

- Polynomials are continuous everywhere.

- Graphs of polynomials have no sharp corners or points of vertical tangency, since polynomials are differentiable everywhere.

- The graph of a nonconstant polynomial eventually increases or decreases without bound as $x \to +\infty$ or as $x \to -\infty$, since the limit of a nonconstant polynomial as $x \to +\infty$ or as $x \to -\infty$ is $\pm\infty$ (see the subsection in Section 2.3 entitled Limits of Polynomials as $x \to \pm\infty$).

- The graph of a polynomial of degree n has at most n x-intercepts, at most $n - 1$ relative extrema, and at most $n - 2$ inflection points.

The last property is a consequence of the fact that the x-intercepts, relative extrema, and inflection points occur at real roots of $p(x) = 0$, $p'(x) = 0$, and $p''(x) = 0$, respectively, so if $p(x)$ has degree n greater than 1, then $p'(x)$ has degree $n - 1$ and $p''(x)$ has degree $n - 2$. Thus, for example, the graph of a quadratic polynomial has at most two x-intercepts, one relative extremum, and no inflection points; and the graph of a cubic polynomial has at most three x-intercepts, two relative extrema, and one inflection point.

FOR THE READER. For each of the graphs in Figure 5.3.1, count the number of x-intercepts, relative extrema, and inflection points, and confirm that your count is consistent with the degree of the polynomial.

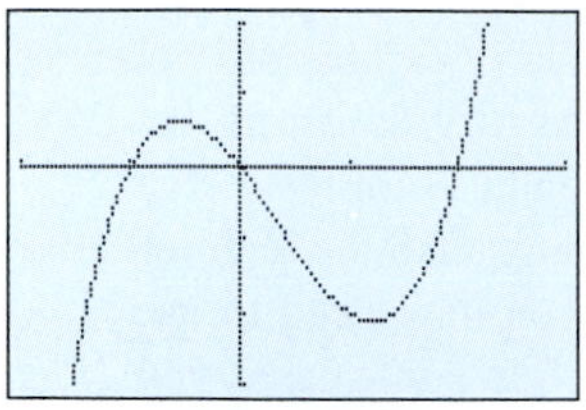

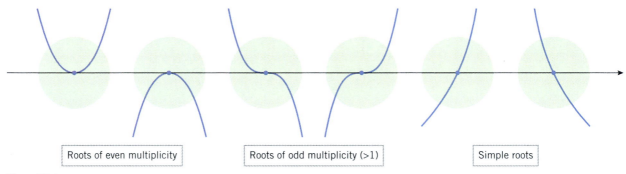

$[-2, 3] \times [-3, 2]$
xScl $= 1$, yScl $= 1$

$$y = x^3 - x^2 - 2x$$

Figure 5.3.2

Example 1   Figure 5.3.2 shows the graph of

$$y = x^3 - x^2 - 2x$$

produced on a graphing calculator. Confirm that the graph is not missing any significant features.

Solution.   We can be confident that the graph exhibits all the significant features of the polynomial because the polynomial has degree 3, and three roots, two relative extrema, and one inflection point are accounted for. Moreover, the graph indicates the correct behavior as $x \to +\infty$ and as $x \to -\infty$, since

$$\lim_{x \to +\infty} (x^3 - x^2 - 2x) = \lim_{x \to +\infty} x^3 = +\infty$$

$$\lim_{x \to -\infty} (x^3 - x^2 - 2x) = \lim_{x \to -\infty} x^3 = -\infty$$ ◄

· ·

GEOMETRIC IMPLICATIONS OF MULTIPLICITY

A root $x = r$ of a polynomial $p(x)$ has ***multiplicity m*** if $(x-r)^m$ divides $p(x)$ but $(x-r)^{m+1}$ does not. A root of multiplicity 1 is called a ***simple root***. There is a close relationship between the multiplicity of a root of a polynomial and the behavior of the graph in the vicinity of the root. This relationship, stated below, is illustrated in Figure 5.3.3.

| Roots of even multiplicity | Roots of odd multiplicity (>1) | Simple roots |

Figure 5.3.3

5.3.1   THE GEOMETRIC IMPLICATIONS OF MULTIPLICITY.   *Suppose that $p(x)$ is a polynomial with a root of multiplicity m at $x = r$.*

(a)   *If m is even, then the graph of $y = p(x)$ is tangent to the x-axis at $x = r$, does not cross the x-axis there, and does not have an inflection point there.*

(b)   *If m is odd and greater than 1, then the graph is tangent to the x-axis at $x = r$, crosses the x-axis there, and also has an inflection point there.*

(c)   *If m = 1 (so that the root is simple), then the graph is not tangent to the x-axis at $x = r$, crosses the x-axis there, and may or may not have an inflection point there.*

Example 2   Make a conjecture about the behavior of the graph of

$$y = x^3(3x - 4)(x + 2)^2$$

in the vicinity of its x-intercepts, and test your conjecture by generating the graph.

Solution.   The x-intercepts occur at $x = 0$, $x = \frac{4}{3}$, and $x = -2$. The root $x = 0$ has multiplicity 3, which is odd, so at that point the graph should be tangent to the x-axis, cross the x-axis, and have an inflection point there. The root $x = -2$ has multiplicity 2, which is even, so the graph should be tangent to but not cross the x-axis there. The root $x = \frac{4}{3}$ is simple, so at that point the curve should cross the x-axis without being tangent to it. All of this is consistent with the graph in Figure 5.3.4.   ◄

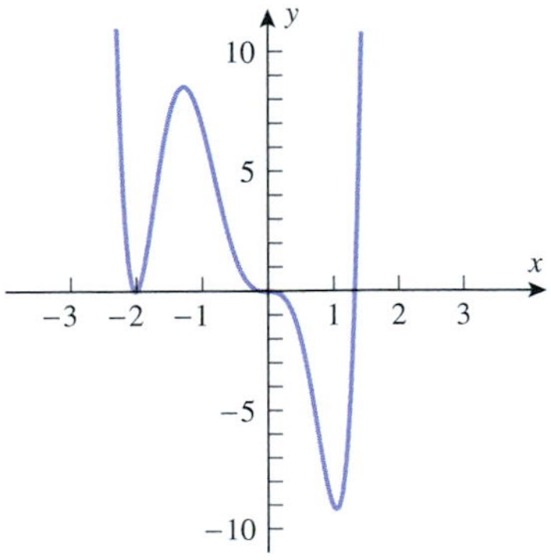

$$y = x^3(3x - 4)(x + 2)^2$$

Figure 5.3.4

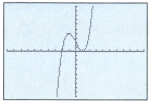

[−10, 10] × [−10, 10]
xScl = 1, yScl = 1

$y = x^3 - 3x + 2$

Figure 5.3.5

Example 3   Generate or sketch a graph of the equation

$$y = x^3 - 3x + 2 = (x + 2)(x - 1)^2$$

and identify the exact locations of the intercepts, relative extrema, and inflection points.

Solution.   Figure 5.3.5 shows a graph of the given equation produced with a graphing utility. Since the polynomial has degree 3, all roots, relative extrema, and inflection points are accounted for in the graph: there are three roots (a simple negative root and a positive root of multiplicity 2), and there are two relative extrema and one inflection point. The following analysis will identify the exact locations of the intercepts, relative extrema, and inflection points.

- *x-intercepts:* Setting $y = 0$ yields roots at $x = -2$ and at $x = 1$.
- *y-intercept:* Setting $x = 0$ yields $y = 2$.
- *Behavior as $x \to +\infty$ and as $x \to -\infty$:* The graph in Figure 5.3.5 suggests that the graph increases without bound as $x \to +\infty$ and decreases without bound as $x \to -\infty$. This is confirmed by the limits

$$\lim_{x \to +\infty} (x^3 - 3x + 2) = \lim_{x \to +\infty} x^3 = +\infty$$

$$\lim_{x \to -\infty} (x^3 - 3x + 2) = \lim_{x \to -\infty} x^3 = -\infty$$

- *Derivatives:*

$$\frac{dy}{dx} = 3x^2 - 3 = 3(x - 1)(x + 1)$$

$$\frac{d^2y}{dx^2} = 6x$$

- *Intervals of increase and decrease; relative extrema; concavity:* Figure 5.3.6 shows the sign pattern of the first and second derivatives and what they imply about the graph shape.

Figure 5.3.7 shows the graph labeled with the coordinates of the intercepts, relative extrema, and inflection point.   ◄

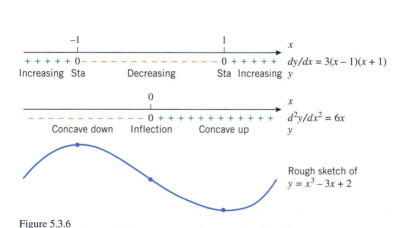

Figure 5.3.6

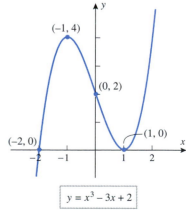

$y = x^3 - 3x + 2$

Figure 5.3.7

GRAPHING RATIONAL FUNCTIONS

Rational functions (ratios of polynomials) are more complicated to graph than polynomials because they may have discontinuities and asymptotes.

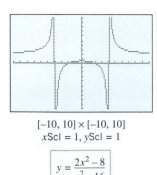

[−10, 10] × [−10, 10]
xScl = 1, yScl = 1

$$y = \frac{2x^2 - 8}{x^2 - 16}$$

Figure 5.3.8

Example 4 Generate or sketch a graph of the equation

$$y = \frac{2x^2 - 8}{x^2 - 16}$$

and identify the exact locations of the intercepts, relative extrema, inflection points, and asymptotes.

Solution. Figure 5.3.8 shows a calculator-generated graph of the equation in the window $[-10, 10] \times [-10, 10]$. The figure suggests that the graph is symmetric about the y-axis and has two vertical asymptotes and a horizontal asymptote. The figure also suggests that there is a relative maximum at $x = 0$ and two x-intercepts. There do not seem to be any inflection points. The following analysis will identify the exact locations of the key features of the graph.

- *Symmetries:* Replacing x by $-x$ does not change the equation, so the graph is symmetric about the y-axis.

- *x-intercepts:* Setting $y = 0$ yields the x-intercepts $x = -2$ and $x = 2$.

- *y-intercept:* Setting $x = 0$ yields the y-intercept $y = \frac{1}{2}$.

- *Vertical asymptotes:* Setting $x^2 - 16 = 0$ yields the solutions $x = -4$ and $x = 4$. Since neither solution is a root of $2x^2 - 8$, the graph has vertical asymptotes at $x = -4$ and $x = 4$.

- *Horizontal asymptotes:* The limits

$$\lim_{x \to +\infty} \frac{2x^2 - 8}{x^2 - 16} = \lim_{x \to +\infty} \frac{2 - (8/x^2)}{1 - (16/x^2)} = 2$$

$$\lim_{x \to -\infty} \frac{2x^2 - 8}{x^2 - 16} = \lim_{x \to -\infty} \frac{2 - (8/x^2)}{1 - (16/x^2)} = 2$$

yield the horizontal asymptote $y = 2$.

The set of values where x-intercepts or vertical asymptotes occur is $\{-4, -2, 2, 4\}$. These values divide the x-axis into the open intervals

$$(-\infty, -4), \quad (-4, -2), \quad (-2, 2), \quad (2, 4), \quad (4, +\infty)$$

Over each of these intervals, y cannot change sign (why?). We can find the sign of y on each interval by choosing an arbitrary test value in the interval and evaluating $y = f(x)$ at the test value (Table 5.3.1).

Table 5.3.1

INTERVAL	TEST VALUE	$y = \dfrac{2x^2 - 8}{x^2 - 16}$	SIGN OF y
$(-\infty, -4)$	$x = -5$	$y = 14/3$	$+$
$(-4, -2)$	$x = -3$	$y = -10/7$	$-$
$(-2, 2)$	$x = 0$	$y = 1/2$	$+$
$(2, 4)$	$x = 3$	$y = -10/7$	$-$
$(4, +\infty)$	$x = 5$	$y = 14/3$	$+$

The information in Table 5.3.1 is consistent with Figure 5.3.8, so we can be certain that the calculator graph has not missed any sign changes. The next step is to use the first and second derivatives to determine whether the calculator graph has missed any relative extrema or changes in concavity.

- *Derivatives:*

$$\frac{dy}{dx} = \frac{(x^2-16)(4x) - (2x^2-8)(2x)}{(x^2-16)^2} = -\frac{48x}{(x^2-16)^2}$$

$$\frac{d^2y}{dx^2} = \frac{48(16+3x^2)}{(x^2-16)^3} \quad \text{(verify)}$$

- *Intervals of increase and decrease; relative extrema:* A sign analysis of dy/dx yields

$$
\begin{array}{c}
\begin{array}{ccccc}
 & -4 & 0 & 4 & \\
\hline
\end{array} \quad x \\
\begin{array}{ccccc}
+ + + + +\infty + + + + +0 - - - - -\infty - - - - - & \text{Sign of } dy/dx \\
\text{Incr \ Undef \ Incr \ Sta \ Decr \ Undef \ Decr} & y
\end{array}
\end{array}
$$

Thus, the graph is increasing on the intervals $(-\infty, -4)$ and $(-4, 0]$; and it is decreasing on the intervals $[0, 4)$ and $(4, +\infty)$. There is a relative maximum at $x = 0$.

- *Concavity:* A sign analysis of d^2y/dx^2 yields

$$
\begin{array}{c}
\begin{array}{ccc}
 & -4 & 4 \\
\hline
\end{array} \quad x \\
\begin{array}{ccc}
+ + + + +\infty - - - - - - - - - - -\infty + + + + + & \text{Sign of } d^2y/dx^2 \\
\text{Concave} \quad \text{Concave} \quad \text{Concave} & y \\
\text{up} \qquad \text{down} \qquad \text{up}
\end{array}
\end{array}
$$

There are changes in concavity at the vertical asymptotes, $x = -4$ and $x = 4$, but there are no inflection points.

This analysis confirms that our calculator-generated graph exhibited all important features of the rational function. Figure 5.3.9 shows a graph of the equation with the asymptotes, intercepts, and relative maximum identified. ◀

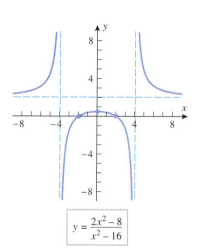

$$y = \frac{2x^2-8}{x^2-16}$$

Figure 5.3.9

Example 5   Generate or sketch a graph of

$$y = \frac{x^2-1}{x^3}$$

and identify the exact locations of all asymptotes, intercepts, relative extrema, and inflection points.

Solution.   Figure 5.3.10*a* shows a calculator-generated graph of the given equation in the window $[-10, 10] \times [-10, 10]$, and Figure 5.3.10*b* shows a second version of the graph that gives more detail in the vicinity of the x-axis. These figures suggest that the graph is symmetric about the origin. They also suggest that there are two relative extrema, two inflection points, two x-intercepts, a vertical asymptote at $x = 0$, and a horizontal asymptote at $y = 0$. The following analysis will identify the exact locations of all the key features and will determine whether the calculator-generated graphs in Figure 5.3.10 have missed any of these features.

- *Symmetries:* Replacing x by $-x$ and y by $-y$ yields an equation that simplifies back to the original equation, so the graph is symmetric about the origin.
- *x-intercepts:* Setting $y = 0$ yields the x-intercepts $x = -1$ and $x = 1$.
- *y-intercept:* Setting $x = 0$ leads to a division by zero, so that there is no y-intercept.
- *Vertical asymptotes:* Setting $x^3 = 0$ yields the solution $x = 0$. This is not a root of $x^2 - 1$, so $x = 0$ is a vertical asymptote.

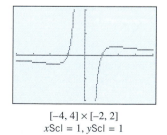

$[-10, 10] \times [-10, 10]$
xScl = 1, yScl = 1

(*a*)

$[-4, 4] \times [-2, 2]$
xScl = 1, yScl = 1

(*b*)

$$y = \frac{x^2-1}{x^3}$$

Figure 5.3.10

- *Horizontal asymptotes:* The limits

$$\lim_{x \to +\infty} \frac{x^2 - 1}{x^3} = \lim_{x \to +\infty} \frac{\dfrac{1}{x} - \dfrac{1}{x^3}}{1} = 0$$

$$\lim_{x \to -\infty} \frac{x^2 - 1}{x^3} = \lim_{x \to -\infty} \frac{\dfrac{1}{x} - \dfrac{1}{x^3}}{1} = 0$$

yield the horizontal asymptote $y = 0$.

- *Derivatives:*

$$\frac{dy}{dx} = \frac{x^3(2x) - (x^2 - 1)(3x^2)}{\left(x^3\right)^2} = \frac{3 - x^2}{x^4}$$

$$\frac{d^2y}{dx^2} = \frac{x^4(-2x) - (3 - x^2)(4x^3)}{\left(x^4\right)^2} = \frac{2(x^2 - 6)}{x^5}$$

- *Intervals of increase and decrease; relative extrema:*

This analysis reveals a relative minimum at $x = -\sqrt{3}$ and a relative maximum at $x = \sqrt{3}$.

- *Concavity:*

This analysis reveals that changes in concavity occur at the vertical asymptote $x = 0$ and at the inflection points at $x = -\sqrt{6}$ and at $x = \sqrt{6}$.

Figure 5.3.11 shows a table of coordinate values at the relative extrema and inflection points together with a graph of the equation on which we have emphasized these points. ◀

Suppose that the numerator polynomial of a rational function $f(x)$ has degree greater than the degree of the denominator polynomial $d(x)$. Then by division we can write

$$f(x) = q(x) + \frac{r(x)}{d(x)}$$

where $q(x)$ and $r(x)$ are polynomials and the degree of $r(x)$ is less than that of $d(x)$. In this case, $f(x)$ will be asymptotic to the quotient polynomial $q(x)$; that is,

$$\lim_{x \to -\infty} [f(x) - q(x)] = 0 \quad \text{and} \quad \lim_{x \to +\infty} [f(x) - q(x)] = 0$$

(see the end of Exercise Set 2.3). Exercises 66–72 at the end of this section deal with the instance of an *oblique* asymptote, where the numerator has degree one more than the degree of the denominator. Example 6 illustrates an instance where the difference in degree is two.

Example 6   Generate or sketch a graph of $y = \dfrac{x^3 - x^2 - 8}{x - 1}$.

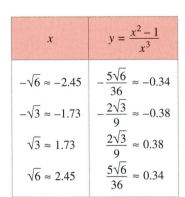

x	$y = \dfrac{x^2 - 1}{x^3}$
$-\sqrt{6} \approx -2.45$	$-\dfrac{5\sqrt{6}}{36} \approx -0.34$
$-\sqrt{3} \approx -1.73$	$-\dfrac{2\sqrt{3}}{9} \approx -0.38$
$\sqrt{3} \approx 1.73$	$\dfrac{2\sqrt{3}}{9} \approx 0.38$
$\sqrt{6} \approx 2.45$	$\dfrac{5\sqrt{6}}{36} \approx 0.34$

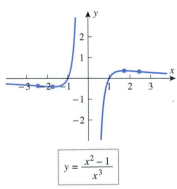

Figure 5.3.11

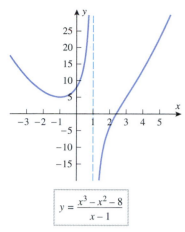

$$y = \frac{x^3 - x^2 - 8}{x - 1}$$

Figure 5.3.12

Solution. Figure 5.3.12 shows a computer-generated graph of

$$f(x) = \frac{x^3 - x^2 - 8}{x - 1}$$

to which an asymptote has been added for clarity. Note that

$$f(x) = x^2 - \frac{8}{x - 1}$$

so $f(x) \approx x^2$ [since $8/(x - 1) \approx 0$] as $x \to -\infty$ and as $x \to +\infty$. Thus, we would expect the graph to be concave up for large values of x, but the vertical asymptote at $x = 1$ indicates that $f(x)$ should be concave down in an interval just to the right of 1, so there should be an inflection point to the right of $x = 1$. Also, our sketch indicates a relative minimum to the left of $x = 1$. To determine the locations of these features we proceed as follows.

- *Symmetries:* There are no symmetries about a vertical axis or about a point.

- *x-intercepts:* Setting $y = 0$ leads to solving the equation $x^3 - x^2 - 8 = 0$. From Figure 5.3.12 it appears there is one solution in the interval $[2, 3]$. Using a solver yields $x \approx 2.39486$.

- *y-intercepts:* Setting $x = 0$ yields the y-intercept $y = 8$.

- *Vertical asymptotes:* Setting $x = 1$ would produce a nonzero numerator and a zero denominator for $f(x)$, so $x = 1$ is a vertical asymptote.

- *Horizontal asymptotes:* There are no horizontal asymptotes; however, as noted,

$$f(x) = x^2 - \frac{8}{x - 1}$$

so

$$\lim_{x \to -\infty} [f(x) - x^2] = \lim_{x \to -\infty} -\frac{8}{x - 1} = 0 \quad \text{and} \quad \lim_{x \to +\infty} [f(x) - x^2] = 0$$

Thus, $f(x)$ is asymptotic to $y = x^2$ as $x \to -\infty$ and as $x \to +\infty$.

- *Derivatives:*

$$f'(x) = \frac{d}{dx}\left[x^2 - \frac{8}{x - 1}\right] = 2x + \frac{8}{(x - 1)^2}$$

$$f''(x) = \frac{d}{dx}\left[2x + \frac{8}{(x - 1)^2}\right] = 2 - \frac{16}{(x - 1)^3}$$

- *Intervals of increase and decrease; relative extrema:* $f'(x) = 0$ when

$$2x = -\frac{8}{(x - 1)^2}$$

or when $2(x^3 - 2x^2 + x + 4) = 2(x + 1)(x^2 - 3x + 4) = 0$. The only real solution to this equation is $x = -1$.

	-1		1		
$- - - - -0$	$+ + + + +$	∞	$+ + + + + + + + + + + +$		Sign of dy/dx
Decr	Sta	Incr	Undef	Incr	y

The analysis reveals a relative minimum $f(-1) = 5$ at $x = -1$.

- *Concavity:* $f''(x) = 0$ when

$$2 = \frac{16}{(x - 1)^3}$$

or when $(x - 1)^3 = 8$. Then $x - 1 = 2$, so $x = 3$.

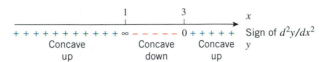

The analysis reveals an inflection point at $x = 3$. The coordinates of the inflection point are $(3, 5)$.

Figure 5.3.13 shows a graph of $y = f(x)$ with the relative minimum and inflection point highlighted and the asymptotes indicated. ◀

GRAPHS WITH VERTICAL TANGENTS AND CUSPS

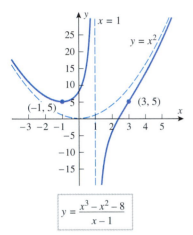

$$y = \frac{x^3 - x^2 - 8}{x - 1}$$

Figure 5.3.13

Figure 5.3.14 shows four curve elements that are commonly found in graphs of functions that involve radicals or fractional exponents. In all four cases, the function is not differentiable at x_0 because the secant line through $(x_0, f(x_0))$ and $(x, f(x))$ approaches a vertical position as x approaches x_0 from either side. Thus, in each case, the curve has a vertical tangent line at $(x_0, f(x_0))$.

It can be shown that the graph of a function f has a vertical tangent line at $(x_0, f(x_0))$ if f is continuous at x_0 and $f'(x)$ approaches either $+\infty$ or $-\infty$ as $x \to x_0^+$ and as $x \to x_0^-$. Furthermore, in the case where $f'(x)$ approaches $+\infty$ from one side and $-\infty$ from the other side, the function f is said to have a *cusp* at x_0.

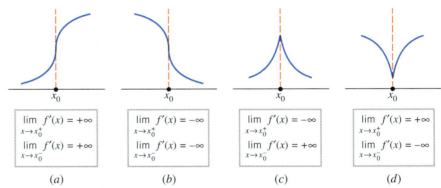

(a) (b) (c) (d)

Figure 5.3.14

REMARK. It is important to observe that vertical tangent lines occur only at points of nondifferentiability, whereas nonvertical tangent lines occur at points of differentiability.

Example 7 Generate or sketch a graph of $y = (x - 4)^{2/3}$.

Solution. Figure 5.3.15 shows a computer-generated graph of the equation $y = (x - 4)^{2/3}$. (As suggested in the discussion preceding Exercise 29 of Section 1.3, we had to trick the computer into producing the left branch by graphing the equation $y = |x - 4|^{2/3}$.) To locate the important features of this graph, we let $f(x) = (x - 4)^{2/3}$ and proceed as follows.

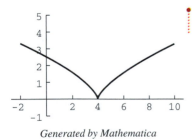

Generated by Mathematica

$$y = (x - 4)^{2/3}$$

Figure 5.3.15

- *Symmetries:* There are no symmetries about the coordinate axes or the origin (verify). However, the graph of $y = (x - 4)^{2/3}$ is symmetric about the line $x = 4$, since it is a translation (four units to the right) of the graph of $y = x^{2/3}$, which is symmetric about the y-axis.

- *x-intercepts:* Setting $y = 0$ yields the x-intercept $x = 4$.

- *y-intercepts:* Setting $x = 0$ yields the y-intercept $y = \sqrt[3]{16}$.

- *Vertical asymptotes:* None, since $f(x) = (x - 4)^{2/3}$ is continuous everywhere.

- *Horizontal asymptotes:* None, since

$$\lim_{x \to +\infty} (x-4)^{2/3} = +\infty \quad \text{and} \quad \lim_{x \to -\infty} (x-4)^{2/3} = +\infty$$

- *Derivatives:*

$$\frac{dy}{dx} = f'(x) = \frac{2}{3}(x-4)^{-1/3} = \frac{2}{3(x-4)^{1/3}}$$

$$\frac{d^2 y}{dx^2} = f''(x) = -\frac{2}{9}(x-4)^{-4/3} = -\frac{2}{9(x-4)^{4/3}}$$

- *Relative extrema; concavity:* There is a critical number at $x = 4$, since f is not differentiable there; and by the first derivative test there is a relative minimum at that critical number, since $f'(x) < 0$ if $x < 4$ and $f'(x) > 0$ if $x > 4$. Since $f''(x) < 0$ if $x \neq 4$, the graph is concave down for $x < 4$ and for $x > 4$.

- *Vertical tangent lines:* There is a vertical tangent line and cusp at $x = 4$ of the type in Figure 5.3.14d since $f(x) = (x-4)^{2/3}$ is continuous at $x = 4$ and

$$\lim_{x \to 4^+} f'(x) = \lim_{x \to 4^+} \frac{2}{3(x-4)^{1/3}} = +\infty$$

$$\lim_{x \to 4^-} f'(x) = \lim_{x \to 4^-} \frac{2}{3(x-4)^{1/3}} = -\infty$$

Combining the preceding information with a sign analysis of the first and second derivatives yields Figure 5.3.16. This confirms that the computer-generated graph in Figure 5.3.15 exhibited the important features of the graph. ◀

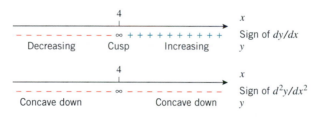

Figure 5.3.16

Example 8 Generate or sketch a graph of $y = 6x^{1/3} + 3x^{4/3}$.

Solution. Figure 5.3.17 shows a computer-generated graph of the equation. Once again, we had to call on the discussion preceding Exercise 29 of Section 1.3 to trick the computer into graphing a portion of the graph over the negative x-axis. (See if you can figure out how to do this.) To find the important features of this graph, we let

$$f(x) = 6x^{1/3} + 3x^{4/3} = 3x^{1/3}(2 + x)$$

and proceed as follows.

- *Symmetries:* There are no symmetries about the coordinate axes or the origin (verify).
- *x-intercepts:* Setting $y = 3x^{1/3}(2 + x) = 0$ yields the x-intercepts $x = 0$ and $x = -2$.
- *y-intercept:* Setting $x = 0$ yields the y-intercept $y = 0$.

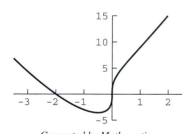

Generated by Mathematica

$$y = 6x^{1/3} + 3x^{4/3}$$

Figure 5.3.17

- *Vertical asymptotes:* None, since $f(x) = 6x^{1/3} + 3x^{4/3}$ is continuous everywhere.
- *Horizontal asymptotes:* None, since

$$\lim_{x \to +\infty} (6x^{1/3} + 3x^{4/3}) = \lim_{x \to +\infty} 3x^{1/3}(2 + x) = +\infty$$

$$\lim_{x \to -\infty} (6x^{1/3} + 3x^{4/3}) = \lim_{x \to -\infty} 3x^{1/3}(2 + x) = +\infty$$

- *Derivatives:*

$$\frac{dy}{dx} = f'(x) = 2x^{-2/3} + 4x^{1/3} = 2x^{-2/3}(1 + 2x) = \frac{2(2x + 1)}{x^{2/3}}$$

$$\frac{d^2y}{dx^2} = f''(x) = -\frac{4}{3}x^{-5/3} + \frac{4}{3}x^{-2/3} = \frac{4}{3}x^{-5/3}(-1 + x) = \frac{4(x - 1)}{3x^{5/3}}$$

- *Relative extrema; vertical tangent lines; concavity:* The critical numbers are $x = 0$ and $x = -\frac{1}{2}$. From the first derivative test and the sign analysis of dy/dx in Figure 5.3.18, there is a relative minimum at $x = -\frac{1}{2}$. There is a point of vertical tangency at $x = 0$, since

$$\lim_{x \to 0^+} f'(x) = \lim_{x \to 0^+} \frac{2(2x + 1)}{x^{2/3}} = +\infty$$

$$\lim_{x \to 0^-} f'(x) = \lim_{x \to 0^-} \frac{2(2x + 1)}{x^{2/3}} = +\infty$$

From the sign analysis of d^2y/dx^2 in Figure 5.3.18, the graph is concave up for $x < 0$, concave down for $0 < x < 1$, and concave up again for $x > 1$. There are inflection points at $(0, 0)$ and $(1, 9)$.

Combining the preceding information with a sign analysis of the first and second derivatives yields the graph shape shown in Figure 5.3.18.

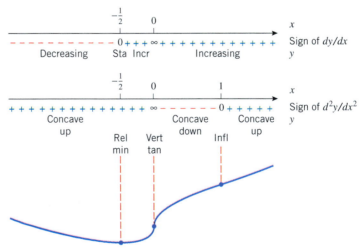

Figure 5.3.18

This confirms that the computer-generated graph in Figure 5.3.15 exhibits most of the important features of the graph, except for the fact that it did not reveal the very subtle inflection point at $x = 1$. In this case the artistic rendering of the curve in Figure 5.3.18 emphasizes the subtleties of the graph shape more effectively than the computer-generated graph. ◀

Example 9 Generate or sketch a graph of $y = e^{-x^2/2}$ and identify the exact locations of all relative extrema and inflection points.

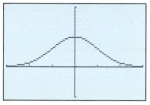

[−3, 3] × [−1, 2]
xScl = 1, yScl = 1

Figure 5.3.19

Solution. Figure 5.3.19 shows a calculator-generated graph of $y = e^{-x^2/2}$ in the window $[-3, 3] \times [-1, 2]$. This figure suggests that the graph is symmetric about the y-axis and has a relative maximum at $x = 0$, a horizontal asymptote $y = 0$, and two inflection points. The following analysis will identify the exact locations of these features.

- *Symmetries:* Replacing x by $-x$ does not change the equation, so the graph is symmetric about the y-axis.

- *x-intercepts:* Setting $y = 0$ leads to the equation $e^{-x^2/2} = 0$, which has no solutions since all powers of e have positive values. Thus, there are no x-intercepts.

- *y-intercepts:* Setting $x = 0$ yields the y-intercept $y = 1$.

- *Vertical asymptotes:* There are no vertical asymptotes since $e^{-x^2/2}$ is defined and continuous on $(-\infty, +\infty)$.

- *Horizontal asymptotes:* Since $-x^2/2 \to -\infty$ as $x \to -\infty$ or $x \to +\infty$, it follows from Formula (12) of Section 4.2 that

$$\lim_{x \to -\infty} e^{-x^2/2} = \lim_{x \to +\infty} e^{-x^2/2} = 0$$

Thus, $e^{-x^2/2}$ is asymptotic to $y = 0$ as $x \to -\infty$ and as $x \to +\infty$.

- *Derivatives:*

$$\frac{dy}{dx} = e^{-x^2/2} \frac{d}{dx}\left[-\frac{x^2}{2}\right] = -xe^{-x^2/2}$$

$$\frac{d^2y}{dx^2} = -x\frac{d}{dx}\left[e^{-x^2/2}\right] + e^{-x^2/2}\frac{d}{dx}[-x]$$

$$= x^2 e^{-x^2/2} - e^{-x^2/2} = (x^2 - 1)e^{-x^2/2}$$

- *Intervals of increase and decrease; relative extrema:* Since $e^{-x^2/2} > 0$ for all x, the sign of $dy/dx = -xe^{-x^2/2}$ is the same as the sign of $-x$.

```
              0
   ───────────┼──────────────→  x
   + + + + + + 0 − − − − − −     Sign of dy/dx
   Increasing  Sta  Decreasing   y
```

The analysis reveals a relative maximum $e^0 = 1$ at $x = 0$.

- *Concavity:* Since $e^{-x^2/2} > 0$ for all x, the sign of $d^2y/dx^2 = (x^2 - 1)e^{-x^2/2}$ is the same as the sign of $x^2 - 1$.

```
          −1              1
   ────────┼───────────────┼─────────→  x
   + + + + + 0 − − − − − − 0 + + + + + +   Sign of d²y/dx²
   Concave  Infl  Concave  Infl  Concave   y
     up            down            up
```

Thus, the inflection points occur at $x = -1$ and at $x = 1$. These inflection points are $(-1, e^{-1/2}) \approx (-1, 0.61)$ and $(1, e^{-1/2}) \approx (1, 0.61)$.

Our analysis confirms that the calculator-generated graph in Figure 5.3.19 reveals all of the essential features of the curve $y = e^{-x^2/2}$. ◀

EXERCISE SET 5.3 〰 Graphing Utility

In Exercises 1–10, give a graph of the polynomial and label the coordinates of the stationary points and inflection points. Check your work with a graphing utility.

〰 **1.** $x^2 - 2x - 3$

〰 **2.** $1 + x - x^2$

〰 **3.** $x^3 - 3x + 1$

〰 **4.** $2x^3 - 3x^2 + 12x + 9$

〰 **5.** $x^4 + 2x^3 - 1$

〰 **6.** $x^4 - 2x^2 - 12$

〰 **7.** $3x^5 - 5x^3$

〰 **8.** $3x^4 + 4x^3$

9. $x(x-1)^3$ **10.** $x^5 + 5x^4$

In Exercises 11–22, give a graph of the rational function and label the coordinates of the stationary points and inflection points. Show the horizontal and vertical asymptotes, and label them with their equations. Check your work with a graphing utility.

11. $\dfrac{2x}{x-3}$ **12.** $\dfrac{x}{x^2-1}$ **13.** $\dfrac{x^2}{x^2-1}$

14. $\dfrac{x^2-1}{x^2+1}$ **15.** $x^2 - \dfrac{1}{x}$ **16.** $\dfrac{2x^2-1}{x^2}$

17. $\dfrac{x^3-1}{x^3+1}$ **18.** $\dfrac{8}{4-x^2}$ **19.** $\dfrac{x-1}{x^2-4}$

20. $\dfrac{x+3}{x^2-4}$ **21.** $\dfrac{x+2}{x^2-4}$ **22.** $\dfrac{x^2-1}{x^3-1}$

In Exercises 23–26, the graph of the rational function crosses its horizontal asymptote. Give a graph of the function and label the coordinates of the stationary points and inflection points. Show the horizontal and vertical asymptotes, and label the point(s) where the graph crosses a horizontal asymptote. Check your work with a graphing utility.

23. $\dfrac{(x-1)^2}{x^2}$ **24.** $\dfrac{3x^2-4x-4}{x^2}$

25. $4 + \dfrac{x-1}{x^4}$ **26.** $2 + \dfrac{3}{x} - \dfrac{1}{x^3}$

27. In each part, match the function with graphs I–VI without using a graphing utility, and then use a graphing utility to generate the graphs.

(a) $x^{1/3}$ (b) $x^{1/4}$ (c) $x^{1/5}$

(d) $x^{2/5}$ (e) $x^{4/3}$ (f) $x^{-1/3}$

I
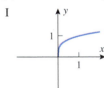

II
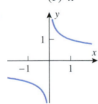

III
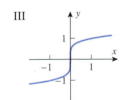

IV
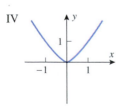

V
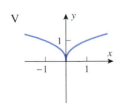

VI
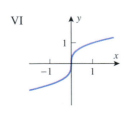

Figure Ex-27

28. Sketch the general shape of the graph of $y = x^{1/n}$, and then explain in words what happens to the shape of the graph as n increases if
(a) n is a positive even integer
(b) n is a positive odd integer.

In Exercises 29–36, give a graph of the function and identify the locations of all critical numbers and inflection points. Check your work with a graphing utility.

29. $\sqrt{x^2-1}$ **30.** $\sqrt[3]{x^2-4}$

31. $2x + 3x^{2/3}$ **32.** $4x - 3x^{4/3}$

33. $x\sqrt{3-x}$ **34.** $4x^{1/3} - x^{4/3}$

35. $\dfrac{8(\sqrt{x}-1)}{x}$ **36.** $\dfrac{1+\sqrt{x}}{1-\sqrt{x}}$

In Exercises 37–42, give a graph of the function and identify the locations of all relative extrema and inflection points. Check your work with a graphing utility.

37. $x + \sin x$ **38.** $x - \cos x$

39. $\sin x + \cos x$ **40.** $\sqrt{3}\cos x + \sin x$

41. $\sin^2 x, \quad 0 \le x \le 2\pi$

42. $x\tan x, \quad -\pi/2 < x < \pi/2$

Using L'Hôpital's rule (Section 4.5) we can verify that
$$\lim_{x \to +\infty} \frac{e^x}{x} = +\infty, \quad \lim_{x \to +\infty} \frac{x}{e^x} = 0, \quad \lim_{x \to -\infty} xe^x = 0$$
In Exercises 43–52: (a) Use these results, as necessary, to find the limits of $f(x)$ as $x \to +\infty$ and as $x \to -\infty$. (b) Give a graph of $f(x)$ and identify all relative extrema, inflection points, and asymptotes (as appropriate). Check your work with a graphing utility.

43. $f(x) = xe^x$ **44.** $f(x) = xe^{-2x}$

45. $f(x) = x^2 e^{-2x}$ **46.** $f(x) = x^2 e^{2x}$

47. $f(x) = xe^{x^2}$ **48.** $f(x) = e^{-1/x^2}$

49. $f(x) = \dfrac{e^x}{x}$ **50.** $f(x) = xe^{-x}$

51. $f(x) = x^2 e^{1-x}$ **52.** $f(x) = x^3 e^{x-1}$

Using L'Hôpital's rule (Section 4.5) we can verify that
$$\lim_{x \to +\infty} \frac{\ln x}{x^r} = 0, \quad \lim_{x \to +\infty} \frac{x^r}{\ln x} = +\infty, \quad \lim_{x \to 0^+} x^r \ln x = 0$$
for any positive real number r. In Exercises 53–58: (a) Use these results, as necessary, to find the limits of $f(x)$ as $x \to +\infty$ and as $x \to 0^+$. (b) Give a graph of $f(x)$ and identify all relative extrema, inflection points, and asymptotes (as appropriate). Check your work with a graphing utility.

53. $f(x) = x \ln x$ **54.** $f(x) = x^2 \ln x$

55. $f(x) = \dfrac{\ln x}{x^2}$ **56.** $f(x) = \dfrac{\ln x}{\sqrt{x}}$

57. $f(x) = x^2 \ln(2x)$ **58.** $f(x) = \ln(x^2 + 1)$

59. In each part: (i) Make a conjecture about the behavior of the graph in the vicinity of its x-intercepts. (ii) Make a rough sketch of the graph based on your conjecture and the limits of the polynomials as $x \to +\infty$ and as $x \to -\infty$. (iii) Compare your sketch to the graph generated with a graphing utility.

  (a) $y = x(x - 1)(x + 1)$ (b) $y = x^2(x - 1)^2(x + 1)^2$
  (c) $y = x^2(x - 1)^2(x + 1)^3$ (d) $y = x(x - 1)^5(x + 1)^4$

60. Sketch the graph of $y = (x - a)^m (x - b)^n$ for the stated values of m and n, assuming that $a < b$ (six graphs in total).

  (a) $m = 1, \ n = 1, 2, 3$ (b) $m = 2, \ n = 2, 3$
  (c) $m = 3, \ n = 3$

61. In each part, make a rough sketch of the graph using asymptotes and appropriate limits but no derivatives. Compare your sketch to that generated with a graphing utility.

  (a) $y = \dfrac{3x^2 - 8}{x^2 - 4}$ (b) $y = \dfrac{x^2 + 2x}{x^2 - 1}$

  (c) $y = \dfrac{2x - x^2}{x^2 + x - 2}$ (d) $y = \dfrac{x^2}{x^2 - x - 2}$

62. (a) Sketch the graph of

$$y = \frac{1}{(x - a)(x - b)}$$

    assuming that $a \neq b$.

  (b) Prove that if $a \neq b$, then the function

$$f(x) = \frac{1}{(x - a)(x - b)}$$

    is symmetric about the line $x = (a + b)/2$.

63. Consider the family of curves $y = xe^{-bx}$ ($b > 0$).
  (a) Use a graphing utility to generate some members of this family.
  (b) Discuss the effect of varying b on the shape of the graph, and discuss the locations of the relative extrema and inflection points.

64. Consider the family of curves $y = e^{-bx^2}$ ($b > 0$).
  (a) Use a graphing utility to generate some members of this family.
  (b) Discuss the effect of varying b on the shape of the graph, and discuss the locations of the relative extrema and inflection points.

65. (a) Determine whether the following limits exist, and if so, find them:

$$\lim_{x \to +\infty} e^x \cos x, \qquad \lim_{x \to -\infty} e^x \cos x$$

  (b) Sketch the graphs of $y = e^x$, $y = -e^x$, and $y = e^x \cos x$ in the same coordinate system, and label any points of intersection.
  (c) Use a graphing utility to generate some members of the family $y = e^{ax} \cos bx$ ($a > 0$ and $b > 0$), and discuss the effect of varying a and b on the shape of the curve.

66. (**Oblique Asymptotes**) If a rational function $P(x)/Q(x)$ is such that the degree of the numerator exceeds the degree of the denominator by *one*, then the graph of $P(x)/Q(x)$ will have an ***oblique asymptote***, that is, an asymptote that is neither vertical nor horizontal. To see why, we perform the division of $P(x)$ by $Q(x)$ to obtain

$$\frac{P(x)}{Q(x)} = (ax + b) + \frac{R(x)}{Q(x)}$$

where $ax + b$ is the quotient and $R(x)$ is the remainder. Use the fact that the degree of the remainder $R(x)$ is less than the degree of the divisor $Q(x)$ to help prove

$$\lim_{x \to +\infty} \left[\frac{P(x)}{Q(x)} - (ax + b) \right] = 0$$

$$\lim_{x \to -\infty} \left[\frac{P(x)}{Q(x)} - (ax + b) \right] = 0$$

As illustrated in the accompanying figure, these results tell us that the graph of the equation $y = P(x)/Q(x)$ "approaches" the line (an oblique asymptote) $y = ax + b$ as $x \to +\infty$ or as $x \to -\infty$.

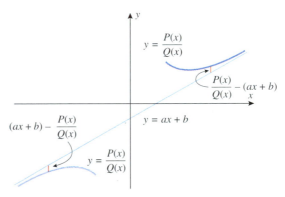

Figure Ex-66

In Exercises 67–71, sketch the graph of the rational function. Show all vertical, horizontal, and oblique asymptotes (see Exercise 66).

67. $\dfrac{x^2 - 2}{x}$ **68.** $\dfrac{x^2 - 2x - 3}{x + 2}$ **69.** $\dfrac{(x - 2)^3}{x^2}$

70. $\dfrac{4 - x^3}{x^2}$ **71.** $x + 1 - \dfrac{1}{x} - \dfrac{1}{x^2}$

72. Find all values of x where the graph of

$$y = \frac{2x^3 - 3x + 4}{x^2}$$

crosses its oblique asymptote. (See Exercise 66.)

73. Let $f(x) = (x^3 + 1)/x$. Show that the graph of $y = f(x)$ approaches the curve $y = x^2$ asymptotically. Sketch the graph of $y = f(x)$ showing this asymptotic behavior.

74. Let $f(x) = (2 + 3x - x^3)/x$. Show that $y = f(x)$ approaches the curve $y = 3 - x^2$ asymptotically. Sketch the graph of $y = f(x)$ showing this asymptotic behavior.

75. A rectangular plot of land is to be fenced off so that the area enclosed will be 400 ft². Let L be the length of fencing needed and x the length of one side of the rectangle. Show that $L = 2x + 800/x$ for $x > 0$, and sketch the graph of L versus x for $x > 0$.

76. A box with a square base and open top is to be made from sheet metal so that its volume is 500 in³. Let S be the area of the surface of the box and x the length of a side of the square base. Show that $S = x^2 + 2000/x$ for $x > 0$, and sketch the graph of S versus x for $x > 0$.

77. The accompanying figure shows a computer-generated graph of the polynomial $y = 0.1x^5(x - 1)$ using a viewing window of $[-2, 2.5] \times [-1, 5]$. Show that the choice of the vertical scale caused the computer to miss important features of the graph. Find the features that were missed and make your own sketch of the graph that shows the missing features.

78. The accompanying figure shows a computer-generated graph of the polynomial $y = 0.1x^5(x + 1)^2$ using a viewing window of $[-2, 1.5] \times [-0.2, 0.2]$. Show that the choice of the vertical scale caused the computer to miss important features of the graph. Find the features that were missed and make your own sketch of the graph that shows the missing features.

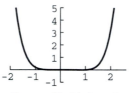

Generated by Mathematica

Figure Ex-77

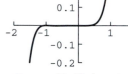

Generated by Mathematica

Figure Ex-78

5.4 RECTILINEAR MOTION (MOTION ALONG A LINE)

In Section 1.5 we discussed the motion of a particle moving with constant velocity along a line, and in Section 3.1 we discussed the motion of a particle moving with variable velocity along a line. In this section we will continue to investigate situations in which a particle may move back and forth with variable velocity along a line. Some examples are a piston moving up and down in a cylinder, a buoy bobbing up and down in the waves, or an object attached to a vibrating spring.

TERMINOLOGY

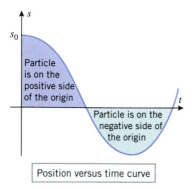

Figure 5.4.1

In this section we will assume that a point representing some object is allowed to move in either direction along a coordinate line. This is called ***rectilinear motion***. The coordinate line might be an x-axis, a y-axis, or an axis that is inclined at some angle. To avoid being specific, we will denote the coordinate line as the s-axis. We will assume that units are chosen for measuring distance and time and that we begin observing the particle at time $t = 0$. As the particle moves along the s-axis, its coordinate is some function of the elapsed time t, say $s = s(t)$. We call $s(t)$ the ***position function*** of the particle, and we call the graph of s versus t the ***position versus time curve***.

Figure 5.4.1 shows a typical position versus time curve for a particle in rectilinear motion. We can tell from that graph that the coordinate of the particle at time $t = 0$ is s_0, and we can tell from the sign of s when the particle is on the negative or the positive side of the origin as it moves along the coordinate line.

Example 1 Figure 5.4.2a shows the position versus time curve for a jackrabbit moving along an s-axis. In words, describe how the position of the rabbit changes with time.

Solution. The rabbit is at $s = -3$ at time $t = 0$. It moves in the positive direction until time $t = 4$, since s is increasing. At time $t = 4$ the rabbit is at position $s = 3$. At that time it turns around and travels in the negative direction until time $t = 7$, since s is decreasing. At time $t = 7$ the rabbit is at position $s = -1$, and it remains stationary thereafter, since s is constant for $t > 7$. This is illustrated in Figure 5.4.2b. ◄

INSTANTANEOUS VELOCITY

We stated in Section 3.1 that the instantaneous velocity of a particle at any time can be interpreted as the slope of the position versus time curve of the particle at that time. Since

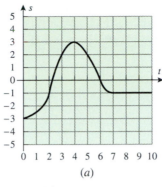

Figure 5.4.2

the slope of this curve is also given by the derivative of the position function for the particle, we make the following formal definition of the velocity function.

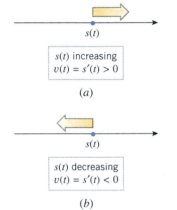

Figure 5.4.3

> **5.4.1**   DEFINITION.   If $s(t)$ is the position function of a particle moving on a coordinate line, then the *instantaneous velocity* of the particle at time t is defined by
>
> $$v(t) = s'(t) = \frac{ds}{dt} \qquad (1)$$

Since the instantaneous velocity at a given time is equal to the slope of the position versus time curve at that time, the sign of the velocity tells us which way the particle is moving—a positive velocity means that s is increasing with time, so the particle is moving in the positive direction; a negative velocity means that s is decreasing with time, so the particle is moving in the negative direction (Figure 5.4.3). For example, in Figure 5.4.2 the rabbit is moving in the positive direction between times $t = 0$ and $t = 4$ and is moving in the negative direction between times $t = 4$ and $t = 7$.

SPEED VERSUS VELOCITY

Recall from our discussion of uniform rectilinear motion in Section 1.5 that there is a distinction between the terms *speed* and *velocity*—speed describes how fast an object is moving without regard to direction, whereas velocity describes how fast it is moving and in what direction. Mathematically, we define the *instantaneous speed* of a particle to be the absolute value of its instantaneous velocity; that is,

$$\begin{bmatrix} \text{instantaneous} \\ \text{speed at} \\ \text{time } t \end{bmatrix} = |v(t)| = \left| \frac{ds}{dt} \right| \qquad (2)$$

For example, if two particles on the same coordinate line have velocities $v = 5$ m/s and $v = -5$ m/s, respectively, then the particles are moving in opposite directions, but they both have a speed of $|v| = 5$ m/s.

Example 2   Let $s(t) = t^3 - 6t^2$ be the position function of a particle moving along an s-axis, where s is in meters and t is in seconds. Find the instantaneous velocity and speed, and show the graphs of position, velocity, and speed versus time.

Solution.   From (1) and (2), the instantaneous velocity and speed are given by

$$v(t) = \frac{ds}{dt} = 3t^2 - 12t \quad \text{and} \quad |v(t)| = |3t^2 - 12t|$$

The graphs of position, velocity, and speed versus time are shown in Figure 5.4.4. Observe that velocity and speed both have units of meters per second (m/s), since s is in meters (m) and time is in seconds (s). ◀

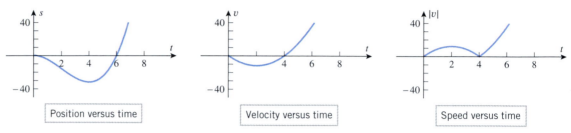

Position versus time

Velocity versus time

Speed versus time

Figure 5.4.4

The graphs in Figure 5.4.4 provide a wealth of visual information about the motion of the particle. For example, the position versus time curve tells us that the particle is on the negative side of the origin for $0 < t < 6$, is on the positive side of the origin for $t > 6$, and is at the origin at times $t = 0$ and $t = 6$. The velocity versus time curve tells us that the particle is moving in the negative direction if $0 < t < 4$, is moving in the positive direction if $t > 4$, and is momentarily stopped at times $t = 0$ and $t = 4$ (the velocity is zero at those times). The speed versus time curve tells us that the speed of the particle is increasing for $0 < t < 2$, decreasing for $2 < t < 4$, and increasing again for $t > 4$.

ACCELERATION

In rectilinear motion, the rate at which the velocity of a particle changes with time is called its *acceleration*. More precisely, we make the following definition.

5.4.2 DEFINITION. If $s(t)$ is the position function of a particle moving on a coordinate line, then the instantaneous acceleration of the particle at time t is defined by

$$a(t) = v'(t) = \frac{dv}{dt} \tag{3}$$

or alternatively, since $v(t) = s'(t)$,

$$a(t) = s''(t) = \frac{d^2s}{dt^2} \tag{4}$$

Example 3 Let $s(t) = t^3 - 6t^2$ be the position function of a particle moving along an s-axis, where s is in meters and t is in seconds. Find the instantaneous acceleration $a(t)$, and show the graph of acceleration versus time.

Solution. From Example 2, the instantaneous velocity of the particle is $v(t) = 3t^2 - 12t$, so the instantaneous acceleration is

$$a(t) = \frac{dv}{dt} = 6t - 12$$

and the acceleration versus time curve is the line shown in Figure 5.4.5. Note that in this example the acceleration has units of m/s^2, since v is in meters per second (m/s) and time is in seconds (s). ◀

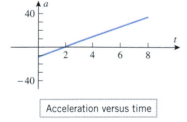

Acceleration versus time

Figure 5.4.5

SPEEDING UP AND SLOWING DOWN

We will say that a particle in rectilinear motion is *speeding up* when its instantaneous speed is increasing and is *slowing down* when its instantaneous speed is decreasing. In everyday language an object that is speeding up is said to be "accelerating" and an object that is slowing down is said to be "decelerating"; thus, one might expect that a particle in rectilinear motion will be speeding up when its instantaneous acceleration is positive and slowing down when it is negative. Although this is true for a particle moving in the positive direction, it is *not* true for a particle moving in the negative direction—a particle with negative velocity is speeding up when its acceleration is negative and slowing down when its acceleration is positive. This is because a positive acceleration implies an increasing velocity, and increasing a negative velocity decreases its absolute value; similarly, a negative acceleration implies a decreasing velocity, and decreasing a negative velocity increases its absolute value.

The following statement, which we will ask you to prove in Exercise 39, summarizes these informal ideas.

> **5.4.3** INTERPRETING THE SIGN OF ACCELERATION. *A particle in rectilinear motion is speeding up when its velocity and acceleration have the same sign and slowing down when they have opposite signs.*

FOR THE READER. For a particle in rectilinear motion, what is happening when $v(t) = 0$? When $a(t) = 0$?

Example 4 In Examples 2 and 3 we found the velocity versus time curve and the acceleration versus time curve for a particle with position function $s(t) = t^3 - 6t^2$. Use those curves to determine when the particle is speeding up and slowing down, and confirm that your results are consistent with the speed versus time curve obtained in Example 2.

Solution. Over the time interval $0 < t < 2$ the velocity and acceleration are negative, so the particle is speeding up. This is consistent with the speed versus time curve, since the speed is increasing over this time interval. Over the time interval $2 < t < 4$ the velocity is negative and the acceleration is positive, so the particle is slowing down. This is also consistent with the speed versus time curve, since the speed is decreasing over this time interval. Finally, on the time interval $t > 4$ the velocity and acceleration are positive, so the particle is speeding up, which again is consistent with the speed versus time curve. ◀

ANALYZING THE POSITION VERSUS TIME CURVE

The position versus time curve contains all of the significant information about the position and velocity of a particle in rectilinear motion:

- If $s(t) > 0$, the particle is on the positive side of the s-axis.
- If $s(t) < 0$, the particle is on the negative side of the s-axis.
- The slope of the curve at any time is equal to the instantaneous velocity at that time.
- Where the curve has positive slope, the velocity is positive and the particle is moving in the positive direction.
- Where the curve has negative slope, the velocity is negative and the particle is moving in the negative direction.
- Where the slope of the curve is zero, the velocity is zero, and the particle is momentarily stopped.

Information about the acceleration of a particle in rectilinear motion can also be deduced from the position versus time curve by examining its concavity. To see why this is so, observe that the position versus time curve will be concave up on intervals where $s''(t) > 0$, and it will be concave down on intervals where $s''(t) < 0$. But we know from (4) that $s''(t)$ is the instantaneous acceleration, so that on intervals where the position versus time curve is concave up the particle has a positive acceleration, and on intervals where it is concave down the particle has a negative acceleration.

Table 5.4.1 summarizes our observations about the position versus time curve.

Table 5.4.1

POSITION VERSUS TIME CURVE	CHARACTERISTICS OF THE CURVE AT $t = t_0$	BEHAVIOR OF THE PARTICLE AT TIME $t = t_0$
	• $s(t_0) > 0$ • Curve has positive slope. • Curve is concave down.	• Particle is on the positive side of the origin. • Particle is moving in the positive direction. • Velocity is decreasing. • Particle is slowing down.
	• $s(t_0) > 0$ • Curve has negative slope. • Curve is concave down.	• Particle is on the positive side of the origin. • Particle is moving in the negative direction. • Velocity is decreasing. • Particle is speeding up.
	• $s(t_0) < 0$ • Curve has negative slope. • Curve is concave up.	• Particle is on the negative side of the origin. • Particle is moving in the negative direction. • Velocity is increasing. • Particle is slowing down.
	• $s(t_0) > 0$ • Curve has zero slope. • Curve is concave down.	• Particle is on the positive side of the origin. • Particle is momentarily stopped. • Velocity is decreasing.

Example 5 Use the position versus time curve in Figure 5.4.2 to determine when the jackrabbit in Example 1 is speeding up and slowing down.

Solution. From $t = 0$ to $t = 2$, the acceleration and velocity are positive, so the rabbit is speeding up. From $t = 2$ to $t = 4$, the acceleration is negative and the velocity is positive, so the rabbit is slowing down. At $t = 4$, the velocity is zero, so the rabbit has momentarily stopped. From $t = 4$ to $t = 6$, the acceleration is negative and the velocity is negative, so the rabbit is speeding up. From $t = 6$ to $t = 7$, the acceleration is positive and the velocity is negative, so the rabbit is slowing down. Thereafter, the velocity is zero, so the rabbit has stopped. ◄

Example 6 Suppose that the position function of a particle moving on a coordinate line is given by $s(t) = 2t^3 - 21t^2 + 60t + 3$. Analyze the motion of the particle for $t \geq 0$.

Solution. The velocity and acceleration at time t are

$$v(t) = s'(t) = 6t^2 - 42t + 60 = 6(t - 2)(t - 5)$$
$$a(t) = v'(t) = 12t - 42 = 12\left(t - \tfrac{7}{2}\right)$$

At each instant we can determine the direction of motion from the sign of $v(t)$ and whether the particle is speeding up or slowing down from the signs of $v(t)$ and $a(t)$ together (Figures 5.4.6a and 5.4.6b). The motion of the particle is described schematically by the curved line in Figure 5.4.6c. At time $t = 0$ the particle is at $s(0) = 3$ moving right with velocity $v(0) = 60$, but slowing down with acceleration $a(0) = -42$. The particle continues moving right until time $t = 2$, when it stops at $s(2) = 55$, reverses direction, and begins to speed up with an acceleration of $a(2) = -18$. At time $t = \tfrac{7}{2}$ the particle begins to slow down, but continues moving left until time $t = 5$, when it stops at $s(5) = 28$, reverses direction again,

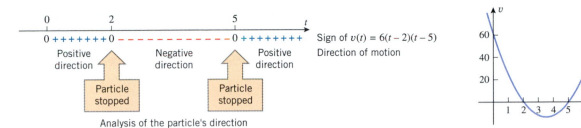

Figure 5.4.6a

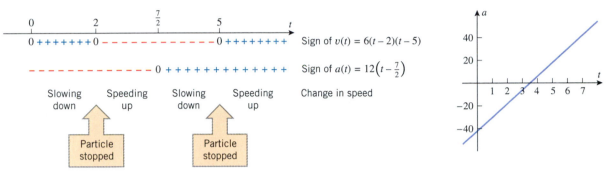

Figure 5.4.6b

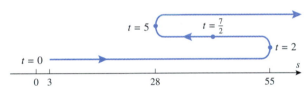

Figure 5.4.6c

and begins to speed up with acceleration $a(5) = 18$. The particle then continues moving right thereafter with increasing speed.   ◀

REMARK.   The curved line in Figure 5.4.6c is descriptive only. The actual path of the particle is back and forth on the coordinate line.

FOR THE READER.   Figure 5.4.7a shows the graph of the position function $s(t)$ for the particle in Example 6, and Figure 5.4.7b shows the graphs of position, velocity, and acceleration superimposed in one figure. Describe how the signs and slopes of the velocity and acceleration curves relate to the shape of the graph of the position function.

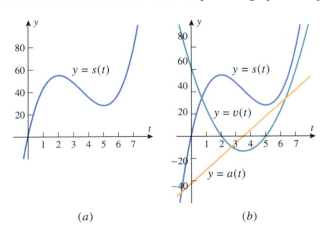

Figure 5.4.7   (a)   (b)

FREE-FALL MOTION

s-axis

s

Height

Earth

Figure 5.4.8

We will now discuss how some of the ideas in this section can be applied to the study of *free-fall motion*, which is the motion that occurs when an object near the Earth is imparted some initial vertical velocity (up or down), and thereafter moves on a vertical line. In modeling free-fall motion it is assumed that the only force acting on the object is the Earth's gravity and that the object stays sufficiently close to the Earth's surface so that the gravitational force is constant. In particular, air resistance and the gravitational pull of other celestial bodies are neglected.

In our study of free-fall motion, we will ignore the physical size of the object by treating it as a particle, and we will assume that the object moves along an *s*-axis whose origin is at the surface of the Earth and whose positive direction is up. With this convention, the *s*-coordinate of the particle is the height of the particle above the Earth's surface (Figure 5.4.8). The following result will be derived later using calculus and some basic principles of physics.

5.4.4   THE FREE-FALL MODEL.   Suppose that at time $t = 0$ an object at a height of s_0 above the Earth's surface is imparted an upward or downward velocity of v_0 and thereafter moves vertically subject only to the force of the Earth's gravity. If the positive direction of the *s*-axis is up, and if the origin is at the surface of the Earth, then at any time t the height $s = s(t)$ of the object is given by the formula

$$s = s_0 + v_0 t - \tfrac{1}{2} g t^2 \tag{5}$$

where g is a constant, called the *acceleration due to gravity*. In this text we will use the following approximations for g, depending on the units of measurement:

$g = 9.8 \text{ m/s}^2$ [distance in meters and time in seconds]

$g = 32 \text{ ft/s}^2$ [distance in feet and time in seconds]

It follows from (5) that the instantaneous velocity and acceleration of an object in free-fall motion are

$$v = \frac{ds}{dt} = v_0 - gt \tag{6}$$

$$a = \frac{dv}{dt} = -g \tag{7}$$

REMARK.   Because we have chosen the positive direction of the *s*-axis to be up, a positive velocity implies an upward motion and a negative velocity a downward motion. Thus, it makes sense that instantaneous acceleration $-g$ is negative, since an upward-moving object has positive velocity and negative acceleration, which implies that it is slowing down; and a downward-moving object has negative velocity and negative acceleration, which implies that it is speeding up. (It is a little confusing that the positive constant g is called the *acceleration due to gravity* in 5.4.4, given that the instantaneous acceleration is actually the negative constant $-g$. This mismatch in terminology is caused by the upward orientation of the *s*-axis in Figure 5.4.8; had we chosen the positive direction to be down, then the instantaneous acceleration would have turned out to be g. However, our orientation has the advantage of allowing us to interpret s as the height of the object.)

Example 7   Nolan Ryan, one of the fastest baseball pitchers of all time, was capable of throwing a baseball 150 ft/s (over 102 mi/h). During his career, he had the opportunity to pitch in the Houston Astrodome, home to the Houston Astros Baseball Team from 1965 to 1999. The Astrodome was an indoor stadium with a ceiling 208 ft high. Could Nolan Ryan have hit the ceiling of the Astrodome if he were capable of giving a baseball an upward velocity of 100 ft/s from a height of 7 ft?

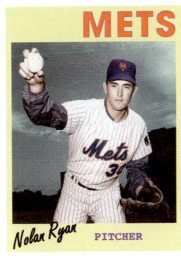

Nolan Ryan's rookie baseball card

Solution. Taking $g = 32$ ft/s^2, $v_0 = 100$ ft/s, and $s_0 = 7$ ft in (5) and (6) yields the equations

$$s = 7 + 100t - 16t^2 \quad \text{and} \quad v = 100 - 32t \tag{8–9}$$

whose graphs are shown in Figure 5.4.9. It is evident from the graph of s versus t that the maximum height of the baseball is less than 208 ft, so Ryan could not have hit the ceiling. However, let us go a step further and determine exactly how high the ball will go. The maximum height s occurs at the stationary point obtained by solving the equation $ds/dt = 0$. However, $ds/dt = v$, which means that the maximum height occurs when $v = 0$, which from (9) can be expressed as

$$100 - 32t = 0 \tag{10}$$

Solving this equation yields $t = 25/8$. To find the height s at this time we substitute this value of t in (8), from which we obtain

$$s = 7 + 100(25/8) - 16(25/8)^2 = 163.25 \text{ ft}$$

which is roughly 45 ft short of hitting the ceiling.   ◄

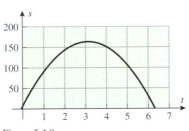

Figure 5.4.9

> **REMARK.** Equation (10) can also be deduced by physical reasoning: The ball is moving up when the velocity is positive and moving down when the velocity is negative, so it makes sense that the velocity is zero when the ball reaches its peak.

EXERCISE SET 5.4   ∿ Graphing Utility

1. The graphs of three position functions are shown in the accompanying figure. In each case determine the signs of the velocity and acceleration, then determine whether the particle is speeding up or slowing down.

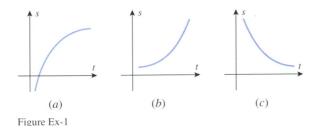

   (a)   (b)   (c)

   Figure Ex-1

2. The graphs of three velocity functions are shown in the accompanying figure. In each case determine the sign of the acceleration, then determine whether the particle is speeding up or slowing down.

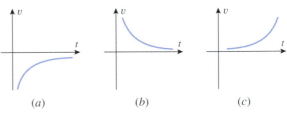

   (a)   (b)   (c)

   Figure Ex-2

3. The position function of a particle moving on a horizontal x-axis is shown in Figure Ex-3 (next page).
   (a) Is the particle moving left or right at time t_0?
   (b) Is the acceleration positive or negative at time t_0?
   (c) Is the particle speeding up or slowing down at time t_0?
   (d) Is the particle speeding up or slowing down at time t_1?

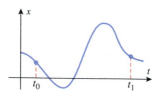

Figure Ex-3

4. For the graphs in the accompanying figure, match the position functions with their corresponding velocity functions.

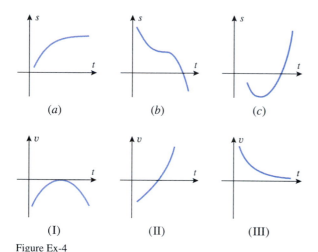

(a)   (b)   (c)

(I)   (II)   (III)

Figure Ex-4

5. Sketch a reasonable graph of s versus t for a mouse that is trapped in a narrow corridor (an s-axis with the positive direction to the right) and scurries back and forth as follows. It runs right with a constant speed of 1.2 m/s for awhile, then gradually slows down to 0.6 m/s, then quickly speeds up to 2.0 m/s, then gradually slows to a stop but immediately reverses direction and quickly speeds up to 1.2 m/s.

6. The accompanying figure shows the graph of s versus t for an ant that moves along a narrow vertical pipe (an s-axis with the positive direction up).
(a) When, if ever, is the ant above the origin?
(b) When, if ever, does the ant have velocity zero?
(c) When, if ever, is the ant moving down the pipe?

7. The accompanying figure shows the graph of velocity versus time for a particle moving along a coordinate line. Make a rough sketch of the graphs of speed versus time and acceleration versus time.

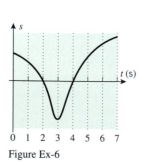

Figure Ex-6

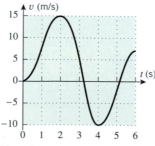

Figure Ex-7

8. The accompanying figure shows the position versus time graph for an elevator that ascends 40 m from one stop to the next.
(a) Estimate the velocity when the elevator is halfway up.
(b) Sketch rough graphs of the velocity versus time curve and the acceleration versus time curve.

9. The accompanying figure shows the velocity versus time graph for a test run on a classic Grand Prix GTP. Using this graph, estimate
(a) the acceleration at 60 mi/h (in units of ft/s²)
(b) the time at which the maximum acceleration occurs.
[Data from *Car and Driver Magazine*, October 1990.]

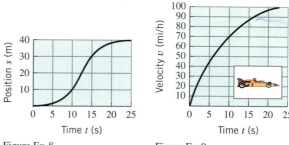

Figure Ex-8   Figure Ex-9

10. Let $s(t) = \sin(\pi t/4)$ be the position function of a particle moving along a coordinate line, where s is in meters and t is in seconds.
(a) Make a table showing the position, velocity, and acceleration to two decimal places at times $t = 1, 2, 3, 4$, and 5.
(b) At each of the times in part (a), determine whether the particle is stopped; if it is not, state its direction of motion.
(c) At each of the times in part (a), determine whether the particle is speeding up, slowing down, or neither.

In Exercises 11–14, the position function of a particle moving along a coordinate line is given, where s is in feet and t is in seconds.
(a) Find the velocity and acceleration functions.
(b) Find the position, velocity, speed, and acceleration at time $t = 1$.
(c) At what times is the particle stopped?
(d) When is the particle speeding up? Slowing down?
(e) Find the total distance traveled by the particle from time $t = 0$ to time $t = 5$.

11. $s(t) = t^3 - 6t^2, \quad t \geq 0$

12. $s(t) = t^4 - 4t + 2, \quad t \geq 0$

13. $s(t) = 3\cos(\pi t/2), \quad 0 \leq t \leq 5$

14. $s(t) = \dfrac{t}{t^2 + 4}, \quad t \geq 0$

15. Let $s(t) = t/(t^2 + 5)$ be the position function of a particle moving along a coordinate line, where s is in meters and t is in seconds. Use a graphing utility to generate the graphs of $s(t)$, $v(t)$, and $a(t)$ for $t \geq 0$, and use those graphs where needed.

(a) Use the appropriate graph to make a rough estimate of the time at which the particle first reverses the direction of its motion; and then find the time exactly.

(b) Find the exact position of the particle when it first reverses the direction of its motion.

(c) Use the appropriate graphs to make a rough estimate of the time intervals on which the particle is speeding up and on which it is slowing down; and then find those time intervals exactly.

 16. Let $s(t) = t/e^t$ be the position function of a particle moving along a coordinate line, where s is in meters and t is in seconds. Use a graphing utility to generate the graphs of $s(t)$, $v(t)$, and $a(t)$ for $t \geq 0$, and use those graphs where needed.

(a) Use the appropriate graph to make a rough estimate of the time at which the particle first reverses the direction of its motion; and then find the time exactly.

(b) Find the exact position of the particle when it first reverses the direction of its motion.

(c) Use the appropriate graphs to make a rough estimate of the time intervals on which the particle is speeding up and on which it is slowing down; and then find those time intervals exactly.

In Exercises 17–22, the position function of a particle moving along a coordinate line is given. Use the method of Example 6 to analyze the motion of the particle for $t \geq 0$, and give a schematic picture of the motion (as in Figure 5.4.6).

---

17. $s = -3t + 2$

18. $s = t^3 - 6t^2 + 9t + 1$

19. $s = t^3 - 9t^2 + 24t$

20. $s = t + \dfrac{9}{t+1}$

21. $s = \begin{cases} \cos t, & 0 \leq t \leq 2\pi \\ 1, & t > 2\pi \end{cases}$

22. $s = \sqrt{t}(4 - 4t + 2t^2)$

23. Let $s(t) = 5t^2 - 22t$ be the position function of a particle moving along a coordinate line, where s is in feet and t is in seconds.

(a) Find the maximum speed of the particle during the time interval $1 \leq t \leq 3$.

(b) When, during the time interval $1 \leq t \leq 3$, is the particle farthest from the origin? What is its position at that instant?

24. Let $s = 100/(t^2 + 12)$ be the position function of a particle moving along a coordinate line, where s is in feet and t is in seconds. Find the maximum speed of the particle for $t \geq 0$, and find the direction of motion of the particle when it has its maximum speed.

In Exercises 25–29, assume that the free-fall model applies and that the positive direction is up, so that Formulas (5), (6), and (7) can be used. In those problems stating that an object is "dropped" or "released from rest," you should interpret that to mean that the initial velocity of the object is zero. Take $g = 32$ ft/s^2 or $g = 9.8$ m/s^2, depending on the units.

---

25. A wrench is accidentally dropped at the top of an elevator shaft in a tall building.

(a) How many meters does the wrench fall in 1.5 s?

(b) What is the velocity of the wrench at that time?

(c) How long does it take for the wrench to reach a speed of 12 m/s?

(d) How long does it take for the wrench to fall 100 m?

26. In 1939, Joe Sprinz of the San Francisco Seals Baseball Club attempted to catch a ball dropped from a blimp at a height of 800 ft (for the purpose of breaking the record for catching a ball dropped from the greatest height set the preceding year by members of the Cleveland Indians).

(a) How long does it take for a ball to drop 800 ft?

(b) What is the velocity of a ball in miles per hour after an 800-ft drop (88 ft/s $= 60$ mi/h)?

[*Note:* As a practical matter, it is unrealistic to ignore wind resistance in this problem; however, even with the slowing effect of wind resistance, the impact of the ball slammed Sprinz's glove hand into his face, fractured his upper jaw in 12 places, broke five teeth, and knocked him unconscious. He dropped the ball!]

27. A projectile is launched upward from ground level with an initial speed of 60 m/s.

(a) How long does it take for the projectile to reach its highest point?

(b) How high does the projectile go?

(c) How long does it take for the projectile to drop back to the ground from its highest point?

(d) What is the speed of the projectile when it hits the ground?

28. (a) Use the results in Exercise 27 to make a conjecture about the relationship between the initial and final speeds of a projectile that is launched upward from ground level and returns to ground level.

(b) Prove your conjecture.

29. In Example 7, how fast would Nolan Ryan have to throw a ball upward from a height of 7 feet in order to hit the ceiling of the Astrodome?

30. The free-fall formulas (5) and (6) can be combined and rearranged in various useful ways. Derive the following variations of those formulas.

(a) $v^2 = v_0^2 - 2g(s - s_0)$ (b) $s = s_0 + \frac{1}{2}(v_0 + v)t$

(c) $s = s_0 + vt + \frac{1}{2}gt^2$

31. A rock, dropped from an unknown height, strikes the ground with a speed of 24 m/s. Use the formula in part (a) of Exercise 30 to find the unknown height.

32. A rock thrown downward with an unknown initial velocity from a height of 1000 ft reaches the ground in 5 s. Use the formula in part (c) of Exercise 30 to find the velocity of the rock when it hits the ground.

33. (a) A ball is thrown upward from a height s_0 with an initial velocity of v_0. Use the formula in part (a) of Exercise 30 to show that the maximum height of the ball is $s_{max} = s_0 + v_0^2/2g$.

(b) Use this result to solve Exercise 29.

34. Let $s = t^3 - 6t^2 + 1$.

    (a) Find s and v when $a = 0$.

    (b) Find s and a when $v = 0$.

35. Let $s = \sqrt{2t^2 + 1}$ be the position function of a particle moving along a coordinate line.

    (a) Use a graphing utility to generate the graph of v versus t, and make a conjecture about the velocity of the particle as $t \to +\infty$.

    (b) Check your conjecture by finding $\lim\limits_{t \to +\infty} v$.

36. (a) Use the chain rule to show that for a particle in rectilinear motion $a = v(dv/ds)$.

    (b) Let $s = \sqrt{3t + 7}, \, t \geq 0$. Find a formula for v in terms of s and use the equation in part (a) to find the acceleration when $s = 5$.

37. Suppose that the position functions of two particles, P_1 and P_2, in motion along the same line are

$$s_1 = \tfrac{1}{2}t^2 - t + 3 \quad \text{and} \quad s_2 = -\tfrac{1}{4}t^2 + t + 1$$

respectively, for $t \geq 0$.

    (a) Prove that P_1 and P_2 do not collide.

    (b) How close can P_1 and P_2 get to one another?

    (c) During what intervals of time are they moving in opposite directions?

38. Let $s_A = 15t^2 + 10t + 20$ and $s_B = 5t^2 + 40t, \, t \geq 0$, be the position functions of cars A and B that are moving along parallel straight lanes of a highway.

    (a) How far is car A ahead of car B when $t = 0$?

    (b) At what instants of time are the cars next to one another?

    (c) At what instant of time do they have the same velocity? Which car is ahead at this instant?

39. Prove that a particle is speeding up if the velocity and acceleration have the same sign, and slowing down if they have opposite signs. [*Hint:* Let $r(t) = |v(t)|$ and find $r'(t)$ using the chain rule.]

5.5 ABSOLUTE MAXIMA AND MINIMA

At the beginning of Section 5.2 we observed that if the graph of a function f is viewed as a two-dimensional mountain range (Figure 5.2.1), then the relative maxima and minima correspond to the tops of the hills and the bottoms of the valleys; that is, they are the high and low points in their immediate vicinity. In this section we will be concerned with the more encompassing problem of finding the highest and lowest points over the entire mountain range, that is, we will be looking for the top of the highest hill and the bottom of the deepest valley. In mathematical terms, we will be looking for the largest and smallest values of a function over an interval.

ABSOLUTE EXTREMA

We will be concerned here with finding the largest and smallest values of a function over a finite or infinite interval I. We begin with some terminology.

> **5.5.1 DEFINITION.** A function f is said to have an ***absolute maximum*** on an interval I at x_0 if $f(x_0)$ is the largest value of f on I; that is, $f(x_0) \geq f(x)$ for all x in the domain of f that are in I. Similarly, f is said to have an ***absolute minimum*** on I at x_0 if $f(x_0)$ is the smallest value of f on I; that is, $f(x_0) \leq f(x)$ for all x in the domain of f that are in I. If f has either an absolute maximum or absolute minimum on I at x_0, then f is said to have an ***absolute extremum*** on I at x_0.

As illustrated in Figure 5.5.1, there is no guarantee that a function f will have absolute extrema on a given interval.

EXISTENCE OF ABSOLUTE EXTREMA

The remainder of this section will focus on the following problem.

> **5.5.2 PROBLEM.**
> (a) Determine whether a function f has any absolute extrema on a given interval I.
> (b) If there are absolute extrema, determine where they occur and what the absolute maximum and minimum values are.

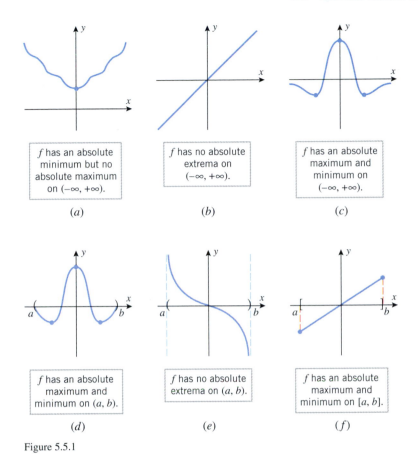

Figure 5.5.1

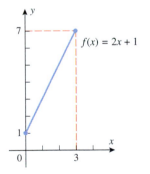

Figure 5.5.2

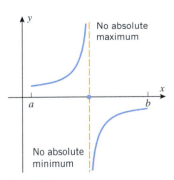

Figure 5.5.3

Parts (*a*)–(*e*) of Figure 5.5.1 show that a continuous function may or may not have absolute maxima or minima on an infinite interval or on a finite open interval. However, the following theorem shows that a continuous function must have both an absolute maximum and an absolute minimum on every *finite closed* interval [see part (*f*) of Figure 5.5.1].

5.5.3 THEOREM (*Extreme-Value Theorem*). *If a function f is continuous on a finite closed interval* [*a, b*], *then f has both an absolute maximum and an absolute minimum on* [*a, b*].

FOR THE READER. Although the proof of this theorem is too difficult to include here, you should be able to convince yourself of its validity with a little experimentation—try graphing various continuous functions over the interval [0, 1], and convince yourself that there is no way to avoid having a highest and lowest point on the graph. As a physical analogy, if you imagine the graph to be a roller coaster track starting at $x = 0$ and ending at $x = 1$, the roller coaster will have to pass through a highest point and a lowest point during the trip.

The function $f(x) = 2x + 1$ is continuous everywhere, so the Extreme-Value Theorem guarantees that $f(x)$ has both an absolute maximum and an absolute minimum on every finite closed interval. For example, on the interval [0, 3], the absolute minimum occurs at $x = 0$ and the absolute maximum occurs at $x = 3$. The absolute minimum and maximum values for $f(x)$ on [0, 3] are $f(0) = 1$ and $f(3) = 7$, respectively (Figure 5.5.2).

The hypotheses of the Extreme-Value Theorem are essential. Figure 5.5.3 shows the graph of a function that is defined on a closed interval [*a, b*] but fails to be continuous on that interval. This function has neither an absolute maximum nor an absolute minimum on the interval $[a, b]$. If f is continuous on an interval that is not both closed and finite, then we could encounter situations such as those in Figure 5.5.1.

To illustrate further, consider again the function $f(x) = 2x + 1$, but now for values of x in the half-open interval $[0, 3)$. The function f has an absolute minimum value of 1 at $x = 0$ in the interval $[0, 3)$. However, for any number x_0 in $[0, 3)$ that we might choose as a candidate for the location of an absolute maximum, we can find another number, say $x_1 = (x_0 + 3)/2$, also in $[0, 3)$, with $f(x_1) > f(x_0)$ (Figure 5.5.4). Thus, for any particular value of $f(x)$ on $[0, 3)$, we can find a larger value of the function on this interval; that is, f does not attain an absolute maximum value on $[0, 3)$.

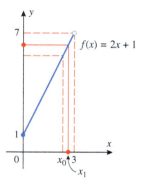

Figure 5.5.4

FINDING ABSOLUTE EXTREMA ON FINITE CLOSED INTERVALS

The Extreme-Value Theorem is an example of what mathematicians call an *existence theorem*. Such theorems state conditions under which certain objects exist, in this case absolute extrema. However, knowing that an object exists and finding it are two separate things. We will now address methods for determining the locations of absolute extrema under the conditions of the Extreme-Value Theorem.

If f is continuous on the finite closed interval $[a, b]$, then the absolute extrema of f can occur either at the endpoints of the interval or inside on the open interval (a, b). If the absolute extrema happen to fall inside, then the following theorem tells us that they must occur at critical numbers of f.

5.5.4 THEOREM. *If f has an absolute extremum on an open interval (a, b), then it must occur at a critical number of f.*

Proof. If f has an absolute maximum on (a, b) at x_0, then $f(x_0)$ is also a relative maximum for f; for if $f(x_0)$ is the largest value of f on all of (a, b), then $f(x_0)$ is certainly the largest value for f in the immediate vicinity of x_0. Thus, x_0 is a critical number of f by Theorem 5.2.2. The proof for absolute minima is similar. ∎

• REMARK.   Theorem 5.5.4 is also valid for functions on infinite open intervals.

It follows from this theorem, that if f is continuous on the finite closed interval $[a, b]$, then the absolute extrema occur either at the endpoints of the interval or at critical numbers inside the interval (Figure 5.5.5). Thus, we can use the following procedure to find the absolute extrema of a continuous function on a finite closed interval $[a, b]$.

A Procedure for Finding the Absolute Extrema of a Continuous Function f on a Finite Closed Interval $[a, b]$.

Step 1.   Find the critical numbers of f in (a, b).

Step 2.   Evaluate f at all the critical numbers and at the endpoints a and b.

Step 3.   The largest of the values in Step 2 is the absolute maximum value of f on $[a, b]$ and the smallest value is the absolute minimum.

Example 1   Find the absolute maximum and minimum values of $f(x) = 2x^3 - 15x^2 + 36x$ on the interval $[1, 5]$, and determine where these values occur.

Solution.   Since f is continuous and differentiable everywhere, the absolute extrema must occur either at endpoints of the interval or at solutions to the equation $f'(x) = 0$ in the open interval $(1, 5)$. The equation $f'(x) = 0$ can be written as

$$6x^2 - 30x + 36 = 6(x^2 - 5x + 6) = 6(x - 2)(x - 3) = 0$$

Figure 5.5.5

Absolute maximum occurs at an endpoint.

Absolute maximum occurs in the open interval (a, b) at a value x_0 where $f'(x_0) = 0$.

Absolute maximum occurs in the open interval (a, b) at a value x_0 where f is not differentiable.

Thus, there are stationary points at $x = 2$ and at $x = 3$. Evaluating f at the endpoints, at $x = 2$, and at $x = 3$ yields

$$f(1) = 2(1)^3 - 15(1)^2 + 36(1) = 23$$
$$f(2) = 2(2)^3 - 15(2)^2 + 36(2) = 28$$
$$f(3) = 2(3)^3 - 15(3)^2 + 36(3) = 27$$
$$f(5) = 2(5)^3 - 15(5)^2 + 36(5) = 55$$

from which we conclude that an absolute minimum of f on $[1, 5]$ is 23, occurring at $x = 1$, and the absolute maximum of f on $[1, 5]$ is 55, occurring at $x = 5$. This is consistent with the graph of f in Figure 5.5.6. ◄

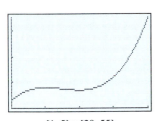

$[1, 5] \times [20, 55]$
xScl $= 1$, yScl $= 10$

$y = 2x^3 - 15x^2 + 36x$

Figure 5.5.6

Example 2 Find the absolute extrema of $f(x) = 6x^{4/3} - 3x^{1/3}$ on the interval $[-1, 1]$, and determine where these values occur.

Solution. Note that f is continuous everywhere and therefore the Extreme-Value Theorem guarantees that f has a maximum and a minimum value in the interval $[-1, 1]$. Differentiating, we obtain

$$f'(x) = 8x^{1/3} - x^{-2/3} = x^{-2/3}(8x - 1) = \frac{8x - 1}{x^{2/3}}$$

Table 5.5.1

x	-1	0	$\frac{1}{8}$	1
$f(x)$	9	0	$-\frac{9}{8}$	3

Thus, $f'(x) = 0$ at $x = \frac{1}{8}$, and $f'(x)$ is undefined at $x = 0$. Evaluating f at these critical numbers and endpoints yields Table 5.5.1, from which we conclude that an absolute minimum value of $-\frac{9}{8}$ occurs at $x = \frac{1}{8}$, and an absolute maximum value of 9 occurs at $x = -1$. ◄

ABSOLUTE EXTREMA ON INFINITE INTERVALS

We observed earlier that a continuous function may or may not have absolute extrema on an infinite interval (see Figure 5.5.1). However, certain conclusions about the existence of absolute extrema of a continuous function f on $(-\infty, +\infty)$ can be drawn from the behavior of $f(x)$ as $x \to -\infty$ and as $x \to +\infty$ (Table 5.5.2).

Example 3 What can you say about the existence of absolute extrema on $(-\infty, +\infty)$ for polynomials?

Solution. If $p(x)$ is a polynomial of odd degree, then

$$\lim_{x \to +\infty} p(x) \quad \text{and} \quad \lim_{x \to -\infty} p(x) \tag{1}$$

have opposite signs (one is $+\infty$ and the other is $-\infty$), so there are no absolute extrema. On the other hand, if $p(x)$ has even degree, then the limits in (1) have the same sign (both $+\infty$ or both $-\infty$). If the leading coefficient is positive, then both limits are $+\infty$, and there is an

Table 5.5.2

LIMITS	$\lim\limits_{x \to -\infty} f(x) = +\infty$ $\lim\limits_{x \to +\infty} f(x) = +\infty$	$\lim\limits_{x \to -\infty} f(x) = -\infty$ $\lim\limits_{x \to +\infty} f(x) = -\infty$	$\lim\limits_{x \to -\infty} f(x) = -\infty$ $\lim\limits_{x \to +\infty} f(x) = +\infty$	$\lim\limits_{x \to -\infty} f(x) = +\infty$ $\lim\limits_{x \to +\infty} f(x) = -\infty$
CONCLUSION IF f **IS CONTINUOUS EVERYWHERE**	f has an absolute minimum but no absolute maximum on $(-\infty, +\infty)$.	f has an absolute maximum but no absolute minimum on $(-\infty, +\infty)$.	f has neither an absolute maximum nor an absolute minimum on $(-\infty, +\infty)$.	f has neither an absolute maximum nor an absolute minimum on $(-\infty, +\infty)$.
GRAPH				

absolute minimum but no absolute maximum; if the leading coefficient is negative, then both limits are $-\infty$, and there is an absolute maximum but no absolute minimum. ◀

Example 4 Determine by inspection whether $p(x) = 3x^4 + 4x^3$ has any absolute extrema. If so, find them and state where they occur.

Solution. Since $p(x)$ has even degree and the leading coefficient is positive, $p(x) \to +\infty$ as $x \to \pm\infty$. Thus, there is an absolute minimum but no absolute maximum. From Theorem 5.5.4 [applied to the interval $(-\infty, +\infty)$], the absolute minimum must occur at a critical number of p. Since p is differentiable everywhere, we can find all critical numbers by solving the equation $p'(x) = 0$. This equation is

$$12x^3 + 12x^2 = 12x^2(x + 1) = 0$$

from which we conclude that the critical numbers are $x = 0$ and $x = -1$. Evaluating p at these critical numbers yields

$$p(0) = 0 \quad \text{and} \quad p(-1) = -1$$

Therefore, p has an absolute minimum of -1 at $x = -1$ (Figure 5.5.7). ◀

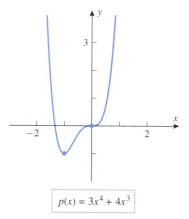

$p(x) = 3x^4 + 4x^3$

Figure 5.5.7

ABSOLUTE EXTREMA ON OPEN INTERVALS

We know that a continuous function may or may not have absolute extrema on an open interval. However, certain conclusions about the existence of absolute extrema of a continuous function f on a finite open interval (a, b) can be drawn from the behavior of $f(x)$ as $x \to a^+$ and as $x \to b^-$ (Table 5.5.3). Similar conclusions can be drawn for intervals of the form $(-\infty, b)$ or $(a, +\infty)$.

Table 5.5.3

LIMITS	$\lim\limits_{x \to a^+} f(x) = +\infty$ $\lim\limits_{x \to b^-} f(x) = +\infty$	$\lim\limits_{x \to a^+} f(x) = -\infty$ $\lim\limits_{x \to b^-} f(x) = -\infty$	$\lim\limits_{x \to a^+} f(x) = -\infty$ $\lim\limits_{x \to b^-} f(x) = +\infty$	$\lim\limits_{x \to a^+} f(x) = +\infty$ $\lim\limits_{x \to b^-} f(x) = -\infty$
CONCLUSION IF f **IS CONTINUOUS ON** (a, b)	f has an absolute minimum but no absolute maximum on (a, b).	f has an absolute maximum but no absolute minimum on (a, b).	f has neither an absolute maximum nor an absolute minimum on (a, b).	f has neither an absolute maximum nor an absolute minimum on (a, b).
GRAPH				

Example 5   Determine whether the function

$$f(x) = \frac{1}{x^2 - x}$$

has any absolute extrema on the interval $(0, 1)$. If so, find them and state where they occur.

Solution.   Since f is continuous on the interval $(0, 1)$ and

$$\lim_{x \to 0^+} f(x) = \lim_{x \to 0^+} \frac{1}{x^2 - x} = \lim_{x \to 0^+} \frac{1}{x(x - 1)} = -\infty$$

$$\lim_{x \to 1^-} f(x) = \lim_{x \to 1^-} \frac{1}{x^2 - x} = \lim_{x \to 1^-} \frac{1}{x(x - 1)} = -\infty$$

the function f has an absolute maximum but no absolute minimum on the interval $(0, 1)$. By Theorem 5.5.4 the absolute maximum must occur at a critical number of f in the interval $(0, 1)$. We have

$$f'(x) = -\frac{2x - 1}{\left(x^2 - x\right)^2}$$

so the only solution of the equation $f'(x) = 0$ is $x = \frac{1}{2}$. Although f is not differentiable at $x = 0$ or at $x = 1$, these values are doubly disqualified since they are neither in the domain of f nor in the interval $(0, 1)$. Thus, the absolute maximum occurs at $x = \frac{1}{2}$, and this absolute maximum is

$$f\left(\tfrac{1}{2}\right) = \frac{1}{\left(\tfrac{1}{2}\right)^2 - \tfrac{1}{2}} = -4$$

(Figure 5.5.8).   ◀

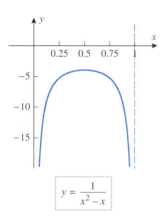

$$y = \frac{1}{x^2 - x}$$

Figure 5.5.8

....

ABSOLUTE EXTREMA OF FUNCTIONS WITH ONE RELATIVE EXTREMUM

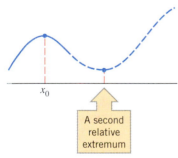

A second relative extremum

Figure 5.5.9

If a continuous function has only one relative extremum on a finite or infinite interval I, then that relative extremum must of necessity also be an absolute extremum. To understand why this is so, suppose that f has a relative maximum at x_0 in an interval I, and there are no other relative extrema of f on I. If $f(x_0)$ is *not* the absolute maximum of f on I, then the graph of f has to make an upward turn somewhere on I to rise above $f(x_0)$. However, this cannot happen because in the process of making an upward turn it would produce a second relative extremum on I (Figure 5.5.9). Thus, $f(x_0)$ must be the absolute maximum as well as a relative maximum. This idea is captured in the following theorem, which we state without proof.

5.5.5   THEOREM.   *Suppose that f is continuous and has exactly one relative extremum on an interval I, say at x_0.*

(a)   If f has a relative minimum at x_0, then $f(x_0)$ is the absolute minimum of f on I.

(b)   If f has a relative maximum at x_0, then $f(x_0)$ is the absolute maximum of f on I.

This theorem is often helpful in situations where other methods are difficult or tedious to apply.

Example 6   Find the absolute extrema of the function $f(x) = e^{(x^3 - 3x^2)}$ on the interval

(a) $(-\infty, +\infty)$ (b) $(0, +\infty)$

Solution (a).   Note that $f(x) > 0$ for all x and that

$$\lim_{x \to -\infty} f(x) = 0 \quad \text{and} \quad \lim_{x \to +\infty} f(x) = +\infty$$

Thus, the range of f on $(-\infty, +\infty)$ is $(0, +\infty)$ and there are no absolute extrema on the interval $(-\infty, +\infty)$.

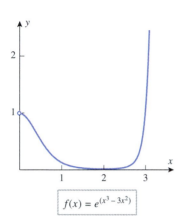

$f(x) = e^{(x^3 - 3x^2)}$

Figure 5.5.10

Solution (b). Since $\lim_{x \to +\infty} f(x) = +\infty$, we know that f cannot have an absolute maximum on the interval $(0, +\infty)$. However, the limit

$$\lim_{x \to 0^+} f(x) = e^0 = 1$$

is not infinite, so there is a possibility that f may have an absolute minimum on this interval. In this case it would have to occur at a stationary point, which suggests that we look for solutions of the equation $f'(x) = 0$. But,

$$f'(x) = e^{(x^3 - 3x^2)}(3x^2 - 6x) = 3x(x - 2)e^{(x^3 - 3x^2)}$$

so f has critical numbers $x = 0$ and $x = 2$. However, the only critical number inside the interval $(0, +\infty)$ is $x = 2$. Thus, Theorem 5.5.5 is applicable here. Since

$$f''(x) = e^{(x^3 - 3x^2)}(3x^2 - 6x)^2 + e^{(x^3 - 3x^2)}(6x - 6) = [(3x^2 - 6x)^2 + (6x - 6)]e^{(x^3 - 3x^2)}$$

we have

$$f''(2) = [0 + 6]e^{-4} = 6e^{-4} > 0$$

so a relative minimum occurs at $x = 2$ by the second derivative test. Thus, $f(x)$ has an absolute minimum at $x = 2$, and this absolute minimum is $f(2) = e^{-4} \approx 0.0183$ (Figure 5.5.10). ◀

ABSOLUTE EXTREMA AND PARAMETRIC CURVES

Suppose that a curve C is given parametrically by the equations

$$x = f(t), \quad y = g(t) \qquad (a \leq t \leq b)$$

where f and g are *continuous* on the finite closed interval $[a, b]$. It follows from the Extreme-Value Theorem that $f(t)$ and $g(t)$ have absolute maxima and absolute minima for $a \leq t \leq b$; this means that a particle moving along the curve cannot move away from the origin indefinitely—there must be a smallest and largest x-coordinate and a smallest and largest y-coordinate. Geometrically, the entire curve is contained within a box determined by these smallest and largest coordinates.

Example 7 Suppose that the equations of motion for a paper airplane during its first 10 seconds of flight are

$$x = t - 3\sin t, \quad y = 4 - 3\cos t \qquad (0 \leq t \leq 10)$$

What are the highest and lowest points in the trajectory, and when is the airplane at those points?

Solution. The trajectory, pictured in Figure 5.5.11, is shown in more detail in Figure 1.8.2. We want to find the absolute maximum and minimum values of y over the time interval $[0, 10]$ and the values of t for which these absolute extrema occur. The absolute extrema must occur either at the endpoints of the closed interval $[0, 10]$ or at critical numbers in the open interval $(0, 10)$. To find the critical numbers, we must solve the equation $dy/dt = 0$,

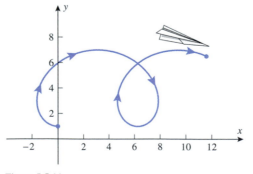

Figure 5.5.11

which is

$$3 \sin t = 0$$

Thus, there are critical numbers in the interval $(0, 10)$ at $t = \pi, 2\pi,$ and 3π. Evaluating $y = 4 - 3 \cos t$ at the endpoints and the critical numbers yields

$$y = 4 - 3 \cos 0 = 4 - 3 = 1$$
$$y = 4 - 3 \cos \pi = 4 - (-3) = 7$$
$$y = 4 - 3 \cos 2\pi = 4 - 3 = 1$$
$$y = 4 - 3 \cos 3\pi = 4 - (-3) = 7$$
$$y = 4 - 3 \cos 10 \approx 6.517$$

Thus, a high point of $y = 7$ is reached at times $t = \pi$ and $t = 3\pi$, and a low point of $y = 1$ is reached at times $t = 0$ and $t = 2\pi$. This is consistent with Figure 1.8.2.   ◄

EXERCISE SET 5.5   ∿ Graphing Utility   c CAS

In Exercises 1–2, use the graph to find x-coordinates of the relative extrema and absolute extrema of f on $[0, 7]$.

1.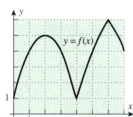

2.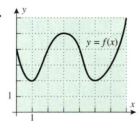

3. In each part, sketch the graph of a continuous function f with the stated properties on the interval $[0, 10]$.
(a) f has an absolute minimum at $x = 0$ and an absolute maximum at $x = 10$.
(b) f has an absolute minimum at $x = 2$ and an absolute maximum at $x = 7$.
(c) f has relative minima at $x = 1$ and $x = 8$, has relative maxima at $x = 3$ and $x = 7$, has an absolute minimum at $x = 5$, and has an absolute maximum at $x = 10$.

4. In each part, sketch the graph of a continuous function f with the stated properties on the interval $(-\infty, +\infty)$.
(a) f has no relative extrema or absolute extrema.
(b) f has an absolute minimum at $x = 0$ but no absolute maximum.
(c) f has an absolute maximum at $x = -5$ and an absolute minimum at $x = 5$.

In Exercises 5–14, find the absolute maximum and minimum values of f on the given closed interval, and state where those values occur.

5. $f(x) = 4x^2 - 4x + 1$; $[0, 1]$
6. $f(x) = 8x - x^2$; $[0, 6]$
7. $f(x) = (x - 1)^3$; $[0, 4]$

8. $f(x) = 2x^3 - 3x^2 - 12x$; $[-2, 3]$
9. $f(x) = \dfrac{3x}{\sqrt{4x^2 + 1}}$; $[-1, 1]$
10. $f(x) = \left(x^2 + x\right)^{2/3}$; $[-2, 3]$
11. $f(x) = x - \tan x$; $[-\pi/4, \pi/4]$
12. $f(x) = \sin x - \cos x$; $[0, \pi]$
13. $f(x) = 1 + |9 - x^2|$; $[-5, 1]$
14. $f(x) = |6 - 4x|$; $[-3, 3]$

In Exercises 15–22, find the absolute maximum and minimum values of f, if any, on the given interval, and state where those values occur.

15. $f(x) = x^2 - 3x - 1$; $(-\infty, +\infty)$
16. $f(x) = 3 - 4x - 2x^2$; $(-\infty, +\infty)$
17. $f(x) = 4x^3 - 3x^4$; $(-\infty, +\infty)$
18. $f(x) = x^4 + 4x$; $(-\infty, +\infty)$
19. $f(x) = x^3 - 3x - 2$; $(-\infty, +\infty)$
20. $f(x) = x^3 - 9x + 1$; $(-\infty, +\infty)$
21. $f(x) = \dfrac{x^2}{x + 1}$; $(-5, -1)$ **22.** $f(x) = \dfrac{x + 3}{x - 3}$; $[-5, 5]$

In Exercises 23–36, use a graphing utility to estimate the absolute maximum and minimum values of f, if any, on the stated interval, and then use calculus methods to find the exact values.

∿ **23.** $f(x) = \left(x^2 - 1\right)^2$; $(-\infty, +\infty)$
∿ **24.** $f(x) = (x - 1)^2(x + 2)^2$; $(-\infty, +\infty)$
∿ **25.** $f(x) = x^{2/3}(20 - x)$; $[-1, 20]$
∿ **26.** $f(x) = \dfrac{x}{x^2 + 2}$; $[-1, 4]$

27. $f(x) = 1 + \dfrac{1}{x}$; $(0, +\infty)$

28. $f(x) = \dfrac{x}{x^2 + 1}$; $[0, +\infty)$

29. $f(x) = 2\sec x - \tan x$; $[0, \pi/4]$

30. $f(x) = \sin^2 x + \cos x$; $[-\pi, \pi]$

31. $f(x) = x^3 e^{-2x}$; $[1, 4]$ **32.** $f(x) = \dfrac{\ln(2x)}{x}$; $[1, e]$

33. $f(x) = \dfrac{x}{2} + \ln(x^2 + 1)$; $[-4, 0]$

34. $f(x) = (x^2 - 1)e^x$; $[-2, 2]$

35. $f(x) = \sin(\cos x)$; $[0, 2\pi]$

36. $f(x) = \cos(\sin x)$; $[0, \pi]$

37. Find the absolute maximum and minimum values of

$$f(x) = \begin{cases} 4x - 2, & x < 1 \\ (x - 2)(x - 3), & x \geq 1 \end{cases}$$

on $\left[\frac{1}{2}, \frac{7}{2}\right]$.

38. Let $f(x) = x^2 + px + q$. Find the values of p and q such that $f(1) = 3$ is an extreme value of f on $[0, 2]$. Is this value a maximum or minimum?

If f is a periodic function, then the locations of all absolute extrema on the interval $(-\infty, +\infty)$ can be obtained by finding the locations of the absolute extrema for one period and using the periodicity to locate the rest. Use this idea in Exercises 39 and 40 to find the absolute maximum and minimum values of the function, and state the x-values at which they occur.

---

39. $f(x) = 2\sin 2x + \sin 4x$ **40.** $f(x) = 3\cos\dfrac{x}{3} + 2\cos\dfrac{x}{2}$

One way of proving that $f(x) \leq g(x)$ for all x in a given interval is to show that $0 \leq g(x) - f(x)$ for all x in the interval; and one way of proving the latter inequality is to show that the absolute minimum value of $g(x) - f(x)$ on the interval is nonnegative. Use this idea to prove the inequalities in Exercises 41 and 42.

---

41. Prove that $\sin x \leq x$ for all x in the interval $[0, 2\pi]$.

42. Prove that $\cos x \geq 1 - (x^2/2)$ for all x in the interval $[0, 2\pi]$.

43. What is the smallest possible slope for a tangent to the graph of the equation $y = x^3 - 3x^2 + 5x$?

44. (a) Show that

$$f(x) = \dfrac{64}{\sin x} + \dfrac{27}{\cos x}$$

has a minimum value but no maximum value on the interval $(0, \pi/2)$.

(b) Find the minimum value.

c 45. Show that the absolute minimum value of

$$f(x) = x^2 + \dfrac{16x^2}{(8 - x)^2}, \quad x > 8$$

occurs at $x = 4(2 + \sqrt[3]{2})$ by using a CAS to find $f'(x)$ and to solve the equation $f'(x) = 0$.

46. The concentration $C(t)$ of a drug in the bloodstream t hours after it has been injected is commonly modeled by an equation of the form

$$C(t) = \dfrac{K(e^{-bt} - e^{-at})}{a - b}$$

where $K > 0$ and $a > b > 0$.

(a) At what time does the maximum concentration occur?

(b) Let $K = 1$ for simplicity, and use a graphing utility to check your result in part (a) by graphing $C(t)$ for various values of a and b.

47. It can be proved that if f is differentiable on (a, b) and L is a line that does not intersect the curve $y = f(x)$ over an interval (a, b), then the points at which the curve is closest to or farthest from the line L, if any, occur at points where the tangent line to the curve is parallel to L (see the accompanying figure). Use this result to find the points on the graph of $y = -x^2$, $-1 \leq x \leq 1.5$, that are closest to and farthest from the line $y = 2 - x$.

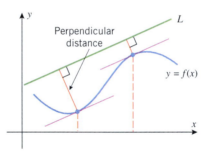

Figure Ex-47

48. Use the idea discussed in Exercise 47 to find the coordinates of all points on the graph of $y = x^3$, $-1 \leq x \leq 1$, closest to and farthest from the line $y = \frac{4}{3}x - 1$.

49. Suppose that the equations of motion of a paper airplane during the first 12 seconds of flight are

$$x = t - 2\sin t, \quad y = 2 - 2\cos t \quad (0 \leq t \leq 12)$$

What are the highest and lowest points in the trajectory, and when is the airplane at those points?

50. The accompanying figure shows the path of a fly whose equations of motion are

$$x = \dfrac{\cos t}{2 + \sin t}, \quad y = 3 + \sin(2t) - 2\sin^2 t \quad (0 \leq t \leq 2\pi)$$

(a) How high and low does it fly?

(b) How far left and right of the origin does it fly?

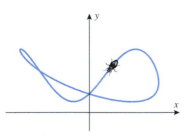

Figure Ex-50

51. Let $f(x) = ax^2 + bx + c$, where $a > 0$. Prove that $f(x) \geq 0$ for all x if and only if $b^2 - 4ac \leq 0$. [*Hint: Find the minimum of $f(x)$.*]

52. Prove Theorem 5.5.4 in the case where the extreme value is a minimum.

5.6 APPLIED MAXIMUM AND MINIMUM PROBLEMS

In this section we will show how the methods discussed in the last section can be used to solve various applied optimization problems.

CLASSIFICATION OF OPTIMIZATION PROBLEMS

The applied optimization problems that we will consider in this section fall into the following two categories:

- Problems that reduce to maximizing or minimizing a continuous function over a finite closed interval.

- Problems that reduce to maximizing or minimizing a continuous function over an infinite interval or a finite interval that is not closed.

For problems of the first type the Extreme-Value Theorem (5.5.3) guarantees that the problem has a solution, and we know that the solution can be obtained by examining the values of the function at the critical numbers and at the endpoints. However, for problems of the second type there may or may not be a solution. If the function is continuous and has exactly one relative extremum of the appropriate type on the interval, then Theorem 5.5.5 guarantees the existence of a solution and provides a method for finding it. In cases where this theorem is not applicable some ingenuity may be required to solve the problem.

PROBLEMS INVOLVING FINITE CLOSED INTERVALS

In his *On a Method for the Evaluation of Maxima and Minima*, the seventeenth century French mathematician Pierre de Fermat[*] solved an optimization problem very similar to

---

[*] PIERRE DE FERMAT (1601–1665). Fermat, the son of a successful French leather merchant, was a lawyer who practiced mathematics as a hobby. He received a Bachelor of Civil Laws degree from the University of Orleans in 1631 and subsequently held various government positions, including a post as councillor to the Toulouse parliament. Although he was apparently financially successful, confidential documents of that time suggest that his performance in office and as a lawyer was poor, perhaps because he devoted so much time to mathematics. Throughout his life, Fermat fought all efforts to have his mathematical results published. He had the unfortunate habit of scribbling his work in the margins of books and often sent his results to friends without keeping copies for himself. As a result, he never received credit for many major achievements until his name was raised from obscurity in the mid-nineteenth century. It is now known that Fermat, simultaneously and independently of Descartes, developed analytic geometry. Unfortunately, Descartes and Fermat argued bitterly over various problems so that there was never any real cooperation between these two great geniuses.

Fermat solved many fundamental calculus problems. He obtained the first procedure for differentiating polynomials, and solved many important maximization, minimization, area, and tangent problems. His work served to inspire Isaac Newton. Fermat is best known for his work in number theory, the study of properties of and relationships between whole numbers. He was the first mathematician to make substantial contributions to this field after the ancient Greek mathematician Diophantus. Unfortunately, none of Fermat's contemporaries appreciated his work in this area, a fact that eventually pushed Fermat into isolation and obscurity in later life. In addition to his work in calculus and number theory, Fermat was one of the founders of probability theory and made major contributions to the theory of optics. Outside mathematics, Fermat was a classical scholar of some note, was fluent in French, Italian, Spanish, Latin, and Greek, and he composed a considerable amount of Latin poetry.

One of the great mysteries of mathematics is shrouded in Fermat's work in number theory. In the margin of a book by Diophantus, Fermat scribbled that for integer values of n greater than 2, the equation $x^n + y^n = z^n$ has no nonzero integer solutions for x, y, and z. He stated, "I have discovered a truly marvelous proof of this, which however the margin is not large enough to contain." This result, which became known as "Fermat's last theorem," appeared to be true, but its proof evaded the greatest mathematical geniuses for 300 years until Professor Andrew Wiles of Princeton University presented a proof in June 1993 in a dramatic series of three lectures that drew international media attention (see *New York Times*, June 27, 1993). As it turned out, that proof had a serious gap that he and Richard Taylor fixed and published in 1995. A prize of 100,000 German marks was offered in 1908 for the solution, but it is worthless today because of inflation.

the one posed in our first example. Fermat's work on such optimization problems prompted the French mathematician Laplace to proclaim Fermat the "true inventor of the differential calculus." Although this honor must still reside with Newton and Leibniz, it is the case that Fermat developed procedures that anticipated parts of differential calculus.

Example 1   A garden is to be laid out in a rectangular area and protected by a chicken wire fence. What is the largest possible area of the garden if only 100 running feet of chicken wire is available for the fence?

Solution.   Let

$$x = \text{length of the rectangle (ft)}$$
$$y = \text{width of the rectangle (ft)}$$
$$A = \text{area of the rectangle (ft}^2)$$

Then

$$A = xy \tag{1}$$

Since the perimeter of the rectangle is 100 ft, the variables x and y are related by the equation

$$2x + 2y = 100 \quad \text{or} \quad y = 50 - x \tag{2}$$

(See Figure 5.6.1.) Substituting (2) in (1) yields

$$A = x(50 - x) = 50x - x^2 \tag{3}$$

Because x represents a length it cannot be negative, and because the two sides of length x cannot have a combined length exceeding the total perimeter of 100 ft, the variable x must satisfy

$$0 \leq x \leq 50 \tag{4}$$

Thus, we have reduced the problem to that of finding the value (or values) of x in $[0, 50]$, for which A is maximum. Since A is a polynomial in x, it is continuous on $[0, 50]$, and so the maximum must occur at an endpoint of this interval or at a critical number.

From (3) we obtain

$$\frac{dA}{dx} = 50 - 2x$$

Setting $dA/dx = 0$ we obtain

$$50 - 2x = 0$$

or $x = 25$. Thus, the maximum occurs at one of the values

$$x = 0, \quad x = 25, \quad x = 50$$

Substituting these values in (3) yields Table 5.6.1, which tells us that the maximum area of 625 ft² occurs at $x = 25$, which is consistent with the graph of (3) in Figure 5.6.2. From (2) the corresponding value of y is 25, so the rectangle of perimeter 100 ft with greatest area is a square with sides of length 25 ft.   ◄

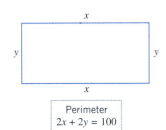

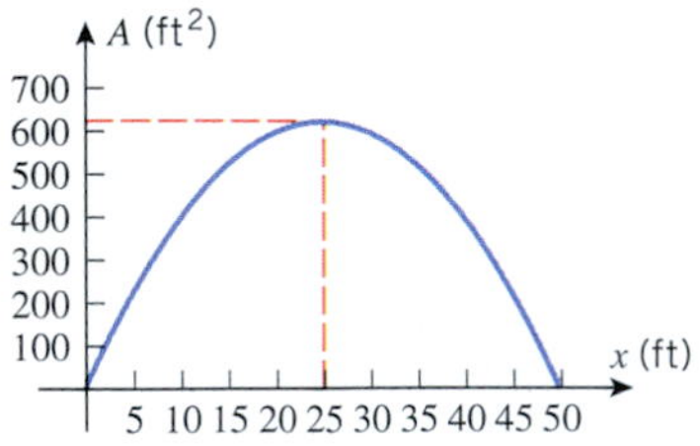

Figure 5.6.1

Perimeter
2x + 2y = 100

Table 5.6.1

x	0	25	50
A	0	625	0

Figure 5.6.2

REMARK.   In this example we included $x = 0$ and $x = 50$ as possible values for x, even though both values lead to rectangles with two sides of length zero. Whether or not these values should be allowed will depend on our objective in the problem. If we view this purely as a mathematical problem, then there is nothing wrong with allowing sides of length zero. However, if we view this as an applied problem in which the rectangle will be formed from physical material, then these values should be excluded.

Example 1 illustrates the following five-step procedure that can be used for solving many applied maximum and minimum problems.

Step 1.   Draw an appropriate figure and label the quantities relevant to the problem.

Step 2.   Find a formula for the quantity to be maximized or minimized.

Step 3.   Using the conditions stated in the problem to eliminate variables, express the quantity to be maximized or minimized as a function of one variable.

Step 4.   Find the interval of possible values for this variable from the physical restrictions in the problem.

Step 5.   If applicable, use the techniques of the preceding section to obtain the maximum or minimum.

Example 2   An open box is to be made from a 16-inch by 30-inch piece of cardboard by cutting out squares of equal size from the four corners and bending up the sides (Figure 5.6.3). What size should the squares be to obtain a box with the largest volume?

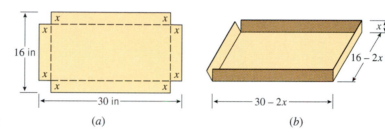

Figure 5.6.3   (a)   (b)

Solution.   For emphasis, we explicitly list the steps of the five-step problem-solving procedure given above as an outline for the solution of this problem. (In later examples we will follow these guidelines implicitly.)

- *Step 1:* Figure 5.6.3a illustrates the cardboard piece with squares removed from its corners. Let

  x = length (in inches) of the sides of the squares to be cut out

  V = volume (in cubic inches) of the resulting box

- *Step 2:* Because we are removing a square of side x from each corner, the resulting box will have dimensions $16 - 2x$ by $30 - 2x$ by x (Figure 5.6.3b). Since the volume of a box is the product of its dimensions, we have

$$V = (16 - 2x)(30 - 2x)x = 480x - 92x^2 + 4x^3 \tag{5}$$

- *Step 3:* Note that our expression for volume is already in terms of the single variable x.

- *Step 4:* The variable x in (5) is subject to certain restrictions. Because x represents a length, it cannot be negative, and because the width of the cardboard is 16 inches, we cannot cut out squares whose sides are more than 8 inches long. Thus, the variable x in (5) must satisfy

  $0 \leq x \leq 8$

and hence we have reduced our problem to finding the value (or values) of x in the interval [0, 8] for which (5) is a maximum.

- *Step 5:* From (5) we obtain

$$\frac{dV}{dx} = 480 - 184x + 12x^2 = 4(120 - 46x + 3x^2)$$
$$= 4(x - 12)(3x - 10)$$

Table 5.6.2

x	0	$\frac{10}{3}$	8
V	0	$\frac{19600}{27} \approx 726$	0

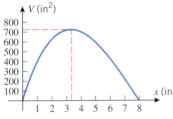

Figure 5.6.4

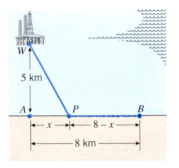

Figure 5.6.5

Setting $dV/dx = 0$ yields

$$x = \tfrac{10}{3} \quad \text{and} \quad x = 12$$

Since $x = 12$ falls outside the interval $[0, 8]$, the maximum value of V occurs either at the critical number $x = \frac{10}{3}$ or at the endpoints $x = 0, x = 8$. Substituting these values into (5) yields Table 5.6.2, which tells us that the greatest possible volume $V = \frac{19600}{27}$ in$^3 \approx 726$ in^3 occurs when we cut out squares whose sides have length $\frac{10}{3}$ inches. This is consistent with the graph of (5) shown in Figure 5.6.4. ◀

In Example 2 of Section 1.1 we used approximate graphical methods to solve a problem of piping oil from an offshore well to a point on the shore with minimal cost. We will now show how to solve that problem exactly using calculus.

Example 3 Figure 5.6.5 shows an offshore oil well located at a point W that is 5 km from the closest point A on a straight shoreline. Oil is to be piped from W to a shore point B that is 8 km from A by piping it on a straight line under water from W to some shore point P between A and B and then on to B via pipe along the shoreline. If the cost of laying pipe is \$1,000,000/km under water and \$500,000/km over land, where should the point P be located to minimize the cost of laying the pipe?

Solution. Let

$x =$ distance (in kilometers) between A and P

$c =$ cost (in millions of dollars) for the entire pipeline

From Figure 5.6.5 the length of pipe under water is the distance between W and P. By the Theorem of Pythagoras, that length is

$$\sqrt{x^2 + 25} \tag{6}$$

Also from Figure 5.6.5, the length of pipe over land is the distance between P and B, which is

$$8 - x \tag{7}$$

From (6) and (7) it follows that the total cost c (in millions of dollars) for the pipeline is

$$c = 1(\sqrt{x^2 + 25}) + \tfrac{1}{2}(8 - x) = \sqrt{x^2 + 25} + \tfrac{1}{2}(8 - x) \tag{8}$$

Because the distance between A and B is 8 km, the distance x between A and P must satisfy

$$0 \le x \le 8$$

We have thus reduced our problem to finding the value (or values) of x in the interval $[0, 8]$ for which (8) is a minimum. Since c is a continuous function of x on the closed interval $[0, 8]$, we can use the methods developed in the preceding section to find the minimum. From (8) we obtain

$$\frac{dc}{dx} = \frac{x}{\sqrt{x^2 + 25}} - \frac{1}{2}$$

Setting $dc/dx = 0$ and solving for x yields

$$\frac{x}{\sqrt{x^2 + 25}} = \frac{1}{2} \tag{9}$$

$$x^2 = \frac{1}{4}(x^2 + 25)$$

$$x = \pm \frac{5}{\sqrt{3}}$$

The number $-5/\sqrt{3}$ is not a solution of (9) and must be discarded, leaving $x = 5/\sqrt{3}$ as the only critical number. Since this number lies in the interval $[0, 8]$, the minimum must occur at one of the values

$$x = 0, \quad x = 5/\sqrt{3}, \quad x = 8$$

Substituting these values into (8) yields Table 5.6.3, which tells us that the least possible cost of the pipeline (to the nearest dollar) is $c = \$8,330,127$, and this occurs when the point P is located at a distance of $5/\sqrt{3} \approx 2.89$ km from A. This is consistent with the graph in Figure 1.1.9c.   ◄

Table 5.6.3

x	0	$\frac{5}{\sqrt{3}}$	8
c	9	$\frac{10}{\sqrt{3}} + \left(4 - \frac{5}{2\sqrt{3}}\right) \approx 8.330127$	$\sqrt{89} \approx 9.433981$

⋮ FOR THE READER.   If you have a CAS, use it to check all of the computations in this example. Specifically, differentiate c with respect to x, solve the equation $dc/dx = 0$, and perform all of the numerical calculations.

Example 4   Find the radius and height of the right circular cylinder of largest volume that can be inscribed in a right circular cone with radius 6 inches and height 10 inches (Figure 5.6.6a).

Solution.   Let

r = radius (in inches) of the cylinder

h = height (in inches) of the cylinder

V = volume (in cubic inches) of the cylinder

The formula for the volume of the inscribed cylinder is

$$V = \pi r^2 h \tag{10}$$

To eliminate one of the variables in (10) we need a relationship between r and h. Using similar triangles (Figure 5.6.6b) we obtain

$$\frac{10 - h}{r} = \frac{10}{6} \quad \text{or} \quad h = 10 - \tfrac{5}{3}r \tag{11}$$

Substituting (11) into (10) we obtain

$$V = \pi r^2 \left(10 - \tfrac{5}{3}r\right) = 10\pi r^2 - \tfrac{5}{3}\pi r^3 \tag{12}$$

which expresses V in terms of r alone. Because r represents a radius it cannot be negative, and because the radius of the inscribed cylinder cannot exceed the radius of the cone, the variable r must satisfy

$$0 \le r \le 6$$

Thus, we have reduced the problem to that of finding the value (or values) of r in [0, 6] for which (12) is a maximum. Since V is a continuous function of r on [0, 6], the methods developed in the preceding section apply.

From (12) we obtain

$$\frac{dV}{dr} = 20\pi r - 5\pi r^2 = 5\pi r (4 - r)$$

Setting $dV/dr = 0$ gives

$$5\pi r (4 - r) = 0$$

so $r = 0$ and $r = 4$ are critical numbers. Since these lie in the interval [0, 6], the maximum must occur at one of the values

$$r = 0, \quad r = 4, \quad r = 6$$

Substituting these values into (12) yields Table 5.6.4, which tells us that the maximum

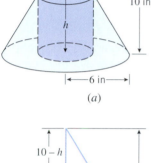

(a)

(b)

Figure 5.6.6

Table 5.6.4

r	0	4	6
V	0	$\frac{160}{3}\pi$	0

volume $V = \frac{160}{3}\pi \approx 168$ in^3 occurs when the inscribed cylinder has radius 4 in. When $r = 4$ it follows from (11) that $h = \frac{10}{3}$. Thus, the inscribed cylinder of largest volume has radius $r = 4$ in and height $h = \frac{10}{3}$ in.   ◀

PROBLEMS INVOLVING INTERVALS THAT ARE NOT BOTH FINITE AND CLOSED

Example 5   A closed cylindrical can is to hold 1 liter (1000 cm^3) of liquid. How should we choose the height and radius to minimize the amount of material needed to manufacture the can?

Solution.   Let

h = height (in cm) of the can

r = radius (in cm) of the can

S = surface area (in cm^2) of the can

Assuming there is no waste or overlap, the amount of material needed for manufacture will be the same as the surface area of the can. Since the can consists of two circular disks of radius r and a rectangular sheet with dimensions h by $2\pi r$ (Figure 5.6.7), the surface area will be

$$S = 2\pi r^2 + 2\pi r h \tag{13}$$

Since S depends on two variables, r and h, we will look for some condition in the problem that will allow us to express one of these variables in terms of the other. For this purpose, observe that the volume of the can is 1000 cm^3, so it follows from the formula $V = \pi r^2 h$ for the volume of a cylinder that

$$1000 = \pi r^2 h \quad \text{or} \quad h = \frac{1000}{\pi r^2} \tag{14--15}$$

Substituting (15) in (13) yields

$$S = 2\pi r^2 + \frac{2000}{r} \tag{16}$$

Thus, we have reduced the problem to finding a value of r in the interval $(0, +\infty)$ for which S is minimum. Since S is a continuous function of r on the interval $(0, +\infty)$ and

$$\lim_{r \to 0^+} \left(2\pi r^2 + \frac{2000}{r} \right) = +\infty \quad \text{and} \quad \lim_{r \to +\infty} \left(2\pi r^2 + \frac{2000}{r} \right) = +\infty$$

the analysis in Table 5.5.3 implies that S does have a minimum on the interval $(0, +\infty)$. Since this minimum must occur at a critical number, we calculate

$$\frac{dS}{dr} = 4\pi r - \frac{2000}{r^2} \tag{17}$$

Setting $dS/dr = 0$ gives

$$r = \frac{10}{\sqrt[3]{2\pi}} \approx 5.4 \tag{18}$$

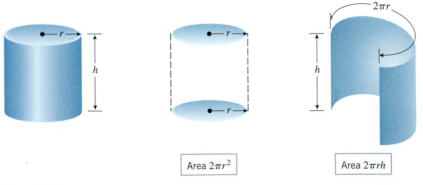

Area $2\pi r^2$ Area $2\pi r h$

Figure 5.6.7

Since (18) is the only critical number in the interval $(0, +\infty)$, this value of r yields the minimum value of S. From (15) the value of h corresponding to this r is

$$h = \frac{1000}{\pi(10/\sqrt[3]{2\pi})^2} = \frac{20}{\sqrt[3]{2\pi}} = 2r$$

It is not an accident here that the minimum occurs when the height of the can is equal to the diameter of its base (Exercise 27).

Second Solution. The conclusion that a minimum occurs at the value of r in (18) can be deduced from Theorem 5.5.5 and the second derivative test by noting that

$$\frac{d^2 S}{dr^2} = 4\pi + \frac{4000}{r^3}$$

is positive if $r > 0$ and hence is positive if $r = 10/\sqrt[3]{2\pi}$. This implies that a relative minimum, and therefore a minimum, occurs at the critical number $r = 10/\sqrt[3]{2\pi}$.

Third Solution. An alternative justification that the critical number $r = 10/\sqrt[3]{2\pi}$ corresponds to a minimum for S is to view the graph of S versus r (Figure 5.6.8). ◀

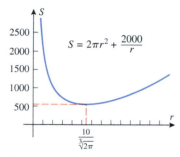

$S = 2\pi r^2 + \dfrac{2000}{r}$

Figure 5.6.8

REMARK. Note that S has no maximum on $(0, +\infty)$. Thus, had we asked for the dimensions of the can requiring the maximum amount of material for its manufacture, there would have been no solution to the problem. Optimization problems with no solution are sometimes called ***ill posed***.

Example 6 Find a point on the curve $y = x^2$ that is closest to the point $(18, 0)$.

Solution. The distance L between $(18, 0)$ and an arbitrary point (x, y) on the curve $y = x^2$ (Figure 5.6.9) is given by

$$L = \sqrt{(x - 18)^2 + (y - 0)^2}$$

Since (x, y) lies on the curve, x and y satisfy $y = x^2$; thus,

$$L = \sqrt{(x - 18)^2 + x^4} \tag{19}$$

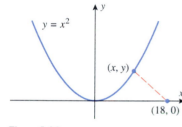

$y = x^2$

(x, y)

$(18, 0)$

Figure 5.6.9

Because there are no restrictions on x, the problem reduces to finding a value of x in $(-\infty, +\infty)$ for which (19) is a minimum. The distance L and the square of the distance L^2 are minimized at the same value (see Exercise 60). Thus, the minimum value of L in (19) and the minimum value of

$$S = L^2 = (x - 18)^2 + x^4 \tag{20}$$

occur at the same x-value.

From (20),

$$\frac{dS}{dx} = 2(x - 18) + 4x^3 = 4x^3 + 2x - 36 \tag{21}$$

so that the critical numbers satisfy $4x^3 + 2x - 36 = 0$ or, equivalently,

$$2x^3 + x - 18 = 0 \tag{22}$$

To solve for x we will begin by checking the divisors of -18 to see whether the polynomial on the left side has any integer roots (see Appendix F). These divisors are $\pm 1, \pm 2, \pm 3, \pm 6, \pm 9,$ and ± 18. A check of these values shows that $x = 2$ is a root, so that $x - 2$ is a factor of the polynomial. After dividing the polynomial by this factor we can rewrite (22) as

$$(x - 2)(2x^2 + 4x + 9) = 0$$

Thus, the remaining solutions of (22) satisfy the quadratic equation

$$2x^2 + 4x + 9 = 0$$

But this equation has no real solutions (using the quadratic formula), so that $x = 2$ is the only critical number of S. To determine the nature of this critical number we will use the second derivative test. From (21),

$$\frac{d^2 S}{dx^2} = 12x^2 + 2, \quad \text{so} \quad \frac{d^2 S}{dx^2}\bigg|_{x=2} = 50 > 0$$

which shows that a relative minimum occurs at $x = 2$. Since $x = 2$ is the only relative extremum for L, it follows from Theorem 5.5.5 that an absolute minimum value of L also occurs at $x = 2$. Thus, the point on the curve $y = x^2$ closest to $(18, 0)$ is

$$(x, y) = (x, x^2) = (2, 4) \qquad \blacktriangleleft$$

AN APPLICATION TO ECONOMICS

Three functions of importance to an economist or a manufacturer are

$C(x)$ = total cost of producing x units of a product during some time period

$R(x)$ = total revenue from selling x units of the product during the time period

$P(x)$ = total profit obtained by selling x units of the product during the time period

These are called, respectively, the *cost function*, *revenue function*, and *profit function*. If all units produced are sold, then these are related by

$$P(x) = R(x) - C(x) \tag{23}$$

[profit] = [revenue] – [cost]

The total cost $C(x)$ of producing x units can be expressed as a sum

$$C(x) = a + M(x) \tag{24}$$

where a is a constant, called *overhead*, and $M(x)$ is a function representing *manufacturing cost*. The overhead, which includes such fixed costs as rent and insurance, does not depend on x; it must be paid even if nothing is produced. On the other hand, the manufacturing cost $M(x)$, which includes such items as cost of materials and labor, depends on the number of items manufactured. It is shown in economics that with suitable simplifying assumptions, $M(x)$ can be expressed in the form

$$M(x) = bx + cx^2$$

where b and c are constants. Substituting this in (24) yields

$$C(x) = a + bx + cx^2 \tag{25}$$

If a manufacturing firm can sell all the items it produces for p dollars apiece, then its total revenue $R(x)$ (in dollars) will be

$$R(x) = px \tag{26}$$

and its total profit $P(x)$ (in dollars) will be

$$P(x) = [\text{total revenue}] - [\text{total cost}] = R(x) - C(x) = px - C(x)$$

Thus, if the cost function is given by (25),

$$P(x) = px - (a + bx + cx^2) \tag{27}$$

Depending on such factors as number of employees, amount of machinery available, economic conditions, and competition, there will be some upper limit ℓ on the number of items a manufacturer is capable of producing and selling. Thus, during a fixed time period the variable x in (27) will satisfy

$$0 \le x \le \ell$$

By determining the value or values of x in $[0, \ell]$ that maximize (27), the firm can determine how many units of its product must be manufactured and sold to yield the greatest profit. This is illustrated in the following numerical example.

Example 7   A liquid form of penicillin manufactured by a pharmaceutical firm is sold in bulk at a price of $200 per unit. If the total production cost (in dollars) for x units is

$$C(x) = 500{,}000 + 80x + 0.003x^2$$

and if the production capacity of the firm is at most 30,000 units in a specified time, how many units of penicillin must be manufactured and sold in that time to maximize the profit?

Solution.   Since the total revenue for selling x units is $R(x) = 200x$, the profit $P(x)$ on x units will be

$$P(x) = R(x) - C(x) = 200x - (500{,}000 + 80x + 0.003x^2) \qquad (28)$$

Since the production capacity is at most 30,000 units, x must lie in the interval $[0, 30{,}000]$. From (28)

$$\frac{dP}{dx} = 200 - (80 + 0.006x) = 120 - 0.006x$$

Setting $dP/dx = 0$ gives

$$120 - 0.006x = 0 \quad \text{or} \quad x = 20{,}000$$

Since this critical number lies in the interval $[0, 30{,}000]$, the maximum profit must occur at one of the values

$$x = 0, \quad x = 20{,}000, \quad \text{or} \quad x = 30{,}000$$

Substituting these values in (28) yields Table 5.6.5, which tells us that the maximum profit $P = \$700{,}000$ occurs when $x = 20{,}000$ units are manufactured and sold in the specified time.   ◄

Table 5.6.5

x	0	20,000	30,000
$P(x)$	−500,000	700,000	400,000

MARGINAL ANALYSIS

Economists call $P'(x)$, $R'(x)$, and $C'(x)$ the ***marginal profit***, ***marginal revenue***, and ***marginal cost***, respectively; and they interpret these quantities as the *additional* profit, revenue, and cost that result from producing and selling one additional unit of the product when the production and sales levels are at x units. These interpretations follow from the local linear approximations of the profit, revenue, and cost functions. For example, it follows from Formula (2) of Section 3.8 that when the production and sales levels are at x units the local linear approximation of the profit function is

$$P(x + \Delta x) \approx P(x) + P'(x)\Delta x$$

Thus, if $\Delta x = 1$ (one additional unit produced and sold), this formula implies

$$P(x + 1) \approx P(x) + P'(x)$$

and hence the *additional* profit that results from producing and selling one additional unit can be approximated as

$$P(x + 1) - P(x) \approx P'(x)$$

A BASIC PRINCIPLE OF ECONOMICS

It follows from (23) that $P'(x) = 0$ has the same solution as $C'(x) = R'(x)$, and this implies that the maximum profit must occur where the marginal revenue is equal to the marginal cost; that is:

> *The maximum profit occurs where the cost of manufacturing and selling an additional unit of a product is approximately equal to the revenue generated by the additional unit.*

In Example 7, the maximum profit occurs when $x = 20{,}000$ units. Note that

$$C(20{,}001) - C(20{,}000) = \$200.003 \quad \text{and} \quad R(20{,}001) - R(20{,}000) = \$200$$

which is consistent with this basic economic principle.

EXERCISE SET 5.6

1. Express the number 10 as a sum of two nonnegative numbers whose product is as large as possible.

2. How should two nonnegative numbers be chosen so that their sum is 1 and the sum of their squares is
   (a) as large as possible
   (b) as small as possible?

3. Find a number in the closed interval $\left[\frac{1}{2}, \frac{3}{2}\right]$ such that the sum of the number and its reciprocal is
   (a) as small as possible
   (b) as large as possible.

4. A rectangular field is to be bounded by a fence on three sides and by a straight stream on the fourth side. Find the dimensions of the field with maximum area that can be enclosed with 1000 feet of fence.

5. A rectangular plot of land is to be fenced in using two kinds of fencing. Two opposite sides will use heavy-duty fencing selling for $3 a foot, while the remaining two sides will use standard fencing selling for $2 a foot. What are the dimensions of the rectangular plot of greatest area that can be fenced in at a cost of $6000?

6. A rectangle is to be inscribed in a right triangle having sides of length 6 in, 8 in, and 10 in. Find the dimensions of the rectangle with greatest area assuming the rectangle is positioned as in the accompanying figure.

7. Solve the problem in Exercise 6 assuming the rectangle is positioned as in the accompanying figure.

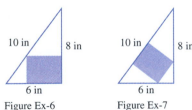

10 in   8 in   10 in   8 in

6 in   6 in
Figure Ex-6   Figure Ex-7

8. A rectangle has its two lower corners on the x-axis and its two upper corners on the curve $y = 16 - x^2$. For all such rectangles, what are the dimensions of the one with largest area?

9. Find the dimensions of the rectangle with maximum area that can be inscribed in a circle of radius 10.

10. Find the dimensions of the rectangle of greatest area that can be inscribed in a semicircle of radius R as shown in the accompanying figure.

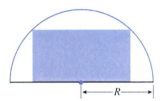

$\longleftarrow R \longrightarrow$

Figure Ex-10

11. A rectangular area of 3200 ft^2 is to be fenced off. Two opposite sides will use fencing costing $1 per foot and the remaining sides will use fencing costing $2 per foot. Find the dimensions of the rectangle of least cost.

12. Show that among all rectangles with perimeter p, the square has the maximum area.

13. Show that among all rectangles with area A, the square has the minimum perimeter.

14. A wire of length 12 in can be bent into a circle, bent into a square, or cut into two pieces to make both a circle and a square. How much wire should be used for the circle if the total area enclosed by the figure(s) is to be
    (a) a maximum   (b) a minimum?

15. Suppose that the number of bacteria in a culture at time t is given by $N = 5000(25 + te^{-t/20})$.
    (a) Find the largest and smallest number of bacteria in the culture during the time interval $0 \le t \le 100$.
    (b) At what time during the time interval in part (a) is the number of bacteria decreasing most rapidly?

16. A church window consisting of a rectangle topped by a semicircle is to have a perimeter p. Find the radius of the semicircle if the area of the window is to be maximum.

17. A sheet of cardboard 12 in square is used to make an open box by cutting squares of equal size from the four corners and folding up the sides. What size squares should be cut to obtain a box with largest possible volume?

18. A square sheet of cardboard of side k is used to make an open box by cutting squares of equal size from the four corners and folding up the sides. What size squares should be

cut from the corners to obtain a box with largest possible volume?

19. An open box is to be made from a 3-ft by 8-ft rectangular piece of sheet metal by cutting out squares of equal size from the four corners and bending up the sides. Find the maximum volume that the box can have.

20. A closed rectangular container with a square base is to have a volume of 2250 in^3. The material for the top and bottom of the container will cost \$2 per in^2, and the material for the sides will cost \$3 per in^2. Find the dimensions of the container of least cost.

21. A closed rectangular container with a square base is to have a volume of 2000 cm^3. It costs twice as much per square centimeter for the top and bottom as it does for the sides. Find the dimensions of the container of least cost.

22. A container with square base, vertical sides, and open top is to be made from 1000 ft^2 of material. Find the dimensions of the container with greatest volume.

23. A rectangular container with two square sides and an open top is to have a volume of V cubic units. Find the dimensions of the container with minimum surface area.

24. Find the dimensions of the right circular cylinder of largest volume that can be inscribed in a sphere of radius R.

25. Find the dimensions of the right circular cylinder of greatest surface area that can be inscribed in a sphere of radius R.

26. Show that the right circular cylinder of greatest volume that can be inscribed in a right circular cone has volume that is $\frac{4}{9}$ the volume of the cone (Figure Ex-26).

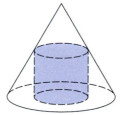

Figure Ex-26

27. A closed, cylindrical can is to have a volume of V cubic units. Show that the can of minimum surface area is achieved when the height is equal to the diameter of the base.

28. A closed cylindrical can is to have a surface area of S square units. Show that the can of maximum volume is achieved when the height is equal to the diameter of the base.

29. A cylindrical can, open at the top, is to hold 500 cm^3 of liquid. Find the height and radius that minimize the amount of material needed to manufacture the can.

30. A soup can in the shape of a right circular cylinder of radius r and height h is to have a prescribed volume V. The top and bottom are cut from squares as shown in the accompanying figure. If the shaded corners are wasted, but there is

no other waste, find the ratio r/h for the can requiring the least material (including waste).

31. A box-shaped wire frame consists of two identical wire squares whose vertices are connected by four straight wires of equal length (Figure Ex-31). If the frame is to be made from a wire of length L, what should the dimensions be to obtain a box of greatest volume?

Figure Ex-30   Figure Ex-31

32. Suppose that the sum of the surface areas of a sphere and a cube is a constant.
  (a) Show that the sum of their volumes is smallest when the diameter of the sphere is equal to the length of an edge of the cube.
  (b) When will the sum of their volumes be greatest?

33. Find the height and radius of the cone of slant height L whose volume is as large as possible.

34. A cone is made from a circular sheet of radius R by cutting out a sector and gluing the cut edges of the remaining piece together (Figure Ex-34). What is the maximum volume attainable for the cone?

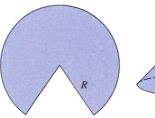

Figure Ex-34

35. A cone-shaped paper drinking cup is to hold 10 cm^3 of water. Find the height and radius of the cup that will require the least amount of paper.

36. Find the dimensions of the isosceles triangle of least area that can be circumscribed about a circle of radius R.

37. Find the height and radius of the right circular cone with least volume that can be circumscribed about a sphere of radius R.

38. A trapezoid is inscribed in a semicircle of radius 2 so that one side is along the diameter (Figure Ex-38 on p. 352). Find the maximum possible area for the trapezoid. [*Hint:* Express the area of the trapezoid in terms of θ.]

39. A drainage channel is to be made so that its cross section is a trapezoid with equally sloping sides (Figure Ex-39 on p. 352). If the sides and bottom all have a length of 5 ft, how should the angle θ ($0 \le \theta \le \pi/2$) be chosen to yield the greatest cross-sectional area of the channel?

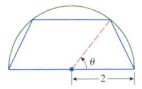

Figure Ex-38

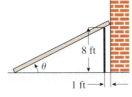

Figure Ex-39

40. A lamp is suspended above the center of a round table of radius r. How high above the table should the lamp be placed to achieve maximum illumination at the edge of the table? [Assume that the illumination I is directly proportional to the cosine of the angle of incidence ϕ of the light rays and inversely proportional to the square of the distance l from the light source (Figure Ex-40).]

41. A plank is used to reach over a fence 8 ft high to support a wall that is 1 ft behind the fence (Figure Ex-41). What is the length of the shortest plank that can be used? [*Hint:* Express the length of the plank in terms of the angle θ shown in the figure.]

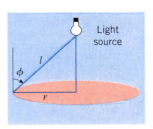

Figure Ex-40   Figure Ex-41

42. A commercial cattle ranch currently allows 20 steers per acre of grazing land; on the average its steers weigh 2000 lb at market. Estimates by the Agriculture Department indicate that the average market weight per steer will be reduced by 50 lb for each additional steer added per acre of grazing land. How many steers per acre should be allowed in order for the ranch to get the largest possible total market weight for its cattle?

43. (a) A chemical manufacturer sells sulfuric acid in bulk at a price of $100 per unit. If the daily total production cost in dollars for x units is

$$C(x) = 100{,}000 + 50x + 0.0025x^2$$

and if the daily production capacity is at most 7000 units, how many units of sulfuric acid must be manufactured and sold daily to maximize the profit?

(b) Would it benefit the manufacturer to expand the daily production capacity?

(c) Use marginal analysis to approximate the effect on profit if daily production could be increased from 7000 to 7001 units.

44. A firm determines that x units of its product can be sold daily at p dollars per unit, where

$$x = 1000 - p$$

The cost of producing x units per day is

$$C(x) = 3000 + 20x$$

(a) Find the revenue function $R(x)$.

(b) Find the profit function $P(x)$.

(c) Assuming that the production capacity is at most 500 units per day, determine how many units the company must produce and sell each day to maximize the profit.

(d) Find the maximum profit.

(e) What price per unit must be charged to obtain the maximum profit?

45. In a certain chemical manufacturing process, the daily weight y of defective chemical output depends on the total weight x of all output according to the empirical formula

$$y = 0.01x + 0.00003x^2$$

where x and y are in pounds. If the profit is $100 per pound of nondefective chemical produced and the loss is $20 per pound of defective chemical produced, how many pounds of chemical should be produced daily to maximize the total daily profit?

46. The cost c (in dollars per hour) to run an ocean liner at a constant speed v (in miles per hour) is given by $c = a + bv^n$, where a, b, and n are positive constants with $n > 1$. Find the speed needed to make the cheapest 3000-mi run.

47. Two particles, A and B, are in motion in the xy-plane. Their coordinates at each instant of time t ($t \ge 0$) are given by $x_A = t$, $y_A = 2t$, $x_B = 1 - t$, and $y_B = t$. Find the minimum distance between A and B.

48. Follow the directions of Exercise 47, with $x_A = t$, $y_A = t^2$, $x_B = 2t$, and $y_B = 2$.

49. Prove that $(1, 0)$ is the closest point on the curve $x^2 + y^2 = 1$ to $(2, 0)$.

50. Find all points on the curve $y = \sqrt{x}$ for $0 \le x \le 3$ that are closest to, and at the greatest distance from, the point $(2, 0)$.

51. Find all points on the curve $x^2 - y^2 = 1$ closest to $(0, 2)$.

52. Find a point on the curve $x = 2y^2$ closest to $(0, 9)$.

53. Find the coordinates of the point P on the curve

$$y = \frac{1}{x^2} \quad (x > 0)$$

where the segment of the tangent line at P that is cut off by the coordinate axes has its shortest length.

54. Find the x-coordinate of the point P on the parabola

$$y = 1 - x^2 \quad (0 < x \le 1)$$

where the triangle that is enclosed by the tangent line at P and the coordinate axes has the smallest area.

55. Where on the curve $y = \left(1 + x^2\right)^{-1}$ does the tangent line have the greatest slope?

56. A man is floating in a rowboat 1 mile from the (straight) shoreline of a large lake. A town is located on the shoreline 1 mile from the point on the shoreline closest to the man. As suggested in the accompanying figure, he intends to row in a straight line to some point P on the shoreline and then walk the remaining distance to the town. To what point should he row in order to reach his destination in the least time if
(a) he can walk 5 mi/h and row 3 mi/h;
(b) he can walk 5 mi/h and row 4 mi/h?

57. A pipe of negligible diameter is to be carried horizontally around a corner from a hallway 8 ft wide into a hallway 4 ft wide (Figure Ex-57). What is the maximum length that the pipe can have? [An interesting discussion of this problem in the case where the diameter of the pipe is not neglected is given by Norman Miller in the *American Mathematical Monthly*, Vol. 56, 1949, pp. 177–179.]

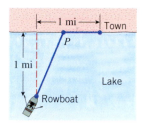

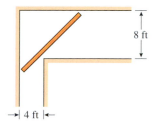

Figure Ex-56 | Figure Ex-57

58. If an unknown physical quantity x is measured n times, the measurements $x_1, x_2, \ldots, x_n$ often vary because of uncontrollable factors such as temperature, atmospheric pressure, and so forth. Thus, a scientist is often faced with the problem of using n different observed measurements to obtain an estimate $\bar{x}$ of an unknown quantity x. One method for making such an estimate is based on the **least squares principle**, which states that the estimate $\bar{x}$ should be chosen to minimize

$$s = (x_1 - \bar{x})^2 + (x_2 - \bar{x})^2 + \cdots + (x_n - \bar{x})^2$$

which is the sum of the squares of the deviations between the estimate $\bar{x}$ and the measured values. Show that the estimate resulting from the least squares principle is

$$\bar{x} = \frac{1}{n}(x_1 + x_2 + \cdots + x_n)$$

that is, $\bar{x}$ is the arithmetic average of the observed values.

59. Suppose that the intensity of a point light source is directly proportional to the strength of the source and inversely proportional to the square of the distance from the source. Two point light sources with strengths of S and $8S$ are separated by a distance of 90 cm. Where on the line segment between the two sources is the intensity a minimum?

60. Prove: If $f(x) \geq 0$ on an interval I and if $f(x)$ has a maximum value on I at x_0, then $\sqrt{f(x)}$ also has a maximum value at x_0. Similarly for minimum values. [*Hint:* Use the fact that $\sqrt{x}$ is an increasing function on the interval $[0, +\infty)$.]

61. Given points $A(2, 1)$ and $B(5, 4)$, find the point P in the interval $[2, 5]$ on the x-axis that maximizes angle APB.

62. The lower edge of a painting, 10 ft in height, is 2 ft above an observer's eye level. Assuming that the best view is obtained when the angle subtended at the observer's eye by the painting is maximum, how far from the wall should the observer stand?

63. *Fermat's* (biography on p. 341) *principle* in optics states that light traveling from one point to another follows that path for which the total travel time is minimum. In a uniform medium, the paths of "minimum time" and "shortest distance" turn out to be the same, so that light, if unobstructed, travels along a straight line. Assume that we have a light source, a flat mirror, and an observer in a uniform medium. If a light ray leaves the source, bounces off the mirror, and travels on to the observer, then its path will consist of two line segments, as shown in Figure Ex-63. According to Fermat's principle, the path will be such that the total travel time t is minimum or, since the medium is uniform, the path will be such that the total distance traveled from A to P to B is as small as possible. Assuming the minimum occurs when $dt/dx = 0$, show that the light ray will strike the mirror at the point P where the "angle of incidence" θ_1 equals the "angle of reflection" θ_2.

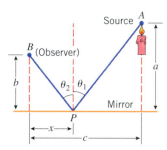

Figure Ex-63

64. Fermat's principle (Exercise 63) also explains why light rays traveling between air and water undergo bending (refraction). Imagine that we have two uniform media (such as air and water) and a light ray traveling from a source A in one medium to an observer B in the other medium (Figure Ex-64 on p. 354). It is known that light travels at a constant speed in a uniform medium, but more slowly in a dense medium (such as water) than in a thin medium (such as air). Consequently, the path of shortest time from A to B is not necessarily a straight line, but rather some broken line path A to P to B allowing the light to take greatest advantage of its higher speed through the thin medium. *Snell's* [*] (biography on p. 354) *law of refraction* states that the path of the light ray will be such that

$$\frac{\sin\theta_1}{v_1} = \frac{\sin\theta_2}{v_2}$$

where v_1 is the speed of light in the first medium, v_2 is the speed of light in the second medium, and θ_1 and θ_2 are the angles shown in Figure Ex-64. Show that this follows from the assumption that the path of minimum time occurs when $dt/dx = 0$.

65. A farmer wants to walk at a constant rate from her barn to a straight river, fill her pail, and carry it to her house in the least time.
    (a) Explain how this problem relates to Fermat's principle and the light-reflection problem in Exercise 63.
    (b) Use the result of Exercise 63 to describe geometrically the best path for the farmer to take.
    (c) Use part (b) to determine where the farmer should fill her pail if her house and barn are located as in Figure Ex-65.

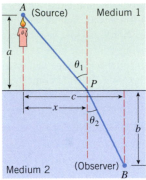

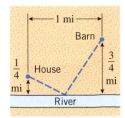

Figure Ex-64

Figure Ex-65

5.7 NEWTON'S METHOD

In Section 2.5 we showed how to approximate the roots of an equation $f(x) = 0$ by using the Intermediate-Value Theorem and also by zooming in on the x-intercepts of $y = f(x)$ with a graphing utility. In this section we will study a technique, called Newton's Method, that is usually more efficient than either of those methods. Newton's Method is the technique used by many commercial and scientific computer programs for finding roots.

NEWTON'S METHOD

In beginning algebra one learns that the solution of a first-degree equation $ax + b = 0$ is given by the formula $x = -b/a$, and the solutions of a second-degree equation

$$ax^2 + bx + c = 0$$

are given by the quadratic formula. Formulas also exist for the solutions of all third- and fourth-degree equations, although they are too complicated to be of practical use. In 1826 it was shown by the Norwegian mathematician Niels Henrik Abel[†] that it is impossible to

---
[*] WILLEBRORD VAN ROIJEN SNELL (1591–1626). Dutch mathematician. Snell, who succeeded his father to the post of Professor of Mathematics at the University of Leiden in 1613, is most famous for the result of light refraction that bears his name. Although this phenomenon was studied as far back as the ancient Greek astronomer Ptolemy, until Snell's work the relationship was incorrectly thought to be $\theta_1/v_1 = \theta_2/v_2$. Snell's law was published by Descartes in 1638 without giving proper credit to Snell. Snell also discovered a method for determining distances by triangulation that founded the modern technique of mapmaking.

[†] NIELS HENRIK ABEL (1802–1829). Norwegian mathematician. Abel was the son of a poor Lutheran minister and a remarkably beautiful mother from whom he inherited strikingly good looks. In his brief life of 26 years Abel lived in virtual poverty and suffered a succession of adversities; yet he managed to prove major results that altered the mathematical landscape forever. At the age of thirteen he was sent away from home to a school whose better days had long passed. By a stroke of luck the school had just hired a teacher named Bernt Michael Holmboe, who quickly discovered that Abel had extraordinary mathematical ability. Together, they studied the calculus texts of Euler and works of Newton and the later French mathematicians. By the time he graduated, Abel was familiar with most of the great mathematical literature. In 1820 his father died, leaving the family in dire financial straits. Abel was able to enter the University of Christiania in Oslo only because he was granted a free room and several professors supported him directly from their salaries. The University had no advanced courses in mathematics, so Abel took a preliminary degree in 1822 and then continued to study mathematics on his own. In 1824 he published at his own expense the proof that it is impossible to solve the general fifth-degree polynomial equation algebraically. With the hope that this landmark paper would lead to his recognition and acceptance by the European mathematical community, Abel sent the paper to the great German mathematician Gauss, who casually declared it to be a "monstrosity" and tossed it aside. However, in 1826 Abel's paper on the fifth-degree equation and other work was published in the first issue of a new journal, founded by his friend, Leopold Crelle. In the summer of 1826 he completed a landmark work on transcendental functions, which he submitted to the French Academy of

construct a similar formula for the solutions of a *general* fifth-degree equation or higher. Thus, for a *specific* fifth-degree polynomial equation such as

$$x^5 - 9x^4 + 2x^3 - 5x^2 + 17x - 8 = 0$$

it may be difficult or impossible to find exact values for all of the solutions. Similar difficulties occur for nonpolynomial equations such as

$$x - \cos x = 0$$

For such equations the solutions are generally approximated in some way, often by the method we will now discuss.

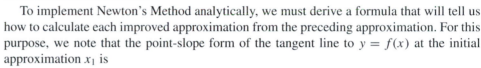

Figure 5.7.1

Suppose that we are trying to find a root r of the equation $f(x) = 0$, and suppose that by some method we are able to obtain an initial rough estimate, x_1, of r, say by generating the graph of $y = f(x)$ with a graphing utility and examining the x-intercept. If $f(x_1) = 0$, then $r = x_1$. If $f(x_1) \neq 0$, then we consider an easier problem, that of finding a root to a linear equation. The best linear approximation to $y = f(x)$ near $x = x_1$ is given by the tangent line to the graph of f at x_1, so it might be reasonable to expect that the x-intercept to this tangent line provides an improved approximation to r. Call this intercept x_2 (Figure 5.7.1). We can now treat x_2 in the same way we did x_1. If $f(x_2) = 0$, then $r = x_2$. If $f(x_2) \neq 0$, then construct the tangent line to the graph of f at x_2, and take x_3 to be the x-intercept of this tangent line. Continuing in this way we can generate a succession of values $x_1, x_2, x_3, x_4, \ldots$ that will usually approach r. This procedure for approximating r is called ***Newton's Method***.

To implement Newton's Method analytically, we must derive a formula that will tell us how to calculate each improved approximation from the preceding approximation. For this purpose, we note that the point-slope form of the tangent line to $y = f(x)$ at the initial approximation x_1 is

$$y - f(x_1) = f'(x_1)(x - x_1) \tag{1}$$

If $f'(x_1) \neq 0$, then this line is not parallel to the x-axis and consequently it crosses the x-axis at some point $(x_2, 0)$. Substituting the coordinates of this point in (1) yields

$$-f(x_1) = f'(x_1)(x_2 - x_1)$$

Solving for x_2 we obtain

$$x_2 = x_1 - \frac{f(x_1)}{f'(x_1)} \tag{2}$$

The next approximation can be obtained more easily. If we view x_2 as the starting approximation and x_3 the new approximation, we can simply apply (2) with x_2 in place of x_1 and x_3 in place of x_2. This yields

$$x_3 = x_2 - \frac{f(x_2)}{f'(x_2)} \tag{3}$$

provided $f'(x_2) \neq 0$. In general, if x_n is the nth approximation, then it is evident from the pattern in (2) and (3) that the improved approximation x_{n+1} is given by

> **Newton's Method**
>
> $$x_{n+1} = x_n - \frac{f(x_n)}{f'(x_n)}, \quad n = 1, 2, 3, \ldots$$ (4)

Example 1   Use Newton's Method to approximate the real solutions of

$$x^3 - x - 1 = 0$$

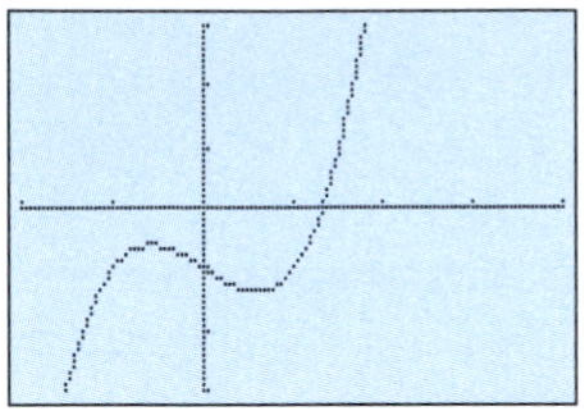

Solution.   Let $f(x) = x^3 - x - 1$, so $f'(x) = 3x^2 - 1$ and (4) becomes

$$x_{n+1} = x_n - \frac{x_n^3 - x_n - 1}{3x_n^2 - 1}$$ (5)

From the graph of f in Figure 5.7.2, we see that the given equation has only one real solution. This solution lies between 1 and 2 because $f(1) = -1 < 0$ and $f(2) = 5 > 0$. We will use $x_1 = 1.5$ as our first approximation ($x_1 = 1$ or $x_1 = 2$ would also be reasonable choices).

Letting $n = 1$ in (5) and substituting $x_1 = 1.5$ yields

$$x_2 = 1.5 - \frac{(1.5)^3 - 1.5 - 1}{3(1.5)^2 - 1} \approx 1.34782609$$ (6)

(We used a calculator that displays nine digits.) Next, we let $n = 2$ in (5) and substitute x_2 to obtain

$$x_3 = x_2 - \frac{x_2^3 - x_2 - 1}{3x_2^2 - 1} \approx 1.32520040$$ (7)

If we continue this process until two identical approximations are generated in succession, we obtain

$$x_1 = 1.5$$
$$x_2 \approx 1.34782609$$
$$x_3 \approx 1.32520040$$
$$x_4 \approx 1.32471817$$
$$x_5 \approx 1.32471796$$
$$x_6 \approx 1.32471796$$

At this stage there is no need to continue further because we have reached the display accuracy limit of our calculator, and all subsequent approximations that the calculator generates will likely be the same. Thus, the solution is approximately $x \approx 1.32471796$. ◀

REMARK.   Many calculators and computer programs calculate internally with more digits than they display. Thus, where possible, you should use stored calculated values rather than displayed values from intermediate calculations. For example, the value of x_2 used in (7) should be the stored value, not (6).

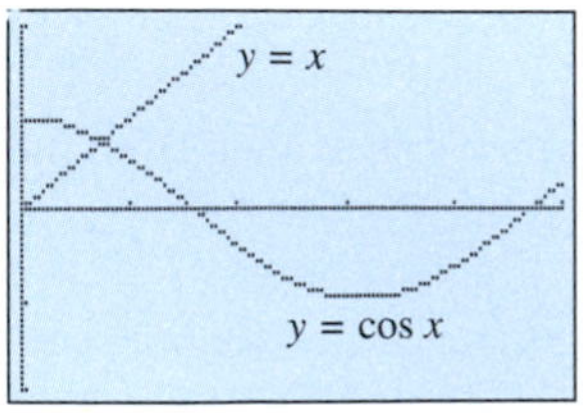

Example 2   It is evident from Figure 5.7.3 that if x is in radians, then the equation

$$\cos x = x$$

has a solution between 0 and 1. Use Newton's Method to approximate it.

Solution.   Rewrite the equation as

$$x - \cos x = 0$$

and apply (4) with $f(x) = x - \cos x$. Since $f'(x) = 1 + \sin x$, (4) becomes

$$x_{n+1} = x_n - \frac{x_n - \cos x_n}{1 + \sin x_n}$$ (8)

[−2, 4] × [−3, 3]
xScl = 1, yScl = 1

$y = x^3 - x - 1$

Figure 5.7.2

$y = x$

$y = \cos x$

[0, 5] × [−2, 2]
xScl = 1, yScl = 1

Figure 5.7.3

From Figure 5.7.3, the solution seems closer to $x = 1$ than $x = 0$, so we will use $x_1 = 1$ (radian) as our initial approximation. Letting $n = 1$ in (8) and substituting $x_1 = 1$ yields

$$x_2 = 1 - \frac{1 - \cos 1}{1 + \sin 1} \approx 0.750363868$$

Next, letting $n = 2$ in (8) and substituting this value of x_2 yields

$$x_3 = x_2 - \frac{x_2 - \cos x_2}{1 + \sin x_2} \approx 0.739112891$$

If we continue this process until two identical approximations are generated in succession, we obtain

$$x_1 = 1$$
$$x_2 \approx 0.750363868$$
$$x_3 \approx 0.739112891$$
$$x_4 \approx 0.739085133$$
$$x_5 \approx 0.739085133$$

Thus, to the accuracy limit of our calculator, the solution of the equation $\cos x = x$ is $x \approx 0.739085133$. ◄

SOME DIFFICULTIES WITH NEWTON'S METHOD

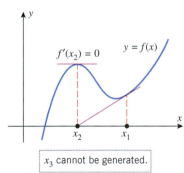

x_3 cannot be generated.

Figure 5.7.4

When Newton's Method works, the approximations usually converge toward the solution with dramatic speed. However, there are situations in which the method fails. For example, if $f'(x_n) = 0$ for some n, then (4) involves a division by zero, making it impossible to generate x_{n+1}. However, this is to be expected because the tangent line to $y = f(x)$ is parallel to the x-axis where $f'(x_n) = 0$, and hence this tangent line does not cross the x-axis to generate the next approximation (Figure 5.7.4).

Newton's Method can fail for other reasons as well; sometimes it may overlook the root you are trying to find and converge to a different root, and sometimes it may fail to converge altogether. For example, consider the equation

$$x^{1/3} = 0$$

which has $x = 0$ as its only solution, and try to approximate this solution by Newton's Method with a starting value of $x_0 = 1$. Letting $f(x) = x^{1/3}$, Formula (4) becomes

$$x_{n+1} = x_n - \frac{(x_n)^{1/3}}{\frac{1}{3}(x_n)^{-2/3}} = x_n - 3x_n = -2x_n$$

Beginning with $x_1 = 1$, the successive values generated by this formula are

$$x_1 = 1, \quad x_2 = -2, \quad x_3 = 4, \quad x_4 = -8, \ldots$$

which obviously do not converge to $x = 0$. Figure 5.7.5 illustrates what is happening geometrically in this situation.

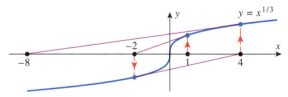

Figure 5.7.5

To learn more about the conditions under which Newton's Method converges and for a discussion of error questions, you should consult a book on numerical analysis. For a more in-depth discussion of Newton's Method and its relationship to contemporary studies of chaos and fractals, you may want to read the article, "Newton's Method and Fractal Patterns," by Phillip Straffin, which appears in *Applications of Calculus*, MAA Notes, Vol. 3, No. 29, 1993, published by the Mathematical Association of America.

EXERCISE SET 5.7 ~ Graphing Utility

In this exercise set express your answer with as many decimal digits as your calculating utility can display, but use the procedure in the remark following Example 1.

1. Approximate $\sqrt{2}$ by applying Newton's Method to the equation $x^2 - 2 = 0$.

2. Approximate $\sqrt{7}$ by applying Newton's Method to the equation $x^2 - 7 = 0$.

3. Approximate $\sqrt[3]{6}$ by applying Newton's Method to the equation $x^3 - 6 = 0$.

4. To what equation would you apply Newton's Method to approximate the nth root of a?

In Exercises 5–8, the equation has one real solution. Approximate it by Newton's Method.

5. $x^3 - x + 3 = 0$

6. $x^3 + x - 1 = 0$

7. $x^5 + x^4 - 5 = 0$

8. $x^5 - x + 1 = 0$

In Exercises 9–14, use a graphing utility to determine how many solutions the equation has, and then use Newton's Method to approximate the solution that satisfies the stated condition.

~ 9. $x^4 + x - 3 = 0$; $x < 0$

~ 10. $x^5 - 5x^3 - 2 = 0$; $x > 0$

~ 11. $2 \sin x = x$; $x > 0$

~ 12. $\sin x = x^2$; $x > 0$

~ 13. $x - \tan x = 0$; $\pi/2 < x < 3\pi/2$

~ 14. $1 + e^x \cos x = 0$; $0 < x < \pi$

In Exercises 15–20, use a graphing utility to determine the number of times the curves intersect; and then apply Newton's Method, where needed, to approximate the x-coordinates of all intersections.

~ 15. $y = x^3$ and $y = \frac{1}{2}x - 1$

~ 16. $y = \sin x$ and $y = x^3 - 2x^2 + 1$

~ 17. $y = x^2$ and $y = \sqrt{2x + 1}$

~ 18. $y = \frac{1}{8}x^3 + 1$ and $y = \cos 2x$

~ 19. $y = 1$ and $y = e^x \cos x$; $0 < x < \pi$

~ 20. $y = e^{-x}$ and $y = \ln x$

21. The *mechanic's rule* for approximating square roots states that $\sqrt{a} \approx x_{n+1}$, where
$$x_{n+1} = \frac{1}{2}\left(x_n + \frac{a}{x_n}\right), \quad n = 1, 2, 3, \ldots$$
and x_1 is any positive approximation to $\sqrt{a}$.

(a) Apply Newton's Method to
$$f(x) = x^2 - a$$
to derive the mechanic's rule.

(b) Use the mechanic's rule to approximate $\sqrt{10}$.

22. Many calculators compute reciprocals using the approximation $1/a \approx x_{n+1}$, where
$$x_{n+1} = x_n(2 - ax_n), \quad n = 1, 2, 3, \ldots$$
and x_1 is an initial approximation to $1/a$. This formula makes it possible to perform divisions using multiplications and subtractions, which is a faster procedure than dividing directly.

(a) Apply Newton's Method to
$$f(x) = \frac{1}{x} - a$$
to derive this approximation.

(b) Use the formula to approximate $\frac{1}{17}$.

23. Use Newton's Method to find the absolute minimum of
$$f(x) = \tfrac{1}{4}x^4 + x^2 + 5x$$

24. Use Newton's Method to find the absolute maximum of $f(x) = x \sin x$ on the interval $[0, \pi]$.

25. For the function $f(x) = e^x/(1 + x^2)$, use Newton's Method to approximate the x-coordinates of the inflection points to two decimal places.

26. Use Newton's Method to find the absolute maximum of
$$f(x) = (1 - 2x) \tan^{-1} x$$

27. Use Newton's Method to find the coordinates of the point on the parabola $y = x^2$ that is closest to the point $(1, 0)$.

28. Use Newton's Method to find the dimensions of the rectangle of largest area that can be inscribed under the curve $y = \cos x$ for $0 \le x \le \pi/2$, as shown in the accompanying figure.

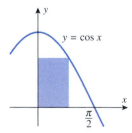

$y = \cos x$

Figure Ex-28

29. (a) Show that on a circle of radius r, the central angle θ that subtends an arc whose length is 1.5 times the length L of its chord satisfies the equation $\theta = 3 \sin(\theta/2)$ (Figure Ex-29 on p. 359).

(b) Use Newton's Method to approximate θ.

30. A *segment* of a circle is the region enclosed by an arc and its chord (Figure Ex-30 on p. 359). If r is the radius of the circle and θ the angle subtended at the center of the circle, then it can be shown that the area A of the segment

is $A = \frac{1}{2}r^2(\theta - \sin\theta)$, where θ is in radians. Find the value of θ for which the area of the segment is one-fourth the area of the circle. Give θ to the nearest degree.

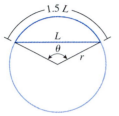

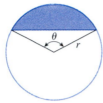

Figure Ex-29   Figure Ex-30

In Exercises 31 and 32, use Newton's Method to approximate all real values of y satisfying the given equation for the indicated value of x.

31. $xy^4 + x^3 y = 1$; $x = 1$

32. $xy - \cos\left(\frac{1}{2}xy\right) = 0$; $x = 2$

33. An *annuity* is a sequence of equal payments that are paid or received at regular time intervals. For example, you may want to deposit equal amounts at the end of each year into an interest-bearing account for the purpose of accumulating a lump sum at some future time. If, at the end of each year, interest of $i \times 100\%$ on the account balance for that year is added to the account, then the account is said to pay $i \times 100\%$ interest, *compounded annually*. It can be shown that if payments of Q dollars are deposited at the end of each year into an account that pays $i \times 100\%$ compounded annually, then at the time when the nth payment and the accrued interest for the past year are deposited, the amount $S(n)$ in the account is given by the formula

$$S(n) = \frac{Q}{i}[(1+i)^n - 1]$$

Suppose that you can invest \$5000 in an interest-bearing account at the end of each year, and your objective is to have \$250,000 on the 25th payment. What annual compound interest rate must the account pay for you to achieve your goal? [*Hint:* Show that the interest rate i satisfies the equation $50i = (1+i)^{25} - 1$, and solve it using Newton's Method.]

34. (a) Use a graphing utility to generate the graph of

$$f(x) = \frac{x}{x^2 + 1}$$

and use it to explain what happens if you apply Newton's Method with a starting value of $x_1 = 2$. Check your conclusion by computing $x_2, x_3, x_4,$ and x_5.

(b) Use the graph generated in part (a) to explain what happens if you apply Newton's Method with a starting value of $x_1 = 0.5$. Check your conclusion by computing $x_2, x_3, x_4,$ and x_5.

35. (a) Apply Newton's Method to the function $f(x) = x^2 + 1$ with a starting value of $x_1 = 0.5$, and determine if the values of $x_2, \ldots, x_{10}$ appear to converge.

(b) Explain what is happening.

5.8 ROLLE'S THEOREM; MEAN-VALUE THEOREM

In this section we will discuss a result called the Mean-Value Theorem. This theorem has so many important consequences that it is regarded as one of the major principles in calculus.

ROLLE'S THEOREM

We will begin with a special case of the Mean-Value Theorem, called Rolle's Theorem, in honor of the mathematician Michel Rolle.* This theorem states the geometrically obvious

---

* MICHEL ROLLE (1652-1719), French mathematician. Rolle, the son of a shopkeeper, received only an elementary education. He married early and as a young man struggled hard to support his family on the meager wages of a transcriber for notaries and attorneys. In spite of his financial problems and minimal education, Rolle studied algebra and Diophantine analysis (a branch of number theory) on his own. Rolle's fortune changed dramatically in 1682 when he published an elegant solution of a difficult, unsolved problem in Diophantine analysis. The public recognition of his achievement led to a patronage under minister Louvois, a job as an elementary mathematics teacher, and eventually to a short-term administrative post in the Ministry of War. In 1685 he joined the Académie des Sciences in a low-level position for which he received no regular salary until 1699. He stayed there until he died of apoplexy in 1719.

While Rolle's forté was always Diophantine analysis, his most important work was a book on the algebra of equations, called *Traité d'algèbre*, published in 1690. In that book Rolle firmly established the notation $\sqrt[n]{a}$ [earlier written as $\sqrt{n}\,a$] for the nth root of a, and proved a polynomial version of the theorem that today bears his name. (Rolle's Theorem was named by Giusto Bellavitis in 1846.) Ironically, Rolle was one of the most vocal early antagonists of calculus. He strove intently to demonstrate that it gave erroneous results and was based on unsound reasoning. He quarreled so vigorously on the subject that the Académie des Sciences was forced to intervene on several occasions. Among his several achievements, Rolle helped advance the currently accepted size order for negative numbers. Descartes, for example, viewed -2 as smaller than -5. Rolle preceded most of his contemporaries by adopting the current convention in 1691.

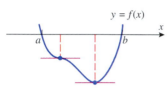

Figure 5.8.1

fact that if the graph of a differentiable function intersects the x-axis at two places, a and b, then somewhere between a and b there must be at least one place where the tangent line is horizontal (Figure 5.8.1). The precise statement of the theorem is as follows:

5.8.1   THEOREM (*Rolle's Theorem*).   *Let f be differentiable on (a, b) and continuous on $[a, b]$. If $f(a) = f(b) = 0$, then there is at least one number c in (a, b) such that $f'(c) = 0$.*

Proof.   Either $f(x)$ is equal to zero for all x in $[a, b]$ or it is not. If it is, then $f'(x) = 0$ for all x in (a, b), since f is constant on (a, b). Thus, for any c in (a, b)

$$f'(c) = 0$$

If $f(x)$ is not equal to zero for all x in $[a, b]$, then there must be a value of x in (a, b) where $f(x) > 0$ or $f(x) < 0$. We will consider the first case and leave the second as an exercise.

   Since f is continuous on $[a, b]$, it follows from the Extreme-Value Theorem (5.5.3) that f has a maximum value at some number c in $[a, b]$. Since $f(a) = f(b) = 0$ and $f(x) > 0$ somewhere in (a, b), the number c cannot be an endpoint; it must lie in (a, b). By hypothesis, f is differentiable everywhere on (a, b). In particular, it is differentiable at c so that $f'(c) = 0$ by Theorem 5.5.4.   ∎

Example 1   The function $f(x) = \sin x$ has roots at $x = 0$ and $x = 2\pi$. Verify the hypotheses and conclusion of Rolle's Theorem for $f(x) = \sin x$ on $[0, 2\pi]$.

Solution.   Since f is continuous and differentiable everywhere, it is differentiable on $(0, 2\pi)$ and continuous on $[0, 2\pi]$. Thus, Rolle's Theorem guarantees that there is at least one number c in the interval $(0, 2\pi)$ such that $f'(c) = 0$. Since $f'(x) = \cos x$, we can find c by solving the equation $\cos c = 0$ on the interval $(0, 2\pi)$. This yields two values for c, namely $c_1 = \pi/2$ and $c_2 = 3\pi/2$ (Figure 5.8.2).   ◄

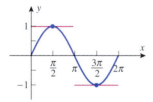

Figure 5.8.2

• REMARK.   In the preceding example, we were able to find the exact values of c because the equation $f'(c) = 0$ was easy to solve. However, if this equation cannot be solved, then you may not be able to find precise values of c, even though you know they exist. This will rarely cause problems because usually one is more interested in knowing that the values of c exist than in finding them.

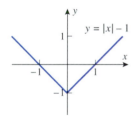

Figure 5.8.3

   The hypotheses in Rolle's Theorem are critical—if f fails to be differentiable at even one place in the interval, then the conclusion may not hold. For example, the function $f(x) = |x| - 1$ has roots at $x = \pm 1$, yet there is no horizontal tangent line to the graph of f over the interval $(-1, 1)$ (Figure 5.8.3).

THE MEAN-VALUE THEOREM

Rolle's Theorem is a special case of the ***Mean-Value Theorem***, which states that between any two points A and B on the graph of a differentiable function, there must be at least one place where the tangent line to the curve is parallel to the secant line joining A and B (Figure 5.8.4).

   Noting that the slope of the secant line joining $A(a, f(a))$ and $B(b, f(b))$ is

$$\frac{f(b) - f(a)}{b - a}$$

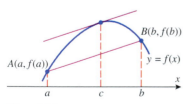

Figure 5.8.4

and the slope of the tangent at c is $f'(c)$, the Mean-Value Theorem can be stated precisely as follows.

5.8.2   THEOREM (*Mean-Value Theorem*).   *Let f be differentiable on (a, b) and continuous on $[a, b]$. Then there is at least one number c in (a, b) such that*

$$f'(c) = \frac{f(b) - f(a)}{b - a} \qquad (1)$$

**VELOCITY INTERPRETATION OF
THE MEAN-VALUE THEOREM**

There is a nice interpretation of the Mean-Value Theorem in the situation where $x = f(t)$ is the position versus time curve for a car moving along a straight road. In this case, the right side of (1) is the average velocity of the car over the time interval from $a \le t \le b$, and the left side is the instantaneous velocity at time $t = c$. Thus, the Mean-Value Theorem implies that at least once during the time interval the instantaneous velocity must equal the average velocity. This agrees with our real-world experience—if the average velocity for a trip is 40 mi/h, then sometime during the trip the speedometer has to read 40 mi/h.

Example 2   You are driving on a straight highway on which the speed limit is 55 mi/h. At 8:05 A.M. a police car clocks your velocity at 50 mi/h and at 8:10 A.M. a second police car posted 5 mi down the road clocks your velocity at 55 mi/h. Explain why the police have a right to charge you with a speeding violation.

Solution.   You traveled 5 mi in 5 min $\left(= \frac{1}{12} \text{ h}\right)$, so your average velocity was 60 mi/h. However, the Mean-Value Theorem guarantees the police that your instantaneous velocity was 60 mi/h at least once over the 5-mi section of highway. ◄

**PROOF OF THE MEAN-VALUE
THEOREM**

Motivation for the Proof of Theorem 5.8.2.   Figure 5.8.4 suggests that (1) will hold (i.e., the tangent line will be parallel to the secant line) at a number c where the vertical distance between the curve and the secant line is maximum. Thus, to prove the Mean-Value Theorem it is natural to begin by looking for a formula for the vertical distance $v(x)$ between the curve $y = f(x)$ and the secant line joining $(a, f(a))$ and $(b, f(b))$.

Proof of Theorem 5.8.2.   Since the two-point form of the equation of the secant line joining $(a, f(a))$ and $(b, f(b))$ is

$$y - f(a) = \frac{f(b) - f(a)}{b - a}(x - a)$$

or equivalently,

$$y = \frac{f(b) - f(a)}{b - a}(x - a) + f(a)$$

the difference $v(x)$ between the height of the graph of f and the height of the secant line is

$$v(x) = f(x) - \left[\frac{f(b) - f(a)}{b - a}(x - a) + f(a) \right] \tag{2}$$

Since $f(x)$ is continuous on $[a, b]$ and differentiable on (a, b), so is $v(x)$. Moreover,

$$v(a) = 0 \quad \text{and} \quad v(b) = 0$$

so that $v(x)$ satisfies the hypotheses of Rolle's Theorem on the interval $[a, b]$. Thus, there is a number c in (a, b) such that $v'(c) = 0$. But from Equation (2)

$$v'(x) = f'(x) - \frac{f(b) - f(a)}{b - a}$$

so

$$v'(c) = f'(c) - \frac{f(b) - f(a)}{b - a}$$

Since $v'(c) = 0$, we have

$$f'(c) = \frac{f(b) - f(a)}{b - a}$$

Example 3

(a)   Generate the graph of $f(x) = (x^3/4) + 1$ over the interval $[0, 2]$, and use it to determine the number of tangent lines to the graph of f over the interval $(0, 2)$ that are parallel to the secant line joining the endpoints of the graph.

(b)   Show that f satisfies the hypotheses of the Mean-Value Theorem on the interval $[0, 2]$, and find all values of c in the interval $(0, 2)$ whose existence is guaranteed by the Mean-Value Theorem. Confirm that these values of c are consistent with your graph in part (a).

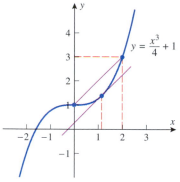

Figure 5.8.5

Solution (a). The graph of f in Figure 5.8.5 suggests that there is only one tangent line over the interval $(0, 2)$ that is parallel to the secant line joining the endpoints.

Solution (b). The function f is continuous and differentiable everywhere because it is a polynomial. In particular, f is continuous on $[0, 2]$ and differentiable on $(0, 2)$, so the hypotheses of the Mean-Value Theorem are satisfied with $a = 0$ and $b = 2$. But

$$f(a) = f(0) = 1, \quad f(b) = f(2) = 3$$

$$f'(x) = \frac{3x^2}{4}, \qquad f'(c) = \frac{3c^2}{4}$$

so in this case Equation (1) becomes

$$\frac{3c^2}{4} = \frac{3-1}{2-0} \quad \text{or} \quad 3c^2 = 4$$

which has the two solutions $c = \pm 2/\sqrt{3} \approx \pm 1.15$. However, only the positive solution lies in the interval $(0, 2)$; this value of c is consistent with Figure 5.8.5. ◄

CONSEQUENCES OF THE MEAN-VALUE THEOREM

We stated at the beginning of this section that the Mean-Value Theorem is the starting point for many important results in calculus. As an example of this, we will use it to prove Theorem 5.1.2, which was one of our fundamental tools for analyzing graphs of functions.

5.1.2 THEOREM (*Revisited*).   *Let f be a function that is continuous on a closed interval $[a, b]$ and differentiable on the open interval (a, b).*

(a)   If $f'(x) > 0$ for every value of x in (a, b), then f is increasing on $[a, b]$.

(b)   If $f'(x) < 0$ for every value of x in (a, b), then f is decreasing on $[a, b]$.

(c)   If $f'(x) = 0$ for every value of x in (a, b), then f is constant on $[a, b]$.

Proof (a). Suppose that x_1 and x_2 are numbers in $[a, b]$ such that $x_1 < x_2$. We must show that $f(x_1) < f(x_2)$. Because the hypotheses of the Mean-Value Theorem are satisfied on the entire interval $[a, b]$, they are satisfied on the subinterval $[x_1, x_2]$. Thus, there is some number c in the open interval (x_1, x_2) such that

$$f'(c) = \frac{f(x_2) - f(x_1)}{x_2 - x_1}$$

or equivalently,

$$f(x_2) - f(x_1) = f'(c)(x_2 - x_1) \tag{3}$$

Since c is in the open interval (x_1, x_2), it follows that $a < c < b$; thus, $f'(c) > 0$. However, $x_2 - x_1 > 0$ since we assumed that $x_1 < x_2$. It follows from (3) that $f(x_2) - f(x_1) > 0$ or, equivalently, $f(x_1) < f(x_2)$, which is what we were to prove. The proofs of parts (b) and (c) are similar and are left as exercises. ∎

THE CONSTANT DIFFERENCE THEOREM

We know from our earliest study of derivatives that the derivative of a constant is zero. Part (c) of Theorem 5.1.2 is the converse of that result; that is, a function whose derivative is zero on an interval must be constant on that interval. If we apply this to the difference of two functions, we obtain the following useful theorem.

5.8.3 THEOREM (*The Constant Difference Theorem*).   *If f and g are continuous on a closed interval $[a, b]$, and if $f'(x) = g'(x)$ for all x in the open interval (a, b), then f and g differ by a constant on $[a, b]$; that is, there is a constant k such that $f(x) - g(x) = k$ for all x in $[a, b]$.*

Proof. Let $h(x) = f(x) - g(x)$. Then for every x in (a, b)

$$h'(x) = f'(x) - g'(x) = 0$$

Thus, $h(x) = f(x) - g(x)$ is constant on $[a, b]$ by Theorem 5.1.2(c). ∎

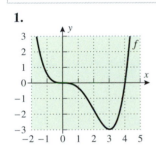

• REMARK. This theorem remains true if the closed interval $[a, b]$ is replaced by a finite or infinite interval (a, b), $[a, b)$, or $(a, b]$, provided f and g are differentiable on (a, b) and continuous on the entire interval.

The Constant Difference Theorem has a simple geometric interpretation—it tells us that if f and g have the same derivative on an interval, then there is a constant k such that $f(x) = g(x) + k$ for each x in the interval; that is, the graphs of f and g can be obtained from one another by a vertical translation (Figure 5.8.6).

If $f'(x) = g'(x)$ on an interval, then the graphs of f and g are vertical translations of one another.

Figure 5.8.6

EXERCISE SET 5.8  Graphing Utility

In Exercises 1 and 2, use the graph of f to find an interval $[a, b]$ on which Rolle's Theorem applies, and find all values of c in that interval that satisfy the conclusion of the theorem.

1.   **2.**

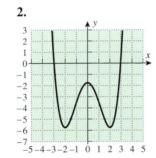

In Exercises 3–8, verify that the hypotheses of Rolle's Theorem are satisfied on the given interval, and find all values of c in that interval that satisfy the conclusion of the theorem.

3. $f(x) = x^2 - 6x + 8$; $[2, 4]$

4. $f(x) = x^3 - 3x^2 + 2x$; $[0, 2]$

5. $f(x) = \cos x$; $[\pi/2, 3\pi/2]$

6. $f(x) = \dfrac{x^2 - 1}{x - 2}$; $[-1, 1]$

7. $f(x) = \frac{1}{2}x - \sqrt{x}$; $[0, 4]$

8. $f(x) = \dfrac{1}{x^2} - \dfrac{4}{3x} + \dfrac{1}{3}$; $[1, 3]$

9. Use the graph of f in the accompanying figure to estimate all values of c that satisfy the conclusion of the Mean-Value Theorem on the interval $[0, 8]$.

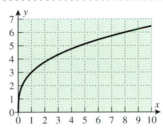

Figure Ex-9

10. Use the graph of f in Exercise 9 to estimate all values of c that satisfy the conclusion of the Mean-Value Theorem on the interval $[0, 4]$.

In Exercises 11–16, verify that the hypotheses of the Mean-Value Theorem are satisfied on the given interval, and find all values of c in that interval that satisfy the conclusion of the theorem.

11. $f(x) = x^2 + x$; $[-4, 6]$

12. $f(x) = x^3 + x - 4$; $[-1, 2]$

13. $f(x) = \sqrt{x + 1}$; $[0, 3]$

14. $f(x) = x + \dfrac{1}{x}$; $[3, 4]$

15. $f(x) = \sqrt{25 - x^2}$; $[-5, 3]$

16. $f(x) = \dfrac{1}{x - 1}$; $[2, 5]$

17. (a) Find an interval $[a, b]$ on which

$$f(x) = x^4 + x^3 - x^2 + x - 2$$

satisfies the hypotheses of Rolle's Theorem.

(b) Generate the graph of $f'(x)$, and use it to make rough estimates of all values of c in the interval obtained in part (a) that satisfy the conclusion of Rolle's Theorem.

(c) Use Newton's Method to improve on the rough estimates obtained in part (b).

18. Let $f(x) = x^3 + 4x$.

(a) Find the equation of the secant line through the points $(-2, f(-2))$ and $(1, f(1))$.

(b) Show that there is only one number c in the interval $(-2, 1)$ that satisfies the conclusion of the Mean-Value Theorem for the secant line in part (a).

(c) Find the equation of the tangent line to the graph of f at the point $(c, f(c))$.

(d) Use a graphing utility to generate the secant line in part (a) and the tangent line in part (c) in the same coordinate system, and confirm visually that the two lines seem parallel.

19. Let $f(x) = \tan x$.

(a) Show that there is no number c in the interval $(0, \pi)$ such that $f'(c) = 0$, even though $f(0) = f(\pi) = 0$.

(b) Explain why the result in part (a) does not violate Rolle's Theorem.

20. Let $f(x) = x^{2/3}$, $a = -1$, and $b = 8$.

(a) Show that there is no number c in (a, b) such that

$$f'(c) = \frac{f(b) - f(a)}{b - a}$$

(b) Explain why the result in part (a) does not violate the Mean-Value Theorem.

21. (a) Show that if f is differentiable on $(-\infty, +\infty)$, and if $y = f(x)$ and $y = f'(x)$ are graphed in the same coordinate system, then between any two x-intercepts of f there is at least one x-intercept of f'.

(b) Give some examples that illustrate this.

22. Review Definitions 3.1.3 and 3.1.4 of average and instantaneous rate of change of y with respect to x, and use the Mean-Value Theorem to show that if f is differentiable on $(-\infty, +\infty)$, then for any interval $[x_0, x_1]$ there is at least one number in (x_0, x_1) where the instantaneous rate of change of y with respect to x is equal to the average rate of change over the interval.

In Exercises 23–25, use the result of Exercise 22.

23. An automobile travels 4 mi along a straight road in 5 min. Show that the speedometer reads exactly 48 mi/h at least once during the trip.

24. At 11 A.M. on a certain morning the outside temperature was $76°$ F. At 11 P.M. that evening it had dropped to $52°$ F.

(a) Show that at some instant during this period the temperature was decreasing at the rate of $2°$F/h.

(b) Suppose that you know that the temperature reached a high of $88°$ F sometime between 11 A.M. and 11 P.M. Show that at some instant during this period the temperature was decreasing at a rate greater than $3°$F/h.

25. Suppose that two runners in a 100-m dash finish in a tie. Show that they had the same velocity at least once during the race.

26. Use the fact that

$$\frac{d}{dx}(x^6 - 2x^2 + x) = 6x^5 - 4x + 1$$

to show that the equation $6x^5 - 4x + 1 = 0$ has at least one solution in the interval $(0, 1)$.

27. (a) Use the Constant Difference Theorem (5.8.3) to show that if $f'(x) = g'(x)$ for all x in the interval $(-\infty, +\infty)$, and if f and g have the same value at some number x_0, then $f(x) = g(x)$ for all x in $(-\infty, +\infty)$.

(b) Use the result in part (a) to confirm the trigonometric identity $\sin^2 x + \cos^2 x = 1$.

28. (a) Use the Constant Difference Theorem (5.8.3) to show that if $f'(x) = g'(x)$ for all x in $(-\infty, +\infty)$, and if $f(x_0) - g(x_0) = c$ at some number x_0, then

$$f(x) - g(x) = c$$

for all x in $(-\infty, +\infty)$.

(b) Use the result in part (a) to show that the function

$$h(x) = (x - 1)^3 - (x^2 + 3)(x - 3)$$

is constant for all x in $(-\infty, +\infty)$, and find the constant.

(c) Check the result in part (b) by multiplying out and simplifying the formula for $h(x)$.

29. Let $g(x) = x^3 - 4x + 6$. Find $f(x)$ so that $f'(x) = g'(x)$ and $f(1) = 2$.

30. Let $g(x) = \tan^{-1} x$. Find $f(x)$ so that $f'(x) = g'(x)$ and $f(1) = 2$.

31. (a) Use the Mean-Value Theorem to show that if f is differentiable on an interval I, and if $|f'(x)| \leq M$ for all values of x in I, then

$$|f(x) - f(y)| \leq M|x - y|$$

for all values of x and y in I.

(b) Use the result in part (a) to show that

$$|\sin x - \sin y| \leq |x - y|$$

for all real values of x and y.

32. (a) Use the Mean-Value Theorem to show that if f is differentiable on an open interval I, and if $|f'(x)| \geq M$ for all values of x in I, then

$$|f(x) - f(y)| \geq M|x - y|$$

for all values of x and y in I.

(b) Use the result in part (a) to show that

$$|\tan x - \tan y| \geq |x - y|$$

for all values of x and y in the interval $(-\pi/2, \pi/2)$.

(c) Use the result in part (b) to show that

$$|\tan x + \tan y| \geq |x + y|$$

for all values of x and y in the interval $(-\pi/2, \pi/2)$.

33. (a) Use the Mean-Value Theorem to show that

$$\sqrt{y} - \sqrt{x} < \frac{y - x}{2\sqrt{x}}$$

if $0 < x < y$.
(b) Use the result in part (a) to show that if $0 < x < y$, then $\sqrt{xy} < \frac{1}{2}(x + y)$.

34. Show that if f is differentiable on an open interval I and $f'(x) \neq 0$ on I, the equation $f(x) = 0$ can have at most one real root in I.

35. Use the result in Exercise 34 to show the following:
(a) The equation $x^3 + 4x - 1 = 0$ has exactly one real root.

(b) If $b^2 - 3ac < 0$ and if $a \neq 0$, then the equation

$$ax^3 + bx^2 + cx + d = 0$$

has exactly one real root.

36. Use the inequality $\frac{1}{6}\sqrt{3} < 0.29$ to prove that

$$1.71 < \sqrt{3} < 1.75$$

[*Hint:* Let $f(x) = \sqrt{x}$, $a = 3$, and $b = 4$ in the Mean-Value Theorem.]

37. Use the Mean-Value Theorem to prove that

$$\frac{x}{1 + x^2} < \tan^{-1} x < x \quad (x > 0)$$

38. (a) Show that if f and g are functions for which

$$f'(x) = g(x) \quad \text{and} \quad g'(x) = f(x)$$

for all x, then $f^2(x) - g^2(x)$ is a constant.
(b) Show that the function $f(x) = \frac{1}{2}(e^x + e^{-x})$ and the function $g(x) = \frac{1}{2}(e^x - e^{-x})$ have this property.

39. (a) Show that if f and g are functions for which

$$f'(x) = g(x) \quad \text{and} \quad g'(x) = -f(x)$$

for all x, then $f^2(x) + g^2(x)$ is a constant.
(b) Give an example of functions f and g with this property.

40. Let f and g be continuous on $[a, b]$ and differentiable on (a, b). Prove: If $f(a) = g(a)$ and $f(b) = g(b)$, then there is a number c in (a, b) such that $f'(c) = g'(c)$.

41. Illustrate the result in Exercise 40 by drawing an appropriate picture.

42. (a) Prove: If $f''(x) > 0$ for all x in (a, b), then $f'(x) = 0$ at most once in (a, b).
(b) Give a geometric interpretation of the result in (a).

43. (a) Prove part (b) of Theorem 5.1.2.
(b) Prove part (c) of Theorem 5.1.2.

44. Use the Mean-Value Theorem to prove the following result: Let f be continuous at x_0 and suppose that $\lim_{x \to x_0} f'(x)$ exists. Then f is differentiable at x_0, and

$$f'(x_0) = \lim_{x \to x_0} f'(x)$$

[*Hint:* The derivative $f'(x_0)$ is given by

$$f'(x_0) = \lim_{x \to x_0} \frac{f(x) - f(x_0)}{x - x_0}$$

provided this limit exists.]

45. Let

$$f(x) = \begin{cases} 3x^2, & x \leq 1 \\ ax + b, & x > 1 \end{cases}$$

Find the values of a and b so that f will be differentiable at $x = 1$.

46. (a) Let

$$f(x) = \begin{cases} x^2, & x \leq 0 \\ x^2 + 1, & x > 0 \end{cases}$$

Show that

$$\lim_{x \to 0^-} f'(x) = \lim_{x \to 0^+} f'(x)$$

but that $f'(0)$ does not exist.
(b) Let

$$f(x) = \begin{cases} x^2, & x \leq 0 \\ x^3, & x > 0 \end{cases}$$

Show that $f'(0)$ exists but $f''(0)$ does not.

47. Use the Mean-Value Theorem to prove the following result, alluded to in Section 5.3: The graph of a function f has a vertical tangent line at $(x_0, f(x_0))$ if f is continuous at x_0 and $f'(x)$ approaches either $+\infty$ or $-\infty$ as $x \to x_0^+$ and as $x \to x_0^-$.

SUPPLEMENTARY EXERCISES

 Graphing Utility   [c] CAS

1. (a) If $x_1 < x_2$, what relationship must hold between $f(x_1)$ and $f(x_2)$ if f is increasing on an interval containing x_1 and x_2? Decreasing? Constant?
(b) What condition on f' ensures that f is increasing on an interval $[a, b]$? Decreasing? Constant?

2. (a) What condition on f' ensures that f is concave up on an open interval I? Concave down?
(b) What condition on f'' ensures that f is concave up on an open interval I? Concave down?
(c) In words, what is an inflection point of f?

3. (a) Where on the graph of $y = f(x)$ would you expect y to be increasing or decreasing most rapidly with respect to x?
   (b) In words, what is a relative extremum?
   (c) State a procedure for determining where the relative extrema of f occur.

4. Determine whether the statement is true or false. If it is false, give an example for which the statement fails.
   (a) If f has a relative maximum at x_0, then $f(x_0)$ is the largest value that $f(x)$ can have.
   (b) If $f(x_0)$ is the largest value for f on the interval (a, b), then f has a relative maximum at x_0.
   (c) A function f has a relative extremum at each of its critical numbers.

5. (a) According to the first derivative test, what conditions ensure that f has a relative maximum at x_0? A relative minimum?
   (b) According to the second derivative test, what conditions ensure that f has a relative maximum at x_0? A relative minimum?

6. In each part, sketch a continuous curve $y = f(x)$ with the stated properties.
   (a) $f(2) = 4$, $f'(2) = 1$, $f''(x) < 0$ for $x < 2$, $f''(x) > 0$ for $x > 2$
   (b) $f(2) = 4$, $f''(x) > 0$ for $x < 2$, $f''(x) < 0$ for $x > 2$, and $\lim_{x \to 2^-} f'(x) = +\infty$, $\lim_{x \to 2^+} f'(x) = +\infty$
   (c) $f(2) = 4$, $f''(x) < 0$ for $x \neq 2$, and $\lim_{x \to 2^-} f'(x) = 1$, $\lim_{x \to 2^+} f'(x) = -1$

7. In each part, find all critical numbers, and use the first derivative test to classify them as relative maxima, relative minima, or neither.
   (a) $f(x) = x^{1/3}(x - 7)^2$
   (b) $f(x) = 2\sin x - \cos 2x$, $0 \leq x \leq 2\pi$
   (c) $f(x) = 3x - (x - 1)^{3/2}$

8. In each part, find all critical numbers, and use the second derivative test (where possible) to classify them as relative maxima, relative minima, or neither.
   (a) $f(x) = x^{-1/2} + \frac{1}{9}x^{1/2}$
   (b) $f(x) = x^2 + 8/x$
   (c) $f(x) = \sin^2 x - \cos x$, $0 \leq x \leq 2\pi$

In Exercises 9–16, give a graph of f, and identify the limits as $x \to \pm\infty$, as well as locations of all relative extrema, inflection points, and asymptotes (as appropriate).

---

9. $f(x) = x^4 - 3x^3 + 3x^2 + 1$

10. $f(x) = x^5 - 4x^4 + 4x^3$

11. $f(x) = \tan(x^2 + 1)$ 12. $f(x) = x - \cos x$

13. $f(x) = \dfrac{x^2}{x^2 + 2x + 5}$ 14. $f(x) = \dfrac{25 - 9x^2}{x^3}$

15. $f(x) = \begin{cases} \frac{1}{2}x^2, & x \leq 0 \\ -x^2, & x > 0 \end{cases}$

16. $f(x) = (1 + x)^{2/3}(3 - x)^{1/3}$

When using a graphing utility, important features of a graph may be missed if the viewing window is not chosen appropriately. This is illustrated in Exercises 17 and 18.

---

17. (a) Generate the graph of $f(x) = \frac{1}{3}x^3 - \frac{1}{400}x$ over the interval $[-5, 5]$, and make a conjecture about the locations and nature of all critical numbers.
    (b) Find the exact locations of all the critical numbers, and classify them as relative maxima, relative minima, or neither.
    (c) Confirm the results in part (b) by graphing f over an appropriate interval.

18. (a) Generate the graph of
    $$f(x) = \frac{1}{5}x^5 - \frac{7}{8}x^4 + \frac{1}{3}x^3 + \frac{7}{2}x^2 - 6x$$
    over the interval $[-5, 5]$, and make a conjecture about the locations and nature of all critical numbers.
    (b) Find the exact locations of all the critical numbers, and classify them as relative maxima, relative minima, or neither.
    (c) Confirm the results in part (b) by graphing portions of f over appropriate intervals. [*Note:* It will not be possible to find a single window in which all of the critical numbers are clearly visible.]

19. (a) Use a graphing utility to generate the graphs of $y = x$ and $y = (x^3 - 8)/(x^2 + 1)$ together over the interval $[-5, 5]$, and make a conjecture about the relationship between the two graphs.
    (b) Use Exercise 66 of Section 5.3 to confirm your conjecture in part (a).

20. In parts (a)–(d), the graph of a polynomial with degree at most 6 is given. Find equations for polynomials that produce graphs with these shapes, and check your answers with a graphing utility.

(a)
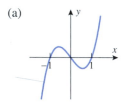

(b)
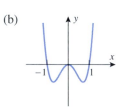

(c)
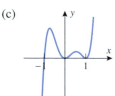

(d)
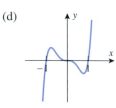

21. Find the equations of the tangent lines at all inflection points of the graph of

$$f(x) = x^4 - 6x^3 + 12x^2 - 8x + 3$$

22. Use implicit differentiation to show that a function defined implicitly by $\sin x + \cos y = 2y$ has a critical number whenever $\cos x = 0$. Then use either the first or second derivative test to classify these critical numbers as relative maxima or minima.

23. Let

$$f(x) = \frac{2x^3 + x^2 - 15x + 7}{(2x - 1)(3x^2 + x - 1)}$$

Graph $y = f(x)$, and find the equations of all horizontal and vertical asymptotes. Explain why there is no vertical asymptote at $x = \frac{1}{2}$, even though the denominator of f is zero at that point.

c 24. Let

$$f(x) = \frac{x^5 - x^4 - 3x^3 + 2x + 4}{x^7 - 2x^6 - 3x^5 + 6x^4 + 4x - 8}$$

(a) Use a CAS to factor the numerator and denominator of f, and use the results to determine the locations of all vertical asymptotes.

(b) Confirm that your answer is consistent with the graph of f.

25. For a general quadratic polynomial

$$f(x) = ax^2 + bx + c \quad (a \neq 0)$$

find conditions on a, b, and c to ensure that f is always increasing or always decreasing on $[0, +\infty)$.

26. For the general cubic polynomial

$$f(x) = ax^3 + bx^2 + cx + d \quad (a \neq 0)$$

find conditions on a, b, c, and d to ensure that f is always increasing or always decreasing on $(-\infty, +\infty)$.

~ 27. In each part, approximate the coordinates (x, y) of the relative extrema, and confirm that your answers are consistent with the graph of f.

(a) $f(x) = x^2 - \sin x$

(b) $f(x) = \sqrt{x^4 + 1} - \sqrt{x^2 + 1}$

(c) $f(x) = \dfrac{x}{x^2 - \sin x + 1}$

28. Approximate to six decimal places the largest value of k such that the function $f(x) = 1 + 2x + x^3 - x^4$ is increasing on $(-\infty, k]$.

29. (a) Can an object in rectilinear motion reverse direction if its acceleration is constant? Justify your answer using a velocity versus time curve.

(b) Can an object in rectilinear motion have increasing speed and decreasing acceleration? Justify your answer using a velocity versus time curve.

~ 30. Suppose that the position function of a particle in rectilinear motion is given by the formula $s(t) = t/(t^2 + 5)$ for $t \geq 0$.

(a) Use a graphing utility to generate the position, velocity, and acceleration versus time curves.

(b) Use the appropriate graph to make a rough estimate of the time when the particle reverses direction, and then find that time exactly.

(c) Find the position, velocity, and acceleration at the instant when the particle reverses direction.

(d) Use the appropriate graphs to make rough estimates of the time intervals on which the particle is speeding up and the time intervals on which it is slowing down, and then find those time intervals exactly.

(e) When does the particle have its maximum and minimum velocities?

31. A basketball player, standing near the basket to grab a rebound, jumps 76.0 cm vertically.

(a) How much time does the player spend in the top 15.0 cm of the jump and how much time in the bottom 15.0 cm?

(b) In words, explain why basketball players seem to be suspended in air when they jump.

32. (a) Suppose that an object is released from rest from the top of a high building. Assuming that a free-fall model applies and that time is in seconds and distance is in meters, make a table that shows the distance traveled by the object and its speed to one decimal place at 1-second increments from $t = 0$ to $t = 4$.

(b) Confirm that doubling the elapsed time doubles the velocity, and explain why this happens.

(c) Confirm that doubling the elapsed time increases the distance traveled by a factor of 4, and explain why this happens.

c 33. Suppose that the position function of a particle in rectilinear motion is given by the formula

$$s(t) = \frac{t^2 + 1}{t^4 + 1}, \quad t \geq 0$$

(a) Use a CAS to find simplified formulas for the velocity $v(t)$ and the acceleration $a(t)$.

(b) Graph the position, velocity, and acceleration versus time curves.

(c) Use the appropriate graph to make a rough estimate of the time at which the particle is farthest from the origin and its distance from the origin at that time.

(d) Use the appropriate graph to make a rough estimate of the time interval during which the particle is moving in the positive direction.

(e) Use the appropriate graphs to make rough estimates of the time intervals during which the particle is speeding up and the time intervals during which it is slowing down.

(f) Use the appropriate graph to make a rough estimate of the maximum speed of the particle and the time at which the maximum speed occurs.

34. Is it true or false that a particle in rectilinear motion is speeding up when its velocity is increasing and slowing down when its velocity is decreasing? Justify your answer.

35. (a) What inequality must $f(x)$ satisfy for the function f to have an absolute maximum on an interval I at x_0?
(b) What inequality must $f(x)$ satisfy for f to have an absolute minimum on I at x_0?
(c) What is the difference between an absolute extremum and a relative extremum?

36. According to the Extreme-Value Theorem, what conditions on a function f and an interval I guarantee that f will have both an absolute maximum and an absolute minimum on I?

37. In each part, determine whether the statement is true or false, and justify your answer.
(a) If f is differentiable on the open interval (a, b), and if f has an absolute extremum on that interval, then it must occur at a stationary point of f.
(b) If f is continuous on the open interval (a, b), and if f has an absolute extremum on that interval, then it must occur at a stationary point of f.

38. Suppose that f is continuous on the closed interval $[a, b]$ and differentiable on the open interval (a, b), and suppose that $f(a) = f(b)$. Is it true or false that f must have at least one stationary point in (a, b)? Justify your answer.

39. In each part, find the absolute minimum m and the absolute maximum M of f on the given interval (if they exist), and state where the absolute extrema occur.
(a) $f(x) = 1/x$; $[-2, -1]$
(b) $f(x) = x^3 - x^4$; $\left[-1, \frac{3}{2}\right]$
(c) $f(x) = x^2(x - 2)^{1/3}$; $(0, 3]$
(d) $f(x) = e^x/x^2$; $(0, +\infty)$

40. In each part, find the absolute minimum m and the absolute maximum M of f on the given interval (if they exist), and state where the absolute extrema occur.
(a) $f(x) = 2x/(x^2 + 3)$; $(0, 2]$
(b) $f(x) = 2x^5 - 5x^4 + 7$; $(-1, 3)$
(c) $f(x) = -|x^2 - 2x|$; $[1, 3]$
(d) $f(x) = x^x$; $(0, +\infty)$

41. Draw an appropriate picture, and describe the basic idea of Newton's Method without using any formulas.

42. Use Newton's Method to approximate all three solutions of $x^3 - 4x + 1 = 0$.

43. Use Newton's Method to approximate the smallest positive solution of $\sin x + \cos x = 0$.

44. Suppose that f is an increasing function on $[a, b]$ and that x_0 is a number in (a, b). Prove that if f is differentiable at x_0, then $f'(x_0) \geq 0$.

45. In each part, determine whether all of the hypotheses of Rolle's Theorem are satisfied on the stated interval. If not,

state which hypotheses fail; if so, find all values of c guaranteed in the conclusion of the theorem.
(a) $f(x) = \sqrt{4 - x^2}$ on $[-2, 2]$
(b) $f(x) = x^{2/3} - 1$ on $[-1, 1]$
(c) $f(x) = \sin(x^2)$ on $[0, \sqrt{\pi}\,]$

46. In each part, determine whether all of the hypotheses of the Mean-Value Theorem are satisfied on the stated interval. If not, state which hypotheses fail; if so, find all values of c guaranteed in the conclusion of the theorem.
(a) $f(x) = |x - 1|$ on $[-2, 2]$
(b) $f(x) = \dfrac{x + 1}{x - 1}$ on $[2, 3]$
(c) $f(x) = \begin{cases} 3 - x^2 & \text{if } x \leq 1 \\ 2/x & \text{if } x > 1 \end{cases}$ on $[0, 2]$

47. A church window consists of a blue semicircular section surmounting a clear rectangular section as shown in the accompanying figure. The blue glass lets through half as much light per unit area as the clear glass. Find the radius r of the window that admits the most light if the perimeter of the entire window is to be P feet.

48. Find the dimensions of the rectangle of maximum area that can be inscribed inside the ellipse $(x/4)^2 + (y/3)^2 = 1$ (see the accompanying figure).

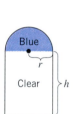

Figure Ex-47   Figure Ex-48

c 49. Let
$$f(x) = \frac{x^3 + 2}{x^4 + 1}$$
(a) Generate the graph of $y = f(x)$, and use the graph to make rough estimates of the coordinates of the absolute extrema.
(b) Use a CAS to solve the equation $f'(x) = 0$ and then use it to make more accurate approximations of the coordinates in part (a).

c 50. As shown in the accompanying figure, suppose that a boat enters the river at the point $(1, 0)$ and maintains a heading toward the origin. As a result of the strong current, the boat follows the path
$$y = \frac{x^{10/3} - 1}{2x^{2/3}}$$
where x and y are in miles.
(a) Graph the path taken by the boat.

(b) Can the boat reach the origin? If not, discuss its fate and find how close it comes to the origin.

(c) What is the velocity of the boat in the x-direction at the instant when it is closest to the origin if the velocity in the y-direction is -4 mi/h at this instant?

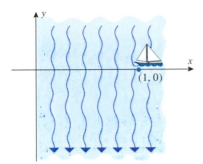

Figure Ex-50

51. According to *Kepler's law*, the planets in our solar system move in elliptical orbits around the Sun. If a planet's closest approach to the Sun occurs at time $t = 0$, then the distance r from the center of the planet to the center of the Sun at some later time t can be determined from the equation

$$r = a(1 - e \cos \phi)$$

where a is the average distance between centers, e is a positive constant that measures the "flatness" of the elliptical orbit, and ϕ is the solution of *Kepler's equation*1

$$\frac{2\pi t}{T} = \phi - e \sin \phi$$

in which T is the time it takes for one complete orbit of the planet. Estimate the distance from the Earth to the Sun when $t = 90$ days. [First find ϕ from Kepler's equation, and then use this value of ϕ to find the distance. Use $a = 150 \times 10^6$ km, $e = 0.0167$, and $T = 365$ days.]

52. Using the formulas in Exercise 51, find the distance from the planet Mars to the Sun when $t = 1$ year. For Mars use $a = 228 \times 10^6$ km, $e = 0.0934$, and $T = 1.88$ years.

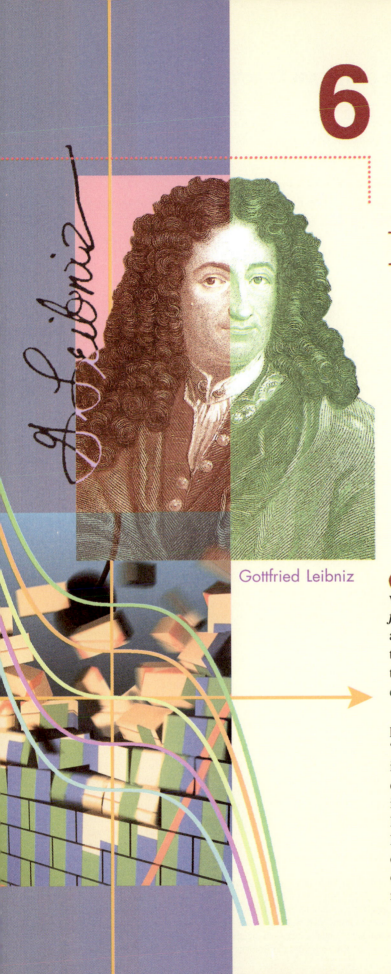

6

INTEGRATION

Gottfried Leibniz

Traditionally, that portion of calculus concerned with finding tangent lines and rates of change is called *differential calculus* and that portion concerned with finding areas is called *integral calculus*. However, we will see in this chapter that the two problems are so closely related that the distinction between differential and integral calculus is often hard to discern.

In this chapter we will begin with an overview of the problem of finding areas—we will discuss what the term "area" means, and we will outline two approaches to defining and calculating areas. Following this overview, we will discuss the "Fundamental Theorem of Calculus," which is the theorem that relates the problems of finding tangent lines and areas, and we will discuss techniques for calculating areas. Finally, we will use the ideas in this chapter to continue our study of rectilinear motion, to examine some consequences of the chain rule in integral calculus, and to reexamine the concept of the natural logarithm function.

6.1 AN OVERVIEW OF THE AREA PROBLEM

In this introductory section we will consider the problem of calculating areas of plane regions with curvilinear boundaries. All of the results in this section will be reexamined in more detail later in this chapter, so our purpose here is simply to introduce the fundamental concepts.

Figure 6.1.1

The main goal of this chapter is to study the following major problem of calculus:

> **6.1.1** THE AREA PROBLEM. Given a function f that is continuous and nonnegative on an interval $[a, b]$, find the area between the graph of f and the interval $[a, b]$ on the x-axis (Figure 6.1.1).

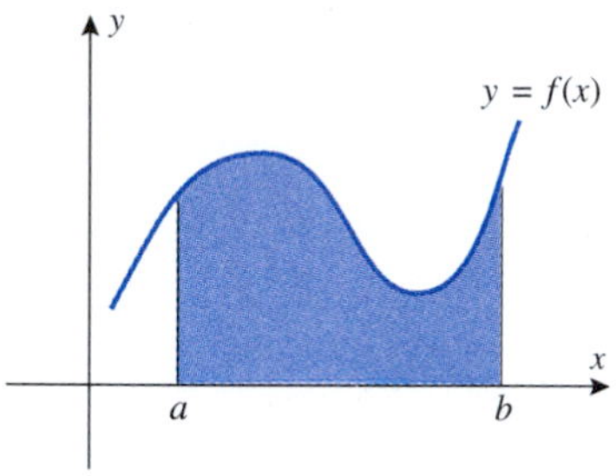

Of course, from a strictly logical point of view, we should first provide a precise definition of the term *area* before discussing methods for calculating areas. However, in this section we will treat the concept of area intuitively, postponing a more formal definition until Section 6.4.

Formulas for the areas of plane regions with straight-line boundaries (squares, rectangles, triangles, trapezoids, etc.) were well known in many early civilizations. On the other hand, obtaining formulas for regions with curvilinear boundaries (a circle being the simplest case) caused problems for early mathematicians. The first real progress on such problems was made by the Greek mathematician, Archimedes,[*] who obtained the areas of regions bounded by arcs of circles, parabolas, spirals, and various other curves by ingenious use of a procedure later known as the *method of exhaustion*. That method, when applied to a circle of radius r, consists of inscribing a succession of regular polygons in the circle and allowing the number of sides n to increase indefinitely (Figure 6.1.2). As n increases, the polygons tend to "exhaust" the region inside the circle, and the areas of those polygons become better and better approximations to the exact area of the circle.

---

[*] ARCHIMEDES (287 B.C.–212 B.C.). Greek mathematician and scientist. Born in Syracuse, Sicily, Archimedes was the son of the astronomer Pheidias and possibly related to Heiron II, king of Syracuse. Most of the facts about his life come from the Roman biographer, Plutarch, who inserted a few tantalizing pages about him in the massive biography of the Roman soldier, Marcellus. In the words of one writer, "the account of Archimedes is slipped like a tissue-thin shaving of ham in a bull-choking sandwich."

Archimedes ranks with Newton and Gauss as one of the three greatest mathematicians who ever lived, and he is certainly the greatest mathematician of antiquity. His mathematical work is so modern in spirit and technique that it is barely distinguishable from that of a seventeenth-century mathematician, yet it was all done without benefit of algebra or a convenient number system. Among his mathematical achievements, Archimedes developed a general method (exhaustion) for finding areas and volumes, and he used the method to find areas bounded by parabolas and spirals and to find volumes of cylinders, paraboloids, and segments of spheres. He gave a procedure for approximating π and bounded its value between $3\frac{10}{71}$ and $3\frac{1}{7}$. In spite of the limitations of the Greek numbering system, he devised methods for finding square roots and invented a method based on the Greek myriad (10,000) for representing numbers as large as 1 followed by 80 million billion zeros.

Of all his mathematical work, Archimedes was most proud of his discovery of the method for finding the volume of a sphere—he showed that the volume of a sphere is two-thirds the volume of the smallest cylinder that can contain it. At his request, the figure of a sphere and cylinder was engraved on his tombstone.

In addition to mathematics, Archimedes worked extensively in mechanics and hydrostatics. Nearly every schoolchild knows Archimedes as the absent-minded scientist who, on realizing that a floating object displaces its weight of liquid, leaped from his bath and ran naked through the streets of Syracuse shouting, "Eureka, Eureka!"—(meaning, "I have found it!"). Archimedes actually created the discipline of hydrostatics and used it to find equilibrium positions for various floating bodies. He laid down the fundamental postulates of mechanics, discovered the laws of levers, and calculated centers of gravity for various flat surfaces and solids. In the excitement of discovering the mathematical laws of the lever, he is said to have declared, "Give me a place to stand and I will move the earth."

Although Archimedes was apparently more interested in pure mathematics than its applications, he was an engineering genius. During the second Punic war, when Syracuse was attacked by the Roman fleet under the command of Marcellus, it was reported by Plutarch that Archimedes' military inventions held the fleet at bay for three years. He invented super catapults that showered the Romans with rocks weighing a quarter ton or more,

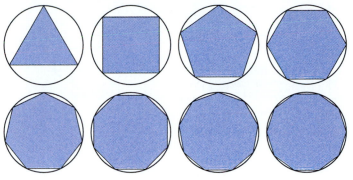

Figure 6.1.2

Table 6.1.1

n	$A(n)$
100	3.13952597647
200	3.14107590781
300	3.14136298250
400	3.14146346236
500	3.14150997084
600	3.14153523487
700	3.14155046835
800	3.14156035548
900	3.14156713408
1000	3.14157198278
2000	3.14158748588
3000	3.14159035683
4000	3.14159136166
5000	3.14159182676
6000	3.14159207940
7000	3.14159223174
8000	3.14159233061
9000	3.14159239839
10000	3.14159244688

To see how this works numerically, let $A(n)$ denote the area of a regular n-sided polygon inscribed in a circle of radius 1. Table 6.1.1 shows the values of $A(n)$ for various choices of n. Note that for large values of n the area $A(n)$ appears to be close to π (square units), as one would expect. This suggests that for a circle of radius 1, the method of exhaustion is equivalent to an equation of the form

$$\lim_{n \to \infty} A(n) = \pi$$

However, Greek mathematicians were very suspicious of the concept of "infinity" and intentionally avoided explanations that referred to the "limiting behavior" of some quantity. As a consequence, obtaining exact answers by the classical method of exhaustion was a cumbersome procedure. In our discussion of the area problem, we will consider a more modern version of the method of exhaustion that explicitly incorporates the notion of a limit. Because our approach uses a collection of rectangles to "exhaust" an area, we will refer to it as the *rectangle method*.

THE RECTANGLE METHOD FOR FINDING AREAS

There are two basic methods for finding the area of the region having the form shown in Figure 6.1.1—the *rectangle method* and the *antiderivative method*. The idea behind the rectangle method is as follows:

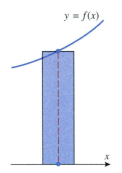

$y = f(x)$

x

Figure 6.1.3

- Divide the interval $[a, b]$ into n equal subintervals, and over each subinterval construct a rectangle that extends from the x-axis to any point on the curve $y = f(x)$ that is above the subinterval; the particular point does not matter—it can be above the center, above an endpoint, or above any other point in the subinterval. In Figure 6.1.3 it is above the center.

- For each n, the total area of the rectangles can be viewed as an *approximation* to the exact area under the curve over the interval $[a, b]$. Moreover, it is evident intuitively that as n increases these approximations will get better and better and will approach the exact area as a limit (Figure 6.1.4).

Later, this procedure will serve both as a mathematical definition and a method of computation—we will *define* the area under $y = f(x)$ over the interval $[a, b]$ as the limit of the areas of the approximating rectangles, and we will use the method itself to approximate this area.

and fearsome mechanical devices with iron "beaks and claws" that reached over the city walls, grasped the ships, and spun them against the rocks. After the first repulse, Marcellus called Archimedes a "geometrical Briareus (a hundred-armed mythological monster) who uses our ships like cups to ladle water from the sea."

  Eventually the Roman army was victorious and contrary to Marcellus' specific orders the 75-year-old Archimedes was killed by a Roman soldier. According to one report of the incident, the soldier cast a shadow across the sand in which Archimedes was working on a mathematical problem. When the annoyed Archimedes yelled, "Don't disturb my circles," the soldier flew into a rage and cut the old man down.

  With his death the Greek gift of mathematics passed into oblivion, not to be fully resurrected again until the sixteenth century. Unfortunately, there is no known accurate likeness or statue of this great man.

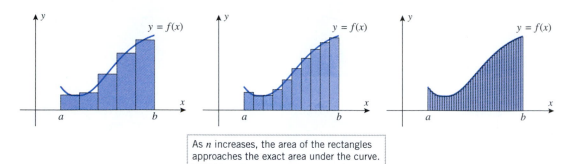

As *n* increases, the area of the rectangles approaches the exact area under the curve.

Figure 6.1.4

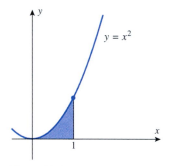

Figure 6.1.5

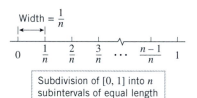

Subdivision of [0, 1] into *n* subintervals of equal length

Figure 6.1.6

To illustrate this idea, we will use the rectangle method to approximate the area under the curve $y = x^2$ over the interval $[0, 1]$ (Figure 6.1.5). We will begin by dividing the interval $[0, 1]$ into *n* equal subintervals, from which it follows that each subinterval has length $1/n$; the endpoints of the subintervals occur at

$$0, \quad \frac{1}{n}, \quad \frac{2}{n}, \quad \frac{3}{n}, \ldots, \quad \frac{n-1}{n}, \quad 1$$

(Figure 6.1.6). We want to construct a rectangle over each of these subintervals whose height is the value of the function $f(x) = x^2$ at some number in the subinterval. To be specific, let us use the right endpoints, in which case the heights of our rectangles will be

$$\left(\frac{1}{n}\right)^2, \quad \left(\frac{2}{n}\right)^2, \quad \left(\frac{3}{n}\right)^2, \ldots, \quad 1^2$$

and since each rectangle has a base of width $1/n$, the total area A_n of the *n* rectangles will be

$$A_n = \left[\left(\frac{1}{n}\right)^2 + \left(\frac{2}{n}\right)^2 + \left(\frac{3}{n}\right)^2 + \cdots + 1^2\right]\left(\frac{1}{n}\right) \tag{1}$$

For example, if $n = 4$, then the total area of the four approximating rectangles would be

$$A_4 = \left[\left(\tfrac{1}{4}\right)^2 + \left(\tfrac{2}{4}\right)^2 + \left(\tfrac{3}{4}\right)^2 + 1^2\right]\left(\tfrac{1}{4}\right) = \tfrac{15}{32} = 0.46875$$

Table 6.1.2 shows the result of evaluating (1) on a computer for some increasingly large values of *n*. These computations suggest that the exact area is close to $\tfrac{1}{3}$. In Section 6.4 we will prove that this area is exactly $\tfrac{1}{3}$ by showing that

$$\lim_{n \to \infty} A_n = \tfrac{1}{3}$$

Table 6.1.2

n	4	10	100	1000	10,000	100,000
A_n	0.468750	0.385000	0.338350	0.333834	0.333383	0.333338

Equation (1) may be written more concisely by using *sigma notation*, which is discussed in Section 6.4 in detail. [Sigma (Σ) is an uppercase letter in the Greek alphabet used to denote sums.] With sigma notation, the sum

$$\left(\frac{1}{n}\right)^2 + \left(\frac{2}{n}\right)^2 + \left(\frac{3}{n}\right)^2 + \cdots + 1^2$$

may be expressed simply as

$$\sum_{k=1}^{n} \left(\frac{k}{n}\right)^2$$

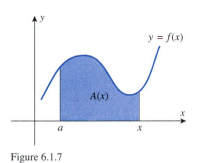

Figure 6.1.7

This notation tells us to form the sum of the terms that result when we substitute successive integers for k in the expression $(k/n)^2$, starting with $k = 1$ and ending with $k = n$. Each value of a positive integer n then determines a value of the sum. For example, if $n = 4$, then

$$\sum_{k=1}^{4} \left(\frac{k}{4}\right)^2 = \left(\frac{1}{4}\right)^2 + \left(\frac{2}{4}\right)^2 + \left(\frac{3}{4}\right)^2 + \left(\frac{4}{4}\right)^2 = \frac{30}{16} = \frac{15}{8}$$

In general, using sigma notation we write

$$A_n = \frac{1}{n} \sum_{k=1}^{n} \left(\frac{k}{n}\right)^2$$

FOR THE READER.   Many calculating utilities perform automatic summations for expressions that involve some version of the sigma notation. If your calculating utility performs such summations, use it to verify the value of A_{100} given in Table 6.1.2. (Otherwise, use it to confirm A_{10}.)

THE ANTIDERIVATIVE METHOD FOR FINDING AREAS

Despite the intuitive appeal of the rectangle method, the limits involved can be evaluated directly only in certain special cases. For this reason, work on the area problem remained at a rudimentary level until the latter half of the seventeenth century. Two results that were to prove to be a major breakthrough in the area problem were discovered by mathematicians Isaac Barrow and Isaac Newton in Great Britain, and Gottfried Leibniz in Germany. These results appeared, without fanfare, as Proposition 11 of Lecture X and Proposition 19 of Lecture XI in Isaac Barrow's *Lectiones geometricae*. Each of the two results can be used to solve the area problem.

The solution based on Proposition 11 was preferred by Isaac Newton and provides us with a paradoxically effective *indirect* approach to the area problem. According to this line of argument, to find the area under the curve in Figure 6.1.1, one should first consider the seemingly *harder* problem of finding the area $A(x)$ between the graph of f and the interval $[a, x]$, where x denotes an arbitrary number in $[a, b]$ (Figure 6.1.7). If one can discover a formula for the area function $A(x)$, then the area under the curve from a to b can be obtained simply by substituting $x = b$ into this formula.

This may seem to be a surprising approach to the area problem. After all, why should the problem of determining the area $A(x)$ for *every* x in the interval $[a, b]$ be more tractable than the problem of computing a *single* value $A(b)$? However, the basis for this approach is the observation that although the area function $A(x)$ may be difficult to compute, its *derivative* $A'(x)$ is easy to find. To illustrate, let us consider some examples of area functions $A(x)$ that *can* be computed from simple geometry.

Example 1   For each of the functions f, find the area $A(x)$ between the graph of f and the interval $[a, x] = [-1, x]$, and find the derivative $A'(x)$ of this area function.

(a) $f(x) = 2$ (b) $f(x) = x + 1$ (c) $f(x) = 2x + 3$

Solution (a).   From Figure 6.1.8a we see that

$$A(x) = 2(x - (-1)) = 2(x + 1) = 2x + 2$$

is the area of a rectangle of height 2 and base $x + 1$. For this area function,

$$A'(x) = 2 = f(x)$$

Solution (b).   From Figure 6.1.8b we see that

$$A(x) = \frac{1}{2}(x + 1)(x + 1) = \frac{x^2}{2} + x + \frac{1}{2}$$

is the area of an isosceles right triangle with base and height equal to $x + 1$. For this area function,

$$A'(x) = x + 1 = f(x)$$

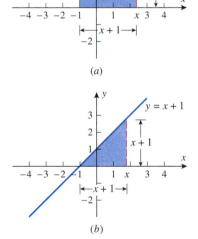

Figure 6.1.8

Solution (c). Recall that the formula for the area of a trapezoid is $A = \frac{1}{2}(b + b')h$, where b and b' denote the lengths of the parallel sides of the trapezoid, and the altitude h denotes the distance between the parallel sides. From Figure 6.1.8c we see that

$$A(x) = \tfrac{1}{2}((2x + 3) + 1)(x - (-1)) = x^2 + 3x + 2$$

is the area of a trapezoid with parallel sides of lengths 1 and $2x + 3$ and with altitude $x - (-1) = x + 1$. For this area function,

$$A'(x) = 2x + 3 = f(x) \qquad\qquad ◄$$

Note that in every case in Example 1,

$$A'(x) = f(x) \tag{2}$$

That is, *the derivative of the area function $A(x)$ is the function whose graph forms the upper boundary of the region.* We will show in Section 6.6 that Equation (2) is valid not simply for linear functions such as those in Example 1, but for any continuous function. Thus, to find the area function $A'(x)$, we can look instead for a (particular) function whose *derivative* is $f(x)$. This is called an ***antidifferentiation*** problem because we are trying to find $A(x)$ by "undoing" a differentiation. Whereas earlier in the text we were concerned with the process of differentiation, we will now also be concerned with the process of antidifferentiation.

To see how this *antiderivative method* applies to a specific example, let us return to the problem of finding the area between the graph of $f(x) = x^2$ and the interval $[0, 1]$. If we let $A(x)$ denote the area between the graph of f and the interval $[0, x]$, then (2) tells us that $A'(x) = f(x) = x^2$. By simple guesswork, we see that one function whose derivative is $f(x) = x^2$ is $\frac{1}{3}x^3$. It then follows from Theorem 5.8.3 that $A(x) = \frac{1}{3}x^3 + C$ for some constant C. This is where the decision to solve the area problem for a general right-hand endpoint helps. If we consider the case $x = 0$, then the interval $[0, x]$ reduces to a single point. If we agree that the area above a single point should be taken as zero, then it follows that

$$0 = A(0) = \tfrac{1}{3}0^3 + C = 0 + C = C \quad \text{or} \quad C = 0$$

Therefore, $A(x) = \frac{1}{3}x^3$ and the area between the graph of f and the interval $[0, 1]$ is $A(1) = \frac{1}{3}$. Note that this conclusion agrees with our numerical estimates in Table 6.1.2.

Although the antiderivative method provides us with a convenient solution to the area problem, it appears to have little to do with the rectangle method. It would be nice to have a solution that more clearly elucidates the connection between the operation of summing areas of rectangles on the one hand and the operation of antidifferentiation on the other. Fortunately, the solution to the area problem based on Barrow's Proposition 19 reveals just this connection. In addition, it allows us to formulate in modern language the approach to the area problem preferred by Leibniz. We will provide this solution in Section 6.6 (Theorem 6.6.1), as well as develop a modern version of Barrow's Proposition 11 (Theorem 6.6.3). Together, these two approaches to the area problem comprise what is now known as the *Fundamental Theorem of Calculus.*

INTEGRAL CALCULUS

We see that the rectangle method and the use of antidifferentiation provide us with quite different approaches to the area problem. The rectangle method is a frontal assault on the problem, whereas antidifferentiation is more in the form of a sneak attack. In this chapter we will carefully study both approaches to the problem.

In Sections 6.2 and 6.3 we will begin to develop some techniques for the process of antidifferentiation, a process that is also known as *integration.* Later, in Section 6.5 we will discuss a more general version of the rectangle method known as the *Riemann sum.* In much the same way that area can be interpreted as a "limit" using the rectangle method, we will define the *definite integral* as a "limit" of Riemann sums.

The definite integral and antidifferentiation are the twin pillars on which integral calculus rests. Both are important. The definite integral is generally the means by which problems in integral calculus are recognized and formulated. For example, in addition to the area problem, the problems of computing the volume of a solid, finding the arc length of a curve, and determining the work done in pumping water out of a tank are all examples of problems that may be solved by means of a definite integral. On the other hand, it can be difficult to obtain exact solutions to such problems by direct computation of a definite integral. Fortunately, in many cases of interest, the Fundamental Theorem of Calculus will allow us to evaluate a definite integral by means of antidifferentiation. Much of the power of integral calculus lies in the two-pronged approach of the definite integral and antidifferentiation.

EXERCISE SET 6.1

In Exercises 1–8, estimate the area between the graph of the function f and the interval $[a, b]$. Use an approximation scheme with n rectangles similar to our treatment of $f(x) = x^2$ in this section. If your calculating utility will perform automatic summations, estimate the specified area using $n = 10, 50$, and 100 rectangles. Otherwise, estimate this area using $n = 2, 5$, and 10 rectangles.

1. $f(x) = \sqrt{x}$; $[a, b] = [0, 1]$
2. $f(x) = \dfrac{1}{x + 1}$; $[a, b] = [0, 1]$
3. $f(x) = \sin x$; $[a, b] = [0, \pi]$
4. $f(x) = \cos x$; $[a, b] = [0, \pi/2]$
5. $f(x) = \dfrac{1}{x}$; $[a, b] = [1, 2]$
6. $f(x) = \cos x$; $[a, b] = [-\pi/2, \pi/2]$
7. $f(x) = \sqrt{1 - x^2}$; $[a, b] = [0, 1]$
8. $f(x) = \sqrt{1 - x^2}$; $[a, b] = [-1, 1]$

In Exercises 9–14, use simple area formulas from geometry to find the area function $A(x)$ that gives the area between the graph of the specified function f and the interval $[a, x]$. Confirm that $A'(x) = f(x)$ in every case.

9. $f(x) = 3$; $[a, x] = [1, x]$
10. $f(x) = 5$; $[a, x] = [2, x]$
11. $f(x) = 2x + 2$; $[a, x] = [0, x]$
12. $f(x) = 3x - 3$; $[a, x] = [1, x]$
13. $f(x) = 2x + 2$; $[a, x] = [1, x]$
14. $f(x) = 3x - 3$; $[a, x] = [2, x]$
15. How do the area functions in Exercises 11 and 13 compare? Explain.
16. Let $f(x)$ denote a *linear function* that is nonnegative on the interval $[a, b]$. For each value of x in $[a, b]$, define $A(x)$ to be the area between the graph of f and the interval $[a, x]$.
   (a) Prove that $A(x) = \frac{1}{2}[f(a) + f(x)](x - a)$.
   (b) Use part (a) to verify that $A'(x) = f(x)$.
17. Let A denote the area between the graph of $f(x) = \sqrt{x}$ and the interval $[0, 1]$, and let B denote the area between the graph of $f(x) = x^2$ and the interval $[0, 1]$. Explain geometrically why $A + B = 1$.
18. Let A denote the area between the graph of $f(x) = 1/x$ and the interval $[1, 2]$, and let B denote the area between the graph of f and the interval $[\frac{1}{2}, 1]$. Explain geometrically why $A = B$.

6.2 THE INDEFINITE INTEGRAL; INTEGRAL CURVES AND DIRECTION FIELDS

In the last section we saw the potential for antidifferentiation to play an important role in finding exact areas. In this section we will develop some fundamental results about antidifferentiation that will ultimately lead us to systematic procedures for solving many antiderivative problems.

6.2.1 DEFINITION. A function F is called an ***antiderivative*** of a function f on a given interval I if $F'(x) = f(x)$ for all x in the interval.

For example, the function $F(x) = \frac{1}{3}x^3$ is an antiderivative of $f(x) = x^2$ on the interval $(-\infty, +\infty)$ because for each x in this interval

$$F'(x) = \frac{d}{dx}\left[\tfrac{1}{3}x^3\right] = x^2 = f(x)$$

However, $F(x) = \frac{1}{3}x^3$ is not the only antiderivative of f on this interval. If we add any constant C to $\frac{1}{3}x^3$, then the function $G(x) = \frac{1}{3}x^3 + C$ is also an antiderivative of f on $(-\infty, +\infty)$, since

$$G'(x) = \frac{d}{dx}\left[\tfrac{1}{3}x^3 + C\right] = x^2 + 0 = f(x)$$

In general, once any single antiderivative is known, other antiderivatives can be obtained by adding constants to the known antiderivative. Thus,

$$\tfrac{1}{3}x^3, \quad \tfrac{1}{3}x^3 + 2, \quad \tfrac{1}{3}x^3 - 5, \quad \tfrac{1}{3}x^3 + \sqrt{2}$$

are all antiderivatives of $f(x) = x^2$.

It is reasonable to ask if there are antiderivatives of a function f that cannot be obtained by adding some constant to a known antiderivative F. The answer is *no*—once a single antiderivative of f on an interval I is known, all other antiderivatives on that interval are obtainable by adding constants to the known antiderivative. This is so because Theorem 5.8.3 tells us that if two functions are differentiable on an open interval I such that their derivatives are equal on I, then the functions differ by a constant on I. The following theorem summarizes these observations.

> **6.2.2 THEOREM.** *If $F(x)$ is any antiderivative of $f(x)$ on an interval I, then for any constant C the function $F(x) + C$ is also an antiderivative on that interval. Moreover, each antiderivative of $f(x)$ on the interval I can be expressed in the form $F(x) + C$ by choosing the constant C appropriately.*

THE INDEFINITE INTEGRAL

The process of finding antiderivatives is called *antidifferentiation* or *integration*. Thus, if

$$\frac{d}{dx}[F(x)] = f(x) \tag{1}$$

then *integrating* (or *antidifferentiating*) the function $f(x)$ produces an antiderivative of the form $F(x) + C$. To emphasize this process, Equation (1) is recast using *integral notation*,

$$\int f(x)\,dx = F(x) + C \tag{2}$$

where C is understood to represent an arbitrary constant. It is important to note that (1) and (2) are just different notations to express the same fact. For example,

$$\int x^2\,dx = \tfrac{1}{3}x^3 + C \quad \text{is equivalent to} \quad \frac{d}{dx}\left[\tfrac{1}{3}x^3\right] = x^2$$

Note that if we differentiate an antiderivative of $f(x)$, we obtain $f(x)$ back again. Thus,

$$\frac{d}{dx}\left[\int f(x)\,dx\right] = f(x) \tag{3}$$

The expression $\int f(x)\,dx$ is called an *indefinite integral*. The adjective "indefinite" emphasizes that the result of antidifferentiation is a "generic" function, descibed only up to a constant summand. The "elongated s" that appears on the left side of (2) is called an *integral sign*,[*] the function $f(x)$ is called the *integrand*, and the constant C is called the *constant of integration*. Equation (2) should be read as:

> *The integral of $f(x)$ with respect to x is equal to $F(x)$ plus a constant.*

The differential symbol, dx, in the differentiation and antidifferentiation operations

$$\frac{d}{dx}[\ ] \quad \text{and} \quad \int [\ ]\,dx$$

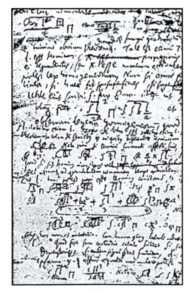

Extract from the manuscript of Leibniz dated October 29, 1675 in which the integral sign first appeared.

---

[*] This notation was devised by Leibniz. In his early papers Leibniz used the notation "omn." (an abbreviation for the Latin word "omnes") to denote integration. Then on October 29, 1675 he wrote, "It will be useful to write $\int$ for omn., thus $\int \ell$ for omn. ℓ ...." Two or three weeks later he refined the notation further and wrote $\int [\ ]\,dx$ rather than $\int$ alone. This notation is so useful and so powerful that its development by Leibniz must be regarded as a major milestone in the history of mathematics and science.

serves to identify the independent variable. If an independent variable other than x is used, say t, then the notation must be adjusted appropriately. Thus,

$$\frac{d}{dt}[F(t)] = f(t) \quad \text{and} \quad \int f(t)\,dt = F(t) + C$$

are equivalent statements.

Example 1

DERIVATIVE FORMULA	EQUIVALENT INTEGRATION FORMULA
$\dfrac{d}{dx}[x^3] = 3x^2$	$\displaystyle\int 3x^2\,dx = x^3 + C$
$\dfrac{d}{dx}[\sqrt{x}] = \dfrac{1}{2\sqrt{x}}$	$\displaystyle\int \dfrac{1}{2\sqrt{x}}\,dx = \sqrt{x} + C$
$\dfrac{d}{dt}[\tan t] = \sec^2 t$	$\displaystyle\int \sec^2 t\,dt = \tan t + C$
$\dfrac{d}{du}[u^{3/2}] = \frac{3}{2}u^{1/2}$	$\displaystyle\int \frac{3}{2}u^{1/2}\,du = u^{3/2} + C$

◀

For simplicity, the dx is sometimes absorbed into the integrand. For example,

$$\int 1\,dx \quad \text{can be written as} \quad \int dx$$

$$\int \frac{1}{x^2}\,dx \quad \text{can be written as} \quad \int \frac{dx}{x^2}$$

The integral sign and differential serve as delimiters, flanking the integrand on the left and right, respectively. In particular, we do *not* write $\int dx f(x)$ when we intend $\int f(x)\,dx$.

INTEGRATION FORMULAS

Integration is essentially educated guesswork—given the derivative f of a function F, one tries to guess what the function F is. However, many basic integration formulas can be obtained directly from their companion differentiation formulas. Some of the most important are given in Table 6.2.1.

FOR THE READER.   Recall from Section 4.4 that
$$\frac{d}{dx}[\sec^{-1} x] = \frac{1}{|x|\sqrt{x^2 - 1}}$$
Use this to verify Formula 14 in Table 6.2.1.

Example 2   The second integration formula in Table 6.2.1 will be easier to remember if you express it in words:

To integrate a power of x (other than -1), add 1 to the exponent and divide by the new exponent.

Here are some examples:

$$\int x^2\,dx = \frac{x^3}{3} + C \qquad \boxed{r = 2}$$

$$\int x^3\,dx = \frac{x^4}{4} + C \qquad \boxed{r = 3}$$

$$\int \frac{1}{x^5}\,dx = \int x^{-5}\,dx = \frac{x^{-5+1}}{-5+1} + C = -\frac{1}{4x^4} + C \qquad \boxed{r = -5}$$

$$\int \sqrt{x}\,dx = \int x^{\frac{1}{2}}\,dx = \frac{x^{\frac{1}{2}+1}}{\frac{1}{2}+1} + C = \frac{2}{3}x^{\frac{3}{2}} + C = \frac{2}{3}(\sqrt{x})^3 + C \qquad \boxed{r = \frac{1}{2}}$$

Table 6.2.1

DIFFERENTIATION FORMULA	INTEGRATION FORMULA
1. $\dfrac{d}{dx}[x] = 1$	$\displaystyle\int dx = x + C$
2. $\dfrac{d}{dx}\left[\dfrac{x^{r+1}}{r+1}\right] = x^r \quad (r \neq -1)$	$\displaystyle\int x^r\, dx = \dfrac{x^{r+1}}{r+1} + C \quad (r \neq -1)$
3. $\dfrac{d}{dx}[\sin x] = \cos x$	$\displaystyle\int \cos x\, dx = \sin x + C$
4. $\dfrac{d}{dx}[-\cos x] = \sin x$	$\displaystyle\int \sin x\, dx = -\cos x + C$
5. $\dfrac{d}{dx}[\tan x] = \sec^2 x$	$\displaystyle\int \sec^2 x\, dx = \tan x + C$
6. $\dfrac{d}{dx}[-\cot x] = \csc^2 x$	$\displaystyle\int \csc^2 x\, dx = -\cot x + C$
7. $\dfrac{d}{dx}[\sec x] = \sec x \tan x$	$\displaystyle\int \sec x \tan x\, dx = \sec x + C$
8. $\dfrac{d}{dx}[-\csc x] = \csc x \cot x$	$\displaystyle\int \csc x \cot x\, dx = -\csc x + C$
9. $\dfrac{d}{dx}[e^x] = e^x$	$\displaystyle\int e^x\, dx = e^x + C$
10. $\dfrac{d}{dx}\left[\dfrac{b^x}{\ln b}\right] = b^x \quad (0 < b,\, b \neq 1)$	$\displaystyle\int b^x\, dx = \dfrac{b^x}{\ln b} + C \quad (0 < b,\, b \neq 1)$
11. $\dfrac{d}{dx}[\ln |x|] = \dfrac{1}{x}$	$\displaystyle\int \dfrac{1}{x}\, dx = \ln |x| + C$
12. $\dfrac{d}{dx}[\tan^{-1} x] = \dfrac{1}{1 + x^2}$	$\displaystyle\int \dfrac{1}{1 + x^2}\, dx = \tan^{-1} x + C$
13. $\dfrac{d}{dx}[\sin^{-1} x] = \dfrac{1}{\sqrt{1 - x^2}}$	$\displaystyle\int \dfrac{1}{\sqrt{1 - x^2}}\, dx = \sin^{-1} x + C$
14. $\dfrac{d}{dx}[\sec^{-1} |x|] = \dfrac{1}{x\sqrt{x^2 - 1}}$	$\displaystyle\int \dfrac{1}{x\sqrt{x^2 - 1}}\, dx = \sec^{-1} |x| + C$

The case $r = -1$ does not fit this pattern. However, note from Formula 11 in Table 6.2.1 that

$$\int x^{-1}\, dx = \int \frac{1}{x}\, dx = \ln |x| + C \qquad \blacktriangleleft$$

PROPERTIES OF THE INDEFINITE INTEGRAL

Our first properties of antiderivatives follow directly from the simple constant factor, sum, and difference rules for derivatives.

6.2.3 THEOREM. *Suppose that $F(x)$ and $G(x)$ are antiderivatives of $f(x)$ and $g(x)$, respectively, and that c is a constant. Then:*

(a) *A constant factor can be moved through an integral sign; that is,*

$$\int cf(x)\, dx = cF(x) + C$$

(b) *An antiderivative of a sum is the sum of the antiderivatives; that is,*

$$\int [f(x) + g(x)]\, dx = F(x) + G(x) + C$$

(c) *An antiderivative of a difference is the difference of the antiderivatives; that is,*

$$\int [f(x) - g(x)]\, dx = F(x) - G(x) + C$$

Proof. In general, to establish the validity of an equation of the form

$$\int h(x)\,dx = H(x) + C$$

one must show that

$$\frac{d}{dx}[H(x)] = h(x)$$

We are given that $F(x)$ and $G(x)$ are antiderivatives of $f(x)$ and $g(x)$, respectively, so we know that

$$\frac{d}{dx}[F(x)] = f(x) \quad \text{and} \quad \frac{d}{dx}[G(x)] = g(x)$$

Thus,

$$\frac{d}{dx}[cF(x)] = c\frac{d}{dx}[F(x)] = cf(x)$$

$$\frac{d}{dx}[F(x) + G(x)] = \frac{d}{dx}[F(x)] + \frac{d}{dx}[G(x)] = f(x) + g(x)$$

$$\frac{d}{dx}[F(x) - G(x)] = \frac{d}{dx}[F(x)] - \frac{d}{dx}[G(x)] = f(x) - g(x)$$

which proves the three statements of the theorem. ∎

In practice, the results of Theorem 6.2.3 are summarized by the following formulas:

$$\int cf(x)\,dx = c\int f(x)\,dx \tag{4}$$

$$\int [f(x) + g(x)]\,dx = \int f(x)\,dx + \int g(x)\,dx \tag{5}$$

$$\int [f(x) - g(x)]\,dx = \int f(x)\,dx - \int g(x)\,dx \tag{6}$$

However, these equations must be applied carefully to avoid errors and unnecessary complexities arising from the constants of integration. For example, if you were to use (4) to integrate $0x$ by writing

$$\int 0x\,dx = 0\int x\,dx = 0\left(\frac{x^2}{2} + C\right) = 0$$

then you will have erroneously lost the constant of integration, and if you use (4) to integrate $2x$ by writing

$$\int 2x\,dx = 2\int x\,dx = 2\left(\frac{x^2}{2} + C\right) = x^2 + 2C$$

then you will have an unnecessarily complicated form of the arbitrary constant. Similarly, if you use (5) to integrate $1 + x$ by writing

$$\int (1 + x)\,dx = \int 1\,dx + \int x\,dx = (x + C_1) + \left(\frac{x^2}{2} + C_2\right) = x + \frac{x^2}{2} + C_1 + C_2$$

then you will have two arbitrary constants when one will suffice. These three kinds of problems are caused by introducing constants of integration too soon and can be avoided by inserting the constant of integration in the final result, rather than in intermediate computations.

Example 3 Evaluate

(a) $\displaystyle\int 4\cos x\,dx$ (b) $\displaystyle\int (x + x^2)\,dx$

Solution (a). Since $F(x) = \sin x$ is an antiderivative for $f(x) = \cos x$ (Table 6.2.1), we obtain

$$\int \underbrace{4\cos x\, dx}_{(4)} = 4\int \cos x\, dx = 4\sin x + C$$

Solution (b). From Table 6.2.1 we obtain

$$\int \underbrace{(x + x^2)\, dx}_{(5)} = \int x\, dx + \int x^2\, dx = \frac{x^2}{2} + \frac{x^3}{3} + C \qquad \blacktriangleleft$$

Parts (b) and (c) of Theorem 6.2.3 can be extended to more than two functions, which in combination with part (a) results in the following general formula:

$$\int [c_1 f_1(x) + c_2 f_2(x) + \cdots + c_n f_n(x)]\, dx$$
$$= c_1 \int f_1(x)\, dx + c_2 \int f_2(x)\, dx + \cdots + c_n \int f_n(x)\, dx \qquad (7)$$

Example 4

$$\int (3x^6 - 2x^2 + 7x + 1)\, dx = 3\int x^6\, dx - 2\int x^2\, dx + 7\int x\, dx + \int 1\, dx$$
$$= \frac{3x^7}{7} - \frac{2x^3}{3} + \frac{7x^2}{2} + x + C \qquad \blacktriangleleft$$

Sometimes it is useful to rewrite an integrand in a different form before performing the integration.

Example 5 Evaluate

(a) $\displaystyle\int \frac{\cos x}{\sin^2 x}\, dx$ (b) $\displaystyle\int \frac{t^2 - 2t^4}{t^4}\, dt$ (c) $\displaystyle\int \frac{x^2}{x^2 + 1}\, dx$

Solution (a).

$$\int \frac{\cos x}{\sin^2 x}\, dx = \int \frac{1}{\sin x}\frac{\cos x}{\sin x}\, dx = \int \csc x \cot x\, dx = -\csc x + C$$

> Formula 8 in Table 6.2.1

Solution (b).

$$\int \frac{t^2 - 2t^4}{t^4}\, dt = \int \left(\frac{1}{t^2} - 2\right) dt = \int (t^{-2} - 2)\, dt$$
$$= \frac{t^{-1}}{-1} - 2t + C = -\frac{1}{t} - 2t + C$$

Solution (c). Using Formula 12 in Table 6.2.1,

$$\int \frac{x^2}{x^2 + 1}\, dx = \int \left(\frac{x^2 + 1}{x^2 + 1} - \frac{1}{x^2 + 1}\right) dx$$
$$= \int \left(1 - \frac{1}{x^2 + 1}\right) dx = x - \tan^{-1} x + C \qquad \blacktriangleleft$$

INTEGRAL CURVES

Graphs of antiderivatives of a function f are called *integral curves* of f. We know from Theorem 6.2.2 that if $y = F(x)$ is any integral curve of $f(x)$, then all other integral curves are vertical translations of this curve, since they have equations of the form $y = F(x) + C$.

For example, $y = \frac{1}{3}x^3$ is one integral curve for $f(x) = x^2$, so all the other integral curves have equations of the form $y = \frac{1}{3}x^3 + C$; conversely, the graph of any equation of this form is an integral curve (Figure 6.2.1).

In many problems one is interested in finding a function whose derivative satisfies specified conditions. The following example illustrates a geometric problem of this type.

Example 6   Suppose that a point moves along some unknown curve $y = f(x)$ in the xy-plane in such a way that at each point (x, y) on the curve, the tangent line has slope x^2. Find an equation for the curve given that it passes through the point $(2, 1)$.

Solution.   We know that $dy/dx = x^2$, so

$$y = \int x^2\,dx = \tfrac{1}{3}x^3 + C$$

Since the curve passes through $(2, 1)$, a specific value for C can be found by using the fact that $y = 1$ if $x = 2$. Substituting these values in the above equation yields

$$1 = \tfrac{1}{3}(2^3) + C \quad \text{or} \quad C = -\tfrac{5}{3}$$

so the curve is $y = \tfrac{1}{3}x^3 - \tfrac{5}{3}$.   ◀

Observe that in this example the requirement that the unknown curve pass through the point $(2, 1)$ enabled us to determine a specific value for the constant of integration, thereby isolating the single integral curve $y = \frac{1}{3}x^3 - \frac{5}{3}$ from the family $y = \frac{1}{3}x^3 + C$ (Figure 6.2.2).

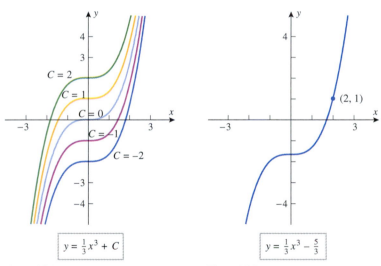

Figure 6.2.1   Figure 6.2.2

INTEGRATION FROM THE VIEWPOINT OF DIFFERENTIAL EQUATIONS

We will now consider another way of looking at integration that will be useful in our later work. Suppose that $f(x)$ is a known function and we are interested in finding a function $F(x)$ such that $y = F(x)$ satisfies the equation

$$\frac{dy}{dx} = f(x) \tag{8}$$

The solutions of this equation are the antiderivatives of $f(x)$, and we know that these can be obtained by integrating $f(x)$. For example, the solutions of the equation

$$\frac{dy}{dx} = x^2 \tag{9}$$

are

$$y = \int x^2\,dx = \frac{x^3}{3} + C$$

Equation (8) is called a ***differential equation*** because it involves a derivative of an unknown function. Differential equations are different from the kinds of equations we have encountered so far in that the unknown is a *function* and not a *number* as in an equation such as $x^2 + 5x - 6 = 0$.

Sometimes we will not be interested in finding all of the solutions of (8), but rather we will want only the solution whose integral curve passes through a specified point (x_0, y_0). For example, in Example 6 we solved (9) for the integral curve that passed through the point $(2, 1)$.

For simplicity, it is common in the study of differential equations to denote a solution of $dy/dx = f(x)$ as $y(x)$ rather than $F(x)$, as earlier. With this notation, the problem of finding a function $y(x)$ whose derivative is $f(x)$ and whose integral curve passes through the point (x_0, y_0) is expressed as

$$\frac{dy}{dx} = f(x), \quad y(x_0) = y_0 \tag{10}$$

This is called an ***initial-value problem***, and the requirement that $y(x_0) = y_0$ is called the ***initial condition*** for the problem.

Example 7 Solve the initial-value problem

$$\frac{dy}{dx} = \cos x, \quad y(0) = 1$$

Solution. The solution of the differential equation is

$$y = \int \cos x \, dx = \sin x + C \tag{11}$$

The initial condition $y(0) = 1$ implies that $y = 1$ if $x = 0$; substituting these values in (11) yields

$$1 = \sin(0) + C \quad \text{or} \quad C = 1$$

Thus, the solution of the initial-value problem is $y = \sin x + 1$. ◀

DIRECTION FIELDS

If we interpret dy/dx as the slope of a tangent line, then at a point (x, y) on an integral curve of the equation $dy/dx = f(x)$, the slope of the tangent line is $f(x)$. What is interesting about this is that the slopes of the tangent lines to the integral curves can be obtained without actually solving the differential equation. For example, if

$$\frac{dy}{dx} = \sqrt{x^2 + 1}$$

then we know without solving the equation that at the point where $x = 1$ the tangent line to an integral curve has slope $\sqrt{1^2 + 1} = \sqrt{2}$; and more generally, at a point where $x = a$, the tangent line to an integral curve has slope $\sqrt{a^2 + 1}$.

A geometric description of the integral curves of a differential equation $dy/dx = f(x)$ can be obtained by choosing a rectangular grid of points in the xy-plane, calculating the slopes of the tangent lines to the integral curves at the gridpoints, and drawing small portions of the tangent lines at those points. The resulting picture, which is called a ***direction field*** or ***slope field*** for the equation, shows the "direction" of the integral curves at the gridpoints. With sufficiently many gridpoints it is often possible to visualize the integral curves themselves; for example, Figure 6.2.3a shows a direction field for the differential equation $dy/dx = x^2$, and Figure 6.2.3b shows that same field with the integral curves imposed on it—the more gridpoints that are used, the more completely the direction field reveals the shape of the integral curves. However, the amount of computation can be considerable, so computers are usually used when direction fields with many gridpoints are needed.

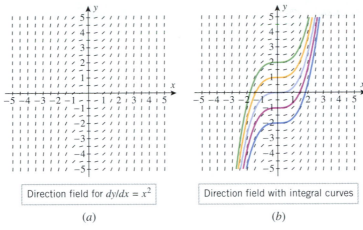

Direction field for $dy/dx = x^2$

(*a*)

Direction field with integral curves

(*b*)

Figure 6.2.3

EXERCISE SET 6.2 ⬜ Graphing Utility

1. In each part, confirm that the formula is correct, and state a corresponding integration formula.

(a) $\dfrac{d}{dx}[\sqrt{1+x^2}] = \dfrac{x}{\sqrt{1+x^2}}$

(b) $\dfrac{d}{dx}[xe^x] = (x+1)e^x$

2. In each part, confirm that the stated formula is correct by differentiating.

(a) $\displaystyle\int x \sin x \, dx = \sin x - x \cos x + C$

(b) $\displaystyle\int \dfrac{dx}{(1-x^2)^{3/2}} = \dfrac{x}{\sqrt{1-x^2}} + C$

In Exercises 3–6, find the derivative and state a corresponding integration formula.

3. $\dfrac{d}{dx}[\sqrt{x^3+5}]$

4. $\dfrac{d}{dx}\left[\dfrac{x}{x^2+3}\right]$

5. $\dfrac{d}{dx}[\sin(2\sqrt{x}\,)]$

6. $\dfrac{d}{dx}[\sin x - x \cos x]$

In Exercises 7 and 8, evaluate the integral by rewriting the integrand appropriately, if required, and then apply Formula 2 in Table 6.2.1.

7. (a) $\displaystyle\int x^8 \, dx$ (b) $\displaystyle\int x^{5/7} \, dx$ (c) $\displaystyle\int x^3 \sqrt{x} \, dx$

8. (a) $\displaystyle\int \sqrt[3]{x^2} \, dx$ (b) $\displaystyle\int \dfrac{1}{x^6} \, dx$ (c) $\displaystyle\int x^{-7/8} \, dx$

In Exercises 9–12, evaluate each integral by applying Theorem 6.2.3 and Formula 2 in Table 6.2.1 appropriately.

9. (a) $\displaystyle\int \dfrac{1}{2x^3} \, dx$ (b) $\displaystyle\int (u^3 - 2u + 7) \, du$

10. $\displaystyle\int (x^{2/3} - 4x^{-1/5} + 4) \, dx$

11. $\displaystyle\int (x^{-3} + \sqrt{x} - 3x^{1/4} + x^2) \, dx$

12. $\displaystyle\int \left(\dfrac{7}{y^{3/4}} - \sqrt[3]{y} + 4\sqrt{y}\right) dy$

In Exercises 13–32, evaluate the integral, and check your answer by differentiating.

13. $\displaystyle\int x(1+x^3) \, dx$ **14.** $\displaystyle\int (2+y^2)^2 \, dy$

15. $\displaystyle\int x^{1/3}(2-x)^2 \, dx$ **16.** $\displaystyle\int (1+x^2)(2-x) \, dx$

17. $\displaystyle\int \dfrac{x^5 + 2x^2 - 1}{x^4} \, dx$ **18.** $\displaystyle\int \dfrac{1 - 2t^3}{t^3} \, dt$

19. $\displaystyle\int \left[\dfrac{2}{x} + 3e^x\right] dx$ **20.** $\displaystyle\int \left[\dfrac{1}{2t} - \sqrt{2}e^t\right] dt$

21. $\displaystyle\int [4\sin x + 2\cos x] \, dx$

22. $\displaystyle\int [4\sec^2 x + \csc x \cot x] \, dx$

23. $\displaystyle\int \sec x (\sec x + \tan x) \, dx$ **24.** $\displaystyle\int \sec x (\tan x + \cos x) \, dx$

25. $\displaystyle\int \dfrac{\sec \theta}{\cos \theta} \, d\theta$ **26.** $\displaystyle\int \dfrac{dy}{\csc y}$

27. $\displaystyle\int \dfrac{\sin x}{\cos^2 x} \, dx$ **28.** $\displaystyle\int \left[\phi + \dfrac{2}{\sin^2 \phi}\right] d\phi$

29. $\displaystyle\int [1 + \sin^2 \theta \csc \theta] \, d\theta$ **30.** $\displaystyle\int \dfrac{\sin 2x}{\cos x} \, dx$

31. $\displaystyle\int \left[\dfrac{1}{2\sqrt{1-x^2}} - \dfrac{3}{1+x^2}\right] dx$

32. $\displaystyle\int \left[\frac{4}{x\sqrt{x^2-1}} + \frac{1+x+x^3}{1+x^2} \right] dx$

33. Evaluate the integral

$$\int \frac{1}{1+\sin x}\, dx$$

by multiplying the numerator and denominator by an appropriate expression.

34. Use the double-angle formula $\cos 2x = 2\cos^2 x - 1$ to evaluate the integral

$$\int \frac{1}{1+\cos 2x}\, dx$$

35. (a) Use a graphing utility to generate a slope field for the differential equation $dy/dx = x$ in the region $-5 \le x \le 5$ and $-5 \le y \le 5$.
(b) Graph some representative integral curves of the function $f(x) = x$.
(c) Find an equation for the integral curve that passes through the point $(4, 7)$.

36. (a) Use a graphing utility to generate a slope field for the differential equation $dy/dx = e^x/2$ in the region $-1 \le x \le 4$ and $-1 \le y \le 4$.
(b) Graph some representative integral curves of the function $f(x) = e^x/2$ for $x > 0$.
(c) Find an equation for the integral curve that passes through the point $(0, 1)$.

37. Use a graphing utility to generate some representative integral curves of the function $f(x) = 5x^4 - \sec^2 x$ over the interval $(-\pi/2, \pi/2)$.

38. Use a graphing utility to generate some representative integral curves of the function $f(x) = (x-1)/x$ over the interval $(0, 5)$.

39. Suppose that a point moves along a curve $y = f(x)$ in the xy-plane in such a way that at each point (x, y) on the curve the tangent line has slope $-\sin x$. Find an equation for the curve, given that it passes through the point $(0, 2)$.

40. Suppose that a point moves along a curve $y = f(x)$ in the xy-plane in such a way that at each point (x, y) on the curve the tangent line has slope $(x+1)^2$. Find an equation for the curve, given that it passes through the point $(-2, 8)$.

In Exercises 41–44, solve the initial-value problems.

41. (a) $\dfrac{dy}{dx} = \sqrt[3]{x}, \quad y(1) = 2$

(b) $\dfrac{dy}{dt} = \sin t + 1, \quad y\!\left(\dfrac{\pi}{3}\right) = \dfrac{1}{2}$

(c) $\dfrac{dy}{dx} = \dfrac{x+1}{\sqrt{x}}, \quad y(1) = 0$

42. (a) $\dfrac{dy}{dx} = \dfrac{1}{(2x)^3}, \quad y(1) = 0$

(b) $\dfrac{dy}{dt} = \sec^2 t - \sin t, \quad y\!\left(\dfrac{\pi}{4}\right) = 1$

(c) $\dfrac{dy}{dx} = x^2\sqrt{x^3}, \quad y(0) = 0$

43. (a) $\dfrac{dy}{dx} = 4e^x, \quad y(0) = 1$

(b) $\dfrac{dy}{dt} = \dfrac{1}{t}, \quad y(-1) = 5$

44. (a) $\dfrac{dy}{dt} = \dfrac{3}{\sqrt{1-t^2}}, \quad y\!\left(\dfrac{\sqrt{3}}{2}\right) = 0$

(b) $\dfrac{dy}{dx} = \dfrac{x^2-1}{x^2+1}, \quad y(1) = \dfrac{\pi}{2}$

45. Find the general form of a function whose second derivative is $\sqrt{x}$. [*Hint:* Solve the equation $f''(x) = \sqrt{x}$ for $f(x)$ by integrating both sides twice.]

46. Find a function f such that $f''(x) = x + \cos x$ and such that $f(0) = 1$ and $f'(0) = 2$. [*Hint:* Integrate both sides of the equation twice.]

In Exercises 47–49, find an equation of the curve that satisfies the given conditions.

47. At each point (x, y) on the curve the slope is $2x + 1$; the curve passes through the point $(-3, 0)$.

48. At each point (x, y) on the curve the slope equals the square of the distance between the point and the y-axis; the point $(-1, 2)$ is on the curve.

49. At each point (x, y) on the curve, y satisfies the condition $d^2y/dx^2 = 6x$; the line $y = 5 - 3x$ is tangent to the curve at the point where $x = 1$.

50. Suppose that a uniform metal rod 50 cm long is insulated laterally, and the temperatures at the exposed ends are maintained at $25°C$ and $85°C$, respectively. Assume that an x-axis is chosen as in the accompanying figure and that the temperature $T(x)$ satisfies the equation

$$\frac{d^2T}{dx^2} = 0$$

Find $T(x)$ for $0 \le x \le 50$.

Figure Ex-50

51. (a) Show that

$$F(x) = \tfrac{1}{6}(3x+4)^2 \quad \text{and} \quad G(x) = \tfrac{3}{2}x^2 + 4x$$

differ by a constant by showing that they are antiderivatives of the same function.
(b) Find the constant C such that $F(x) - G(x) = C$ by evaluating $F(x)$ and $G(x)$ at a particular value of x.
(c) Check your answer in part (b) by simplifying the expression $F(x) - G(x)$ algebraically.

52. Follow the directions of Exercise 51 with

$$F(x) = \frac{x^2}{x^2+5} \quad \text{and} \quad G(x) = -\frac{5}{x^2+5}.$$

In Exercises 53 and 54, use a trigonometric identity to help evaluate the integral.

53. $\displaystyle\int \tan^2 x\, dx$

54. $\displaystyle\int \cot^2 x\, dx$

55. Use the identities $\cos 2\theta = 1 - 2\sin^2\theta = 2\cos^2\theta - 1$ to help evaluate the integrals

(a) $\displaystyle\int \sin^2(x/2)\, dx$

(b) $\displaystyle\int \cos^2(x/2)\, dx$

56. Let F and G be the functions defined piecewise by

$$F(x) = \begin{cases} x, & x > 0 \\ -x, & x < 0 \end{cases} \quad \text{and} \quad G(x) = \begin{cases} x + 2, & x > 0 \\ -x + 3, & x < 0 \end{cases}$$

(a) Show that F and G have the same derivative.

(b) Show that $G(x) \neq F(x) + C$ for any constant C.

(c) Do parts (a) and (b) violate Theorem 6.2.2? Explain.

57. The speed of sound in air at $0°C$ (or 273 K on the Kelvin scale) is 1087 ft/s, but the speed v increases as the temperature T rises. Experimentation has shown that the rate of change of v with respect to T is

$$\frac{dv}{dT} = \frac{1087}{2\sqrt{273}} T^{-1/2}$$

where v is in feet per second and T is in kelvins (K). Find a formula that expresses v as a function of T.

6.3 INTEGRATION BY SUBSTITUTION

---

*In this section we will study a technique, called **substitution**, that can often be used to transform complicated integration problems into simpler ones.*

---

u-SUBSTITUTION

The method of substitution can be motivated by examining the chain rule from the viewpoint of antidifferentiation. For this purpose, suppose that F is an antiderivative of f and that g is a differentiable function. The chain rule implies that the derivative of $F(g(x))$ can be expressed as

$$\frac{d}{dx}[F(g(x))] = F'(g(x))g'(x)$$

which we can write in integral form as

$$\int F'(g(x))g'(x)\, dx = F(g(x)) + C \tag{1}$$

or since F is an antiderivative of f,

$$\int f(g(x))g'(x)\, dx = F(g(x)) + C \tag{2}$$

For our purposes it will be useful to let $u = g(x)$ and to write $du/dx = g'(x)$ in the differential form $du = g'(x)\, dx$. With this notation (1) can be expressed as

$$\int f(u)\, du = F(u) + C \tag{3}$$

The process of evaluating an integral of form (2) by converting it into form (3) with the substitution

$$u = g(x) \quad \text{and} \quad du = g'(x)\, dx$$

is called the **method of u-substitution**. Here our emphasis is *not* on the interpretation of the expression $du = g'(x)\, dx$ as a function of dx as was done in Section 3.8. Instead, the differential notation serves primarily as a useful "bookkeeping" device for the method of u-substitution. The following example illustrates how the method works.

Example 1 Evaluate $\displaystyle\int (x^2 + 1)^{50} \cdot 2x\, dx$.

Solution. If we let $u = x^2 + 1$, then $du/dx = 2x$, which implies that $du = 2x\, dx$. Thus, the given integral can be written as

$$\int (x^2 + 1)^{50} \cdot 2x\, dx = \int u^{50}\, du = \frac{u^{51}}{51} + C = \frac{(x^2 + 1)^{51}}{51} + C \quad \blacktriangleleft$$

It is important to realize that in the method of u-substitution you have control over the choice of u, but once you make that choice you have no control over the resulting expression for du. Thus, in the last example we *chose* $u = x^2 + 1$ but $du = 2x\,dx$ was *computed*. Fortunately, our choice of u, combined with the computed du, worked out perfectly to produce an integral involving u that was easy to evaluate. However, in general, the method of u-substitution will fail if the chosen u and the computed du cannot be used to produce an integrand in which no expressions involving x remain, or if you cannot evaluate the resulting integral. Thus, for example, the substitution $u = x^2$, $du = 2x\,dx$ will not work for the integral

$$\int 2x \sin x^4\,dx$$

because this substitution results in the integral

$$\int \sin u^2\,du$$

which still cannot be evaluated in terms of familiar functions.

In general, there are no hard and fast rules for choosing u, and in some problems no choice of u will work. In such cases other methods need to be used, some of which will be discussed later. Making appropriate choices for u will come with experience, but you may find the following *guidelines*, combined with a mastery of the basic integrals in Table 6.2.1, helpful.

> **Step 1.** Look for some composition $f(g(x))$ within the integrand for which the substitution
>
> $$u = g(x), \quad du = g'(x)\,dx$$
>
> produces an integral that is expressed entirely in terms of u and du. This may or may not be possible.
>
> **Step 2.** If you are successful in Step 1, then try to evaluate the resulting integral in terms of u. Again, this may or may not be possible.
>
> **Step 3.** If you are successful in Step 2, then replace u by $g(x)$ to express your final answer in terms of x.

EASY TO RECOGNIZE SUBSTITUTIONS

The easiest substitutions occur when the integrand is the derivative of a known function, except for a constant added to or subtracted from the independent variable.

Example 2

$$\int \sin(x+9)\,dx = \int \sin u\,du = -\cos u + C = -\cos(x+9) + C$$

$$u = x + 9$$
$$du = 1 \cdot dx = dx$$

$$\int (x-8)^{23}\,dx = \int u^{23}\,du = \frac{u^{24}}{24} + C = \frac{(x-8)^{24}}{24} + C \qquad \blacktriangleleft$$

$$u = x - 8$$
$$du = 1 \cdot dx = dx$$

Another easy u-substitution occurs when the integrand is the derivative of a known function, except for a constant that multiplies or divides the independent variable. The following example illustrates two ways to evaluate such integrals.

Example 3 Evaluate $\displaystyle\int \cos 5x\,dx$.

Solution.

$$\int \cos 5x \, dx = \int (\cos u) \cdot \frac{1}{5} du = \frac{1}{5} \int \cos u \, du = \frac{1}{5} \sin u + C = \frac{1}{5} \sin 5x + C$$

$$\boxed{\begin{array}{l} u = 5x \\ du = 5\,dx \text{ or } dx = \frac{1}{5}\,du \end{array}}$$

Alternative Solution. There is a variation of the preceding method that some people prefer. The substitution $u = 5x$ requires $du = 5\,dx$. If there were a factor of 5 in the integrand, then we could group the 5 and dx together to form the du required by the substitution. Since there is no factor of 5, we will insert one and compensate by putting a factor of $\frac{1}{5}$ in front of the integral. The computations are as follows:

$$\int \cos 5x \, dx = \frac{1}{5} \int \cos 5x \cdot 5\,dx = \frac{1}{5} \int \cos u \, du = \frac{1}{5} \sin u + C = \frac{1}{5} \sin 5x + C \quad \blacktriangleleft$$

$$\boxed{\begin{array}{l} u = 5x \\ du = 5\,dx \end{array}}$$

More generally, if the integrand is a composition of the form $f(ax + b)$, where $f(x)$ is an easy to integrate function, then the substitution $u = ax + b$, $du = a\,dx$ will work.

Example 4

$$\int \frac{dx}{\left(\frac{1}{3}x - 8\right)^5} = \int \frac{3\,du}{u^5} = 3 \int u^{-5} \, du = -\frac{3}{4} u^{-4} + C = -\frac{3}{4} \left(\frac{1}{3}x - 8\right)^{-4} + C \quad \blacktriangleleft$$

$$\boxed{\begin{array}{l} u = \frac{1}{3}x - 8 \\ du = \frac{1}{3}\,dx \text{ or } dx = 3\,du \end{array}}$$

Example 5   Evaluate $\displaystyle\int \frac{dx}{1 + 3x^2}$.

Solution. Substituting

$$u = \sqrt{3}x, \quad du = \sqrt{3}\,dx$$

yields

$$\int \frac{dx}{1 + 3x^2} = \frac{1}{\sqrt{3}} \int \frac{du}{1 + u^2} = \frac{1}{\sqrt{3}} \tan^{-1} u + C = \frac{1}{\sqrt{3}} \tan^{-1}(\sqrt{3}x) + C \quad \blacktriangleleft$$

With the help of Theorem 6.2.3, a complicated integral can sometimes be computed by expressing it as a sum of simpler integrals.

Example 6

$$\int \left(\frac{1}{x} + \sec^2 \pi x\right) dx = \int \frac{dx}{x} + \int \sec^2 \pi x \, dx = \ln|x| + \int \sec^2 \pi x \, dx$$

$$= \ln|x| + \frac{1}{\pi} \int \sec^2 u \, du$$

$$\boxed{\begin{array}{l} u = \pi x \\ du = \pi\,dx \text{ or } dx = \frac{1}{\pi}\,du \end{array}}$$

$$= \ln|x| + \frac{1}{\pi} \tan u + C = \ln|x| + \frac{1}{\pi} \tan \pi x + C \quad \blacktriangleleft$$

The next four examples illustrate a substitution $u = g(x)$ where $g(x)$ is a nonlinear function.

Example 7 Evaluate $\int \sin^2 x \cos x \, dx$.

Solution. If we let $u = \sin x$, then

$$\frac{du}{dx} = \cos x, \quad \text{so} \quad du = \cos x \, dx$$

Thus,

$$\int \sin^2 x \cos x \, dx = \int u^2 \, du = \frac{u^3}{3} + C = \frac{\sin^3 x}{3} + C \qquad \blacktriangleleft$$

Example 8 Evaluate $\int \dfrac{e^{\sqrt{x}}}{\sqrt{x}} \, dx$.

Solution. If we let $u = \sqrt{x}$, then

$$\frac{du}{dx} = \frac{1}{2\sqrt{x}}, \quad \text{so} \quad du = \frac{1}{2\sqrt{x}} \, dx \quad \text{or} \quad 2\,du = \frac{1}{\sqrt{x}} \, dx$$

Thus,

$$\int \frac{e^{\sqrt{x}}}{\sqrt{x}} \, dx = \int 2e^u \, du = 2 \int e^u \, du = 2e^u + C = 2e^{\sqrt{x}} + C \qquad \blacktriangleleft$$

Example 9 Evaluate $\int t^4 \sqrt[3]{3 - 5t^5} \, dt$.

Solution.

$$\int t^4 \sqrt[3]{3 - 5t^5} \, dt = -\frac{1}{25} \int \sqrt[3]{u} \, du = -\frac{1}{25} \int u^{1/3} \, du$$

$$u = 3 - 5t^5$$
$$du = -25t^4 \, dt \quad \text{or} \quad -\tfrac{1}{25} \, du = t^4 \, dt$$

$$= -\frac{1}{25} \frac{u^{4/3}}{4/3} + C = -\frac{3}{100} \left(3 - 5t^5\right)^{4/3} + C \qquad \blacktriangleleft$$

Example 10 Evaluate $\int \dfrac{e^x}{\sqrt{1 - e^{2x}}} \, dx$.

Solution. Substituting

$$u = e^x, \quad du = e^x \, dx$$

yields

$$\int \frac{e^x}{\sqrt{1 - e^{2x}}} \, dx = \int \frac{du}{\sqrt{1 - u^2}} = \sin^{-1} u + C = \sin^{-1}(e^x) + C \qquad \blacktriangleleft$$

LESS APPARENT SUBSTITUTIONS

The method of substitution is relatively straightforward, provided the integrand contains an easily recognized composition $f(g(x))$ and the remainder of the integrand is a constant multiple of $g'(x)$. If this is not the case, the method may still apply but can require more computation.

Example 11 Evaluate $\int x^2 \sqrt{x - 1} \, dx$.

Solution. The composition $\sqrt{x - 1}$ suggests the substitution

$$u = x - 1 \quad \text{so that} \quad du = dx \qquad (4)$$

From the first equality in (4)

$$x^2 = (u+1)^2 = u^2 + 2u + 1$$

so that

$$\int x^2 \sqrt{x-1}\, dx = \int (u^2 + 2u + 1)\sqrt{u}\, du = \int (u^{5/2} + 2u^{3/2} + u^{1/2})\, du$$

$$= \tfrac{2}{7} u^{7/2} + \tfrac{4}{5} u^{5/2} + \tfrac{2}{3} u^{3/2} + C$$

$$= \tfrac{2}{7}(x-1)^{7/2} + \tfrac{4}{5}(x-1)^{5/2} + \tfrac{2}{3}(x-1)^{3/2} + C \qquad \blacktriangleleft$$

Example 12   Evaluate $\displaystyle\int \cos^3 x\, dx$.

Solution.   The only compositions in the integrand that suggest themselves are

$$\cos^3 x = (\cos x)^3 \quad \text{and} \quad \cos^2 x = (\cos x)^2$$

However, neither the substitution $u = \cos x$ nor the substitution $u = \cos^2 x$ work (verify). In this case, an appropriate substitution is not suggested by the composition contained in the integrand. On the other hand, note from Equation (2) that the derivative $g'(x)$ appears as a factor in the integrand. This suggests that we write

$$\int \cos^3 x\, dx = \int \cos^2 x \cos x\, dx$$

and solve the equation $du = \cos x\, dx$ for $u = \sin x$. Since $\sin^2 x + \cos^2 x = 1$, we then have

$$\int \cos^3 x\, dx = \int \cos^2 x \cos x\, dx = \int (1 - \sin^2 x) \cos x\, dx = \int (1 - u^2)\, du$$

$$= u - \frac{u^3}{3} + C = \sin x - \frac{1}{3}\sin^3 x + C \qquad \blacktriangleleft$$

Example 13   Evaluate $\displaystyle\int \frac{dx}{a^2 + x^2}\, dx$, where $a \neq 0$ is a constant.

Solution.   Some simple algebra and an appropriate u-substitution will allow us to use Formula 12 in Table 6.2.1.

$$\int \frac{dx}{a^2 + x^2} = \frac{1}{a}\int \frac{dx/a}{1 + (x/a)^2} \underset{\substack{u = x/a \\ du = dx/a}}{=} \frac{1}{a}\int \frac{du}{1 + u^2} = \frac{1}{a}\tan^{-1} u + C = \frac{1}{a}\tan^{-1}\frac{x}{a} + C \qquad \blacktriangleleft$$

The method of Example 13 leads to the following generalizations of Formulas 12, 13, and 14 in Table 6.2.1 for $a > 0$:

$$\int \frac{du}{a^2 + u^2} = \frac{1}{a}\tan^{-1}\frac{u}{a} + C \tag{5}$$

$$\int \frac{du}{\sqrt{a^2 - u^2}} = \sin^{-1}\frac{u}{a} + C \tag{6}$$

$$\int \frac{du}{u\sqrt{u^2 - a^2}} = \frac{1}{a}\sec^{-1}\left|\frac{u}{a}\right| + C \tag{7}$$

Example 14 Evaluate $\displaystyle\int \frac{dx}{\sqrt{2 - x^2}}$.

Solution. Applying (6) with $u = x$ and $a = \sqrt{2}$ yields

$$\int \frac{dx}{\sqrt{2 - x^2}} = \sin^{-1} \frac{x}{\sqrt{2}} + C$$ ◀

....
INTEGRATION USING COMPUTER ALGEBRA SYSTEMS

The advent of computer algebra systems has made it possible to evaluate many kinds of integrals that would be laborious to evaluate by hand. For example, *Derive*, running on a handheld calculator, evaluated the integral

$$\int \frac{5x^2}{(1 + x)^{1/3}} dx = \frac{3(x + 1)^{2/3}(5x^2 - 6x + 9)}{8} + C$$

in about a second. The computer algebra system *Mathematica*, running on a personal computer, required even less time to evaluate this same integral. However, just as one would not want to rely on a calculator to compute $2 + 2$, so one would not want to use a CAS to integrate a simple function such as $f(x) = x^2$. Thus, even if you have a CAS, you will want to develop a reasonable level of competence in evaluating basic integrals. Moreover, the mathematical techniques that we will introduce for evaluating basic integrals are precisely the techniques that computer algebra systems use to evaluate more complicated integrals.

FOR THE READER. If you have a CAS, use it to calculate the integrals in the examples of this section. If your CAS produces a form of the answer that is different from the one in the text, then confirm algebraically that the two answers agree. Your CAS has various commands for simplifying answers. Explore the effect of using the CAS to simplify the expressions it produces for the integrals.

EXERCISE SET 6.3   ⟋ Graphing Utility   🅒 CAS
....

In Exercises 1–6, evaluate the integrals by making the indicated substitutions.

1. (a) $\displaystyle\int 2x\,(x^2 + 1)^{23}\, dx$; $u = x^2 + 1$

(b) $\displaystyle\int \cos^3 x \sin x\, dx$; $u = \cos x$

(c) $\displaystyle\int \frac{1}{\sqrt{x}} \sin \sqrt{x}\, dx$; $u = \sqrt{x}$

(d) $\displaystyle\int \frac{3x\, dx}{\sqrt{4x^2 + 5}}$; $u = 4x^2 + 5$

2. (a) $\displaystyle\int \sec^2(4x + 1)\, dx$; $u = 4x + 1$

(b) $\displaystyle\int y\sqrt{1 + 2y^2}\, dy$; $u = 1 + 2y^2$

(c) $\displaystyle\int \sqrt{\sin \pi\theta} \cos \pi\theta\, d\theta$; $u = \sin \pi\theta$

(d) $\displaystyle\int (2x + 7)(x^2 + 7x + 3)^{4/5}\, dx$; $u = x^2 + 7x + 3$

3. (a) $\displaystyle\int \cot x \csc^2 x\, dx$; $u = \cot x$

(b) $\displaystyle\int (1 + \sin t)^9 \cos t\, dt$; $u = 1 + \sin t$

(c) $\displaystyle\int \cos 2x\, dx$; $u = 2x$

(d) $\displaystyle\int x \sec^2 x^2\, dx$; $u = x^2$

4. (a) $\displaystyle\int x^2\sqrt{1 + x}\, dx$; $u = 1 + x$

(b) $\displaystyle\int [\csc(\sin x)]^2 \cos x\, dx$; $u = \sin x$

(c) $\displaystyle\int \sin(x - \pi)\, dx$; $u = x - \pi$

(d) $\displaystyle\int \frac{5x^4}{(x^5 + 1)^2}\, dx$; $u = x^5 + 1$

5. (a) $\displaystyle\int \frac{dx}{x \ln x}$; $u = \ln x$ (b) $\displaystyle\int e^{-5x}\, dx$; $u = -5x$

(c) $\displaystyle\int \frac{\sin 3\theta}{1 + \cos 3\theta}\, d\theta$; $u = 1 + \cos 3\theta$

(d) $\displaystyle\int \frac{e^x}{1 + e^x}\, dx$; $u = 1 + e^x$

6. (a) $\displaystyle\int \frac{x^2\, dx}{1 + x^6}$; $u = x^3$

(b) $\displaystyle\int \frac{dx}{x\sqrt{1 - (\ln x)^2}}$; $u = \ln x$

(c) $\displaystyle\int \frac{dx}{x\sqrt{9x^2 - 1}}$; $u = 3x$

(d) $\displaystyle\int \frac{dx}{\sqrt{x}(1 + x)}$; $u = \sqrt{x}$

In Exercises 7–48, evaluate the integrals by making appropriate substitutions.

7. $\int x\left(2-x^2\right)^3 dx$

8. $\int (3x-1)^5 dx$

9. $\int \cos 8x\, dx$

10. $\int \sin 3x\, dx$

11. $\int \sec 4x \tan 4x\, dx$

12. $\int \sec^2 5x\, dx$

13. $\int e^{2x} dx$

14. $\int \dfrac{dx}{2x}$

15. $\int \dfrac{dx}{\sqrt{1-4x^2}}$

16. $\int \dfrac{dx}{1+16x^2}$

17. $\int t\sqrt{7t^2+12}\, dt$

18. $\int \dfrac{x}{\sqrt{4-5x^2}}\, dx$

19. $\int \dfrac{x^2}{\sqrt{x^3+1}}\, dx$

20. $\int \dfrac{1}{(1-3x)^2}\, dx$

21. $\int \dfrac{x}{(4x^2+1)^3}\, dx$

22. $\int x\cos(3x^2)\, dx$

23. $\int e^{\sin x}\cos x\, dx$

24. $\int x^3 e^{x^4}\, dx$

25. $\int x^2 e^{-2x^3}\, dx$

26. $\int \dfrac{e^x+e^{-x}}{e^x-e^{-x}}\, dx$

27. $\int \dfrac{e^x}{1+e^{2x}}\, dx$

28. $\int \dfrac{t}{t^4+1}\, dt$

29. $\int \dfrac{\sin(5/x)}{x^2}\, dx$

30. $\int \dfrac{\sec^2(\sqrt{x})}{\sqrt{x}}\, dx$

31. $\int x^2 \sec^2(x^3)\, dx$

32. $\int \cos^3 2t \sin 2t\, dt$

33. $\int \sin^5 3t \cos 3t\, dt$

34. $\int \dfrac{\sin 2\theta}{(5+\cos 2\theta)^3}\, d\theta$

35. $\int \cos 4\theta \sqrt{2-\sin 4\theta}\, d\theta$

36. $\int \tan^3 5x \sec^2 5x\, dx$

37. $\int \dfrac{\sec^2 x\, dx}{\sqrt{1-\tan^2 x}}$

38. $\int \dfrac{\sin\theta}{\cos^2\theta+1}\, d\theta$

39. $\int \sec^3 2x \tan 2x\, dx$

40. $\int [\sin(\sin\theta)]\cos\theta\, d\theta$

41. $\int \dfrac{dx}{e^x}$

42. $\int \sqrt{e^x}\, dx$

43. $\int \dfrac{e^{\sqrt{y+1}}}{\sqrt{y+1}}\, dy$

44. $\int \dfrac{dy}{\sqrt{y}e^{\sqrt{y}}}$

45. $\int x\sqrt{x-3}\, dx$

46. $\int \dfrac{y\, dy}{\sqrt{y+1}}$

47. $\int \sin^3 2\theta\, d\theta$

48. $\int \sec^4 3\theta\, d\theta$

[*Hint:* Apply a trigonometric identity.]

In Exercises 49–52, evaluate each integral by first modifying the form of the integrand and then making an appropriate substitution, if needed.

49. $\int \dfrac{t+1}{t}\, dt$

50. $\int e^{2\ln x}\, dx$

51. $\int [\ln(e^x)+\ln(e^{-x})]\, dx$

52. $\int \cot x\, dx$

In Exercises 53 and 54, use Formulas (5), (6), and (7) to evaluate the integrals.

53. (a) $\int \dfrac{dx}{\sqrt{9-x^2}}$ (b) $\int \dfrac{dx}{5+x^2}$ (c) $\int \dfrac{dx}{x\sqrt{x^2-\pi}}$

54. (a) $\int \dfrac{e^x}{4+e^{2x}}\, dx$ (b) $\int \dfrac{dx}{\sqrt{9-4x^2}}$ (c) $\int \dfrac{dy}{y\sqrt{5y^2-3}}$

In Exercises 55–57, evaluate the integrals assuming that n is a positive integer and $b\neq 0$.

55. $\int (a+bx)^n\, dx$

56. $\int \sqrt[n]{a+bx}\, dx$

57. $\int \sin^n(a+bx)\cos(a+bx)\, dx$

c **58.** Use a CAS to check the answers you obtained in Exercises 55–57. If the answer produced by the CAS does not match yours, show that the two answers are equivalent. [*Suggestion: Mathematica* users may find it helpful to apply the Simplify command to the answer.]

59. (a) Evaluate the integral $\int \sin x \cos x\, dx$ by two methods: first by letting $u=\sin x$, then by letting $u=\cos x$.
(b) Explain why the two apparently different answers obtained in part (a) are really equivalent.

60. (a) Evaluate $\int (5x-1)^2\, dx$ by two methods: first square and integrate, then let $u=5x-1$.
(b) Explain why the two apparently different answers obtained in part (a) are really equivalent.

In Exercises 61–64, solve the initial-value problems.

61. $\dfrac{dy}{dx}=\sqrt{3x+1}$; $y(1)=5$

62. $\dfrac{dy}{dx}=6-5\sin 2x$; $y(0)=3$

63. $\dfrac{dy}{dt}=2e^{-t}$, $y(1)=3-\dfrac{2}{e}$

64. $\dfrac{dy}{dt}=\dfrac{1}{100+4t^2}$, $y(-5)=\dfrac{3\pi}{80}$

65. Use a graphing utility to generate some typical integral curves of $f(x)=x/\sqrt{x^2+1}$ over the interval $(-5,5)$.

66. Use a graphing utility to generate some typical integral curves of $f(x)=x/(x^2+1)$ over the interval $(-5,5)$.

67. Find a function f such that the slope of the tangent line at a point (x,y) on the curve $y=f(x)$ is $\sqrt{3x+1}$, and the curve passes through the point $(0,1)$.

68. A population of frogs is estimated to be 100,000 at the beginning of the year 2000. Suppose that the rate of growth of the population $p(t)$ (in thousands) after t years is given by $p'(t)=(4+0.15t)^{3/2}$. Estimate the projected population at the beginning of the year 2005.

69. Derive integration Formula (6).

70. Derive integration Formula (7).

6.4 SIGMA NOTATION; AREA AS A LIMIT

Recall from the informal discussion in Section 6.1 that if a function f is continuous and nonnegative on an interval $[a, b]$, then the "rectangle method" provides us with one approach to computing the area between the graph of f and the interval $[a, b]$. We begin this section with a discussion of a notation to represent lengthy sums in a concise form. Then we will discuss the rectangle method in more detail, both as a means for defining and for computing the area under a curve. In particular, we will show that such an area may be interpreted as a limit.

SIGMA NOTATION

The notation we will discuss is called **sigma notation** or **summation notation** because it uses the uppercase Greek letter Σ (sigma) to denote various kinds of sums. To illustrate how this notation works, consider the sum

$$1^2 + 2^2 + 3^2 + 4^2 + 5^2$$

in which each term is of the form k^2, where k is one of the integers from 1 to 5. In sigma notation this sum can be written as

$$\sum_{k=1}^{5} k^2$$

which is read "the summation of k^2, where k runs from 1 to 5." The notation tells us to form the sum of the terms that result when we substitute successive integers for k in the expression k^2, starting with $k = 1$ and ending with $k = 5$.

More generally, if $f(k)$ is a function of k, and if m and n are integers such that $m \leq n$, then

$$\sum_{k=m}^{n} f(k) \tag{1}$$

denotes the sum of the terms that result when we substitute successive integers for k, starting with $k = m$ and ending with $k = n$ (Figure 6.4.1).

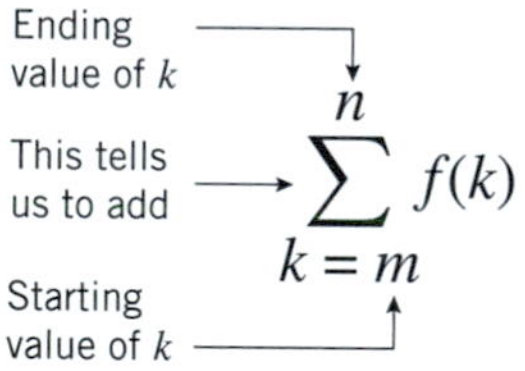

Ending value of k ——→ n

This tells us to add ——→ $\displaystyle\sum_{k=m} f(k)$

Starting value of k ——→ $k = m$

Figure 6.4.1

Example 1

$$\sum_{k=4}^{8} k^3 = 4^3 + 5^3 + 6^3 + 7^3 + 8^3$$

$$\sum_{k=1}^{5} 2k = 2 \cdot 1 + 2 \cdot 2 + 2 \cdot 3 + 2 \cdot 4 + 2 \cdot 5 = 2 + 4 + 6 + 8 + 10$$

$$\sum_{k=0}^{5} (2k + 1) = 1 + 3 + 5 + 7 + 9 + 11$$

$$\sum_{k=0}^{5} (-1)^k (2k + 1) = 1 - 3 + 5 - 7 + 9 - 11$$

$$\sum_{k=-3}^{1} k^3 = (-3)^3 + (-2)^3 + (-1)^3 + 0^3 + 1^3 = -27 - 8 - 1 + 0 + 1$$

$$\sum_{k=1}^{3} k \sin\left(\frac{k\pi}{5}\right) = \sin\frac{\pi}{5} + 2\sin\frac{2\pi}{5} + 3\sin\frac{3\pi}{5} \qquad \blacktriangleleft$$

The numbers m and n in (1) are called, respectively, the **lower** and **upper limits of summation**; and the letter k is called the **index of summation**. It is not essential to use k as the index of summation; any letter not reserved for another purpose will do. For example,

$$\sum_{i=1}^{6} \frac{1}{i}, \quad \sum_{j=1}^{6} \frac{1}{j}, \quad \text{and} \quad \sum_{n=1}^{6} \frac{1}{n}$$

all denote the sum

$$1 + \frac{1}{2} + \frac{1}{3} + \frac{1}{4} + \frac{1}{5} + \frac{1}{6}$$

If the upper and lower limits of summation are the same, then the "sum" in (1) reduces to a single term. For example,

$$\sum_{k=2}^{2} k^3 = 2^3 \quad \text{and} \quad \sum_{i=1}^{1} \frac{1}{i+2} = \frac{1}{1+2} = \frac{1}{3}$$

In the sums

$$\sum_{i=1}^{5} 2 \quad \text{and} \quad \sum_{j=0}^{2} x^3$$

the expression to the right of the Σ sign does not involve the index of summation. In such cases, we take all the terms in the sum to be the same, with one term for each allowable value of the summation index. Thus,

$$\sum_{i=1}^{5} 2 = 2+2+2+2+2 \quad \text{and} \quad \sum_{j=0}^{2} x^3 = x^3 + x^3 + x^3$$

CHANGING THE LIMITS OF SUMMATION

A sum can be written in more than one way using sigma notation with different limits of summation and correspondingly different summands. For example,

$$\sum_{i=1}^{5} 2i = 2+4+6+8+10 = \sum_{j=0}^{4}(2j+2) = \sum_{k=3}^{7}(2k-4)$$

On occasion we will want to change the sigma notation for a given sum to a sigma notation with different limits of summation.

Example 2 Express

$$\sum_{k=3}^{7} 5^{k-2}$$

in sigma notation so that the lower limit of summation is 0 rather than 3.

Solution.

$$\sum_{k=3}^{7} 5^{k-2} = 5^1 + 5^2 + 5^3 + 5^4 + 5^5$$
$$= 5^{0+1} + 5^{1+1} + 5^{2+1} + 5^{3+1} + 5^{4+1}$$
$$= \sum_{j=0}^{4} 5^{j+1} = \sum_{k=0}^{4} 5^{k+1} \quad \blacktriangleleft$$

PROPERTIES OF SUMS

When stating general properties of sums it is often convenient to use a subscripted letter such as a_k in place of the function notation $f(k)$. For example,

$$\sum_{k=1}^{5} a_k = a_1 + a_2 + a_3 + a_4 + a_5 = \sum_{j=1}^{5} a_j = \sum_{k=-1}^{3} a_{k+2}$$
$$\sum_{k=1}^{n} a_k = a_1 + a_2 + \cdots + a_n = \sum_{j=1}^{n} a_j = \sum_{k=-1}^{n-2} a_{k+2}$$

Our first properties provide some basic rules for manipulating sums.

6.4.1 THEOREM.

(a) $\displaystyle\sum_{k=1}^{n} ca_k = c\sum_{k=1}^{n} a_k$ (*if c does not depend on k*)

(b) $\displaystyle\sum_{k=1}^{n}(a_k + b_k) = \sum_{k=1}^{n} a_k + \sum_{k=1}^{n} b_k$

(c) $\displaystyle\sum_{k=1}^{n}(a_k - b_k) = \sum_{k=1}^{n} a_k - \sum_{k=1}^{n} b_k$

We will prove parts (*a*) and (*b*) and leave part (*c*) as an exercise.

Proof (a).

$$\sum_{k=1}^{n} ca_k = ca_1 + ca_2 + \cdots + ca_n = c(a_1 + a_2 + \cdots + a_n) = c\sum_{k=1}^{n} a_k$$

Proof (b).

$$\sum_{k=1}^{n}(a_k + b_k) = (a_1 + b_1) + (a_2 + b_2) + \cdots + (a_n + b_n)$$

$$= (a_1 + a_2 + \cdots + a_n) + (b_1 + b_2 + \cdots + b_n) = \sum_{k=1}^{n} a_k + \sum_{k=1}^{n} b_k \quad ∎$$

Restating Theorem 6.4.1 in words:

> (a) *A constant factor can be moved through a sigma sign.*
> (b) *Sigma distributes across sums.*
> (c) *Sigma distributes across differences.*

SUMMATION FORMULAS

6.4.2 THEOREM.

(a) $\displaystyle\sum_{k=1}^{n} k = 1 + 2 + \cdots + n = \frac{n(n+1)}{2}$

(b) $\displaystyle\sum_{k=1}^{n} k^2 = 1^2 + 2^2 + \cdots + n^2 = \frac{n(n+1)(2n+1)}{6}$

(c) $\displaystyle\sum_{k=1}^{n} k^3 = 1^3 + 2^3 + \cdots + n^3 = \left[\frac{n(n+1)}{2}\right]^2$

We will prove parts (*a*) and (*b*) and leave part (*c*) as an exercise.

Proof (a). Writing

$$\sum_{k=1}^{n} k$$

two ways, with summands in increasing order and in decreasing order, and then adding, we obtain

$$\sum_{k=1}^{n} k = \quad 1 \quad + \quad 2 \quad + \quad 3 \quad + \cdots + (n-2) + (n-1) + \quad n$$

$$\sum_{k=1}^{n} k = \quad n \quad + (n-1) + (n-2) + \cdots + \quad 3 \quad + \quad 2 \quad + \quad 1$$

$$2\sum_{k=1}^{n} k = (n+1) + (n+1) + (n+1) + \cdots + (n+1) + (n+1) + (n+1)$$

$$= n(n+1)$$

Thus,

$$\sum_{k=1}^{n} k = \frac{n(n+1)}{2}$$

Proof (b). Note that

$$(k+1)^3 - k^3 = k^3 + 3k^2 + 3k + 1 - k^3 = 3k^2 + 3k + 1$$

So,

$$\sum_{k=1}^{n}[(k+1)^3 - k^3] = \sum_{k=1}^{n}(3k^2 + 3k + 1) \tag{2}$$

Writing out the left side of (2) with the index running *down* from $k = n$ to $k = 1$, we have

$$\sum_{k=1}^{n}[(k+1)^3 - k^3] = [(n+1)^3 - n^3] + \cdots + [4^3 - 3^3] + [3^3 - 2^3] + [2^3 - 1^3]$$

$$= (n+1)^3 - 1 \tag{3}$$

Combining (3) and (2), and expanding the right side of (2) by using Theorem 6.4.1 and part (*a*) of this theorem yields

$$(n+1)^3 - 1 = 3\sum_{k=1}^{n} k^2 + 3\sum_{k=1}^{n} k + \sum_{k=1}^{n} 1$$

$$= 3\sum_{k=1}^{n} k^2 + 3\frac{n(n+1)}{2} + n$$

So,

$$3\sum_{k=1}^{n} k^2 = [(n+1)^3 - 1] - 3\frac{n(n+1)}{2} - n$$

$$= (n+1)^3 - 3(n+1)\left(\frac{n}{2}\right) - (n+1)$$

$$= \frac{n+1}{2}[2(n+1)^2 - 3n - 2]$$

$$= \frac{n+1}{2}[2n^2 + n] = \frac{n(n+1)(2n+1)}{2}$$

Thus,

$$\sum_{k=1}^{n} k^2 = \frac{n(n+1)(2n+1)}{6}$$

REMARK. The sum in (3) is an example of a ***telescoping sum***, since the cancellation of each of the two parts of an interior summand with parts of its neighboring summands allows the entire sum to collapse like a telescope.

Example 3 Evaluate $\displaystyle\sum_{k=1}^{30} k(k+1)$.

Solution.

$$\sum_{k=1}^{30} k(k+1) = \sum_{k=1}^{30}(k^2+k) = \sum_{k=1}^{30}k^2 + \sum_{k=1}^{30}k$$

$$= \frac{30(31)(61)}{6} + \frac{30(31)}{2} = 9920 \qquad \boxed{\text{Theorem } 6.4.2(a), (b)} \qquad \blacktriangleleft$$

In formulas such as

$$\sum_{k=1}^{n} k = \frac{n(n+1)}{2} \quad \text{or} \quad 1+2+\cdots+n = \frac{n(n+1)}{2}$$

the left side of the equality is said to express the sum in **open form** and the right side is said to express it in **closed form**. The open form indicates the summands and the closed form is an explicit formula for the sum.

Example 4 Express $\displaystyle\sum_{k=1}^{n}(3+k)^2$ in closed form.

Solution.

$$\sum_{k=1}^{n}(3+k)^2 = 4^2 + 5^2 + \cdots + (3+n)^2$$

$$= [1^2 + 2^2 + 3^3 + 4^2 + 5^2 + \cdots + (3+n)^2] - [1^2 + 2^2 + 3^2]$$

$$= \left(\sum_{k=1}^{3+n} k^2\right) - 14$$

$$= \frac{(3+n)(4+n)(7+2n)}{6} - 14 = \frac{1}{6}(73n + 21n^2 + 2n^3) \qquad \blacktriangleleft$$

FOR THE READER. Your numerical calculating utility probably provides some way of evaluating sums that can be expressed in sigma notation. Check your documentation to find out how to do this, and then use your utility to confirm that the numerical result obtained in Example 3 is correct. If you have access to a CAS, it provides some method for finding closed forms for sums such as those in Theorem 6.4.2. Use your CAS to confirm the formulas in that theorem, and then find closed forms for

$$\sum_{k=1}^{n} k^4 \quad \text{and} \quad \sum_{k=1}^{n} k^5$$

A DEFINITION OF AREA

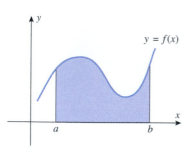

Figure 6.4.2

Suppose that f is a continuous function that is nonnegative on an interval $[a, b]$, and let R denote the region that is bounded below by the x-axis, bounded on the sides by the vertical lines $x = a$ and $x = b$, and bounded above by the curve $y = f(x)$ (Figure 6.4.2). Recall from the informal discussion in Section 6.1 that the "rectangle method" provides us with one approach to computing the area between the graph of f and the interval $[a, b]$. Our goal now is to define formally what we mean by the area of R. We will work from the definition of the area of a rectangle as the product of its length and width. Define the area of a region decomposed into a finite collection of rectangles to be the sum of the areas of those rectangles. To define the area of the region R, we will use these definitions and the rectangle method of Section 6.1. The basic idea is as follows:

- Divide the interval $[a, b]$ into n equal subintervals.

- Over each subinterval construct a rectangle whose height is the value of f at any point in the subinterval.

- The union of these rectangles forms a region R_n whose area can be regarded as an approximation to the "area" A of the region R.

- Repeat the process using more and more subdivisions.
- Define the area of R to be the "limit" of the areas of the approximating regions R_n as n is made larger and larger without bound. We can express this idea symbolically as

$$A = \text{area}(R) = \lim_{n \to +\infty} [\text{area}(R_n)] \tag{4}$$

REMARK.   There is a difference in interpretation between writing $\lim_{n \to +\infty}$ and writing $\lim_{x \to +\infty}$, where n represents a positive integer and x has no such restriction. Equation (4) should be interpreted to mean that by choosing the positive integer n sufficiently large, we can make area(R_n) as close to A as desired. Later we will study limits of the type $\lim_{n \to +\infty}$ in detail, but for now suffice it to say that the computational techniques we have used for limits of type $\lim_{x \to +\infty}$ will also work for $\lim_{n \to +\infty}$.

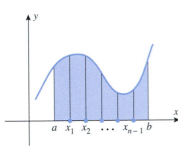

Figure 6.4.3

To make all of this more precise, it will be helpful to capture this procedure in mathematical notation. For this purpose, suppose that we divide the interval $[a, b]$ into n subintervals by inserting $n - 1$ equally spaced points between a and b, say

$$x_1, x_2, \ldots, x_{n-1}$$

(Figure 6.4.3). Each of these subintervals has width $(b - a)/n$, which it is customary to denote by

$$\Delta x = \frac{b - a}{n}$$

In each subinterval we need to choose an x-value at which to evaluate the function f to determine the height of a rectangle over the interval. If we denote those x-values by

$$x_1^*, x_2^*, \ldots, x_n^*$$

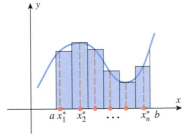

Figure 6.4.4

(Figure 6.4.4), then the areas of the rectangles constructed over these intervals will be

$$f(x_1^*)\Delta x, \quad f(x_2^*)\Delta x, \ldots, \quad f(x_n^*)\Delta x$$

(Figure 6.4.5), and the total area of the region R_n will be

$$\text{area}(R_n) = f(x_1^*)\Delta x + f(x_2^*)\Delta x + \cdots + f(x_n^*)\Delta x$$

With this notation (4) can be expressed as

$$A = \lim_{n \to +\infty} \sum_{k=1}^{n} f(x_k^*)\Delta x$$

which suggests the following definition of the area of the region R.

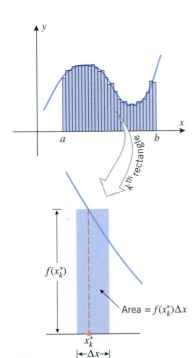

Figure 6.4.5

6.4.3 DEFINITION (*Area Under a Curve*).   If the function f is continuous on $[a, b]$ and if $f(x) \geq 0$ for all x in $[a, b]$, then the *area* under the curve $y = f(x)$ over the interval $[a, b]$ is defined by

$$A = \lim_{n \to +\infty} \sum_{k=1}^{n} f(x_k^*)\Delta x \tag{5}$$

In (5) the values of $x_1^*, x_2^*, \ldots, x_n^*$ may be chosen in many different ways, so it is conceivable that different choices of these values might produce different values of A. Were this to happen, then Definition 6.4.3 would not be an acceptable definition of area. Fortunately, this does not happen; it is proved in advanced courses that when f is continuous (as we have assumed), the same value of A results no matter how the x_k^* are chosen. In practice they are chosen in some systematic fashion, some common choices being

- the left endpoint of each subinterval;
- the right endpoint of each subinterval;
- the midpoint of each subinterval.

If, as shown in Figure 6.4.6, the subinterval $[a, b]$ is divided by $x_1, x_2, x_3, \ldots, x_{n-1}$ into n equal parts each of length $\Delta x = (b - a)/n$, and if we let $x_0 = a$ and $x_n = b$, then

$$x_k = a + k\Delta x \quad \text{for } k = 0, 1, 2, \ldots, n$$

Thus,

$$x_k^* = x_{k-1} = a + (k - 1)\Delta x \qquad \boxed{\text{Left endpoint}} \qquad (6)$$

$$x_k^* = x_k = a + k\Delta x \qquad \boxed{\text{Right endpoint}} \qquad (7)$$

$$x_k^* = \tfrac{1}{2}(x_{k-1} + x_k) = a + \left(k - \tfrac{1}{2}\right)\Delta x \qquad \boxed{\text{Midpoint}} \qquad (8)$$

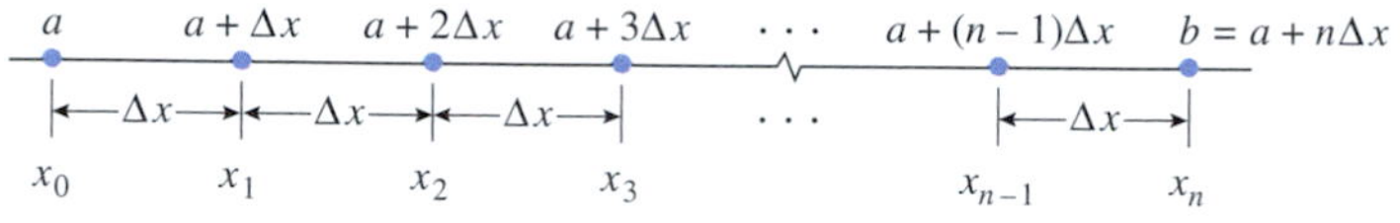

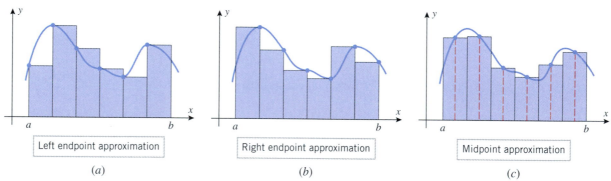

Figure 6.4.6

NUMERICAL APPROXIMATIONS OF AREA

We would expect from Definition 6.4.3 that for each of the choices (6), (7), and (8), the sum

$$\sum_{k=1}^{n} f(x_k^*)\Delta x = \Delta x \sum_{k=1}^{n} f(x_k^*) = \Delta x[f(x_1^*) + f(x_2^*) + \cdots + f(x_n^*)] \qquad (9)$$

would yield a good approximation to the area A, provided n is a large positive integer. According to which of these three options is used in choosing the x_k^*, we refer to Formula (9) as the **left endpoint approximation**, the **right endpoint approximation**, or the **midpoint approximation** of the exact area (Figure 6.4.7).

Left endpoint approximation	Right endpoint approximation	Midpoint approximation
(a)	(b)	(c)

Figure 6.4.7

Example 5   Find the left endpoint, right endpoint, and midpoint approximations of the area under the curve $y = 9 - x^2$ over the interval $[0, 3]$ with $n = 10$, $n = 20$, and $n = 50$ (Figure 6.4.8).

Solution.   Details of the computations for the case $n = 10$ are shown to six decimal places in Table 6.4.1 and the results of all computations are given in Table 6.4.2.   ◀

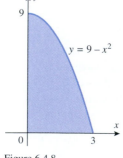

$y = 9 - x^2$

Figure 6.4.8

REMARK.   We will show below that the exact area under $y = 9 - x^2$ over the interval $[0, 3]$ is 18 (i.e., 18 square units), so that in the preceding example the midpoint approximation is more accurate than either of the endpoint approximations. This can also be seen geometrically from the approximating rectangles: Since the graph of $y = 9 - x^2$ is decreasing over the interval $[0, 3]$, each left endpoint approximation overestimates the area, each right endpoint approximation underestimates the area, and each midpoint approximation falls between the overestimate and the underestimate (Figure 6.4.9). This is consistent with the values in Table 6.4.2. Later in the text we will investigate the error that results when an area is approximated by the midpoint rule.

Table 6.4.1

$n = 10, \Delta x = (b - a)/n = (3 - 0)/10 = 0.3$

	LEFT ENDPOINT APPROXIMATION		RIGHT ENDPOINT APPROXIMATION		MIDPOINT APPROXIMATION	
k	x_k^*	$9 - (x_k^*)^2$	x_k^*	$9 - (x_k^*)^2$	x_k^*	$9 - (x_k^*)^2$
1	0.0	9.000000	0.3	8.910000	0.15	8.977500
2	0.3	8.910000	0.6	8.640000	0.45	8.797500
3	0.6	8.640000	0.9	8.190000	0.75	8.437500
4	0.9	8.190000	1.2	7.560000	1.05	7.897500
5	1.2	7.560000	1.5	6.750000	1.35	7.177500
6	1.5	6.750000	1.8	5.760000	1.65	6.277500
7	1.8	5.760000	2.1	4.590000	1.95	5.197500
8	2.1	4.590000	2.4	3.240000	2.25	3.937500
9	2.4	3.240000	2.7	1.710000	2.55	2.497500
10	2.7	1.710000	3.0	0.000000	2.85	0.877500
		64.350000		55.350000		60.075000

$$\Delta x \sum_{k=1}^{n} f(x_k^*)$$

(0.3)(64.350000)	(0.3)(55.350000)	(0.3)(60.075000)
= 19.305000	= 16.605000	= 18.022500

Table 6.4.2

n	LEFT ENDPOINT APPROXIMATION	RIGHT ENDPOINT APPROXIMATION	MIDPOINT APPROXIMATION
10	19.305000	16.605000	18.022500
20	18.663750	17.313750	18.005625
50	18.268200	17.728200	18.000900

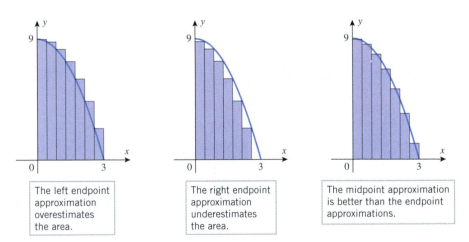

The left endpoint approximation overestimates the area.

The right endpoint approximation underestimates the area.

The midpoint approximation is better than the endpoint approximations.

Figure 6.4.9

COMPUTING THE EXACT VALUE OF AN AREA

Although numerical approximations of area are useful, we will often wish to compute the *exact* value of some area. In certain cases this can be done by explicitly evaluating the limit in Definition 6.4.3.

Example 6 Use Definition 6.4.3 with x_k^* as the right endpoint of each subinterval to find the area between the graph of $f(x) = x^2$ and the interval [0, 1].

Solution. We have

$$\Delta x = \frac{b - a}{n} = \frac{1 - 0}{n} = \frac{1}{n}$$

and from (7)

$$x_k^* = a + k\Delta x = \frac{k}{n}$$

so that

$$\sum_{k=1}^{n} f(x_k^*)\Delta x = \sum_{k=1}^{n} (x_k^*)^2 \Delta x = \sum_{k=1}^{n} \left(\frac{k}{n}\right)^2 \frac{1}{n} = \frac{1}{n^3} \sum_{k=1}^{n} k^2$$

$$= \frac{1}{n^3} \left[\frac{n(n+1)(2n+1)}{6}\right] = \frac{1}{3} + \frac{1}{2n} + \frac{1}{6n^2}$$

Therefore,

$$A = \lim_{n \to +\infty} \sum_{k=1}^{n} f(x_k^*)\Delta x = \lim_{n \to +\infty} \left(\frac{1}{3} + \frac{1}{2n} + \frac{1}{6n^2}\right) = \frac{1}{3}$$

(Note that this conclusion agrees with the numerical evidence we collected in Table 6.1.2.) ◀

In the solution to Example 6 we made use of one of the "closed form" summation formulas from Theorem 6.4.2. The next result collects some consequences of Theorem 6.4.2 that can facilitate computations of area using Definition 6.4.3.

6.4.4 THEOREM.

(*a*) $\displaystyle \lim_{n \to +\infty} \frac{1}{n} \sum_{k=1}^{n} 1 = 1$ (*b*) $\displaystyle \lim_{n \to +\infty} \frac{1}{n^2} \sum_{k=1}^{n} k = \frac{1}{2}$

(*c*) $\displaystyle \lim_{n \to +\infty} \frac{1}{n^3} \sum_{k=1}^{n} k^2 = \frac{1}{3}$ (*d*) $\displaystyle \lim_{n \to +\infty} \frac{1}{n^4} \sum_{k=1}^{n} k^3 = \frac{1}{4}$

The proof of Theorem 6.4.4 is left as an exercise for the reader.

Example 7 Use Definition 6.4.3 with x_k^* as the midpoint of each subinterval to find the area under the parabola $y = f(x) = 9 - x^2$ and over the interval $[0, 3]$.

Solution. Each subinterval will have length

$$\Delta x = \frac{b - a}{n} = \frac{3 - 0}{n} = \frac{3}{n}$$

and from (8)

$$x_k^* = a + \left(k - \frac{1}{2}\right)\Delta x = \left(k - \frac{1}{2}\right)\left(\frac{3}{n}\right)$$

Thus,

$$f(x_k^*)\Delta x = [9 - (x_k^*)^2]\Delta x = \left[9 - \left(k - \frac{1}{2}\right)^2 \left(\frac{3}{n}\right)^2\right]\left(\frac{3}{n}\right)$$

$$= \left[9 - \left(k^2 - k + \frac{1}{4}\right)\left(\frac{9}{n^2}\right)\right]\left(\frac{3}{n}\right)$$

$$= \frac{27}{n} - \frac{27}{n^3}k^2 + \frac{27}{n^3}k - \frac{27}{4n^3}$$

and

$$\sum_{k=1}^{n} f(x_k^*)\Delta x = \sum_{k=1}^{n}\left(\frac{27}{n} - \frac{27}{n^3}k^2 + \frac{27}{n^3}k - \frac{27}{4n^3}\right)$$

$$= 27\left[\frac{1}{n}\sum_{k=1}^{n}1 - \frac{1}{n^3}\sum_{k=1}^{n}k^2 + \frac{1}{n}\left(\frac{1}{n^2}\sum_{k=1}^{n}k\right) - \frac{1}{4n^2}\left(\frac{1}{n}\sum_{k=1}^{n}1\right)\right]$$

Therefore,

$$A = \lim_{n\to+\infty}\sum_{k=1}^{n} f(x_k^*)\Delta x$$

$$= \lim_{n\to+\infty}27\left[\frac{1}{n}\sum_{k=1}^{n}1 - \frac{1}{n^3}\sum_{k=1}^{n}k^2 + \frac{1}{n}\left(\frac{1}{n^2}\sum_{k=1}^{n}k\right) - \frac{1}{4n^2}\left(\frac{1}{n}\sum_{k=1}^{n}1\right)\right]$$

$$= 27\left[1 - \frac{1}{3} + 0\cdot\frac{1}{2} - 0\cdot 1\right] = 18$$

where we used Theorem 6.4.4 to compute the limits as $n\to+\infty$ of the expressions

$$\frac{1}{n^j}\sum_{k=1}^{n}k^{j-1}\quad \text{for } j = 1, 2, 3$$ ◀

NET SIGNED AREA

In Definition 6.4.3 we assumed that f is continuous and nonnegative on the interval $[a, b]$. If f is continuous and attains both positive and negative values on $[a, b]$, then the limit

$$\lim_{n\to+\infty}\sum_{k=1}^{n} f(x_k^*)\Delta x \tag{10}$$

no longer represents the area between the curve $y = f(x)$ and the interval $[a, b]$ on the x-axis; rather, it represents a difference of areas—the area of the region that is above the interval $[a, b]$ and below the curve $y = f(x)$ minus the area of the region that is below the interval $[a, b]$ and above the curve $y = f(x)$. We call this the *net signed area* between the graph of $y = f(x)$ and the interval $[a, b]$. For example, in Figure 6.4.10a, the net signed area between the curve $y = f(x)$ and the interval $[a, b]$ is

$$(A_I + A_{III}) - A_{II} = \left[\text{area above } [a, b]\right] - \left[\text{area below } [a, b]\right]$$

To explain why the limit in (10) represents this net signed area, let us subdivide the interval $[a, b]$ in Figure 6.4.10a into n equal subintervals and examine the terms in the sum

$$\sum_{k=1}^{n} f(x_k^*)\Delta x \tag{11}$$

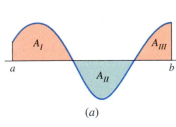

(a)

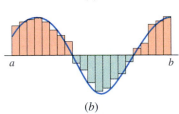

(b)

Figure 6.4.10

If $f(x_k^*)$ is positive, then the product $f(x_k^*)\Delta x$ represents the area of the rectangle with height $f(x_k^*)$ and base Δx (the pink rectangles in Figure 6.4.10b). However, if $f(x_k^*)$ is negative, then the product $f(x_k^*)\Delta x$ is the *negative* of the area of the rectangle with height $|f(x_k^*)|$ and base Δx (the green rectangles in Figure 6.4.10b). Thus, (11) represents the total area of the pink rectangles minus the total area of the green rectangles. As n increases, the pink rectangles fill out the regions with areas A_I and A_{III} and the green rectangles fill out the region with area A_{II}, which explains why the limit in (10) represents the signed area between $y = f(x)$ and the interval $[a, b]$. We formalize this in the following definition.

> **6.4.5 DEFINITION (*Net Signed Area*).** If the function f is continuous on $[a, b]$, then the *net signed area* A between $y = f(x)$ and the interval $[a, b]$ is defined by
>
> $$A = \lim_{n\to+\infty}\sum_{k=1}^{n} f(x_k^*)\Delta x$$

As with Definition 6.4.3, it can be shown that for a continuous function this limit always exists (independently of the choice of the numbers x_k^*). The net signed area between the curve $y = f(x)$ and $[a, b]$ can be positive, negative, or zero; it is positive when there is more area above the interval than below, negative when there is more area below than above, and zero when the areas above and below are equal.

Example 8   Use Definition 6.4.5 with x_k^* as the left endpoint of each subinterval to find the net signed area between the graph of $y = f(x) = x - 1$ and the interval $[0, 2]$.

Solution.   Each subinterval will have length

$$\Delta x = \frac{b - a}{n} = \frac{2 - 0}{n} = \frac{2}{n}$$

and from (6)

$$x_k^* = a + (k - 1)\Delta x = (k - 1)\left(\frac{2}{n}\right)$$

Thus,

$$f(x_k^*)\Delta x = (x_k^* - 1)\Delta x = \left[(k - 1)\left(\frac{2}{n}\right) - 1\right]\left(\frac{2}{n}\right) = \left(\frac{4}{n^2}\right)k - \frac{4}{n^2} - \frac{2}{n}$$

and

$$\sum_{k=1}^{n} f(x_k^*)\Delta x = \sum_{k=1}^{n}\left[\left(\frac{4}{n^2}\right)k - \frac{4}{n^2} - \frac{2}{n}\right]$$

$$= 4\left(\frac{1}{n^2}\sum_{k=1}^{n}k\right) - \frac{4}{n}\left(\frac{1}{n}\sum_{k=1}^{n}1\right) - 2\left(\frac{1}{n}\sum_{k=1}^{n}1\right)$$

Therefore,

$$A = \lim_{n \to +\infty}\sum_{k=1}^{n} f(x_k^*)\Delta x = \lim_{n \to +\infty}\left[4\left(\frac{1}{n^2}\sum_{k=1}^{n}k\right) - \frac{4}{n}\left(\frac{1}{n}\sum_{k=1}^{n}1\right) - 2\left(\frac{1}{n}\sum_{k=1}^{n}1\right)\right]$$

$$= 4\left(\frac{1}{2}\right) - 0 \cdot 1 - 2(1) = 0$$

Since the net signed area is zero, the area A_1 below the graph of f and above the interval $[0, 2]$ must equal the area A_2 above the graph of f and below the interval $[0, 2]$. This conclusion agrees with the graph of f shown in Figure 6.4.11. ◄

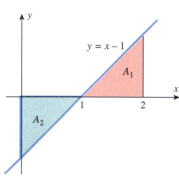

Figure 6.4.11

EXERCISE SET 6.4   ⒸCAS

1. Evaluate

(a) $\displaystyle\sum_{k=1}^{3} k^3$ (b) $\displaystyle\sum_{j=2}^{6}(3j - 1)$ (c) $\displaystyle\sum_{i=-4}^{1}(i^2 - i)$

(d) $\displaystyle\sum_{n=0}^{5} 1$ (e) $\displaystyle\sum_{k=0}^{4}(-2)^k$ (f) $\displaystyle\sum_{n=1}^{6}\sin n\pi$.

2. Evaluate

(a) $\displaystyle\sum_{k=1}^{4} k\sin\frac{k\pi}{2}$ (b) $\displaystyle\sum_{j=0}^{5}(-1)^j$ (c) $\displaystyle\sum_{i=7}^{20}\pi^2$

(d) $\displaystyle\sum_{m=3}^{5} 2^{m+1}$ (e) $\displaystyle\sum_{n=1}^{6}\sqrt{n}$ (f) $\displaystyle\sum_{k=0}^{10}\cos k\pi$.

In Exercises 3–8, write each expression in sigma notation, but do not evaluate.

3. $1 + 2 + 3 + \cdots + 10$

4. $3 \cdot 1 + 3 \cdot 2 + 3 \cdot 3 + \cdots + 3 \cdot 20$

5. $2 + 4 + 6 + 8 + \cdots + 20$ 6. $1 + 3 + 5 + 7 + \cdots + 15$

7. $1 - 3 + 5 - 7 + 9 - 11$ 8. $1 - \frac{1}{2} + \frac{1}{3} - \frac{1}{4} + \frac{1}{5}$

9. (a) Express the sum of the even integers from 2 to 100 in sigma notation.

   (b) Express the sum of the odd integers from 1 to 99 in sigma notation.

10. Express in sigma notation.
   (a) $a_1 - a_2 + a_3 - a_4 + a_5$
   (b) $-b_0 + b_1 - b_2 + b_3 - b_4 + b_5$
   (c) $a_0 + a_1x + a_2x^2 + \cdots + a_nx^n$
   (d) $a^5 + a^4b + a^3b^2 + a^2b^3 + ab^4 + b^5$

In Exercises 11–16, use Theorem 6.4.2 to evaluate the sums, and check your answers using the summation feature of a calculating utility.

11. $\displaystyle\sum_{k=1}^{100} k$ **12.** $\displaystyle\sum_{k=1}^{100}(7k+1)$ **13.** $\displaystyle\sum_{k=1}^{20} k^2$

14. $\displaystyle\sum_{k=4}^{20} k^2$ **15.** $\displaystyle\sum_{k=1}^{30} k(k-2)(k+2)$ **16.** $\displaystyle\sum_{k=1}^{6}(k-k^3)$

In Exercises 17–20, express the sums in closed form.

17. $\displaystyle\sum_{k=1}^{n} \frac{3k}{n}$ **18.** $\displaystyle\sum_{k=1}^{n-1} \frac{k^2}{n}$ **19.** $\displaystyle\sum_{k=1}^{n-1} \frac{k^3}{n^2}$

20. $\displaystyle\sum_{k=1}^{n}\left(\frac{5}{n}-\frac{2k}{n}\right)$

c **21.** For each of the sums that you obtained in Exercises 17–20, use a CAS to check your answer. If the answer produced by the CAS does not match your own, show that the two answers are equivalent.

22. Solve the equation $\displaystyle\sum_{k=1}^{n} k = 465$.

In Exercises 23–26, express the function of n in closed form and then find the limit.

23. $\displaystyle\lim_{n\to+\infty} \frac{1+2+3+\cdots+n}{n^2}$

24. $\displaystyle\lim_{n\to+\infty} \frac{1^2+2^2+3^2+\cdots+n^2}{n^3}$

25. $\displaystyle\lim_{n\to+\infty}\sum_{k=1}^{n} \frac{5k}{n^2}$ **26.** $\displaystyle\lim_{n\to+\infty}\sum_{k=1}^{n-1} \frac{2k^2}{n^3}$

27. Express $1+2+2^2+2^3+2^4+2^5$ in sigma notation with
(a) $j=0$ as the lower limit of summation
(b) $j=1$ as the lower limit of summation
(c) $j=2$ as the lower limit of summation.

28. Express

$$\sum_{k=5}^{9} k2^{k+4}$$

in sigma notation with
(a) $k=1$ as the lower limit of summation
(b) $k=13$ as the upper limit of summation.

In Exercises 29–32, divide the interval $[a, b]$ into $n = 4$ subintervals of equal length, and then compute

$$\sum_{k=1}^{4} f(x_k^*)\Delta x$$

with x_k^* as (a) the left endpoint of each subinterval, (b) the midpoint of each subinterval, and (c) the right endpoint of each subinterval.

29. $f(x) = 3x + 1$; $a = 2, b = 6$

30. $f(x) = 1/x$; $a = 1, b = 9$

31. $f(x) = \cos x$; $a = 0, b = \pi$

32. $f(x) = 2x - x^2$; $a = -1, b = 3$

In Exercises 33–36, use a calculating utility with summation capabilities or a CAS to obtain an approximate value for the area between the curve and the specified interval with $n = 10, 20$, and 50 subintervals by using the (a) left endpoint, (b) right endpoint, and (c) midpoint approximations. (If you do not have access to such a utility, then just do the case $n = 10$.)

c **33.** $y = 1/x$; $[1, 2]$ **c** **34.** $y = 1/x^2$; $[1, 3]$

c **35.** $y = \sqrt{x}$; $[0, 4]$ **c** **36.** $y = \sin x$; $[0, \pi/2]$

In Exercises 37–42, use Definition 6.4.3 with x_k^* as the *right* endpoint of each subinterval to find the area under the curve $y = f(x)$ over the interval $[a, b]$.

37. $y = \frac{1}{2}x$; $a = 1, b = 4$

38. $y = 5 - x$; $a = 0, b = 5$

39. $y = 9 - x^2$; $a = 0, b = 3$

40. $y = 4 - \frac{1}{4}x^2$; $a = 0, b = 3$

41. $y = x^3$; $a = 2, b = 6$

42. $y = 1 - x^3$; $a = -3, b = -1$

In Exercises 43–46, use Definition 6.4.5 with x_k^* as the *left* endpoint of each subinterval to find the area under the curve $y = f(x)$ over the interval $[a, b]$.

43. The function f and interval $[a, b]$ of Exercise 37.

44. The function f and interval $[a, b]$ of Exercise 38.

45. The function f and interval $[a, b]$ of Exercise 39.

46. The function f and interval $[a, b]$ of Exercise 40.

In Exercises 47 and 48, use Definition 6.4.3 with x_k^* as the *midpoint* of each subinterval to find the area under the curve $y = f(x)$ over the interval $[a, b]$.

47. The function $f(x) = x^2$; $a = 0, b = 1$

48. The function $f(x) = x^2$; $a = -1, b = 1$

In Exercises 49–52, use Definition 6.4.5 with x_k^* as the *right* endpoint of each subinterval to find the net signed area between the curve $y = f(x)$ and the interval $[a, b]$.

49. $y = x$; $a = -1, b = 1$. Verify your answer with a simple geometric argument.

50. $y = x$; $a = -1, b = 2$. Verify your answer with a simple geometric argument.

51. $y = x^2 - 1$; $a = 0, b = 2$ **52.** $y = x^3$; $a = -1, b = 1$

53. Use Definition 6.4.3 with x_k^* as the left endpoint of each subinterval to find the area under the graph of $y = mx$ and over the interval $[a, b]$, where $m > 0$ and $a \geq 0$.

54. Use Definition 6.4.5 with x_k^* as the right endpoint of each subinterval to find the net signed area between the graph of $y = mx$ and the interval $[a, b]$.

55. (a) Show that the area under the graph of $y = x^3$ and over the interval $[0, b]$ is $b^4/4$.
   (b) Find a formula for the area under $y = x^3$ over the interval $[a, b]$, where $a \geq 0$.

56. Find the area between the graph of $y = \sqrt{x}$ and the interval $[0, 1]$. [*Hint:* Use the result of Exercise 17 of Section 6.1.]

57. An artist wants to create a rough triangular design using uniform square tiles glued edge to edge. She places n tiles in a row to form the base of the triangle and then makes each successive row two tiles shorter than the preceding row. Find a formula for the number of tiles used in the design. [*Hint:* Your answer will depend on whether n is even or odd.]

58. An artist wants to create a sculpture by gluing together uniform spheres. She creates a rough rectangular base that has 50 spheres along one edge and 30 spheres along the other. She then creates successive layers by gluing spheres in the grooves of the preceding layer. How many spheres will there be in the sculpture?

59. By writing out the sums, determine whether the following are valid identities.

(a) $\displaystyle \int \left[\sum_{i=1}^{n} f_i(x) \right] dx = \sum_{i=1}^{n} \left[\int f_i(x)\, dx \right]$

(b) $\displaystyle \frac{d}{dx} \left[\sum_{i=1}^{n} f_i(x) \right] = \sum_{i=1}^{n} \left[\frac{d}{dx} [f_i(x)] \right]$

60. Which of the following are valid identities?

(a) $\displaystyle \sum_{i=1}^{n} a_i b_i = \sum_{i=1}^{n} a_i \sum_{i=1}^{n} b_i$ (b) $\displaystyle \sum_{i=1}^{n} \frac{a_i}{b_i} = \sum_{i=1}^{n} a_i \left/ \sum_{i=1}^{n} b_i \right.$

(c) $\displaystyle \sum_{i=1}^{n} a_i^2 = \left(\sum_{i=1}^{n} a_i \right)^2$

61. Prove part (*c*) of Theorem 6.4.1.

62. Prove part (*c*) of Theorem 6.4.2. [*Hint:* Begin with the difference $(k+1)^4 - k^4$ and follow the steps used to prove part (*b*) of the theorem.]

63. Prove Theorem 6.4.4.

6.5 THE DEFINITE INTEGRAL

In this section we will introduce the concept of a "definite integral," which will link the concept of area to other important concepts such as length, volume, density, probability, and work.

RIEMANN SUMS AND THE DEFINITE INTEGRAL

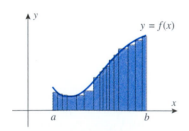

Figure 6.5.1

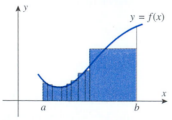

Figure 6.5.2

In our definition of net signed area (Definition 6.4.5), we assumed that for each positive number n, the interval $[a, b]$ was subdivided into n subintervals of equal length to create bases for the approximating rectangles. For some functions it may be more convenient to use rectangles with different widths (see Exercise 33); however, if we are to "exhaust" an area with rectangles of different widths, then it is important that successive subdivisions be constructed in such a way that the widths of the rectangles approach zero as n increases (Figure 6.5.1). Thus, we must preclude the kind of situation that occurs in Figure 6.5.2 in which the right half of the interval is never subdivided. If this kind of subdivision were allowed, the error in the approximation would not approach zero as n increased.

A ***partition*** of the interval $[a, b]$ is a collection of numbers

$$a = x_0 < x_1 < x_2 < \cdots < x_{n-1} < x_n = b$$

that divides $[a, b]$ into n subintervals of lengths

$$\Delta x_1 = x_1 - x_0, \quad \Delta x_2 = x_2 - x_1, \quad \Delta x_3 = x_3 - x_2, \ldots, \quad \Delta x_n = x_n - x_{n-1}$$

The partition is said to be ***regular*** provided the subintervals all have the same length

$$\Delta x_k = \Delta x = \frac{b - a}{n}$$

For a regular partition, the widths of the approximating rectangles approach zero as n is made large. Since this need not be the case for a general partition, we need some way to measure the "size" of these widths. One approach is to let max Δx_k denote the largest of the subinterval widths. The magnitude max Δx_k is called the **mesh size** of the partition. For example, Figure 6.5.3 shows a partition of the interval $[0, 6]$ into four subintervals with a mesh size of 2.

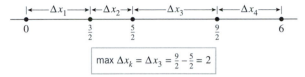

$$\max \Delta x_k = \Delta x_3 = \tfrac{9}{2} - \tfrac{5}{2} = 2$$

Figure 6.5.3

If we are to generalize Definition 6.4.5 so that it allows for unequal subinterval widths, we must replace the constant length Δx by the variable length Δx_k. When this is done the sum

$$\sum_{k=1}^{n} f(x_k^*)\Delta x \quad \text{is replaced by} \quad \sum_{k=1}^{n} f(x_k^*)\Delta x_k$$

We also need to replace the expression $n \to +\infty$ by an expression that guarantees us that the lengths of all subintervals approach zero. We will use the expression $\max \Delta x_k \to 0$ for this purpose. (Some writers use the symbol $\|\Delta\|$ rather than $\max \Delta x_k$ for the mesh size of the partition, in which case $\max \Delta x_k \to 0$ would be replaced by $\|\Delta\| \to 0$.) Based on our intuitive concept of area, we would then expect the net signed area A between the graph of f and the interval $[a, b]$ to satisfy the equation

$$A = \lim_{\max \Delta x_k \to 0} \sum_{k=1}^{n} f(x_k^*)\Delta x_k$$

(We will see in a moment that this is the case.) The limit that appears in this expression is one of the fundamental concepts of integral calculus and forms the basis for the following definition.

6.5.1 DEFINITION. A function f is said to be **integrable** on a finite closed interval $[a, b]$ if the limit

$$\lim_{\max \Delta x_k \to 0} \sum_{k=1}^{n} f(x_k^*)\Delta x_k$$

exists and does not depend on the choice of partitions or on the choice of the numbers x_k^* in the subintervals. When this is the case we denote the limit by the symbol

$$\int_a^b f(x)\,dx = \lim_{\max \Delta x_k \to 0} \sum_{k=1}^{n} f(x_k^*)\Delta x_k$$

which is called the **definite integral** of f from a to b. The numbers a and b are called the **lower limit of integration** and the **upper limit of integration**, respectively, and $f(x)$ is called the **integrand**.

The notation used for the definite integral deserves some comment. Historically, the expression "$f(x)\,dx$" was interpreted to be the "infinitesimal area" of a rectangle with height $f(x)$ and "infinitesimal" width dx. By "summing" these infinitesimal areas, the entire area under the curve was obtained. The integral symbol "$\int$" is an "elongated s" that was used to indicate this summation. For us, the integral symbol "$\int$" and the symbol "dx" can serve as reminders that the definite integral is actually a limit of a *summation* as $\Delta x_k \to 0$.

The sum that appears in Definition 6.5.1 is called a ***Riemann*** [*] ***sum***, and the definite integral is sometimes called the ***Riemann integral*** in honor of the German mathematician Bernhard Riemann who formulated many of the basic concepts of integral calculus. (The reason for the similarity in notation between the definite integral and the indefinite integral will become clear in the next section, where we will establish a link between the two types of "integration.")

The limit that appears in Definition 6.5.1 is somewhat different from the kinds of limits discussed in Chapter 2. Loosely phrased, the expression

$$\lim_{\max \Delta x_k \to 0} \sum_{k=1}^{n} f(x_k^*) \Delta x_k = L$$

is intended to convey the idea that we can force the Riemann sums to be as close as we please to L, regardless of how the values of x_k^* are chosen, by making the mesh size of the partition sufficiently small. Although it is possible to give a more formal definition of this limit, we will simply rely on intuitive arguments when applying Definition 6.5.1.

Example 1   Use Definition 6.5.1 to show that if $f(x) = C$ is a constant function, then

$$\int_a^b f(x)\,dx = C(b - a)$$

Solution.   Since $f(x) = C$ is constant, it follows that no matter how the values of x_k^* are chosen,

$$\sum_{k=1}^{n} f(x_k^*) \Delta x_k = \sum_{k=1}^{n} C \Delta x_k = C \sum_{k=1}^{n} \Delta x_k = C(b - a)$$

Since every Riemann sum has the same value $C(b - a)$, it follows that

$$\lim_{\max \Delta x_k \to 0} \sum_{k=1}^{n} f(x_k^*) \Delta x_k = \lim_{\max \Delta x_k \to 0} C(b - a) = C(b - a) \qquad \blacktriangleleft$$

Note that in Definition 6.5.1, we do *not* assume that the function f is necessarily continuous on the interval $[a, b]$.

Example 2   Define a function f on the interval $[0, 1]$ by $f(x) = 1$ if $0 < x \leq 1$ and $f(0) = 0$. Use Definition 6.5.1 to show that

$$\int_0^1 f(x)\,dx = 1$$

---

[*] GEORG FRIEDRICH BERNHARD RIEMANN (1826–1866). German mathematician. Bernhard Riemann, as he is commonly known, was the son of a Protestant minister. He received his elementary education from his father and showed brilliance in arithmetic at an early age. In 1846 he enrolled at Göttingen University to study theology and philology, but he soon transferred to mathematics. He studied physics under W. E. Weber and mathematics under Carl Friedrich Gauss, whom some people consider to be the greatest mathematician who ever lived. In 1851 Riemann received his Ph.D. under Gauss, after which he remained at Göttingen to teach. In 1862, one month after his marriage, Riemann suffered an attack of pleuritis, and for the remainder of his life was an extremely sick man. He finally succumbed to tuberculosis in 1866 at age 39.

An interesting story surrounds Riemann's work in geometry. For his introductory lecture prior to becoming an associate professor, Riemann submitted three possible topics to Gauss. Gauss surprised Riemann by choosing the topic Riemann liked the least, the foundations of geometry. The lecture was like a scene from a movie. The old and failing Gauss, a giant in his day, watching intently as his brilliant and youthful protégé skillfully pieced together portions of the old man's own work into a complete and beautiful system. Gauss is said to have gasped with delight as the lecture neared its end, and on the way home he marveled at his student's brilliance. Gauss died shortly thereafter. The results presented by Riemann that day eventually evolved into a fundamental tool that Einstein used some 50 years later to develop relativity theory.

In addition to his work in geometry, Riemann made major contributions to the theory of complex functions and mathematical physics. The notion of the definite integral, as it is presented in most basic calculus courses, is due to him. Riemann's early death was a great loss to mathematics, for his mathematical work was brilliant and of fundamental importance.

Solution. We first note that since

$$\lim_{x \to 0^+} f(x) = \lim_{x \to 0^+} 1 = 1 \neq 0 = f(0)$$

f is *not* continuous on the interval $[0, 1]$. Consider any partition of $[0, 1]$ and any choice of the x_k^* corresponding to this partition. Then either $x_1^* = 0$ or it does not. If not, then

$$\sum_{k=1}^{n} f(x_k^*)\Delta x_k = \sum_{k=1}^{n} \Delta x_k = 1$$

On the other hand, if $x_1^* = 0$, then $f(x_1^*) = f(0) = 0$ and

$$\sum_{k=1}^{n} f(x_k^*)\Delta x_k = \sum_{k=2}^{n} \Delta x_k = -\Delta x_1 + \sum_{k=1}^{n} \Delta x_k = 1 - \Delta x_1$$

In either case we see that the *difference* between the Riemann sum

$$\sum_{k=1}^{n} f(x_k^*)\Delta x_k$$

and 1 is at most Δx_1. Since Δx_1 approaches zero as max $\Delta x_k \to 0$, it follows that

$$\int_0^1 f(x)\,dx = 1 \qquad \blacktriangleleft$$

Although Example 2 shows that a function does not have to be continuous on an interval to be integrable on that interval, we will be interested primarily in the definite integrals of continuous functions. Our earlier discussion of net signed area suggests that a function that is continuous on an interval should also be integrable on that interval. This is the content of the next result, which we state without proof.

6.5.2 THEOREM. *If a function f is continuous on an interval $[a, b]$, then f is integrable on $[a, b]$.*

We can use Theorem 6.5.2 to clarify the connection between the definite integral and net signed area. Suppose that f is a continuous function on an interval $[a, b]$. Recall that in Section 6.4 we defined the net signed area A between the graph of f and the interval $[a, b]$ to be given by the limit

$$A = \lim_{n \to +\infty} \sum_{k=1}^{n} f(x_k^*)\Delta x$$

On the other hand, it follows from Theorem 6.5.2 and Definition 6.5.1 that we can use *regular* partitions of $[a, b]$ to compute the definite integral of f over $[a, b]$ as the limit

$$\int_a^b f(x)\,dx = \lim_{n \to +\infty} \sum_{k=1}^{n} f(x_k^*)\Delta x$$

Since the two limits are the same, we conclude that

$$A = \lim_{n \to +\infty} \sum_{k=1}^{n} f(x_k^*)\Delta x = \int_a^b f(x)\,dx = \lim_{\max \Delta x_k \to 0} \sum_{k=1}^{n} f(x_k^*)\Delta x_k$$

In other words, the definite integral of a continuous function f from a to b may always be interpreted as the net signed area between the graph of f and the interval $[a, b]$. Of course, if f is nonnegative, this is simply the area beneath the graph of f and above the interval $[a, b]$. It follows that our area computations in Section 6.4 may be reformulated as computations of particular definite integrals. For example, we showed that the area between the graph of $f(x) = 9 - x^2$ and the interval $[0, 3]$ is 18 square units. Equivalently, this computation shows us that

$$\int_0^3 (9 - x^2)\, dx = 18$$

Fortunately, there are often effective and efficient methods for evaluating definite integrals that do not require the explicit evaluation of limits. (We will have more to say about this in Section 6.6.) In the simplest cases, definite integrals can be calculated using formulas from plane geometry to compute signed areas.

Example 3 Sketch the region whose area is represented by the definite integral, and evaluate the integral using an appropriate formula from geometry.

(a) $\displaystyle\int_1^4 2\, dx$ (b) $\displaystyle\int_{-1}^2 (x + 2)\, dx$ (c) $\displaystyle\int_0^1 \sqrt{1 - x^2}\, dx$

Solution (a). The graph of the integrand is the horizontal line $y = 2$, so the region is a rectangle of height 2 extending over the interval from 1 to 4 (Figure 6.5.4a). Thus,

$$\int_1^4 2\, dx = (\text{area of rectangle}) = 2(3) = 6$$

Solution (b). The graph of the integrand is the line $y = x + 2$, so the region is a trapezoid whose base extends from $x = -1$ to $x = 2$ (Figure 6.5.4b). Thus,

$$\int_{-1}^2 (x + 2)\, dx = (\text{area of trapezoid}) = \tfrac{1}{2}(1 + 4)(3) = \tfrac{15}{2}$$

Solution (c). The graph of $y = \sqrt{1 - x^2}$ is the upper semicircle of radius 1, centered at the origin, so the region is the right quarter-circle extending from $x = 0$ to $x = 1$ (Figure 6.5.4c). Thus,

$$\int_0^1 \sqrt{1 - x^2}\, dx = (\text{area of quarter-circle}) = \frac{1}{4}\pi(1^2) = \frac{\pi}{4} \qquad \blacktriangleleft$$

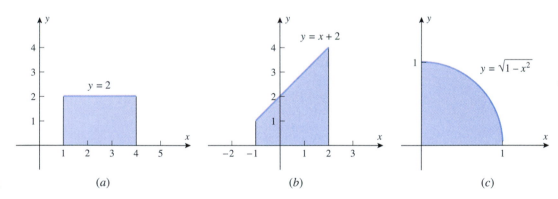

Figure 6.5.4 (a) (b) (c)

Example 4 Evaluate

(a) $\displaystyle\int_0^2 (x - 1)\, dx$ (b) $\displaystyle\int_0^1 (x - 1)\, dx$

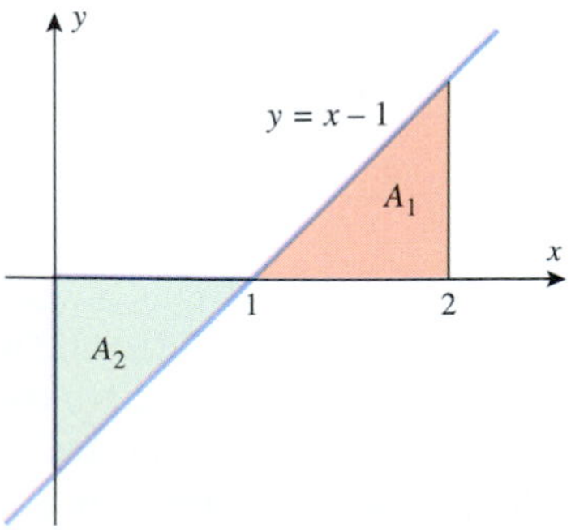

Solution. The graph of $y = x - 1$ is shown in Figure 6.5.5, and we leave it for you to verify that the shaded triangular regions both have area $\tfrac{1}{2}$. Over the interval $[0, 2]$ the net signed area is $A_1 - A_2 = \tfrac{1}{2} - \tfrac{1}{2} = 0$, and over the interval $[0, 1]$ the net signed area is $-A_2 = -\tfrac{1}{2}$. Thus,

$$\int_0^2 (x - 1)\, dx = 0 \quad \text{and} \quad \int_0^1 (x - 1)\, dx = -\tfrac{1}{2}$$

Figure 6.5.5

(Recall that in Example 8 of Section 6.4, we used Definition 6.4.5 to show that the net signed area between the graph of $y = x - 1$ and the interval $[0, 2]$ is 0.) ◄

PROPERTIES OF THE DEFINITE INTEGRAL

It is assumed in Definition 6.5.1 that $[a, b]$ is a finite closed interval with $a < b$, and hence the upper limit of integration in the definite integral is greater than the lower limit of integration. However, it will be convenient to extend this definition to allow for cases in which the upper and lower limits of integration are equal or the lower limit of integration is greater than the upper limit of integration. For this purpose we make the following special definitions.

6.5.3 DEFINITION.

(a) If a is in the domain of f, we define

$$\int_a^a f(x)\,dx = 0$$

(b) If f is integrable on $[a, b]$, then we define

$$\int_b^a f(x)\,dx = -\int_a^b f(x)\,dx$$

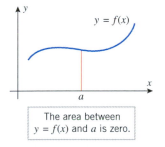

The area between $y = f(x)$ and a is zero.

Figure 6.5.6

REMARK. Part (a) of this definition is consistent with the intuitive idea that the area between a point on the x-axis and a curve $y = f(x)$ should be zero (Figure 6.5.6). Part (b) of the definition is simply a useful convention; it states that interchanging the limits of integration reverses the sign of the integral.

Example 5

(a) $\displaystyle\int_1^1 x^2\,dx = 0$

(b) $\displaystyle\int_1^0 \sqrt{1 - x^2}\,dx = -\int_0^1 \sqrt{1 - x^2}\,dx = -\frac{\pi}{4}$ ◄

Example 3(c)

Because definite integrals are defined as limits, they inherit many of the properties of limits. For example, we know that constants can be moved through limit signs and that the limit of a sum or difference is the sum or difference of the limits. Thus, you should not be surprised by the following theorem, which we state without formal proof.

6.5.4 THEOREM. *If f and g are integrable on $[a, b]$ and if c is a constant, then cf, $f + g$, and $f - g$ are integrable on $[a, b]$ and*

(a) $\displaystyle\int_a^b cf(x)\,dx = c\int_a^b f(x)\,dx$

(b) $\displaystyle\int_a^b [f(x) + g(x)]\,dx = \int_a^b f(x)\,dx + \int_a^b g(x)\,dx$

(c) $\displaystyle\int_a^b [f(x) - g(x)]\,dx = \int_a^b f(x)\,dx - \int_a^b g(x)\,dx$

Part (*b*) of this theorem can be extended to more than two functions. More precisely,

$$\int_a^b [f_1(x) + f_2(x) + \cdots + f_n(x)] \, dx$$

$$= \int_a^b f_1(x) \, dx + \int_a^b f_2(x) \, dx + \cdots + \int_a^b f_n(x) \, dx$$

Some properties of definite integrals can be motivated by interpreting the integral as an area. For example, if f is continuous and nonnegative on the interval $[a, b]$, and if c is a point between a and b, then the area under $y = f(x)$ over the interval $[a, b]$ can be split into two parts and expressed as the area under the graph from a to c plus the area under the graph from c to b (Figure 6.5.7), that is,

$$\int_a^b f(x) \, dx = \int_a^c f(x) \, dx + \int_c^b f(x) \, dx$$

This is a special case of the following theorem about definite integrals, which we state without proof.

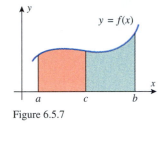
Figure 6.5.7

6.5.5   THEOREM.   *If f is integrable on a closed interval containing the three numbers a, b, and c, then*

$$\int_a^b f(x) \, dx = \int_a^c f(x) \, dx + \int_c^b f(x) \, dx$$

no matter how the numbers are ordered.

The following theorem, which we state without formal proof, can also be motivated by interpreting definite integrals as areas.

6.5.6   THEOREM.

(*a*)   *If f is integrable on $[a, b]$ and $f(x) \geq 0$ for all x in $[a, b]$, then*

$$\int_a^b f(x) \, dx \geq 0$$

(*b*)   *If f and g are integrable on $[a, b]$ and $f(x) \geq g(x)$ for all x in $[a, b]$, then*

$$\int_a^b f(x) \, dx \geq \int_a^b g(x) \, dx$$

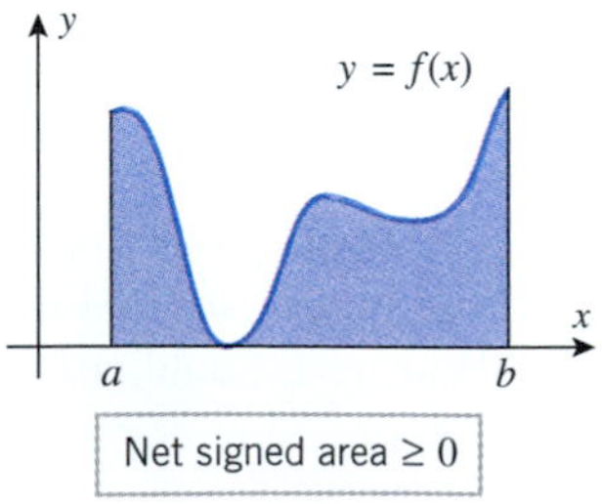

Geometrically, part (*a*) of this theorem states the obvious fact that if f is nonnegative on $[a, b]$, then the net signed area between the graph of f and the interval $[a, b]$ is also nonnegative (Figure 6.5.8). Part (*b*) has its simplest interpretation when f and g are nonnegative on $[a, b]$, in which case the theorem states that if the graph of f does not go below the graph of g, then the area under the graph of f is at least as large as the area under the graph of g (Figure 6.5.9).

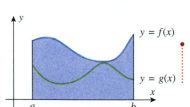

REMARK.   Part (*b*) of this theorem states that one can integrate both sides of the inequality $f(x) \geq g(x)$ without altering the sense of the inequality. We also note that in the case where $b > a$, both parts of the theorem remain true if $\geq$ is replaced by $\leq$, $>$, or $<$ throughout.

Example 6   Evaluate

$$\int_0^1 (5 - 3\sqrt{1 - x^2}) \, dx$$

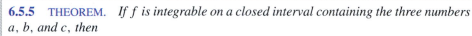
y = f(x)

Net signed area ≥ 0

Figure 6.5.8

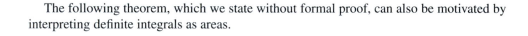
y = f(x)

y = g(x)

Area under $f \geq$ area under g

Figure 6.5.9

Solution. From parts (*a*) and (*c*) of Theorem 6.5.4 we can write

$$\int_0^1 (5 - 3\sqrt{1 - x^2})\, dx = \int_0^1 5\, dx - \int_0^1 3\sqrt{1 - x^2}\, dx = \int_0^1 5\, dx - 3\int_0^1 \sqrt{1 - x^2}\, dx$$

The first integral can be interpreted as the area of a rectangle of height 5 and base 1, so its value is 5, and from Example 3 the value of the second integral is $\pi/4$. Thus,

$$\int_0^1 (5 - 3\sqrt{1 - x^2})\, dx = 5 - 3\left(\frac{\pi}{4}\right) = 5 - \frac{3\pi}{4} \qquad \blacktriangleleft$$

DISCONTINUITIES AND INTEGRABILITY

The problem of determining when functions with discontinuities are integrable is quite complex and beyond the scope of this text. However, there are a few basic results about integrability that are important to know; we begin with a definition.

6.5.7 DEFINITION. A function f that is defined on an interval I is said to be **bounded** on I if there is a positive number M such that

$$-M \le f(x) \le M$$

for all x in the interval I. Geometrically, this means that the graph of f over the interval I lies between the lines $y = -M$ and $y = M$.

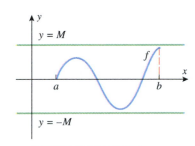

f is bounded on $[a, b]$.

Figure 6.5.10

For example, a continuous function f is bounded on *every* finite closed interval because the Extreme-Value Theorem (5.5.3) implies that f has an absolute maximum and an absolute minimum on the interval; hence, its graph will lie between the line $y = -M$ and $y = M$, provided we make M large enough (Figure 6.5.10). In contrast, a function that has a vertical asymptote inside of an interval is not bounded on that interval because its graph over the interval cannot be made to lie between the lines $y = -M$ and $y = M$, no matter how large we make the value of M (Figure 6.5.11).

The following theorem, which we state without proof, provides some facts about integrability for functions with discontinuities.

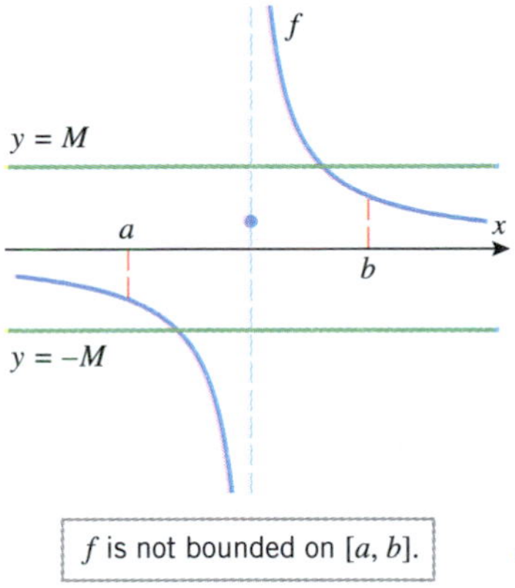

6.5.8 THEOREM. *Let f be a function that is defined on the finite closed interval $[a, b]$.*

(a) If f has finitely many discontinuities in $[a, b]$ but is bounded on $[a, b]$, then f is integrable on $[a, b]$.

(b) If f is not bounded on $[a, b]$, then f is not integrable on $[a, b]$.

---

f is not bounded on $[a, b]$.

Figure 6.5.11

FOR THE READER. Sketch the graph of a function over the interval $[0, 1]$ that has the properties stated in part (*a*) of this theorem.

EXERCISE SET 6.5

In Exercises 1–4, find the value of

(a) $\displaystyle\sum_{k=1}^{n} f(x_k^*)\Delta x_k$ (b) $\max \Delta x_k$.

1. $f(x) = x + 1$; $a = 0, b = 4$; $n = 3$;
   $\Delta x_1 = 1, \Delta x_2 = 1, \ \Delta x_3 = 2$;
   $x_1^* = \frac{1}{3}, x_2^* = \frac{3}{2}, x_3^* = 3$

2. $f(x) = \cos x$; $a = 0, b = 2\pi$; $n = 4$;
   $\Delta x_1 = \pi/2, \Delta x_2 = 3\pi/4, \ \Delta x_3 = \pi/2, \Delta x_4 = \pi/4$;
   $x_1^* = \pi/4, x_2^* = \pi, x_3^* = 3\pi/2, x_4^* = 7\pi/4$

3. $f(x) = 4 - x^2$; $a = -3, b = 4$; $n = 4$;
   $\Delta x_1 = 1, \Delta x_2 = 2, \Delta x_3 = 1, \Delta x_4 = 3$;
   $x_1^* = -\frac{5}{2}, x_2^* = -1, x_3^* = \frac{1}{4}, x_4^* = 3$

4. $f(x) = x^3$; $a = -3, b = 3$; $n = 4$;
   $\Delta x_1 = 2, \Delta x_2 = 1, \Delta x_3 = 1, \Delta x_4 = 2$;
   $x_1^* = -2, x_2^* = 0, x_3^* = 0, x_4^* = 2$

In Exercises 5–8, use the given values of a and b to express the following limits as definite integrals. (Do not evaluate the integrals.)

5. $\lim\limits_{\max \Delta x_k \to 0} \sum\limits_{k=1}^{n} (x_k^*)^2 \Delta x_k;\ a = -1, b = 2$

6. $\lim\limits_{\max \Delta x_k \to 0} \sum\limits_{k=1}^{n} (x_k^*)^3 \Delta x_k;\ a = 1, b = 2$

7. $\lim\limits_{\max \Delta x_k \to 0} \sum\limits_{k=1}^{n} 4x_k^*(1 - 3x_k^*) \Delta x_k;\ a = -3, b = 3$

8. $\lim\limits_{\max \Delta x_k \to 0} \sum\limits_{k=1}^{n} (\sin^2 x_k^*) \Delta x_k;\ a = 0, b = \pi/2$

In Exercises 9 and 10, use Definition 6.5.1 to express the integrals as limits of Riemann sums. Do not try to evaluate the integrals.

9. (a) $\displaystyle\int_1^2 2x\,dx$ **(b)** $\displaystyle\int_0^1 \frac{x}{x+1}\,dx$

10. (a) $\displaystyle\int_1^2 \sqrt{x}\,dx$ **(b)** $\displaystyle\int_{-\pi/2}^{\pi/2} (1 + \cos x)\,dx$

In Exercises 11–14, sketch the region whose signed area is represented by the definite integral, and evaluate the integral using an appropriate formula from geometry, where needed.

11. (a) $\displaystyle\int_0^3 x\,dx$ **(b)** $\displaystyle\int_{-2}^{-1} x\,dx$

(c) $\displaystyle\int_{-1}^4 x\,dx$ **(d)** $\displaystyle\int_{-5}^5 x\,dx$

12. (a) $\displaystyle\int_0^2 \left(1 - \tfrac{1}{2}x\right)dx$ **(b)** $\displaystyle\int_{-1}^1 \left(1 - \tfrac{1}{2}x\right)dx$

(c) $\displaystyle\int_2^3 \left(1 - \tfrac{1}{2}x\right)dx$ **(d)** $\displaystyle\int_0^3 \left(1 - \tfrac{1}{2}x\right)dx$

13. (a) $\displaystyle\int_0^5 2\,dx$ **(b)** $\displaystyle\int_0^\pi \cos x\,dx$

(c) $\displaystyle\int_{-1}^2 |2x - 3|\,dx$ **(d)** $\displaystyle\int_{-1}^1 \sqrt{1 - x^2}\,dx$

14. (a) $\displaystyle\int_{-10}^{-5} 6\,dx$ **(b)** $\displaystyle\int_{-\pi/3}^{\pi/3} \sin x\,dx$

(c) $\displaystyle\int_0^3 |x - 2|\,dx$ **(d)** $\displaystyle\int_0^2 \sqrt{4 - x^2}\,dx$

15. Use the areas shown in the accompanying figure to find

(a) $\displaystyle\int_a^b f(x)\,dx$ **(b)** $\displaystyle\int_b^c f(x)\,dx$

(c) $\displaystyle\int_a^c f(x)\,dx$ **(d)** $\displaystyle\int_a^d f(x)\,dx.$

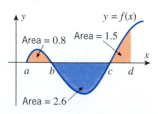

Figure Ex-15

16. In each part, evaluate the integral, given that
$$f(x) = \begin{cases} 2x, & x \le 1 \\ 2, & x > 1 \end{cases}$$

(a) $\displaystyle\int_0^1 f(x)\,dx$ **(b)** $\displaystyle\int_{-1}^1 f(x)\,dx$

(c) $\displaystyle\int_1^{10} f(x)\,dx$ **(d)** $\displaystyle\int_{1/2}^5 f(x)\,dx$

17. Find $\displaystyle\int_{-1}^2 [f(x) + 2g(x)]\,dx$ if
$$\int_{-1}^2 f(x)\,dx = 5 \quad \text{and} \quad \int_{-1}^2 g(x)\,dx = -3$$

18. Find $\displaystyle\int_1^4 [3f(x) - g(x)]\,dx$ if
$$\int_1^4 f(x)\,dx = 2 \quad \text{and} \quad \int_1^4 g(x)\,dx = 10$$

19. Find $\displaystyle\int_1^5 f(x)\,dx$ if
$$\int_0^1 f(x)\,dx = -2 \quad \text{and} \quad \int_0^5 f(x)\,dx = 1$$

20. Find $\displaystyle\int_3^{-2} f(x)\,dx$ if
$$\int_{-2}^1 f(x)\,dx = 2 \quad \text{and} \quad \int_1^3 f(x)\,dx = -6$$

In Exercises 21 and 22, use Theorem 6.5.4 and appropriate formulas from geometry to evaluate the integrals.

21. (a) $\displaystyle\int_0^1 (x + 2\sqrt{1 - x^2})\,dx$ **(b)** $\displaystyle\int_{-1}^3 (4 - 5x)\,dx$

22. (a) $\displaystyle\int_{-3}^0 (2 + \sqrt{9 - x^2})\,dx$ **(b)** $\displaystyle\int_{-2}^2 (1 - 3|x|)\,dx$

In Exercises 23 and 24, use Theorem 6.5.6 to determine whether the value of the integral is positive or negative.

23. (a) $\displaystyle\int_2^3 \frac{\sqrt{x}}{1 - x}\,dx$ **(b)** $\displaystyle\int_0^4 \frac{x^2}{3 - \cos x}\,dx$

24. (a) $\displaystyle\int_{-3}^{-1} \frac{x^4}{\sqrt{3 - x}}\,dx$ **(b)** $\displaystyle\int_{-2}^2 \frac{x^3 - 9}{|x| + 1}\,dx$

In Exercises 25 and 26, evaluate the integrals by completing the square and applying appropriate formulas from geometry.

25. $\displaystyle\int_0^{10} \sqrt{10x - x^2}\,dx$ **26.** $\displaystyle\int_0^3 \sqrt{6x - x^2}\,dx$

In Exercises 27 and 28, evaluate the limit over the interval $[a, b]$ by expressing it as a definite integral and applying an appropriate formula from geometry.

27. $\displaystyle\lim_{\max \Delta x_k \to 0} \sum_{k=1}^{n} (3x_k^* + 1)\Delta x_k; \ a = 0, b = 1$

28. $\displaystyle\lim_{\max \Delta x_k \to 0} \sum_{k=1}^{n} \sqrt{4 - (x_k^*)^2} \, \Delta x_k; \ a = -2, b = 2$

29. In each part, use Theorems 6.5.2 and 6.5.8 to determine whether the function f is integrable on the interval $[-1, 1]$.
    (a) $f(x) = \cos x$
    (b) $f(x) = \begin{cases} x/|x|, & x \neq 0 \\ 0, & x = 0 \end{cases}$
    (c) $f(x) = \begin{cases} 1/x^2, & x \neq 0 \\ 0, & x = 0 \end{cases}$
    (d) $f(x) = \begin{cases} \sin 1/x, & x \neq 0 \\ 0, & x = 0 \end{cases}$

30. It can be shown that every interval contains both rational and irrational numbers. Accepting this to be so, do you believe that the function
$$f(x) = \begin{cases} 1 & \text{if} \quad x \text{ is rational} \\ 0 & \text{if} \quad x \text{ is irrational} \end{cases}$$
is integrable on a closed interval $[a, b]$? Explain your reasoning.

31. It can be shown that the limit in Definition 6.5.1 has all of the limit properties stated in Theorem 2.2.2. Accepting this to be so, show that
    (a) $\displaystyle\int_a^b cf(x)\,dx = c\int_a^b f(x)\,dx$
    (b) $\displaystyle\int_a^b [f(x) + g(x)]\,dx = \int_a^b f(x)\,dx + \int_a^b g(x)\,dx$

32. Find the smallest and largest values that the Riemann sum
$$\sum_{k=1}^{3} f(x_k^*)\Delta x_k$$

can have on the interval $[0, 4]$ if $f(x) = x^2 - 3x + 4$ and $\Delta x_1 = 1$, $\Delta x_2 = 2$, $\Delta x_3 = 1$.

33. The function $f(x) = \sqrt{x}$ is continuous on $[0, 4]$ and therefore integrable on this interval. Evaluate
$$\int_0^4 \sqrt{x}\,dx$$
by using Definition 6.5.1. Use subintervals of unequal length given by the partition
$$0 < 4(1)^2/n^2 < 4(2)^2/n^2 < \cdots < 4(n-1)^2/n^2 < 4$$
and let x_k^* be the right endpoint of the kth subinterval.

34. Suppose that f is defined on the interval $[a, b]$ and that $f(x) = 0$ for $a < x \leq b$. Use Definition 6.5.1 to prove that
$$\int_a^b f(x)\,dx = 0$$

35. Suppose that g is a continuous function on the interval $[a, b]$ and that f is a function defined on $[a, b]$ with $f(x) = g(x)$ for $a < x \leq b$. Prove that
$$\int_a^b f(x)\,dx = \int_a^b g(x)\,dx$$
[*Hint:* Write
$$\int_a^b f(x)\,dx = \int_a^b [(f(x) - g(x)) + g(x)]\,dx$$
and use the result of Exercise 34 along with Theorem 6.5.4(b).]

36. Define the function f by $f(x) = 1/x$, $x \neq 0$ and $f(0) = 0$. It follows from Theorem 6.5.8(b) that f is not integrable on the interval $[0, 1]$. Prove this to be the case by applying Definition 6.5.1. [*Hint:* Argue that no matter how small the mesh size is for a partition of $[0, 1]$, there will always be a choice of x_1^* that will make the Riemann sum in Definition 6.5.1 as large as we like.]

6.6 THE FUNDAMENTAL THEOREM OF CALCULUS

In this section we will establish two basic relationships between definite and indefinite integrals that together constitute a result called the Fundamental Theorem of Calculus. One part of this theorem will relate the rectangle and antiderivative methods for calculating areas, and the second part will provide a powerful method for evaluating definite integrals using antiderivatives.

THE FUNDAMENTAL THEOREM OF CALCULUS

As in earlier sections, let us begin by assuming that f is nonnegative and continuous on an interval $[a, b]$, in which case the area A under the graph of f over the interval $[a, b]$ is represented by the definite integral
$$A = \int_a^b f(x)\,dx \tag{1}$$
(Figure 6.6.1).

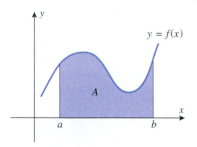

Figure 6.6.1

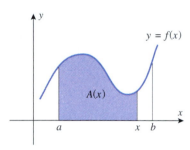

Figure 6.6.2

Recall that our discussion of the antiderivative method in Section 6.1 suggested that if $A(x)$ is the area under the graph of f from a to x (Figure 6.6.2), then

- $A'(x) = f(x)$
- $A(a) = 0$ The area under the curve from a to a is the area above the single point a, and hence is zero.
- $A(b) = A$ The area under the curve from a to b is A.

The formula $A'(x) = f(x)$ states that $A(x)$ is an antiderivative of $f(x)$, which implies that every other antiderivative of $f(x)$ on $[a, b]$ can be obtained by adding a constant to $A(x)$. Accordingly, let

$$F(x) = A(x) + C$$

be any antiderivative of $f(x)$, and consider what happens when we subtract $F(a)$ from $F(b)$:

$$F(b) - F(a) = [A(b) + C] - [A(a) + C] = A(b) - A(a) = A - 0 = A$$

Hence (1) can be expressed as

$$\int_a^b f(x)\,dx = F(b) - F(a)$$

In words, this equation states:

> *The definite integral can be evaluated by finding any antiderivative of the integrand and then subtracting the value of this antiderivative at the lower limit of integration from its value at the upper limit of integration.*

Although our evidence for this result assumed that f is nonnegative on $[a, b]$, this assumption is not essential.

6.6.1   THEOREM (*The Fundamental Theorem of Calculus, Part 1*).   *If f is continuous on $[a, b]$ and F is any antiderivative of f on $[a, b]$, then*

$$\int_a^b f(x)\,dx = F(b) - F(a) \tag{2}$$

Proof.   Let $x_1, x_2, \ldots, x_{n-1}$ be any numbers in $[a, b]$ such that

$$a < x_1 < x_2 < \cdots < x_{n-1} < b$$

These values divide $[a, b]$ into n subintervals

$$[a, x_1], [x_1, x_2], \ldots, [x_{n-1}, b] \tag{3}$$

whose lengths, as usual, we denote by

$$\Delta x_1, \Delta x_2, \ldots, \Delta x_n$$

By hypothesis, $F'(x) = f(x)$ for all x in $[a, b]$, so F satisfies the hypotheses of the Mean-Value Theorem (5.8.2) on each subinterval in (3). Hence, we can find numbers $x_1^*, x_2^*, \ldots, x_n^*$ in the respective subintervals in (3) such that

$$F(x_1) - F(a) = F'(x_1^*)(x_1 - a) = f(x_1^*)\Delta x_1$$
$$F(x_2) - F(x_1) = F'(x_2^*)(x_2 - x_1) = f(x_2^*)\Delta x_2$$
$$F(x_3) - F(x_2) = F'(x_3^*)(x_3 - x_2) = f(x_3^*)\Delta x_3$$
$$\vdots$$
$$F(b) - F(x_{n-1}) = F'(x_n^*)(b - x_{n-1}) = f(x_n^*)\Delta x_n$$

Adding the preceding equations yields

$$F(b) - F(a) = \sum_{k=1}^{n} f(x_k^*)\Delta x_k \qquad (4)$$

Let us now increase n in such a way that max $\Delta x_k \to 0$. Since f is assumed to be continuous, the right side of (4) approaches $\int_a^b f(x)\, dx$ by Theorem 6.5.2 and Definition 6.5.1. However, the left side of (4) is independent of n; that is, the left side of (4) remains constant as n increases. Thus,

$$F(b) - F(a) = \lim_{\max \Delta x_k \to 0} \sum_{k=1}^{n} f(x_k^*)\Delta x_k = \int_a^b f(x)\, dx \qquad \blacksquare$$

It is standard to denote the difference $F(b) - F(a)$ as

$$F(x)\Big]_a^b = F(b) - F(a) \quad \text{or} \quad \left[F(x)\right]_a^b = F(b) - F(a)$$

For example, using the first of these notations we can express (2) as

$$\int_a^b f(x)\, dx = F(x)\Bigg]_a^b \qquad (5)$$

Example 1 Evaluate $\int_1^2 x\, dx$.

Solution. The function $F(x) = \frac{1}{2}x^2$ is an antiderivative of $f(x) = x$; thus, from (2)

$$\int_1^2 x\, dx = \frac{1}{2}x^2\Bigg]_1^2 = \frac{1}{2}(2)^2 - \frac{1}{2}(1)^2 = 2 - \frac{1}{2} = \frac{3}{2} \qquad \blacktriangleleft$$

Example 2 In Example 5 of Section 6.4 we approximated the area under the graph of $y = 9 - x^2$ over the interval $[0, 3]$ using left endpoint, right endpoint, and midpoint approximations, all of which produced an approximation of roughly 18 (square units). In Example 7 of that section we used Definition 6.4.3 to prove that the exact area A is indeed 18. We can now solve this problem more quickly using the Fundamental Theorem of Calculus:

$$A = \int_0^3 (9 - x^2)\, dx = 9x - \frac{x^3}{3}\Bigg]_0^3 = \left(27 - \frac{27}{3}\right) - 0 = 18 \qquad \blacktriangleleft$$

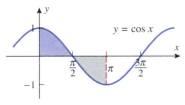

Figure 6.6.3

Example 3

(a) Find the area under the curve $y = \cos x$ over the interval $[0, \pi/2]$ (Figure 6.6.3).

(b) Make a conjecture about the value of the integral

$$\int_0^\pi \cos x\, dx$$

and confirm your conjecture using the Fundamental Theorem of Calculus.

Solution (a). Since $\cos x \geq 0$ over the interval $[0, \pi/2]$, the area A under the curve is

$$A = \int_0^{\pi/2} \cos x\, dx = \sin x\Bigg]_0^{\pi/2} = \sin\frac{\pi}{2} - \sin 0 = 1$$

Solution (b). The given integral can be interpreted as the signed area between the graph of $y = \cos x$ and the interval $[0, \pi]$. The graph in Figure 6.6.3 suggests that over the interval $[0, \pi]$ the portion of area above the x-axis is the same as the portion of area below the x-axis, so we conjecture that the signed area is zero; this implies that the value of the integral is zero. This is confirmed by the computations

$$\int_0^\pi \cos x \, dx = \sin x \Big]_0^\pi = \sin \pi - \sin 0 = 0 \qquad \blacktriangleleft$$

THE RELATIONSHIP BETWEEN DEFINITE AND INDEFINITE INTEGRALS

Observe that in the preceding examples we did not include a constant of integration in the antiderivatives. In general, when applying the Fundamental Theorem of Calculus there is no need to include a constant of integration because it will drop out anyway. To see that this is so, let F be any antiderivative of the integrand on $[a, b]$, and let C be any constant; then

$$\int_a^b f(x) \, dx = F(x) + C \Big]_a^b = [F(b) + C] - [F(a) + C] = F(b) - F(a)$$

Thus, for purposes of evaluating a definite integral we can omit the constant of integration in

$$\int_a^b f(x) \, dx = F(x) + C \Big]_a^b$$

and express (5) as

$$\int_a^b f(x) \, dx = \left[\int f(x) \, dx \right]_a^b \qquad (6)$$

which relates the definite and indefinite integrals.

Example 4

$$\int_1^9 \sqrt{x} \, dx = \int \sqrt{x} \, dx \Big]_1^9 = \int x^{1/2} \, dx \Big]_1^9 = \frac{2}{3} x^{3/2} \Big]_1^9 = \frac{2}{3}(27 - 1) = \frac{52}{3} \qquad \blacktriangleleft$$

REMARK. Usually, we will dispense with the step of displaying the indefinite integral explicitly and write the antiderivative immediately, as in our first three examples.

Example 5 Table 6.2.1 will be helpful for the following computations.

$$\int_4^9 x^2 \sqrt{x} \, dx = \int_4^9 x^{5/2} \, dx = \frac{2}{7} x^{7/2} \Big]_4^9 = \frac{2}{7}(2187 - 128) = \frac{4118}{7} = 588\frac{2}{7}$$

$$\int_0^{\pi/2} \frac{\sin x}{5} \, dx = -\frac{\cos x}{5} \Big]_0^{\pi/2} = -\frac{1}{5} \left[\cos\left(\frac{\pi}{2}\right) - \cos 0 \right] = -\frac{1}{5}[0 - 1] = \frac{1}{5}$$

$$\int_0^{\pi/3} \sec^2 x \, dx = \tan x \Big]_0^{\pi/3} = \tan\left(\frac{\pi}{3}\right) - \tan 0 = \sqrt{3} - 0 = \sqrt{3}$$

$$\int_0^{\ln 3} 5e^x \, dx = 5e^x \Big]_0^{\ln 3} = 5[e^{\ln 3} - e^0] = 5[3 - 1] = 10$$

$$\int_{-e}^{-1} \frac{1}{x} \, dx = \ln|x| \Big]_{-e}^{-1} = \ln|-1| - \ln|-e| = 0 - 1 = -1$$

$$\int_{-1/2}^{1/2} \frac{1}{\sqrt{1-x^2}} \, dx = \sin^{-1} x \Big]_{-1/2}^{1/2} = \sin^{-1}\left(\frac{1}{2}\right) - \sin^{-1}\left(-\frac{1}{2}\right) = \frac{\pi}{6} - \left(-\frac{\pi}{6}\right) = \frac{\pi}{3} \qquad \blacktriangleleft$$

WARNING. The requirements in the Fundamental Theorem of Calculus that f be continuous on $[a, b]$ and that F be an antiderivative for f over the entire interval $[a, b]$ are important to keep in mind. Disregarding these assumptions will likely lead to incorrect results. For example, the function $f(x) = 1/x^2$ fails on two counts to be continuous at $x = 0$: $f(x)$ is not defined at $x = 0$ and $\lim_{x \to 0} f(x)$ does not exist. Thus, the Fundamental Theorem of Calculus should not be used to integrate f on any interval that contains $x = 0$. However, if we ignore this and blindly apply Formula (2) over the interval $[-1, 1]$, we might think that

$$\cdot \text{WRONG} \cdot \text{WRONG} \cdot \text{WRONG} \cdot \text{WRONG} \cdot \text{W}$$
$$\int_{-1}^{1} \frac{1}{x^2}\, dx = -\frac{1}{x}\Bigg]_{-1}^{1} = -[1 - (-1)] = -2$$
$$\text{G} \cdot \text{WRONG} \cdot \text{WRONG} \cdot \text{WRONG} \cdot \text{WRONG} \cdot$$

This answer is clearly ridiculous, since $f(x) = 1/x^2$ is a nonnegative function and hence cannot possibly produce a negative definite integral. Indeed, even if we were to extend f to be defined at 0, say by setting

$$f(x) = \begin{cases} 1/x^2, & x \neq 0 \\ 0, & x = 0 \end{cases}$$

f would still be unbounded on any interval containing $x = 0$, so Theorem 6.5.8(*b*) tells us that f is not even integrable across any such interval.

FOR THE READER. If you have a CAS, read the documentation on evaluating definite integrals, and then check the results in the preceding examples.

The Fundamental Theorem of Calculus can be applied without modification to definite integrals in which the lower limit of integration is greater than or equal to the upper limit of integration.

Example 6

$$\int_{1}^{1} x^2\, dx = \frac{x^3}{3}\Bigg]_{1}^{1} = \frac{1}{3} - \frac{1}{3} = 0$$

$$\int_{4}^{0} x\, dx = \frac{x^2}{2}\Bigg]_{4}^{0} = \frac{0}{2} - \frac{16}{2} = -8$$

The latter result is consistent with the result that would be obtained by first reversing the limits of integration in accordance with Definition 6.5.3(b):

$$\int_{4}^{0} x\, dx = -\int_{0}^{4} x\, dx = -\frac{x^2}{2}\Bigg]_{0}^{4} = -\left[\frac{16}{2} - \frac{0}{2}\right] = -8 \qquad \blacktriangleleft$$

To integrate a continuous function that is defined piecewise on an interval $[a, b]$, split this interval into subintervals at the breakpoints of the function, and integrate separately over each subinterval in accordance with Theorem 6.5.5.

Example 7 Evaluate $\int_{0}^{6} f(x)\, dx$ if

$$f(x) = \begin{cases} x^2, & x < 2 \\ 3x - 2, & x \geq 2 \end{cases}$$

Solution. From Theorem 6.5.5

$$\int_{0}^{6} f(x)\, dx = \int_{0}^{2} f(x)\, dx + \int_{2}^{6} f(x)\, dx = \int_{0}^{2} x^2\, dx + \int_{2}^{6} (3x - 2)\, dx$$

$$= \frac{x^3}{3}\Bigg]_{0}^{2} + \left[\frac{3x^2}{2} - 2x\right]_{2}^{6} = \left(\frac{8}{3} - 0\right) + (42 - 2) = \frac{128}{3} \qquad \blacktriangleleft$$

Example 8   Evaluate $\displaystyle\int_{-1}^{2} |x|\, dx$.

Solution.   Since $|x| = x$ when $x \geq 0$ and $|x| = -x$ when $x \leq 0$,

$$\int_{-1}^{2} |x|\, dx = \int_{-1}^{0} |x|\, dx + \int_{0}^{2} |x|\, dx$$

$$= \int_{-1}^{0} (-x)\, dx + \int_{0}^{2} x\, dx$$

$$= -\frac{x^2}{2}\Bigg]_{-1}^{0} + \frac{x^2}{2}\Bigg]_{0}^{2} = \frac{1}{2} + 2 = \frac{5}{2} \qquad \blacktriangleleft$$

DUMMY VARIABLES

To evaluate a definite integral using the Fundamental Theorem of Calculus, one needs to be able to find an antiderivative of the integrand; thus, it is important to know what kinds of functions have antiderivatives. It is our next objective to show that all continuous functions have antiderivatives, but to do this we will need some preliminary results.

Formula (6) shows that there is a close relationship between the integrals

$$\int_{a}^{b} f(x)\, dx \quad \text{and} \quad \int f(x)\, dx$$

However, the definite and indefinite integrals differ in some important ways. For one thing, the two integrals are different kinds of objects—the definite integral is a *number* (the net signed area between the graph of $y = f(x)$ and the interval $[a, b]$), whereas the indefinite integral is a *function*, or more accurately a set of functions [the antiderivatives of $f(x)$]. However, the two types of integrals also differ in the role played by the variable of integration. In an indefinite integral, the variable of integration is "passed through" to the antiderivative in the sense that integrating a function of x produces a function of x, integrating a function of t produces a function of t, and so forth. For example,

$$\int x^2\, dx = \frac{x^3}{3} + C \quad \text{and} \quad \int t^2\, dt = \frac{t^3}{3} + C$$

In contrast, the variable of integration in a definite integral is not passed through to the end result, since the end result is a number. Thus, integrating a function of x over an interval and integrating the same function of t over the same interval of integration produce the same value for the integral. For example,

$$\int_{1}^{3} x^2\, dx = \frac{x^3}{3}\Bigg]_{x=1}^{3} = \frac{27}{3} - \frac{1}{3} = \frac{26}{3} \quad \text{and} \quad \int_{1}^{3} t^2\, dt = \frac{t^3}{3}\Bigg]_{t=1}^{3} = \frac{27}{3} - \frac{1}{3} = \frac{26}{3}$$

However, this latter result should not be surprising, since the area under the graph of the curve $y = f(x)$ over an interval $[a, b]$ on the x-axis is the same as the area under the graph of the curve $y = f(t)$ over the interval $[a, b]$ on the t-axis (Figure 6.6.4).

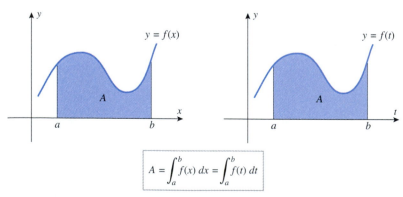

$$A = \int_{a}^{b} f(x)\, dx = \int_{a}^{b} f(t)\, dt$$

Figure 6.6.4

Because the variable of integration in a definite integral plays no role in the end result, it is often referred to as a ***dummy variable***. In summary:

Whenever you find it convenient to change the letter used for the variable of integration in a definite integral, you can do so without changing the value of the integral.

THE MEAN-VALUE THEOREM FOR INTEGRALS

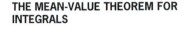

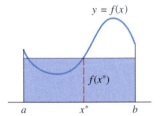

Figure 6.6.5

To reach our goal of showing that continuous functions have antiderivatives, we will need to develop a basic property of definite integrals, known as the *Mean-Value Theorem for Integrals*. In the next section we will use this theorem to extend the familiar idea of "average value" so that it applies to continuous functions, but here we will need it as a tool for developing other results.

Let f be a continuous nonnegative function on $[a, b]$, and let m and M be the minimum and maximum values of $f(x)$ on this interval. Consider the rectangles of heights m and M over the interval $[a, b]$ (Figure 6.6.5). It is clear geometrically from this figure that the area

$$A = \int_a^b f(x)\, dx$$

under $y = f(x)$ is at least as large as the area of the rectangle of height m and no larger than the area of the rectangle of height M. It seems reasonable, therefore, that there is a rectangle over the interval $[a, b]$ of some appropriate height $f(x^*)$ between m and M whose area is precisely A; that is,

$$\int_a^b f(x)\, dx = f(x^*)(b - a)$$

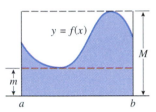

Figure 6.6.6

(Figure 6.6.6). This is a special case of the following result.

6.6.2 THEOREM (*The Mean-Value Theorem for Integrals*). *If f is continuous on a closed interval $[a, b]$, then there is at least one number x^* in $[a, b]$ such that*

$$\int_a^b f(x)\, dx = f(x^*)(b - a) \tag{7}$$

Proof. By the Extreme-Value Theorem (5.5.3), f assumes a maximum value M and a minimum value m on $[a, b]$. Thus, for all x in $[a, b]$,

$$m \le f(x) \le M$$

and from Theorem 6.5.6(b)

$$\int_a^b m\, dx \le \int_a^b f(x)\, dx \le \int_a^b M\, dx$$

or

$$m(b - a) \le \int_a^b f(x)\, dx \le M(b - a) \tag{8}$$

or

$$m \le \frac{1}{b - a} \int_a^b f(x)\, dx \le M$$

This implies that

$$\frac{1}{b - a} \int_a^b f(x)\, dx \tag{9}$$

is a number between m and M, and since $f(x)$ assumes the values m and M on $[a, b]$, it follows from the Intermediate-Value Theorem (2.5.8) that $f(x)$ must assume the value (9) at some x^* in $[a, b]$; that is,

$$\frac{1}{b-a} \int_a^b f(x)\,dx = f(x^*) \quad \text{or} \quad \int_a^b f(x)\,dx = f(x^*)(b-a)$$ ∎

Example 9 Since $f(x) = x^2$ is continuous on the interval $[1, 4]$, the Mean-Value Theorem for Integrals guarantees that there is a number x^* in $[1, 4]$ such that

$$\int_1^4 x^2\,dx = f(x^*)(4 - 1) = (x^*)^2(4 - 1) = 3(x^*)^2$$

But

$$\int_1^4 x^2\,dx = \frac{x^3}{3}\Bigg]_1^4 = 21$$

so that

$$3(x^*)^2 = 21 \quad \text{or} \quad (x^*)^2 = 7 \quad \text{or} \quad x^* = \pm\sqrt{7}$$

Thus, $x^* = \sqrt{7} \approx 2.65$ is the number in the interval $[1, 4]$ whose existence is guaranteed by the Mean-Value Theorem for Integrals. ◀

PART 2 OF THE FUNDAMENTAL THEOREM OF CALCULUS

In Section 6.1 we suggested that if f is continuous and nonnegative on $[a, b]$, and if $A(x)$ is the area under the graph of $y = f(x)$ over the interval $[a, x]$ (Figure 6.2.2), then $A'(x) = f(x)$. But $A(x)$ can be expressed as the definite integral

$$A(x) = \int_a^x f(t)\,dt$$

(where we have used t rather than x as the variable of integration to avoid confusion with the x that appears as the upper limit of integration). Thus, the relationship $A'(x) = f(x)$ can be expressed as

$$\frac{d}{dx}\left[\int_a^x f(t)\,dt\right] = f(x)$$

This is a special case of the following more general result, which applies even if f has negative values.

6.6.3 THEOREM (*The Fundamental Theorem of Calculus, Part 2*). *If f is continuous on an interval I, then f has an antiderivative on I. In particular, if a is any number in I, then the function F defined by*

$$F(x) = \int_a^x f(t)\,dt$$

is an antiderivative of f on I; that is, $F'(x) = f(x)$ for each x in I, or in an alternative notation

$$\frac{d}{dx}\left[\int_a^x f(t)\,dt\right] = f(x) \tag{10}$$

Proof. We will show first that $F(x)$ is defined at each x in the interval I. If $x > a$ and x is in the interval I, then Theorem 6.5.2 applied to the interval $[a, x]$ and the continuity of f on I ensure that $F(x)$ is defined; and if x is in the interval I and $x \leq a$, then Definition 6.5.3 combined with Theorem 6.5.2 ensures that $F(x)$ is defined. Thus, $F(x)$ is defined for all x in I.

Next we will show that $F'(x) = f(x)$ for each x in the interval I. If x is not an endpoint of I, then it follows from the definition of a derivative that

$$F'(x) = \lim_{w \to x} \frac{F(w) - F(x)}{w - x}$$

$$= \lim_{w \to x} \left(\frac{1}{w - x} \left[\int_a^w f(t)\,dt - \int_a^x f(t)\,dt \right] \right)$$

$$= \lim_{w \to x} \left(\frac{1}{w - x} \left[\int_a^w f(t)\,dt + \int_x^a f(t)\,dt \right] \right)$$

$$= \lim_{w \to x} \left(\frac{1}{w - x} \int_x^w f(t)\,dt \right) \tag{11}$$

Applying the Mean-Value Theorem for Integrals (6.6.2) to $\int_x^w f(t)\,dt$, we obtain

$$\frac{1}{w - x} \int_x^w f(t)\,dt = \frac{1}{w - x}[f(t^*) \cdot (w - x)] = f(t^*) \tag{12}$$

where t^* is some number between x and w. Because t^* is between x and w, it follows that $t^* \to x$ as $w \to x$. Thus $f(t^*) \to f(x)$ as $w \to x$, since f is assumed continuous at x. Therefore, it follows from (11) and (12) that

$$F'(x) = \lim_{w \to x} \left(\frac{1}{w - x} \int_x^w f(t)\,dt \right) = \lim_{w \to x} f(t^*) = f(x)$$

If x is an endpoint of the interval I, then the two-sided limits in the proof must be replaced by the appropriate one-sided limits, but otherwise the arguments are identical. ∎

In words, Formula (10) states:

If a definite integral has a variable upper limit of integration, a constant lower limit of integration, and a continuous integrand, then the derivative of the integral with respect to its upper limit is equal to the integrand evaluated at the upper limit.

Example 10   Find

$$\frac{d}{dx} \left[\int_1^x t^3\,dt \right]$$

by applying Part 2 of the Fundamental Theorem of Calculus, and then confirm the result by performing the integration and then differentiating.

Solution.   The integrand is a continuous function, so from (10)

$$\frac{d}{dx} \left[\int_1^x t^3\,dt \right] = x^3$$

Alternatively, evaluating the integral and then differentiating yields

$$\int_1^x t^3\,dt = \frac{t^4}{4} \Bigg]_{t=1}^x = \frac{x^4}{4} - \frac{1}{4}, \quad \frac{d}{dx}\left[\frac{x^4}{4} - \frac{1}{4} \right] = x^3$$

so the two methods for differentiating the integral agree.   ◄

Example 11   Since

$$f(x) = \frac{\sin x}{x}$$

is continuous on any interval that does not contain the origin, it follows from (10) that on

the interval $(0, +\infty)$ we have

$$\frac{d}{dx}\left[\int_1^x \frac{\sin t}{t}\, dt\right] = \frac{\sin x}{x}$$

Unlike the preceding example, there is no way to evaluate the integral in terms of familiar functions, so Formula (10) provides the only simple method for finding the derivative. ◀

DIFFERENTIATION AND INTEGRATION ARE INVERSE PROCESSES

The two parts of the Fundamental Theorem of Calculus, when taken together, tell us that differentiation and integration are inverse processes in the sense that each undoes the effect of the other. To see why this is so, note that Part 1 of the Fundamental Theorem of Calculus (6.6.1) implies that

$$\int_a^x f'(t)\, dt = f(x) - f(a)$$

which tells us that if the value of $f(a)$ is known, then the function f can be recovered from its derivative f' by integrating. Conversely, Part 2 of the Fundamental Theorem of Calculus (6.6.3) states that

$$\frac{d}{dx}\left[\int_a^x f(t)\, dt\right] = f(x)$$

which tells us that the function f can be recovered from its integral by differentiating. Thus, differentiation and integration can be viewed as inverse processes.

It is common to treat parts 1 and 2 of the Fundamental Theorem of Calculus as a single theorem, and refer to it simply as the *Fundamental Theorem of Calculus*. This theorem ranks as one of the greatest discoveries in the history of science, and its formulation by Newton and Leibniz is generally regarded to be the "discovery of calculus."

EXERCISE SET 6.6   ~ Graphing Utility   C CAS

1. In each part, use a definite integral to find the area of the region, and check your answer using an appropriate formula from geometry.

(a)   (b)   (c)

2. In each part, use a definite integral to find the area under the curve $y = f(x)$ over the stated interval, and check your answer using an appropriate formula from geometry.
   (a) $f(x) = x$; $[0, 5]$
   (b) $f(x) = 5$; $[3, 9]$
   (c) $f(x) = x + 3$; $[-1, 2]$

In Exercises 3–8, find the area under the curve $y = f(x)$ over the stated interval.

3. $f(x) = x^3$; $[2, 3]$
4. $f(x) = x^4$; $[-1, 1]$
5. $f(x) = \sqrt{x}$; $[1, 9]$
6. $f(x) = x^{-3/5}$; $[1, 4]$
7. $f(x) = e^x$; $[1, 3]$
8. $f(x) = \dfrac{1}{x}$; $[1, 5]$

In Exercises 9–27, evaluate the integrals using Part 1 of the Fundamental Theorem of Calculus.

9. $\displaystyle\int_{-3}^0 (x^2 - 4x + 7)\, dx$

10. $\displaystyle\int_{-1}^2 x(1 + x^3)\, dx$

11. $\displaystyle\int_1^3 \frac{1}{x^2}\, dx$

12. $\displaystyle\int_1^2 \frac{1}{x^6}\, dx$

13. $\displaystyle\int_4^9 2x\sqrt{x}\, dx$

14. $\displaystyle\int_1^8 (5x^{2/3} - 4x^{-2})\, dx$

15. $\displaystyle\int_{-\pi/2}^{\pi/2} \sin\theta\, d\theta$

16. $\displaystyle\int_0^{\pi/4} \sec^2\theta\, d\theta$

17. $\displaystyle\int_{-\pi/4}^{\pi/4} \cos x\, dx$

18. $\displaystyle\int_0^1 (x - \sec x \tan x)\, dx$

19. $\displaystyle\int_{\ln 2}^3 5e^x\, dx$

20. $\displaystyle\int_{1/2}^1 \frac{1}{2x}\, dx$

21. $\displaystyle\int_0^{1/\sqrt{2}} \frac{dx}{\sqrt{1 - x^2}}$

22. $\displaystyle\int_{-1}^1 \frac{dx}{1 + x^2}$

23. $\displaystyle\int_{\sqrt{2}}^2 \frac{dx}{x\sqrt{x^2 - 1}}$

24. $\displaystyle\int_{-\sqrt{2}}^{-2/\sqrt{3}} \frac{dx}{x\sqrt{x^2 - 1}}$

25. $\displaystyle\int_1^4 \left(\frac{3}{\sqrt{t}} - 5\sqrt{t} - t^{-3/2}\right) dt$

26. $\displaystyle\int_4^9 (4y^{-1/2} + 2y^{1/2} + y^{-5/2})\,dy$

27. $\displaystyle\int_{\pi/6}^{\pi/2} \left(x + \frac{2}{\sin^2 x}\right)dx$

c 28. Use a CAS to evaluate the integral

$$\int_a^{4a} (a^{1/2} - x^{1/2})\,dx$$

and check the answer by hand.

In Exercises 29–32, use Theorem 6.5.5 to evaluate the given integrals.

29. (a) $\displaystyle\int_0^2 |2x - 3|\,dx$ **(b)** $\displaystyle\int_0^{3\pi/4} |\cos x|\,dx$

30. (a) $\displaystyle\int_{-1}^2 \sqrt{2 + |x|}\,dx$ **(b)** $\displaystyle\int_0^{\pi/2} \left|\tfrac{1}{2} - \sin x\right|\,dx$

31. (a) $\displaystyle\int_{-1}^1 |e^x - 1|\,dx$ **(b)** $\displaystyle\int_1^4 \frac{|2-x|}{x}\,dx$

32. (a) $\displaystyle\int_{-3}^3 \left|x^2 - 1 - \frac{15}{x^2+1}\right|dx$

(b) $\displaystyle\int_0^{\sqrt{3}/2} \left|\frac{1}{\sqrt{1-x^2}} - \sqrt{2}\right|dx$

c 33. (a) CAS programs provide methods for entering functions that are defined piecewise. Check your documentation to see how this is done, and then use the CAS to evaluate

$$\int_0^2 f(x)\,dx, \quad \text{where} \quad f(x) = \begin{cases} x, & x \le 1 \\ x^2, & x > 1 \end{cases}$$

Use Theorem 6.5.5 to check the answer by hand.
(b) Find a formula for an antiderivative F of f on the interval $[0, 4]$ and verify that

$$\int_0^2 f(x)\,dx = F(2) - F(0)$$

c 34. (a) Use a CAS to evaluate

$$\int_0^4 f(x)\,dx, \quad \text{where} \quad f(x) = \begin{cases} \sqrt{x}, & 0 \le x < 1 \\ 1/x^2, & x \ge 1 \end{cases}$$

Use Theorem 6.5.5 to check the answer by hand.
(b) Find a formula for an antiderivative F of f on the interval $[0, 4]$ and verify that

$$\int_0^4 f(x)\,dx = F(4) - F(0)$$

In Exercises 35–38, use a calculating utility to find the midpoint approximation of the integral using $n = 20$ subintervals, and then find the exact value of the integral using Part 1 of the Fundamental Theorem of Calculus.

35. $\displaystyle\int_1^3 \frac{1}{x^2}\,dx$ **36.** $\displaystyle\int_0^{\pi/2} \sin x\,dx$

37. $\displaystyle\int_{-1}^1 \sec^2 x\,dx$ **38.** $\displaystyle\int_1^3 \frac{1}{x}\,dx$

39. Find the area under the curve $y = x^2 + 1$ over the interval $[0, 3]$. Make a sketch of the region.

40. Find the area that is above the x-axis, but below the curve $y = (1 - x)(x - 2)$. Make a sketch of the region.

41. Find the area under the curve $y = 3\sin x$ over the interval $[0, 2\pi/3]$. Sketch the region.

42. Find the area below the interval $[-2, -1]$, but above the curve $y = x^3$. Make a sketch of the region.

43. A student wants to find the area enclosed by the graphs of $y = 1/\sqrt{1 - x^2}$, $y = 0$, $x = 0$, and $x = 0.8$.
(a) Show that the exact area is $\sin^{-1} 0.8$.
(b) The student uses a calculator to approximate the result in part (a) to two decimal places and obtains an incorrect answer of 53.13. What was the student's error? Find the correct approximation.

∿ 44. (a) Use a graphing utility to generate the graph of

$$f(x) = \frac{1}{100}(x + 2)(x + 1)(x - 3)(x - 5)$$

and use the graph to make a conjecture about the sign of the integral

$$\int_{-2}^5 f(x)\,dx$$

(b) Check your conjecture by evaluating the integral.

45. (a) Let f be an odd function; that is, $f(-x) = -f(x)$. Invent a theorem that makes a statement about the value of an integral of the form

$$\int_{-a}^a f(x)\,dx$$

(b) Confirm that your theorem works for the integrals

$$\int_{-1}^1 x^3\,dx \quad \text{and} \quad \int_{-\pi/2}^{\pi/2} \sin x\,dx$$

(c) Let f be an even function; that is, $f(-x) = f(x)$. Invent a theorem that makes a statement about the relationship between the integrals

$$\int_{-a}^a f(x)\,dx \quad \text{and} \quad \int_0^a f(x)\,dx$$

(d) Confirm that your theorem works for the integrals

$$\int_{-1}^1 x^2\,dx \quad \text{and} \quad \int_{-\pi/2}^{\pi/2} \cos x\,dx$$

c 46. Use the theorem you invented in Exercise 45(a) to evaluate the integral

$$\int_{-5}^5 \frac{x^7 - x^5 + x}{x^4 + x^2 + 7}\,dx$$

and check your answer with a CAS.

47. Define $F(x)$ by

$$F(x) = \int_1^x (t^3 + 1)\,dt$$

(a) Use Part 2 of the Fundamental Theorem of Calculus to find $F'(x)$.
(b) Check the result in part (a) by first integrating and then differentiating.

48. Define $F(x)$ by

$$F(x) = \int_{\pi/4}^{x} \cos 2t\, dt$$

(a) Use Part 2 of the Fundamental Theorem of Calculus to find $F'(x)$.
(b) Check the result in part (a) by first integrating and then differentiating.

In Exercises 49–52, use Part 2 of the Fundamental Theorem of Calculus to find the derivatives.

49. (a) $\dfrac{d}{dx} \int_1^x \sin(\sqrt{t})\, dt$ (b) $\dfrac{d}{dx} \int_0^x e^{t^2}\, dt$

50. (a) $\dfrac{d}{dx} \int_0^x \dfrac{dt}{1+\sqrt{t}}$ (b) $\dfrac{d}{dx} \int_1^x \ln t\, dt$

51. $\dfrac{d}{dx} \int_x^0 \dfrac{t}{\cos t}\, dt$ [*Hint:* Use Definition 6.5.3(b).]

52. $\dfrac{d}{du} \int_0^u |x|\, dx$

53. Let $F(x) = \int_2^x \sqrt{3t^2+1}\, dt$. Find
(a) $F(2)$ (b) $F'(2)$ (c) $F''(2)$.

54. Let $F(x) = \int_{\sqrt{3}}^x \tan^{-1} t\, dt$. Find
(a) $F(\sqrt{3})$ (b) $F'(\sqrt{3})$ (c) $F''(\sqrt{3})$.

55. Let $F(x) = \int_0^x \dfrac{t-3}{t^2+7}\, dt$ for $-\infty < x < +\infty$.
(a) Find the value of x where F attains its minimum value.
(b) Find intervals over which F is only increasing or only decreasing.
(c) Find open intervals over which F is only concave up or only concave down.

56. Use the plotting and numerical integration commands of a CAS to generate the graph of the function F in Exercise 55 over the interval $-20 \le x \le 20$, and confirm that the graph is consistent with the results obtained in that exercise.

57. (a) Over what open interval does the formula
$$F(x) = \int_1^x \dfrac{dt}{t}$$
represent an antiderivative of $f(x) = 1/x$?
(b) Find a point where the graph of F crosses the x-axis.

58. (a) Over what open interval does the formula
$$F(x) = \int_1^x \dfrac{1}{t^2-9}\, dt$$
represent an antiderivative of
$$f(x) = \dfrac{1}{x^2-9}?$$

(b) Find a point where the graph of F crosses the x-axis.

In Exercises 59 and 60, find all values of x^* in the stated interval that satisfy Equation (7) in the Mean-Value Theorem for Integrals (6.6.2), and explain what these numbers represent.

59. (a) $f(x) = \sqrt{x}$; $[0, 9]$
(b) $f(x) = 3x^2 + 2x + 1$; $[-1, 2]$

60. (a) $f(x) = \sin x$; $[-\pi, \pi]$ (b) $f(x) = 1/x^2$; $[1, 3]$

It was shown in the proof of the Mean-Value Theorem for Integrals (6.6.2) that if f is continuous on $[a, b]$, and if $m \le f(x) \le M$ on $[a, b]$, then

$$m(b-a) \le \int_a^b f(x)\, dx \le M(b-a)$$

[see (8)]. These inequalities make it possible to obtain bounds on the size of a definite integral from bounds on the size of its integrand. This is illustrated in Exercises 61 and 62.

61. Find the maximum and minimum values of $\sqrt{x^3+2}$ for $0 \le x \le 3$, and use these values to find bounds on the value of the integral
$$\int_0^3 \sqrt{x^3+2}\, dx$$

62. Find values of m and M such that $m \le x \sin x \le M$ for $0 \le x \le \pi$, and use these values to find bounds on the value of the integral
$$\int_0^\pi x \sin x\, dx$$

63. Prove:
(a) $[cF(x)]_a^b = c[F(x)]_a^b$
(b) $[F(x) + G(x)]_a^b = F(x)]_a^b + G(x)]_a^b$
(c) $[F(x) - G(x)]_a^b = F(x)]_a^b - G(x)]_a^b$.

64. Prove the Mean-Value Theorem for Integrals (Theorem 6.6.2) by applying the Mean-Value Theorem (5.8.2) to an antiderivative F for f.

In Exercises 65 and 66, evaluate each limit by interpreting it as a Riemann sum in which the given interval is divided into n subintervals of equal width.

65. $\displaystyle\lim_{n \to +\infty} \sum_{k=1}^n \dfrac{\pi}{4n} \sec^2\left(\dfrac{\pi k}{4n}\right)$; $\left[0, \dfrac{\pi}{4}\right]$

66. $\displaystyle\lim_{n \to +\infty} \sum_{k=1}^n \dfrac{n}{n^2+k^2}$; $[0, 1]$

6.7 RECTILINEAR MOTION REVISITED; AVERAGE VALUE

In Section 5.4 we used the derivative to define the notions of instantaneous velocity and acceleration for a particle moving along a line. In this section we will resume the study of such motion using the tools of integration. We will also investigate the general problem of integrating a rate of change, and we will show how the definite integral can be used to define the average value of a continuous function. More applications of integration will be given in Chapter 7.

FINDING POSITION AND VELOCITY BY INTEGRATION

Recall from Definitions 5.4.1 and 5.4.2 that if $s(t)$ is the position function of a particle moving on a coordinate line, then the instantaneous velocity and acceleration of the particle are given by the formulas

$$v(t) = s'(t) = \frac{ds}{dt} \quad \text{and} \quad a(t) = v'(t) = \frac{dv}{dt} = \frac{d^2s}{dt^2}$$

It follows from these formulas that $s(t)$ is an antiderivative of $v(t)$ and $v(t)$ is an antiderivative of $a(t)$; that is,

$$s(t) = \int v(t)\,dt \qquad \text{and} \qquad v(t) = \int a(t)\,dt \qquad (1\text{--}2)$$

Thus, if the velocity of a particle is known, then its position function can be obtained from (1) by integration, provided there is sufficient additional information to determine the constant of integration. In particular, we can determine the constant of integration if we know the position s_0 of the particle at some time t_0, since this information determines a unique antiderivative $s(t)$ (Figure 6.7.1). Similarly, if the acceleration function of the particle is known, then its velocity function can be obtained from (2) by integration if we know the velocity v_0 of the particle at some time t_0 (Figure 6.7.2).

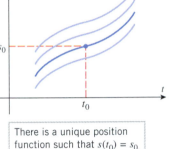

There is a unique position function such that $s(t_0) = s_0$.

Figure 6.7.1

Example 1 Find the position function of a particle that moves with velocity $v(t) = \cos \pi t$ along a coordinate line, assuming that the particle has coordinate $s = 4$ at time $t = 0$.

Solution. The position function is

$$s(t) = \int v(t)\,dt = \int \cos \pi t\,dt = \frac{1}{\pi}\sin \pi t + C$$

Since $s = 4$ when $t = 0$, it follows that

$$4 = s(0) = \frac{1}{\pi}\sin 0 + C = C$$

Thus,

$$s(t) = \frac{1}{\pi}\sin \pi t + 4 \qquad \blacktriangleleft$$

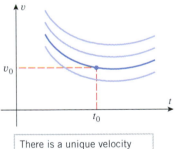

There is a unique velocity function such that $v(t_0) = v_0$.

Figure 6.7.2

UNIFORMLY ACCELERATED MOTION

One of the most important cases of rectilinear motion occurs when a particle has constant acceleration. We call this ***uniformly accelerated motion***.

We will show that if a particle moves with constant acceleration along an s-axis, and if the position and velocity of the particle are known at some point in time, say when $t = 0$, then it is possible to derive formulas for the position $s(t)$ and the velocity $v(t)$ at any time t. To see how this can be done, suppose that the particle has constant acceleration

$$a(t) = a \qquad (3)$$

and

$$s = s_0 \quad \text{when} \quad t = 0 \qquad (4)$$
$$v = v_0 \quad \text{when} \quad t = 0 \qquad (5)$$

where s_0 and v_0 are known. We call (4) and (5) the ***initial conditions*** for the motion.

With (3) as a starting point, we can integrate $a(t)$ to obtain $v(t)$, and we can integrate $v(t)$ to obtain $s(t)$, using an initial condition in each case to determine the constant of integration. The computations are as follows:

$$v(t) = \int a(t)\, dt = \int a\, dt = at + C_1 \tag{6}$$

To determine the constant of integration C_1 we apply initial condition (5) to this equation to obtain

$$v_0 = v(0) = a \cdot 0 + C_1 = C_1$$

Substituting this in (6) and putting the constant term first yields

$$v(t) = v_0 + at$$

Since v_0 is constant, it follows that

$$s(t) = \int v(t)\, dt = \int (v_0 + at)\, dt = v_0 t + \tfrac{1}{2} a t^2 + C_2 \tag{7}$$

To determine the constant C_2 we apply initial condition (4) to this equation to obtain

$$s_0 = s(0) = v_0 \cdot 0 + \tfrac{1}{2} a \cdot 0 + C_2 = C_2$$

Substituting this in (7) and putting the constant term first yields

$$s(t) = s_0 + v_0 t + \tfrac{1}{2} a t^2$$

In summary, we have the following result.

6.7.1 UNIFORMLY ACCELERATED MOTION. *If a particle moves with constant acceleration a along an s-axis, and if the position and velocity at time $t = 0$ are s_0 and v_0, respectively, then the position and velocity functions of the particle are*

$$s(t) = s_0 + v_0 t + \tfrac{1}{2} a t^2 \tag{8}$$

$$v(t) = v_0 + at \tag{9}$$

FOR THE READER. How can you tell from the velocity versus time curve whether a particle moving along a line has uniformly accelerated motion?

Example 2 Suppose that an intergalactic spacecraft uses a sail and the "solar wind" to produce a constant acceleration of 0.032 m/s². Assuming that the spacecraft has a velocity of 10,000 m/s when the sail is first raised, how far will the spacecraft travel in 1 hour, and what will its velocity be at the end of this hour?

Solution. In this problem the choice of a coordinate axis is at our discretion, so we will choose it to make the computations as simple as possible. Accordingly, let us introduce an s-axis whose positive direction is in the direction of motion, and let us take the origin to coincide with the position of the spacecraft at the time $t = 0$ when the sail is raised. Thus, the Formulas (8) and (9) for uniformly accelerated motion apply with

$$s_0 = s(0) = 0, \quad v_0 = v(0) = 10{,}000, \quad \text{and} \quad a = 0.032$$

Since 1 hour corresponds to $t = 3600$ s, it follows from (8) that in 1 hour the spacecraft travels a distance of

$$s(3600) = 10{,}000(3600) + \tfrac{1}{2}(0.032)(3600)^2 \approx 36{,}200{,}000 \text{ m}$$

and it follows from (9) that after 1 hour its velocity is

$$v(3600) = 10{,}000 + (0.032)(3600) \approx 10{,}100 \text{ m/s}$$

◀

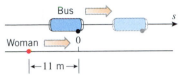

Figure 6.7.3

Example 3   A bus has stopped to pick up riders, and a woman is running at a constant velocity of 5 m/s to catch it. When she is 11 m behind the front door the bus pulls away with a constant acceleration of 1 m/s^2. From that point in time, how long will it take for the woman to reach the front door of the bus if she keeps running with a velocity of 5 m/s?

Solution.   As shown in Figure 6.7.3, choose the s-axis so that the bus and the woman are moving in the positive direction, and the front door of the bus is at the origin at the time $t = 0$ when the bus begins to pull away. To catch the bus at some later time t, the woman will have to cover a distance $s_w(t)$ that is equal to 11 m plus the distance $s_b(t)$ traveled by the bus; that is, the woman will catch the bus when

$$s_w(t) = s_b(t) + 11 \tag{10}$$

Since the woman has a constant velocity of 5 m/s, the distance she travels in t seconds is $s_w(t) = 5t$. Thus, (10) can be written as

$$s_b(t) = 5t - 11 \tag{11}$$

Since the bus has a constant acceleration of $a = 1$ m/s^2, and since $s_0 = v_0 = 0$ at time $t = 0$ (why?), it follows from (8) that

$$s_b(t) = \tfrac{1}{2}t^2$$

Substituting this equation into (11) and reorganizing the terms yields the quadratic equation

$$\tfrac{1}{2}t^2 - 5t + 11 = 0 \quad \text{or} \quad t^2 - 10t + 22 = 0$$

Solving this equation for t using the quadratic formula yields two solutions:

$$t = 5 - \sqrt{3} \approx 3.3 \quad \text{and} \quad t = 5 + \sqrt{3} \approx 6.7$$

(verify). Thus, the woman can reach the door at two different times, $t = 3.3$ s and $t = 6.7$ s. The reason that there are two solutions can be explained as follows: When the woman first reaches the door, she is running faster than the bus and can run past it if the driver does not see her. However, as the bus speeds up, it eventually catches up to her, and she has another chance to flag it down. ◀

THE FREE-FALL MODEL

In Section 5.4 we discussed the free-fall model of motion near the surface of the Earth with the promise that we would derive Formula (5) of that section later in the text; we will now show how to do this. As stated in 5.4.4 and illustrated in Figure 5.4.8, we will assume that the object moves on an s-axis whose origin is at the surface of the Earth and whose positive direction is up; and we will assume that the position and velocity of the object at time $t = 0$ are s_0 and v_0, respectively.

It is a fact of physics that a particle moving on a vertical line near the Earth's surface and subject only to the force of the Earth's gravity moves with essentially constant acceleration. The magnitude of this constant, denoted by the letter g, is approximately 9.8 m/s^2 or 32 ft/s^2, depending on whether distance is measured in meters or feet.*

Recall that a particle is speeding up when its velocity and acceleration have the same sign and is slowing down when they have opposite signs. Thus, because we have chosen the positive direction to be up, it follows that the acceleration $a(t)$ of a particle in free fall is negative for all values of t. To see that this is so, observe that an upward-moving particle (positive velocity) is slowing down, so its acceleration must be negative; and a downward-moving particle (negative velocity) is speeding up, so its acceleration must also be negative. Thus, we conclude that

$$a(t) = -g$$

and hence it follows from (8) and (9) that the position and velocity functions of an object

---

*Strictly speaking, the constant g varies with the latitude and the distance from the Earth's center. However, for motion at a fixed latitude and near the surface of the Earth, the assumption of a constant g is satisfactory for many applications.

in free fall are

$$s(t) = s_0 + v_0 t - \tfrac{1}{2} g t^2 \qquad\qquad\qquad (12)$$

$$v(t) = v_0 - g t \qquad\qquad\qquad (13)$$

FOR THE READER.   Had we chosen the positive direction of the s-axis to be down, then the acceleration would have been $a(t) = g$ (why?). How would this have affected Formulas (12) and (13)?

Example 4   A ball is hit directly upward with an initial velocity of 49 m/s and is struck at a point that is 1 m above the ground. Assuming that the free-fall model applies, how high will the ball travel?

Solution.   Since distance is in meters, we take $g = 9.8$ m/s^2. Initially, we have $s_0 = 1$ and $v_0 = 49$, so from (12) and (13)

$$v(t) = -9.8t + 49$$
$$s(t) = -4.9t^2 + 49t + 1$$

The ball will rise until $v(t) = 0$, that is, until $-9.8t + 49 = 0$ or $t = 5$. At this instant the height above the ground will be

$$s(5) = -4.9(5)^2 + 49(5) + 1 = 123.5 \text{ m} \qquad\qquad \blacktriangleleft$$

Example 5   A penny is released from rest near the top of the Empire State Building at a point that is 1250 ft above the ground (Figure 6.7.4). Assuming that the free-fall model applies, how long does it take for the penny to hit the ground, and what is its speed at the time of impact?

Solution.   Since distance is in feet, we take $g = 32$ ft/s^2. Initially, we have $s_0 = 1250$ and $v_0 = 0$, so from (12)

$$s(t) = -16t^2 + 1250 \qquad\qquad\qquad (14)$$

Impact occurs when $s(t) = 0$. Solving this equation for t, we obtain

$$-16t^2 + 1250 = 0$$
$$t^2 = \frac{1250}{16} = \frac{625}{8}$$
$$t = \pm \frac{25}{\sqrt{8}} \approx \pm 8.8 \text{ s}$$

Since $t \geq 0$, we can discard the negative solution and conclude that it takes $25/\sqrt{8} \approx 8.8$ s for the penny to hit the ground. To obtain the velocity at the time of impact, we substitute $t = 25/\sqrt{8}$, $v_0 = 0$, and $g = 32$ in (13) to obtain

$$v\left(\frac{25}{\sqrt{8}}\right) = 0 - 32\left(\frac{25}{\sqrt{8}}\right) = -200\sqrt{2} \approx -282.8 \text{ ft/s}$$

Thus, the speed at the time of impact is

$$\left| v\left(\frac{25}{\sqrt{8}}\right) \right| = 200\sqrt{2} \approx 282.8 \text{ ft/s}$$

which is more than 192 mi/h. $\blacktriangleleft$

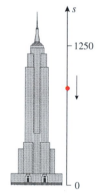

Figure 6.7.4

INTEGRATING RATES OF CHANGE

The Fundamental Theorem of Calculus

$$\int_a^b f(x)\,dx = F(b) - F(a) \qquad\qquad\qquad (15)$$

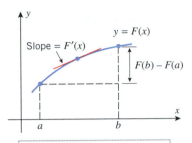

Integrating the slope of $y = F(x)$ over the interval $[a, b]$ produces the change $F(b) - F(a)$ in the value of $F(x)$.

Figure 6.7.5

has a useful interpretation that can be seen by rewriting it in a slightly different form. Since F is an antiderivative of f on the interval $[a, b]$, we can use the relationship $F'(x) = f(x)$ to rewrite (15) as

$$\int_a^b F'(x)\, dx = F(b) - F(a) \tag{16}$$

In this formula we can view $F'(x)$ as the rate of change of $F(x)$ with respect to x, and we can view $F(b) - F(a)$ as the *change* in the value of $F(x)$ as x increases from a to b (Figure 6.7.5). Thus, we have the following useful principle.

> **6.7.2** INTEGRATING A RATE OF CHANGE. Integrating the rate of change of $F(x)$ with respect to x over an interval $[a, b]$ produces the change in the value of $F(x)$ that occurs as x increases from a to b.

Here are some examples of this idea:

- If $P(t)$ is a population (e.g., plants, animals, or people) at time t, then $P'(t)$ is the rate at which the population is changing at time t, and

$$\int_{t_1}^{t_2} P'(t)\, dt = P(t_2) - P(t_1)$$

is the change in the population between times t_1 and t_2.

- If $A(t)$ is the area of an oil spill at time t, then $A'(t)$ is the rate at which the area of the spill is changing at time t, and

$$\int_{t_1}^{t_2} A'(t)\, dt = A(t_2) - A(t_1)$$

is the change in the area of the spill between times t_1 and t_2.

- If $P'(x)$ is the marginal profit that results from producing and selling x units of a product (see Section 5.6), then

$$\int_{x_1}^{x_2} P'(x)\, dx = P(x_2) - P(x_1)$$

is the change in the profit that results when the production level increases from x_1 units to x_2 units.

DISPLACEMENT IN RECTILINEAR MOTION

As another application of (16), suppose that $s(t)$ and $v(t)$ are the position and velocity functions of a particle moving on a coordinate line. Since $v(t)$ is the rate of change of $s(t)$ with respect to t, it follows from the principle in 6.7.2 that integrating $v(t)$ over an interval $[t_0, t_1]$ will produce the change in the value of $s(t)$ as t increases from t_0 to t_1; that is,

$$\int_{t_0}^{t_1} v(t)\, dt = \int_{t_0}^{t_1} s'(t)\, dt = s(t_1) - s(t_0) \tag{17}$$

The expression $s(t_1) - s(t_0)$ in this formula is called the *displacement* or *change in position* of the particle over the time interval $[t_0, t_1]$. For a particle moving horizontally, the displacement is positive if the final position of the particle is to the right of its initial position, negative if it is to the left of its initial position, and zero if it coincides with the initial position (Figure 6.7.6).

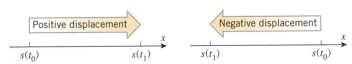

Figure 6.7.6

REMARK.   In physical problems it is important to associate the correct units with definite integrals. In general, the units for the definite integral

$$\int_a^b f(x)\, dx$$

will be units of $f(x)$ times units of x. This is because the definite integral is a limit of Riemann sums each of whose terms is a product of the form $f(x) \cdot \Delta x$. For example, if time is measured in seconds (s) and velocity is measured in meters per second (m/s), then integrating velocity over a time interval will produce a result whose units are in meters, since m/s $\times$ s $=$ m. Note that this is consistent with Formula (17), since displacement has units of length.

DISTANCE TRAVELED IN RECTILINEAR MOTION

In general, the displacement of a particle is not the same as the distance traveled by the particle. For example, a particle that travels 100 units in the positive direction and then 100 units in the negative direction travels a distance of 200 units but has a displacement of zero, since it returns to its starting position. The only case in which the displacement and the distance traveled are the same occurs when the particle moves in the positive direction without reversing the direction of its motion.

FOR THE READER.   What is the relationship between the displacement of a particle and the distance it travels if the particle moves in the negative direction without reversing the direction of motion?

From (17), integrating the velocity function of a particle over a time interval yields the displacement of a particle over that time interval. In contrast, to find the *total distance traveled* by the particle over the time interval (the distance traveled in the positive direction plus the distance traveled in the negative direction), we must integrate the *absolute value* of the velocity function; that is, we must integrate the speed:

$$\begin{bmatrix} \text{total distance} \\ \text{traveled during} \\ \text{time interval} \\ [t_0, t_1] \end{bmatrix} = \int_{t_0}^{t_1} |v(t)|\, dt \tag{18}$$

Example 6   Suppose that a particle moves on a coordinate line so that its velocity at time t is $v(t) = t^2 - 2t$ m/s.

(a)   Find the displacement of the particle during the time interval $0 \le t \le 3$.
(b)   Find the distance traveled by the particle during the time interval $0 \le t \le 3$.

Solution (a). From (17) the displacement is

$$\int_0^3 v(t)\, dt = \int_0^3 (t^2 - 2t)\, dt = \left[\frac{t^3}{3} - t^2 \right]_0^3 = 0$$

Thus, the particle is at the same position at time $t = 3$ as at $t = 0$.

Solution (b). The velocity can be written as $v(t) = t^2 - 2t = t(t - 2)$, from which we see that $v(t) \le 0$ for $0 \le t \le 2$ and $v(t) \ge 0$ for $2 \le t \le 3$. Thus, it follows from (18) that the distance traveled is

$$\int_0^3 |v(t)|\, dt = \int_0^2 -v(t)\, dt + \int_2^3 v(t)\, dt$$

$$= \int_0^2 -(t^2 - 2t)\, dt + \int_2^3 (t^2 - 2t)\, dt$$

$$= -\left[\frac{t^3}{3} - t^2 \right]_0^2 + \left[\frac{t^3}{3} - t^2 \right]_2^3 = \frac{4}{3} + \frac{4}{3} = \frac{8}{3} \text{ m}$$

◄

ANALYZING THE VELOCITY VERSUS TIME CURVE

In Section 5.4 we showed how to use the position versus time curve to obtain information about the behavior of a particle moving on a coordinate line (Table 5.4.1). Similarly, there is valuable information that can be obtained from the *velocity versus time curve*. For example, the integral in (17) can be interpreted geometrically as the net signed area between the graph of $v(t)$ and the interval $[t_0, t_1]$, and it can be interpreted physically as the displacement of the particle over this interval. Thus, we have the following result.

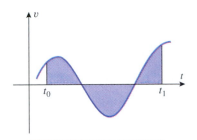

> **6.7.3** FINDING DISPLACEMENT FROM THE VELOCITY VERSUS TIME CURVE.   For a particle in rectilinear motion, the net signed area between the velocity versus time curve and an interval $[t_0, t_1]$ on the t-axis represents the displacement of the particle over that time interval (Figure 6.7.7).

The net signed area is the displacement of the particle during the interval $[t_0, t_1]$.

Figure 6.7.7

Example 7   Figure 6.7.8 shows three velocity versus time curves for a particle in rectilinear motion along a horizontal line. In each case, find the displacement of the particle over the time interval $0 \leq t \leq 4$, and explain what it tells you about the motion of the particle.

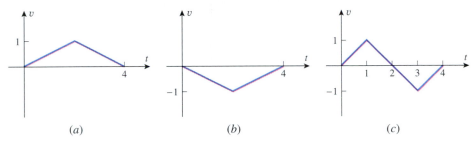

(a)   (b)   (c)

Figure 6.7.8

Solution.   In part (a) of Figure 6.7.8 the net signed area under the curve is 2, so the particle is 2 units to the right of its starting point at the end of the time period. In part (b) the net signed area under the curve is -2, so the particle is 2 units to the left of its starting point at the end of the time period. In part (c) the net signed area under the curve is 0, so the particle is back at its starting point at the end of the time period.   ◀

By replacing the concept of net signed area with that of "total area," we can also interpret geometrically the total distance traveled by a particle in rectilinear motion. If $f(x)$ is a continuous function on an interval $[a, b]$, then we define the ***total area*** between the curve $y = f(x)$ and the interval to be the integral of $|f(x)|$ over the interval $[a, b]$. Geometrically, the total area is the area of the region that is between the graph of f and the x-axis.

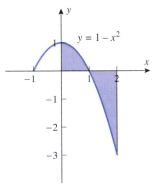

Figure 6.7.9

Example 8   Find the total area between the curve $y = 1 - x^2$ and the x-axis over the interval $[0, 2]$ (Figure 6.7.9).

Solution.   The area A is given by

$$A = \int_0^2 |1 - x^2| \, dx = \int_0^1 (1 - x^2) \, dx + \int_1^2 -(1 - x^2) \, dx$$

$$= \left[x - \frac{x^3}{3} \right]_0^1 - \left[x - \frac{x^3}{3} \right]_1^2$$

$$= \frac{2}{3} - \left(-\frac{4}{3} \right) = 2 \qquad ◀$$

From (18), integrating the speed $|v(t)|$ over a time interval $[t_0, t_1]$ produces the distance traveled by the particle during the time interval. However, we can also interpret the integral in (18) as the total area between the velocity versus time curve and the interval $[t_0, t_1]$ on the t-axis. Thus, we have the following result.

6.7.4 FINDING DISTANCE TRAVELED FROM THE VELOCITY VERSUS TIME CURVE. For a particle in rectilinear motion, the total area between the velocity versus time curve and an interval $[t_0, t_1]$ on the t-axis represents the distance traveled by the particle over that time interval.

Example 9 For each of the velocity versus time curves in Figure 6.7.8 find the total distance traveled by the particle over the time interval $0 \le t \le 4$.

Solution. In all three parts of Figure 6.7.8 the total area between the curve and the interval $[0, 4]$ is 2, so the particle travels a distance of 2 units during the time period in all three cases, even though the displacement is different in each case, as discussed in Example 7. ◀

AVERAGE VALUE OF A CONTINUOUS FUNCTION

In scientific work, numerical information is often summarized by computing some sort of *average* or *mean* value of the observed data. There are various kinds of averages, but the most common is the **arithmetic mean** or **arithmetic average**, which is formed by adding the data and dividing by the number of data points. Thus, the arithmetic average $\bar{a}$ of n numbers $a_1, a_2, \ldots, a_n$ is

$$\bar{a} = \frac{1}{n}(a_1 + a_2 + \cdots + a_n) = \frac{1}{n}\sum_{k=1}^{n} a_k$$

In the case where the a_k's are values of a function f, say,

$$a_1 = f(x_1), a_2 = f(x_2), \ldots, a_n = f(x_n)$$

then the arithmetic average $\bar{a}$ of these function values is

$$\bar{a} = \frac{1}{n}\sum_{k=1}^{n} f(x_k)$$

We will now show how to extend this concept so that we can compute not only the arithmetic average of finitely many function values but an average of *all* values of $f(x)$ as x varies over a closed interval $[a, b]$. For this purpose recall the Mean-Value Theorem for Integrals (6.6.2), which states that if f is continuous on the interval $[a, b]$, then there is at least one number x^* in this interval such that

$$\int_a^b f(x)\,dx = f(x^*)(b - a)$$

The quantity

$$f(x^*) = \frac{1}{b - a}\int_a^b f(x)\,dx \tag{19}$$

will be our candidate for the average value of f over the interval $[a, b]$. To explain what motivates this, divide the interval $[a, b]$ into n subintervals of equal length

$$\Delta x = \frac{b - a}{n} \tag{20}$$

and choose arbitrary numbers $x_1^*, x_2^*, \ldots, x_n^*$ in successive subintervals. Then the arithmetic average of the values $f(x_1^*), f(x_2^*), \ldots, f(x_n^*)$ is

$$\text{ave} = \frac{1}{n}[f(x_1^*) + f(x_2^*) + \cdots + f(x_n^*)]$$

or from (20)

$$\text{ave} = \frac{1}{b - a}[f(x_1^*)\Delta x + f(x_2^*)\Delta x + \cdots + f(x_n^*)\Delta x] = \frac{1}{b - a}\sum_{k=1}^{n} f(x_k^*)\Delta x$$

Taking the limit as $n \to +\infty$ yields

$$\lim_{n \to +\infty} \frac{1}{b - a}\sum_{k=1}^{n} f(x_k^*)\Delta x = \frac{1}{b - a}\int_a^b f(x)\,dx$$

Since this equation describes what happens when we compute the average of "more and more" values of $f(x)$, we are led to the following definition.

6.7.5 DEFINITION.   If f is continuous on $[a, b]$, then the ***average value*** (or ***mean value***) of f on $[a, b]$ is defined to be

$$f_{\text{ave}} = \frac{1}{b-a} \int_a^b f(x)\, dx \tag{21}$$

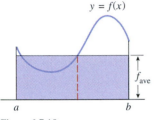

$y = f(x)$

f_{ave}

a b

Figure 6.7.10

• REMARK.   When f is nonnegative on $[a, b]$, the quantity f_{ave} has a simple geometric interpretation, which can be seen by writing (21) as

$$f_{\text{ave}} \cdot (b - a) = \int_a^b f(x)\, dx$$

The left side of this equation is the area of a rectangle with a height of f_{ave} and base of length $b - a$, and the right side is the area under $y = f(x)$ over $[a, b]$. Thus, f_{ave} is the height of a rectangle constructed over the interval $[a, b]$, whose area is the same as the area under the graph of f over that interval (Figure 6.7.10). Note also that the Mean-Value Theorem, when expressed in form (21), ensures that there is always at least one number x^* in $[a, b]$ at which the value of f is equal to the average value of f over the interval.

Example 10   Find the average value of the function $f(x) = \sqrt{x}$ over the interval $[1, 4]$, and find all numbers in the interval at which the value of f is the same as the average.

Solution.

$$f_{\text{ave}} = \frac{1}{b-a} \int_a^b f(x)\, dx = \frac{1}{4-1} \int_1^4 \sqrt{x}\, dx = \frac{1}{3} \left[\frac{2x^{3/2}}{3} \right]_1^4$$

$$= \frac{1}{3} \left[\frac{16}{3} - \frac{2}{3} \right] = \frac{14}{9} \approx 1.6$$

The x-values at which $f(x) = \sqrt{x}$ is the same as the average satisfy $\sqrt{x} = 14/9$, from which we obtain $x = 196/81 \approx 2.4$ (Figure 6.7.11).   ◀

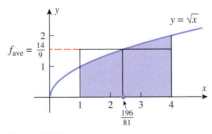

$y = \sqrt{x}$

$f_{\text{ave}} = \frac{14}{9}$

Figure 6.7.11

AVERAGE VELOCITY REVISITED

In Section 3.1 we considered the motion of a particle moving along a coordinate line, and we motivated the concept of instantaneous velocity by viewing it as the limit of average velocities over smaller and smaller time intervals. That discussion led us to conclude that the average velocity of the particle over a time interval could be interpreted as the slope of a secant line of the position versus time curve (Figure 3.1.6). We will now show that the same result is true if Definition 6.7.5 is used to compute the average velocity.

For this purpose, suppose that $s(t)$ and $v(t)$ are the position and velocity functions of such a particle, and let us use Formula (21) to calculate the average velocity of the particle over a time interval $[t_0, t_1]$. This yields

$$v_{\text{ave}} = \frac{1}{t_1 - t_0} \int_{t_0}^{t_1} v(t)\, dt = \frac{1}{t_1 - t_0} \int_{t_0}^{t_1} s'(t)\, dt = \frac{s(t_1) - s(t_0)}{t_1 - t_0}$$

Thus, *the average velocity over a time interval is the displacement divided by the elapsed time.* Geometrically, this is the slope of the secant line shown in Figure 6.7.12. Thus, the discussion of average velocity in Section 3.1 is consistent with Definition 6.7.5.

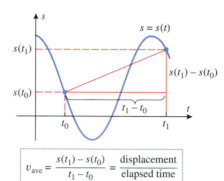

Figure 6.7.12

EXERCISE SET 6.7   ~ Graphing Utility   [C] CAS

1. (a) If $h'(t)$ is the rate of change of a child's height measured in inches per year, what does the integral $\int_0^{10} h'(t)\, dt$ represent, and what are its units?

   (b) If $r'(t)$ is the rate of change of the radius of a spherical balloon measured in centimeters per second, what does the integral $\int_1^2 r'(t)\, dt$ represent, and what are its units?

   (c) If $H(t)$ is the rate of change of the speed of sound with respect to temperature measured in ft/s per °F, what does the integral $\int_{32}^{100} H(t)\, dt$ represent, and what are its units?

   (d) If $v(t)$ is the velocity of a particle in rectilinear motion, measured in cm/h, what does the integral $\int_{t_1}^{t_2} v(t)\, dt$ represent, and what are its units?

2. (a) Suppose that sludge is emptied into a river at the rate of $V(t)$ gallons per minute, starting at time $t = 0$. Write an integral that represents the total volume of sludge that is emptied into the river during the first hour.

   (b) Suppose that the tangent line to a curve $y = f(x)$ has slope $m(x)$ at x. What does the integral $\int_{x_1}^{x_2} m(x)\, dx$ represent?

3. In each part, the velocity versus time curve is given for a particle moving along a line. Use the curve to find the displacement and the distance traveled by the particle over the time interval $0 \le t \le 3$.

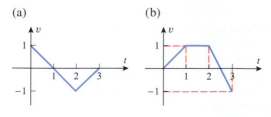

4. Sketch a velocity versus time curve for a particle that travels a distance of 5 units along a coordinate line during the time interval $0 \le t \le 10$ and has a displacement of 0 units.

5. The accompanying figure shows the acceleration versus time curve for a particle moving along a coordinate line. If the initial velocity of the particle is 20 m/s, estimate

   (a) the velocity at time $t = 4$ s

   (b) the velocity at time $t = 6$ s.

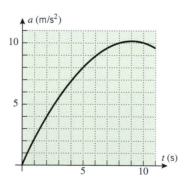

Figure Ex-5

6. Determine whether the particle in Exercise 5 is speeding up or slowing down at times $t = 4$ s and $t = 6$ s.

In Exercises 7–10, a particle moves along an s-axis. Use the given information to find the position function of the particle.

7. (a) $v(t) = t^3 - 2t^2 + 1$; $s(0) = 1$

   (b) $a(t) = 4\cos 2t$; $v(0) = -1$; $s(0) = -3$

8. (a) $v(t) = 1 + \sin t$; $s(0) = -3$

   (b) $a(t) = t^2 - 3t + 1$; $v(0) = 0$; $s(0) = 0$

9. (a) $v(t) = 2t - 3$; $s(1) = 5$
   (b) $a(t) = \cos t$; $v(\pi/2) = 2$; $s(\pi/2) = 0$

10. (a) $v(t) = t^{2/3}$; $s(8) = 0$
   (b) $a(t) = \sqrt{t}$; $v(4) = 1$; $s(4) = -5$

In Exercises 11–14, a particle moves with a velocity of $v(t)$ m/s along an s-axis. Find the displacement and the distance traveled by the particle during the given time interval.

11. (a) $v(t) = \sin t$; $0 \le t \le \pi/2$
   (b) $v(t) = \cos t$; $\pi/2 \le t \le 2\pi$

12. (a) $v(t) = 2t - 4$; $0 \le t \le 6$
   (b) $v(t) = |t - 3|$; $0 \le t \le 5$

13. (a) $v(t) = t^3 - 3t^2 + 2t$; $0 \le t \le 3$
   (b) $v(t) = \sqrt{t} - 2$; $0 \le t \le 3$

14. (a) $v(t) = \frac{1}{2} - (1/t^2)$; $1 \le t \le 3$
   (b) $v(t) = 3/\sqrt{t}$; $4 \le t \le 9$

In Exercises 15–18, a particle moves with acceleration $a(t)$ m/s² along an s-axis and has velocity v_0 m/s at time $t = 0$. Find the displacement and the distance traveled by the particle during the given time interval.

15. $a(t) = -2$; $v_0 = 3$; $1 \le t \le 4$

16. $a(t) = t - 2$; $v_0 = 0$; $1 \le t \le 5$

17. $a(t) = 1/\sqrt{5t + 1}$; $v_0 = 2$; $0 \le t \le 3$

18. $a(t) = \sin t$; $v_0 = 1$; $\pi/4 \le t \le \pi/2$

19. In each part use the given information to find the position, velocity, speed, and acceleration at time $t = 1$.
   (a) $v = \sin \frac{1}{2}\pi t$; $s = 0$ when $t = 0$
   (b) $a = -3t$; $s = 1$ and $v = 0$ when $t = 0$

20. The accompanying figure shows the velocity versus time curve over the time interval $1 \le t \le 5$ for a particle moving along a horizontal coordinate line.
   (a) What can you say about the sign of the acceleration over the time interval?
   (b) When is the particle speeding up? Slowing down?
   (c) What can you say about the location of the particle at time $t = 5$ relative to its location at time $t = 1$? Explain your reasoning.

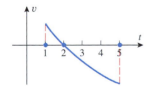

Figure Ex-20

In Exercises 21–26, sketch the curve and find the total area between the curve and the given interval on the x-axis.

21. $y = x^2 - 1$; $[0, 3]$

22. $y = \sin x$; $[0, 3\pi/2]$

23. $y = \sqrt{x + 1} - 1$; $[-1, 1]$ **24.** $y = \dfrac{x^2 - 1}{x^2}$; $\left[\frac{1}{2}, 2\right]$

25. $y = e^x - 1$; $[-1, 1]$ **26.** $y = \dfrac{x - 1}{x}$; $\left[\frac{1}{2}, 2\right]$

27. Suppose that a particle moves along a line so that its velocity v at time t is given by
$$v(t) = \begin{cases} 5t, & 0 \le t < 1 \\ 6\sqrt{t} - \dfrac{1}{t}, & 1 \le t \end{cases}$$
where t is in seconds and v is in centimeters per second (cm/s). Estimate the time(s) at which the particle is 4 cm from its starting position.

28. Suppose that a particle moves along a line so that its velocity v at time t is given by
$$v(t) = \frac{3}{t^2 + 1} - 0.5t, \quad t \ge 0$$
where t is in seconds and v is in centimeters per second (cm/s). Estimate the time(s) at which the particle is 2 cm from its starting position.

29. Suppose that the velocity function of a particle moving along an s-axis is $v(t) = 20t^2 - 100t + 50$ ft/s and that the particle is at the origin at time $t = 0$. Use a graphing utility to generate the graphs of $s(t)$, $v(t)$, and $a(t)$ for the first 6 s of motion.

30. Suppose that the acceleration function of a particle moving along an s-axis is $a(t) = 4t - 30$ m/s and that the position and velocity at time $t = 0$ are $s_0 = -5$ m and $v_0 = 3$ m/s. Use a graphing utility to generate the graphs of $s(t)$, $v(t)$, and $a(t)$ for the first 25 s of motion.

31. Let the velocity function for a particle that is at the origin initially and moves along an s-axis be $v(t) = 0.5 - t \sin t$.
   (a) Generate the velocity versus time curve, and use it to make a conjecture about the sign of the displacement over the time interval $0 \le t \le 5$.
   (b) Use a CAS to find the displacement.

32. Let the velocity function for a particle that is at the origin initially and moves along an s-axis be $v(t) = 0.5 - t \cos \pi t$.
   (a) Generate the velocity versus time curve, and use it to make a conjecture about the sign of the displacement over the time interval $0 \le t \le 1$.
   (b) Use a CAS to find the displacement.

33. Let the velocity function for a particle that is at the origin initially and moves along an s-axis be $v(t) = 0.5 - te^{-t}$.
   (a) Generate the velocity versus time curve, and use it to make a conjecture about the sign of the displacement over the time interval $0 \le t \le 5$.
   (b) Use a CAS to find the displacement.

34. Let the velocity function for a particle that is at the origin initially and moves along an s-axis be $v(t) = t \ln(t + 0.1)$.
   (a) Generate the velocity versus time curve, and use it to make a conjecture about the sign of the displacement over the time interval $0 \le t \le 1$.
   (b) Use a CAS to find the displacement.

35. Suppose that at time $t = 0$ a particle is at the origin of an x-axis and has a velocity of $v_0 = 25$ cm/s. For the first 4 s thereafter it has no acceleration, and then it is acted on by a retarding force that produces a constant negative acceleration of $a = -10 \text{ cm/s}^2$.

(a) Sketch the acceleration versus time curve over the interval $0 \le t \le 12$.

(b) Sketch the velocity versus time curve over the time interval $0 \le t \le 12$.

(c) Find the x-coordinate of the particle at times $t = 8$ s and $t = 12$ s.

(d) What is the maximum x-coordinate of the particle over the time interval $0 \le t \le 12$?

36. Formulas (8) and (9) for uniformly accelerated motion can be rearranged in various useful ways. For simplicity, let $s = s(t)$ and $v = v(t)$, and derive the following variations of those formulas.

(a) $a = \dfrac{v^2 - v_0^2}{2(s - s_0)}$ (b) $t = \dfrac{2(s - s_0)}{v_0 + v}$

(c) $s = s_0 + vt - \frac{1}{2}at^2$ [Note how this differs from (8).]

Exercises 37–44 involve uniformly accelerated motion. In these exercises assume that the object is moving in the positive direction of a coordinate line, and apply Formulas (8) and (9) or those from Exercise 36, as appropriate. In some of these problems you will need the fact that $88 \text{ ft/s} = 60 \text{ mi/h}$.

37. (a) An automobile traveling on a straight road decelerates uniformly from 55 mi/h to 25 mi/h in 30 s. Find its acceleration in ft/s^2.

(b) A bicycle rider traveling on a straight path accelerates uniformly from rest to 30 km/h in 1 min. Find his acceleration in km/s^2.

38. A car traveling 60 mi/h along a straight road decelerates at a constant rate of 10 ft/s^2.

(a) How long will it take until the speed is 45 mi/h?

(b) How far will the car travel before coming to a stop?

39. Spotting a police car, you hit the brakes on your new Porsche to reduce your speed from 90 mi/h to 60 mi/h at a constant rate over a distance of 200 ft.

(a) Find the acceleration in ft/s^2.

(b) How long does it take for you to reduce your speed to 55 mi/h?

(c) At the acceleration obtained in part (a), how long would it take for you to bring your Porsche to a complete stop from 90 mi/h?

40. A particle moving along a straight line is accelerating at a constant rate of 3 m/s^2. Find the initial velocity if the particle moves 40 m in the first 4 s.

41. A motorcycle, starting from rest, speeds up with a constant acceleration of 2.6 m/s^2. After it has traveled 120 m, it slows down with a constant acceleration of -1.5 m/s^2 until it attains a speed of 12 m/s. What is the distance traveled by the motorcycle at that point?

42. A sprinter in a 100-m race explodes out of the starting block with an acceleration of 4.0 m/s^2, which she sustains for 2.0 s. Her acceleration then drops to zero for the rest of race.

(a) What is her time for the race?

(b) Make a graph of her distance from the starting block versus time.

43. A car that has stopped at a toll booth leaves the booth with a constant acceleration of 2 ft/s^2. At the time the car leaves the booth it is 5000 ft behind a truck traveling with a constant velocity of 50 ft/s. How long will it take for the car to catch the truck, and how far will the car be from the toll booth at that time?

44. In the final sprint of a rowing race the challenger is rowing at a constant speed of 12 m/s. At the point where the leader is 100 m from the finish line and the challenger is 15 m behind, the leader is rowing at 8 m/s but starts accelerating at a constant 0.5 m/s^2. Who wins?

In Exercises 45–54, assume that a free-fall model applies. Solve these exercises by applying Formulas (12) and (13) or, if appropriate, use those from Exercise 36 with $a = -g$. In these exercises take $g = 32 \text{ ft/s}^2$ or $g = 9.8 \text{ m/s}^2$, depending on the units.

45. A projectile is launched vertically upward from ground level with an initial velocity of 112 ft/s.

(a) Find the velocity at $t = 3$ s and $t = 5$ s.

(b) How high will the projectile rise?

(c) Find the speed of the projectile when it hits the ground.

46. A projectile fired downward from a height of 112 ft reaches the ground in 2 s. What is its initial velocity?

47. A projectile is fired vertically upward from ground level with an initial velocity of 16 ft/s.

(a) How long will it take for the projectile to hit the ground?

(b) How long will the projectile be moving upward?

48. A rock is dropped from the top of the Washington Monument, which is 555 ft high.

(a) How long will it take for the rock to hit the ground?

(b) What is the speed of the rock at impact?

49. A helicopter pilot drops a package when the helicopter is 200 ft above the ground and rising at a speed of 20 ft/s.

(a) How long will it take for the package to hit the ground?

(b) What will be its speed at impact?

50. A stone is thrown downward with an initial speed of 96 ft/s from a height of 112 ft.

(a) How long will it take for the stone to hit the ground?

(b) What will be its speed at impact?

51. A projectile is fired vertically upward with an initial velocity of 49 m/s from a tower 150 m high.

(a) How long will it take for the projectile to reach its maximum height?

(b) What is the maximum height?

(c) How long will it take for the projectile to pass its starting point on the way down?

(d) What is the velocity when it passes the starting point on the way down?

(e) How long will it take for the projectile to hit the ground?

(f) What will be its speed at impact?

52. A man drops a stone from a bridge. What is the height of the bridge if
 (a) the stone hits the water 4 s later
 (b) the sound of the splash reaches the man 4 s later? [Take 1080 ft/s as the speed of sound.]

53. In the final stages of a Moon landing, a lunar module fires its retrorockets and descends to a height of $h = 5$ m above the lunar surface (Figure Ex-53). At that point the retrorockets are cut off, and the module goes into free fall. Given that the Moon's gravity is $1/6$ of the Earth's, find the speed of the module when it touches the lunar surface.

Figure Ex-53

54. Given that the Moon's gravity is $1/6$ of the Earth's, how much faster would a projectile have to be launched upward from the surface of the Earth than from the surface of the Moon to reach a height of 1000 ft?

In Exercises 55–60, find the average value of the function over the given interval.

55. $f(x) = 3x$; $[1, 3]$ **56.** $f(x) = x^2$; $[-1, 2]$

57. $f(x) = \sin x$; $[0, \pi]$ **58.** $f(x) = \cos x$; $[0, \pi]$

59. $f(x) = 1/x$; $[1, e]$ **60.** $f(x) = e^x$; $[-1, \ln 5]$

61. (a) Find f_{ave} of $f(x) = x^2$ over $[0, 2]$.
 (b) Find a number x^* in $[0, 2]$ such that $f(x^*) = f_{ave}$.
 (c) Sketch the graph of $f(x) = x^2$ over $[0, 2]$ and construct a rectangle over the interval whose area is the same as the area under the graph of f over the interval.

62. (a) Find f_{ave} of $f(x) = 2x$ over $[0, 4]$.
 (b) Find a number x^* in $[0, 4]$ such that $f(x^*) = f_{ave}$.
 (c) Sketch the graph of $f(x) = 2x$ over $[0, 4]$ and construct a rectangle over the interval whose area is the same as the area under the graph of f over the interval.

63. (a) Suppose that the velocity function of a particle moving along a coordinate line is $v(t) = 3t^3 + 2$. Find the average velocity of the particle over the time interval $1 \leq t \leq 4$ by integrating.
 (b) Suppose that the position function of a particle moving along a coordinate line is $s(t) = 6t^2 + t$. Find the average velocity of the particle over the time interval $1 \leq t \leq 4$ algebraically.

64. (a) Suppose that the acceleration function of a particle moving along a coordinate line is $a(t) = t + 1$. Find the av-

erage acceleration of the particle over the time interval $0 \leq t \leq 5$ by integrating.
 (b) Suppose that the velocity function of a particle moving along a coordinate line is $v(t) = \cos t$. Find the average acceleration of the particle over the time interval $0 \leq t \leq \pi/4$ algebraically.

65. Water is run at a constant rate of 1 ft^3/min to fill a cylindrical tank of radius 3 ft and height 5 ft. Assuming that the tank is empty initially, make a conjecture about the average weight of the water in the tank over the time period required to fill it, and then check your conjecture by integrating. [Take the weight density of water to be 62.4 lb/ft^3.]

66. (a) The temperature of a 10-m-long metal bar is $15°C$ at one end and $30°C$ at the other end. Assuming that the temperature increases linearly from the cooler end to the hotter end, what is the average temperature of the bar?
 (b) Explain why there must be a point on the bar where the temperature is the same as the average, and find it.

67. (a) Suppose that a reservoir supplies water to an industrial park at a constant rate of $r = 4$ gallons per minute (gal/min) between 8:30 A.M. and 9:00 A.M. How much water does the reservoir supply during that time period?
 (b) Suppose that one of the industrial plants increases its water consumption between 9:00 A.M. and 10:00 A.M. and that the rate at which the reservoir supplies water increases linearly, as shown in the accompanying figure. How much water does the reservoir supply during that 1-hour time period?
 (c) Suppose that from 10:00 A.M. to 12 noon the rate at which the reservoir supplies water is given by the formula $r(t) = 10 + \sqrt{t}$ gal/min, where t is the time (in minutes) since 10:00 A.M. How much water does the reservoir supply during that 2-hour time period?

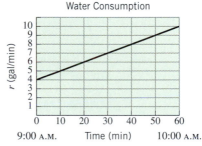

Water Consumption

Figure Ex-67

68. A traffic engineer monitors the rate at which cars enter the main highway during the afternoon rush hour. From her data she estimates that between 4:30 P.M. and 5:30 P.M. the rate $R(t)$ at which cars enter the highway is given by the formula $R(t) = 100(1 - 0.0001t^2)$ cars per minute, where t is the time (in minutes) since 4:30 P.M.

 (a) When does the peak traffic flow into the highway occur?
 (b) Estimate the number of cars that enter the highway during the rush hour.

69. (a) Prove: If f is continuous on $[a, b]$, then

$$\int_a^b [f(x) - f_{\text{ave}}]\, dx = 0$$

(b) Does there exist a constant $c \neq f_{\text{ave}}$ such that

$$\int_a^b [f(x) - c]\, dx = 0?$$

6.8 EVALUATING DEFINITE INTEGRALS BY SUBSTITUTION

In this section we will discuss two methods for evaluating definite integrals in which a substitution is required.

TWO METHODS FOR MAKING SUBSTITUTIONS IN DEFINITE INTEGRALS

Recall from Section 6.3 that indefinite integrals of the form

$$\int f(g(x))g'(x)\, dx$$

can sometimes be evaluated by making the u-substitution

$$u = g(x), \quad du = g'(x)\, dx \tag{1}$$

which converts the integral to the form

$$\int f(u)\, du$$

To apply this method to a definite integral of the form

$$\int_a^b f(g(x))g'(x)\, dx$$

we need to account for the effect that the substitution has on the x-limits of integration. There are two ways of doing this.

Method 1 First evaluate the indefinite integral

$$\int f(g(x))g'(x)\, dx$$

by substitution, and then use the relationship

$$\int_a^b f(g(x))g'(x)\, dx = \left[\int f(g(x))g'(x)\, dx \right]_a^b$$

to evaluate the definite integral. This procedure does not require any modification of the x-limits of integration.

Method 2 Make the substitution (1) directly in the definite integral, and then use the relationship $u = g(x)$ to replace the x-limits, $x = a$ and $x = b$, by corresponding u-limits, $u = g(a)$ and $u = g(b)$. This produces a new definite integral

$$\int_{g(a)}^{g(b)} f(u)\, du$$

that is expressed entirely in terms of u.

Example 1 Use the two methods above to evaluate $\displaystyle\int_0^2 x(x^2 + 1)^3\, dx$.

Solution by Method 1. If we let

$$u = x^2 + 1 \quad \text{so that} \quad du = 2x\, dx \tag{2}$$

then we obtain

$$\int x(x^2+1)^3\,dx = \frac{1}{2}\int u^3\,du = \frac{u^4}{8}+C = \frac{(x^2+1)^4}{8}+C$$

Thus,

$$\int_0^2 x(x^2+1)^3\,dx = \left[\int x(x^2+1)^3\,dx\right]_{x=0}^2 = \frac{(x^2+1)^4}{8}\Bigg]_{x=0}^2$$

$$= \frac{625}{8} - \frac{1}{8} = 78$$

Solution by Method 2. If we make the substitution $u = x^2+1$ in (2), then

if $x = 0, \quad u = 1$

if $x = 2, \quad u = 5$

Thus,

$$\int_0^2 x(x^2+1)^3\,dx = \frac{1}{2}\int_1^5 u^3\,du = \frac{u^4}{8}\Bigg]_{u=1}^5 = \frac{625}{8} - \frac{1}{8} = 78$$

which agrees with the result obtained by Method 1.   ◀

The following theorem states precise conditions under which Method 2 can be used.

6.8.1   THEOREM.   *If g' is continuous on $[a,b]$ and f is continuous on an interval containing the values of $g(x)$ for $a \le x \le b$, then*

$$\int_a^b f(g(x))g'(x)\,dx = \int_{g(a)}^{g(b)} f(u)\,du$$

Proof.   Since f is continuous on an interval containing the values of $g(x)$ for $a \le x \le b$, it follows that f has an antiderivative F on that interval. If we let $u = g(x)$, then the chain rule implies that

$$\frac{d}{dx}F(g(x)) = \frac{d}{dx}F(u) = \frac{dF}{du}\frac{du}{dx} = f(u)\frac{du}{dx} = f(g(x))g'(x)$$

for each x in $[a,b]$. Thus, $F(g(x))$ is an antiderivative of $f(g(x))g'(x)$ on $[a,b]$. Therefore, by Part 1 of the Fundamental Theorem of Calculus (Theorem 6.6.1)

$$\int_a^b f(g(x))g'(x)\,dx = F(g(x))\Bigg]_a^b = F(g(b)) - F(g(a)) = \int_{g(a)}^{g(b)} f(u)\,du \qquad ■$$

The choice of methods for evaluating definite integrals by substitution is generally a matter of taste, but in the following examples we will use the second method, since the idea is new.

Example 2   Evaluate

(a) $\displaystyle\int_0^{\pi/8} \sin^5 2x \cos 2x\,dx$ (b) $\displaystyle\int_2^5 (2x-5)(x-3)^9\,dx$

Solution (a).   Let

$$u = \sin 2x \quad \text{so that} \quad du = 2\cos 2x\,dx \quad \left(\text{or } \tfrac{1}{2}\,du = \cos 2x\,dx\right)$$

With this substitution,

$$
\begin{aligned}
&\text{if} \quad x = 0, \quad u = \sin(0) = 0 \\
&\text{if} \quad x = \pi/8, \quad u = \sin(\pi/4) = 1/\sqrt{2}
\end{aligned}
$$

so

$$
\int_0^{\pi/8} \sin^5 2x \cos 2x \, dx = \frac{1}{2} \int_0^{1/\sqrt{2}} u^5 \, du = \frac{1}{2} \cdot \frac{u^6}{6} \Bigg]_{u=0}^{1/\sqrt{2}}
$$

$$
= \frac{1}{2} \left[\frac{1}{6(\sqrt{2})^6} - 0 \right] = \frac{1}{96}
$$

Solution (b). Let

$$
u = x - 3 \quad \text{so that} \quad du = dx
$$

This leaves a factor of $2x + 5$ unresolved in the integrand. However,

$$
x = u + 3, \quad \text{so} \quad 2x - 5 = 2(u + 3) - 5 = 2u + 1
$$

With this substitution,

$$
\begin{aligned}
&\text{if} \quad x = 2, \quad u = 2 - 3 = -1 \\
&\text{if} \quad x = 5, \quad u = 5 - 3 = 2
\end{aligned}
$$

so

$$
\int_2^5 (2x - 5)(x - 3)^9 \, dx = \int_{-1}^2 (2u + 1)u^9 \, du = \int_{-1}^2 (2u^{10} + u^9) \, du
$$

$$
= \left[\frac{2u^{11}}{11} + \frac{u^{10}}{10} \right]_{u=-1}^2 = \left(\frac{2^{12}}{11} + \frac{2^{10}}{10} \right) - \left(-\frac{2}{11} + \frac{1}{10} \right)
$$

$$
= \frac{52{,}233}{110} \approx 474.8
$$

◀

Example 3 Evaluate

(a) $\displaystyle\int_0^{3/4} \frac{dx}{1 - x}$ (b) $\displaystyle\int_0^{\ln 3} e^x (1 + e^x)^{1/2} \, dx$

Solution (a). Let

$$
u = 1 - x \quad \text{so that} \quad du = -dx
$$

With this substitution,

$$
\begin{aligned}
&\text{if} \quad x = 0, \quad u = 1 \\
&\text{if} \quad x = \tfrac{3}{4}, \quad u = \tfrac{1}{4}
\end{aligned}
$$

Thus,

$$
\int_0^{3/4} \frac{dx}{1 - x} = -\int_1^{1/4} \frac{du}{u} = -\ln|u| \Bigg]_{u=1}^{1/4}
$$

$$
= -\left[\ln\left(\frac{1}{4} \right) - \ln(1) \right] = \ln 4
$$

Solution (b). Make the u-substitution

$$
u = 1 + e^x, \quad du = e^x \, dx
$$

and change the x-limits of integration ($x = 0$, $x = \ln 3$) to the u-limits

$$
u = 1 + e^0 = 2, \quad u = 1 + e^{\ln 3} = 1 + 3 = 4
$$

This yields

$$\int_{0}^{\ln 3} e^{x}(1+e^{x})^{1/2}\,dx = \int_{2}^{4} u^{1/2}\,du = \frac{2}{3}u^{3/2}\Big]_{2}^{4} = \frac{2}{3}[4^{3/2} - 2^{3/2}] = \frac{16 - 4\sqrt{2}}{3} \quad \blacktriangleleft$$

NEWTON'S LAW OF COOLING

Example 4   A glass of lemonade with a temperature of $40°$F is left to sit in a room whose temperature is a constant $70°$F. Using a principle of physics, called **Newton's Law of Cooling**, one can show that if the temperature of the lemonade reaches $52°$F in 1 hour, then the temperature T of the lemonade as a function of the elapsed time t is modeled by the equation

$$T = 70 - 30e^{-0.5t}$$

where T is in $°$F and t is in hours. The graph of this equation, shown in Figure 6.8.1, conforms to our everyday experience that the temperature of the lemonade gradually approaches the temperature of the room. Find the average temperature T_{ave} of the lemonade over the first 5 hours.

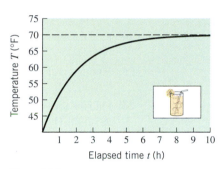

Figure 6.8.1

Solution.   From Definition 6.7.5 the average value of T over the time interval [0, 5] is

$$T_{\text{ave}} = \frac{1}{5}\int_{0}^{5}(70 - 30e^{-0.5t})\,dt \tag{3}$$

To evaluate this integral, we make the substitution

$$u = -0.5t \quad \text{so that} \quad du = -0.5\,dt \quad [\text{or } dt = -2\,du]$$

With this substitution,

$$\text{if} \quad t = 0, \quad u = 0$$
$$\text{if} \quad t = 5, \quad u = (-0.5)5 = -2.5$$

Thus, (3) can be expressed as

$$T_{\text{ave}} = \frac{1}{5}\int_{0}^{-2.5}(70 - 30e^{u})(-2)\,du = -\frac{2}{5}\int_{0}^{-2.5}(70 - 30e^{u})\,du$$

$$= -\frac{2}{5}\left[70u - 30e^{u}\right]_{u=0}^{-2.5} = -\frac{2}{5}\left[(-175 - 30e^{-2.5}) - (-30)\right]$$

$$= 58 + 12e^{-2.5} \approx 59°\text{F} \quad \blacktriangleleft$$

REMARK.   Observe that the u-substitution in this example produced an integral in which the upper u-limit of integration was smaller than the lower u-limit of integration. In our computations we left the limits of integration in that order, but we could have reversed the order to put the larger limit on top and compensated by reversing the sign of the integral in accordance with Definition 6.5.3(b). The choice of procedures is a matter of taste; both produce the same result (verify).

EXERCISE SET 6.8 C CAS

In Exercises 1–4, express the integral in terms of the variable u, but do not evaluate it.

1. (a) $\displaystyle\int_0^2 (x+1)^7\,dx;\ u = x+1$

(b) $\displaystyle\int_{-1}^2 x\sqrt{8-x^2}\,dx;\ u = 8-x^2$

(c) $\displaystyle\int_{-1}^1 \sin(\pi\theta)\,d\theta;\ u = \pi\theta$

(d) $\displaystyle\int_0^3 (x+2)(x-3)^{20}\,dx;\ u = x-3$

2. (a) $\displaystyle\int_{-1}^4 (5-2x)^8\,dx;\ u = 5-2x$

(b) $\displaystyle\int_{-\pi/3}^{2\pi/3} \frac{\sin x}{\sqrt{2+\cos x}}\,dx;\ u = 2+\cos x$

(c) $\displaystyle\int_0^{\pi/4} \tan^2 x\sec^2 x\,dx;\ u = \tan x$

(d) $\displaystyle\int_0^1 x^3\sqrt{x^2+3}\,dx;\ u = x^2+3$

3. (a) $\displaystyle\int_0^1 e^{2x-1}\,dx;\ u = 2x-1$

(b) $\displaystyle\int_e^{e^2} \frac{\ln x}{x}\,dx;\ u = \ln x$

4. (a) $\displaystyle\int_1^{\sqrt 3} \frac{\sqrt{\tan^{-1} x}}{1+x^2}\,dx;\ u = \tan^{-1} x$

(b) $\displaystyle\int_1^{\sqrt e} \frac{dx}{x\sqrt{1-(\ln x)^2}};\ u = \ln x$

In Exercises 5–18, evaluate the definite integral two ways: first by a u-substitution in the definite integral and then by a u-substitution in the corresponding indefinite integral.

5. $\displaystyle\int_0^1 (2x+1)^4\,dx$

6. $\displaystyle\int_1^2 (4x-2)^3\,dx$

7. $\displaystyle\int_{-1}^0 (1-2x)^3\,dx$

8. $\displaystyle\int_1^2 (4-3x)^8\,dx$

9. $\displaystyle\int_0^8 x\sqrt{1+x}\,dx$

10. $\displaystyle\int_{-5}^0 x\sqrt{4-x}\,dx$

11. $\displaystyle\int_0^{\pi/2} 4\sin(x/2)\,dx$

12. $\displaystyle\int_0^{\pi/6} 2\cos 3x\,dx$

13. $\displaystyle\int_{-2}^{-1} \frac{x}{(x^2+2)^3}\,dx$

14. $\displaystyle\int_{1-\pi}^{1+\pi} \sec^2\left(\tfrac14 x - \tfrac14\right)\,dx$

15. $\displaystyle\int_{-\ln 3}^{\ln 3} \frac{e^x}{e^x+4}\,dx$

16. $\displaystyle\int_0^{\ln 5} e^x(3-4e^x)\,dx$

17. $\displaystyle\int_1^3 \frac{dx}{\sqrt x\,(x+1)}$

18. $\displaystyle\int_{\ln 2}^{\ln(2/\sqrt 3)} \frac{e^{-x}\,dx}{\sqrt{1-e^{-2x}}}$

In Exercises 19–22, evaluate the definite integral by expressing it in terms of u and evaluating the resulting integral using a formula from geometry.

19. $\displaystyle\int_0^{5/3} \sqrt{25-9x^2}\,dx;\ u = 3x$

20. $\displaystyle\int_0^2 x\sqrt{16-x^4}\,dx;\ u = x^2$

21. $\displaystyle\int_{\pi/3}^{\pi/2} \sin\theta\sqrt{1-4\cos^2\theta}\,d\theta;\ u = 2\cos\theta$

22. $\displaystyle\int_{e^{-6}}^{e^6} \frac{\sqrt{36-(\ln x)^2}}{x}\,dx;\ u = \ln x$

23. Find the area under the curve $y = \sin\pi x$ over the interval $[0, 1]$.

24. Find the area under the curve $y = 3\cos 2x$ over the interval $[0, \pi/8]$.

25. Find the area under the curve $y = 1/(x+5)^2$ over the interval $[3, 7]$.

26. Find the area under the curve $y = 1/(3x+1)^2$ over the interval $[0, 1]$.

27. Find the area of the region enclosed by the graphs of $y = 1/\sqrt{1-9x^2}$, $y = 0$, $x = 0$, and $x = 1/6$.

28. Find the area of the region enclosed by the graphs of $y = \sin^{-1} x$, $x = 0$, and $y = \pi/2$.

29. Find the average value of
$$f(x) = \frac{x}{(5x^2+1)^2}$$
over the interval $[0, 2]$.

30. Find the average value of $f(x) = \sec^2\pi x$ over the interval $\left[-\tfrac14, \tfrac14\right]$.

31. Find the average value of $f(x) = e^{-2x}$ over the interval $[0, 4]$.

32. Find the average value of $f(x) = e^{3x}/(1+e^{6x})$ over the interval $[-(\ln 3)/6, 0]$.

In Exercises 33–52, evaluate the integrals by any method.

33. $\displaystyle\int_0^1 \frac{dx}{\sqrt{3x+1}}$

34. $\displaystyle\int_1^2 \sqrt{5x-1}\,dx$

35. $\displaystyle\int_{-1}^1 \frac{x^2\,dx}{\sqrt{x^3+9}}$

36. $\displaystyle\int_{-1}^0 6t^2(t^3+1)^{19}\,dt$

37. $\displaystyle\int_1^3 \frac{x+2}{\sqrt{x^2+4x+7}}\,dx$

38. $\displaystyle\int_1^2 \frac{dx}{x^2-6x+9}$

39. $\displaystyle\int_{-3\pi/4}^{\pi/4} \sin x\cos x\,dx$

40. $\displaystyle\int_0^{\pi/4} \sqrt{\tan x}\sec^2 x\,dx$

41. $\displaystyle\int_0^{\sqrt\pi} 5x\cos(x^2)\,dx$

42. $\displaystyle\int_{\pi^2}^{4\pi^2} \frac{1}{\sqrt x}\sin\sqrt x\,dx$

43. $\displaystyle\int_{\pi/12}^{\pi/9} \sec^2 3\theta\,d\theta$

44. $\displaystyle\int_0^{\pi/2} \sin^2 3\theta\cos 3\theta\,d\theta$

45. $\displaystyle\int_0^1 \frac{y^2\,dy}{\sqrt{4-3y}}$

46. $\displaystyle\int_{-1}^4 \frac{x\,dx}{\sqrt{5+x}}$

47. $\displaystyle\int_0^e \frac{dx}{x+e}$

48. $\displaystyle\int_1^{\sqrt{2}} xe^{-x^2}\,dx$

49. $\displaystyle\int_0^1 \frac{x}{\sqrt{4-3x^4}}\,dx$

50. $\displaystyle\int_1^2 \frac{1}{\sqrt{x}\sqrt{4-x}}\,dx$

51. $\displaystyle\int_0^{2/\sqrt{3}} \frac{1}{4+9x^2}\,dx$

52. $\displaystyle\int_1^{\sqrt{2}} \frac{x}{3+x^4}\,dx$

C **53.** (a) Use a CAS to find the exact value of the integral

$$\int_0^{\pi/6} \sin^4 x \cos^3 x\,dx$$

(b) Confirm the exact value by hand calculation.
[*Hint:* Use the identity $\cos^2 x = 1 - \sin^2 x$.]

C **54.** (a) Use a CAS to find the exact value of the integral

$$\int_{-\pi/4}^{\pi/4} \tan^4 x\,dx$$

(b) Confirm the exact value by hand calculation.
[*Hint:* Use the identity $1 + \tan^2 x = \sec^2 x$.]

55. (a) Find $\displaystyle\int_0^1 f(3x+1)\,dx$ if $\displaystyle\int_1^4 f(x)\,dx = 5$.

(b) Find $\displaystyle\int_0^3 f(3x)\,dx$ if $\displaystyle\int_0^9 f(x)\,dx = 5$.

(c) Find $\displaystyle\int_{-2}^0 xf(x^2)\,dx$ if $\displaystyle\int_0^4 f(x)\,dx = 1$.

56. Given that m and n are positive integers, show that

$$\int_0^1 x^m(1-x)^n\,dx = \int_0^1 x^n(1-x)^m\,dx$$

by making a substitution. Do not attempt to evaluate the integrals.

57. Given that n is a positive integer, show that

$$\int_0^{\pi/2} \sin^n x\,dx = \int_0^{\pi/2} \cos^n x\,dx$$

by using a trigonometric identity and making a substitution. Do not attempt to evaluate the integrals.

58. Given that n is a positive integer, evaluate the integral

$$\int_0^1 x(1-x)^n\,dx$$

59. Suppose that at time $t = 0$ there are 750 bacteria in a growth medium and the bacteria population $y(t)$ grows at the rate $y'(t) = 802.137e^{1.528t}$ bacteria per hour. How many bacteria will there be in 12 hours?

60. Suppose that the value of a yacht in dollars after t years of use is $V(t) = 275{,}000e^{-0.17t}$. What is the average value of the yacht over its first 10 years of use?

61. Suppose that a particle moving along a coordinate line has velocity $v(t) = 25 + 10e^{-0.05t}$ ft/s.
(a) What is the distance traveled by the particle from time $t = 0$ to time $t = 10$?

(b) Does the term $10e^{-0.05t}$ have much effect on the distance traveled by the particle over that time interval? Explain your reasoning.

62. Find a positive value of k such that the area under the graph of $y = e^{2x}$ over the interval $[0, k]$ is 3 square units.

∿ **63.** Estimate the value of k ($k > 0$) so that the region enclosed by $y = 1/(1+kx^2)$, $y = 0$, $x = 0$, and $x = 2$ has an area of 0.6 square unit.

C **64.** (a) Find the limit

$$\lim_{n \to +\infty} \sum_{k=1}^n \frac{\sin(k\pi/n)}{n}$$

by evaluating an appropriate definite integral over the interval $[0, 1]$.

(b) Check your answer to part (a) by evaluating the limit directly with a CAS.

65. Electricity is supplied to homes in the form of **alternating current**, which means that the voltage has a sinusoidal waveform described by an equation of the form

$$V = V_p \sin(2\pi ft)$$

(see the accompanying figure). In this equation, V_p is called the **peak voltage** or **amplitude** of the current, f is called its **frequency**, and $1/f$ is called its **period**. The voltages V and V_p are measured in volts (V), the time t is measured in seconds (s), and the frequency is measured in hertz (Hz) or sometimes in cycles per second. (A **cycle** is the electrical term for one period of the waveform.) Most alternating-current voltmeters read what is called the **rms** or **root-mean-square** value of V. By definition, this is the square root of the average value of V^2 over one period.
(a) Show that

$$V_{\text{rms}} = \frac{V_p}{\sqrt{2}}$$

[*Hint:* Compute the average over the cycle from $t = 0$ to $t = 1/f$, and use the identity $\sin^2\theta = \frac{1}{2}(1 - \cos 2\theta)$ to help evaluate the integral.]

(b) In the United States, electrical outlets supply alternating current with an rms voltage of 120 V at a frequency of 60 Hz. What is the peak voltage at such an outlet?

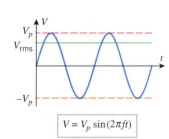

$$V = V_p \sin(2\pi ft)$$

Figure Ex-65

66. Show that if f and g are continuous functions, then

$$\int_0^t f(t-x)g(x)\,dx = \int_0^t f(x)g(t-x)\,dx$$

67. (a) Let $I = \int_0^a \dfrac{f(x)}{f(x) + f(a-x)}\, dx$. Show that $I = a/2$.

   [*Hint:* Let $u = a - x$, and then note the difference between the resulting integrand and 1.]

   (b) Use the result of part (a) to find

   $$\int_0^3 \frac{\sqrt{x}}{\sqrt{x} + \sqrt{3-x}}\, dx$$

   (c) Use the result of part (a) to find

   $$\int_0^{\pi/2} \frac{\sin x}{\sin x + \cos x}\, dx$$

68. Let $I = \int_{-1}^1 \dfrac{1}{1+x^2}\, dx$. Show that the substitution $x = 1/u$ results in

   $$I = -\int_{-1}^1 \frac{1}{1+u^2}\, du = -I$$

   so $2I = 0$, which implies that $I = 0$. However, this is impossible since the integrand of the given integral is positive over the interval of integration. Where is the error?

69. (a) Prove that if f is an odd function, then

   $$\int_{-a}^a f(x)\, dx = 0$$

   and give a geometric explanation of this result.
   [*Hint:* One way to prove that a quantity q is zero is to show that $q = -q$.]

   (b) Prove that if f is an even function, then

   $$\int_{-a}^a f(x)\, dx = 2\int_0^a f(x)\, dx$$

   and give a geometric explanation of this result. [*Hint:* Split the interval of integration from $-a$ to a into two parts at 0.]

70. Evaluate

   (a) $\displaystyle\int_{-1}^1 x\sqrt{\cos(x^2)}\, dx$

   (b) $\displaystyle\int_0^{\pi} \sin^8 x \cos^5 x\, dx$.

   [*Hint:* Use the substitution $u = x - (\pi/2)$.]

6.9 LOGARITHMIC FUNCTIONS FROM THE INTEGRAL POINT OF VIEW

In Section 4.2 we discussed natural logarithms from the viewpoint of exponents; that is, we took $y = \ln x$ to mean that $e^y = x$. In this section we will show that $\ln x$ can also be expressed as an integral with a variable upper limit. This integral representation of $\ln x$ is important mathematically because it provides a convenient way of establishing properties such as differentiability and continuity. However, it is also important in applications because it provides a way of recognizing when integral solutions of problems can be expressed as natural logarithms.

EXPONENTS

Our work earlier in Chapter 4 was built on the somewhat shaky foundation of extending our definition of exponential expressions b^x ($b > 0$) to allow for exponents that could be any real number. The process started by defining integer exponents by

$$b^0 = 1, \quad b^1 = b, \quad b^2 = b \cdot b, \quad b^3 = b \cdot b \cdot b, \ldots, \quad b^{-1} = \frac{1}{b}, \quad b^{-2} = \frac{1}{b^2}, \ldots$$

Rational exponents were defined as solutions to equations involving integer exponents:

   $b^{p/q}$ is the (positive) solution to $x^q = b^p$

For example, $2^{3.1}$ is the (positive) solution to $x^{10} = 2^{31}$. We claimed that this could be extended to irrational exponents via approximations using rational exponents. For example, it was argued that 2^{π} could be defined as the limiting value of the sequence

$$2^3, \quad 2^{3.1}, \quad 2^{3.14}, \quad 2^{3.141}, \quad 2^{3.1415}, \quad 2^{3.14159}, \ldots$$

where the exponents are successive terminating decimal approximations of π. We then claimed that the resulting exponential function $y = b^x$ is continuous on $(-\infty, +\infty)$ and has the familiar properties of exponents:

$$b^0 = 1, \quad b^{-p} = \frac{1}{b^p}, \quad b^{p+q} = b^p b^q, \quad b^{p-q} = \frac{b^p}{b^q}, \quad (b^p)^q = b^{pq}$$

We further claimed that (for $b > 0, b \neq 1$) $f(x) = b^x$ is a one-to-one function, so it has an inverse function that we named $\log_b x$. We also claimed that

$$\lim_{v \to +\infty} \left(1 + \frac{1}{v}\right)^v = e \quad \text{and} \quad \lim_{v \to -\infty} \left(1 + \frac{1}{v}\right)^v = e$$

which allowed us to use these limits to define e and to find derivatives of exponential and logarithmic functions:

$$\frac{d}{dx}[b^x] = b^x \log_e b \quad \text{and} \quad \frac{d}{dx}[\log_b x] = \frac{1}{x \log_e b}$$

In particular, defining $\ln x = \log_e x$, we have

$$\frac{d}{dx}[e^x] = e^x \quad \text{and} \quad \frac{d}{dx}[\ln x] = \frac{1}{x}$$

Now, for $x > 0$ we have

$$\int_1^x \frac{1}{t}\,dt = \ln t \Big]_1^x = \ln x - \ln 1 = \ln x \tag{1}$$

This relates the natural logarithm function $\ln x$ to a definite integral of a continuous function, an expression for which we have developed a precise definition.

FORMAL DEFINITION OF ln x

A rigorous approach to logarithmic and exponential functions uses (1) as a starting point to define $\ln x$ and defines the natural exponential function as the inverse function for $\ln x$. The challenge is then to demonstrate the consistency of these definitions with our familiar properties for logarithms and exponents.

6.9.1   DEFINITION.   The ***natural logarithm*** of x is denoted by $\ln x$ and is defined by the integral

$$\ln x = \int_1^x \frac{1}{t}\,dt, \quad x > 0 \tag{2}$$

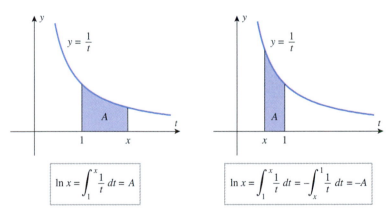

$$\ln x = \int_1^x \frac{1}{t}\,dt = A$$

$$\ln x = \int_1^x \frac{1}{t}\,dt = -\int_x^1 \frac{1}{t}\,dt = -A$$

Figure 6.9.1

Geometrically, $\ln x$ is the area under the curve $y = 1/t$ from $t = 1$ to $t = x$ when $x > 1$, and $\ln x$ is the negative of the area under the curve $y = 1/t$ from $t = x$ to $t = 1$ when $0 < x < 1$ (Figure 6.9.1). Since $1/t > 0$ for $t > 0$, $\ln x$ will be an increasing function on $(0, +\infty)$. Moreover, if $x = 1$, then $\ln x = 0$, since the upper and lower limits of (2) are the same. All of this is consistent with the computer-generated graph of $y = \ln x$ in Figure 4.2.4.

FOR THE READER.   Review Theorem 6.5.8, and then explain why x is required to be positive in Definition 6.9.1.

For specific values of x, the value of $\ln x$ can be approximated numerically by approximating the definite integral in (2), say by using the midpoint approximation that was discussed in Section 6.4.

Example 1   Approximate $\ln 2$ using the midpoint approximation with $n = 10$.

Solution.   From (2), the exact value of $\ln 2$ is represented by the integral

$$\ln 2 = \int_1^2 \frac{1}{t}\, dt$$

The midpoint rule is given in Formulas (8) and (9) of Section 6.4. Expressed in terms of t, the latter formula is

$$\int_a^b f(t)\, dt \approx \Delta t \sum_{k=1}^n f(t_k^*)$$

where Δt is the common width of the subintervals and $t_1^*, t_2^*, \ldots, t_n^*$ are the midpoints. In this case we have 10 subintervals, so $\Delta t = (2 - 1)/10 = 0.1$. The computations to six decimal places are shown in Table 6.9.1. By comparison, a calculator set to display six decimal places gives $\ln 2 \approx 0.693147$, so the magnitude of the error in the midpoint approximation is about 0.000311. Greater accuracy in the midpoint approximation can be obtained by increasing n. For example, the midpoint approximation with $n = 100$ yields $\ln 2 \approx 0.693144$, which is correct to five decimal places.   ◀

PROPERTIES OF ln x

Definition 6.9.1 is not only useful for approximating values of $\ln x$; it is the key to establishing many of the fundamental properties of the natural logarithm. For example, by Part 2 of the Fundamental Theorem of Calculus (Theorem 6.6.3), we have

$$\frac{d}{dx}[\ln x] = \frac{1}{x} \quad (x > 0) \tag{3}$$

In particular, the natural logarithm function is differentiable on $(0, +\infty)$, so we also have that $\ln x$ is continuous on $(0, +\infty)$.

We can use (3) to establish that our definition for $\ln x$ satisfies the expected logarithm properties.

Table 6.9.1

$$n = 10$$
$$\Delta t = (b - a)/n = (2 - 1)/10 = 0.1$$

k	t_k^*	$1/t_k^*$
1	1.05	0.952381
2	1.15	0.869565
3	1.25	0.800000
4	1.35	0.740741
5	1.45	0.689655
6	1.55	0.645161
7	1.65	0.606061
8	1.75	0.571429
9	1.85	0.540541
10	1.95	0.512821
		6.928355

$$\Delta t \sum_{k=1}^n f(t_k^*) \approx (0.1)(6.928355)$$
$$\approx 0.692836$$

6.9.2   THEOREM.   *For any positive numbers a and c and any rational number r:*

(a) $\ln ac = \ln a + \ln c$ (b) $\ln \dfrac{1}{c} = -\ln c$

(c) $\ln \dfrac{a}{c} = \ln a - \ln c$ (d) $\ln a^r = r \ln a$

Proof (a).   Treating a as a constant, consider the function $f(x) = \ln(ax)$. Then

$$f'(x) = \frac{1}{ax} \cdot \frac{d}{dx}(ax) = \frac{1}{ax} \cdot a = \frac{1}{x}$$

Thus, $\ln ax$ and $\ln x$ have the same derivative on $(0, +\infty)$, so these functions must differ by a constant on this interval. That is, there is a constant k such that

$$\ln ax - \ln x = k \tag{4}$$

on $(0, +\infty)$. Substituting $x = 1$ into this equation we conclude that $\ln a = k$ (verify). Thus, (4) can be written as

$$\ln ax - \ln x = \ln a$$

Setting $x = c$ establishes that

$$\ln ac - \ln c = \ln a \quad \text{or} \quad \ln ac = \ln a + \ln c$$

Proofs (b) and (c). Part (b) follows immediately from part (a) by substituting $1/c$ for a (verify). Then

$$\ln \frac{a}{c} = \ln \left(a \cdot \frac{1}{c} \right) = \ln a + \ln \frac{1}{c} = \ln a - \ln c$$

Proof (d). Since

$$\frac{d}{dx}[\ln x^r] = \frac{1}{x^r} \cdot \frac{d}{dx}[x^r] = \frac{1}{x^r} \cdot rx^{r-1} = \frac{r}{x}$$

and

$$\frac{d}{dx}[r \ln x] = r \cdot \frac{d}{dx}[\ln x] = \frac{r}{x}$$

the functions $\ln x^r$ and $r \ln x$ have the same derivative on $(0, +\infty)$. Thus, there is a constant k such that

$$\ln x^r - r \ln x = k$$

Substituting $x = 1$ into this equation we conclude that $k = 0$ (verify), so

$$\ln x^r - r \ln x = 0 \quad \text{or} \quad \ln x^r = r \ln x$$

Setting $x = a$ completes the proof. ■

The function $\ln x$ is defined and increasing for x in the interval $(0, +\infty)$. Now, for any integer N, if $x > 2^N$, then

$$\ln x > \ln 2^N = N \ln 2$$

by Theorem 6.9.2(d). Since

$$\ln 2 = \int_1^2 \frac{1}{t} \, dt > 0$$

$N \ln 2$ can be made arbitrarily large by choosing N appropriately, so

$$\lim_{x \to +\infty} \ln x = +\infty$$

Furthermore, by observing that $v = 1/x \to +\infty$ as $x \to 0^+$, we can use the preceding limit and Theorem 6.9.2(b) to conclude that

$$\lim_{x \to 0^+} \ln x = \lim_{v \to +\infty} \ln \frac{1}{v} = \lim_{v \to +\infty} (-\ln v) = -\infty$$

These results are summarized in the following theorem.

6.9.3 THEOREM.

(a) *The domain of $\ln x$ is $(0, +\infty)$.*

(b) $\displaystyle \lim_{x \to 0^+} \ln x = -\infty$ *and* $\displaystyle \lim_{x \to +\infty} \ln x = +\infty$

(c) *The range of $\ln x$ is $(-\infty, +\infty)$.*

DEFINITION OF e^x

In Section 4.2 we introduced e informally as the value of a limit, although we did not have the mathematical tools to prove the existence of this limit. We now give a precise definition of the number e and confirm that it matches the desired limit.

Since $\ln x$ is increasing and continuous on $(0, +\infty)$ with range $(-\infty, +\infty)$, there is exactly one (positive) solution to the equation $\ln x = 1$. We *define* e to be the unique solution to

$\ln x = 1$, so

$$\ln e = 1 \tag{5}$$

Furthermore, if x is any real number, there is a unique positive solution y to $\ln y = x$, so for irrational values of x we *define* e^x to be this solution. That is, when x is irrational, e^x is defined by

$$\ln e^x = x \tag{6}$$

Note that for rational values of x, we also have $\ln e^x = x \ln e = x$ from Theorem 6.9.2(*d*). Moreover, it follows immediately that $e^{\ln x} = x$ for any $x > 0$. Thus, (6) defines the exponential function for all real values of x as the inverse of the natural logarithm function.

6.9.4 DEFINITION. The inverse of the natural logarithm function $\ln x$ is denoted by e^x and is called the *natural exponential function*.

We can now establish the differentiability of e^x, confirm that

$$\frac{d}{dx}[e^x] = e^x$$

and verify the limits in Formulas (3)–(5) of Section 4.2.

6.9.5 THEOREM. *The natural exponential function e^x is differentiable on $(-\infty, +\infty)$ and its derivative is*

$$\frac{d}{dx}[e^x] = e^x$$

Proof. Because $\ln x$ is differentiable and

$$\frac{d}{dx}[\ln x] = \frac{1}{x} > 0$$

for all x in $(0, +\infty)$, it follows from Corollary 4.1.7, with $f(x) = \ln x$ and $f^{-1}(x) = e^x$, that e^x is differentiable on $(-\infty, +\infty)$ and its derivative is

$$\frac{d}{dx}\underbrace{[e^x]}_{f^{-1}(x)} = \underbrace{\frac{1}{1/e^x}}_{f'(f^{-1}(x))} = e^x \qquad \blacksquare$$

6.9.6 THEOREM.

(*a*) $\displaystyle\lim_{x \to 0} (1 + x)^{1/x} = e$ $\qquad$ (*b*) $\displaystyle\lim_{x \to +\infty} \left(1 + \frac{1}{x}\right)^x = e$ $\qquad$ (*c*) $\displaystyle\lim_{x \to -\infty} \left(1 + \frac{1}{x}\right)^x = e$

Proof. We will prove part (*a*); the proofs of parts (*b*) and (*c*) follow from this limit and are left as exercises. We first observe that

$$\frac{d}{dx}[\ln(x + 1)]\Big|_{x=0} = \frac{1}{x + 1} \cdot 1\Big|_{x=0} = 1$$

However, using the definition of the derivative, we obtain

$$\frac{d}{dx}[\ln(x + 1)]\Big|_{x=0} = \lim_{w \to 0} \frac{\ln(w + 1) - \ln(0 + 1)}{w - 0}$$

$$= \lim_{w \to 0}\left[\frac{1}{w} \cdot \ln(w + 1)\right] = \lim_{w \to 0}[\ln(w + 1)^{1/w}]$$

Thus,

$$1 = \lim_{w \to 0} [\ln(w+1)^{1/w}], \quad \text{so} \quad e = e^{\left(\lim_{w \to 0} [\ln(w+1)^{1/w}]\right)}$$

Since e^x is continuous on $(-\infty, +\infty)$, we can move the limit symbol through the function symbol, and once more using the inverse relationship between e^x and $\ln x$, we obtain

$$e = \lim_{w \to 0} e^{[\ln(w+1)^{1/w}]} = \lim_{w \to 0} (w+1)^{1/w}$$

which establishes the limit in part (a). ■

IRRATIONAL EXPONENTS

Recall from Theorem 6.9.2(d) that if $a > 0$ and r is a rational number, then $\ln a^r = r \ln a$. Then $a^r = e^{\ln a^r} = e^{r \ln a}$ for any positive value of a and any rational number r. But the expression $e^{r \ln a}$ makes sense for *any* real number r, whether rational or irrational, so it is a good candidate to give meaning to a^r for any real number r.

6.9.7   DEFINITION.   If $a > 0$ and r is a real number, a^r is defined by

$$a^r = e^{r \ln a} \tag{7}$$

With this definition it can be shown that the standard algebraic properties of exponents, such as

$$a^p a^q = a^{p+q}, \quad \frac{a^p}{a^q} = a^{p-q}, \quad (a^p)^q = a^{pq}, \quad (a^p)(b^p) = (ab)^p$$

hold for any real values of $a, b, p,$ and q, where a and b are positive. In addition, using (7) for a real exponent r, we can define the power function x^r whose domain consists of all positive real numbers and, for a positive base b, we can define the **base b exponential function b^x** whose domain consists of all real numbers.

6.9.8   THEOREM.

(a) *For any real number r, the power function x^r is differentiable on $(0, +\infty)$ and its derivative is*

$$\frac{d}{dx}[x^r] = rx^{r-1}$$

(b) *For $b > 0$ and $b \neq 1$, the base b exponential function b^x is differentiable on $(-\infty, +\infty)$ and its derivative is*

$$\frac{d}{dx}[b^x] = b^x \ln b$$

Proof.   The differentiability of $x^r = e^{r \ln x}$ and $b^x = e^{x \ln b}$ on their domains follows from the differentiability of $\ln x$ on $(0, +\infty)$ and of e^x on $(-\infty, +\infty)$:

$$\frac{d}{dx}[x^r] = \frac{d}{dx}[e^{r \ln x}] = e^{r \ln x} \cdot \frac{d}{dx}[r \ln x] = x^r \cdot \frac{r}{x} = rx^{r-1}$$

$$\frac{d}{dx}[b^x] = \frac{d}{dx}[e^{x \ln b}] = e^{x \ln b} \cdot \frac{d}{dx}[x \ln b] = b^x \ln b \qquad ■$$

GENERAL LOGARITHMS

We note that for $b > 0$ and $b \neq 1$, the function b^x is one-to-one, and so has an inverse function. Using the definition of b^x, we can solve $y = b^x$ for x as a function of y:

$$y = b^x = e^{x \ln b}$$
$$\ln y = \ln(e^{x \ln b}) = x \ln b$$
$$\frac{\ln y}{\ln b} = x$$

Thus, the inverse function for b^x is $(\ln x)/(\ln b)$.

6.9.9 DEFINITION. For $b > 0$ and $b \neq 1$, the **base b logarithm** function, denoted $\log_b x$, is defined by

$$\log_b x = \frac{\ln x}{\ln b} \tag{8}$$

It follows immediately from this definition that $\log_b x$ is the inverse function for b^x and satisfies the properties in Theorem 4.2.2. Furthermore, $\log_b x$ is differentiable on $(0, +\infty)$, and its derivative is

$$\frac{d}{dx}[\log_b x] = \frac{1}{x \ln b}$$

As a final note of consistency, we observe that $\log_e x = \ln x$.

FUNCTIONS DEFINED BY INTEGRALS

The functions we have dealt with thus far in this text are called **elementary functions**; they include polynomial, rational, power, exponential, logarithmic, and trigonometric functions, and all other functions that can be obtained from these by addition, subtraction, multiplication, division, root extraction, and composition.

However, there are many important functions that do not fall into this category. Such functions occur in many ways, but they commonly arise in the course of solving initial-value problems of the form

$$\frac{dy}{dx} = f(x), \quad y(x_0) = y_0 \tag{9}$$

Recall from Example 7 of Section 6.2 and the discussion preceding it that the basic method for solving (9) is to integrate $f(x)$, and then use the initial condition to determine the constant of integration. It can be proved that if f is continuous, then (9) has a unique solution and that this procedure produces it. However, there is another approach: Instead of solving each initial-value problem individually, we can find a general formula for the solution of (9), and then apply that formula to solve specific problems. We will now show that

$$y(x) = y_0 + \int_{x_0}^{x} f(t)\, dt \tag{10}$$

is a formula for the solution of (9). To confirm that this is so we must show that $dy/dx = f(x)$ and that $y(x_0) = y_0$. The computations are as follows:

$$\frac{dy}{dx} = \frac{d}{dx}\left[y_0 + \int_{x_0}^{x} f(t)\, dt\right] = 0 + f(x) = f(x)$$

$$y(x_0) = y_0 + \int_{x_0}^{x_0} f(t)\, dt = y_0 + 0 = y_0$$

Example 2 In Example 7 of Section 6.2 we showed that the solution of the initial-value problem

$$\frac{dy}{dx} = \cos x, \quad y(0) = 1$$

is $y(x) = 1 + \sin x$. This initial-value problem can also be solved by applying Formula (10) with $f(x) = \cos x$, $x_0 = 0$, and $y_0 = 1$. This yields

$$y(x) = 1 + \int_0^x \cos t\, dt = 1 + \left[\sin t\right]_{t=0}^{x} = 1 + \sin x \quad \blacktriangleleft$$

In the last example we were able to perform the integration in Formula (10) and express the solution of the initial-value problem as an elementary function. However, sometimes this will not be possible, in which case the solution of the initial-value problem must be left in terms of an "unevaluated" integral. For example, from (10), the solution of the initial-value problem

$$\frac{dy}{dx} = e^{-x^2}, \quad y(0) = 1$$

is

$$y(x) = 1 + \int_0^x e^{-t^2}\, dt$$

However, it can be shown that there is no way to express the integral in this solution as an elementary function. Thus, we have encountered a *new* function, which we regard to be *defined* by the integral. A close relative of this function, known as the ***error function***, plays an important role in probability and statistics; it is denoted by $\mathrm{erf}(x)$ and is defined as

$$\mathrm{erf}(x) = \frac{2}{\sqrt{\pi}} \int_0^x e^{-t^2}\, dt \tag{11}$$

Indeed, many of the most important functions in science and engineering are defined as integrals that have special names and notations associated with them. For example, the functions defined by

$$S(x) = \int_0^x \sin\left(\frac{\pi t^2}{2}\right) dt \quad \text{and} \quad C(x) = \int_0^x \cos\left(\frac{\pi t^2}{2}\right) dt \tag{12–13}$$

are called the ***Fresnel sine and cosine functions***, respectively, in honor of the French physicist Augustin Fresnel (1788–1827), who first encountered them in his study of diffraction of light waves.

EVALUATING AND GRAPHING FUNCTIONS DEFINED BY INTEGRALS

The following values of $S(1)$ and $C(1)$ were produced by a CAS that has a built-in algorithm for approximating definite integrals:

$$S(1) = \int_0^1 \sin\left(\frac{\pi t^2}{2}\right) dt \approx 0.438259, \qquad C(1) = \int_0^1 \cos\left(\frac{\pi t^2}{2}\right) dt \approx 0.779893$$

To generate graphs of functions defined by integrals, computer programs choose a set of x-values in the domain, approximate the integral for each of those values, and then plot the resulting points. Thus, there is a lot of computation involved in generating such graphs, since each plotted point requires the approximation of an integral. The graphs of the Fresnel functions in Figure 6.9.2 were generated in this way using a CAS.

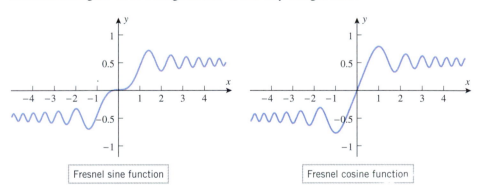

Figure 6.9.2

Fresnel sine function

Fresnel cosine function

REMARK.   Although it required a considerable amount of computation to generate the graphs of the Fresnel functions, the derivatives of $S(x)$ and $C(x)$ are easy to obtain using Part 2 of the Fundamental Theorem of Calculus (6.6.3); they are

$$S'(x) = \sin\left(\frac{\pi x^2}{2}\right) \quad \text{and} \quad C'(x) = \cos\left(\frac{\pi x^2}{2}\right) \tag{14–15}$$

These derivatives can be used to determine the locations of the relative extrema and inflection points and to investigate other properties of $S(x)$ and $C(x)$.

Various applications can lead to integrals in which one or both of the limits of integration is a function of x. Some examples are

$$\int_x^1 \sqrt{\sin t}\, dt, \qquad \int_{x^2}^{\sin x} \sqrt{t^3 + 1}\, dt, \qquad \int_{\ln x}^{\pi} \frac{dt}{t^7 - 8}$$

We will complete this section by showing how to differentiate integrals of the form

$$\int_a^{g(x)} f(t)\, dt \tag{16}$$

where a is constant. Derivatives of other kinds of integrals with functions as limits of integration will be discussed in the exercises.

To differentiate (16) we can view the integral as a composition $F(g(x))$, where

$$F(x) = \int_a^x f(t)\, dt$$

If we now apply the chain rule, we obtain

$$\frac{d}{dx}\left[\int_a^{g(x)} f(t)\, dt\right] = \frac{d}{dx}[F(g(x))] = F'(g(x))g'(x) = f(g(x))g'(x)$$

$$\boxed{\text{Theorem 6.6.3}}$$

Thus,

$$\frac{d}{dx}\left[\int_a^{g(x)} f(t)\, dt\right] = f(g(x))g'(x) \tag{17}$$

In words:

> *To differentiate an integral with a constant lower limit and a function as the upper limit, substitute the upper limit into the integrand, and multiply by the derivative of the upper limit.*

Example 3

$$\frac{d}{dx}\left[\int_1^{\sin x} (1 - t^2)\, dt\right] = (1 - \sin^2 x)\cos x = \cos^3 x \qquad \blacktriangleleft$$

The connection between natural logarithms and integrals was made in the middle of the seventeenth century in the course of investigating areas under the curve $y = 1/t$. The problem being considered was to find values of $t_1, t_2, t_3, \ldots, t_n, \ldots$ for which the areas $A_1, A_2, A_3, \ldots, A_n, \ldots$ in Figure 6.9.3a would be equal. Through the combined work of

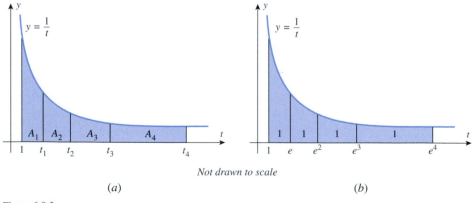

Not drawn to scale

(a)   (b)

Figure 6.9.3

Isaac Newton, the Belgian Jesuit priest, Gregory of St. Vincent (1584–1667), and Gregory's student, Alfons A. de Sarasa (1618–1667), it was shown that by taking the points to be

$$t_1 = e, \quad t_2 = e^2, \quad t_3 = e^3, \dots, \quad t_n = e^n, \dots$$

each of the areas would be 1 (Figure 6.9.3b). Thus, in modern integral notation

$$\int_1^{e^n} \frac{1}{t}\, dt = n$$

which can be expressed as

$$\int_1^{e^n} \frac{1}{t}\, dt = \ln(e^n)$$

By comparing the upper limit of the integral and the expression inside the logarithm, it is a natural leap to the more general result

$$\int_1^x \frac{1}{t}\, dt = \ln x$$

which today we take as the formal definition of the natural logarithm.

EXERCISE SET 6.9 ~ Graphing Utility C CAS

1. Sketch the curve $y = 1/t$, and shade a region under the curve whose area is
   (a) $\ln 2$ (b) $-\ln 0.5$ (c) 2.

2. Sketch the curve $y = 1/t$, and shade two different regions under the curve whose areas are $\ln 1.5$.

3. Given that $\ln a = 2$ and $\ln c = 5$, find
   (a) $\displaystyle\int_1^{ac} \frac{1}{t}\, dt$ (b) $\displaystyle\int_1^{1/c} \frac{1}{t}\, dt$
   (c) $\displaystyle\int_1^{a/c} \frac{1}{t}\, dt$ (d) $\displaystyle\int_1^{a^3} \frac{1}{t}\, dt$.

4. Given that $\ln a = 9$, find
   (a) $\displaystyle\int_1^{\sqrt{a}} \frac{1}{t}\, dt$ (b) $\displaystyle\int_1^{2a} \frac{1}{t}\, dt$
   (c) $\displaystyle\int_1^{2/a} \frac{1}{t}\, dt$ (d) $\displaystyle\int_2^{a} \frac{1}{t}\, dt$.

5. Approximate $\ln 5$ using the midpoint rule with $n = 10$, and estimate the magnitude of the error by comparing your answer to that produced directly by a calculating utility.

6. Approximate $\ln 3$ using the midpoint rule with $n = 20$, and estimate the magnitude of the error by comparing your answer to that produced directly by a calculating utility.

7. Simplify the expression and state the values of x for which your simplification is valid.
   (a) $e^{-\ln x}$ (b) $e^{\ln x^2}$
   (c) $\ln\left(e^{-x^2}\right)$ (d) $\ln(1/e^x)$
   (e) $\exp(3\ln x)$ (f) $\ln(xe^x)$
   (g) $\ln\left(e^{x - \sqrt[3]{x}}\right)$ (h) $e^{x - \ln x}$

8. (a) Let $f(x) = e^{-2x}$. Find the simplest exact value of the function $f(\ln 3)$.
   (b) Let $f(x) = e^x + 3e^{-x}$. Find the simplest exact value of the function $f(\ln 2)$.

In Exercises 9 and 10, express the given quantity as a power of e.

9. (a) 3^π (b) $2^{\sqrt{2}}$

10. (a) π^{-x} (b) $x^{2x}, \quad x > 0$

In Exercises 11 and 12, find the limits by making appropriate substitutions in the limits given in Theorem 6.9.6.

11. (a) $\displaystyle\lim_{x \to +\infty} \left(1 + \frac{1}{x}\right)^{2x}$ (b) $\displaystyle\lim_{x \to 0} (1 + 2x)^{1/x}$

12. (a) $\displaystyle\lim_{x \to +\infty} \left(1 + \frac{1}{3x}\right)^x$ (b) $\displaystyle\lim_{x \to 0} (1 + x)^{1/(3x)}$

In Exercises 13 and 14, find $g'(x)$ using Part 2 of the Fundamental Theorem of Calculus, and check your answer by evaluating the integral and then differentiating.

13. $g(x) = \displaystyle\int_1^x (t^2 - t)\, dt$ 14. $g(x) = \displaystyle\int_\pi^x (1 - \cos t)\, dt$

In Exercises 15 and 16, find the derivative using Formula (17), and check your answer by evaluating the integral and then differentiating.

15. (a) $\dfrac{d}{dx}\displaystyle\int_1^{x^3}\dfrac{1}{t}\,dt$ (b) $\dfrac{d}{dx}\displaystyle\int_1^{\ln x}e^t\,dt$

16. (a) $\dfrac{d}{dx}\displaystyle\int_{-1}^{x^2}\sqrt{t+1}\,dt$ (b) $\dfrac{d}{dx}\displaystyle\int_{\pi}^{1/x}\sin t\,dt$

17. Let $F(x)=\displaystyle\int_0^x\dfrac{\cos t}{t^2+3}\,dt$. Find

    (a) $F(0)$ (b) $F'(0)$ (c) $F''(0)$.

18. Let $F(x)=\displaystyle\int_2^x\sqrt{3t^2+1}\,dt$. Find

    (a) $F(2)$ (b) $F'(2)$ (c) $F''(2)$.

C   **19.** (a) Use Formula (17) to find

$$\dfrac{d}{dx}\int_1^{x^2}t\sqrt{1+t}\,dt$$

    (b) Use a CAS to evaluate the integral and differentiate the resulting function.

    (c) Use the simplification command of the CAS, if necessary, to confirm that answers in parts (a) and (b) are the same.

20. Show that

    (a) $\dfrac{d}{dx}\left[\displaystyle\int_x^a f(t)\,dt\right]=-f(x)$

    (b) $\dfrac{d}{dx}\left[\displaystyle\int_{g(x)}^a f(t)\,dt\right]=-f(g(x))g'(x).$

> In Exercises 21 and 22, use the results in Exercise 20 to find the derivative.

21. (a) $\dfrac{d}{dx}\displaystyle\int_x^1\sin(t^2)\,dt$ (b) $\dfrac{d}{dx}\displaystyle\int_{\tan x}^3\dfrac{t^2}{1+t^2}\,dt$

22. (a) $\dfrac{d}{dx}\displaystyle\int_x^0(t^2+1)^{40}\,dt$ (b) $\dfrac{d}{dx}\displaystyle\int_{1/x}^\pi\cos^3 t\,dt$

23. Find

$$\dfrac{d}{dx}\left[\int_{3x}^{x^2}\dfrac{t-1}{t^2+1}\,dt\right]$$

by writing

$$\int_{3x}^{x^2}\dfrac{t-1}{t^2+1}\,dt=\int_{3x}^0\dfrac{t-1}{t^2+1}\,dt+\int_0^{x^2}\dfrac{t-1}{t^2+1}\,dt$$

24. Use Exercise 20(b) and the idea in Exercise 23 to show that

$$\dfrac{d}{dx}\int_{h(x)}^{g(x)}f(t)\,dt=f(g(x))g'(x)-f(h(x))h'(x)$$

25. Use the result obtained in Exercise 24 to perform the following differentiations:

    (a) $\dfrac{d}{dx}\displaystyle\int_{x^2}^{x^3}\sin^2 t\,dt$ (b) $\dfrac{d}{dx}\displaystyle\int_{-x}^x\dfrac{1}{1+t}\,dt.$

26. Prove that the function

$$F(x)=\int_x^{3x}\dfrac{1}{t}\,dt$$

is constant on the interval $(0,+\infty)$ by using Exercise 24 to find $F'(x)$. What is that constant?

27. Let $F(x)=\displaystyle\int_0^x f(t)\,dt$, where f is the function whose graph is shown in the accompanying figure.

    (a) Find $F(0)$, $F(3)$, $F(5)$, $F(7)$, and $F(10)$.

    (b) On what subintervals of the interval $[0, 10]$ is F increasing? Decreasing?

    (c) Where does F have its maximum value? Its minimum value?

    (d) Sketch the graph of F.

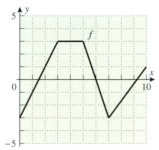

Figure Ex-27

28. Use the appropriate values found in part (a) of Exercise 27 to find the average value of f over the interval $[0, 10]$.

> In Exercises 29 and 30, express $F(x)$ in a piecewise form that does not involve an integral.

29. $F(x)=\displaystyle\int_{-1}^x|t|\,dt$

30. $F(x)=\displaystyle\int_0^x f(t)\,dt$, where $f(x)=\begin{cases}x, & 0\le x\le 2\\ 2, & x>2\end{cases}$

> In Exercises 31–34, use Formula (10) to solve the initial-value problem.

31. $\dfrac{dy}{dx}=\sqrt[3]{x}$; $y(1)=2$ **32.** $\dfrac{dy}{dx}=\dfrac{x+1}{\sqrt{x}}$; $y(1)=0$

33. $\dfrac{dy}{dx}=\sec^2 x-\sin x$; $y(\pi/4)=1$

34. $\dfrac{dy}{dx}=xe^{x^2}$; $y(0)=0$

35. Suppose that at time $t=0$ there are P_0 individuals who have disease X, and suppose that a certain model for the spread of the disease predicts that the disease will spread at the rate of $r(t)$ individuals per day. Write a formula for the number of individuals who will have disease X after x days.

36. Suppose that $v(t)$ is the velocity function of a particle moving along an s-axis. Write a formula for the coordinate of the particle at time T if the particle is at s_1 at time $t=1$.

37. The accompanying figure shows the graphs of $y=f(x)$ and $y=\displaystyle\int_0^x f(t)\,dt$. Determine which graph is which, and explain your reasoning.

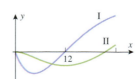

Figure Ex-37

38. (a) Make a conjecture about the value of the limit

$$\lim_{k\to 0}\int_1^b t^{k-1}\,dt \quad (b>0)$$

(b) Check your conjecture by evaluating the integral and finding the limit. [*Hint:* Interpret the limit as the definition of the derivative of an exponential function.]

39. Let $F(x)=\int_0^x f(t)\,dt$, where f is the function graphed in the accompanying figure.

(a) Where do the relative minima of F occur?

(b) Where do the relative maxima of F occur?

(c) Where does the absolute maximum of F on the interval $[0, 5]$ occur?

(d) Where does the absolute minimum of F on the interval $[0, 5]$ occur?

(e) Where is F concave up? Concave down?

(f) Sketch the graph of F.

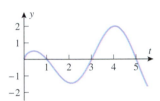

Figure Ex-39

[c] **40.** CAS programs have commands for working with most of the important nonelementary functions. Check your CAS documentation for information about the error function erf(x) [see Formula (11)], and then complete the following.

(a) Generate the graph of erf(x).

(b) Use the graph to make a conjecture about the existence and location of any relative maxima and minima of erf(x).

(c) Check your conjecture in part (b) using the derivative of erf(x).

(d) Use the graph to make a conjecture about the existence and location of any inflection points of erf(x).

(e) Check your conjecture in part (d) using the second derivative of erf(x).

(f) Use the graph to make a conjecture about the existence of horizontal asymptotes of erf(x).

(g) Check your conjecture in part (f) by using the CAS to find the limits of erf(x) as $x\to\pm\infty$.

41. The Fresnel sine and cosine functions $S(x)$ and $C(x)$ were defined in Formulas (12) and (13) and graphed in Figure 6.9.2. Their derivatives were given in Formulas (14) and (15).

(a) At what points does $C(x)$ have relative minima? Relative maxima?

(b) Where do the inflection points of $C(x)$ occur?

(c) Confirm that your answers in parts (a) and (b) are consistent with the graph of $C(x)$.

42. Find the limit

$$\lim_{h\to 0}\frac{1}{h}\int_x^{x+h}\ln t\,dt$$

43. Find a function f and a number a such that

$$2+\int_a^x f(t)\,dt = e^{3x}$$

44. (a) Give a geometric argument to show that

$$\frac{1}{x+1} < \int_x^{x+1}\frac{1}{t}\,dt < \frac{1}{x}, \quad x>0$$

(b) Use the result in part (a) to prove that

$$\frac{1}{x+1} < \ln\left(1+\frac{1}{x}\right) < \frac{1}{x}, \quad x>0$$

(c) Use the result in part (b) to prove that

$$e^{x/(x+1)} < \left(1+\frac{1}{x}\right)^x < e, \quad x>0$$

and hence that

$$\lim_{x\to+\infty}\left(1+\frac{1}{x}\right)^x = e$$

(d) Use the inequality in part (c) to prove that

$$\left(1+\frac{1}{x}\right)^x < e < \left(1+\frac{1}{x}\right)^{x+1}, \quad x>0$$

45. Use a graphing utility to generate the graph of

$$y = \left(1+\frac{1}{x}\right)^{x+1} - \left(1+\frac{1}{x}\right)^x$$

in the window $[0, 100]\times[0, 0.2]$, and use that graph and part (d) of Exercise 44 to make a rough estimate of the error in the approximation

$$e \approx \left(1+\frac{1}{50}\right)^{50}$$

46. Prove: If f is continuous on an open interval I and a is any point in I, then

$$F(x) = \int_a^x f(t)\,dt$$

is continuous on I.

C CAS

1. Write a paragraph that describes the *rectangle method* for defining the area under a curve $y = f(x)$ over an interval $[a, b]$.

2. What is an *integral curve* of a function f? How are two integral curves of a function f related?

3. The *definite integral* of f over the interval $[a, b]$ is defined as the limit

$$\int_a^b f(x)\,dx = \lim_{\max \Delta x_k \to 0} \sum_{k=1}^n f(x_k^*)\Delta x_k$$

Explain what the various symbols on the right side of this equation mean.

4. State the two parts of the Fundamental Theorem of Calculus, and explain what is meant by the phrase "differentiation and integration are inverse processes."

5. Derive the formulas for the position and velocity functions of a particle that moves with uniformly accelerated motion along a coordinate line.

6. (a) Devise a procedure for finding upper and lower estimates of the area of the region in the accompanying figure (in cm²).
   (b) Use your procedure to find upper and lower estimates of the area.
   (c) Improve on the estimates you obtained in part (b).

Figure Ex-6

7. Suppose that

$$\int_0^1 f(x)\,dx = \tfrac{1}{2}, \quad \int_1^2 f(x)\,dx = \tfrac{1}{4},$$

$$\int_0^3 f(x)\,dx = -1, \quad \int_0^1 g(x)\,dx = 2$$

In each part, use this information to evaluate the given integral, if possible. If there is not enough information to evaluate the integral, then say so.

(a) $\displaystyle\int_0^2 f(x)\,dx$ (b) $\displaystyle\int_1^3 f(x)\,dx$ (c) $\displaystyle\int_2^3 5f(x)\,dx$

(d) $\displaystyle\int_1^0 g(x)\,dx$ (e) $\displaystyle\int_0^1 g(2x)\,dx$ (f) $\displaystyle\int_0^1 [g(x)]^2\,dx$

8. In each part, use the information in Exercise 7 to evaluate the given integral. If there is not enough information to evaluate the integral, then say so.

(a) $\displaystyle\int_0^1 [f(x) + g(x)]\,dx$ (b) $\displaystyle\int_0^1 f(x)g(x)\,dx$

(c) $\displaystyle\int_0^1 \frac{f(x)}{g(x)}\,dx$ (d) $\displaystyle\int_0^1 [4g(x) - 3f(x)]\,dx$

9. In each part, evaluate the integral. Where appropriate, you may use a geometric formula.

(a) $\displaystyle\int_{-1}^1 (1 + \sqrt{1 - x^2})\,dx$

(b) $\displaystyle\int_0^3 (x\sqrt{x^2 + 1} - \sqrt{9 - x^2})\,dx$

(c) $\displaystyle\int_0^1 x\sqrt{1 - x^4}\,dx$

10. Evaluate the integral $\int_0^1 |2x - 1|\,dx$, and sketch the region whose area it represents.

11. One of the numbers π, $\pi/2$, $35\pi/128$, $1 - \pi$ is the correct value of the integral

$$\int_0^\pi \sin^8 x\,dx$$

Use the accompanying graph of $y = \sin^8 x$ and a logical process of elimination to find the correct value. [Do not attempt to evaluate the integral.]

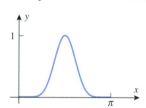

Figure Ex-11

12. Evaluate $\displaystyle\int \frac{e^{2x}}{e^x + 3}\,dx$

    [*Hint:* Divide $e^x + 3$ into e^{2x}.]

13. Give a convincing geometric argument to show that

$$\int_1^e \ln x\,dx + \int_0^1 e^x\,dx = e$$

14. In each part, find the limit by interpreting it as a limit of Riemann sums in which the interval $[0, 1]$ is divided into n subintervals of equal length.

(a) $\displaystyle\lim_{n \to +\infty} \frac{\sqrt{1} + \sqrt{2} + \sqrt{3} + \cdots + \sqrt{n}}{n^{3/2}}$

(b) $\displaystyle\lim_{n \to +\infty} \frac{1^4 + 2^4 + 3^4 + \cdots + n^4}{n^5}$

(c) $\displaystyle\lim_{n \to +\infty} \frac{e^{1/n} + e^{2/n} + e^{3/n} + \cdots + e^{n/n}}{n}$

15. (a) Divide the interval $[1, 2]$ into 5 subintervals of equal length, and use appropriate Riemann sums to show that

$$0.2\left[\frac{1}{1.2} + \frac{1}{1.4} + \frac{1}{1.6} + \frac{1}{1.8} + \frac{1}{2.0}\right] < \ln 2$$

$$< 0.2\left[\frac{1}{1.0} + \frac{1}{1.2} + \frac{1}{1.4} + \frac{1}{1.6} + \frac{1}{1.8}\right]$$

(b) Show that if the interval $[1, 2]$ is divided into n subintervals of equal length, then

$$\sum_{k=1}^{n} \frac{1}{n+k} < \ln 2 < \sum_{k=0}^{n-1} \frac{1}{n+k}$$

(c) Show that the difference between the two sums in part (b) is $1/(2n)$, and use this result to show that the sums in part (a) approximate $\ln 2$ with an error of at most 0.1.

(d) How large must n be to ensure that the sums in part (b) approximate $\ln 2$ to three decimal places?

16. The accompanying figure shows the direction field for a differential equation $dy/dx = f(x)$. Which of the following functions is most likely to be $f(x)$?

$$\sqrt{x}, \quad \sin x, \quad x^4, \quad x$$

Explain your reasoning.

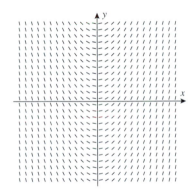

Figure Ex-16

17. In each part, confirm the stated equality.

(a) $1 \cdot 2 + 2 \cdot 3 + \cdots + n(n+1) = \frac{1}{3}n(n+1)(n+2)$

(b) $\lim\limits_{n \to +\infty} \sum\limits_{k=1}^{n-1}\left(\frac{9}{n} - \frac{k}{n^2}\right) = \frac{17}{2}$

(c) $\sum\limits_{i=1}^{3}\left(\sum\limits_{j=1}^{2}(i+j)\right) = 21$

18. Express

$$\sum_{k=4}^{18} k(k-3)$$

in sigma notation with

(a) $k = 0$ as the lower limit of summation

(b) $k = 5$ as the lower limit of summation.

19. The accompanying figure shows a square that is n units by n units that has been subdivided into a one-unit square and $n - 1$ "L-shaped" regions. Use this figure to show that the sum of the first n consecutive positive odd integers is n^2.

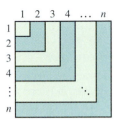

Figure Ex-19

20. Derive the result of Exercise 19 by writing

$$1 + 3 + 5 + \cdots + 2n - 1 = \sum_{k=1}^{n}(2k - 1)$$

When part of each term of a sum cancels part of the next term, leaving only portions of the first and last terms at the end, the sum is said to *telescope*. In Exercises 21–24, evaluate the telescoping sum.

21. $\sum\limits_{k=5}^{17}(3^k - 3^{k-1})$

22. $\sum\limits_{k=1}^{50}\left(\frac{1}{k} - \frac{1}{k+1}\right)$

23. $\sum\limits_{k=2}^{20}\left(\frac{1}{k^2} - \frac{1}{(k-1)^2}\right)$

24. $\sum\limits_{k=1}^{100}(2^{k+1} - 2^k)$

25. (a) Show that

$$\frac{1}{1 \cdot 3} + \frac{1}{3 \cdot 5} + \cdots + \frac{1}{(2n-1)(2n+1)} = \frac{n}{2n+1}$$

$$\left[Hint: \frac{1}{(2n-1)(2n+1)} = \frac{1}{2}\left(\frac{1}{2n-1} - \frac{1}{2n+1}\right).\right]$$

(b) Use the result in part (a) to find

$$\lim_{n \to +\infty} \sum_{k=1}^{n} \frac{1}{(2k-1)(2k+1)}$$

26. (a) Show that

$$\frac{1}{1 \cdot 2} + \frac{1}{2 \cdot 3} + \frac{1}{3 \cdot 4} + \cdots + \frac{1}{n(n+1)} = \frac{n}{n+1}$$

$$\left[Hint: \frac{1}{n(n+1)} = \frac{1}{n} - \frac{1}{n+1}.\right]$$

(b) Use the result in part (a) to find

$$\lim_{n \to +\infty} \sum_{k=1}^{n} \frac{1}{k(k+1)}$$

27. Let $\bar{x}$ denote the arithmetic average of the n numbers $x_1, x_2, \ldots, x_n$. Use Theorem 6.4.1 to prove that

$$\sum_{i=1}^{n}(x_i - \bar{x}) = 0$$

28. Let
$$S = \sum_{k=0}^{n} ar^k$$
Show that $S - rS = a - ar^{n+1}$ and hence that
$$\sum_{k=0}^{n} ar^k = \frac{a - ar^{n+1}}{1 - r} \quad (r \neq 1)$$
(A sum of this form is called a **geometric sum**.)

29. In each part, rewrite the sum, if necessary, so that the lower limit is 0, and then use the formula derived in Exercise 28 to evaluate the sum. Check your answers using the summation feature of a calculating utility.

(a) $\displaystyle\sum_{k=1}^{20} 3^k$ (b) $\displaystyle\sum_{k=5}^{30} 2^k$ (c) $\displaystyle\sum_{k=0}^{100} (-1)^{k+1} \frac{1}{2^k}$

c 30. In each part, make a conjecture about the limit by using a CAS to evaluate the sum for $n = 10$, 20, and 50; and then check your conjecture by using the formula in Exercise 28 to express the sum in closed form, and then finding the limit exactly.

(a) $\displaystyle\lim_{n \to +\infty} \sum_{k=0}^{n} \frac{1}{2^k}$ (b) $\displaystyle\lim_{n \to +\infty} \sum_{k=1}^{n} \left(\frac{3}{4}\right)^k$

31. (a) Show that the substitutions $u = \sec x$ and $u = \tan x$ produce different values for the integral
$$\int \sec^2 x \tan x \, dx$$
(b) Explain why both are correct.

32. Use the two substitutions in Exercise 31 to evaluate the definite integral
$$\int_0^{\pi/4} \sec^2 x \tan x \, dx$$
and confirm that they produce the same result.

33. Evaluate the integral
$$\int \sqrt{1 + x^{-2/3}} \, dx$$
by making the substitution $u = 1 + x^{2/3}$.

34. (a) Express the equation
$$\int_a^b [f_1(x) + f_2(x) + \cdots + f_n(x)] \, dx$$
$$= \int_a^b f_1(x) \, dx + \int_a^b f_2(x) \, dx + \cdots + \int_a^b f_n(x) \, dx$$
in sigma notation.
(b) If $c_1, c_2, \ldots, c_n$ are constants and $f_1, f_2, \ldots, f_n$ are integrable functions on $[a, b]$, do you think it is always true that
$$\int_a^b \left(\sum_{k=1}^{n} c_k f_k(x)\right) dx = \sum_{k=1}^{n} \left[c_k \int_a^b f_k(x) \, dx \right]?$$
Explain your reasoning.

35. The accompanying figure shows five points on the graph of an unknown function f. Devise a strategy for using the known points to approximate the area A under the graph of $y = f(x)$ over the interval $[1, 5]$. Describe your strategy, and use it to approximate A.

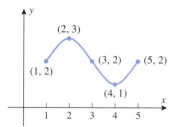

Figure Ex-35

36. Find the left endpoint, right endpoint, and midpoint approximations of the area under the curve $y = e^x$ over the interval $[0, 5]$ using $n = 5$ subintervals.

37. For $f(x) = 1/x$, find all values of x^* in the interval $[1, e]$ that satisfy Equation (7) in the Mean-Value Theorem for Integrals (6.6.2), and explain what these numbers represent.

38. Find the area under the graph of $f(x) = 5x - x^2$ over the interval $[0, 5]$ using Definition 6.4.3 with x_k^* as the *left* endpoint of each subinterval.

In Exercises 39 and 40, use a calculating utility to find the left endpoint, right endpoint, and midpoint approximations to the area under the curve $y = f(x)$ over the stated interval using $n = 10$ subintervals.

39. $y = \ln x$; $[1, 2]$ **40.** $y = e^x$; $[0, 1]$

41. Express the limit as a definite integral over $[0, 1]$, and then evaluate the limit by evaluating the integral.
$$\lim_{\max \Delta x_k \to 0} \sum_{k=1}^{n} e^{x_k^*} \Delta x_k$$

42. Use Part 2 of the Fundamental Theorem of Calculus (6.6.3) to find the derivative.

(a) $\displaystyle\frac{d}{dx} \int_0^x e^{t^2} \, dt$ (b) $\displaystyle\frac{d}{dx} \int_1^x \ln t \, dt$

43. Find an integral formula for the antiderivative of $1/(1 + e^x)$ on the interval $(-\infty, +\infty)$ whose value at $x = 1$ is (a) 0 and (b) 2.

c 44. Let $\displaystyle F(x) = \int_0^x \frac{t^2 - 3}{t^4 + 7} \, dt$.

(a) Find the intervals on which F is increasing. Decreasing.
(b) Find the open intervals on which F is concave up. Concave down.
(c) Find the x-values, if any, at which the function F has absolute extrema.
(d) Use a CAS to graph F, and confirm that the results in parts (a), (b), and (c) are consistent with the graph.

45. Prove that the function

$$F(x) = \int_0^x \frac{1}{1+t^2}\, dt + \int_0^{1/x} \frac{1}{1+t^2}\, dt$$

is constant on the interval $(0, +\infty)$.

46. What is the natural domain of the function

$$F(x) = \int_1^x \frac{1}{t^2 - 9}\, dt?$$

Explain your reasoning.

47. In each part, determine the values of x for which $F(x)$ is positive, negative, or zero without performing the integration; explain your reasoning.

(a) $F(x) = \int_1^x \frac{t^4}{t^2 + 3}\, dt$ (b) $F(x) = \int_{-1}^x \sqrt{4 - t^2}\, dt$

48. Find a formula (defined piecewise) for the upper boundary of the trapezoid shown in the accompanying figure, and then integrate that function to derive the formula for the area of the trapezoid given on the inside front cover of this text.

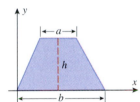

Figure Ex-48

49. The velocity of a particle moving along an s-axis is measured at 5-s intervals for 40 s, and the velocity function is modeled by a smooth curve. The curve and the data points are shown in the accompanying figure.
(a) Does the particle have constant acceleration? Explain your reasoning.
(b) Is there any 15-s time interval during which the acceleration is constant? Explain your reasoning.
(c) Estimate the average velocity of the particle over the 40-s time period.
(d) Estimate the distance traveled by the particle from time $t = 0$ to time $t = 40$.
(e) Is the particle ever slowing down during the 40-s time period? Explain your reasoning.
(f) Is there sufficient information for you to determine the s-coordinate of the particle at time $t = 10$? If so, find it. If not, explain what additional information you need.

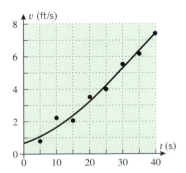

Figure Ex-49

50. Suppose that a tumor grows at the rate of $r(t) = t/7$ grams per week. When, during the second 26 weeks of growth, is the weight of the tumor the same as its average weight during that period?

In Exercises 51–59, evaluate the integrals by hand, and check your answers with a CAS if you have one.

51. $\displaystyle\int \frac{\cos 3x}{\sqrt{5 + 2\sin 3x}}\, dx$ **52.** $\displaystyle\int \frac{\sqrt{3 + \sqrt{x}}}{\sqrt{x}}\, dx$

53. $\displaystyle\int \frac{x^2}{(ax^3 + b)^2}\, dx$ **54.** $\displaystyle\int x\sec^2(ax^2)\, dx$

55. $\displaystyle\int_{-2}^{-1}\left(u^{-4} + 3u^{-2} - \frac{1}{u^5}\right) du$

56. $\displaystyle\int_0^1 \sin^2(\pi x)\cos(\pi x)\, dx$ **57.** $\displaystyle\int_e^{e^2} \frac{dx}{x\ln x}$

58. $\displaystyle\int_0^1 \frac{dx}{\sqrt{e^x}}$ **59.** $\displaystyle\int_0^{\ln\sqrt{2}} \frac{1 + \cos(e^{-2x})}{e^{2x}}\, dx$

C **60.** In number theory, $\pi(n)$ denotes the number of prime numbers that are less than or equal to the positive integer n. For example, it can be shown with the help of a computer that $\pi(100{,}000) = 9592$; that is, there are 9592 prime numbers that are less than or equal to 100,000. There are two useful approximations to $\pi(n)$ that are appropriate for large values of n:

$$\pi(n) \approx \frac{n}{\ln n} \quad \text{and} \quad \pi(n) \approx \int_2^n \frac{1}{\ln t}\, dt$$

Use a CAS to determine which of these approximations produces the better estimate of $\pi(100{,}000)$.

C **61.** Use a CAS to approximate the area of the region in the first quadrant that lies below the curve $y = x + x^2 - x^3$ and above the x-axis.

C **62.** In each part, use a CAS to solve the initial-value problem.

(a) $\dfrac{dy}{dx} = x^2\cos 3x; \ y(\pi/2) = -1$

(b) $\dfrac{dy}{dx} = \dfrac{x^3}{(4 + x^2)^{3/2}}; \ y(0) = -2$

C **63.** In each part, use a CAS, where needed, to solve for k.

(a) $\displaystyle\int_1^k (x^3 - 2x - 1)\, dx = 0, \quad k > 1$

(b) $\displaystyle\int_0^k (x^2 + \sin 2x)\, dx = 3, \quad k \geq 0$

C **64.** Use a CAS to approximate the largest and smallest values of the integral

$$\int_{-1}^x \frac{t}{\sqrt{2 + t^3}}\, dt$$

for $1 \leq x \leq 3$.

c **65.** The function J_0 defined by

$$J_0(x) = \frac{1}{\pi} \int_0^\pi \cos(x \sin t)\, dt$$

is called the **Bessel function of order zero**.

(a) Use a CAS to graph the equation $y = J_0(x)$ over the interval $0 \leq x \leq 8$.
(b) Estimate $J_0(1)$.
(c) Estimate the smallest positive zero of $J_0(x)$.

EXPANDING THE CALCULUS HORIZON

Blammo the Human Cannonball

*Blammo the Human Cannonball will be fired from a cannon and hopes to land in a small net at the opposite end of the circus arena. Your job as Blammo's manager is to do the mathematical calculations that will allow Blammo to perform his death-defying act safely. The methods that you will use are from the field of **ballistics** (the study of projectile motion).*

The Problem

Blammo's cannon has a **muzzle velocity** of 35 m/s, which means that Blammo will leave the muzzle with that velocity. The muzzle opening will be 5 m above the ground, and Blammo's objective is to land in a net that is also 5 m above the ground and that extends a distance of 10 m between 90 m and 100 m from the cannon opening (Figure 1). Your mathematical problem is to determine the **elevation angle** α of the cannon (the angle from the horizontal to the cannon barrel) that will make Blammo land in the net.

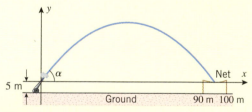

Figure 1

Modeling Assumptions

Blammo's trajectory will be determined by his initial velocity, the elevation angle of the cannon, and the forces that act on him after he leaves the muzzle. We will assume that the only force acting on Blammo after he leaves the muzzle is the downward force of the Earth's gravity. In particular, we will ignore the effect of air resistance. It will be convenient to introduce the xy-coordinate system shown in Figure 1 and to assume that Blammo is at the origin at time $t = 0$. We will also assume that Blammo's motion can be decomposed into two independent components, a horizontal component parallel to the x-axis and a vertical component parallel to the y-axis. We will analyze the horizontal and vertical components of Blammo's motion separately, and then we will combine the information to obtain a complete picture of his trajectory.

Blammo's Equations of Motion

We will denote the position and velocity functions for Blammo's horizontal component of motion by $x(t)$ and $v_x(t)$, and we will denote the position and velocity functions for his vertical component of motion by $y(t)$ and $v_y(t)$.

Since the only force acting on Blammo after he leaves the muzzle is the downward force of the Earth's gravity, there are no horizontal forces to alter his initial horizontal velocity $v_x(0)$. Thus, Blammo will have a constant velocity of $v_x(0)$ in the x-direction; this implies that

$$x(t) = v_x(0)t \tag{1}$$

In the y-direction Blammo is acted on only by the downward force of the Earth's gravity. Thus, his motion in this direction is governed by the free-fall model; hence, from (12) in Section 6.7 his vertical position function is

$$y(t) = y(0) + v_y(0)t - \tfrac{1}{2}gt^2$$

Taking $g = 9.8 \text{ m/s}^2$, and using the fact that $y(0) = 0$, this equation can be written as

$$y(t) = v_y(0)t - 4.9t^2 \tag{2}$$

....
Exercise 1   At time $t = 0$ Blammo's velocity is 35 m/s, and this velocity is directed at an angle α with the horizontal. It is a fact of physics that the initial velocity components $v_x(0)$ and $v_y(0)$ can be obtained geometrically from the muzzle velocity and the angle of elevation using the triangle shown in Figure 2. We will justify this later in the text, but for now use this fact to show that Equations (1) and (2) can be expressed as

$$x(t) = (35\cos\alpha)t$$
$$y(t) = (35\sin\alpha)t - 4.9t^2$$

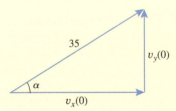

Figure 2

....
Exercise 2

(a) Use the result in Exercise 1 to find the velocity functions $v_x(t)$ and $v_y(t)$ in terms of the elevation angle α.

(b) Find the time t at which Blammo is at his maximum height above the x-axis, and show that this maximum height (in meters) is

$$y_{\max} = 62.5\sin^2\alpha$$

....
Exercise 3   The equations obtained in Exercise 1 can be viewed as parametric equations for Blammo's trajectory. Show, by eliminating the parameter t, that if $0 < \alpha < \pi/2$, then Blammo's trajectory is given by the equation

$$y = (\tan\alpha)x - \frac{0.004}{\cos^2\alpha}x^2$$

Explain why Blammo's trajectory is a parabola.

Finding the Elevation Angle

Define Blammo's *horizontal range* R to be the horizontal distance he travels until he returns to the height of the muzzle opening ($y = 0$). Your objective is to find elevation angles that will make the horizontal range fall between 90 m and 100 m, thereby ensuring that Blammo lands in the net (Figure 3).

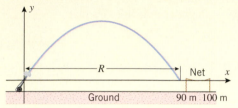

Figure 3

Exercise 4   Use a graphing utility and either the parametric equations obtained in Exercise 1 or the single equation obtained in Exercise 3 to generate Blammo's trajectories, taking elevation angles at increments of 10° from 15° to 85°. In each case, determine visually whether Blammo lands in the net.

Exercise 5   Find the time required for Blammo to return to his starting height ($y = 0$), and use that result to show that Blammo's range R is given by the formula

$$R = 125 \sin 2\alpha$$

Exercise 6

(a) Use the result in Exercise 5 to find two elevation angles that will allow Blammo to hit the midpoint of the net 95 m away.

(b) The tent is 55 m high. Explain why the larger elevation angle cannot be used.

Exercise 7   How much can the smaller elevation angle in Exercise 6 vary and still have Blammo hit the net between 90 m and 100 m?

Blammo's Shark Trick

Blammo is to be fired from 5 m above ground level with a muzzle velocity of 35 m/s over a flaming wall that is 20 m high and past a 5-m-high shark pool (Figure 4). To make the feat impressive, the pool will be made as long as possible. Your job as Blammo's manager is to determine the length of the pool, how far to place the cannon from the wall, and what elevation angle to use to ensure that Blammo clears the pool.

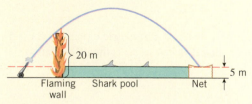

Figure 4

Exercise 8   Prepare a written presentation of the problem and your solution of it that is at an appropriate level for an engineer, physicist, or mathematician to read. Your presentation should contain the following elements: an explanation of all notation, a list and description of all formulas that will be used, a diagram that shows the orientation of any coordinate systems that will be used, a description of any assumptions you make to solve the problem, graphs that you think will enhance the presentation, and a clear step-by-step explanation of your solution.

Module by: *John Rickert, Rose-Hulman Institute of Technology*
Howard Anton, Drexel University

7

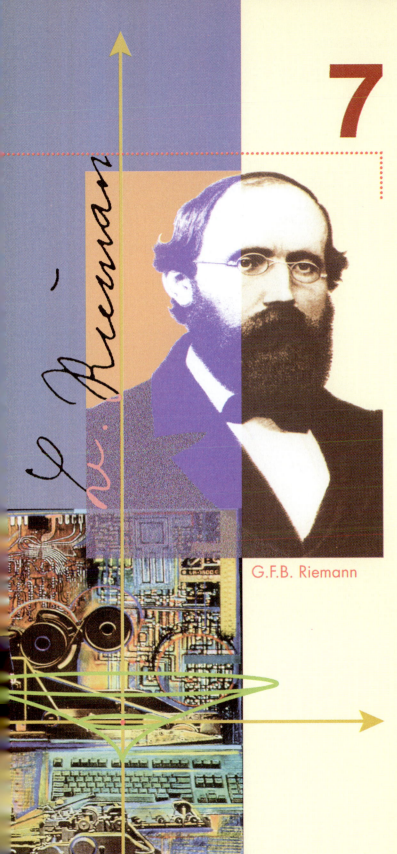

G.F.B. Riemann

APPLICATIONS OF THE DEFINITE INTEGRAL IN GEOMETRY, SCIENCE, AND ENGINEERING

*I*n the last chapter we introduced the definite integral as the limit of Riemann sums in the context of finding areas. However, Riemann sums and definite integrals have applications that extend far beyond the area problem. In this chapter we will show how Riemann sums and definite integrals arise in such problems as finding the volume and surface area of a solid, finding the length of a plane curve, calculating the work done by a force, finding the pressure and force exerted by a fluid on a submerged object, and finding properties of suspended cables.

Although these problems are diverse, the required calculations can all be approached by the same procedure that we used to find areas—breaking the required calculation into "small parts," making an approximation that is good because the part is small, adding the approximations from the parts to produce a Riemann sum that approximates the entire quantity to be calculated, and then taking the limit of the Riemann sums to produce an exact result.

7.1 AREA BETWEEN TWO CURVES

In the last chapter we showed how to find the area between a curve $y = f(x)$ and an interval on the x-axis. Here we will show how to find the area between two curves.

A REVIEW OF RIEMANN SUMS

Before we consider the problem of finding the area between two curves it will be helpful to review the basic principle that underlies the calculation of area as a definite integral. Recall that if f is continuous and nonnegative on $[a, b]$, then the definite integral for the area A under $y = f(x)$ over the interval $[a, b]$ is obtained in four steps (Figure 7.1.1):

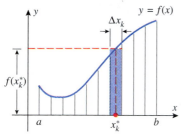

Figure 7.1.1

- Divide the interval $[a, b]$ into n subintervals, and use those subintervals to divide the area under the curve $y = f(x)$ into n strips.

- Assuming that the width of the kth strip is Δx_k, approximate the area of that strip by the area of a rectangle of width Δx_k and height $f(x_k^*)$, where x_k^* is a number in the kth subinterval.

- Add the approximate areas of the strips to approximate the entire area A by the Riemann sum:

$$A \approx \sum_{k=1}^{n} f(x_k^*)\Delta x_k$$

- Take the limit of the Riemann sums as the number of subintervals increases and their widths approach zero. This causes the error in the approximations to approach zero and produces the following definite integral for the exact area A:

$$A = \lim_{\max \Delta x_k \to 0} \sum_{k=1}^{n} f(x_k^*)\Delta x_k = \int_a^b f(x)\, dx$$

Observe the effect that the limit process has on the various parts of the Riemann sum:

- The quantity x_k^* in the Riemann sum becomes the variable x in the definite integral.
- The interval width Δx_k in the Riemann sum becomes the dx in the definite integral.
- The interval $[a, b]$ is implicit in the Riemann sum as the aggregate of the subintervals with widths $\Delta x_1, \ldots, \Delta x_n$, but $[a, b]$ is explicitly represented by the upper and lower limits of integration in the definite integral.

AREA BETWEEN $y = f(x)$ AND $y = g(x)$

We will now consider the following extension of the area problem.

7.1.1 FIRST AREA PROBLEM. Suppose that f and g are continuous functions on an interval $[a, b]$ and

$$f(x) \geq g(x) \quad \text{for} \quad a \leq x \leq b$$

[This means that the curve $y = f(x)$ lies above the curve $y = g(x)$ and that the two can touch but not cross.] Find the area A of the region bounded above by $y = f(x)$, below by $y = g(x)$, and on the sides by the lines $x = a$ and $x = b$ (Figure 7.1.2a).

To solve this problem we divide the interval $[a, b]$ into n subintervals, which has the effect of subdividing the region into n strips (Figure 7.1.2b). If we assume that the width of the kth strip is Δx_k, then the area of the strip can be approximated by the area of a rectangle of width Δx_k and height $f(x_k^*) - g(x_k^*)$, where x_k^* is a number in the kth subinterval. Adding these approximations yields the following Riemann sum that approximates the area A:

$$A \approx \sum_{k=1}^{n} [f(x_k^*) - g(x_k^*)]\Delta x_k$$

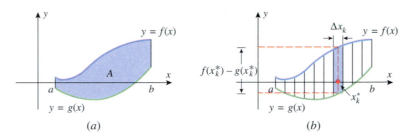

Figure 7.1.2   (a)   (b)

Taking the limit as n increases and the widths of the subintervals approach zero yields the following definite integral for the area A between the curves:

$$A = \lim_{\max \Delta x_k \to 0} \sum_{k=1}^{n} [f(x_k^*) - g(x_k^*)] \Delta x_k = \int_a^b [f(x) - g(x)]\, dx$$

In summary, we have the following result:

7.1.2 AREA FORMULA. If f and g are continuous functions on the interval $[a, b]$, and if $f(x) \geq g(x)$ for all x in $[a, b]$, then the area of the region bounded above by $y = f(x)$, below by $y = g(x)$, on the left by the line $x = a$, and on the right by the line $x = b$ is

$$A = \int_a^b [f(x) - g(x)]\, dx \qquad (1)$$

In the case where f and g are *nonnegative* on the interval $[a, b]$, the formula

$$A = \int_a^b [f(x) - g(x)]\, dx = \int_a^b f(x)\, dx - \int_a^b g(x)\, dx$$

states that the area A between the curves can be obtained by subtracting the area under $y = g(x)$ from the area under $y = f(x)$ (Figure 7.1.3).

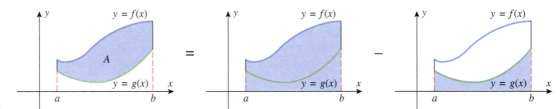

Figure 7.1.3

When the region is complicated, it may require some careful thought to determine the integrand and limits of integration in (1). Here is a systematic procedure that you can follow to set up this formula.

Step 1. Sketch the region and then draw a vertical line segment through the region at an arbitrary point x on the x-axis, connecting the top and bottom boundaries (Figure 7.1.4a).

Step 2. The y-coordinate of the top endpoint of the line segment sketched in Step 1 will be $f(x)$, the bottom one $g(x)$, and the length of the line segment will be $f(x) - g(x)$. This is the integrand in (1).

Step 3. To determine the limits of integration, imagine moving the line segment left and then right. The leftmost position at which the line segment intersects the region is $x = a$ and the rightmost is $x = b$ (Figures 7.1.4b and 7.1.4c).

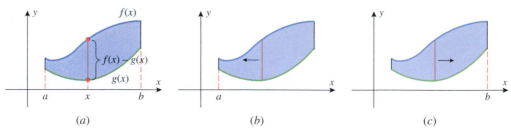

Figure 7.1.4

• REMARK.   It is not necessary to make an extremely accurate sketch in Step 1; the only
   purpose of the sketch is to determine which curve is the upper boundary and which is the
   lower boundary.

• REMARK.   There is a useful way of thinking about this procedure: If you view the vertical
   line segment as the "cross section" of the region at the point x, then Formula (1) states that
   the area between the curves is obtained by integrating the length of the cross section over
   the interval from a to b.

Example 1   Find the area of the region bounded above by $y = x + 6$, bounded below by
$y = x^2$, and bounded on the sides by the lines $x = 0$ and $x = 2$.

Solution.   The region and a cross section are shown in Figure 7.1.5. The cross section
extends from $g(x) = x^2$ on the bottom to $f(x) = x + 6$ on the top. If the cross section is
moved through the region, then its leftmost position will be $x = 0$ and its rightmost position
will be $x = 2$. Thus, from (1)

$$A = \int_0^2 [(x + 6) - x^2]\,dx = \left[\frac{x^2}{2} + 6x - \frac{x^3}{3}\right]_0^2 = \frac{34}{3} - 0 = \frac{34}{3} \quad \blacktriangleleft$$

It is possible that the upper and lower boundaries of a region may intersect at one or
both endpoints, in which case the sides of the region will be points, rather than vertical
line segments (Figure 7.1.6). When that occurs you will have to determine the points of
intersection to obtain the limits of integration.

Example 2   Find the area of the region that is enclosed between the curves $y = x^2$ and
$y = x + 6$.

Solution.   A sketch of the region (Figure 7.1.7) shows that the lower boundary is $y = x^2$
and the upper boundary is $y = x + 6$. At the endpoints of the region, the upper and lower
boundaries have the same y-coordinates; thus, to find the endpoints we equate

$$y = x^2 \quad \text{and} \quad y = x + 6 \tag{2}$$

This yields

$$x^2 = x + 6 \quad \text{or} \quad x^2 - x - 6 = 0 \quad \text{or} \quad (x + 2)(x - 3) = 0$$

from which we obtain

$$x = -2 \quad \text{and} \quad x = 3$$

Although the y-coordinates of the endpoints are not essential to our solution, they may be
obtained from (2) by substituting $x = -2$ and $x = 3$ in either equation. This yields $y = 4$
and $y = 9$, so the upper and lower boundaries intersect at $(-2, 4)$ and $(3, 9)$.
From (1) with $f(x) = x + 6$, $g(x) = x^2$, $a = -2$, and $b = 3$, we obtain the area

$$A = \int_{-2}^3 [(x + 6) - x^2]\,dx = \left[\frac{x^2}{2} + 6x - \frac{x^3}{3}\right]_{-2}^3 = \frac{27}{2} - \left(-\frac{22}{3}\right) = \frac{125}{6} \quad \blacktriangleleft$$

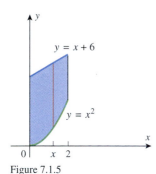

Figure 7.1.5

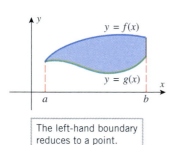

Both side boundaries
reduce to points.

The left-hand boundary
reduces to a point.

Figure 7.1.6

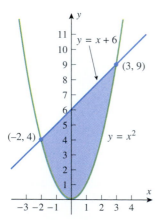

Figure 7.1.7

It is possible for the upper or lower boundary of a region to consist of two or more different curves, in which case it will be necessary to subdivide the region into smaller pieces in order to apply Formula (1). This is illustrated in the next example.

Example 3   Find the area of the region enclosed by $x = y^2$ and $y = x - 2$.

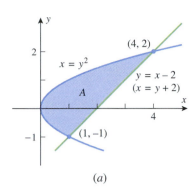

(a)

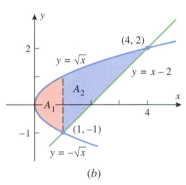

(b)

Figure 7.1.8

Solution.   To make an accurate sketch of the region, we need to know where the curves $x = y^2$ and $y = x - 2$ intersect. In Example 2 we found intersections by equating the expressions for y. Here it is easier to rewrite the latter equation as $x = y + 2$ and equate the expressions for x, namely

$$x = y^2 \quad \text{and} \quad x = y + 2 \tag{3}$$

This yields

$$y^2 = y + 2 \quad \text{or} \quad y^2 - y - 2 = 0 \quad \text{or} \quad (y + 1)(y - 2) = 0$$

from which we obtain $y = -1$, $y = 2$. Substituting these values in either equation in (3) we see that the corresponding x-values are $x = 1$ and $x = 4$, respectively, so the points of intersection are $(1, -1)$ and $(4, 2)$ (Figure 7.1.8a).

To apply Formula (1), the equations of the boundaries must be written so that y is expressed explicitly as a function of x. The upper boundary can be written as $y = \sqrt{x}$ (rewrite $x = y^2$ as $y = \pm\sqrt{x}$ and choose the $+$ for the upper portion of the curve). The lower portion of the boundary consists of two parts: $y = -\sqrt{x}$ for $0 \leq x \leq 1$ and $y = x - 2$ for $1 \leq x \leq 4$ (Figure 7.1.8b). Because of this change in the formula for the lower boundary, it is necessary to divide the region into two parts and find the area of each part separately.

From (1) with $f(x) = \sqrt{x}$, $g(x) = -\sqrt{x}$, $a = 0$, and $b = 1$, we obtain

$$A_1 = \int_0^1 [\sqrt{x} - (-\sqrt{x})]\, dx = 2 \int_0^1 \sqrt{x}\, dx = 2\left[\frac{2}{3}x^{3/2}\right]_0^1 = \frac{4}{3} - 0 = \frac{4}{3}$$

From (1) with $f(x) = \sqrt{x}$, $g(x) = x - 2$, $a = 1$, and $b = 4$, we obtain

$$A_2 = \int_1^4 [\sqrt{x} - (x - 2)]\, dx = \int_1^4 (\sqrt{x} - x + 2)\, dx$$

$$= \left[\frac{2}{3}x^{3/2} - \frac{1}{2}x^2 + 2x\right]_1^4 = \left(\frac{16}{3} - 8 + 8\right) - \left(\frac{2}{3} - \frac{1}{2} + 2\right) = \frac{19}{6}$$

Thus, the area of the entire region is

$$A = A_1 + A_2 = \frac{4}{3} + \frac{19}{6} = \frac{9}{2} \qquad \blacktriangleleft$$

FOR THE READER.   It is assumed in Formula (1) that $f(x) \geq g(x)$ for all x in the interval $[a, b]$. What do you think that the integral represents if this condition is not satisfied, that is, the graphs of f and g cross one another over the interval? Explain your reasoning, and give an example to support your conclusion. Using definite integrals, write an expression for the area between the graphs of f and g in your example.

Example 4   Figure 7.1.9 shows velocity versus time curves for two race cars that move along a straight track, starting from rest at the same line. What does the area A between the curves over the interval $0 \leq t \leq T$ represent?

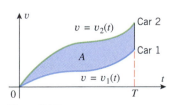

Figure 7.1.9

Solution.   From (1)

$$A = \int_0^T [v_2(t) - v_1(t)]\, dt = \int_0^T v_2(t)\, dt - \int_0^T v_1(t)\, dt$$

But from 6.7.4, the first integral is the distance traveled by car 2 during the time interval, and the second integral is the distance traveled by car 1. Thus, A is the distance by which car 2 is ahead of car 1 at time T. $\qquad \blacktriangleleft$

REVERSING THE ROLES OF x AND y

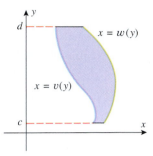

Figure 7.1.10

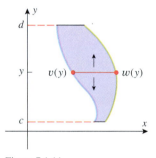

Figure 7.1.11

Sometimes it is possible to avoid splitting a region into parts by integrating with respect to y rather than x. We will now show how this can be done.

7.1.3 SECOND AREA PROBLEM. Suppose that w and v are continuous functions of y on an interval $[c, d]$ and that

$$w(y) \geq v(y) \quad \text{for} \quad c \leq y \leq d$$

[This means that the curve $x = w(y)$ lies to the right of the curve $x = v(y)$ and that the two can touch but not cross.] Find the area A of the region bounded on the left by $x = v(y)$, on the right by $x = w(y)$, and above and below by the lines $y = d$ and $y = c$ (Figure 7.1.10).

Proceeding as in the derivation of (1), but with the roles of x and y reversed, leads to the following analog of 7.1.2.

7.1.4 AREA FORMULA. If w and v are continuous functions and if $w(y) \geq v(y)$ for all y in $[c, d]$, then the area of the region bounded on the left by $x = v(y)$, on the right by $x = w(y)$, below by $y = c$, and above by $y = d$ is

$$A = \int_c^d [w(y) - v(y)] \, dy \qquad (4)$$

The guiding principle in applying this formula is the same as with (1): The integrand in (4) can be viewed as the length of the horizontal cross section at an arbitrary point y on the y-axis, in which case Formula (4) states that the area can be obtained by integrating the length of the horizontal cross section over the interval $[c, d]$ on the y-axis (Figure 7.1.11).

In Example 3, where we integrated with respect to x to find the area of the region enclosed by $x = y^2$ and $y = x - 2$, we had to split the region into parts and evaluate two integrals. In the next example we will see that by integrating with respect to y no splitting of the region is necessary.

Example 5 Find the area of the region enclosed by $x = y^2$ and $y = x - 2$, integrating with respect to y.

Solution. From Figure 7.1.8 the left boundary is $x = y^2$, the right boundary is $y = x - 2$, and the region extends over the interval $-1 \leq y \leq 2$. However, to apply (4) the equations for the boundaries must be written so that x is expressed explicitly as a function of y. Thus, we rewrite $y = x - 2$ as $x = y + 2$. It now follows from (4) that

$$A = \int_{-1}^{2} [(y + 2) - y^2] \, dy = \left[\frac{y^2}{2} + 2y - \frac{y^3}{3} \right]_{-1}^{2} = \frac{9}{2}$$

which agrees with the result obtained in Example 3. ◄

REMARK. The choice between Formulas (1) and (4) is generally dictated by the shape of the region, and one would usually choose the formula that requires the least amount of splitting. However, if the integral(s) resulting by one method are difficult to evaluate, then the other method might be preferable, even if it requires more splitting.

up a Riemann sum whose limit is the volume of the entire solid. We will give the details shortly, but first we need to discuss how to find the volume of a solid whose cross sections do not vary in size and shape (i.e., are congruent).

One of the simplest examples of a solid with congruent cross sections is a right circular cylinder of radius r, since all cross sections taken perpendicular to the central axis are circular regions of radius r. The volume V of a right circular cylinder of radius r and height h can be expressed in terms of the height and the area of a cross section as

$$V = \pi r^2 h = [\text{area of a cross section}] \times [\text{height}] \tag{1}$$

This is a special case of a more general volume formula that applies to solids called *right cylinders*. A **right cylinder** is a solid that is generated when a plane region is translated along a line or **axis** that is perpendicular to the region (Figure 7.2.3). The distance h that the region is translated is called the **height** or sometimes the **width** of the cylinder, and each cross section is a duplicate of the translated region. We will assume that the volume V of a right cylinder with cross-sectional area A and height h is given by

$$V = A \cdot h = [\text{area of a cross section}] \times [\text{height}] \tag{2}$$

(Figure 7.2.4). Note that this is consistent with Formula (1) for the volume of a right circular cylinder. We now have all of the tools required to solve the following problem.

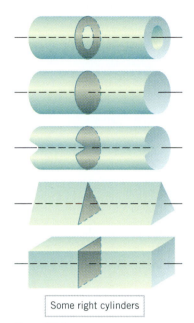

Some right cylinders

Figure 7.2.3

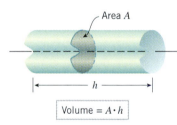

Area A

h

Volume = $A \cdot h$

Figure 7.2.4

> **7.2.1**   PROBLEM.   Let S be a solid that extends along the x-axis and is bounded on the left and right, respectively, by the planes that are perpendicular to the x-axis at $x = a$ and $x = b$ (Figure 7.2.5a). Find the volume V of the solid, assuming that its cross-sectional area $A(x)$ is known at each x in the interval $[a, b]$.

To solve this problem we divide the interval $[a, b]$ into n subintervals, which has the effect of dividing the solid into n slabs (Figure 7.2.5b).

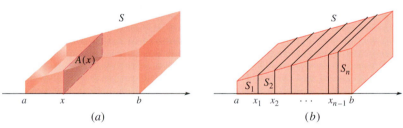

Figure 7.2.5

If we assume that the width of the kth slab is Δx_k, then the volume of the slab can be approximated by the volume of a right cylinder of width (height) Δx_k and cross-sectional area $A(x_k^*)$, where x_k^* is a number in the kth subinterval (Figure 7.2.6). Adding these approximations yields the following Riemann sum that approximates the volume V:

$$V \approx \sum_{k=1}^{n} A(x_k^*) \Delta x_k$$

Taking the limit as n increases and the widths of the subintervals approach zero yields the definite integral

$$V = \lim_{\max \Delta x_k \to 0} \sum_{k=1}^{n} A(x_k^*) \Delta x_k = \int_{a}^{b} A(x)\, dx$$

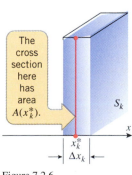

The cross section here has area $A(x_k^*)$.

S_k

x_k^*

Δx_k

Figure 7.2.6

In summary, we have the following result:

7.2.2   VOLUME FORMULA.   Let S be a solid bounded by two parallel planes perpendicular to the x-axis at $x = a$ and $x = b$. If, for each x in $[a, b]$, the cross-sectional area of S perpendicular to the x-axis is $A(x)$, then the volume of the solid is

$$V = \int_a^b A(x)\, dx \qquad (3)$$

provided $A(x)$ is integrable.

There is a similar result for cross sections perpendicular to the y-axis.

7.2.3   VOLUME FORMULA.   Let S be a solid bounded by two parallel planes perpendicular to the y-axis at $y = c$ and $y = d$. If, for each y in $[c, d]$, the cross-sectional area of S perpendicular to the y-axis is $A(y)$, then the volume of the solid is

$$V = \int_c^d A(y)\, dy \qquad (4)$$

provided $A(y)$ is integrable.

In words, these formulas state:

> *The volume of a solid can be obtained by integrating the cross-sectional area from one end of the solid to the other.*

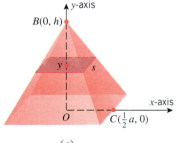

(a)

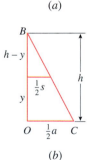

(b)

Figure 7.2.7

Example 1   Derive the formula for the volume of a right pyramid whose altitude is h and whose base is a square with sides of length a.

Solution.   As illustrated in Figure 7.2.7a, we introduce a rectangular coordinate system in which the y-axis passes through the apex and is perpendicular to the base, and the x-axis passes through the base and is parallel to a side of the base.

At any y in the interval $[0, h]$ on the y-axis, the cross section perpendicular to the y-axis is a square. If s denotes the length of a side of this square, then by similar triangles (Figure 7.2.7b)

$$\frac{\frac{1}{2}s}{\frac{1}{2}a} = \frac{h - y}{h} \quad \text{or} \quad s = \frac{a}{h}(h - y)$$

Thus, the area $A(y)$ of the cross section at y is

$$A(y) = s^2 = \frac{a^2}{h^2}(h - y)^2$$

and by (4) the volume is

$$V = \int_0^h A(y)\, dy = \int_0^h \frac{a^2}{h^2}(h - y)^2\, dy = \frac{a^2}{h^2} \int_0^h (h - y)^2\, dy$$

$$= \frac{a^2}{h^2}\left[-\frac{1}{3}(h - y)^3 \right]_{y=0}^{h} = \frac{a^2}{h^2}\left[0 + \frac{1}{3}h^3 \right] = \frac{1}{3}a^2 h$$

That is, the volume is $\frac{1}{3}$ of the area of the base times the altitude.   ◀

SOLIDS OF REVOLUTION

A *solid of revolution* is a solid that is generated by revolving a plane region about a line that lies in the same plane as the region; the line is called the *axis of revolution*. Many familiar solids are of this type (Figure 7.2.8).

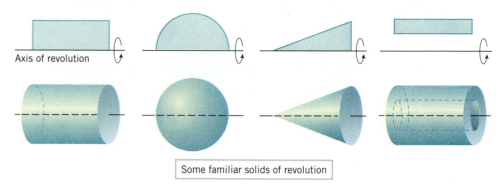

Axis of revolution

Some familiar solids of revolution

Figure 7.2.8

VOLUMES BY DISKS PERPENDICULAR TO THE x-AXIS

We will be interested in the following general problem:

> **7.2.4** PROBLEM. Let f be continuous and nonnegative on $[a, b]$, and let R be the region that is bounded above by $y = f(x)$, below by the x-axis, and on the sides by the lines $x = a$ and $x = b$ (Figure 7.2.9a). Find the volume of the solid of revolution that is generated by revolving the region R about the x-axis.

We can solve this problem by slicing. For this purpose, observe that the cross section of the solid taken perpendicular to the x-axis at the point x is a circular disk of radius $f(x)$ (Figure 7.2.9b). The area of this region is

$$A(x) = \pi[f(x)]^2$$

Thus, from (3) the volume of the solid is

$$V = \int_a^b \pi[f(x)]^2 \, dx \qquad (5)$$

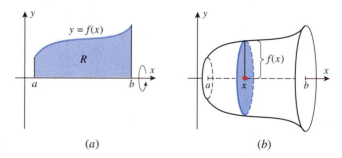

Figure 7.2.9   (a)   (b)

Because the cross sections are disk shaped, the application of this formula is called the *method of disks*.

Example 2 Find the volume of the solid that is obtained when the region under the curve $y = \sqrt{x}$ over the interval $[1, 4]$ is revolved about the x-axis (Figure 7.2.10).

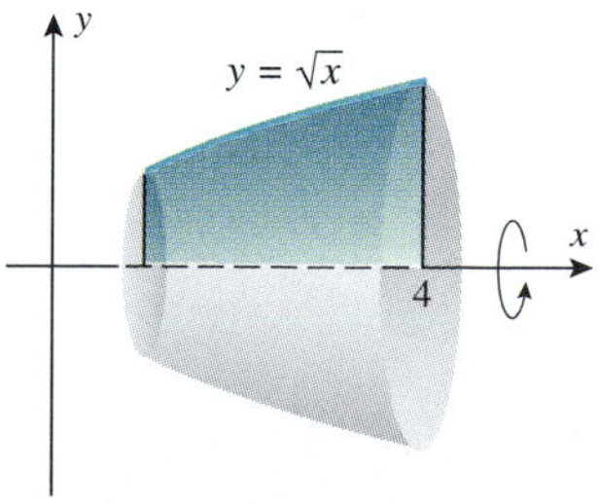

Solution. From (5), the volume is

$$V = \int_a^b \pi[f(x)]^2 \, dx = \int_1^4 \pi x \, dx = \frac{\pi x^2}{2} \bigg]_1^4 = 8\pi - \frac{\pi}{2} = \frac{15\pi}{2} \qquad \blacktriangleleft$$

y

$y = \sqrt{x}$

x

4

Figure 7.2.10

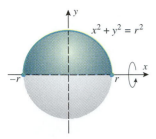

Figure 7.2.11

Example 3 Derive the formula for the volume of a sphere of radius r.

Solution. As indicated in Figure 7.2.11, a sphere of radius r can be generated by revolving the upper semicircular disk enclosed between the x-axis and

$$x^2 + y^2 = r^2$$

about the x-axis. Since the upper half of this circle is the graph of $y = f(x) = \sqrt{r^2 - x^2}$, it follows from (5) that the volume of the sphere is

$$V = \int_a^b \pi[f(x)]^2\, dx = \int_{-r}^r \pi(r^2 - x^2)\, dx = \pi\left[r^2 x - \frac{x^3}{3}\right]_{-r}^r = \frac{4}{3}\pi r^3 \qquad \blacktriangleleft$$

**VOLUMES BY WASHERS
PERPENDICULAR TO THE x-AXIS**

Not all solids of revolution have solid interiors; some have holes or channels that create interior surfaces, as in the last part of Figure 7.2.8. Thus, we will be interested in problems of the following type.

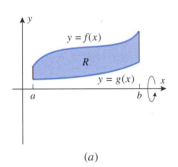

(a)

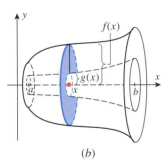

(b)

Figure 7.2.12

7.2.5 PROBLEM. Let f and g be continuous and nonnegative on $[a, b]$, and suppose that $f(x) \geq g(x)$ for all x in the interval $[a, b]$. Let R be the region that is bounded above by $y = f(x)$, below by $y = g(x)$, and on the sides by the lines $x = a$ and $x = b$ (Figure 7.2.12a). Find the volume of the solid of revolution that is generated by revolving the region R about the x-axis.

We can solve this problem by slicing. For this purpose, observe that the cross section of the solid taken perpendicular to the x-axis at the point x is the annular or "washer-shaped" region with inner radius $g(x)$ and outer radius $f(x)$ (Figure 7.2.12b); hence its area is

$$A(x) = \pi[f(x)]^2 - \pi[g(x)]^2 = \pi([f(x)]^2 - [g(x)]^2)$$

Thus, from (3) the volume of the solid is

$$V = \int_a^b \pi([f(x)]^2 - [g(x)]^2)\, dx \qquad (6)$$

Because the cross sections are washer shaped, the application of this formula is called the *method of washers*.

Example 4 Find the volume of the solid generated when the region between the graphs of the equations $f(x) = \frac{1}{2} + x^2$ and $g(x) = x$ over the interval $[0, 2]$ is revolved about the x-axis (Figure 7.2.13).

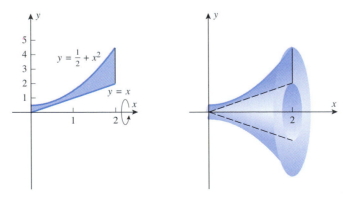

Unequal scales on axes

Figure 7.2.13

Solution. From (6) the volume is

$$V = \int_a^b \pi([f(x)]^2 - [g(x)]^2)\,dx = \int_0^2 \pi\left(\left[\tfrac{1}{2} + x^2\right]^2 - x^2\right)\,dx$$

$$= \int_0^2 \pi\left(\frac{1}{4} + x^4\right)\,dx = \pi\left[\frac{x}{4} + \frac{x^5}{5}\right]_0^2 = \frac{69\pi}{10} \qquad \blacktriangleleft$$

VOLUMES BY DISKS AND WASHERS PERPENDICULAR TO THE *y*-AXIS

The methods of disks and washers have analogs for regions that are revolved about the *y*-axis (Figures 7.2.14 and 7.2.15). Using the method of slicing and Formula (4), you should have no trouble deducing the following formulas for the volumes of the solids in the figures.

$$V = \int_c^d \pi[u(y)]^2\,dy \qquad\qquad V = \int_c^d \pi([w(y)]^2 - [v(y)]^2)\,dy \qquad\qquad (7\text{--}8)$$

Disks   Washers

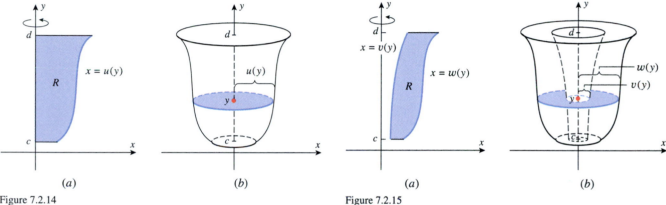

(a)   (b)   (a)   (b)

Figure 7.2.14   Figure 7.2.15

Example 5   Find the volume of the solid generated when the region enclosed by $y = \sqrt{x}$, $y = 2$, and $x = 0$ is revolved about the *y*-axis (Figure 7.2.16).

Solution. The cross sections taken perpendicular to the *y*-axis are disks, so we will apply (7). But first we must rewrite $y = \sqrt{x}$ as $x = y^2$. Thus, from (7) with $u(y) = y^2$, the volume is

$$V = \int_c^d \pi[u(y)]^2\,dy = \int_0^2 \pi y^4\,dy = \frac{\pi y^5}{5}\Bigg]_0^2 = \frac{32\pi}{5} \qquad \blacktriangleleft$$

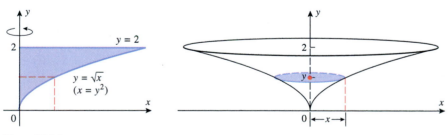

Figure 7.2.16

EXERCISE SET 7.2 ☐C CAS

In Exercises 1–4, find the volume of the solid that results when the shaded region is revolved about the indicated axis.

1.

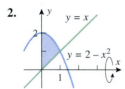
$y = \sqrt{3 - x}$

2.

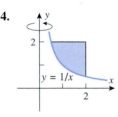
$y = x$
$y = 2 - x^2$

3.
$y = 3 - 2x$

4.
$y = 1/x$

In Exercises 5–16, find the volume of the solid that results when the region enclosed by the given curves is revolved about the x-axis.

5. $y = x^2$, $x = 0$, $x = 2$, $y = 0$

6. $y = \sec x$, $x = \pi/4$, $x = \pi/3$, $y = 0$

7. $y = \sqrt{\cos x}$, $x = \pi/4$, $x = \pi/2$, $y = 0$

8. $y = x^2$, $y = x^3$ **9.** $y = \sqrt{25 - x^2}$, $y = 3$

10. $y = 9 - x^2$, $y = 0$ **11.** $x = \sqrt{y}$, $x = y/4$

12. $y = \sin x$, $y = \cos x$, $x = 0$, $x = \pi/4$. [*Hint:* Use the identity $\cos 2x = \cos^2 x - \sin^2 x$.]

13. $y = e^x$, $y = 0$, $x = 0$, $x = \ln 3$

14. $y = e^{-2x}$, $y = 0$, $x = 0$, $x = 1$

15. $y = \dfrac{1}{\sqrt{4 + x^2}}$, $x = -2$, $x = 2$, $y = 0$

16. $y = \dfrac{e^{3x}}{\sqrt{1 + e^{6x}}}$, $x = 0$, $x = 1$, $y = 0$

In Exercises 17–26, find the volume of the solid that results when the region enclosed by the given curves is revolved about the y-axis.

17. $y = x^3$, $x = 0$, $y = 1$ **18.** $x = 1 - y^2$, $x = 0$

19. $x = \sqrt{1 + y}$, $x = 0$, $y = 3$

20. $y = x^2 - 1$, $x = 2$, $y = 0$

21. $x = \csc y$, $y = \pi/4$, $y = 3\pi/4$, $x = 0$

22. $y = x^2$, $x = y^2$ **23.** $x = y^2$, $x = y + 2$

24. $x = 1 - y^2$, $x = 2 + y^2$, $y = -1$, $y = 1$

25. $y = \ln x$, $x = 0$, $y = 0$, $y = 1$

26. $y = \sqrt{\dfrac{1 - x^2}{x^2}}$ ($x > 0$), $x = 0$, $y = 0$, $y = 2$

27. Find the volume of the solid that results when the region above the x-axis and below the ellipse
$$\frac{x^2}{a^2} + \frac{y^2}{b^2} = 1 \quad (a > 0, b > 0)$$
is revolved about the x-axis.

28. Let V be the volume of the solid that results when the region enclosed by $y = 1/x$, $y = 0$, $x = 2$, and $x = b$ $(0 < b < 2)$ is revolved about the x-axis. Find the value of b for which $V = 3$.

29. Find the volume of the solid generated when the region enclosed by $y = \sqrt{x + 1}$, $y = \sqrt{2x}$, and $y = 0$ is revolved about the x-axis. [*Hint:* Split the solid into two parts.]

30. Find the volume of the solid generated when the region enclosed by $y = \sqrt{x}$, $y = 6 - x$, and $y = 0$ is revolved about the x-axis. [*Hint:* Split the solid into two parts.]

31. Find the volume of the solid that results when the region enclosed by $y = \sqrt{x}$, $y = 0$, and $x = 9$ is revolved about the line $x = 9$.

32. Find the volume of the solid that results when the region in Exercise 31 is revolved about the line $y = 3$.

33. Find the volume of the solid that results when the region enclosed by $x = y^2$ and $x = y$ is revolved about the line $y = -1$.

34. Find the volume of the solid that results when the region in Exercise 33 is revolved about the line $x = -1$.

35. A nose cone for a space reentry vehicle is designed so that a cross section, taken x ft from the tip and perpendicular to the axis of symmetry, is a circle of radius $\frac{1}{4}x^2$ ft. Find the volume of the nose cone given that its length is 20 ft.

36. A certain solid is 1 ft high, and a horizontal cross section taken x ft above the bottom of the solid is an annulus of inner radius x^2 and outer radius $\sqrt{x}$. Find the volume of the solid.

37. Find the volume of the solid whose base is the region bounded between the curves $y = x$ and $y = x^2$, and whose cross sections perpendicular to the x-axis are squares.

38. The base of a certain solid is the region enclosed by $y = \sqrt{x}$, $y = 0$, and $x = 4$. Every cross section perpendicular to the x-axis is a semicircle with its diameter across the base. Find the volume of the solid.

39. Find the volume of the solid whose base is enclosed by the circle $x^2 + y^2 = 1$ and whose cross sections taken perpendicular to the base are

(a) semicircles   (b) squares

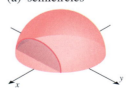

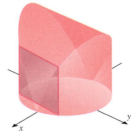

(c) equilateral triangles.

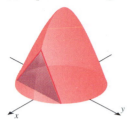

40. Derive the formula for the volume of a right circular cone with radius r and height h.

In Exercises 41–44, use a CAS to estimate the volume of the solid that results when the region enclosed by the curves is revolved about the stated axis.

C **41.** $y = \sin^8 x$, $y = 2x/\pi$, $x = 0$, $x = \pi/2$; x-axis

C **42.** $y = \pi^2 \sin x \cos^3 x$, $y = 4x^2$, $x = 0$, $x = \pi/4$; x-axis

C **43.** $y = e^x$, $x = 1$, $y = 1$; y-axis

C **44.** $y = x\sqrt{\tan^{-1} x}$, $y = x$; x-axis

45. The accompanying figure shows a ***spherical cap*** of radius ρ and height h cut from a sphere of radius r. Show that the volume V of the spherical cap can be expressed as
(a) $V = \frac{1}{3}\pi h^2 (3r - h)$ (b) $V = \frac{1}{6}\pi h (3\rho^2 + h^2)$.

Figure Ex-45

46. If fluid enters a hemispherical bowl with a radius of 10 ft at a rate of $\frac{1}{2}$ ft^3/min, how fast will the fluid be rising when the depth is 5 ft? [*Hint:* See Exercise 45.]

47. The accompanying figure shows the dimensions of a small lightbulb at 10 equally spaced points.
(a) Use formulas from geometry to make a rough estimate of the volume enclosed by the glass portion of the bulb.
(b) Use the average of left and right endpoint approximations to approximate the volume.

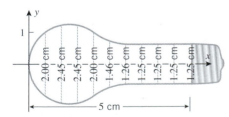

Figure Ex-47

48. Use the result in Exercise 45 to find the volume of the solid that remains when a hole of radius $r/2$ is drilled through the

center of a sphere of radius r, and then check your answer by integrating.

49. As shown in the accompanying figure, a cocktail glass with a bowl shaped like a hemisphere of diameter 8 cm contains a cherry with a diameter of 2 cm. If the glass is filled to a depth of h cm, what is the volume of liquid it contains? [*Hint:* First consider the case where the cherry is partially submerged, then the case where it is totally submerged.]

Figure Ex-49

50. Find the volume of the torus that results when the region enclosed by the circle of radius r with center at $(h, 0)$, $h > r$, is revolved about the y-axis. [*Hint:* Use an appropriate formula from plane geometry to help evaluate the definite integral.]

51. A wedge is cut from a right circular cylinder of radius r by two planes, one perpendicular to the axis of the cylinder and the other making an angle θ with the first. Find the volume of the wedge by slicing perpendicular to the y-axis as shown in the accompanying figure.

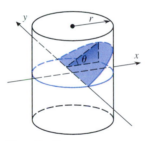

Figure Ex-51

52. Find the volume of the wedge described in Exercise 51 by slicing perpendicular to the x-axis.

53. Two right circular cylinders of radius r have axes that intersect at right angles. Find the volume of the solid common to the two cylinders. [*Hint:* One-eighth of the solid is sketched in the accompanying figure.]

54. In 1635 Bonaventura Cavalieri, a student of Galileo, stated the following result, called ***Cavalieri's principle***: *If two solids have the same height, and if the areas of their cross sections taken parallel to and at equal distances from their bases are always equal, then the solids have the same volume.* Use this result to find the volume of the oblique cylinder in the accompanying figure.

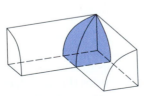

Figure Ex-53

Figure Ex-54

7.3 VOLUMES BY CYLINDRICAL SHELLS

The methods for computing volumes that have been discussed so far depend on our ability to compute the cross-sectional area of the solid and to integrate that area across the solid. In this section we will develop another method for finding volumes that may be applicable when the cross-sectional area cannot be found or the integration is too difficult.

CYLINDRICAL SHELLS

In this section we will be interested in the following problem:

7.3.1 PROBLEM. Let f be continuous and nonnegative on $[a, b]$, and let R be the region that is bounded above by $y = f(x)$, below by the x-axis, and on the sides by the lines $x = a$ and $x = b$. Find the volume V of the solid of revolution S that is generated by revolving the region R about the y-axis (Figure 7.3.1).

Sometimes problems of this type can be solved by the method of disks or washers perpendicular to the y-axis, but when that method is not applicable or the resulting integral is difficult, the *method of cylindrical shells*, which we will discuss here, will often work.

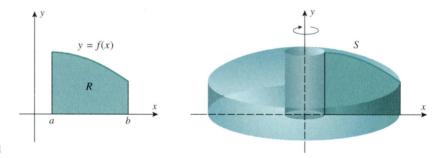

Figure 7.3.1

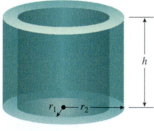

Figure 7.3.2

A *cylindrical shell* is a solid enclosed by two concentric right circular cylinders (Figure 7.3.2). The volume V of a cylindrical shell with inner radius r_1, outer radius r_2, and height h can be written as

$$V = [\text{area of cross section}] \cdot [\text{height}] = (\pi r_2^2 - \pi r_1^2)h$$
$$= \pi(r_2 + r_1)(r_2 - r_1)h = 2\pi \cdot \left[\tfrac{1}{2}(r_1 + r_2)\right] \cdot h \cdot (r_2 - r_1)$$

But $\frac{1}{2}(r_1 + r_2)$ is the average radius of the shell and $r_2 - r_1$ is its thickness, so

$$V = 2\pi \cdot [\text{average radius}] \cdot [\text{height}] \cdot [\text{thickness}] \qquad (1)$$

We will now show how this formula can be used to solve the problem posed above. The underlying idea is to divide the interval $[a, b]$ into n subintervals, thereby subdividing the region R into n strips, $R_1, R_2, \ldots, R_n$ (Figure 7.3.3a). When the region R is revolved about the y-axis, these strips generate "tube-like" solids $S_1, S_2, \ldots, S_n$ that are nested one inside the other and together comprise the entire solid S (Figure 7.3.3b). Thus, the volume V of the solid can be obtained by adding together the volumes of the tubes; that is,

$$V = V(S_1) + V(S_2) + \cdots + V(S_n)$$

As a rule, the tubes will have curved upper surfaces, so there will be no simple formulas for their volumes. However, if the strips are thin, then we can approximate each strip by a rectangle (Figure 7.3.4a). These rectangles, when revolved about the y-axis, will produce cylindrical shells whose volumes closely approximate the volumes generated by the original strips (Figure 7.3.4b). We will show that by adding the volumes of the cylindrical shells we can obtain a Riemann sum that approximates the volume V, and by taking the limit of the Riemann sums we can obtain an integral for the exact volume V.

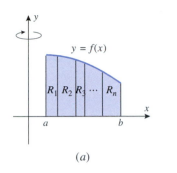

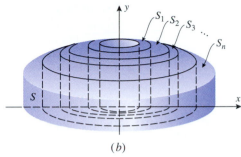

Figure 7.3.3

(a)

(b)

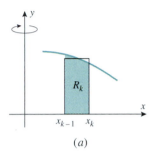

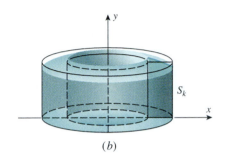

Figure 7.3.4

(a)

(b)

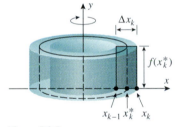

Figure 7.3.5

To implement this idea, suppose that the kth strip extends from x_{k-1} to x_k and that the width of this strip is

$$\Delta x_k = x_k - x_{k-1}$$

If we let x_k^* be the *midpoint* of the interval $[x_{k-1}, x_k]$, and if we construct a rectangle of height $f(x_k^*)$ over the interval, then revolving this rectangle about the y-axis produces a cylindrical shell of height $f(x_k^*)$, average radius x_k^*, and thickness Δx_k (Figure 7.3.5). From (1), the volume V_k of this cylindrical shell is

$$V_k = 2\pi x_k^* f(x_k^*) \Delta x_k$$

Adding the volumes of the n cylindrical shells yields the following Riemann sum that approximates the volume V:

$$V \approx \sum_{k=1}^{n} 2\pi x_k^* f(x_k^*) \Delta x_k$$

Taking the limit as n increases and the widths of the subintervals approach zero yields the definite integral

$$V = \lim_{\max \Delta x_k \to 0} \sum_{k=1}^{n} 2\pi x_k^* f(x_k^*) \Delta x_k = \int_a^b 2\pi x f(x)\, dx$$

In summary, we have the following result.

7.3.2 **VOLUME BY CYLINDRICAL SHELLS ABOUT THE y-AXIS.** Let f be continuous and nonnegative on $[a, b]$, and let R be the region that is bounded above by $y = f(x)$, below by the x-axis, and on the sides by the lines $x = a$ and $x = b$. Then the volume V of the solid of revolution that is generated by revolving the region R about the y-axis is given by

$$V = \int_a^b 2\pi x f(x)\, dx \qquad (2)$$

Example 1 Use cylindrical shells to find the volume of the solid generated when the region enclosed between $y = \sqrt{x}, x = 1, x = 4$, and the x-axis is revolved about the y-axis (Figure 7.3.6).

Solution. Since $f(x) = \sqrt{x}, a = 1$, and $b = 4$, Formula (2) yields

$$V = \int_1^4 2\pi x \sqrt{x}\, dx = 2\pi \int_1^4 x^{3/2}\, dx = \left[2\pi \cdot \frac{2}{5} x^{5/2} \right]_1^4 = \frac{4\pi}{5}[32 - 1] = \frac{124\pi}{5} \quad \blacktriangleleft$$

VARIATIONS OF THE METHOD OF CYLINDRICAL SHELLS

The method of cylindrical shells is applicable in a variety of situations that do not fit the conditions required by Formula (2). For example, the region may be enclosed between two curves, or the axis of revolution may be some line other than the y-axis. However, rather than develop a separate formula for every possible situation, we will give a general way of thinking about the method of cylindrical shells that can be adapted to each new situation as it arises.

For this purpose, we will need to reexamine the integrand in Formula (2): At each x in the interval $[a, b]$, the vertical line segment from the x-axis to the curve $y = f(x)$ can be viewed as the cross section of the region R at x (Figure 7.3.7a). When the region R is revolved about the y-axis, the cross section at x sweeps out the *surface* of a right circular cylinder of height $f(x)$ and radius x (Figure 7.3.7b). The area of this surface is

$$2\pi x f(x)$$

(Figure 7.3.7c), which is the integrand in (2). Thus, Formula (2) can be viewed informally in the following way.

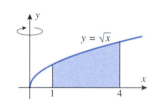

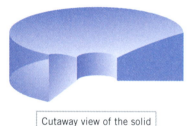

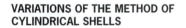
Cutaway view of the solid

Figure 7.3.6

> **7.3.3** **AN INFORMAL VIEWPOINT ABOUT CYLINDRICAL SHELLS.** The volume V of a solid of revolution that is generated by revolving a region R about an axis can be obtained by integrating the area of the surface generated by an arbitrary cross section of R taken parallel to the axis of revolution.

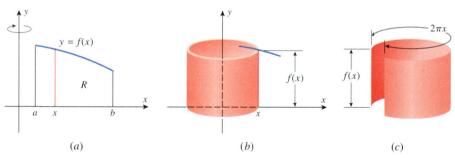

(a)   (b)   (c)

Figure 7.3.7

The following examples illustrate how to apply this result in situations where Formula (2) is not applicable.

Example 2 Use cylindrical shells to find the volume of the solid generated when the region R in the first quadrant enclosed between $y = x$ and $y = x^2$ is revolved about the y-axis (Figure 7.3.8).

Solution. At each x in $[0, 1]$ the cross section of R parallel to the y-axis generates a cylindrical surface of height $x - x^2$ and radius x. Since the area of this surface is

$$2\pi x (x - x^2)$$

the volume of the solid is

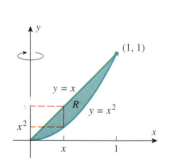

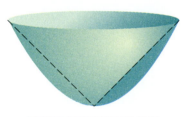

This solid looks like a bowl with a cone-shaped interior.

Figure 7.3.8

$$V = \int_0^1 2\pi x(x - x^2)\, dx = 2\pi \int_0^1 (x^2 - x^3)\, dx = 2\pi \left[\frac{x^3}{3} - \frac{x^4}{4} \right]_0^1 = 2\pi \left[\frac{1}{3} - \frac{1}{4} \right] = \frac{\pi}{6} \quad \blacktriangleleft$$

FOR THE READER. The volume in this example can also be obtained by the method of washers. Confirm that the volume produced by that method agrees with the volume obtained by cylindrical shells.

Example 3 Use cylindrical shells to find the volume of the solid generated when the region R under $y = x^2$ over the interval $[0, 2]$ is revolved about the x-axis (Figure 7.3.9).

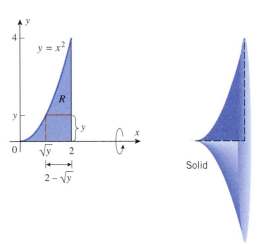

Figure 7.3.9

Solution. At each y in the interval $0 \leq y \leq 4$, the cross section of R parallel to the x-axis generates a cylindrical surface of height $2 - \sqrt{y}$ and radius y. Since the area of this surface is $2\pi y(2 - \sqrt{y})$, the volume of the solid is

$$V = \int_0^4 2\pi y(2 - \sqrt{y})\, dy = 2\pi \int_0^4 (2y - y^{3/2})\, dy = 2\pi \left[y^2 - \frac{2}{5}y^{5/2} \right]_0^4 = \frac{32\pi}{5} \quad \blacktriangleleft$$

FOR THE READER. The volume in this example can also be obtained by the method of disks. Confirm that the volume produced by that method agrees with the volume obtained by cylindrical shells.

EXERCISE SET 7.3 🇨 CAS

In Exercises 1–4, use cylindrical shells to find the volume of the solid generated when the shaded region is revolved about the indicated axis.

1.
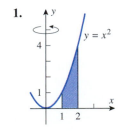

2.

3.

4.

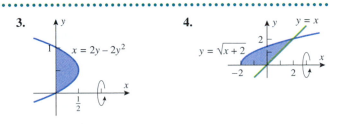

In Exercises 5–12, use cylindrical shells to find the volume of the solid generated when the region enclosed by the given curves is revolved about the y-axis.

5. $y = x^3$, $x = 1$, $y = 0$

6. $y = \sqrt{x}, \ x = 4, \ x = 9, \ y = 0$

7. $y = 1/x, \ y = 0, \ x = 1, \ x = 3$

8. $y = \cos(x^2), \ x = 0, \ x = \frac{1}{2}\sqrt{\pi}, \ y = 0$

9. $y = 2x - 1, \ y = -2x + 3, \ x = 2$

10. $y = 2x - x^2, \ y = 0$

11. $y = \dfrac{1}{x^2 + 1}, \ x = 0, \ x = 1, \ y = 0$

12. $y = e^{x^2}, \ x = 1, \ x = \sqrt{3}, \ y = 0$

> In Exercises 13–16, use cylindrical shells to find the volume of the solid generated when the region enclosed by the given curves is revolved about the x-axis.

13. $y^2 = x, \ y = 1, \ x = 0$

14. $x = 2y, \ y = 2, \ y = 3, \ x = 0$

15. $y = x^2, \ x = 1, \ y = 0$

16. $xy = 4, \ x + y = 5$

c **17.** Use a CAS to find the volume of the solid generated when the region enclosed by $y = \sin x$ and $y = 0$ for $0 \le x \le \pi$ is revolved about the y-axis.

c **18.** Use a CAS to find the volume of the solid generated when the region enclosed by $y = \cos x$, $y = 0$, and $x = 0$ for $0 \le x \le \pi/2$ is revolved about the y-axis.

19. (a) Use cylindrical shells to find the volume of the solid that is generated when the region under the curve
$$y = x^3 - 3x^2 + 2x$$
over $[0, 1]$ is revolved about the y-axis.
(b) For this problem, is the method of cylindrical shells easier or harder than the method of slicing discussed in the last section? Explain.

20. Use cylindrical shells to find the volume of the solid that is generated when the region that is enclosed by $y = 1/x^3$, $x = 1$, $x = 2$, $y = 0$ is revolved about the line $x = -1$.

21. Use cylindrical shells to find the volume of the solid that is generated when the region that is enclosed by $y = x^3$, $y = 1$, $x = 0$ is revolved about the line $y = 1$.

22. Let R_1 and R_2 be regions of the form shown in the accompanying figure. Use cylindrical shells to find a formula for the volume of the solid that results when
(a) region R_1 is revolved about the y-axis
(b) region R_2 is revolved about the x-axis.

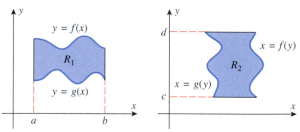

Figure Ex-22

23. Use cylindrical shells to find the volume of the cone generated when the triangle with vertices $(0, 0)$, $(0, r)$, $(h, 0)$, where $r > 0$ and $h > 0$, is revolved about the x-axis.

24. The region enclosed between the curve $y^2 = kx$ and the line $x = \frac{1}{4}k$ is revolved about the line $x = \frac{1}{2}k$. Use cylindrical shells to find the volume of the resulting solid. (Assume $k > 0$.)

25. A round hole of radius a is drilled through the center of a solid sphere of radius r. Use cylindrical shells to find the volume of the portion removed. (Assume $r > a$.)

26. Use cylindrical shells to find the volume of the torus obtained by revolving the circle $x^2 + y^2 = a^2$ about the line $x = b$, where $b > a > 0$. [*Hint:* It may help in the integration to think of an integral as an area.]

27. Let V_x and V_y be the volumes of the solids that result when the region enclosed by $y = 1/x$, $y = 0$, $x = \frac{1}{2}$, and $x = b$ $(b > \frac{1}{2})$ is revolved about the x-axis and y-axis, respectively. Is there a value of b for which $V_x = V_y$?

28. (a) Find the volume V of the solid generated when the region bounded by $y = 1/(1 + x^4)$, $y = 0$, $x = 1$, and $x = b$ $(b > 1)$ is revolved about the y-axis.
(b) Find $\lim\limits_{b \to +\infty} V$.

7.4 LENGTH OF A PLANE CURVE

In this section we will consider the problem of finding the length of a plane curve.

ARC LENGTH

Although formulas for lengths of circular arcs appear in early historical records, very little was known about the lengths of more general curves until the mid-seventeenth century. About that time formulas were discovered for a few specific curves such as the length of an arch of a cycloid. However, such basic problems as finding the length of an ellipse defied the mathematicians of that period, and almost no progress was made on the general problem of finding lengths of curves until the advent of calculus in the next century.

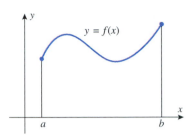

Figure 7.4.1

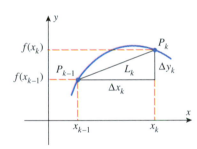

Figure 7.4.2

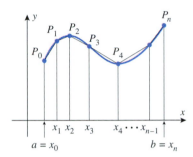

Figure 7.4.3

Our first objective in this section is to *define* what we mean by the length (also called the *arc length*) of a plane curve $y = f(x)$ over an interval $[a, b]$ (Figure 7.4.1). Once that is done we will be able to focus on computational matters. To avoid some complications that would otherwise occur, we will impose the requirement that f' be continuous on $[a, b]$, in which case we will say that $y = f(x)$ is a **smooth curve** on $[a, b]$ (or that f is a **smooth function** on $[a, b]$).

We will be concerned with the following problem:

7.4.1 ARC LENGTH PROBLEM. Suppose that $y = f(x)$ is a smooth curve on the interval $[a, b]$. Define and find a formula for the arc length L of the curve $y = f(x)$ over the interval $[a, b]$.

The basic idea for defining arc length is to break up the curve into small segments, approximate the curve segments by line segments, add the lengths of the line segments to form a Riemann sum that approximates the arc length L, and take the limit of the Riemann sums to obtain an integral for L.

To implement this idea, divide the interval $[a, b]$ into n subintervals by inserting numbers $x_1, x_2, \ldots, x_{n-1}$ between $a = x_0$ and $b = x_n$. As shown in Figure 7.4.2, let $P_0, P_1, \ldots, P_n$ be the points on the curve with x-coordinates $a = x_0, x_1, x_2, \ldots, x_{n-1}, b = x_n$ and join these points with straight line segments. These line segments form a **polygonal path** that we can regard as an approximation to the curve $y = f(x)$. As suggested by Figure 7.4.3, the length L_k of the kth line segment in the polygonal path is

$$L_k = \sqrt{(\Delta x_k)^2 + (\Delta y_k)^2} = \sqrt{(\Delta x_k)^2 + [f(x_k) - f(x_{k-1})]^2} \tag{1}$$

If we now add the lengths of these line segments, we obtain the following approximation to the length L of the curve

$$L \approx \sum_{k=1}^{n} L_k = \sum_{k=1}^{n} \sqrt{(\Delta x_k)^2 + [f(x_k) - f(x_{k-1})]^2} \tag{2}$$

To put this in the form of a Riemann sum we will apply the Mean-Value Theorem (5.8.2). This theorem implies that there is a number x_k^* between x_{k-1} and x_k such that

$$\frac{f(x_k) - f(x_{k-1})}{x_k - x_{k-1}} = f'(x_k^*) \quad \text{or} \quad f(x_k) - f(x_{k-1}) = f'(x_k^*)\Delta x_k$$

and hence we can rewrite (2) as

$$L \approx \sum_{k=1}^{n} \sqrt{1 + [f'(x_k^*)]^2}\, \Delta x_k$$

Thus, taking the limit as n increases and the widths of the subintervals approach zero yields the following integral that defines the arc length L:

$$L = \lim_{\max \Delta x_k \to 0} \sum_{k=1}^{n} \sqrt{1 + [f'(x_k^*)]^2}\, \Delta x_k = \int_a^b \sqrt{1 + [f'(x)]^2}\, dx$$

In summary, we have the following definition:

7.4.2 DEFINITION. If $y = f(x)$ is a smooth curve on the interval $[a, b]$, then the arc length L of this curve over $[a, b]$ is defined as

$$L = \int_a^b \sqrt{1 + [f'(x)]^2}\, dx \tag{3}$$

This result provides both a definition and a formula for computing arc lengths. Where convenient, (3) can also be expressed as

$$L = \int_a^b \sqrt{1 + [f'(x)]^2}\,dx = \int_a^b \sqrt{1 + \left(\frac{dy}{dx}\right)^2}\,dx \qquad (4)$$

Moreover, for a curve expressed in the form $x = g(y)$, where g' is continuous on $[c, d]$, the arc length L from $y = c$ to $y = d$ can be expressed as

$$L = \int_c^d \sqrt{1 + [g'(y)]^2}\,dy = \int_c^d \sqrt{1 + \left(\frac{dx}{dy}\right)^2}\,dy \qquad (5)$$

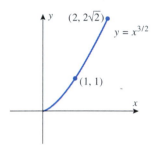

Figure 7.4.4

Example 1   Find the arc length of the curve $y = x^{3/2}$ from $(1, 1)$ to $(2, 2\sqrt{2})$ (Figure 7.4.4) in two ways: (a) using Formula (4) and (b) using Formula (5).

Solution (a).   Since

$$\frac{dy}{dx} = \frac{3}{2}x^{1/2}$$

and since the curve extends from $x = 1$ to $x = 2$, it follows from (4) that

$$L = \int_1^2 \sqrt{1 + \frac{9}{4}x}\,dx$$

To evaluate this integral we make the u-substitution

$$u = 1 + \frac{9}{4}x, \quad du = \frac{9}{4}\,dx$$

and then change the x-limits of integration $(x = 1, x = 2)$ to the corresponding u-limits $\left(u = \frac{13}{4}, u = \frac{22}{4}\right)$:

$$L = \frac{4}{9}\int_{13/4}^{22/4} u^{1/2}\,du = \frac{8}{27}u^{3/2}\Big]_{13/4}^{22/4} = \frac{8}{27}\left[\left(\frac{22}{4}\right)^{3/2} - \left(\frac{13}{4}\right)^{3/2}\right]$$

$$= \frac{22\sqrt{22} - 13\sqrt{13}}{27} \approx 2.09$$

Solution (b).   To apply Formula (5) we must first rewrite the equation $y = x^{3/2}$ so that x is expressed as a function of y. This yields $x = y^{2/3}$ and

$$\frac{dx}{dy} = \frac{2}{3}y^{-1/3}$$

Since the curve extends from $y = 1$ to $y = 2\sqrt{2}$, it follows from (5) that

$$L = \int_1^{2\sqrt{2}} \sqrt{1 + \frac{4}{9}y^{-2/3}}\,dy = \frac{1}{3}\int_1^{2\sqrt{2}} y^{-1/3}\sqrt{9y^{2/3} + 4}\,dy$$

To evaluate this integral we make the u-substitution

$$u = 9y^{2/3} + 4, \quad du = 6y^{-1/3}\,dy$$

and change the y-limits of integration $(y = 1, y = 2\sqrt{2})$ to the corresponding u-limits $(u = 13, u = 22)$. This gives

$$L = \frac{1}{18}\int_{13}^{22} u^{1/2}\,du = \frac{1}{27}u^{3/2}\Big]_{13}^{22} = \frac{1}{27}[(22)^{3/2} - (13)^{3/2}] = \frac{22\sqrt{22} - 13\sqrt{13}}{27}$$

This result agrees with that in part (a); however, the integration here is more tedious. In problems where there is a choice between using (4) or (5), it is often the case that one of the formulas leads to a simpler integral than the other.   ◄

ARC LENGTH OF PARAMETRIC CURVES

The following result provides a formula for finding the arc length of a curve from parametric equations for the curve. Its derivation is similar to that of Formula (3) and will be omitted.

7.4.3 ARC LENGTH FORMULA FOR PARAMETRIC CURVES. If no segment of the curve represented by the parametric equations

$$x = x(t), \quad y = y(t) \quad (a \leq t \leq b)$$

is traced more than once as t increases from a to b, and if dx/dt and dy/dt are continuous functions for $a \leq t \leq b$, then the arc length L of the curve is given by

$$L = \int_a^b \sqrt{\left(\frac{dx}{dt}\right)^2 + \left(\frac{dy}{dt}\right)^2}\, dt \tag{6}$$

REMARK. Note that Formulas (4) and (5) are special cases of (6). For example, Formula (4) can be obtained from (6) by writing $y = f(x)$ parametrically as $x = t$, $y = f(t)$; similarly, Formula (5) can be obtained from (6) by writing $x = g(y)$ parametrically as $x = g(t)$, $y = t$. We leave the details as exercises.

Example 2 Use (6) to find the circumference of a circle of radius a from the parametric equations

$$x = a\cos t, \quad y = a\sin t \quad (0 \leq t \leq 2\pi)$$

Solution.

$$L = \int_0^{2\pi} \sqrt{\left(\frac{dx}{dt}\right)^2 + \left(\frac{dy}{dt}\right)^2}\, dt = \int_0^{2\pi} \sqrt{(-a\sin t)^2 + (a\cos t)^2}\, dt$$

$$= \int_0^{2\pi} a\, dt = at\Big]_0^{2\pi} = 2\pi a \qquad \blacktriangleleft$$

FINDING ARC LENGTH BY NUMERICAL METHODS

As a rule, the integrals that arise in calculating arc length tend to be impossible to evaluate in terms of elementary functions, so it will often be necessary to approximate the integral using a numerical method such as the midpoint approximation (discussed in Section 6.4) or some other comparable method. Examples 1 and 2 are rare exceptions.

Example 3 From (4), the arc length of $y = \sin x$ from $x = 0$ to $x = \pi$ is given by the integral

$$L = \int_0^{\pi} \sqrt{1 + (\cos x)^2}\, dx$$

This integral cannot be evaluated in terms of elementary functions; however, using a calculating utility with a numerical integration capability yields the approximation $L \approx 3.8202$.
$\blacktriangleleft$

FOR THE READER. In Figure 7.4.5, the scale on both axes is 2 centimeters per unit. Confirm that the result in Example 3 is reasonable by laying a piece of string as closely as possible along the curve in the figure and measuring its length in centimeters.

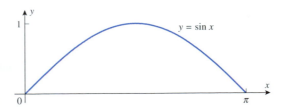

Figure 7.4.5

> **FOR THE READER.** Computer algebra systems and some scientific calculators have commands for evaluating integrals numerically, and some scientific calculators have built-in commands for approximating arc lengths. If you have a scientific calculator with one of these capabilities or a CAS, read the documentation, and then use your calculator or CAS to check the result in Example 3.

EXERCISE SET 7.4 Graphing Utility   **c** CAS

1. Use the Theorem of Pythagoras to find the length of the line segment $y = 2x$ from $(1, 2)$ to $(2, 4)$, and confirm that the value is consistent with the length computed using
   (a) Formula (4)   (b) Formula (5).

2. Use the Theorem of Pythagoras to find the length of the line segment $x = t$, $y = 5t$ $(0 \le t \le 1)$, and confirm that the value is consistent with the length computed using Formula (6).

In Exercises 3–8, find the exact arc length of the curve over the stated interval.

3. $y = 3x^{3/2} - 1$ from $x = 0$ to $x = 1$

4. $x = \frac{1}{3}(y^2 + 2)^{3/2}$ from $y = 0$ to $y = 1$

5. $y = x^{2/3}$ from $x = 1$ to $x = 8$

6. $y = (x^6 + 8)/(16x^2)$ from $x = 2$ to $x = 3$

7. $24xy = y^4 + 48$ from $y = 2$ to $y = 4$

8. $x = \frac{1}{8}y^4 + \frac{1}{4}y^{-2}$ from $y = 1$ to $y = 4$

In Exercises 9–14, find the exact arc length of the parametric curve without eliminating the parameter.

9. $x = \frac{1}{3}t^3$, $y = \frac{1}{2}t^2$ $(0 \le t \le 1)$

10. $x = (1 + t)^2$, $y = (1 + t)^3$ $(0 \le t \le 1)$

11. $x = \cos 2t$, $y = \sin 2t$ $(0 \le t \le \pi/2)$

12. $x = \cos t + t \sin t$, $y = \sin t - t \cos t$ $(0 \le t \le \pi)$

13. $x = e^t \cos t$, $y = e^t \sin t$ $(0 \le t \le \pi/2)$

14. $x = e^t(\sin t + \cos t)$, $y = e^t(\cos t - \sin t)$ $(1 \le t \le 4)$

In Exercises 15 and 16, express the exact arc length of the curve over the given interval as an integral that has been simplified to eliminate the radical, and then evaluate the integral using a CAS.

c 15. $y = \ln(\sec x)$ from $x = 0$ to $x = \pi/4$

c 16. $y = \ln(\sin x)$ from $x = \pi/4$ to $x = \pi/2$

c 17. (a) Recall from Section 1.8 that a cycloid is the path traced by a point on the rim of a wheel that rolls along a line (Figure 1.8.13). Use the parametric equations in Formula (9) of that section to show that the length L of one arch of a cycloid is given by the integral

$$L = a \int_0^{2\pi} \sqrt{2(1 - \cos\theta)}\, d\theta$$

(b) Use a CAS to show that L is eight times the radius of the wheel (see the accompanying figure).

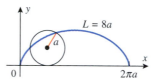

Figure Ex-17

 18. It was stated in Exercise 41 of Section 1.8 that the curve given parametrically by the equations

$$x = a\cos^3\phi, \quad y = a\sin^3\phi$$

is called a *four-cusped hypocycloid* (also called an **astroid**).
   (a) Use a graphing utility to generate the graph in the case where $a = 1$, so that it is traced exactly once.
   (b) Find the exact arc length of the curve in part (a).

19. Consider the curve $y = x^{2/3}$.
   (a) Sketch the portion of the curve between $x = -1$ and $x = 8$.
   (b) Explain why Formula (4) cannot be used to find the arc length of the curve sketched in part (a).
   (c) Find the arc length of the curve sketched in part (a).

20. Derive Formulas (4) and (5) from Formula (6) by choosing appropriate parametrizations of the curves.

In Exercises 21 and 22, use the midpoint approximation with $n = 20$ subintervals to approximate the arc length of the curve over the given interval.

21. $y = x^2$ from $x = 0$ to $x = 2$

22. $x = \sin y$ from $y = 0$ to $y = \pi$

c 23. Use a CAS or a scientific calculator with numerical integration capabilities to approximate the arc lengths in Exercises 21 and 22.

24. Let $y = f(x)$ be a smooth curve on the closed interval $[a, b]$. Prove that if there are nonnegative numbers m and M such that $m \le f'(x) \le M$ for all x in $[a, b]$, then the arc length L of $y = f(x)$ over the interval $[a, b]$ satisfies the inequalities

$$(b - a)\sqrt{1 + m^2} \le L \le (b - a)\sqrt{1 + M^2}$$

25. Use the result of Exercise 24 to show that the arc length L of $y = \sin x$ over the interval $0 \le x \le \pi/4$ satisfies

$$\frac{\pi}{4}\sqrt{\frac{3}{2}} \le L \le \frac{\pi}{4}\sqrt{2}$$

26. Show that the total arc length of the ellipse $x = a \cos t$, $y = b \sin t$, $0 \le t \le 2\pi$ for $a > b > 0$ is given by

$$4a \int_0^{\pi/2} \sqrt{1 - k^2 \cos^2 t}\, dt$$

where $k = \sqrt{a^2 - b^2}/a$.

c 27. (a) Show that the total arc length of the ellipse

$$x = 2 \cos t, \quad y = \sin t \quad (0 \le t \le 2\pi)$$

is given by

$$4 \int_0^{\pi/2} \sqrt{1 + 3 \sin^2 t}\, dt$$

(b) Use a CAS or a scientific calculator with numerical integration capabilities to approximate the arc length in part (a). Round your answer to two decimal places.

(c) Suppose that the parametric equations in part (a) describe the path of a particle moving in the xy-plane, where t is time in seconds and x and y are in centimeters. Use a CAS or a scientific calculator with numerical integration capabilities to approximate the distance traveled by the particle from $t = 1.5$ s to $t = 4.8$ s. Round your answer to two decimal places.

c 28. A basketball player makes a successful shot from the free throw line. Suppose that the path of the ball from the moment of release to the moment it enters the hoop is described by

$$y = 2.15 + 2.09x - 0.41x^2, \quad 0 \le x \le 4.6$$

where x is the horizontal distance (in meters) from the point of release, and y is the vertical distance (in meters) above the floor. Use a CAS or a scientific calculator with numerical integration capabilities to approximate the distance the ball travels from the moment it is released to the moment it enters the hoop. Round your answer to two decimal places.

c 29. Find a positive value of k (to two decimal places) such that the curve $y = k \sin x$ has an arc length of $L = 5$ units over the interval from $x = 0$ to $x = \pi$. [*Hint:* Find an integral for the arc length L in terms of k, and then use a CAS or a scientific calculator with a numeric integration capability to find integer values of k at which the values of $L - 5$ have opposite signs. Complete the solution by using the Intermediate-Value Theorem (2.5.8) to approximate the value of k to two decimal places.]

7.5 AREA OF A SURFACE OF REVOLUTION

In this section we will consider the problem of finding the area of a surface that is generated by revolving a plane curve about a line.

SURFACE AREA

A *surface of revolution* is a surface that is generated by revolving a plane curve about an axis that lies in the same plane as the curve. For example, the surface of a sphere can be generated by revolving a semicircle about its diameter, and the lateral surface of a right circular cylinder can be generated by revolving a line segment about an axis that is parallel to it (Figure 7.5.1).

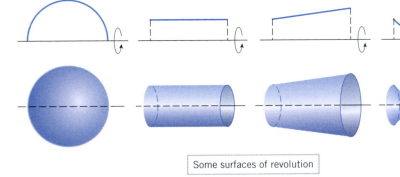

Some surfaces of revolution

Figure 7.5.1

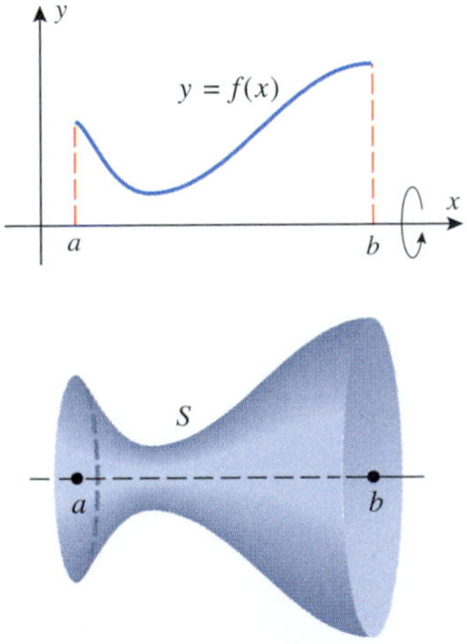

In this section we will be concerned with the following problem:

7.5.1   SURFACE AREA PROBLEM.   Suppose that f is a smooth, nonnegative function on $[a, b]$ and that a surface of revolution is generated by revolving the portion of the curve $y = f(x)$ between $x = a$ and $x = b$ about the x-axis (Figure 7.5.2). Define what is meant by the *area S* of the surface, and find a formula for computing it.

Figure 7.5.2

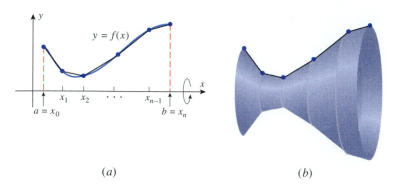

Figure 7.5.3   (*a*)   (*b*)

To motivate an appropriate definition for the area S of a surface of revolution, we will decompose the surface into small sections whose areas can be approximated by elementary formulas, add the approximations of the areas of the sections to form a Riemann sum that approximates S, and then take the limit of the Riemann sums to obtain an integral for the exact value of S.

To implement this idea, divide the interval $[a, b]$ into n subintervals by inserting numbers $x_1, x_2, \ldots, x_{n-1}$ between $a = x_0$ and $b = x_n$. As illustrated in Figure 7.5.3*a*, the corresponding points on the graph of f define a polygonal path that approximates the curve $y = f(x)$ over the interval $[a, b]$. When this polygonal path is revolved about the x-axis, it generates a surface consisting of n parts, each of which is a frustum of a right circular cone (Figure 7.5.3*b*). Thus, the area of each part of the approximating surface can be obtained from the formula

$$S = \pi(r_1 + r_2)l \tag{1}$$

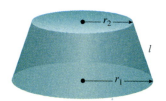

Figure 7.5.4

for the lateral area S of a frustum of slant height l and base radii r_1 and r_2 (Figure 7.5.4). As suggested by Figure 7.5.5, the kth frustum has radii $f(x_{k-1})$ and $f(x_k)$ and height Δx_k. Its slant height is the length L_k of the kth line segment in the polygonal path, which from Formula (1) of Section 7.4 is

$$L_k = \sqrt{(\Delta x_k)^2 + [f(x_k) - f(x_{k-1})]^2}$$

Thus, the lateral area S_k of the kth frustum is

$$S_k = \pi[f(x_{k-1}) + f(x_k)]\sqrt{(\Delta x_k)^2 + [f(x_k) - f(x_{k-1})]^2}$$

If we add these areas, we obtain the following approximation to the area S of the entire surface:

$$S \approx \sum_{k=1}^{n} \pi[f(x_{k-1}) + f(x_k)]\sqrt{(\Delta x_k)^2 + [f(x_k) - f(x_{k-1})]^2} \tag{2}$$

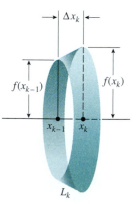

Figure 7.5.5

To put this in the form of a Riemann sum we will apply the Mean-Value Theorem (5.8.2). This theorem implies that there is a number x_k^* between x_{k-1} and x_k such that

$$\frac{f(x_k) - f(x_{k-1})}{x_k - x_{k-1}} = f'(x_k^*) \quad \text{or} \quad f(x_k) - f(x_{k-1}) = f'(x_k^*)\Delta x_k$$

and hence we can rewrite (2) as

$$S \approx \sum_{k=1}^{n} \pi[f(x_{k-1}) + f(x_k)]\sqrt{1 + [f'(x_k^*)]^2}\,\Delta x_k \tag{3}$$

However, this is not yet a Riemann sum because it involves the variables x_{k-1} and x_k. To eliminate these variables from the expression, observe that the average value of the numbers $f(x_{k-1})$ and $f(x_k)$ lies between these numbers, so the continuity of f and the Intermediate-Value Theorem (2.5.8) imply that there is a number x_k^{**} between x_{k-1} and x_k such that

$$\tfrac{1}{2}[f(x_{k-1}) + f(x_k)] = f(x_k^{**})$$

Thus, (2) can be expressed as

$$S \approx \sum_{k=1}^{n} 2\pi f(x_k^{**})\sqrt{1 + [f'(x_k^*)]^2}\,\Delta x_k$$

Although this expression is close to a Riemann sum in form, it is not a true Riemann sum because it involves two variables x_k^* and x_k^{**}, rather than x_k^* alone. However, it is proved in advanced calculus courses that this has no effect on the limit because of the continuity of f. Thus, we can assume that $x_k^{**} = x_k^*$ when taking the limit, and this suggests that S can be defined as

$$S = \lim_{\max \Delta x_k \to 0} \sum_{k=1}^{n} 2\pi f(x_k^{**})\sqrt{1 + [f'(x_k^*)]^2}\,\Delta x_k = \int_a^b 2\pi f(x)\sqrt{1 + [f'(x)]^2}\,dx$$

In summary, we have the following definition:

7.5.2   DEFINITION.   If f is a smooth, nonnegative function on $[a, b]$, then the surface area S of the surface of revolution that is generated by revolving the portion of the curve $y = f(x)$ between $x = a$ and $x = b$ about the x-axis is defined as

$$S = \int_a^b 2\pi f(x)\sqrt{1 + [f'(x)]^2}\,dx$$

This result provides both a definition and a formula for computing surface areas. Where convenient, this formula can also be expressed as

$$S = \int_a^b 2\pi f(x)\sqrt{1 + [f'(x)]^2}\,dx = \int_a^b 2\pi y\sqrt{1 + \left(\frac{dy}{dx}\right)^2}\,dx \qquad (4)$$

Moreover, if g is nonnegative and $x = g(y)$ is a smooth curve on the interval $[c, d]$, then the area of the surface that is generated by revolving the portion of a curve $x = g(y)$ between $y = c$ and $y = d$ about the y-axis can be expressed as

$$S = \int_c^d 2\pi g(y)\sqrt{1 + [g'(y)]^2}\,dy = \int_c^d 2\pi x\sqrt{1 + \left(\frac{dx}{dy}\right)^2}\,dy \qquad (5)$$

Example 1   Find the area of the surface that is generated by revolving the portion of the curve $y = x^3$ between $x = 0$ and $x = 1$ about the x-axis (Figure 7.5.6).

Solution.   Since $y = x^3$, we have $dy/dx = 3x^2$, and hence from (4) the surface area S is

$$S = \int_0^1 2\pi y\sqrt{1 + \left(\frac{dy}{dx}\right)^2}\,dx$$

$$= \int_0^1 2\pi x^3\sqrt{1 + (3x^2)^2}\,dx$$

$$= 2\pi \int_0^1 x^3(1 + 9x^4)^{1/2}\,dx$$

$$= \frac{2\pi}{36}\int_1^{10} u^{1/2}\,du \qquad \boxed{\begin{array}{l} u = 1 + 9x^4 \\ du = 36x^3\,dx \end{array}}$$

$$= \frac{2\pi}{36}\cdot\frac{2}{3}u^{3/2}\Big]_{u=1}^{10} = \frac{\pi}{27}(10^{3/2} - 1) \approx 3.56 \qquad \blacktriangleleft$$

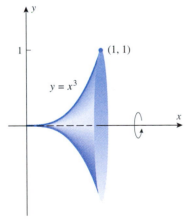

$y = x^3$

$(1, 1)$

Figure 7.5.6

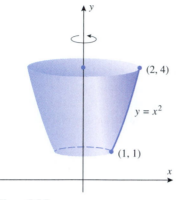

Figure 7.5.7

Example 2 Find the area of the surface that is generated by revolving the portion of the curve $y = x^2$ between $x = 1$ and $x = 2$ about the y-axis (Figure 7.5.7).

Solution. Because the curve is revolved about the y-axis we will apply Formula (5). Toward this end, we rewrite $y = x^2$ as $x = \sqrt{y}$ and observe that the y-values corresponding to $x = 1$ and $x = 2$ are $y = 1$ and $y = 4$. Since $x = \sqrt{y}$, we have $dx/dy = 1/(2\sqrt{y})$, and hence from (5) the surface area S is

$$S = \int_1^4 2\pi x \sqrt{1 + \left(\frac{dx}{dy}\right)^2}\, dy$$

$$= \int_1^4 2\pi \sqrt{y} \sqrt{1 + \left(\frac{1}{2\sqrt{y}}\right)^2}\, dy$$

$$= \pi \int_1^4 \sqrt{4y + 1}\, dy$$

$$= \frac{\pi}{4} \int_5^{17} u^{1/2}\, du \qquad \boxed{\begin{array}{l} u = 4y + 1 \\ du = 4\,dy \end{array}}$$

$$= \frac{\pi}{4} \cdot \frac{2}{3} u^{3/2} \Big]_{u=5}^{17} = \frac{\pi}{6}(17^{3/2} - 5^{3/2}) \approx 30.85 \qquad \blacktriangleleft$$

EXERCISE SET 7.5　[c] CAS

In Exercises 1–4, find the area of the surface generated by revolving the given curve about the x-axis.

1. $y = 7x,\ 0 \le x \le 1$

2. $y = \sqrt{x},\ 1 \le x \le 4$

3. $y = \sqrt{4 - x^2},\ -1 \le x \le 1$

4. $x = \sqrt[3]{y},\ 1 \le y \le 8$

In Exercises 5–8, find the area of the surface generated by revolving the given curve about the y-axis.

5. $x = 9y + 1,\ 0 \le y \le 2$

6. $x = y^3,\ 0 \le y \le 1$

7. $x = \sqrt{9 - y^2},\ -2 \le y \le 2$

8. $x = 2\sqrt{1 - y},\ -1 \le y \le 0$

In Exercises 9–12, use a CAS to find the exact area of the surface generated by revolving the curve about the stated axis.

[c] **9.** $y = \sqrt{x} - \frac{1}{3}x^{3/2},\ 1 \le x \le 3$; x-axis

[c] **10.** $y = \frac{1}{3}x^3 + \frac{1}{4}x^{-1},\ 1 \le x \le 2$; x-axis

[c] **11.** $8xy^2 = 2y^6 + 1,\ 1 \le y \le 2$; y-axis

[c] **12.** $x = \sqrt{16 - y},\ 0 \le y \le 15$; y-axis

In Exercises 13–16, use a CAS or a calculator with numerical integration capabilities to approximate the area of the surface generated by revolving the curve about the stated axis. Round your answer to two decimal places.

[c] **13.** $y = \sin x,\ 0 \le x \le \pi$; x-axis

[c] **14.** $x = \tan y,\ 0 \le y \le \pi/4$; y-axis

[c] **15.** $y = e^x,\ 0 \le x \le 1$; x-axis

[c] **16.** $y = e^x,\ 1 \le y \le e$; y-axis

17. Use Formula (4) to show that the lateral area S of a right circular cone with height h and base radius r is

$$S = \pi r \sqrt{r^2 + h^2}$$

18. Show that the area of the surface of a sphere of radius r is $4\pi r^2$. [*Hint:* Revolve the semicircle $y = \sqrt{r^2 - x^2}$ about the x-axis.]

19. (a) The figure in Exercise 45 of Section 7.2 shows a spherical cap of height h cut from a sphere of radius r. Show that the surface area S of the cap is $S = 2\pi r h$. [*Hint:* Revolve an appropriate portion of the circle $x^2 + y^2 = r^2$ about the y-axis.]

(b) The portion of a sphere that is cut by two parallel planes is called a *zone*. Use the result in part (a) to show that the surface area of a zone depends on the radius of the sphere and the distance between the planes, but not on the location of the zone.

Exercises 20–26 require the formulas developed in the following discussion: If $x'(t)$ and $y'(t)$ are continuous functions and if no segment of the curve

$$x = x(t), \quad y = y(t) \quad (a \le t \le b)$$

is traced more than once, then it can be shown that the area of the surface generated by revolving this curve about the x-axis is

$$S = \int_a^b 2\pi y(t) \sqrt{[x'(t)]^2 + [y'(t)]^2}\, dt \tag{A}$$

and the area of the surface generated by revolving the curve about the y-axis is

$$S = \int_a^b 2\pi x(t) \sqrt{[x'(t)]^2 + [y'(t)]^2}\, dt \tag{B}$$

20. Derive Formulas (4) and (5) from Formulas (A) and (B) above by choosing appropriate parametrizations for the curves $y = f(x)$ and $x = g(y)$.

21. Find the area of the surface generated by revolving the parametric curve $x = t^2$, $y = 2t$, $0 \le t \le 4$ about the x-axis.

C 22. Use a CAS to find the area of the surface generated by revolving the parametric curve

$$x = \cos^2 t, \quad y = 5 \sin t \quad (0 \le t \le \pi/2)$$

about the x-axis.

23. Find the area of the surface generated by revolving the parametric curve $x = t$, $y = 2t^2$, $0 \le t \le 1$ about the y-axis.

24. Find the area of the surface generated by revolving the parametric curve $x = \cos^2 t$, $y = \sin^2 t$, $0 \le t \le \pi/2$ about the y-axis.

25. By revolving the semicircle

$$x = r \cos t, \quad y = r \sin t \quad (0 \le t \le \pi)$$

about the x-axis, show that the surface area of a sphere of radius r is $4\pi r^2$.

26. The equations

$$x = a\phi - a \sin \phi, \quad y = a - a \cos \phi \quad (0 \le \phi \le 2\pi)$$

represent one arch of a cycloid. Show that the surface area generated by revolving this curve about the x-axis is $S = 64\pi a^2/3$. [*Hint:* Use the identities $\sin^2 \dfrac{\phi}{2} = \dfrac{1 - \cos \phi}{2}$ and $\sin^3 \phi = (1 - \cos^2 \phi) \sin \phi$ to help with the integration.]

27. (a) If a cone of slant height l and base radius r is cut along a lateral edge and laid flat, then as shown in the accompanying figure it becomes a sector of a circle of radius l. Use the formula $A = \frac{1}{2} l^2 \theta$ for the area of a sector with radius l and central angle θ (in radians) to show that the lateral surface area of the cone is $\pi r l$.

(b) Use the result in part (a) to obtain Formula (1) for the lateral surface area of a frustum.

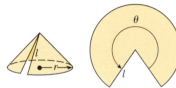

Figure Ex-27

28. Assume that $y = f(x)$ is a smooth curve on the interval $[a, b]$ and assume that $f(x) \ge 0$ for $a \le x \le b$. Derive a formula for the surface area generated when the curve $y = f(x)$, $a \le x \le b$, is revolved about the line $y = -k$ $(k > 0)$.

29. Let $y = f(x)$ be a smooth curve on the interval $[a, b]$ and assume that $f(x) \ge 0$ for $a \le x \le b$. By the Extreme-Value Theorem (5.5.3), the function f has a maximum value K and a minimum value k on $[a, b]$. Prove: If L is the arc length of the curve $y = f(x)$ between $x = a$ and $x = b$ and if S is the area of the surface that is generated by revolving this curve about the x-axis, then

$$2\pi k L \le S \le 2\pi K L$$

30. Let $y = f(x)$ be a smooth curve on $[a, b]$ and assume that $f(x) \ge 0$ for $a \le x \le b$. Let A be the area under the curve $y = f(x)$ between $x = a$ and $x = b$ and let S be the area of the surface obtained when this section of curve is revolved about the x-axis.

(a) Prove that $2\pi A \le S$.

(b) For what functions f is $2\pi A = S$?

7.6 WORK

In this section we will use the integration tools developed in the preceding chapter to study some of the basic principles of "work," which is one of the fundamental concepts in physics and engineering.

THE ROLE OF WORK IN PHYSICS AND ENGINEERING

In this section we will be concerned with two related concepts, *work* and *energy*. To put these ideas in a familiar setting, when you push a stalled car for a certain distance you are performing work, and the effect of your work is to make the car move. The energy of motion caused by the work is called the *kinetic energy* of the car. The exact connection between work and kinetic energy is governed by a principle of physics, called the *work–energy relationship*. Although we will touch on this idea in this section, a detailed study of the relationship between work and energy will be left for courses in physics and engineering. Our primary goal here will be to explain the role of integration in the study of work.

WORK DONE BY A CONSTANT FORCE APPLIED IN THE DIRECTION OF MOTION

When a stalled car is pushed, the speed that the car attains depends on the force F with which it is pushed and the distance d over which that force is applied (Figure 7.6.1). Thus, force and distance are the ingredients of work in the following definition.

> **7.6.1 DEFINITION.** If a constant force of magnitude F is applied in the direction of motion of an object, and if that object moves a distance d, then we define the **work** W performed by the force on the object to be
>
> $$W = F \cdot d \tag{1}$$

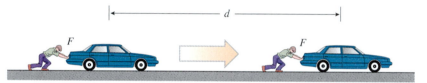

Figure 7.6.1

FOR THE READER. If you push against an immovable object, such as a brick wall, you may tire yourself out, but you will perform no work. Why?

Common units for measuring force are newtons (N) in the International System of Units (SI), dynes (dyn) in the centimeter-gram-second (CGS) system, and pounds (lb) in the British Engineering (BE) system. One newton is the force required to give a mass of 1 kg an acceleration of 1 m/s^2, one dyne is the force required to give a mass of 1 g an acceleration of 1 cm/s^2, and one pound of force is the force required to give a mass of 1 slug an acceleration of 1 ft/s^2.

It follows from Definition 7.6.1 that work has units of force times distance. The most common units of work are newton-meters (N·m), dyne-centimeters (dyn·cm), and foot-pounds (ft·lb). As indicated in Table 7.6.1, one newton-meter is also called a *joule* (J), and one dyne-centimeter is also called an *erg*. One foot-pound is approximately 1.36 J.

Table 7.6.1

SYSTEM	FORCE	×	DISTANCE	=	WORK
SI	newton (N)		meter (m)		joule (J)
CGS	dyne (dyn)		centimeter (cm)		erg
BE	pound (lb)		foot (ft)		foot-pound (ft·lb)

CONVERSION FACTORS:
$1 \text{ N} = 10^5 \text{ dyn} \approx 0.225 \text{ lb}$ $1 \text{ lb} \approx 4.45 \text{ N}$
$1 \text{ J} = 10^7 \text{ erg} \approx 0.738 \text{ ft·lb}$ $1 \text{ ft·lb} \approx 1.36 \text{ J} = 1.36 \times 10^7 \text{ erg}$

Example 1 An object moves 5 ft along a line while subjected to a constant force of 100 lb in its direction of motion. The work done is

$$W = F \cdot d = 100 \cdot 5 = 500 \text{ ft·lb}$$

An object moves 25 m along a line while subjected to a constant force of 4 N in its direction of motion. The work done is

$$W = F \cdot d = 4 \cdot 25 = 100 \text{ N·m} = 100 \text{ J} \qquad \blacktriangleleft$$

Vasili Alexeev lifting a record-breaking 562 lb in the 1976 Olympics

Example 2   In the 1976 Olympics, Vasili Alexeev astounded the world by lifting a record-breaking 562 lb from the floor to above his head (about 2 m). Equally astounding was the feat of strongman Paul Anderson, who in 1957 braced himself on the floor and used his back to lift 6270 lb of lead and automobile parts a distance of 1 cm. Who did more work?

Solution.   To lift an object one must apply sufficient force to overcome the gravitational force that the Earth exerts on that object. The force that the Earth exerts on an object is that object's weight; thus, in performing their feats, Alexeev applied a force of 562 lb over a distance of 2 m and Anderson applied a force of 6270 lb over a distance of 1 cm. Pounds are units in the BE system, meters are units in SI, and centimeters are units in the CGS system. We will need to decide on the measurement system we want to use and be consistent. Let us agree to use SI and express the work of the two men in joules. Using the conversion factor in Table 7.6.1 we obtain

$$562 \text{ lb} \approx 562 \text{ lb} \times 4.45 \text{ N/lb} = 2500.9 \text{ N}$$
$$6270 \text{ lb} \approx 6270 \text{ lb} \times 4.45 \text{ N/lb} = 27{,}901.5 \text{ N}$$

Using these values and the fact that 1 cm = 0.01 m we obtain

$$\text{Alexeev's work} = (2500.9 \text{ N}) \times (2 \text{ m}) \approx 5002 \text{ J}$$
$$\text{Anderson's work} = (27{,}901.5 \text{ N}) \times (0.01 \text{ m}) \approx 279 \text{ J}$$

Therefore, even though Anderson's lift required a tremendous upward force, it was applied over such a short distance that Alexeev did more work.   ◀

WORK DONE BY A VARIABLE FORCE APPLIED IN THE DIRECTION OF MOTION

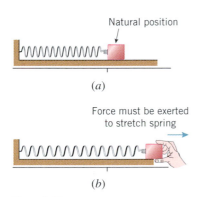

Figure 7.6.2

Many important problems are concerned with finding the work done by a *variable* force that is applied in the direction of motion. For example, Figure 7.6.2a shows a spring in its natural state (neither compressed nor stretched). If we want to pull the block horizontally (Figure 7.6.2b), then we would have to apply more and more force to the block to overcome the increasing force of the stretching spring. Thus, our next objective is to define what is meant by the work performed by a variable force and to find a formula for computing it. This will require calculus.

7.6.2   PROBLEM.   Suppose that an object moves in the positive direction along a coordinate line while subjected to a variable force $F(x)$ that is applied in the direction of motion. Define what is meant by the *work* W performed by the force on the object as the object moves from $x = a$ to $x = b$, and find a formula for computing the work.

The basic idea for solving this problem is to break up the interval $[a, b]$ into subintervals that are sufficiently small that the force does not vary much on each subinterval. This will allow us to treat the force as constant on each subinterval and to approximate the work on each subinterval using Formula (1). By adding the approximations to the work on the subintervals, we will obtain a Riemann sum that approximates the work W over the entire interval, and by taking the limit of the Riemann sums we will obtain an integral for W.

To implement this idea, divide the interval $[a, b]$ into n subintervals by inserting numbers $x_1, x_2, \ldots, x_{n-1}$ between $a = x_0$ and $b = x_n$. We can use Formula (1) to approximate the work W_k done in the kth subinterval by choosing any number x_k^* in this interval and regarding the force to have a constant value $F(x_k^*)$ throughout the interval. Since the width of the kth subinterval is $x_k - x_{k-1} = \Delta x_k$, this yields the approximation

$$W_k \approx F(x_k^*)\Delta x_k$$

Adding these approximations yields the following Riemann sum that approximates the work

W done over the entire interval:

$$W \approx \sum_{k=1}^{n} F(x_k^*)\Delta x_k$$

Taking the limit as n increases and the widths of the subintervals approach zero yields the definite integral

$$W = \lim_{\max \Delta x_k \to 0} \sum_{k=1}^{n} F(x_k^*)\Delta x_k = \int_a^b F(x)\,dx$$

In summary, we have the following result:

7.6.3 DEFINITION. Suppose that an object moves in the positive direction along a coordinate line over the interval $[a, b]$ while subjected to a variable force $F(x)$ that is applied in the direction of motion. Then we define the **work** W performed by the force on the object to be

$$W = \int_a^b F(x)\,dx \tag{2}$$

Hooke's law [Robert Hooke (1635–1703), English physicist] states that under appropriate conditions a spring that is stretched x units beyond its natural length pulls back with a force

$$F(x) = kx$$

where k is a constant (called the **spring constant** or **spring stiffness**). The value of k depends on such factors as the thickness of the spring and the material used in its composition. Since $k = F(x)/x$, the constant k has units of force per unit length.

Example 3 A spring exerts a force of 5 N when stretched 1 m beyond its natural length.

(a) Find the spring constant k.

(b) How much work is required to stretch the spring 1.8 m beyond its natural length?

Solution (a). From Hooke's law,

$$F(x) = kx$$

From the data, $F(x) = 5$ N when $x = 1$ m, so $5 = k \cdot 1$. Thus, the spring constant is $k = 5$ newtons per meter (N/m). This means that the force $F(x)$ required to stretch the spring x meters is

$$F(x) = 5x \tag{3}$$

Natural position
of spring

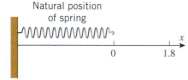

0   1.8

Figure 7.6.3

Solution (b). Place the spring along a coordinate line as shown in Figure 7.6.3. We want to find the work W required to stretch the spring over the interval from $x = 0$ to $x = 1.8$. From (2) and (3) the work W required is

$$W = \int_a^b F(x)\,dx = \int_0^{1.8} 5x\,dx = \left. \frac{5x^2}{2} \right]_0^{1.8} = 8.1 \text{ J} \quad \blacktriangleleft$$

Example 4 An astronaut's *weight* (or more precisely, *Earth weight*) is the force exerted on the astronaut by the Earth's gravity. As the astronaut moves upward into space, the gravitational pull of the Earth decreases, and hence so does his or her weight. We will show later in the text that if the Earth is assumed to be a sphere of radius 4000 mi, then an astronaut who weighs 150 lb on Earth will have a weight of

$$w(x) = \frac{2,400,000,000}{x^2} \text{ lb}, \quad x \geq 4000$$

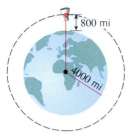

Figure 7.6.4

at a distance of x mi from the Earth's center. Use this formula to determine the work in foot-pounds required to lift the astronaut to a point that is 800 mi above the surface of the Earth (Figure 7.6.4).

Solution. Since the Earth has a radius of 4000 mi, the astronaut is lifted from a point that is 4000 mi from the Earth's center to a point that is 4800 mi from the Earth's center. Thus, from (2), the work W required to lift the astronaut is

$$
\begin{aligned}
W &= \int_{4000}^{4800} \frac{2,400,000,000}{x^2}\, dx \\
&= -\frac{2,400,000,000}{x}\Bigg]_{4000}^{4800} \\
&= -500,000 + 600,000 \\
&= 100,000 \text{ mile-pounds} \\
&= (100,000 \text{ mi·lb}) \times (5280 \text{ ft/mi}) \\
&= 5.28 \times 10^8 \text{ ft·lb} \qquad \blacktriangleleft
\end{aligned}
$$

CALCULATING WORK FROM BASIC PRINCIPLES

Some problems cannot be solved by mechanically substituting into formulas, and one must return to basic principles to obtain solutions. This is illustrated in the next example.

Example 5 A conical water tank of radius 10 ft and height 30 ft is filled with water to a depth of 15 ft (Figure 7.6.5a). How much work is required to pump all of the water out through a hole in the top of the tank?

Solution. Our strategy will be to divide the water into thin layers, approximate the work required to move each layer to the top of the tank, add the approximations for the layers to obtain a Riemann sum that approximates the total work, and then take the limit of the Riemann sums to produce an integral for the total work.

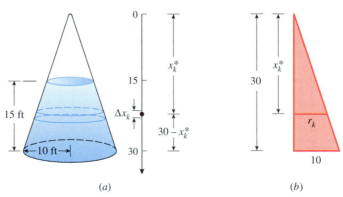

(a)   (b)

Figure 7.6.5

To implement this idea, introduce an x-axis as shown in Figure 7.6.5a, and divide the water into n layers with Δx_k denoting the thickness of the kth layer. This division induces a partition of the interval $[15, 30]$ into n subintervals. Although the upper and lower surfaces of the kth layer are at different distances from the top, the difference will be small if the layer is thin, and we can reasonably assume that the entire layer is concentrated at a single point x_k^* (Figure 7.6.5a). Thus, the work W_k required to move the kth layer to the top of the tank is approximately

$$W_k \approx F_k x_k^* \qquad (4)$$

where F_k is the force required to lift the kth layer. But the force required to lift the kth layer is the force needed to overcome gravity, and this is the same as the weight of the layer. If the layer is very thin, we can approximate the volume of the kth layer with the volume of a cylinder of height Δx_k and radius r_k, where (by similar triangles)

$$\frac{r_k}{x_k^*} = \frac{10}{30} = \frac{1}{3}$$

or equivalently, $r_k = x_k^*/3$ (Figure 7.6.5b). Therefore, the volume of the kth layer of water is approximately

$$\pi r_k^2 \Delta x_k = \pi (x_k^*/3)^2 \Delta x_k = \frac{\pi}{9}(x_k^*)^2 \Delta x_k$$

Since the weight density of water is 62.4 lb/ft³, it follows that

$$F_k \approx \frac{62.4\pi}{9}(x_k^*)^2 \Delta x_k$$

Thus, from (4)

$$W_k \approx \left(\frac{62.4\pi}{9}(x_k^*)^2 \Delta x_k \right) x_k^* = \frac{62.4\pi}{9}(x_k^*)^3 \Delta x_k$$

and hence the work W required to move all n layers has the approximation

$$W = \sum_{k=1}^{n} W_k \approx \sum_{k=1}^{n} \frac{62.4\pi}{9}(x_k^*)^3 \Delta x_k$$

To find the *exact* value of the work we take the limit as max $\Delta x_k \to 0$. This yields

$$W = \lim_{\max \Delta x_k \to 0} \sum_{k=1}^{n} \frac{62.4\pi}{9}(x_k^*)^3 \Delta x_k = \int_{15}^{30} \frac{62.4\pi}{9} x^3 \, dx$$

$$= \frac{62.4\pi}{9} \left(\frac{x^4}{4} \right) \Bigg]_{15}^{30} = 1{,}316{,}250\pi \approx 4{,}135{,}000 \text{ ft·lb} \quad \blacktriangleleft$$

THE WORK–ENERGY RELATIONSHIP

When you see an object in motion, you can be certain that somehow work has been expended to create that motion. For example, when you drop a stone from a building, the stone gathers speed because the force of the Earth's gravity is performing work on it, and when a hockey player strikes a puck with a hockey stick, the work performed on the puck during the brief period of contact with the stick creates the enormous speed of the puck across the ice. However, experience shows that the speed obtained by an object depends not only on the amount of work done, but also on the mass of the object. For example, the work required to throw a 5-oz baseball 50 mi/h would accelerate a 10-lb bowling ball to less than 9 mi/h.

Using the method of substitution for definite integrals, we will derive a simple equation that relates the work done on an object to the object's mass and velocity. Furthermore, this equation will allow us to motivate an appropriate definition for the "energy of motion" of an object. As in Definition 7.6.3, we will assume that an object moves in the positive direction along a coordinate line over the interval $[a, b]$ while subjected to a force $F(x)$ that is applied in the direction of motion. We let $x = x(t)$, $v = v(t) = x'(t)$, and $v'(t)$ denote the respective position, velocity, and acceleration of the object at time t. It follows from Newton's Second Law of Motion that

$$F(x(t)) = mv'(t)$$

where m is the mass of the object. Assume that

$$x(t_0) = a \quad \text{and} \quad x(t_1) = b$$

with

$$v(t_0) = v_i \quad \text{and} \quad v(t_1) = v_f$$

the initial and final velocities of the object, respectively. Then

$$W = \int_a^b F(x)\,dx = \int_{x(t_0)}^{x(t_1)} F(x)\,dx$$

$$= \int_{t_0}^{t_1} F(x(t))x'(t)\,dt \qquad \boxed{\text{By Theorem 6.8.1 with } x = x(t), dx = x'(t)\,dt}$$

$$= \int_{t_0}^{t_1} mv'(t)v(t)\,dt = \int_{t_0}^{t_1} mv(t)v'(t)\,dt$$

$$= \int_{v(t_0)}^{v(t_1)} mv\,dv \qquad \boxed{\text{By Theorem 6.8.1 with } v = v(t), dv = v'(t)\,dt}$$

$$= \int_{v_i}^{v_f} mv\,dv = \tfrac{1}{2}mv^2\Big|_{v_i}^{v_f} = \tfrac{1}{2}mv_f^2 - \tfrac{1}{2}mv_i^2$$

We see from the equation

$$W = \tfrac{1}{2}mv_f^2 - \tfrac{1}{2}mv_i^2 \tag{5}$$

that the work done on the object is equal to the change in the quantity $\tfrac{1}{2}mv^2$ from its initial value to its final value. We will refer to Equation (5) as the **work–energy relationship**. If we define the "energy of motion" or **kinetic energy** of our object to be given by

$$K = \tfrac{1}{2}mv^2 \tag{6}$$

then Equation (5) tells us that the work done on an object is equal to the *change* in the object's kinetic energy. Loosely speaking, we may think of work done on an object as being "transformed" into kinetic energy of the object. The units of kinetic energy are the same as the units of work. For example, in SI kinetic energy is measured in joules (J).

Example 6 A space probe of mass $m = 5.00 \times 10^4$ kg travels in deep space subjected only to the force of its own engine. Starting at a time when the speed of the probe is $v = 1.10 \times 10^4$ m/s, the engine is fired continuously over a distance of 2.50×10^6 m with a constant force of 4.00×10^5 N in the direction of motion. What is the final speed of the probe?

Solution. Since the force applied by the engine is constant and in the direction of motion, the work W expended by the engine on the probe is

$$W = \text{force} \times \text{distance} = (4.00 \times 10^5 \text{ N}) \times (2.50 \times 10^6 \text{ m}) = 1.00 \times 10^{12} \text{ J}$$

From (5), the final kinetic energy $K_f = \tfrac{1}{2}mv_f^2$ of the probe can be expressed in terms of the work W and the initial kinetic energy $K_i = \tfrac{1}{2}mv_i^2$ as

$$K_f = W + K_i$$

Thus, from the known mass and initial speed we have

$$K_f = (1.00 \times 10^{12} \text{ J}) + \tfrac{1}{2}(5.00 \times 10^4 \text{ kg})(1.10 \times 10^4 \text{ m/s})^2 = 4.025 \times 10^{12} \text{ J}$$

The final kinetic energy is $K_f = \tfrac{1}{2}mv_f^2$, so the final speed of the probe is

$$v_f = \sqrt{\frac{2K_f}{m}} = \sqrt{\frac{2(4.025 \times 10^{12})}{5.00 \times 10^4}} \approx 1.27 \times 10^4 \text{ m/s} \qquad \blacktriangleleft$$

EXERCISE SET 7.6

1. Find the work done when
   (a) a constant force of 30 lb in the positive x-direction moves an object from $x = -2$ to $x = 5$ ft
   (b) a variable force of $F(x) = 1/x^2$ lb in the positive x-direction moves an object from $x = 1$ to $x = 6$ ft.

2. A variable force $F(x)$ in the positive x-direction is graphed in the accompanying figure. Find the work done by the force on a particle that moves from $x = 0$ to $x = 5$.

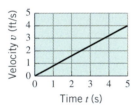

Figure Ex-2

3. A constant force of 10 lb in the positive x-direction is applied to a particle whose velocity versus time curve is shown in the accompanying figure. Find the work done by the force on the particle from time $t = 0$ to $t = 5$.

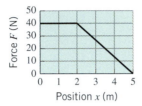

Figure Ex-3

4. A spring whose natural length is 15 cm exerts a force of 45 N when stretched to a length of 20 cm.
   (a) Find the spring constant (in newtons/meter).
   (b) Find the work that is done in stretching the spring 3 cm beyond its natural length.
   (c) Find the work done in stretching the spring from a length of 20 cm to a length of 25 cm.

5. A spring exerts a force of 100 N when it is stretched 0.2 m beyond its natural length. How much work is required to stretch the spring 0.8 m beyond its natural length?

6. Assume that a force of 6 N is required to compress a spring from a natural length of 4 m to a length of $3\frac{1}{2}$ m. Find the work required to compress the spring from its natural length to a length of 2 m. (Hooke's law applies to compression as well as extension.)

7. Assume that 10 ft·lb of work is required to stretch a spring 1 ft beyond its natural length. What is the spring constant?

8. A cylindrical tank of radius 5 ft and height 9 ft is two-thirds filled with water. Find the work required to pump all the water over the upper rim.

9. Solve Exercise 8 assuming that the tank is two-thirds filled with a liquid that weighs ρ lb/ft³.

10. A cone-shaped water reservoir is 20 ft in diameter across the top and 15 ft deep. If the reservoir is filled to a depth of 10 ft, how much work is required to pump all the water to the top of the reservoir?

11. The vat shown in the accompanying figure contains water to a depth of 2 m. Find the work required to pump all the water to the top of the vat. [Use 9810 N/m³ as the weight density of water.]

12. The cylindrical tank shown in the accompanying figure is filled with a liquid weighing 50 lb/ft³. Find the work required to pump all the liquid to a level 1 ft above the top of the tank.

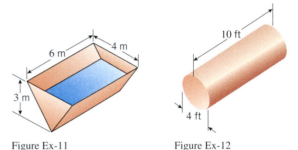

Figure Ex-11   Figure Ex-12

13. A swimming pool is built in the shape of a rectangular parallelepiped 10 ft deep, 15 ft wide, and 20 ft long.
    (a) If the pool is filled to 1 ft below the top, how much work is required to pump all the water into a drain at the top edge of the pool?
    (b) A one-horsepower motor can do 550 ft·lb of work per second. What size motor is required to empty the pool in 1 hour?

14. How much work is required to fill the swimming pool in Exercise 13 to 1 ft below the top if the water is pumped in through an opening located at the bottom of the pool?

15. A 100-ft length of steel chain weighing 15 lb/ft is dangling from a pulley. How much work is required to wind the chain onto the pulley?

16. A 3-lb bucket containing 20 lb of water is hanging at the end of a 20-ft rope that weighs 4 oz/ft. The other end of the rope is attached to a pulley. How much work is required to wind the length of rope onto the pulley, assuming that the rope is wound onto the pulley at a rate of 2 ft/s and that as the bucket is being lifted, water leaks from the bucket at a rate of 0.5 lb/s?

17. A rocket weighing 3 tons is filled with 40 tons of liquid fuel. In the initial part of the flight, fuel is burned off at a constant rate of 2 tons per 1000 ft of vertical height. How much work is done in lifting the rocket to 3000 ft?

18. It follows from Coulomb's law in physics that two like electrostatic charges repel each other with a force inversely proportional to the square of the distance between them. Suppose that two charges A and B repel with a force of k newtons when they are positioned at points $A(-a, 0)$ and $B(a, 0)$, where a is measured in meters. Find the work W required to move charge A along the x-axis to the origin if charge B remains stationary.

19. It is a law of physics that the gravitational force exerted by the Earth on an object above the Earth's surface varies inversely as the square of its distance from the Earth's center. Thus, an object's weight $w(x)$ is related to its distance x from the Earth's center by a formula of the form

$$w(x) = \frac{k}{x^2}$$

where k is a constant of proportionality that depends on the mass of the object.

(a) Use this fact and the assumption that the Earth is a sphere of radius 4000 mi to obtain the formula for $w(x)$ in Example 4.

(b) Find a formula for the weight $w(x)$ of a satellite that is x mi from the Earth's surface if its weight on Earth is 6000 lb.

(c) How much work is required to lift the satellite from the surface of the Earth to an orbital position that is 1000 mi high?

20. (a) The formula $w(x) = k/x^2$ in Exercise 19 is applicable to all celestial bodies. Assuming that the Moon is a sphere of radius 1080 mi, find the force that the Moon exerts on an astronaut who is x mi from the surface of the Moon if her weight on the Moon's surface is 20 lb.

(b) How much work is required to lift the astronaut to a point that is 10.8 mi above the Moon's surface?

21. The Yamanashi Maglev Test Line in Japan that runs between Sakaigawa and Akiyama is currently testing magnetic levitation (MAGLEV) trains that are designed to levitate inches above powerful magnetic fields. Suppose that a MAGLEV train has a mass of $m = 4.00 \times 10^5$ kg and that starting at a time when the train has a speed of 20 m/s the engine applies a force of 6.40×10^5 N in the direction of motion over a distance of 3.00×10^3 m. Use the work–energy relationship (5) to find the final speed of the train.

22. Assume that a Mars probe of mass $m = 2.00 \times 10^3$ kg is subjected only to the force of its own engine. Starting at a time when the speed of the probe is $v = 1.00 \times 10^4$ m/s, the engine is fired continuously over a distance of 2.00×10^5 m with a constant force of 2.00×10^5 N in the direction of motion. Use the work–energy relationship (5) to find the final speed of the probe.

23. On August 10, 1972 a meteorite with an estimated mass of 4×10^6 kg and an estimated speed of 15 km/s skipped across the atmosphere above the western United States and Canada but fortunately did not hit the Earth.

(a) Assuming that the meteorite had hit the Earth with a speed of 15 km/s, what would have been its change in kinetic energy in joules (J)?

(b) Express the energy as a multiple of the explosive energy of 1 megaton of TNT, which is 4.2×10^{15} J.

(c) The energy associated with the Hiroshima atomic bomb was 13 kilotons of TNT. To how many such bombs would the meteorite impact have been equivalent?

7.7 FLUID PRESSURE AND FORCE

In this section we will use the integration tools developed in the preceding chapter to study the pressures and forces exerted by fluids on submerged objects.

WHAT IS A FLUID?

A *fluid* is a substance that flows to conform to the boundaries of any container in which it is placed. Fluids include *liquids*, such as water, oil, and mercury, as well as *gases*, such as helium, oxygen, and air. The study of fluids falls into two categories: *fluid statics* (the study of fluids at rest) and *fluid dynamics* (the study of fluids in motion). In this section we will be concerned only with fluid statics; toward the end of this text we will investigate problems in fluid dynamics.

THE CONCEPT OF PRESSURE

The effect that a force has on an object depends on how that force is spread over the surface of the object. For example, when you walk on soft snow with boots, the weight of your body crushes the snow and you sink into it. However, if you put on a pair of skis to spread the weight of your body over a greater surface area, then the weight of your body has less of a crushing effect on the snow, and you are able to glide across the surface. The concept that accounts for both the magnitude of a force and the area over which it is applied is called *pressure*.

> **7.7.1 DEFINITION.** If a force of magnitude F is applied to a surface of area A, then we define the ***pressure*** P exerted by the force on the surface to be
>
> $$P = \frac{F}{A} \tag{1}$$

It follows from this definition that pressure has units of force per unit area. The most common units of pressure are newtons per square meter (N/m^2) in SI and pounds per square inch (lb/in^2) or pounds per square foot (lb/ft^2) in the BE system. As indicated in Table 7.7.1, one newton per square meter is called a *pascal*[*] (Pa). A pressure of 1 Pa is quite small ($1\ Pa = 1.45 \times 10^{-4}\ lb/in^2$), so in countries using SI, tire pressure gauges are usually calibrated in kilopascals (kPa), which is 1000 pascals.

Table 7.7.1

SYSTEM	FORCE	÷	AREA	=	PRESSURE
SI	newton (N)		square meter (m^2)		pascal (Pa)
BE	pound (lb)		square foot (ft^2)		lb/ft^2
BE	pound (lb)		square inch (in^2)		lb/in^2 (psi)

CONVERSION FACTORS:
$1\ Pa \approx 1.45 \times 10^{-4}\ lb/in^2 \approx 2.09 \times 10^{-2}\ lb/ft^2$
$1\ lb/in^2 \approx 6.89 \times 10^3\ Pa$ $\qquad$ $1\ lb/ft^2 \approx 47.9\ Pa$

In this section we will be interested in pressures and forces on objects submerged in fluids. Pressures themselves have no directional characteristics, but the forces that they create always act perpendicular to the face of the submerged object. Thus, in Figure 7.7.1 the water pressure creates horizontal forces on the sides of the tank, vertical forces on the bottom of the tank, and forces that vary in direction, so as to be perpendicular to the different parts of the swimmer's body.

Example 1 Referring to Figure 7.7.1, suppose that the back of the swimmer's hand has a surface area of $8.4 \times 10^{-3}\ m^2$ and that the pressure acting on it is $5.1 \times 10^4\ Pa$ (a realistic value near the bottom of a deep diving pool). Find the force that acts on the swimmer's hand.

Fluid forces always act perpendicular to the surface of a submerged object.

Figure 7.7.1

[*] BLAISE PASCAL (1623–1662). French mathematician and scientist. Pascal's mother died when he was three years old and his father, a highly educated magistrate, personally provided the boy's early education. Although Pascal showed an inclination for science and mathematics, his father refused to tutor him in those subjects until he mastered Latin and Greek. Pascal's sister and primary biographer claimed that he independently discovered the first thirty-two propositions of Euclid without ever reading a book on geometry. (However, it is generally agreed that the story is apocryphal.) Nevertheless, the precocious Pascal published a highly respected essay on conic sections by the time he was sixteen years old. Descartes, who read the essay, thought it so brilliant that he could not believe that it was written by such a young man. By age 18 his health began to fail and until his death he was in frequent pain. However, his creativity was unimpaired.

Pascal's contributions to physics include the discovery that air pressure decreases with altitude and the principle of fluid pressure that bears his name. However, the originality of his work is questioned by some historians. Pascal made major contributions to a branch of mathematics called "projective geometry," and he helped to develop probability theory through a series of letters with Fermat.

In 1646, Pascal's health problems resulted in a deep emotional crisis that led him to become increasingly concerned with religious matters. Although born a Catholic, he converted to a religious doctrine called Jansenism and spent most of his final years writing on religion and philosophy.

Solution. From (1), the force F is

$$F = PA = (5.1 \times 10^4 \text{ N/m}^2)(8.4 \times 10^{-3} \text{ m}^2) \approx 4.3 \times 10^2 \text{ N}$$

This is quite a large force (nearly 100 lb in the BE system). ◄

Scuba divers know that the deeper they dive, the greater the pressure and the forces that they feel on their bodies. This sense of pressure and force is caused by the weight of the water and air above—the deeper the diver goes, the greater the weight above and hence the greater the pressure and force that he or she feels.

To calculate pressures and forces on submerged objects, we need to know something about the characteristics of the fluids in which they are submerged. For simplicity, we will assume that the fluids under consideration are *homogeneous*, by which we mean that any two samples of the fluid with the same volume have the same mass. It follows from this assumption that the mass per unit volume is a constant δ that depends on the physical characteristics of the fluid but not on the size or location of the sample; we call

$$\delta = \frac{m}{V} \tag{2}$$

the ***mass density*** of the fluid. Sometimes it is more convenient to work with weight per unit volume than with mass per unit volume. Thus, we define the ***weight density*** ρ of a fluid to be

$$\rho = \frac{w}{V} \tag{3}$$

where w is the weight of a fluid sample of volume V. Thus, if the weight density of a fluid is known, then the weight w of a fluid sample of volume V can be computed from the formula $w = \rho V$. Table 7.7.2 shows some typical weight densities.

To calculate fluid pressures and forces we will need to make use of an experimental observation. Suppose that a flat surface of area A is submerged in a homogeneous fluid of weight density ρ such that the entire surface lies between depths h_1 and h_2, where $h_1 \leq h_2$ (Figure 7.7.2). Experiments show that on both sides of the surface, the fluid exerts a force that is perpendicular to the surface and whose magnitude F satisfies the inequalities

$$\rho h_1 A \leq F \leq \rho h_2 A \tag{4}$$

Thus, it follows from (1) that the pressure $P = F/A$ on a given side of the surface satisfies the inequalities

$$\rho h_1 \leq P \leq \rho h_2 \tag{5}$$

Note that it is now a straightforward matter to calculate fluid force and pressure on a flat surface that is submerged *horizontally* at depth h, for then $h = h_1 = h_2$ and inequalities (4) and (5) become the *equalities*

$$F = \rho h A \tag{6}$$

and

$$P = \rho h \tag{7}$$

Example 2 Find the fluid pressure and force on the top of a flat circular plate of radius 2 m that is submerged horizontally in water at a depth of 6 m (Figure 7.7.3).

FLUID DENSITY

Table 7.7.2

WEIGHT DENSITIES

SI	N/m^3
Machine oil	4,708
Gasoline	6,602
Fresh water	9,810
Seawater	10,045
Mercury	133,416

BE SYSTEM	lb/ft^3
Machine oil	30.0
Gasoline	42.0
Fresh water	62.4
Seawater	64.0
Mercury	849.0

All densities are affected by variations in temperature and pressure. Weight densities are also affected by variations in g.

FLUID PRESSURE

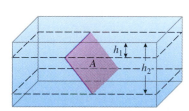

Figure 7.7.2

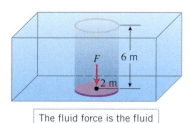

The fluid force is the fluid pressure times the area.

Figure 7.7.3

Solution. Since the weight density of water is $\rho = 9810 \text{ N/m}^3$, it follows from (7) that the fluid pressure is

$$P = \rho h = (9810)(6) = 58{,}860 \text{ Pa}$$

and it follows from (6) that the fluid force is

$$F = \rho h A = \rho h (\pi r^2) = (9810)(6)(4\pi) = 235{,}440\pi \approx 739{,}700 \text{ N} \quad \blacktriangleleft$$

FLUID FORCE ON A VERTICAL SURFACE

It was easy to calculate the fluid force on the horizontal plate in Example 2 because each point on the plate was at the same depth. The problem of finding the fluid force on a vertical surface is more complicated because the depth, and hence the pressure, is not constant over the surface. To find the fluid force on a vertical surface we will need calculus.

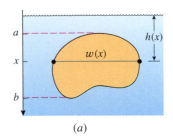

(a)

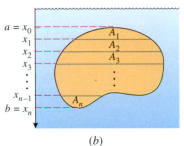

(b)

> **7.7.2 PROBLEM.** Suppose that a flat surface is immersed vertically in a fluid of weight density ρ and that the submerged portion of the surface extends from $x = a$ to $x = b$ along an x-axis whose positive direction is down (Figure 7.7.4a). For $a \le x \le b$, suppose that $w(x)$ is the width of the surface and that $h(x)$ is the depth of the point x. Define what is meant by the *fluid force F* on the surface, and find a formula for computing it.

The basic idea for solving this problem is to divide the surface into horizontal strips whose areas may be approximated by areas of rectangles. These area approximations, along with inequalities (4), will allow us to create a Riemann sum that approximates the total force on the surface. By taking a limit of Riemann sums we will then obtain an integral for F.

To implement this idea, we divide the interval $[a, b]$ into n subintervals by inserting the numbers $x_1, x_2, \ldots, x_{n-1}$ between $a = x_0$ and $b = x_n$. This has the effect of dividing the surface into n strips of area $A_k, k = 1, 2, \ldots, n$ (Figure 7.7.4b). It follows from (4) that the force F_k on the kth strip satisfies the inequalities

$$\rho h(x_{k-1}) A_k \le F_k \le \rho h(x_k) A_k$$

or equivalently,

$$h(x_{k-1}) \le \frac{F_k}{\rho A_k} \le h(x_k)$$

Since the depth function $h(x)$ increases linearly, there must exist a number x_k^* between x_{k-1} and x_k such that

$$h(x_k^*) = \frac{F_k}{\rho A_k}$$

or equivalently,

$$F_k = \rho h(x_k^*) A_k$$

Figure 7.7.4

We now approximate the area A_k of the kth strip of the surface by the area of a rectangle of width $w(x_k^*)$ and height $\Delta x_k = x_k - x_{k-1}$ (Figure 7.7.4c). It follows that F_k may be approximated as

$$F_k = \rho h(x_k^*) A_k \approx \rho h(x_k^*) \cdot \underbrace{w(x_k^*) \Delta x_k}_{\text{Area of rectangle}}$$

Adding these approximations yields the following Riemann sum that approximates the total force F on the surface:

$$F = \sum_{k=1}^{n} F_k \approx \sum_{k=1}^{n} \rho h(x_k^*) w(x_k^*) \Delta x_k$$

Taking the limit as n increases and the widths of the subintervals approach zero yields the definite integral

$$F = \lim_{\max \Delta x_k \to 0} \sum_{k=1}^{n} \rho h(x_k^*) w(x_k^*) \Delta x_k = \int_a^b \rho h(x) w(x) \, dx$$

In summary, we have the following result:

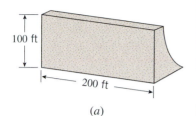

(a)

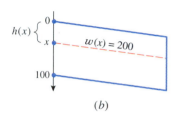

(b)

Figure 7.7.5

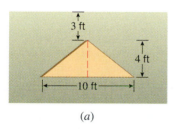

(a)

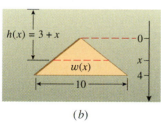

(b)

Figure 7.7.6

7.7.3 DEFINITION. Suppose that a flat surface is immersed vertically in a fluid of weight density ρ and that the submerged portion of the surface extends from $x = a$ to $x = b$ along an x-axis whose positive direction is down (Figure 7.7.4a). For $a \leq x \leq b$, suppose that $w(x)$ is the width of the surface and that $h(x)$ is the depth of the point x. Then we define the **fluid force** F on the surface to be

$$F = \int_a^b \rho h(x) w(x)\, dx \qquad (8)$$

Example 3 The face of a dam is a vertical rectangle of height 100 ft and width 200 ft (Figure 7.7.5a). Find the total fluid force exerted on the face when the water surface is level with the top of the dam.

Solution. Introduce an x-axis with its origin at the water surface as shown in Figure 7.7.5b. At a point x on this axis, the width of the dam in feet is $w(x) = 200$ and the depth in feet is $h(x) = x$. Thus, from (8) with $\rho = 62.4$ lb/ft^3 (the weight density of water) we obtain as the total force on the face

$$F = \int_0^{100} (62.4)(x)(200)\, dx = 12{,}480 \int_0^{100} x\, dx = 12{,}480 \left.\frac{x^2}{2}\right]_0^{100} = 62{,}400{,}000 \text{ lb}$$

◄

Example 4 A plate in the form of an isosceles triangle with base 10 ft and altitude 4 ft is submerged vertically in machine oil as shown in Figure 7.7.6a. Find the fluid force F against the plate surface if the oil has weight density $\rho = 30$ lb/ft^3.

Solution. Introduce an x-axis as shown in Figure 7.7.6b. By similar triangles, the width of the plate, in feet, at a depth of $h(x) = (3 + x)$ ft satisfies

$$\frac{w(x)}{10} = \frac{x}{4}, \quad \text{so} \quad w(x) = \frac{5}{2}x$$

Thus, it follows from (8) that the force on the plate is

$$F = \int_a^b \rho h(x) w(x)\, dx = \int_0^4 (30)(3 + x)\left(\frac{5}{2}x\right) dx$$

$$= 75 \int_0^4 (3x + x^2)\, dx = 75 \left[\frac{3x^2}{2} + \frac{x^3}{3}\right]_0^4 = 3400 \text{ lb}$$

◄

EXERCISE SET 7.7

In this exercise set, refer to Table 7.7.2 for weight densities of fluids, when needed.

1. A flat rectangular plate is submerged horizontally in water.

 (a) Find the force (in lb) and the pressure (in lb/ft^2) on the top surface of the plate if its area is 100 ft^2 and the surface is at a depth of 5 ft.

 (b) Find the force (in N) and the pressure (in Pa) on the top surface of the plate if its area is 25 m^2 and the surface is at a depth of 10 m.

2. (a) Find the force (in N) on the deck of a sunken ship if its area is 160 m^2 and the pressure acting on it is 6.0×10^5 Pa.

 (b) Find the force (in lb) on a diver's face mask if its area is 60 in^2 and the pressure acting on it is 100 lb/in^2.

In Exercises 3–8, the flat surfaces shown are submerged vertically in water. Find the fluid force against the surface.

3.
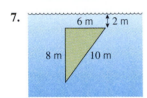
2 ft

4 ft

4.
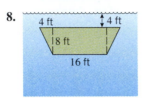
1 m

2 m

4 m

5.
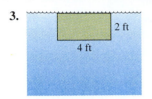
←10 m→

6.
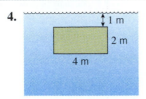
←4 ft→

4 ft   4 ft

7.
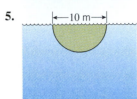
6 m   2 m

8 m   10 m

8.
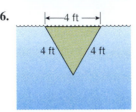
4 ft   4 ft

8 ft

16 ft

9. Suppose that a flat surface is immersed vertically in a fluid of weight density ρ. If ρ is doubled, is the force on the plate also doubled? Explain your reasoning.

10. An oil tank is shaped like a right circular cylinder of diameter 4 ft. Find the total fluid force against one end when the axis is horizontal and the tank is half filled with oil of weight density 50 lb/ft^3.

11. A square plate of side a feet is dipped in a liquid of weight density ρ lb/ft^3. Find the fluid force on the plate if a vertex is at the surface and a diagonal is perpendicular to the surface.

Formula (8) gives the fluid force on a flat surface immersed vertically in a fluid. More generally, if a flat surface is immersed so that it makes an angle of $0 \leq \theta < \pi/2$ with the vertical, then the fluid force on the surface is given by

$$F = \int_a^b \rho h(x) w(x) \sec \theta \, dx$$

Use this formula in Exercises 12–15.

12. Derive the formula given above for the fluid force on a flat surface immersed at an angle in a fluid.

13. The accompanying figure shows a rectangular swimming pool whose bottom is an inclined plane. Find the fluid force on the bottom when the pool is filled to the top.

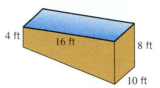
4 ft   16 ft   8 ft

10 ft

Figure Ex-13

14. By how many feet should the water in the pool of Exercise 13 be lowered in order for the force on the bottom to be reduced by a factor of $\frac{1}{2}$?

15. The accompanying figure shows a dam whose face is an inclined rectangle. Find the fluid force on the face when the water is level with the top of this dam.

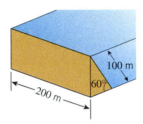
100 m

200 m   60°

Figure Ex-15

16. An observation window on a submarine is a square with 2-ft sides. Using ρ_0 for the weight density of seawater, find the fluid force on the window when the submarine has descended so that the window is vertical and its top is at a depth of h feet.

17. (a) Show: If the submarine in Exercise 16 descends vertically at a constant rate, then the fluid force on the window increases at a constant rate.

(b) At what rate is the force on the window increasing if the submarine is descending vertically at 20 ft/min?

18. (a) Let $D = D_a$ denote a disk of radius a submerged in a fluid of weight density ρ such that the center of D is h units below the surface of the fluid. For each value of r in the interval $(0, a]$, let D_r denote the disk of radius r that is concentric with D. Select a side of the disk D and define $P(r)$ to be the fluid pressure on the chosen side of D_r. Use (5) to prove that

$$\lim_{r \to 0^+} P(r) = \rho h$$

(b) Explain why the result in part (a) may be interpreted to mean that *fluid pressure at a given depth is the same in all directions*. (This statement is one version of a result known as *Pascal's Principle*.)

7.8 HYPERBOLIC FUNCTIONS AND HANGING CABLES

In this section we will study certain combinations of e^x and e^{-x}, called "hyperbolic functions." These functions, which arise in various engineering applications, have many properties in common with the trigonometric functions. This similarity is somewhat surprising, since there is little on the surface to suggest that there should be any relationship between exponential and trigonometric functions. This is because the relationship occurs within the context of complex numbers, a topic which we will leave for more advanced courses.

DEFINITIONS OF HYPERBOLIC FUNCTIONS

To introduce the hyperbolic functions, observe that the function e^x can be expressed in the following way as the sum of an even function and an odd function:

$$e^x = \underbrace{\frac{e^x + e^{-x}}{2}}_{\text{Even}} + \underbrace{\frac{e^x - e^{-x}}{2}}_{\text{Odd}}$$

These functions are sufficiently important that there are names and notation associated with them: the odd function is called the *hyperbolic sine* of x and the even function is called the *hyperbolic cosine* of x. They are denoted by

$$\sinh x = \frac{e^x - e^{-x}}{2} \quad \text{and} \quad \cosh x = \frac{e^x + e^{-x}}{2}$$

where sinh is pronounced "cinch" and cosh rhymes with "gosh." From these two building blocks we can create four more functions to produce the following set of six *hyperbolic functions*.

7.8.1 DEFINITION.

Hyperbolic sine	$\sinh x = \dfrac{e^x - e^{-x}}{2}$	
Hyperbolic cosine	$\cosh x = \dfrac{e^x + e^{-x}}{2}$	
Hyperbolic tangent	$\tanh x = \dfrac{\sinh x}{\cosh x}$	$= \dfrac{e^x - e^{-x}}{e^x + e^{-x}}$
Hyperbolic cotangent	$\coth x = \dfrac{\cosh x}{\sinh x}$	$= \dfrac{e^x + e^{-x}}{e^x - e^{-x}}$
Hyperbolic secant	$\text{sech } x = \dfrac{1}{\cosh x}$	$= \dfrac{2}{e^x + e^{-x}}$
Hyperbolic cosecant	$\text{csch } x = \dfrac{1}{\sinh x}$	$= \dfrac{2}{e^x - e^{-x}}$

REMARK. The terms "tanh," "sech," and "csch" are pronounced "tanch," "seech," and "coseech," respectively.

Example 1

$$\sinh 0 = \frac{e^0 - e^{-0}}{2} = \frac{1 - 1}{2} = 0$$

$$\cosh 0 = \frac{e^0 + e^{-0}}{2} = \frac{1 + 1}{2} = 1$$

$$\sinh 2 = \frac{e^2 - e^{-2}}{2} \approx 3.6269$$

◄

GRAPHS OF THE HYPERBOLIC FUNCTIONS

The graphs of the hyperbolic functions, which are shown in Figure 7.8.1, can be generated with a graphing utility, but it is worthwhile to observe that the general shape of the graph of $y = \cosh x$ can be obtained by sketching the graphs of $y = \frac{1}{2}e^x$ and $y = \frac{1}{2}e^{-x}$ separately and adding the corresponding y-coordinates [see part (a) of the figure]. Similarly, the general shape of the graph of $y = \sinh x$ can be obtained by sketching the graphs of $y = \frac{1}{2}e^x$ and $y = -\frac{1}{2}e^{-x}$ separately and adding corresponding y-coordinates [see part (b) of the figure].

The design of the Gateway Arch near St. Louis is based on an inverted hyperbolic cosine curve.

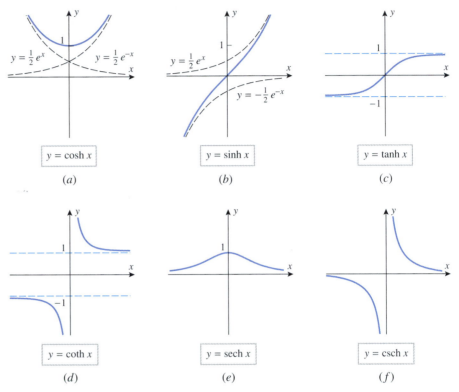

Figure 7.8.1

Observe that $\sinh x$ has a domain of $(-\infty, +\infty)$ and a range of $(-\infty, +\infty)$, whereas $\cosh x$ has a domain of $(-\infty, +\infty)$ and a range of $[1, +\infty)$. Observe also that $y = \frac{1}{2}e^x$ and $y = \frac{1}{2}e^{-x}$ are *curvilinear asymptotes* for $y = \cosh x$ in the sense that the graph of $y = \cosh x$ gets closer and closer to the graph of $y = \frac{1}{2}e^x$ as $x \to +\infty$ and gets closer and closer to the graph of $y = \frac{1}{2}e^{-x}$ as $x \to -\infty$. (See Exercise Set 2.3.) Similarly, $y = \frac{1}{2}e^x$ is a curvilinear asymptote for $y = \sinh x$ as $x \to +\infty$ and $y = -\frac{1}{2}e^{-x}$ is a curvilinear asymptote as $x \to -\infty$. Other properties of the hyperbolic functions are explored in the exercises.

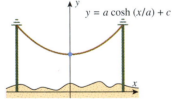

Figure 7.8.2

HANGING CABLES AND OTHER APPLICATIONS

Hyperbolic functions arise in vibratory motions inside elastic solids and more generally in many problems where mechanical energy is gradually absorbed by a surrounding medium. They also occur when a homogeneous, flexible cable is suspended between two points, as with a telephone line hanging between two poles. Such a cable forms a curve, called a *catenary* (from the Latin *catena*, meaning "chain"). If, as in Figure 7.8.2, a coordinate system is introduced so that the low point of the cable lies on the y-axis, then it can be shown using principles of physics that the cable has an equation of the form

$$y = a \cosh\left(\frac{x}{a}\right) + c$$

HYPERBOLIC IDENTITIES

The hyperbolic functions satisfy various identities that are similar to identities for trigonometric functions. The most fundamental of these is

$$\cosh^2 x - \sinh^2 x = 1 \tag{1}$$

which can be proved by writing

$$\cosh^2 x - \sinh^2 x = \left(\frac{e^x + e^{-x}}{2}\right)^2 - \left(\frac{e^x - e^{-x}}{2}\right)^2$$

$$= \tfrac{1}{4}(e^{2x} + 2e^0 + e^{-2x}) - \tfrac{1}{4}(e^{2x} - 2e^0 + e^{-2x})$$

$$= 1$$

Other hyperbolic identities can be derived in a similar manner or, alternatively, by performing algebraic operations on known identities. For example, if we divide (1) by $\cosh^2 x$, we obtain

$$1 - \tanh^2 x = \operatorname{sech}^2 x$$

and if we divide (1) by $\sinh^2 x$, we obtain

$$\coth^2 x - 1 = \operatorname{csch}^2 x$$

The following theorem summarizes some of the more useful hyperbolic identities. The proofs of those not already obtained are left as exercises.

7.8.2 THEOREM.

$\cosh x + \sinh x = e^x$	$\sinh(x + y) = \sinh x \cosh y + \cosh x \sinh y$
$\cosh x - \sinh x = e^{-x}$	$\cosh(x + y) = \cosh x \cosh y + \sinh x \sinh y$
$\cosh^2 x - \sinh^2 x = 1$	$\sinh(x - y) = \sinh x \cosh y - \cosh x \sinh y$
$1 - \tanh^2 x = \operatorname{sech}^2 x$	$\cosh(x - y) = \cosh x \cosh y - \sinh x \sinh y$
$\coth^2 x - 1 = \operatorname{csch}^2 x$	$\sinh 2x = 2 \sinh x \cosh x$
$\cosh(-x) = \cosh x$	$\cosh 2x = \cosh^2 x + \sinh^2 x$
$\sinh(-x) = -\sinh x$	$\cosh 2x = 2 \sinh^2 x + 1$
	$\cosh 2x = 2 \cosh^2 x - 1$

WHY THEY ARE CALLED HYPERBOLIC FUNCTIONS

Recall that the parametric equations

$$x = \cos t, \quad y = \sin t \qquad (0 \le t \le 2\pi)$$

represent the unit circle $x^2 + y^2 = 1$ (Figure 7.8.3a), as may be seen by writing

$$x^2 + y^2 = \cos^2 t + \sin^2 t = 1$$

If $0 \le t \le 2\pi$, then the parameter t can be interpreted as the angle in radians from the positive x-axis to the point $(\cos t, \sin t)$ or, alternatively, as twice the shaded area of the sector in Figure 7.8.3a (verify). Analogously, the parametric equations

$$x = \cosh t, \quad y = \sinh t \qquad (-\infty < t < +\infty)$$

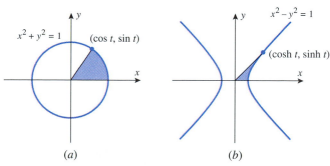

(a) $\qquad\qquad$ (b)

Figure 7.8.3

represent a portion of the curve $x^2 - y^2 = 1$, as may be seen by writing

$$x^2 - y^2 = \cosh^2 t - \sinh^2 t = 1$$

and observing that $x = \cosh t > 0$. This curve, which is shown in Figure 7.8.3b, is the right half of a larger curve called the **unit hyperbola**; this is the reason why the functions in this section are called *hyperbolic* functions. It can be shown that if $t \geq 0$, then the parameter t can be interpreted as twice the shaded area in Figure 7.8.3b. (We omit the details.)

DERIVATIVE AND INTEGRAL FORMULAS

Derivative formulas for $\sinh x$ and $\cosh x$ can be obtained by expressing these functions in terms of e^x and e^{-x}:

$$\frac{d}{dx}[\sinh x] = \frac{d}{dx}\left[\frac{e^x - e^{-x}}{2}\right] = \frac{e^x + e^{-x}}{2} = \cosh x$$

$$\frac{d}{dx}[\cosh x] = \frac{d}{dx}\left[\frac{e^x + e^{-x}}{2}\right] = \frac{e^x - e^{-x}}{2} = \sinh x$$

Derivatives of the remaining hyperbolic functions can be obtained by expressing them in terms of sinh and cosh and applying appropriate identities. For example,

$$\frac{d}{dx}[\tanh x] = \frac{d}{dx}\left[\frac{\sinh x}{\cosh x}\right] = \frac{\cosh x \dfrac{d}{dx}[\sinh x] - \sinh x \dfrac{d}{dx}[\cosh x]}{\cosh^2 x}$$

$$= \frac{\cosh^2 x - \sinh^2 x}{\cosh^2 x} = \frac{1}{\cosh^2 x} = \operatorname{sech}^2 x$$

The following theorem provides a complete list of the generalized derivative formulas and corresponding integration formulas for the hyperbolic functions.

7.8.3 THEOREM.

$$\frac{d}{dx}[\sinh u] = \cosh u \frac{du}{dx} \qquad \int \cosh u \, du = \sinh u + C$$

$$\frac{d}{dx}[\cosh u] = \sinh u \frac{du}{dx} \qquad \int \sinh u \, du = \cosh u + C$$

$$\frac{d}{dx}[\tanh u] = \operatorname{sech}^2 u \frac{du}{dx} \qquad \int \operatorname{sech}^2 u \, du = \tanh u + C$$

$$\frac{d}{dx}[\coth u] = -\operatorname{csch}^2 u \frac{du}{dx} \qquad \int \operatorname{csch}^2 u \, du = -\coth u + C$$

$$\frac{d}{dx}[\operatorname{sech} u] = -\operatorname{sech} u \tanh u \frac{du}{dx} \qquad \int \operatorname{sech} u \tanh u \, du = -\operatorname{sech} u + C$$

$$\frac{d}{dx}[\operatorname{csch} u] = -\operatorname{csch} u \coth u \frac{du}{dx} \qquad \int \operatorname{csch} u \coth u \, du = -\operatorname{csch} u + C$$

Example 2

$$\frac{d}{dx}[\cosh(x^3)] = \sinh(x^3) \cdot \frac{d}{dx}[x^3] = 3x^2 \sinh(x^3)$$

$$\frac{d}{dx}[\ln(\tanh x)] = \frac{1}{\tanh x} \cdot \frac{d}{dx}[\tanh x] = \frac{\operatorname{sech}^2 x}{\tanh x}$$

◄

Example 3

$$\int \sinh^5 x \cosh x \, dx = \tfrac{1}{6} \sinh^6 x + C \qquad \boxed{\begin{array}{l} u = \sinh x \\ du = \cosh x \, dx \end{array}}$$

$$\int \tanh x \, dx = \int \frac{\sinh x}{\cosh x} dx$$

$$= \ln |\cosh x| + C \qquad \boxed{\begin{array}{l} u = \cosh x \\ du = \sinh x \, dx \end{array}}$$

$$= \ln(\cosh x) + C$$

We were justified in dropping the absolute value signs since $\cosh x > 0$ for all x. ◄

Example 4 Find the length of the catenary $y = 10 \cosh(x/10)$ from $x = -10$ to $x = 10$ (Figure 7.8.4).

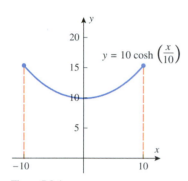

$y = 10 \cosh\left(\frac{x}{10}\right)$

Figure 7.8.4

Solution. From Formula (4) of Section 7.4, the length L of the catenary is

$$L = \int_{-10}^{10} \sqrt{1 + \left(\frac{dy}{dx}\right)^2} \, dx$$

$$= 2 \int_0^{10} \sqrt{1 + \left(\frac{dy}{dx}\right)^2} \, dx \qquad \boxed{\begin{array}{l} \text{By symmetry} \\ \text{about the } y\text{-axis} \end{array}}$$

$$= 2 \int_0^{10} \sqrt{1 + \sinh^2 \left(\frac{x}{10}\right)} \, dx$$

$$= 2 \int_0^{10} \cosh \left(\frac{x}{10}\right) \, dx \qquad \boxed{\begin{array}{l} \text{By (1) and the fact} \\ \text{that } \cosh x > 0 \end{array}}$$

$$= 20 \sinh \left(\frac{x}{10}\right) \Big]_0^{10}$$

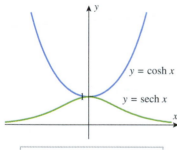

$y = \cosh x$

$y = \sech x$

With the restriction that $x \geq 0$, the curves $y = \cosh x$ and $y = \sech x$ pass the horizontal line test.

Figure 7.8.5

$$= 20[\sinh 1 - \sinh 0] = 20 \sinh 1 = 20 \left(\frac{e - e^{-1}}{2}\right) \approx 23.50 \qquad ◄$$

REMARK. Computer algebra systems, such as *Mathematica*, *Maple*, and *Derive* have built-in capabilities for evaluating hyperbolic functions directly, but some calculators do not. However, if you need to evaluate a hyperbolic function on a calculator, you can do so by expressing it in terms of exponential functions, as in this example.

INVERSES OF HYPERBOLIC FUNCTIONS

Referring to Figure 7.8.1, it is evident that the graphs of $\sinh x$, $\tanh x$, $\coth x$, and $\csch x$ pass the horizontal line test, but the graphs of $\cosh x$ and $\sech x$ do not. In the latter case restricting x to be nonnegative makes the functions invertible (Figure 7.8.5). The graphs of the six inverse hyperbolic functions in Figure 7.8.6 were obtained by reflecting the graphs of the hyperbolic functions (with the appropriate restrictions) about the line $y = x$.

Table 7.8.1 summarizes the basic properties of the inverse hyperbolic functions. You should confirm that the domains and ranges listed in this table agree with the graphs in Figure 7.8.6.

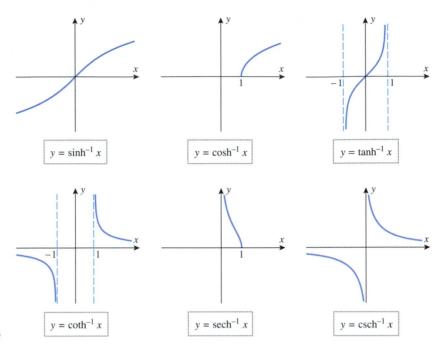

Figure 7.8.6

Table 7.8.1

FUNCTION	DOMAIN	RANGE	BASIC RELATIONSHIPS
$\sinh^{-1} x$	$(-\infty, +\infty)$	$(-\infty, +\infty)$	$\sinh^{-1}(\sinh x) = x \quad \text{if} \quad -\infty < x < +\infty$ $\sinh(\sinh^{-1} x) = x \quad \text{if} \quad -\infty < x < +\infty$
$\cosh^{-1} x$	$[1, +\infty)$	$[0, +\infty)$	$\cosh^{-1}(\cosh x) = x \quad \text{if} \quad x \geq 0$ $\cosh(\cosh^{-1} x) = x \quad \text{if} \quad x \geq 1$
$\tanh^{-1} x$	$(-1, 1)$	$(-\infty, +\infty)$	$\tanh^{-1}(\tanh x) = x \quad \text{if} \quad -\infty < x < +\infty$ $\tanh(\tanh^{-1} x) = x \quad \text{if} \quad -1 < x < 1$
$\coth^{-1} x$	$(-\infty, -1) \cup (1, +\infty)$	$(-\infty, 0) \cup (0, +\infty)$	$\coth^{-1}(\coth x) = x \quad \text{if} \quad x < 0 \text{ or } x > 0$ $\coth(\coth^{-1} x) = x \quad \text{if} \quad x < -1 \text{ or } x > 1$
$\text{sech}^{-1} x$	$(0, 1]$	$[0, +\infty)$	$\text{sech}^{-1}(\text{sech } x) = x \quad \text{if} \quad x \geq 0$ $\text{sech}(\text{sech}^{-1} x) = x \quad \text{if} \quad 0 < x \leq 1$
$\text{csch}^{-1} x$	$(-\infty, 0) \cup (0, +\infty)$	$(-\infty, 0) \cup (0, +\infty)$	$\text{csch}^{-1}(\text{csch } x) = x \quad \text{if} \quad x < 0 \text{ or } x > 0$ $\text{csch}(\text{csch}^{-1} x) = x \quad \text{if} \quad x < 0 \text{ or } x > 0$

LOGARITHMIC FORMS OF INVERSE HYPERBOLIC FUNCTIONS

Because the hyperbolic functions are expressible in terms of e^x, it should not be surprising that the inverse hyperbolic functions are expressible in terms of natural logarithms; the next theorem shows that this is so.

7.8.4 THEOREM. *The following relationships hold for all x in the domains of the stated inverse hyperbolic functions:*

$$\sinh^{-1} x = \ln(x + \sqrt{x^2 + 1}) \qquad \cosh^{-1} x = \ln(x + \sqrt{x^2 - 1})$$

$$\tanh^{-1} x = \frac{1}{2} \ln\left(\frac{1 + x}{1 - x}\right) \qquad \coth^{-1} x = \frac{1}{2} \ln\left(\frac{x + 1}{x - 1}\right)$$

$$\text{sech}^{-1} x = \ln\left(\frac{1 + \sqrt{1 - x^2}}{x}\right) \qquad \text{csch}^{-1} x = \ln\left(\frac{1}{x} + \frac{\sqrt{1 + x^2}}{|x|}\right)$$

We will show how to derive the first formula in this theorem, and leave the rest as exercises. The basic idea is to write the equation $x = \sinh y$ in terms of exponential functions and solve this equation for y as a function of x. This will produce the equation $y = \sinh^{-1} x$ with $\sinh^{-1} x$ expressed in terms of natural logarithms. Expressing $x = \sinh y$ in terms of exponentials yields

$$x = \sinh y = \frac{e^y - e^{-y}}{2}$$

which can be rewritten as

$$e^y - 2x - e^{-y} = 0$$

Multiplying this equation through by e^y we obtain

$$e^{2y} - 2xe^y - 1 = 0$$

and applying the quadratic formula yields

$$e^y = \frac{2x \pm \sqrt{4x^2 + 4}}{2} = x \pm \sqrt{x^2 + 1}$$

Since $e^y > 0$, the solution involving the minus sign is extraneous and must be discarded. Thus,

$$e^y = x + \sqrt{x^2 + 1}$$

Taking natural logarithms yields

$$y = \ln(x + \sqrt{x^2 + 1}) \quad \text{or} \quad \sinh^{-1} x = \ln(x + \sqrt{x^2 + 1})$$

Example 5

$$\sinh^{-1} 1 = \ln(1 + \sqrt{1^2 + 1}) = \ln(1 + \sqrt{2}) \approx 0.8814$$

$$\tanh^{-1}\left(\frac{1}{2}\right) = \frac{1}{2} \ln\left(\frac{1 + \frac{1}{2}}{1 - \frac{1}{2}}\right) = \frac{1}{2} \ln 3 \approx 0.5493 \qquad \blacktriangleleft$$

DERIVATIVES AND INTEGRALS INVOLVING INVERSE HYPERBOLIC FUNCTIONS

Theorem 4.1.6 can be used to establish the differentiability of the inverse hyperbolic functions (we omit the details), and formulas for the derivatives can be obtained from Theorem 7.8.4. For example,

$$\frac{d}{dx}[\sinh^{-1} x] = \frac{d}{dx}[\ln(x + \sqrt{x^2 + 1})] = \frac{1}{x + \sqrt{x^2 + 1}}\left(1 + \frac{x}{\sqrt{x^2 + 1}}\right)$$

$$= \frac{\sqrt{x^2 + 1} + x}{(x + \sqrt{x^2 + 1})(\sqrt{x^2 + 1})} = \frac{1}{\sqrt{x^2 + 1}}$$

This computation leads to two integral formulas, a formula that involves $\sinh^{-1} x$ and an equivalent formula that involves logarithms:

$$\int \frac{dx}{\sqrt{x^2 + 1}} = \sinh^{-1} x + C = \ln(x + \sqrt{x^2 + 1}) + C$$

FOR THE READER. The derivative of $\sinh^{-1} x$ can also be obtained by letting $y = \sinh^{-1} x$ and differentiating the equation $x = \sinh y$ implicitly. Try it.

The following two theorems list the generalized derivative formulas and corresponding integration formulas for the inverse hyperbolic functions. Some of the proofs appear as exercises.

7.8.5 THEOREM.

$$\frac{d}{dx}(\sinh^{-1} u) = \frac{1}{\sqrt{1+u^2}}\frac{du}{dx}$$

$$\frac{d}{dx}(\cosh^{-1} u) = \frac{1}{\sqrt{u^2-1}}\frac{du}{dx}, \quad u > 1$$

$$\frac{d}{dx}(\tanh^{-1} u) = \frac{1}{1-u^2}\frac{du}{dx}, \quad |u| < 1$$

$$\frac{d}{dx}(\coth^{-1} u) = \frac{1}{1-u^2}\frac{du}{dx}, \quad |u| > 1$$

$$\frac{d}{dx}(\operatorname{sech}^{-1} u) = -\frac{1}{u\sqrt{1-u^2}}\frac{du}{dx}, \quad 0 < u < 1$$

$$\frac{d}{dx}(\operatorname{csch}^{-1} u) = -\frac{1}{|u|\sqrt{1+u^2}}\frac{du}{dx}, \quad u \neq 0$$

7.8.6 THEOREM. *If $a > 0$, then*

$$\int \frac{du}{\sqrt{a^2+u^2}} = \sinh^{-1}\left(\frac{u}{a}\right) + C \quad or \quad \ln(u + \sqrt{u^2+a^2}) + C$$

$$\int \frac{du}{\sqrt{u^2-a^2}} = \cosh^{-1}\left(\frac{u}{a}\right) + C \quad or \quad \ln(u + \sqrt{u^2-a^2}) + C, \quad u > a$$

$$\int \frac{du}{a^2-u^2} = \begin{cases} \dfrac{1}{a}\tanh^{-1}\left(\dfrac{u}{a}\right) + C, & |u| < a \\[2ex] \dfrac{1}{a}\coth^{-1}\left(\dfrac{u}{a}\right) + C, & |u| > a \end{cases} \quad or \quad \frac{1}{2a}\ln\left|\frac{a+u}{a-u}\right| + C, \quad |u| \neq a$$

$$\int \frac{du}{u\sqrt{a^2-u^2}} = -\frac{1}{a}\operatorname{sech}^{-1}\left|\frac{u}{a}\right| + C \quad or \quad -\frac{1}{a}\ln\left(\frac{a+\sqrt{a^2-u^2}}{|u|}\right) + C, \quad 0 < |u| < a$$

$$\int \frac{du}{u\sqrt{a^2+u^2}} = -\frac{1}{a}\operatorname{csch}^{-1}\left|\frac{u}{a}\right| + C \quad or \quad -\frac{1}{a}\ln\left(\frac{a+\sqrt{a^2+u^2}}{|u|}\right) + C, \quad u \neq 0$$

Example 6 Evaluate $\displaystyle\int \frac{dx}{\sqrt{4x^2-9}}, x > \frac{3}{2}$.

Solution. Let $u = 2x$. Thus, $du = 2\,dx$ and

$$\int \frac{dx}{\sqrt{4x^2-9}} = \frac{1}{2}\int \frac{2\,dx}{\sqrt{4x^2-9}} = \frac{1}{2}\int \frac{du}{\sqrt{u^2-3^2}}$$

$$= \frac{1}{2}\cosh^{-1}\left(\frac{u}{3}\right) + C = \frac{1}{2}\cosh^{-1}\left(\frac{2x}{3}\right) + C$$

Alternatively, we can use the logarithmic equivalent of $\cosh^{-1}(2x/3)$,

$$\cosh^{-1}\left(\frac{2x}{3}\right) = \ln(2x + \sqrt{4x^2-9}) - \ln 3$$

(verify), and express the answer as

$$\int \frac{dx}{\sqrt{4x^2-9}} = \frac{1}{2}\ln(2x + \sqrt{4x^2-9}) + C$$

◀

EXERCISE SET 7.8 ~ Graphing Utility [c] CAS

In Exercises 1 and 2, approximate the expression to four decimal places.

1. (a) $\sinh 3$ (b) $\cosh(-2)$ (c) $\tanh(\ln 4)$
(d) $\sinh^{-1}(-2)$ (e) $\cosh^{-1} 3$ (f) $\tanh^{-1} \frac{3}{4}$

2. (a) $\operatorname{csch}(-1)$ (b) $\operatorname{sech}(\ln 2)$ (c) $\coth 1$
(d) $\operatorname{sech}^{-1}\frac{1}{2}$ (e) $\coth^{-1} 3$ (f) $\operatorname{csch}^{-1}(-\sqrt{3})$

3. In each part, find the exact numerical value of the expression.
(a) $\sinh(\ln 3)$ (b) $\cosh(-\ln 2)$
(c) $\tanh(2\ln 5)$ (d) $\sinh(-3\ln 2)$

4. In each part, rewrite the expression as a ratio of polynomials.
(a) $\cosh(\ln x)$ (b) $\sinh(\ln x)$
(c) $\tanh(2\ln x)$ (d) $\cosh(-\ln x)$

5. In each part, a value for one of the hyperbolic functions is given at an unspecified positive number x_0. Use appropriate identities to find the exact values of the remaining five hyperbolic functions at x_0.
(a) $\sinh x_0 = 2$ (b) $\cosh x_0 = \frac{5}{4}$ (c) $\tanh x_0 = \frac{4}{5}$

6. Obtain the derivative formulas for $\operatorname{csch} x$, $\operatorname{sech} x$, and $\coth x$ from the derivative formulas for $\sinh x$, $\cosh x$, and $\tanh x$.

7. Find the derivatives of $\sinh^{-1} x$, $\cosh^{-1} x$, and $\tanh^{-1} x$ by differentiating the equations $x = \sinh y$, $x = \cosh y$, and $x = \tanh y$ implicitly.

[c] **8.** Use a CAS to find the derivatives of $\sinh^{-1} x$, $\cosh^{-1} x$, $\tanh^{-1} x$, $\coth^{-1} x$, $\operatorname{sech}^{-1} x$, and $\operatorname{csch}^{-1} x$, and confirm that your answers are consistent with those in Theorem 7.8.5.

In Exercises 9–28, find dy/dx.

9. $y = \sinh(4x - 8)$ **10.** $y = \cosh(x^4)$

11. $y = \coth(\ln x)$ **12.** $y = \ln(\tanh 2x)$

13. $y = \operatorname{csch}(1/x)$ **14.** $y = \operatorname{sech}(e^{2x})$

15. $y = \sqrt{4x + \cosh^2(5x)}$ **16.** $y = \sinh^3(2x)$

17. $y = x^3 \tanh^2(\sqrt{x})$ **18.** $y = \sinh(\cos 3x)$

19. $y = \sinh^{-1}\left(\frac{1}{3}x\right)$ **20.** $y = \sinh^{-1}(1/x)$

21. $y = \ln(\cosh^{-1} x)$ **22.** $y = \cosh^{-1}(\sinh^{-1} x)$

23. $y = \dfrac{1}{\tanh^{-1} x}$ **24.** $y = (\coth^{-1} x)^2$

25. $y = \cosh^{-1}(\cosh x)$ **26.** $y = \sinh^{-1}(\tanh x)$

27. $y = e^x \operatorname{sech}^{-1}\sqrt{x}$ **28.** $y = (1 + x\operatorname{csch}^{-1} x)^{10}$

[c] **29.** Use a CAS to find the derivatives in Example 2. If the answers produced by the CAS do not match those in the text,

then use appropriate identities to show that the answers are equivalent.

[c] **30.** For each of the derivatives you obtained in Exercises 9–28, use a CAS to check your answer. If the answer produced by the CAS does not match your own, show that the two answers are equivalent.

In Exercises 31–46, evaluate the integrals.

31. $\displaystyle\int \sinh^6 x \cosh x\, dx$ **32.** $\displaystyle\int \cosh(2x - 3)\, dx$

33. $\displaystyle\int \sqrt{\tanh x}\ \operatorname{sech}^2 x\, dx$ **34.** $\displaystyle\int \operatorname{csch}^2(3x)\, dx$

35. $\displaystyle\int \tanh x\, dx$ **36.** $\displaystyle\int \coth^2 x \operatorname{csch}^2 x\, dx$

37. $\displaystyle\int_{\ln 2}^{\ln 3} \tanh x \operatorname{sech}^3 x\, dx$ **38.** $\displaystyle\int_0^{\ln 3} \frac{e^x - e^{-x}}{e^x + e^{-x}}\, dx$

39. $\displaystyle\int \frac{dx}{\sqrt{1 + 9x^2}}$ **40.** $\displaystyle\int \frac{dx}{\sqrt{x^2 - 2}}\quad (x > \sqrt{2})$

41. $\displaystyle\int \frac{dx}{\sqrt{1 - e^{2x}}}\quad (x < 0)$ **42.** $\displaystyle\int \frac{\sin\theta\, d\theta}{\sqrt{1 + \cos^2\theta}}$

43. $\displaystyle\int \frac{dx}{x\sqrt{1 + 4x^2}}$ **44.** $\displaystyle\int \frac{dx}{\sqrt{9x^2 - 25}}\quad (x > 5/3)$

45. $\displaystyle\int_0^{1/2} \frac{dx}{1 - x^2}$ **46.** $\displaystyle\int_0^{\sqrt{3}} \frac{dt}{\sqrt{t^2 + 1}}$

[c] **47.** For each of the integrals you evaluated in Exercises 31–46, use a CAS to check your answer. If the answer produced by the CAS does not match your own, show that the two answers are equivalent.

~ **48.** Use a graphing utility to generate the graphs of $\sinh x$, $\cosh x$, and $\tanh x$ by expressing these functions in terms of e^x and e^{-x}. If your graphing utility can graph the hyperbolic functions directly, then generate the graphs that way as well.

49. Find the area enclosed by $y = \sinh 2x$, $y = 0$, and $x = \ln 3$.

50. Find the volume of the solid that is generated when the region enclosed by $y = \operatorname{sech} x$, $y = 0$, $x = 0$, and $x = \ln 2$ is revolved about the x-axis.

51. Find the volume of the solid that is generated when the region enclosed by $y = \cosh 2x$, $y = \sinh 2x$, $x = 0$, and $x = 5$ is revolved about the x-axis.

~ **52.** Approximate the positive value of the constant a such that the area enclosed by $y = \cosh ax$, $y = 0$, $x = 0$, and $x = 1$ is 2 square units. Express your answer to at least five decimal places.

53. Find the arc length of $y = \cosh x$ between $x = 0$ and $x = \ln 2$.

54. Find the arc length of the catenary $y = a \cosh(x/a)$ between $x = 0$ and $x = x_1$ ($x_1 > 0$).

55. Prove that $\sinh x$ is an odd function of x and that $\cosh x$ is an even function of x, and check that this is consistent with the graphs in Figure 7.8.1.

In Exercises 56 and 57, prove the identities.

56. (a) $\cosh x + \sinh x = e^x$
(b) $\cosh x - \sinh x = e^{-x}$
(c) $\sinh(x + y) = \sinh x \cosh y + \cosh x \sinh y$
(d) $\sinh 2x = 2 \sinh x \cosh x$
(e) $\cosh(x + y) = \cosh x \cosh y + \sinh x \sinh y$
(f) $\cosh 2x = \cosh^2 x + \sinh^2 x$
(g) $\cosh 2x = 2 \sinh^2 x + 1$
(h) $\cosh 2x = 2 \cosh^2 x - 1$

57. (a) $1 - \tanh^2 x = \operatorname{sech}^2 x$
(b) $\tanh(x + y) = \dfrac{\tanh x + \tanh y}{1 + \tanh x \tanh y}$
(c) $\tanh 2x = \dfrac{2 \tanh x}{1 + \tanh^2 x}$

58. Prove:
(a) $\cosh^{-1} x = \ln(x + \sqrt{x^2 - 1})$, $x \geq 1$
(b) $\tanh^{-1} x = \dfrac{1}{2} \ln\left(\dfrac{1 + x}{1 - x}\right)$, $-1 < x < 1$.

59. Use Exercise 58 to obtain the derivative formulas for $\cosh^{-1} x$ and $\tanh^{-1} x$.

60. Prove:
$$\operatorname{sech}^{-1} x = \cosh^{-1}(1/x), \quad 0 < x \leq 1$$
$$\coth^{-1} x = \tanh^{-1}(1/x), \quad |x| > 1$$
$$\operatorname{csch}^{-1} x = \sinh^{-1}(1/x), \quad x \neq 0$$

61. Use Exercise 60 to express the integral
$$\int \frac{du}{1 - u^2}$$
entirely in terms of $\tanh^{-1}$.

62. Show that
(a) $\dfrac{d}{dx}[\operatorname{sech}^{-1}|x|] = -\dfrac{1}{x\sqrt{1 - x^2}}$
(b) $\dfrac{d}{dx}[\operatorname{csch}^{-1}|x|] = -\dfrac{1}{x\sqrt{1 + x^2}}$.

63. Find the limits, and confirm that they are consistent with the graphs in Figures 7.8.1 and 7.8.6.
(a) $\lim\limits_{x \to +\infty} \sinh x$
(b) $\lim\limits_{x \to -\infty} \sinh x$
(c) $\lim\limits_{x \to +\infty} \tanh x$
(d) $\lim\limits_{x \to -\infty} \tanh x$
(e) $\lim\limits_{x \to +\infty} \sinh^{-1} x$
(f) $\lim\limits_{x \to 1^-} \tanh^{-1} x$

64. In each part, find the limit.
(a) $\lim\limits_{x \to +\infty} (\cosh^{-1} x - \ln x)$
(b) $\lim\limits_{x \to +\infty} \dfrac{\cosh x}{e^x}$

65. Use the first and second derivatives to show that the graph of $y = \tanh^{-1} x$ is always increasing and has an inflection point at the origin.

66. The integration formulas for $1/\sqrt{u^2 - a^2}$ in Theorem 7.8.6 are valid for $u > a$. Show that the following formula is valid for $u < -a$:
$$\int \frac{du}{\sqrt{u^2 - a^2}} = -\cosh^{-1}\left(-\frac{u}{a}\right) + C \quad \text{or} \quad \ln\left|u + \sqrt{u^2 - a^2}\right| + C$$

67. Show that $(\sinh x + \cosh x)^n = \sinh nx + \cosh nx$.

68. Show that
$$\int_{-a}^{a} e^{tx}\, dx = \frac{2 \sinh at}{t}$$

69. A cable is suspended between two poles as shown in Figure 7.8.2. Assume that the equation of the curve formed by the cable is $y = a \cosh(x/a)$, where a is a positive constant. Suppose that the x-coordinates of the points of support are $x = -b$ and $x = b$, where $b > 0$.
(a) Show that the length L of the cable is given by
$$L = 2a \sinh \frac{b}{a}$$
(b) Show that the sag S (the vertical distance between the highest and lowest points on the cable) is given by
$$S = a \cosh \frac{b}{a} - a$$

Exercises 70 and 71 refer to the hanging cable described in Exercise 69.

70. Assuming that the cable is 120 ft long and the poles are 100 ft apart, approximate the sag in the cable by approximating a. Express your final answer to the nearest tenth of a foot. [*Hint:* First let $u = 50/a$.]

71. Assuming that the poles are 400 ft apart and the sag in the cable is 30 ft, approximate the length of the cable by approximating a. Express your final answer to the nearest tenth of a foot. [*Hint:* First let $u = 200/a$.]

72. Figure Ex-72 (next page) shows a person pulling a boat by holding a rope of length a attached to the bow and walking along the edge of a dock. If we assume that the rope is always tangent to the curve traced by the bow of the boat, then this curve, which is called a *tractrix*, has the property that the segment of the tangent line between the curve and the y-axis has a constant length a. It can be proved that the equation of this tractrix is
$$y = a \operatorname{sech}^{-1} \frac{x}{a} - \sqrt{a^2 - x^2}$$
(a) Show that to move the bow of the boat to a point (x, y), the person must walk a distance
$$D = a \operatorname{sech}^{-1} \frac{x}{a}$$
from the origin.

(b) If the rope has a length of 15 m, how far must the person walk from the origin to bring the boat 10 m from the dock? Round your answer to two decimal places.

(c) Find the distance traveled by the bow along the tractrix as it moves from its initial position to the point where it is 5 m from the dock.

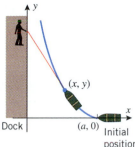

Figure Ex-72

SUPPLEMENTARY EXERCISES

 Graphing Utility   [C] CAS

1. State an integral formula for finding the arc length of a smooth curve $y = f(x)$ over an interval $[a, b]$, and use Riemann sums to derive the formula.

2. Describe the method of slicing for finding volumes, and use that method to derive an integral formula for finding volumes by the method of disks.

3. State an integral formula for finding a volume by the method of cylindrical shells, and use Riemann sums to derive the formula.

4. State an integral formula for the work W done by a variable force $F(x)$ applied in the direction of motion to an object moving from $x = a$ to $x = b$, and use Riemann sums to derive the formula.

5. State an integral formula for the fluid force F exerted on a vertical flat surface immersed in a fluid of weight density ρ, and use Riemann sums to derive the formula.

6. Let R be the region in the first quadrant enclosed by $y = x^2$, $y = 2 + x$, and $x = 0$. In each part, set up, but *do not evaluate*, an integral or a sum of integrals that will solve the problem.
   (a) Find the area of R by integrating with respect to x.
   (b) Find the area of R by integrating with respect to y.
   (c) Find the volume of the solid generated by revolving R about the x-axis by integrating with respect to x.
   (d) Find the volume of the solid generated by revolving R about the x-axis by integrating with respect to y.
   (e) Find the volume of the solid generated by revolving R about the y-axis by integrating with respect to x.
   (f) Find the volume of the solid generated by revolving R about the y-axis by integrating with respect to y.

7. (a) Set up a sum of definite integrals that represents the total shaded area between the curves $y = f(x)$ and $y = g(x)$ in the accompanying figure.
   (b) Find the total area enclosed between $y = x^3$ and $y = x$ over the interval $[-1, 2]$.

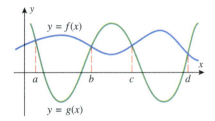

Figure Ex-7

8. Let C be the curve $27x - y^3 = 0$ between $y = 0$ and $y = 2$. In each part, set up, but *do not evaluate*, an integral or a sum of integrals that solves the problem.
   (a) Find the area of the surface generated by revolving C about the y-axis by integrating with respect to x.
   (b) Find the area of the surface generated by revolving C about the y-axis by integrating with respect to y.
   (c) Find the area of the surface generated by revolving C about the line $y = -2$ by integrating with respect to y.

9. Find the arc length in the second quadrant of the curve $x^{2/3} + y^{2/3} = a^{2/3}$ from $x = -a$ to $x = -\frac{1}{8}a$, where $a > 0$.

10. As shown in the accompanying figure, a cathedral dome is designed with three semicircular supports of radius r so that each horizontal cross section is a regular hexagon. Show that the volume of the dome is $r^3\sqrt{3}$.

11. As shown in the accompanying figure, a cylindrical hole is drilled all the way through the center of a sphere. Show that the volume of the remaining solid depends only on the length L of the hole, not on the size of the sphere.

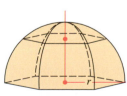

Figure Ex-10

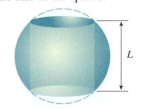

Figure Ex-11

12. A football has the shape of the solid generated by revolving the region bounded between the x-axis and the parabola $y = 4R(x^2 - \frac{1}{4}L^2)/L^2$ about the x-axis. Find its volume.

[c] 13. As shown in the accompanying figure, a horizontal beam with dimensions 2 in × 6 in × 16 ft is fixed at both ends and is subjected to a uniformly distributed load of 120 lb/ft. As a result of the load, the centerline of the beam undergoes a deflection that is described by

$$y = -1.67 \times 10^{-8}(x^4 - 2Lx^3 + L^2x^2)$$

($0 \le x \le 192$), where $L = 192$ inches is the length of the unloaded beam, x is the horizontal distance along the beam measured in inches from the left end, and y is the deflection of the centerline in inches.

(a) Graph y versus x for $0 \le x \le 192$.

(b) Find the maximum deflection of the centerline.

(c) Use a CAS or a calculator with a numerical integration capability to find the length of the centerline of the loaded beam. Round your answer to two decimal places.

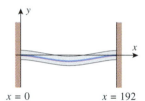

x = 0   x = 192   Figure Ex-13

[c] 14. A golfer makes a successful chip shot to the green. Suppose that the path of the ball from the moment it is struck to the moment it hits the green is described by

$$y = 12.54x - 0.41x^2$$

where x is the horizontal distance (in yards) from the point where the ball is struck, and y is the vertical distance (in yards) above the fairway. Use a CAS or a calculating utility with a numerical integration capability to find the distance the ball travels from the moment it is struck to the moment it hits the green. Assume that the fairway and green are at the same level and round your answer to two decimal places.

[c] 15. Use a CAS to find the area of the surface generated by revolving the parametric curve $x = e^t \cos t$, $y = e^t \sin t$, $0 \le t \le \pi/2$ about the x-axis.

[c] 16. Consider the region enclosed by $y = \sin^{-1} x$, $y = 0$, and $x = 1$. Find the volume of the solid generated by revolving the region about the x-axis using

(a) disks   (b) cylindrical shells.

17. (a) A spring exerts a force of 0.5 N when stretched 0.25 m beyond its natural length. Assuming that Hooke's law applies, how much work was performed in stretching the spring to this length?

(b) How far beyond its natural length can the spring be stretched with 25 J of work?

18. A boat is anchored so that the anchor is 150 ft below the surface of the water. In the water, the anchor weighs 2000 lb

and the chain weighs 30 lb/ft. How much work is required to raise the anchor to the surface?

19. In each part, set up, but *do not evaluate*, an integral that solves the problem.

(a) Find the fluid force exerted on a side of a box that has a 3-m-square base and is filled to a depth of 1 m with a liquid of weight density ρ N/m³.

(b) Find the fluid force exerted by a liquid of weight density ρ lb/ft³ on a face of the vertical plate shown in part (a) of the accompanying figure.

(c) Find the fluid force exerted on the parabolic dam in part (b) of the accompanying figure by water that extends to the top of the dam.

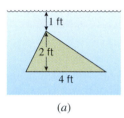   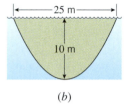

(a)   (b)

Figure Ex-19

20. Show that for any constant a, the function $y = \sinh(ax)$ satisfies the equation $y'' = a^2 y$.

21. In each part, prove the identity.

(a) $\cosh 3x = 4\cosh^3 x - 3\cosh x$

(b) $\cosh \frac{1}{2}x = \sqrt{\frac{1}{2}(\cosh x + 1)}$

(c) $\sinh \frac{1}{2}x = \pm\sqrt{\frac{1}{2}(\cosh x - 1)}$

22. Suppose that a hollow tube rotates with a constant angular velocity of ω rad/s about a horizontal axis at one end of the tube, as shown in the accompanying figure. Assume that an object is free to slide without friction in the tube while the tube is rotating. Let r be the distance from the object to the pivot point at time $t \ge 0$, and assume that the object is at rest and $r = 0$ when $t = 0$. It can be shown that if the tube is horizontal at time $t = 0$ and rotating as shown in the figure, then

$$r = \frac{g}{2\omega^2}[\sinh(\omega t) - \sin(\omega t)]$$

during the period that the object is in the tube. Assume that t is in seconds and r is in meters, and use $g = 9.8$ m/s² and $\omega = 2$ rad/s.

(a) Graph r versus t for $0 \le t \le 1$.

(b) Assuming that the tube has a length of 1 m, approximately how long does it take for the object to reach the end of the tube?

(c) Use the result of part (b) to approximate dr/dt at the instant that the object reaches the end of the tube.

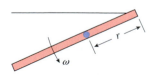

Figure Ex-22

23. The design of the Gateway Arch in St. Louis, Missouri, by architect Eero Saarinan was implemented using equations provided by Dr. Hannskarl Badel. The equation used for the centerline of the arch was

$$y = 693.8597 - 68.7672\cosh(0.0100333x)\ \text{ft}$$

for x between -299.2239 and 299.2239.

(a) Use a graphing utility to graph the centerline of the arch.
(b) Find the length of the centerline to four decimal places.
(c) For what values of x is the height of the arch 100 ft? Round your answers to four decimal places.
(d) Approximate, to the nearest degree, the acute angle that the tangent line to the centerline makes with the ground at the ends of the arch.

Exercises 24–26 lead to equations that cannot be solved exactly. Use any method you choose to approximate the solutions of those equations, and round your answers to two decimal places.

24. Find the area of the region enclosed by the curves $y = x^2 - 1$ and $y = 2\sin x$.

25. Referring to the accompanying figure, find the value of k so that the areas of the shaded regions are equal. [*Note:* This exercise is based on Problem A1 of the Fifty-Fourth Annual William Lowell Putnam Mathematical Competition.]

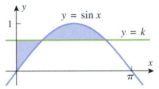

Figure Ex-25

[C] **26.** Consider the region to the left of the vertical line $x = k$ $(0 < k < \pi)$ and between the curve $y = \sin x$ and the x-axis. Use a CAS to find the value of k so that the solid generated by revolving the region about the y-axis has a volume of 8 cubic units.

8

PRINCIPLES OF INTEGRAL EVALUATION

Augustin Cauchy

*I*n earlier chapters we obtained many basic integration formulas from the corresponding differentiation formulas. For example, knowing that the derivative of $\sin x$ is $\cos x$ enabled us to deduce that the integral of $\cos x$ is $\sin x$. Subsequently, we expanded our integration repertoire by introducing the method of u-substitution. That method enabled us to integrate many functions by transforming the integrand of an unfamiliar integral into a familiar form. However, u-substitution alone is not adequate to handle the wide variety of integrals that arise in applications, so additional integration techniques are still needed. In this chapter we will discuss some of those techniques, and we will provide a more systematic procedure for attacking unfamiliar integrals. We will talk more about numerical approximations of definite integrals, and we will explore the idea of integrating over infinite intervals.

8.1 AN OVERVIEW OF INTEGRATION METHODS

In this section we will give a brief overview of methods for evaluating integrals, and we will review the integration formulas that were discussed in earlier sections.

There are three basic approaches for evaluating unfamiliar integrals:

- **Technology**—CAS programs such as *Mathematica*, *Maple*, and *Derive* are capable of evaluating extremely complicated integrals, and for both the computer and handheld calculator such programs are increasingly available.

- **Tables**—Prior to the development of CAS programs, scientists relied heavily on tables to evaluate difficult integrals arising in applications. Such tables were compiled over many years, incorporating the skills and experience of many people. One such table appears in the endpapers of this text, but more comprehensive tables appear in various reference books such as the *CRC Standard Mathematical Tables and Formulae*, CRC Press, Inc., 1996.

- **Transformation Methods**—Transformation methods are methods for converting unfamiliar integrals into familiar integrals. These include u-substitution, algebraic manipulation of the integrand, and other methods that we will discuss in this chapter.

None of the three methods is perfect; for example, CAS programs often encounter integrals that they cannot evaluate and they sometimes produce answers that are excessively complicated, tables are not exhaustive and hence may not include a particular integral of interest, and transformation methods rely on human ingenuity that may prove to be inadequate in difficult problems.

   In this chapter we will focus on transformation methods and tables, so it will *not be necessary* to have a CAS such as *Mathematica*, *Maple*, or *Derive*. However, if you have a CAS, then you can use it to confirm the results in the examples, and there are exercises that are designed to be solved with a CAS. If you have a CAS, keep in mind that many of the algorithms that it uses are based on the methods we will discuss here, so an understanding of these methods will help you to use your technology in a more informed way.

The following is a list of basic integrals that we have encountered thus far:

CONSTANTS, POWERS, EXPONENTIALS

1. $\displaystyle\int du = u + C$ **2.** $\displaystyle\int a\,du = a\int du = au + C$

3. $\displaystyle\int u^r\,du = \frac{u^{r+1}}{r+1} + C,\ r \neq -1$ **4.** $\displaystyle\int \frac{du}{u} = \ln|u| + C$

5. $\displaystyle\int e^u\,du = e^u + C$ **6.** $\displaystyle\int b^u\,du = \frac{b^u}{\ln b} + C,\ b > 0,\ b \neq 1$

TRIGONOMETRIC FUNCTIONS

7. $\displaystyle\int \sin u\,du = -\cos u + C$ **8.** $\displaystyle\int \cos u\,du = \sin u + C$

9. $\displaystyle\int \sec^2 u\,du = \tan u + C$ **10.** $\displaystyle\int \csc^2 u\,du = -\cot u + C$

11. $\displaystyle\int \sec u \tan u\,du = \sec u + C$ **12.** $\displaystyle\int \csc u \cot u\,du = -\csc u + C$

13. $\displaystyle\int \tan u\,du = -\ln|\cos u| + C$ **14.** $\displaystyle\int \cot u\,du = \ln|\sin u| + C$

HYPERBOLIC FUNCTIONS

15. $\int \sinh u \, du = \cosh u + C$ **16.** $\int \cosh u \, du = \sinh u + C$

17. $\int \operatorname{sech}^2 u \, du = \tanh u + C$ **18.** $\int \operatorname{csch}^2 u \, du = -\coth u + C$

19. $\int \operatorname{sech} u \tanh u \, du = -\operatorname{sech} u + C$ **20.** $\int \operatorname{csch} u \coth u \, du = -\operatorname{csch} u + C$

ALGEBRAIC FUNCTIONS ($a > 0$)

21. $\int \dfrac{du}{\sqrt{a^2 - u^2}} = \sin^{-1} \dfrac{u}{a} + C \quad (|u| < a)$

22. $\int \dfrac{du}{a^2 + u^2} = \dfrac{1}{a} \tan^{-1} \dfrac{u}{a} + C$

23. $\int \dfrac{du}{u\sqrt{u^2 - a^2}} = \dfrac{1}{a} \sec^{-1} \left|\dfrac{u}{a}\right| + C \quad (0 < a < |u|)$

24. $\int \dfrac{du}{\sqrt{a^2 + u^2}} = \ln(u + \sqrt{u^2 + a^2}) + C$

25. $\int \dfrac{du}{\sqrt{u^2 - a^2}} = \ln\left|u + \sqrt{u^2 - a^2}\right| + C \quad (0 < a < |u|)$

26. $\int \dfrac{du}{a^2 - u^2} = \dfrac{1}{2a} \ln\left|\dfrac{a + u}{a - u}\right| + C$

27. $\int \dfrac{du}{u\sqrt{a^2 - u^2}} = -\dfrac{1}{a} \ln\left|\dfrac{a + \sqrt{a^2 - u^2}}{u}\right| + C \quad (0 < |u| < a)$

28. $\int \dfrac{du}{u\sqrt{a^2 + u^2}} = -\dfrac{1}{a} \ln\left|\dfrac{a + \sqrt{a^2 + u^2}}{u}\right| + C$

REMARK. Formula 25 is a generalization of a result in Theorem 7.8.6. Readers who did not cover Section 7.8 can ignore Formulas 24–28 for now, since we will develop other methods for obtaining them in this chapter.

EXERCISE SET 8.1

Review: Without looking at the text, complete the following integration formulas and then check your results by referring to the list of formulas at the beginning of this section.

Constants, Powers, Exponentials

$\int du =$ $\int a \, du =$

$\int u^r \, du =$ $\int \dfrac{du}{u} =$

$\int e^u \, du =$ $\int b^u \, du =$

Trigonometric Functions

$\int \sin u \, du =$ $\int \cos u \, du =$

$\int \sec^2 u \, du =$ $\int \csc^2 u \, du =$

$\int \sec u \tan u \, du =$ $\int \csc u \cot u \, du =$

$\int \tan u \, du =$ $\int \cot u \, du =$

Algebraic Functions

$$\int \frac{du}{\sqrt{1-u^2}} = \qquad \int \frac{du}{1+u^2} =$$

$$\int \frac{du}{u\sqrt{u^2-1}} = \qquad \int \frac{du}{\sqrt{1+u^2}} =$$

$$\int \frac{du}{\sqrt{u^2-1}} = \qquad \int \frac{du}{1-u^2} =$$

$$\int \frac{du}{u\sqrt{1-u^2}} = \qquad \int \frac{du}{u\sqrt{1+u^2}} =$$

Hyperbolic Functions

$$\int \sinh u \, du = \qquad \int \cosh u \, du =$$

$$\int \operatorname{sech}^2 u \, du = \qquad \int \operatorname{csch}^2 u \, du =$$

$$\int \operatorname{sech} u \tanh u \, du =$$

$$\int \operatorname{csch} u \coth u \, du =$$

In Exercises 1–30, evaluate the integrals by making appropriate *u*-substitutions and applying the formulas reviewed in this section.

1. $\int (3-2x)^3 \, dx$

2. $\int \sqrt{4+9x} \, dx$

3. $\int x \sec^2(x^2) \, dx$

4. $\int 4x \tan(x^2) \, dx$

5. $\int \frac{\sin 3x}{2 + \cos 3x} \, dx$

6. $\int \frac{1}{4 + 9x^2} \, dx$

7. $\int e^x \sinh(e^x) \, dx$

8. $\int \frac{\sec(\ln x) \tan(\ln x)}{x} \, dx$

9. $\int e^{\cot x} \csc^2 x \, dx$

10. $\int \frac{x}{\sqrt{1-x^4}} \, dx$

11. $\int \cos^5 7x \sin 7x \, dx$

12. $\int \frac{\cos x}{\sin x \sqrt{\sin^2 x + 1}} \, dx$

13. $\int \frac{e^x}{\sqrt{4 + e^{2x}}} \, dx$

14. $\int \frac{e^{\tan^{-1} x}}{1 + x^2} \, dx$

15. $\int \frac{e^{\sqrt{x-2}}}{\sqrt{x-2}} \, dx$

16. $\int (3x+1) \cot(3x^2 + 2x) \, dx$

17. $\int \frac{\cosh \sqrt{x}}{\sqrt{x}} \, dx$

18. $\int \frac{dx}{x \ln x}$

19. $\int \frac{dx}{\sqrt{x} \, 3^{\sqrt{x}}}$

20. $\int \sec(\sin \theta) \tan(\sin \theta) \cos \theta \, d\theta$

21. $\int \frac{\operatorname{csch}^2(2/x)}{x^2} \, dx$

22. $\int \frac{dx}{\sqrt{x^2 - 3}}$

23. $\int \frac{e^{-x}}{4 - e^{-2x}} \, dx$

24. $\int \frac{\cos(\ln x)}{x} \, dx$

25. $\int \frac{e^x}{\sqrt{1 - e^{2x}}} \, dx$

26. $\int \frac{\sinh(x^{-1/2})}{x^{3/2}} \, dx$

27. $\int \frac{x}{\sec(x^2)} \, dx$

28. $\int \frac{e^x}{\sqrt{4 - e^{2x}}} \, dx$

29. $\int x 4^{-x^2} \, dx$

30. $\int 2^{\pi x} \, dx$

8.2 INTEGRATION BY PARTS

In this section we will discuss an integration technique that is essentially an anti-derivative formulation of the formula for differentiating a product of two functions.

THE PRODUCT RULE AND INTEGRATION BY PARTS

We saw in Section 6.3 that the *u*-substitution method of integration is based on the chain rule for differentiation. In this section we will examine a method of integration that is based on the *product rule* for differentiation. To motivate the general formula, we will consider the problem of evaluating $\int x \cos x \, dx$. Our approach to this problem will be by means of a two-step process. The first step is to choose a function whose derivative is the sum of *two* functions, *one* of which is $x \cos x$. For example, the function $x \sin x$ has this property, since by the product rule

$$\frac{d}{dx}(x \sin x) = x \cos x + \sin x$$

(Note that $x \sin x$ may be obtained from $x \cos x$ by integrating the $\cos x$ "part" of $x \cos x$ while leaving the x "part" alone.) The second step in evaluating $\int x \cos x \, dx$ is to subtract from our chosen function an antiderivative for the "extra" function that is produced by the product rule. What results will then be an antiderivative for $x \cos x$. For example, from the function $x \sin x$, we would need to subtract an antiderivative of $\sin x$. Since $-\cos x$ is an antiderivative of $\sin x$, we conclude that

$$x \sin x - (-\cos x) = x \sin x + \cos x$$

is an antiderivative of $x \cos x$. Indeed, this conclusion is easily verified since

$$\frac{d}{dx}(x \sin x + \cos x) = x \cos x + \sin x - \sin x = x \cos x$$

It follows that

$$\int x \cos x \, dx = x \sin x + \cos x + C$$

This two-step process is an illustration of a method of integration known as ***integration by parts***. More generally, suppose that we wish to evaluate an integral of the form $\int f(x)g(x)\,dx$. If $G(x)$ is an antiderivative of $g(x)$, then by the product rule for derivatives, the function $f(x)G(x)$ satisfies the equation

$$\frac{d}{dx}(f(x)G(x)) = f(x)g(x) + f'(x)G(x)$$

Consequently, if we subtract an antiderivative for $f'(x)G(x)$ from the function $f(x)G(x)$, the result will be an antiderivative for $f(x)g(x)$. We may express this conclusion symbolically by writing

$$\int f(x)g(x)\,dx = f(x)G(x) - \int f'(x)G(x)\,dx \tag{1}$$

which is one version of the integration by parts formula. By using this formula we can sometimes reduce a difficult integration problem to an easier one.

In practice, it is usual to rewrite (1) by letting

$$u = f(x), \quad du = f'(x)\,dx$$
$$v = G(x), \quad dv = G'(x)\,dx = g(x)\,dx$$

This yields the following alternative form for (1):

$$\int u\,dv = uv - \int v\,du \tag{2}$$

To illustrate the use of Formula (2) we will reevaluate $\int x \cos x \, dx$. The first step is to make a choice of u and dv. We will let $u = x$ and $dv = \cos x \, dx$ from which it follows that $du = dx$ and $v = \sin x$. Then, from Formula (2)

$$\int x \cos x \, dx = x \sin x - \int \sin x \, dx$$

$$= x \sin x - (-\cos x) + C = x \sin x + \cos x + C$$

REMARK. In the calculation of $v = \sin x$ from $dv = \cos x \, dx$, we omitted a constant of integration. Had we included a constant of integration and written $v = \sin x + C_1$, the constant C_1 would have eventually canceled out [Exercise 62(a)]. This is always the case in integration by parts [Exercise 62(b)], and it is common to omit consideration of a constant of integration when going from dv to v. However, in certain cases a clever choice of a constant of integration can simplify the computation of $\int v \, du$ [Exercises 63–65].

REMARK.   To use integration by parts successfully, the choice of u and dv must be made so that the new integral is easier than the original. For example, if we decided above to let

$$u = \cos x, \quad dv = x \, dx, \quad du = -\sin x \, dx, \quad v = \frac{x^2}{2}$$

then we would have obtained

$$\int x \cos x \, dx = \frac{x^2}{2} \cos x - \int \frac{x^2}{2} (-\sin x) \, dx = \frac{x^2}{2} \cos x + \frac{1}{2} \int x^2 \sin x \, dx$$

For this choice of u and dv, the new integral is actually more complicated than the original. In general there are no hard and fast rules for choosing u and dv; it is mainly a matter of experience that comes from lots of practice.

For the case in which the integrand is the product of different "types" of functions, an interesting mnemonic device was suggested by Herbert Kasube in his article "A Technique for Integration by Parts" (*American Mathematical Monthly*, Vol. 90, 1983, pp. 210–211). In this article the author suggests the use of the acronym LIATE, which is short for **l**ogarithmic, **i**nverse trigonometric, **a**lgebraic, **t**rigonometric, and **e**xponential. According to the author, when the integrand of an integration by parts problem consists of the product of two different types of functions, we should let u designate the function that appears first in LIATE, and let dv denote the rest. For example, since the integrand of $\int x \cos x \, dx$ is the product of the algebraic function x with the trigonometric function $\cos x$, we should let $u = x$ and $dv = \cos x \, dx$, which agrees with our choice in the reevaluation of this integral. Although LIATE does not always produce the correct choice of u and dv, it does work much of the time.

Example 1   Evaluate $\int x e^x \, dx$.

Solution.   In this case the integrand is the product of the algebraic function x with the exponential function e^x. According to LIATE we should let

$$u = x \quad \text{and} \quad dv = e^x \, dx$$

so that

$$du = dx \quad \text{and} \quad v = e^x$$

Thus, from (2)

$$\int x e^x \, dx = \int u \, dv = uv - \int v \, du = x e^x - \int e^x \, dx = x e^x - e^x + C \qquad \blacktriangleleft$$

In some cases there is only one reasonable choice of u and dv.

Example 2   Evaluate $\int \ln x \, dx$.

Solution.   One choice is to let $u = 1$ and $dv = \ln x \, dx$. But with this choice finding v is equivalent to evaluating $\int \ln x \, dx$ and we have gained nothing. Therefore, the only reasonable choice is to let $u = \ln x$ and $dv = dx$, so that $du = (1/x) \, dx$ and $v = x$. Thus, from (2)

$$\int \ln x \, dx = \int u \, dv = uv - \int v \, du = x \ln x - \int dx = x \ln x - x + C \qquad \blacktriangleleft$$

REPEATED INTEGRATION BY PARTS

It is sometimes necessary to use integration by parts more than once in the same problem.

Example 3   Evaluate $\int x^2 e^{-x} \, dx$.

Solution.   Let

$$u = x^2, \quad dv = e^{-x} \, dx, \quad du = 2x \, dx, \quad v = -e^{-x}$$

so that from (2)

$$\int x^2 e^{-x}\, dx = \int u\, dv = uv - \int v\, du = x^2(-e^{-x}) - \int -e^{-x}(2x)\, dx$$

$$= -x^2 e^{-x} + 2\int xe^{-x}\, dx$$

The last integral is similar to the original except that we have replaced x^2 by x. Another integration by parts applied to $\int xe^{-x}\, dx$ will complete the problem. We let

$$u = x, \quad dv = e^{-x}\, dx, \quad du = dx, \quad v = -e^{-x}$$

so that

$$\int xe^{-x}\, dx = x(-e^{-x}) - \int -e^{-x}\, dx = -xe^{-x} + \int e^{-x}\, dx = -xe^{-x} - e^{-x} + C$$

Since $-xe^{-x} - e^{-x}$ is an antiderivative for xe^{-x}, it follows that

$$\int x^2 e^{-x}\, dx = -x^2 e^{-x} + 2\int xe^{-x}\, dx = -x^2 e^{-x} + 2(-xe^{-x} - e^{-x}) + C$$

$$= -(x^2 + 2x + 2)e^{-x} + C \qquad \blacktriangleleft$$

Note that the integrand in Example 3 is of the form $p(x)q(x)$, where $p(x) = x^2$ is a *polynomial* and $q(x) = e^{-x}$ is a function that can be *repeatedly integrated*. For integrands of this form, repeated integration by parts can be done more efficiently by means of a procedure known as **tabular integration by parts**. The procedure depends on the fact that repeated differentiation of a polynomial eventually results in 0. Since the method is easier to illustrate than to describe, we will show how tabular integration by parts may be used to evaluate the integral in Example 3. The first step is to create the following table:

REPEATED DIFFERENTIATION		REPEATED ANTIDIFFERENTIATION
x^2	+	e^{-x}
$2x$	−	$-e^{-x}$
2	+	e^{-x}
0		$-e^{-x}$

The entries in the left column of the table are obtained by starting with $p(x) = x^2$ and repeatedly *differentiating* until 0 results. The entries in the right column are obtained by starting with $q(x) = e^{-x}$ and repeatedly *integrating* until an entry is opposite the 0 in the left column. The diagonal segments shown in the table are alternately labeled with + and − signs. To evaluate $\int x^2 e^{-x}\, dx$, we sum the products of the entries joined by a diagonal, incorporating the sign of the corresponding diagonal into each product. It follows that

$$\int x^2 e^{-x}\, dx = -x^2 e^{-x} - 2xe^{-x} - 2e^{-x} + C = -(x^2 + 2x + 2)e^{-x} + C$$

which agrees with our result in Example 3.

A second example should make the procedure clear.

Example 4 In Example 11 of Section 6.3 we evaluated $\int x^2 \sqrt{x-1}\, dx$ using u-substitution. Evaluate this integral using tabular integration by parts.

Solution. The integrand is the product of a polynomial $p(x) = x^2$ and a function

$$q(x) = \sqrt{x-1} = (x-1)^{1/2}$$

that can be repeatedly integrated. First we form the table:

REPEATED DIFFERENTIATION		REPEATED ANTIDIFFERENTIATION
x^2	$+$	$(x-1)^{1/2}$
$2x$	$-$	$\frac{2}{3}(x-1)^{3/2}$
2	$+$	$\frac{4}{15}(x-1)^{5/2}$
0		$\frac{8}{105}(x-1)^{7/2}$

Then it follows that

$$\int x^2\sqrt{x-1}\,dx = \tfrac{2}{3}x^2(x-1)^{3/2} - \tfrac{8}{15}x(x-1)^{5/2} + \tfrac{16}{105}(x-1)^{7/2} + C$$

We leave it for the reader to show that this solution is equivalent to that of Example 11 in Section 6.3. ◀

The next illustration of repeated integration by parts deserves special attention.

Example 5 Evaluate $\displaystyle\int e^x \cos x\,dx$.

Solution. Let

$$u = e^x, \quad dv = \cos x\,dx, \quad du = e^x\,dx, \quad v = \sin x$$

Thus,

$$\int e^x \cos x\,dx = \int u\,dv = uv - \int v\,du = e^x \sin x - \int e^x \sin x\,dx \qquad (3)$$

Since the integral $\int e^x \sin x\,dx$ is similar in form to the original integral $\int e^x \cos x\,dx$, it seems that nothing has been accomplished. However, let us integrate this new integral by parts. We let

$$u = e^x, \quad dv = \sin x\,dx, \quad du = e^x\,dx, \quad v = -\cos x$$

Thus,

$$\int e^x \sin x\,dx = \int u\,dv = uv - \int v\,du = -e^x \cos x + \int e^x \cos x\,dx$$

Together with Equation (3) this yields

$$\int e^x \cos x\,dx = e^x \sin x + e^x \cos x - \int e^x \cos x\,dx \qquad (4)$$

It appears that we are going in circles since our original integral has now reappeared on the right side of this equation. However, at this point it is helpful to remind ourselves of the *meaning* of Equation (4). Equation (4) is a symbolic way of stating that if $F(x)$ is any antiderivative of $e^x \cos x$, then the function $e^x \sin x + e^x \cos x - F(x)$ is also an antiderivative of $e^x \cos x$. In other words,

$$e^x \cos x = \frac{d}{dx}[e^x \sin x + e^x \cos x - F(x)] = \frac{d}{dx}[e^x \sin x + e^x \cos x] - F'(x)$$

$$= \frac{d}{dx}[e^x \sin x + e^x \cos x] - e^x \cos x$$

Equivalently,

$$2e^x \cos x = \frac{d}{dx}[e^x \sin x + e^x \cos x]$$

or

$$e^x \cos x = \frac{1}{2}\frac{d}{dx}[e^x \sin x + e^x \cos x] = \frac{d}{dx}\left[\frac{1}{2}(e^x \sin x + e^x \cos x)\right]$$

[Note that this last equation may also be verified by direct computation of the derivative of $\frac{1}{2}(e^x \sin x + e^x \cos x)$.] It follows that

$$\int e^x \cos x \, dx = \tfrac{1}{2}(e^x \sin x + e^x \cos x) + C \tag{5}$$

We can also obtain Equation (5) directly from Equation (4) by an informal argument. The idea is to "solve" Equation (4) for $\int e^x \cos x \, dx$, adding the (necessary) constant of integration only at the very end. That is, from Equation (4) we obtain

$$2 \int e^x \cos x \, dx = e^x \sin x + e^x \cos x$$

or

$$\int e^x \cos x \, dx = \tfrac{1}{2}(e^x \sin x + e^x \cos x)$$

Since the left side of this equation is an indefinite integral, we need a constant of integration C on the right side. Adding this C to the right side results in Equation (5). Although informal arguments such as these can save time, they must be used with care (Exercise 66).

INTEGRATION BY PARTS FOR DEFINITE INTEGRALS

For definite integrals the formula corresponding to (2) is

$$\int_a^b u \, dv = uv \Big]_a^b - \int_a^b v \, du \tag{6}$$

REMARK. It is important to keep in mind that the variables u and v in this formula are functions of x and that the limits of integration in (6) are limits on the variable x. Sometimes it is helpful to emphasize this by writing (6) as

$$\int_{x=a}^{x=b} u \, dv = uv \Big]_{x=a}^{x=b} - \int_{x=a}^{x=b} v \, du \tag{7}$$

The next example illustrates how integration by parts can be used to integrate the inverse trigonometric functions.

Example 6 Evaluate $\displaystyle\int_0^1 \tan^{-1} x \, dx$.

Solution. Let

$$u = \tan^{-1} x, \quad dv = dx, \quad du = \frac{1}{1 + x^2} \, dx, \quad v = x$$

Thus,

$$\int_0^1 \tan^{-1} x \, dx = \int_0^1 u \, dv = uv \Big]_0^1 - \int_0^1 v \, du$$

> The limits of integration refer to x; that is, $x = 0$ and $x = 1$.

$$= x \tan^{-1} x \Big]_0^1 - \int_0^1 \frac{x}{1 + x^2} \, dx$$

But

$$\int_0^1 \frac{x}{1 + x^2} \, dx = \frac{1}{2} \int_0^1 \frac{2x}{1 + x^2} \, dx = \frac{1}{2} \ln(1 + x^2) \Big]_0^1 = \frac{1}{2} \ln 2$$

so

$$\int_0^1 \tan^{-1} x \, dx = x \tan^{-1} x \Big]_0^1 - \frac{1}{2} \ln 2 = \left(\frac{\pi}{4} - 0\right) - \frac{1}{2} \ln 2 = \frac{\pi}{4} - \ln\sqrt{2} \qquad \blacktriangleleft$$

REDUCTION FORMULAS

Integration by parts can be used to derive *reduction formulas* for integrals. These are formulas that express an integral involving a power of a function in terms of an integral that involves a *lower* power of that function. For example, if n is a positive integer and $n \geq 2$, then integration by parts can be used to obtain the reduction formulas

$$\int \sin^n x \, dx = -\frac{1}{n} \sin^{n-1} x \cos x + \frac{n-1}{n} \int \sin^{n-2} x \, dx \qquad (8)$$

$$\int \cos^n x \, dx = \frac{1}{n} \cos^{n-1} x \sin x + \frac{n-1}{n} \int \cos^{n-2} x \, dx \qquad (9)$$

To illustrate how such formulas can be obtained, let us derive (9). We begin by writing $\cos^n x$ as $\cos^{n-1} x \cdot \cos x$ and letting

$$u = \cos^{n-1} x \qquad\qquad dv = \cos x \, dx$$
$$du = (n-1) \cos^{n-2} x(-\sin x) \, dx \qquad v = \sin x$$
$$= -(n-1) \cos^{n-2} x \sin x \, dx$$

so that

$$\int \cos^n x \, dx = \int \cos^{n-1} x \cos x \, dx = \int u \, dv = uv - \int v \, du$$

$$= \cos^{n-1} x \sin x + (n-1) \int \sin^2 x \cos^{n-2} x \, dx$$

$$= \cos^{n-1} x \sin x + (n-1) \int (1 - \cos^2 x) \cos^{n-2} x \, dx$$

$$= \cos^{n-1} x \sin x + (n-1) \int \cos^{n-2} x \, dx - (n-1) \int \cos^n x \, dx$$

We now appeal to an informal argument and "solve" for $\int \cos^n x \, dx$. (See our comments following Example 5.) Transposing the last term on the right to the left side yields

$$n \int \cos^n x \, dx = \cos^{n-1} x \sin x + (n-1) \int \cos^{n-2} x \, dx$$

from which (9) follows.

Reduction formulas (8) and (9) reduce the exponent of sine (or cosine) by 2. Thus, if the formulas are applied repeatedly, the exponent can eventually be reduced to 0 if n is even or 1 if n is odd, at which point the integration can be completed. We will discuss this method in more detail in the next section, but for now, here is an example that illustrates how reduction formulas work.

Example 7 Evaluate $\int \cos^4 x \, dx$.

Solution. From (9) with $n = 4$

$$\int \cos^4 x \, dx = \frac{1}{4} \cos^3 x \sin x + \frac{3}{4} \int \cos^2 x \, dx \qquad \boxed{\text{Now apply (9) with } n = 2.}$$

$$= \frac{1}{4} \cos^3 x \sin x + \frac{3}{4} \left(\frac{1}{2} \cos x \sin x + \frac{1}{2} \int dx \right)$$

$$= \frac{1}{4} \cos^3 x \sin x + \frac{3}{8} \cos x \sin x + \frac{3}{8} x + C \qquad \blacktriangleleft$$

EXERCISE SET 8.2

In Exercises 1–40, evaluate the integral.

1. $\int xe^{-x}\,dx$

2. $\int xe^{3x}\,dx$

3. $\int x^2 e^x\,dx$

4. $\int x^2 e^{-2x}\,dx$

5. $\int x\sin 2x\,dx$

6. $\int x\cos 3x\,dx$

7. $\int x^2\cos x\,dx$

8. $\int x^2\sin x\,dx$

9. $\int \sqrt{x}\ln x\,dx$

10. $\int x\ln x\,dx$

11. $\int (\ln x)^2\,dx$

12. $\int \frac{\ln x}{\sqrt{x}}\,dx$

13. $\int \ln(2x+3)\,dx$

14. $\int \ln(x^2+4)\,dx$

15. $\int \sin^{-1}x\,dx$

16. $\int \cos^{-1}(2x)\,dx$

17. $\int \tan^{-1}(2x)\,dx$

18. $\int x\tan^{-1}x\,dx$

19. $\int e^x\sin x\,dx$

20. $\int e^{2x}\cos 3x\,dx$

21. $\int e^{ax}\sin bx\,dx$

22. $\int e^{-3\theta}\sin 5\theta\,d\theta$

23. $\int \sin(\ln x)\,dx$

24. $\int \cos(\ln x)\,dx$

25. $\int x\sec^2 x\,dx$

26. $\int x\tan^2 x\,dx$

27. $\int x^3 e^{x^2}\,dx$

28. $\int \frac{xe^x}{(x+1)^2}\,dx$

29. $\int_0^1 xe^{-5x}\,dx$

30. $\int_0^2 xe^{2x}\,dx$

31. $\int_1^e x^2\ln x\,dx$

32. $\int_{\sqrt{e}}^e \frac{\ln x}{x^2}\,dx$

33. $\int_{-2}^2 \ln(x+3)\,dx$

34. $\int_0^{1/2} \sin^{-1}x\,dx$

35. $\int_2^4 \sec^{-1}\sqrt{\theta}\,d\theta$

36. $\int_1^2 x\sec^{-1}x\,dx$

37. $\int_0^{\pi/2} x\sin 4x\,dx$

38. $\int_0^\pi (x+x\cos x)\,dx$

39. $\int_1^3 \sqrt{x}\tan^{-1}\sqrt{x}\,dx$

40. $\int_0^2 \ln(x^2+1)\,dx$

41. In each part, evaluate the integral by making a u-substitution and then integrating by parts.

(a) $\int e^{\sqrt{x}}\,dx$

(b) $\int \cos\sqrt{x}\,dx$

42. Prove that tabular integration by parts gives the correct answer for

$$\int p(x)q(x)\,dx$$

where $p(x)$ is any *quadratic polynomial* and $q(x)$ is any function that can be repeatedly integrated.

In Exercises 43–46, evaluate the integral using tabular integration by parts.

43. $\int (3x^2-x+2)e^{-x}\,dx$

44. $\int (x^2+x+1)\sin x\,dx$

45. $\int 8x^4\cos 2x\,dx$

46. $\int x^3\sqrt{2x+1}\,dx$

47. (a) Find the area of the region enclosed by $y=\ln x$, the line $x=e$, and the x-axis.
(b) Find the volume of the solid generated when the region in part (a) is revolved about the x-axis.

48. Find the area of the region between $y=x\sin x$ and $y=x$ for $0\le x\le \pi/2$.

49. Find the volume of the solid generated when the region between $y=\sin x$ and $y=0$ for $0\le x\le\pi$ is revolved about the y-axis.

50. Find the volume of the solid generated when the region enclosed between $y=\cos x$ and $y=0$ for $0\le x\le \pi/2$ is revolved about the y-axis.

51. A particle moving along the x-axis has velocity function $v(t)=t^2e^{-t}$. How far does the particle travel from time $t=0$ to $t=5$?

52. The study of sawtooth waves in electrical engineering leads to integrals of the form

$$\int_{-\pi/\omega}^{\pi/\omega} t\sin(k\omega t)\,dt$$

where k is an integer and ω is a nonzero constant. Evaluate the integral.

53. Use reduction formula (8) to evaluate
(a) $\int \sin^3 x\,dx$
(b) $\int_0^{\pi/4} \sin^4 x\,dx.$

54. Use reduction formula (9) to evaluate
(a) $\int \cos^5 x\,dx$
(b) $\int_0^{\pi/2} \cos^6 x\,dx.$

55. Derive reduction formula (8).

56. In each part, use integration by parts or other methods to derive the reduction formula.

(a) $\int \sec^n x\,dx = \frac{\sec^{n-2}x\tan x}{n-1}+\frac{n-2}{n-1}\int \sec^{n-2}x\,dx$

(b) $\int \tan^n x\,dx = \frac{\tan^{n-1}x}{n-1}-\int \tan^{n-2}x\,dx$

(c) $\int x^n e^x\,dx = x^n e^x - n\int x^{n-1}e^x\,dx$

In Exercises 57 and 58, use the reduction formulas in Exercise 56 to evaluate the integrals.

57. (a) $\displaystyle\int \tan^4 x \, dx$ (b) $\displaystyle\int \sec^4 x \, dx$ (c) $\displaystyle\int x^3 e^x \, dx$

58. (a) $\displaystyle\int x^2 e^{3x} \, dx$ (b) $\displaystyle\int_0^1 x e^{-\sqrt{x}} \, dx$

[*Hint:* First make a substitution.]

59. Let f be a function whose second derivative is continuous on $[-1, 1]$. Show that

$$\int_{-1}^{1} x f''(x) \, dx = f'(1) + f'(-1) - f(1) + f(-1)$$

60. Recall from Theorem 4.1.4 and the discussion preceding it that if $f'(x) > 0$, then the function f is increasing and has an inverse function. The purpose of this problem is to show that if this condition is satisfied and if f' is continuous, then a definite integral of f^{-1} can be expressed in terms of a definite integral of f.

(a) Use integration by parts to show that

$$\int_a^b f(x) \, dx = bf(b) - af(a) - \int_a^b x f'(x) \, dx$$

(b) Use the result in part (a) to show that if $y = f(x)$, then

$$\int_a^b f(x) \, dx = bf(b) - af(a) - \int_{f(a)}^{f(b)} f^{-1}(y) \, dy$$

(c) Show that if we let $\alpha = f(a)$ and $\beta = f(b)$, then the result in part (b) can be written as

$$\int_\alpha^\beta f^{-1}(x) \, dx = \beta f^{-1}(\beta) - \alpha f^{-1}(\alpha) - \int_{f^{-1}(\alpha)}^{f^{-1}(\beta)} f(x) \, dx$$

61. In each part, use the result in Exercise 60 to obtain the equation, and then confirm that the equation is correct by performing the integrations.

(a) $\displaystyle\int_0^{1/2} \sin^{-1} x \, dx = \frac{1}{2} \sin^{-1}\left(\frac{1}{2}\right) - \int_0^{\pi/6} \sin x \, dx$

(b) $\displaystyle\int_e^{e^2} \ln x \, dx = (2e^2 - e) - \int_1^2 e^x \, dx$

62. (a) In the integral $\int x \cos x \, dx$, let

$$u = x, \qquad dv = \cos x \, dx,$$
$$du = dx, \qquad v = \sin x + C_1$$

Show that the constant C_1 cancels out, thus giving the same solution obtained by omitting C_1.

(b) Show that in general

$$uv - \int v \, du = u(v + C_1) - \int (v + C_1) \, du$$

thereby justifying the omission of the constant of integration when calculating v in integration by parts.

63. Evaluate $\int \ln(x + 1) \, dx$ using integration by parts. Simplify the computation of $\int v \, du$ by introducing a constant of integration $C_1 = 1$ when going from dv to v.

64. Evaluate $\int \ln(2x + 3) \, dx$ using integration by parts. Simplify the computation of $\int v \, du$ by introducing a constant of integration $C_1 = \frac{3}{2}$ when going from dv to v. Compare your solution with your answer to Exercise 13.

65. Evaluate $\int x \tan^{-1} x \, dx$ using integration by parts. Simplify the computation of $\int v \, du$ by introducing a constant of integration $C_1 = \frac{1}{2}$ when going from dv to v.

66. What equation results if integration by parts is applied to the integral

$$\int \frac{1}{x \ln x} \, dx$$

with the choices

$$u = \frac{1}{\ln x} \quad \text{and} \quad dv = \frac{1}{x} \, dx?$$

In what sense is this equation true? In what sense is it false?

8.3 TRIGONOMETRIC INTEGRALS

In the last section we derived reduction formulas for integrating positive integer powers of sine, cosine, tangent, and secant. In this section we will show how to work with those reduction formulas, and we will discuss methods for integrating other kinds of integrals that involve trigonometric functions.

INTEGRATING POWERS OF SINE AND COSINE

We begin by recalling two reduction formulas from the preceding section.

$$\int \sin^n x \, dx = -\frac{1}{n} \sin^{n-1} x \cos x + \frac{n-1}{n} \int \sin^{n-2} x \, dx \qquad (1)$$

$$\int \cos^n x \, dx = \frac{1}{n} \cos^{n-1} x \sin x + \frac{n-1}{n} \int \cos^{n-2} x \, dx \qquad (2)$$

In the case where $n = 2$, these formulas yield

$$\int \sin^2 x \, dx = -\frac{1}{2}\sin x \cos x + \frac{1}{2}\int dx = \frac{1}{2}x - \frac{1}{2}\sin x \cos x + C \tag{3}$$

$$\int \cos^2 x \, dx = \frac{1}{2}\cos x \sin x + \frac{1}{2}\int dx = \frac{1}{2}x + \frac{1}{2}\sin x \cos x + C \tag{4}$$

Alternative forms of these integration formulas can be derived from the trigonometric identities

$$\sin^2 x = \tfrac{1}{2}(1 - \cos 2x) \quad \text{and} \quad \cos^2 x = \tfrac{1}{2}(1 + \cos 2x) \tag{5–6}$$

which follow from the double-angle formulas

$$\cos 2x = 1 - 2\sin^2 x \quad \text{and} \quad \cos 2x = 2\cos^2 x - 1$$

These identities yield

$$\int \sin^2 x \, dx = \frac{1}{2}\int (1 - \cos 2x)\, dx = \frac{1}{2}x - \frac{1}{4}\sin 2x + C \tag{7}$$

$$\int \cos^2 x \, dx = \frac{1}{2}\int (1 + \cos 2x)\, dx = \frac{1}{2}x + \frac{1}{4}\sin 2x + C \tag{8}$$

Observe that the antiderivatives in Formulas (3) and (4) involve both sines and cosines, whereas those in (7) and (8) involve sines alone. However, the apparent discrepancy is easy to resolve by using the identity

$$\sin 2x = 2 \sin x \cos x$$

to rewrite (7) and (8) in forms (3) and (4), or conversely.

In the case where $n = 3$, the reduction formulas for integrating $\sin^3 x$ and $\cos^3 x$ yield

$$\int \sin^3 x \, dx = -\frac{1}{3}\sin^2 x \cos x + \frac{2}{3}\int \sin x \, dx = -\frac{1}{3}\sin^2 x \cos x - \frac{2}{3}\cos x + C \tag{9}$$

$$\int \cos^3 x \, dx = \frac{1}{3}\cos^2 x \sin x + \frac{2}{3}\int \cos x \, dx = \frac{1}{3}\cos^2 x \sin x + \frac{2}{3}\sin x + C \tag{10}$$

If desired, Formula (9) can be expressed in terms of cosines alone by using the identity $\sin^2 x = 1 - \cos^2 x$, and Formula (10) can be expressed in terms of sines alone by using the identity $\cos^2 x = 1 - \sin^2 x$. We leave it for you to do this and confirm that

$$\int \sin^3 x \, dx = \tfrac{1}{3}\cos^3 x - \cos x + C \tag{11}$$

$$\int \cos^3 x \, dx = \sin x - \tfrac{1}{3}\sin^3 x + C \tag{12}$$

FOR THE READER. When asked to integrate $\sin^3 x$ and $\cos^3 x$, the *Maple* CAS produces forms (11) and (12). However, the *Mathematica* CAS produces

$$\int \sin^3 x \, dx = -\tfrac{3}{4}\cos x + \tfrac{1}{12}\cos 3x + C$$

$$\int \cos^3 x \, dx = \tfrac{3}{4}\sin x + \tfrac{1}{12}\sin 3x + C$$

See if you can reconcile *Mathematica*'s results with (11) and (12).

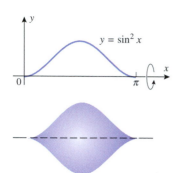

Figure 8.3.1

We leave it as an exercise to obtain the following formulas by first applying the reduction formulas, and then using appropriate trigonometric identities.

$$\int \sin^4 x \, dx = \tfrac{3}{8}x - \tfrac{1}{4}\sin 2x + \tfrac{1}{32}\sin 4x + C \tag{13}$$

$$\int \cos^4 x \, dx = \tfrac{3}{8}x + \tfrac{1}{4}\sin 2x + \tfrac{1}{32}\sin 4x + C \tag{14}$$

Example 1 Find the volume V of the solid that is obtained when the region under the curve $y = \sin^2 x$ over the interval $[0, \pi]$ is revolved about the x-axis (Figure 8.3.1).

Solution. Using the method of disks, Formula (5) of Section 7.2 yields

$$V = \int_0^\pi \pi \sin^4 x \, dx = \pi \left[\tfrac{3}{8}x - \tfrac{1}{4}\sin 2x + \tfrac{1}{32}\sin 4x \right]_0^\pi = \tfrac{3}{8}\pi^2 \qquad \blacktriangleleft$$

INTEGRATING PRODUCTS OF SINES AND COSINES

If m and n are positive integers, then the integral

$$\int \sin^m x \cos^n x \, dx$$

can be evaluated by one of the three procedures stated in Table 8.3.1, depending on whether m and n are odd or even.

Table 8.3.1

$\int \sin^m x \cos^n x \, dx$	PROCEDURE	RELEVANT IDENTITIES
n odd	• Split off a factor of $\cos x$. • Apply the relevant identity. • Make the substitution $u = \sin x$.	$\cos^2 x = 1 - \sin^2 x$
m odd	• Split off a factor of $\sin x$. • Apply the relevant identity. • Make the substitution $u = \cos x$.	$\sin^2 x = 1 - \cos^2 x$
$\left\{\begin{array}{l} m \text{ even} \\ n \text{ even} \end{array}\right.$	• Use the relevant identities to reduce the powers on $\sin x$ and $\cos x$.	$\left\{\begin{array}{l} \sin^2 x = \tfrac{1}{2}(1 - \cos 2x) \\ \cos^2 x = \tfrac{1}{2}(1 + \cos 2x) \end{array}\right.$

Example 2 Evaluate

(a) $\displaystyle\int \sin^4 x \cos^5 x \, dx$ (b) $\displaystyle\int \sin^4 x \cos^4 x \, dx$

Solution (a). Since $n = 5$ is odd, we will follow the first procedure in Table 8.3.1:

$$\int \sin^4 x \cos^5 x \, dx = \int \sin^4 x \cos^4 x \cos x \, dx$$

$$= \int \sin^4 x (1 - \sin^2 x)^2 \cos x \, dx$$

$$= \int u^4 (1 - u^2)^2 \, du$$

$$= \int (u^4 - 2u^6 + u^8) \, du$$

$$= \tfrac{1}{5}u^5 - \tfrac{2}{7}u^7 + \tfrac{1}{9}u^9 + C$$

$$= \tfrac{1}{5}\sin^5 x - \tfrac{2}{7}\sin^7 x + \tfrac{1}{9}\sin^9 x + C$$

Solution (b). Since $m = n = 4$, both exponents are even, so we will follow the third procedure in Table 8.3.1:

$$\int \sin^4 x \cos^4 x \, dx = \int (\sin^2 x)^2 (\cos^2 x)^2 \, dx$$

$$= \int \left(\frac{1}{2}[1 - \cos 2x] \right)^2 \left(\frac{1}{2}[1 + \cos 2x] \right)^2 dx$$

$$= \frac{1}{16} \int (1 - \cos^2 2x)^2 \, dx$$

$$= \frac{1}{16} \int \sin^4 2x \, dx \qquad \boxed{\begin{array}{l} \text{Note that this can be obtained more} \\ \text{directly from the original integral using} \\ \text{the identity } \sin x \cos x = \frac{1}{2} \sin 2x. \end{array}}$$

$$= \frac{1}{32} \int \sin^4 u \, du \qquad \boxed{\begin{array}{l} u = 2x \\ du = 2\,dx \text{ or } dx = \frac{1}{2}\,du \end{array}}$$

$$= \frac{1}{32} \left(\frac{3}{8}u - \frac{1}{4}\sin 2u + \frac{1}{32}\sin 4u \right) + C \qquad \boxed{\text{Formula (13)}}$$

$$= \frac{3}{128}x - \frac{1}{128}\sin 4x + \frac{1}{1024}\sin 8x + C \qquad \blacktriangleleft$$

Integrals of the form

$$\int \sin mx \cos nx \, dx, \quad \int \sin mx \sin nx \, dx, \quad \int \cos mx \cos nx \, dx \qquad (15)$$

can be found by using the trigonometric identities

$$\sin \alpha \cos \beta = \tfrac{1}{2}[\sin(\alpha - \beta) + \sin(\alpha + \beta)] \qquad (16)$$

$$\sin \alpha \sin \beta = \tfrac{1}{2}[\cos(\alpha - \beta) - \cos(\alpha + \beta)] \qquad (17)$$

$$\cos \alpha \cos \beta = \tfrac{1}{2}[\cos(\alpha - \beta) + \cos(\alpha + \beta)] \qquad (18)$$

to express the integrand as a sum or difference of sines and cosines.

Example 3 Evaluate $\int \sin 7x \cos 3x \, dx$.

Solution. Using (16) yields

$$\int \sin 7x \cos 3x \, dx = \frac{1}{2} \int (\sin 4x + \sin 10x) \, dx = -\frac{1}{8}\cos 4x - \frac{1}{20}\cos 10x + C \qquad \blacktriangleleft$$

INTEGRATING POWERS OF TANGENT AND SECANT

The procedures for integrating powers of tangent and secant closely parallel those for sine and cosine. The idea is to use the following reduction formulas (which were derived in Exercise 56 of Section 8.2) to reduce the exponent in the integrand until the resulting integral can be evaluated:

$$\int \tan^n x \, dx = \frac{\tan^{n-1} x}{n - 1} - \int \tan^{n-2} x \, dx \qquad (19)$$

$$\int \sec^n x \, dx = \frac{\sec^{n-2} x \tan x}{n - 1} + \frac{n - 2}{n - 1} \int \sec^{n-2} x \, dx \qquad (20)$$

In the case where n is odd, the exponent can be reduced to 1, leaving us with the problem of integrating $\tan x$ or $\sec x$. These integrals are given by

$$\int \tan x \, dx = \ln |\sec x| + C \tag{21}$$

$$\int \sec x \, dx = \ln |\sec x + \tan x| + C \tag{22}$$

Formula (21) can be obtained by writing

$$\int \tan x \, dx = \int \frac{\sin x}{\cos x} \, dx$$

$$= -\ln |\cos x| + C \qquad \boxed{\begin{array}{l} u = \cos x \\ du = -\sin x \, dx \end{array}}$$

$$= \ln |\sec x| + C \qquad \boxed{\ln |\cos x| = -\ln \dfrac{1}{|\cos x|}}$$

Formula (22) requires a trick. We write

$$\int \sec x \, dx = \int \sec x \left(\frac{\sec x + \tan x}{\sec x + \tan x} \right) dx = \int \frac{\sec^2 x + \sec x \tan x}{\sec x + \tan x} \, dx$$

$$= \ln |\sec x + \tan x| + C \qquad \boxed{\begin{array}{l} u = \sec x + \tan x \\ du = (\sec^2 x + \sec x \tan x) \, dx \end{array}}$$

The following basic integrals occur frequently and are worth noting:

$$\int \tan^2 x \, dx = \tan x - x + C \tag{23}$$

$$\int \sec^2 x \, dx = \tan x + C \tag{24}$$

Formula (24) is already known to us, since the derivative of $\tan x$ is $\sec^2 x$. Formula (23) can be obtained by applying reduction formula (19) with $n = 2$ (verify) or, alternatively, by using the identity

$$1 + \tan^2 x = \sec^2 x$$

to write

$$\int \tan^2 x \, dx = \int (\sec^2 x - 1) \, dx = \tan x - x + C$$

The formulas

$$\int \tan^3 x \, dx = \tfrac{1}{2} \tan^2 x - \ln |\sec x| + C \tag{25}$$

$$\int \sec^3 x \, dx = \tfrac{1}{2} \sec x \tan x + \tfrac{1}{2} \ln |\sec x + \tan x| + C \tag{26}$$

can be deduced from (21), (22), and reduction formulas (19) and (20) as follows:

$$\int \tan^3 x \, dx = \frac{1}{2} \tan^2 x - \int \tan x \, dx = \frac{1}{2} \tan^2 x - \ln |\sec x| + C$$

$$\int \sec^3 x \, dx = \frac{1}{2} \sec x \tan x + \frac{1}{2} \int \sec x \, dx = \frac{1}{2} \sec x \tan x + \frac{1}{2} \ln |\sec x + \tan x| + C$$

INTEGRATING PRODUCTS OF TANGENTS AND SECANTS

If m and n are positive integers, then the integral

$$\int \tan^m x \sec^n x \, dx$$

can be evaluated by one of the three procedures stated in Table 8.3.2, depending on whether m and n are odd or even.

Table 8.3.2

$\int \tan^m x \sec^n x \, dx$	PROCEDURE	RELEVANT IDENTITIES
n even	• Split off a factor of $\sec^2 x$. • Apply the relevant identity. • Make the substitution $u = \tan x$.	$\sec^2 x = \tan^2 x + 1$
m odd	• Split off a factor of $\sec x \tan x$. • Apply the relevant identity. • Make the substitution $u = \sec x$.	$\tan^2 x = \sec^2 x - 1$
$\begin{cases} m \text{ even} \\ n \text{ odd} \end{cases}$	• Use the relevant identities to reduce the integrand to powers of $\sec x$ alone. • Then use the reduction formula for powers of $\sec x$.	$\tan^2 x = \sec^2 x - 1$

Example 4 Evaluate

(a) $\displaystyle\int \tan^2 x \sec^4 x \, dx$ (b) $\displaystyle\int \tan^3 x \sec^3 x \, dx$ (c) $\displaystyle\int \tan^2 x \sec x \, dx$

Solution (a). Since $n = 4$ is even, we will follow the first procedure in Table 8.3.2:

$$\int \tan^2 x \sec^4 x \, dx = \int \tan^2 x \sec^2 x \sec^2 x \, dx$$

$$= \int \tan^2 x (\tan^2 x + 1) \sec^2 x \, dx$$

$$= \int u^2 (u^2 + 1) \, du$$

$$= \tfrac{1}{5} u^5 + \tfrac{1}{3} u^3 + C = \tfrac{1}{5} \tan^5 x + \tfrac{1}{3} \tan^3 x + C$$

Solution (b). Since $m = 3$ is odd, we will follow the second procedure in Table 8.3.2:

$$\int \tan^3 x \sec^3 x \, dx = \int \tan^2 x \sec^2 x (\sec x \tan x) \, dx$$

$$= \int (\sec^2 x - 1) \sec^2 x (\sec x \tan x) \, dx$$

$$= \int (u^2 - 1) u^2 \, du$$

$$= \tfrac{1}{5} u^5 - \tfrac{1}{3} u^3 + C = \tfrac{1}{5} \sec^5 x - \tfrac{1}{3} \sec^3 x + C$$

Solution (c). Since $m = 2$ is even and $n = 1$ is odd, we will follow the third procedure in Table 8.3.2:

$$\int \tan^2 x \sec x \, dx = \int (\sec^2 x - 1) \sec x \, dx$$

$$= \int \sec^3 x \, dx - \int \sec x \, dx \qquad \boxed{\text{See (26) and (22).}}$$

$$= \tfrac{1}{2} \sec x \tan x + \tfrac{1}{2} \ln |\sec x + \tan x| - \ln |\sec x + \tan x| + C$$

$$= \tfrac{1}{2} \sec x \tan x - \tfrac{1}{2} \ln |\sec x + \tan x| + C$$

◀

AN ALTERNATIVE METHOD FOR INTEGRATING POWERS OF SINE, COSINE, TANGENT, AND SECANT

The methods in Tables 8.3.1 and 8.3.2 can sometimes be applied if $m = 0$ or $n = 0$ to integrate positive integer powers of sine, cosine, tangent, and secant without reduction formulas. For example, instead of using the reduction formula to integrate $\sin^3 x$, we can apply the second procedure in Table 8.3.1.

$$\int \sin^3 x\, dx = \int (\sin^2 x)\sin x\, dx$$

$$= \int (1 - \cos^2 x)\sin x\, dx \qquad \begin{array}{l} u = \cos x \\ du = -\sin x\, dx \end{array}$$

$$= -\int (1 - u^2)\, du$$

$$= \tfrac{1}{3}u^3 - u + C = \tfrac{1}{3}\cos^3 x - \cos x + C$$

which agrees with (11).

REMARK.   With the aid of the identity $1 + \cot^2 x = \csc^2 x$ the techniques in Table 8.3.2 can be adapted to treat integrals of the form

$$\int \cot^m x \csc^n x\, dx$$

Also, there are reduction formulas for powers of cosecant and cotangent that are analogous to Formulas (19) and (20).

MERCATOR'S MAP OF THE WORLD

The integral of $\sec x$ plays an important role in the design of navigational maps for charting nautical and aeronautical courses. Sailors and pilots usually chart their courses along paths with constant compass headings; for example, the course might be $30°$ northeast or $135°$ southwest. Except for courses that are parallel to the equator or run due north or south, a course with constant compass heading spirals around the Earth toward one of the poles (as in Figure 8.3.2a). In 1569 the Flemish mathematician and geographer Gerhard Kramer (1512–1594) (better known by the Latin name Mercator) devised a world map, called the *Mercator projection*, in which spirals of constant compass headings appear as straight lines. This was extremely important because it enabled sailors to determine compass headings between two points by connecting them with a straight line on a map (Figure 8.3.2b).

A flight with constant compass heading from New York City to Moscow as it appears on a globe

(a)

A flight with constant compass heading from New York City to Moscow as it appears on a Mercator projection

(b)

Figure 8.3.2

If the Earth is assumed to be a sphere of radius 4000 mi, then the lines of latitude at $1°$ increments are equally spaced about 70 mi apart (why?). However, in the Mercator projection, the lines of latitude become wider apart toward the poles, so that two widely spaced latitude lines near the poles may be actually the same distance apart on the Earth as two closely spaced latitude lines near the equator. It can be proved that on a Mercator map in which the equatorial line has length L, the vertical distance D_β on the map between the equator (latitude $0°$) and the line of latitude $\beta°$ is

$$D_\beta = \frac{L}{2\pi} \int_0^{\beta\pi/180} \sec x \, dx \qquad (27)$$

(see Exercises 59 and 60).

EXERCISE SET 8.3

In Exercises 1–52, evaluate the integral.

1. $\displaystyle\int \cos^5 x \sin x \, dx$

2. $\displaystyle\int \sin^4 3x \cos 3x \, dx$

3. $\displaystyle\int \sin ax \cos ax \, dx$

4. $\displaystyle\int \cos^2 3x \, dx$

5. $\displaystyle\int \sin^2 5\theta \, d\theta$

6. $\displaystyle\int \cos^3 at \, dt$

7. $\displaystyle\int \cos^5 \theta \, d\theta$

8. $\displaystyle\int \sin^3 x \cos^3 x \, dx$

9. $\displaystyle\int \sin^2 2t \cos^3 2t \, dt$

10. $\displaystyle\int \sin^3 2x \cos^2 2x \, dx$

11. $\displaystyle\int \sin^2 x \cos^2 x \, dx$

12. $\displaystyle\int \sin^2 x \cos^4 x \, dx$

13. $\displaystyle\int \sin x \cos 2x \, dx$

14. $\displaystyle\int \sin 3\theta \cos 2\theta \, d\theta$

15. $\displaystyle\int \sin x \cos(x/2) \, dx$

16. $\displaystyle\int \cos^{1/5} x \sin x \, dx$

17. $\displaystyle\int_0^{\pi/4} \cos^3 x \, dx$

18. $\displaystyle\int_0^{\pi/2} \sin^2 \frac{x}{2} \cos^2 \frac{x}{2} \, dx$

19. $\displaystyle\int_0^{\pi/3} \sin^4 3x \cos^3 3x \, dx$

20. $\displaystyle\int_{-\pi}^{\pi} \cos^2 5\theta \, d\theta$

21. $\displaystyle\int_0^{\pi/6} \sin 2x \cos 4x \, dx$

22. $\displaystyle\int_0^{2\pi} \sin^2 kx \, dx$

23. $\displaystyle\int \sec^2(3x+1) \, dx$

24. $\displaystyle\int \tan 5x \, dx$

25. $\displaystyle\int e^{-2x} \tan(e^{-2x}) \, dx$

26. $\displaystyle\int \cot 3x \, dx$

27. $\displaystyle\int \sec 2x \, dx$

28. $\displaystyle\int \frac{\sec(\sqrt{x})}{\sqrt{x}} \, dx$

29. $\displaystyle\int \tan^2 x \sec^2 x \, dx$

30. $\displaystyle\int \tan^5 x \sec^4 x \, dx$

31. $\displaystyle\int \tan^3 4x \sec^4 4x \, dx$

32. $\displaystyle\int \tan^4 \theta \sec^4 \theta \, d\theta$

33. $\displaystyle\int \sec^5 x \tan^3 x \, dx$

34. $\displaystyle\int \tan^5 \theta \sec \theta \, d\theta$

35. $\displaystyle\int \tan^4 x \sec x \, dx$

36. $\displaystyle\int \tan^2 \frac{x}{2} \sec^3 \frac{x}{2} \, dx$

37. $\displaystyle\int \tan 2t \sec^3 2t \, dt$

38. $\displaystyle\int \tan x \sec^5 x \, dx$

39. $\displaystyle\int \sec^4 x \, dx$

40. $\displaystyle\int \sec^5 x \, dx$

41. $\displaystyle\int \tan^4 x \, dx$

42. $\displaystyle\int \tan^3 4x \, dx$

43. $\displaystyle\int \sqrt{\tan x} \sec^4 x \, dx$

44. $\displaystyle\int \tan x \sec^{3/2} x \, dx$

45. $\displaystyle\int_0^{\pi/6} \tan^2 2x \, dx$

46. $\displaystyle\int_0^{\pi/6} \sec^3 \theta \tan \theta \, d\theta$

47. $\displaystyle\int_0^{\pi/2} \tan^5 \frac{x}{2} \, dx$

48. $\displaystyle\int_0^{1/4} \sec \pi x \tan \pi x \, dx$

49. $\displaystyle\int \cot^3 x \csc^3 x \, dx$

50. $\displaystyle\int \cot^2 3t \sec 3t \, dt$

51. $\displaystyle\int \cot^3 x \, dx$

52. $\displaystyle\int \csc^4 x \, dx$

53. Let m, n be distinct nonnegative integers. Use Formulas (16)–(18) to prove:

(a) $\displaystyle\int_0^{2\pi} \sin mx \cos nx \, dx = 0$

(b) $\displaystyle\int_0^{2\pi} \cos mx \cos nx \, dx = 0$

(c) $\displaystyle\int_0^{2\pi} \sin mx \sin nx \, dx = 0.$

54. Evaluate the integrals in Exercise 53 when m and n denote the *same* nonnegative integer.

55. Find the arc length of the curve $y = \ln(\cos x)$ over the interval $[0, \pi/4]$.

56. Find the volume of the solid generated when the region enclosed by $y = \tan x$, $y = 1$, and $x = 0$ is revolved about the x-axis.

57. Find the volume of the solid that results when the region enclosed by $y = \cos x$, $y = \sin x$, $x = 0$, and $x = \pi/4$ is revolved about the x-axis.

58. The region bounded below by the x-axis and above by the portion of $y = \sin x$ from $x = 0$ to $x = \pi$ is revolved about the x-axis. Find the volume of the resulting solid.

59. Use Formula (27) to show that if the length of the equatorial line on a Mercator projection is L, then the vertical distance D between the latitude lines at $\alpha°$ and $\beta°$ on the same side of the equator (where $\alpha < \beta$) is

$$D = \frac{L}{2\pi} \ln \left| \frac{\sec \beta° + \tan \beta°}{\sec \alpha° + \tan \alpha°} \right|$$

60. Suppose that the equator has a length of 100 cm on a Mercator projection. In each part, use the result in Exercise 59 to answer the question.
  (a) What is the vertical distance on the map between the equator and the line at $25°$ north latitude?
  (b) What is the vertical distance on the map between New Orleans, Louisiana, at $30°$ north latitude and Winnepeg, Canada, at $50°$ north latitude?

61. (a) Show that

$$\int \csc x \, dx = -\ln |\csc x + \cot x| + C$$

  (b) Show that the result in part (a) can also be written as

$$\int \csc x \, dx = \ln |\csc x - \cot x| + C$$

and

$$\int \csc x \, dx = \ln \left| \tan \tfrac{1}{2} x \right| + C$$

62. Rewrite $\sin x + \cos x$ in the form

$$A \sin(x + \phi)$$

and use your result together with Exercise 61 to evaluate

$$\int \frac{dx}{\sin x + \cos x}$$

63. Use the method of Exercise 62 to evaluate

$$\int \frac{dx}{a \sin x + b \cos x} \quad (a, b \text{ not both zero})$$

64. (a) Use Formula (8) in Section 8.2 to show that

$$\int_0^{\pi/2} \sin^n x \, dx = \frac{n-1}{n} \int_0^{\pi/2} \sin^{n-2} x \, dx \quad (n \geq 2)$$

  (b) Use this result to derive the **Wallis sine formulas**:

$$\int_0^{\pi/2} \sin^n x \, dx = \frac{\pi}{2} \cdot \frac{1 \cdot 3 \cdot 5 \cdots (n-1)}{2 \cdot 4 \cdot 6 \cdots n} \quad \left(\begin{array}{c} n \text{ even} \\ \text{and} \geq 2 \end{array} \right)$$

$$\int_0^{\pi/2} \sin^n x \, dx = \frac{2 \cdot 4 \cdot 6 \cdots (n-1)}{3 \cdot 5 \cdot 7 \cdots n} \quad \left(\begin{array}{c} n \text{ odd} \\ \text{and} \geq 3 \end{array} \right)$$

65. Use the Wallis formulas in Exercise 64 to evaluate
  (a) $\displaystyle\int_0^{\pi/2} \sin^3 x \, dx$ (b) $\displaystyle\int_0^{\pi/2} \sin^4 x \, dx$
  (c) $\displaystyle\int_0^{\pi/2} \sin^5 x \, dx$ (d) $\displaystyle\int_0^{\pi/2} \sin^6 x \, dx$.

66. Use Formula (9) in Section 8.2 and the method of Exercise 64 to derive the **Wallis cosine formulas**:

$$\int_0^{\pi/2} \cos^n x \, dx = \frac{\pi}{2} \cdot \frac{1 \cdot 3 \cdot 5 \cdots (n-1)}{2 \cdot 4 \cdot 6 \cdots n} \quad \left(\begin{array}{c} n \text{ even} \\ \text{and} \geq 2 \end{array} \right)$$

$$\int_0^{\pi/2} \cos^n x \, dx = \frac{2 \cdot 4 \cdot 6 \cdots (n-1)}{3 \cdot 5 \cdot 7 \cdots n} \quad \left(\begin{array}{c} n \text{ odd} \\ \text{and} \geq 3 \end{array} \right)$$

8.4 TRIGONOMETRIC SUBSTITUTIONS

In this section we will discuss a method for evaluating integrals containing radicals by making substitutions involving trigonometric functions. We will also show how integrals containing quadratic polynomials can sometimes be evaluated by completing the square.

THE METHOD OF TRIGONOMETRIC SUBSTITUTION

To start, we will be concerned with integrals that contain expressions of the form

$$\sqrt{a^2 - x^2}, \quad \sqrt{x^2 + a^2}, \quad \sqrt{x^2 - a^2}$$

in which a is a positive constant. The basic idea for evaluating such integrals is to make a substitution for x that will eliminate the radical. For example, to eliminate the radical in the expression $\sqrt{a^2 - x^2}$, we can make the substitution

$$x = a \sin \theta, \quad -\pi/2 \leq \theta \leq \pi/2 \tag{1}$$

which yields

$$\sqrt{a^2 - x^2} = \sqrt{a^2 - a^2 \sin^2 \theta} = \sqrt{a^2(1 - \sin^2 \theta)}$$

$$= a\sqrt{\cos^2 \theta} = a|\cos \theta| = a \cos \theta \qquad \boxed{\cos \theta \geq 0 \text{ since } -\pi/2 \leq \theta \leq \pi/2}$$

The restriction on θ in (1) serves two purposes—it enables us to replace $|\cos \theta|$ by $\cos \theta$ to simplify the calculations, and it also ensures that the substitutions can be rewritten as $\theta = \sin^{-1}(x/a)$, if needed.

Example 1   Evaluate $\displaystyle\int \frac{dx}{x^2\sqrt{4 - x^2}}$.

Solution.   To eliminate the radical we make the substitution

$$x = 2\sin\theta, \quad dx = 2\cos\theta\, d\theta$$

This yields

$$\int \frac{dx}{x^2\sqrt{4 - x^2}} = \int \frac{2\cos\theta\, d\theta}{(2\sin\theta)^2\sqrt{4 - 4\sin^2\theta}}$$

$$= \int \frac{2\cos\theta\, d\theta}{(2\sin\theta)^2(2\cos\theta)} = \frac{1}{4}\int \frac{d\theta}{\sin^2\theta}$$

$$= \frac{1}{4}\int \csc^2\theta\, d\theta = -\frac{1}{4}\cot\theta + C \tag{2}$$

At this point we have completed the integration; however, because the original integral was expressed in terms of x, it is desirable to express $\cot\theta$ in terms of x as well. This can be done using trigonometric identities, but the expression can also be obtained by writing the substitution $x = 2\sin\theta$ as $\sin\theta = x/2$ and representing it geometrically as in Figure 8.4.1. From that figure we obtain

$$\cot\theta = \frac{\sqrt{4 - x^2}}{x}$$

Substituting this in (2) yields

$$\int \frac{dx}{x^2\sqrt{4 - x^2}} = -\frac{1}{4}\frac{\sqrt{4 - x^2}}{x} + C \qquad \blacktriangleleft$$

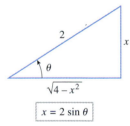

$$x = 2\sin\theta$$

Figure 8.4.1

Example 2   Evaluate $\displaystyle\int_1^{\sqrt{2}} \frac{dx}{x^2\sqrt{4 - x^2}}$.

Solution.   There are two possible approaches: we can make the substitution in the indefinite integral (as in Example 1) and then evaluate the definite integral using the x-limits of integration, or we can make the substitution in the definite integral and convert the x-limits to the corresponding θ-limits.

Method 1.   Using the result from Example 1 with the x-limits of integration yields

$$\int_1^{\sqrt{2}} \frac{dx}{x^2\sqrt{4 - x^2}} = -\frac{1}{4}\left[\frac{\sqrt{4 - x^2}}{x}\right]_1^{\sqrt{2}} = -\frac{1}{4}[1 - \sqrt{3}] = \frac{\sqrt{3} - 1}{4}$$

Method 2.   The substitution $x = 2\sin\theta$ can be expressed as $x/2 = \sin\theta$ or $\theta = \sin^{-1}(x/2)$, so the θ-limits that correspond to $x = 1$ and $x = \sqrt{2}$ are

$$x = 1: \quad \theta = \sin^{-1}(1/2) = \pi/6$$

$$x = \sqrt{2}: \quad \theta = \sin^{-1}(\sqrt{2}/2) = \pi/4$$

Thus, from (2) in Example 1 we obtain

$$\int_1^{\sqrt{2}} \frac{dx}{x^2\sqrt{4-x^2}} = -\frac{1}{4}\Big[\cot\theta\Big]_{\pi/6}^{\pi/4} = -\frac{1}{4}[1-\sqrt{3}\,] = \frac{\sqrt{3}-1}{4}$$ ◄

Example 3   Find the area of the ellipse

$$\frac{x^2}{a^2} + \frac{y^2}{b^2} = 1$$

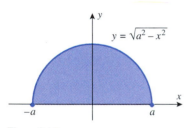

$$\frac{x^2}{a^2} + \frac{y^2}{b^2} = 1$$

Figure 8.4.2

Solution.   Because the ellipse is symmetric about both axes, its area A is four times the area in the first quadrant (Figure 8.4.2). If we solve the equation of the ellipse for y in terms of x, we obtain

$$y = \pm\frac{b}{a}\sqrt{a^2-x^2}$$

where the positive square root gives the equation of the upper half. Thus, the area A is given by

$$A = 4\int_0^a \frac{b}{a}\sqrt{a^2-x^2}\,dx = \frac{4b}{a}\int_0^a \sqrt{a^2-x^2}\,dx$$

To evaluate this integral, we will make the substitution $x = a\sin\theta$ $(dx = a\cos\theta\,d\theta)$ and convert the x-limits of integration to θ-limits. Since the substitution can be expressed as $\theta = \sin^{-1}(x/a)$, the θ-limits of integration are

$$x = 0:\quad \theta = \sin^{-1}(0) = 0$$

$$x = a:\quad \theta = \sin^{-1}(1) = \pi/2$$

Thus, we obtain

$$A = \frac{4b}{a}\int_0^a \sqrt{a^2-x^2}\,dx = \frac{4b}{a}\int_0^{\pi/2} a\cos\theta \cdot a\cos\theta\,d\theta$$

$$= 4ab\int_0^{\pi/2}\cos^2\theta\,d\theta = 4ab\int_0^{\pi/2}\frac{1}{2}(1+\cos 2\theta)\,d\theta$$

$$= 2ab\Big[\theta + \frac{1}{2}\sin 2\theta\Big]_0^{\pi/2} = 2ab\Big[\frac{\pi}{2} - 0\Big] = \pi ab$$ ◄

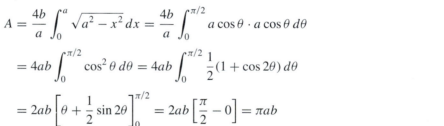

$$y = \sqrt{a^2 - x^2}$$

Figure 8.4.3

REMARK.   In the special case where $a = b$, the ellipse becomes a circle of radius a, and the area formula becomes $A = \pi a^2$, as expected. It is worth noting that

$$\int_{-a}^a \sqrt{a^2-x^2}\,dx = \tfrac{1}{2}\pi a^2 \tag{3}$$

since this integral represents the area of the upper semicircle (Figure 8.4.3).

FOR THE READER.   If you have a calculating utility with a numerical integration capability, use it and Formula (3) to approximate π to three decimal places.

Thus far, we have focused on using the substitution $x = a\sin\theta$ to evaluate integrals involving radicals of the form $\sqrt{a^2-x^2}$. Table 8.4.1 summarizes this method and describes some other substitutions of this type.

Example 4   Find the arc length of the curve $y = x^2/2$ from $x = 0$ to $x = 1$ (Figure 8.4.4).

Solution.   From Formula (4) of Section 7.4 the arc length L of the curve is

$$L = \int_0^1 \sqrt{1 + \Big(\frac{dy}{dx}\Big)^2}\,dx = \int_0^1 \sqrt{1+x^2}\,dx$$

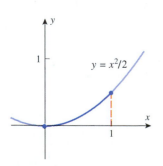

$$y = x^2/2$$

Figure 8.4.4

The integrand involves a radical of the form $\sqrt{a^2+x^2}$ with $a = 1$, so from Table 8.4.1 we

Table 8.4.1

EXPRESSION IN THE INTEGRAND	SUBSTITUTION	RESTRICTION ON θ	SIMPLIFICATION
$\sqrt{a^2 - x^2}$	$x = a \sin\theta$	$-\pi/2 \le \theta \le \pi/2$	$a^2 - x^2 = a^2 - a^2 \sin^2\theta = a^2 \cos^2\theta$
$\sqrt{a^2 + x^2}$	$x = a \tan\theta$	$-\pi/2 < \theta < \pi/2$	$a^2 + x^2 = a^2 + a^2 \tan^2\theta = a^2 \sec^2\theta$
$\sqrt{x^2 - a^2}$	$x = a \sec\theta$	$\begin{cases} 0 \le \theta < \pi/2 & (\text{if } x \ge a) \\ \pi/2 < \theta \le \pi & (\text{if } x \le -a) \end{cases}$	$x^2 - a^2 = a^2 \sec^2\theta - a^2 = a^2 \tan^2\theta$

make the substitution

$$x = \tan\theta, \quad -\pi/2 < \theta < \pi/2$$

$$\frac{dx}{d\theta} = \sec^2\theta \quad \text{or} \quad dx = \sec^2\theta \, d\theta$$

Since this substitution can be expressed as $\theta = \tan^{-1} x$, the θ-limits of integration that correspond to the x-limits, $x = 0$ and $x = 1$, are

$$x = 0: \quad \theta = \tan^{-1} 0 = 0$$
$$x = 1: \quad \theta = \tan^{-1} 1 = \pi/4$$

Thus,

$$L = \int_0^1 \sqrt{1 + x^2}\, dx = \int_0^{\pi/4} \sqrt{1 + \tan^2\theta}\, \sec^2\theta\, d\theta$$

$$= \int_0^{\pi/4} \sqrt{\sec^2\theta}\, \sec^2\theta\, d\theta$$

$$= \int_0^{\pi/4} |\sec\theta|\sec^2\theta\, d\theta$$

$$= \int_0^{\pi/4} \sec^3\theta\, d\theta \qquad \boxed{\sec\theta > 0 \text{ since } -\pi/2 < \theta < \pi/2}$$

$$= \left[\tfrac{1}{2}\sec\theta\tan\theta + \tfrac{1}{2}\ln|\sec\theta + \tan\theta| \right]_0^{\pi/4} \qquad \boxed{\begin{array}{l}\text{Formula (26)}\\\text{of Section 8.3}\end{array}}$$

$$= \tfrac{1}{2}[\sqrt{2} + \ln(\sqrt{2} + 1)] \approx 1.148 \qquad \blacktriangleleft$$

Example 5 Evaluate $\displaystyle\int \frac{\sqrt{x^2 - 25}}{x}\, dx$, assuming that $x \ge 5$.

Solution. The integrand involves a radical of the form $\sqrt{x^2 - a^2}$ with $a = 5$, so from Table 8.4.1 we make the substitution

$$x = 5\sec\theta, \quad 0 \le \theta < \pi/2$$

$$\frac{dx}{d\theta} = 5\sec\theta\tan\theta \quad \text{or} \quad dx = 5\sec\theta\tan\theta\, d\theta$$

Thus,

$$\int \frac{\sqrt{x^2 - 25}}{x}\, dx = \int \frac{\sqrt{25\sec^2\theta - 25}}{5\sec\theta}(5\sec\theta\tan\theta)\, d\theta$$

$$= \int \frac{5|\tan\theta|}{5\sec\theta}(5\sec\theta\tan\theta)\, d\theta$$

$$= 5\int \tan^2\theta\, d\theta \qquad \boxed{\begin{array}{l}\tan\theta \ge 0 \text{ since}\\0 \le \theta < \pi/2\end{array}}$$

$$= 5\int (\sec^2\theta - 1)\, d\theta = 5\tan\theta - 5\theta + C$$

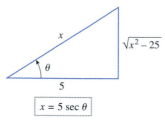

$$x = 5 \sec \theta$$

Figure 8.4.5

To express the solution in terms of x, we will represent the substitution $x = 5 \sec \theta$ geometrically by the triangle in Figure 8.4.5, from which we obtain

$$\tan \theta = \frac{\sqrt{x^2 - 25}}{5}$$

From this and the fact that the substitution can be expressed as $\theta = \sec^{-1}(x/5)$, we obtain

$$\int \frac{\sqrt{x^2 - 25}}{x}\, dx = \sqrt{x^2 - 25} - 5 \sec^{-1}\left(\frac{x}{5}\right) + C \qquad \blacktriangleleft$$

• •

INTEGRALS INVOLVING
$ax^2 + bx + c$

Integrals that involve a quadratic expression $ax^2 + bx + c$, where $a \neq 0$ and $b \neq 0$, can often be evaluated by first completing the square, then making an appropriate substitution. The following examples illustrate this idea.

Example 6   Evaluate $\displaystyle\int \frac{x}{x^2 - 4x + 8}\, dx$.

Solution.   Completing the square yields

$$x^2 - 4x + 8 = (x^2 - 4x + 4) + 8 - 4 = (x - 2)^2 + 4$$

Thus, the substitution

$$u = x - 2, \quad du = dx$$

yields

$$\int \frac{x}{x^2 - 4x + 8}\, dx = \int \frac{x}{(x - 2)^2 + 4}\, dx = \int \frac{u + 2}{u^2 + 4}\, du$$

$$= \int \frac{u}{u^2 + 4}\, du + 2 \int \frac{du}{u^2 + 4}$$

$$= \frac{1}{2} \int \frac{2u}{u^2 + 4}\, du + 2 \int \frac{du}{u^2 + 4}$$

$$= \frac{1}{2} \ln(u^2 + 4) + 2 \left(\frac{1}{2}\right) \tan^{-1} \frac{u}{2} + C$$

$$= \frac{1}{2} \ln[(x - 2)^2 + 4] + \tan^{-1}\left(\frac{x - 2}{2}\right) + C \qquad \blacktriangleleft$$

Example 7   Evaluate $\displaystyle\int \frac{dx}{\sqrt{5 - 4x - 2x^2}}$.

Solution.   Completing the square yields

$$5 - 4x - 2x^2 = 5 - 2(x^2 + 2x) = 5 - 2(x^2 + 2x + 1) + 2$$
$$= 5 - 2(x + 1)^2 + 2 = 7 - 2(x + 1)^2$$

Thus,

$$\int \frac{dx}{\sqrt{5 - 4x - 2x^2}} = \int \frac{dx}{\sqrt{7 - 2(x + 1)^2}}$$

$$= \int \frac{du}{\sqrt{7 - 2u^2}} \qquad \boxed{\begin{array}{l} u = x + 1 \\ du = dx \end{array}}$$

$$= \frac{1}{\sqrt{2}} \int \frac{du}{\sqrt{(7/2) - u^2}}$$

$$= \frac{1}{\sqrt{2}} \sin^{-1}\left(\frac{u}{\sqrt{7/2}}\right) + C \qquad \boxed{\begin{array}{l} \text{Formula 21, Section 8.1 with} \\ a = \sqrt{7/2} \end{array}}$$

$$= \frac{1}{\sqrt{2}} \sin^{-1}(\sqrt{2/7}(x + 1)) + C \qquad \blacktriangleleft$$

EXERCISE SET 8.4 [C] CAS

In Exercises 1–26, evaluate the integral.

1. $\displaystyle\int \sqrt{4 - x^2}\, dx$

2. $\displaystyle\int \sqrt{1 - 4x^2}\, dx$

3. $\displaystyle\int \frac{x^2}{\sqrt{9 - x^2}}\, dx$

4. $\displaystyle\int \frac{dx}{x^2\sqrt{16 - x^2}}$

5. $\displaystyle\int \frac{dx}{(4 + x^2)^2}$

6. $\displaystyle\int \frac{x^2}{\sqrt{5 + x^2}}\, dx$

7. $\displaystyle\int \frac{\sqrt{x^2 - 9}}{x}\, dx$

8. $\displaystyle\int \frac{dx}{x^2\sqrt{x^2 - 16}}$

9. $\displaystyle\int \frac{x^3}{\sqrt{2 - x^2}}\, dx$

10. $\displaystyle\int x^3\sqrt{5 - x^2}\, dx$

11. $\displaystyle\int \frac{dx}{x^2\sqrt{4x^2 - 9}}$

12. $\displaystyle\int \frac{\sqrt{1 + t^2}}{t}\, dt$

13. $\displaystyle\int \frac{dx}{(1 - x^2)^{3/2}}$

14. $\displaystyle\int \frac{dx}{x^2\sqrt{x^2 + 25}}$

15. $\displaystyle\int \frac{dx}{\sqrt{x^2 - 1}}$

16. $\displaystyle\int \frac{dx}{1 + 2x^2 + x^4}$

17. $\displaystyle\int \frac{dx}{(9x^2 - 1)^{3/2}}$

18. $\displaystyle\int \frac{x^2}{\sqrt{x^2 - 25}}\, dx$

19. $\displaystyle\int e^x\sqrt{1 - e^{2x}}\, dx$

20. $\displaystyle\int \frac{\cos\theta}{\sqrt{2 - \sin^2\theta}}\, d\theta$

21. $\displaystyle\int_0^4 x^3\sqrt{16 - x^2}\, dx$

22. $\displaystyle\int_0^{1/3} \frac{dx}{(4 - 9x^2)^2}$

23. $\displaystyle\int_{\sqrt{2}}^2 \frac{dx}{x^2\sqrt{x^2 - 1}}$

24. $\displaystyle\int_{\sqrt{2}}^2 \frac{\sqrt{2x^2 - 4}}{x}\, dx$

25. $\displaystyle\int_1^3 \frac{dx}{x^4\sqrt{x^2 + 3}}$

26. $\displaystyle\int_0^3 \frac{x^3}{(3 + x^2)^{5/2}}\, dx$

27. The integral

$$\int \frac{x}{x^2 + 4}\, dx$$

can be evaluated either by a trigonometric substitution or by the substitution $u = x^2 + 4$. Do it both ways and show that the results are equivalent.

28. The integral

$$\int \frac{x^2}{x^2 + 4}\, dx$$

can be evaluated either by a trigonometric substitution or by algebraically rewriting the numerator of the integrand as $(x^2 + 4) - 4$. Do it both ways and show that the results are equivalent.

29. Find the arc length of the curve $y = \ln x$ from $x = 1$ to $x = 2$.

30. Find the arc length of the curve $y = x^2$ from $x = 0$ to $x = 1$.

31. Find the area of the surface generated when the curve in Exercise 30 is revolved about the x-axis.

32. Find the volume of the solid generated when the region enclosed by $x = y(1 - y^2)^{1/4}$, $y = 0$, $y = 1$, and $x = 0$ is revolved about the y-axis.

In Exercise 33, the trigonometric substitutions $x = a\sec\theta$ and $x = a\tan\theta$ lead to difficult integrals; for such integrals it is sometimes possible to use the **hyperbolic substitutions**

$$x = a\sinh u \text{ for integrals involving } \sqrt{x^2 + a^2}$$

$$x = a\cosh u \text{ for integrals involving } \sqrt{x^2 - a^2},\, x \geq a$$

These substitutions are useful because in each case the hyperbolic identity

$$a^2\cosh^2 u - a^2\sinh^2 u = a^2$$

removes the radical.

33. (a) Evaluate

$$\int \frac{dx}{\sqrt{x^2 + 9}}$$

using the hyperbolic substitution that is suggested above.

(b) Evaluate the integral in part (a) by a trigonometric substitution and show that the results in parts (a) and (b) agree.

(c) Use a hyperbolic substitution to evaluate

$$\int \sqrt{x^2 - 1}\, dx, \quad x \geq 1$$

34. In Example 3 we found the area of an ellipse by making the substitution $x = a\sin\theta$ in the required integral. Find the area by making the substitution $x = a\cos\theta$, and discuss any restrictions on θ that are needed.

In Exercises 35–46, evaluate the integral.

35. $\displaystyle\int \frac{dx}{x^2 - 4x + 13}$

36. $\displaystyle\int \frac{dx}{\sqrt{2x - x^2}}$

37. $\displaystyle\int \frac{dx}{\sqrt{8 + 2x - x^2}}$

38. $\displaystyle\int \frac{dx}{16x^2 + 16x + 5}$

39. $\displaystyle\int \frac{dx}{\sqrt{x^2 - 6x + 10}}$

40. $\displaystyle\int \frac{x}{x^2 + 6x + 10}\, dx$

41. $\displaystyle\int \sqrt{3 - 2x - x^2}\, dx$

42. $\displaystyle\int \frac{e^x}{\sqrt{1 + e^x + e^{2x}}}\, dx$

43. $\displaystyle\int \frac{dx}{2x^2 + 4x + 7}$

44. $\displaystyle\int \frac{2x + 3}{4x^2 + 4x + 5}\, dx$

45. $\displaystyle\int_1^2 \frac{dx}{\sqrt{4x - x^2}}$

46. $\displaystyle\int_0^1 \sqrt{x(4 - x)}\, dx$

In Exercises 47 and 48, there is a good chance that your CAS will not be able to evaluate the integral as stated. If this is so, make a substitution that converts the integral into one that your CAS can evaluate.

C **47.** $\displaystyle\int \cos x \sin x \sqrt{1 - \sin^4 x} \, dx$

C **48.** $\displaystyle\int (x \cos x + \sin x) \sqrt{1 + x^2 \sin^2 x} \, dx$

8.5 INTEGRATING RATIONAL FUNCTIONS BY PARTIAL FRACTIONS

Recall that a rational function is a ratio of two polynomials. In this section we will give a general method for integrating rational functions that is based on the idea of decomposing a rational function into a sum of simple rational functions that can be integrated by the methods studied in earlier sections.

PARTIAL FRACTIONS

In algebra one learns to combine two or more fractions into a single fraction by finding a common denominator. For example,

$$\frac{2}{x - 4} + \frac{3}{x + 1} = \frac{2(x + 1) + 3(x - 4)}{(x - 4)(x + 1)} = \frac{5x - 10}{x^2 - 3x - 4} \tag{1}$$

However, for purposes of integration, the left side of (1) is preferable to the right side since each of the terms is easy to integrate:

$$\int \frac{5x - 10}{x^2 - 3x - 4} \, dx = \int \frac{2}{x - 4} \, dx + \int \frac{3}{x + 1} \, dx = 2 \ln |x - 4| + 3 \ln |x + 1| + C$$

Thus, it is desirable to have some method that will enable us to obtain the left side of (1), starting with the right side. To illustrate how this can be done, we begin by noting that on the left side the numerators are constants and the denominators are the factors of the denominator on the right side. Thus, to find the left side of (1), starting from the right side, we could factor the denominator of the right side and look for constants A and B such that

$$\frac{5x - 10}{(x - 4)(x + 1)} = \frac{A}{x - 4} + \frac{B}{x + 1} \tag{2}$$

One way to find the constants A and B is to multiply (2) through by $(x - 4)(x + 1)$ to clear fractions. This yields

$$5x - 10 = A(x + 1) + B(x - 4) \tag{3}$$

This relationship holds for all x, so it holds in particular if $x = 4$ or $x = -1$. Substituting $x = 4$ in (3) makes the second term on the right drop out and yields the equation $10 = 5A$ or $A = 2$; and substituting $x = -1$ in (3) makes the first term on the right drop out and yields the equation $-15 = -5B$ or $B = 3$. Substituting these values in (2) we obtain

$$\frac{5x - 10}{(x - 4)(x + 1)} = \frac{2}{x - 4} + \frac{3}{x + 1} \tag{4}$$

which agrees with (1).

A second method for finding the constants A and B is to multiply out the right side of (3) and collect like powers of x to obtain

$$5x - 10 = (A + B)x + (A - 4B)$$

Since the polynomials on the two sides are identical, their corresponding coefficients must be the same. Equating the corresponding coefficients on the two sides yields the following system of equations in the unknowns A and B:

$$A + B = 5$$
$$A - 4B = -10$$

Solving this system yields $A = 2$ and $B = 3$ as before (verify).

The terms on the right side of (4) are called *partial fractions* of the expression on the left side because they each constitute *part* of that expression. To find those partial fractions we first had to make a guess about their form, and then we had to find the unknown constants. Our next objective is to extend this idea to general rational functions. For this purpose, suppose that $P(x)/Q(x)$ is a *proper rational function*, by which we mean that the degree of the numerator is less than the degree of the denominator. There is a theorem in advanced algebra which states that every proper rational function can be expressed as a sum

$$\frac{P(x)}{Q(x)} = F_1(x) + F_2(x) + \cdots + F_n(x)$$

where $F_1(x), F_2(x), \ldots, F_n(x)$ are rational functions of the form

$$\frac{A}{(ax+b)^k} \quad \text{or} \quad \frac{Ax+B}{(ax^2+bx+c)^k}$$

in which the denominators are factors of $Q(x)$. The sum is called the *partial fraction decomposition* of $P(x)/Q(x)$, and the terms are called *partial fractions*. As in our opening example, there are two parts to finding a partial fraction decomposition: determining the exact form of the decomposition and finding the unknown constants.

FINDING THE FORM OF A PARTIAL FRACTION DECOMPOSITION

The first step in finding the form of the partial fraction decomposition of a proper rational function $P(x)/Q(x)$ is to factor $Q(x)$ completely into linear and irreducible quadratic factors, and then collect all repeated factors so that $Q(x)$ is expressed as a product of *distinct* factors of the form

$$(ax+b)^m \quad \text{and} \quad (ax^2+bx+c)^m$$

From these factors we can determine the form of the partial fraction decomposition using two rules that we will now discuss.

LINEAR FACTORS

If all of the factors of $Q(x)$ are linear, then the partial fraction decomposition of $P(x)/Q(x)$ can be determined by using the following rule:

> **LINEAR FACTOR RULE.** For each factor of the form $(ax+b)^m$, the partial fraction decomposition contains the following sum of m partial fractions:
>
> $$\frac{A_1}{ax+b} + \frac{A_2}{(ax+b)^2} + \cdots + \frac{A_m}{(ax+b)^m}$$
>
> where $A_1, A_2, \ldots, A_m$ are constants to be determined. In the case where $m = 1$, only the first term in the sum appears.

Example 1 Evaluate $\displaystyle\int \frac{dx}{x^2+x-2}$.

Solution. The integrand is a proper rational function that can be written as

$$\frac{1}{x^2+x-2} = \frac{1}{(x-1)(x+2)}$$

The factors $x - 1$ and $x + 2$ are both linear and appear to the first power, so each contributes one term to the partial fraction decomposition by the linear factor rule. Thus, the decomposition has the form

$$\frac{1}{(x-1)(x+2)} = \frac{A}{x-1} + \frac{B}{x+2} \tag{5}$$

where A and B are constants to be determined. Multiplying this expression through by $(x-1)(x+2)$ yields

$$1 = A(x+2) + B(x-1) \tag{6}$$

As discussed earlier, there are two methods for finding A and B: we can substitute values of x that are chosen to make terms on the right drop out, or we can multiply out on the right and equate corresponding coefficients on the two sides to obtain a system of equations that can be solved for A and B. We will use the first approach.

Setting $x = 1$ makes the second term in (6) drop out and yields $1 = 3A$ or $A = \frac{1}{3}$; and setting $x = -2$ makes the first term in (6) drop out and yields $1 = -3B$ or $B = -\frac{1}{3}$. Substituting these values in (5) yields the partial fraction decomposition

$$\frac{1}{(x-1)(x+2)} = \frac{\frac{1}{3}}{x-1} + \frac{-\frac{1}{3}}{x+2}$$

The integration can now be completed as follows:

$$\int \frac{dx}{(x-1)(x+2)} = \frac{1}{3}\int \frac{dx}{x-1} - \frac{1}{3}\int \frac{dx}{x+2}$$

$$= \frac{1}{3}\ln|x-1| - \frac{1}{3}\ln|x+2| + C = \frac{1}{3}\ln\left|\frac{x-1}{x+2}\right| + C \qquad \blacktriangleleft$$

If the factors of $Q(x)$ are linear and none are repeated, as in the last example, then the recommended method for finding the constants in the partial fraction decomposition is to substitute appropriate values of x to make terms drop out. However, if some of the linear factors are repeated, then it will not be possible to find all of the constants in this way. In this case the recommended procedure is to find as many constants as possible by substitution and then find the rest by equating coefficients. This is illustrated in the next example.

Example 2   Evaluate $\displaystyle\int \frac{2x+4}{x^3-2x^2}\,dx$.

Solution.   The integrand can be rewritten as

$$\frac{2x+4}{x^3-2x^2} = \frac{2x+4}{x^2(x-2)}$$

Although x^2 is a quadratic factor, it is *not* irreducible since $x^2 = xx$. Thus, by the linear factor rule, x^2 introduces two terms (since $m = 2$) of the form

$$\frac{A}{x} + \frac{B}{x^2}$$

and the factor $x - 2$ introduces one term (since $m = 1$) of the form

$$\frac{C}{x-2}$$

so the partial fraction decomposition is

$$\frac{2x+4}{x^2(x-2)} = \frac{A}{x} + \frac{B}{x^2} + \frac{C}{x-2} \tag{7}$$

Multiplying by $x^2(x-2)$ yields

$$2x+4 = Ax(x-2) + B(x-2) + Cx^2 \tag{8}$$

which, after multiplying out and collecting like powers of x, becomes

$$2x+4 = (A+C)x^2 + (-2A+B)x - 2B \tag{9}$$

Setting $x = 0$ in (8) makes the first and third terms drop out and yields $B = -2$, and setting $x = 2$ in (8) makes the first and second terms drop out and yields $C = 2$ (verify). However, there is no substitution in (8) that produces A directly, so we look to Equation (9) to find this value. This can be done by equating the coefficients of x^2 on the two sides to obtain

$$A + C = 0 \quad \text{or} \quad A = -C = -2$$

Substituting the values $A = -2$, $B = -2$, and $C = 2$ in (7) yields the partial fraction decomposition

$$\frac{2x + 4}{x^2(x - 2)} = \frac{-2}{x} + \frac{-2}{x^2} + \frac{2}{x - 2}$$

Thus,

$$\int \frac{2x + 4}{x^2(x - 2)} \, dx = -2 \int \frac{dx}{x} - 2 \int \frac{dx}{x^2} + 2 \int \frac{dx}{x - 2}$$

$$= -2 \ln |x| + \frac{2}{x} + 2 \ln |x - 2| + C = 2 \ln \left| \frac{x - 2}{x} \right| + \frac{2}{x} + C \quad \blacktriangleleft$$

QUADRATIC FACTORS

If some of the factors of $Q(x)$ are irreducible quadratics, then the contribution of those factors to the partial fraction decomposition of $P(x)/Q(x)$ can be determined from the following rule:

> **QUADRATIC FACTOR RULE.** For each factor of the form $(ax^2 + bx + c)^m$, the partial fraction decomposition contains the following sum of m partial fractions:
>
> $$\frac{A_1 x + B_1}{ax^2 + bx + c} + \frac{A_2 x + B_2}{(ax^2 + bx + c)^2} + \cdots + \frac{A_m x + B_m}{(ax^2 + bx + c)^m}$$
>
> where $A_1, A_2, \ldots, A_m, B_1, B_2, \ldots, B_m$ are constants to be determined. In the case where $m = 1$, only the first term in the sum appears.

Example 3 Evaluate $\displaystyle\int \frac{x^2 + x - 2}{3x^3 - x^2 + 3x - 1} \, dx.$

Solution. The denominator in the integrand can be factored by grouping:

$$\frac{x^2 + x - 2}{3x^3 - x^2 + 3x - 1} = \frac{x^2 + x - 2}{x^2(3x - 1) + (3x - 1)} = \frac{x^2 + x - 2}{(3x - 1)(x^2 + 1)}$$

By the linear factor rule, the factor $3x - 1$ introduces one term; namely

$$\frac{A}{3x - 1}$$

and by the quadratic factor rule, the factor $x^2 + 1$ introduces one term; namely

$$\frac{Bx + C}{x^2 + 1}$$

Thus, the partial fraction decomposition is

$$\frac{x^2 + x - 2}{(3x - 1)(x^2 + 1)} = \frac{A}{3x - 1} + \frac{Bx + C}{x^2 + 1} \tag{10}$$

Multiplying by $(3x - 1)(x^2 + 1)$ yields

$$x^2 + x - 2 = A(x^2 + 1) + (Bx + C)(3x - 1) \tag{11}$$

We could find A by substituting $x = \frac{1}{3}$ to make the last term drop out, and then find the rest of the constants by equating corresponding coefficients. However, in this case it is just as easy to find *all* of the constants by equating coefficients and solving the resulting system. For this purpose we multiply out the right side of (11) and collect like terms:

$$x^2 + x - 2 = (A + 3B)x^2 + (-B + 3C)x + (A - C)$$

Equating corresponding coefficients gives

$$A + 3B \qquad = \quad 1$$
$$- \quad B + 3C = \quad 1$$
$$A \qquad - \quad C = -2$$

To solve this system, subtract the third equation from the first to eliminate A. Then use the resulting equation together with the second equation to solve for B and C. Finally, determine A from the first or third equation. This yields (verify)

$$A = -\tfrac{7}{5}, \quad B = \tfrac{4}{5}, \quad C = \tfrac{3}{5}$$

Thus, (10) becomes

$$\frac{x^2 + x - 2}{(3x - 1)(x^2 + 1)} = \frac{-\tfrac{7}{5}}{3x - 1} + \frac{\tfrac{4}{5}x + \tfrac{3}{5}}{x^2 + 1}$$

and

$$\int \frac{x^2 + x - 2}{(3x - 1)(x^2 + 1)}\, dx = -\frac{7}{5}\int \frac{dx}{3x - 1} + \frac{4}{5}\int \frac{x}{x^2 + 1}\, dx + \frac{3}{5}\int \frac{dx}{x^2 + 1}$$

$$= -\frac{7}{15}\ln|3x - 1| + \frac{2}{5}\ln(x^2 + 1) + \frac{3}{5}\tan^{-1}x + C \qquad \blacktriangleleft$$

FOR THE READER. Computer algebra systems have built-in capabilities for finding partial fraction decompositions. If you have a CAS, read the documentation on partial fraction decompositions, and use your CAS to find the decompositions in Examples 1, 2, and 3.

Example 4 Evaluate $\displaystyle\int \frac{3x^4 + 4x^3 + 16x^2 + 20x + 9}{(x + 2)(x^2 + 3)^2}\, dx$.

Solution. Observe that the integrand is a proper rational function since the numerator has degree 4 and the denominator has degree 5. Thus, the method of partial fractions is applicable. By the linear factor rule, the factor $x + 2$ introduces the single term

$$\frac{A}{x + 2}$$

and by the quadratic factor rule, the factor $(x^2 + 3)^2$ introduces two terms (since $m = 2$):

$$\frac{Bx + C}{x^2 + 3} + \frac{Dx + E}{(x^2 + 3)^2}$$

Thus, the partial fraction decomposition of the integrand is

$$\frac{3x^4 + 4x^3 + 16x^2 + 20x + 9}{(x + 2)(x^2 + 3)^2} = \frac{A}{x + 2} + \frac{Bx + C}{x^2 + 3} + \frac{Dx + E}{(x^2 + 3)^2} \qquad (12)$$

Multiplying by $(x + 2)(x^2 + 3)^2$ yields

$$3x^4 + 4x^3 + 16x^2 + 20x + 9$$
$$= A(x^2 + 3)^2 + (Bx + C)(x^2 + 3)(x + 2) + (Dx + E)(x + 2) \qquad (13)$$

which, after multiplying out and collecting like powers of x, becomes

$$3x^4 + 4x^3 + 16x^2 + 20x + 9$$
$$= (A + B)x^4 + (2B + C)x^3 + (6A + 3B + 2C + D)x^2$$
$$+ (6B + 3C + 2D + E)x + (9A + 6C + 2E) \qquad (14)$$

Equating corresponding coefficients in (14) yields the following system of five linear equations in five unknowns:

$$\begin{aligned} A + B &= 3 \\ 2B + C &= 4 \\ 6A + 3B + 2C + D &= 16 \\ 6B + 3C + 2D + E &= 20 \\ 9A + 6C + 2E &= 9 \end{aligned} \qquad (15)$$

Efficient methods for solving systems of linear equations such as this are studied in a branch of mathematics called *linear algebra*; those methods are outside the scope of this text. However, as a practical matter most linear systems of any size are solved by computer, and most computer algebra systems have commands that in many cases can solve linear systems exactly. In this particular case we can simplify the work by first substituting $x = -2$ in (13), which yields $A = 1$. Substituting this known value of A in (15) yields the simpler system

$$\begin{aligned} B &= 2 \\ 2B + C &= 4 \\ 3B + 2C + D &= 10 \\ 6B + 3C + 2D + E &= 20 \\ 6C + 2E &= 0 \end{aligned} \qquad (16)$$

This system can be solved by starting at the top and working down, first substituting $B = 2$ in the second equation to get $C = 0$, then substituting the known values of B and C in the third equation to get $D = 4$, and so forth. This yields

$$A = 1, \quad B = 2, \quad C = 0, \quad D = 4, \quad E = 0$$

Thus, (12) becomes

$$\frac{3x^4 + 4x^3 + 16x^2 + 20x + 9}{(x+2)(x^2+3)^2} = \frac{1}{x+2} + \frac{2x}{x^2+3} + \frac{4x}{(x^2+3)^2}$$

and so

$$\int \frac{3x^4 + 4x^3 + 16x^2 + 20x + 9}{(x+2)(x^2+3)^2}\, dx$$

$$= \int \frac{dx}{x+2} + \int \frac{2x}{x^2+3}\, dx + 4\int \frac{x}{(x^2+3)^2}\, dx$$

$$= \ln|x+2| + \ln(x^2+3) - \frac{2}{x^2+3} + C \qquad \blacktriangleleft$$

INTEGRATING IMPROPER RATIONAL FUNCTIONS

Although the method of partial fractions only applies to proper rational functions, an improper rational function can be integrated by performing a long division and expressing the function as the quotient plus the remainder over the divisor. The remainder over the divisor will be a proper rational function, which can then be decomposed into partial fractions. This idea is illustrated in the following example:

Example 5 Evaluate $\displaystyle\int \frac{3x^4 + 3x^3 - 5x^2 + x - 1}{x^2 + x - 2}\, dx.$

Solution. The integrand is an improper rational function since the numerator has degree 4 and the denominator has degree 2. Thus, we first perform the long division

$$\begin{array}{r} 3x^2 + 1 \\ x^2 + x - 2 \overline{\big)\ 3x^4 + 3x^3 - 5x^2 + x - 1} \\ \underline{3x^4 + 3x^3 - 6x^2} \\ x^2 + x - 1 \\ \underline{x^2 + x - 2} \\ 1 \end{array}$$

It follows that the integrand can be expressed as

$$\frac{3x^4 + 3x^3 - 5x^2 + x - 1}{x^2 + x - 2} = (3x^2 + 1) + \frac{1}{x^2 + x - 2}$$

and hence

$$\int \frac{3x^4 + 3x^3 - 5x^2 + x - 1}{x^2 + x - 2}\,dx = \int (3x^2 + 1)\,dx + \int \frac{dx}{x^2 + x - 2}$$

The second integral on the right now involves a proper rational function and can thus be evaluated by a partial fraction decomposition. Using the result of Example 1 we obtain

$$\int \frac{3x^4 + 3x^3 - 5x^2 + x - 1}{x^2 + x - 2}\,dx = x^3 + x + \frac{1}{3}\ln\left|\frac{x-1}{x+2}\right| + C \quad \blacktriangleleft$$

CONCLUDING REMARKS

There are some cases in which the method of partial fractions is inappropriate. For example, it would be illogical to use partial fractions to perform the integration

$$\int \frac{3x^2 + 2}{x^3 + 2x - 8}\,dx = \ln|x^3 + 2x - 8| + C$$

since the substitution $u = x^3 + 2x - 8$ is more direct. Similarly, the integration

$$\int \frac{2x - 1}{x^2 + 1}\,dx = \int \frac{2x}{x^2 + 1}\,dx - \int \frac{dx}{x^2 + 1} = \ln(x^2 + 1) - \tan^{-1}x + C$$

requires only a little algebra since the integrand is already in partial-fraction form.

EXERCISE SET 8.5   C CAS

In Exercises 1–8, write out the form of the partial fraction decomposition. (Do not find the numerical values of the coefficients.)

1. $\dfrac{3x - 1}{(x - 2)(x + 5)}$ **2.** $\dfrac{5}{x(x^2 - 9)}$

3. $\dfrac{2x - 3}{x^3 - x^2}$ **4.** $\dfrac{x^2}{(x + 2)^3}$

5. $\dfrac{1 - 5x^2}{x^3(x^2 + 1)}$ **6.** $\dfrac{2x}{(x - 1)(x^2 + 5)}$

7. $\dfrac{4x^3 - x}{(x^2 + 5)^2}$ **8.** $\dfrac{1 - 3x^4}{(x - 2)(x^2 + 1)^2}$

In Exercises 9–32, evaluate the integral.

9. $\displaystyle\int \frac{dx}{x^2 + 3x - 4}$ **10.** $\displaystyle\int \frac{dx}{x^2 + 8x + 7}$

11. $\displaystyle\int \frac{11x + 17}{2x^2 + 7x - 4}\,dx$ **12.** $\displaystyle\int \frac{5x - 5}{3x^2 - 8x - 3}\,dx$

13. $\displaystyle\int \frac{2x^2 - 9x - 9}{x^3 - 9x}\,dx$ **14.** $\displaystyle\int \frac{dx}{x(x^2 - 1)}$

15. $\displaystyle\int \frac{x^2 + 2}{x + 2}\,dx$ **16.** $\displaystyle\int \frac{x^2 - 4}{x - 1}\,dx$

17. $\displaystyle\int \frac{3x^2 - 10}{x^2 - 4x + 4}\,dx$ **18.** $\displaystyle\int \frac{x^2}{x^2 - 3x + 2}\,dx$

19. $\displaystyle\int \frac{x^5 + 2x^2 + 1}{x^3 - x}\,dx$ **20.** $\displaystyle\int \frac{2x^5 - x^3 - 1}{x^3 - 4x}\,dx$

21. $\displaystyle\int \frac{2x^2 + 3}{x(x - 1)^2}\,dx$ **22.** $\displaystyle\int \frac{3x^2 - x + 1}{x^3 - x^2}\,dx$

23. $\displaystyle\int \frac{x^2 + x - 16}{(x + 1)(x - 3)^2}\,dx$ **24.** $\displaystyle\int \frac{2x^2 - 2x - 1}{x^3 - x^2}\,dx$

25. $\displaystyle\int \frac{x^2}{(x + 2)^3}\,dx$ **26.** $\displaystyle\int \frac{2x^2 + 3x + 3}{(x + 1)^3}\,dx$

27. $\displaystyle\int \frac{2x^2 - 1}{(4x - 1)(x^2 + 1)}\,dx$ **28.** $\displaystyle\int \frac{dx}{x^3 + x}$

29. $\displaystyle\int \frac{x^3 + 3x^2 + x + 9}{(x^2 + 1)(x^2 + 3)}\,dx$ **30.** $\displaystyle\int \frac{x^3 + x^2 + x + 2}{(x^2 + 1)(x^2 + 2)}\,dx$

31. $\displaystyle\int \frac{x^3 - 3x^2 + 2x - 3}{x^2 + 1}\,dx$

32. $\displaystyle\int \frac{x^4 + 6x^3 + 10x^2 + x}{x^2 + 6x + 10}\,dx$

In Exercises 33 and 34, evaluate the integral by making a substitution that converts the integrand to a rational function.

33. $\displaystyle\int \frac{\cos\theta}{\sin^2\theta + 4\sin\theta - 5}\,d\theta$ **34.** $\displaystyle\int \frac{e^t}{e^{2t} - 4}\,dt$

35. Find the volume of the solid generated when the region enclosed by $y = x^2/(9 - x^2)$, $y = 0$, $x = 0$, and $x = 2$ is revolved about the x-axis.

36. Find the area of the region under the curve $y = 1/(1 + e^x)$, over the interval $[-\ln 5, \ln 5]$. [*Hint:* Make a substitution that converts the integrand to a rational function.]

In Exercises 37 and 38, use a CAS to evaluate the integral in two ways: (i) integrate directly; (ii) use the CAS to find the partial fraction decomposition and integrate the decomposition. Integrate by hand to check the results.

[c] 37. $\displaystyle\int \frac{x^2 + 1}{(x^2 + 2x + 3)^2}\, dx$

[c] 38. $\displaystyle\int \frac{x^5 + x^4 + 4x^3 + 4x^2 + 4x + 4}{(x^2 + 2)^3}\, dx$

In Exercises 39 and 40, integrate by hand and check your answers using a CAS.

[c] 39. $\displaystyle\int \frac{dx}{x^4 - 3x^3 - 7x^2 + 27x - 18}$

[c] 40. $\displaystyle\int \frac{dx}{16x^3 - 4x^2 + 4x - 1}$

41. Show that

$$\int_0^1 \frac{x}{x^4 + 1}\, dx = \frac{\pi}{8}$$

42. Use partial fractions to derive the integration formula

$$\int \frac{1}{a^2 - x^2}\, dx = \frac{1}{2a} \ln \left| \frac{a + x}{a - x} \right| + C$$

8.6 USING TABLES OF INTEGRALS AND COMPUTER ALGEBRA SYSTEMS

In this section we will discuss how to integrate using tables, and we will address some of the issues that relate to using computer algebra systems for integration. Readers who are not using computer algebra systems can skip that material with no problem.

INTEGRAL TABLES

Tables of integrals are useful for eliminating tedious hand computation. The endpapers of this text contain a relatively brief table of integrals that we will refer to as the ***Endpaper Integral Table***; more comprehensive tables are published in standard reference books such as the *CRC Standard Mathematical Tables and Formulae*, CRC Press, Inc., 1996.

All integral tables have their own scheme for classifying integrals according to the form of the integrand. For example, the Endpaper Integral Table classifies the integrals into 15 categories; *Basic Functions, Reciprocals of Basic Functions, Powers of Trigonometric Functions, Products of Trigonometric Functions*, and so forth. The first step in working with tables is to read through the classifications so that you understand the classification scheme and know where to look in the table for integrals of different types.

PERFECT MATCHES

If you are lucky, the integral you are attempting to evaluate will match up perfectly with one of the forms in the table. However, when looking for matches you may have to make an adjustment for the variable of integration. For example, the integral

$$\int x^2 \sin x\, dx$$

is a perfect match with Formula (46) in the Endpaper Integral Table, except for the letter used for the variable of integration. Thus, to apply Formula (46) to the given integral we need to change the variable of integration in the formula from u to x. With that minor modification we obtain

$$\int x^2 \sin x\, dx = 2x \sin x + (2 - x^2) \cos x + C$$

Here are some more examples of perfect matches:

Example 1 Use the Endpaper Integral Table to evaluate

(a) $\displaystyle\int \sin 7x \cos 2x \, dx$ (b) $\displaystyle\int x^2\sqrt{7+3x}\,dx$

(c) $\displaystyle\int \frac{\sqrt{2-x^2}}{x}\,dx$ (d) $\displaystyle\int (x^3+7x+1)\sin \pi x\,dx$

Solution (a). The integrand can be classified as a product of trigonometric functions. Thus, from Formula (40) with $m = 7$ and $n = 2$ we obtain

$$\int \sin 7x \cos 2x \, dx = -\frac{\cos 9x}{18} - \frac{\cos 5x}{10} + C$$

Solution (b). The integrand can be classified as a power of x multiplying $\sqrt{a+bx}$. Thus, from Formula (103) with $a = 7$ and $b = 3$ we obtain

$$\int x^2\sqrt{7+3x}\,dx = \frac{2}{2835}(135x^2 - 252x + 392)(7+3x)^{3/2} + C$$

Solution (c). The integrand can be classified as a power of x dividing $\sqrt{a^2-x^2}$. Thus, from Formula (79) with $a = \sqrt{2}$ we obtain

$$\int \frac{\sqrt{2-x^2}}{x}\,dx = \sqrt{2-x^2} - \sqrt{2}\ln\left|\frac{\sqrt{2}+\sqrt{2-x^2}}{x}\right| + C$$

Solution (d). The integrand can be classified as a polynomial multiplying a trigonometric function. Thus, we apply Formula (58) with $p(x) = x^3+7x+1$ and $a = \pi$. The successive nonzero derivatives of $p(x)$ are

$$p'(x) = 3x^2 + 7, \quad p''(x) = 6x, \quad p'''(x) = 6$$

and hence

$$\int (x^3+7x+1)\sin \pi x\,dx$$

$$= -\frac{x^3+7x+1}{\pi}\cos \pi x + \frac{3x^2+7}{\pi^2}\sin \pi x + \frac{6x}{\pi^3}\cos \pi x - \frac{6}{\pi^4}\sin \pi x + C \quad \blacktriangleleft$$

MATCHES REQUIRING SUBSTITUTIONS

Sometimes an integral that does not match any table entry can be made to match by making an appropriate substitution. Here are some examples.

Example 2 Use the Endpaper Integral Table to evaluate $\displaystyle\int \sqrt{x-4x^2}\,dx$.

Solution. The integrand does not match any of the forms in the table precisely. It comes closest to matching Formula (112), but it misses because of the factor of 4 multiplying x^2 inside the radical. However, if we make the substitution

$$u = 2x, \quad du = 2\,dx$$

then the $4x^2$ will become a u^2, and the transformed integral will be

$$\int \sqrt{x-4x^2}\,dx = \frac{1}{2}\int \sqrt{\tfrac{1}{2}u - u^2}\,du$$

which matches Formula (112) with $a = \frac{1}{4}$. Thus, we obtain

$$\int \sqrt{x-4x^2}\,dx = \frac{1}{2}\left[\frac{u-\frac{1}{4}}{2}\sqrt{\tfrac{1}{2}u-u^2} + \frac{1}{32}\sin^{-1}\left(\frac{u-\frac{1}{4}}{\frac{1}{4}}\right)\right] + C$$

$$= \frac{1}{2}\left[\frac{2x-\frac{1}{4}}{2}\sqrt{x-4x^2} + \frac{1}{32}\sin^{-1}\left(\frac{2x-\frac{1}{4}}{\frac{1}{4}}\right)\right] + C$$

$$= \frac{8x-1}{16}\sqrt{x-4x^2} + \frac{1}{64}\sin^{-1}(8x-1) + C \quad \blacktriangleleft$$

Example 3   Use the Endpaper Integral Table to evaluate

(a) $\displaystyle\int e^{\pi x} \sin^{-1}(e^{\pi x})\,dx$ (b) $\displaystyle\int x\sqrt{x^2 - 4x + 5}\,dx$

Solution (a). The integrand does not even come close to matching any of the forms in the table. However, a little thought suggests the substitution

$$u = e^{\pi x}, \quad du = \pi e^{\pi x}\,dx$$

from which we obtain

$$\int e^{\pi x} \sin^{-1}(e^{\pi x})\,dx = \frac{1}{\pi}\int \sin^{-1} u\,du$$

The integrand is now a basic function, and Formula (7) yields

$$\int e^{\pi x} \sin^{-1}(e^{\pi x})\,dx = \frac{1}{\pi}[u \sin^{-1} u + \sqrt{1 - u^2}] + C$$

$$= \frac{1}{\pi}[e^{\pi x} \sin^{-1}(e^{\pi x}) + \sqrt{1 - e^{2\pi x}}] + C$$

Solution (b). Again, the integrand does not closely match any of the forms in the table. However, a little thought suggests that it may be possible to bring the integrand closer to the form $x\sqrt{x^2 + a^2}$ by completing the square to eliminate the term involving x inside the radical. Doing this yields

$$\int x\sqrt{x^2 - 4x + 5}\,dx = \int x\sqrt{(x^2 - 4x + 4) + 1}\,dx = \int x\sqrt{(x - 2)^2 + 1}\,dx \quad (1)$$

At this point we are closer to the form $x\sqrt{x^2 + a^2}$, but we are not quite there because of the $(x - 2)^2$ rather than x^2 inside the radical. However, we can resolve that problem with the substitution

$$u = x - 2, \quad du = dx$$

With this substitution we have $x = u + 2$, so (1) can be expressed in terms of u as

$$\int x\sqrt{x^2 - 4x + 5}\,dx = \int (u + 2)\sqrt{u^2 + 1}\,du = \int u\sqrt{u^2 + 1}\,du + 2\int \sqrt{u^2 + 1}\,du$$

The first integral on the right is now a perfect match with Formula (84) with $a = 1$, and the second is a perfect match with Formula (72) with $a = 1$. Thus, applying these formulas we obtain

$$\int x\sqrt{x^2 - 4x + 5}\,dx = \left[\frac{1}{3}(u^2 + 1)^{3/2}\right] + 2\left[\frac{u}{2}\sqrt{u^2 + 1} + \frac{1}{2}\ln(u + \sqrt{u^2 + 1})\right] + C$$

If we now replace u by $x - 2$ (in which case $u^2 + 1 = x^2 - 4x + 5$), we obtain

$$\int x\sqrt{x^2 - 4x + 5}\,dx = \tfrac{1}{3}(x^2 - 4x + 5)^{3/2} + (x - 2)\sqrt{x^2 - 4x + 5}$$

$$+ \ln(x - 2 + \sqrt{x^2 - 4x + 5}) + C$$

Although correct, this form of the answer has an unnecessary mixture of radicals and fractional exponents. If desired, we can "clean up" the answer by writing

$$(x^2 - 4x + 5)^{3/2} = (x^2 - 4x + 5)\sqrt{x^2 - 4x + 5}$$

from which it follows that (verify)

$$\int x\sqrt{x^2 - 4x + 5}\,dx = \tfrac{1}{3}(x^2 - x - 1)\sqrt{x^2 - 4x + 5}$$

$$+ \ln(x - 2 + \sqrt{x^2 - 4x + 5}) + C \quad \blacktriangleleft$$

MATCHES REQUIRING REDUCTION FORMULAS

In cases where the entry in an integral table is a reduction formula, that formula will have to be applied first to reduce the given integral to a form in which it can be evaluated.

Example 4   Use the Endpaper Integral Table to evaluate $\displaystyle\int \frac{x^3}{\sqrt{1+x}}\,dx$.

Solution.   The integrand can be classified as a power of x multiplying the reciprocal of $\sqrt{a+bx}$. Thus, from Formula (107) with $a=1$, $b=1$, and $n=3$, followed by Formula (106), we obtain

$$\int \frac{x^3}{\sqrt{1+x}}\,dx = \frac{2x^3\sqrt{1+x}}{7} - \frac{6}{7}\int \frac{x^2}{\sqrt{1+x}}\,dx$$

$$= \frac{2x^3\sqrt{1+x}}{7} - \frac{6}{7}\left[\frac{2}{15}(3x^2-4x+8)\sqrt{1+x}\right]+C$$

$$= \left(\frac{2x^3}{7} - \frac{12x^2}{35} + \frac{16x}{35} - \frac{32}{35}\right)\sqrt{1+x}+C \qquad \blacktriangleleft$$

MATCHES REQUIRING SPECIAL SUBSTITUTIONS

The Endpaper Integral Table has numerous entries involving an exponent of $3/2$ or involving square roots (exponent $1/2$), but it has no entries with other fractional exponents. However, integrals involving fractional powers of x can often be simplified by making the substitution $u=x^{1/n}$ in which n is the least common multiple of the denominators of the exponents. Here are some examples.

Example 5   Evaluate

(a) $\displaystyle\int \frac{\sqrt{x}}{1+\sqrt[3]{x}}\,dx$ (b) $\displaystyle\int \frac{dx}{2+2\sqrt{x}}$ (c) $\displaystyle\int \sqrt{1+e^x}\,dx$

Solution (a).   The integrand contains $x^{1/2}$ and $x^{1/3}$, so we make the substitution $u=x^{1/6}$, from which we obtain

$$x=u^6, \quad dx=6u^5\,du$$

Thus,

$$\int \frac{\sqrt{x}}{1+\sqrt[3]{x}}\,dx = \int \frac{(u^6)^{1/2}}{1+(u^6)^{1/3}}(6u^5)\,du = 6\int \frac{u^8}{1+u^2}\,du$$

By long division

$$\frac{u^8}{1+u^2} = u^6 - u^4 + u^2 - 1 + \frac{1}{1+u^2}$$

from which it follows that

$$\int \frac{\sqrt{x}}{1+\sqrt[3]{x}}\,dx = 6\int \left(u^6 - u^4 + u^2 - 1 + \frac{1}{1+u^2}\right)du$$

$$= \tfrac{6}{7}u^7 - \tfrac{6}{5}u^5 + 2u^3 - 6u + 6\tan^{-1}u + C$$

$$= \tfrac{6}{7}x^{7/6} - \tfrac{6}{5}x^{5/6} + 2x^{1/2} - 6x^{1/6} + 6\tan^{-1}(x^{1/6}) + C$$

Solution (b).   The integrand contains $x^{1/2}$ but does not match any of the forms in the Endpaper Integral Table. Thus, we make the substitution $u=x^{1/2}$, from which we obtain

$$x=u^2, \quad dx=2u\,du$$

Making this substitution yields

$$\int \frac{dx}{2+2\sqrt{x}} = \int \frac{2u}{2+2u}\,du$$

$$= \int \left(1 - \frac{1}{1+u}\right)du \qquad \boxed{\text{Long division}}$$

$$= u - \ln|1+u| + C$$

$$= \sqrt{x} - \ln(1+\sqrt{x}) + C \qquad \boxed{\text{Absolute value not needed}}$$

Solution (c). Again, the integral does not match any of the forms in the Endpaper Integral Table. However, the integrand contains $(1 + e^x)^{1/2}$, which is analogous to the situation in part (b), except that here it is $1 + e^x$ rather than x that is raised to the $1/2$ power. This suggests the substitution $u = (1 + e^x)^{1/2}$, from which we obtain (verify)

$$x = \ln(u^2 - 1), \quad dx = \frac{2u}{u^2 - 1}\,du$$

Thus,

$$\int \sqrt{1 + e^x}\,dx = \int u\left(\frac{2u}{u^2 - 1}\right)du$$

$$= \int \frac{2u^2}{u^2 - 1}\,du$$

$$= \int \left(2 + \frac{2}{u^2 - 1}\right)du \qquad \boxed{\text{Long division}}$$

$$= 2u + \int \left(\frac{1}{u - 1} - \frac{1}{u + 1}\right)du \qquad \boxed{\text{Partial fractions}}$$

$$= 2u + \ln|u - 1| - \ln|u + 1| + C$$

$$= 2u + \ln\left|\frac{u - 1}{u + 1}\right| + C$$

$$= 2\sqrt{1 + e^x} + \ln\left[\frac{\sqrt{1 + e^x} - 1}{\sqrt{1 + e^x} + 1}\right] + C \qquad \boxed{\begin{array}{l}\text{Absolute value}\\\text{not needed}\end{array}} \qquad \blacktriangleleft$$

Functions that consist of finitely many sums, differences, quotients, and products of $\sin x$ and $\cos x$ are called *rational functions of sin x and cos x*. Some examples are

$$\frac{\sin x + 3\cos^2 x}{\cos x + 4\sin x}, \qquad \frac{\sin x}{1 + \cos x - \cos^2 x}, \qquad \frac{3\sin^5 x}{1 + 4\sin x}$$

The Endpaper Integral Table gives a few formulas for integrating rational functions of $\sin x$ and $\cos x$ under the heading *Reciprocals of Basic Functions*. For example, it follows from Formula (18) that

$$\int \frac{1}{1 + \sin x}\,dx = \tan x - \sec x + C \tag{2}$$

However, since the integrand is a rational function of $\sin x$, it may be desirable in a particular application to express the value of the integral in terms of $\sin x$ and $\cos x$ and rewrite (2) as

$$\int \frac{1}{1 + \sin x}\,dx = \frac{\sin x - 1}{\cos x} + C$$

Many rational functions of $\sin x$ and $\cos x$ can be evaluated by an ingenious method that was discovered by the mathematician Karl Weierstrass (see p. 140 for biography). The idea is to make the substitution

$$u = \tan(x/2), \quad -\pi/2 < x/2 < \pi/2$$

from which it follows that

$$x = 2\tan^{-1} u, \quad dx = \frac{2}{1 + u^2}\,du$$

To implement this substitution we need to express $\sin x$ and $\cos x$ in terms of u. For this purpose we will use the identities

$$\sin x = 2\sin(x/2)\cos(x/2) \tag{3}$$

$$\cos x = \cos^2(x/2) - \sin^2(x/2) \tag{4}$$

and the following relationships suggested by Figure 8.6.1:

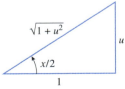

Figure 8.6.1

$$\sin(x/2) = \frac{u}{\sqrt{1 + u^2}} \quad \text{and} \quad \cos(x/2) = \frac{1}{\sqrt{1 + u^2}}$$

Substituting these expressions in (3) and (4) yields

$$\sin x = 2\left(\frac{u}{\sqrt{1+u^2}}\right)\left(\frac{1}{\sqrt{1+u^2}}\right) = \frac{2u}{1+u^2}$$

$$\cos x = \left(\frac{1}{\sqrt{1+u^2}}\right)^2 - \left(\frac{u}{\sqrt{1+u^2}}\right)^2 = \frac{1-u^2}{1+u^2}$$

In summary, we have shown that the substitution $u = \tan(x/2)$ can be implemented in a rational function of $\sin x$ and $\cos x$ by letting

$$\sin x = \frac{2u}{1+u^2}, \qquad \cos x = \frac{1-u^2}{1+u^2}, \qquad dx = \frac{2}{1+u^2}\,du \tag{5}$$

Example 6 Evaluate $\displaystyle\int \frac{dx}{1-\sin x + \cos x}$.

Solution. The integrand is a rational function of $\sin x$ and $\cos x$ that does not match any of the formulas in the Endpaper Integral Table, so we make the substitution $u = \tan(x/2)$. Thus, from (5) we obtain

$$\int \frac{dx}{1-\sin x + \cos x} = \int \frac{\dfrac{2\,du}{1+u^2}}{1 - \left(\dfrac{2u}{1+u^2}\right) + \left(\dfrac{1-u^2}{1+u^2}\right)}$$

$$= \int \frac{2\,du}{(1+u^2) - 2u + (1-u^2)}$$

$$= \int \frac{du}{1-u} = -\ln|1-u| + C = -\ln|1 - \tan(x/2)| + C \qquad \blacktriangleleft$$

REMARK. The substitution $u = \tan(x/2)$ will convert any rational function of $\sin x$ and $\cos x$ to an ordinary rational function of u. However, the method can lead to cumbersome partial fraction decompositions, so it may be worthwhile to explore the existence of simpler methods when hand computation is to be used.

INTEGRATING WITH COMPUTER ALGEBRA SYSTEMS

Integration tables are rapidly giving way to computerized integration using computer algebra systems. However, as with many powerful tools, a knowledgeable operator is an important component of the system.

Sometimes computer algebra systems do not produce the most general form of the indefinite integral. For example, the integral formula

$$\int \frac{dx}{x-1} = \ln|x-1| + C$$

which can be obtained by inspection or by using the substitution $u = x-1$, is valid for $x > 1$ or for $x < 1$. However, *Mathematica, Maple, Derive*, and the computer algebra systems used by the Texas Instruments TI-89 and Hewlett-Packard HP-49 calculators evaluate this integral as[*]

$$\ln(-1+x), \quad \ln(x-1), \quad \ln(x-1), \quad \ln(|x-1|), \quad \ln(x-1)$$
$$\quad\;\text{\small\textit{Mathematica}} \qquad \text{\small\textit{Maple}} \qquad \text{\small\textit{Derive}} \qquad \text{\small TI-89} \qquad \text{\small HP-49}$$

Observe that none of the systems include the constant of integration—the answer produced is a particular antiderivative and not the most general antiderivative (indefinite integral).

---

[*]Results produced by *Mathematica, Maple, Derive*, the TI-89, and the HP-49 may vary depending on the version of the software that is used.

Observe also that only the TI-89 includes the absolute value signs; consequently, the antiderivatives produced in this instance by the other systems are valid only for $x > 1$. All systems, however, are able to recover to correctly calculate the definite integral

$$\int_0^{1/2} \frac{dx}{x-1} = -\ln 2$$

Now let us examine how these systems handle the integral

$$\int x\sqrt{x^2-4x+5}\,dx = \tfrac{1}{3}(x^2-x-1)\sqrt{x^2-4x+5}$$
$$+ \ln(x-2+\sqrt{x^2-4x+5}) \tag{6}$$

which we obtained in Example 3(b) (with the constant of integration included). *Derive*, the TI-89, and the HP-49 produce this result in slightly different algebraic forms, but *Maple* produces the result

$$\int x\sqrt{x^2-4x+5}\,dx = \tfrac{1}{3}(x^2-4x+5)^{3/2} + \tfrac{1}{2}(2x-4)\sqrt{x^2-4x+5} + \sinh^{-1}(x-2)$$

This can be rewritten as (6) by expressing the fractional exponent in radical form and expressing $\sinh^{-1}(x-2)$ in logarithmic form using Theorem 7.8.4 (verify). *Mathematica* produces the result

$$\int x\sqrt{x^2-4x+5}\,dx = \tfrac{1}{3}(x^2-x-1)\sqrt{x^2-4x+5} - \sinh^{-1}(2-x)$$

which can be rewritten in form (6) by using Theorem 7.8.4 together with the identity $\sinh^{-1}(-x) = -\sinh^{-1} x$ (verify).

Computer algebra systems can sometimes produce inconvenient or unnatural answers to integration problems. For example, the systems mentioned above produce the following results when asked to integrate $(x+1)^7$:

$$\frac{(x+1)^8}{8},$$
Mathematica, Maple, Derive, TI-89

$$\tfrac{1}{8}x^8 + x^7 + \tfrac{7}{2}x^6 + 7x^5 + \tfrac{35}{4}x^4 + 7x^3 + \tfrac{7}{2}x^2 + x$$
HP-49

The answers produced by the majority of these systems are in keeping with the hand computation

$$\int (x+1)^7\,dx = \frac{(x+1)^8}{8} + C$$

that uses the substitution $u = x+1$, whereas the answer produced by the HP-49 appears to be based on expanding $(x+1)^7$ and integrating term by term.

FOR THE READER. If you expand the expression $\tfrac{1}{8}(x+1)^8$, you will discover that it contains a summand $\tfrac{1}{8}$ that does not appear in the HP-49 result. What is the explanation?

In Example 2(a) of Section 8.3 we showed that

$$\int \sin^4 x \cos^5 x\,dx = \tfrac{1}{5}\sin^5 x - \tfrac{2}{7}\sin^7 x + \tfrac{1}{9}\sin^9 x + C$$

This is the answer produced by the HP-49. In contrast, *Mathematica* integrates this as

$$\tfrac{3}{128}\sin x - \tfrac{1}{192}\sin 3x - \tfrac{1}{320}\sin 5x + \tfrac{1}{1792}\sin 7x + \tfrac{1}{2304}\sin 9x$$

and *Maple*, *Derive*, and the TI-89 essentially integrate it as

$$-\tfrac{1}{9}\sin^3 x \cos^6 x - \tfrac{1}{21}\sin x \cos^6 x + \tfrac{1}{105}\cos^4 x \sin x + \tfrac{4}{315}\cos^2 x \sin x + \tfrac{8}{315}\sin x$$

Although these three results look quite different, they can be obtained from one another using appropriate trigonometric identities.

COMPUTER ALGEBRA SYSTEMS HAVE LIMITATIONS

A computer algebra system combines a set of integration rules (such as substitution) with a library of functions that it can use to construct antiderivatives. Such libraries contain elementary functions, such as polynomials, rational functions, trigonometric functions, as well as various nonelementary functions that arise in engineering, physics, and other applied fields. Just as our Endpaper Integral Table has only 121 indefinite integrals, these libraries are not exhaustive of all possible integrands. If the system cannot manipulate the integrand to a form matching one in its library, the program will give some indication that it cannot evaluate the integral. For example, when asked to evaluate the integral

$$\int (1 + \ln x)\sqrt{1 + (x \ln x)^2}\, dx \tag{7}$$

all of the systems mentioned above respond by displaying some form of the unevaluated integral as an answer, indicating that they could not perform the integration.

FOR THE READER. Sometimes integrals that cannot be evaluated by a CAS in their given form can be evaluated by first rewriting them in a different form or by making a substitution. Make a u-substitution in (7) that will enable you to evaluate the integral with your CAS.

Sometimes computer algebra systems respond by expressing an integral in terms of another integral. For example, if you try to integrate e^{x^2} using *Mathematica*, *Maple*, or *Derive*, you will obtain an expression involving erf (which stands for **error function**). The function erf(x) is defined as

$$\text{erf}(x) = \frac{2}{\sqrt{\pi}} \int_0^x e^{-t^2}\, dt$$

so all three programs essentially rewrite the given integral in terms of a closely related integral. Indeed, this is what we did in integrating $1/x$, since the natural logarithm function is (formally) defined as

$$\ln x = \int_1^x \frac{1}{t}\, dt$$

(see Section 6.9).

Example 7 A particle moves along an x-axis in such a way that its velocity $v(t)$ at time t is

$$v(t) = 30\cos^7 t \sin^4 t \quad (t \geq 0)$$

Graph the position versus time curve for the particle, given that the particle is at $x = 1$ when $t = 0$.

Solution. Since $dx/dt = v(t)$ and $x = 1$ when $t = 0$, the position function $x(t)$ is given by

$$x(t) = 1 + \int_0^t v(s)\, ds$$

Some computer algebra systems will allow this expression to be entered directly into a command for plotting functions, but it is often more efficient to perform the integration first. The authors' integration utility yields

$$x = \int 30\cos^7 t \sin^4 t\, dt$$
$$= -\tfrac{30}{11}\sin^{11} t + 10\sin^9 t - \tfrac{90}{7}\sin^7 t + 6\sin^5 t + C$$

where we have added the required constant of integration. Using the initial condition $x(0) = 1$, we substitute the values $x = 1$ and $t = 0$ into this equation to find that $C = 1$, so

$$x(t) = -\tfrac{30}{11}\sin^{11} t + 10\sin^9 t - \tfrac{90}{7}\sin^7 t + 6\sin^5 t + 1 \quad (t \geq 0)$$

The graph of x versus t is shown in Figure 8.6.2. ◀

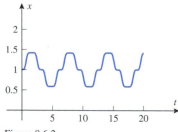

Figure 8.6.2

EXERCISE SET 8.6 [c] CAS

In Exercises 1–24:
(a) Use the Endpaper Integral Table to evaluate the integral.
(b) If you have a CAS, use it to evaluate the integral, and then confirm that the result is equivalent to the one that you found in part (a).

1. $\displaystyle\int \frac{3x}{4x-1}\,dx$

2. $\displaystyle\int \frac{x}{(2-3x)^2}\,dx$

3. $\displaystyle\int \frac{1}{x(2x+5)}\,dx$

4. $\displaystyle\int \frac{1}{x^2(1-5x)}\,dx$

5. $\displaystyle\int x\sqrt{2x-3}\,dx$

6. $\displaystyle\int \frac{x}{\sqrt{2-x}}\,dx$

7. $\displaystyle\int \frac{1}{x\sqrt{4-3x}}\,dx$

8. $\displaystyle\int \frac{1}{x\sqrt{3x-4}}\,dx$

9. $\displaystyle\int \frac{1}{5-x^2}\,dx$

10. $\displaystyle\int \frac{1}{x^2-9}\,dx$

11. $\displaystyle\int \sqrt{x^2-3}\,dx$

12. $\displaystyle\int \frac{\sqrt{x^2+5}}{x^2}\,dx$

13. $\displaystyle\int \frac{x^2}{\sqrt{x^2+4}}\,dx$

14. $\displaystyle\int \frac{1}{x^2\sqrt{x^2-2}}\,dx$

15. $\displaystyle\int \sqrt{9-x^2}\,dx$

16. $\displaystyle\int \frac{\sqrt{4-x^2}}{x^2}\,dx$

17. $\displaystyle\int \frac{\sqrt{3-x^2}}{x}\,dx$

18. $\displaystyle\int \frac{1}{x\sqrt{6x-x^2}}\,dx$

19. $\displaystyle\int \sin 3x \sin 2x\,dx$

20. $\displaystyle\int \sin 2x \cos 5x\,dx$

21. $\displaystyle\int x^3 \ln x\,dx$

22. $\displaystyle\int \frac{\ln x}{\sqrt{x}}\,dx$

23. $\displaystyle\int e^{-2x} \sin 3x\,dx$

24. $\displaystyle\int e^x \cos 2x\,dx$

In Exercises 25–36:
(a) Make the indicated u-substitution, and then use the Endpaper Integral Table to evaluate the integral.
(b) If you have a CAS, use it to evaluate the integral, and then confirm that the result is equivalent to the one that you found in part (a).

25. $\displaystyle\int \frac{e^{4x}}{(4-3e^{2x})^2}\,dx, \quad u=e^{2x}$

26. $\displaystyle\int \frac{\cos 2x}{(\sin 2x)(3-\sin 2x)}\,dx, \quad u=\sin 2x$

27. $\displaystyle\int \frac{1}{\sqrt{x}(9x+4)}\,dx, \quad u=3\sqrt{x}$

28. $\displaystyle\int \frac{\cos 4x}{9+\sin^2 4x}\,dx, \quad u=\sin 4x$

29. $\displaystyle\int \frac{1}{\sqrt{9x^2-4}}\,dx, \quad u=3x$

30. $\displaystyle\int x\sqrt{2x^4+3}\,dx, \quad u=\sqrt{2}x^2$

31. $\displaystyle\int \frac{x^5}{\sqrt{5-9x^4}}\,dx, \quad u=3x^2$

32. $\displaystyle\int \frac{1}{x^2\sqrt{3-4x^2}}\,dx, \quad u=2x$

33. $\displaystyle\int \frac{\sin^2(\ln x)}{x}\,dx, \quad u=\ln x$

34. $\displaystyle\int e^{-2x}\cos^2(e^{-2x})\,dx, \quad u=e^{-2x}$

35. $\displaystyle\int xe^{-2x}\,dx, \quad u=-2x$

36. $\displaystyle\int \ln(5x-1)\,dx, \quad u=5x-1$

In Exercises 37–48:
(a) Make an appropriate u-substitution, and then use the Endpaper Integral Table to evaluate the integral.
(b) If you have a CAS, use it to evaluate the integral (no substitution), and then confirm that the result is equivalent to that in part (a).

37. $\displaystyle\int \frac{\sin 3x}{(\cos 3x)(\cos 3x+1)^2}\,dx$

38. $\displaystyle\int \frac{\ln x}{x\sqrt{4\ln x-1}}\,dx$

39. $\displaystyle\int \frac{x}{16x^4-1}\,dx$

40. $\displaystyle\int \frac{e^x}{3-4e^{2x}}\,dx$

41. $\displaystyle\int e^x\sqrt{3-4e^{2x}}\,dx$

42. $\displaystyle\int \frac{\sqrt{4-9x^2}}{x^2}\,dx$

43. $\displaystyle\int \sqrt{5x-9x^2}\,dx$

44. $\displaystyle\int \frac{1}{x\sqrt{x-5x^2}}\,dx$

45. $\displaystyle\int x\sin 3x\,dx$

46. $\displaystyle\int \cos\sqrt{x}\,dx$

47. $\displaystyle\int e^{-\sqrt{x}}\,dx$

48. $\displaystyle\int x\ln(2-3x^2)\,dx$

In Exercises 49–52:
(a) Complete the square, make an appropriate u-substitution, and then use the Endpaper Integral Table to evaluate the integral.
(b) If you have a CAS, use it to evaluate the integral (no substitution or square completion), and then confirm that the result is equivalent to that in part (a).

49. $\displaystyle\int \frac{1}{x^2+4x-5}\,dx$

50. $\displaystyle\int \sqrt{3-2x-x^2}\,dx$

51. $\displaystyle\int \frac{x}{\sqrt{5+4x-x^2}}\,dx$

52. $\displaystyle\int \frac{x}{x^2+6x+13}\,dx$

In Exercises 53–66:
(a) Make an appropriate u-substitution of the form $u = x^{1/n}$, $u = (x + a)^{1/n}$, or $u = x^n$, and then use the Endpaper Integral Table to evaluate the integral.
(b) If you have a CAS, use it to evaluate the integral, and then confirm that the result is equivalent to the one that you found in part (a).

53. $\displaystyle\int x\sqrt{x-2}\,dx$

54. $\displaystyle\int \frac{x}{\sqrt{x+1}}\,dx$

55. $\displaystyle\int x^5\sqrt{x^3+1}\,dx$

56. $\displaystyle\int \frac{1}{x\sqrt{x^3-1}}\,dx$

57. $\displaystyle\int \frac{dx}{\sqrt{x}+\sqrt[3]{x}}$

58. $\displaystyle\int \frac{dx}{x-x^{3/5}}$

59. $\displaystyle\int \frac{dx}{x(1-x^{1/4})}$

60. $\displaystyle\int \frac{x^{2/3}}{x+1}\,dx$

61. $\displaystyle\int \frac{dx}{x^{1/2}-x^{1/3}}$

62. $\displaystyle\int \frac{1+\sqrt{x}}{1-\sqrt{x}}\,dx$

63. $\displaystyle\int \frac{x^3}{\sqrt{1+x^2}}\,dx$

64. $\displaystyle\int \frac{x}{(x+3)^{1/5}}\,dx$

65. $\displaystyle\int \sin\sqrt{x}\,dx$

66. $\displaystyle\int e^{\sqrt{x}}\,dx$

In Exercises 67–72:
(a) Make u-substitution (5) to convert the integrand to a rational function of u, and then use the Endpaper Integral Table to evaluate the integral.
(b) If you have a CAS, use it to evaluate the integral (no substitution), and then confirm that the result is equivalent to that in part (a).

67. $\displaystyle\int \frac{dx}{1+\sin x+\cos x}$

68. $\displaystyle\int \frac{dx}{2+\sin x}$

69. $\displaystyle\int \frac{d\theta}{1-\cos\theta}$

70. $\displaystyle\int \frac{dx}{4\sin x-3\cos x}$

71. $\displaystyle\int \frac{\cos x}{2-\cos x}\,dx$

72. $\displaystyle\int \frac{dx}{\sin x+\tan x}$

In Exercises 73 and 74, use any method to solve for x.

73. $\displaystyle\int_2^x \frac{1}{t(4-t)}\,dt = 0.5,\ 2 < x < 4$

74. $\displaystyle\int_1^x \frac{1}{t\sqrt{2t-1}}\,dt = 1,\ x > \tfrac{1}{2}$

In Exercises 75–78, use any method to find the area of the region enclosed by the curves.

75. $y = \sqrt{25-x^2},\ y = 0,\ x = 0,\ x = 4$

76. $y = \sqrt{9x^2-4},\ y = 0,\ x = 2$

77. $y = \dfrac{1}{25-16x^2},\ y = 0,\ x = 0,\ x = 1$

78. $y = \sqrt{x}\ln x,\ y = 0,\ x = 4$

In Exercises 79–82, use any method to find the volume of the solid generated when the region enclosed by the curves is revolved about the y-axis.

79. $y = \cos x,\ y = 0,\ x = 0,\ x = \pi/2$

80. $y = \sqrt{x-4},\ y = 0,\ x = 8$

81. $y = e^{-x},\ y = 0,\ x = 0,\ x = 3$

82. $y = \ln x,\ y = 0,\ x = 5$

In Exercises 83 and 84, use any method to find the arc length of the curve.

83. $y = 2x^2,\ 0 \le x \le 2$

84. $y = 3\ln x,\ 1 \le x \le 3$

In Exercises 85 and 86, use any method to find the area of the surface generated by revolving the curve about the x-axis.

85. $y = \sin x,\ 0 \le x \le \pi$

86. $y = 1/x,\ 1 \le x \le 4$

In Exercises 87 and 88, information is given about the motion of a particle moving along a coordinate line.
(a) Use a CAS to find the position function of the particle for $t \ge 0$. You may approximate the constants of integration, where necessary.
(b) Graph the position versus time curve.

c **87.** $v(t) = 20\cos^6 t\,\sin^3 t,\ s(0) = 2$

c **88.** $a(t) = e^{-t}\sin 2t\,\sin 4t,\ v(0) = 0,\ s(0) = 10$

89. (a) Use the substitution $u = \tan(x/2)$ to show that
$$\int \sec x\,dx = \ln\left|\frac{1+\tan(x/2)}{1-\tan(x/2)}\right| + C$$
and confirm that this is consistent with Formula (22) of Section 8.3.
(b) Use the result in part (a) to show that
$$\int \sec x\,dx = \ln\left|\tan\left(\frac{\pi}{4}+\frac{x}{2}\right)\right| + C$$

90. Use the substitution $u = \tan(x/2)$ to show that
$$\int \csc x\,dx = \frac{1}{2}\ln\left[\frac{1-\cos x}{1+\cos x}\right] + C$$
and confirm that this is consistent with the result in Exercise 61(a) of Section 8.3.

91. Find a substitution that can be used to integrate rational functions of $\sinh x$ and $\cosh x$ and use your substitution to evaluate
$$\int \frac{dx}{2\cosh x+\sinh x}$$
without expressing the integrand in terms of e^x and e^{-x}.

8.7 NUMERICAL INTEGRATION; SIMPSON'S RULE

Our usual procedure for evaluating a definite integral is to find an antiderivative of the integrand and apply the Fundamental Theorem of Calculus. However, if an anti- derivative of the integrand cannot be found, then we must settle for a numerical approximation of the integral. In Section 6.4 we discussed three procedures for ap- proximating areas using Riemann sums—left endpoint approximation, right endpoint approximation, and midpoint approximation. In this section we will adapt those ideas to approximating general definite integrals, and we will discuss some new approxima- tion methods that often provide more accuracy with less computation.

A REVIEW OF RIEMANN SUM APPROXIMATIONS

Recall from Section 6.5 that the definite integral of a continuous function f over an interval $[a, b]$ may be computed as

$$\int_a^b f(x)\,dx = \lim_{n \to +\infty} \sum_{k=1}^n f(x_k^*)\Delta x$$

where the sum that appears on the right side is called a Riemann sum. In this formula, the interval $[a, b]$ is divided into n subintervals of width $\Delta x = (b - a)/n$, and x_k^* denotes an arbitrary point in the kth subinterval. It follows that as n increases the Riemann sum will eventually be a good approximation to the integral, which we denote by writing

$$\int_a^b f(x)\,dx \approx \sum_{k=1}^n f(x_k^*)\Delta x$$

or, equivalently,

$$\int_a^b f(x)\,dx \approx \Delta x \left[f(x_1^*) + f(x_2^*) + \cdots + f(x_n^*) \right]$$

In this section we will denote the values of f at the endpoints of the subintervals by

$$y_0 = f(a), \quad y_1 = f(x_1), \quad y_2 = f(x_2), \ldots, \quad y_{n-1} = f(x_{n-1}), \quad y_n = f(b)$$

and we will denote the values of f at the midpoints of the subintervals by

$$y_{m_1}, y_{m_2}, \ldots, y_{m_n}$$

(Figure 8.7.1).

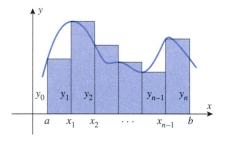 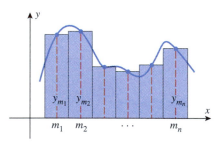

Figure 8.7.1

With this notation the left endpoint, right endpoint, and midpoint approximations dis- cussed in Section 6.4 can be expressed as shown in Table 8.7.1.

TRAPEZOIDAL APPROXIMATION

The left-hand and right-hand endpoint approximations are rarely used in applications; how- ever, if we take the average of the left-hand and right-hand endpoint approximations, we obtain a result, called the *trapezoidal approximation*, which is commonly used:

> **Trapezoidal Approximation**
>
> $$\int_a^b f(x)\,dx \approx \left(\frac{b - a}{2n}\right) [y_0 + 2y_1 + \cdots + 2y_{n-1} + y_n] \tag{1}$$

Table 8.7.1

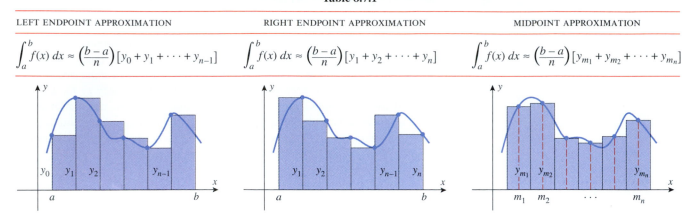

LEFT ENDPOINT APPROXIMATION	RIGHT ENDPOINT APPROXIMATION	MIDPOINT APPROXIMATION
$$\int_a^b f(x)\,dx \approx \left(\frac{b-a}{n}\right)[y_0 + y_1 + \cdots + y_{n-1}]$$	$$\int_a^b f(x)\,dx \approx \left(\frac{b-a}{n}\right)[y_1 + y_2 + \cdots + y_n]$$	$$\int_a^b f(x)\,dx \approx \left(\frac{b-a}{n}\right)[y_{m_1} + y_{m_2} + \cdots + y_{m_n}]$$

The name *trapezoidal approximation* can be explained by considering the case in which $f(x) \geq 0$ on $[a, b]$, so that $\int_a^b f(x)\,dx$ represents the area under $f(x)$ over $[a, b]$. Geometrically, the trapezoidal approximation formula results if we approximate this area by the sum of the trapezoidal areas shown in Figure 8.7.2 (Exercise 43).

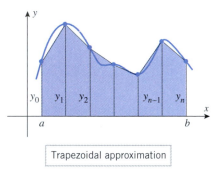

Trapezoidal approximation

Figure 8.7.2

Example 1 In Table 8.7.2 we have approximated

$$\ln 2 = \int_1^2 \frac{1}{x}\,dx$$

using the midpoint approximation and the trapezoidal approximation. In each case we used $n = 10$ subdivisions of the interval $[1, 2]$, so that

$$\underbrace{\frac{b-a}{n} = \frac{2-1}{10} = 0.1}_{\text{Midpoint}} \quad \text{and} \quad \underbrace{\frac{b-a}{2n} = \frac{2-1}{20} = 0.05}_{\text{Trapezoidal}} \qquad \blacktriangleleft$$

> REMARK. In Example 1 we rounded the numerical values to nine places to the right of the decimal point; we will follow this procedure throughout this section. If your calculator cannot produce this many places, then you will have to make the appropriate adjustments. What is important here is that you understand the principles involved.

COMPARISON OF THE MIDPOINT AND TRAPEZOIDAL APPROXIMATIONS

The value of $\ln 2$ rounded to nine decimal places is

$$\ln 2 = \int_1^2 \frac{1}{x}\,dx \approx 0.693147181 \qquad (2)$$

so that the midpoint approximation in Example 1 produced a more accurate result than the

Table 8.7.2

Midpoint Approximation			Trapezoidal Approximation				
	MIDPOINT			ENDPOINT		MULTIPLIER	
i	m_i	$y_{m_i} = f(m_i) = 1/m_i$	i	x_i	$y_i = f(x_i) = 1/x_i$	w_i	$w_i y_i$
1	1.05	0.952380952	0	1.0	1.000000000	1	1.000000000
2	1.15	0.869565217	1	1.1	0.909090909	2	1.818181818
3	1.25	0.800000000	2	1.2	0.833333333	2	1.666666667
4	1.35	0.740740741	3	1.3	0.769230769	2	1.538461538
5	1.45	0.689655172	4	1.4	0.714285714	2	1.428571429
6	1.55	0.645161290	5	1.5	0.666666667	2	1.333333333
7	1.65	0.606060606	6	1.6	0.625000000	2	1.250000000
8	1.75	0.571428571	7	1.7	0.588235294	2	1.176470588
9	1.85	0.540540541	8	1.8	0.555555556	2	1.111111111
10	1.95	0.512820513	9	1.9	0.526315789	2	1.052631579
		6.928353603	10	2.0	0.500000000	1	0.500000000
							13.875428063

$$\int_1^2 \frac{1}{x}\,dx \approx (0.1)(6.928353603) \approx 0.692835360$$

$$\int_1^2 \frac{1}{x}\,dx \approx (0.05)(13.875428063) \approx 0.693771403$$

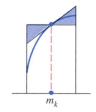

The shaded triangles have equal areas.

Figure 8.7.3

trapezoidal approximation (verify). To see why this should be so, we need to look at the midpoint approximation from another viewpoint. [For simplicity in the explanations, we will assume that $f(x) \geq 0$, but the conclusions will be true without this assumption.] For differentiable functions, the midpoint approximation is sometimes called the *tangent line approximation* because over each subinterval the area of the rectangle used in the midpoint approximation is equal to the area of the trapezoid whose upper boundary is the tangent line to $y = f(x)$ at the midpoint of the interval (Figure 8.7.3). The equality of these areas follows from the fact that the shaded triangles in Figure 8.7.3 are congruent.

In this section we will denote the midpoint and trapezoidal approximations of $\int_a^b f(x)\,dx$ with n subintervals by M_n and T_n, respectively, and we will denote the errors in these approximations by

$$|E_M| = \left| \int_a^b f(x)\,dx - M_n \right| \quad \text{and} \quad |E_T| = \left| \int_a^b f(x)\,dx - T_n \right|$$

In Figure 8.7.4a we have isolated a subinterval of $[a, b]$ on which the graph of a function f is concave down, and we have shaded the areas that represent the errors in the midpoint and trapezoidal approximations over the subinterval. In Figure 8.7.4b we show a succession of four illustrations which make it evident that the error from the midpoint approximation is less than that from the trapezoidal approximation. If the graph of f were concave up, analogous figures would lead to the same conclusion. (This argument, due to Frank Buck, appeared in *The College Mathematics Journal*, Vol. 16, No. 1, 1985.)

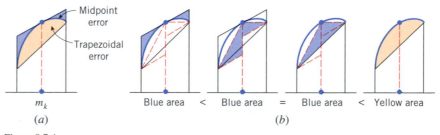

Figure 8.7.4

Figure 8.7.4a also suggests that on a subinterval where the graph is concave down, the midpoint approximation is larger than the value of the integral and the trapezoidal approximation is smaller. On an interval where the graph is concave up it is the other way around. In summary, we have the following result, which we state without formal proof:

8.7.1   THEOREM.   *Let f be continuous on $[a, b]$, and let $|E_M|$ and $|E_T|$ be the absolute errors that result from the midpoint and trapezoidal approximations of $\int_a^b f(x)\,dx$ using n subintervals.*

(a)   *If the graph of f is either concave up or concave down on (a, b), then $|E_M| < |E_T|$, that is, the error from the midpoint approximation is less than that from the trapezoidal approximation.*

(b)   *If the graph of f is concave down on (a, b), then*

$$T_n < \int_a^b f(x)\,dx < M_n$$

(c)   *If the graph of f is concave up on (a, b), then*

$$M_n < \int_a^b f(x)\,dx < T_n$$

Example 2   As observed earlier and illustrated in Table 8.7.3, the midpoint approximation of

$$\int_1^2 \frac{1}{x}\,dx = \ln 2$$

in Example 1 is more accurate than the trapezoidal approximation when partitioning $[1, 2]$ into $n = 10$ subintervals. This is consistent with part (a) of Theorem 8.7.1, since $f(x) = 1/x$ is continuous on $[1, 2]$ and concave up on $(1, 2)$. Moreover, $M_{10} < \ln 2 < T_{10}$, as predicted by part (c) of Theorem 8.7.1.   ◀

Table 8.7.3

$\ln 2$ (NINE DECIMAL PLACES)	APPROXIMATION	DIFFERENCE
0.693147181	$T_{10} \approx 0.693771403$	$E_T = \ln 2 - T_{10} \approx -0.000624222$
0.693147181	$M_{10} \approx 0.692835360$	$E_M = \ln 2 - M_{10} \approx 0.000311821$

Example 3   In Table 8.7.4 we have approximated

$$\sin 1 = \int_0^1 \cos x\,dx$$

using the midpoint and trapezoidal approximations with $n = 5$ subdivisions of the interval $[0, 1]$. (As before, the numerical values are rounded to nine decimal places.) Note that $f(x) = \cos x$ is continuous on $[0, 1]$ and concave down on $(0, 1)$. Thus, Theorem 8.7.1(a)

Table 8.7.4

$\sin 1$ (NINE DECIMAL PLACES)	APPROXIMATION	DIFFERENCE
0.841470985	$T_5 \approx 0.838664210$	$E_T = \sin 1 - T_5 \approx 0.002806775$
0.841470985	$M_5 \approx 0.842875074$	$E_M = \sin 1 - M_5 \approx -0.001404089$

guarantees that $|E_M| < |E_T|$, as shown in Table 8.7.4. Also, $T_5 < \sin 1 < M_5$, as predicted by Theorem 8.7.1(b).

Table 8.7.5 shows approximations for

$$\sin 3 = \int_0^3 \cos x \, dx$$

using the midpoint and trapezoidal approximations with $n = 10$ subdivisions of the interval $[0, 3]$. Note that $|E_M| < |E_T|$ and $T_{10} < \sin 3 < M_{10}$, although these results are not guaranteed by Theorem 8.7.1 since $f(x) = \cos x$ changes concavity on the interval $(0, 3)$. ◀

Table 8.7.5

sin 3 (NINE DECIMAL PLACES)	APPROXIMATION	DIFFERENCE
0.141120008	$T_{10} \approx 0.140060017$	$E_T = \sin 3 - T_{10} \approx 0.001059991$
0.141120008	$M_{10} \approx 0.141650601$	$E_M = \sin 3 - M_{10} \approx -0.000530592$

WARNING. Do not conclude that the midpoint approximation is always better than the trapezoidal approximation; for some values of n, the trapezoidal approximation can be more accurate over an interval on which the function changes concavity.

SIMPSON'S RULE

Over an interval on which the integrand does not change concavity, Theorem 8.7.1 guarantees that a definite integral is better approximated by the midpoint approximation than by the trapezoidal approximation and that the value of the definite integral lies between these two approximations. The numerical evidence in Tables 8.7.3 and 8.7.4 (and even in Table 8.7.5, despite the change in concavity of the integrand over the interval) reveals that $E_T \approx -2E_M$ in these instances. This suggests that

$$3\int_a^b f(x) \, dx = 2\int_a^b f(x) \, dx + \int_a^b f(x) \, dx$$

$$= 2(M_n + E_M) + (T_n + E_T)$$

$$= (2M_n + T_n) + (2E_M + E_T) \approx 2M_n + T_n$$

That is,

$$\int_a^b f(x) \, dx \approx \tfrac{1}{3}(2M_n + T_n)$$

Table 8.7.6 displays the approximations $\frac{1}{3}(2M_n + T_n)$ corresponding to the data in Tables 8.7.3 to 8.7.5. Thus, with little extra effort, we have much improved approximations for these definite integrals.

Table 8.7.6

CALCULATOR VALUE (NINE DECIMAL PLACES)	DEFINITE INTEGRAL APPROXIMATION	DIFFERENCE
$\ln 2 \approx 0.693147181$	$\int_1^2 (1/x) \, dx \approx \frac{1}{3}(2M_{10} + T_{10}) \approx 0.693147375$	-0.000000194
$\sin 1 \approx 0.841470985$	$\int_0^1 \cos x \, dx \approx \frac{1}{3}(2M_5 + T_5) \approx 0.841471453$	-0.000000468
$\sin 3 \approx 0.141120008$	$\int_0^3 \cos x \, dx \approx \frac{1}{3}(2M_{10} + T_{10}) \approx 0.141120406$	-0.000000398

Using the midpoint and trapezoidal approximation formulas in Table 8.7.1 and Formula (1), we can derive a similar formula for this approximation. For convenience, we partition the interval $[a, b]$ into $2n$ subintervals, each of length $(b - a)/(2n)$. As before, label the endpoints of these subintervals by $a = x_0, x_1, x_2, \ldots, x_{2n} = b$. Then $x_0, x_2, x_4, \ldots, x_{2n}$ define a partition of $[a, b]$ into n equal subintervals, and the midpoints of these subintervals are $x_1, x_3, x_5, \ldots, x_{2n-1}$, respectively. Using $y_i = f(x_i)$, we have

$$M_n = \left(\frac{b-a}{n}\right)[y_1 + y_3 + \cdots + y_{2n-1}] = \left(\frac{b-a}{2n}\right)[2y_1 + 2y_3 + \cdots + 2y_{2n-1}]$$

$$T_n = \left(\frac{b-a}{2n}\right)[y_0 + 2y_2 + 2y_4 + \cdots + 2y_{2n-2} + y_{2n}]$$

Now define S_{2n} by

$$S_{2n} = \frac{1}{3}(2M_n + T_n)$$

$$= \frac{1}{3}\left(\frac{b-a}{2n}\right)[y_0 + 4y_1 + 2y_2 + 4y_3 + 2y_4 + \cdots + 2y_{2n-2} + 4y_{2n-1} + y_{2n}] \qquad (3)$$

The approximation

$$\int_a^b f(x)\,dx \approx S_{2n}$$

as given in (3) is known as **Simpson's**[*] **rule**. We denote the absolute error in this approximation by

$$|E_S| = \left|\int_a^b f(x)\,dx - S_{2n}\right|$$

Example 4 Table 8.7.6 shows the Simpson's rule approximations

$$S_{20} = \tfrac{1}{3}(2M_{10} + T_{10}), \quad S_{10} = \tfrac{1}{3}(2M_5 + T_5), \quad \text{and} \quad S_{20} = \tfrac{1}{3}(2M_{10} + T_{10})$$

for the definite integrals

$$\int_1^2 \frac{1}{x}\,dx, \quad \int_0^1 \cos x\,dx, \quad \text{and} \quad \int_0^3 \cos x\,dx$$

respectively.

In Table 8.7.7 we have approximated

$$\ln 2 = \int_1^2 \frac{1}{x}\,dx$$

using (3) for Simpson's rule, where the interval $[1, 2]$ is partitioned into $2n = 10$ subintervals. Thus,

$$\frac{1}{3}\left(\frac{b-a}{2n}\right) = \frac{1}{3}\left(\frac{2-1}{10}\right) = \frac{1}{30}$$

---

[*]THOMAS SIMPSON (1710–1761). English mathematician. Simpson was the son of a weaver. He was trained to follow in his father's footsteps and had little formal education in his early life. His interest in science and mathematics was aroused in 1724, when he witnessed an eclipse of the Sun and received two books from a peddler, one on astrology and the other on arithmetic. Simpson quickly absorbed their contents and soon became a successful local fortune teller. His improved financial situation enabled him to give up weaving and marry his landlady, an older woman. Then in 1733 some mysterious "unfortunate incident" forced him to move. He settled in Derby, where he taught in an evening school and worked at weaving during the day. In 1736 he moved to London and published his first mathematical work in a periodical called the *Ladies' Diary* (of which he later became the editor). In 1737 he published a successful calculus textbook that enabled him to give up weaving completely and concentrate on textbook writing and teaching. His fortunes improved further in 1740 when one Robert Heath accused him of plagiarism. The publicity was marvelous, and Simpson proceeded to dash off a succession of best-selling textbooks: *Algebra* (ten editions plus translations), *Geometry* (twelve editions plus translations), *Trigonometry* (five editions plus translations), and numerous others. It is interesting to note that Simpson did not discover the rule that bears his name. It was a well-known result by Simpson's time.

Table 8.7.7   Simpson's Rule

	ENDPOINT		MULTIPLIER	
i	x_i	$y_i = f(x_i) = 1/x_i$	w_i	$w_i y_i$
0	1.0	1.000000000	1	1.000000000
1	1.1	0.909090909	4	3.636363636
2	1.2	0.833333333	2	1.666666667
3	1.3	0.769230769	4	3.076923077
4	1.4	0.714285714	2	1.428571429
5	1.5	0.666666667	4	2.666666667
6	1.6	0.625000000	2	1.250000000
7	1.7	0.588235294	4	2.352941176
8	1.8	0.555555556	2	1.111111111
9	1.9	0.526315789	4	2.105263158
10	2.0	0.500000000	1	0.500000000
				20.794506921

$$\int_1^2 \frac{1}{x}\,dx \approx \left(\tfrac{1}{30}\right)(20.794506921) \approx 0.693150231$$

Then

$$|E_S| = \left| \int_1^2 \frac{1}{x}\,dx - S_{10} \right|$$

$$= |\ln 2 - S_{10}| \approx |0.693147181 - 0.693150231| = 0.000003050$$

By contrast, $M_5 \approx 0.691907886$ and $T_5 \approx 0.695634921$ have absolute errors

$$|E_M| \approx 0.001239295 \quad \text{and} \quad |E_T| \approx 0.002487740$$

respectively, so S_{10} is a much more accurate approximation of $\ln 2$ than either M_5 or T_5.
◀

GEOMETRIC INTERPRETATION OF SIMPSON'S RULE

Both the midpoint and trapezoidal approximations for a definite integral are obtained by approximating a segment of the curve $y = f(x)$ by a linear segment (Figure 8.7.5). Formula (3) for Simpson's rule can be obtained by approximating a segment of the curve $y = f(x)$ by a segment of a quadratic function $y = Ax^2 + Bx + C$, thus capturing some sense of the concavity of the function.

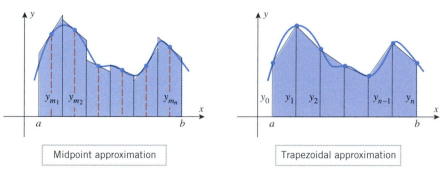

Midpoint approximation   Trapezoidal approximation

Figure 8.7.5

For this interpretation of Simpson's rule we start with the observation that for

$$a \le X_0 < X_2 \le b$$

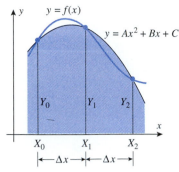

Figure 8.7.6

there is a unique function $g(x)$ of the form

$$g(x) = Ax^2 + Bx + C$$

such that

$$g(X_0) = f(X_0), \quad g(X_2) = f(X_2), \quad \text{and} \quad g(X_1) = f(X_1)$$

where $X_1 = (X_0 + X_2)/2$ (Figure 8.7.6). That is, we approximate $f(x)$ on $[X_0, X_2]$ by fitting a polynomial $g(x)$ of degree at most 2 to the points on the graph of $y = f(x)$ corresponding to $x = X_0, X_1,$ and X_2. We then use $\int_{X_0}^{X_2} g(x)\,dx$ to approximate $\int_{X_0}^{X_2} f(x)\,dx$.
Setting

$$\Delta x = \frac{X_2 - X_0}{2}$$

so that

$$X_2 = X_0 + 2\Delta x, \quad Y_0 = f(X_0), \quad Y_1 = f(X_1), \quad \text{and} \quad Y_2 = f(X_2)$$

the key result to establish is

$$\int_{X_0}^{X_2} g(x)\,dx = \int_{X_0}^{X_2} (Ax^2 + Bx + C)\,dx = \frac{\Delta x}{3}[Y_0 + 4Y_1 + Y_2] \tag{4}$$

We verify (4) by working from both ends to arrive at a common middle. Starting with the expression $Y_0 + 4Y_1 + Y_2$ on the right side of Equation (4),

$$Y_0 + 4Y_1 + Y_2$$

$$= g(X_0) + 4g(X_1) + g(X_2)$$

$$= A[X_0^2 + 4X_1^2 + X_2^2] + B[X_0 + 4X_1 + X_2] + C[1 + 4 + 1]$$

$$= A\left[X_0^2 + 4\left(\frac{X_0 + X_2}{2}\right)^2 + X_2^2\right] + B\left[X_0 + 4\left(\frac{X_0 + X_2}{2}\right) + X_2\right] + 6C$$

$$= A[X_0^2 + (X_0 + X_2)^2 + X_2^2] + B[3X_0 + 3X_2] + 6C$$

$$= 2A[X_0^2 + X_0X_2 + X_2^2] + 3B[X_0 + X_2] + 6C \tag{5}$$

Furthermore,

$$\int_{X_0}^{X_2} g(x)\,dx = \int_{X_0}^{X_2} (Ax^2 + Bx + C)\,dx = \frac{A}{3}x^3 + \frac{B}{2}x^2 + Cx\Bigg]_{X_0}^{X_2}$$

$$= \frac{A}{3}(X_2^3 - X_0^3) + \frac{B}{2}(X_2^2 - X_0^2) + C(X_2 - X_0)$$

$$= \left(\frac{X_2 - X_0}{3}\right)\left[A(X_2^2 + X_2X_0 + X_0^2) + \frac{3B}{2}(X_2 + X_0) + 3C\right]$$

$$= \left(\frac{2\Delta x}{3}\right)\left[A(X_2^2 + X_2X_0 + X_0^2) + \frac{3B}{2}(X_2 + X_0) + 3C\right]$$

$$= \frac{\Delta x}{3}[2A(X_2^2 + X_2X_0 + X_0^2) + 3B(X_2 + X_0) + 6C] \tag{6}$$

Substituting (5) into (6) gives (4).

Using the partition $a = x_0, x_1, x_2, \ldots, x_{2n} = b$ of the interval $[a, b]$ into $2n$ subintervals, each of width

$$\Delta x = \frac{b - a}{2n}$$

and applying (4) to the subintervals $[x_0, x_2], [x_2, x_4], \ldots, [x_{2n-2}, x_{2n}]$, we can now derive

Simpson's rule in (3) as the integral of a piecewise-quadratic approximation to $f(x)$:

$$\int_{a=x_0}^{b=x_{2n}} f(x)\,dx$$

$$= \int_{x_0}^{x_2} f(x)\,dx + \int_{x_2}^{x_4} f(x)\,dx + \cdots + \int_{x_{2n-2}}^{x_{2n}} f(x)\,dx$$

$$\approx \frac{\Delta x}{3}[y_0 + 4y_1 + y_2] + \frac{\Delta x}{3}[y_2 + 4y_3 + y_4] + \cdots + \frac{\Delta x}{3}[y_{2n-2} + 4y_{2n-1} + y_{2n}]$$

$$= \frac{\Delta x}{3}[y_0 + 4y_1 + 2y_2 + 4y_3 + 2y_4 + \cdots + 2y_{2n-2} + 4y_{2n-1} + y_{2n}]$$

$$= S_{2n}$$

ERROR ESTIMATES

With all the methods studied in this section, there are two sources of error: the *intrinsic* or *truncation error* due to the approximation formula, and the *roundoff error* introduced in the calculations. In general, increasing n reduces the truncation error but increases the roundoff error, since more computations are required for larger n. In practical applications, it is important to know how large n must be taken to ensure that a specified degree of accuracy is obtained. The analysis of roundoff error is complicated and will not be considered here. However, the following theorems, which are proved in books on numerical analysis, provide upper bounds on the truncation errors in the midpoint, trapezoidal, and Simpson's rule approximations.

8.7.2 THEOREM (*Midpoint and Trapezoidal Error Estimates*).   *If f'' is continuous on $[a, b]$ and if K_2 is the maximum value of $|f''(x)|$ on $[a, b]$, then for n subintervals of $[a, b]$*

$$(a) \quad |E_M| = \left| \int_a^b f(x)\,dx - M_n \right| \leq \frac{(b-a)^3 K_2}{24n^2} \tag{7}$$

$$(b) \quad |E_T| = \left| \int_a^b f(x)\,dx - T_n \right| \leq \frac{(b-a)^3 K_2}{12n^2} \tag{8}$$

8.7.3 THEOREM (*Simpson Error Estimate*).   *If $f^{(4)}$ is continuous on $[a, b]$ and if K_4 is the maximum value of $|f^{(4)}(x)|$ on $[a, b]$, then for $2n$ subintervals of $[a, b]$*

$$|E_S| = \left| \int_a^b f(x)\,dx - S_{2n} \right| \leq \frac{(b-a)^5 K_4}{180(2n)^4} \tag{9}$$

Example 5   Find an upper bound on the absolute error that results from approximating

$$\ln 2 = \int_1^2 \frac{1}{x}\,dx$$

using (a) the midpoint approximation M_{10} with $n = 10$ subintervals, (b) the trapezoidal approximation T_{10} with $n = 10$ subintervals, and (c) Simpson's rule S_{10} with $2n = 10$ subintervals.

Solution.   We will apply Formulas (7), (8), and (9) with

$$f(x) = \frac{1}{x}, \quad a = 1, \quad \text{and} \quad b = 2$$

For (7) and (8) we use $n = 10$; for (9) we use $2n = 10$, or $n = 5$. We have

$$f'(x) = -\frac{1}{x^2}, \quad f''(x) = \frac{2}{x^3}, \quad f'''(x) = -\frac{6}{x^4}, \quad f^{(4)}(x) = \frac{24}{x^5}$$

Thus,

$$|f''(x)| = \left|\frac{2}{x^3}\right| = \frac{2}{x^3}, \quad |f^{(4)}(x)| = \left|\frac{24}{x^5}\right| = \frac{24}{x^5} \tag{10--11}$$

where we have dropped the absolute values because $f''(x)$ and $f^{(4)}(x)$ have positive values for $1 \leq x \leq 2$. Since (10) and (11) are continuous and decreasing on $[1, 2]$, both functions have their maximum values at $x = 1$; for (10) this maximum value is 2 and for (11) the maximum value is 24. Thus we can take $K_2 = 2$ in (7) and (8), and $K_4 = 24$ in (9). This yields

$$|E_M| \leq \frac{(b-a)^3 K_2}{24n^2} = \frac{1^3 \cdot 2}{24 \cdot 10^2} \approx 0.000833333$$

$$|E_T| \leq \frac{(b-a)^3 K_2}{12n^2} = \frac{1^3 \cdot 2}{12 \cdot 10^2} \approx 0.001666667$$

$$|E_S| \leq \frac{(b-a)^5 K_4}{180(2n)^4} = \frac{1^5 \cdot 24}{180 \cdot 10^4} \approx 0.000013333 \qquad \blacktriangleleft$$

Note that the error bounds calculated in the preceding example are consistent with the values of E_M, E_T, and E_S calculated in Examples 2 and 4. In fact, these errors are considerably smaller in absolute value than the upper bounds of Example 5. It is quite common that the actual errors in the approximations M_n, T_n, and S_{2n} are substantially smaller than the upper bounds given in Theorems 8.7.2 and 8.7.3.

Example 6   How many subintervals should be used in approximating

$$\ln 2 = \int_1^2 \frac{1}{x}\, dx$$

by Simpson's rule for five decimal-place accuracy?

Solution.   To obtain five decimal-place accuracy, we must choose the number of subintervals so that

$$|E_S| \leq 0.000005 = 5 \times 10^{-6}$$

From (9), this can be achieved by taking $2n$ in Simpson's rule to satisfy

$$\frac{(b-a)^5 K_4}{180(2n)^4} \leq 5 \times 10^{-6}$$

Taking $a = 1$, $b = 2$, and $K_4 = 24$ (found in Example 5) in this inequality yields

$$\frac{1^5 \cdot 24}{180 \cdot (2n)^4} \leq 5 \times 10^{-6}$$

which, on taking reciprocals, can be rewritten as

$$(2n)^4 \geq \frac{2 \times 10^6}{75} \quad \text{or} \quad n^4 \geq \frac{10^4}{6}$$

Thus,

$$n \geq \frac{10}{\sqrt[4]{6}} \approx 6.389$$

Since n must be an integer, the smallest value of n that satisfies this requirement is $n = 7$, or $2n = 14$. Thus, the approximation S_{14} using 14 subintervals will produce five decimal-place accuracy. $\blacktriangleleft$

REMARK.   In cases where it is difficult to find the values of K_2 and K_4 in Formulas (7), (8), and (9), these constants may be replaced by any larger constants. For example, suppose that a constant K can be easily found with the certainty that $|f''(x)| < K$ on the interval. Then $K_2 \le K$ and

$$|E_T| \le \frac{(b-a)^3 K_2}{12n^2} \le \frac{(b-a)^3 K}{12n^2} \tag{12}$$

so the right side of (12) is also an upper bound on the value of $|E_T|$. Using K, however, will likely increase the computed value of n needed for a given error tolerance. Many applications involve the resolution of competing practical issues, here illustrated through the trade-off between the convenience of finding a crude bound for $|f''(x)|$ versus the efficiency of using the smallest possible n for a desired accuracy.

Example 7   How many subintervals should be used in approximating

$$\int_0^1 \cos(x^2)\, dx$$

by the midpoint approximation for three decimal-place accuracy?

Solution.   To obtain three decimal-place accuracy, we must choose n so that

$$|E_M| \le 0.0005 = 5 \times 10^{-4} \tag{13}$$

From (7) with $f(x) = \cos(x^2)$, $a = 0$, and $b = 1$, an upper bound on $|E_M|$ is given by

$$|E_M| \le \frac{K_2}{24n^2} \tag{14}$$

where $|K_2|$ is the maximum value of $|f''(x)|$ on the interval $[0, 1]$. But,

$$f'(x) = -2x \sin(x^2)$$
$$f''(x) = -4x^2 \cos(x^2) - 2\sin(x^2) = -[4x^2 \cos(x^2) + 2\sin(x^2)]$$

so that

$$|f''(x)| = |4x^2 \cos(x^2) + 2\sin(x^2)| \tag{15}$$

It would be tedious to look for the maximum value of this function on the interval $[0, 1]$. For x in $[0, 1]$, it is easy to see that each of the expressions x^2, $\cos(x^2)$, and $\sin(x^2)$ is bounded in absolute value by 1, so $|4x^2 \cos(x^2) + 2\sin(x^2)| \le 4 + 2 = 6$ on $[0, 1]$. We can improve on this by using a graphing utility to sketch $|f''(x)|$, as shown in Figure 8.7.7. It is evident from the graph that

$$|f''(x)| < 4 \quad \text{for} \quad 0 \le x \le 1$$

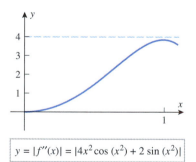

$y = |f''(x)| = |4x^2 \cos(x^2) + 2\sin(x^2)|$

Figure 8.7.7

Thus, it follows from (14) that

$$|E_M| \le \frac{K_2}{24n^2} < \frac{4}{24n^2} = \frac{1}{6n^2}$$

and hence we can satisfy (13) by choosing n so that

$$\frac{1}{6n^2} < 5 \times 10^{-4}$$

which, on taking reciprocals, can be written as

$$n^2 > \frac{10^4}{30} \quad \text{or} \quad n > \frac{10^2}{\sqrt{30}} \approx 18.257$$

The smallest integer value of n satisfying this inequality is $n = 19$. Thus, the midpoint approximation M_{19} using 19 subintervals will produce three decimal-place accuracy.   ◄

A COMPARISON OF THE THREE METHODS

Of the three methods studied in this section, Simpson's rule generally produces more accurate results than the midpoint or trapezoidal approximations for an equivalent amount of effort. To make this plausible, let us express (7), (8), and (9) in terms of the subinterval width

$$\Delta x = \frac{b-a}{n} \quad \text{for } M_n \text{ and } T_n$$

and

$$\Delta x = \frac{b-a}{2n} \quad \text{for } S_{2n}$$

We obtain

$$|E_M| \le \frac{1}{24} K_2 (b-a)(\Delta x)^2 \tag{16}$$

$$|E_T| \le \frac{1}{12} K_2 (b-a)(\Delta x)^2 \tag{17}$$

$$|E_S| \le \frac{1}{180} K_4 (b-a)(\Delta x)^4 \tag{18}$$

(verify). Thus, for Simpson's rule the upper bound on the absolute error is proportional to $(\Delta x)^4$, whereas the upper bound on the absolute error for the midpoint and trapezoidal approximations is proportional to $(\Delta x)^2$. Thus, reducing the interval width by a factor of 10, for example, reduces the error bound by a factor of 100 for the midpoint and trapezoidal approximations but reduces the error bound by a factor of 10,000 for Simpson's rule. This suggests that, as n increases, the accuracy of Simpson's rule improves much more rapidly than that of the other approximations.

As a final note, observe that if $f(x)$ is a polynomial of degree 3 or less, then we have $f^{(4)}(x) = 0$ for all x, so $K_4 = 0$ in (9) and consequently $|E_S| = 0$. Thus, Simpson's rule gives exact results for polynomials of degree 3 or less. Similarly, the midpoint and trapezoidal approximations give exact results for polynomials of degree 1 or less. (You should also be able to see that this is so geometrically.)

EXERCISE SET 8.7 $\boxed{\text{C}}$ CAS

In Exercises 1–6, use $n = 10$ subintervals to approximate the integral by (a) the midpoint approximation, (b) the trapezoidal approximation, and use $2n = 10$ subintervals to approximate the integral by (c) Simpson's rule. In each case, find the exact value of the integral and approximate the absolute error. Express your answers to at least four decimal places.

1. $\displaystyle\int_0^3 \sqrt{x+1}\,dx$ **2.** $\displaystyle\int_1^4 \frac{1}{\sqrt{x}}\,dx$ **3.** $\displaystyle\int_0^\pi \sin x\,dx$

4. $\displaystyle\int_0^1 \cos x\,dx$ **5.** $\displaystyle\int_1^3 e^{-x}\,dx$ **6.** $\displaystyle\int_{-1}^1 \frac{1}{2x+3}\,dx$

In Exercises 7–12, use inequalities (7), (8), and (9) to find upper bounds on the errors in parts (a), (b), and (c) of the indicated exercise.

7. Exercise 1   **8.** Exercise 2   **9.** Exercise 3

10. Exercise 4   **11.** Exercise 5   **12.** Exercise 6

In Exercises 13–18, use inequalities (7), (8), and (9) to find a number n of subintervals for (a) the midpoint approximation and (b) the trapezoidal approximation to ensure that the absolute error of the approximation will be less than the given value. Also, (c) find a number $2n$ of subintervals to ensure that the absolute error for the Simpson's rule approximation will be less than the given value.

13. Exercise 1; 5×10^{-4} **14.** Exercise 2; 5×10^{-4}

15. Exercise 3; 10^{-3} **16.** Exercise 4; 10^{-3}

17. Exercise 5; 10^{-6} **18.** Exercise 6; 10^{-6}

In Exercises 19 and 20, find a function $g(x)$ of the form

$$g(x) = Ax^2 + Bx + C$$

whose graph contains the points $(X_0, f(X_0))$, $(X_1, f(X_1))$, and $(X_2, f(X_2))$, for the given function $f(x)$ and the given values X_0, X_1, and X_2. Verify that

$$\int_{X_0}^{X_2} g(x)\,dx = \frac{\Delta x}{3}[f(X_0) + 4f(X_1) + f(X_2)]$$

where $\Delta x = (X_2 - X_0)/2$ as asserted with Formula (4).

19. $f(x) = \dfrac{1}{x}$; $X_0 = 2$, $X_1 = 3$, $X_2 = 4$

20. $f(x) = \cos^2(\pi x)$; $X_0 = 0$, $X_1 = \frac{1}{6}$, $X_2 = \frac{1}{3}$

In Exercises 21–26, approximate the integral using Simpson's rule with $2n = 10$ subintervals, and compare your answer to that produced by a calculating utility with a numerical integration capability. Express your answers to at least four decimal places.

21. $\displaystyle\int_0^1 e^{-x^2}\, dx$

22. $\displaystyle\int_0^2 \dfrac{x}{\sqrt{1+x^3}}\, dx$

23. $\displaystyle\int_1^2 \sqrt{1+x^3}\, dx$

24. $\displaystyle\int_0^\pi \dfrac{1}{2-\sin x}\, dx$

25. $\displaystyle\int_0^2 \sin(x^2)\, dx$

26. $\displaystyle\int_1^3 \sqrt{\ln x}\, dx$

In Exercises 27 and 28, the exact value of the integral is π (verify). Use $n = 10$ subintervals to approximate the integral by (a) the midpoint approximation and (b) the trapezoidal approximation, and use $2n = 10$ subintervals to approximate the integral by (c) Simpson's rule. Estimate the absolute error, and express your answers to at least four decimal places.

27. $\displaystyle\int_0^1 \dfrac{4}{1+x^2}\, dx$

28. $\displaystyle\int_0^2 \sqrt{4-x^2}\, dx$

29. In Example 6 we showed that taking $2n = 14$ subdivisions ensures that the approximation of

$$\ln 2 = \int_1^2 \dfrac{1}{x}\, dx$$

by Simpson's rule is accurate to five decimal places. Confirm this by comparing the approximation of $\ln 2$ produced by Simpson's rule with $2n = 14$ to the value produced directly by your calculating utility.

30. In parts (a) and (b), determine whether an approximation of the integral by the trapezoidal rule would be less than or would be greater than the exact value of the integral.

(a) $\displaystyle\int_1^2 e^{-x^2}\, dx$ (b) $\displaystyle\int_0^{0.5} e^{-x^2}\, dx$

In Exercises 31 and 32, find a value for n to ensure that the absolute error in approximating the integral by the midpoint rule will be less than 10^{-4}.

31. $\displaystyle\int_0^2 x \sin x\, dx$

32. $\displaystyle\int_0^1 e^{\cos x}\, dx$

In Exercises 33 and 34, show that inequalities (7) and (8) are of no value in finding an upper bound on the absolute error that results from approximating the integral using either the midpoint approximation or the trapezoidal approximation.

33. $\displaystyle\int_0^1 \sqrt{x}\, dx$

34. $\displaystyle\int_0^1 \sin\sqrt{x}\, dx$

In Exercises 35 and 36, use Simpson's rule with $2n = 10$ subintervals to approximate the length of the curve. Express your answers to at least four decimal places.

35. $y = \sin x$, $0 \le x \le \pi$ **36.** $y = 1/x$, $1 \le x \le 3$

Numerical integration methods can be used in problems where only measured or experimentally determined values of the integrand are available. In Exercises 37–42, use Simpson's rule to estimate the value of the relevant integral.

37. A graph of the speed v versus time t for a test run of an Infiniti G20 automobile is shown in the accompanying figure. Estimate the speeds at $t = 0, 5, 10, 15$, and 20 s from the graph, convert to ft/s using 1 mi/h $= 22/15$ ft/s, and use these speeds to approximate the number of feet traveled during the first 20 s. Round your answer to the nearest foot. [*Hint*: Distance traveled $= \int_0^{20} v(t)\, dt$.] [Data from *Road and Track*, October 1990.]

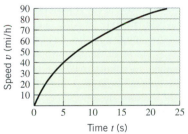

Figure Ex-37

38. A graph of the acceleration a versus time t for an object moving on a straight line is shown in the accompanying figure. Estimate the accelerations at $t = 0, 1, 2, \ldots, 8$ s from the graph and use them to approximate the change in velocity from $t = 0$ to $t = 8$ s. Round your answer to the nearest tenth cm/s. [*Hint*: Change in velocity $= \int_0^8 a(t)\, dt$.]

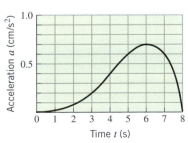

Figure Ex-38

39. The accompanying table gives the speeds, in miles per second, at various times for a test rocket that was fired upward from the surface of the Earth. Use these values to approximate the number of miles traveled during the first 180 s. Round your answer to the nearest tenth of a mile. [*Hint*: Distance traveled $= \int_0^{180} v(t)\, dt$.]

40. The accompanying table gives the speeds of a bullet at various distances from the muzzle of a rifle. Use these values to approximate the number of seconds for the bullet to travel 1800 ft. Express your answer to the nearest hundredth of a second. [*Hint:* If v is the speed of the bullet and x is the distance traveled, then $v = dx/dt$ so that $dt/dx = 1/v$ and $t = \int_0^{1800} (1/v)\, dx$.]

TIME t (s)	SPEED v (mi/s)	DISTANCE x (ft)	SPEED v (ft/s)
0	0.00	0	3100
30	0.03	300	2908
60	0.08	600	2725
90	0.16	900	2549
120	0.27	1200	2379
150	0.42	1500	2216
180	0.65	1800	2059

Table Ex-39   Table Ex-40

41. Measurements of a pottery shard recovered from an archaeological dig reveal that the shard came from a pot with a flat bottom and circular cross sections (see the accompanying figure). The figure shows interior radius measurements of the shard made every 4 cm from the bottom of the pot to the top. Use those values to approximate the interior volume of the pot to the nearest tenth of a liter (1 L = 1000 cm^3). [*Hint:* Use 6.2.3 (volume by cross sections) to set up an appropriate integral for the volume.]

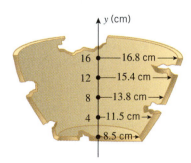

Figure Ex-41

42. Engineers want to construct a straight and level road 600 ft long and 75 ft wide by making a vertical cut through an intervening hill (see the accompanying figure). Heights of the hill above the centerline of the proposed road, as obtained at various points from a contour map of the region, are shown in the accompanying figure. To estimate the construction costs, the engineers need to know the volume of earth that must be removed. Approximate this volume, rounded to the nearest cubic foot. [*Hint:* First, set up an integral for the cross-sectional area of the cut along the centerline of the road, then assume that the height of the hill does not vary between the centerline and edges of the road.]

HORIZONTAL DISTANCE x (ft)	HEIGHT h (ft)
0	0
100	7
200	16
300	24
400	25
500	16
600	0

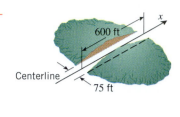

Figure Ex-42

43. Derive the trapezoidal rule by summing the areas of the trapezoids in Figure 8.7.2.

44. Let f be a function that is positive, continuous, decreasing, and concave down on the interval $[a, b]$. Assuming that $[a, b]$ is subdivided into n equal subintervals, arrange the following approximations of $\int_a^b f(x)\, dx$ in order of increasing value: left endpoint, right endpoint, midpoint, and trapezoidal.

C 45. Let $f(x) = \cos(x^2)$.
(a) Use a CAS to approximate the maximum value of $|f''(x)|$ on the interval $[0, 1]$.
(b) How large must n be in the midpoint approximation of $\int_0^1 f(x)\, dx$ to ensure that the absolute error is less than 5×10^{-4}? Compare your result with that obtained in Example 7.
(c) Estimate the integral using the midpoint approximation with the value of n obtained in part (b).

C 46. Let $f(x) = \sqrt{1 + x^3}$.
(a) Use a CAS to approximate the maximum value of $|f''(x)|$ on the interval $[0, 1]$.
(b) How large must n be in the trapezoidal approximation of $\int_0^1 f(x)\, dx$ to ensure that the absolute error is less than 10^{-3}?
(c) Estimate the integral using the trapezoidal approximation with the value of n obtained in part (b).

C 47. Let $f(x) = \cos(x^2)$.
(a) Use a CAS to approximate the maximum value of $|f^{(4)}(x)|$ on the interval $[0, 1]$.
(b) How large must the value of n be in the approximation of $\int_0^1 f(x)\, dx$ by Simpson's rule to ensure that the absolute error is less than 10^{-4}?
(c) Estimate the integral using Simpson's rule with the value of n obtained in part (b).

C 48. Let $f(x) = \sqrt{1 + x^3}$.
(a) Use a CAS to approximate the maximum value of $|f^{(4)}(x)|$ on the interval $[0, 1]$.
(b) How large must the value of n be in the approximation of $\int_0^1 f(x)\, dx$ by Simpson's rule to ensure that the absolute error is less than 10^{-5}?
(c) Estimate the integral using Simpson's rule with the value of n obtained in part (b).

8.8   IMPROPER INTEGRALS

Up to now we have focused on definite integrals with continuous integrands and finite intervals of integration. In this section we will extend the concept of a definite integral to include infinite intervals of integration and integrands that become infinite within the interval of integration.

IMPROPER INTEGRALS

It is assumed in the definition of the definite integral

$$\int_a^b f(x)\,dx$$

that $[a, b]$ is a finite interval and that the limit that defines the integral exists; that is, the function f is integrable. We observed in Theorems 6.5.2 and 6.5.8 that continuous functions are integrable, as are bounded functions with finitely many points of discontinuity. We also observed in Theorem 6.5.8 that functions that are not bounded on the interval of integration are not integrable. Thus, for example, a function with a vertical asymptote within the interval of integration would not be integrable.

Our main objective in this section is to extend the concept of a definite integral to allow for infinite intervals of integration and integrands with vertical asymptotes within the interval of integration. We will call the vertical asymptotes *infinite discontinuities*, and we will call integrals with infinite intervals of integration or infinite discontinuities within the interval of integration *improper integrals*. Here are some examples:

- Improper integrals with infinite intervals of integration:

$$\int_1^{+\infty} \frac{dx}{x^2}, \quad \int_{-\infty}^0 e^x\,dx, \quad \int_{-\infty}^{+\infty} \frac{dx}{1+x^2}$$

- Improper integrals with infinite discontinuities in the interval of integration:

$$\int_{-3}^3 \frac{dx}{x^2}, \quad \int_1^2 \frac{dx}{x-1}, \quad \int_0^{\pi} \tan x\,dx$$

- Improper integrals with infinite discontinuities and infinite intervals of integration:

$$\int_0^{+\infty} \frac{dx}{\sqrt{x}}, \quad \int_{-\infty}^{+\infty} \frac{dx}{x^2-9}, \quad \int_1^{+\infty} \sec x\,dx$$

INTEGRALS OVER INFINITE INTERVALS

To motivate a reasonable definition for improper integrals of the form

$$\int_a^{+\infty} f(x)\,dx$$

let us begin with the case where f is continuous and nonnegative on $[a, +\infty)$, so we can think of the integral as the area under the curve $y = f(x)$ over the interval $[a, +\infty)$ (Figure 8.8.1). At first, you might be inclined to argue that this area is infinite because the region has infinite extent. However, such an argument would be based on vague intuition rather than precise mathematical logic, since the concept of area has only been defined over intervals of *finite extent*. Thus, before we can make any reasonable statements about the area of the region in Figure 8.8.1, we need to begin by defining what we mean by the area of this region. For that purpose, it will help to focus on a specific example.

Suppose we are interested in the area A of the region that lies below the curve $y = 1/x^2$ and above the interval $[1, +\infty)$ on the x-axis. Instead of trying to find the entire area at once, let us begin by calculating the portion of the area that lies above a finite interval $[1, \ell]$,

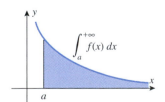

Figure 8.8.1

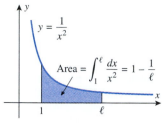

$y = \dfrac{1}{x^2}$

Area $= \displaystyle\int_1^\ell \dfrac{dx}{x^2} = 1 - \dfrac{1}{\ell}$

Figure 8.8.2

where $\ell > 1$ is arbitrary. That area is

$$\int_1^\ell \frac{dx}{x^2} = -\frac{1}{x}\bigg]_1^\ell = 1 - \frac{1}{\ell}$$

(Figure 8.8.2). If we now allow ℓ to increase so that $\ell \to +\infty$, then the portion of the area over the interval $[1, \ell]$ will begin to fill out the area over the entire interval $[1, +\infty)$ (Figure 8.8.3), and hence we can reasonably define the area A under $y = 1/x^2$ over the interval $[1, +\infty)$ to be

$$A = \int_1^{+\infty} \frac{dx}{x^2} = \lim_{\ell \to +\infty} \int_1^\ell \frac{dx}{x^2} = \lim_{\ell \to +\infty} \left(1 - \frac{1}{\ell}\right) = 1 \qquad (1)$$

Thus, the area has a finite value of 1 and is not infinite as we first conjectured.

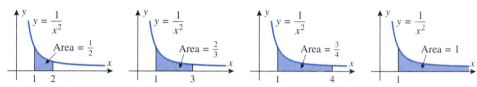

Figure 8.8.3

With the preceding discussion as our guide, we make the following definition (which is applicable to functions with both positive and negative values):

> **8.8.1**   DEFINITION.   The *improper integral of f over the interval* $[a, +\infty)$ is defined as
>
> $$\int_a^{+\infty} f(x)\,dx = \lim_{\ell \to +\infty} \int_a^\ell f(x)\,dx$$
>
> In the case where the limit exists, the improper integral is said to **converge**, and the limit is defined to be the value of the integral. In the case where the limit does not exist, the improper integral is said to **diverge**, and it is not assigned a value.

If f is nonnegative on $[a, +\infty)$ and the improper integral converges, then the value of the integral is regarded to be the area under the graph of f over the interval $[a, +\infty)$; and if the integral diverges, then the area under the graph of f over the interval $[a, +\infty)$ is regarded to be infinite.

Example 1   Evaluate

(a) $\displaystyle\int_1^{+\infty} \frac{dx}{x^3}$ (b) $\displaystyle\int_1^{+\infty} \frac{dx}{x}$

Solution (a). Following the definition, we replace the infinite upper limit by a finite upper limit ℓ, and then take the limit of the resulting integral. This yields

$$\int_1^{+\infty} \frac{dx}{x^3} = \lim_{\ell \to +\infty} \int_1^\ell \frac{dx}{x^3} = \lim_{\ell \to +\infty} \left[-\frac{1}{2x^2}\right]_1^\ell = \lim_{\ell \to +\infty} \left(\frac{1}{2} - \frac{1}{2\ell^2}\right) = \frac{1}{2}$$

Solution (b).

$$\int_1^{+\infty} \frac{dx}{x} = \lim_{\ell \to +\infty} \int_1^\ell \frac{dx}{x} = \lim_{\ell \to +\infty} \left[\ln x\right]_1^\ell = \lim_{\ell \to +\infty} \ln \ell = +\infty$$

In this case the integral diverges and hence has no value. ◀

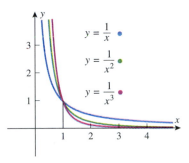

Figure 8.8.4

Because the functions $1/x^3$, $1/x^2$, and $1/x$ are nonnegative over the interval $[1, +\infty)$, it follows from (1) and the last example that over this interval the area under $y = 1/x^3$ is $\frac{1}{2}$, the area under $y = 1/x^2$ is 1, and the area under $y = 1/x$ is infinite. However, on the surface the graphs of the three functions seem very much alike (Figure 8.8.4), and there is nothing to suggest why one of the areas should be infinite and the other two finite. One explanation is that $1/x^3$ and $1/x^2$ approach zero more rapidly than $1/x$ as $x \to +\infty$, so that the area over the interval $[1, \ell]$ accumulates less rapidly under the curves $y = 1/x^3$ and $y = 1/x^2$ than under $y = 1/x$ as $\ell \to +\infty$, and the difference is just enough that the first two areas are finite and the third is infinite.

Example 2 For what values of p does the integral $\displaystyle\int_1^{+\infty} \frac{dx}{x^p}$ converge?

Solution. We know from the preceding example that the integral diverges if $p = 1$, so let us assume that $p \neq 1$. In this case we have

$$\int_1^{+\infty} \frac{dx}{x^p} = \lim_{\ell \to +\infty} \int_1^{\ell} x^{-p}\, dx = \lim_{\ell \to +\infty} \frac{x^{1-p}}{1-p} \bigg]_1^{\ell} = \lim_{\ell \to +\infty} \left[\frac{\ell^{1-p}}{1-p} - \frac{1}{1-p} \right]$$

If $p > 1$, then the exponent $1 - p$ is negative and $\ell^{1-p} \to 0$ as $\ell \to +\infty$; and if $p < 1$, then the exponent $1 - p$ is positive and $\ell^{1-p} \to +\infty$ as $\ell \to +\infty$. Thus, the integral converges if $p > 1$ and diverges otherwise. In the convergent case the value of the integral is

$$\int_1^{+\infty} \frac{dx}{x^p} = \left[0 - \frac{1}{1-p} \right] = \frac{1}{p-1} \quad (p > 1) \qquad \blacktriangleleft$$

The following theorem summarizes this result:

8.8.2 THEOREM.

$$\int_1^{+\infty} \frac{dx}{x^p} = \begin{cases} \dfrac{1}{p-1} & \text{if } p > 1 \\[2mm] \text{diverges} & \text{if } p \leq 1 \end{cases}$$

Example 3 Evaluate $\displaystyle\int_0^{+\infty} (1 - x)e^{-x}\, dx$.

Solution. Integrating by parts with $u = 1 - x$ and $dv = e^{-x}\, dx$ yields

$$\int (1 - x)e^{-x}\, dx = -e^{-x}(1-x) - \int e^{-x}\, dx = -e^{-x} + xe^{-x} + e^{-x} + C = xe^{-x} + C$$

Thus,

$$\int_0^{+\infty} (1 - x)e^{-x}\, dx = \lim_{\ell \to +\infty} \left[xe^{-x} \right]_0^{\ell} = \lim_{\ell \to +\infty} \frac{\ell}{e^{\ell}}$$

The limit is an indeterminate form of type ∞/∞, so we will apply L'Hôpital's rule by differentiating the numerator and denominator with respect to ℓ. This yields

$$\int_0^{+\infty} (1 - x)e^{-x}\, dx = \lim_{\ell \to +\infty} \frac{1}{e^{\ell}} = 0$$

An explanation of why this integral is zero can be obtained by interpreting the integral as the net signed area between the graph of $y = (1 - x)e^{-x}$ and the interval $[0, +\infty)$ (Figure 8.8.5). $\blacktriangleleft$

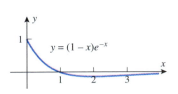

$y = (1 - x)e^{-x}$

The net signed area between the graph and the interval $[0, +\infty)$ is zero.

Figure 8.8.5

We also make the following definition:

8.8.3   DEFINITION.   The *improper integral of f over the interval* $(-\infty, b]$ is defined as

$$\int_{-\infty}^{b} f(x)\,dx = \lim_{k \to -\infty} \int_{k}^{b} f(x)\,dx \tag{2}$$

The integral is said to *converge* if the limit exists and *diverge* if it does not. The *improper integral of f over the interval* $(-\infty, +\infty)$ is defined as

$$\int_{-\infty}^{+\infty} f(x)\,dx = \int_{-\infty}^{c} f(x)\,dx + \int_{c}^{+\infty} f(x)\,dx \tag{3}$$

where c is any real number. The improper integral is said to *converge* if both terms converge and *diverge* if either term diverges.

REMARK.   In this definition, if f is nonnegative on the interval of integration, then the improper integral is regarded to be the area under the graph of f over that interval; the area has a finite value if the integral converges and is infinite if it diverges. We also note that in (3) it is usual to choose $c = 0$, but the choice does not matter; it can be proved that neither the convergence nor the value of the integral depends on the choice of c.

Example 4   Evaluate $\displaystyle\int_{-\infty}^{+\infty} \frac{dx}{1 + x^2}$.

Solution.   We will evaluate the integral by choosing $c = 0$ in (3). With this value for c we obtain

$$\int_{0}^{+\infty} \frac{dx}{1 + x^2} = \lim_{\ell \to +\infty} \int_{0}^{\ell} \frac{dx}{1 + x^2} = \lim_{\ell \to +\infty} \left[\tan^{-1} x\right]_{0}^{\ell} = \lim_{\ell \to +\infty} (\tan^{-1} \ell) = \frac{\pi}{2}$$

$$\int_{-\infty}^{0} \frac{dx}{1 + x^2} = \lim_{k \to -\infty} \int_{k}^{0} \frac{dx}{1 + x^2} = \lim_{k \to -\infty} \left[\tan^{-1} x\right]_{k}^{0} = \lim_{k \to -\infty} (-\tan^{-1} k) = \frac{\pi}{2}$$

Thus, the integral converges and its value is

$$\int_{-\infty}^{+\infty} \frac{dx}{1 + x^2} = \int_{-\infty}^{0} \frac{dx}{1 + x^2} + \int_{0}^{+\infty} \frac{dx}{1 + x^2} = \frac{\pi}{2} + \frac{\pi}{2} = \pi$$

Since the integrand is nonnegative on the interval $(-\infty, +\infty)$, the integral represents the area of the region shown in Figure 8.8.6.   ◀

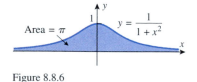

Area = π $y = \dfrac{1}{1 + x^2}$

Figure 8.8.6

INTEGRALS WHOSE INTEGRANDS HAVE INFINITE DISCONTINUITIES

Next we will consider improper integrals whose integrands have infinite discontinuities. We will start with the case where the interval of integration is a finite interval $[a, b]$ and the infinite discontinuity occurs at the right-hand endpoint.

To motivate an appropriate definition for such an integral let us consider the case where f is nonnegative on $[a, b]$, so we can interpret the improper integral $\int_{a}^{b} f(x)\,dx$ as the area of the region in Figure 8.8.7a. The problem of finding the area of this region is complicated by the fact that it extends indefinitely in the positive y-direction. However, instead of trying to find the entire area at once, we can proceed indirectly by calculating the portion of the area over the interval $[a, \ell]$ and then letting ℓ approach b to fill out the area of the entire region (Figure 8.8.7b). Motivated by this idea, we make the following definition:

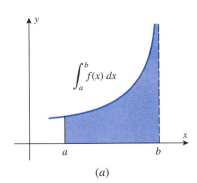

$$\int_a^b f(x)\, dx$$

(a)

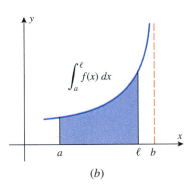

$$\int_a^\ell f(x)\, dx$$

(b)

Figure 8.8.7

8.8.4 DEFINITION. If f is continuous on the interval $[a, b]$, except for an infinite discontinuity at b, then the *improper integral of f over the interval $[a, b]$* is defined as

$$\int_a^b f(x)\, dx = \lim_{\ell \to b^-} \int_a^\ell f(x)\, dx \tag{4}$$

In the case where the limit exists, the improper integral is said to *converge*, and the limit is defined to be the value of the integral. In the case where the limit does not exist, the improper integral is said to *diverge*, and it is not assigned a value.

Example 5 Evaluate $\displaystyle\int_0^1 \frac{dx}{\sqrt{1 - x}}$.

Solution. The integral is improper because the integrand approaches $+\infty$ as x approaches the upper limit 1 from the left. From (4),

$$\int_0^1 \frac{dx}{\sqrt{1 - x}} = \lim_{\ell \to 1^-} \int_0^\ell \frac{dx}{\sqrt{1 - x}} = \lim_{\ell \to 1^-} \left[-2\sqrt{1 - x} \right]_0^\ell$$

$$= \lim_{\ell \to 1^-} [-2\sqrt{1 - \ell} + 2] = 2 \quad \blacktriangleleft$$

Improper integrals with an infinite discontinuity at the left-hand endpoint or inside the interval of integration are defined as follows.

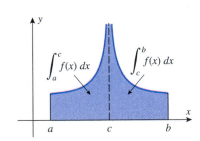

$$\int_a^c f(x)\, dx \qquad \int_c^b f(x)\, dx$$

Figure 8.8.8

8.8.5 DEFINITION. If f is continuous on the interval $[a, b]$, except for an infinite discontinuity at a, then the *improper integral of f over the interval $[a, b]$* is defined as

$$\int_a^b f(x)\, dx = \lim_{k \to a^+} \int_k^b f(x)\, dx \tag{5}$$

The integral is said to *converge* if the limit exists and *diverge* if it does not. If f is continuous on the interval $[a, b]$, except for an infinite discontinuity at a number c in (a, b), then the *improper integral of f over the interval $[a, b]$* is defined as

$$\int_a^b f(x)\, dx = \int_a^c f(x)\, dx + \int_c^b f(x)\, dx \tag{6}$$

The improper integral is said to *converge* if both terms converge and *diverge* if either term diverges (Figure 8.8.8).

Example 6 Evaluate

(a) $\displaystyle\int_1^2 \frac{dx}{1 - x}$ (b) $\displaystyle\int_1^4 \frac{dx}{(x - 2)^{2/3}}$ (c) $\displaystyle\int_0^{+\infty} \frac{dx}{\sqrt{x}(x + 1)}$

Solution (a). The integral is improper because the integrand approaches $-\infty$ as x approaches the lower limit 1 from the right (Figure 8.8.9). From Definition 8.8.5 we obtain

$$\int_1^2 \frac{dx}{1 - x} = \lim_{k \to 1^+} \int_k^2 \frac{dx}{1 - x} = \lim_{k \to 1^+} \left[-\ln|1 - x| \right]_k^2$$

$$= \lim_{k \to 1^+} \left[-\ln|-1| + \ln|1 - k| \right] = \lim_{k \to 1^+} \ln|1 - k| = -\infty$$

so the integral diverges.

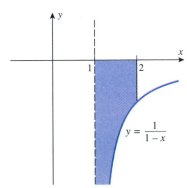

$$y = \frac{1}{1 - x}$$

Figure 8.8.9

Solution (b). The integral is improper because the integrand approaches $+\infty$ at $x = 2$, which is inside the interval of integration. From Definition 8.8.5 we obtain

$$\int_1^4 \frac{dx}{(x-2)^{2/3}} = \int_1^2 \frac{dx}{(x-2)^{2/3}} + \int_2^4 \frac{dx}{(x-2)^{2/3}} \tag{7}$$

But

$$\int_1^2 \frac{dx}{(x-2)^{2/3}} = \lim_{\ell \to 2^-} \int_1^\ell \frac{dx}{(x-2)^{2/3}} = \lim_{\ell \to 2^-} [3(\ell - 2)^{1/3} - 3(1-2)^{1/3}] = 3$$

$$\int_2^4 \frac{dx}{(x-2)^{2/3}} = \lim_{k \to 2^+} \int_k^4 \frac{dx}{(x-2)^{2/3}} = \lim_{k \to 2^+} [3(4-2)^{1/3} - 3(k-2)^{1/3}] = 3\sqrt[3]{2}$$

Thus, from (7)

$$\int_1^4 \frac{dx}{(x-2)^{2/3}} = 3 + 3\sqrt[3]{2}$$

Solution (c). This integral is improper for two reasons—the interval of integration is infinite, and there is an infinite discontinuity at $x = 0$. To evaluate this integral we will split the interval of integration at a convenient point, say $x = 1$, and write

$$\int_0^{+\infty} \frac{dx}{\sqrt{x}(x+1)} = \int_0^1 \frac{dx}{\sqrt{x}(x+1)} + \int_1^{+\infty} \frac{dx}{\sqrt{x}(x+1)}$$

The integrand in these two improper integrals does not match any of the forms in the Endpaper Integral Table, but the radical suggests the substitution $x = u^2$, $dx = 2u\, du$, from which we obtain

$$\int \frac{dx}{\sqrt{x}(x+1)} = \int \frac{2u\, du}{u(u^2+1)} = 2 \int \frac{du}{u^2+1}$$
$$= 2\tan^{-1} u + C = 2\tan^{-1} \sqrt{x} + C$$

Thus,

$$\int_0^{+\infty} \frac{dx}{\sqrt{x}(x+1)} = 2 \lim_{k \to 0^+} \left[\tan^{-1} \sqrt{x}\right]_k^1 + 2 \lim_{\ell \to +\infty} \left[\tan^{-1} \sqrt{x}\right]_1^\ell$$
$$= 2\left[\frac{\pi}{4} - 0\right] + 2\left[\frac{\pi}{2} - \frac{\pi}{4}\right] = \pi \qquad \blacktriangleleft$$

WARNING. It is sometimes tempting to apply the Fundamental Theorem of Calculus directly to an improper integral without taking the appropriate limits. To illustrate what can go wrong with this procedure, suppose we ignore the fact that the integral

$$\int_0^2 \frac{dx}{(x-1)^2} \tag{8}$$

is improper and write

$$\int_0^2 \frac{dx}{(x-1)^2} = -\frac{1}{x-1}\Big]_0^2 = -1 - (1) = -2$$

This result is clearly nonsense because the integrand is never negative and consequently the integral cannot be negative! To evaluate (8) correctly we should write

$$\int_0^2 \frac{dx}{(x-1)^2} = \int_0^1 \frac{dx}{(x-1)^2} + \int_1^2 \frac{dx}{(x-1)^2}$$

But

$$\int_0^1 \frac{dx}{(x-1)^2} = \lim_{\ell \to 1^-} \int_0^\ell \frac{dx}{(x-1)^2} = \lim_{\ell \to 1^-} \left[-\frac{1}{\ell-1} - 1\right] = +\infty$$

so that (8) diverges.

THE APPLICATION OF IMPROPER INTEGRALS TO ARC LENGTH AND SURFACE AREA

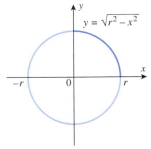

Figure 8.8.10

In Definitions 7.4.2 and 7.5.2 for arc length and surface area we required the function f to be smooth (continuous first derivative) to ensure the integrability in the resulting formula. However, smoothness is overly restrictive since some of the most basic formulas in geometry involve functions that are not smooth but lead to convergent improper integrals. Accordingly, let us agree to extend the definitions of arc length and surface area to allow functions that are not smooth, but for which the resulting integral in the formula converges.

Example 7 Derive the formula for the circumference of a circle of radius r.

Solution. For convenience, let us assume that the circle is centered at the origin, in which case its equation is $x^2 + y^2 = r^2$. We will find the arc length of the portion of the circle that lies in the first quadrant and then multiply by 4 to obtain the total circumference (Figure 8.8.10).

Since the equation of the upper semicircle is $y = \sqrt{r^2 - x^2}$, it follows from Formula (4) of Section 7.4 that the circumference C is

$$C = 4 \int_0^r \sqrt{1 + (dy/dx)^2}\, dx = 4 \int_0^r \sqrt{1 + \left(-\frac{x}{\sqrt{r^2 - x^2}}\right)^2}\, dx$$

$$= 4r \int_0^r \frac{dx}{\sqrt{r^2 - x^2}}$$

This integral is improper because of the infinite discontinuity at $x = r$, and hence we evaluate it by writing

$$C = 4r \lim_{\ell \to r^-} \int_0^\ell \frac{dx}{\sqrt{r^2 - x^2}}$$

$$= 4r \lim_{\ell \to r^-} \left[\sin^{-1}\left(\frac{x}{r}\right)\right]_0^\ell \qquad \boxed{\begin{array}{l}\text{Formula (77) in the}\\\text{Endpaper Integral Table}\end{array}}$$

$$= 4r \lim_{\ell \to r^-} \left[\sin^{-1}\left(\frac{\ell}{r}\right) - \sin^{-1} 0\right]$$

$$= 4r[\sin^{-1} 1 - \sin^{-1} 0] = 4r\left(\frac{\pi}{2} - 0\right) = 2\pi r \qquad \blacktriangleleft$$

EXERCISE SET 8.8 ~ Graphing Utility [C] CAS

1. In each part, determine whether the integral is improper, and if so, explain why.

   (a) $\displaystyle\int_1^5 \frac{dx}{x - 3}$ (b) $\displaystyle\int_1^5 \frac{dx}{x + 3}$ (c) $\displaystyle\int_0^1 \ln x\, dx$

   (d) $\displaystyle\int_1^{+\infty} e^{-x}\, dx$ (e) $\displaystyle\int_{-\infty}^{+\infty} \frac{dx}{\sqrt[3]{x - 1}}$ (f) $\displaystyle\int_0^{\pi/4} \tan x\, dx$

2. In each part, determine all values of p for which the integral is improper.

   (a) $\displaystyle\int_0^1 \frac{dx}{x^p}$ (b) $\displaystyle\int_1^2 \frac{dx}{x - p}$ (c) $\displaystyle\int_0^1 e^{-px}\, dx$

   In Exercises 3–30, evaluate the integrals that converge.

3. $\displaystyle\int_0^{+\infty} e^{-x}\, dx$

4. $\displaystyle\int_{-1}^{+\infty} \frac{x}{1 + x^2}\, dx$

5. $\displaystyle\int_4^{+\infty} \frac{2}{x^2 - 1}\, dx$

6. $\displaystyle\int_0^{+\infty} xe^{-x^2}\, dx$

7. $\displaystyle\int_e^{+\infty} \frac{1}{x \ln^3 x}\, dx$

8. $\displaystyle\int_2^{+\infty} \frac{1}{x\sqrt{\ln x}}\, dx$

9. $\displaystyle\int_{-\infty}^0 \frac{dx}{(2x - 1)^3}$

10. $\displaystyle\int_{-\infty}^2 \frac{dx}{x^2 + 4}$

11. $\displaystyle\int_{-\infty}^0 e^{3x}\, dx$

12. $\displaystyle\int_{-\infty}^0 \frac{e^x\, dx}{3 - 2e^x}$

13. $\displaystyle\int_{-\infty}^{+\infty} x^3\, dx$

14. $\displaystyle\int_{-\infty}^{+\infty} \frac{x}{\sqrt{x^2 + 2}}\, dx$

15. $\displaystyle\int_{-\infty}^{+\infty} \frac{x}{(x^2 + 3)^2}\, dx$

16. $\displaystyle\int_{-\infty}^{+\infty} \frac{e^{-t}}{1 + e^{-2t}}\, dt$

17. $\displaystyle\int_3^4 \frac{dx}{(x - 3)^2}$

18. $\displaystyle\int_0^8 \frac{dx}{\sqrt[3]{x}}$

19. $\displaystyle\int_0^{\pi/2} \tan x\, dx$

20. $\displaystyle\int_0^9 \frac{dx}{\sqrt{9-x}}$

21. $\displaystyle\int_0^1 \frac{dx}{\sqrt{1-x^2}}$

22. $\displaystyle\int_{-3}^1 \frac{x\,dx}{\sqrt{9-x^2}}$

23. $\displaystyle\int_0^{\pi/6} \frac{\cos x}{\sqrt{1-2\sin x}}\,dx$

24. $\displaystyle\int_0^{\pi/4} \frac{\sec^2 x}{1-\tan x}\,dx$

25. $\displaystyle\int_0^3 \frac{dx}{x-2}$

26. $\displaystyle\int_{-2}^2 \frac{dx}{x^2}$

27. $\displaystyle\int_{-1}^8 x^{-1/3}\,dx$

28. $\displaystyle\int_0^4 \frac{dx}{(x-2)^{2/3}}$

29. $\displaystyle\int_0^{+\infty} \frac{1}{x^2}\,dx$

30. $\displaystyle\int_1^{+\infty} \frac{dx}{x\sqrt{x^2-1}}$

In Exercises 31–34, make the u-substitution and evaluate the resulting definite integral.

31. $\displaystyle\int_0^{+\infty} \frac{e^{-\sqrt{x}}}{\sqrt{x}}\,dx;\ u=\sqrt{x}$ [*Note:* $u\to +\infty$ as $x\to +\infty$.]

32. $\displaystyle\int_0^{+\infty} \frac{dx}{\sqrt{x}(x+4)};\ u=\sqrt{x}$

33. $\displaystyle\int_0^{+\infty} \frac{e^{-x}}{\sqrt{1-e^{-x}}}\,dx;\ u=1-e^{-x}$
  [*Note:* $u\to 1$ as $x\to +\infty$.]

34. $\displaystyle\int_0^{+\infty} \frac{e^{-x}}{\sqrt{1-e^{-2x}}}\,dx;\ u=e^{-x}$

In Exercises 35 and 36, express the improper integral as a limit, and then evaluate that limit with a CAS. Confirm the answer by evaluating the integral directly with the CAS.

c 35. $\displaystyle\int_0^{+\infty} e^{-x}\cos x\,dx$

c 36. $\displaystyle\int_0^{+\infty} xe^{-3x}\,dx$

c 37. In each part, try to evaluate the integral exactly with a CAS. If your result is not a simple numerical answer, then use the CAS to find a numerical approximation of the integral.

(a) $\displaystyle\int_{-\infty}^{+\infty} \frac{1}{x^8+x+1}\,dx$

(b) $\displaystyle\int_0^{+\infty} \frac{1}{\sqrt{1+x^3}}\,dx$

(c) $\displaystyle\int_1^{+\infty} \frac{\ln x}{e^x}\,dx$

(d) $\displaystyle\int_1^{+\infty} \frac{\sin x}{x^2}\,dx$

c 38. In each part, confirm the result with a CAS.

(a) $\displaystyle\int_0^{+\infty} \frac{\sin x}{\sqrt{x}}\,dx = \sqrt{\frac{\pi}{2}}$

(b) $\displaystyle\int_{-\infty}^{+\infty} e^{-x^2}\,dx = \sqrt{\pi}$

(c) $\displaystyle\int_0^1 \frac{\ln x}{1+x}\,dx = -\frac{\pi^2}{12}$

39. Find the length of the curve $y=(4-x^{2/3})^{3/2}$ over the interval $[0,8]$.

40. Find the length of the curve $y=\sqrt{9-x^2}$ over the interval $[0,3]$.

In Exercises 41 and 42, use L'Hôpital's rule to help evaluate the improper integral.

41. $\displaystyle\int_0^1 \ln x\,dx$

42. $\displaystyle\int_1^{+\infty} \frac{\ln x}{x^2}\,dx$

43. Find the area of the region between the x-axis and the curve $y=e^{-3x}$ for $x\ge 0$.

44. Find the area of the region between the x-axis and the curve $y=8/(x^2-4)$ for $x\ge 3$.

45. Suppose that the region between the x-axis and the curve $y=e^{-x}$ for $x\ge 0$ is revolved about the x-axis.
(a) Find the volume of the solid that is generated.
(b) Find the surface area of the solid.

46. Suppose that f and g are continuous functions and that
$$0\le f(x)\le g(x)$$
if $x\ge a$. Give a reasonable informal argument using areas to explain why the following results are true.
(a) If $\int_a^{+\infty} f(x)\,dx$ diverges, then $\int_a^{+\infty} g(x)\,dx$ diverges.
(b) If $\int_a^{+\infty} g(x)\,dx$ converges, then $\int_a^{+\infty} f(x)\,dx$ converges and $\int_a^{+\infty} f(x)\,dx \le \int_a^{+\infty} g(x)\,dx$.
[*Note:* The results in this exercise are sometimes called *comparison tests* for improper integrals.]

In Exercises 47–51, use the results in Exercise 46.

47. (a) Confirm graphically and algebraically that $e^{-x^2}\le e^{-x}$ if $x\ge 1$.
(b) Evaluate the integral
$$\int_1^{+\infty} e^{-x}\,dx$$
(c) What does the result obtained in part (b) tell you about the integral
$$\int_1^{+\infty} e^{-x^2}\,dx?$$

48. (a) Confirm graphically and algebraically that
$$\frac{1}{2x+1}\le \frac{e^x}{2x+1}\quad (x\ge 0)$$
(b) Evaluate the integral
$$\int_0^{+\infty} \frac{dx}{2x+1}$$
(c) What does the result obtained in part (b) tell you about the integral
$$\int_0^{+\infty} \frac{e^x}{2x+1}\,dx?$$

49. Let R be the region to the right of $x=1$ that is bounded by the x-axis and the curve $y=1/x$. When this region is revolved about the x-axis it generates a solid whose surface is known as *Gabriel's Horn* (for reasons that should be clear from the accompanying figure). Show that the solid has a finite volume but its surface has an infinite area. [*Note:* It has been suggested that if one could saturate the interior of the solid with paint and allow it to seep through to the surface, then one could paint an infinite surface with a finite amount of paint! What do you think?]

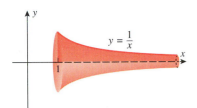

$y = \dfrac{1}{x}$

Figure Ex-49

50. In each part, use Exercise 46 to determine whether the integral converges or diverges. If it converges, then use part (b) of that exercise to find an upper bound on the value of the integral.

(a) $\displaystyle\int_{2}^{+\infty} \frac{\sqrt{x^3 + 1}}{x}\, dx$ (b) $\displaystyle\int_{2}^{+\infty} \frac{x}{x^5 + 1}\, dx$

(c) $\displaystyle\int_{0}^{+\infty} \frac{xe^x}{2x + 1}\, dx$

51. Show that

$$\lim_{x \to +\infty} \frac{\int_{0}^{2x} \sqrt{1 + t^3}\, dt}{x^{5/2}}$$

is an indeterminate form of type ∞/∞, and then use L'Hôpital's rule to find the limit.

52. (a) Give a reasonable informal argument, based on areas, that explains why the integrals

$$\int_{0}^{+\infty} \sin x\, dx \quad\text{and}\quad \int_{0}^{+\infty} \cos x\, dx$$

diverge.

(b) Show that $\displaystyle\int_{0}^{+\infty} \frac{\cos \sqrt{x}}{\sqrt{x}}\, dx$ diverges.

53. In electromagnetic theory, the magnetic potential at a point on the axis of a circular coil is given by

$$u = \frac{2\pi N I r}{k} \int_{a}^{+\infty} \frac{dx}{(r^2 + x^2)^{3/2}}$$

where N, I, r, k, and a are constants. Find u.

C **54.** The *average speed*, $\bar{v}$, of the molecules of an ideal gas is given by

$$\bar{v} = \frac{4}{\sqrt{\pi}} \left(\frac{M}{2RT}\right)^{3/2} \int_{0}^{+\infty} v^3 e^{-Mv^2/(2RT)}\, dv$$

and the *root-mean-square speed*, v_{rms}, by

$$v_{\text{rms}}^2 = \frac{4}{\sqrt{\pi}} \left(\frac{M}{2RT}\right)^{3/2} \int_{0}^{+\infty} v^4 e^{-Mv^2/(2RT)}\, dv$$

where v is the molecular speed, T is the gas temperature, M is the molecular weight of the gas, and R is the gas constant.

(a) Use a CAS to show that

$$\int_{0}^{+\infty} x^3 e^{-a^2 x^2}\, dx = \frac{1}{2a^4}, \quad a > 0$$

and use this result to show that $\bar{v} = \sqrt{8RT/\pi M}$.

(b) Use a CAS to show that

$$\int_{0}^{+\infty} x^4 e^{-a^2 x^2}\, dx = \frac{3\sqrt{\pi}}{8a^5}, \quad a > 0$$

and use this result to show that $v_{\text{rms}} = \sqrt{3RT/M}$.

55. In Exercise 19 of Section 7.6, we determined the work required to lift a 6000-lb satellite to an orbital position that is 1000 mi above the Earth's surface. The ideas discussed in that exercise will be needed here.

(a) Find a definite integral that represents the work required to lift a 6000-lb satellite to a position ℓ miles above the Earth's surface.

(b) Find a definite integral that represents the work required to lift a 6000-lb satellite an "infinite distance" above the Earth's surface. Evaluate the integral. [*Note:* The result obtained here is sometimes called the work required to "escape" the Earth's gravity.]

A *transform* is a formula that converts or "transforms" one function into another. Transforms are used in applications to convert a difficult problem into an easier problem whose solution can then be used to solve the original difficult problem. The *Laplace transform* of a function $f(t)$, which plays an important role in the study of differential equations, is denoted by $\mathcal{L}\{f(t)\}$ and is defined by

$$\mathcal{L}\{f(t)\} = \int_{0}^{+\infty} e^{-st} f(t)\, dt$$

In this formula s is treated as a constant in the integration process; thus, the Laplace transform has the effect of transforming $f(t)$ into a function of s. Use this formula in Exercises 56 and 57.

---

56. Show that

(a) $\mathcal{L}\{1\} = \dfrac{1}{s}, \ s > 0$ (b) $\mathcal{L}\{e^{2t}\} = \dfrac{1}{s - 2}, \ s > 2$

(c) $\mathcal{L}\{\sin t\} = \dfrac{1}{s^2 + 1}, \ s > 0$

(d) $\mathcal{L}\{\cos t\} = \dfrac{s}{s^2 + 1}, \ s > 0$.

57. In each part, find the Laplace transform.

(a) $f(t) = t, \ s > 0$ (b) $f(t) = t^2, \ s > 0$

(c) $f(t) = \begin{cases} 0, & t < 3 \\ 1, & t \geq 3 \end{cases}, \ s > 0$

C **58.** Later in the text, we will show that

$$\int_{0}^{+\infty} e^{-x^2}\, dx = \tfrac{1}{2}\sqrt{\pi}$$

Confirm that this is reasonable by using a CAS or a calculator with a numerical integration capability.

59. Use the result in Exercise 58 to show that

(a) $\displaystyle\int_{-\infty}^{+\infty} e^{-ax^2}\, dx = \sqrt{\frac{\pi}{a}}, \ a > 0$

(b) $\dfrac{1}{\sqrt{2\pi}\sigma} \displaystyle\int_{-\infty}^{+\infty} e^{-x^2/2\sigma^2}\, dx = 1, \ \sigma > 0$.

A convergent improper integral over an infinite interval can be approximated by first replacing the infinite limit(s) of integration by finite limit(s), then using a numerical integration technique, such as Simpson's rule, to approximate the integral with finite limit(s). This technique is illustrated in Exercises 60 and 61.

60. Suppose that the integral in Exercise 58 is approximated by first writing it as

$$\int_0^{+\infty} e^{-x^2}\,dx = \int_0^K e^{-x^2}\,dx + \int_K^{+\infty} e^{-x^2}\,dx$$

then dropping the second term, and then applying Simpson's rule to the integral

$$\int_0^K e^{-x^2}\,dx$$

The resulting approximation has two sources of error: the error from Simpson's rule and the error

$$E = \int_K^{+\infty} e^{-x^2}\,dx$$

that results from discarding the second term. We call E the **truncation error**.

(a) Approximate the integral in Exercise 58 by applying Simpson's rule with $2n = 10$ subdivisions to the integral

$$\int_0^3 e^{-x^2}\,dx$$

Round your answer to four decimal places and compare it to $\frac{1}{2}\sqrt{\pi}$ rounded to four decimal places.

(b) Use the result that you obtained in Exercise 46 and the fact that $e^{-x^2} \le \frac{1}{3}xe^{-x^2}$ for $x \ge 3$ to show that the truncation error for the approximation in part (a) satisfies $0 < E < 2.1 \times 10^{-5}$.

61. (a) It can be shown that

$$\int_0^{+\infty} \frac{1}{x^6+1}\,dx = \frac{\pi}{3}$$

Approximate this integral by applying Simpson's rule with $2n = 20$ subdivisions to the integral

$$\int_0^4 \frac{1}{x^6+1}\,dx$$

Round your answer to three decimal places and compare it to $\pi/3$ rounded to three decimal places.

(b) Use the result that you obtained in Exercise 46 and the fact that $1/(x^6+1) < 1/x^6$ for $x \ge 4$ to show that the truncation error for the approximation in part (a) satisfies $0 < E < 2 \times 10^{-4}$.

62. For what values of p does $\displaystyle\int_0^{+\infty} e^{px}\,dx$ converge?

63. Show that $\displaystyle\int_0^1 \frac{dx}{x^p}$ converges if $p < 1$ and diverges if $p \ge 1$.

c **64.** It is sometimes possible to convert an improper integral into a "proper" integral having the same value by making an appropriate substitution. Evaluate the following integral by making the indicated substitution, and investigate what happens if you evaluate the integral directly using a CAS.

$$\int_0^1 \sqrt{\frac{1+x}{1-x}}\,dx;\quad u = \sqrt{1-x}$$

In Exercises 65 and 66, transform the given improper integral into a proper integral by making the stated u-substitution, then approximate the proper integral by Simpson's rule with $2n = 10$ subdivisions. Round your answer to three decimal places.

65. $\displaystyle\int_0^1 \frac{\cos x}{\sqrt{x}}\,dx;\quad u = \sqrt{x}$

66. $\displaystyle\int_0^1 \frac{\sin x}{\sqrt{1-x}}\,dx;\quad u = \sqrt{1-x}$

SUPPLEMENTARY EXERCISES

c CAS

1. Consider the following methods for evaluating integrals: u-substitution, integration by parts, partial fractions, reduction formulas, and trigonometric substitutions. In each part, state the approach that you would try first to evaluate the integral. If none of them seems appropriate, then say so. You need not evaluate the integral.

(a) $\displaystyle\int x\sin x\,dx$

(b) $\displaystyle\int \cos x\sin x\,dx$

(c) $\displaystyle\int \tan^7 x\,dx$

(d) $\displaystyle\int \tan^7 x\sec^2 x\,dx$

(e) $\displaystyle\int \frac{3x^2}{x^3+1}\,dx$

(f) $\displaystyle\int \frac{3x^2}{(x+1)^3}\,dx$

(g) $\displaystyle\int \tan^{-1} x\,dx$

(h) $\displaystyle\int \sqrt{4-x^2}\,dx$

(i) $\displaystyle\int x\sqrt{4-x^2}\,dx$

2. Consider the following trigonometric substitutions:

$$x = 3\sin\theta, \quad x = 3\tan\theta, \quad x = 3\sec\theta$$

In each part, state the substitution that you would try first to evaluate the integral. If none seems appropriate, then state a trigonometric substitution that you would use. You need not evaluate the integral.

(a) $\displaystyle\int \sqrt{9+x^2}\,dx$

(b) $\displaystyle\int \sqrt{9-x^2}\,dx$

(c) $\int \sqrt{1 - 9x^2}\, dx$ (d) $\int \sqrt{x^2 - 9}\, dx$

(e) $\int \sqrt{9 + 3x^2}\, dx$ (f) $\int \sqrt{1 + (9x)^2}\, dx$

3. (a) What condition must a rational function satisfy for the method of partial fractions to be applicable directly?
   (b) If the condition in part (a) is not satisfied, what must you do if you want to use partial fractions?

4. What is an improper integral?

5. In each part, find the number of the formula in the Endpaper Integral Table that you would apply to evaluate the integral. You need not evaluate the integral.

   (a) $\int \sin 7x \cos 9x\, dx$ (b) $\int (x^7 - x^5)e^{9x}\, dx$

   (c) $\int x\sqrt{x - x^2}\, dx$ (d) $\int \dfrac{dx}{x\sqrt{4x + 3}}$

   (e) $\int x^9 \pi^x\, dx$ (f) $\int \dfrac{3x - 1}{2 + x^2}\, dx$

6. Evaluate the integral $\displaystyle\int_0^1 \dfrac{x^3}{\sqrt{x^2 + 1}}\, dx$ using
   (a) integration by parts
   (b) the substitution $u = \sqrt{x^2 + 1}$.

7. In each part, evaluate the integral by making an appropriate substitution and applying a reduction formula.

   (a) $\int \sin^4 2x\, dx$ (b) $\int x \cos^5(x^2)\, dx$

8. Consider the integral $\displaystyle\int \dfrac{1}{x^3 - x}\, dx$.
   (a) Evaluate the integral using the substitution $x = \sec\theta$. For what values of x is your result valid?
   (b) Evaluate the integral using the substitution $x = \sin\theta$. For what values of x is your result valid?
   (c) Evaluate the integral using the method of partial fractions. For what values of x is your result valid?

9. (a) Evaluate the integral

   $$\int \dfrac{1}{\sqrt{2x - x^2}}\, dx$$

   three ways: using the substitution $u = \sqrt{x}$, using the substitution $u = \sqrt{2 - x}$, and completing the square.
   (b) Show that the answers in part (a) are equivalent.

10. Find the area of the region that is enclosed by the curves $y = (x - 3)/(x^3 + x^2)$, $y = 0$, $x = 1$, and $x = 2$.

11. Sketch the region whose area is $\displaystyle\int_0^{+\infty} \dfrac{dx}{1 + x^2}$, and use your sketch to show that

    $$\int_0^{+\infty} \dfrac{dx}{1 + x^2} = \int_0^1 \sqrt{\dfrac{1 - y}{y}}\, dy$$

12. Find the area that is enclosed between the x-axis and the curve $y = (\ln x - 1)/x^2$ for $x \geq e$.

13. Find the volume of the solid that is generated when the region between the x-axis and the curve $y = e^{-x}$ for $x \geq 0$ is revolved about the y-axis.

14. Find a positive value of a that satisfies the equation

    $$\int_0^{+\infty} \dfrac{1}{x^2 + a^2}\, dx = 1$$

In Exercises 15–30, evaluate the integral.

15. $\displaystyle\int \sqrt{\cos\theta}\, \sin\theta\, d\theta$ 16. $\displaystyle\int_0^{\pi/4} \tan^7\theta\, d\theta$

17. $\displaystyle\int x \tan^2(x^2) \sec^2(x^2)\, dx$ 18. $\displaystyle\int_{-1/\sqrt{2}}^{1/\sqrt{2}} (1 - 2x^2)^{3/2}\, dx$

19. $\displaystyle\int \dfrac{dx}{(3 + x^2)^{3/2}}$

20. $\displaystyle\int \dfrac{\cos\theta}{\sin^2\theta - 6\sin\theta + 12}\, d\theta$

21. $\displaystyle\int \dfrac{x + 3}{\sqrt{x^2 + 2x + 2}}\, dx$ 22. $\displaystyle\int \dfrac{\sec^2\theta}{\tan^3\theta - \tan^2\theta}\, d\theta$

23. $\displaystyle\int \dfrac{dx}{(x - 1)(x + 2)(x - 3)}$ 24. $\displaystyle\int \dfrac{dx}{x(x^2 + x + 1)}$

25. $\displaystyle\int_4^8 \dfrac{\sqrt{x - 4}}{x}\, dx$ 26. $\displaystyle\int_0^9 \dfrac{\sqrt{x}}{x + 9}\, dx$

27. $\displaystyle\int \dfrac{1}{\sqrt{e^x + 1}}\, dx$ 28. $\displaystyle\int_0^{\ln 2} \sqrt{e^x - 1}\, dx$

29. $\displaystyle\int_a^{+\infty} \dfrac{x\, dx}{(x^2 + 1)^2}$

30. $\displaystyle\int_0^{+\infty} \dfrac{dx}{a^2 + b^2 x^2}, \quad a, b > 0$

Some integrals that can be evaluated by hand cannot be evaluated by all computer algebra systems. In Exercises 31–34, evaluate the integral by hand, and determine if it can be evaluated on your CAS.

[C] 31. $\displaystyle\int \dfrac{x^3}{\sqrt{1 - x^8}}\, dx$

[C] 32. $\displaystyle\int (\cos^{32} x \sin^{30} x - \cos^{30} x \sin^{32} x)\, dx$

[C] 33. $\displaystyle\int \sqrt{x - \sqrt{x^2 - 4}}\, dx.$ [Hint: $\frac{1}{2}(\sqrt{x + 2} - \sqrt{x - 2})^2 =?$]

[C] 34. $\displaystyle\int \dfrac{1}{x^{10} + x}\, dx.$ [Hint: Rewrite the denominator as $x^{10}(1 + x^{-9})$.]

[C] 35. Let

$$f(x) = \dfrac{-2x^5 + 26x^4 + 15x^3 + 6x^2 + 20x + 43}{x^6 - x^5 - 18x^4 - 2x^3 - 39x^2 - x - 20}$$

(a) Use a CAS to factor the denominator, and then write down the form of the partial fraction decomposition. You need not find the values of the constants.
(b) Check your answer in part (a) by using the CAS to find the partial fraction decomposition of f.
(c) Integrate f by hand, and then check your answer by integrating with the CAS.

36. The **Gamma function**, $\Gamma(x)$, is defined as

$$\Gamma(x) = \int_0^{+\infty} t^{x-1} e^{-t}\, dt$$

It can be shown that this improper integral converges if and only if $x > 0$.

(a) Find $\Gamma(1)$.

(b) Prove: $\Gamma(x+1) = x\Gamma(x)$ for all $x > 0$. [*Hint:* Use integration by parts.]

(c) Use the results in parts (a) and (b) to find $\Gamma(2)$, $\Gamma(3)$, and $\Gamma(4)$; and then make a conjecture about $\Gamma(n)$ for positive integer values of n.

(d) Show that $\Gamma\left(\frac{1}{2}\right) = \sqrt{\pi}$. [*Hint:* See Exercise 58 of Section 8.8.]

(e) Use the results obtained in parts (b) and (d) to show that $\Gamma\left(\frac{3}{2}\right) = \frac{1}{2}\sqrt{\pi}$ and $\Gamma\left(\frac{5}{2}\right) = \frac{3}{4}\sqrt{\pi}$.

37. Refer to the Gamma function defined in Exercise 36 to show that

(a) $\displaystyle\int_0^1 (\ln x)^n\, dx = (-1)^n \Gamma(n+1), \quad n > 0.$

[*Hint:* Let $t = -\ln x$.]

(b) $\displaystyle\int_0^{+\infty} e^{-x^n}\, dx = \Gamma\left(\frac{n+1}{n}\right), \quad n > 0.$

[*Hint:* Let $t = x^n$. Use the result in Exercise 36(b).]

c **38.** A **simple pendulum** consists of a mass that swings in a vertical plane at the end of a massless rod of length L, as shown in the accompanying figure. Suppose that a simple pendulum is displaced through an angle θ_0 and released from rest. It can be shown that in the absence of friction, the time T required for the pendulum to make one complete back-and-forth swing, called the **period**, is given by

$$T = \sqrt{\frac{8L}{g}} \int_0^{\theta_0} \frac{1}{\sqrt{\cos\theta - \cos\theta_0}}\, d\theta \qquad (1)$$

where $\theta = \theta(t)$ is the angle the pendulum makes with the vertical at time t. The improper integral in (1) is difficult to evaluate numerically. By a substitution outlined below it can be shown that the period can be expressed as

$$T = 4\sqrt{\frac{L}{g}} \int_0^{\pi/2} \frac{1}{\sqrt{1 - k^2 \sin^2\phi}}\, d\phi \qquad (2)$$

where $k = \sin(\theta_0/2)$. The integral in (2) is called a **complete elliptic integral of the first kind** and is more easily evaluated by numerical methods.

(a) Obtain (2) from (1) by substituting

$$\cos\theta = 1 - 2\sin^2(\theta/2)$$
$$\cos\theta_0 = 1 - 2\sin^2(\theta_0/2)$$
$$k = \sin(\theta_0/2)$$

and then making the change of variable

$$\sin\phi = \sin(\theta/2)/\sin(\theta_0/2) = \sin(\theta/2)/k$$

(b) Use (2) and the numerical integration capability of your CAS to estimate the period of a simple pendulum for which $L = 1.5$ ft, $\theta_0 = 20°$, and $g = 32$ ft/s^2.

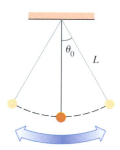

Figure Ex-38

EXPANDING THE CALCULUS HORIZON

Railroad Design

Y*our company has a contract to construct a track bed for a railroad line between towns A and B shown on the contour map in Figure 1. The bed can be created by cutting trenches through the surface or by using some combination of trenches and tunnels. As chief engineer, your assignment is to analyze the costs of trenches and tunnels and to propose a design strategy for minimizing the total construction cost.*

Engineering Requirements

The Transportation Board submits the following engineering requirements to your company:

- The track bed is to be straight and 10 m wide. The grade is to increase at a constant rate from the existing elevation of 100 m at town A to an elevation of 110 m at point M and then decrease at a constant rate to the existing elevation of 88 m at town B.

- From town *A* to point *M* and from point *N* to town *B* the track bed is to be created by excavating a trench whose vertical cross sections are trapezoids with the dimensions shown in Figure 2.
- Between points *M* and *N* your company must decide whether to excavate a trench of the type in Figure 2 or to excavate a tunnel whose vertical cross sections have the dimensions shown in Figure 3.

CONTOUR MAP
ELEVATIONS IN METERS

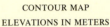

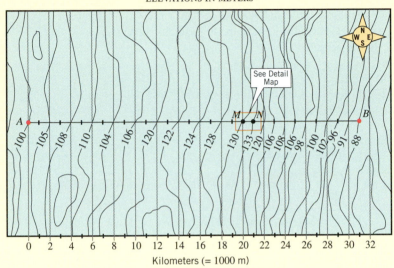

Figure 1

DETAIL MAP
PROPOSED TUNNEL CONSTRUCTION

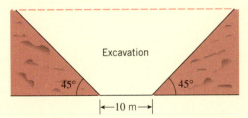

Figure 2

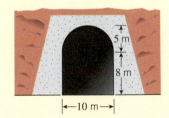

Figure 3

Cost Factors

Surface excavation of railbeds is performed using bulldozers, hydraulic excavators (backhoes), loading tractors, and other specialized equipment. Typically, the excavated dirt is piled at the side of the tracks to form sloped embankments, and the excavation cost is estimated from the volume of dirt to be removed and piled.

Tunnels in rock are often excavated by drilling shafts and inserting boring machines (called *moles*) to loosen and remove rock and dirt. Tunnels in soft ground are often excavated by starting at the tunnel face and using bucket or rotary excavators housed inside of shields. As the excavator progresses, tunnel liners are inserted behind it to support the earth and prevent cave-ins. Dirt removal is performed using conveyors or sometimes using railcars (called *muck cars*) that run on specially constructed tracks. Ventilation and air compression are other factors that add to the excavation cost of tunnels. In general, the excavation cost for a tunnel can be estimated from two components, the total volume of dirt to be removed and a cost that increases with the distance to the tunnel opening.

Make the following cost assumptions:

- The excavation and dirt-piling cost for a trench is $4.00 per cubic meter.
- The drilling and dirt-piling cost for a tunnel is $8.00 per cubic meter, and the costs involved in moving a load of dirt inside the tunnel a distance of 1 m toward the entrance along the track line is $0.06 per cubic meter.

Cost Analysis of Trenches

Assume that variations in elevation are negligible for short distances at right angles to the track, so that the cross sections of the dirt to be excavated always have the trapezoidal shape shown in Figure 2 (straight horizontal edges at the surface).

Exercise 1   Complete Table 1, and then use the table and Simpson's rule with $2n = 10$ to approximate the cost of a trench from town A to point M.

Table 1

DISTANCE x FROM TOWN A (m)	TERRAIN ELEVATION (m)	TRACK ELEVATION (m)	DEPTH OF CUT (m)	CROSS-SECTIONAL AREA $f(x)$ OF CUT (m^2)
0	100	100	0	0
2,000	105	101	4	56
4,000				
6,000				
8,000				
10,000				
12,000				
14,000				
16,000				
18,000				
20,000				

Exercise 2   As in Exercise 1, use Simpson's rule with $2n = 10$ to approximate the cost of constructing a trench from (a) point M to point N, and (b) point N to town B.

Exercise 3   Find the total cost of the project if a trench is used along the entire line from town A to town B.

Cost Analysis of a Tunnel

Exercise 4

(a) Find the volume of dirt that must be removed from the tunnel, and calculate the drilling and dirt-piling cost.

(b) Find an integral for the cost of moving all of the dirt inside the tunnel to the tunnel entrance. [*Suggestion:* Use Riemann sums.]

(c) Find the total cost of excavating the tunnel.

••••••••••
Exercise 5   Find the total cost of the project using a trench from town A to point M, a tunnel from point M to point N, and a trench from point N to town B. Compare the cost to that obtained in Exercise 3 and state which method is cheaper.

••

Module by: *C. Lynn Kiaer, Rose-Hulman Institute of Technology*
               David Ryeburn, Simon Fraser University
               Howard Anton, Drexel University
               Peter Dunn, Railroad Construction Company, Inc., Paterson, NJ

9

MATHEMATICAL MODELING WITH DIFFERENTIAL EQUATIONS

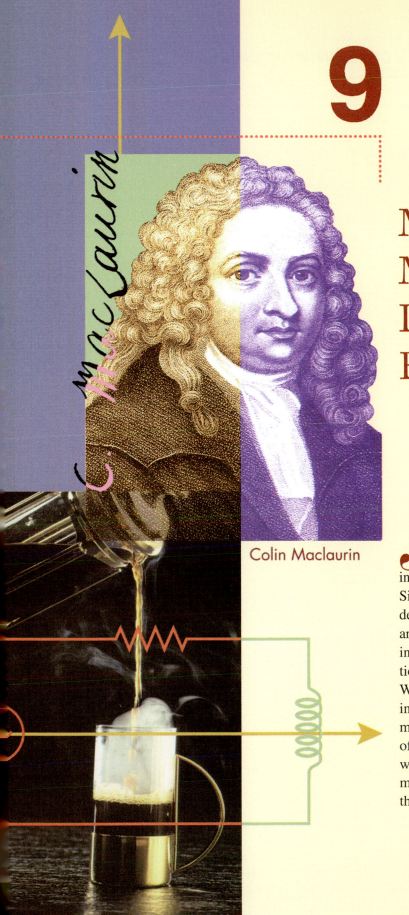

Colin Maclaurin

*M*any of the principles in science and engineering concern relationships between changing quantities. Since rates of change are represented mathematically by derivatives, it should not be surprising that such principles are often expressed in terms of differential equations. We introduced the concept of a differential equation in Section 6.2, but in this chapter we will go into more detail. We will discuss some important mathematical models that involve differential equations, and we will discuss some methods for solving and approximating solutions of some of the basic types of differential equations. However, we will only be able to touch the surface of this topic, leaving many important topics in differential equations to courses that are devoted completely to the subject.

9.1 FIRST-ORDER DIFFERENTIAL EQUATIONS AND APPLICATIONS

In this section we will introduce some basic terminology and concepts concerning differential equations. We will also discuss methods for solving certain basic types of differential equations, and we will give some applications of our work.

TERMINOLOGY

Recall from Section 6.2 that a ***differential equation*** is an equation involving one or more derivatives of an unknown function. In this section we will denote the unknown function by $y = y(x)$ unless the differential equation arises from an applied problem involving time, in which case we will denote it by $y = y(t)$. The ***order*** of a differential equation is the order of the highest derivative that it contains. Here are some examples:

DIFFERENTIAL EQUATION	ORDER
$\dfrac{dy}{dx} = 3y$	1
$\dfrac{d^2y}{dx^2} - 6\dfrac{dy}{dx} + 8y = 0$	2
$\dfrac{d^3y}{dx^3} - t\dfrac{dy}{dt} + (t^2 - 1)y = e^t$	3
$y' - y = e^{2x}$	1
$y'' + y' = \cos t$	2

In the last two equations the derivatives of y are expressed in "prime" notation. You will usually be able to tell from the equation itself or the context in which it arises whether to interpret y' as dy/dx or as dy/dt.

SOLUTIONS OF DIFFERENTIAL EQUATIONS

A function $y = y(x)$ is a ***solution*** of a differential equation on an open interval I if the equation is satisfied identically on I when y and its derivatives are substituted into the equation. For example, $y = e^{2x}$ is a solution of the differential equation

$$\frac{dy}{dx} - y = e^{2x} \tag{1}$$

on the interval $I = (-\infty, +\infty)$, since substituting y and its derivative into the left side of this equation yields

$$\frac{dy}{dx} - y = \frac{d}{dx}[e^{2x}] - e^{2x} = 2e^{2x} - e^{2x} = e^{2x}$$

for all real values of x. However, this is not the only solution on I; for example, the function

$$y = Ce^x + e^{2x} \tag{2}$$

is also a solution for every real value of the constant C, since

$$\frac{dy}{dx} - y = \frac{d}{dx}[Ce^x + e^{2x}] - (Ce^x + e^{2x}) = (Ce^x + 2e^{2x}) - (Ce^x + e^{2x}) = e^{2x}$$

After developing some techniques for solving equations such as (1), we will be able to show that *all* solutions of (1) on $I = (-\infty, +\infty)$ can be obtained by substituting values for the constant C in (2). On a given interval I, a solution of a differential equation from which all solutions on I can be derived by substituting values for arbitrary constants is called a ***general solution*** of the equation on I. Thus (2) is a general solution of (1) on the interval $I = (-\infty, +\infty)$.

REMARK. Usually, the general solution of an *n*th-order differential equation on an interval will contain *n* arbitrary constants. Although we will not prove this, it makes sense intuitively because *n* integrations are needed to recover a function from its *n*th derivative, and each integration introduces an arbitrary constant. For example, (2) has one arbitrary constant, which is consistent with the fact that it is the general solution of the *first-order* equation (1).

The graph of a solution of a differential equation is called an ***integral curve*** for the equation, so the general solution of a differential equation produces a family of integral curves corresponding to the different possible choices for the arbitrary constants. For example, Figure 9.1.1 shows some integral curves for (1), which were obtained by assigning values to the arbitrary constant in (2).

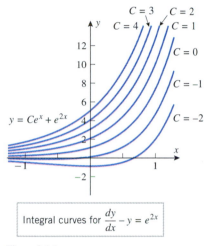

Integral curves for $\dfrac{dy}{dx} - y = e^{2x}$

Figure 9.1.1

INITIAL-VALUE PROBLEMS

When an applied problem leads to a differential equation, there are usually conditions in the problem that determine specific values for the arbitrary constants. As a rule of thumb, it requires *n* conditions to determine values for all *n* arbitrary constants in the general solution of an *n*th-order differential equation (one condition for each constant). For a first-order equation, the single arbitrary constant can be determined by specifying the value of the unknown function $y(x)$ at an arbitrary *x*-value x_0, say $y(x_0) = y_0$. This is called an ***initial condition***, and the problem of solving a first-order equation subject to an initial condition is called a ***first-order initial-value problem***. Geometrically, the initial condition $y(x_0) = y_0$ has the effect of isolating the integral curve that passes through the point (x_0, y_0) from the complete family of integral curves.

Example 1 The solution of the initial-value problem

$$\frac{dy}{dx} - y = e^{2x}, \quad y(0) = 3$$

can be obtained by substituting the initial condition $x = 0$, $y = 3$ in the general solution (2) to find C. We obtain

$$3 = Ce^0 + e^0 = C + 1$$

Thus, $C = 2$, and the solution of the initial-value problem, which is obtained by substituting this value of C in (2), is

$$y = 2e^x + e^{2x}$$

Geometrically, this solution is realized as the integral curve in Figure 9.1.1 that passes through the point $(0, 3)$. ◄

FIRST-ORDER LINEAR EQUATIONS

The simplest first-order equations are those that can be written in the form

$$\frac{dy}{dx} = q(x) \tag{3}$$

Such equations can often be solved by integration. For example, if

$$\frac{dy}{dx} = x^3 \tag{4}$$

then

$$y = \int x^3 \, dx = \frac{x^4}{4} + C$$

is the general solution of (4) on the interval $I = (-\infty, +\infty)$. More generally, a first-order differential equation is called **linear** if it is expressible in the form

$$\frac{dy}{dx} + p(x)y = q(x) \tag{5}$$

Equation (3) is the special case of (5) that results when the function $p(x)$ is identically 0. Some other examples of first-order linear differential equations are

$$\frac{dy}{dx} + x^2 y = e^x, \qquad \frac{dy}{dx} + (\sin x)y + x^3 = 0, \qquad \frac{dy}{dx} + 5y = 2$$

$$\underbrace{p(x) = x^2, \, q(x) = e^x} \qquad \underbrace{p(x) = \sin x, \, q(x) = -x^3} \qquad \underbrace{p(x) = 5, \, q(x) = 2}$$

Let us assume that the functions $p(x)$ and $q(x)$ are both *continuous* on some common open interval I. We will now describe a procedure for finding a general solution to (5) on I. From the Fundamental Theorem of Calculus (Theorem 6.6.3) it follows that $p(x)$ has an antiderivative $P = P(x)$ on I. That is, there exists a differentiable function $P(x)$ on I such that $dP/dx = p(x)$. Consequently, the function $\mu = \mu(x)$ defined by $\mu = e^{P(x)}$ is differentiable on I with

$$\frac{d\mu}{dx} = \frac{d}{dx}\left(e^{P(x)}\right) = \frac{dP}{dx}e^{P(x)} = \mu p(x)$$

Suppose now that $y = y(x)$ is a solution to (5) on I. Then

$$\frac{d}{dx}(\mu y) = \mu\frac{dy}{dx} + \frac{d\mu}{dx}y = \mu\frac{dy}{dx} + \mu p(x)y = \mu\left(\frac{dy}{dx} + p(x)y\right) = \mu q(x)$$

That is, the function μy is an antiderivative (or integral) of the known function $\mu q(x)$. For this reason, the function $\mu = e^{P(x)}$ is known as an **integrating factor** for Equation (5). On the other hand, the function $\mu q(x)$ is continuous on I and therefore possesses an antiderivative $H(x)$. It then follows from Theorem 6.2.2 that $\mu y = H(x) + C$ for some constant C or, equivalently, that

$$y = \frac{1}{\mu}[H(x) + C] \tag{6}$$

Conversely, it is straightforward to check that for any choice of C, Equation (6) defines a solution to (5) on I [Exercise 58(a)]. We conclude that a general solution to (5) on I is given by (6). Since

$$\int \mu q(x) \, dx = H(x) + C$$

this general solution can be expressed as

$$y = \frac{1}{\mu}\int \mu q(x) \, dx \tag{7}$$

We will refer to this process for solving Equation (5) as **the method of integrating factors**.

Example 2 Solve the differential equation

$$\frac{dy}{dx} - y = e^{2x}$$

Solution. This is a first-order linear differential equation with the functions $p(x) = -1$ and $q(x) = e^{2x}$ that are both continuous on the interval $I = (-\infty, +\infty)$. Thus, we can choose $P(x) = -x$, with $\mu = e^{-x}$, and $\mu q(x) = e^{-x}e^{2x} = e^x$ so that the general solution to this equation on I is given by

$$y = \frac{1}{\mu} \int \mu q(x)\,dx = \frac{1}{e^{-x}} \int e^x\,dx = e^x[e^x + C] = e^{2x} + Ce^x$$

Note that this solution is in agreement with Equation (2) discussed earlier. ◄

It is not necessary to memorize Equation (7) to apply the method of integrating factors; you need only remember the integrating factor $\mu = e^{P(x)}$ and the steps used to derive Equation (7).

Example 3 Solve the initial-value problem

$$x\frac{dy}{dx} - y = x, \quad y(1) = 2$$

Solution. This differential equation can be written in the form of (5) by dividing through by x. This yields

$$\frac{dy}{dx} - \frac{1}{x}y = 1 \tag{8}$$

where $q(x) = 1$ is continuous on $(-\infty, +\infty)$ and $p(x) = -1/x$ is continuous on $(-\infty, 0)$ and $(0, +\infty)$. Since we need $p(x)$ and $q(x)$ to be continuous on a common interval, and since our initial condition presumes a solution for $x = 1$, we will find the general solution of (8) on the interval $(0, +\infty)$. On this interval

$$\int \frac{1}{x}\,dx = \ln x + C$$

so that we can take $P(x) = -\ln x$ with $\mu = e^{P(x)} = e^{-\ln x} = 1/x$ the corresponding integrating factor. Multiplying both sides of Equation (8) by this integrating factor yields

$$\frac{1}{x}\frac{dy}{dx} - \frac{1}{x^2}y = \frac{1}{x}$$

or

$$\frac{d}{dx}\left[\frac{1}{x}y\right] = \frac{1}{x}$$

Therefore, on the interval $(0, +\infty)$,

$$\frac{1}{x}y = \int \frac{1}{x}\,dx = \ln x + C$$

from which it follows that

$$y = x\ln x + Cx \tag{9}$$

The initial condition $y(1) = 2$ requires that $y = 2$ if $x = 1$. Substituting these values into (9) and solving for C yields $C = 2$ (verify), so the solution of the initial-value problem is

$$y = x\ln x + 2x$$ ◄

The result of Example 3 illustrates an important property of first-order linear initial-value problems: Given any x_0 in I and any value of y_0, there will always *exist* a solution $y = y(x)$ to (5) on I with $y(x_0) = y_0$; furthermore, this solution will be *unique* [Exercise 58(b)]. Such existence and uniqueness results need not hold for nonlinear equations (Exercise 60).

Solving a first-order linear differential equation involves only the integration of functions of x. We will now consider a collection of equations whose solutions require integration of functions of y as well. A first-order **separable** differential equation is one that can be written in the form

$$h(y)\frac{dy}{dx} = g(x) \tag{10}$$

For example, the equation

$$(4y - \cos y)\frac{dy}{dx} = 3x^2$$

is a separable equation with

$$h(y) = 4y - \cos y \quad \text{and} \quad g(x) = 3x^2$$

We will assume that the functions $h(y)$ and $g(x)$ both possess antiderivatives in their respective variables y and x. That is, there exists a differentiable function $H(y)$ with $dH/dy = h(y)$ and there exists a differentiable function $G(x)$ with $dG/dx = g(x)$.

Suppose now that $y = y(x)$ is a solution to (10) on an open interval I. Then it follows from the chain rule that

$$\frac{d}{dx}[H(y)] = \frac{dH}{dy}\frac{dy}{dx} = h(y)\frac{dy}{dx} = g(x)$$

In other words, the function $H(y(x))$ is an antiderivative of $g(x)$ on the interval I. By Theorem 6.2.2, there must exist a constant C such that $H(y(x)) = G(x) + C$ on I. Equivalently, the solution $y = y(x)$ to (10) is defined *implicitly* by the equation

$$H(y) = G(x) + C \tag{11}$$

Conversely, suppose that for some choice of C a differentiable function $y = y(x)$ is defined implicitly by (11), Then $y(x)$ will be a solution to (10) (Exercise 59). We conclude that every solution to (10) will be given implicitly by Equation (11) for some appropriate choice of C.

We can express Equation (11) symbolically by writing

$$\int h(y)\,dy = \int g(x)\,dx \tag{12}$$

Informally, we first "multiply" both sides of Equation (10) by dx to "separate" the variables into the equation $h(y)\,dy = g(x)\,dx$. Integrating both sides of this equation then gives Equation (12). This process is called the method of **separation of variables**. Although separation of variables provides us with a convenient way of recovering Equation (11), it must be interpreted with care. For example, the constant C in Equation (11) is often *not* arbitrary; some choices of C may yield solutions, and others may not. Furthermore, even when solutions do exist, their domains can vary in unexpected ways with the choice of C. It is for reasons such as these that we will not refer to a "general" solution of a separable equation.

In some cases Equation (11) can be solved to yield explicit solutions to (10).

Example 4   Solve the differential equation

$$\frac{dy}{dx} = -4xy^2$$

and then solve the initial-value problem

$$\frac{dy}{dx} = -4xy^2, \quad y(0) = 1$$

Solution.   For $y \neq 0$ we can write this equation in the form of (10) as

$$\frac{1}{y^2}\frac{dy}{dx} = -4x$$

Separating variables and integrating yields

$$\frac{1}{y^2}\,dy = -4x\,dx$$

$$\int \frac{1}{y^2}\,dy = \int -4x\,dx$$

which is a symbolic expression of the equation

$$-\frac{1}{y} = -2x^2 + C$$

Solving for y as a function of x, we obtain

$$y = \frac{1}{2x^2 - C}$$

The initial condition $y(0) = 1$ requires that $y = 1$ when $x = 0$. Substituting these values into our solution yields $C = -1$ (verify). Thus, a solution to the initial-value problem is

$$y = \frac{1}{2x^2 + 1}$$

Some integral curves and our solution of the initial-value problem are graphed in Figure 9.1.2.   ◀

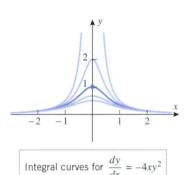

Integral curves for $\dfrac{dy}{dx} = -4xy^2$

Figure 9.1.2

One aspect of our solution to Example 4 deserves special comment. Had the initial condition been $y(0) = 0$ instead of $y(0) = 1$, the method we used would have failed to yield a solution to the resulting initial-value problem (Exercise 39). This is due to the fact that we assumed $y \neq 0$ in order to rewrite the equation $dy/dx = -4xy^2$ in the form

$$\frac{1}{y^2}\frac{dy}{dx} = -4x$$

It is important to be aware of such assumptions when manipulating a differential equation algebraically.

As a second example, consider the first-order linear equation $dy/dx - 3y = 0$. Using the method of integrating factors, it is easy to see that the general solution of this equation is $y = Ce^{3x}$ (verify). On the other hand, we can also apply the method of separation of variables to this differential equation. For $y \neq 0$ the equation can be written in the form

$$\frac{1}{y}\frac{dy}{dx} = 3$$

Separating the variables and integrating yields

$$\int \frac{dy}{y} = \int 3\,dx$$

$$\ln|y| = 3x + c$$

$$|y| = e^{3x+c} = e^c e^{3x} \qquad \boxed{\text{We have used } c \text{ as the constant of integration here to reserve } C \text{ for the constant in the final result.}}$$

$$y = \pm e^c e^{3x} = Ce^{3x} \qquad \boxed{\text{Letting } C = \pm e^c}$$

This appears to be the same solution that we obtained using the method of integrating factors. However, the careful reader may have observed that the constant $C = \pm e^c$ is not truly arbitrary, since $C = 0$ is not an allowable value. Thus, separation of variables missed the solution $y = 0$, which the method of integrating factors did not. The problem occurred because we had to divide by y to separate the variables. (Exercises 7 and 8 ask you to compare the two methods with some other first-order linear equations.)

It is often not possible to solve Equation (11) for y as an explicit function of x. In such cases, it is common to refer to Equation (11) as a "solution" to (10).

Example 5   Solve the initial-value problem

$$(4y - \cos y)\frac{dy}{dx} - 3x^2 = 0, \quad y(0) = 0$$

Solution.   We can write this equation in the form of (10) as

$$(4y - \cos y)\frac{dy}{dx} = 3x^2$$

Separating variables and integrating yields

$$(4y - \cos y)\,dy = 3x^2\,dx$$

$$\int (4y - \cos y)\,dy = \int 3x^2\,dx$$

which is a symbolic expression of the equation

$$2y^2 - \sin y = x^3 + C \tag{13}$$

Equation (13) defines solutions of the differential equation implicitly; it cannot be solved explicitly for y as a function of x.

For the initial-value problem, the initial condition $y(0) = 0$ requires that $y = 0$ if $x = 0$. Substituting these values into (13) to determine the constant of integration yields $C = 0$ (verify). Thus, the solution of the initial-value problem is

$$2y^2 - \sin y = x^3 \qquad\blacktriangleleft$$

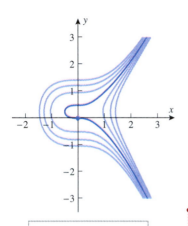

Integral curves for
$$(4y - \cos y)\frac{dy}{dx} - 3x^2 = 0$$

Figure 9.1.3

FOR THE READER.   Some computer algebra systems can graph implicit equations. For example, Figure 9.1.3 shows the graphs of (13) for $C = 0, \pm 1, \pm 2$, and ± 3, with emphasis on the solution of the initial-value problem. If you have a CAS that can graph implicit equations, read the documentation on graphing them and try to duplicate this figure. Also, try to determine which values of C produce which curves.

APPLICATIONS IN GEOMETRY

We conclude this section with some applications of first-order differential equations.

Example 6   Find a curve in the xy-plane that passes through $(0, 3)$ and whose tangent line at a point (x, y) has slope $2x/y^2$.

Solution.   Since the slope of the tangent line is dy/dx, we have

$$\frac{dy}{dx} = \frac{2x}{y^2} \tag{14}$$

and, since the curve passes through $(0, 3)$, we have the initial condition

$$y(0) = 3 \tag{15}$$

Equation (14) is separable and can be written as

$$y^2\,dy = 2x\,dx$$

so

$$\int y^2\,dy = \int 2x\,dx \quad \text{or} \quad \tfrac{1}{3}y^3 = x^2 + C$$

It follows from the initial condition (15) that $y = 3$ if $x = 0$. Substituting these values into the last equation yields $C = 9$ (verify), so the equation of the desired curve is

$$\tfrac{1}{3}y^3 = x^2 + 9 \quad \text{or} \quad y = (3x^2 + 27)^{1/3} \qquad\blacktriangleleft$$

MIXING PROBLEMS

In a typical mixing problem, a tank is filled to a specified level with a solution that contains a known amount of some soluble substance (say salt). The thoroughly stirred solution is allowed to drain from the tank at a known rate, and at the same time a solution with a known concentration of the soluble substance is added to the tank at a known rate that may or may not differ from the draining rate. As time progresses, the amount of the soluble substance in the tank will generally change, and the usual mixing problem seeks to determine the amount of the substance in the tank at a specified time. This type of problem serves as a model for many kinds of problems: discharge and filtration of pollutants in a river, injection and absorption of medication in the bloodstream, and migrations of species into and out of an ecological system, for example.

Example 7   At time $t = 0$, a tank contains 4 lb of salt dissolved in 100 gal of water. Suppose that brine containing 2 lb of salt per gallon of brine is allowed to enter the tank at a rate of 5 gal/min and that the mixed solution is drained from the tank at the same rate (Figure 9.1.4). Find the amount of salt in the tank after 10 minutes.

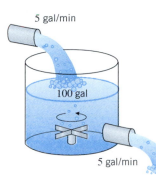
5 gal/min

100 gal

5 gal/min

Figure 9.1.4

Solution.   Let $y(t)$ be the amount of salt (in pounds) after t minutes. We are given that $y(0) = 4$, and we want to find $y(10)$. We will begin by finding a differential equation that is satisfied by $y(t)$. To do this, observe that dy/dt, which is the rate at which the amount of salt in the tank changes with time, can be expressed as

$$\frac{dy}{dt} = \text{rate in} - \text{rate out} \tag{16}$$

where *rate in* is the rate at which salt enters the tank and *rate out* is the rate at which salt leaves the tank. But the rate at which salt enters the tank is

$$\text{rate in} = (2 \text{ lb/gal}) \cdot (5 \text{ gal/min}) = 10 \text{ lb/min}$$

Since brine enters and drains from the tank at the same rate, the volume of brine in the tank stays constant at 100 gal. Thus, after t minutes have elapsed, the tank contains $y(t)$ lb of salt per 100 gal of brine, and hence the rate at which salt leaves the tank at that instant is

$$\text{rate out} = \left(\frac{y(t)}{100} \text{ lb/gal} \right) \cdot (5 \text{ gal/min}) = \frac{y(t)}{20} \text{ lb/min}$$

Therefore, (16) can be written as

$$\frac{dy}{dt} = 10 - \frac{y}{20} \quad \text{or} \quad \frac{dy}{dt} + \frac{y}{20} = 10$$

which is a first-order linear differential equation satisfied by $y(t)$. Since we are given that $y(0) = 4$, the function $y(t)$ can be obtained by solving the initial-value problem

$$\frac{dy}{dt} + \frac{y}{20} = 10, \quad y(0) = 4$$

The integrating factor for the differential equation is

$$\mu = e^{t/20}$$

If we multiply the differential equation through by μ, then we obtain

$$\frac{d}{dt}(e^{t/20}y) = 10e^{t/20}$$

$$e^{t/20}y = \int 10e^{t/20}dt = 200e^{t/20} + C$$

$$y(t) = 200 + Ce^{-t/20} \tag{17}$$

The initial condition states that $y = 4$ when $t = 0$. Substituting these values into (17) and solving for C yields $C = -196$ (verify), so

$$y(t) = 200 - 196e^{-t/20} \tag{18}$$

Thus, at time $t = 10$ the amount of salt in the tank is

$$y(10) = 200 - 196e^{-0.5} \approx 81.1 \text{ lb}$$
◀

FOR THE READER.   Figure 9.1.5 shows the graph of (18). Observe that $y(t) \to 200$ as $t \to +\infty$, which means that over an extended period of time the amount of salt in the tank tends toward 200 lb. Give an informal physical argument to explain why this result is to be expected.

A MODEL OF FREE-FALL MOTION RETARDED BY AIR RESISTANCE

In Section 5.4 we considered the free-fall model of an object moving along a vertical axis near the surface of the Earth. It was assumed in that model that there is no air resistance and that the only force acting on the object is the Earth's gravity. Our goal here is to find a model that takes air resistance into account. For this purpose we make the following assumptions:

- The object moves along a vertical s-axis whose origin is at the surface of the Earth and whose positive direction is up (Figure 5.4.8).
- At time $t = 0$ the height of the object is s_0 and the velocity is v_0.
- The only forces on the object are the force $F_G = -mg$ of the Earth's gravity acting down and the force F_R of air resistance acting opposite to the direction of motion. The force F_R is called the **drag force**.

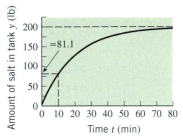

Figure 9.1.5

We will also need the following result from physics:

9.1.1   NEWTON'S SECOND LAW OF MOTION.   If an object with mass m is subjected to a force F, then the object undergoes an acceleration a that satisfies the equation

$$F = ma \tag{19}$$

In the case of free-fall motion retarded by air resistance, the net force acting on the object is

$$F_G + F_R = -mg + F_R$$

and the acceleration is d^2s/dt^2, so Newton's second law implies that

$$-mg + F_R = m\frac{d^2s}{dt^2} \tag{20}$$

Experimentation has shown that the force F_R of air resistance depends on the shape of the object and its speed—the greater the speed, the greater the drag force. There are many possible models for air resistance, but one of the most basic assumes that the drag force F_R is proportional to the velocity of the object, that is,

$$F_R = -cv$$

where c is a positive constant that depends on the object's shape and properties of the air.* (The minus sign ensures that the drag force is opposite to the direction of motion.) Substituting this in (20) and writing d^2s/dt^2 as dv/dt, we obtain

$$-mg - cv = m\frac{dv}{dt}$$

or on dividing by m and rearranging we obtain

$$\frac{dv}{dt} + \frac{c}{m}v = -g$$

which is a first-order linear differential equation in the unknown function $v = v(t)$ with $p(t) = c/m$ and $q(t) = -g$ [see (5)]. For a specific object, the coefficient c can be determined experimentally, so we can assume that m, g, and c are known constants. Thus,

---

*Other common models assume that $F_R = -cv^2$ or, more generally, $F_R = -cv^p$ for some value of p.

the velocity function $v = v(t)$ can be obtained by solving the initial-value problem

$$\frac{dv}{dt} + \frac{c}{m}v = -g, \quad v(0) = v_0 \tag{21}$$

Once the velocity function is found, the position function $s = s(t)$ can be obtained by solving the initial-value problem

$$\frac{ds}{dt} = v(t), \quad s(0) = s_0 \tag{22}$$

In Exercise 47 we will ask you to solve (21) and show that

$$v(t) = e^{-ct/m}\left(v_0 + \frac{mg}{c}\right) - \frac{mg}{c} \tag{23}$$

Note that

$$\lim_{t \to +\infty} v(t) = -\frac{mg}{c} \tag{24}$$

(verify). Thus, the speed $|v(t)|$ does not increase indefinitely, as in free fall; rather, because of the air resistance, it approaches a finite limiting speed v_τ given by

$$v_\tau = \left|-\frac{mg}{c}\right| = \frac{mg}{c} \tag{25}$$

This is called the **terminal speed** of the object, and (24) is called its **terminal velocity**.

> REMARK. Intuition suggests that near the limiting velocity, the velocity $v(t)$ changes very slowly; that is, $dv/dt \approx 0$. Thus, it should not be surprising that the limiting velocity can be obtained informally from (21) by setting $dv/dt = 0$ in the differential equation and solving for v. This yields
>
> $$v = -\frac{mg}{c}$$
>
> which agrees with (24).

EXERCISE SET 9.1   ~ Graphing Utility   [c] CAS

1. Confirm that $y = 2e^{x^3/3}$ is a solution of the initial-value problem $y' = x^2 y$, $y(0) = 2$.

2. Confirm that $y = \frac{1}{4}x^4 + 2\cos x + 1$ is a solution of the initial-value problem $y' = x^3 - 2\sin x$, $y(0) = 3$.

In Exercises 3 and 4, state the order of the differential equation, and confirm that the functions in the given family are solutions.

3. (a) $(1 + x)\dfrac{dy}{dx} = y$; $y = c(1 + x)$

    (b) $y'' + y = 0$; $y = c_1 \sin t + c_2 \cos t$

4. (a) $2\dfrac{dy}{dx} + y = x - 1$; $y = ce^{-x/2} + x - 3$

    (b) $y'' - y = 0$; $y = c_1 e^t + c_2 e^{-t}$

In Exercises 5 and 6, use implicit differentiation to confirm that the equation defines implicit solutions of the differential equation.

5. $\ln y = xy + C$; $\dfrac{dy}{dx} = \dfrac{y^2}{1 - xy}$

6. $x^2 + xy^2 = C$; $2x + y^2 + 2xy\dfrac{dy}{dx} = 0$

The first-order linear equations in Exercises 7 and 8 can be rewritten as first-order separable equations. Solve the equations using both the method of integrating factors and the method of separation of variables, and determine whether the solutions produced are the same.

7. (a) $\dfrac{dy}{dx} + 3y = 0$ (b) $\dfrac{dy}{dt} - 2y = 0$

8. (a) $\dfrac{dy}{dx} - 4xy = 0$ (b) $\dfrac{dy}{dt} + y = 0$

In Exercises 9–14, solve the differential equation by the method of integrating factors.

9. $\dfrac{dy}{dx} + 3y = e^{-2x}$

10. $\dfrac{dy}{dx} + 2xy = x$

11. $y' + y = \cos(e^x)$

12. $2\dfrac{dy}{dx} + 4y = 1$

13. $(x^2 + 1)\dfrac{dy}{dx} + xy = 0$

14. $\dfrac{dy}{dx} + y - \dfrac{1}{1 + e^x} = 0$

In Exercises 15–24, solve the differential equation by separation of variables. Where reasonable, express the family of solutions as explicit functions of x.

---

15. $\dfrac{dy}{dx} = \dfrac{y}{x}$

16. $\dfrac{dy}{dx} = (1 + y^2)x^2$

17. $\dfrac{\sqrt{1 + x^2}}{1 + y}\dfrac{dy}{dx} = -x$

18. $(1 + x^4)\dfrac{dy}{dx} = \dfrac{x^3}{y}$

19. $(1 + y^2)y' = e^x y$

20. $y' = -xy$

21. $e^{-y}\sin x - y'\cos^2 x = 0$

22. $y' - (1 + x)(1 + y^2) = 0$

23. $\dfrac{dy}{dx} - \dfrac{y^2 - y}{\sin x} = 0$

24. $3\tan y - \dfrac{dy}{dx}\sec x = 0$

25. In each part, find the solution of the differential equation

$$x\dfrac{dy}{dx} + y = x$$

that satisfies the initial condition.

(a) $y(1) = 2$ (b) $y(-1) = 2$

26. In each part, find the solution of the differential equation

$$\dfrac{dy}{dx} = xy$$

that satisfies the initial condition.

(a) $y(0) = 1$ (b) $y(0) = \frac{1}{2}$

In Exercises 27–32, solve the initial-value problem by any method.

---

27. $\dfrac{dy}{dx} - xy = x,\quad y(0) = 3$

28. $\dfrac{dy}{dt} + y = 2,\quad y(0) = 1$

29. $y' = \dfrac{4x^2}{y + \cos y},\quad y(1) = \pi$

30. $y' - xe^y = 2e^y,\quad y(0) = 0$

31. $\dfrac{dy}{dt} = \dfrac{2t + 1}{2y - 2},\quad y(0) = -1$

32. $y'\cosh x + y\sinh x = \cosh^2 x,\quad y(0) = \frac{1}{4}$

33. (a) Sketch some typical integral curves of the differential equation $y' = y/2x$.

(b) Find an equation for the integral curve that passes through the point $(2, 1)$.

34. (a) Sketch some typical integral curves of the differential equation $y' = -x/y$.

(b) Find an equation for the integral curve that passes through the point $(3, 4)$.

In Exercises 35 and 36, solve the differential equation, and then use a graphing utility to generate five integral curves for the equation.

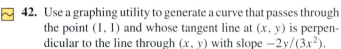

35. $(x^2 + 4)\dfrac{dy}{dx} + xy = 0$ **36.** $y' + 2y - 3e^t = 0$

If you have a CAS that can graph implicit equations, solve the differential equations in Exercises 37 and 38, and then use the CAS to generate five integral curves for the equation.

---

37. $y' = \dfrac{x^2}{1 - y^2}$ **38.** $y' = \dfrac{y}{1 + y^2}$

39. Suppose that the initial condition in Example 4 had been $y(0) = 0$. Show that none of the solutions generated in Example 4 satisfy this initial condition, and then solve the initial-value problem

$$\dfrac{dy}{dx} = -4xy^2,\quad y(0) = 0$$

Why does the method of Example 4 fail to produce this particular solution?

40. Find all ordered pairs (x_0, y_0) such that if the initial condition in Example 4 is replaced by $y(x_0) = y_0$, the solution of the resulting initial-value problem is defined for all real numbers.

41. Find an equation of a curve with x-intercept 2 whose tangent line at any point (x, y) has slope xe^y.

42. Use a graphing utility to generate a curve that passes through the point $(1, 1)$ and whose tangent line at (x, y) is perpendicular to the line through (x, y) with slope $-2y/(3x^2)$.

43. At time $t = 0$, a tank contains 25 ounces of salt dissolved in 50 gal of water. Then brine containing 4 ounces of salt per gallon of brine is allowed to enter the tank at a rate of 2 gal/min and the mixed solution is drained from the tank at the same rate.

(a) How much salt is in the tank at an arbitrary time t?

(b) How much salt is in the tank after 25 min?

44. A tank initially contains 200 gal of pure water. Then at time $t = 0$ brine containing 5 lb of salt per gallon of brine is allowed to enter the tank at a rate of 10 gal/min and the mixed solution is drained from the tank at the same rate.

(a) How much salt is in the tank at an arbitrary time t?

(b) How much salt is in the tank after 30 min?

45. A tank with a 1000-gal capacity initially contains 500 gal of water that is polluted with 50 lb of particulate matter. At time $t = 0$, pure water is added at a rate of 20 gal/min and the mixed solution is drained off at a rate of 10 gal/min. How much particulate matter is in the tank when it reaches the point of overflowing?

46. The water in a polluted lake initially contains 1 lb of mercury salts per 100,000 gal of water. The lake is circular with diameter 30 m and uniform depth 3 m. Polluted water is pumped from the lake at a rate of 1000 gal/h and is replaced with fresh water at the same rate. Construct a table that shows the amount of mercury in the lake (in lb) at the end of each hour over a 12-hour period. Discuss any assumptions you made. [Use 264 gal/m³.]

47. (a) Use the method of integrating factors to derive solution (23) to the initial-value problem (21). [*Note:* Keep in mind that c, m, and g are constants.]

(b) Show that (23) can be expressed in terms of the terminal speed (25) as

$$v(t) = e^{-gt/v_\tau}(v_0 + v_\tau) - v_\tau$$

(c) Show that if $s(0) = s_0$, then the position function of the object can be expressed as

$$s(t) = s_0 - v_\tau t + \frac{v_\tau}{g}(v_0 + v_\tau)(1 - e^{-gt/v_\tau})$$

48. Suppose a fully equipped sky diver weighing 240 lb has a terminal speed of 120 ft/s with a closed parachute and 24 ft/s with an open parachute. Suppose further that this sky diver is dropped from an airplane at an altitude of 10,000 ft, falls for 25 s with a closed parachute, and then falls the rest of the way with an open parachute.

(a) Assuming that the sky diver's initial vertical velocity is zero, use Exercise 47 to find the sky diver's vertical velocity and height at the time the parachute opens. [Take $g = 32$ ft/s².]

(b) Use a calculating utility to find a numerical solution for the total time that the sky diver is in the air.

49. The accompanying figure is a schematic diagram of a basic RL series electrical circuit that contains a power source with a time-dependent voltage of $V(t)$ volts (V), a resistor with a constant resistance of R ohms (Ω), and an inductor with a constant inductance of L henrys (H). If you don't know anything about electrical circuits, don't worry; all you need to know is that electrical theory states that a current of $I(t)$ amperes (A) flows through the circuit where $I(t)$ satisfies the differential equation

$$L\frac{dI}{dt} + RI = V(t)$$

(a) Find $I(t)$ if $R = 10\,\Omega$, $L = 4\,$H, V is a constant 12 V, and $I(0) = 0$ A.

(b) What happens to the current over a long period of time?

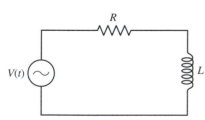

Figure Ex-49

50. Find $I(t)$ for the electrical circuit in Exercise 49 if $R = 6\,\Omega$, $L = 3\,$H, $V(t) = 3\sin t$ V, and $I(0) = 15$ A.

51. A rocket, fired upward from rest at time $t = 0$, has an initial mass of m_0 (including its fuel). Assuming that the fuel is consumed at a constant rate k, the mass m of the rocket, while fuel is being burned, will be given by $m = m_0 - kt$. It can be shown that if air resistance is neglected and the fuel gases are expelled at a constant speed c relative to the rocket, then the velocity v of the rocket will satisfy the equation

$$m\frac{dv}{dt} = ck - mg$$

where g is the acceleration due to gravity.

(a) Find $v(t)$ keeping in mind that the mass m is a function of t.

(b) Suppose that the fuel accounts for 80% of the initial mass of the rocket and that all of the fuel is consumed in 100 s. Find the velocity of the rocket in meters per second at the instant the fuel is exhausted. [Take $g = 9.8$ m/s² and $c = 2500$ m/s.]

52. A bullet of mass m, fired straight up with an initial velocity of v_0, is slowed by the force of gravity and a drag force of air resistance kv^2, where g is the constant acceleration due to gravity and k is a positive constant. As the bullet moves upward, its velocity v satisfies the equation

$$m\frac{dv}{dt} = -(kv^2 + mg)$$

(a) Show that if $x = x(t)$ is the height of the bullet above the barrel opening at time t, then

$$mv\frac{dv}{dx} = -(kv^2 + mg)$$

(b) Express x in terms of v given that $x = 0$ when $v = v_0$.

(c) Assuming that

$$v_0 = 988 \text{ m/s}, \quad g = 9.8 \text{ m/s}^2$$
$$m = 3.56 \times 10^{-3} \text{ kg}, \quad k = 7.3 \times 10^{-6} \text{ kg/m}$$

use the result in part (b) to find out how high the bullet rises. [*Hint:* Find the velocity of the bullet at its highest point.]

The following discussion is needed for Exercises 53 and 54. Suppose that a tank containing a liquid is vented to the air at the top and has an outlet at the bottom through which the liquid can drain. It follows from *Torricelli's law* in physics that if the outlet is opened at time $t = 0$, then at each instant the depth of the liquid $h(t)$ and the area $A(h)$ of the liquid's surface are related by

$$A(h)\frac{dh}{dt} = -k\sqrt{h}$$

where k is a positive constant that depends on such factors as the viscosity of the liquid and the cross-sectional area of the outlet. Use this result in Exercises 53 and 54, assuming that h is in feet, $A(h)$ is in square feet, and t is in seconds.

53. Suppose that the cylindrical tank in the accompanying figure is filled to a depth of 4 feet at time $t = 0$ and that the constant in Torricelli's law is $k = 0.025$.
  (a) Find $h(t)$.
  (b) How many minutes will it take for the tank to drain completely?

54. Follow the directions of Exercise 53 for the cylindrical tank in the accompanying figure, assuming that the tank is filled to a depth of 4 feet at time $t = 0$ and that the constant in Torricelli's law is $k = 0.025$.

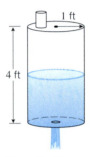 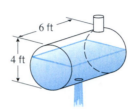

Figure Ex-53   Figure Ex-54

55. Suppose that a particle moving along the x-axis encounters a resisting force that results in an acceleration of $a = dv/dt = -0.04v^2$. Given that $x = 0$ cm and $v = 50$ cm/s at time $t = 0$, find the velocity v and position x as a function of t for $t \geq 0$.

56. Suppose that a particle moving along the x-axis encounters a resisting force that results in an acceleration of $a = dv/dt = -0.02\sqrt{v}$. Given that $x = 0$ cm and $v = 9$ cm/s at time $t = 0$, find the velocity v and position x as a function of t for $t \geq 0$.

57. Find an initial-value problem whose solution is

$$y = \cos x + \int_0^x e^{-t^2}\, dt$$

58. (a) Prove that if C is an arbitrary constant, then any function $y = y(x)$ defined by Equation (6) will be a solution to (5) on the interval I.
  (b) Consider the initial-value problem

$$\frac{dy}{dx} + p(x)y = q(x), \quad y(x_0) = y_0$$

where the functions $p(x)$ and $q(x)$ are both continuous on some open interval I. Using the general solution for a first-order linear equation, prove that this initial-value problem has a unique solution on I.

59. Use implicit differentiation to prove that any differentiable function defined implicitly by Equation (11) will be a solution to (10).

60. (a) Prove that solutions need not be unique for nonlinear initial-value problems by finding two solutions to

$$y\frac{dy}{dx} = x, \quad y(0) = 0$$

  (b) Prove that solutions need not exist for nonlinear initial-value problems by showing that there is no solution for

$$y\frac{dy}{dx} = -x, \quad y(0) = 0$$

61. In our derivation of Equation (6) we did not consider the possibility of a solution $y = y(x)$ to (5) that was defined on an open subset $I_1 \subseteq I$, $I_1 \neq I$. Prove that there was no loss of generality in our analysis by showing that any such solution must extend to a solution to (5) on the entire interval I.

9.2 DIRECTION FIELDS; EULER'S METHOD

In this section we will reexamine the concept of a direction field, and we will discuss a method for approximating solutions of first-order equations numerically. Numerical approximations are important in cases where the differential equation cannot be solved exactly.

FUNCTIONS OF TWO VARIABLES

We will be concerned here with first-order equations that are expressed with the derivative by itself on one side of the equation. For example,

$$y' = x^3 \quad \text{and} \quad y' = \sin(xy)$$

The first of these equations involves only x on the right side, so it has the form $y' = f(x)$. However, the second equation involves both x and y on the right side, so it has the form $y' = f(x, y)$, where the symbol $f(x, y)$ stands for a function of the two variables x and y. Later in the text we will study functions of two variables in more depth, but for now it will suffice to think of $f(x, y)$ as a formula that produces a unique output when values of x and

y are given as inputs. For example, if

$$f(x, y) = x^2 + 3y$$

and if the inputs are $x = 2$ and $y = -4$, then the output is

$$f(2, -4) = 2^2 + 3(-4) = 4 - 12 = -8$$

> **REMARK.** In applied problems involving time, it is usual to use t as the independent variable, in which case we would be concerned with equations of the form $y' = f(t, y)$, where $y' = dy/dt$.

DIRECTION FIELDS

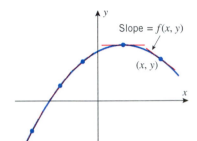

At each point (x, y) on an integral curve of $y' = f(x, y)$, the tangent line has slope $f(x, y)$.

Figure 9.2.1

In Section 6.2 we introduced the concept of a direction field in the context of differential equations of the form $y' = f(x)$; the same principles apply to differential equations of the form

$$y' = f(x, y)$$

To see why this is so, let us review the basic idea. If we interpret y' as the slope of a tangent line, then the differential equation states that at each point (x, y) on an integral curve, the slope of the tangent line is equal to the value of f at that point (Figure 9.2.1). For example, suppose that $f(x, y) = y - x$, in which case we have the differential equation

$$y' = y - x \tag{1}$$

A geometric description of the set of integral curves can be obtained by choosing a rectangular grid of points in the xy-plane, calculating the slopes of the tangent lines to the integral curves at the gridpoints, and drawing small segments of the tangent lines at those points. The resulting picture is called a ***direction field*** or a ***slope field*** for the differential equation because it shows the "direction" or "slope" of the integral curves at the gridpoints. The more gridpoints that are used, the better the description of the integral curves. For example, Figure 9.2.2 shows two direction fields for (1)—the first was obtained by hand calculation using the 49 gridpoints shown in the accompanying table, and the second, which gives a clearer picture of the integral curves, was obtained using 625 gridpoints and a CAS.

VALUES OF $f(x, y) = y - x$

	$y = -3$	$y = -2$	$y = -1$	$y = 0$	$y = 1$	$y = 2$	$y = 3$
$x = -3$	0	1	2	3	4	5	6
$x = -2$	−1	0	1	2	3	4	5
$x = -1$	−2	−1	0	1	2	3	4
$x = 0$	−3	−2	−1	0	1	2	3
$x = 1$	−4	−3	−2	−1	0	1	2
$x = 2$	−5	−4	−3	−2	−1	0	1
$x = 3$	−6	−5	−4	−3	−2	−1	0

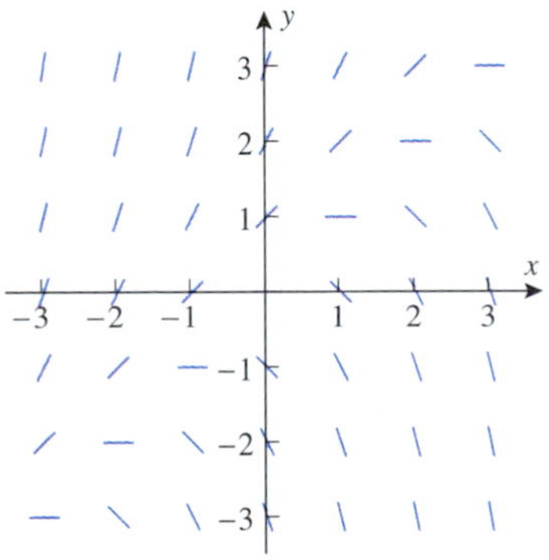

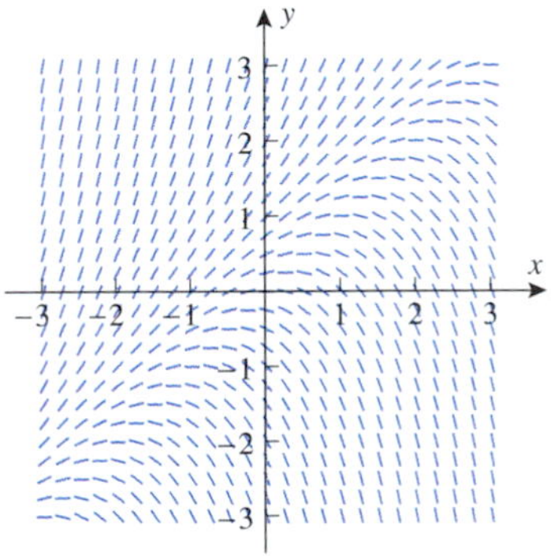

Figure 9.2.2

It so happens that Equation (1) can be solved exactly, since it can be written as

$$y' - y = -x$$

which, by comparison with Equation (5) in Section 9.1, is a first-order linear equation with $p(x) = -1$ and $q(x) = -x$. We leave it for you to use the method of integrating factors to show that the general solution of this equation is

$$y = x + 1 + Ce^x \tag{2}$$

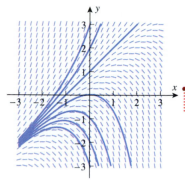

Figure 9.2.3

Figure 9.2.3 shows some of the integral curves superimposed on the direction field. Observe, however, that it was not necessary to have the general solution to construct the direction field. Indeed, direction fields are important precisely because they can be constructed in cases where the differential equation cannot be solved exactly.

FOR THE READER.   Confirm that the first direction field in Figure 9.2.2 is consistent with the values in the accompanying table.

Example 1   In Example 7 of Section 9.1 we considered a mixing problem in which the amount of salt $y(t)$ in a tank at time t was shown to satisfy the differential equation

$$\frac{dy}{dt} + \frac{y}{20} = 10$$

which can be rewritten as

$$y' = 10 - \frac{y}{20} \tag{3}$$

We subsequently found the general solution of this equation to be

$$y(t) = 200 + Ce^{-t/20} \tag{4}$$

and then we found the value of the arbitrary constant C from the initial condition in the problem [the known amount of salt $y(0)$ at time $t = 0$]. However, it follows from (4) that

$$\lim_{t \to +\infty} y(t) = 200$$

for all values of C, so regardless of the amount of salt that is present in the tank initially, the amount of salt in the tank will eventually begin to stabilize at 200 lb. This can also be seen geometrically from the direction field for (3) shown in Figure 9.2.4. This direction field suggests that if the amount of salt present in the tank is greater than 200 lb initially, then the amount of salt will decrease steadily over time toward a limiting value of 200 lb; and if it is less than 200 lb initially, then it will increase steadily toward a limiting value of 200 lb. The direction field also suggests that if the amount present initially is exactly 200 lb, then the amount of salt in the tank will stay constant at 200 lb. This can also be seen from (4), since $C = 0$ in this case (verify).   ◀

Observe that for the direction field shown in Figure 9.2.4 the tangent segments along any horizontal line are parallel. This occurs because the differential equation has the form $y' = f(y)$ with t absent from the right side [see (3)]. Thus, for a fixed y the slope y' does not change as time varies. Because of this time independence of slope, differential equations of the form $y' = f(y)$ are said to be **autonomous** (from the Greek word *autonomous*, meaning "independent").

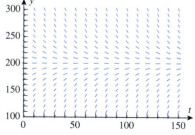

Figure 9.2.4

EULER'S METHOD

Our next objective is to develop a method for approximating the solution of an initial-value problem of the form

$$y' = f(x, y), \quad y(x_0) = y_0$$

We will not attempt to approximate $y(x)$ for all values of x; rather, we will choose some small increment Δx and focus on approximating the values of $y(x)$ at a succession of

x-values spaced Δx units apart, starting from x_0. We will denote these x-values by

$$x_1 = x_0 + \Delta x, \quad x_2 = x_1 + \Delta x, \quad x_3 = x_2 + \Delta x, \quad x_4 = x_3 + \Delta x, \ldots$$

and we will denote the approximations of $y(x)$ at these points by

$$y_1 \approx y(x_1), \quad y_2 \approx y(x_2), \quad y_3 \approx y(x_3), \quad y_4 \approx y(x_4), \ldots$$

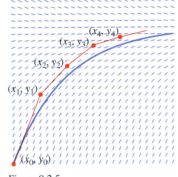

The technique that we will describe for obtaining these approximations is called *Euler's Method*. Although there are better approximation methods available, many of them use Euler's Method as a starting point, so the underlying concepts are important to understand.

The basic idea behind Euler's Method is to start at the known initial point (x_0, y_0) and draw a line segment in the direction determined by the direction field until we reach the point (x_1, y_1) with x-coordinate $x_1 = x_0 + \Delta x$ (Figure 9.2.5). If Δx is small, then it is reasonable to expect that this line segment will not deviate much from the integral curve $y = y(x)$, and thus y_1 should closely approximate $y(x_1)$. To obtain the subsequent approximations, we repeat the process using the direction field as a guide at each step. Starting at the endpoint (x_1, y_1), we draw a line segment determined by the direction field until we reach the point (x_2, y_2) with x-coordinate $x_2 = x_1 + \Delta x$, and from that point we draw a line segment determined by the direction field to the point (x_3, y_3) with x-coordinate $x_3 = x_2 + \Delta x$, and so forth. As indicated in Figure 9.2.5, this procedure produces a polygonal path that tends to follow the integral curve closely, so it is reasonable to expect that the y-values $y_2, y_3, y_4, \ldots$ will closely approximate $y(x_2), y(x_3), y(x_4), \ldots$.

Figure 9.2.5

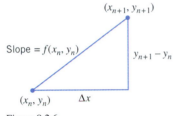

Figure 9.2.6

To explain how the approximations $y_1, y_2, y_3, \ldots$ can be computed, let us focus on a typical line segment. As indicated in Figure 9.2.6, assume that we have found the point (x_n, y_n), and we are trying to determine the next point (x_{n+1}, y_{n+1}), where $x_{n+1} = x_n + \Delta x$. Since the slope of the line segment joining the points is determined by the direction field at the starting point, the slope is $f(x_n, y_n)$, and hence

$$\frac{y_{n+1} - y_n}{x_{n+1} - x_n} = \frac{y_{n+1} - y_n}{\Delta x} = f(x_n, y_n)$$

which we can rewrite as

$$y_{n+1} = y_n + f(x_n, y_n)\Delta x$$

This formula, which is the heart of Euler's Method, tells us how to use each approximation to compute the next approximation.

Euler's Method

To approximate the solution of the initial-value problem

$$y' = f(x, y), \quad y(x_0) = y_0$$

proceed as follows:

Step 1. Choose a nonzero number Δx to serve as an *increment* or *step size* along the x-axis, and let

$$x_1 = x_0 + \Delta x, \quad x_2 = x_1 + \Delta x, \quad x_3 = x_2 + \Delta x, \ldots$$

Step 2. Compute successively

$$y_1 = y_0 + f(x_0, y_0)\Delta x$$
$$y_2 = y_1 + f(x_1, y_1)\Delta x$$
$$y_3 = y_2 + f(x_2, y_2)\Delta x$$
$$\vdots$$
$$y_{n+1} = y_n + f(x_n, y_n)\Delta x$$

The numbers $y_1, y_2, y_3, \ldots$ in these equations are the approximations of $y(x_1), y(x_2), y(x_3), \ldots$.

Example 2   Use Euler's Method with a step size of 0.1 to make a table of approximate values of the solution of the initial-value problem

$$y' = y - x, \quad y(0) = 2 \tag{5}$$

over the interval $0 \le x \le 1$.

Solution.   In this problem we have $f(x, y) = y - x$, $x_0 = 0$, and $y_0 = 2$. Moreover, since the step size is 0.1, the x-values at which the approximate values will be obtained are

$$x_1 = 0.1, \quad x_2 = 0.2, \quad x_3 = 0.3, \dots, \quad x_9 = 0.9, \quad x_{10} = 1$$

The first three approximations are

$$y_1 = y_0 + f(x_0, y_0)\Delta x = 2 + (2 - 0)(0.1) = 2.2$$

$$y_2 = y_1 + f(x_1, y_1)\Delta x = 2.2 + (2.2 - 0.1)(0.1) = 2.41$$

$$y_3 = y_2 + f(x_2, y_2)\Delta x = 2.41 + (2.41 - 0.2)(0.1) = 2.631$$

Here is a way of organizing all 10 approximations rounded to five decimal places:

EULER'S METHOD FOR $y' = y - x$, $y(0) = 2$ WITH $\Delta x = 0.1$

n	x_n	y_n	$f(x_n, y_n)\Delta x$	$y_{n+1} = y_n + f(x_n, y_n)\Delta x$
0	0	2.00000	0.20000	2.20000
1	0.1	2.20000	0.21000	2.41000
2	0.2	2.41000	0.22100	2.63100
3	0.3	2.63100	0.23310	2.86410
4	0.4	2.86410	0.24641	3.11051
5	0.5	3.11051	0.26105	3.37156
6	0.6	3.37156	0.27716	3.64872
7	0.7	3.64872	0.29487	3.94359
8	0.8	3.94359	0.31436	4.25795
9	0.9	4.25795	0.33579	4.59374
10	1.0	4.59374	—	—

Observe that each entry in the last column becomes the next entry in the third column.   ◀

ACCURACY OF EULER'S METHOD

It follows from (5) and the initial condition $y(0) = 2$ that the exact solution of the initial-value problem in Example 2 is

$$y = x + 1 + e^x$$

Thus, in this case we can compare the approximate values of $y(x)$ produced by Euler's Method with decimal approximations of the exact values (Table 9.2.1). In Table 9.2.1 the **absolute error** is calculated as

$$|\text{exact value} - \text{approximation}|$$

and the **percentage error** as

$$\frac{|\text{exact value} - \text{approximation}|}{|\text{exact value}|} \times 100\%$$

REMARK.   As a rough rule of thumb, the absolute error in an approximation produced by Euler's Method is proportional to the step size; thus, reducing the step size by half reduces the absolute error (and hence the percentage error) by roughly half. However, reducing the step size also increases the amount of computation, thereby increasing the potential for roundoff error. We will leave a detailed study of error issues for courses in differential equations or numerical analysis.

Table 9.2.1

x	EXACT SOLUTION	EULER APPROXIMATION	ABSOLUTE ERROR	PERCENTAGE ERROR
0	2.00000	2.00000	0.00000	0.00
0.1	2.20517	2.20000	0.00517	0.23
0.2	2.42140	2.41000	0.01140	0.47
0.3	2.64986	2.63100	0.01886	0.71
0.4	2.89182	2.86410	0.02772	0.96
0.5	3.14872	3.11051	0.03821	1.21
0.6	3.42212	3.37156	0.05056	1.48
0.7	3.71375	3.64872	0.06503	1.75
0.8	4.02554	3.94359	0.08195	2.04
0.9	4.35960	4.25795	0.10165	2.33
1.0	4.71828	4.59374	0.12454	2.64

EXERCISE SET 9.2 ~ Graphing Utility c CAS

1. Sketch the direction field for $y' = xy/8$ at the gridpoints (x, y), where $x = 0, 1, \ldots, 4$ and $y = 0, 1, \ldots, 4$.

2. Sketch the direction field for $y' + y = 2$ at the gridpoints (x, y), where $x = 0, 1, \ldots, 4$ and $y = 0, 1, \ldots, 4$.

3. A direction field for the differential equation $y' = 1 - y$ is shown in the accompanying figure. In each part, sketch the graph of the solution that satisfies the initial condition.
   (a) $y(0) = -1$ (b) $y(0) = 1$ (c) $y(0) = 2$

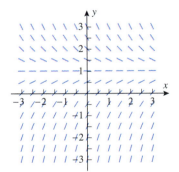

Figure Ex-3

~ 4. Solve the initial-value problems in Exercise 3, and use a graphing utility to confirm that the integral curves for these solutions are consistent with the sketches you obtained from the direction field.

5. A direction field for the differential equation $y' = 2y - x$ is shown in the accompanying figure. In each part, sketch the graph of the solution that satisfies the initial condition.
   (a) $y(1) = 1$ (b) $y(0) = -1$ (c) $y(-1) = 0$

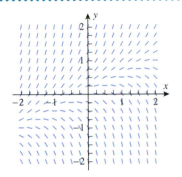

Figure Ex-5

~ 6. Solve the initial-value problems in Exercise 5, and use a graphing utility to confirm that the integral curves for these solutions are consistent with the sketches you obtained from the direction field.

7. Use the direction field in Exercise 3 to make a conjecture about the behavior of the solutions of $y' = 1 - y$ as $x \to +\infty$, and confirm your conjecture by examining the general solution of the equation.

8. Use the direction field in Exercise 5 to make a conjecture about the effect of y_0 on the behavior of the solution of the initial-value problem $y' = 2y - x$, $y(0) = y_0$ as $x \to +\infty$, and check your conjecture by examining the solution of the initial-value problem.

9. In each part, match the differential equation with the direction field (see next page), and explain your reasoning.
   (a) $y' = 1/x$ (b) $y' = 1/y$ (c) $y' = e^{-x^2}$
   (d) $y' = y^2 - 1$ (e) $y' = \dfrac{x + y}{x - y}$
   (f) $y' = (\sin x)(\sin y)$

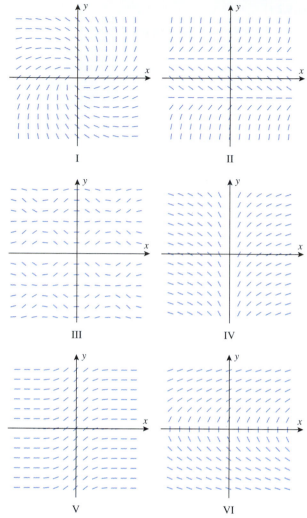

I   II

III   IV

V   VI

Figure Ex-9

C **10.** If you have a CAS or a graphing utility that can generate direction fields, read the documentation on how to do it and check your answers in Exercise 9 by generating the direction fields for the differential equations.

11. (a) Use Euler's Method with a step size of $\Delta x = 0.2$ to approximate the solution of the initial-value problem

$$y' = x + y, \quad y(0) = 1$$

over the interval $0 \le x \le 1$.

(b) Solve the initial-value problem exactly, and calculate the error and the percentage error in each of the approximations in part (a).

(c) Sketch the exact solution and the approximate solution together.

12. It was stated at the end of this section that reducing the step size in Euler's Method by half reduces the error in each approximation by about half. Confirm that the error in $y(1)$ is reduced by about half if a step size of $\Delta x = 0.1$ is used in Exercise 11.

In Exercises 13–16, use Euler's Method with the given step size Δx to approximate the solution of the initial-value problem over the stated interval. Present your answer as a table and as a graph.

13. $dy/dx = \sqrt{y}$, $y(0) = 1$, $0 \le x \le 4$, $\Delta x = 0.5$

14. $dy/dx = x - y^2$, $y(0) = 1$, $0 \le x \le 2$, $\Delta x = 0.25$

15. $dy/dt = \sin y$, $y(0) = 1$, $0 \le t \le 2$, $\Delta x = 0.5$

16. $dy/dt = e^{-y}$, $y(0) = 0$, $0 \le t \le 1$, $\Delta x = 0.1$

17. Consider the initial-value problem

$$y' = \cos 2\pi t, \quad y(0) = 1$$

Use Euler's Method with five steps to approximate $y(1)$.

18. (a) Show that the solution of the initial-value problem $y' = e^{-x^2}$, $y(0) = 0$ is

$$y(x) = \int_0^x e^{-t^2} \, dt$$

(b) Use Euler's Method with $\Delta x = 0.05$ to approximate the value of

$$y(1) = \int_0^1 e^{-t^2} \, dt$$

and compare the answer to that produced by a calculating utility with a numerical integration capability.

19. The accompanying figure shows a direction field for the differential equation $y' = -x/y$.

(a) Use the direction field to estimate $y\left(\frac{1}{2}\right)$ for the solution that satisfies the given initial condition $y(0) = 1$.

(b) Compare your estimate to the exact value of $y\left(\frac{1}{2}\right)$.

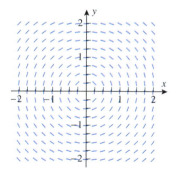

Figure Ex-19

20. Consider the initial-value problem

$$\frac{dy}{dx} = \frac{\sqrt{y}}{2}, \quad y(0) = 1$$

(a) Use Euler's Method with step sizes of $\Delta x = 0.2, 0.1$, and 0.05 to obtain three approximations of $y(1)$.

(b) Plot the three approximations versus Δx, and make a conjecture about the exact value of $y(1)$. Explain your reasoning.

(c) Check your conjecture by finding $y(1)$ exactly.

9.3 MODELING WITH FIRST-ORDER DIFFERENTIAL EQUATIONS

Since many of the fundamental laws of the physical and social sciences involve rates of change, it should not be surprising that such laws are modeled by differential equations. In this section we will discuss the general idea of modeling with differential equations, and we will investigate some important models that can be applied to population growth, carbon dating, medicine, and ecology.

POPULATION GROWTH

One of the simplest models of population growth is based on the observation that when populations (people, plants, bacteria, and fruit flies, for example) are not constrained by environmental limitations, they tend to grow at a rate that is proportional to the size of the population—the larger the population, the more rapidly it grows.

To translate this principle into a mathematical model, suppose that $y = y(t)$ denotes the population at time t. At each point in time, the rate of increase of the population with respect to time is dy/dt, so the assumption that the rate of growth is proportional to the population is described by the differential equation

$$\frac{dy}{dt} = ky \tag{1}$$

where k is a positive constant of proportionality that can usually be determined experimentally. Thus, if the population is known at some point in time, say $y = y_0$ at time $t = 0$, then a general formula for the population $y(t)$ can be obtained by solving the initial-value problem

$$\frac{dy}{dt} = ky, \quad y(0) = y_0$$

PHARMACOLOGY

When a drug (say, penicillin or aspirin) is administered to an individual, it enters the bloodstream and then is absorbed by the body over time. Medical research has shown that the amount of a drug that is present in the bloodstream tends to decrease at a rate that is proportional to the amount of the drug present—the more of the drug that is present in the bloodstream, the more rapidly it is absorbed by the body.

To translate this principle into a mathematical model, suppose that $y = y(t)$ is the amount of the drug present in the bloodstream at time t. At each point in time, the rate of change in y with respect to t is dy/dt, so the assumption that the rate of decrease is proportional to the amount y in the bloodstream translates into the differential equation

$$\frac{dy}{dt} = -ky \tag{2}$$

where k is a positive constant of proportionality that depends on the drug and can be determined experimentally. The negative sign is required because y decreases with time. Thus, if the initial dosage of the drug is known, say $y = y_0$ at time $t = 0$, then a general formula for $y(t)$ can be obtained by solving the initial-value problem

$$\frac{dy}{dt} = -ky, \quad y(0) = y_0$$

SPREAD OF DISEASE

Suppose that a disease begins to spread in a population of L individuals. Logic suggests that at each point in time the rate at which the disease spreads will depend on how many individuals are already affected and how many are not—as more individuals are affected, the opportunity to spread the disease tends to increase, but at the same time there are fewer individuals who are not affected, so the opportunity to spread the disease tends to decrease. Thus, there are two conflicting influences on the rate at which the disease spreads.

To translate this into a mathematical model, suppose that $y = y(t)$ is the number of individuals who have the disease at time t, so of necessity the number of individuals who do not have the disease at time t is $L - y$. As the value of y increases, the value of $L - y$

decreases, so the conflicting influences of the two factors on the rate of spread dy/dt are taken into account by the differential equation

$$\frac{dy}{dt} = ky(L - y)$$

where k is a positive constant of proportionality that depends on the nature of the disease and the behavior patterns of the individuals and can be determined experimentally. Thus, if the number of affected individuals is known at some point in time, say $y = y_0$ at time $t = 0$, then a general formula for $y(t)$ can be obtained by solving the initial-value problem

$$\frac{dy}{dt} = ky(L - y), \quad y(0) = y_0 \tag{3}$$

INHIBITED POPULATION GROWTH

The population growth model that we discussed at the beginning of this section was predicated on the assumption that the population $y = y(t)$ is not constrained by the environment. For this reason, it is sometimes called the **uninhibited growth model**. However, in the real world this assumption is usually not valid—populations generally grow within ecological systems that can only support a certain number of individuals; the number L of such individuals is called the **carrying capacity** of the system. Thus, when $y > L$, the population exceeds the capacity of the ecological system and tends to decrease toward L; when $y < L$, the population is below the capacity of the ecological system and tends to increase toward L; and when $y = L$, the population is in balance with the capacity of the ecological system and tends to remain stable.

To translate this into a mathematical model, we must look for a differential equation in which $y > 0$, $L > 0$, and

$$\frac{dy}{dt} < 0 \quad \text{if} \quad \frac{y}{L} > 1$$

$$\frac{dy}{dt} > 0 \quad \text{if} \quad \frac{y}{L} < 1$$

$$\frac{dy}{dt} = 0 \quad \text{if} \quad \frac{y}{L} = 1$$

Moreover, logic suggests that when the population is far below the carrying capacity (i.e., $y/L \approx 0$), then the environmental constraints should have little effect, and the growth rate should behave very much like the uninhibited model. Thus, we want

$$\frac{dy}{dt} \approx ky \quad \text{if} \quad \frac{y}{L} \approx 0$$

A simple differential equation that meets all of these requirements is

$$\frac{dy}{dt} = k \left(1 - \frac{y}{L}\right) y$$

where k is a positive constant of proportionality. Thus, if k and L can be determined experimentally, and if the population is known at some point in time, say $y(0) = y_0$, then a general formula for the population $y(t)$ can be determined by solving the initial-value problem

$$\frac{dy}{dt} = k \left(1 - \frac{y}{L}\right) y, \quad y(0) = y_0 \tag{4}$$

This theory of population growth is due to the Belgian mathematician, P. F. Verhulst (1804–1849), who introduced it in 1838 and described it as "logistic growth."[*] Thus, the differential equation in (4) is called the **logistic differential equation**, and the growth model described by (4) is called the **logistic model** or the **inhibited growth model**.

---

[*] Verhulst's model fell into obscurity for nearly a hundred years because he did not have sufficient census data to test its validity. However, interest in the model was revived in the 1930s when biologists used it successfully to describe the growth of fruit fly and flour beetle populations. Verhulst himself used the model to predict that an upper limit on Belgium's population would be approximately 9,400,000. In 1998 the population was about 10,175,000.

REMARK. Observe that the differential equation in (3) can be expressed as

$$\frac{dy}{dt} = kL\left(1 - \frac{y}{L}\right)y$$

which is a logistic equation with kL rather than k as the constant of proportionality. Thus, this model for the spread of disease is also a logistic or inhibited growth model.

EXPONENTIAL GROWTH AND DECAY MODELS

Equations (1) and (2) are examples of a general class of models called *exponential models*. In general, exponential models arise in situations where a quantity increases or decreases at a rate that is proportional to the amount of the quantity present. More precisely, we make the following definition:

9.3.1 DEFINITION. A quantity $y = y(t)$ is said to have an ***exponential growth model*** if it increases at a rate that is proportional to the amount of the quantity present, and it is said to have an ***exponential decay model*** if it decreases at a rate that is proportional to the amount of the quantity present. Thus, for an exponential growth model, the quantity $y(t)$ satisfies an equation of the form

$$\frac{dy}{dt} = ky \quad (k > 0) \tag{5}$$

and for an exponential decay model, the quantity $y(t)$ satisfies an equation of the form

$$\frac{dy}{dt} = -ky \quad (k > 0) \tag{6}$$

The constant k is called the ***growth constant*** or the ***decay constant***, as appropriate.

Equations (5) and (6) are first-order linear equations, since they can be rewritten as

$$\frac{dy}{dt} - ky = 0 \quad \text{and} \quad \frac{dy}{dt} + ky = 0$$

both of which have the form of Equation (5) in Section 9.1 (but with t rather than x as the independent variable); in the first equation we have $p(t) = -k$ and $q(t) = 0$, and in the second we have $p(t) = k$ and $q(t) = 0$.

To illustrate how these equations can be solved, suppose that a quantity $y = y(t)$ has an exponential growth model and we know the amount of the quantity at some point in time, say $y = y_0$ when $t = 0$. Thus, a general formula for $y(t)$ can be obtained by solving the initial-value problem

$$\frac{dy}{dt} - ky = 0, \quad y(0) = y_0$$

Multiplying the differential equation through by the integrating factor

$$\mu = e^{-kt}$$

yields

$$\frac{d}{dt}(e^{-kt}y) = 0$$

and then integrating with respect to t yields

$$e^{-kt}y = C \quad \text{or} \quad y = Ce^{kt}$$

The initial condition implies that $y = y_0$ when $t = 0$, from which it follows that $C = y_0$ (verify). Thus, the solution of the initial-value problem is

$$y = y_0 e^{kt} \tag{7}$$

We leave it for you to show that if $y = y(t)$ has an exponential decay model, and if $y(0) = y_0$, then

$$y = y_0 e^{-kt} \tag{8}$$

INTERPRETING THE GROWTH AND DECAY CONSTANTS

The significance of the constant k in Formulas (7) and (8) can be understood by reexamining the differential equations that gave rise to these formulas. For example, in the case of the exponential growth model, Equation (5) can be rewritten as

$$k = \frac{dy/dt}{y}$$

which states that the growth rate as a fraction of the entire population remains constant over time, and this constant is k. For this reason, k is called the ***relative growth rate*** of the population. It is usual to express the relative growth rate as a percentage. Thus, a relative growth rate of 3% per unit of time in an exponential growth model means that $k = 0.03$. Similarly, the constant k in an exponential decay model is called the ***relative decay rate***.

REMARK.   It is standard practice in applications to call the relative growth rate the *growth rate*, even though it is not really correct (the growth rate is dy/dt). However, the practice is so common that we will follow it here.

Example 1   According to United Nations data, the world population in 1998 was approximately 5.9 billion and growing at a rate of about 1.33% per year. Assuming an exponential growth model, estimate the world population at the beginning of the year 2023.

Solution.   We assume that the population at the beginning of 1998 was 5.9 billion and let

t = time elapsed from the beginning of 1998 (in years)

y = world population (in billions)

Since the beginning of 1998 corresponds to $t = 0$, it follows from the given data that

$y_0 = y(0) = 5.9$ (billion)

Since the growth rate is 1.33% ($k = 0.0133$), it follows from (7) that the world population at time t will be

$$y(t) = y_0 e^{kt} = 5.9 e^{0.0133t} \qquad (9)$$

Since the beginning of the year 2023 corresponds to an elapsed time of $t = 25$ years ($2023 - 1998 = 25$), it follows from (9) that the world population by the year 2023 will be

$$y(25) = 5.9 e^{0.0133(25)} \approx 8.2$$

which is a population of approximately 8.2 billion.   ◀

REMARK.   In this example, the growth rate was given, so there was no need to calculate it. If the growth rate or decay rate in an exponential model is unknown, then it can be calculated using the initial condition and the value of y at one other point in time (Exercise 34).

DOUBLING TIME AND HALF-LIFE

If a quantity y has an exponential growth model, then the time required for the original size to double is called the ***doubling time***, and if y has an exponential decay model, then the time required for the original size to reduce by half is called the ***half-life***. As it turns out, doubling time and half-life depend only on the growth or decay rate and not on the amount present initially. To see why this is so, suppose that $y = y(t)$ has an exponential growth model

$$y = y_0 e^{kt} \qquad (10)$$

and let T denote the amount of time required for y to double in size. Thus, at time $t = T$ the value of y will be $2y_0$, and hence from (10)

$$2y_0 = y_0 e^{kT} \quad \text{or} \quad e^{kT} = 2$$

Taking the natural logarithm of both sides yields $kT = \ln 2$, which implies that the doubling time is

$$T = \frac{1}{k} \ln 2 \qquad (11)$$

We leave it as an exercise to show that Formula (11) also gives the half-life of an exponential decay model. Observe that this formula does not involve the initial amount y_0, so that in an exponential growth or decay model, the quantity y doubles (or reduces by half) every T units (Figure 9.3.1).

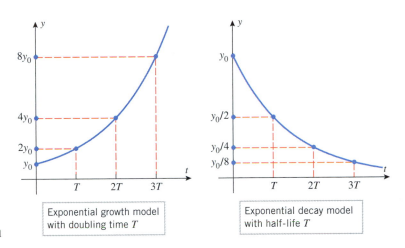

Figure 9.3.1

Exponential growth model with doubling time T

Exponential decay model with half-life T

Example 2 It follows from (11) that with a continued growth rate of 1.33% per year, the doubling time for the world population will be

$$T = \frac{1}{0.0133} \ln 2 \approx 52.116$$

or approximately 52 years. Thus, with a continued 1.33% annual growth rate the population of 5.9 billion in 1998 will double to 11.8 billion by the year 2050 and will double again to 23.6 billion by 2102. ◄

RADIOACTIVE DECAY

It is a fact of physics that radioactive elements disintegrate spontaneously in a process called *radioactive decay*. Experimentation has shown that the rate of disintegration is proportional to the amount of the element present, which implies that the amount $y = y(t)$ of a radioactive element present as a function of time has an exponential decay model.

Every radioactive element has a specific half-life; for example, the half-life of radioactive carbon-14 is about 5730 years. Thus, from (11), the decay constant for this element is

$$k = \frac{1}{T} \ln 2 = \frac{\ln 2}{5730} \approx 0.000121$$

and this implies that if there are y_0 units of carbon-14 present at time $t = 0$, then the number of units present after t years will be approximately

$$y(t) = y_0 e^{-0.000121t} \qquad (12)$$

Example 3 If 100 grams of radioactive carbon-14 are stored in a cave for 1000 years, how many grams will be left at that time?

Solution. From (12) with $y_0 = 100$ and $t = 1000$, we obtain

$$y(1000) = 100e^{-0.000121(1000)} = 100e^{-0.121} \approx 88.6$$

Thus, about 88.6 grams will be left. ◄

CARBON DATING

When the nitrogen in the Earth's upper atmosphere is bombarded by cosmic radiation, the radioactive element carbon-14 is produced. This carbon-14 combines with oxygen to form carbon dioxide, which is ingested by plants, which in turn are eaten by animals. In this way all living plants and animals absorb quantities of radioactive carbon-14. In 1947 the American nuclear scientist W. F. Libby[*] proposed the theory that the percentage of carbon-14 in the atmosphere and in living tissues of plants is the same. When a plant or animal dies, the carbon-14 in the tissue begins to decay. Thus, the age of an artifact that contains plant or animal material can be estimated by determining what percentage of its original carbon-14 content remains. Various procedures, called *carbon dating* or *carbon-14 dating*, have been developed for measuring this percentage.

Example 4 In 1988 the Vatican authorized the British Museum to date a cloth relic known as the Shroud of Turin, possibly the burial shroud of Jesus of Nazareth. This cloth, which first surfaced in 1356, contains the negative image of a human body that was widely believed to be that of Jesus. The report of the British Museum showed that the fibers in the cloth contained between 92% and 93% of their original carbon-14. Use this information to estimate the age of the shroud.

The Shroud of Turin

Solution. From (12), the fraction of the original carbon-14 that remains after t years is

$$\frac{y(t)}{y_0} = e^{-0.000121t}$$

Taking the natural logarithm of both sides and solving for t, we obtain

$$t = -\frac{1}{0.000121} \ln\left(\frac{y(t)}{y_0}\right)$$

Thus, taking $y(t)/y_0$ to be 0.93 and 0.92, we obtain

$$t = -\frac{1}{0.000121} \ln(0.93) \approx 600$$

$$t = -\frac{1}{0.000121} \ln(0.92) \approx 689$$

This means that when the test was done in 1988, the shroud was between 600 and 689 years old, thereby placing its origin between 1299 A.D. and 1388 A.D. Thus, if one accepts the validity of carbon-14 dating, the Shroud of Turin cannot be the burial shroud of Jesus of Nazareth. ◄

LOGISTIC MODELS

Recall that the logistic model of population growth in an ecological system with carrying capacity L is determined by initial-value problem (4). To illustrate how this initial-value problem can be solved for $y(t)$, let us focus on the differential equation

$$\frac{dy}{dt} = k\left(1 - \frac{y}{L}\right)y \tag{13}$$

Note that the constant functions $y = 0$ and $y = L$ are particular solutions of (13). To find nonconstant solutions, it will be convenient to rewrite Equation (13) as

$$\frac{dy}{dt} = \frac{k}{L}(L - y)y = \frac{k}{L}y(L - y)$$

This equation is separable, since it can be rewritten in differential form as

$$\frac{L}{y(L - y)}\,dy = k\,dt$$

Integrating both sides yields the equation

$$\int \frac{L}{y(L - y)}\,dy = \int k\,dt$$

---

[*] W. F. Libby, "Radiocarbon Dating," *American Scientist*, Vol. 44, 1956, pp. 98–112.

Using partial fractions on the left side, we can rewrite this equation as (verify)

$$\int \left(\frac{1}{y} + \frac{1}{L - y} \right) dy = \int k \, dt$$

Integrating and rearranging the form of the result, we obtain

$$\ln |y| - \ln |L - y| = kt + C$$

$$\ln \left| \frac{y}{L - y} \right| = kt + C$$

$$\left| \frac{y}{L - y} \right| = e^{kt + C}$$

$$\left| \frac{L - y}{y} \right| = e^{-kt - C} = e^{-C} e^{-kt}$$

$$\frac{L - y}{y} = \pm e^{-C} e^{-kt}$$

$$\frac{L}{y} - 1 = Ae^{-kt} \quad \text{(where } A = \pm e^{-C}\text{)}$$

Solving this equation for y yields (verify)

$$y = \frac{L}{1 + Ae^{-kt}} \tag{14}$$

As the final step, we want to use the initial condition in (4) to determine the constant A. But the initial condition implies that $y = y_0$ if $t = 0$, so from (14)

$$y_0 = \frac{L}{1 + A}$$

from which we obtain

$$A = \frac{L - y_0}{y_0}$$

Thus, the solution of the initial-value problem (4) is

$$y = \frac{L}{1 + \left(\dfrac{L - y_0}{y_0} \right) e^{-kt}}$$

which can be rewritten more simply as

$$y = \frac{y_0 L}{y_0 + (L - y_0)e^{-kt}} \tag{15}$$

Note that the constant solutions of (13) are also given in (15); they correspond to the initial conditions $y_0 = 0$ and $y_0 = L$.

The graph of (15) has one of four general shapes, depending on the relationship between the initial population y_0 and the carrying capacity L (Figure 9.3.2).

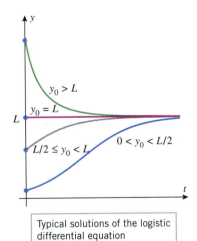

$y_0 > L$

$y_0 = L$

$L/2 \leq y_0 < L$

$0 < y_0 < L/2$

Typical solutions of the logistic differential equation

Figure 9.3.2

Example 5 Figure 9.3.3 shows the graph of a population $y = y(t)$ with a logistic growth model. Estimate the values of y_0, L, and k, and use the estimates to deduce a formula for y as a function of t.

Solution. The graph suggests that the carrying capacity is $L = 5$, and the population at time $t = 0$ is $y_0 = 1$. Thus, from (15), the equation has the form

$$y = \frac{5}{1 + 4e^{-kt}} \tag{16}$$

where k must still be determined. However, the graph passes through the point $(1, 2)$, which tells us that $y = 2$ if $t = 1$. Substituting these values in (16) yields

$$2 = \frac{5}{1 + 4e^{-k}}$$

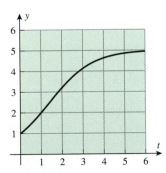

Figure 9.3.3

Solving for k we obtain (verify)

$$k = \ln \tfrac{8}{3} \approx 0.98$$

and substituting this in (16) yields

$$y = \frac{5}{1 + 4e^{-0.98t}}$$

◀

EXERCISE SET 9.3 ⌇ Graphing Utility

1. (a) Suppose that a quantity $y = y(t)$ increases at a rate that is proportional to the square of the amount present, and suppose that at time $t = 0$, the amount present is y_0. Find an initial-value problem whose solution is $y(t)$.

   (b) Suppose that a quantity $y = y(t)$ decreases at a rate that is proportional to the square of the amount present, and suppose that at a time $t = 0$, the amount present is y_0. Find an initial-value problem whose solution is $y(t)$.

2. (a) Suppose that a quantity $y = y(t)$ changes in such a way that $dy/dt = k\sqrt{y}$, where $k > 0$. Describe how y changes in words.

   (b) Suppose that a quantity $y = y(t)$ changes in such a way that $dy/dt = -ky^3$, where $k > 0$. Describe how y changes in words.

3. (a) Suppose that a particle moves along an s-axis in such a way that its velocity $v(t)$ is always half of $s(t)$. Find a differential equation whose solution is $s(t)$.

   (b) Suppose that an object moves along an s-axis in such a way that its acceleration $a(t)$ is always twice the velocity. Find a differential equation whose solution is $s(t)$.

4. Suppose that a body moves along an s-axis through a resistive medium in such a way that the velocity $v = v(t)$ decreases at a rate that is twice the square of the velocity.

   (a) Find a differential equation whose solution is the velocity $v(t)$.

   (b) Find a differential equation whose solution is the position $s(t)$.

5. Suppose that an initial population of 10,000 bacteria grows exponentially at a rate of 1% per hour and that $y = y(t)$ is the number of bacteria present t hours later.

   (a) Find an initial-value problem whose solution is $y(t)$.

   (b) Find a formula for $y(t)$.

   (c) How long does it take for the initial population of bacteria to double?

   (d) How long does it take for the population of bacteria to reach 45,000?

6. A cell of the bacterium *E. coli* divides into two cells every 20 minutes when placed in a nutrient culture. Let $y = y(t)$ be the number of cells that are present t minutes after a single cell is placed in the culture. Assume that the growth of the bacteria is approximated by a continuous exponential growth model.

   (a) Find an initial-value problem whose solution is $y(t)$.

   (b) Find a formula for $y(t)$.

   (c) How many cells are present after 2 hours?

   (d) How long does it take for the number of cells to reach 1,000,000?

7. Radon-222 is a radioactive gas with a half-life of 3.83 days. This gas is a health hazard because it tends to get trapped in the basements of houses, and many health officials suggest that homeowners seal their basements to prevent entry of the gas. Assume that 5.0×10^7 radon atoms are trapped in a basement at the time it is sealed and that $y(t)$ is the number of atoms present t days later.

   (a) Find an initial-value problem whose solution is $y(t)$.

   (b) Find a formula for $y(t)$.

   (c) How many atoms will be present after 30 days?

   (d) How long will it take for 90% of the original quantity of gas to decay?

8. Polonium-210 is a radioactive element with a half-life of 140 days. Assume that 10 milligrams of the element are placed in a lead container and that $y(t)$ is the number of milligrams present t days later.

   (a) Find an initial-value problem whose solution is $y(t)$.

   (b) Find a formula for $y(t)$.

   (c) How many milligrams will be present after 10 weeks?

   (d) How long will it take for 70% of the original sample to decay?

9. Suppose that 100 fruit flies are placed in a breeding container that can support at most 5000 flies. Assuming that the population grows exponentially at a rate of 2% per day, how long will it take for the container to reach capacity?

10. Suppose that the town of Grayrock had a population of 10,000 in 1987 and a population of 12,000 in 1997. Assuming an exponential growth model, in what year will the population reach 20,000?

11. A scientist wants to determine the half-life of a certain radioactive substance. She determines that in exactly 5 days a 10.0-milligram sample of the substance decays to 3.5 milligrams. Based on these data, what is the half-life?

12. Suppose that 40% of a certain radioactive substance decays in 5 years.

    (a) What is the half-life of the substance in years?

    (b) Suppose that a certain quantity of this substance is stored in a cave. What percentage of it will remain after t years?

13. In each part, find an exponential growth model $y = y_0 e^{kt}$ that satisfies the stated conditions.
   (a) $y_0 = 2$; doubling time $T = 5$
   (b) $y(0) = 5$; growth rate 1.5%
   (c) $y(1) = 1$; $y(10) = 100$
   (d) $y(1) = 1$; doubling time $T = 5$

14. In each part, find an exponential decay model $y = y_0 e^{-kt}$ that satisfies the stated conditions.
   (a) $y_0 = 10$; half-life $T = 5$
   (b) $y(0) = 10$; decay rate 1.5%
   (c) $y(1) = 100$; $y(10) = 1$
   (d) $y(1) = 10$; half-life $T = 5$

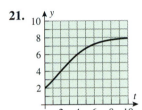 **15.** (a) Make a conjecture about the effect on the graphs of $y = y_0 e^{kt}$ and $y = y_0 e^{-kt}$ of varying k and keeping y_0 fixed. Confirm your conjecture with a graphing utility.
   (b) Make a conjecture about the effect on the graphs of $y = y_0 e^{kt}$ and $y = y_0 e^{-kt}$ of varying y_0 and keeping k fixed. Confirm your conjecture with a graphing utility.

16. (a) What effect does increasing y_0 and keeping k fixed have on the doubling time or half-life of an exponential model? Justify your answer.
   (b) What effect does increasing k and keeping y_0 fixed have on the doubling time and half-life of an exponential model? Justify your answer.

17. (a) There is a trick, called the **Rule of 70**, that can be used to get a quick estimate of the doubling time or half-life of an exponential model. According to this rule, the doubling time or half-life is roughly 70 divided by the percentage growth or decay rate. For example, we showed in Example 2 that with a continued growth rate of 1.33% per year the world population would double every 52 years. This result agrees with the Rule of 70, since $70/1.33 \approx 52.6$. Explain why this rule works.
   (b) Use the Rule of 70 to estimate the doubling time of a population that grows exponentially at a rate of 1% per year.
   (c) Use the Rule of 70 to estimate the half-life of a population that decreases exponentially at a rate of 3.5% per hour.
   (d) Use the Rule of 70 to estimate the growth rate that would be required for a population growing exponentially to double every 10 years.

18. Find a formula for the tripling time of an exponential growth model.

19. In 1950, a research team digging near Folsom, New Mexico, found charred bison bones along with some leaf-shaped projectile points (called the "Folsom points") that had been made by a Paleo-Indian hunting culture. It was clear from the evidence that the bison had been cooked and eaten by the makers of the points, so that carbon-14 dating of the bones made it possible for the researchers to determine when the hunters roamed North America. Tests showed that the bones contained between 27% and 30% of their original carbon-14. Use this information to show that the hunters lived roughly between 9000 B.C. and 8000 B.C.

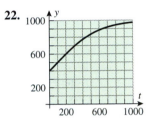 **20.** (a) Use a graphing utility to make a graph of p_{rem} versus t, where p_{rem} is the percentage of carbon-14 that remains in an artifact after t years.
   (b) Use the graph to estimate the percentage of carbon-14 that would have to have been present in the 1988 test of the Shroud of Turin for it to have been the burial shroud of Jesus. [See Example 4.]

In Exercises 21 and 22, the graph of a logistic model
$$y = \frac{y_0 L}{y_0 + (L - y_0)e^{-kt}}$$
is shown. Estimate y_0, L, and k.

21.

22.

23. Suppose that the growth of a population $y = y(t)$ is given by the logistic equation
$$y = \frac{60}{5 + 7e^{-t}}$$
   (a) What is the population at time $t = 0$?
   (b) What is the carrying capacity L?
   (c) What is the constant k?
   (d) When does the population reach half of the carrying capacity?
   (e) Find an initial-value problem whose solution is $y(t)$.

24. Suppose that the growth of a population $y = y(t)$ is given by the logistic equation
$$y = \frac{1000}{1 + 999e^{-0.9t}}$$
   (a) What is the population at time $t = 0$?
   (b) What is the carrying capacity L?
   (c) What is the constant k?
   (d) When does the population reach 75% of the carrying capacity?
   (e) Find an initial-value problem whose solution is $y(t)$.

25. Suppose that a population $y(t)$ grows in accordance with the logistic model
$$\frac{dy}{dt} = 10(1 - 0.1y)y$$
   (a) What is the carrying capacity?
   (b) What is the value of k?
   (c) For what value of y is the population growing most rapidly?

26. Suppose that a population $y(t)$ grows in accordance with the logistic model

$$\frac{dy}{dt} = 50y - 0.001y^2$$

(a) What is the carrying capacity?
(b) What is the value of k?
(c) For what value of y is the population growing most rapidly?

27. Suppose that a college residence hall houses 1000 students. Following the semester break, 20 students in the hall return with the flu, and 5 days later 35 students have the flu.
(a) Use model (4) to set up an initial-value problem whose solution is the number of students who will have had the flu t days after the return from the break. [*Note:* The differential equation in this case will involve a constant of proportionality.]
(b) Solve the initial-value problem, and use the given data to find the constant of proportionality.
(c) Make a table that illustrates how the flu spreads day to day over a 2-week period.
(d) Use a graphing utility to generate a graph that illustrates how the flu spreads over a 2-week period.

28. It has been observed experimentally that at a constant temperature the rate of change of the atmospheric pressure p with respect to the altitude h above sea level is proportional to the pressure.
(a) Assuming that the pressure at sea level is p_0, find an initial-value problem whose solution is $p(h)$. [*Note:* The differential equation in this case will involve a constant of proportionality.]
(b) Find a formula for $p(h)$ in atmospheres (atm) if the pressure at sea level is 1 atm and the pressure at 5000 ft above sea level is 0.83 atm.

Newton's Law of Cooling states that the rate at which the temperature of a cooling object decreases and the rate at which a warming object increases are proportional to the difference between the temperature of the object and the temperature of the surrounding medium. Use this result in Exercises 29–32.

29. A cup of water with a temperature of $95°C$ is placed in a room with a constant temperature $21°C$.
(a) Assuming that Newton's Law of Cooling applies, set up and solve an initial-value problem whose solution is the temperature of the water t minutes after it is placed in the room. [*Note:* The differential equation will involve a constant of proportionality.]
(b) How many minutes will it take for the water to reach a temperature of $51°C$ if it cools to $85°C$ in 1 minute?

30. A glass of lemonade with a temperature of $40°F$ is placed in a room with a constant temperature of $70°F$, and 1 hour later its temperature is $52°F$. We stated in Example 4 of Section 6.8 that t hours after the lemonade is placed in the room its

temperature is approximated by $T = 70 - 30e^{-0.5t}$. Confirm this using Newton's Law of Cooling and the method used in Exercise 29.

31. The great detective Sherlock Holmes and his assistant Dr. Watson are discussing the murder of actor Cornelius McHam. McHam was shot in the head, and his understudy, Barry Moore, was found standing over the body with the murder weapon in hand. Let's listen in.

Watson: Open-and-shut case Holmes—Moore is the murderer.

Holmes: Not so fast Watson—you are forgetting Newton's Law of Cooling!

Watson: Huh?

Holmes: Elementary my dear Watson—Moore was found standing over McHam at 10:06 P.M., at which time the coroner recorded a body temperature of $77.9°F$ and noted that the room thermostat was set to $72°F$. At 11:06 P.M. the coroner took another reading and recorded a body temperature of $75.6°F$. Since McHam's normal temperature is $98.6°F$, and since Moore was on stage between 6:00 P.M. and 8:00 P.M., Moore is obviously innocent.

Watson: Huh?

Holmes: Sometimes you are so dull Watson. Ask any calculus student to figure it out for you.

Watson: Hrrumph....

32. Suppose that at time $t = 0$ an object with temperature T_0 is placed in a room with constant temperature T_a. If $T_0 < T_a$, then the temperature of the object will increase, and if $T_0 > T_a$, then the temperature will decrease. Assuming that Newton's Law of Cooling applies, show that in both cases the temperature $T(t)$ at time t is given by

$$T(t) = T_a + (T_0 - T_a)e^{-kt}$$

where k is a positive constant.

33. (a) Show that if $b > 1$, then the equation $y = y_0 b^t$ can be expressed as $y = y_0 e^{kt}$ for some positive constant k. [*Note:* This shows that if $b > 1$, and if y grows in accordance with the equation $y = y_0 b^t$, then y has an exponential growth model.]
(b) Show that if $0 < b < 1$, then the equation $y = y_0 b^t$ can be expressed as $y = y_0 e^{-kt}$ for some positive constant k. [*Note:* This shows that if $0 < b < 1$, and if y decays in accordance with the equation $y = y_0 b^t$, then y has an exponential decay model.]
(c) Express $y = 4(2^t)$ in the form $y = y_0 e^{kt}$.
(d) Express $y = 4(0.5^t)$ in the form $y = y_0 e^{-kt}$.

34. Suppose that a quantity y has an exponential growth model $y = y_0 e^{kt}$ or an exponential decay model $y = y_0 e^{-kt}$, and it is known that $y = y_1$ if $t = t_1$. In each case find a formula for k in terms of y_0, y_1, and t_1, assuming that $t_1 \neq 0$.

9.4 SECOND-ORDER LINEAR HOMOGENEOUS DIFFERENTIAL EQUATIONS; THE VIBRATING SPRING

In this section we will show how to solve an important collection of second-order differential equations. As an application, we will study the motion of a vibrating spring.

SECOND-ORDER LINEAR HOMOGENEOUS DIFFERENTIAL EQUATIONS WITH CONSTANT COEFFICIENTS

A *second-order linear differential equation* is one of the form

$$\frac{d^2y}{dx^2} + p(x)\frac{dy}{dx} + q(x)y = r(x) \tag{1}$$

or in alternative notation,

$$y'' + p(x)y' + q(x)y = r(x)$$

If $r(x)$ is identically 0, then (1) reduces to

$$\frac{d^2y}{dx^2} + p(x)\frac{dy}{dx} + q(x)y = 0$$

which is called the second-order linear **homogeneous** differential equation.

In order to discuss the solutions to a second-order linear homogeneous differential equation, it will be useful to introduce some terminology. Two functions f and g are said to be **linearly dependent** if one is a *constant* multiple of the other. If neither is a constant multiple of the other, then they are called **linearly independent**. Thus,

$$f(x) = \sin x \quad \text{and} \quad g(x) = 3\sin x$$

are linearly dependent, but

$$f(x) = x \quad \text{and} \quad g(x) = x^2$$

are linearly independent. The following theorem is central to the study of second-order linear homogeneous differential equations.

9.4.1   THEOREM. *Consider the homogeneous equation*

$$\frac{d^2y}{dx^2} + p(x)\frac{dy}{dx} + q(x)y = 0 \tag{2}$$

where the functions $p(x)$ and $q(x)$ are continuous on some common open interval I. Then there exist linearly independent solutions $y_1(x)$ and $y_2(x)$ to (2) on I. Furthermore, given any such pair of linearly independent solutions $y_1(x)$ and $y_2(x)$, a general solution of (2) on I is given by

$$y(x) = c_1 y_1(x) + c_2 y_2(x) \tag{3}$$

That is, every solution of (2) on I can be obtained from (3) by choosing appropriate values of the constants c_1 and c_2; conversely, (3) is a solution of (2) for all choices of c_1 and c_2.

A complete proof of this theorem is best left for a course in differential equations. (Readers interested in portions of the argument are referred to Chapter 3 of *Elementary Differential Equations*, 7th ed., John Wiley & Sons, New York, 2001, by William E. Boyce and Richard C. DiPrima.)

We will restrict our attention to second-order linear homogeneous equations of the form

$$\frac{d^2y}{dx^2} + p\frac{dy}{dx} + qy = 0 \tag{4}$$

where p and q are *constants*. Since the constant functions $p(x) = p$ and $q(x) = q$ are continuous on $I = (-\infty, +\infty)$, it follows from Theorem 9.4.1 that to determine a general solution to (4) we need only find two linearly independent solutions $y_1(x)$ and $y_2(x)$ on I. The general solution will then be given by $y(x) = c_1 y_1(x) + c_2 y_2(x)$, where c_1 and c_2 are arbitrary constants.

We will start by looking for solutions to (4) of the form $y = e^{mx}$. This is motivated by the fact that the first and second derivatives of this function are multiples of y, suggesting that a solution of (4) might result by choosing m appropriately. To find such an m, we substitute

$$y = e^{mx}, \quad \frac{dy}{dx} = me^{mx}, \quad \frac{d^2 y}{dx^2} = m^2 e^{mx} \tag{5}$$

into (4) to obtain

$$(m^2 + pm + q)e^{mx} = 0 \tag{6}$$

which is satisfied if and only if

$$m^2 + pm + q = 0 \tag{7}$$

since $e^{mx} \neq 0$ for every x.

Equation (7), which is called the *auxiliary equation* for (4), can be obtained from (4) by replacing $d^2 y/dx^2$ by m^2, dy/dx by $m \ (= m^1)$, and y by $1 \ (= m^0)$. The solutions, m_1 and m_2, of the auxiliary equation can be obtained by factoring or by the quadratic formula. These solutions are

$$m_1 = \frac{-p + \sqrt{p^2 - 4q}}{2}, \quad m_2 = \frac{-p - \sqrt{p^2 - 4q}}{2} \tag{8}$$

Depending on whether $p^2 - 4q$ is positive, zero, or negative, these roots will be distinct and real, equal and real, or complex conjugates.[*] We will consider each of these cases separately.

DISTINCT REAL ROOTS

If m_1 and m_2 are distinct real roots, then (4) has the two solutions

$$y_1 = e^{m_1 x}, \quad y_2 = e^{m_2 x}$$

Neither of the functions $e^{m_1 x}$ and $e^{m_2 x}$ is a constant multiple of the other (Exercise 29), so the general solution of (4) in this case is

$$y(x) = c_1 e^{m_1 x} + c_2 e^{m_2 x} \tag{9}$$

Example 1   Find the general solution of $y'' - y' - 6y = 0$.

Solution.   The auxiliary equation is

$$m^2 - m - 6 = 0 \quad \text{or equivalently,} \quad (m + 2)(m - 3) = 0$$

so its roots are $m = -2, m = 3$. Thus, from (9) the general solution of the differential equation is

$$y = c_1 e^{-2x} + c_2 e^{3x}$$

where c_1 and c_2 are arbitrary constants.   ◀

EQUAL REAL ROOTS

If m_1 and m_2 are equal real roots, say $m_1 = m_2 \ (= m)$, then the auxiliary equation yields only one solution of (4):

$$y_1(x) = e^{mx}$$

We will now show that

$$y_2(x) = xe^{mx} \tag{10}$$

is a second linearly independent solution. To see that this is so, note that $p^2 - 4q = 0$ in

---

[*]Recall that the complex solutions of a polynomial equation, and in particular of a quadratic equation, occur as conjugate pairs $a + bi$ and $a - bi$.

(8) since the roots are equal. Thus,

$$m = m_1 = m_2 = -p/2$$

and (10) becomes

$$y_2(x) = xe^{(-p/2)x}$$

Differentiating yields

$$y_2'(x) = \left(1 - \frac{p}{2}x\right)e^{(-p/2)x} \quad \text{and} \quad y_2''(x) = \left(\frac{p^2}{4}x - p\right)e^{-(p/2)x}$$

so

$$y_2''(x) + py_2'(x) + qy_2(x) = \left[\left(\frac{p^2}{4}x - p\right) + p\left(1 - \frac{p}{2}x\right) + qx\right]e^{(-p/2)x}$$

$$= \left[-\frac{p^2}{4} + q\right]xe^{(-p/2)x} \tag{11}$$

But $p^2 - 4q = 0$ implies that $(-p^2/4) + q = 0$, so (11) becomes

$$y_2''(x) + py_2'(x) + qy_2(x) = 0$$

which tells us that $y_2(x)$ is a solution of (4). It can be shown that

$$y_1(x) = e^{mx} \quad \text{and} \quad y_2(x) = xe^{mx}$$

are linearly independent (Exercise 29), so the general solution of (4) in this case is

$$y = c_1 e^{mx} + c_2 x e^{mx} \tag{12}$$

Example 2 Find the general solution of $y'' - 8y' + 16y = 0$.

Solution. The auxiliary equation is

$$m^2 - 8m + 16 = 0 \quad \text{or equivalently,} \quad (m - 4)^2 = 0$$

so $m = 4$ is the only root. Thus, from (12) the general solution of the differential equation is

$$y = c_1 e^{4x} + c_2 x e^{4x} \qquad \blacktriangleleft$$

COMPLEX ROOTS

If the auxiliary equation has complex roots $m_1 = a + bi$ and $m_2 = a - bi$, then both $y_1(x) = e^{ax}\cos bx$ and $y_2(x) = e^{ax}\sin bx$ are linearly independent solutions of (4) and

$$y = e^{ax}(c_1\cos bx + c_2\sin bx) \tag{13}$$

is the general solution. The proof is discussed in the exercises (Exercise 30).

Example 3 Find the general solution of $y'' + y' + y = 0$.

Solution. The auxiliary equation $m^2 + m + 1 = 0$ has roots

$$m_1 = \frac{-1 + \sqrt{1 - 4}}{2} = -\frac{1}{2} + \frac{\sqrt{3}}{2}i$$

$$m_2 = \frac{-1 - \sqrt{1 - 4}}{2} = -\frac{1}{2} - \frac{\sqrt{3}}{2}i$$

Thus, from (13) with $a = -1/2$ and $b = \sqrt{3}/2$, the general solution of the differential equation is

$$y = e^{-x/2}\left(c_1\cos\frac{\sqrt{3}}{2}x + c_2\sin\frac{\sqrt{3}}{2}x\right) \qquad \blacktriangleleft$$

$\bullet$

INITIAL-VALUE PROBLEMS

When a physical problem leads to a second-order differential equation, there are usually two conditions in the problem that determine specific values for the two arbitrary constants in the general solution of the equation. Conditions that specify the value of the solution $y(x)$ and its derivative $y'(x)$ at $x = x_0$ are called *initial conditions*. A second-order differential equation with initial conditions is called a *second-order initial-value problem*.

Example 4   Solve the initial-value problem

$$y'' - y = 0, \quad y(0) = 1, \quad y'(0) = 0$$

Solution. We must first solve the differential equation. The auxiliary equation

$$m^2 - 1 = 0$$

has distinct real roots $m_1 = 1$, $m_2 = -1$, so from (9) the general solution is

$$y(x) = c_1 e^x + c_2 e^{-x} \tag{14}$$

and the derivative of this solution is

$$y'(x) = c_1 e^x - c_2 e^{-x} \tag{15}$$

Substituting $x = 0$ in (14) and (15) and using the initial conditions $y(0) = 1$ and $y'(0) = 0$ yields the system of equations

$$c_1 + c_2 = 1$$
$$c_1 - c_2 = 0$$

Solving this system yields $c_1 = \frac{1}{2}$, $c_2 = \frac{1}{2}$, so from (14) the solution of the initial-value problem is

$$y(x) = \tfrac{1}{2}e^x + \tfrac{1}{2}e^{-x} = \cosh x \qquad \blacktriangleleft$$

The following summary is included as a ready reference for the solution of second-order homogeneous linear differential equations with constant coefficients.

Summary

EQUATION: $y'' + py' + qy = 0$

AUXILIARY EQUATION: $m^2 + pm + q = 0$

CASE	GENERAL SOLUTION
Distinct real roots m_1, m_2 of the auxiliary equation	$y = c_1 e^{m_1 x} + c_2 e^{m_2 x}$
Equal real roots $m_1 = m_2 \ (= m)$ of the auxiliary equation	$y = c_1 e^{mx} + c_2 x e^{mx}$
Complex roots $m_1 = a + bi$, $m_2 = a - bi$ of the auxiliary equation	$y = e^{ax}(c_1 \cos bx + c_2 \sin bx)$

$\bullet$

VIBRATIONS OF SPRINGS

We conclude this section with an engineering model that leads to a second-order differential equation of type (4).

As shown in Figure 9.4.1, consider a block of mass M that is suspended from a vertical spring and allowed to settle into an *equilibrium position*. Assume that the block is then set into vertical vibratory motion by pulling or pushing on it and releasing it at time $t = 0$. We will be interested in finding a mathematical model that describes the vibratory motion of the block over time.

To translate this problem into mathematical form, we introduce a vertical y-axis whose positive direction is up and whose origin is at the connection of the spring to the block when the block is in equilibrium (Figure 9.4.2). Our goal is to find the coordinate $y = y(t)$ of the top of the block as a function of time. For this purpose we will need Newton's Second Law

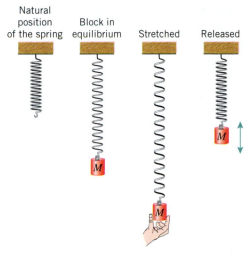

Figure 9.4.1

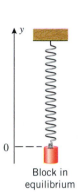

Figure 9.4.2

of Motion, which we will write as

$$F = Ma$$

rather than $F = ma$, as in Formula (19) of Section 9.1. This is to avoid a conflict with the letter "m" in the auxiliary equation. We will also need the following two results from physics:

9.4.2 HOOKE'S LAW.   If a spring is stretched (or compressed) ℓ units beyond its natural position, then it pulls (or pushes) with a force of magnitude

$$F = k\ell$$

where k is a positive constant, called the ***spring constant***. This constant, which is measured in units of force per unit length, depends on such factors as the thickness of the spring and its composition. The force exerted by the spring is called the ***restoring force***.

9.4.3 WEIGHT.   The gravitational force exerted by the Earth on an object is called the object's ***weight*** (or more precisely, its ***Earth weight***). It follows from Newton's Second Law of Motion that an object with mass M has a weight w of magnitude Mg, where g is the acceleration due to gravity. However, if the positive direction is up, as we are assuming here, then the force of the Earth's gravity is in the negative direction, so

$$w = -Mg$$

The weight of an object is measured in units of force.

The motion of the block in Figure 9.4.1 will depend on how far it is stretched or compressed initially and the forces that act on it while it moves. In our model we will assume that there are only two such forces: its weight w and the restoring force F_s of the spring. In particular, we will ignore such forces as air resistance, internal frictional forces in the spring, forces due to movement of the spring support, and so forth. With these assumptions, the model is called the ***simple harmonic model*** and the motion of the block is called ***simple harmonic motion***.

Our goal is to produce a differential equation whose solution gives the position function $y(t)$ of the block as a function of time. We will do this by determining the net force $F(t)$ acting on the block at a general time t and then applying Newton's Second Law of Motion.

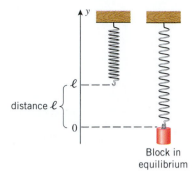

Figure 9.4.3

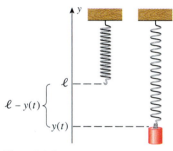

Figure 9.4.4

Since the only forces acting on the block are its weight $w = -Mg$ and the restoring force F_s of the spring, and since the acceleration of the block at time t is $y''(t)$, it follows from Newton's Second Law of Motion that

$$F_s(t) - Mg = My''(t) \tag{16}$$

To express $F_s(t)$ in terms of $y(t)$, we will begin by examining the forces on the block when it is in its equilibrium position. In this position the downward force of the weight is perfectly balanced by the upward restoring force of the spring, so that the sum of these two forces must be zero. Thus, if we assume that the spring constant is k and that the spring is stretched a distance of ℓ units beyond its natural length when the block is in equilibrium (Figure 9.4.3), then

$$k\ell - Mg = 0 \tag{17}$$

Now let us examine the restoring force acting on the block when the connection point has coordinate $y(t)$. At this point the end of the spring is displaced $\ell - y(t)$ units from its natural position (Figure 9.4.4), so Hooke's law implies that the restoring force is

$$F_s(t) = k(\ell - y(t)) = k\ell - ky(t)$$

which from (17) can be rewritten as

$$F_s(t) = Mg - ky(t)$$

Substituting this in (16) and canceling the Mg terms yields

$$-ky(t) = My''(t)$$

which we can rewrite as the homogeneous equation

$$y''(t) + \left(\frac{k}{M}\right) y(t) = 0 \tag{18}$$

The auxiliary equation for (18) is

$$m^2 + \frac{k}{M} = 0$$

which has imaginary roots $m_1 = \sqrt{k/M}\,i$, $m_2 = -\sqrt{k/M}\,i$ (since k and M are positive). It follows that the general solution of (18) is

$$y(t) = c_1 \cos\left(\sqrt{\frac{k}{M}}\,t\right) + c_2 \sin\left(\sqrt{\frac{k}{M}}\,t\right) \tag{19}$$

• **FOR THE READER.** Confirm that the functions in family (19) are solutions of (18).

To determine the constants c_1 and c_2 in (19) we will take as our initial conditions the position and velocity at time $t = 0$. Specifically, we will ask you to show in Exercise 40 that if the position of the block at time $t = 0$ is y_0, and if the initial velocity of the block is zero (i.e., it is *released* from rest), then

$$y(t) = y_0 \cos\left(\sqrt{\frac{k}{M}}\,t\right) \tag{20}$$

This formula describes a periodic vibration with an amplitude of $|y_0|$, a period T given by

$$T = \frac{2\pi}{\sqrt{k/M}} = 2\pi\sqrt{M/k} \tag{21}$$

and a frequency f given by

$$f = \frac{1}{T} = \frac{\sqrt{k/M}}{2\pi} \tag{22}$$

(Figure 9.4.5).

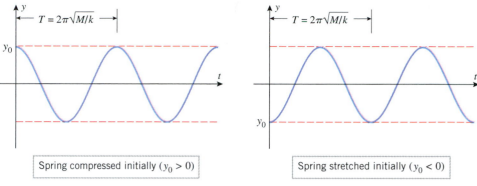

Spring compressed initially ($y_0 > 0$)   Spring stretched initially ($y_0 < 0$)

Figure 9.4.5

Example 5   Suppose that the block in Figure 9.4.2 stretches the spring 0.2 m in equilibrium. Suppose also that the block is pulled 0.5 m below its equilibrium position and released at time $t = 0$.

(a)   Find the position function $y(t)$ of the block.

(b)   Find the amplitude, period, and frequency of the vibration.

Solution (a).   The appropriate formula is (20). Although we are not given the mass M of the block or the spring constant k, it does not matter because we can use the equilibrium condition (17) to find the ratio k/M without having values for k and M. Specifically, we are given that in equilibrium the block stretches the spring $\ell = 0.2$ m, and we know that $g = 9.8$ m/s^2. Thus, (17) implies that

$$\frac{k}{M} = \frac{g}{\ell} = \frac{9.8}{0.2} = 49 \text{ s}^{-2} \tag{23}$$

Substituting this in (20) yields

$$y(t) = y_0 \cos 7t$$

where y_0 is the coordinate of the block at time $t = 0$. However, we are given that the block is initially 0.5 m *below* the equilibrium position, so $y_0 = -0.5$ and hence the position function of the block is $y(t) = -0.5 \cos 7t$.

Solution (b).   The amplitude of the vibration is

$$\text{amplitude } = |y_0| = |-0.5| = 0.5 \text{ m}$$

and from (21), (22), and (23) the period and frequency are

$$\text{period} = T = 2\pi\sqrt{\frac{M}{k}} = 2\pi\sqrt{\frac{1}{49}} = \frac{2\pi}{7} \text{ s}, \quad \text{frequency} = f = \frac{1}{T} = \frac{7}{2\pi} \text{ Hz} \quad \blacktriangleleft$$

EXERCISE SET 9.4   ⌁ Graphing Utility   **c** CAS

• •

1. Verify that the following are solutions of the differential equation $y'' - y' - 2y = 0$ by substituting these functions into the equation.
   (a) e^{2x} and e^{-x}
   (b) $c_1 e^{2x} + c_2 e^{-x}$ (c_1, c_2 constants)

2. Verify that the following are solutions of the differential equation $y'' + 4y' + 4y = 0$ by substituting these functions into the equation.
   (a) e^{-2x} and xe^{-2x}
   (b) $c_1 e^{-2x} + c_2 xe^{-2x}$ (c_1, c_2 constants)

In Exercises 3–16, find the general solution of the differential equation.

3. $y'' + 3y' - 4y = 0$

4. $y'' + 6y' + 5y = 0$

5. $y'' - 2y' + y = 0$

6. $y'' + 6y' + 9y = 0$

7. $y'' + 5y = 0$

8. $y'' + y = 0$

9. $\dfrac{d^2y}{dx^2} - \dfrac{dy}{dx} = 0$

10. $\dfrac{d^2y}{dx^2} + 3\dfrac{dy}{dx} = 0$

11. $\dfrac{d^2y}{dt^2} + 4\dfrac{dy}{dt} + 4y = 0$

12. $\dfrac{d^2y}{dt^2} - 10\dfrac{dy}{dt} + 25y = 0$

13. $\dfrac{d^2y}{dx^2} - 4\dfrac{dy}{dx} + 13y = 0$

14. $\dfrac{d^2y}{dx^2} - 6\dfrac{dy}{dx} + 25y = 0$

15. $8y'' - 2y' - y = 0$

16. $9y'' - 6y' + y = 0$

In Exercises 17–22, solve the initial-value problem.

17. $y'' + 2y' - 3y = 0;$ $\quad y(0) = 1,$ $y'(0) = 5$

18. $y'' - 6y' - 7y = 0;$ $\quad y(0) = 5,$ $y'(0) = 3$

19. $y'' - 6y' + 9y = 0;$ $\quad y(0) = 2,$ $y'(0) = 1$

20. $y'' + 4y' + y = 0;$ $\quad y(0) = 5,$ $y'(0) = 4$

21. $y'' + 4y' + 5y = 0;$ $\quad y(0) = -3,$ $y'(0) = 0$

22. $y'' - 6y' + 13y = 0;$ $\quad y(0) = -1,$ $y'(0) = 1$

23. In each part find a second-order linear homogeneous differential equation with constant coefficients that has the given functions as solutions.

(a) $y_1 = e^{5x},$ $y_2 = e^{-2x}$ (b) $y_1 = e^{4x},$ $y_2 = xe^{4x}$

(c) $y_1 = e^{-x}\cos 4x,$ $y_2 = e^{-x}\sin 4x$

24. Show that if e^x and e^{-x} are solutions of a second-order linear homogeneous differential equation, then so are $\cosh x$ and $\sinh x$.

25. Find all values of k for which the differential equation $y'' + ky' + ky = 0$ has a general solution of the given form.

(a) $y = c_1 e^{ax} + c_2 e^{bx}$ (b) $y = c_1 e^{ax} + c_2 x e^{ax}$

(c) $y = c_1 e^{ax}\cos bx + c_2 e^{ax}\sin bx$

26. The equation

$$x^2\frac{d^2y}{dx^2} + px\frac{dy}{dx} + qy = 0 \quad (x > 0)$$

where p and q are constants, is called *Euler's equidimensional equation*. Show that the substitution $x = e^z$ transforms this equation into the equation

$$\frac{d^2y}{dz^2} + (p-1)\frac{dy}{dz} + qy = 0$$

27. Use the result in Exercise 26 to find the general solution of

(a) $x^2\dfrac{d^2y}{dx^2} + 3x\dfrac{dy}{dx} + 2y = 0$ $\quad (x > 0)$

(b) $x^2\dfrac{d^2y}{dx^2} - x\dfrac{dy}{dx} - 2y = 0$ $\quad (x > 0)$.

28. Let $y(x)$ be a solution of $y'' + py' + qy = 0$. Prove: If p and q are positive constants, then $\lim_{x \to +\infty} y(x) = 0$.

29. Prove that the following functions are linearly independent.

(a) $y_1 = e^{m_1 x},$ $y_2 = e^{m_2 x}$ $(m_1 \neq m_2)$

(b) $y_1 = e^{mx},$ $y_2 = xe^{mx}$

30. Prove: If the auxiliary equation of

$$y'' + py' + qy = 0$$

has complex roots $a + bi$ and $a - bi$, then the general solution of this differential equation is

$$y(x) = e^{ax}(c_1 \cos bx + c_2 \sin bx)$$

[*Hint:* Using substitution, verify that $y_1 = e^{ax}\cos bx$ and $y_2 = e^{ax}\sin bx$ are solutions of the differential equation. Then prove that y_1 and y_2 are linearly independent.]

31. Suppose that the auxiliary equation of the equation $y'' + py' + qy = 0$ has distinct real roots μ and m.

(a) Show that the function

$$g_\mu(x) = \frac{e^{\mu x} - e^{mx}}{\mu - m}$$

is a solution of the differential equation.

(b) Use L'Hôpital's rule to show that

$$\lim_{\mu \to m} g_\mu(x) = xe^{mx}$$

[*Note:* Can you see how the result in part (b) makes it plausible that the function $y(x) = xe^{mx}$ is a solution of $y'' + py' + qy = 0$ when m is a repeated root of the auxiliary equation?]

32. Consider the problem of solving the differential equation

$$y'' + \lambda y = 0$$

subject to the conditions $y(0) = 0$, $y(\pi) = 0$.

(a) Show that if $\lambda \leq 0$, then $y = 0$ is the only solution.

(b) Show that if $\lambda > 0$, then the solution is

$$y = c\sin\sqrt{\lambda}x$$

where c is an arbitrary constant, if

$$\lambda = 1, 2^2, 3^2, 4^2, \ldots$$

and the only solution is $y = 0$ otherwise.

Exercises 33–38 involve vibrations of the block pictured in Figure 9.4.1. Assume that the y-axis is as shown in Figure 9.4.2 and that the simple harmonic model applies.

33. Suppose that the block has a mass of 1 kg, the spring constant is $k = 0.25$ N/m, and the block is pushed 0.3 m above its equilibrium position and released at time $t = 0$.

(a) Find the position function $y(t)$ of the block.

(b) Find the period and frequency of the vibration.

(c) Sketch the graph of $y(t)$.

(d) At what time does the block first pass through the equilibrium position?

(e) At what time does the block first reach its maximum distance below the equilibrium position?

34. Suppose that the block has a weight of 64 lb, the spring constant is $k = 0.25$ lb/ft, and the block is pushed 1 ft above its equilibrium position and released at time $t = 0$.
  (a) Find the position function $y(t)$ of the block.
  (b) Find the period and frequency of the vibration.
  (c) Sketch the graph of $y(t)$.
  (d) At what time does the block first pass through the equilibrium position?
  (e) At what time does the block first reach its maximum distance below the equilibrium position?

35. Suppose that the block stretches the spring 0.05 m in equilibrium, and the block is pulled 0.12 m below the equilibrium position and released at time $t = 0$.
  (a) Find the position function $y(t)$ of the block.
  (b) Find the period and frequency of the vibration.
  (c) Sketch the graph of $y(t)$.
  (d) At what time does the block first pass through the equilibrium position?
  (e) At what time does the block first reach its maximum distance above the equilibrium position?

36. Suppose that the block stretches the spring 0.5 ft in equilibrium, and is pulled 1.5 ft below the equilibrium position and released at time $t = 0$.
  (a) Find the position function $y(t)$ of the block.
  (b) Find the period and frequency of the vibration.
  (c) Sketch the graph of $y(t)$.
  (d) At what time does the block first pass through the equilibrium position?
  (e) At what time does the block first reach its maximum distance above the equilibrium position?

37. (a) For what values of y would you expect the block in Exercise 36 to have its maximum speed? Confirm your answer to this question mathematically.
  (b) For what values of y would you expect the block to have its minimum speed? Confirm your answer to this question mathematically.

38. Suppose that the block weighs w pounds and vibrates with a period of 3 s when it is pulled below the equilibrium position and released. Suppose also that if the process is repeated with an additional 4 lb of weight, then the period is 5 s.
  (a) Find the spring constant. (b) Find w.

39. As shown in the accompanying figure, suppose that a toy cart of mass M is attached to a wall by a spring with spring constant k, and let a horizontal x-axis be introduced with its origin at the connection point of the spring and cart when the cart is in equilibrium. Suppose that the cart is pulled or pushed horizontally to a point x_0 and then released at time $t = 0$. Find an initial-value problem whose solution is the position function of the cart, and state any assumptions you have made.

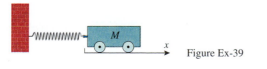

Figure Ex-39

40. Use the initial position $y(0) = y_0$ and the initial velocity $v(0) = 0$ to find the constants c_1 and c_2 in (19).

The accompanying figure shows a mass–spring system in which an object of mass M is suspended by a spring and linked to a piston that moves in a *dashpot* containing a viscous fluid. If there are no external forces acting on the system, then the object is said to have **free motion** and the motion of the object is completely determined by the displacement and velocity of the object at time $t = 0$, the stiffness of the spring as measured by the spring constant k, and the viscosity of the fluid in the dashpot as measured by a *damping constant* c. Mathematically, the displacement $y = y(t)$ of the object from its equilibrium position is the solution of an initial-value problem of the form

$$y'' + Ay' + By = 0, \quad y(0) = y_0, \quad y'(0) = v_0$$

where the coefficient A is determined by M and c and the coefficient B is determined by M and k. In our derivation of Equation (21) we considered only motion in which the coefficient A is zero and in which the object is released from rest, that is, $v_0 = 0$. In Exercises 41–45, you are asked to consider initial-value problems for which both the coefficient A and the initial velocity v_0 are nonzero.

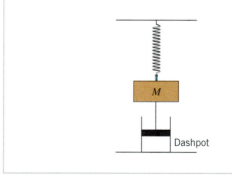

Dashpot

41. (a) Solve the initial-value problem $y'' + 2.4y' + 1.44y = 0$, $y(0) = 1$, $y'(0) = 2$ and graph $y = y(t)$ on the interval $[0, 5]$.
  (b) Find the maximum distance above the equilibrium position attained by the object.
  (c) The graph of $y(t)$ suggests that the object does not pass through the equilibrium position. Show that this is so.

42. (a) Solve the initial-value problem $y'' + 5y' + 2y = 0$, $y(0) = 1/2$, $y'(0) = -4$ and graph $y = y(t)$ on the interval $[0, 5]$.
  (b) Find the maximum distance below the equilibrium position attained by the object.
  (c) The graph of $y(t)$ suggests that the object passes through the equilibrium position exactly once. With what speed does the object pass through the equilibrium position?

43. (a) Solve the initial-value problem $y'' + y' + 5y = 0$, $y(0) = 1$, $y'(0) = -3.5$ and graph $y = y(t)$ on the interval $[0, 8]$.

(b) Find the maximum distance below the equilibrium position attained by the object.

(c) Find the velocity of the object when it passes through the equilibrium position the first time.

(d) Find, by inspection, the acceleration of the object when it passes through the equilibrium position the first time. [*Hint:* Examine the differential equation and use the result in part (c).]

44. (a) Solve the initial-value problem $y'' + y' + 3y = 0$, $y(0) = -2$, $y'(0) = v_0$.

(b) Find the largest positive value of v_0 for which the object will rise no higher than 1 unit above the equilibrium position. [*Hint:* Use a trial-and-error strategy. Estimate v_0 to the nearest hundredth.]

(c) Graph the solution of the initial-value problem on the interval $[0, 8]$ using the value of v_0 obtained in part (b).

45. (a) Solve the initial-value problem $y'' + 3.5y' + 3y = 0$, $y(0) = 1$, $y'(0) = v_0$.

(b) Use the result in part (a) to find the solutions for $v_0 = 2$, $v_0 = -1$, and $v_0 = -4$ and graph all three solutions on the interval $[0, 4]$ in the same coordinate system.

(c) Discuss the effect of the initial velocity on the motion of the object.

46. Consider the first-order linear homogeneous equation

$$\frac{dy}{dx} + p(x)y = 0$$

where $p(x)$ is a continuous function on some open interval I. By analogy to the results of Theorem 9.4.1, we might expect the general solution of this equation to be of the form

$$y = cy_1(x)$$

where $y_1(x)$ is a solution of the equation on the interval I and c is an arbitrary constant. Prove this to be the case.

SUPPLEMENTARY EXERCISES

c CAS

1. We have seen that the general solution of a first-order linear equation involves a single arbitrary constant and that the general solution of a second-order linear differential equation involves two arbitrary constants. Give an informal explanation of why one might expect the number of arbitrary constants to equal the order of the equation.

2. Write a paragraph that describes Euler's Method.

3. (a) List the steps in the method of integrating factors for solving first-order linear differential equations.

(b) What would you do if you had to solve an important initial-value problem involving a first-order linear differential equation whose integrating factor could not be obtained because of the complexity of the integration?

4. Which of the following differential equations are separable?

(a) $\dfrac{dy}{dx} = f(x)g(y)$ (b) $\dfrac{dy}{dx} = \dfrac{f(x)}{g(y)}$

(c) $\dfrac{dy}{dx} = f(x) + g(y)$ (d) $\dfrac{dy}{dx} = \sqrt{f(x)g(y)}$

5. Classify the following first-order differential equations as separable, linear, both, or neither.

(a) $\dfrac{dy}{dx} - 3y = \sin x$ (b) $\dfrac{dy}{dx} + xy = x$

(c) $y\dfrac{dy}{dx} - x = 1$ (d) $\dfrac{dy}{dx} + xy^2 = \sin(xy)$

6. Determine whether the methods of integrating factors and separation of variables produce the same solution of the differential equation

$$\frac{dy}{dx} - 4xy = x$$

7. Consider the model $dy/dt = ky(L - y)$ for the spread of a disease, where $k > 0$ and $0 < y \leq L$. For what value of y is the disease spreading most rapidly, and at what rate is it spreading?

8. (a) Show that if a quantity $y = y(t)$ has an exponential model, and if $y(t_1) = y_1$ and $y(t_2) = y_2$, then the doubling time or the half-life T is

$$T = \left| \frac{(t_2 - t_1) \ln 2}{\ln(y_2/y_1)} \right|$$

(b) In a certain 1-hour period the number of bacteria in a colony increases by 25%. Assuming an exponential growth model, what is the doubling time for the colony?

9. Assume that a spherical meteoroid burns up at a rate that is proportional to its surface area. Given that the radius is originally 4 m and 1 min later its radius is 3 m, find a formula for the radius as a function of time.

10. A tank contains 1000 gal of fresh water. At time $t = 0$ min, brine containing 5 ounces of salt per gallon of brine is poured into the tank at a rate of 10 gal/min, and the mixed solution is drained from the tank at the same rate. After 15 min that process is stopped and fresh water is poured into the tank at the rate of 5 gal/min, and the mixed solution is drained from the tank at the same rate. Find the amount of salt in the tank at time $t = 30$ min.

11. Suppose that a room containing 1200 ft³ of air is free of carbon monoxide. At time $t = 0$ cigarette smoke containing 4% carbon monoxide is introduced at the rate of 0.1 ft³/min, and the well-circulated mixture is vented from the room at the same rate.

(a) Find a formula for the percentage of carbon monoxide in the room at time t.

(b) Extended exposure to air containing 0.012% carbon monoxide is considered dangerous. How long will it take to reach this level? [This is based on a problem from William E. Boyce and Richard C. DiPrima, *Elementary Differential Equations*, 7th ed., John Wiley & Sons, New York, 2001.]

In Exercises 12–16, solve the initial-value problem.

---

12. $y' = 1 + y^2$, $y(0) = 1$

13. $y' = \dfrac{y^5}{x(1 + y^4)}$, $y(1) = 1$

14. $xy' + 2y = 4x^2$, $y(1) = 2$

15. $y' = 4y^2 \sec^2 2x$, $y(\pi/8) = 1$

16. $y' = 6 - 5y + y^2$, $y(0) = \ln 2$

c 17. (a) Solve the initial-value problem
$$y' - y = x \sin 3x, \quad y(0) = 1$$
by the method of integrating factors, using a CAS to perform any difficult integrations.

(b) Use the CAS to solve the initial-value problem directly, and confirm that the answer is consistent with that obtained in part (a).

(c) Graph the solution.

c 18. Use a CAS to derive Formula (23) of Section 9.1 by solving initial-value problem (21).

19. (a) It is currently accepted that the half-life of carbon-14 might vary ±40 years from its nominal value of 5730 years. Does this variation make it possible that the Shroud of Turin dates to the time of Jesus of Nazareth? [See Example 4 of Section 9.3.]

(b) Review the subsection of Section 3.8 entitled Error Propagation in Applications, and then estimate the percentage error that results in the computed age of an artifact from an $r\%$ error in the half-life of carbon-14.

20. (a) Use Euler's Method with a step-size of $\Delta x = 0.1$ to approximate the solution of the initial-value problem
$$y' = 1 + 5t - y, \quad y(1) = 5$$
over the interval $[1, 2]$.

(b) Find the percentage error in the values computed.

21. Find the general solution of each differential equation.
(a) $y'' - 3y' + 2y = 0$ (b) $4y'' - 4y' + y = 0$
(c) $y'' + y' + 2y = 0$

22. (a) Sketch the integral curve of $2yy' = 1$ that passes through the point $(0, 1)$ and the integral curve that passes through the point $(0, -1)$.

(b) Sketch the integral curve of $y' = -2xy^2$ that passes through the point $(0, 1)$.

23. Suppose that a herd of 19 deer is moved to a small island whose estimated carrying capacity is 95 deer, and assume that the population has a logistic growth model.
(a) Given that 1 year later the population is 25, how long will it take for the deer population to reach 80% of the island's carrying capacity?

(b) Find an initial-value problem whose solution gives the deer population as a function of time.

c 24. If the block in Figure 9.4.1 is displaced y_0 units from its equilibrium position and given an initial velocity of v_0, rather than being released with an initial velocity of 0, then its position function $y(t)$ given in Equation (19) of Section 9.4 must satisfy the initial conditions $y(0) = y_0$ and $y'(0) = v_0$.
(a) Show that
$$y(t) = y_0 \cos\left(\sqrt{\frac{k}{M}}\, t\right) + v_0 \sqrt{\frac{M}{k}} \sin\left(\sqrt{\frac{k}{M}}\, t\right)$$

(b) Suppose that a block with a mass of 1 kg stretches the spring 0.5 m in equilibrium. Use a graphing utility to graph the position function of the block if it is set in motion by pulling it down 1 m and imparting it an initial upward velocity of 0.25 m/s.

(c) What is the maximum displacement of the block from the equilibrium position?

25. A block attached to a vertical spring is displaced from its equilibrium position and released, thereby causing it to vibrate with amplitude $|y_0|$ and period T.
(a) Show that the velocity of the block has maximum magnitude $2\pi|y_0|/T$ and that the maximum occurs when the block is at its equilibrium position.

(b) Show that the acceleration of the block has maximum magnitude $4\pi^2|y_0|/T^2$ and that the maximum occurs when the block is at a top or bottom point of its motion.

26. Suppose that P dollars is invested at an annual interest rate of $r \times 100\%$. If the accumulated interest is credited to the account at the end of the year, then the interest is said to be *compounded annually*; if it is credited at the end of each 6-month period, then it is said to be *compounded semiannually*; and if it is credited at the end of each 3-month period, then it is said to be *compounded quarterly*. The more frequently the interest is compounded, the better it is for the investor since more of the interest is itself earning interest.
(a) Show that if interest is compounded n times a year at equally spaced intervals, then the value A of the investment after t years is
$$A = P\left(1 + \frac{r}{n}\right)^{nt}$$

(b) One can imagine interest to be compounded each day, each hour, each minute, and so forth. Carried to the limit one can conceive of interest compounded at each instant of time; this is called *continuous compounding*. Thus, from part (a), the value A of P dollars after t years when invested at an annual rate of $r \times 100\%$, compounded continuously, is

$$A = \lim_{n \to +\infty} P \left(1 + \frac{r}{n}\right)^{nt}$$

Use the fact that $\lim_{x \to 0} (1 + x)^{1/x} = e$ to prove that $A = Pe^{rt}$.

(c) Use the result in part (b) to show that money invested at continuous compound interest increases at a rate proportional to the amount present.

27. (a) If $1000 is invested at 8% per year compounded continuously (Exercise 26), what will the investment be worth after 5 years?

(b) If it is desired that an investment at 8% per year compounded continuously should have a value of $10,000 after 10 years, how much should be invested now?

(c) How long does it take for an investment at 8% per year compounded continuously to double in value?

28. Prove Theorem 9.4.1 in the special case where $q(x)$ is identically zero.

29. Assume that the motion of a block of mass M is governed by the simple harmonic model (18) in Section 9.4. Define the *potential energy* of the block at time t to be $\frac{1}{2}k[y(t)]^2$, and define the *kinetic energy* of the block at time t to be $\frac{1}{2}M[y'(t)]^2$. Prove that the sum of the potential energy of the block and the kinetic energy of the block is constant.

10

INFINITE SERIES

Brook Taylor

*I*n this chapter we will be concerned with *infinite series*, which are sums that involve infinitely many terms. Infinite series play a fundamental role in both mathematics and science—they are used, for example, to approximate trigonometric functions and logarithms, to solve differential equations, to evaluate difficult integrals, to create new functions, and to construct mathematical models of physical laws. Since it is impossible to add up infinitely many numbers directly, one goal will be to define exactly what we mean by the sum of an infinite series. However, unlike finite sums, it turns out that not all infinite series actually have a sum, so we will need to develop tools for determining which infinite series have sums and which do not. Once the basic ideas have been developed we will begin to apply our work; we will show how infinite series are used to evaluate such quantities as $\sin 17°$ and $\ln 5$, how they are used to create functions, and finally, how they are used to model physical laws.

10.1 MACLAURIN AND TAYLOR POLYNOMIAL APPROXIMATIONS

In Chapter 3 we used a tangent line to the graph of a function to obtain a linear approximation to the function near the point of tangency. In this section we will see how to improve such local approximations by using polynomials. We conclude the section by obtaining a bound on the error in these approximations. We have placed this section here for those who want an early discussion of Maclaurin and Taylor polynomials. If desired, this section can be delayed and used as a prelude to Section 10.8.

LOCAL QUADRATIC APPROXIMATIONS

Recall from Formula (1) in Section 3.8 that the local linear approximation of a function f at x_0 is

$$f(x) \approx f(x_0) + f'(x_0)(x - x_0) \tag{1}$$

In this formula, the approximating function

$$p(x) = f(x_0) + f'(x_0)(x - x_0)$$

is a first-degree polynomial satisfying $p(x_0) = f(x_0)$ and $p'(x_0) = f'(x_0)$ (verify). Thus, the local linear approximation of f at x_0 has the property that its value and the values of its first derivative match those of f at x_0.

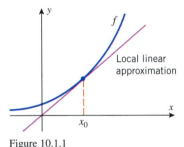

Figure 10.1.1

If the graph of a function f has a pronounced "bend" at x_0, then we can expect that the accuracy of the local linear approximation of f at x_0 will decrease rapidly as we progress away from x_0 (Figure 10.1.1). One way to deal with this problem is to approximate the function f at x_0 by a polynomial p of degree 2 with the property that the value of p and the values of its first two derivatives match those of f at x_0. This ensures that the graphs of f and p not only have the same tangent line at x_0, but they also bend in the same direction at x_0 (both concave up or concave down). As a result, we can expect that the graph of p will remain close to the graph of f over a larger interval around x_0 than the graph of the local linear approximation. The polynomial p is called the *local quadratic approximation of f at $x = x_0$.*

To illustrate this idea, let us try to find a formula for the local quadratic approximation of a function f at $x = 0$. This approximation has the form

$$f(x) \approx c_0 + c_1 x + c_2 x^2 \tag{2}$$

where c_0, c_1, and c_2 must be chosen so that the values of

$$p(x) = c_0 + c_1 x + c_2 x^2$$

and its first two derivatives match those of f at 0. Thus, we want

$$p(0) = f(0), \quad p'(0) = f'(0), \quad p''(0) = f''(0) \tag{3}$$

But the values of $p(0)$, $p'(0)$, and $p''(0)$ are as follows:

$$p(x) = c_0 + c_1 x + c_2 x^2 \qquad p(0) = c_0$$
$$p'(x) = c_1 + 2c_2 x \qquad p'(0) = c_1$$
$$p''(x) = 2c_2 \qquad p''(0) = 2c_2$$

Thus, it follows from (3) that

$$c_0 = f(0), \quad c_1 = f'(0), \quad c_2 = \frac{f''(0)}{2}$$

and substituting these in (2) yields the following formula for the local quadratic approximation of f at $x = 0$:

$$f(x) \approx f(0) + f'(0)x + \frac{f''(0)}{2}x^2 \tag{4}$$

REMARK.   Observe that with $x_0 = 0$, Formula (1) becomes

$$f(x) \approx f(0) + f'(0)x \tag{5}$$

and hence the linear part of the local quadratic approximation of f at 0 is the local linear approximation of f at 0.

Example 1   Find the local linear and quadratic approximations of e^x at $x = 0$, and graph e^x and the two approximations together.

Solution.   If we let $f(x) = e^x$, then $f'(x) = f''(x) = e^x$; and hence

$$f(0) = f'(0) = f''(0) = e^0 = 1$$

Thus, from (4) the local quadratic approximation of e^x at $x = 0$ is

$$e^x \approx 1 + x + \frac{x^2}{2}$$

and the local linear approximation (which is the linear part of the local quadratic approximation) is

$$e^x \approx 1 + x$$

The graphs of e^x and the two approximations are shown in Figure 10.1.2. As expected, the local quadratic approximation is more accurate than the local linear approximation near $x = 0$.   ◄

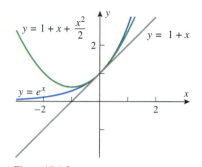

$y = 1 + x + \dfrac{x^2}{2}$

$y = 1 + x$

$y = e^x$

Figure 10.1.2

MACLAURIN POLYNOMIALS

It is natural to ask whether one can improve on the accuracy of a local quadratic approximation by using a polynomial of degree 3. Specifically, one might look for a polynomial of degree 3 with the property that its value and the values of its first three derivatives match those of f at a point; and if this provides an improvement in accuracy, why not go on to polynomials of even higher degree? Thus, we are led to consider the following general problem.

> **10.1.1   PROBLEM.**   Given a function f that can be differentiated n times at $x = x_0$, find a polynomial p of degree n with the property that the value of p and the values of its first n derivatives match those of f at x_0.

We will begin by solving this problem in the case where $x_0 = 0$. Thus, we want a polynomial

$$p(x) = c_0 + c_1 x + c_2 x^2 + c_3 x^3 + \cdots + c_n x^n \tag{6}$$

such that

$$f(0) = p(0), \quad f'(0) = p'(0), \quad f''(0) = p''(0), \ldots, \quad f^{(n)}(0) = p^{(n)}(0) \tag{7}$$

But

$$\begin{aligned}
p(x) &= c_0 + c_1 x + c_2 x^2 + c_3 x^3 + \cdots + c_n x^n \\
p'(x) &= c_1 + 2c_2 x + 3c_3 x^2 + \cdots + nc_n x^{n-1} \\
p''(x) &= 2c_2 + 3 \cdot 2c_3 x + \cdots + n(n-1)c_n x^{n-2} \\
p'''(x) &= 3 \cdot 2c_3 + \cdots + n(n-1)(n-2)c_n x^{n-3} \\
&\vdots \\
p^{(n)}(x) &= n(n-1)(n-2) \cdots (1)c_n
\end{aligned}$$

Thus, to satisfy (7) we must have[*]

$$f(0) = p(0) = c_0$$
$$f'(0) = p'(0) = c_1$$
$$f''(0) = p''(0) = 2c_2 = 2!c_2$$
$$f'''(0) = p'''(0) = 3 \cdot 2c_3 = 3!c_3$$
$$\vdots$$
$$f^{(n)}(0) = p^{(n)}(0) = n(n-1)(n-2)\cdots(1)c_n = n!c_n$$

which yields the following values for the coefficients of $p(x)$:

$$c_0 = f(0), \quad c_1 = f'(0), \quad c_2 = \frac{f''(0)}{2!}, \quad c_3 = \frac{f'''(0)}{3!}, \ldots, \quad c_n = \frac{f^{(n)}(0)}{n!}$$

The polynomial that results by using these coefficients in (6) is called the *nth Maclaurin[†]
polynomial for f*.

> **10.1.2 DEFINITION.** If f can be differentiated n times at 0, then we define the ***nth
> Maclaurin polynomial for f*** to be
>
> $$p_n(x) = f(0) + f'(0)x + \frac{f''(0)}{2!}x^2 + \frac{f'''(0)}{3!}x^3 + \cdots + \frac{f^{(n)}(0)}{n!}x^n \qquad (8)$$
>
> This polynomial has the property that its value and the values of its first n derivatives
> match the values of f and its first n derivatives at $x = 0$.

REMARK. Observe that $p_1(x)$ is the local linear approximation of f at 0 and $p_2(x)$ is the
local quadratic approximation of f at $x = 0$.

Example 2 Find the Maclaurin polynomials p_0, p_1, p_2, p_3, and p_n for e^x.

Solution. Let $f(x) = e^x$. Thus,

$$f'(x) = f''(x) = f'''(x) = \cdots = f^{(n)}(x) = e^x$$

and

$$f(0) = f'(0) = f''(0) = f'''(0) = \cdots = f^{(n)}(0) = e^0 = 1$$

---

[*]Recall that if n is a positive integer, then the symbol $n!$ (read "n factorial") denotes the product of the first n
positive integers; that is,

$$n! = 1 \cdot 2 \cdot 3 \cdots n \quad \text{or equivalently,} \quad n! = n(n-1)(n-2)\cdots 1$$

Moreover, it is agreed by convention that $0! = 1$.

[†]COLIN MACLAURIN (1698–1746). Scottish mathematician. Maclaurin's father, a minister, died when the boy
was only six months old, and his mother when he was nine years old. He was then raised by an uncle who
was also a minister. Maclaurin entered Glasgow University as a divinity student, but transferred to mathematics
after one year. He received his Master's degree at age 17 and, in spite of his youth, began teaching at Marischal
College in Aberdeen, Scotland. He met Isaac Newton during a visit to London in 1719 and from that time
on became Newton's disciple. During that era, some of Newton's analytic methods were bitterly attacked by
major mathematicians and much of Maclaurin's important mathematical work resulted from his efforts to defend
Newton's ideas geometrically. Maclaurin's work, *A Treatise of Fluxions* (1742), was the first systematic formulation
of Newton's methods. The treatise was so carefully done that it was a standard of mathematical rigor in calculus
until the work of Cauchy in 1821. Maclaurin was also an outstanding experimentalist; he devised numerous
ingenious mechanical devices, made important astronomical observations, performed actuarial computations for
insurance societies, and helped to improve maps of the islands around Scotland.

Therefore,

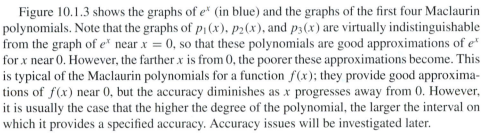

$$p_0(x) = f(0) = 1$$

$$p_1(x) = f(0) + f'(0)x = 1 + x$$

$$p_2(x) = f(0) + f'(0)x + \frac{f''(0)}{2!}x^2 = 1 + x + \frac{x^2}{2!} = 1 + x + \frac{1}{2}x^2$$

$$p_3(x) = f(0) + f'(0)x + \frac{f''(0)}{2!}x^2 + \frac{f'''(0)}{3!}x^3$$

$$= 1 + x + \frac{x^2}{2!} + \frac{x^3}{3!} = 1 + x + \frac{1}{2}x^2 + \frac{1}{6}x^3$$

$$p_n(x) = f(0) + f'(0)x + \frac{f''(0)}{2!}x^2 + \cdots + \frac{f^{(n)}(0)}{n!}x^n$$

$$= 1 + x + \frac{x^2}{2!} + \cdots + \frac{x^n}{n!}$$ ◄

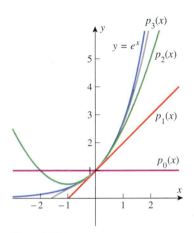

Figure 10.1.3

Figure 10.1.3 shows the graphs of e^x (in blue) and the graphs of the first four Maclaurin polynomials. Note that the graphs of $p_1(x)$, $p_2(x)$, and $p_3(x)$ are virtually indistinguishable from the graph of e^x near $x = 0$, so that these polynomials are good approximations of e^x for x near 0. However, the farther x is from 0, the poorer these approximations become. This is typical of the Maclaurin polynomials for a function $f(x)$; they provide good approximations of $f(x)$ near 0, but the accuracy diminishes as x progresses away from 0. However, it is usually the case that the higher the degree of the polynomial, the larger the interval on which it provides a specified accuracy. Accuracy issues will be investigated later.

TAYLOR POLYNOMIALS

Up to now we have focused on approximating a function f in the vicinity of $x = 0$. Now we will consider the more general case of approximating f in the vicinity of an arbitrary domain value x_0. The basic idea is the same as before; we want to find an nth-degree polynomial p with the property that its value and the values of its first n derivatives match those of f at x_0. However, rather than expressing $p(x)$ in powers of x, it will simplify the computations if we express it in powers of $x - x_0$; that is,

$$p(x) = c_0 + c_1(x - x_0) + c_2(x - x_0)^2 + \cdots + c_n(x - x_0)^n \qquad (9)$$

We will leave it as an exercise for you to imitate the computations used in the case where $x_0 = 0$ to show that

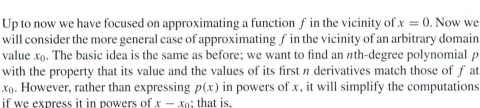

$$c_0 = f(x_0), \quad c_1 = f'(x_0), \quad c_2 = \frac{f''(x_0)}{2!}, \quad c_3 = \frac{f'''(x_0)}{3!}, \ldots, \quad c_n = \frac{f^{(n)}(x_0)}{n!}$$

Substituting these values in (9) we obtain a polynomial called the *nth Taylor*[*] *polynomial about $x = x_0$ for f*.

---

[*] BROOK TAYLOR (1685–1731). English mathematician. Taylor was born of well-to-do parents. Musicians and artists were entertained frequently in the Taylor home, which undoubtedly had a lasting influence on young Brook. In later years, Taylor published a definitive work on the mathematical theory of perspective and obtained major mathematical results about the vibrations of strings. There also exists an unpublished work, *On Musick*, that was intended to be part of a joint paper with Isaac Newton. Taylor's life was scarred with unhappiness, illness, and tragedy. Because his first wife was not rich enough to suit his father, the two men argued bitterly and parted ways. Subsequently, his wife died in childbirth. Then, after he remarried, his second wife also died in childbirth, though his daughter survived. Taylor's most productive period was from 1714 to 1719, during which time he wrote on a wide range of subjects—magnetism, capillary action, thermometers, perspective, and calculus. In his final years, Taylor devoted his writing efforts to religion and philosophy. According to Taylor, the results that bear his name were motivated by coffeehouse conversations about works of Newton on planetary motion and works of Halley ("Halley's comet") on roots of polynomials. Unfortunately, Taylor's writing style was so terse and hard to understand that he never received credit for many of his innovations.

10.1.3 DEFINITION. If f can be differentiated n times at x_0, then we define the **nth Taylor polynomial for f about $x = x_0$** to be

$$p_n(x) = f(x_0) + f'(x_0)(x - x_0) + \frac{f''(x_0)}{2!}(x - x_0)^2$$

$$+ \frac{f'''(x_0)}{3!}(x - x_0)^3 + \cdots + \frac{f^{(n)}(x_0)}{n!}(x - x_0)^n \quad (10)$$

REMARK. Observe that the Maclaurin polynomials are special cases of the Taylor polynomials; that is, the nth-order Maclaurin polynomial is the nth-order Taylor polynomial about $x = 0$. Observe also that $p_1(x)$ is the local linear approximation of f at $x = x_0$ and $p_2(x)$ is the local quadratic approximation of f at $x = x_0$.

Example 3 Find the first four Taylor polynomials for $\ln x$ about $x = 2$.

Solution. Let $f(x) = \ln x$. Thus,

$$
\begin{array}{ll}
f(x) = \ln x & f(2) = \ln 2 \\
f'(x) = 1/x & f'(2) = 1/2 \\
f''(x) = -1/x^2 & f''(2) = -1/4 \\
f'''(x) = 2/x^3 & f'''(2) = 1/4
\end{array}
$$

Substituting in (10) with $x_0 = 2$ yields

$$p_0(x) = f(2) = \ln 2$$

$$p_1(x) = f(2) + f'(2)(x - 2) = \ln 2 + \tfrac{1}{2}(x - 2)$$

$$p_2(x) = f(2) + f'(2)(x - 2) + \frac{f''(2)}{2!}(x - 2)^2 = \ln 2 + \tfrac{1}{2}(x - 2) - \tfrac{1}{8}(x - 2)^2$$

$$p_3(x) = f(2) + f'(2)(x - 2) + \frac{f''(2)}{2!}(x - 2)^2 + \frac{f'''(2)}{3!}(x - 2)^3$$

$$= \ln 2 + \tfrac{1}{2}(x - 2) - \tfrac{1}{8}(x - 2)^2 + \tfrac{1}{24}(x - 2)^3$$

The graph of $\ln x$ (in blue) and its first four Taylor polynomials about $x = 2$ are shown in Figure 10.1.4. As expected, these polynomials produce their best approximations of $\ln x$ near 2. ◀

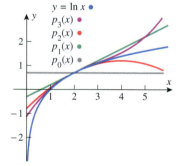

Figure 10.1.4

SIGMA NOTATION FOR TAYLOR AND MACLAURIN POLYNOMIALS

Frequently, we will want to express Formula (10) in sigma notation. To do this, we use the notation $f^{(k)}(x_0)$ to denote the kth derivative of f at $x = x_0$, and we make the convention that $f^{(0)}(x_0)$ denotes $f(x_0)$. This enables us to write

$$\sum_{k=0}^{n} \frac{f^{(k)}(x_0)}{k!}(x - x_0)^k = f(x_0) + f'(x_0)(x - x_0)$$

$$+ \frac{f''(x_0)}{2!}(x - x_0)^2 + \cdots + \frac{f^{(n)}(x_0)}{n!}(x - x_0)^n \quad (11)$$

In particular, we can write the nth-order Maclaurin polynomial for $f(x)$ as

$$\sum_{k=0}^{n} \frac{f^{(k)}(0)}{k!}x^k = f(0) + f'(0)x + \frac{f''(0)}{2!}x^2 + \cdots + \frac{f^{(n)}(0)}{n!}x^n \quad (12)$$

Example 4 Find the nth Maclaurin polynomials for

(a) $\sin x$ (b) $\cos x$ (c) $\dfrac{1}{1 - x}$

Solution (a). In the Maclaurin polynomials for $\sin x$, only the odd powers of x appear explicitly. To see this, let $f(x) = \sin x$; thus,

$$f(x) = \sin x \qquad f(0) = 0$$
$$f'(x) = \cos x \qquad f'(0) = 1$$
$$f''(x) = -\sin x \qquad f''(0) = 0$$
$$f'''(x) = -\cos x \qquad f'''(0) = -1$$

Since $f^{(4)}(x) = \sin x = f(x)$, the pattern $0, 1, 0, -1$ will repeat as we evaluate successive derivatives at 0. Therefore, the successive Maclaurin polynomials for $\sin x$ are

$$p_0(x) = 0$$
$$p_1(x) = 0 + x$$
$$p_2(x) = 0 + x + 0$$
$$p_3(x) = 0 + x + 0 - \frac{x^3}{3!}$$
$$p_4(x) = 0 + x + 0 - \frac{x^3}{3!} + 0$$
$$p_5(x) = 0 + x + 0 - \frac{x^3}{3!} + 0 + \frac{x^5}{5!}$$
$$p_6(x) = 0 + x + 0 - \frac{x^3}{3!} + 0 + \frac{x^5}{5!} + 0$$
$$p_7(x) = 0 + x + 0 - \frac{x^3}{3!} + 0 + \frac{x^5}{5!} + 0 - \frac{x^7}{7!}$$

Because of the zero terms, each even-order Maclaurin polynomial [after $p_0(x)$] is the same as the preceding odd-order Maclaurin polynomial. That is,

$$p_{2k+1}(x) = p_{2k+2}(x) = x - \frac{x^3}{3!} + \frac{x^5}{5!} - \frac{x^7}{7!} + \cdots + (-1)^k \frac{x^{2k+1}}{(2k+1)!} \quad (k = 0, 1, 2, \ldots)$$

The graphs of $\sin x$, $p_1(x)$, $p_3(x)$, $p_5(x)$, and $p_7(x)$ are shown in Figure 10.1.5.

Solution (b). In the Maclaurin polynomials for $\cos x$, only the even powers of x appear explicitly; the computations are similar to those in part (a). The reader should be able to show that

$$p_0(x) = p_1(x) = 1$$
$$p_2(x) = p_3(x) = 1 - \frac{x^2}{2!}$$
$$p_4(x) = p_5(x) = 1 - \frac{x^2}{2!} + \frac{x^4}{4!}$$
$$p_6(x) = p_7(x) = 1 - \frac{x^2}{2!} + \frac{x^4}{4!} - \frac{x^6}{6!}$$

In general, the Maclaurin polynomials for $\cos x$ are given by

$$p_{2k}(x) = p_{2k+1}(x) = 1 - \frac{x^2}{2!} + \frac{x^4}{4!} - \frac{x^6}{6!} + \cdots + (-1)^k \frac{x^{2k}}{(2k)!} \quad (k = 0, 1, 2, \ldots)$$

The graphs of $\cos x$, $p_0(x)$, $p_2(x)$, $p_4(x)$, and $p_6(x)$ are shown in Figure 10.1.6.

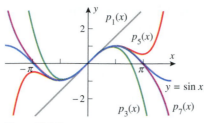

Figure 10.1.5   Figure 10.1.6

Solution (c). Let $f(x) = 1/(1 - x)$. The values of f and its first k derivatives at $x = 0$ are as follows:

$$f(x) = \frac{1}{1 - x} \qquad f(0) = 1 = 0!$$

$$f'(x) = \frac{1}{(1 - x)^2} \qquad f'(0) = 1 = 1!$$

$$f''(x) = \frac{2}{(1 - x)^3} \qquad f''(0) = 2 = 2!$$

$$f'''(x) = \frac{3 \cdot 2}{(1 - x)^4} \qquad f'''(0) = 3!$$

$$f^{(4)}(x) = \frac{4 \cdot 3 \cdot 2}{(1 - x)^5} \qquad f^{(4)}(0) = 4!$$

$$\vdots \qquad\qquad\qquad \vdots$$

$$f^{(k)}(x) = \frac{k!}{(1 - x)^{k+1}} \qquad f^{(k)}(0) = k!$$

Thus, substituting $f^{(k)}(0) = k!$ into Formula (12) yields the nth Maclaurin polynomial for $1/(1 - x)$:

$$p_n(x) = \sum_{k=0}^{n} x^k = 1 + x + x^2 + \cdots + x^n \quad (n = 0, 1, 2, \ldots) \qquad \blacktriangleleft$$

Example 5   Find the nth Taylor polynomial for $1/x$ about $x = 1$.

Solution.   Let $f(x) = 1/x$. The computations are similar to those in part (c) of Example 4. We leave it for you to show that

$$f(1) = 1, \quad f'(1) = -1, \quad f''(1) = 2!, \quad f'''(1) = -3!,$$

$$f^{(4)}(1) = 4!, \ldots, \quad f^{(k)}(1) = (-1)^k k!$$

Thus, substituting $f^{(k)}(1) = (-1)^k k!$ into Formula (11) with $x_0 = 1$ yields the nth Taylor polynomial for $1/x$:

$$\sum_{k=0}^{n} (-1)^k (x - 1)^k = 1 - (x - 1) + (x - 1)^2 - (x - 1)^3 + \cdots + (-1)^n (x - 1)^n \qquad \blacktriangleleft$$

FOR THE READER.   CAS programs have commands for generating Taylor polynomials of any specified degree. If you have a CAS, read the documentation to determine how this is done, and then use the CAS to confirm the computations in the examples in this section.

THE nTH REMAINDER

The nth Taylor polynomial p_n for a function f about $x = x_0$ has been introduced as a tool to obtain good approximations to values of $f(x)$ for x near x_0. We now develop a method to forecast how good these approximations will be.

It is convenient to develop a notation for the error in using $p_n(x)$ to approximate $f(x)$, so we define $R_n(x)$ to be the difference between $f(x)$ and its nth Taylor polynomial. That is,

$$R_n(x) = f(x) - p_n(x) = f(x) - \sum_{k=0}^{n} \frac{f^{(k)}(x_0)}{k!}(x - x_0)^k \tag{13}$$

This can also be written as

$$f(x) = p_n(x) + R_n(x) = \sum_{k=0}^{n} \frac{f^{(k)}(x_0)}{k!}(x - x_0)^k + R_n(x) \tag{14}$$

which is called *Taylor's formula with remainder*.

Finding a bound for $R_n(x)$ gives an indication of the accuracy of the approximation $p_n(x) \approx f(x)$. The following theorem, which is proved in Appendix G, provides such a bound.

10.1.4   THEOREM (*The Remainder Estimation Theorem*).   *If the function f can be differentiated $n + 1$ times on an interval I containing the number x_0, and if M is an upper bound for $|f^{(n+1)}(x)|$ on I, that is, $|f^{(n+1)}(x)| \leq M$ for all x in I, then*

$$|R_n(x)| \leq \frac{M}{(n + 1)!}|x - x_0|^{n+1} \tag{15}$$

for all x in I.

Example 6   Use an nth Maclaurin polynomial for e^x to approximate e to five decimal-place accuracy.

Solution.   We note first that the exponential function e^x has derivatives of all orders for every real number x. From Example 2, the nth Maclaurin polynomial for e^x is

$$\sum_{k=0}^{n} \frac{x^k}{k!} = 1 + x + \frac{x^2}{2!} + \cdots + \frac{x^n}{n!}$$

from which we have

$$e = e^1 \approx \sum_{k=0}^{n} \frac{1^k}{k!} = 1 + 1 + \frac{1}{2!} + \cdots + \frac{1}{n!}$$

Thus, our problem is to determine how many terms to include in a Maclaurin polynomial for e^x to achieve five decimal-place accuracy; that is, we want to choose n so that the absolute value of the nth remainder at $x = 1$ satisfies

$$|R_n(1)| \leq 0.000005$$

To determine n we use the Remainder Estimation Theorem with $f(x) = e^x$, $x = 1$, $x_0 = 0$, and I being the interval $[0, 1]$. In this case it follows from Formula (15) that

$$|R_n(1)| \leq \frac{M}{(n + 1)!} \tag{16}$$

where M is an upper bound on the value of $f^{(n+1)}(x) = e^x$ for x in the interval $[0, 1]$. However, e^x is an increasing function, so its maximum value on the interval $[0, 1]$ occurs at $x = 1$; that is, $e^x \leq e$ on this interval. Thus, we can take $M = e$ in (16) to obtain

$$|R_n(1)| \leq \frac{e}{(n + 1)!} \tag{17}$$

Unfortunately, this inequality is not very useful because it involves e, which is the very quantity we are trying to approximate. However, if we accept that $e < 3$, then we can replace (17) with the following less precise, but more easily applied, inequality:

$$|R_n(1)| \le \frac{3}{(n+1)!}$$

Thus, we can achieve five decimal-place accuracy by choosing n so that

$$\frac{3}{(n+1)!} \le 0.000005 \quad \text{or} \quad (n+1)! \ge 600{,}000$$

Since $9! = 362{,}880$ and $10! = 3{,}628{,}800$, the smallest value of n that meets this criterion is $n = 9$. Thus, to five decimal-place accuracy

$$e \approx 1 + 1 + \frac{1}{2!} + \frac{1}{3!} + \frac{1}{4!} + \frac{1}{5!} + \frac{1}{6!} + \frac{1}{7!} + \frac{1}{8!} + \frac{1}{9!} \approx 2.71828$$

As a check, a calculator's twelve-digit representation of e is $e \approx 2.71828182846$, which agrees with the preceding approximation when rounded to five decimal places. ◄

EXERCISE SET 10.1   Graphing Utility   CAS

1. In each part, find the local quadratic approximation of f at $x = x_0$, and use that approximation to find the local linear approximation of f at x_0.
   - (a) $f(x) = e^{-x}$; $x_0 = 0$
   - (b) $f(x) = \cos x$; $x_0 = 0$
   - (c) $f(x) = \sin x$; $x_0 = \pi/2$
   - (d) $f(x) = \sqrt{x}$; $x_0 = 1$

2. In each part, use a CAS to find the local quadratic approximation of f at $x = x_0$, and use that approximation to find the local linear approximation of f at $x = x_0$.
   - (a) $f(x) = e^{\sin x}$; $x_0 = 0$
   - (b) $f(x) = \sqrt{x}$; $x_0 = 9$
   - (c) $f(x) = \sec^{-1} x$; $x_0 = 2$
   - (d) $f(x) = \sin^{-1} x$; $x_0 = 0$

3. (a) Find the local quadratic approximation of $\sqrt{x}$ at $x_0 = 1$.
   - (b) Use the result obtained in part (a) to approximate $\sqrt{1.1}$, and compare your approximation to that produced directly by your calculating utility. [See Example 1 of Section 3.8.]

4. (a) Find the local quadratic approximation of $\cos x$ at $x_0 = 0$.
   - (b) Use the result obtained in part (a) to approximate $\cos 2°$, and compare the approximation to that produced directly by your calculating utility.

5. Use an appropriate local quadratic approximation to approximate $\tan 61°$, and compare the result to that produced directly by your calculating utility.

6. Use an appropriate local quadratic approximation to approximate $\sqrt{36.03}$, and compare the result to that produced directly by your calculating utility.

In Exercises 7–16, find the Maclaurin polynomials of orders $n = 0, 1, 2, 3$, and 4, and then find the nth Maclaurin polynomials for the function in sigma notation.

7. e^{-x} 8. e^{ax} 9. $\cos \pi x$

10. $\sin \pi x$ 11. $\ln(1 + x)$ 12. $\dfrac{1}{1+x}$

13. $\cosh x$ 14. $\sinh x$ 15. $x \sin x$

16. xe^x

In Exercises 17–24, find the Taylor polynomials of orders $n = 0, 1, 2, 3$, and 4 about $x = x_0$, and then find the nth Taylor polynomials for the function in sigma notation.

17. e^x; $x_0 = 1$ 18. e^{-x}; $x_0 = \ln 2$

19. $\dfrac{1}{x}$; $x_0 = -1$ 20. $\dfrac{1}{x+2}$; $x_0 = 3$

21. $\sin \pi x$; $x_0 = \dfrac{1}{2}$ 22. $\cos x$; $x_0 = \dfrac{\pi}{2}$

23. $\ln x$; $x_0 = 1$ 24. $\ln x$; $x_0 = e$

25. (a) Find the third Maclaurin polynomial for
$$f(x) = 1 + 2x - x^2 + x^3$$
   - (b) Find the third Taylor polynomial about $x = 1$ for
$$f(x) = 1 + 2(x - 1) - (x - 1)^2 + (x - 1)^3$$

26. (a) Find the nth Maclaurin polynomial for
$$f(x) = c_0 + c_1 x + c_2 x^2 + \cdots + c_n x^n$$
   - (b) Find the nth Taylor polynomial about $x = 1$ for
$$f(x) = c_0 + c_1(x - 1) + c_2(x - 1)^2 + \cdots + c_n(x - 1)^n$$

In Exercises 27–30, find the first four distinct Taylor polynomials about $x = x_0$, and use a graphing utility to graph the given function and the Taylor polynomials on the same screen.

 27. $f(x) = e^{-2x}$; $x_0 = 0$ **28.** $f(x) = \sin x$; $x_0 = \pi/2$

 29. $f(x) = \cos x$; $x_0 = \pi$ **30.** $\ln(x + 1)$; $x_0 = 0$

31. Use the method of Example 6 to approximate $\sqrt{e}$ to four decimal-place accuracy, and check your work by comparing your answer to that produced directly by your calculating utility. [*Suggestion:* Write $\sqrt{e}$ as $e^{0.5}$.]

32. Use the method of Example 6 to approximate $1/e$ to three decimal-place accuracy, and check your work by comparing your answer to that produced directly by your calculating utility.

33. Which of the functions graphed in the following figure is most likely to have $p(x) = 1 - x + 2x^2$ as its second-order Maclaurin polynomial? Explain your reasoning.

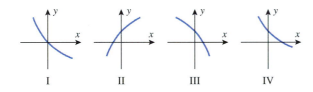

I   II   III   IV

34. Suppose that the values of a function f and its first three derivatives at $x = 1$ are

$$f(1) = 2, \quad f'(1) = -3, \quad f''(1) = 0, \quad f'''(1) = 6$$

Find as many Taylor polynomials for f as you can about $x = 1$.

35. Show that the nth Taylor polynomial for $\sinh x$ about $x = \ln 4$ is

$$\sum_{k=0}^{n} \frac{16 - (-1)^k}{8k!} (x - \ln 4)^k$$

36. (a) The accompanying figure shows a sector of radius r and central angle 2α. Assuming that the angle α is small, use the local quadratic approximation of $\cos \alpha$ at $\alpha = 0$ to show that $x \approx r\alpha^2/2$.

(b) Assuming that the Earth is a sphere of radius 4000 mi, use the result in part (a) to approximate the maximum amount by which a 100-mi arc along the equator will diverge from its chord.

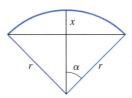

Figure Ex-36

 37. Let $p_1(x)$ and $p_2(x)$ be the local linear and local quadratic approximations of $f(x) = e^{\sin x}$ at $x = 0$.

(a) Use a graphing utility to generate the graphs of $f(x)$, $p_1(x)$, and $p_2(x)$ on the same screen for $-1 \leq x \leq 1$.

(b) Construct a table of values of $f(x)$, $p_1(x)$, and $p_2(x)$ for $x = -1.00, -0.75, -0.50, -0.25, 0, 0.25, 0.50, 0.75, 1.00$. Round the values to three decimal places.

(c) Generate the graph of $|f(x) - p_1(x)|$, and use the graph to determine an interval on which $p_1(x)$ approximates $f(x)$ with an error of at most ± 0.01. [*Suggestion:* Review the discussion relating to Figure 3.8.5.]

(d) Generate the graph of $|f(x) - p_2(x)|$, and use the graph to determine an interval on which $p_2(x)$ approximates $f(x)$ with an error of at most ± 0.01.

 38. (a) Find an interval $[0, b]$ over which e^x can be approximated by $1 + x + (x^2/2!)$ to three decimal-place accuracy throughout the interval.

(b) Check your answer in part (a) by graphing

$$\left| e^x - \left(1 + x + \frac{x^2}{2!} \right) \right|$$

over the interval you obtained.

 39. (a) Use the Remainder Estimation Theorem to find an interval containing $x = 0$ over which $\sin x$ can be approximated by $x - (x^3/3!)$ to three decimal-place accuracy throughout the interval.

(b) Check your answer in part (a) by graphing

$$\left| \sin x - \left(x - \frac{x^3}{3!} \right) \right|$$

over the interval you obtained.

10.2 SEQUENCES

In everyday language, the term "sequence" means a succession of things in a definite order—chronological order, size order, or logical order, for example. In mathematics, the term "sequence" is commonly used to denote a succession of numbers whose order is determined by a rule or a function. In this section, we will develop some of the basic ideas concerning sequences of numbers.

Stated informally, an ***infinite sequence***, or more simply a ***sequence***, is an unending succession of numbers, called ***terms***. It is understood that the terms have a definite order; that is, there is a first term a_1, a second term a_2, a third term a_3, a fourth term a_4, and so forth. Such a sequence would typically be written as

$$a_1, a_2, a_3, a_4, \ldots$$

where the dots are used to indicate that the sequence continues indefinitely. Some specific examples are

$$1, 2, 3, 4, \ldots, \qquad 1, \tfrac{1}{2}, \tfrac{1}{3}, \tfrac{1}{4}, \ldots,$$
$$2, 4, 6, 8, \ldots, \qquad 1, -1, 1, -1, \ldots$$

Each of these sequences has a definite pattern that makes it easy to generate additional terms if we assume that those terms follow the same pattern as the displayed terms. However, such patterns can be deceiving, so it is better to have a rule or formula for generating the terms. One way of doing this is to look for a function that relates each term in the sequence to its term number. For example, in the sequence

$$2, 4, 6, 8, \ldots$$

each term is twice the term number; that is, the nth term in the sequence is given by the formula $2n$. We denote this by writing the sequence as

$$2, 4, 6, 8, \ldots, 2n, \ldots$$

We call the function $f(n) = 2n$ the *general term* of this sequence. Now, if we want to know a specific term in the sequence, we need only substitute its term number in the formula for the general term. For example, the 37th term in the sequence is $2 \cdot 37 = 74$.

Example 1 In each part, find the general term of the sequence.

(a) $\tfrac{1}{2}, \tfrac{2}{3}, \tfrac{3}{4}, \tfrac{4}{5}, \ldots$ (b) $\tfrac{1}{2}, \tfrac{1}{4}, \tfrac{1}{8}, \tfrac{1}{16}, \ldots$

(c) $\tfrac{1}{2}, -\tfrac{2}{3}, \tfrac{3}{4}, -\tfrac{4}{5}, \ldots$ (d) $1, 3, 5, 7, \ldots$

Solution (a). In Table 10.2.1, the four known terms have been placed below their term numbers, from which we see that the numerator is the same as the term number and the denominator is one greater than the term number. This suggests that the nth term has numerator n and denominator $n + 1$, as indicated in the table. Thus, the sequence can be expressed as

$$\frac{1}{2}, \frac{2}{3}, \frac{3}{4}, \frac{4}{5}, \ldots, \frac{n}{n + 1}, \ldots$$

Solution (b). In Table 10.2.2, the denominators of the four known terms have been expressed as powers of 2 and the first four terms have been placed below their term numbers, from which we see that the exponent in the denominator is the same as the term number. This suggests that the denominator of the nth term is 2^n, as indicated in the table. Thus, the sequence can be expressed as

$$\frac{1}{2}, \frac{1}{4}, \frac{1}{8}, \frac{1}{16}, \ldots, \frac{1}{2^n}, \ldots$$

Table 10.2.1

TERM NUMBER	1	2	3	4	⋯	n	⋯
TERM	$\tfrac{1}{2}$	$\tfrac{2}{3}$	$\tfrac{3}{4}$	$\tfrac{4}{5}$	⋯	$\tfrac{n}{n+1}$	⋯

Table 10.2.2

TERM NUMBER	1	2	3	4	⋯	n	⋯
TERM	$\tfrac{1}{2}$	$\tfrac{1}{2^2}$	$\tfrac{1}{2^3}$	$\tfrac{1}{2^4}$	⋯	$\tfrac{1}{2^n}$	⋯

Solution (c). This sequence is identical to that in part (a), except for the alternating signs. Thus, the nth term in the sequence can be obtained by multiplying the nth term in part (a) by $(-1)^{n+1}$. This factor produces the correct alternating signs, since its successive values, starting with $n = 1$, are $1, -1, 1, -1, \ldots$. Thus, the sequence can be written as

$$\frac{1}{2}, -\frac{2}{3}, \frac{3}{4}, -\frac{4}{5}, \ldots, (-1)^{n+1}\frac{n}{n+1}, \ldots$$

Solution (d). In Table 10.2.3, the four known terms have been placed below their term numbers, from which we see that each term is one less than twice its term number. This suggests that the nth term in the sequence is $2n - 1$, as indicated in the table. Thus, the sequence can be expressed as

$$1, 3, 5, 7, \ldots, 2n - 1, \ldots \quad \blacktriangleleft$$

Table 10.2.3

TERM NUMBER	1	2	3	4	$\cdots$	n	$\cdots$
TERM	1	3	5	7	$\cdots$	$2n - 1$	$\cdots$

FOR THE READER. Consider the sequence whose general term is

$$f(n) = \tfrac{1}{3}(3 - 5n + 6n^2 - n^3)$$

Calculate the first three terms, and make a conjecture about the fourth term. Check your conjecture by calculating the fourth term. What message does this convey?

When the general term of a sequence

$$a_1, a_2, a_3, \ldots, a_n, \ldots \tag{1}$$

is known, there is no need to write out the initial terms, and it is common to write only the general term enclosed in braces. Thus, (1) might be written as

$$\{a_n\}_{n=1}^{+\infty}$$

For example, here are the four sequences in Example 1 expressed in brace notation.

SEQUENCE	BRACE NOTATION
$\dfrac{1}{2}, \dfrac{2}{3}, \dfrac{3}{4}, \dfrac{4}{5}, \ldots, \dfrac{n}{n+1}, \ldots$	$\left\{\dfrac{n}{n+1}\right\}_{n=1}^{+\infty}$
$\dfrac{1}{2}, \dfrac{1}{4}, \dfrac{1}{8}, \dfrac{1}{16}, \ldots, \dfrac{1}{2^n}, \ldots$	$\left\{\dfrac{1}{2^n}\right\}_{n=1}^{+\infty}$
$\dfrac{1}{2}, -\dfrac{2}{3}, \dfrac{3}{4}, -\dfrac{4}{5}, \ldots, (-1)^{n+1}\dfrac{n}{n+1}, \ldots$	$\left\{(-1)^{n+1}\dfrac{n}{n+1}\right\}_{n=1}^{+\infty}$
$1, 3, 5, 7, \ldots, 2n - 1, \ldots$	$\{2n - 1\}_{n=1}^{+\infty}$

The letter n in (1) is called the ***index*** for the sequence. It is not essential to use n for the index; any letter not reserved for another purpose can be used. For example, we might view the general term of the sequence $a_1, a_2, a_3, \ldots$ to be the kth term, in which case we would denote this sequence as $\{a_k\}_{k=1}^{+\infty}$. Moreover, it is not essential to start the index at 1; sometimes it is more convenient to start it at 0 (or some other integer). For example, consider the sequence

$$1, \frac{1}{2}, \frac{1}{2^2}, \frac{1}{2^3}, \ldots$$

One way to write this sequence is

$$\left\{ \frac{1}{2^{n-1}} \right\}_{n=1}^{+\infty}$$

However, the general term will be simpler if we think of the initial term in the sequence as the zeroth term, in which case we can write the sequence as

$$\left\{ \frac{1}{2^n} \right\}_{n=0}^{+\infty}$$

REMARK.   In general discussions that involve sequences in which the specific terms and the starting point for the index are not important, it is common to write $\{a_n\}$ rather than $\{a_n\}_{n=1}^{+\infty}$ or $\{a_n\}_{n=0}^{+\infty}$. Moreover, we can distinguish between different sequences by using different letters for their general terms; thus, $\{a_n\}$, $\{b_n\}$, and $\{c_n\}$ denote three different sequences.

We began this section by describing a sequence as an unending succession of numbers. Although this conveys the general idea, it is not a satisfactory mathematical definition because it relies on the term "succession," which is itself an undefined term. To motivate a precise definition, consider the sequence

$$2, 4, 6, 8, \ldots, 2n, \ldots$$

If we denote the general term by $f(n) = 2n$, then we can write this sequence as

$$f(1), f(2), f(3), \ldots, f(n), \ldots$$

which is a "list" of values of the function

$$f(n) = 2n, \quad n = 1, 2, 3, \ldots$$

whose domain is the set of positive integers. This suggests the following definition.

10.2.1   DEFINITION.   A *sequence* is a function whose domain is a set of integers. Specifically, we will regard the expression $\{a_n\}_{n=1}^{+\infty}$ to be an alternative notation for the function $f(n) = a_n, n = 1, 2, 3, \ldots$.

GRAPHS OF SEQUENCES

Since sequences are functions, it makes sense to talk about the graph of a sequence. For example, the graph of the sequence $\{1/n\}_{n=1}^{+\infty}$ is the graph of the equation

$$y = \frac{1}{n}, \quad n = 1, 2, 3, \ldots$$

Because the right side of this equation is defined only for positive integer values of n, the graph consists of a succession of isolated points (Figure 10.2.1a). This is in distinction to the graph of

$$y = \frac{1}{x}, \quad x \geq 1$$

which is a continuous curve (Figure 10.2.1b).

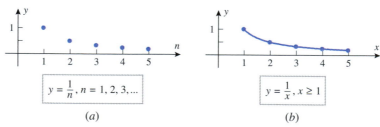

$$y = \frac{1}{n}, n = 1, 2, 3, \ldots$$

(a)

$$y = \frac{1}{x}, x \geq 1$$

(b)

Figure 10.2.1

Since sequences are functions, we can inquire about their limits. However, because a sequence $\{a_n\}$ is only defined for integer values of n, the only limit that makes sense is the limit of a_n as $n \to +\infty$. In Figure 10.2.2 we have shown the graphs of four sequences, each of which behaves differently as $n \to +\infty$:

- The terms in the sequence $\{n + 1\}$ increase without bound.

- The terms in the sequence $\{(-1)^{n+1}\}$ oscillate between -1 and 1.

- The terms in the sequence $\{n/(n + 1)\}$ increase toward a "limiting value" of 1.

- The terms in the sequence $\left\{1 + \left(-\frac{1}{2}\right)^n\right\}$ also tend toward a "limiting value" of 1, but do so in an oscillatory fashion.

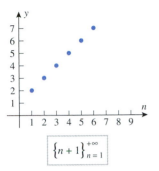
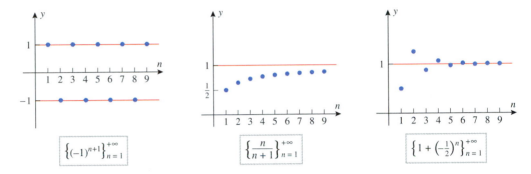

Figure 10.2.2

Informally speaking, the limit of a sequence $\{a_n\}$ is intended to describe how a_n behaves as $n \to +\infty$. To be more specific, we will say that *a sequence $\{a_n\}$ approaches a limit L if the terms in the sequence eventually become arbitrarily close to L.* Geometrically, this means that for any positive number ϵ there is a point in the sequence after which all terms lie between the lines $y = L - \epsilon$ and $y = L + \epsilon$ (Figure 10.2.3).

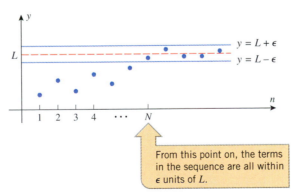

From this point on, the terms in the sequence are all within ϵ units of L.

Figure 10.2.3

The following definition makes these ideas precise.

10.2.2 DEFINITION. A sequence $\{a_n\}$ is said to **converge** to the **limit** L if given any $\epsilon > 0$, there is a positive integer N such that $|a_n - L| < \epsilon$ for $n \geq N$. In this case we write

$$\lim_{n \to +\infty} a_n = L$$

A sequence that does not converge to some finite limit is said to **diverge**.

Example 2 The first two sequences in Figure 10.2.2 diverge, and the second two converge to 1; that is,

$$\lim_{n \to +\infty} \frac{n}{n+1} = 1 \quad \text{and} \quad \lim_{n \to +\infty} \left(1 + \left(-\tfrac{1}{2}\right)^n\right) = 1 \qquad \blacktriangleleft$$

FOR THE READER. How would you define

$$\lim_{n \to +\infty} a_n = +\infty \quad \text{and} \quad \lim_{n \to +\infty} a_n = -\infty?$$

The following theorem, which we state without proof, shows that the familiar properties of limits apply to sequences. This theorem ensures that the algebraic techniques used to find limits of the form $\lim_{x \to +\infty}$ can also be used for limits of the form $\lim_{n \to +\infty}$.

10.2.3 THEOREM. *Suppose that the sequences $\{a_n\}$ and $\{b_n\}$ converge to limits L_1 and L_2, respectively, and c is a constant. Then*

(a) $\displaystyle \lim_{n \to +\infty} c = c$

(b) $\displaystyle \lim_{n \to +\infty} c a_n = c \lim_{n \to +\infty} a_n = cL_1$

(c) $\displaystyle \lim_{n \to +\infty} (a_n + b_n) = \lim_{n \to +\infty} a_n + \lim_{n \to +\infty} b_n = L_1 + L_2$

(d) $\displaystyle \lim_{n \to +\infty} (a_n - b_n) = \lim_{n \to +\infty} a_n - \lim_{n \to +\infty} b_n = L_1 - L_2$

(e) $\displaystyle \lim_{n \to +\infty} (a_n b_n) = \lim_{n \to +\infty} a_n \cdot \lim_{n \to +\infty} b_n = L_1 L_2$

(f) $\displaystyle \lim_{n \to +\infty} \left(\frac{a_n}{b_n}\right) = \frac{\displaystyle \lim_{n \to +\infty} a_n}{\displaystyle \lim_{n \to +\infty} b_n} = \frac{L_1}{L_2} \quad (\text{if } L_2 \neq 0)$

Example 3 In each part, determine whether the sequence converges or diverges. If it converges, find the limit.

(a) $\left\{ \dfrac{n}{2n+1} \right\}_{n=1}^{+\infty}$

(b) $\left\{ (-1)^{n+1} \dfrac{n}{2n+1} \right\}_{n=1}^{+\infty}$

(c) $\left\{ (-1)^{n+1} \dfrac{1}{n} \right\}_{n=1}^{+\infty}$

(d) $\{8 - 2n\}_{n=1}^{+\infty}$

Solution (a). Dividing numerator and denominator by n yields

$$\lim_{n \to +\infty} \frac{n}{2n+1} = \lim_{n \to +\infty} \frac{1}{2 + 1/n} = \frac{\displaystyle \lim_{n \to +\infty} 1}{\displaystyle \lim_{n \to +\infty} (2 + 1/n)} = \frac{\displaystyle \lim_{n \to +\infty} 1}{\displaystyle \lim_{n \to +\infty} 2 + \lim_{n \to +\infty} 1/n}$$

$$= \frac{1}{2+0} = \frac{1}{2}$$

Thus, the sequence converges to $\tfrac{1}{2}$.

Solution (b). This sequence is the same as that in part (a), except for the factor of $(-1)^{n+1}$, which oscillates between $+1$ and -1. Thus, the terms in this sequence oscillate between positive and negative values, with the odd-numbered terms being identical to those in part (a) and the even-numbered terms being the negatives of those in part (a). Since the sequence in part (a) has a limit of $\tfrac{1}{2}$, it follows that the odd-numbered terms in this sequence approach $\tfrac{1}{2}$, and the even-numbered terms approach $-\tfrac{1}{2}$. Therefore, this sequence has no limit—it diverges.

Solution (c). Since $\lim_{n \to +\infty} 1/n = 0$, the product $(-1)^{n+1}(1/n)$ oscillates between positive and negative values, with the odd-numbered terms approaching 0 through positive

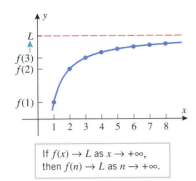

If $f(x) \to L$ as $x \to +\infty$,
then $f(n) \to L$ as $n \to +\infty$.

(a)

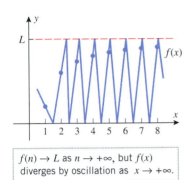

$f(n) \to L$ as $n \to +\infty$, but $f(x)$
diverges by oscillation as $x \to +\infty$.

(b)

Figure 10.2.4

values and the even-numbered terms approaching 0 through negative values. Thus,

$$\lim_{n \to +\infty} (-1)^{n+1} \frac{1}{n} = 0$$

so the sequence converges to 0.

Solution (d). $\displaystyle\lim_{n \to +\infty} (8 - 2n) = -\infty$, so the sequence $\{8 - 2n\}_{n=1}^{+\infty}$ diverges. ◄

If the general term of a sequence is $f(n)$, and if we replace n by x, where x can vary over the entire interval $[1, +\infty)$, then the values of $f(n)$ can be viewed as "sample values" of $f(x)$ taken at the positive integers. Thus, if $f(x) \to L$ as $x \to +\infty$, then it must also be true that $f(n) \to L$ as $n \to +\infty$ (Figure 10.2.4*a*). However, the converse is not true; that is, one cannot infer that $f(x) \to L$ as $x \to +\infty$ from the fact that $f(n) \to L$ as $n \to +\infty$ (Figure 10.2.4*b*).

Example 4   In each part, determine whether the sequence converges, and if so, find its limit.

(a) $1, \dfrac{1}{2}, \dfrac{1}{2^2}, \dfrac{1}{2^3}, \ldots, \dfrac{1}{2^n}, \ldots$ (b) $1, 2, 2^2, 2^3, \ldots, 2^n, \ldots$

Solution. Replacing n by x in the first sequence produces the power function $(1/2)^x$, and replacing n by x in the second sequence produces the power function 2^x. Now recall that if $0 < b < 1$, then $b^x \to 0$ as $x \to +\infty$, and if $b > 1$, then $b^x \to +\infty$ as $x \to +\infty$ (Figure 4.2.1). Thus,

$$\lim_{n \to +\infty} \frac{1}{2^n} = 0 \quad \text{and} \quad \lim_{n \to +\infty} 2^n = +\infty$$ ◄

Example 5   Find the limit of the sequence $\left\{ \dfrac{n}{e^n} \right\}_{n=1}^{+\infty}$.

Solution. The expression n/e^n is an indeterminate form of type ∞/∞ as $n \to +\infty$, so L'Hôpital's rule is indicated. However, we cannot apply this rule directly to n/e^n because the functions n and e^n have been defined here only at the positive integers, and hence are not differentiable functions. To circumvent this problem we extend the domains of these functions to all real numbers, here implied by replacing n by x, and apply L'Hôpital's rule to the limit of the quotient x/e^x. This yields

$$\lim_{x \to +\infty} \frac{x}{e^x} = \lim_{x \to +\infty} \frac{1}{e^x} = 0$$

from which we can conclude that

$$\lim_{n \to +\infty} \frac{n}{e^n} = 0$$ ◄

Example 6   Show that $\displaystyle\lim_{n \to +\infty} \sqrt[n]{n} = 1$.

Solution.

$$\lim_{n \to +\infty} \sqrt[n]{n} = \lim_{n \to +\infty} n^{1/n} = \lim_{n \to +\infty} e^{(1/n)\ln n} = e^0 = 1 \qquad \boxed{\begin{array}{l}\text{By L'Hôpital's rule applied}\\ \text{to } (1/x)\ln x\end{array}}$$ ◄

Sometimes the even-numbered and odd-numbered terms of a sequence behave sufficiently differently that it is desirable to investigate their convergence separately. The following theorem, whose proof is omitted, is helpful for that purpose.

> **10.2.4** THEOREM. *A sequence converges to a limit L if and only if the sequences of even-numbered terms and odd-numbered terms both converge to L.*

Example 7 The sequence

$$\frac{1}{2}, \frac{1}{3}, \frac{1}{2^2}, \frac{1}{3^2}, \frac{1}{2^3}, \frac{1}{3^3}, \dots$$

converges to 0, since the even-numbered terms and the odd-numbered terms both converge to 0, and the sequence

$$1, \tfrac{1}{2}, 1, \tfrac{1}{3}, 1, \tfrac{1}{4}, \dots$$

diverges, since the odd-numbered terms converge to 1 and the even-numbered terms converge to 0. ◄

THE SQUEEZING THEOREM FOR SEQUENCES

The following theorem, which we state without proof, is an adaptation of the Squeezing Theorem (2.6.2) to sequences. This theorem will be useful for finding limits of sequences that cannot be obtained directly.

10.2.5 THEOREM (*The Squeezing Theorem for Sequences*). *Let $\{a_n\}$, $\{b_n\}$, and $\{c_n\}$ be sequences such that*

$$a_n \le b_n \le c_n \quad (\text{for all values of } n \text{ beyond some index } N)$$

If the sequences $\{a_n\}$ and $\{c_n\}$ have a common limit L as $n \to +\infty$, then $\{b_n\}$ also has the limit L as $n \to +\infty$.

Table 10.2.4

n	$\dfrac{n!}{n^n}$
1	1.0000000000
2	0.5000000000
3	0.2222222222
4	0.0937500000
5	0.0384000000
6	0.0154320988
7	0.0061198990
8	0.0024032593
9	0.0009366567
10	0.0003628800
11	0.0001399059
12	0.0000537232

Example 8 Use numerical evidence to make a conjecture about the limit of the sequence*

$$\left\{ \frac{n!}{n^n} \right\}_{n=1}^{+\infty}$$

and then confirm that your conjecture is correct.

Solution. Table 10.2.4, which was obtained with a calculating utility, suggests that the limit of the sequence may be 0. To confirm this we need to examine the limit of

$$a_n = \frac{n!}{n^n}$$

as $n \to +\infty$. Although this is an indeterminate form of type ∞/∞, L'Hôpital's rule is not helpful because we have no definition of $x!$ for values of x that are not integers. However, let us write out some of the initial terms and the general term in the sequence:

$$a_1 = 1, \quad a_2 = \frac{1 \cdot 2}{2 \cdot 2}, \quad a_3 = \frac{1 \cdot 2 \cdot 3}{3 \cdot 3 \cdot 3}, \dots, \quad a_n = \frac{1 \cdot 2 \cdot 3 \cdots n}{n \cdot n \cdot n \cdots n}, \dots$$

We can rewrite the general term as

$$a_n = \frac{1}{n} \left(\frac{2 \cdot 3 \cdots n}{n \cdot n \cdots n} \right)$$

from which it is evident that

$$0 \le a_n \le \frac{1}{n}$$

However, the two outside expressions have a limit of 0 as $n \to +\infty$; thus, the Squeezing Theorem for Sequences implies that $a_n \to 0$ as $n \to +\infty$, which confirms our conjecture. ◄

The following theorem is often useful for finding the limit of a sequence with both positive and negative terms—it states that if the sequence $\{|a_n|\}$ that is obtained by taking the absolute value of each term in the sequence $\{a_n\}$ converges to 0, then $\{a_n\}$ also converges to 0.

---

*The symbol $n!$ (read "n factorial") is defined on p. 640.

> **10.2.6** THEOREM. *If* $\lim\limits_{n \to +\infty} |a_n| = 0$, *then* $\lim\limits_{n \to +\infty} a_n = 0$.

Proof. Depending on the sign of a_n, either $a_n = |a_n|$ or $a_n = -|a_n|$. Thus, in all cases we have

$$-|a_n| \le a_n \le |a_n|$$

However, the limit of the two outside terms is 0, and hence the limit of a_n is 0 by the Squeezing Theorem for Sequences. ■

Example 9 Consider the sequence

$$1, \ -\frac{1}{2}, \ \frac{1}{2^2}, \ -\frac{1}{2^3}, \dots, \ (-1)^n \frac{1}{2^n}, \dots$$

If we take the absolute value of each term, we obtain the sequence

$$1, \ \frac{1}{2}, \ \frac{1}{2^2}, \ \frac{1}{2^3}, \dots, \ \frac{1}{2^n}, \dots$$

which, as shown in Example 4, converges to 0. Thus, from Theorem 10.2.6 we have

$$\lim_{n \to +\infty} \left[(-1)^n \frac{1}{2^n} \right] = 0$$ ◄

SEQUENCES DEFINED RECURSIVELY

Some sequences do not arise from a formula for the general term, but rather from a formula or set of formulas that specify how to generate each term in the sequence from terms that precede it; such sequences are said to be defined *recursively*, and the defining formulas are called *recursion formulas*. A good example is the mechanic's rule for approximating square roots. In Exercise 21 of Section 5.7 you were asked to show that

$$x_1 = 1, \quad x_{n+1} = \frac{1}{2}\left(x_n + \frac{a}{x_n}\right) \tag{2}$$

describes the sequence produced by Newton's Method to approximate $\sqrt{a}$ as a root of the function $f(x) = x^2 - a$. Table 10.2.5 shows the first five terms in an application of the mechanic's rule to approximate $\sqrt{2}$.

Table 10.2.5

n	$x_1 = 1, \quad x_{n+1} = \frac{1}{2}\left(x_n + \frac{2}{x_n}\right)$	DECIMAL APPROXIMATION
	$x_1 = 1$ (Starting value)	1.00000000000
1	$x_2 = \frac{1}{2}\left[1 + \frac{2}{1}\right] = \frac{3}{2}$	1.50000000000
2	$x_3 = \frac{1}{2}\left[\frac{3}{2} + \frac{2}{3/2}\right] = \frac{17}{12}$	1.41666666667
3	$x_4 = \frac{1}{2}\left[\frac{17}{12} + \frac{2}{17/12}\right] = \frac{577}{408}$	1.41421568627
4	$x_5 = \frac{1}{2}\left[\frac{577}{408} + \frac{2}{577/408}\right] = \frac{665,857}{470,832}$	1.41421356237
5	$x_6 = \frac{1}{2}\left[\frac{665,857}{470,832} + \frac{2}{665,857/470,832}\right] = \frac{886,731,088,897}{627,013,566,048}$	1.41421356237

It would take us too far afield to investigate the convergence of sequences defined recursively, but we will conclude this section with a useful technique that can sometimes be used to compute limits of such sequences.

Example 10   Assuming that the sequence in Table 10.2.5 converges, show that the limit is $\sqrt{2}$.

Solution.   Assume that $x_n \to L$, where L is to be determined. Since $n + 1 \to +\infty$ as $n \to +\infty$, it is also true that $x_{n+1} \to L$ as $n \to +\infty$. Thus, if we take the limit of the expression

$$x_{n+1} = \frac{1}{2}\left(x_n + \frac{2}{x_n}\right)$$

as $n \to +\infty$, we obtain

$$L = \frac{1}{2}\left(L + \frac{2}{L}\right)$$

which can be rewritten as $L^2 = 2$. The negative solution of this equation is extraneous because $x_n > 0$ for all n, so $L = \sqrt{2}$.   ◀

EXERCISE SET 10.2   ~ Graphing Utility   C CAS

1. In each part, find a formula for the general term of the sequence, starting with $n = 1$.

(a) $1, \dfrac{1}{3}, \dfrac{1}{9}, \dfrac{1}{27}, \ldots$ (b) $1, -\dfrac{1}{3}, \dfrac{1}{9}, -\dfrac{1}{27}, \ldots$

(c) $\dfrac{1}{2}, \dfrac{3}{4}, \dfrac{5}{6}, \dfrac{7}{8}, \ldots$ (d) $\dfrac{1}{\sqrt{\pi}}, \dfrac{4}{\sqrt[3]{\pi}}, \dfrac{9}{\sqrt[4]{\pi}}, \dfrac{16}{\sqrt[5]{\pi}}, \ldots$

2. In each part, find two formulas for the general term of the sequence, one starting with $n = 1$ and the other with $n = 0$.

(a) $1, -r, r^2, -r^3, \ldots$ (b) $r, -r^2, r^3, -r^4, \ldots$

3. (a) Write out the first four terms of the sequence $\{1 + (-1)^n\}$, starting with $n = 0$.

(b) Write out the first four terms of the sequence $\{\cos n\pi\}$, starting with $n = 0$.

(c) Use the results in parts (a) and (b) to express the general term of the sequence $4, 0, 4, 0, \ldots$ in two different ways, starting with $n = 0$.

4. In each part, find a formula for the general term using factorials and starting with $n = 1$.

(a) $1 \cdot 2, \, 1 \cdot 2 \cdot 3 \cdot 4, \, 1 \cdot 2 \cdot 3 \cdot 4 \cdot 5 \cdot 6,$
$1 \cdot 2 \cdot 3 \cdot 4 \cdot 5 \cdot 6 \cdot 7 \cdot 8, \ldots$

(b) $1, \, 1 \cdot 2 \cdot 3, \, 1 \cdot 2 \cdot 3 \cdot 4 \cdot 5, \, 1 \cdot 2 \cdot 3 \cdot 4 \cdot 5 \cdot 6 \cdot 7, \ldots$

In Exercises 5–22, write out the first five terms of the sequence, determine whether the sequence converges, and if so find its limit.

5. $\left\{\dfrac{n}{n+2}\right\}_{n=1}^{+\infty}$ **6.** $\left\{\dfrac{n^2}{2n+1}\right\}_{n=1}^{+\infty}$ **7.** $\{2\}_{n=1}^{+\infty}$

8. $\left\{\ln\left(\dfrac{1}{n}\right)\right\}_{n=1}^{+\infty}$ **9.** $\left\{\dfrac{\ln n}{n}\right\}_{n=1}^{+\infty}$ **10.** $\left\{n \sin \dfrac{\pi}{n}\right\}_{n=1}^{+\infty}$

11. $\{1 + (-1)^n\}_{n=1}^{+\infty}$ **12.** $\left\{\dfrac{(-1)^{n+1}}{n^2}\right\}_{n=1}^{+\infty}$

13. $\left\{(-1)^n \dfrac{2n^3}{n^3+1}\right\}_{n=1}^{+\infty}$ **14.** $\left\{\dfrac{n}{2^n}\right\}_{n=1}^{+\infty}$

15. $\left\{\dfrac{(n+1)(n+2)}{2n^2}\right\}_{n=1}^{+\infty}$ **16.** $\left\{\dfrac{\pi^n}{4^n}\right\}_{n=1}^{+\infty}$

17. $\left\{\cos \dfrac{3}{n}\right\}_{n=1}^{+\infty}$ **18.** $\left\{\cos \dfrac{\pi n}{2}\right\}_{n=1}^{+\infty}$

19. $\{n^2 e^{-n}\}_{n=1}^{+\infty}$ **20.** $\{\sqrt{n^2 + 3n} - n\}_{n=1}^{+\infty}$

21. $\left\{\left(\dfrac{n+3}{n+1}\right)^n\right\}_{n=1}^{+\infty}$ **22.** $\left\{\left(1 - \dfrac{2}{n}\right)^n\right\}_{n=1}^{+\infty}$

In Exercises 23–30, find the general term of the sequence, starting with $n = 1$, determine whether the sequence converges, and if so find its limit.

23. $\dfrac{1}{2}, \dfrac{3}{4}, \dfrac{5}{6}, \dfrac{7}{8}, \ldots$ **24.** $0, \dfrac{1}{2^2}, \dfrac{2}{3^2}, \dfrac{3}{4^2}, \ldots$

25. $\dfrac{1}{3}, \dfrac{1}{9}, \dfrac{1}{27}, \dfrac{1}{81}, \ldots$ **26.** $-1, 2, -3, 4, -5, \ldots$

27. $\left(1 - \dfrac{1}{2}\right), \left(\dfrac{1}{2} - \dfrac{1}{3}\right), \left(\dfrac{1}{3} - \dfrac{1}{4}\right), \left(\dfrac{1}{4} - \dfrac{1}{5}\right), \ldots$

28. $3, \dfrac{3}{2}, \dfrac{3}{2^2}, \dfrac{3}{2^3}, \ldots$

29. $(\sqrt{2} - \sqrt{3}), (\sqrt{3} - \sqrt{4}), (\sqrt{4} - \sqrt{5}), \ldots$

30. $\dfrac{1}{3^5}, -\dfrac{1}{3^6}, \dfrac{1}{3^7}, -\dfrac{1}{3^8}, \ldots$

31. (a) Starting with $n = 1$, write out the first six terms of the sequence $\{a_n\}$, where

$$a_n = \begin{cases} 1, & \text{if } n \text{ is odd} \\ n, & \text{if } n \text{ is even} \end{cases}$$

(b) Starting with $n = 1$, and considering the even and odd terms separately, find a formula for the general term of the sequence

$$1, \frac{1}{2^2}, 3, \frac{1}{2^4}, 5, \frac{1}{2^6}, \ldots$$

(c) Starting with $n = 1$, and considering the even and odd terms separately, find a formula for the general term of the sequence

$$1, \frac{1}{3}, \frac{1}{3}, \frac{1}{5}, \frac{1}{5}, \frac{1}{7}, \frac{1}{7}, \frac{1}{9}, \frac{1}{9}, \ldots$$

(d) Determine whether the sequences in parts (a), (b), and (c) converge. For those that do, find the limit.

32. For what positive values of b does the sequence $b, 0, b^2, 0, b^3, 0, b^4, \ldots$ converge? Justify your answer.

C 33. (a) Use numerical evidence to make a conjecture about the limit of the sequence $\{\sqrt[n]{n^3}\}_{n=2}^{+\infty}$.
(b) Use a CAS to confirm your conjecture.

C 34. (a) Use numerical evidence to make a conjecture about the limit of the sequence $\{\sqrt[n]{3^n + n^3}\}_{n=2}^{+\infty}$.
(b) Use a CAS to confirm your conjecture.

35. Assuming that the sequence given in Formula (2) of this section converges, use the method of Example 10 to show that the limit of this sequence is $\sqrt{a}$.

36. Consider the sequence

$$a_1 = \sqrt{6}$$
$$a_2 = \sqrt{6 + \sqrt{6}}$$
$$a_3 = \sqrt{6 + \sqrt{6 + \sqrt{6}}}$$
$$a_4 = \sqrt{6 + \sqrt{6 + \sqrt{6 + \sqrt{6}}}}$$
$$\vdots$$

(a) Find a recursion formula for a_{n+1}.
(b) Assuming that the sequence converges, use the method of Example 10 to find the limit.

37. Consider the sequence $\{a_n\}_{n=1}^{+\infty}$, where

$$a_n = \frac{1}{n^2} + \frac{2}{n^2} + \cdots + \frac{n}{n^2}$$

(a) Find $a_1, a_2, a_3,$ and a_4.
(b) Use numerical evidence to make a conjecture about the limit of the sequence.
(c) Confirm your conjecture by expressing a_n in closed form and calculating the limit.

38. Follow the directions in Exercise 37 with

$$a_n = \frac{1^2}{n^3} + \frac{2^2}{n^3} + \cdots + \frac{n^2}{n^3}$$

In Exercises 39 and 40, use numerical evidence to make a conjecture about the limit of the sequence, and then use the Squeezing Theorem for Sequences (Theorem 10.2.5) to confirm that your conjecture is correct.

39. $\displaystyle\lim_{n \to +\infty} \frac{\sin^2 n}{n}$

40. $\displaystyle\lim_{n \to +\infty} \left(\frac{1+n}{2n}\right)^n$

41. (a) A bored student enters the number 0.5 in a calculator display and then repeatedly computes the square of the number in the display. Taking $a_0 = 0.5$, find a formula for the general term of the sequence $\{a_n\}$ of numbers that appear in the display.
(b) Try this with a calculator and make a conjecture about the limit of a_n.
(c) Confirm your conjecture by finding the limit of a_n.
(d) For what values of a_0 will this procedure produce a convergent sequence?

42. Let

$$f(x) = \begin{cases} 2x, & 0 \le x < 0.5 \\ 2x - 1, & 0.5 \le x < 1 \end{cases}$$

Does the sequence $f(0.2), f(f(0.2)), f(f(f(0.2))), \ldots$ converge? Justify your reasoning.

43. (a) Use a graphing utility to generate the graph of the equation $y = (2^x + 3^x)^{1/x}$, and then use the graph to make a conjecture about the limit of the sequence

$$\{(2^n + 3^n)^{1/n}\}_{n=1}^{+\infty}$$

(b) Confirm your conjecture by calculating the limit.

44. Consider the sequence $\{a_n\}_{n=1}^{+\infty}$ whose nth term is

$$a_n = \frac{1}{n} \sum_{k=1}^{n} \frac{1}{1 + (k/n)}$$

Show that $\lim_{n \to +\infty} a_n = \ln 2$ by interpreting a_n as the Riemann sum of a definite integral.

45. Let a_n be the average value of $f(x) = 1/x$ over the interval $[1, n]$. Determine whether the sequence $\{a_n\}$ converges, and if so find its limit.

46. The sequence whose terms are $1, 1, 2, 3, 5, 8, 13, 21, \ldots$ is called the **Fibonacci sequence** in honor of Leonardo ("Fibonacci") da Pisa (c. 1170–1250). This sequence has the property that after starting with two 1's, each term is the sum of the preceding two.
(a) Denoting the sequence by $\{a_n\}$ and starting with $a_1 = 1$ and $a_2 = 1$, show that

$$\frac{a_{n+2}}{a_{n+1}} = 1 + \frac{a_n}{a_{n+1}} \quad \text{if } n \ge 1$$

(b) Give a reasonable informal argument to show that if the sequence $\{a_{n+1}/a_n\}$ converges to some limit L, then the sequence $\{a_{n+2}/a_{n+1}\}$ must also converge to L.
(c) Assuming that the sequence $\{a_{n+1}/a_n\}$ converges, show that its limit is $(1 + \sqrt{5})/2$.

47. If we accept the fact that the sequence $\{1/n\}_{n=1}^{+\infty}$ converges to the limit $L = 0$, then according to Definition 10.2.2, for every $\epsilon > 0$, there exists a positive integer N such that $|a_n - L| = |(1/n) - 0| < \epsilon$ when $n \geq N$. In each part, find the smallest possible value of N for the given value of ϵ.

(a) $\epsilon = 0.5$ (b) $\epsilon = 0.1$ (c) $\epsilon = 0.001$

48. If we accept the fact that the sequence

$$\left\{ \frac{n}{n+1} \right\}_{n=1}^{+\infty}$$

converges to the limit $L = 1$, then according to Definition 10.2.2, for every $\epsilon > 0$ there exists an integer N such that

$$|a_n - L| = \left| \frac{n}{n+1} - 1 \right| < \epsilon$$

when $n \geq N$. In each part, find the smallest value of N for the given value of ϵ.

(a) $\epsilon = 0.25$ (b) $\epsilon = 0.1$ (c) $\epsilon = 0.001$

49. Use Definition 10.2.2 to prove that

(a) the sequence $\{1/n\}_{n=1}^{+\infty}$ converges to 0

(b) the sequence $\left\{ \dfrac{n}{n+1} \right\}_{n=1}^{+\infty}$ converges to 1.

50. Find $\lim_{n \to +\infty} r^n$, where r is a real number. [*Hint:* Consider the cases $|r| < 1$, $|r| > 1$, $r = 1$, and $r = -1$ separately.]

10.3 MONOTONE SEQUENCES

There are many situations in which it is important to know whether a sequence converges, but the value of the limit is not relevant to the problem at hand. In this section we will study several techniques that can be used to determine whether a sequence converges.

TERMINOLOGY

We begin with some terminology.

10.3.1   DEFINITION.   A sequence $\{a_n\}_{n=1}^{+\infty}$ is called

strictly increasing if $a_1 < a_2 < a_3 < \cdots < a_n < \cdots$

increasing if $a_1 \leq a_2 \leq a_3 \leq \cdots \leq a_n \leq \cdots$

strictly decreasing if $a_1 > a_2 > a_3 > \cdots > a_n > \cdots$

decreasing if $a_1 \geq a_2 \geq a_3 \geq \cdots \geq a_n \geq \cdots$

In words, a sequence is strictly increasing if each term is larger than its predecessor, increasing if each term is the same as or larger than its predecessor, strictly decreasing if each term is smaller than its predecessor, and decreasing if each term is the same as or smaller than its predecessor. It follows that every strictly increasing sequence is increasing (but not conversely), and every strictly decreasing sequence is decreasing (but not conversely). A sequence that is either strictly increasing or strictly decreasing is called *strictly monotone*, and a sequence that is either increasing or decreasing is called *monotone*.

Example 1

SEQUENCE	DESCRIPTION
$\dfrac{1}{2}, \dfrac{2}{3}, \dfrac{3}{4}, \ldots, \dfrac{n}{n+1}, \ldots$	Strictly increasing
$1, \dfrac{1}{2}, \dfrac{1}{3}, \ldots, \dfrac{1}{n}, \ldots$	Strictly decreasing
$1, 1, 2, 2, 3, 3, \ldots$	Increasing; not strictly increasing
$1, 1, \dfrac{1}{2}, \dfrac{1}{2}, \dfrac{1}{3}, \dfrac{1}{3}, \ldots$	Decreasing; not strictly decreasing
$1, -\dfrac{1}{2}, \dfrac{1}{3}, -\dfrac{1}{4}, \ldots, (-1)^{n+1}\dfrac{1}{n}, \ldots$	Neither increasing nor decreasing

The first and second sequences are strictly monotone, and the third and fourth sequences are monotone but not strictly monotone. The fifth sequence is not monotone. ◄

• FOR THE READER. Can a sequence be both increasing and decreasing? Explain.

····································

TESTING FOR MONOTONICITY

In order for a sequence to be strictly increasing, *all* pairs of successive terms, a_n and a_{n+1}, must satisfy $a_n < a_{n+1}$ or, equivalently, $a_{n+1} - a_n > 0$. More generally, monotone sequences can be classified as follows:

DIFFERENCE BETWEEN SUCCESSIVE TERMS	CLASSIFICATION
$a_{n+1} - a_n > 0$	Strictly increasing
$a_{n+1} - a_n < 0$	Strictly decreasing
$a_{n+1} - a_n \geq 0$	Increasing
$a_{n+1} - a_n \leq 0$	Decreasing

Frequently, one can *guess* whether a sequence is monotone or strictly monotone by writing out some of the initial terms. However, to be certain that the guess is correct, one must give a precise mathematical argument. The following example illustrates one method for doing this.

Example 2 Show that

$$\frac{1}{2}, \frac{2}{3}, \frac{3}{4}, \ldots, \frac{n}{n+1}, \ldots$$

is a strictly increasing sequence.

Solution. The pattern of the initial terms suggests that the sequence is strictly increasing. To prove that this is so, let

$$a_n = \frac{n}{n+1}$$

We can obtain a_{n+1} by replacing n by $n+1$ in this formula. This yields

$$a_{n+1} = \frac{n+1}{(n+1)+1} = \frac{n+1}{n+2}$$

Thus, for $n \geq 1$

$$a_{n+1} - a_n = \frac{n+1}{n+2} - \frac{n}{n+1} = \frac{n^2 + 2n + 1 - n^2 - 2n}{(n+1)(n+2)} = \frac{1}{(n+1)(n+2)} > 0$$

which proves that the sequence is strictly increasing. ◄

If a_n and a_{n+1} are any successive terms in a strictly increasing sequence, then $a_n < a_{n+1}$. If the terms in the sequence are all positive, then we can divide both sides of this inequality by a_n to obtain $1 < a_{n+1}/a_n$ or, equivalently, $a_{n+1}/a_n > 1$. More generally, monotone sequences with *positive* terms can be classified as follows:

RATIO OF SUCCESSIVE TERMS	CONCLUSION
$a_{n+1}/a_n > 1$	Strictly increasing
$a_{n+1}/a_n < 1$	Strictly decreasing
$a_{n+1}/a_n \geq 1$	Increasing
$a_{n+1}/a_n \leq 1$	Decreasing

Example 3 Show that the sequence in Example 2 is strictly increasing by examining the ratio of successive terms.

Solution. As shown in the solution of Example 2,

$$a_n = \frac{n}{n+1} \quad \text{and} \quad a_{n+1} = \frac{n+1}{n+2}$$

Thus,

$$\frac{a_{n+1}}{a_n} = \frac{(n+1)/(n+2)}{n/(n+1)} = \frac{n+1}{n+2} \cdot \frac{n+1}{n} = \frac{n^2 + 2n + 1}{n^2 + 2n} \tag{1}$$

Since the numerator in (1) exceeds the denominator, it follows that $a_{n+1}/a_n > 1$ for $n \geq 1$. This proves that the sequence is strictly increasing. ◀

The following example illustrates still a third technique for determining whether a sequence is strictly monotone.

Example 4 In Examples 2 and 3 we proved that the sequence

$$\frac{1}{2}, \frac{2}{3}, \frac{3}{4}, \ldots, \frac{n}{n+1}, \ldots$$

is strictly increasing by considering the difference and ratio of successive terms. Alternatively, we can proceed as follows. Let

$$f(x) = \frac{x}{x+1}$$

so that the nth term in the given sequence is $a_n = f(n)$. The function f is increasing for $x \geq 1$ since

$$f'(x) = \frac{(x+1)(1) - x(1)}{(x+1)^2} = \frac{1}{(x+1)^2} > 0$$

Thus,

$$a_n = f(n) < f(n+1) = a_{n+1}$$

which proves that the given sequence is strictly increasing. ◀

In general, if $f(n) = a_n$ is the nth term of a sequence, and if f is differentiable for $x \geq 1$, then we have the following results:

DERIVATIVE OF f FOR $x \geq 1$	CONCLUSION FOR THE SEQUENCE WITH $a_n = f(n)$
$f'(x) > 0$	Strictly increasing
$f'(x) < 0$	Strictly decreasing
$f'(x) \geq 0$	Increasing
$f'(x) \leq 0$	Decreasing

PROPERTIES THAT HOLD EVENTUALLY

Sometimes a sequence will behave erratically at first and then settle down into a definite pattern. For example, the sequence

$$9, -8, -17, 12, 1, 2, 3, 4, \ldots \tag{2}$$

is strictly increasing from the fifth term on, but the sequence as a whole cannot be classified as strictly increasing because of the erratic behavior of the first four terms. To describe such sequences, we introduce the following terminology.

> **10.3.2 DEFINITION.** If discarding finitely many terms from the beginning of a sequence produces a sequence with a certain property, then the original sequence is said to have that property *eventually*.

For example, although we cannot say that sequence (2) is strictly increasing, we can say that it is eventually strictly increasing.

Example 5 Show that the sequence $\left\{\dfrac{10^n}{n!}\right\}_{n=1}^{+\infty}$ is eventually strictly decreasing.

Solution. We have

$$a_n = \frac{10^n}{n!} \quad \text{and} \quad a_{n+1} = \frac{10^{n+1}}{(n+1)!}$$

so

$$\frac{a_{n+1}}{a_n} = \frac{10^{n+1}/(n+1)!}{10^n/n!} = \frac{10^{n+1}n!}{10^n(n+1)!} = 10\frac{n!}{(n+1)n!} = \frac{10}{n+1} \tag{3}$$

From (3), $a_{n+1}/a_n < 1$ for all $n \geq 10$, so the sequence is eventually strictly decreasing. ◄

AN INTUITIVE VIEW OF CONVERGENCE

Informally stated, the convergence or divergence of a sequence does not depend on the behavior of its *initial terms*, but rather on how the terms behave *eventually*. For example, the sequence

$$3, \ -9, \ -13, \ 17, \ 1, \ \frac{1}{2}, \ \frac{1}{3}, \ \frac{1}{4}, \dots$$

eventually behaves like the sequence

$$1, \ \frac{1}{2}, \ \frac{1}{3}, \dots, \ \frac{1}{n}, \dots$$

and hence has a limit of 0.

CONVERGENCE OF MONOTONE SEQUENCES

The following two theorems, whose proofs are discussed at the end of this section, show that a monotone sequence either converges or becomes infinite—divergence by oscillation cannot occur.

> **10.3.3 THEOREM.** *If a sequence $\{a_n\}$ is eventually increasing, then there are two possibilities:*
>
> *(a) There is a constant M, called an **upper bound** for the sequence, such that $a_n \leq M$ for all n, in which case the sequence converges to a limit L satisfying $L \leq M$.*
>
> *(b) No upper bound exists, in which case $\lim\limits_{n \to +\infty} a_n = +\infty$.*

> **10.3.4 THEOREM.** *If a sequence $\{a_n\}$ is eventually decreasing, then there are two possibilities:*
>
> *(a) There is a constant M, called a **lower bound** for the sequence, such that $a_n \geq M$ for all n, in which case the sequence converges to a limit L satisfying $L \geq M$.*
>
> *(b) No lower bound exists, in which case $\lim\limits_{n \to +\infty} a_n = -\infty$.*

Note that these results do not give a method for obtaining limits; they tell us only whether a limit exists.

Example 6   Show that the sequence $\left\{ \dfrac{10^n}{n!} \right\}_{n=1}^{+\infty}$ converges and find its limit.

Solution.   We showed in Example 5 that the sequence is eventually strictly decreasing. Since all terms in the sequence are positive, it is bounded below by $M = 0$, and hence Theorem 10.3.4 guarantees that it converges to a nonnegative limit L. However, the limit is not evident directly from the formula $10^n/n!$ for the nth term, so we will need some ingenuity to obtain it.

Recall from Formula (3) of Example 5 that successive terms in the given sequence are related by the recursion formula

$$a_{n+1} = \frac{10}{n+1} a_n \tag{4}$$

where $a_n = 10^n/n!$. We will take the limit as $n \to +\infty$ of both sides of (4) and use the fact that

$$\lim_{n \to +\infty} a_{n+1} = \lim_{n \to +\infty} a_n = L$$

We obtain

$$L = \lim_{n \to +\infty} a_{n+1} = \lim_{n \to +\infty} \left(\frac{10}{n+1} a_n \right) = \lim_{n \to +\infty} \frac{10}{n+1} \lim_{n \to +\infty} a_n = 0 \cdot L = 0$$

so that

$$L = \lim_{n \to +\infty} \frac{10^n}{n!} = 0 \qquad \blacktriangleleft$$

REMARK.   In the exercises we will show that the technique illustrated in this example can be adapted to obtain the limit

$$\lim_{n \to +\infty} \frac{x^n}{n!} = 0 \tag{5}$$

for any real value of x (Exercise 26). This result will be useful in our later work.

THE COMPLETENESS AXIOM

In this text we have accepted the familiar properties of real numbers without proof, and indeed, we have not even attempted to define the term *real number*. Although this is sufficient for many purposes, it was recognized by the late nineteenth century that the study of limits and functions in calculus requires a precise axiomatic formulation of the real numbers analogous to the axiomatic development of Euclidean geometry. Although we will not attempt to pursue this development, we will need to discuss one of the axioms about real numbers in order to prove Theorems 10.3.3 and 10.3.4. But first we will introduce some terminology.

If S is a nonempty set of real numbers, then we call u an ***upper bound*** for S if u is greater than or equal to every number in S, and we call ℓ a ***lower bound*** for S if ℓ is smaller than or equal to every number in S. For example, if S is the set of numbers in the interval $(1, 3)$, then $u = 4$, 10, and 100 are upper bounds for S and $\ell = -10$, 0, and $\frac{1}{2}$ are lower bounds for S. Observe also that $u = 3$ is the smallest of all upper bounds and $\ell = 1$ is the largest of all lower bounds. The existence of a smallest upper bound and a greatest lower bound for S is not accidental; it is a consequence of the following axiom.

10.3.5   AXIOM (*The Completeness Axiom*).   *If a nonempty set S of real numbers has an upper bound, then it has a smallest upper bound (called the **least upper bound**), and if a nonempty set S of real numbers has a lower bound, then it has a largest lower bound (called the **greatest lower bound**).*

Proof of Theorem 10.3.3.

(a) We will prove the result for increasing sequences, and leave it for the reader to adapt the argument to sequences that are eventually increasing. Assume there exists a number M such that $a_n \leq M$ for $n = 1, 2, \ldots$. Then M is an upper bound for the set of terms in the sequence. By the Completeness Axiom there is a least upper bound for the terms, call it L. Now let ϵ be any positive number. Since L is the least upper bound for the terms, $L - \epsilon$ is not an upper bound for the terms, which means that there is at least one term a_N such that

$$a_N > L - \epsilon$$

Moreover, since $\{a_n\}$ is an increasing sequence, we must have

$$a_n \geq a_N > L - \epsilon \tag{6}$$

when $n \geq N$. But a_n cannot exceed L since L is an upper bound for the terms. This observation together with (6) tells us that $L \geq a_n > L - \epsilon$ for $n \geq N$, so all terms from the Nth on are within ϵ units of L. This is exactly the requirement to have

$$\lim_{n \to +\infty} a_n = L$$

Finally, $L \leq M$ since M is an upper bound for the terms and L is the least upper bound. This proves part (a).

(b) If there is no number M such that $a_n \leq M$ for $n = 1, 2, \ldots$, then no matter how large we choose M, there is a term a_N such that

$$a_N > M$$

and, since the sequence is increasing,

$$a_n \geq a_N > M$$

when $n \geq N$. Thus, the terms in the sequence become arbitrarily large as n increases. That is,

$$\lim_{n \to +\infty} a_n = +\infty$$

The proof of Theorem 10.3.4 will be omitted since it is similar to that of 10.3.3.

EXERCISE SET 10.3

In Exercises 1–6, use $a_{n+1} - a_n$ to show that the given sequence $\{a_n\}$ is strictly increasing or strictly decreasing.

1. $\left\{\dfrac{1}{n}\right\}_{n=1}^{+\infty}$ **2.** $\left\{1 - \dfrac{1}{n}\right\}_{n=1}^{+\infty}$ **3.** $\left\{\dfrac{n}{2n+1}\right\}_{n=1}^{+\infty}$

4. $\left\{\dfrac{n}{4n-1}\right\}_{n=1}^{+\infty}$ **5.** $\{n - 2^n\}_{n=1}^{+\infty}$ **6.** $\{n - n^2\}_{n=1}^{+\infty}$

In Exercises 7–12, use a_{n+1}/a_n to show that the given sequence $\{a_n\}$ is strictly increasing or strictly decreasing.

7. $\left\{\dfrac{n}{2n+1}\right\}_{n=1}^{+\infty}$ **8.** $\left\{\dfrac{2^n}{1+2^n}\right\}_{n=1}^{+\infty}$ **9.** $\{ne^{-n}\}_{n=1}^{+\infty}$

10. $\left\{\dfrac{10^n}{(2n)!}\right\}_{n=1}^{+\infty}$ **11.** $\left\{\dfrac{n^n}{n!}\right\}_{n=1}^{+\infty}$ **12.** $\left\{\dfrac{5^n}{2^{(n^2)}}\right\}_{n=1}^{+\infty}$

In Exercises 13–18, use differentiation to show that the sequence is strictly increasing or strictly decreasing.

13. $\left\{\dfrac{n}{2n+1}\right\}_{n=1}^{+\infty}$ **14.** $\left\{3 - \dfrac{1}{n}\right\}_{n=1}^{+\infty}$

15. $\left\{\dfrac{1}{n + \ln n}\right\}_{n=1}^{+\infty}$ **16.** $\{ne^{-2n}\}_{n=1}^{+\infty}$

17. $\left\{\dfrac{\ln(n+2)}{n+2}\right\}_{n=1}^{+\infty}$ **18.** $\{\tan^{-1} n\}_{n=1}^{+\infty}$

In Exercises 19–24, use any method to show that the given sequence is eventually strictly increasing or eventually strictly decreasing.

19. $\{2n^2 - 7n\}_{n=1}^{+\infty}$ **20.** $\{n^3 - 4n^2\}_{n=1}^{+\infty}$

21. $\left\{\dfrac{n}{n^2 + 10}\right\}_{n=1}^{+\infty}$

22. $\left\{n + \dfrac{17}{n}\right\}_{n=1}^{+\infty}$

23. $\left\{\dfrac{n!}{3^n}\right\}_{n=1}^{+\infty}$

24. $\{n^5 e^{-n}\}_{n=1}^{+\infty}$

25. (a) Suppose that $\{a_n\}$ is a monotone sequence such that $1 \le a_n \le 2$. Must the sequence converge? If so, what can you say about the limit?

(b) Suppose that $\{a_n\}$ is a monotone sequence such that $a_n \le 2$. Must the sequence converge? If so, what can you say about the limit?

26. The goal in this exercise is to prove Formula (5) in this section. The case where $x = 0$ is obvious, so we will focus on the case where $x \ne 0$.

(a) Let $a_n = |x|^n / n!$. Show that

$$a_{n+1} = \frac{|x|}{n + 1} a_n$$

(b) Show that the sequence $\{a_n\}$ is eventually strictly decreasing.

(c) Show that the sequence $\{a_n\}$ converges.

(d) Use the results in parts (a) and (c) to show that $a_n \to 0$ as $n \to +\infty$.

(e) Obtain Formula (5) from the result in part (d).

27. Let $\{a_n\}$ be the sequence defined recursively by $a_1 = \sqrt{2}$ and $a_{n+1} = \sqrt{2 + a_n}$ for $n \ge 1$.

(a) List the first three terms of the sequence.

(b) Show that $a_n < 2$ for $n \ge 1$.

(c) Show that $a_{n+1}^2 - a_n^2 = (2 - a_n)(1 + a_n)$ for $n \ge 1$.

(d) Use the results in parts (b) and (c) to show that $\{a_n\}$ is a strictly increasing sequence. [*Hint:* If x and y are positive real numbers such that $x^2 - y^2 > 0$, then it follows by factoring that $x - y > 0$.]

(e) Show that $\{a_n\}$ converges and find its limit L.

28. Let $\{a_n\}$ be the sequence defined recursively by $a_1 = 1$ and $a_{n+1} = \frac{1}{2}[a_n + (3/a_n)]$ for $n \ge 1$.

(a) Show that $a_n \ge \sqrt{3}$ for $n \ge 2$. [*Hint:* What is the minimum value of $\frac{1}{2}[x + (3/x)]$ for $x > 0$?]

(b) Show that $\{a_n\}$ is eventually decreasing. [*Hint:* Examine $a_{n+1} - a_n$ or a_{n+1}/a_n and use the result in part (a).]

(c) Show that $\{a_n\}$ converges and find its limit L.

29. (a) Compare appropriate areas in the accompanying figure to deduce the following inequalities for $n \ge 2$:

$$\int_1^n \ln x \, dx < \ln n! < \int_1^{n+1} \ln x \, dx$$

(b) Use the result in part (a) to show that

$$\frac{n^n}{e^{n-1}} < n! < \frac{(n+1)^{n+1}}{e^n}, \qquad n > 1$$

(c) Use the Squeezing Theorem for Sequences (Theorem 10.2.5) and the result in part (b) to show that

$$\lim_{n \to +\infty} \frac{\sqrt[n]{n!}}{n} = \frac{1}{e}$$

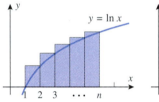 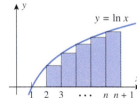

Figure Ex-29

30. Use the left inequality in Exercise 29(b) to show that

$$\lim_{n \to +\infty} \sqrt[n]{n!} = +\infty$$

10.4 INFINITE SERIES

The purpose of this section is to discuss sums that contain infinitely many terms. The most familiar examples of such sums occur in the decimal representations of real numbers. For example, when we write $\frac{1}{3}$ in the decimal form $\frac{1}{3} = 0.3333\ldots$, we mean

$$\frac{1}{3} = 0.3 + 0.03 + 0.003 + 0.0003 + \cdots$$

which suggests that the decimal representation of $\frac{1}{3}$ can be viewed as a sum of infinitely many real numbers.

SUMS OF INFINITE SERIES

Our first objective is to define what is meant by the "sum" of infinitely many real numbers. We begin with some terminology.

10.4.1 DEFINITION. An ***infinite series*** is an expression that can be written in the form

$$\sum_{k=1}^{\infty} u_k = u_1 + u_2 + u_3 + \cdots + u_k + \cdots$$

The numbers $u_1, u_2, u_3, \ldots$ are called the ***terms*** of the series.

Since it is impossible to add infinitely many numbers together directly, sums of infinite series are defined and computed by an indirect limiting process. To motivate the basic idea, consider the decimal

$$0.3333\ldots \tag{1}$$

This can be viewed as the infinite series

$$0.3 + 0.03 + 0.003 + 0.0003 + \cdots$$

or, equivalently,

$$\frac{3}{10} + \frac{3}{10^2} + \frac{3}{10^3} + \frac{3}{10^4} + \cdots \tag{2}$$

Since (1) is the decimal expansion of $\frac{1}{3}$, any reasonable definition for the sum of an infinite series should yield $\frac{1}{3}$ for the sum of (2). To obtain such a definition, consider the following sequence of (finite) sums:

$$s_1 = \frac{3}{10} = 0.3$$

$$s_2 = \frac{3}{10} + \frac{3}{10^2} = 0.33$$

$$s_3 = \frac{3}{10} + \frac{3}{10^2} + \frac{3}{10^3} = 0.333$$

$$s_4 = \frac{3}{10} + \frac{3}{10^2} + \frac{3}{10^3} + \frac{3}{10^4} = 0.3333$$

$$\vdots$$

The sequence of numbers $s_1, s_2, s_3, s_4, \ldots$ can be viewed as a succession of approximations to the "sum" of the infinite series, which we want to be $\frac{1}{3}$. As we progress through the sequence, more and more terms of the infinite series are used, and the approximations get better and better, suggesting that the desired sum of $\frac{1}{3}$ might be the *limit* of this sequence of approximations. To see that this is so, we must calculate the limit of the general term in the sequence of approximations, namely

$$s_n = \frac{3}{10} + \frac{3}{10^2} + \cdots + \frac{3}{10^n} \tag{3}$$

The problem of calculating

$$\lim_{n \to +\infty} s_n = \lim_{n \to +\infty} \left(\frac{3}{10} + \frac{3}{10^2} + \cdots + \frac{3}{10^n} \right)$$

is complicated by the fact that both the last term and the number of terms in the sum change with n. It is best to rewrite such limits in a closed form in which the number of terms does not vary, if possible. (See the discussion of closed form and open form following Example 3 in Section 6.4.) To do this, we multiply both sides of (3) by $\frac{1}{10}$ to obtain

$$\frac{1}{10} s_n = \frac{3}{10^2} + \frac{3}{10^3} + \cdots + \frac{3}{10^n} + \frac{3}{10^{n+1}} \tag{4}$$

and then subtract (4) from (3) to obtain

$$s_n - \frac{1}{10} s_n = \frac{3}{10} - \frac{3}{10^{n+1}}$$

$$\frac{9}{10} s_n = \frac{3}{10}\left(1 - \frac{1}{10^n}\right)$$

$$s_n = \frac{1}{3}\left(1 - \frac{1}{10^n}\right)$$

Since $1/10^n \to 0$ as $n \to +\infty$, it follows that

$$\lim_{n \to +\infty} s_n = \lim_{n \to +\infty} \frac{1}{3}\left(1 - \frac{1}{10^n}\right) = \frac{1}{3}$$

which we denote by writing

$$\frac{1}{3} = \frac{3}{10} + \frac{3}{10^2} + \frac{3}{10^3} + \cdots + \frac{3}{10^n} + \cdots$$

Motivated by the preceding example, we are now ready to define the general concept of the "sum" of an infinite series

$$u_1 + u_2 + u_3 + \cdots + u_k + \cdots$$

We begin with some terminology: Let s_n denote the sum of the initial terms of the series, up to and including the term with index n. Thus,

$$s_1 = u_1$$
$$s_2 = u_1 + u_2$$
$$s_3 = u_1 + u_2 + u_3$$
$$\vdots$$
$$s_n = u_1 + u_2 + u_3 + \cdots + u_n = \sum_{k=1}^{n} u_k$$

The number s_n is called the **nth partial sum** of the series and the sequence $\{s_n\}_{n=1}^{+\infty}$ is called the **sequence of partial sums**.

> WARNING.   In everyday language the words "sequence" and "series" are often used interchangeably. However, this is not so in mathematics—mathematically, a sequence is a *succession* and a series is a *sum*. It is essential that you keep this distinction in mind.

As n increases, the partial sum $s_n = u_1 + u_2 + \cdots + u_n$ includes more and more terms of the series. Thus, if s_n tends toward a limit as $n \to +\infty$, it is reasonable to view this limit as the sum of *all* the terms in the series. This suggests the following definition.

10.4.2   DEFINITION.   Let $\{s_n\}$ be the sequence of partial sums of the series

$$u_1 + u_2 + u_3 + \cdots + u_k + \cdots$$

If the sequence $\{s_n\}$ converges to a limit S, then the series is said to **converge** to S, and S is called the **sum** of the series. We denote this by writing

$$S = \sum_{k=1}^{\infty} u_k$$

If the sequence of partial sums diverges, then the series is said to **diverge**. A divergent series has no sum.

> REMARK.   Sometimes it will be desirable to start the summation index in an infinite series at $k = 0$ rather than $k = 1$, in which case we will view u_0 as the zeroth term and $s_0 = u_0$ as the zeroth partial sum. It can be proved that changing the starting value for the index has no effect on the convergence or divergence of an infinite series.

Example 1   Determine whether the series

$$1 - 1 + 1 - 1 + 1 - 1 + \cdots$$

converges or diverges. If it converges, find the sum.

Solution. It is tempting to conclude that the sum of the series is zero by arguing that the positive and negative terms cancel one another. However, this is *not correct*; the problem is that algebraic operations that hold for finite sums do not carry over to infinite series in all cases. Later, we will discuss conditions under which familiar algebraic operations can be applied to infinite series, but for this example we turn directly to Definition 10.4.2. The partial sums are

$$s_1 = 1$$
$$s_2 = 1 - 1 = 0$$
$$s_3 = 1 - 1 + 1 = 1$$
$$s_4 = 1 - 1 + 1 - 1 = 0$$

and so forth. Thus, the sequence of partial sums is

$$1, 0, 1, 0, 1, 0, \ldots$$

Since this is a divergent sequence, the given series diverges and consequently has no sum. ◄

GEOMETRIC SERIES

In many important series, each term is obtained by multiplying the preceding term by some fixed constant. Thus, if the initial term of the series is a and each term is obtained by multiplying the preceding term by r, then the series has the form

$$\sum_{k=0}^{\infty} ar^k = a + ar + ar^2 + ar^3 + \cdots + ar^k + \cdots \quad (a \neq 0)$$

Such series are called *geometric series*, and the number r is called the *ratio* for the series. Here are some examples:

$$1 + 2 + 4 + 8 + \cdots + 2^k + \cdots \qquad \boxed{a = 1, r = 2}$$

$$\frac{3}{10} + \frac{3}{10^2} + \frac{3}{10^3} + \cdots + \frac{3}{10^k} + \cdots \qquad \boxed{a = \frac{3}{10}, r = \frac{1}{10}}$$

$$\frac{1}{2} - \frac{1}{4} + \frac{1}{8} - \frac{1}{16} + \cdots + (-1)^{k+1}\frac{1}{2^k} + \cdots \qquad \boxed{a = \frac{1}{2}, r = -\frac{1}{2}}$$

$$1 + 1 + 1 + \cdots + 1 + \cdots \qquad \boxed{a = 1, r = 1}$$

$$1 - 1 + 1 - 1 + \cdots + (-1)^{k+1} + \cdots \qquad \boxed{a = 1, r = -1}$$

$$1 + x + x^2 + x^3 + \cdots + x^k + \cdots \qquad \boxed{a = 1, r = x}$$

The following theorem is the fundamental result on convergence of geometric series.

10.4.3 THEOREM. *A geometric series*

$$\sum_{k=0}^{\infty} ar^k = a + ar + ar^2 + \cdots + ar^k + \cdots \quad (a \neq 0)$$

converges if $|r| < 1$ and diverges if $|r| \geq 1$. If the series converges, then the sum is

$$\sum_{k=0}^{\infty} ar^k = \frac{a}{1 - r}$$

Proof. Let us treat the case $|r| = 1$ first. If $r = 1$, then the series is

$$a + a + a + a + \cdots$$

so the nth partial sum is $s_n = (n + 1)a$ and $\lim_{n \to +\infty} s_n = \lim_{n \to +\infty}(n + 1)a = \pm\infty$ (the sign depending on whether a is positive or negative). This proves divergence. If $r = -1$, the series is

$$a - a + a - a + \cdots$$

so the sequence of partial sums is

$$a, 0, a, 0, a, 0, \ldots$$

which diverges.

Now let us consider the case where $|r| \neq 1$. The nth partial sum of the series is

$$s_n = a + ar + ar^2 + \cdots + ar^n \tag{5}$$

Multiplying both sides of (5) by r yields

$$rs_n = ar + ar^2 + \cdots + ar^n + ar^{n+1} \tag{6}$$

and subtracting (6) from (5) gives

$$s_n - rs_n = a - ar^{n+1}$$

or

$$(1 - r)s_n = a - ar^{n+1} \tag{7}$$

Since $r \neq 1$ in the case we are considering, this can be rewritten as

$$s_n = \frac{a - ar^{n+1}}{1 - r} = \frac{a}{1 - r} - \frac{ar^{n+1}}{1 - r} \tag{8}$$

If $|r| < 1$, then $\lim_{n \to +\infty} r^{n+1} = 0$ (can you see why?), so $\{s_n\}$ converges. From (8)

$$\lim_{n \to +\infty} s_n = \frac{a}{1 - r}$$

If $|r| > 1$, then either $r > 1$ or $r < -1$. In the case $r > 1$, $\lim_{n \to +\infty} r^{n+1} = +\infty$, and in the case $r < -1$, r^{n+1} oscillates between positive and negative values that grow in magnitude, so $\{s_n\}$ diverges in both cases. ∎

Example 2   The series

$$\sum_{k=0}^{\infty} \frac{5}{4^k} = 5 + \frac{5}{4} + \frac{5}{4^2} + \cdots + \frac{5}{4^k} + \cdots$$

is a geometric series with $a = 5$ and $r = \frac{1}{4}$. Since $|r| = \frac{1}{4} < 1$, the series converges and the sum is

$$\frac{a}{1 - r} = \frac{5}{1 - \frac{1}{4}} = \frac{20}{3} \qquad \blacktriangleleft$$

Example 3   Find the rational number represented by the repeating decimal

$$0.784784784\ldots$$

Solution.   We can write

$$0.784784784\ldots = 0.784 + 0.000784 + 0.000000784 + \cdots$$

so the given decimal is the sum of a geometric series with $a = 0.784$ and $r = 0.001$. Thus,

$$0.784784784\ldots = \frac{a}{1 - r} = \frac{0.784}{1 - 0.001} = \frac{0.784}{0.999} = \frac{784}{999} \qquad \blacktriangleleft$$

Example 4   In each part, determine whether the series converges, and if so find its sum.

(a) $\displaystyle\sum_{k=1}^{\infty} 3^{2k} 5^{1-k}$ (b) $\displaystyle\sum_{k=0}^{\infty} x^k$

Solution (a). This is a geometric series in a concealed form, since we can rewrite it as

$$\sum_{k=1}^{\infty} 3^{2k} 5^{1-k} = \sum_{k=1}^{\infty} \frac{9^k}{5^{k-1}} = \sum_{k=1}^{\infty} 9 \left(\frac{9}{5} \right)^{k-1}$$

Since $r = \frac{9}{5} > 1$, the series diverges.

Solution (b). The expanded form of the series is

$$\sum_{k=0}^{\infty} x^k = 1 + x + x^2 + \cdots + x^k + \cdots$$

The series is a geometric series with $a = 1$ and $r = x$, so it converges if $|x| < 1$ and diverges otherwise. When the series converges its sum is

$$\sum_{k=0}^{\infty} x^k = \frac{1}{1-x} \qquad \blacktriangleleft$$

TELESCOPING SUMS

Example 5   Determine whether the series

$$\sum_{k=1}^{\infty} \frac{1}{k(k+1)} = \frac{1}{1 \cdot 2} + \frac{1}{2 \cdot 3} + \frac{1}{3 \cdot 4} + \frac{1}{4 \cdot 5} + \cdots$$

converges or diverges. If it converges, find the sum.

Solution. The nth partial sum of the series is

$$s_n = \sum_{k=1}^{n} \frac{1}{k(k+1)} = \frac{1}{1 \cdot 2} + \frac{1}{2 \cdot 3} + \frac{1}{3 \cdot 4} + \cdots + \frac{1}{n(n+1)}$$

To calculate $\lim_{n \to +\infty} s_n$ we will rewrite s_n in closed form. This can be accomplished by using the method of partial fractions to obtain (verify)

$$\frac{1}{k(k+1)} = \frac{1}{k} - \frac{1}{k+1}$$

from which we obtain the telescoping sum

$$s_n = \sum_{k=1}^{n} \left(\frac{1}{k} - \frac{1}{k+1} \right)$$

$$= \left(1 - \frac{1}{2} \right) + \left(\frac{1}{2} - \frac{1}{3} \right) + \left(\frac{1}{3} - \frac{1}{4} \right) + \cdots + \left(\frac{1}{n} - \frac{1}{n+1} \right)$$

$$= 1 + \left(-\frac{1}{2} + \frac{1}{2} \right) + \left(-\frac{1}{3} + \frac{1}{3} \right) + \cdots + \left(-\frac{1}{n} + \frac{1}{n} \right) - \frac{1}{n+1}$$

$$= 1 - \frac{1}{n+1}$$

so

$$\sum_{k=1}^{\infty} \frac{1}{k(k+1)} = \lim_{n \to +\infty} s_n = \lim_{n \to +\infty} \left(1 - \frac{1}{n+1} \right) = 1 \qquad \blacktriangleleft$$

FOR THE READER.   If you have a CAS, read the documentation to determine how to find sums of infinite series; then use the CAS to check the results in Example 5.

• •
HARMONIC SERIES

One of the most important of all diverging series is the *harmonic series*,

$$\sum_{k=1}^{\infty} \frac{1}{k} = 1 + \frac{1}{2} + \frac{1}{3} + \frac{1}{4} + \frac{1}{5} + \cdots$$

which arises in connection with the overtones produced by a vibrating musical string. It is not immediately evident that this series diverges. However, the divergence will become apparent when we examine the partial sums in detail. Because the terms in the series are all positive, the partial sums

$$s_1 = 1, \quad s_2 = 1 + \tfrac{1}{2}, \quad s_3 = 1 + \tfrac{1}{2} + \tfrac{1}{3}, \quad s_4 = 1 + \tfrac{1}{2} + \tfrac{1}{3} + \tfrac{1}{4}, \ldots$$

form a strictly increasing sequence

$$s_1 < s_2 < s_3 < \cdots < s_n < \cdots$$

Thus, by Theorem 10.3.3 we can prove divergence by demonstrating that there is no constant M that is greater than or equal to *every* partial sum. To this end, we will consider some selected partial sums, namely $s_2, s_4, s_8, s_{16}, s_{32}, \ldots$. Note that the subscripts are successive powers of 2, so that these are the partial sums of the form s_{2^n}. These partial sums satisfy the inequalities

$$s_2 = 1 + \tfrac{1}{2} > \tfrac{1}{2} + \tfrac{1}{2} = \tfrac{2}{2}$$

$$s_4 = s_2 + \tfrac{1}{3} + \tfrac{1}{4} > s_2 + \left(\tfrac{1}{4} + \tfrac{1}{4}\right) = s_2 + \tfrac{1}{2} > \tfrac{3}{2}$$

$$s_8 = s_4 + \tfrac{1}{5} + \tfrac{1}{6} + \tfrac{1}{7} + \tfrac{1}{8} > s_4 + \left(\tfrac{1}{8} + \tfrac{1}{8} + \tfrac{1}{8} + \tfrac{1}{8}\right) = s_4 + \tfrac{1}{2} > \tfrac{4}{2}$$

$$s_{16} = s_8 + \tfrac{1}{9} + \tfrac{1}{10} + \tfrac{1}{11} + \tfrac{1}{12} + \tfrac{1}{13} + \tfrac{1}{14} + \tfrac{1}{15} + \tfrac{1}{16}$$

$$> s_8 + \left(\tfrac{1}{16} + \tfrac{1}{16} + \tfrac{1}{16} + \tfrac{1}{16} + \tfrac{1}{16} + \tfrac{1}{16} + \tfrac{1}{16} + \tfrac{1}{16}\right) = s_8 + \tfrac{1}{2} > \tfrac{5}{2}$$

$$\vdots$$

$$s_{2^n} > \frac{n+1}{2}$$

If M is any constant, we can find a positive integer n such that $(n+1)/2 > M$. But for this n

$$s_{2^n} > \frac{n+1}{2} > M$$

so that no constant M is greater than or equal to *every* partial sum of the harmonic series. This proves divergence.

This divergence proof, which predates the discovery of calculus, is due to a French bishop and teacher, Nicole Oresme (1323–1382). This series eventually attracted the interest of Johann and Jakob Bernoulli (p. 94) and led them to begin thinking about the general concept of convergence, which was a new idea at that time.

This is a proof of the divergence of the harmonic series, as it appeared in an appendix of Jakob Bernoulli's posthumous publication, *Ars Conjectandi*, which appeared in 1713.

EXERCISE SET 10.4 [C] CAS
• •

1. In each part, find exact values for the first four partial sums, find a closed form for the nth partial sum, and determine whether the series converges by calculating the limit of the nth partial sum. If the series converges, then state its sum.

(a) $2 + \dfrac{2}{5} + \dfrac{2}{5^2} + \cdots + \dfrac{2}{5^{k-1}} + \cdots$

(b) $\dfrac{1}{4} + \dfrac{2}{4} + \dfrac{2^2}{4} + \cdots + \dfrac{2^{k-1}}{4} + \cdots$

(c) $\dfrac{1}{2 \cdot 3} + \dfrac{1}{3 \cdot 4} + \dfrac{1}{4 \cdot 5} + \cdots + \dfrac{1}{(k+1)(k+2)} + \cdots$

2. In each part, find exact values for the first four partial sums, find a closed form for the nth partial sum, and determine whether the series converges by calculating the limit of the nth partial sum. If the series converges, then state its sum.

(a) $\displaystyle\sum_{k=1}^{\infty} \left(\frac{1}{4}\right)^k$ (b) $\displaystyle\sum_{k=1}^{\infty} 4^{k-1}$ (c) $\displaystyle\sum_{k=1}^{\infty} \left(\frac{1}{k+3} - \frac{1}{k+4}\right)$

In Exercises 3–14, determine whether the series converges, and if so, find its sum.

3. $\displaystyle\sum_{k=1}^{\infty}\left(-\frac{3}{4}\right)^{k-1}$

4. $\displaystyle\sum_{k=1}^{\infty}\left(\frac{2}{3}\right)^{k+2}$

5. $\displaystyle\sum_{k=1}^{\infty}(-1)^{k-1}\frac{7}{6^{k-1}}$

6. $\displaystyle\sum_{k=1}^{\infty}\left(-\frac{3}{2}\right)^{k+1}$

7. $\displaystyle\sum_{k=1}^{\infty}\frac{1}{(k+2)(k+3)}$

8. $\displaystyle\sum_{k=1}^{\infty}\left(\frac{1}{2^k}-\frac{1}{2^{k+1}}\right)$

9. $\displaystyle\sum_{k=1}^{\infty}\frac{1}{9k^2+3k-2}$

10. $\displaystyle\sum_{k=2}^{\infty}\frac{1}{k^2-1}$

11. $\displaystyle\sum_{k=3}^{\infty}\frac{1}{k-2}$

12. $\displaystyle\sum_{k=5}^{\infty}\left(\frac{e}{\pi}\right)^{k-1}$

13. $\displaystyle\sum_{k=1}^{\infty}\frac{4^{k+2}}{7^{k-1}}$

14. $\displaystyle\sum_{k=1}^{\infty}5^{3k}7^{1-k}$

In Exercises 15–20, express the given repeating decimal as a fraction.

15. $0.4444\ldots$

16. $0.9999\ldots$

17. $5.373737\ldots$

18. $0.159159159\ldots$

19. $0.782178217821\ldots$

20. $0.451141414\ldots$

21. A ball is dropped from a height of 10 m. Each time it strikes the ground it bounces vertically to a height that is $\frac{3}{4}$ of the preceding height. Find the total distance the ball will travel if it is assumed to bounce infinitely often.

22. The accompanying figure shows an "infinite staircase" constructed from cubes. Find the total volume of the staircase, given that the largest cube has a side of length 1 and each successive cube has a side whose length is half that of the preceding cube.

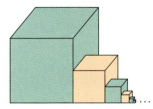

Figure Ex-22

23. In each part, find a closed form for the nth partial sum of the series, and determine whether the series converges. If so, find its sum.

(a) $\ln\dfrac{1}{2}+\ln\dfrac{2}{3}+\ln\dfrac{3}{4}+\cdots+\ln\dfrac{n}{n+1}+\cdots$

(b) $\ln\left(1-\dfrac{1}{4}\right)+\ln\left(1-\dfrac{1}{9}\right)+\ln\left(1-\dfrac{1}{16}\right)+\cdots$

$\qquad\qquad +\ln\left(1-\dfrac{1}{(k+1)^2}\right)+\cdots$

24. Use geometric series to show that

(a) $\displaystyle\sum_{k=0}^{\infty}(-1)^k x^k=\frac{1}{1+x}$ if $-1<x<1$

(b) $\displaystyle\sum_{k=0}^{\infty}(x-3)^k=\frac{1}{4-x}$ if $2<x<4$

(c) $\displaystyle\sum_{k=0}^{\infty}(-1)^k x^{2k}=\frac{1}{1+x^2}$ if $-1<x<1$.

25. In each part, find all values of x for which the series converges, and find the sum of the series for those values of x.

(a) $x-x^3+x^5-x^7+x^9-\cdots$

(b) $\dfrac{1}{x^2}+\dfrac{2}{x^3}+\dfrac{4}{x^4}+\dfrac{8}{x^5}+\dfrac{16}{x^6}+\cdots$

(c) $e^{-x}+e^{-2x}+e^{-3x}+e^{-4x}+e^{-5x}+\cdots$

26. Show: $\displaystyle\sum_{k=1}^{\infty}\frac{\sqrt{k+1}-\sqrt{k}}{\sqrt{k^2+k}}=1.$

27. Show: $\displaystyle\sum_{k=1}^{\infty}\left(\frac{1}{k}-\frac{1}{k+2}\right)=\frac{3}{2}.$

28. Show: $\dfrac{1}{1\cdot3}+\dfrac{1}{2\cdot4}+\dfrac{1}{3\cdot5}+\cdots=\dfrac{3}{4}.$

29. Show: $\dfrac{1}{1\cdot3}+\dfrac{1}{3\cdot5}+\dfrac{1}{5\cdot7}+\cdots=\dfrac{1}{2}.$

30. Show that for all real values of x

$$\sin x-\frac{1}{2}\sin^2 x+\frac{1}{4}\sin^3 x-\frac{1}{8}\sin^4 x+\cdots=\frac{2\sin x}{2+\sin x}$$

31. Let a_1 be any real number, and let $\{a_n\}$ be the sequence defined recursively by

$$a_{n+1}=\tfrac{1}{2}(a_n+1)$$

Make a conjecture about the limit of the sequence, and confirm your conjecture by expressing a_n in terms of a_1 and taking the limit.

32. Recall that a *terminating decimal* is a decimal whose digits are all 0 from some point on ($0.5=0.50000\ldots$, for example). Show that a decimal of the form $0.a_1a_2\ldots a_n 9999\ldots$, where $a_n\neq 9$, can be expressed as a terminating decimal.

33. The great Swiss mathematician Leonhard Euler (biography on p. 11) sometimes reached incorrect conclusions in his pioneering work on infinite series. For example, Euler deduced that

$$\tfrac{1}{2}=1-1+1-1+\cdots$$

and

$$-1=1+2+4+8+\cdots$$

by substituting $x=-1$ and $x=2$ in the formula

$$\frac{1}{1-x}=1+x+x^2+x^3+\cdots$$

What was the problem with his reasoning?

34. As shown in the accompanying figure, suppose that lines L_1 and L_2 form an angle θ, $0<\theta<\pi/2$, at their point of intersection P. A point P_0 is chosen that is on L_1 and a units from P. Starting from P_0 a zig-zag path is constructed by successively going back and forth between L_1 and L_2

along a perpendicular from one line to the other. Find the following sums in terms of θ.

(a) $P_0P_1 + P_1P_2 + P_2P_3 + \cdots$

(b) $P_0P_1 + P_2P_3 + P_4P_5 + \cdots$

(c) $P_1P_2 + P_3P_4 + P_5P_6 + \cdots$

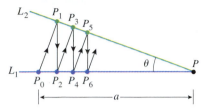

Figure Ex-34

35. As shown in the accompanying figure, suppose that an angle θ is bisected using a straightedge and compass to produce ray R_1, then the angle between R_1 and the initial side is bisected to produce ray R_2. Thereafter, rays R_3, R_4, R_5, ... are constructed in succession by bisecting the angle between the preceding two rays. Show that the sequence of angles that these rays make with the initial side has a limit of $\theta/3$. [This problem is based on *Trisection of an Angle in an Infinite Number of Steps* by Eric Kincannon, which appeared in *The College Mathematics Journal*, Vol. 21, No. 5, November 1990.]

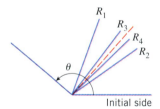

Figure Ex-35

36. In his *Treatise on the Configurations of Qualities and Motions* (written in the 1350s), the French Bishop of Lisieux, Nicole Oresme, used a geometric method to find the sum

of the series

$$\sum_{k=1}^{\infty} \frac{k}{2^k} = \frac{1}{2} + \frac{2}{4} + \frac{3}{8} + \frac{4}{16} + \cdots$$

In part (a) of the accompanying figure, each term in the series is represented by the area of a rectangle, and in part (b) the configuration in part (a) has been divided into rectangles with areas A_1, A_2, A_3, ... Find the sum $A_1 + A_2 + A_3 + \cdots$.

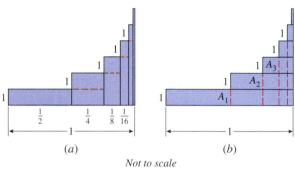

Figure Ex-36

37. (a) See if your CAS can find the sum of the series

$$\sum_{k=1}^{\infty} \frac{6^k}{(3^{k+1} - 2^{k+1})(3^k - 2^k)}$$

(b) Find A and B such that

$$\frac{6^k}{(3^{k+1} - 2^{k+1})(3^k - 2^k)} = \frac{2^k A}{3^k - 2^k} + \frac{2^k B}{3^{k+1} - 2^{k+1}}$$

(c) Use the result in part (b) to find a closed form for the nth partial sum, and then find the sum of the series. [This exercise is adapted from a problem that appeared in the Forty-Fifth Annual William Lowell Putnam Competition.]

38. In each part, use a CAS to find the sum of the series if it converges, and then confirm the result by hand calculation.

(a) $\sum_{k=1}^{\infty} (-1)^{k+1} 2^k 3^{2-k}$ (b) $\sum_{k=1}^{\infty} \frac{3^{3k}}{5^{k-1}}$ (c) $\sum_{k=1}^{\infty} \frac{1}{4k^2 - 1}$

10.5 CONVERGENCE TESTS

In the last section we showed how to find the sum of a series by finding a closed form for the nth partial sum and taking its limit. However, it is relatively rare that one can find a closed form for the nth partial sum of a series, so alternative methods are needed for finding the sum of a series. One possibility is to prove that the series converges, and then to approximate the sum by a partial sum with sufficiently many terms to achieve the desired degree of accuracy. In this section we will develop various tests that can be used to determine whether a given series converges or diverges.

In stating general results about convergence or divergence of series, it is convenient to use the notation $\sum u_k$ as a generic template for a series, thus avoiding the issue of whether the sum begins with $k = 0$ or $k = 1$ or some other value. Indeed, we will see shortly that the starting index value is irrelevant to the issue of convergence. The kth term in an infinite series $\sum u_k$ is called the **general term** of the series. The following theorem establishes a relationship between the limit of the general term and the convergence properties of a series.

10.5.1 THEOREM (*The Divergence Test*).

(a) If $\lim\limits_{k \to +\infty} u_k \neq 0$, then the series $\sum u_k$ diverges.

(b) If $\lim\limits_{k \to +\infty} u_k = 0$, then the series $\sum u_k$ may either converge or diverge.

Proof (a). To prove this result, it suffices to show that if the series converges, then $\lim_{k \to +\infty} u_k = 0$ (why?). We will prove this alternative form of (a).

Let us assume that the series converges. The general term u_k can be written as

$$u_k = s_k - s_{k-1} \tag{1}$$

where s_k is the sum of the terms through u_k and s_{k-1} is the sum of the terms through u_{k-1}. If S denotes the sum of the series, then $\lim_{k \to +\infty} s_k = S$, and since $(k-1) \to +\infty$ as $k \to +\infty$, we also have $\lim_{k \to +\infty} s_{k-1} = S$. Thus, from (1)

$$\lim_{k \to +\infty} u_k = \lim_{k \to +\infty} (s_k - s_{k-1}) = S - S = 0$$

Proof (b). To prove this result, it suffices to produce both a convergent series and a divergent series for which $\lim_{k \to +\infty} u_k = 0$. The following series both have this property:

$$\frac{1}{2} + \frac{1}{2^2} + \cdots + \frac{1}{2^k} + \cdots \quad \text{and} \quad 1 + \frac{1}{2} + \frac{1}{3} + \cdots + \frac{1}{k} + \cdots$$

The first is a convergent geometric series and the second is the divergent harmonic series. ■

The alternative form of part (a) given in the preceding proof is sufficiently important that we state it separately for future reference.

10.5.2 THEOREM. *If the series $\sum u_k$ converges, then $\lim\limits_{k \to +\infty} u_k = 0$.*

Example 1 The series

$$\sum_{k=1}^{\infty} \frac{k}{k+1} = \frac{1}{2} + \frac{2}{3} + \frac{3}{4} + \cdots + \frac{k}{k+1} + \cdots$$

diverges since

$$\lim_{k \to +\infty} \frac{k}{k+1} = \lim_{k \to +\infty} \frac{1}{1 + 1/k} = 1 \neq 0 \qquad \blacktriangleleft$$

WARNING. The converse of Theorem 10.5.2 is false. To prove that a series converges it does not suffice to show that $\lim_{k \to +\infty} u_k = 0$, since this property may hold for divergent as well as convergent series, as we saw in the proof of part (b) of Theorem 10.5.1.

For brevity, the proof of the following result is omitted.

10.5.3   THEOREM.

(a)   *If $\sum u_k$ and $\sum v_k$ are convergent series, then $\sum(u_k + v_k)$ and $\sum(u_k - v_k)$ are convergent series and the sums of these series are related by*

$$\sum_{k=1}^{\infty}(u_k + v_k) = \sum_{k=1}^{\infty}u_k + \sum_{k=1}^{\infty}v_k$$

$$\sum_{k=1}^{\infty}(u_k - v_k) = \sum_{k=1}^{\infty}u_k - \sum_{k=1}^{\infty}v_k$$

(b)   *If c is a nonzero constant, then the series $\sum u_k$ and $\sum cu_k$ both converge or both diverge. In the case of convergence, the sums are related by*

$$\sum_{k=1}^{\infty}cu_k = c\sum_{k=1}^{\infty}u_k$$

(c)   *Convergence or divergence is unaffected by deleting a finite number of terms from a series; in particular, for any positive integer K, the series*

$$\sum_{k=1}^{\infty}u_k = u_1 + u_2 + u_3 + \cdots$$

$$\sum_{k=K}^{\infty}u_k = u_K + u_{K+1} + u_{K+2} + \cdots$$

both converge or both diverge.

REMARK.   Do not read too much into part (c) of this theorem. Although the convergence is not affected when a finite number of terms is deleted from the beginning of a convergent series, the *sum* of a convergent series is changed by the removal of these terms.

Example 2   Find the sum of the series

$$\sum_{k=1}^{\infty}\left(\frac{3}{4^k} - \frac{2}{5^{k-1}}\right)$$

Solution.   The series

$$\sum_{k=1}^{\infty}\frac{3}{4^k} = \frac{3}{4} + \frac{3}{4^2} + \frac{3}{4^3} + \cdots$$

is a convergent geometric series $\left(a = \frac{3}{4}, r = \frac{1}{4}\right)$, and the series

$$\sum_{k=1}^{\infty}\frac{2}{5^{k-1}} = 2 + \frac{2}{5} + \frac{2}{5^2} + \frac{2}{5^3} + \cdots$$

is also a convergent geometric series $\left(a = 2, r = \frac{1}{5}\right)$. Thus, from Theorems 10.5.3(a) and 10.4.3 the given series converges and

$$\sum_{k=1}^{\infty}\left(\frac{3}{4^k} - \frac{2}{5^{k-1}}\right) = \sum_{k=1}^{\infty}\frac{3}{4^k} - \sum_{k=1}^{\infty}\frac{2}{5^{k-1}} = \frac{\frac{3}{4}}{1 - \frac{1}{4}} - \frac{2}{1 - \frac{1}{5}} = -\frac{3}{2} \quad \blacktriangleleft$$

Example 3   Determine whether the following series converge or diverge.

(a) $\displaystyle\sum_{k=1}^{\infty}\frac{5}{k} = 5 + \frac{5}{2} + \frac{5}{3} + \cdots + \frac{5}{k} + \cdots$ (b) $\displaystyle\sum_{k=10}^{\infty}\frac{1}{k} = \frac{1}{10} + \frac{1}{11} + \frac{1}{12} + \cdots$

Solution.   The first series is a constant times the divergent harmonic series, and hence diverges by part (b) of Theorem 10.5.3. The second series results by deleting the first nine terms from the divergent harmonic series, and hence diverges by part (*c*) of Theorem 10.5.3. ◄

THE INTEGRAL TEST

The expressions

$$\sum_{k=1}^{\infty} \frac{1}{k^2} \quad \text{and} \quad \int_{1}^{+\infty} \frac{1}{x^2}\,dx$$

are related in that the integrand in the improper integral results when the index *k* in the general term of the series is replaced by *x* and the limits of summation in the series are replaced by the corresponding limits of integration. The following theorem shows that there is a relationship between the convergence of the series and the integral.

10.5.4 THEOREM (*The Integral Test*).   *Let* $\sum u_k$ *be a series with positive terms, and let* $f(x)$ *be the function that results when k is replaced by x in the general term of the series. If f is decreasing and continuous on the interval* $[a, +\infty)$*, then*

$$\sum_{k=1}^{\infty} u_k \quad \text{and} \quad \int_{a}^{+\infty} f(x)\,dx$$

both converge or both diverge.

Example 4   Use the integral test to determine whether the following series converge or diverge.

(a) $\displaystyle\sum_{k=1}^{\infty} \frac{1}{k}$ (b) $\displaystyle\sum_{k=1}^{\infty} \frac{1}{k^2}$

Solution (a). We already know that this is the divergent harmonic series, so the integral test will simply provide another way of establishing the divergence. If we replace *k* by *x* in the general term $1/k$, we obtain the function $f(x) = 1/x$, which is decreasing and continuous for $x \geq 1$ (as required to apply the integral test with $a = 1$). Since

$$\int_{1}^{+\infty} \frac{1}{x}\,dx = \lim_{\ell \to +\infty} \int_{1}^{\ell} \frac{1}{x}\,dx = \lim_{\ell \to +\infty} [\ln \ell - \ln 1] = +\infty$$

the integral diverges and consequently so does the series.

Solution (b). If we replace *k* by *x* in the general term $1/k^2$, we obtain the function $f(x) = 1/x^2$, which is decreasing and continuous for $x \geq 1$. Since

$$\int_{1}^{+\infty} \frac{1}{x^2}\,dx = \lim_{\ell \to +\infty} \int_{1}^{\ell} \frac{dx}{x^2} = \lim_{\ell \to +\infty} \left[-\frac{1}{x}\right]_{1}^{\ell} = \lim_{\ell \to +\infty} \left[1 - \frac{1}{\ell}\right] = 1$$

the integral converges and consequently the series converges by the integral test with $a = 1$. ◄

REMARK.   In part (b) of the last example, do *not* erroneously conclude that the sum of the series is 1 because the value of the corresponding integral is 1. It can be proved that the sum of the series is actually $\pi^2/6$ and, indeed, the sum of the first two terms alone exceeds 1.

p-**SERIES**

The series in Example 4 are special cases of a class of series called ***p-series*** or ***hyperharmonic series***. A *p*-series is an infinite series of the form

$$\sum_{k=1}^{\infty} \frac{1}{k^p} = 1 + \frac{1}{2^p} + \frac{1}{3^p} + \cdots + \frac{1}{k^p} + \cdots$$

where $p > 0$. Examples of p-series are

$$\sum_{k=1}^{\infty} \frac{1}{k} = 1 + \frac{1}{2} + \frac{1}{3} + \cdots + \frac{1}{k} + \cdots \qquad \boxed{p = 1}$$

$$\sum_{k=1}^{\infty} \frac{1}{k^2} = 1 + \frac{1}{2^2} + \frac{1}{3^2} + \cdots + \frac{1}{k^2} + \cdots \qquad \boxed{p = 2}$$

$$\sum_{k=1}^{\infty} \frac{1}{\sqrt{k}} = 1 + \frac{1}{\sqrt{2}} + \frac{1}{\sqrt{3}} + \cdots + \frac{1}{\sqrt{k}} + \cdots \qquad \boxed{p = \tfrac{1}{2}}$$

The following theorem tells when a p-series converges.

10.5.5   THEOREM (*Convergence of p-Series*).

$$\sum_{k=1}^{\infty} \frac{1}{k^p} = 1 + \frac{1}{2^p} + \frac{1}{3^p} + \cdots + \frac{1}{k^p} + \cdots$$

converges if $p > 1$ and diverges if $0 < p \leq 1$.

Proof.   To establish this result when $p \neq 1$, we will use the integral test.

$$\int_1^{+\infty} \frac{1}{x^p}\, dx = \lim_{\ell \to +\infty} \int_1^\ell x^{-p}\, dx = \lim_{\ell \to +\infty} \frac{x^{1-p}}{1-p}\Bigg]_1^\ell = \lim_{\ell \to +\infty} \left[\frac{\ell^{1-p}}{1-p} - \frac{1}{1-p} \right]$$

If $p > 1$, then $1 - p < 0$, so $\ell^{1-p} \to 0$ as $\ell \to +\infty$. Thus, the integral converges [its value is $-1/(1-p)$] and consequently the series also converges. For $0 < p < 1$, it follows that $1 - p > 0$ and $\ell^{1-p} \to +\infty$ as $\ell \to +\infty$, so the integral and the series diverge. The case $p = 1$ is the harmonic series, which was previously shown to diverge. ∎

Example 5

$$1 + \frac{1}{\sqrt[3]{2}} + \frac{1}{\sqrt[3]{3}} + \cdots + \frac{1}{\sqrt[3]{k}} + \cdots$$

diverges since it is a p-series with $p = \frac{1}{3} < 1$.   ◀

PROOF OF THE INTEGRAL TEST

Before we can prove the integral test, we need a basic result about convergence of series with *nonnegative* terms. If $u_1 + u_2 + u_3 + \cdots + u_k + \cdots$ is such a series, then its sequence of partial sums is increasing, that is,

$$s_1 \leq s_2 \leq s_3 \leq \cdots \leq s_n \leq \cdots$$

Thus, from Theorem 10.3.3 the sequence of partial sums converges to a limit S if and only if it has some upper bound M, in which case $S \leq M$. If no upper bound exists, then the sequence of partial sums diverges. Since convergence of the sequence of partial sums corresponds to convergence of the series, we have the following theorem.

10.5.6   THEOREM.   *If $\sum u_k$ is a series with nonnegative terms, and if there is a constant M such that*

$$s_n = u_1 + u_2 + \cdots + u_n \leq M$$

for every n, then the series converges and the sum S satisfies $S \leq M$. If no such M exists, then the series diverges.

In words, this theorem implies that *a series with nonnegative terms converges if and only if its sequence of partial sums is bounded above.*

Proof of Theorem 10.5.4. We need only show that the series converges when the integral converges and that the series diverges when the integral diverges. For simplicity, we will limit the proof to the case where $a = 1$. Assume that $f(x)$ satisfies the hypotheses of the theorem for $x \geq 1$. Since

$$f(1) = u_1, \ f(2) = u_2, \ \ldots, \ f(n) = u_n, \ \ldots$$

the values of $u_1, u_2, \ldots, u_n, \ldots$ can be interpreted as the areas of the rectangles shown in Figure 10.5.1.

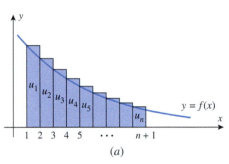

$y = f(x)$

(a)

$y = f(x)$

(b)

Figure 10.5.1

The following inequalities result by comparing the areas under the curve $y = f(x)$ to the areas of the rectangles in Figure 10.5.1 for $n > 1$:

$$\int_1^{n+1} f(x)\,dx < u_1 + u_2 + \cdots + u_n = s_n \qquad \text{Figure 10.5.1}a$$

$$s_n - u_1 = u_2 + u_3 + \cdots + u_n < \int_1^n f(x)\,dx \qquad \text{Figure 10.5.1}b$$

These inequalities can be combined as

$$\int_1^{n+1} f(x)\,dx < s_n < u_1 + \int_1^n f(x)\,dx \qquad (2)$$

If the integral $\int_1^\infty f(x)\,dx$ converges to a finite value L, then from the right-hand inequality in (2)

$$s_n < u_1 + \int_1^n f(x)\,dx < u_1 + \int_1^\infty f(x)\,dx = u_1 + L$$

Thus, each partial sum is less than the finite constant $u_1 + L$, and the series converges by Theorem 10.5.6. On the other hand, if the integral $\int_1^\infty f(x)\,dx$ diverges, then

$$\lim_{n \to +\infty} \int_1^{n+1} f(x)\,dx = +\infty$$

so that from the left-hand inequality in (2), $\lim_{n \to +\infty} s_n = +\infty$. This implies that the series also diverges. ∎

EXERCISE SET 10.5 ~ Graphing Utility C CAS

1. In each part, use Theorem 10.5.3 to find the sum of the series.

(a) $\left(\dfrac{1}{2} + \dfrac{1}{4}\right) + \left(\dfrac{1}{2^2} + \dfrac{1}{4^2}\right) + \cdots + \left(\dfrac{1}{2^k} + \dfrac{1}{4^k}\right) + \cdots$

(b) $\displaystyle\sum_{k=1}^{\infty} \left(\dfrac{1}{5^k} - \dfrac{1}{k(k+1)}\right)$

2. In each part, use Theorem 10.5.3 to find the sum of the series.

(a) $\displaystyle\sum_{k=2}^{\infty} \left[\dfrac{1}{k^2 - 1} - \dfrac{7}{10^{k-1}}\right]$ (b) $\displaystyle\sum_{k=1}^{\infty} \left[7^{-k}3^{k+1} - \dfrac{2^{k+1}}{5^k}\right]$

In Exercises 3 and 4, various p-series are given. In each case, find p and determine whether the series converges.

3. (a) $\displaystyle\sum_{k=1}^{\infty} \frac{1}{k^3}$ (b) $\displaystyle\sum_{k=1}^{\infty} \frac{1}{\sqrt{k}}$ (c) $\displaystyle\sum_{k=1}^{\infty} k^{-1}$ (d) $\displaystyle\sum_{k=1}^{\infty} k^{-2/3}$

4. (a) $\displaystyle\sum_{k=1}^{\infty} k^{-4/3}$ (b) $\displaystyle\sum_{k=1}^{\infty} \frac{1}{\sqrt[4]{k}}$ (c) $\displaystyle\sum_{k=1}^{\infty} \frac{1}{\sqrt[3]{k^5}}$ (d) $\displaystyle\sum_{k=1}^{\infty} \frac{1}{k^\pi}$

> In Exercises 5 and 6, apply the divergence test, and state what it tells you about the series.

5. (a) $\displaystyle\sum_{k=1}^{\infty} \frac{k^2 + k + 3}{2k^2 + 1}$ (b) $\displaystyle\sum_{k=1}^{\infty} \left(1 + \frac{1}{k}\right)^k$

(c) $\displaystyle\sum_{k=1}^{\infty} \cos k\pi$ (d) $\displaystyle\sum_{k=1}^{\infty} \frac{1}{k!}$

6. (a) $\displaystyle\sum_{k=1}^{\infty} \frac{k}{e^k}$ (b) $\displaystyle\sum_{k=1}^{\infty} \ln k$ (c) $\displaystyle\sum_{k=1}^{\infty} \frac{1}{\sqrt{k}}$ (d) $\displaystyle\sum_{k=1}^{\infty} \frac{\sqrt{k}}{\sqrt{k} + 3}$

> In Exercises 7 and 8, confirm that the integral test is applicable, and use it to determine whether the series converges.

7. (a) $\displaystyle\sum_{k=1}^{\infty} \frac{1}{5k + 2}$ (b) $\displaystyle\sum_{k=1}^{\infty} \frac{1}{1 + 9k^2}$

8. (a) $\displaystyle\sum_{k=1}^{\infty} \frac{k}{1 + k^2}$ (b) $\displaystyle\sum_{k=1}^{\infty} \frac{1}{(4 + 2k)^{3/2}}$

> In Exercises 9–24, use any method to determine whether the series converges.

9. $\displaystyle\sum_{k=1}^{\infty} \frac{1}{k + 6}$ **10.** $\displaystyle\sum_{k=1}^{\infty} \frac{3}{5k}$ **11.** $\displaystyle\sum_{k=1}^{\infty} \frac{1}{\sqrt{k + 5}}$

12. $\displaystyle\sum_{k=1}^{\infty} \frac{1}{\sqrt[k]{e}}$ **13.** $\displaystyle\sum_{k=1}^{\infty} \frac{1}{\sqrt[3]{2k - 1}}$ **14.** $\displaystyle\sum_{k=3}^{\infty} \frac{\ln k}{k}$

15. $\displaystyle\sum_{k=1}^{\infty} \frac{k}{\ln(k + 1)}$ **16.** $\displaystyle\sum_{k=1}^{\infty} ke^{-k^2}$ **17.** $\displaystyle\sum_{k=1}^{\infty} \left(1 + \frac{1}{k}\right)^{-k}$

18. $\displaystyle\sum_{k=1}^{\infty} \frac{k^2 + 1}{k^2 + 3}$ **19.** $\displaystyle\sum_{k=1}^{\infty} \frac{\tan^{-1} k}{1 + k^2}$ **20.** $\displaystyle\sum_{k=1}^{\infty} \frac{1}{\sqrt{k^2 + 1}}$

21. $\displaystyle\sum_{k=1}^{\infty} k^2 \sin^2\left(\frac{1}{k}\right)$ **22.** $\displaystyle\sum_{k=1}^{\infty} k^2 e^{-k^3}$

23. $\displaystyle\sum_{k=5}^{\infty} 7k^{-1.01}$ **24.** $\displaystyle\sum_{k=1}^{\infty} \text{sech}^2 k$

> In Exercises 25 and 26, use the integral test to investigate the relationship between the value of p and the convergence of the series.

25. $\displaystyle\sum_{k=2}^{\infty} \frac{1}{k(\ln k)^p}$ **26.** $\displaystyle\sum_{k=3}^{\infty} \frac{1}{k(\ln k)[\ln(\ln k)]^p}$

c 27. Use a CAS to confirm that
$$\sum_{k=1}^{\infty} \frac{1}{k^2} = \frac{\pi^2}{6} \quad \text{and} \quad \sum_{k=1}^{\infty} \frac{1}{k^4} = \frac{\pi^4}{90}$$
and then use these results in each part to find the sum of the series.

(a) $\displaystyle\sum_{k=1}^{\infty} \frac{3k^2 - 1}{k^4}$ (b) $\displaystyle\sum_{k=3}^{\infty} \frac{1}{k^2}$ (c) $\displaystyle\sum_{k=2}^{\infty} \frac{1}{(k-1)^4}$

28. Suppose that the series $\sum u_k$ converges and the series $\sum v_k$ diverges.

(a) Show that the series $\sum (u_k + v_k)$ and $\sum (u_k - v_k)$ both diverge. [*Hint:* Assume that each series converges and use Theorem 10.5.3 to obtain a contradiction.]

(b) Find examples to show that if $\sum u_k$ and $\sum v_k$ both diverge, then the series $\sum (u_k + v_k)$ and $\sum (u_k - v_k)$ may either converge or diverge.

29. In each part, use the results in Exercise 28, if needed, to determine whether the series diverges.

(a) $\displaystyle\sum_{k=1}^{\infty} \left[\left(\frac{2}{3}\right)^{k-1} + \frac{1}{k}\right]$ (b) $\displaystyle\sum_{k=1}^{\infty} \left[\frac{1}{3k + 2} - \frac{1}{k^{3/2}}\right]$

(c) $\displaystyle\sum_{k=2}^{\infty} \left[\frac{1}{k(\ln k)^2} - \frac{1}{k^2}\right]$

> Exercise 30 will show how a partial sum can be used to obtain upper and lower bounds on the sum of the series when the hypotheses of the integral test are satisfied. This result will be needed in Exercises 31–35.

30. (a) Let $\sum_{k=1}^{\infty} u_k$ be a convergent series with positive terms, let $f(x)$ be the function that results when k is replaced by x in the general term of the series, and suppose that f satisfies the hypotheses of the integral test for $x \geq n$ (Theorem 10.5.4). Use an area argument and the accompanying figure (see p. 679) to show that
$$\int_{n+1}^{+\infty} f(x)\, dx < \sum_{k=n+1}^{\infty} u_k < \int_n^{+\infty} f(x)\, dx$$

(b) Show that if S is the sum of the series $\sum_{k=1}^{\infty} u_k$ and s_n is the nth partial sum, then
$$s_n + \int_{n+1}^{+\infty} f(x)\, dx < S < s_n + \int_n^{+\infty} f(x)\, dx$$

31. (a) It was stated in Exercise 27 that
$$\sum_{k=1}^{\infty} \frac{1}{k^2} = \frac{\pi^2}{6}$$

Show that if s_n is the nth partial sum of this series, then
$$s_n + \frac{1}{n + 1} < \frac{\pi^2}{6} < s_n + \frac{1}{n}$$

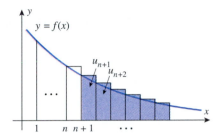

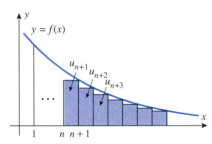

Figure Ex-30

(b) Calculate s_3 exactly, and then use the result in part (a) to show that
$$\frac{29}{18} < \frac{\pi^2}{6} < \frac{61}{36}$$

(c) Use a calculating utility to confirm that the inequalities in part (b) are correct.

(d) Find upper and lower bounds on the error that results if the sum of the series is approximated by the 10th partial sum.

32. In each part, find upper and lower bounds on the error that results if the sum of the series is approximated by the 10th partial sum.

(a) $\displaystyle\sum_{k=1}^{\infty} \frac{1}{(2k+1)^2}$ (b) $\displaystyle\sum_{k=1}^{\infty} \frac{1}{k^2+1}$ (c) $\displaystyle\sum_{k=1}^{\infty} \frac{k}{e^k}$

33. Our objective in this problem is to approximate the sum of the series $\sum_{k=1}^{\infty} 1/k^3$ to two decimal-place accuracy.

(a) Show that if S is the sum of the series and s_n is the nth partial sum, then
$$s_n + \frac{1}{2(n+1)^2} < S < s_n + \frac{1}{2n^2}$$

(b) For two decimal-place accuracy, the error must be less than 0.005 (see Table 2.5.1 on p. 154). We can achieve this by finding an interval of length 0.01 (or less) that contains S and approximating S by the midpoint of that interval. Find the smallest value of n such that the interval containing S in part (a) has a length of 0.01 or less.

(c) Approximate S to two decimal-place accuracy.

34. (a) Use the method of Exercise 33 to approximate the sum of the series $\sum_{k=1}^{\infty} 1/k^4$ to two decimal-place accuracy.

(b) It was stated in Exercise 27 that the sum of this series is $\pi^4/90$. Use a calculating utility to confirm that your answer in part (a) is accurate to two decimal places.

35. We showed in Section 10.4 that the harmonic series $\sum_{k=1}^{\infty} 1/k$ diverges. Our objective in this problem is to demonstrate that although the partial sums of this series approach $+\infty$, they increase extremely slowly.

(a) Use inequality (2) to show that for $n \geq 2$
$$\ln(n+1) < s_n < 1 + \ln n$$

(b) Use the inequalities in part (a) to find upper and lower bounds on the sum of the first million terms in the series.

(c) Show that the sum of the first billion terms in the series is less than 22.

(d) Find a value of n so that the sum of the first n terms is greater than 100.

36. Investigate the relationship between the value of a and the convergence of the series $\sum_{k=1}^{\infty} k^{-\ln a}$.

37. Use a graphing utility to confirm that the integral test applies to the series $\sum_{k=1}^{\infty} k^2 e^{-k}$, and then determine whether the series converges.

38. (a) Show that the integral test applies to the series $\sum_{k=1}^{\infty} 1/(k^3 + 1)$.

(b) Use a CAS and the integral test to confirm that the series converges.

(c) Construct a table of partial sums for $n = 10, 20, 30, \ldots, 100$, showing at least six decimal places.

(d) Based on your table, make a conjecture about the sum of the series to three decimal-place accuracy.

(e) Use part (b) of Exercise 30 to check your conjecture.

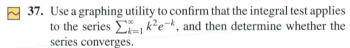

10.6 THE COMPARISON, RATIO, AND ROOT TESTS

In this section we will develop some more basic convergence tests for series with non-negative terms. Later, we will use some of these tests to study the convergence of Taylor series.

THE COMPARISON TEST

We will begin with a test that is useful in its own right and is also the building block for other important convergence tests. The underlying idea of this test is to use the known convergence or divergence of a series to deduce the convergence or divergence of another series.

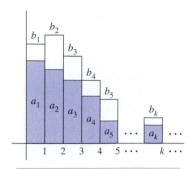

For each rectangle, b_k is the entire area and a_k is the area of the blue portion.

Figure 10.6.1

USING THE COMPARISON TEST

10.6.1   THEOREM (*The Comparison Test*).   *Let $\sum_{k=1}^{\infty} a_k$ and $\sum_{k=1}^{\infty} b_k$ be series with non-negative terms and suppose that*

$$a_1 \leq b_1, \ a_2 \leq b_2, \ a_3 \leq b_3, \ldots, a_k \leq b_k, \ldots$$

(a)   *If the "bigger series" Σb_k converges, then the "smaller series" Σa_k also converges.*

(b)   *If the "smaller series" Σa_k diverges, then the "bigger series" Σb_k also diverges.*

We have left the proof of this theorem for the exercises; however, it is easy to visualize why the theorem is true by interpreting the terms in the series as areas of rectangles (Figure 10.6.1). The comparison test states that if the total area $\sum b_k$ is finite, then the total area $\sum a_k$ must also be finite; and if the total area $\sum a_k$ is infinite, then the total area $\sum b_k$ must also be infinite.

REMARK.   As one would expect, it is not essential in Theorem 10.6.1 that the condition $a_k \leq b_k$ hold for all k, as stated; the conclusions of the theorem remain true if this condition is eventually true.

There are two steps required for using the comparison test to determine whether a series $\sum u_k$ with positive terms converges:

- Guess at whether the series $\sum u_k$ converges or diverges.
- Find a series that proves the guess to be correct. That is, if the guess is divergence, we must find a divergent series whose terms are "smaller" than the corresponding terms of $\sum u_k$, and if the guess is convergence, we must find a convergent series whose terms are "bigger" than the corresponding terms of $\sum u_k$.

In most cases, the series $\sum u_k$ being considered will have its general term u_k expressed as a fraction. To help with the guessing process in the first step, we have formulated two principles that are based on the form of the denominator for u_k. These principles sometimes *suggest* whether a series is likely to converge or diverge. We have called these "informal principles" because they are not intended as formal theorems. In fact, we will not guarantee that they *always* work. However, they work often enough to be useful.

10.6.2   INFORMAL PRINCIPLE.   *Constant summands in the denominator of u_k can usually be deleted without affecting the convergence or divergence of the series.*

10.6.3   INFORMAL PRINCIPLE.   *If a polynomial in k appears as a factor in the numerator or denominator of u_k, all but the leading term in the polynomial can usually be discarded without affecting the convergence or divergence of the series.*

Example 1   Use the comparison test to determine whether the following series converge or diverge.

(a) $\displaystyle\sum_{k=1}^{\infty} \frac{1}{\sqrt{k} - \frac{1}{2}}$ (b) $\displaystyle\sum_{k=1}^{\infty} \frac{1}{2k^2 + k}$

Solution (a).   According to Principle 10.6.2, we should be able to drop the constant in the denominator without affecting the convergence or divergence. Thus, the given series is likely to behave like

$$\sum_{k=1}^{\infty} \frac{1}{\sqrt{k}} \tag{1}$$

which is a divergent p-series $\left(p = \frac{1}{2}\right)$. Thus, we will guess that the given series diverges and try to prove this by finding a divergent series that is "smaller" than the given series. However, series (1) does the trick since

$$\frac{1}{\sqrt{k} - \frac{1}{2}} > \frac{1}{\sqrt{k}} \quad \text{for } k = 1, 2, \dots$$

Thus, we have proved that the given series diverges.

Solution (b). According to Principle 10.6.3, we should be able to discard all but the leading term in the polynomial without affecting the convergence or divergence. Thus, the given series is likely to behave like

$$\sum_{k=1}^{\infty} \frac{1}{2k^2} = \frac{1}{2} \sum_{k=1}^{\infty} \frac{1}{k^2} \tag{2}$$

which converges since it is a constant times a convergent p-series $(p = 2)$. Thus, we will guess that the given series converges and try to prove this by finding a convergent series that is "bigger" than the given series. However, series (2) does the trick since

$$\frac{1}{2k^2 + k} < \frac{1}{2k^2} \quad \text{for } k = 1, 2, \dots$$

Thus, we have proved that the given series converges.   ◀

THE LIMIT COMPARISON TEST

In the last example, Principles 10.6.2 and 10.6.3 provided the guess about convergence or divergence as well as the series needed to apply the comparison test. Unfortunately, it is not always so straightforward to find the series required for comparison, so we will now consider an alternative to the comparison test that is usually easier to apply. The proof is given in Appendix G.

10.6.4   THEOREM (*The Limit Comparison Test*).   *Let* $\sum a_k$ *and* $\sum b_k$ *be series with positive terms and suppose that*

$$\rho = \lim_{k \to +\infty} \frac{a_k}{b_k}$$

If ρ *is finite and* $\rho > 0$, *then the series both converge or both diverge.*

The cases where $\rho = 0$ or $\rho = +\infty$ are discussed in the exercises (Exercise 54).

Example 2   Use the limit comparison test to determine whether the following series converge or diverge.

(a) $\displaystyle\sum_{k=2}^{\infty} \frac{1}{\sqrt{k} - 1}$ (b) $\displaystyle\sum_{k=1}^{\infty} \frac{1}{2k^2 + k}$ (c) $\displaystyle\sum_{k=1}^{\infty} \frac{3k^3 - 2k^2 + 4}{k^7 - k^3 + 2}$

Solution (a). As in Example 1, Principle 10.6.2 suggests that the series is likely to behave like the divergent p-series (1). To prove that the given series diverges, we will apply the limit comparison test with

$$a_k = \frac{1}{\sqrt{k} - 1} \quad \text{and} \quad b_k = \frac{1}{\sqrt{k}}$$

We obtain

$$\rho = \lim_{k \to +\infty} \frac{a_k}{b_k} = \lim_{k \to +\infty} \frac{\sqrt{k}}{\sqrt{k} - 1} = \lim_{k \to +\infty} \frac{1}{1 - \dfrac{1}{\sqrt{k}}} = 1$$

Since ρ is finite and positive, it follows from Theorem 10.6.4 that the given series diverges.

Solution (b). As in Example 1, Principle 10.6.3 suggests that the series is likely to behave like the convergent series (2). To prove that the given series converges, we will apply the limit comparison test with

$$a_k = \frac{1}{2k^2 + k} \quad \text{and} \quad b_k = \frac{1}{2k^2}$$

We obtain

$$\rho = \lim_{k \to +\infty} \frac{a_k}{b_k} = \lim_{k \to +\infty} \frac{2k^2}{2k^2 + k} = \lim_{k \to +\infty} \frac{2}{2 + \dfrac{1}{k}} = 1$$

Since ρ is finite and positive, it follows from Theorem 10.6.4 that the given series converges, which agrees with the conclusion reached in Example 1 using the comparison test.

Solution (c). From Principle 10.6.3, the series is likely to behave like

$$\sum_{k=1}^{\infty} \frac{3k^3}{k^7} = \sum_{k=1}^{\infty} \frac{3}{k^4} \tag{3}$$

which converges since it is a constant times a convergent p-series. Thus, the given series is likely to converge. To prove this, we will apply the limit comparison test to series (3) and the given series. We obtain

$$\rho = \lim_{k \to +\infty} \frac{\dfrac{3k^3 - 2k^2 + 4}{k^7 - k^3 + 2}}{\dfrac{3}{k^4}} = \lim_{k \to +\infty} \frac{3k^7 - 2k^6 + 4k^4}{3k^7 - 3k^3 + 6} = 1$$

Since ρ is finite and nonzero, it follows from Theorem 10.6.4 that the given series converges, since (3) converges. ◄

THE RATIO TEST

The comparison test and the limit comparison test hinge on first making a guess about convergence and then finding an appropriate series for comparison, both of which can be difficult tasks in cases where Principles 10.6.2 and 10.6.3 cannot be applied. In such cases the next test can often be used, since it works exclusively with the terms of the given series—it requires neither an initial guess about convergence nor the discovery of a series for comparison. Its proof is given in Appendix G.

10.6.5   THEOREM (*The Ratio Test*).   *Let $\sum u_k$ be a series with positive terms and suppose that*

$$\rho = \lim_{k \to +\infty} \frac{u_{k+1}}{u_k}$$

(a) *If $\rho < 1$, the series converges.*

(b) *If $\rho > 1$ or $\rho = +\infty$, the series diverges.*

(c) *If $\rho = 1$, the series may converge or diverge, so that another test must be tried.*

Example 3   Use the ratio test to determine whether the following series converge or diverge.

$$\text{(a)} \sum_{k=1}^{\infty} \frac{1}{k!} \quad \text{(b)} \sum_{k=1}^{\infty} \frac{k}{2^k} \quad \text{(c)} \sum_{k=1}^{\infty} \frac{k^k}{k!} \quad \text{(d)} \sum_{k=3}^{\infty} \frac{(2k)!}{4^k} \quad \text{(e)} \sum_{k=1}^{\infty} \frac{1}{2k - 1}$$

Solution (a). The series converges, since

$$\rho = \lim_{k \to +\infty} \frac{u_{k+1}}{u_k} = \lim_{k \to +\infty} \frac{1/(k+1)!}{1/k!} = \lim_{k \to +\infty} \frac{k!}{(k+1)!} = \lim_{k \to +\infty} \frac{1}{k+1} = 0 < 1$$

Solution (b). The series converges, since

$$\rho = \lim_{k \to +\infty} \frac{u_{k+1}}{u_k} = \lim_{k \to +\infty} \frac{k+1}{2^{k+1}} \cdot \frac{2^k}{k} = \frac{1}{2} \lim_{k \to +\infty} \frac{k+1}{k} = \frac{1}{2} < 1$$

Solution (c). The series diverges, since

$$\rho = \lim_{k \to +\infty} \frac{u_{k+1}}{u_k} = \lim_{k \to +\infty} \frac{(k+1)^{k+1}}{(k+1)!} \cdot \frac{k!}{k^k} = \lim_{k \to +\infty} \frac{(k+1)^k}{k^k} = \lim_{k \to +\infty} \left(1 + \frac{1}{k}\right)^k = e > 1$$

See Theorem 6.9.6(*b*)

Solution (d). The series diverges, since

$$\rho = \lim_{k \to +\infty} \frac{u_{k+1}}{u_k} = \lim_{k \to +\infty} \frac{[2(k+1)]!}{4^{k+1}} \cdot \frac{4^k}{(2k)!} = \lim_{k \to +\infty} \left(\frac{(2k+2)!}{(2k)!} \cdot \frac{1}{4}\right)$$

$$= \frac{1}{4} \lim_{k \to +\infty} (2k+2)(2k+1) = +\infty$$

Solution (e). The ratio test is of no help since

$$\rho = \lim_{k \to +\infty} \frac{u_{k+1}}{u_k} = \lim_{k \to +\infty} \frac{1}{2(k+1) - 1} \cdot \frac{2k-1}{1} = \lim_{k \to +\infty} \frac{2k-1}{2k+1} = 1$$

However, the integral test proves that the series diverges since

$$\int_1^{+\infty} \frac{dx}{2x-1} = \lim_{\ell \to +\infty} \int_1^{\ell} \frac{dx}{2x-1} = \lim_{\ell \to +\infty} \frac{1}{2} \ln(2x-1) \Big]_1^{\ell} = +\infty$$

Both the comparison test and the limit comparison test would also have worked here (verify). ◀

THE ROOT TEST

In cases where it is difficult or inconvenient to find the limit required for the ratio test, the next test is sometimes useful. Since its proof is similar to the proof of the ratio test, we will omit it.

10.6.6 THEOREM (*The Root Test*). *Let $\sum u_k$ be a series with positive terms and suppose that*

$$\rho = \lim_{k \to +\infty} \sqrt[k]{u_k} = \lim_{k \to +\infty} (u_k)^{1/k}$$

(a) *If $\rho < 1$, the series converges.*
(b) *If $\rho > 1$ or $\rho = +\infty$, the series diverges.*
(c) *If $\rho = 1$, the series may converge or diverge, so that another test must be tried.*

Example 4 Use the root test to determine whether the following series converge or diverge.

(a) $\displaystyle\sum_{k=2}^{\infty} \left(\frac{4k-5}{2k+1}\right)^k$ (b) $\displaystyle\sum_{k=1}^{\infty} \frac{1}{(\ln(k+1))^k}$

Solution (a). The series diverges, since

$$\rho = \lim_{k \to +\infty} (u_k)^{1/k} = \lim_{k \to +\infty} \frac{4k-5}{2k+1} = 2 > 1$$

Solution (b). The series converges, since

$$\rho = \lim_{k \to +\infty} (u_k)^{1/k} = \lim_{k \to +\infty} \frac{1}{\ln(k+1)} = 0 < 1$$

◀

EXERCISE SET 10.6　🆒 CAS

In Exercises 1 and 2, make a guess about the convergence or divergence of the series, and confirm your guess using the comparison test.

1. (a) $\displaystyle\sum_{k=1}^{\infty} \frac{1}{5k^2 - k}$　　(b) $\displaystyle\sum_{k=1}^{\infty} \frac{3}{k - \frac{1}{4}}$

2. (a) $\displaystyle\sum_{k=2}^{\infty} \frac{k+1}{k^2 - k}$　　(b) $\displaystyle\sum_{k=1}^{\infty} \frac{2}{k^4 + k}$

3. In each part, use the comparison test to show that the series converges.

(a) $\displaystyle\sum_{k=1}^{\infty} \frac{1}{3^k + 5}$　　(b) $\displaystyle\sum_{k=1}^{\infty} \frac{5 \sin^2 k}{k!}$

4. In each part, use the comparison test to show that the series diverges.

(a) $\displaystyle\sum_{k=1}^{\infty} \frac{\ln k}{k}$　　(b) $\displaystyle\sum_{k=1}^{\infty} \frac{k}{k^{3/2} - \frac{1}{2}}$

In Exercises 5–10, use the limit comparison test to determine whether the series converges.

5. $\displaystyle\sum_{k=1}^{\infty} \frac{4k^2 - 2k + 6}{8k^7 + k - 8}$　　**6.** $\displaystyle\sum_{k=1}^{\infty} \frac{1}{9k + 6}$

7. $\displaystyle\sum_{k=1}^{\infty} \frac{5}{3^k + 1}$　　**8.** $\displaystyle\sum_{k=1}^{\infty} \frac{k(k+3)}{(k+1)(k+2)(k+5)}$

9. $\displaystyle\sum_{k=1}^{\infty} \frac{1}{\sqrt[3]{8k^2 - 3k}}$　　**10.** $\displaystyle\sum_{k=1}^{\infty} \frac{1}{(2k+3)^{17}}$

In Exercises 11–16, use the ratio test to determine whether the series converges. If the test is inconclusive, then say so.

11. $\displaystyle\sum_{k=1}^{\infty} \frac{3^k}{k!}$　　**12.** $\displaystyle\sum_{k=1}^{\infty} \frac{4^k}{k^2}$　　**13.** $\displaystyle\sum_{k=1}^{\infty} \frac{1}{5k}$

14. $\displaystyle\sum_{k=1}^{\infty} k \left(\frac{1}{2} \right)^k$　　**15.** $\displaystyle\sum_{k=1}^{\infty} \frac{k!}{k^3}$　　**16.** $\displaystyle\sum_{k=1}^{\infty} \frac{k}{k^2 + 1}$

In Exercises 17–20, use the root test to determine whether the series converges. If the test is inconclusive, then say so.

17. $\displaystyle\sum_{k=1}^{\infty} \left(\frac{3k+2}{2k-1} \right)^k$　　**18.** $\displaystyle\sum_{k=1}^{\infty} \left(\frac{k}{100} \right)^k$

19. $\displaystyle\sum_{k=1}^{\infty} \frac{k}{5^k}$　　**20.** $\displaystyle\sum_{k=1}^{\infty} (1 - e^{-k})^k$

In Exercises 21–44, use any method to determine whether the series converges.

21. $\displaystyle\sum_{k=0}^{\infty} \frac{7^k}{k!}$　　**22.** $\displaystyle\sum_{k=1}^{\infty} \frac{1}{2k+1}$　　**23.** $\displaystyle\sum_{k=1}^{\infty} \frac{k^2}{5^k}$

24. $\displaystyle\sum_{k=1}^{\infty} \frac{k! 10^k}{3^k}$　　**25.** $\displaystyle\sum_{k=1}^{\infty} k^{50} e^{-k}$　　**26.** $\displaystyle\sum_{k=1}^{\infty} \frac{k^2}{k^3 + 1}$

27. $\displaystyle\sum_{k=1}^{\infty} \frac{\sqrt{k}}{k^3 + 1}$　　**28.** $\displaystyle\sum_{k=1}^{\infty} \frac{4}{2 + 3^k k}$

29. $\displaystyle\sum_{k=1}^{\infty} \frac{1}{\sqrt{k(k+1)}}$　　**30.** $\displaystyle\sum_{k=1}^{\infty} \frac{2 + (-1)^k}{5^k}$

31. $\displaystyle\sum_{k=1}^{\infty} \frac{2 + \sqrt{k}}{(k+1)^3 - 1}$　　**32.** $\displaystyle\sum_{k=1}^{\infty} \frac{4 + |\cos k|}{k^3}$

33. $\displaystyle\sum_{k=1}^{\infty} \frac{1}{1 + \sqrt{k}}$　　**34.** $\displaystyle\sum_{k=1}^{\infty} \frac{k!}{k^k}$　　**35.** $\displaystyle\sum_{k=1}^{\infty} \frac{\ln k}{e^k}$

36. $\displaystyle\sum_{k=1}^{\infty} \frac{k!}{e^{k^2}}$　　**37.** $\displaystyle\sum_{k=0}^{\infty} \frac{(k+4)!}{4! k! 4^k}$　　**38.** $\displaystyle\sum_{k=1}^{\infty} \left(\frac{k}{k+1} \right)^{k^2}$

39. $\displaystyle\sum_{k=1}^{\infty} \frac{1}{4 + 2^{-k}}$　　**40.** $\displaystyle\sum_{k=1}^{\infty} \frac{\sqrt{k} \ln k}{k^3 + 1}$　　**41.** $\displaystyle\sum_{k=1}^{\infty} \frac{\tan^{-1} k}{k^2}$

42. $\displaystyle\sum_{k=1}^{\infty} \frac{5^k + k}{k! + 3}$　　**43.** $\displaystyle\sum_{k=0}^{\infty} \frac{(k!)^2}{(2k)!}$　　**44.** $\displaystyle\sum_{k=1}^{\infty} \frac{(k!)^2 2^k}{(2k+2)!}$

In Exercises 45 and 46, find the general term of the series, and use the ratio test to show that the series converges.

45. $1 + \dfrac{1 \cdot 2}{1 \cdot 3} + \dfrac{1 \cdot 2 \cdot 3}{1 \cdot 3 \cdot 5} + \dfrac{1 \cdot 2 \cdot 3 \cdot 4}{1 \cdot 3 \cdot 5 \cdot 7} + \cdots$

46. $1 + \dfrac{1 \cdot 3}{3!} + \dfrac{1 \cdot 3 \cdot 5}{5!} + \dfrac{1 \cdot 3 \cdot 5 \cdot 7}{7!} + \cdots$

In Exercises 47 and 48, use a CAS to investigate the convergence of the series.

🆒 **47.** $\displaystyle\sum_{k=1}^{\infty} \frac{\ln k}{3^k}$　　🆒 **48.** $\displaystyle\sum_{k=1}^{\infty} \frac{[\pi(k+1)]^k}{k^{k+1}}$

49. (a) Make a conjecture about the convergence of the series $\sum_{k=1}^{\infty} \sin(\pi/k)$ by considering the local linear approximation of $\sin x$ near $x = 0$.

(b) Try to confirm your conjecture using the limit comparison test.

50. (a) Make a conjecture about the convergence of the series

$$\sum_{k=1}^{\infty} \left[1 - \cos \left(\frac{1}{k} \right) \right]$$

by considering the local quadratic approximation of $\cos x$ near $x = 0$.

(b) Try to confirm your conjecture using the limit comparison test.

51. Show that $\ln x < \sqrt{x}$ if $x > 0$, and use this result to investigate the convergence of

(a) $\displaystyle\sum_{k=1}^{\infty} \frac{\ln k}{k^2}$　　(b) $\displaystyle\sum_{k=2}^{\infty} \frac{1}{(\ln k)^2}$

52. For which positive values of α does the series $\sum_{k=1}^{\infty}(\alpha^k/k^\alpha)$ converge?

53. Use Theorem 10.5.6 to prove the comparison test (Theorem 10.6.1).

54. Let $\sum a_k$ and $\sum b_k$ be series with positive terms. Prove:

(a) If $\lim_{k \to +\infty}(a_k/b_k) = 0$ and $\sum b_k$ converges, then $\sum a_k$ converges.

(b) If $\lim_{k \to +\infty}(a_k/b_k) = +\infty$ and $\sum b_k$ diverges, then $\sum a_k$ diverges.

10.7 ALTERNATING SERIES; CONDITIONAL CONVERGENCE

Up to now we have focused exclusively on series with nonnegative terms. In this section we will discuss series that contain both positive and negative terms.

ALTERNATING SERIES

Series whose terms alternate between positive and negative, called ***alternating series***, are of special importance. Some examples are

$$\sum_{k=1}^{\infty}(-1)^{k+1}\frac{1}{k} = 1 - \frac{1}{2} + \frac{1}{3} - \frac{1}{4} + \frac{1}{5} - \cdots$$

$$\sum_{k=1}^{\infty}(-1)^{k}\frac{1}{k} = -1 + \frac{1}{2} - \frac{1}{3} + \frac{1}{4} - \frac{1}{5} + \cdots$$

In general, an alternating series has one of the following two forms:

$$\sum_{k=1}^{\infty}(-1)^{k+1}a_k = a_1 - a_2 + a_3 - a_4 + \cdots \tag{1}$$

$$\sum_{k=1}^{\infty}(-1)^{k}a_k = -a_1 + a_2 - a_3 + a_4 - \cdots \tag{2}$$

where the a_k's are assumed to be positive in both cases.

The following theorem is the key result on convergence of alternating series.

> **10.7.1   THEOREM** (***Alternating Series Test***).   *An alternating series of either form* (1) *or form* (2) *converges if the following two conditions are satisfied:*
>
> (*a*) $a_1 \geq a_2 \geq a_3 \geq \cdots \geq a_k \geq \cdots$
>
> (*b*) $\displaystyle\lim_{k \to +\infty} a_k = 0$

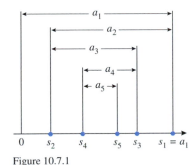

Figure 10.7.1

Proof. We will consider only alternating series of form (1). The idea of the proof is to show that if conditions (*a*) and (*b*) hold, then the sequences of even-numbered and odd-numbered partial sums converge to a common limit S. It will then follow from Theorem 10.2.4 that the entire sequence of partial sums converges to S.

Figure 10.7.1 shows how successive partial sums satisfying conditions (*a*) and (*b*) appear when plotted on a horizontal axis. The even-numbered partial sums

$$s_2, s_4, s_6, s_8, \ldots, s_{2n}, \ldots$$

form an increasing sequence bounded above by a_1, and the odd-numbered partial sums

$$s_1, s_3, s_5, \ldots, s_{2n-1}, \ldots$$

form a decreasing sequence bounded below by 0. Thus, by Theorems 10.3.3 and 10.3.4, the even-numbered partial sums converge to some limit S_E and the odd-numbered partial sums converge to some limit S_O. To complete the proof we must show that $S_E = S_O$. But the $(2n)$-th term in the series is $-a_{2n}$, so that $s_{2n} - s_{2n-1} = -a_{2n}$, which can be written as

$$s_{2n-1} = s_{2n} + a_{2n}$$

However, $2n \to +\infty$ and $2n - 1 \to +\infty$ as $n \to +\infty$, so that

$$S_O = \lim_{n \to +\infty} s_{2n-1} = \lim_{n \to +\infty} (s_{2n} + a_{2n}) = S_E + 0 = S_E$$

which completes the proof. ■

> **REMARK.** As might be expected, it is not essential for condition (*a*) in the alternating series test to hold for all terms; an alternating series will converge if condition (*b*) is true and condition (*a*) holds eventually.

Example 1 Use the alternating series test to show that the following series converge.

$$\text{(a)} \sum_{k=1}^{\infty} (-1)^{k+1} \frac{1}{k} \qquad \text{(b)} \sum_{k=1}^{\infty} (-1)^{k+1} \frac{k+3}{k(k+1)}$$

Solution (a). The two conditions in the alternating series test are satisfied since

$$a_k = \frac{1}{k} > \frac{1}{k+1} = a_{k+1} \quad \text{and} \quad \lim_{k \to +\infty} a_k = \lim_{k \to +\infty} \frac{1}{k} = 0$$

Solution (b). The two conditions in the alternating series test are satisfied since

$$\frac{a_{k+1}}{a_k} = \frac{k+4}{(k+1)(k+2)} \cdot \frac{k(k+1)}{k+3} = \frac{k^2 + 4k}{k^2 + 5k + 6} = \frac{k^2 + 4k}{(k^2 + 4k) + (k+6)} < 1$$

so

$$a_k > a_{k+1}$$

and

$$\lim_{k \to +\infty} a_k = \lim_{k \to +\infty} \frac{k+3}{k(k+1)} = \lim_{k \to +\infty} \frac{\dfrac{1}{k} + \dfrac{3}{k^2}}{1 + \dfrac{1}{k}} = 0 \qquad ◀$$

> **REMARK.** The series in part (*a*) of the last example is called the ***alternating harmonic series***. Observe that this series converges, whereas the harmonic series diverges.

> **REMARK.** If an alternating series violates condition (*b*) of the alternating series test, then the series must diverge by the divergence test (Theorem 10.5.1). However, if condition (*b*) is satisfied, but condition (*a*) is not, the series can either converge or diverge.[*]

APPROXIMATING SUMS OF ALTERNATING SERIES

The following theorem is concerned with the error that results when the sum of an alternating series is approximated by a partial sum.

10.7.2 THEOREM. *If an alternating series satisfies the hypotheses of the alternating series test, and if S is the sum of the series, then:*

(a) *S lies between any two successive partial sums; that is, either*

$$s_n \leq S \leq s_{n+1} \quad \text{or} \quad s_{n+1} \leq S \leq s_n \qquad (3)$$

depending on which partial sum is larger.

(b) *If S is approximated by s_n, then the absolute error $|S - s_n|$ satisfies*

$$|S - s_n| \leq a_{n+1} \qquad (4)$$

Moreover, the sign of the error $S - s_n$ is the same as that of the coefficient of a_{n+1}.

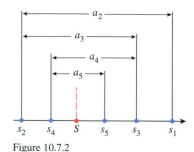

Figure 10.7.2

Proof. We will prove the theorem for series of form (1). Referring to Figure 10.7.2 and keeping in mind our observation in the proof of Theorem 10.7.1 that the odd-numbered

---

[*]The interested reader will find some nice examples in an article by R. Lariviere, "On a Convergence Test for Alternating Series," *Mathematics Magazine*, Vol. 29, 1956, p. 88.

partial sums form a decreasing sequence converging to S and the even-numbered partial sums form an increasing sequence converging to S, we see that successive partial sums oscillate from one side of S to the other in smaller and smaller steps with the odd-numbered partial sums being larger than S and the even-numbered partial sums being smaller than S. Thus, depending on whether n is even or odd, we have

$$s_n \leq S \leq s_{n+1} \quad \text{or} \quad s_{n+1} \leq S \leq s_n$$

which proves (3). Moreover, in either case we have

$$|S - s_n| \leq |s_{n+1} - s_n| \tag{5}$$

But $s_{n+1} - s_n = \pm a_{n+1}$ (the sign depending on whether n is even or odd). Thus, it follows from (5) that $|S - s_n| \leq a_{n+1}$, which proves (4). Finally, since the odd-numbered partial sums are larger than S and the even-numbered partial sums are smaller than S, it follows that $S - s_n$ has the same sign as the coefficient of a_{n+1} (verify). ■

REMARK. In words, inequality (4) states that for a series satisfying the hypotheses of the alternating series test, the magnitude of the error that results from approximating S by s_n is at most that of the first term that is *not* included in the partial sum. Also, note that if $a_1 > a_2 > \cdots > a_k > \cdots$, then inequality (4) can be strengthened to $|s - s_n| < a_{n+1}$.

Example 2 Later in this chapter we will show that the sum of the alternating harmonic series is

$$\ln 2 = 1 - \frac{1}{2} + \frac{1}{3} - \frac{1}{4} + \cdots + (-1)^{k+1}\frac{1}{k} + \cdots$$

(a) Accepting this to be so, find an upper bound on the magnitude of the error that results if $\ln 2$ is approximated by the sum of the first eight terms in the series.

(b) Find a partial sum that approximates $\ln 2$ to one decimal-place accuracy (the nearest tenth).

Solution (a). It follows from (4) that

$$|\ln 2 - s_8| < a_9 = \frac{1}{9} < 0.12 \tag{6}$$

As a check, let us compute s_8 exactly. We obtain

$$s_8 = 1 - \frac{1}{2} + \frac{1}{3} - \frac{1}{4} + \frac{1}{5} - \frac{1}{6} + \frac{1}{7} - \frac{1}{8} = \frac{533}{840}$$

Thus, with the help of a calculator

$$|\ln 2 - s_8| = \left|\ln 2 - \frac{533}{840}\right| \approx 0.059$$

This shows that the error is well under the estimate provided by upper bound (6).

Solution (b). For one decimal-place accuracy, we must choose n so that $|\ln 2 - s_n| \leq 0.05$. However, it follows from (4) that

$$|\ln 2 - s_n| < a_{n+1}$$

so it suffices to choose n so that $a_{n+1} \leq 0.05$.

One way to find n is to use a calculating utility to obtain numerical values for a_1, a_2, a_3, ... until you encounter the first value that is less than or equal to 0.05. If you do this, you will find that it is $a_{20} = 0.05$; this tells us that partial sum s_{19} will provide the desired accuracy. Another way to find n is to solve the inequality

$$\frac{1}{n+1} \leq 0.05$$

algebraically. We can do this by taking reciprocals, reversing the sense of the inequality, and then simplifying to obtain $n \geq 19$. Thus, s_{19} will provide the required accuracy, which is consistent with the previous result.

With the help of a calculating utility, the value of s_{19} is approximately $s_{19} \approx 0.7$ and the value of $\ln 2$ obtained directly is approximately $\ln 2 \approx 0.69$, which agrees with s_{19} when rounded to one decimal place. ◄

REMARK. As this example illustrates, the alternating harmonic series does not provide an efficient way to approximate $\ln 2$, since too much computation is required to achieve reasonable accuracy. Later, we will develop better ways to approximate logarithms.

ABSOLUTE CONVERGENCE

The series

$$1 - \frac{1}{2} - \frac{1}{2^2} + \frac{1}{2^3} + \frac{1}{2^4} - \frac{1}{2^5} - \frac{1}{2^6} + \cdots$$

does not fit in any of the categories studied so far—it has mixed signs, but is not alternating. We will now develop some convergence tests that can be applied to such series.

10.7.3 DEFINITION. A series

$$\sum_{k=1}^{\infty} u_k = u_1 + u_2 + \cdots + u_k + \cdots$$

is said to **converge absolutely** if the series of absolute values

$$\sum_{k=1}^{\infty} |u_k| = |u_1| + |u_2| + \cdots + |u_k| + \cdots$$

converges and is said to **diverge absolutely** if the series of absolute values diverges.

Example 3 Determine whether the following series converge absolutely.

(a) $1 - \dfrac{1}{2} - \dfrac{1}{2^2} + \dfrac{1}{2^3} + \dfrac{1}{2^4} - \dfrac{1}{2^5} - \cdots$ (b) $1 - \dfrac{1}{2} + \dfrac{1}{3} - \dfrac{1}{4} + \dfrac{1}{5} - \cdots$

Solution (a). The series of absolute values is the convergent geometric series

$$1 + \frac{1}{2} + \frac{1}{2^2} + \frac{1}{2^3} + \frac{1}{2^4} + \frac{1}{2^5} + \cdots$$

so the given series converges absolutely.

Solution (b). The series of absolute values is the divergent harmonic series

$$1 + \frac{1}{2} + \frac{1}{3} + \frac{1}{4} + \frac{1}{5} + \cdots$$

so the given series diverges absolutely. ◄

It is important to distinguish between the notions of convergence and absolute convergence. For example, the series in part (b) of Example 3 converges, since it is the alternating harmonic series, yet we demonstrated that it does not converge absolutely. However, the following theorem shows that *if a series converges absolutely, then it converges.*

10.7.4 THEOREM. *If the series*

$$\sum_{k=1}^{\infty} |u_k| = |u_1| + |u_2| + \cdots + |u_k| + \cdots$$

converges, then so does the series

$$\sum_{k=1}^{\infty} u_k = u_1 + u_2 + \cdots + u_k + \cdots$$

Proof. Our proof is based on a trick. We will write the series $\sum u_k$ as

$$\sum_{k=1}^{\infty} u_k = \sum_{k=1}^{\infty} [(u_k + |u_k|) - |u_k|] \tag{7}$$

We are assuming that $\sum |u_k|$ converges, so that if we can show that $\sum (u_k + |u_k|)$ converges, then it will follow from (7) and Theorem 10.5.3(a) that $\sum u_k$ converges. However, the value of $u_k + |u_k|$ is either 0 or $2|u_k|$, depending on the sign of u_k. Thus, in all cases it is true that

$$0 \le u_k + |u_k| \le 2|u_k|$$

But $\sum 2|u_k|$ converges, since it is a constant times the convergent series $\sum |u_k|$; hence $\sum (u_k + |u_k|)$ converges by the comparison test. ∎

Theorem 10.7.4 provides a way of inferring convergence of a series with positive and negative terms from the convergence of a series with nonnegative terms (the series of absolute values). This is important because most of the convergence tests we have developed apply only to series with nonnegative terms.

Example 4 Show that the following series converge.

$$\text{(a) } 1 - \frac{1}{2} - \frac{1}{2^2} + \frac{1}{2^3} + \frac{1}{2^4} - \frac{1}{2^5} - \frac{1}{2^6} + \cdots \qquad \text{(b) } \sum_{k=1}^{\infty} \frac{\cos k}{k^2}$$

Solution (a). Observe that this is not an alternating series because the signs alternate in pairs after the first term. Thus, we have no convergence test that can be applied directly. However, we showed in Example 3(a) that the series converges absolutely, so Theorem 10.7.4 implies that it converges.

Solution (b). With the help of a calculating utility, you will be able to verify that the signs of the terms in this series vary irregularly. Thus, we will test for absolute convergence. The series of absolute values is

$$\sum_{k=1}^{\infty} \left| \frac{\cos k}{k^2} \right|$$

However,

$$\left| \frac{\cos k}{k^2} \right| \le \frac{1}{k^2}$$

But $\sum 1/k^2$ is a convergent p-series ($p = 2$), so the series of absolute values converges by the comparison test. Thus, the given series converges absolutely and hence converges. ◀

CONDITIONAL CONVERGENCE

Although Theorem 10.7.4 is a useful tool for series that converge absolutely, it provides no information about the convergence or divergence of a series that diverges absolutely. For example, consider the two series

$$1 - \frac{1}{2} + \frac{1}{3} - \frac{1}{4} + \cdots + (-1)^{k+1} \frac{1}{k} + \cdots \tag{8}$$

$$-1 - \frac{1}{2} - \frac{1}{3} - \frac{1}{4} - \cdots - \frac{1}{k} - \cdots \tag{9}$$

Both of these series diverge absolutely, since in each case the series of absolute values is the divergent harmonic series

$$1 + \frac{1}{2} + \frac{1}{3} + \cdots + \frac{1}{k} + \cdots$$

However, series (8) converges, since it is the alternating harmonic series, and series (9) diverges, since it is a constant times the divergent harmonic series. As a matter of terminology, a series that converges but diverges absolutely is said to **converge conditionally** (or to be **conditionally convergent**). Thus, (8) is a conditionally convergent series.

THE RATIO TEST FOR ABSOLUTE CONVERGENCE

Although one cannot generally infer convergence or divergence of a series from absolute divergence, the following variation of the ratio test provides a way of deducing divergence from absolute divergence in certain situations. We omit the proof.

10.7.5   THEOREM (*Ratio Test for Absolute Convergence*).   *Let $\sum u_k$ be a series with nonzero terms and suppose that*

$$\rho = \lim_{k \to +\infty} \frac{|u_{k+1}|}{|u_k|}$$

(a)   *If $\rho < 1$, then the series $\sum u_k$ converges absolutely and therefore converges.*

(b)   *If $\rho > 1$ or if $\rho = +\infty$, then the series $\sum u_k$ diverges.*

(c)   *If $\rho = 1$, no conclusion about convergence or absolute convergence can be drawn from this test.*

Example 5   Use the ratio test for absolute convergence to determine whether the series converges.

(a) $\displaystyle\sum_{k=1}^{\infty} (-1)^k \frac{2^k}{k!}$ (b) $\displaystyle\sum_{k=1}^{\infty} (-1)^k \frac{(2k-1)!}{3^k}$

Solution (a). Taking the absolute value of the general term u_k we obtain

$$|u_k| = \left| (-1)^k \frac{2^k}{k!} \right| = \frac{2^k}{k!}$$

Thus,

$$\rho = \lim_{k \to +\infty} \frac{|u_{k+1}|}{|u_k|} = \lim_{k \to +\infty} \frac{2^{k+1}}{(k+1)!} \cdot \frac{k!}{2^k} = \lim_{k \to +\infty} \frac{2}{k+1} = 0 < 1$$

which implies that the series converges absolutely and therefore converges.

Solution (b). Taking the absolute value of the general term u_k we obtain

$$|u_k| = \left| (-1)^k \frac{(2k-1)!}{3^k} \right| = \frac{(2k-1)!}{3^k}$$

Thus,

$$\rho = \lim_{k \to +\infty} \frac{|u_{k+1}|}{|u_k|} = \lim_{k \to +\infty} \frac{[2(k+1)-1]!}{3^{k+1}} \cdot \frac{3^k}{(2k-1)!}$$

$$= \lim_{k \to +\infty} \frac{1}{3} \cdot \frac{(2k+1)!}{(2k-1)!} = \frac{1}{3} \lim_{k \to +\infty} (2k)(2k+1) = +\infty$$

which implies that the series diverges.   ◀

SUMMARY OF CONVERGENCE TESTS

We conclude this section with a summary of convergence tests that can be used for reference.

Summary of Convergence Tests

NAME	STATEMENT	COMMENTS
Divergence Test (10.5.1)	If $\lim\limits_{k \to +\infty} u_k \neq 0$, then $\sum u_k$ diverges.	If $\lim\limits_{k \to +\infty} u_k = 0$, then $\sum u_k$ may or may not converge.
Integral Test (10.5.4)	Let $\sum u_k$ be a series with positive terms, and let $f(x)$ be the function that results when k is replaced by x in the general term of the series. If f is decreasing and continuous for $x \geq a$, then $$\sum_{k=1}^{\infty} u_k \quad \text{and} \quad \int_a^{+\infty} f(x)\, dx$$ both converge or both diverge.	This test only applies to series that have positive terms. Try this test when $f(x)$ is easy to integrate.
Comparison Test (10.6.1)	Let $\sum_{k=1}^{\infty} a_k$ and $\sum_{k=1}^{\infty} b_k$ be series with nonnegative terms such that $$a_1 \leq b_1,\ a_2 \leq b_2, \ldots, a_k \leq b_k, \ldots$$ If $\sum b_k$ converges, then $\sum a_k$ converges, and if $\sum a_k$ diverges, then $\sum b_k$ diverges.	This test only applies to series with nonnegative terms. Try this test as a last resort; other tests are often easier to apply.
Limit Comparison Test (10.6.4)	Let $\sum a_k$ and $\sum b_k$ be series with positive terms such that $$\rho = \lim_{k \to +\infty} \frac{a_k}{b_k}$$ If $0 < \rho < +\infty$, then both series converge or both diverge.	This is easier to apply than the comparison test, but still requires some skill in choosing the series $\sum b_k$ for comparison.
Ratio Test (10.6.5)	Let $\sum u_k$ be a series with positive terms and suppose that $$\rho = \lim_{k \to +\infty} \frac{u_{k+1}}{u_k}$$ (a) Series converges if $\rho < 1$. (b) Series diverges if $\rho > 1$ or $\rho = +\infty$. (c) The test is inconclusive if $\rho = 1$.	Try this test when u_k involves factorials or kth powers.
Root Test (10.6.6)	Let $\sum u_k$ be a series with positive terms such that $$\rho = \lim_{k \to +\infty} \sqrt[k]{u_k}$$ (a) The series converges if $\rho < 1$. (b) The series diverges if $\rho > 1$ or $\rho = +\infty$. (c) The test is inconclusive if $\rho = 1$.	Try this test when u_k involves kth powers.
Alternating Series Test (10.7.1)	If $a_k > 0$ for $k = 1, 2, 3, \ldots$, then the series $$a_1 - a_2 + a_3 - a_4 + \cdots$$ $$-a_1 + a_2 - a_3 + a_4 - \cdots$$ converge if the following conditions hold: (a) $a_1 \geq a_2 \geq a_3 \geq \cdots$ (b) $\lim\limits_{k \to +\infty} a_k = 0$	This test applies only to alternating series.
Ratio Test for Absolute Convergence (10.7.5)	Let $\sum u_k$ be a series with nonzero terms such that $$\rho = \lim_{k \to +\infty} \frac{|u_{k+1}|}{|u_k|}$$ (a) The series converges absolutely if $\rho < 1$. (b) The series diverges if $\rho > 1$ or $\rho = +\infty$. (c) The test is inconclusive if $\rho = 1$.	The series need not have positive terms and need not be alternating to use this test.

EXERCISE SET 10.7 ~Graphing Utility C CAS

In Exercises 1 and 2 show that the series converges by confirming that it satisfies the hypotheses of the alternating series test (Theorem 10.7.1).

1. $\displaystyle\sum_{k=1}^{\infty} \frac{(-1)^{k+1}}{2k+1}$ **2.** $\displaystyle\sum_{k=1}^{\infty} (-1)^{k+1} \frac{k}{3^k}$

In Exercises 3–6, determine whether the alternating series converges, and justify your answer.

3. $\displaystyle\sum_{k=1}^{\infty} (-1)^{k+1} \frac{k+1}{3k+1}$ **4.** $\displaystyle\sum_{k=1}^{\infty} (-1)^{k+1} \frac{k+1}{\sqrt{k}+1}$

5. $\displaystyle\sum_{k=1}^{\infty} (-1)^{k+1} e^{-k}$ **6.** $\displaystyle\sum_{k=3}^{\infty} (-1)^{k} \frac{\ln k}{k}$

In Exercises 7–12, use the ratio test for absolute convergence (Theorem 10.7.5) to determine whether the series converges or diverges. If the test is inconclusive, then say so.

7. $\displaystyle\sum_{k=1}^{\infty} \left(-\frac{3}{5}\right)^k$ **8.** $\displaystyle\sum_{k=1}^{\infty} (-1)^{k+1} \frac{2^k}{k!}$

9. $\displaystyle\sum_{k=1}^{\infty} (-1)^{k+1} \frac{3^k}{k^2}$ **10.** $\displaystyle\sum_{k=1}^{\infty} (-1)^{k} \frac{k}{5^k}$

11. $\displaystyle\sum_{k=1}^{\infty} (-1)^{k} \frac{k^3}{e^k}$ **12.** $\displaystyle\sum_{k=1}^{\infty} (-1)^{k+1} \frac{k^k}{k!}$

In Exercises 13–30, classify the series as absolutely convergent, conditionally convergent, or divergent.

13. $\displaystyle\sum_{k=1}^{\infty} \frac{(-1)^{k+1}}{3k}$ **14.** $\displaystyle\sum_{k=1}^{\infty} \frac{(-1)^{k+1}}{k^{4/3}}$ **15.** $\displaystyle\sum_{k=1}^{\infty} \frac{(-4)^k}{k^2}$

16. $\displaystyle\sum_{k=1}^{\infty} \frac{(-1)^{k+1}}{k!}$ **17.** $\displaystyle\sum_{k=1}^{\infty} \frac{\cos k\pi}{k}$ **18.** $\displaystyle\sum_{k=3}^{\infty} \frac{(-1)^k \ln k}{k}$

19. $\displaystyle\sum_{k=1}^{\infty} (-1)^{k+1} \frac{k+2}{k(k+3)}$ **20.** $\displaystyle\sum_{k=1}^{\infty} \frac{(-1)^{k+1}k^2}{k^3+1}$

21. $\displaystyle\sum_{k=1}^{\infty} \sin \frac{k\pi}{2}$ **22.** $\displaystyle\sum_{k=1}^{\infty} \frac{\sin k}{k^3}$

23. $\displaystyle\sum_{k=2}^{\infty} \frac{(-1)^k}{k \ln k}$ **24.** $\displaystyle\sum_{k=1}^{\infty} \frac{(-1)^k}{\sqrt{k(k+1)}}$

25. $\displaystyle\sum_{k=2}^{\infty} \left(-\frac{1}{\ln k}\right)^k$ **26.** $\displaystyle\sum_{k=1}^{\infty} \frac{(-1)^{k+1}}{\sqrt{k+1}+\sqrt{k}}$

27. $\displaystyle\sum_{k=2}^{\infty} \frac{(-1)^k (k^2+1)}{k^3+2}$ **28.** $\displaystyle\sum_{k=1}^{\infty} \frac{k \cos k\pi}{k^2+1}$

29. $\displaystyle\sum_{k=1}^{\infty} \frac{(-1)^{k+1}k!}{(2k-1)!}$ **30.** $\displaystyle\sum_{k=1}^{\infty} (-1)^{k+1} \frac{3^{2k-1}}{k^2+1}$

In Exercises 31–34, the series satisfies the hypotheses of the alternating series test. For the stated value of n, find an upper bound on the absolute error that results if the sum of the series is approximated by the nth partial sum.

31. $\displaystyle\sum_{k=1}^{\infty} \frac{(-1)^{k+1}}{k}$; $n=7$ **32.** $\displaystyle\sum_{k=1}^{\infty} \frac{(-1)^{k+1}}{k!}$; $n=5$

33. $\displaystyle\sum_{k=1}^{\infty} \frac{(-1)^{k+1}}{\sqrt{k}}$; $n=99$

34. $\displaystyle\sum_{k=1}^{\infty} \frac{(-1)^{k+1}}{(k+1)\ln(k+1)}$; $n=3$

In Exercises 35–38, the series satisfies the hypotheses of the alternating series test. Find a value of n for which the nth partial sum is ensured to approximate the sum of the series to the stated accuracy.

35. $\displaystyle\sum_{k=1}^{\infty} \frac{(-1)^{k+1}}{k}$; $|\text{error}| < 0.0001$

36. $\displaystyle\sum_{k=1}^{\infty} \frac{(-1)^{k+1}}{k!}$; $|\text{error}| < 0.00001$

37. $\displaystyle\sum_{k=1}^{\infty} \frac{(-1)^{k+1}}{\sqrt{k}}$; two decimal places

38. $\displaystyle\sum_{k=1}^{\infty} \frac{(-1)^{k+1}}{(k+1)\ln(k+1)}$; one decimal place

In Exercises 39 and 40, find an upper bound on the absolute error that results if s_{10} is used to approximate the sum of the given *geometric* series. Compute s_{10} rounded to four decimal places and compare this value with the exact sum of the series.

39. $\dfrac{3}{4} - \dfrac{3}{8} + \dfrac{3}{16} - \dfrac{3}{32} + \cdots$ **40.** $1 - \dfrac{2}{3} + \dfrac{4}{9} - \dfrac{8}{27} + \cdots$

In Exercises 41–44, the series satisfies the hypotheses of the alternating series test. Approximate the sum of the series to two decimal-place accuracy.

41. $1 - \dfrac{1}{3!} + \dfrac{1}{5!} - \dfrac{1}{7!} + \cdots$ **42.** $1 - \dfrac{1}{2!} + \dfrac{1}{4!} - \dfrac{1}{6!} + \cdots$

43. $\dfrac{1}{1 \cdot 2} - \dfrac{1}{2 \cdot 2^2} + \dfrac{1}{3 \cdot 2^3} - \dfrac{1}{4 \cdot 2^4} + \cdots$

44. $\dfrac{1}{1^5 + 4 \cdot 1} - \dfrac{1}{3^5 + 4 \cdot 3} + \dfrac{1}{5^5 + 4 \cdot 5} - \dfrac{1}{7^5 + 4 \cdot 7} + \cdots$

C **45.** The purpose of this exercise is to show that the error bound in part (*b*) of Theorem 10.7.2 can be overly conservative in certain cases.

(a) Use a CAS to confirm that
$$\frac{\pi}{4} = 1 - \frac{1}{3} + \frac{1}{5} - \frac{1}{7} + \cdots$$

(b) Use the CAS to show that $|(\pi/4) - s_{26}| < 10^{-2}$.

(c) According to the error bound in part (b) of Theorem 10.7.2, what value of n is required to ensure that $|(\pi/4) - s_n| < 10^{-2}$?

46. Show that the alternating p-series
$$1 - \frac{1}{2^p} + \frac{1}{3^p} - \frac{1}{4^p} + \cdots + (-1)^{k+1}\frac{1}{k^p} + \cdots$$

converges absolutely if $p > 1$, converges conditionally if $0 < p \le 1$, and diverges if $p \le 0$.

> It can be proved that any series that is constructed from an absolutely convergent series by rearranging the terms is absolutely convergent and has the same sum as the original series. Use this fact together with parts (a) and (b) of Theorem 10.5.3 in Exercises 47 and 48.

47. It was stated in Exercise 27 of Section 10.5 that
$$\frac{\pi^2}{6} = 1 + \frac{1}{2^2} + \frac{1}{3^2} + \frac{1}{4^2} + \cdots$$

Use this to show that
$$\frac{\pi^2}{8} = 1 + \frac{1}{3^2} + \frac{1}{5^2} + \frac{1}{7^2} + \cdots$$

48. It was stated in Exercise 27 of Section 10.5 that
$$\frac{\pi^4}{90} = 1 + \frac{1}{2^4} + \frac{1}{3^4} + \frac{1}{4^4} + \cdots$$

Use this to show that
$$\frac{\pi^4}{96} = 1 + \frac{1}{3^4} + \frac{1}{5^4} + \frac{1}{7^4} + \cdots$$

49. It can be proved that the terms of any conditionally convergent series can be rearranged to give either a divergent series or a conditionally convergent series whose sum is any given number S. For example, we stated in Example 2 that
$$\ln 2 = 1 - \frac{1}{2} + \frac{1}{3} - \frac{1}{4} + \frac{1}{5} - \frac{1}{6} + \cdots$$

Show that we can rearrange this series so that its sum is $\frac{1}{2}\ln 2$ by rewriting it as
$$\left(1 - \frac{1}{2} - \frac{1}{4}\right) + \left(\frac{1}{3} - \frac{1}{6} - \frac{1}{8}\right) + \left(\frac{1}{5} - \frac{1}{10} - \frac{1}{12}\right) + \cdots$$

[*Hint:* Add the first two terms in each set of parentheses.]

50. (a) Use a graphing utility to graph
$$f(x) = \frac{4x - 1}{4x^2 - 2x}, \quad x \ge 1$$

(b) Based on your graph, do you think that the series
$$\sum_{k=1}^{\infty}(-1)^{k+1}\frac{4k - 1}{4k^2 - 2k}$$

converges? Explain your reasoning.

51. As illustrated in the accompanying figure, a bug, starting at point A on a 180-cm wire, walks the length of the wire, stops and walks in the opposite direction for half the length of the wire, stops again and walks in the opposite direction for one-third the length of the wire, stops again and walks in the opposite direction for one-fourth the length of the wire, and so forth until it stops for the 1000th time.

(a) Give upper and lower bounds on the distance between the bug and point A when it finally stops. [*Hint:* As stated in Example 2, assume that the sum of the alternating harmonic series is $\ln 2$.]

(b) Give upper and lower bounds on the total distance that the bug has traveled when it finally stops. [*Hint:* Use inequality (2) of Section 10.5.]

Figure Ex-51

52. (a) Prove that if $\sum a_k$ converges absolutely, then $\sum a_k^2$ converges.

(b) Show that the converse of part (a) is false by giving a counterexample.

10.8 MACLAURIN AND TAYLOR SERIES; POWER SERIES

In the last four sections we focused exclusively on series whose terms are numbers. In this section we will introduce Maclaurin and Taylor series, examples of series whose terms are functions. Our primary objective is to develop mathematical tools for the investigation of convergence of Maclaurin and Taylor series.

MACLAURIN AND TAYLOR SERIES

In Section 10.1 we defined the nth Maclaurin polynomial for a function f as
$$\sum_{k=0}^{n}\frac{f^{(k)}(0)}{k!}x^k = f(0) + f'(0)x + \frac{f''(0)}{2!}x^2 + \cdots + \frac{f^{(n)}(0)}{n!}x^n$$

and the nth Taylor polynomial for f about $x = x_0$ as

$$\sum_{k=0}^{n} \frac{f^{(k)}(x_0)}{k!}(x - x_0)^k = f(x_0) + f'(x_0)(x - x_0)$$

$$+ \frac{f''(x_0)}{2!}(x - x_0)^2 + \cdots + \frac{f^{(n)}(x_0)}{n!}(x - x_0)^n$$

Since then we have gone on to consider sums with an infinite number of terms, so it is not a big step to extend the notions of Maclaurin and Taylor polynomials to series by not stopping the summation index at n. Thus, we have the following definition.

10.8.1   DEFINITION.   If f has derivatives of all orders at x_0, then we call the series

$$\sum_{k=0}^{\infty} \frac{f^{(k)}(x_0)}{k!}(x - x_0)^k = f(x_0) + f'(x_0)(x - x_0) + \frac{f''(x_0)}{2!}(x - x_0)^2$$

$$+ \cdots + \frac{f^{(k)}(x_0)}{k!}(x - x_0)^k + \cdots \qquad (1)$$

the *Taylor series for f about x = x_0*. In the special case where $x_0 = 0$, this series becomes

$$\sum_{k=0}^{\infty} \frac{f^{(k)}(0)}{k!}x^k = f(0) + f'(0)x + \frac{f''(0)}{2!}x^2 + \cdots + \frac{f^{(k)}(0)}{k!}x^k + \cdots \qquad (2)$$

in which case we call it the *Maclaurin series for f*.

Note that the nth Maclaurin and Taylor polynomials are the nth partial sums for the corresponding Maclaurin and Taylor series.

Example 1   Find the Maclaurin series for

(a) e^x (b) $\sin x$ (c) $\cos x$ (d) $\dfrac{1}{1-x}$

Solution (a). In Example 2 of Section 10.1 we found that the nth Maclaurin polynomial for e^x is

$$p_n(x) = \sum_{k=0}^{n} \frac{x^k}{k!} = 1 + x + \frac{x^2}{2!} + \cdots + \frac{x^n}{n!}$$

Thus, the Maclaurin series for e^x is

$$\sum_{k=0}^{\infty} \frac{x^k}{k!} = 1 + x + \frac{x^2}{2!} + \cdots + \frac{x^k}{k!} + \cdots$$

Solution (b). In Example 4(a) of Section 10.1 we found that the Maclaurin polynomials for $\sin x$ are given by

$$p_{2k+1}(x) = p_{2k+2}(x) = x - \frac{x^3}{3!} + \frac{x^5}{5!} - \frac{x^7}{7!} + \cdots + (-1)^k \frac{x^{2k+1}}{(2k+1)!} \quad (k = 0, 1, 2, \ldots)$$

Thus, the Maclaurin series for $\sin x$ is

$$\sum_{k=0}^{\infty} (-1)^k \frac{x^{2k+1}}{(2k+1)!} = x - \frac{x^3}{3!} + \frac{x^5}{5!} - \frac{x^7}{7!} + \cdots + (-1)^k \frac{x^{2k+1}}{(2k+1)!} + \cdots$$

Solution (c). In Example 4(b) of Section 10.1 we found that the Maclaurin polynomials for $\cos x$ are given by

$$p_{2k}(x) = p_{2k+1}(x) = 1 - \frac{x^2}{2!} + \frac{x^4}{4!} - \frac{x^6}{6!} + \cdots + (-1)^k \frac{x^{2k}}{(2k)!} \quad (k = 0, 1, 2, \ldots)$$

Thus, the Maclaurin series for $\cos x$ is

$$\sum_{k=0}^{\infty}(-1)^k\frac{x^{2k}}{(2k)!} = 1 - \frac{x^2}{2!} + \frac{x^4}{4!} - \frac{x^6}{6!} + \cdots + (-1)^k\frac{x^{2k}}{(2k)!} + \cdots$$

Solution (d). In Example 4(c) of Section 10.1 we found that the nth Maclaurin polynomial for $1/(1-x)$ is

$$p_n(x) = \sum_{k=0}^{n} x^k = 1 + x + x^2 + \cdots + x^n \quad (n = 0, 1, 2, \ldots)$$

Thus, the Maclaurin series for $1/(1-x)$ is

$$\sum_{k=0}^{\infty} x^k = 1 + x + x^2 + \cdots + x^k + \cdots \qquad \blacktriangleleft$$

Example 2 Find the Taylor series for $1/x$ about $x = 1$.

Solution. In Example 5 of Section 10.1 we found that the nth Taylor polynomial for $1/x$ about $x = 1$ is

$$\sum_{k=0}^{n}(-1)^k(x-1)^k = 1 - (x-1) + (x-1)^2 - (x-1)^3 + \cdots + (-1)^n(x-1)^n$$

Thus, the Taylor series for $1/x$ about $x = 1$ is

$$\sum_{k=0}^{\infty}(-1)^k(x-1)^k = 1 - (x-1) + (x-1)^2 - (x-1)^3 + \cdots + (-1)^k(x-1)^k + \cdots \qquad \blacktriangleleft$$

POWER SERIES IN x

Maclaurin and Taylor series differ from the series that we have considered in the last four sections in that their terms are not merely constants, but instead involve a variable. These are examples of *power series*, which we now define.

If $c_0, c_1, c_2, \ldots$ are constants and x is a variable, then a series of the form

$$\sum_{k=0}^{\infty} c_k x^k = c_0 + c_1 x + c_2 x^2 + \cdots + c_k x^k + \cdots \tag{3}$$

is called a ***power series in x***. Some examples are

$$\sum_{k=0}^{\infty} x^k = 1 + x + x^2 + x^3 + \cdots$$

$$\sum_{k=0}^{\infty} \frac{x^k}{k!} = 1 + x + \frac{x^2}{2!} + \frac{x^3}{3!} + \cdots$$

$$\sum_{k=0}^{\infty}(-1)^k\frac{x^{2k}}{(2k)!} = 1 - \frac{x^2}{2!} + \frac{x^4}{4!} - \frac{x^6}{6!} + \cdots$$

From Example 1, these are the Maclaurin series for the functions $1/(1-x)$, e^x, and $\cos x$, respectively. Indeed, every Maclaurin series

$$\sum_{k=0}^{\infty} \frac{f^{(k)}(0)}{k!}x^k = f(0) + f'(0)x + \frac{f''(0)}{2!}x^2 + \cdots + \frac{f^{(k)}(0)}{k!}x^k + \cdots$$

is a power series in x.

RADIUS AND INTERVAL OF CONVERGENCE

If a numerical value is substituted for x in a power series $\sum c_k x^k$, then the resulting series of numbers may either converge or diverge. This leads to the problem of determining the set of x-values for which a given power series converges; this is called its ***convergence set***.

Observe that every power series in x converges at $x = 0$, since substituting this value in (3) produces the series

$$c_0 + 0 + 0 + 0 + \cdots + 0 + \cdots$$

whose sum is c_0. In rare cases $x = 0$ may be the only number in the convergence set, but more usually the convergence set is some finite or infinite interval containing $x = 0$. This is the content of the following theorem, whose proof will be omitted.

10.8.2   THEOREM.   *For any power series in x, exactly one of the following is true:*

(a)   *The series converges only for $x = 0$.*

(b)   *The series converges absolutely (and hence converges) for all real values of x.*

(c)   *The series converges absolutely (and hence converges) for all x in some finite open interval $(-R, R)$, and diverges if $x < -R$ or $x > R$. At either of the values $x = R$ or $x = -R$, the series may converge absolutely, converge conditionally, or diverge, depending on the particular series.*

This theorem states that the convergence set for a power series in x is always an interval centered at $x = 0$ (possibly just the value $x = 0$ itself or possibly infinite). For this reason, the convergence set of a power series in x is called the **interval of convergence**. In the case where the convergence set is the single value $x = 0$ we say that the series has **radius of convergence 0**, in the case where the convergence set is $(-\infty, +\infty)$ we say that the series has **radius of convergence $+\infty$**, and in the case where the convergence set extends between $-R$ and R we say that the series has **radius of convergence R** (Figure 10.8.1).

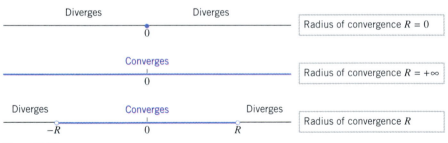

Figure 10.8.1

FINDING THE INTERVAL OF CONVERGENCE

The usual procedure for finding the interval of convergence of a power series is to apply the ratio test for absolute convergence (Theorem 10.7.5). The following example illustrates how this works.

Example 3   Find the interval of convergence and radius of convergence of the following power series.

$$\text{(a) } \sum_{k=0}^{\infty} x^k \quad \text{(b) } \sum_{k=0}^{\infty} \frac{x^k}{k!} \quad \text{(c) } \sum_{k=0}^{\infty} k! x^k \quad \text{(d) } \sum_{k=0}^{\infty} \frac{(-1)^k x^k}{3^k (k+1)}$$

Solution (a). We apply the ratio test for absolute convergence. We have

$$\rho = \lim_{k \to +\infty} \left| \frac{u_{k+1}}{u_k} \right| = \lim_{k \to +\infty} \left| \frac{x^{k+1}}{x^k} \right| = \lim_{k \to +\infty} |x| = |x|$$

so the series converges absolutely if $\rho = |x| < 1$ and diverges if $\rho = |x| > 1$. The test is inconclusive if $|x| = 1$ (i.e., if $x = 1$ or $x = -1$), which means that we will have to investigate convergence at these values separately. At these values the series becomes

$$\sum_{k=0}^{\infty} 1^k = 1 + 1 + 1 + 1 + \cdots \qquad \boxed{x = 1}$$

$$\sum_{k=0}^{\infty} (-1)^k = 1 - 1 + 1 - 1 + \cdots \qquad \boxed{x = -1}$$

both of which diverge; thus, the interval of convergence for the given power series is $(-1, 1)$, and the radius of convergence is $R = 1$.

Solution (b). Applying the ratio test for absolute convergence, we obtain

$$\rho = \lim_{k \to +\infty} \left| \frac{u_{k+1}}{u_k} \right| = \lim_{k \to +\infty} \left| \frac{x^{k+1}}{(k+1)!} \cdot \frac{k!}{x^k} \right| = \lim_{k \to +\infty} \left| \frac{x}{k+1} \right| = 0$$

Since $\rho < 1$ for all x, the series converges absolutely for all x. Thus, the interval of convergence is $(-\infty, +\infty)$ and the radius of convergence is $R = +\infty$.

Solution (c). If $x \neq 0$, then the ratio test for absolute convergence yields

$$\rho = \lim_{k \to +\infty} \left| \frac{u_{k+1}}{u_k} \right| = \lim_{k \to +\infty} \left| \frac{(k+1)! x^{k+1}}{k! x^k} \right| = \lim_{k \to +\infty} |(k+1)x| = +\infty$$

Therefore, the series diverges for all nonzero values of x. Thus, the interval of convergence is the single value $x = 0$ and the radius of convergence is $R = 0$.

Solution (d). Since $|(-1)^k| = |(-1)^{k+1}| = 1$, we obtain

$$\rho = \lim_{k \to +\infty} \left| \frac{u_{k+1}}{u_k} \right| = \lim_{k \to +\infty} \left| \frac{x^{k+1}}{3^{k+1}(k+2)} \cdot \frac{3^k(k+1)}{x^k} \right|$$

$$= \lim_{k \to +\infty} \left[\frac{|x|}{3} \cdot \left(\frac{k+1}{k+2} \right) \right]$$

$$= \frac{|x|}{3} \lim_{k \to +\infty} \left(\frac{1 + (1/k)}{1 + (2/k)} \right) = \frac{|x|}{3}$$

The ratio test for absolute convergence implies that the series converges absolutely if $|x| < 3$ and diverges if $|x| > 3$. The ratio test fails to provide any information when $|x| = 3$, so the cases $x = -3$ and $x = 3$ need separate analyses. Substituting $x = -3$ in the given series yields

$$\sum_{k=0}^{\infty} \frac{(-1)^k(-3)^k}{3^k(k+1)} = \sum_{k=0}^{\infty} \frac{(-1)^k(-1)^k 3^k}{3^k(k+1)} = \sum_{k=0}^{\infty} \frac{1}{k+1}$$

which is the divergent harmonic series $1 + \frac{1}{2} + \frac{1}{3} + \frac{1}{4} + \cdots$. Substituting $x = 3$ in the given series yields

$$\sum_{k=0}^{\infty} \frac{(-1)^k 3^k}{3^k(k+1)} = \sum_{k=0}^{\infty} \frac{(-1)^k}{k+1} = 1 - \frac{1}{2} + \frac{1}{3} - \frac{1}{4} + \cdots$$

which is the conditionally convergent alternating harmonic series. Thus, the interval of convergence for the given series is $(-3, 3]$ and the radius of convergence is $R = 3$. ◀

POWER SERIES IN $x - x_0$

If x_0 is a constant, and if x is replaced by $x - x_0$ in (3), then the resulting series has the form

$$\sum_{k=0}^{\infty} c_k (x - x_0)^k = c_0 + c_1(x - x_0) + c_2(x - x_0)^2 + \cdots + c_k(x - x_0)^k + \cdots$$

This is called a ***power series in*** $x - x_0$. Some examples are

$$\sum_{k=0}^{\infty} \frac{(x-1)^k}{k+1} = 1 + \frac{(x-1)}{2} + \frac{(x-1)^2}{3} + \frac{(x-1)^3}{4} + \cdots \qquad \boxed{x_0 = 1}$$

$$\sum_{k=0}^{\infty} \frac{(-1)^k (x+3)^k}{k!} = 1 - (x+3) + \frac{(x+3)^2}{2!} - \frac{(x+3)^3}{3!} + \cdots \qquad \boxed{x_0 = -3}$$

The first of these is a power series in $x - 1$ and the second is a power series in $x + 3$. Note that a power series in x is a power series in $x - x_0$ in which $x_0 = 0$. More generally, the Taylor series

$$\sum_{k=0}^{\infty} \frac{f^{(k)}(x_0)}{k!} (x - x_0)^k$$

is a power series in $x - x_0$.

The main result on convergence of a power series in $x - x_0$ can be obtained by substituting $x - x_0$ for x in Theorem 10.8.2. This leads to the following theorem.

10.8.3   THEOREM.   *For a power series* $\sum c_k (x - x_0)^k$, *exactly one of the following statements is true:*

(a)   *The series converges only for* $x = x_0$.

(b)   *The series converges absolutely (and hence converges) for all real values of* x.

(c)   *The series converges absolutely (and hence converges) for all x in some finite open interval* $(x_0 - R, x_0 + R)$ *and diverges if* $x < x_0 - R$ *or* $x > x_0 + R$. *At either of the values* $x = x_0 - R$ *or* $x = x_0 + R$, *the series may converge absolutely, converge conditionally, or diverge, depending on the particular series.*

It follows from this theorem that the set of values for which a power series in $x - x_0$ converges is always an interval centered at $x = x_0$; we call this the ***interval of convergence*** (Figure 10.8.2). In part (a) of Theorem 10.8.3 the interval of convergence reduces to the single value $x = x_0$, in which case we say that the series has ***radius of convergence $R = 0$***; in part (b) the interval of convergence is infinite (the entire real line), in which case we say that the series has ***radius of convergence $R = +\infty$***; and in part (c) the interval extends between $x_0 - R$ and $x_0 + R$, in which case we say that the series has ***radius of convergence R***.

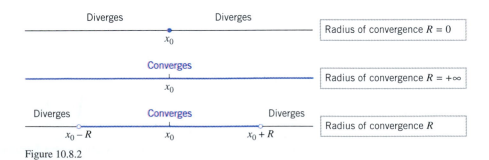

Figure 10.8.2

Example 4   Find the interval of convergence and radius of convergence of the series

$$\sum_{k=1}^{\infty} \frac{(x-5)^k}{k^2}$$

Solution.   We apply the ratio test for absolute convergence.

$$\rho = \lim_{k \to +\infty} \left| \frac{u_{k+1}}{u_k} \right| = \lim_{k \to +\infty} \left| \frac{(x-5)^{k+1}}{(k+1)^2} \cdot \frac{k^2}{(x-5)^k} \right|$$

$$= \lim_{k \to +\infty} \left[|x-5| \left(\frac{k}{k+1} \right)^2 \right]$$

$$= |x-5| \lim_{k \to +\infty} \left(\frac{1}{1+(1/k)} \right)^2 = |x-5|$$

Thus, the series converges absolutely if $|x-5| < 1$, or $-1 < x - 5 < 1$, or $4 < x < 6$. The series diverges if $x < 4$ or $x > 6$.

To determine the convergence behavior at the endpoints $x = 4$ and $x = 6$, we substitute these values in the given series. If $x = 6$, the series becomes

$$\sum_{k=1}^{\infty} \frac{1^k}{k^2} = \sum_{k=1}^{\infty} \frac{1}{k^2} = 1 + \frac{1}{2^2} + \frac{1}{3^2} + \frac{1}{4^2} + \cdots$$

which is a convergent p-series ($p = 2$). If $x = 4$, the series becomes

$$\sum_{k=1}^{\infty} \frac{(-1)^k}{k^2} = -1 + \frac{1}{2^2} - \frac{1}{3^2} + \frac{1}{4^2} - \cdots$$

Since this series converges absolutely, the interval of convergence for the given series is $[4, 6]$. The radius of convergence is $R = 1$ (Figure 10.8.3).   ◀

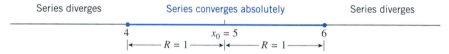

Figure 10.8.3

- FOR THE READER.   It will always be a waste of time to test for convergence at the endpoints of the interval of convergence using the ratio test, since ρ will always be 1 at those points if $\rho = \lim_{n \to +\infty} |a_{n+1}/a_n|$ exists. Explain why this must be so.

FUNCTIONS DEFINED BY POWER SERIES

If a function f is expressed as a power series on some interval, then we say that f is *represented* by the power series on that interval. For example, we saw in Example 4 of Section 10.4 that

$$\frac{1}{1-x} = 1 + x + x^2 + \cdots + x^k + \cdots$$

so that this power series represents the function $1/(1-x)$ on the interval $-1 < x < 1$.

Sometimes new functions actually originate as power series, and the properties of the functions are developed by working with their power series representations. For example, the functions

$$J_0(x) = \sum_{k=0}^{\infty} \frac{(-1)^k x^{2k}}{2^{2k}(k!)^2} = 1 - \frac{x^2}{2^2(1!)^2} + \frac{x^4}{2^4(2!)^2} - \frac{x^6}{2^6(3!)^2} + \cdots \tag{4}$$

and

$$J_1(x) = \sum_{k=0}^{\infty} \frac{(-1)^k x^{2k+1}}{2^{2k+1}(k!)(k+1)!} = \frac{x}{2} - \frac{x^3}{2^3(1!)(2!)} + \frac{x^5}{2^5(2!)(3!)} - \cdots \tag{5}$$

which are called *Bessel functions* in honor of the German mathematician and astronomer

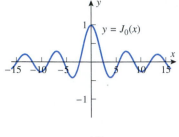

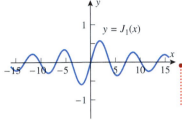

Friedrich Wilhelm Bessel (1784–1846), arise naturally in the study of planetary motion and in various problems that involve heat flow.

To find the domains of these functions, we must determine where their defining power series converge. For example, in the case $J_0(x)$ we have

$$\rho = \lim_{k \to +\infty} \left| \frac{u_{k+1}}{u_k} \right| = \lim_{k \to +\infty} \left| \frac{x^{2(k+1)}}{2^{2(k+1)}[(k+1)!]^2} \cdot \frac{2^{2k}(k!)^2}{x^{2k}} \right|$$

$$= \lim_{k \to +\infty} \left| \frac{x^2}{4(k+1)^2} \right| = 0 < 1$$

so that the series converges for all x; that is, the domain of $J_0(x)$ is $(-\infty, +\infty)$. We leave it as an exercise to show that the power series for $J_1(x)$ also converges for all x.

FOR THE READER.   Many CAS programs have the Bessel functions as part of their libraries. If you have a CAS, read the documentation to determine whether it can graph $J_0(x)$ and $J_1(x)$; if so, generate the graphs shown in Figure 10.8.4.

Figure 10.8.4

EXERCISE SET 10.8   ⌇ Graphing Utility

In Exercises 1–10, find the Maclaurin series for the function in sigma notation.

1. e^{-x} **2.** e^{ax} **3.** $\cos \pi x$ **4.** $\sin \pi x$

5. $\ln(1+x)$ **6.** $\dfrac{1}{1+x}$ **7.** $\cosh x$

8. $\sinh x$ **9.** $x \sin x$ **10.** xe^x

In Exercises 11–18, use sigma notation to write the Taylor series about $x = x_0$ for the given function.

11. e^x; $x_0 = 1$ **12.** e^{-x}; $x_0 = \ln 2$

13. $\dfrac{1}{x}$; $x_0 = -1$ **14.** $\dfrac{1}{x+2}$; $x_0 = 3$

15. $\sin \pi x$; $x_0 = \dfrac{1}{2}$ **16.** $\cos x$; $x_0 = \dfrac{\pi}{2}$

17. $\ln x$; $x_0 = 1$ **18.** $\ln x$; $x_0 = e$

In Exercises 19–22, find the interval of convergence of the power series, and find a familiar function that is represented by the power series on that interval.

19. $1 - x + x^2 - x^3 + \cdots + (-1)^k x^k + \cdots$

20. $1 + x^2 + x^4 + \cdots + x^{2k} + \cdots$

21. $1 + (x-2) + (x-2)^2 + \cdots + (x-2)^k + \cdots$

22. $1 - (x+3) + (x+3)^2 - (x+3)^3 + \cdots + (-1)^k (x+3)^k + \cdots$

23. Suppose that the function f is represented by the power series

$$f(x) = 1 - \frac{x}{2} + \frac{x^2}{4} - \frac{x^3}{8} + \cdots + (-1)^k \frac{x^k}{2^k} + \cdots$$

(a) Find the domain of f.   (b) Find $f(0)$ and $f(1)$.

24. Suppose that the function f is represented by the power series

$$f(x) = 1 - \frac{x-5}{3} + \frac{(x-5)^2}{3^2} - \frac{(x-5)^3}{3^3} + \cdots$$

(a) Find the domain of f.

(b) Find $f(3)$ and $f(6)$.

In Exercises 25–48, find the radius of convergence and the interval of convergence.

25. $\displaystyle\sum_{k=0}^{\infty} \frac{x^k}{k+1}$ **26.** $\displaystyle\sum_{k=0}^{\infty} 3^k x^k$ **27.** $\displaystyle\sum_{k=0}^{\infty} \frac{(-1)^k x^k}{k!}$

28. $\displaystyle\sum_{k=0}^{\infty} \frac{k!}{2^k} x^k$ **29.** $\displaystyle\sum_{k=1}^{\infty} \frac{5^k}{k^2} x^k$ **30.** $\displaystyle\sum_{k=2}^{\infty} \frac{x^k}{\ln k}$

31. $\displaystyle\sum_{k=1}^{\infty} \frac{x^k}{k(k+1)}$ **32.** $\displaystyle\sum_{k=0}^{\infty} \frac{(-2)^k x^{k+1}}{k+1}$

33. $\displaystyle\sum_{k=1}^{\infty} (-1)^{k-1} \frac{x^k}{\sqrt{k}}$ **34.** $\displaystyle\sum_{k=0}^{\infty} \frac{(-1)^k x^{2k}}{(2k)!}$

35. $\displaystyle\sum_{k=0}^{\infty} (-1)^k \frac{x^{2k+1}}{(2k+1)!}$ **36.** $\displaystyle\sum_{k=1}^{\infty} (-1)^k \frac{x^{3k}}{k^{3/2}}$

37. $\displaystyle\sum_{k=0}^{\infty} \frac{3^k}{k!} x^k$ **38.** $\displaystyle\sum_{k=2}^{\infty} (-1)^{k+1} \frac{x^k}{k(\ln k)^2}$

39. $\displaystyle\sum_{k=0}^{\infty} \frac{x^k}{1+k^2}$ **40.** $\displaystyle\sum_{k=0}^{\infty} \frac{(x-3)^k}{2^k}$

41. $\displaystyle\sum_{k=1}^{\infty} (-1)^{k+1} \frac{(x+1)^k}{k}$ **42.** $\displaystyle\sum_{k=0}^{\infty} (-1)^k \frac{(x-4)^k}{(k+1)^2}$

43. $\displaystyle\sum_{k=0}^{\infty} \left(\frac{3}{4}\right)^k (x+5)^k$ **44.** $\displaystyle\sum_{k=1}^{\infty} \frac{(2k+1)!}{k^3} (x-2)^k$

45. $\sum_{k=1}^{\infty}(-1)^k \dfrac{(x+1)^{2k+1}}{k^2+4}$ **46.** $\sum_{k=1}^{\infty}\dfrac{(\ln k)(x-3)^k}{k}$

47. $\sum_{k=0}^{\infty}\dfrac{\pi^k(x-1)^{2k}}{(2k+1)!}$ **48.** $\sum_{k=0}^{\infty}\dfrac{(2x-3)^k}{4^{2k}}$

49. Use the root test to find the interval of convergence of

$$\sum_{k=2}^{\infty}\frac{x^k}{(\ln k)^k}$$

50. Find the domain of the function

$$f(x)=\sum_{k=1}^{\infty}\frac{1\cdot 3\cdot 5\cdots(2k-1)}{(2k-2)!}x^k$$

51. If a function f is represented by a power series on an interval, then the graphs of the partial sums can be used as approximations to the graph of f.
  (a) Use a graphing utility to generate the graph of $1/(1-x)$ together with the graphs of the first four partial sums of its Maclaurin series over the interval $(-1, 1)$.
  (b) In general terms, where are the graphs of the partial sums the most accurate?

52. Show that the power series representation of the Bessel function $J_1(x)$ converges for all x [Formula (5)].

53. Show that if p is a positive integer, then the power series

$$\sum_{k=0}^{\infty}\frac{(pk)!}{(k!)^p}x^k$$

has a radius of convergence of $1/p^p$.

54. Show that if p and q are positive integers, then the power series

$$\sum_{k=0}^{\infty}\frac{(k+p)!}{k!(k+q)!}x^k$$

has a radius of convergence of $+\infty$.

55. (a) Suppose that the power series $\sum c_k(x-x_0)^k$ has radius of convergence R and p is a nonzero constant. What can you say about the radius of convergence of the power series $\sum pc_k(x-x_0)^k$? Explain your reasoning. [*Hint:* See Theorem 10.5.3.]
  (b) Suppose that the power series $\sum c_k(x-x_0)^k$ has a finite radius of convergence R, and the power series $\sum d_k(x-x_0)^k$ has a radius of convergence of $+\infty$. What can you say about the radius of convergence of $\sum(c_k+d_k)(x-x_0)^k$? Explain your reasoning.
  (c) Suppose that the power series $\sum c_k(x-x_0)^k$ has a finite radius of convergence R_1 and the power series $\sum d_k(x-x_0)^k$ has a finite radius of convergence R_2. What can you say about the radius of convergence of $\sum(c_k+d_k)(x-x_0)^k$? Explain your reasoning.

56. Prove: If $\lim_{k\to+\infty}|c_k|^{1/k}=L$, where $L\neq 0$, then $1/L$ is the radius of convergence of the power series $\sum_{k=0}^{\infty}c_kx^k$.

57. Prove: If the power series $\sum_{k=0}^{\infty}c_kx^k$ has radius of convergence R, then the series $\sum_{k=0}^{\infty}c_kx^{2k}$ has radius of convergence $\sqrt{R}$.

58. Prove: If the interval of convergence of the series $\sum_{k=0}^{\infty}c_k(x-x_0)^k$ is $(x_0-R, x_0+R]$, then the series converges conditionally at x_0+R.

10.9 CONVERGENCE OF TAYLOR SERIES; COMPUTATIONAL METHODS

In the last section we introduced power series and intervals of convergence for power series. In this section we focus in particular on Taylor series, and we demonstrate the use of the Remainder Estimation Theorem from Section 10.1 as a tool to determine whether the Taylor series of a function converges to the function on some interval. We will also show how Taylor series can be used to approximate values of trigonometric, exponential, and logarithmic functions.

THE nTH REMAINDER

Recall that the nth Taylor polynomial for a function f about $x=x_0$ has the property that its value and the values of its first n derivatives match those of f at x_0. As n increases, more and more derivatives match up, so it is reasonable to hope that for values of x near x_0 the values of the Taylor polynomials might converge to the value of $f(x)$; that is,

$$\sum_{k=0}^{n}\frac{f^{(k)}(x_0)}{k!}(x-x_0)^k\to f(x)\quad\text{as }n\to+\infty\tag{1}$$

However, the nth Taylor polynomial for f is the nth partial sum of the Taylor series for f, so (1) is equivalent to stating that the Taylor series for f converges at x, and its sum is $f(x)$. Thus, we are led to consider the following problem.

10.9.1 PROBLEM. Given a function f that has derivatives of all orders at $x = x_0$, determine whether there is an open interval containing x_0 such that $f(x)$ is the sum of its Taylor series about $x = x_0$ at each number in the interval; that is,

$$f(x) = \sum_{k=0}^{\infty} \frac{f^{(k)}(x_0)}{k!} (x - x_0)^k \tag{2}$$

for all values of x in the interval.

• FOR THE READER. Show that (2) holds at $x = x_0$, regardless of the function f.

To determine whether (2) holds on some open interval containing x_0, recall the nth remainder for f about $x = x_0$ as given in Formula (13) of Section 10.1,

$$R_n(x) = f(x) - p_n(x) = f(x) - \sum_{k=0}^{n} \frac{f^{(k)}(x_0)}{k!} (x - x_0)^k \tag{3}$$

where $p_n(x)$ is the nth Taylor polynomial for f about $x = x_0$.

One can think of $R_n(x)$ as the error that results at the domain value x when f is approximated by $p_n(x)$. Thus, for a particular value of x, if $p_n(x)$ converges to $f(x)$ as $n \to +\infty$, the error $R_n(x)$ must approach 0; conversely, if $R_n(x) \to 0$ as $n \to +\infty$, then the Taylor polynomials converge to f at x. More precisely:

10.9.2 THEOREM. *The equality*

$$f(x) = \sum_{k=0}^{\infty} \frac{f^{(k)}(x_0)}{k!} (x - x_0)^k$$

holds at a number x if and only if $\lim_{n \to +\infty} R_n(x) = 0$.

ESTIMATING THE nTH REMAINDER

It is relatively rare that one can prove directly that $R_n(x) \to 0$ as $n \to +\infty$. Usually, this is proved indirectly by finding appropriate bounds on $|R_n(x)|$ and applying the Squeezing Theorem for Sequences. The Remainder Estimation Theorem (Theorem 10.1.4) provides a useful bound for this purpose. Recall that this theorem asserts that if M is an upper bound for $|f^{(n+1)}(x)|$ on an interval I containing x_0, then

$$|R_n(x)| \leq \frac{M}{(n+1)!} |x - x_0|^{n+1} \tag{4}$$

for all x in I.

The following example illustrates how the Remainder Estimation Theorem is applied.

Example 1 Show that the Maclaurin series for $\cos x$ converges to $\cos x$ for all x; that is,

$$\cos x = \sum_{k=0}^{\infty} (-1)^k \frac{x^{2k}}{(2k)!} = 1 - \frac{x^2}{2!} + \frac{x^4}{4!} - \frac{x^6}{6!} + \cdots \qquad (-\infty < x < +\infty)$$

Solution. From Theorem 10.9.2 we must show that $R_n(x) \to 0$ for all x as $n \to +\infty$. For this purpose let $f(x) = \cos x$, so that for all x we have

$$f^{(n+1)}(x) = \pm \cos x \quad \text{or} \quad f^{(n+1)}(x) = \pm \sin x$$

In all cases we have

$$|f^{(n+1)}(x)| \leq 1$$

so we can apply (4) with $M = 1$ and $x_0 = 0$ to conclude that

$$0 \le |R_n(x)| \le \frac{|x|^{n+1}}{(n+1)!} \tag{5}$$

However, it follows from Formula (5) of Section 10.3 with $n + 1$ in place of n and $|x|$ in place of x that

$$\lim_{n \to +\infty} \frac{|x|^{n+1}}{(n+1)!} = 0 \tag{6}$$

Thus, it follows from (5) and the Squeezing Theorem for Sequences (Theorem 10.2.5) that $|R_n(x)| \to 0$ as $n \to +\infty$; this implies that $R_n(x) \to 0$ as $n \to +\infty$ by Theorem 10.2.6. Since this is true for all x, we have proved that the Maclaurin series for $\cos x$ converges to $\cos x$ for all x. This is illustrated in Figure 10.9.1, where we can see how successive partial sums approximate the cosine curve more and more closely. ◀

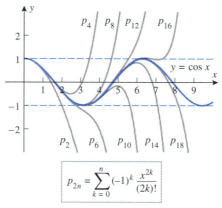

$$P_{2n} = \sum_{k=0}^{n} (-1)^k \frac{x^{2k}}{(2k)!}$$

Figure 10.9.1

REMARK.   The method used in Example 1 can be easily modified to prove that the Taylor series for $\cos x$ about any value $x = x_0$ converges to $\cos x$ for all x, and similarly that the Taylor series for $\sin x$ about any value $x = x_0$ converges to $\sin x$ for all x (Exercises 21 and 22). For reference, there is a list of some of the most important Maclaurin series in Table 10.9.1 at the end of this section.

APPROXIMATING TRIGONOMETRIC FUNCTIONS

In general, to approximate the value of a function f at a number x using a Taylor series, there are two basic questions that must be answered:

- About what domain value x_0 should the Taylor series be expanded?
- How many terms in the series should be used to achieve the desired accuracy?

In response to the first question, x_0 needs to be a number at which the derivatives of f can be evaluated easily, since these values are needed for the coefficients in the Taylor series. Furthermore, if the function f is being evaluated at x, then x_0 should be chosen as close as possible to x, since Taylor series tend to converge more rapidly near x_0. For example, to approximate $\sin 3°$ ($= \pi/60$ radians), it would be reasonable to take $x_0 = 0$, since $\pi/60$ is close to 0 and the derivatives of $\sin x$ are easy to evaluate at 0. On the other hand, to approximate $\sin 85°$ ($= 17\pi/36$ radians), it would be more natural to take $x_0 = \pi/2$, since $17\pi/36$ is close to $\pi/2$ and the derivatives of $\sin x$ are easy to evaluate at $\pi/2$.

In response to the second question posed above, the number of terms required to achieve a specific accuracy needs to be determined on a problem-by-problem basis. The next example gives two methods for doing this.

Example 2 Use the Maclaurin series for $\sin x$ to approximate $\sin 3°$ to five decimal-place accuracy.

Solution. In the Maclaurin series

$$\sin x = \sum_{k=0}^{\infty}(-1)^k \frac{x^{2k+1}}{(2k+1)!} = x - \frac{x^3}{3!} + \frac{x^5}{5!} - \frac{x^7}{7!} + \cdots \tag{7}$$

the angle x is assumed to be in radians (because the differentiation formulas for the trigonometric functions were derived with this assumption). Since $3° = \pi/60$ radians, it follows from (7) that

$$\sin 3° = \sin\frac{\pi}{60} = \left(\frac{\pi}{60}\right) - \frac{(\pi/60)^3}{3!} + \frac{(\pi/60)^5}{5!} - \frac{(\pi/60)^7}{7!} + \cdots \tag{8}$$

We must now determine how many terms in the series are required to achieve five decimal-place accuracy. We will consider two possible approaches, one using the Remainder Estimation Theorem (Theorem 10.1.4) and the other using the fact that (8) satisfies the hypotheses of the alternating series test (Theorem 10.7.1).

Method 1 (The Remainder Estimation Theorem). Since we want to achieve five decimal-place accuracy, our goal is to choose n so that the absolute value of the nth remainder at $x = \pi/60$ does not exceed $0.000005 = 5 \times 10^{-6}$; that is,

$$\left|R_n\left(\frac{\pi}{60}\right)\right| \leq 0.000005 \tag{9}$$

However, if we let $f(x) = \sin x$, then $f^{(n+1)}(x)$ is either $\pm\sin x$ or $\pm\cos x$, and in either case $|f^{(n+1)}(x)| \leq 1$ for all x. Thus, it follows from the Remainder Estimation Theorem with $M = 1$, $x_0 = 0$, and $x = \pi/60$ that

$$\left|R_n\left(\frac{\pi}{60}\right)\right| \leq \frac{|\pi/60|^{n+1}}{(n+1)!}$$

Thus, we can satisfy (9) by choosing n so that

$$\frac{|\pi/60|^{n+1}}{(n+1)!} \leq 0.000005$$

With the help of a calculating utility you can verify that the smallest value of n that meets this criterion is $n = 3$. Thus, to achieve five decimal-place accuracy we need only keep terms up to the third power in (8). This yields

$$\sin 3° \approx \left(\frac{\pi}{60}\right) - \frac{(\pi/60)^3}{3!} \approx 0.05234 \tag{10}$$

(verify). As a check, a calculator gives $\sin 3° \approx 0.05233595624$, which agrees with (10) when rounded to five decimal places.

Method 2 (The Alternating Series Test). We leave it for you to check that (8) satisfies the hypotheses of the alternating series test (Theorem 10.7.1).

Let s_n denote the sum of the terms in (8) up to and including the nth power of $\pi/60$. Since the exponents in the series are odd integers, the integer n must be odd, and the exponent of the first term *not* included in the sum s_n must be $n + 2$. Thus, it follows from part (*b*) of Theorem 10.7.2 that

$$|\sin 3° - s_n| < \frac{(\pi/60)^{n+2}}{(n+2)!}$$

This means that for five decimal-place accuracy we must look for the first positive odd integer n such that

$$\frac{(\pi/60)^{n+2}}{(n+2)!} \leq 0.000005$$

With the help of a calculating utility you can verify that the smallest value of n that meets this criterion is $n = 3$. This agrees with the result obtained above using the Remainder Estimation Theorem and hence leads to approximation (10) as before. ◄

ROUNDOFF AND TRUNCATION ERROR

There are two types of errors that occur when computing with series. The first, called *truncation error*, is the error that results when a series is approximated by a partial sum; and the second, called *roundoff error*, is the error that arises from approximations in numerical computations. For example, in our derivation of (10) we took $n = 3$ to keep the truncation error below 0.000005. However, to evaluate the partial sum we had to approximate π, thereby introducing roundoff error. Had we not exercised some care in choosing this approximation, the roundoff error could easily have degraded the final result.

Methods for estimating and controlling roundoff error are studied in a branch of mathematics called *numerical analysis*. However, as a rule of thumb, to achieve n decimal-place accuracy in a final result, all intermediate calculations must be accurate to at least $n + 1$ decimal places. Thus, in (10) at least six decimal-place accuracy in π is required to achieve the five decimal-place accuracy in the final numerical result. As a practical matter, a good working procedure is to perform all intermediate computations with the maximum number of digits that your calculating utility can handle and then round at the end.

APPROXIMATING EXPONENTIAL FUNCTIONS

Example 3 Show that the Maclaurin series for e^x converges to e^x for all x; that is,

$$e^x = \sum_{k=0}^{\infty} \frac{x^k}{k!} = 1 + x + \frac{x^2}{2!} + \frac{x^3}{3!} + \cdots + \frac{x^k}{k!} + \cdots \qquad (-\infty < x < +\infty)$$

Solution. Let $f(x) = e^x$, so that

$$f^{(n+1)}(x) = e^x$$

We want to show that $R_n(x) \to 0$ as $n \to +\infty$ for all x in the interval $-\infty < x < +\infty$. However, it will be helpful here to consider the cases $x \le 0$ and $x > 0$ separately. If $x \le 0$, then we will take the interval I in the Remainder Estimation Theorem (Theorem 10.1.4) to be $[x, 0]$, and if $x > 0$, then we will take it to be $[0, x]$. Since $f^{(n+1)}(x) = e^x$ is an increasing function, it follows that if c is in the interval $[x, 0]$, then

$$|f^{(n+1)}(c)| \le |f^{(n+1)}(0)| = e^0 = 1$$

and if c is in the interval $[0, x]$, then

$$|f^{(n+1)}(c)| \le |f^{(n+1)}(x)| = e^x$$

Thus, we can apply Theorem 10.1.4 with $M = 1$ in the case where $x \le 0$ and with $M = e^x$ in the case where $x > 0$. This yields

$$0 \le |R_n(x)| \le \frac{|x|^{n+1}}{(n+1)!} \qquad \text{if } x \le 0$$

$$0 \le |R_n(x)| \le e^x \frac{|x|^{n+1}}{(n+1)!} \qquad \text{if } x > 0$$

Thus, in both cases it follows from (6) and the Squeezing Theorem for Sequences that $|R_n(x)| \to 0$ as $n \to +\infty$, which in turn implies that $R_n(x) \to 0$ as $n \to +\infty$. Since this is true for all x, we have proved that the Maclaurin series for e^x converges to e^x for all x. ◄

Since the Maclaurin series for e^x converges to e^x for all x, we can use partial sums of the Maclaurin series to approximate powers of e to arbitrary precision. Recall that in Example 6 of Section 10.1 we were able to use the Remainder Estimation Theorem to determine that evaluating the ninth Maclaurin polynomial for e^x at $x = 1$ yields an approximation for e with five decimal-place accuracy:

$$e \approx 1 + 1 + \frac{1}{2!} + \frac{1}{3!} + \frac{1}{4!} + \frac{1}{5!} + \frac{1}{6!} + \frac{1}{7!} + \frac{1}{8!} + \frac{1}{9!} \approx 2.71828$$

•••••••••••••••••••••••••••••••
APPROXIMATING LOGARITHMS

The Maclaurin series

$$\ln(1+x) = x - \frac{x^2}{2} + \frac{x^3}{3} - \frac{x^4}{4} + \cdots \qquad (-1 < x \leq 1) \tag{11}$$

is the starting point for the approximation of natural logarithms. Unfortunately, the usefulness of this series is limited because of its slow convergence and the restriction $-1 < x \leq 1$. However, if we replace x by $-x$ in this series, we obtain

$$\ln(1-x) = -x - \frac{x^2}{2} - \frac{x^3}{3} - \frac{x^4}{4} - \cdots \qquad (-1 \leq x < 1) \tag{12}$$

and on subtracting (12) from (11) we obtain

$$\ln\left(\frac{1+x}{1-x}\right) = 2\left(x + \frac{x^3}{3} + \frac{x^5}{5} + \frac{x^7}{7} + \cdots\right) \qquad (-1 < x < 1) \tag{13}$$

Series (13), first obtained by James Gregory* in 1668, can be used to compute the natural logarithm of any positive number y by letting

$$y = \frac{1+x}{1-x}$$

or, equivalently,

$$x = \frac{y-1}{y+1} \tag{14}$$

and noting that $-1 < x < 1$. For example, to compute $\ln 2$ we let $y = 2$ in (14), which yields $x = \frac{1}{3}$. Substituting this value in (13) gives

$$\ln 2 = 2\left[\frac{1}{3} + \frac{\left(\frac{1}{3}\right)^3}{3} + \frac{\left(\frac{1}{3}\right)^5}{5} + \frac{\left(\frac{1}{3}\right)^7}{7} + \cdots\right] \tag{15}$$

In Exercise 19 we will ask you to show that five decimal-place accuracy can be achieved using the partial sum with terms up to and including the 13th power of $\frac{1}{3}$. Thus, to five decimal-place accuracy

$$\ln 2 \approx 2\left[\frac{1}{3} + \frac{\left(\frac{1}{3}\right)^3}{3} + \frac{\left(\frac{1}{3}\right)^5}{5} + \frac{\left(\frac{1}{3}\right)^7}{7} + \cdots + \frac{\left(\frac{1}{3}\right)^{13}}{13}\right] \approx 0.69315$$

(verify). As a check, a calculator gives $\ln 2 \approx 0.69314718056$, which agrees with the preceding approximation when rounded to five decimal places.

REMARK. In Example 2 of Section 10.7, we stated without proof that

$$\ln 2 = 1 - \frac{1}{2} + \frac{1}{3} - \frac{1}{4} + \frac{1}{5} - \cdots$$

This result can be obtained letting $x = 1$ in (11). However, this series converges too slowly to be of practical value.

•••••••••••••••••••••••••••••••
APPROXIMATING π

In the next section we will show that

$$\tan^{-1} x = x - \frac{x^3}{3} + \frac{x^5}{5} - \frac{x^7}{7} + \cdots \qquad (-1 \leq x \leq 1) \tag{16}$$

Letting $x = 1$, we obtain

$$\frac{\pi}{4} = \tan^{-1} 1 = 1 - \frac{1}{3} + \frac{1}{5} - \frac{1}{7} + \cdots$$

---

*JAMES GREGORY (1638–1675). Scottish mathematician and astronomer. Gregory, the son of a minister, was famous in his time as the inventor of the Gregorian reflecting telescope, so named in his honor. Although he is not generally ranked with the great mathematicians, much of his work relating to calculus was studied by Leibniz and Newton and undoubtedly influenced some of their discoveries. There is a manuscript, discovered posthumously, which shows that Gregory had anticipated Taylor series well before Taylor.

or

$$\pi = 4\left[1 - \frac{1}{3} + \frac{1}{5} - \frac{1}{7} + \cdots\right]$$

This famous series, obtained by Leibniz in 1674, converges too slowly to be of computational value. A more practical procedure for approximating π uses the identity

$$\frac{\pi}{4} = \tan^{-1}\frac{1}{2} + \tan^{-1}\frac{1}{3} \tag{17}$$

which was derived in Exercise 45 of Section 4.4. By using this identity and series (16) to approximate $\tan^{-1}\frac{1}{2}$ and $\tan^{-1}\frac{1}{3}$, the value of π can be approximated efficiently to any degree of accuracy.

BINOMIAL SERIES

If m is a real number, then the Maclaurin series for $(1 + x)^m$ is called the *binomial series*; it is given by (verify)

$$1 + mx + \frac{m(m-1)}{2!}x^2 + \frac{m(m-1)(m-2)}{3!}x^3 + \cdots + \frac{m(m-1)\cdots(m-k+1)}{k!}x^k + \cdots$$

In the case where m is a nonnegative integer, the function $f(x) = (1 + x)^m$ is a polynomial of degree m, so

$$f^{(m+1)}(0) = f^{(m+2)}(0) = f^{(m+3)}(0) = \cdots = 0$$

and the binomial series reduces to the familiar binomial expansion

$$(1 + x)^m = 1 + mx + \frac{m(m-1)}{2!}x^2 + \frac{m(m-1)(m-2)}{3!}x^3 + \cdots + x^m$$

which is valid for $-\infty < x < +\infty$.

It can be proved that if m is not a nonnegative integer, then the binomial series converges to $(1 + x)^m$ if $|x| < 1$. Thus, for such values of x

$$(1 + x)^m = 1 + mx + \frac{m(m-1)}{2!}x^2 + \cdots + \frac{m(m-1)\cdots(m-k+1)}{k!}x^k + \cdots \tag{18}$$

or in sigma notation,

$$(1 + x)^m = 1 + \sum_{k=1}^{\infty} \frac{m(m-1)\cdots(m-k+1)}{k!}x^k \quad \text{if } |x| < 1 \tag{19}$$

Example 4   Find binomial series for

(a) $\dfrac{1}{(1 + x)^2}$ (b) $\dfrac{1}{\sqrt{1 + x}}$

Solution (a). Since the general term of the binomial series is complicated, you may find it helpful to write out some of the beginning terms of the series, as in Formula (18), to see developing patterns. Substituting $m = -2$ in this formula yields

$$\frac{1}{(1 + x)^2} = (1 + x)^{-2} = 1 + (-2)x + \frac{(-2)(-3)}{2!}x^2$$

$$+ \frac{(-2)(-3)(-4)}{3!}x^3 + \frac{(-2)(-3)(-4)(-5)}{4!}x^4 + \cdots$$

$$= 1 - 2x + \frac{3!}{2!}x^2 - \frac{4!}{3!}x^3 + \frac{5!}{4!}x^4 + \cdots$$

$$= 1 - 2x + 3x^2 - 4x^3 + 5x^4 + \cdots$$

$$= \sum_{k=0}^{\infty} (-1)^k (k + 1)x^k$$

Solution (b). Substituting $m = -\frac{1}{2}$ in (18) yields

$$\frac{1}{\sqrt{1+x}} = 1 - \frac{1}{2}x + \frac{\left(-\frac{1}{2}\right)\left(-\frac{1}{2} - 1\right)}{2!}x^2 + \frac{\left(-\frac{1}{2}\right)\left(-\frac{1}{2} - 1\right)\left(-\frac{1}{2} - 2\right)}{3!}x^3 - \cdots$$

$$= 1 - \frac{1}{2}x + \frac{1 \cdot 3}{2^2 \cdot 2!}x^2 - \frac{1 \cdot 3 \cdot 5}{2^3 \cdot 3!}x^3 + \cdots$$

$$= 1 + \sum_{k=1}^{\infty} (-1)^k \frac{1 \cdot 3 \cdot 5 \cdots (2k-1)}{2^k k!} x^k \qquad \blacktriangleleft$$

For reference, Table 10.9.1 lists the Maclaurin series for some of the most important functions, together with a specification of the intervals over which the Maclaurin series converge to those functions. Some of these results are derived in the exercises and others will be derived in the next section using some special techniques that we will develop.

Table 10.9.1

MACLAURIN SERIES	INTERVAL OF CONVERGENCE
$\dfrac{1}{1-x} = \displaystyle\sum_{k=0}^{\infty} x^k = 1 + x + x^2 + x^3 + \cdots$	$-1 < x < 1$
$\dfrac{1}{1+x^2} = \displaystyle\sum_{k=0}^{\infty} (-1)^k x^{2k} = 1 - x^2 + x^4 - x^6 + \cdots$	$-1 < x < 1$
$e^x = \displaystyle\sum_{k=0}^{\infty} \dfrac{x^k}{k!} = 1 + x + \dfrac{x^2}{2!} + \dfrac{x^3}{3!} + \dfrac{x^4}{4!} + \cdots$	$-\infty < x < +\infty$
$\sin x = \displaystyle\sum_{k=0}^{\infty} (-1)^k \dfrac{x^{2k+1}}{(2k+1)!} = x - \dfrac{x^3}{3!} + \dfrac{x^5}{5!} - \dfrac{x^7}{7!} + \cdots$	$-\infty < x < +\infty$
$\cos x = \displaystyle\sum_{k=0}^{\infty} (-1)^k \dfrac{x^{2k}}{(2k)!} = 1 - \dfrac{x^2}{2!} + \dfrac{x^4}{4!} - \dfrac{x^6}{6!} + \cdots$	$-\infty < x < +\infty$
$\ln(1+x) = \displaystyle\sum_{k=1}^{\infty} (-1)^{k+1} \dfrac{x^k}{k} = x - \dfrac{x^2}{2} + \dfrac{x^3}{3} - \dfrac{x^4}{4} + \cdots$	$-1 < x \leq 1$
$\tan^{-1} x = \displaystyle\sum_{k=0}^{\infty} (-1)^k \dfrac{x^{2k+1}}{2k+1} = x - \dfrac{x^3}{3} + \dfrac{x^5}{5} - \dfrac{x^7}{7} + \cdots$	$-1 \leq x \leq 1$
$\sinh x = \displaystyle\sum_{k=0}^{\infty} \dfrac{x^{2k+1}}{(2k+1)!} = x + \dfrac{x^3}{3!} + \dfrac{x^5}{5!} + \dfrac{x^7}{7!} + \cdots$	$-\infty < x < +\infty$
$\cosh x = \displaystyle\sum_{k=0}^{\infty} \dfrac{x^{2k}}{(2k)!} = 1 + \dfrac{x^2}{2!} + \dfrac{x^4}{4!} + \dfrac{x^6}{6!} + \cdots$	$-\infty < x < +\infty$
$(1+x)^m = 1 + \displaystyle\sum_{k=1}^{\infty} \dfrac{m(m-1)\cdots(m-k+1)}{k!} x^k$	$-1 < x < 1^*$ $(m \neq 0, 1, 2, \ldots)$

*The behavior at the endpoints depends on m: For $m > 0$ the series converges absolutely at both endpoints; for $m \leq -1$ the series diverges at both endpoints; and for $-1 < m < 0$ the series converges conditionally at $x = 1$ and diverges at $x = -1$.

EXERCISE SET 10.9 ~Graphing Utility [c] CAS

1. Use both of the methods given in Example 2 to approximate $\sin 4°$ to five decimal-place accuracy, and check your work by comparing your answer to that produced directly by your calculating utility.

2. Use both of the methods given in Example 2 to approximate $\cos 3°$ to three decimal-place accuracy, and check your work by comparing your answer to that produced directly by your calculating utility.

3. Use the Maclaurin series for $\cos x$ to approximate $\cos 0.1$ to five decimal-place accuracy, and check your work by comparing your answer to that produced directly by your calculating utility.

4. Use the Maclaurin series for $\tan^{-1} x$ to approximate $\tan^{-1} 0.1$ to three decimal-place accuracy, and check your work by comparing your answer to that produced directly by your calculating utility.

5. Use an appropriate Taylor series to approximate $\sin 85°$ to four decimal-place accuracy, and check your work by comparing your answer to that produced directly by your calculating utility.

6. Use a Taylor series to approximate $\cos(-175°)$ to four decimal-place accuracy, and check your work by comparing your answer to that produced directly by your calculating utility.

7. Use the Maclaurin series for $\sinh x$ to approximate $\sinh 0.5$ to three decimal-place accuracy. Check your work by computing $\sinh 0.5$ with a calculating utility.

8. Use the Maclaurin series for $\cosh x$ to approximate $\cosh 0.1$ to three decimal-place accuracy. Check your work by computing $\cosh 0.1$ with a calculating utility.

9. Use the Remainder Estimation Theorem and the method of Example 1 to prove that the Taylor series for $\sin x$ about $x = \pi/4$ converges to $\sin x$ for all x.

10. Use the Remainder Estimation Theorem and the method of Example 3 to prove that the Taylor series for e^x about $x = 1$ converges to e^x for all x.

11. (a) Use Formula (13) in the text to find a series that converges to $\ln 1.25$.
    (b) Approximate $\ln 1.25$ using the first two terms of the series. Round your answer to three decimal places, and compare the result to that produced directly by your calculating utility.

12. (a) Use Formula (13) to find a series that converges to $\ln 3$.
    (b) Approximate $\ln 3$ using the first two terms of the series. Round your answer to three decimal places, and compare the result to that produced directly by your calculating utility.

13. (a) Use the Maclaurin series for $\tan^{-1} x$ to approximate $\tan^{-1} \frac{1}{2}$ and $\tan^{-1} \frac{1}{3}$ to three decimal-place accuracy.

(b) Use the results in part (a) and Formula (17) to approximate π.

(c) Would you be willing to guarantee that your answer in part (b) is accurate to three decimal places? Explain your reasoning.

(d) Compare your answer in part (b) to that produced by your calculating utility.

14. Use an appropriate Taylor series for $\sqrt[3]{x}$ to approximate $\sqrt[3]{28}$ to three decimal-place accuracy, and check your answer by comparing it to that produced directly by your calculating utility.

~ 15. (a) Find an upper bound on the error that can result if $\cos x$ is approximated by $1 - (x^2/2!) + (x^4/4!)$ over the interval $[-0.2, 0.2]$.

(b) Check your answer in part (a) by graphing

$$\left| \cos x - \left(1 - \frac{x^2}{2!} + \frac{x^4}{4!} \right) \right|$$

over the interval.

~ 16. (a) Find an upper bound on the error that can result if $\ln(1 + x)$ is approximated by x over the interval $[-0.01, 0.01]$.

(b) Check your answer in part (a) by graphing

$$|\ln(1 + x) - x|$$

over the interval.

17. Use Formula (18) for the binomial series to obtain the Maclaurin series for

(a) $\dfrac{1}{1 + x}$
(b) $\sqrt[3]{1 + x}$
(c) $\dfrac{1}{(1 + x)^3}$.

18. If m is any real number, and k is a nonnegative integer, then we define the **binomial coefficient**

$\dbinom{m}{k}$ by the formulas $\dbinom{m}{0} = 1$ and

$$\binom{m}{k} = \frac{m(m - 1)(m - 2) \cdots (m - k + 1)}{k!}$$

for $k \geq 1$. Express Formula (18) in the text in terms of binomial coefficients.

19. In this exercise we will use the Remainder Estimation Theorem to determine the number of terms that are required in Formula (15) to approximate $\ln 2$ to five decimal-place accuracy. For this purpose let

$$f(x) = \ln \frac{1 + x}{1 - x} = \ln(1 + x) - \ln(1 - x) \quad (-1 < x < 1)$$

(a) Show that

$$f^{(n+1)}(x) = n! \left[\frac{(-1)^n}{(1 + x)^{n+1}} + \frac{1}{(1 - x)^{n+1}} \right]$$

(b) Use the triangle inequality [Theorem 1.2.2(*d*)] to show that

$$|f^{(n+1)}(x)| \leq n! \left[\frac{1}{(1+x)^{n+1}} + \frac{1}{(1-x)^{n+1}} \right]$$

(c) Since we want to achieve five decimal-place accuracy, our goal is to choose n so that the absolute value of the nth remainder at $x = \frac{1}{3}$ does not exceed the value $0.000005 = 0.5 \times 10^{-5}$; that is, $\left|R_n\left(\frac{1}{3}\right)\right| \leq 0.000005$. Use the Remainder Estimation Theorem to show that this condition will be satisfied if n is chosen so that

$$\frac{M}{(n+1)!} \left(\frac{1}{3}\right)^{n+1} \leq 0.000005$$

where $|f^{(n+1)}(x)| \leq M$ on the interval $\left[0, \frac{1}{3}\right]$.

(d) Use the result in part (b) to show that M can be taken as

$$M = n! \left[1 + \frac{1}{\left(\frac{2}{3}\right)^{n+1}} \right]$$

(e) Use the results in parts (c) and (d) to show that five decimal-place accuracy will be achieved if n satisfies

$$\frac{1}{n+1} \left[\left(\frac{1}{3}\right)^{n+1} + \left(\frac{1}{2}\right)^{n+1} \right] \leq 0.000005$$

and then show that the smallest value of n that satisfies this condition is $n = 13$.

20. Use Formula (13) and the method of Exercise 19 to approximate $\ln\left(\frac{5}{3}\right)$ to five decimal-place accuracy. Then check your work by comparing your answer to that produced directly by your calculating utility.

21. Prove: The Taylor series for $\cos x$ about any value $x = x_0$ converges to $\cos x$ for all x.

22. Prove: The Taylor series for $\sin x$ about any value $x = x_0$ converges to $\sin x$ for all x.

23. (a) In 1706 the British astronomer and mathematician John Machin discovered the following formula for $\pi/4$, called **Machin's formula**:

$$\frac{\pi}{4} = 4\tan^{-1}\frac{1}{5} - \tan^{-1}\frac{1}{239}$$

Use a CAS to approximate $\pi/4$ using Machin's formula to 25 decimal places.

(b) In 1914 the brilliant Indian mathematician Srinivasa Ramanujan (1887–1920) showed that

$$\frac{1}{\pi} = \frac{\sqrt{8}}{9801} \sum_{k=0}^{\infty} \frac{(4k)!(1103 + 26{,}390k)}{(k!)^4 396^{4k}}$$

Use a CAS to compute the first four partial sums in **Ramanujan's formula**.

24. The purpose of this exercise is to show that the Taylor series of a function f may possibly converge to a value different from $f(x)$ for certain x. Let

$$f(x) = \begin{cases} e^{-1/x^2}, & x \neq 0 \\ 0, & x = 0 \end{cases}$$

(a) Use the definition of a derivative to show that $f'(0) = 0$.

(b) With some difficulty it can be shown that $f^{(n)}(0) = 0$ for $n \geq 2$. Accepting this fact, show that the Maclaurin series of f converges for all x, but converges to $f(x)$ only at $x = 0$.

10.10 DIFFERENTIATING AND INTEGRATING POWER SERIES; MODELING WITH TAYLOR SERIES

In this section we will discuss methods for finding power series for derivatives and integrals of functions, and we will discuss some practical methods for finding Taylor series that can be used in situations where it is difficult or impossible to find the series directly.

DIFFERENTIATING POWER SERIES

We begin by considering the following problem:

10.10.1 PROBLEM. Suppose that a function f is represented by a power series on an open interval. How can we use the power series to find the derivative of f on that interval?

The solution to this problem can be motivated by considering the Maclaurin series for $\sin x$:

$$\sin x = x - \frac{x^3}{3!} + \frac{x^5}{5!} - \frac{x^7}{7!} + \cdots \qquad (-\infty < x < +\infty)$$

Of course, we already know that the derivative of $\sin x$ is $\cos x$; however, we are concerned here with using the Maclaurin series to deduce this. The solution is easy—all we need to do is differentiate the Maclaurin series term by term and observe that the resulting series is the Maclaurin series for $\cos x$:

$$\frac{d}{dx}\left[x - \frac{x^3}{3!} + \frac{x^5}{5!} - \frac{x^7}{7!} + \cdots\right] = 1 - 3\frac{x^2}{3!} + 5\frac{x^4}{5!} - 7\frac{x^6}{7!} + \cdots$$

$$= 1 - \frac{x^2}{2!} + \frac{x^4}{4!} - \frac{x^6}{6!} + \cdots = \cos x$$

Here is another example.

$$\frac{d}{dx}[e^x] = \frac{d}{dx}\left[1 + x + \frac{x^2}{2!} + \frac{x^3}{3!} + \frac{x^4}{4!} + \cdots\right]$$

$$= 1 + 2\frac{x}{2!} + 3\frac{x^2}{3!} + 4\frac{x^3}{4!} + \cdots = 1 + x + \frac{x^2}{2!} + \frac{x^3}{3!} + \cdots = e^x$$

• **FOR THE READER.** See whether you can use this method to find the derivative of $\cos x$.

The preceding computations suggest that if a function f is represented by a power series on an open interval, then a power series representation of f' on that interval can be obtained by differentiating the power series for f term by term. This is stated more precisely in the following theorem, which we give without proof.

10.10.2 THEOREM (*Differentiation of Power Series*). *Suppose that a function f is represented by a power series in $x - x_0$ that has a nonzero radius of convergence R; that is,*

$$f(x) = \sum_{k=0}^{\infty} c_k(x - x_0)^k \qquad (x_0 - R < x < x_0 + R)$$

Then:

(a) *The function f is differentiable on the interval $(x_0 - R, x_0 + R)$.*

(b) *If the power series representation for f is differentiated term by term, then the resulting series has radius of convergence R and converges to f' on the interval $(x_0 - R, x_0 + R)$; that is,*

$$f'(x) = \sum_{k=0}^{\infty} \frac{d}{dx}[c_k(x - x_0)^k] \qquad (x_0 - R < x < x_0 + R)$$

This theorem has an important implication about the differentiability of functions that are represented by power series. According to the theorem, the power series for f' has the same radius of convergence as the power series for f, and this means that the theorem can be applied to f' as well as f. However, if we do this, then we conclude that f' is differentiable on the interval $(x_0 - R, x_0 + R)$, and the power series for f'' has the same radius of convergence as the power series for f and f'. We can now repeat this process ad infinitum, applying the theorem successively to f'', f''', ..., $f^{(n)}$, ... to conclude that f has derivatives of all orders on the interval $(x_0 - R, x_0 + R)$. Thus, we have established the following result.

10.10.3 THEOREM. *If a function f can be represented by a power series in $x - x_0$ with a nonzero radius of convergence R, then f has derivatives of all orders on the interval $(x_0 - R, x_0 + R)$.*

In short, it is only the most "well-behaved" functions that can be represented by power series; that is, if a function f does not possess derivatives of all orders on an interval $(x_0 - R, x_0 + R)$, then it cannot be represented by a power series in $x - x_0$ on that interval.

Example 1 In Section 10.8, we showed that the Bessel function $J_0(x)$ is represented by the power series

$$J_0(x) = \sum_{k=0}^{\infty} \frac{(-1)^k x^{2k}}{2^{2k}(k!)^2} \tag{1}$$

with radius of convergence $+\infty$ [see Formula (4) of that section and the related discussion]. Thus, $J_0(x)$ has derivatives of all orders on the interval $(-\infty, +\infty)$, and these can be obtained by differentiating the series term by term. For example, if we write (1) as

$$J_0(x) = 1 + \sum_{k=1}^{\infty} \frac{(-1)^k x^{2k}}{2^{2k}(k!)^2}$$

and differentiate term by term, we obtain

$$J_0'(x) = \sum_{k=1}^{\infty} \frac{(-1)^k (2k) x^{2k-1}}{2^{2k}(k!)^2} = \sum_{k=1}^{\infty} \frac{(-1)^k x^{2k-1}}{2^{2k-1}k!(k-1)!} \qquad \blacktriangleleft$$

REMARK. The computations in this example use some techniques that are worth noting. First, when a power series is expressed in sigma notation, the formula for the general term of the series will often not be of a form that can be used for differentiating the constant term. Thus, if the series has a nonzero constant term, as here, it is usually a good idea to split it off from the summation before differentiating. Second, observe how we simplified the final formula by canceling the factor k from one of the factorials in the denominator. This is a standard simplification technique.

INTEGRATING POWER SERIES

Since the derivative of a function that is represented by a power series can be obtained by differentiating the series term by term, it should not be surprising that an antiderivative of a function represented by a power series can be obtained by integrating the series term by term. For example, we know that $\sin x$ is an antiderivative of $\cos x$. Here is how this result can be obtained by integrating the Maclaurin series for $\cos x$ term by term:

$$\int \cos x \, dx = \int \left[1 - \frac{x^2}{2!} + \frac{x^4}{4!} - \frac{x^6}{6!} + \cdots \right] dx$$

$$= \left[x - \frac{x^3}{3(2!)} + \frac{x^5}{5(4!)} - \frac{x^7}{7(6!)} + \cdots \right] + C$$

$$= \left[x - \frac{x^3}{3!} + \frac{x^5}{5!} - \frac{x^7}{7!} + \cdots \right] + C = \sin x + C$$

The same idea applies to definite integrals. For example, by direct integration we have

$$\int_0^1 \frac{dx}{1+x^2} = \tan^{-1} x \Big]_0^1 = \tan^{-1} 1 - \tan 0 = \frac{\pi}{4} - 0 = \frac{\pi}{4}$$

and we will show later in this section that

$$\frac{\pi}{4} = 1 - \frac{1}{3} + \frac{1}{5} - \frac{1}{7} + \cdots \tag{2}$$

Thus,

$$\int_0^1 \frac{dx}{1+x^2} = 1 - \frac{1}{3} + \frac{1}{5} - \frac{1}{7} + \cdots$$

Here is how this result can be obtained by integrating the Maclaurin series for $1/(1+x^2)$ term by term (see Table 10.9.1):

$$\int_0^1 \frac{dx}{1+x^2} = \int_0^1 [1 - x^2 + x^4 - x^6 + \cdots] \, dx$$

$$= x - \frac{x^3}{3} + \frac{x^5}{5} - \frac{x^7}{7} + \cdots \Big]_0^1 = 1 - \frac{1}{3} + \frac{1}{5} - \frac{1}{7} + \cdots$$

The preceding computations are justified by the following theorem, which we give without proof.

10.10.4 THEOREM (*Integration of Power Series*). *Suppose that a function f is represented by a power series in $x - x_0$ that has a nonzero radius of convergence R; that is,*

$$f(x) = \sum_{k=0}^{\infty} c_k(x - x_0)^k \qquad (x_0 - R < x < x_0 + R)$$

(a) *If the power series representation of f is integrated term by term, then the resulting series has radius of convergence R and converges to an antiderivative for f(x) on the interval $(x_0 - R, x_0 + R)$; that is,*

$$\int f(x)\,dx = \sum_{k=0}^{\infty} \left[\frac{c_k}{k+1}(x - x_0)^{k+1} \right] + C \qquad (x_0 - R < x < x_0 + R)$$

(b) *If α and β are points in the interval $(x_0 - R, x_0 + R)$, and if the power series representation of f is integrated term by term from α to β, then the resulting series converges absolutely on the interval $(x_0 - R, x_0 + R)$ and*

$$\int_{\alpha}^{\beta} f(x)\,dx = \sum_{k=0}^{\infty} \left[\int_{\alpha}^{\beta} c_k(x - x_0)^k\,dx \right]$$

POWER SERIES REPRESENTATIONS MUST BE TAYLOR SERIES

For many functions it is difficult or impossible to find the derivatives that are required to obtain a Taylor series. For example, to find the Maclaurin series for $1/(1 + x^2)$ directly would require some tedious derivative computations (try it). A more practical approach is to substitute $-x^2$ for x in the geometric series

$$\frac{1}{1 - x} = 1 + x + x^2 + x^3 + x^4 + \cdots \qquad (-1 < x < 1)$$

to obtain

$$\frac{1}{1 + x^2} = 1 - x^2 + x^4 - x^6 + x^8 - \cdots$$

However, there are two questions of concern with this procedure:

- Where does the power series that we obtained for $1/(1 + x^2)$ actually converge to $1/(1 + x^2)$?

- How do we know that the power series we have obtained is actually the Maclaurin series for $1/(1 + x^2)$?

The first question is easy to resolve. Since the geometric series converges to $1/(1 - x)$ if $|x| < 1$, the second series will converge to $1/(1 + x^2)$ if $|-x^2| < 1$ or $|x^2| < 1$. However, this is true if and only if $|x| < 1$, so the power series we obtained for the function $1/(1 + x^2)$ converges to this function if $-1 < x < 1$.

The second question is more difficult to answer and leads us to the following general problem.

10.10.5 PROBLEM. Suppose that a function f is represented by a power series in $x - x_0$ that has a nonzero radius of convergence. What relationship exists between the given power series and the Taylor series for f about $x = x_0$?

The answer is that they are the same; and here is the theorem that proves it.

10.10.6 THEOREM. *If a function f is represented by a power series in $x - x_0$ on some open interval containing x_0, then that power series is the Taylor series for f about $x = x_0$.*

Proof. Suppose that

$$f(x) = c_0 + c_1(x - x_0) + c_2(x - x_0)^2 + \cdots + c_k(x - x_0)^k + \cdots$$

for all x in some open interval containing x_0. To prove that this is the Taylor series for f about $x = x_0$, we must show that

$$c_k = \frac{f^{(k)}(x_0)}{k!} \quad \text{for} \quad k = 0, 1, 2, 3, \ldots$$

However, the assumption that the series converges to $f(x)$ on an open interval containing x_0 ensures that it has a nonzero radius of convergence R; hence we can differentiate term by term in accordance with Theorem 10.10.2. Thus,

$$
\begin{aligned}
f(x) &= c_0 + c_1(x - x_0) + c_2(x - x_0)^2 + c_3(x - x_0)^3 + c_4(x - x_0)^4 + \cdots \\
f'(x) &= c_1 + 2c_2(x - x_0) + 3c_3(x - x_0)^2 + 4c_4(x - x_0)^3 + \cdots \\
f''(x) &= 2!c_2 + (3 \cdot 2)c_3(x - x_0) + (4 \cdot 3)c_4(x - x_0)^2 + \cdots \\
f'''(x) &= 3!c_3 + (4 \cdot 3 \cdot 2)c_4(x - x_0) + \cdots \\
&\vdots
\end{aligned}
$$

On substituting $x = x_0$, all the powers of $x - x_0$ drop out, leaving

$$f(x_0) = c_0, \quad f'(x_0) = c_1, \quad f''(x_0) = 2!c_2, \quad f'''(x_0) = 3!c_3, \ldots$$

from which we obtain

$$c_0 = f(x_0), \quad c_1 = f'(x_0), \quad c_2 = \frac{f''(x_0)}{2!}, \quad c_3 = \frac{f'''(x_0)}{3!}, \ldots$$

which shows that the coefficients $c_0, c_1, c_2, c_3, \ldots$ are precisely the coefficients in the Taylor series about x_0 for $f(x)$. ∎

> **REMARK.** This theorem tells us that no matter how we arrive at a power series representation of a function f, be it by substitution, by differentiation, by integration, or by some sort of algebraic manipulation, that series will be the Taylor series for f about $x = x_0$, provided that it converges to f on some open interval containing x_0.

SOME PRACTICAL WAYS TO FIND TAYLOR SERIES

Example 2 Find the Maclaurin series for $\tan^{-1} x$.

Solution. It would be tedious to find the Maclaurin series directly. A better approach is to start with the formula

$$\int \frac{1}{1 + x^2} \, dx = \tan^{-1} x + C$$

and integrate the Maclaurin series

$$\frac{1}{1 + x^2} = 1 - x^2 + x^4 - x^6 + x^8 - \cdots \qquad (-1 < x < 1)$$

term by term. This yields

$$\tan^{-1} x + C = \int \frac{1}{1 + x^2} \, dx = \int [1 - x^2 + x^4 - x^6 + x^8 - \cdots] \, dx$$

or

$$\tan^{-1} x = \left[x - \frac{x^3}{3} + \frac{x^5}{5} - \frac{x^7}{7} + \frac{x^9}{9} - \cdots \right] - C$$

The constant of integration can be evaluated by substituting $x = 0$ and using the condition $\tan^{-1} 0 = 0$. This gives $C = 0$, so that

$$\tan^{-1} x = x - \frac{x^3}{3} + \frac{x^5}{5} - \frac{x^7}{7} + \frac{x^9}{9} - \cdots \qquad (-1 < x < 1) \tag{3}$$

◀

REMARK. Observe that neither Theorem 10.10.2 nor Theorem 10.10.3 addresses what happens at the endpoints of the interval of convergence. However, it can be proved that if the Taylor series for f about $x = x_0$ converges to $f(x)$ for all x in the interval $(x_0 - R, x_0 + R)$, and if the Taylor series converges at the right endpoint $x_0 + R$, then the value that it converges to at that point is the limit of $f(x)$ as $x \to x_0 + R$ from the left; and if the Taylor series converges at the left endpoint $x_0 - R$, then the value that it converges to at that point is the limit of $f(x)$ as $x \to x_0 - R$ from the right.

For example, the Maclaurin series for $\tan^{-1} x$ given in (3) converges at both $x = -1$ and $x = 1$, since the hypotheses of the alternating series test (Theorem 10.7.1) are satisfied at those points. Thus, the continuity of $\tan^{-1} x$ on the interval $[-1, 1]$ implies that at $x = 1$ the Maclaurin series converges to

$$\lim_{x \to 1^-} \tan^{-1} x = \tan^{-1} 1 = \frac{\pi}{4}$$

and at $x = -1$ it converges to

$$\lim_{x \to -1^+} \tan^{-1} x = \tan^{-1}(-1) = -\frac{\pi}{4}$$

This shows that the Maclaurin series for $\tan^{-1} x$ actually converges to $\tan^{-1} x$ on the interval $-1 \leq x \leq 1$. Moreover, the convergence at $x = 1$ establishes Formula (2).

Taylor series provide an alternative to Simpson's rule and other numerical methods for approximating definite integrals.

Example 3 Approximate the integral

$$\int_0^1 e^{-x^2}\, dx$$

to three decimal-place accuracy by expanding the integrand in a Maclaurin series and integrating term by term.

Solution. The simplest way to obtain the Maclaurin series for e^{-x^2} is to replace x by $-x^2$ in the Maclaurin series

$$e^x = 1 + x + \frac{x^2}{2!} + \frac{x^3}{3!} + \frac{x^4}{4!} + \cdots$$

to obtain

$$e^{-x^2} = 1 - x^2 + \frac{x^4}{2!} - \frac{x^6}{3!} + \frac{x^8}{4!} - \cdots$$

Therefore,

$$\int_0^1 e^{-x^2}\, dx = \int_0^1 \left[1 - x^2 + \frac{x^4}{2!} - \frac{x^6}{3!} + \frac{x^8}{4!} - \cdots \right] dx$$

$$= \left[x - \frac{x^3}{3} + \frac{x^5}{5(2!)} - \frac{x^7}{7(3!)} + \frac{x^9}{9(4!)} - \cdots \right]_0^1$$

$$= 1 - \frac{1}{3} + \frac{1}{5 \cdot 2!} - \frac{1}{7 \cdot 3!} + \frac{1}{9 \cdot 4!} - \cdots$$

$$= \sum_{k=0}^{\infty} \frac{(-1)^k}{(2k+1)k!}$$

Since this series clearly satisfies the hypotheses of the alternating series test (Theorem 10.7.1), it follows from Theorem 10.7.2 that if we approximate the integral by s_n (the nth partial sum of the series), then

$$\left| \int_0^1 e^{-x^2}\, dx - s_n \right| < \frac{1}{[2(n+1)+1](n+1)!} = \frac{1}{(2n+3)(n+1)!}$$

Thus, for three decimal-place accuracy we must choose n such that

$$\frac{1}{(2n+3)(n+1)!} \le 0.0005 = 5 \times 10^{-4}$$

With the help of a calculating utility you can show that the smallest value of n that satisfies this condition is $n = 5$. Thus, the value of the integral to three decimal-place accuracy is

$$\int_0^1 e^{-x^2}\, dx \approx 1 - \frac{1}{3} + \frac{1}{5 \cdot 2!} - \frac{1}{7 \cdot 3!} + \frac{1}{9 \cdot 4!} - \frac{1}{11 \cdot 5!} \approx 0.747$$

As a check, a calculator with a built-in numerical integration capability produced the approximation 0.746824, which agrees with our result when rounded to three decimal places. ◀

FOR THE READER.   What advantages does the method in this example have over Simpson's rule? What are its disadvantages?

....
FINDING MACLAURIN SERIES BY MULTIPLICATION AND DIVISION

$$1 - x^2 + \frac{x^4}{2} - \cdots$$
$$\times \quad x - \frac{x^3}{3} + \frac{x^5}{5} - \cdots$$
$$\overline{}$$
$$x - x^3 + \frac{x^5}{2} - \cdots$$
$$\quad - \frac{x^3}{3} + \frac{x^5}{3} - \frac{x^7}{6} + \cdots$$
$$\quad \frac{x^5}{5} - \frac{x^7}{5} + \cdots$$
$$\overline{}$$
$$x - \frac{4}{3}x^3 + \frac{31}{30}x^5 - \cdots$$

The following examples illustrate some algebraic techniques that are sometimes useful for finding Taylor series.

Example 4   Find the first three nonzero terms in the Maclaurin series for the function $f(x) = e^{-x^2} \tan^{-1} x$.

Solution.   Using the series for e^{-x^2} and $\tan^{-1} x$ obtained in Examples 2 and 3 gives

$$e^{-x^2} \tan^{-1} x = \left(1 - x^2 + \frac{x^4}{2} - \cdots \right)\left(x - \frac{x^3}{3} + \frac{x^5}{5} - \cdots \right)$$

Multiplying, as shown in the margin, we obtain

$$e^{-x^2} \tan^{-1} x = x - \frac{4}{3}x^3 + \frac{31}{30}x^5 - \cdots$$

More terms in the series can be obtained by including more terms in the factors. Moreover, one can prove that a series obtained by this method converges at each point in the intersection of the intervals of convergence of the factors (and possibly on a larger interval). Thus, we can be certain that the series we have obtained converges for all x in the interval $-1 \le x \le 1$ (why?). ◀

$$x + \frac{x^3}{3} + \frac{2x^5}{15} + \cdots$$
$$1 - \frac{x^2}{2} + \frac{x^4}{24} - \cdots \,\Big)\, x - \frac{x^3}{6} + \frac{x^5}{120} - \cdots$$
$$\quad x - \frac{x^3}{2} + \frac{x^5}{24} - \cdots$$
$$\overline{}$$
$$\quad \frac{x^3}{3} - \frac{x^5}{30} + \cdots$$
$$\quad \frac{x^3}{3} - \frac{x^5}{6} + \cdots$$
$$\overline{}$$
$$\quad\quad \frac{2x^5}{15} + \cdots$$

FOR THE READER.   If you have a CAS, read the documentation about multiplying polynomials, and then use the CAS to duplicate the result in the last example.

Example 5   Find the first three nonzero terms in the Maclaurin series for $\tan x$.

Solution.   Using the first three terms in the Maclaurin series for $\sin x$ and $\cos x$, we can express $\tan x$ as

$$\tan x = \frac{\sin x}{\cos x} = \frac{x - \dfrac{x^3}{3!} + \dfrac{x^5}{5!} - \cdots}{1 - \dfrac{x^2}{2!} + \dfrac{x^4}{4!} - \cdots}$$

Dividing, as shown in the margin, we obtain

$$\tan x = x + \frac{x^3}{3} + \frac{2x^5}{15} + \cdots$$ ◀

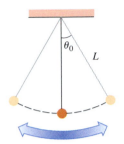

MODELING PHYSICAL LAWS WITH TAYLOR SERIES

Figure 10.10.1

Taylor series provide an important way of modeling physical laws. To illustrate the idea we will consider the problem of modeling the period of a simple pendulum (Figure 10.10.1). As explained in Exercise 38 of the supplementary exercises to Chapter 8, the period T of such a pendulum is given by

$$T = 4\sqrt{\frac{L}{g}} \int_0^{\pi/2} \frac{1}{\sqrt{1 - k^2 \sin^2 \phi}} \, d\phi \qquad (4)$$

where

L = length of the supporting rod

g = acceleration due to gravity

$k = \sin(\theta_0/2)$, where θ_0 is the initial angle of displacement from the vertical

The integral, which is called a ***complete elliptic integral of the first kind***, cannot be expressed in terms of elementary functions and is often approximated by numerical methods. Unfortunately, numerical values are so specific that they often give little insight into general physical principles. However, if we expand the integrand of (4) in a Maclaurin series and integrate term by term, then we can generate an infinite series that can be used to construct various mathematical models for the period T that give a deeper understanding of the behavior of the pendulum.

To obtain the Maclaurin series for the integrand, we will substitute $-k^2 \sin^2 \phi$ for x in the binomial series for $1/\sqrt{1+x}$ that we derived in Example 4 of Section 10.9. If we do this, then we can rewrite (4) as

$$T = 4\sqrt{\frac{L}{g}} \int_0^{\pi/2} \left[1 + \frac{1}{2}k^2 \sin^2 \phi + \frac{1 \cdot 3}{2^2 2!}k^4 \sin^4 \phi + \frac{1 \cdot 3 \cdot 5}{2^3 3!}k^6 \sin^6 \phi + \cdots \right] d\phi \qquad (5)$$

If we integrate term by term, then we can produce a Maclaurin series that converges to the period T. However, one of the most important cases of pendulum motion occurs when the initial displacement is small, in which case all subsequent displacements are small, and we can assume that $k = \sin(\theta_0/2) \approx 0$. In this case we expect the convergence of the Maclaurin series for T to be rapid, and we can approximate the sum of the series by dropping all but the constant term in (5). This yields

$$T = 2\pi \sqrt{\frac{L}{g}} \qquad (6)$$

which is called the ***first-order model*** of T or the model for ***small vibrations***. This model can be improved on by using more terms in the series. For example, if we use the first two terms in the Maclaurin series, we obtain the ***second-order model***

$$T = 2\pi \sqrt{\frac{L}{g} \left(1 + \frac{k^2}{4} \right)} \qquad (7)$$

(verify).

EXERCISE SET 10.10 ⓒ CAS

1. In each part, obtain the Maclaurin series for the function by making an appropriate substitution in the Maclaurin series for $1/(1-x)$. Include the general term in your answer, and state the radius of convergence of the series.

(a) $\dfrac{1}{1+x}$ (b) $\dfrac{1}{1-x^2}$ (c) $\dfrac{1}{1-2x}$ (d) $\dfrac{1}{2-x}$

2. In each part, obtain the Maclaurin series for the function by making an appropriate substitution in the Maclaurin series for $\ln(1+x)$. Include the general term in your answer, and state the radius of convergence of the series.

(a) $\ln(1-x)$ (b) $\ln(1+x^2)$
(c) $\ln(1+2x)$ (d) $\ln(2+x)$

3. In each part, obtain the first four nonzero terms of the Maclaurin series for the function by making an appropriate substitution in one of the binomial series obtained in Example 4 of Section 10.9.

(a) $(2+x)^{-1/2}$ (b) $(1-x^2)^{-2}$

4. (a) Use the Maclaurin series for $1/(1-x)$ to find the Maclaurin series for $1/(a-x)$, where $a \neq 0$, and state the radius of convergence of the series.

(b) Use the binomial series for $1/(1+x)^2$ obtained in Example 4 of Section 10.9 to find the first four nonzero terms in the Maclaurin series for $1/(a+x)^2$, where $a \neq 0$, and state the radius of convergence of the series.

In Exercises 5–8, obtain the first four nonzero terms of the Maclaurin series for the function by making an appropriate substitution in a known Maclaurin series and performing any algebraic operations that are required. State the radius of convergence of the series.

5. (a) $\sin 2x$ (b) e^{-2x} (c) e^{x^2} (d) $x^2 \cos \pi x$

6. (a) $\cos 2x$ (b) $x^2 e^x$ (c) xe^{-x} (d) $\sin(x^2)$

7. (a) $\dfrac{x^2}{1+3x}$ (b) $x \sinh 2x$ (c) $x(1-x^2)^{3/2}$

8. (a) $\dfrac{x}{x-1}$ (b) $3\cosh(x^2)$ (c) $\dfrac{x}{(1+2x)^3}$

In Exercises 9 and 10, find the first four nonzero terms of the Maclaurin series for the function by using an appropriate trigonometric identity or property of logarithms and then substituting in a known Maclaurin series.

9. (a) $\sin^2 x$ (b) $\ln[(1+x^3)^{12}]$

10. (a) $\cos^2 x$ (b) $\ln\left(\dfrac{1-x}{1+x}\right)$

11. (a) Use a known Maclaurin series to find the Taylor series of $1/x$ about $x = 1$ by expressing this function as

$$\frac{1}{x} = \frac{1}{1-(1-x)}$$

(b) Find the interval of convergence of the Taylor series.

12. Use the method of Exercise 11 to find the Taylor series of $1/x$ about $x = x_0$, and state the interval of convergence of the Taylor series.

In Exercises 13 and 14, find the first four nonzero terms of the Maclaurin series for the function by multiplying the Maclaurin series of the factors.

13. (a) $e^x \sin x$ (b) $\sqrt{1+x}\,\ln(1+x)$

14. (a) $e^{-x^2} \cos x$ (b) $(1+x^2)^{4/3}(1+x)^{1/3}$

In Exercises 15 and 16, find the first four nonzero terms of the Maclaurin series for the function by dividing appropriate Maclaurin series.

15. (a) $\sec x \quad \left(= \dfrac{1}{\cos x}\right)$ (b) $\dfrac{\sin x}{e^x}$

16. (a) $\dfrac{\tan^{-1} x}{1+x}$ (b) $\dfrac{\ln(1+x)}{1-x}$

17. Use the Maclaurin series for e^x and e^{-x} to derive the Maclaurin series for $\sinh x$ and $\cosh x$. Include the general terms in your answers and state the radius of convergence of each series.

18. Use the Maclaurin series for $\sinh x$ and $\cosh x$ to obtain the first four nonzero terms in the Maclaurin series for $\tanh x$.

In Exercises 19 and 20, find the first five nonzero terms of the Maclaurin series for the function by using partial fractions and a known Maclaurin series.

19. $\dfrac{4x-2}{x^2-1}$ **20.** $\dfrac{x^3+x^2+2x-2}{x^2-1}$

In Exercises 21 and 22, confirm the derivative formula by differentiating the appropriate Maclaurin series term by term.

21. (a) $\dfrac{d}{dx}[\cos x] = -\sin x$ (b) $\dfrac{d}{dx}[\ln(1+x)] = \dfrac{1}{1+x}$

22. (a) $\dfrac{d}{dx}[\sinh x] = \cosh x$ (b) $\dfrac{d}{dx}[\tan^{-1} x] = \dfrac{1}{1+x^2}$

In Exercises 23 and 24, confirm the integration formula by integrating the appropriate Maclaurin series term by term.

23. (a) $\displaystyle\int e^x \, dx = e^x + C$ (b) $\displaystyle\int \sinh x \, dx = \cosh x + C$

24. (a) $\displaystyle\int \sin x \, dx = -\cos x + C$

(b) $\displaystyle\int \dfrac{1}{1+x} \, dx = \ln(1+x) + C$

25. (a) Use the Maclaurin series for $1/(1-x)$ to find the Maclaurin series for

$$f(x) = \frac{x}{1-x^2}$$

(b) Use the Maclaurin series obtained in part (a) to find $f^{(5)}(0)$ and $f^{(6)}(0)$.

(c) What can you say about the value of $f^{(n)}(0)$?

26. Let $f(x) = x^2 \cos 2x$. Use the method of Exercise 25 to find $f^{(99)}(0)$.

The limit of an indeterminate form as $x \to x_0$ can sometimes be found without using L'Hôpital's rule by expanding the functions involved in Taylor series about $x = x_0$ and taking the limit of the series term by term. Use this method to find the limits in Exercises 27 and 28.

27. (a) $\displaystyle\lim_{x \to 0} \dfrac{\sin x}{x}$ (b) $\displaystyle\lim_{x \to 0} \dfrac{\tan^{-1} x - x}{x^3}$

28. (a) $\lim\limits_{x \to 0} \dfrac{1 - \cos x}{\sin x}$
(b) $\lim\limits_{x \to 0} \dfrac{\ln \sqrt{1 + x} - \sin 2x}{x}$

In Exercises 29–32, use Maclaurin series to approximate the integral to three decimal-place accuracy.

29. $\displaystyle\int_0^1 \sin(x^2)\,dx$

30. $\displaystyle\int_0^{1/2} \tan^{-1}(2x^2)\,dx$

31. $\displaystyle\int_0^{0.2} \sqrt[3]{1 + x^4}\,dx$

32. $\displaystyle\int_0^{1/2} \dfrac{dx}{\sqrt[4]{x^2 + 1}}$

33. (a) Differentiate the Maclaurin series for $1/(1 - x)$, and use the result to show that
$$\sum_{k=1}^{\infty} kx^k = \frac{x}{(1 - x)^2} \quad \text{for } -1 < x < 1$$

(b) Integrate the Maclaurin series for $1/(1 - x)$, and use the result to show that
$$\sum_{k=1}^{\infty} \frac{x^k}{k} = -\ln(1 - x) \quad \text{for } -1 < x < 1$$

(c) Use the result in part (b) to show that
$$\sum_{k=1}^{\infty} (-1)^{k+1} \frac{x^k}{k} = \ln(1 + x) \quad \text{for } -1 < x < 1$$

(d) Show that the series in part (c) converges if $x = 1$.

(e) Use the remark following Example 2 to show that
$$\sum_{k=1}^{\infty} (-1)^{k+1} \frac{x^k}{k} = \ln(1 + x) \quad \text{for } -1 < x \le 1$$

34. In each part, use the results in Exercise 33 to find the sum of the series.

(a) $\displaystyle\sum_{k=1}^{\infty} \frac{k}{3^k} = \frac{1}{3} + \frac{2}{3^2} + \frac{3}{3^3} + \frac{4}{3^4} + \cdots$

(b) $\displaystyle\sum_{k=1}^{\infty} \frac{1}{k(4^k)} = \frac{1}{4} + \frac{1}{2(4^2)} + \frac{1}{3(4^3)} + \frac{1}{4(4^4)} + \cdots$

(c) $\displaystyle\sum_{k=1}^{\infty} (-1)^{k+1} \frac{1}{k} = 1 - \frac{1}{2} + \frac{1}{3} - \frac{1}{4} + \cdots$

35. (a) Use the relationship
$$\int \frac{1}{\sqrt{1 + x^2}}\,dx = \sinh^{-1} x + C$$
to find the first four nonzero terms in the Maclaurin series for $\sinh^{-1} x$.

(b) Express the series in sigma notation.

(c) What is the radius of convergence?

36. (a) Use the relationship
$$\int \frac{1}{\sqrt{1 - x^2}}\,dx = \sin^{-1} x + C$$
to find the first four nonzero terms in the Maclaurin series for $\sin^{-1} x$.

(b) Express the series in sigma notation.

(c) What is the radius of convergence?

37. We showed by Formula (12) of Section 9.3 that if there are y_0 units of radioactive carbon-14 present at time $t = 0$, then the number of units present t years later is
$$y(t) = y_0 e^{-0.000121t}$$

(a) Express $y(t)$ as a Maclaurin series.

(b) Use the first two terms in the series to show that the number of units present after 1 year is approximately $(0.999879)y_0$.

(c) Compare this to the value produced by the formula for $y(t)$.

38. In Section 9.1 we studied the motion of a falling object that has mass m and is retarded by air resistance. We showed that if the initial velocity is v_0 and the drag force F_R is proportional to the velocity, that is, $F_R = -cv$, then the velocity of the object at time t is
$$v(t) = e^{-ct/m}\left(v_0 + \frac{mg}{c}\right) - \frac{mg}{c}$$
where g is the acceleration due to gravity [see Formula (23) of Section 9.1].

(a) Use a Maclaurin series to show that if $ct/m \approx 0$, then the velocity can be approximated as
$$v(t) \approx v_0 - \left(\frac{cv_0}{m} + g\right)t$$

(b) Improve on the approximation in part (a).

39. Suppose that a simple pendulum with a length of $L = 1$ meter is given an initial displacement of $\theta_0 = 5°$ from the vertical.

(a) Approximate the period of the pendulum using Formula (6) for the first-order model. [Take $g = 9.8 \text{ m/s}^2$.]

(b) Approximate the period of the pendulum using Formula (7) for the second-order model.

(c) Use the numerical integration capability of a CAS to approximate the period of the pendulum from Formula (4), and compare it to the values obtained in parts (a) and (b).

40. Use the first three nonzero terms in Formula (5) and the Wallis sine formula in the Endpaper Integral Table (Formula 122) to obtain a model for the period of a simple pendulum.

41. Recall that the gravitational force exerted by the Earth on an object is called the object's *weight* (or more precisely, its *Earth weight*). We noted in statement 9.4.3 that if an object has mass m, then the magnitude of its weight is mg. However, this result presumes that the object is on the surface of the Earth (mean sea level). A more general formula for the magnitude of the gravitational force that the Earth exerts on an object of mass m is
$$F = \frac{mgR^2}{(R + h)^2}$$
where R is the radius of the Earth and h is the height of the object above the Earth's surface.

(a) Use the binomial series for $1/(1 + x)^2$ obtained in Example 4 of Section 10.9 to express F as a Maclaurin series in powers of h/R.

(b) Show that if $h = 0$, then $F = mg$.

(c) Show that if $h/R \approx 0$, then $F \approx mg - (2mgh/R)$.
[*Note:* The quantity $2mgh/R$ can be thought of as a "correction term" for the weight that takes the object's height above the Earth's surface into account.]

(d) If we assume that the Earth is a sphere of radius $R = 4000$ mi at mean sea level, by approximately what percentage does a person's weight change in going from mean sea level to the top of Mt. Everest (29,028 ft)?

42. (a) Show that the Bessel function $J_0(x)$ given by Formula (4) of Section 10.8 satisfies the differential equation

$xy'' + y' + xy = 0$. (This is called the ***Bessel equation of order zero.***)

(b) Show that the Bessel function $J_1(x)$ given by Formula (5) of Section 10.8 satisfies the differential equation $x^2 y'' + xy' + (x^2 - 1)y = 0$. (This is called the ***Bessel equation of order one*.**)

(c) Show that $J_0'(x) = -J_1(x)$.

43. Prove: If the power series $\sum_{k=0}^{\infty} a_k x^k$ and $\sum_{k=0}^{\infty} b_k x^k$ have the same sum on an interval $(-r, r)$, then $a_k = b_k$ for all values of k.

SUPPLEMENTARY EXERCISES

C CAS

1. What is the difference between an infinite sequence and an infinite series?

2. What is meant by the sum of an infinite series?

3. (a) What is a geometric series? Give some examples of convergent and divergent geometric series.

(b) What is a p-series? Give some examples of convergent and divergent p-series.

4. (a) Write down the formula for the Maclaurin series for f in sigma notation.

(b) Write down the formula for the Taylor series for f about $x = x_0$ in sigma notation.

5. State conditions under which an alternating series is guaranteed to converge.

6. (a) What does it mean to say that an infinite series converges absolutely?

(b) What relationship exists between convergence and absolute convergence of an infinite series?

7. If a power series in $x - x_0$ has radius of convergence R, what can you say about the set of x-values at which it converges?

8. State the Remainder Estimation Theorem, and describe some of its uses.

9. Are the following statements true or false? If true, state a theorem to justify your conclusion; if false, then give a counterexample.

(a) If $\sum u_k$ converges, then $u_k \to 0$ as $k \to +\infty$.

(b) If $u_k \to 0$ as $k \to +\infty$, then $\sum u_k$ converges.

(c) If $f(n) = a_n$ for $n = 1, 2, 3, \ldots$, and if $a_n \to L$ as $n \to +\infty$, then $f(x) \to L$ as $x \to +\infty$.

(d) If $f(n) = a_n$ for $n = 1, 2, 3, \ldots$, and if $f(x) \to L$ as $x \to +\infty$, then $a_n \to L$ as $n \to +\infty$.

(e) If $0 < a_n < 1$, then $\{a_n\}$ converges.

(f) If $0 < u_k < 1$, then $\sum u_k$ converges.

(g) If $\sum u_k$ and $\sum v_k$ converge, then $\sum (u_k + v_k)$ diverges.

(h) If $\sum u_k$ and $\sum v_k$ diverge, then $\sum (u_k - v_k)$ converges.

(i) If $0 \le u_k \le v_k$ and $\sum v_k$ converges, then $\sum u_k$ converges.

(j) If $0 \le u_k \le v_k$ and $\sum u_k$ diverges, then $\sum v_k$ diverges.

(k) If an infinite series converges, then it converges absolutely.

(l) If an infinite series diverges absolutely, then it diverges.

10. State whether each of the following is true or false. Justify your answers.

(a) The function $f(x) = x^{1/3}$ has a Maclaurin series.

(b) $1 + \frac{1}{2} - \frac{1}{2} + \frac{1}{3} - \frac{1}{3} + \frac{1}{4} - \frac{1}{4} + \cdots = 1$

(c) $1 + \frac{1}{2} - \frac{1}{2} + \frac{1}{2} - \frac{1}{2} + \frac{1}{2} - \frac{1}{2} + \cdots = 1$

In Exercises 11–14, use any method to determine whether the series converge.

11. (a) $\displaystyle\sum_{k=1}^{\infty} \frac{1}{5^k}$ (b) $\displaystyle\sum_{k=1}^{\infty} \frac{1}{5^k + 1}$ (c) $\displaystyle\sum_{k=1}^{\infty} \frac{9}{\sqrt{k} + 1}$

12. (a) $\displaystyle\sum_{k=1}^{\infty} (-1)^{k+1} \frac{k+4}{k^2+k}$ (b) $\displaystyle\sum_{k=1}^{\infty} (-1)^{k+1} \left(\frac{k+2}{3k-1}\right)^k$

(c) $\displaystyle\sum_{k=1}^{\infty} \frac{k^{-1/2}}{2 + \sin^2 k}$

13. (a) $\displaystyle\sum_{k=1}^{\infty} \frac{1}{k^3 + 2k + 1}$ (b) $\displaystyle\sum_{k=1}^{\infty} \frac{1}{(3+k)^{2/5}}$

(c) $\displaystyle\sum_{k=1}^{\infty} \frac{\cos(1/k)}{k^2}$

14. (a) $\displaystyle\sum_{k=1}^{\infty} \frac{\ln k}{k\sqrt{k}}$ (b) $\displaystyle\sum_{k=1}^{\infty} \frac{k^{4/3}}{8k^2 + 5k + 1}$ (c) $\displaystyle\sum_{k=1}^{\infty} \frac{(-1)^{k+1}}{k^2 + 1}$

15. Find a formula for the exact error that results when the sum of the geometric series $\sum_{k=0}^{\infty} (1/5)^k$ is approximated by the sum of the first 100 terms in the series.

16. Does the series $1 - \frac{2}{3} + \frac{3}{5} - \frac{4}{7} + \frac{5}{9} + \cdots$ converge? Justify your answer.

17. (a) Find the first five Maclaurin polynomials of the function $p(x) = 1 - 7x + 5x^2 + 4x^3$.

(b) Make a general statement about the Maclaurin polynomials of a polynomial of degree n.

18. Use a Maclaurin series and properties of alternating series to show that $|\ln(1+x) - x| \le x^2/2$ if $0 < x < 1$.

19. Show that the approximation

$$\sin x \approx x - \frac{x^3}{3!} + \frac{x^5}{5!}$$

is accurate to four decimal places if $0 \le x \le \pi/4$.

20. Use Maclaurin series to approximate the integral

$$\int_0^1 \frac{1 - \cos x}{x}\,dx$$

to three decimal-place accuracy.

21. It can be proved that

$$\lim_{n \to +\infty} \sqrt[n]{n!} = +\infty \quad \text{and} \quad \lim_{n \to +\infty} \frac{\sqrt[n]{n!}}{n} = \frac{1}{e}$$

In each part, use these limits and the root test to determine whether the series converges.

(a) $\displaystyle\sum_{k=0}^{\infty} \frac{2^k}{k!}$ (b) $\displaystyle\sum_{k=0}^{\infty} \frac{k^k}{k!}$

22. (a) Show that $k^k \ge k!$.

(b) Use the comparison test to show that $\displaystyle\sum_{k=1}^{\infty} k^{-k}$ converges.

(c) Use the root test to show that the series converges.

23. Suppose that $\displaystyle\sum_{k=1}^{n} u_k = 2 - \frac{1}{n}$. Find

(a) u_{100} (b) $\displaystyle\lim_{k \to +\infty} u_k$ (c) $\displaystyle\sum_{k=1}^{\infty} u_k$.

24. In each part, determine whether the series converges; if so, find its sum.

(a) $\displaystyle\sum_{k=1}^{\infty} \left(\frac{3}{2^k} - \frac{2}{3^k}\right)$ (b) $\displaystyle\sum_{k=1}^{\infty} [\ln(k+1) - \ln k]$

(c) $\displaystyle\sum_{k=1}^{\infty} \frac{1}{k(k+2)}$ (d) $\displaystyle\sum_{k=1}^{\infty} [\tan^{-1}(k+1) - \tan^{-1} k]$

25. In each part, find the sum of the series by associating it with some Maclaurin series.

(a) $2 + \dfrac{4}{2!} + \dfrac{8}{3!} + \dfrac{16}{4!} + \cdots$

(b) $\pi - \dfrac{\pi^3}{3!} + \dfrac{\pi^5}{5!} - \dfrac{\pi^7}{7!} + \cdots$

(c) $1 - \dfrac{e^2}{2!} + \dfrac{e^4}{4!} - \dfrac{e^6}{6!} + \cdots$

(d) $1 - \ln 3 + \dfrac{(\ln 3)^2}{2!} - \dfrac{(\ln 3)^3}{3!} + \cdots$

26. Suppose that the sequence $\{a_k\}$ is defined recursively by

$$a_0 = c, \quad a_{k+1} = \sqrt{a_k}$$

Assuming that the sequence converges, find its limit if

(a) $c = \frac{1}{2}$ (b) $c = \frac{3}{2}$.

27. Research has shown that the proportion p of the population with IQs (intelligence quotients) between α and β is approximately

$$p = \frac{1}{16\sqrt{2\pi}} \int_\alpha^\beta e^{-\frac{1}{2}\left(\frac{x-100}{16}\right)^2}\,dx$$

Use the first three terms of an appropriate Maclaurin series to estimate the proportion of the population that has IQs between 100 and 110.

28. Differentiate the Maclaurin series for xe^x and use the result to show that

$$\sum_{k=0}^{\infty} \frac{k+1}{k!} = 2e$$

29. Given: $\dfrac{\pi^2}{6} = 1 + \dfrac{1}{2^2} + \dfrac{1}{3^2} + \dfrac{1}{4^2} + \cdots$.

Show: $\dfrac{\pi^2}{12} = 1 - \dfrac{1}{2^2} + \dfrac{1}{3^2} - \dfrac{1}{4^2} + \cdots$.

30. Let a, b, and p be positive constants. For which values of p does the series $\displaystyle\sum_{k=1}^{\infty} \frac{1}{(a+bk)^p}$ converge?

31. In each part, write out the first four terms of the series, and then find the radius of convergence.

(a) $\displaystyle\sum_{k=1}^{\infty} \frac{1 \cdot 2 \cdot 3 \cdots k}{1 \cdot 4 \cdot 7 \cdots (3k-2)} x^k$

(b) $\displaystyle\sum_{k=1}^{\infty} (-1)^k \frac{1 \cdot 2 \cdot 3 \cdots k}{1 \cdot 3 \cdot 5 \cdots (2k-1)} x^{2k+1}$

32. Find the interval of convergence of

$$\sum_{k=0}^{\infty} \frac{(x - x_0)^k}{b^k} \quad (b > 0)$$

33. Show that the series

$$1 - \frac{x}{2!} + \frac{x^2}{4!} - \frac{x^3}{6!} + \cdots$$

converges to the function

$$f(x) = \begin{cases} \cos\sqrt{x}, & x \ge 0 \\ \cosh\sqrt{-x}, & x < 0 \end{cases}$$

[*Hint:* Use the Maclaurin series for $\cos x$ and $\cosh x$ to obtain series for $\cos\sqrt{x}$, where $x \ge 0$, and $\cosh\sqrt{-x}$, where $x \le 0$.]

34. Prove:
(a) If f is an even function, then all odd powers of x in its Maclaurin series have coefficient 0.
(b) If f is an odd function, then all even powers of x in its Maclaurin series have coefficient 0.

35. In Section 7.6 we defined the kinetic energy K of a particle with mass m and velocity v to be $K = \frac{1}{2}mv^2$ [see Formula (6) of that section]. In this formula the mass m is assumed to be constant, and K is called the ***Newtonian kinetic energy***. However, in Albert Einstein's relativity theory the mass m increases with the velocity and the kinetic energy K is given by the formula

$$K = m_0 c^2 \left[\frac{1}{\sqrt{1 - (v/c)^2}} - 1 \right]$$

in which m_0 is the mass of the particle when its velocity is zero, and c is the speed of light. This is called the ***relativistic kinetic energy***. Use an appropriate binomial series to show that if the velocity is small compared to the speed of light (i.e., $v/c \approx 0$), then the Newtonian and relativistic kinetic energies are in close agreement.

c **36.** If the constant p in the general p-series is replaced by a variable x for $x > 1$, then the resulting function is called

the ***Riemann zeta function*** and is denoted by

$$\zeta(x) = \sum_{k=1}^{\infty} \frac{1}{k^x}$$

(a) Let s_n be the nth partial sum of the series for $\zeta(3.7)$. Find n such that s_n approximates $\zeta(3.7)$ to two decimal-place accuracy, and calculate s_n using this value of n. [*Hint:* Use the right inequality in Exercise 30(b) of Section 10.5 with $f(x) = 1/x^{3.7}$.]

(b) Determine whether your CAS can evaluate the Riemann zeta function directly. If so, compare the value produced by the CAS to the value of s_n obtained in part (a).

11

ANALYTIC GEOMETRY IN CALCULUS

Johannes Kepler

*I*n this chapter we will study aspects of analytic geometry that are important in applications of calculus. We will begin by introducing *polar coordinate systems*, which are used, for example, in tracking the motion of planets and satellites, in identifying the locations of objects from information on radar screens, and in the design of antennas. We will then discuss relationships between curves in polar coordinates and parametric curves in rectangular coordinates, and we will discuss methods for finding areas in polar coordinates and tangent lines to curves given in polar coordinates or parametrically in rectangular coordinates. We will then review the basic properties of parabolas, ellipses, and hyperbolas and discuss these curves in the context of polar coordinates. Finally, we will give some basic applications of our work in astronomy.

11.1 POLAR COORDINATES

Up to now we have specified the location of a point in the plane by means of coordinates relative to two perpendicular coordinate axes. However, sometimes a moving point has a special affinity for some fixed point, such as a planet moving in an orbit under the central attraction of the Sun. In such cases, the path of the particle is best described by its angular direction and its distance from the fixed point. In this section we will discuss a new kind of coordinate system that is based on this idea.

POLAR COORDINATE SYSTEMS

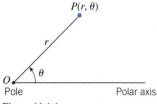

Figure 11.1.1

A *polar coordinate system* in a plane consists of a fixed point O, called the *pole* (or *origin*), and a ray emanating from the pole, called the *polar axis*. In such a coordinate system we can associate with each point P in the plane a pair of *polar coordinates* (r, θ), where r is the distance from P to the pole and θ is an angle from the polar axis to the ray OP (Figure 11.1.1). The number r is called the *radial coordinate* of P and the number θ the *angular coordinate* (or *polar angle*) of P. In Figure 11.1.2, the points $(6, 45°)$, $(5, 120°)$, $(3, 225°)$, and $(4, 330°)$ are plotted in polar coordinate systems. If P is the pole, then $r = 0$, but there is no clearly defined polar angle. We will agree that an arbitrary angle can be used in this case; that is, $(0, \theta)$ are polar coordinates of the pole for all choices of θ.

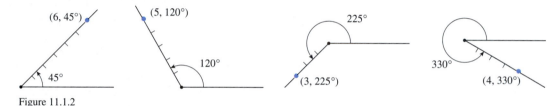

Figure 11.1.2

The polar coordinates of a point are not unique. For example, the polar coordinates

$$(1, 315°), \quad (1, -45°), \quad \text{and} \quad (1, 675°)$$

all represent the same point (Figure 11.1.3). In general, if a point P has polar coordinates (r, θ), then

$$(r, \theta + n \cdot 360°) \quad \text{and} \quad (r, \theta - n \cdot 360°)$$

are also polar coordinates of P for any nonnegative integer n. Thus, every point has infinitely many pairs of polar coordinates.

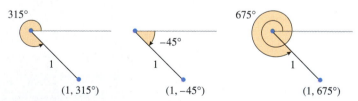

Figure 11.1.3

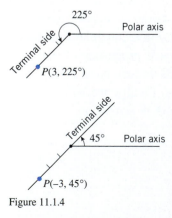

Figure 11.1.4

As defined above, the radial coordinate r of a point P is nonnegative, since it represents the distance from P to the pole. However, it will be convenient to allow for negative values of r as well. To motivate an appropriate definition, consider the point P with polar coordinates $(3, 225°)$. As shown in Figure 11.1.4, we can reach this point by rotating the polar axis through an angle of $225°$ and then moving 3 units from the pole along the terminal side of the angle, or we can reach the point P by rotating the polar axis through an angle of $45°$ and then moving 3 units from the pole along the extension of the terminal side. This suggests that the point $(3, 225°)$ might also be denoted by $(-3, 45°)$, with the minus sign serving to indicate that the point is on the *extension* of the angle's terminal side rather than on the terminal side itself.

In general, the terminal side of the angle $\theta + 180°$ is the extension of the terminal side of θ, so we define negative radial coordinates by agreeing that

$$(-r, \theta) \quad \text{and} \quad (r, \theta + 180°)$$

are polar coordinates of the same point.

FOR THE READER. For many purposes it does not matter whether polar angles are measured in degrees or radians. However, in problems that involve derivatives or integrals they must be measured in radians, since the derivatives of the trigonometric functions were derived under this assumption. Henceforth, we will use radian measure for polar angles, except in certain applications where it is not required and degree measure is more convenient.

RELATIONSHIP BETWEEN POLAR AND RECTANGULAR COORDINATES

Frequently, it will be useful to superimpose a rectangular xy-coordinate system on top of a polar coordinate system, making the positive x-axis coincide with the polar axis. If this is done, then every point P will have both rectangular coordinates (x, y) and polar coordinates (r, θ). As suggested by Figure 11.1.5, these coordinates are related by the equations

$$x = r\cos\theta, \quad y = r\sin\theta \tag{1}$$

These equations are well suited for finding x and y when r and θ are known. However, to find r and θ when x and y are known, it is preferable to use the identities $\sin^2\theta + \cos^2\theta = 1$ and $\tan\theta = \sin\theta / \cos\theta$ to rewrite (1) as

$$r^2 = x^2 + y^2, \quad \tan\theta = \frac{y}{x} \tag{2}$$

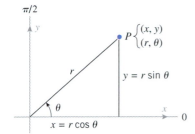

Figure 11.1.5

Example 1 Find the rectangular coordinates of the point P whose polar coordinates are $(6, 2\pi/3)$.

Solution. Substituting the polar coordinates $r = 6$ and $\theta = 2\pi/3$ in (1) yields

$$x = 6\cos\frac{2\pi}{3} = 6\left(-\frac{1}{2}\right) = -3$$

$$y = 6\sin\frac{2\pi}{3} = 6\left(\frac{\sqrt{3}}{2}\right) = 3\sqrt{3}$$

Thus, the rectangular coordinates of P are $(-3, 3\sqrt{3})$ (Figure 11.1.6). ◀

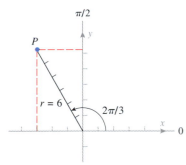

Figure 11.1.6

Example 2 Find polar coordinates of the point P whose rectangular coordinates are $(-2, 2\sqrt{3})$.

Solution. We will find the polar coordinates (r, θ) of P that satisfy the conditions $r > 0$ and $0 \leq \theta < 2\pi$. From the first equation in (2),

$$r^2 = x^2 + y^2 = (-2)^2 + (2\sqrt{3})^2 = 4 + 12 = 16$$

so $r = 4$. From the second equation in (2),

$$\tan\theta = \frac{y}{x} = \frac{2\sqrt{3}}{-2} = -\sqrt{3}$$

From this and the fact that $(-2, 2\sqrt{3})$ lies in the second quadrant, it follows that the angle satisfying the requirement $0 \leq \theta < 2\pi$ is $\theta = 2\pi/3$. Thus, $(4, 2\pi/3)$ are polar coordinates of P. All other polar coordinates of P are expressible in the form

$$\left(4, \frac{2\pi}{3} + 2n\pi\right) \quad \text{or} \quad \left(-4, \frac{5\pi}{3} + 2n\pi\right)$$

where n is an integer. ◀

We will now consider the problem of graphing equations of the form $r = f(\theta)$ in polar coordinates, where θ is assumed to be measured in radians. Some examples of such equations are

$$r = 2\cos\theta, \quad r = \frac{4}{1 - 3\sin\theta}, \quad r = \theta$$

In a rectangular coordinate system the graph of an equation $y = f(x)$ consists of all points whose coordinates (x, y) satisfy the equation. However, in a polar coordinate system, points have infinitely many different pairs of polar coordinates, so that a given point may have some polar coordinates that satisfy the equation $r = f(\theta)$ and others that do not. Taking this into account, we define the **graph of $r = f(\theta)$ in polar coordinates** to consist of all points with *at least one* pair of coordinates (r, θ) that satisfy the equation.

The most elementary way to graph an equation $r = f(\theta)$ in polar coordinates is to plot points. The idea is to choose some typical values of θ, calculate the corresponding values of r, and then plot the resulting pairs (r, θ) in a polar coordinate system. Here are some examples.

Example 3 Sketch the graph of the equation $r = \sin\theta$ in polar coordinates by plotting points.

Solution. Table 11.1.1 shows the coordinates of points on the graph at increments of $\pi/6 (= 30°)$.

Table 11.1.1

θ (RADIANS)	0	$\frac{\pi}{6}$	$\frac{\pi}{3}$	$\frac{\pi}{2}$	$\frac{2\pi}{3}$	$\frac{5\pi}{6}$	π	$\frac{7\pi}{6}$	$\frac{4\pi}{3}$	$\frac{3\pi}{2}$	$\frac{5\pi}{3}$	$\frac{11\pi}{6}$	2π
$r = \sin\theta$	0	$\frac{1}{2}$	$\frac{\sqrt{3}}{2}$	1	$\frac{\sqrt{3}}{2}$	$\frac{1}{2}$	0	$-\frac{1}{2}$	$-\frac{\sqrt{3}}{2}$	-1	$-\frac{\sqrt{3}}{2}$	$-\frac{1}{2}$	0
(r, θ)	$(0,0)$	$\left(\frac{1}{2}, \frac{\pi}{6}\right)$	$\left(\frac{\sqrt{3}}{2}, \frac{\pi}{3}\right)$	$\left(1, \frac{\pi}{2}\right)$	$\left(\frac{\sqrt{3}}{2}, \frac{2\pi}{3}\right)$	$\left(\frac{1}{2}, \frac{5\pi}{6}\right)$	$(0, \pi)$	$\left(-\frac{1}{2}, \frac{7\pi}{6}\right)$	$\left(-\frac{\sqrt{3}}{2}, \frac{4\pi}{3}\right)$	$\left(-1, \frac{3\pi}{2}\right)$	$\left(-\frac{\sqrt{3}}{2}, \frac{5\pi}{3}\right)$	$\left(-\frac{1}{2}, \frac{11\pi}{6}\right)$	$(0, 2\pi)$

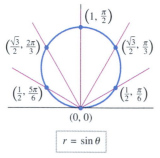

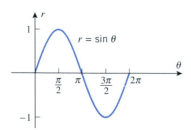

Figure 11.1.7

$r = \sin\theta$

Figure 11.1.8

These points are plotted in Figure 11.1.7. Note, however, that there are 13 points listed in the table but only 6 distinct plotted points. This is because the pairs from $\theta = \pi$ on yield duplicates of the preceding points. For example, $(-1/2, 7\pi/6)$ and $(1/2, \pi/6)$ represent the same point. ◄

Observe that the points in Figure 11.1.7 appear to lie on a circle. We can confirm that this is so by expressing the polar equation $r = \sin\theta$ in terms of x and y. To do this, we multiply the equation through by r to obtain

$$r^2 = r\sin\theta$$

which now allows us to apply Formulas (1) and (2) to rewrite the equation as

$$x^2 + y^2 = y$$

Rewriting this equation as $x^2 + y^2 - y = 0$ and then completing the square yields

$$x^2 + \left(y - \tfrac{1}{2}\right)^2 = \tfrac{1}{4}$$

which is a circle of radius $\frac{1}{2}$ centered at the point $\left(0, \frac{1}{2}\right)$ in the xy-plane.

Just because an equation $r = f(\theta)$ involves the variables r and θ does not mean that it has to be graphed in a polar coordinate system. When useful, this equation can also be graphed in a rectangular coordinate system. For example, Figure 11.1.8 shows the graph of $r = \sin\theta$ in a rectangular θr-coordinate system. This graph can actually help to visualize how the polar graph in Figure 11.1.7 is generated:

- At $\theta = 0$ we have $r = 0$, which corresponds to the pole $(0, 0)$ on the polar graph.
- As θ varies from 0 to $\pi/2$, the value of r increases from 0 to 1, so the point (r, θ) moves along the circle from the pole to the high point at $(1, \pi/2)$.
- As θ varies from $\pi/2$ to π, the value of r decreases from 1 back to 0, so the point (r, θ) moves along the circle from the high point back to the pole.
- As θ varies from π to $3\pi/2$, the values of r are negative, varying from 0 to -1. Thus, the point (r, θ) moves along the circle from the pole to the high point at $(1, \pi/2)$, which is the same as the point $(-1, 3\pi/2)$. This duplicates the motion that occurred for $0 \le \theta \le \pi/2$.
- As θ varies from $3\pi/2$ to 2π, the value of r varies from -1 to 0. Thus, the point (r, θ) moves along the circle from the high point back to the pole, duplicating the motion that occurred for $\pi/2 \le \theta \le \pi$.

Example 4 Sketch the graph of $r = \cos 2\theta$ in polar coordinates.

Solution. Instead of plotting points, we will use the graph of $r = \cos 2\theta$ in rectangular coordinates (Figure 11.1.9) to visualize how the polar graph of this equation is generated. The analysis and the resulting polar graph are shown in Figure 11.1.10. This curve is called a *four-petal rose*. ◀

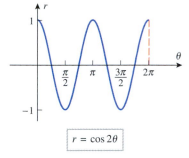

$r = \cos 2\theta$

Figure 11.1.9

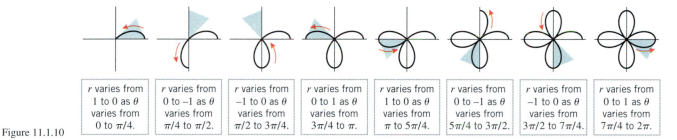

r varies from 1 to 0 as θ varies from 0 to $\pi/4$.	r varies from 0 to -1 as θ varies from $\pi/4$ to $\pi/2$.	r varies from -1 to 0 as θ varies from $\pi/2$ to $3\pi/4$.	r varies from 0 to 1 as θ varies from $3\pi/4$ to π.	r varies from 1 to 0 as θ varies from π to $5\pi/4$.	r varies from 0 to -1 as θ varies from $5\pi/4$ to $3\pi/2$.	r varies from -1 to 0 as θ varies from $3\pi/2$ to $7\pi/4$.	r varies from 0 to 1 as θ varies from $7\pi/4$ to 2π.

Figure 11.1.10

SYMMETRY TESTS

Observe that the polar graph of $r = \cos 2\theta$ in Figure 11.1.10 is symmetric about the x-axis and the y-axis. This symmetry could have been predicted from the following theorem, which is suggested by Figure 11.1.11 (we omit the proof).

11.1.1 THEOREM (*Symmetry Tests*).

(a) *A curve in polar coordinates is symmetric about the x-axis if replacing θ by $-\theta$ in its equation produces an equivalent equation* (Figure 11.1.11a).

(b) *A curve in polar coordinates is symmetric about the y-axis if replacing θ by $\pi - \theta$ in its equation produces an equivalent equation* (Figure 11.1.11b).

(c) *A curve in polar coordinates is symmetric about the origin if replacing θ by $\theta + \pi$, or by replacing r by $-r$ in its equation produces an equivalent equation* (Figure 11.1.11c).

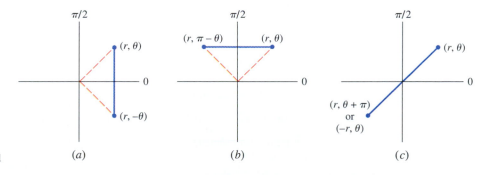

Figure 11.1.11

(a)   (b)   (c)

Example 5 Use Theorem 11.1.1 to confirm that the graph of $r = \cos 2\theta$ in Figure 11.1.10 is symmetric about the x-axis and y-axis.

Solution. To test for symmetry about the x-axis, we replace θ by $-\theta$. This yields

$$r = \cos(-2\theta) = \cos 2\theta$$

Thus, replacing θ by $-\theta$ does not alter the equation.

To test for symmetry about the y-axis, we replace θ by $\pi - \theta$. This yields

$$r = \cos 2(\pi - \theta) = \cos(2\pi - 2\theta) = \cos(-2\theta) = \cos 2\theta$$

Thus, replacing θ by $\pi - \theta$ does not alter the equation. ◄

FOR THE READER. A graph that is symmetric both about the x-axis and about the y-axis is also symmetric about the origin. Use Theorem 11.1.1(c) to verify that the curve of Example 5 is also symmetric about the origin.

Example 6 Sketch the graph of $r = a(1 - \cos \theta)$ in polar coordinates, assuming a to be a positive constant.

Solution. Observe first that replacing θ by $-\theta$ does not alter the equation, so we know in advance that the graph is symmetric about the polar axis. Thus, if we graph the upper half of the curve, then we can obtain the lower half by reflection about the polar axis.

As in our previous examples, we will first graph the equation in rectangular coordinates. This graph, which is shown in Figure 11.1.12a, can be obtained by rewriting the given equation as $r = a - a\cos\theta$, from which we see that the graph in rectangular coordinates can be obtained by first reflecting the graph of $r = a\cos\theta$ about the x-axis to obtain the graph of $r = -a\cos\theta$, and then translating that graph up a units to obtain the graph of $r = a - a\cos\theta$. Now we can see that:

- As θ varies from 0 to $\pi/3$, r increases from 0 to $a/2$.
- As θ varies from $\pi/3$ to $\pi/2$, r increases from $a/2$ to a.
- As θ varies from $\pi/2$ to $2\pi/3$, r increases from a to $3a/2$.
- As θ varies from $2\pi/3$ to π, r increases from $3a/2$ to $2a$.

This produces the polar curve shown in Figure 11.1.12b. The rest of the curve can be obtained by continuing the preceding analysis from π to 2π or, as noted above, by reflecting the portion already graphed about the x-axis (Figure 11.1.12c). This heart-shaped curve is called a **cardioid** (from the Greek word "kardia" for heart). ◄

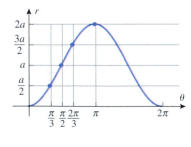

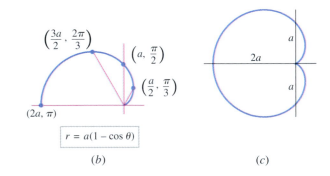

$$r = a(1 - \cos \theta)$$

Figure 11.1.12   (a)   (b)   (c)

Example 7 Sketch the curves

(a) $r = 1$ (b) $\theta = \dfrac{\pi}{4}$ (c) $r = \theta$ $(\theta \geq 0)$

in polar coordinates.

Solution (a). For all values of θ, the point $(1, \theta)$ is 1 unit away from the pole. Thus, the graph is the circle of radius 1 centered at the pole (Figure 11.1.13a).

Solution (b). For all values of r, the point $(r, \pi/4)$ lies on a line that makes an angle of $\pi/4$ with the polar axis (Figure 11.1.13b). Positive values of r correspond to points on the line in the first quadrant and negative values of r to points on the line in the third quadrant. Thus, in absence of any restriction on r, the graph is the entire line. Observe, however, that had we imposed the restriction $r \geq 0$, the graph would have been just the ray in the first quadrant.

Solution (c). Observe that as θ increases, so does r; thus, the graph is a curve that spirals out from the pole as θ increases. A reasonably accurate sketch of the spiral can be obtained by plotting the intersections with the x- and y-axes for values of θ that are multiples of $\pi/2$, keeping in mind that the value of r is always equal to the value of θ (Figure 11.1.13c). ◀

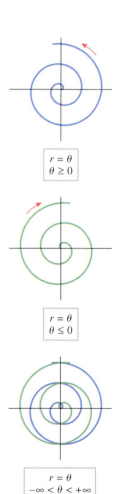

$r = \theta$
$\theta \geq 0$

$r = \theta$
$\theta \leq 0$

$r = \theta$
$-\infty < \theta < +\infty$

Figure 11.1.14

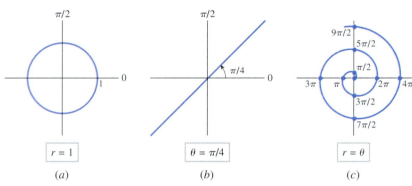

$r = 1$

$\theta = \pi/4$

$r = \theta$

(a)   (b)   (c)

Figure 11.1.13

REMARK.   The spiral in Figure 11.1.13c, which belongs to the family of **Archimedean spirals** $r = a\theta$, coils counterclockwise around the pole because of the restriction $\theta \geq 0$. Had we made the restriction $\theta \leq 0$, the spiral would have coiled clockwise, and had we allowed both positive and negative values of θ, the clockwise and counterclockwise spirals would have been superimposed to form a double Archimedean spiral (Figure 11.1.14).

Example 8   Sketch the graph of $r^2 = 4\cos 2\theta$ in polar coordinates.

Solution.   This equation does not express r as a function of θ, since solving for r in terms of θ yields two functions:

$$r = 2\sqrt{\cos 2\theta} \quad \text{and} \quad r = -2\sqrt{\cos 2\theta}$$

Thus, to graph the equation $r^2 = 4\cos 2\theta$ we will have to graph the two functions separately and then combine those graphs.

We will start with the graph of $r = 2\sqrt{\cos 2\theta}$. Observe first that this equation is not changed if we replace θ by $-\theta$ or if we replace θ by $\pi - \theta$. Thus, the graph is symmetric about the x-axis and the y-axis. This means that the entire graph can be obtained by graphing the portion in the first quadrant, reflecting that portion about the y-axis to obtain the portion in the second quadrant and then reflecting those two portions about the x-axis to obtain the portions in the third and fourth quadrants.

To begin the analysis, we will graph the equation $r = 2\sqrt{\cos 2\theta}$ in rectangular coordinates (see Figure 11.1.15a). Note that there are gaps in that graph over the intervals $\pi/4 < \theta < 3\pi/4$ and $5\pi/4 < \theta < 7\pi/4$ because $\cos 2\theta$ is negative for those values of θ. From this graph we can see that:

- As θ varies from 0 to $\pi/4$, r decreases from 2 to 0.

- As θ varies from $\pi/4$ to $\pi/2$, no points are generated on the polar graph.

This produces the portion of the graph shown in Figure 11.1.15*b*. As noted above, we can complete the graph by a reflection about the y-axis followed by a reflection about the x-axis (11.1.15*c*). The resulting propeller-shaped graph is called a ***lemniscate*** (from the Greek word "lemniscos" for a looped ribbon resembling the number 8). We leave it for you to verify that the equation $r = 2\sqrt{\cos 2\theta}$ has the same graph as $r = -2\sqrt{\cos 2\theta}$, but traced in a diagonally opposite manner. Thus, the graph of the equation $r^2 = 4\cos 2\theta$ consists of two identical superimposed lemniscates. ◀

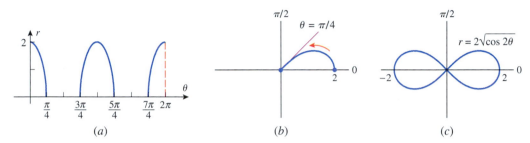

(a) (b) (c)

Figure 11.1.15

FAMILIES OF LINES AND RAYS THROUGH THE POLE

If θ_0 is a fixed angle, then for all values of r the point (r, θ_0) lies on the line that makes an angle of $\theta = \theta_0$ with the polar axis; and, conversely, every point on this line has a pair of polar coordinates of the form (r, θ_0). Thus, the equation $\theta = \theta_0$ represents the line that passes through the pole and makes an angle of θ_0 with the polar axis (Figure 11.1.16*a*). If r is restricted to be nonnegative, then the graph of the equation $\theta = \theta_0$ is the ray that emanates from the pole and makes an angle of θ_0 with the polar axis (Figure 11.1.16*b*). Thus, as θ_0 varies, the equation $\theta = \theta_0$ produces either a family of lines through the pole or a family of rays through the pole, depending on the restrictions on r.

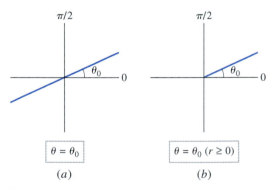

(a) (b)

Figure 11.1.16

FAMILIES OF CIRCLES

We will consider three families of circles in which a is assumed to be a positive constant:

$$r = a \qquad r = 2a\cos\theta \qquad r = 2a\sin\theta \qquad\qquad (3\text{--}5)$$

The equation $r = a$ represents a circle of radius a centered at the pole (Figure 11.1.17*a*). Thus, as a varies, this equation produces a family of circles centered at the pole. For families (4) and (5), recall from plane geometry that a triangle that is inscribed in a circle with a diameter of the circle for a side must be a right triangle. Thus, as indicated in Figures 11.1.17*b* and 11.1.17*c*, the equation $r = 2a\cos\theta$ represents a circle of radius a, centered on the x-axis and tangent to the y-axis at the origin; similarly, the equation $r = 2a\sin\theta$ represents a circle

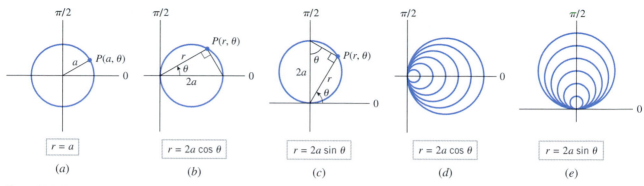

Figure 11.1.17

of radius a, centered on the y-axis and tangent to the x-axis at the origin. Thus, as a varies, Equations (4) and (5) produce the families illustrated in Figures 11.1.17d and 11.1.17e.

REMARK.   Observe that replacing θ by $-\theta$ does not change the equation $r = 2a\cos\theta$, and replacing θ by $\pi - \theta$ does not change the equation $r = 2a\sin\theta$. This explains why the circles in Figure 11.1.17d are symmetric about the x-axis and those in Figure 11.1.17e are symmetric about the y-axis.

FAMILIES OF ROSE CURVES

In polar coordinates, equations of the form

$$r = a\sin n\theta \qquad\qquad r = a\cos n\theta \tag{6-7}$$

in which $a > 0$ and n is a positive integer represent families of flower-shaped curves called *roses* (Figure 11.1.18). The rose consists of n equally spaced petals of radius a if n is odd and $2n$ equally spaced petals of radius a if n is even. It can be shown that a rose with an even number of petals is traced out exactly once as θ varies over the interval $0 \le \theta < 2\pi$ and a rose with an odd number of petals is traced out exactly once as θ varies over the interval $0 \le \theta < \pi$ (Exercise 73). A four-petal rose of radius 1 was graphed in Example 4.

ROSE CURVES

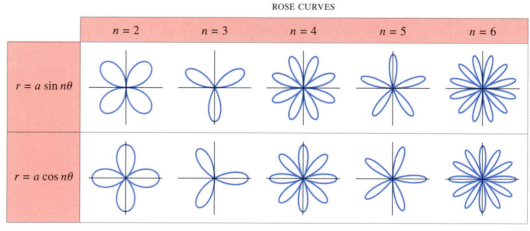

Figure 11.1.18

FOR THE READER.   What do the graphs of the one-petal roses look like?

FAMILIES OF CARDIOIDS AND LIMAÇONS

Equations with any of the four forms

$$r = a \pm b\sin\theta \qquad\qquad r = a \pm b\cos\theta \tag{8-9}$$

in which $a > 0$ and $b > 0$ represent polar curves called *limaçons* (from the Latin word "limax" for a snail-like creature that is commonly called a slug). There are four possible shapes for a limaçon that are determined by the ratio a/b (Figure 11.1.19). If $a = b$ (the case $a/b = 1$), then the limaçon is called a *cardioid* because of its heart-shaped appearance, as noted in Example 6.

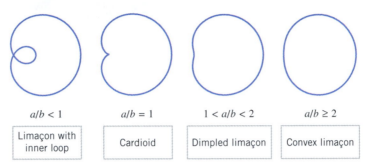

$a/b < 1$	$a/b = 1$	$1 < a/b < 2$	$a/b \geq 2$
Limaçon with inner loop	Cardioid	Dimpled limaçon	Convex limaçon

Figure 11.1.19

Example 9 Figure 11.1.20 shows the family of limaçons $r = a + \cos\theta$ with the constant a varying from 0.25 to 2.50 in steps of 0.25. In keeping with Figure 11.1.19, the limaçons evolve from the loop type to the convex type. As a increases from the starting value of 0.25, the loops get smaller and smaller until the cardioid is reached at $a = 1$. As a increases further, the limaçons evolve through the dimpled type into the convex type. ◄

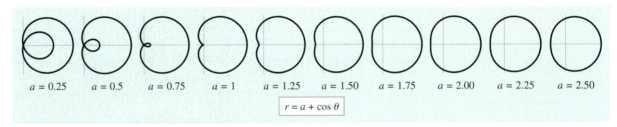

$a = 0.25$	$a = 0.5$	$a = 0.75$	$a = 1$	$a = 1.25$	$a = 1.50$	$a = 1.75$	$a = 2.00$	$a = 2.25$	$a = 2.50$

$r = a + \cos\theta$

Figure 11.1.20

FAMILIES OF SPIRALS

A *spiral* is a curve that coils around a central point. As illustrated in Figure 11.1.14, spirals generally have "left-hand" and "right-hand" versions that coil in opposite directions, depending on the restrictions on the polar angle and the signs of constants that appear in their equations. Some of the more common types of spirals are shown in Figure 11.1.21 for nonnegative values of θ, a, and b.

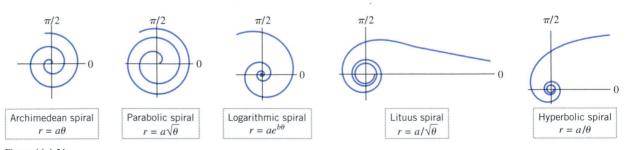

Archimedean spiral $r = a\theta$	Parabolic spiral $r = a\sqrt{\theta}$	Logarithmic spiral $r = ae^{b\theta}$	Lituus spiral $r = a/\sqrt{\theta}$	Hyperbolic spiral $r = a/\theta$

Figure 11.1.21

SPIRALS IN NATURE

Spirals of many kinds occur in nature. For example, the shell of the chambered nautilus (*below*) forms a logarithmic spiral, and a coiled sailor's rope forms an Archimedean spiral. Spirals also occur in flowers, the tusks of certain animals, and in the shapes of galaxies.

The shell of the chambered nautilus reveals a logarithmic spiral. The animal lives in the outermost chamber.

A sailor's coiled rope forms an Archimedean spiral.

GENERATING POLAR CURVES WITH GRAPHING UTILITIES

For polar curves that are too complicated for hand computation, graphing utilities must be used. Although many graphing utilities are capable of graphing polar curves directly, some are not. However, if a graphing utility is capable of graphing parametric equations, then it can be used to graph a polar curve $r = f(\theta)$ by converting this equation to parametric form. This can be done by substituting $f(\theta)$ for r in (1). This yields

$$x = f(\theta)\cos\theta, \quad y = f(\theta)\sin\theta \tag{10}$$

which is a pair of parametric equations for the polar curve in terms of the parameter θ.

Example 10   Express the polar equation

$$r = 2 + \cos\frac{5\theta}{2}$$

parametrically, and generate the polar graph from the parametric equations using a graphing utility.

Solution.   Substituting the given expression for r in $x = r\cos\theta$ and $y = r\sin\theta$ yields the parametric equations

$$x = \left[2 + \cos\frac{5\theta}{2}\right]\cos\theta, \quad y = \left[2 + \cos\frac{5\theta}{2}\right]\sin\theta$$

Next, we need to find an interval over which to vary θ to produce the entire graph. To find such an interval, we will look for the smallest number of complete revolutions that must occur until the value of r begins to repeat. Algebraically, this amounts to finding the smallest positive integer n such that

$$2 + \cos\left(\frac{5(\theta + 2n\pi)}{2}\right) = 2 + \cos\frac{5\theta}{2}$$

or

$$\cos\left(\frac{5\theta}{2} + 5n\pi\right) = \cos\frac{5\theta}{2}$$

For this equality to hold, the quantity $5n\pi$ must be an even multiple of π; the smallest n for which this occurs is $n = 2$. Thus, the entire graph will be traced in two revolutions, which means it can be generated from the parametric equations

$$x = \left[2 + \cos\frac{5\theta}{2}\right]\cos\theta, \quad y = \left[2 + \cos\frac{5\theta}{2}\right]\sin\theta \quad (0 \leq \theta \leq 4\pi)$$

This yields the graph in Figure 11.1.22.   ◄

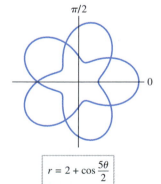

$r = 2 + \cos\dfrac{5\theta}{2}$

Figure 11.1.22

FOR THE READER.   Some graphing utilities require that t be used for the parameter. If this is true of your graphing utility, then you will have to replace θ by t in (10) to generate graphs in polar coordinates. Use a graphing utility to duplicate the curve in Figure 11.1.22.

EXERCISE SET 11.1   ~ Graphing Utility

In Exercises 1 and 2, plot the points in polar coordinates.

1. (a) $(3, \pi/4)$ (b) $(5, 2\pi/3)$ (c) $(1, \pi/2)$
   (d) $(4, 7\pi/6)$ (e) $(-6, -\pi)$ (f) $(-1, 9\pi/4)$

2. (a) $(2, -\pi/3)$ (b) $(3/2, -7\pi/4)$ (c) $(-3, 3\pi/2)$
   (d) $(-5, -\pi/6)$ (e) $(2, 4\pi/3)$ (f) $(0, \pi)$

In Exercises 3 and 4, find the rectangular coordinates of the points whose polar coordinates are given.

3. (a) $(6, \pi/6)$ (b) $(7, 2\pi/3)$ (c) $(-6, -5\pi/6)$
   (d) $(0, -\pi)$ (e) $(7, 17\pi/6)$ (f) $(-5, 0)$

4. (a) $(-8, \pi/4)$ (b) $(7, -\pi/4)$ (c) $(8, 9\pi/4)$
   (d) $(5, 0)$ (e) $(-2, -3\pi/2)$ (f) $(0, \pi)$

5. In each part, a point is given in rectangular coordinates. Find two pairs of polar coordinates for the point, one pair satisfying $r \geq 0$ and $0 \leq \theta < 2\pi$, and the second pair satisfying $r \geq 0$ and $-\pi < \theta \leq \pi$.
   (a) $(-5, 0)$ (b) $(2\sqrt{3}, -2)$ (c) $(0, -2)$
   (d) $(-8, -8)$ (e) $(-3, 3\sqrt{3})$ (f) $(1, 1)$

6. In each part find polar coordinates satisfying the stated conditions for the point whose rectangular coordinates are $(-\sqrt{3}, 1)$.
   (a) $r \geq 0$ and $0 \leq \theta < 2\pi$
   (b) $r \leq 0$ and $0 \leq \theta < 2\pi$
   (c) $r \geq 0$ and $-2\pi < \theta \leq 0$
   (d) $r \leq 0$ and $-\pi < \theta \leq \pi$

In Exercises 7 and 8, use a calculating utility, where needed, to approximate the polar coordinates of the points whose rectangular coordinates are given.

7. (a) $(4, 3)$ (b) $(2, -5)$ (c) $(1, \tan^{-1} 1)$

8. (a) $(-3, 4)$ (b) $(-3, 1.7)$ (c) $\left(2, \sin^{-1} \frac{1}{2}\right)$

In Exercises 9 and 10, identify the curve by transforming the given polar equation to rectangular coordinates.

9. (a) $r = 2$ (b) $r \sin \theta = 4$
   (c) $r = 3 \cos \theta$ (d) $r = \dfrac{6}{3 \cos \theta + 2 \sin \theta}$

10. (a) $r = 5 \sec \theta$ (b) $r = 2 \sin \theta$
   (c) $r = 4 \cos \theta + 4 \sin \theta$ (d) $r = \sec \theta \tan \theta$

In Exercises 11 and 12, express the given equations in polar coordinates.

11. (a) $x = 7$ (b) $x^2 + y^2 = 9$
   (c) $x^2 + y^2 - 6y = 0$ (d) $4xy = 9$

12. (a) $y = -3$ (b) $x^2 + y^2 = 5$
   (c) $x^2 + y^2 + 4x = 0$ (d) $x^2(x^2 + y^2) = y^2$

In Exercises 13–16, a graph is given in a rectangular θr-coordinate system. Sketch the corresponding graph in polar coordinates.

13.
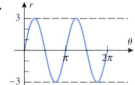

14.
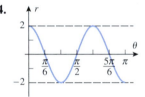

15.
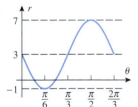

16.
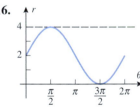

In Exercises 17–20, find an equation for the given polar graph.

17. (a)   (b)   (c)
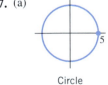
Circle   Circle   Cardioid

18. (a)   (b)   (c)

Limaçon

Circle

Three-petal rose

19. (a) (b) (c)

Four-petal rose   Limaçon   Lemniscate

20. (a) (b) (c)

Cardioid   Five-petal rose   Circle

In Exercises 21–50, sketch the curve in polar coordinates.

21. $\theta = \dfrac{\pi}{6}$

22. $\theta = -\dfrac{3\pi}{4}$

23. $r = 3$

24. $r = 4\sin\theta$

25. $r = 6\cos\theta$

26. $r = 1 + \sin\theta$

27. $2r = \cos\theta$

28. $r - 2 = 2\cos\theta$

29. $r = 3(1 - \sin\theta)$

30. $r = -5 + 5\sin\theta$

31. $r = 4 - 4\cos\theta$

32. $r = 1 + 2\sin\theta$

33. $r = -1 - \cos\theta$

34. $r = 4 + 3\cos\theta$

35. $r = 2 + \sin\theta$

36. $r = 3 - \cos\theta$

37. $r = 3 + 4\cos\theta$

38. $r - 5 = 3\sin\theta$

39. $r = 5 - 2\cos\theta$

40. $r = -3 - 4\sin\theta$

41. $r^2 = 9\cos 2\theta$

42. $r^2 = \sin 2\theta$

43. $r^2 = 16\sin 2\theta$

44. $r = 4\theta \quad (\theta \geq 0)$

45. $r = 4\theta \quad (\theta \leq 0)$

46. $r = 4\theta$

47. $r = \cos 2\theta$

48. $r = 3\sin 2\theta$

49. $r = 9\sin 4\theta$

50. $r = 2\cos 3\theta$

In Exercises 51–55, use a graphing utility to generate the polar graph. Be sure to choose the parameter interval so that a complete graph is generated.

51. $r = \cos\dfrac{\theta}{2}$

52. $r = \sin\dfrac{\theta}{2}$

53. $r = 1 + 2\cos\dfrac{\theta}{4}$

54. $r = 0.5 + \cos\dfrac{\theta}{3}$

55. $r = \cos\dfrac{\theta}{5}$

56. The accompanying figure shows the graph of the "butterfly curve"

$$r = e^{\cos\theta} - 2\cos 4\theta + \sin^3\dfrac{\theta}{4}$$

Generate the complete butterfly with a graphing utility, and state the parameter interval you used.

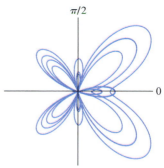

Figure Ex-56

57.  The accompanying figure shows the Archimedean spiral $r = \theta/2$ produced with a graphing calculator.
(a) What interval of values for θ do you think was used to generate the graph?
(b) Duplicate the graph with your own graphing utility.

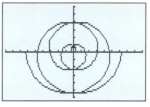

$[-9, 9] \times [-6, 6]$
$x\text{Scl} = 1, y\text{Scl} = 1$

Figure Ex-57

58. The accompanying figure shows graphs of the Archimedean spiral $r = \theta$ and the parabolic spiral $r = \sqrt{\theta}$. Which is which? Explain your reasoning.

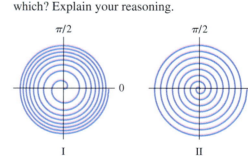

Figure Ex-58

59. (a) Show that if a varies, then the polar equation

$$r = a\sec\theta \quad (-\pi/2 < \theta < \pi/2)$$

describes a family of lines perpendicular to the polar axis.

(b) Show that if b varies, then the polar equation

$$r = b\csc\theta \quad (0 < \theta < \pi)$$

describes a family of lines parallel to the polar axis.

60. Show that if the polar graph of $r = f(\theta)$ is rotated counterclockwise around the origin through an angle α, then $r = f(\theta - \alpha)$ is an equation for the rotated curve. [*Hint:* If (r_0, θ_0) is any point on the original graph, then $(r_0, \theta_0 + \alpha)$ is a point on the rotated graph.]

61. Use the result in Exercise 60 to find an equation for the cardioid $r = 1 + \cos\theta$ after it has been rotated through the given angle, and check your answer with a graphing utility.

(a) $\dfrac{\pi}{4}$ (b) $\dfrac{\pi}{2}$ (c) π (d) $\dfrac{5\pi}{4}$

62. Use the result in Exercise 60 to find an equation for the lemniscate that results when the lemniscate in Example 8 is rotated counterclockwise through an angle of $\pi/2$.

63. Sketch the polar graph of the equation $(r - 1)(\theta - 1) = 0$.

64. (a) Show that if A and B are not both zero, then the graph of the polar equation

$$r = A\sin\theta + B\cos\theta$$

is a circle. Find its radius.

(b) Derive Formulas (4) and (5) from the formula given in part (a).

65. Find the highest point on the cardioid $r = 1 + \cos\theta$.

66. Find the leftmost point on the upper half of the cardioid $r = 1 + \cos\theta$.

67. (a) Show that in a polar coordinate system the distance d between the points (r_1, θ_1) and (r_2, θ_2) is

$$d = \sqrt{r_1^2 + r_2^2 - 2r_1 r_2 \cos(\theta_1 - \theta_2)}$$

(b) Show that if $0 \le \theta_1 < \theta_2 \le \pi$ and if r_1 and r_2 are positive, then the area A of the triangle with vertices $(0, 0)$, (r_1, θ_1), and (r_2, θ_2) is

$$A = \tfrac{1}{2} r_1 r_2 \sin(\theta_2 - \theta_1)$$

(c) Find the distance between the points whose polar coordinates are $(3, \pi/6)$ and $(2, \pi/3)$.

(d) Find the area of the triangle whose vertices in polar coordinates are $(0, 0)$, $(1, 5\pi/6)$, and $(2, \pi/3)$.

68. In the late seventeenth century the Italian astronomer Giovanni Domenico Cassini (1625–1712) introduced the family of curves

$$(x^2 + y^2 + a^2)^2 - b^4 - 4a^2 x^2 = 0 \quad (a > 0, b > 0)$$

in his studies of the relative motions of the Earth and the Sun. These curves, which are called *Cassini ovals*, have one of the three basic shapes shown in the accompanying figure.

(a) Show that if $a = b$, then the polar equation of the Cassini oval is $r^2 = 2a^2 \cos 2\theta$, which is a lemniscate.

(b) Use the formula in Exercise 67(a) to show that the lemniscate in part (a) is the curve traced by a point that moves in such a way that the product of its distances from the polar points $(a, 0)$ and (a, π) is a^2.

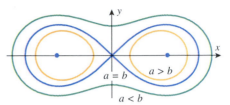

Figure Ex-68

> Vertical and horizontal asymptotes of polar curves can often be detected by investigating the behavior of $x = r\cos\theta$ and $y = r\sin\theta$ as θ varies. This idea is used in Exercises 69–72.

69. Show that the *hyperbolic spiral* $r = 1/\theta$ ($\theta > 0$) has a horizontal asymptote at $y = 1$ by showing that $y \to 1$ and $x \to +\infty$ as $\theta \to 0^+$. Confirm this result by generating the spiral with a graphing utility.

70. Show that the spiral $r = 1/\theta^2$ does not have any horizontal asymptotes.

71. (a) Show that the *kappa curve* $r = 4\tan\theta$ ($0 \le \theta \le 2\pi$) has a vertical asymptote at $x = 4$ by showing that $x \to 4$ and $y \to +\infty$ as $\theta \to \pi/2^-$ and that $x \to 4$ and $y \to -\infty$ as $\theta \to \pi/2^+$.

(b) Use the method in part (a) to show that the kappa curve also has a vertical asymptote at $x = -4$.

(c) Confirm the results in parts (a) and (b) by generating the kappa curve with a graphing utility.

72. Use a graphing utility to make a conjecture about the existence of asymptotes for the *cissoid* $r = 2\sin\theta\tan\theta$, and then confirm your conjecture by calculating appropriate limits.

73. Prove that a rose with an even number of petals is traced out exactly once as θ varies over the interval $0 \le \theta < 2\pi$ and a rose with an odd number of petals is traced out exactly once as θ varies over the interval $0 \le \theta < \pi$.

11.2 TANGENT LINES AND ARC LENGTH FOR PARAMETRIC AND POLAR CURVES

In this section we will derive the formulas required to find slopes, tangent lines, and arc lengths of parametric and polar curves.

TANGENT LINES TO PARAMETRIC CURVES

We will be concerned in this section with curves that are given by parametric equations

$$x = f(t), \quad y = g(t)$$

in which $f(t)$ and $g(t)$ have continuous first derivatives with respect to t. It can be proved that if $dx/dt \ne 0$, then y is a differentiable function of x, in which case the chain rule

implies that

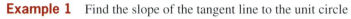

$$\frac{dy}{dx} = \frac{dy/dt}{dx/dt} \tag{1}$$

This formula makes it possible to find dy/dx directly from the parametric equations without eliminating the parameter.

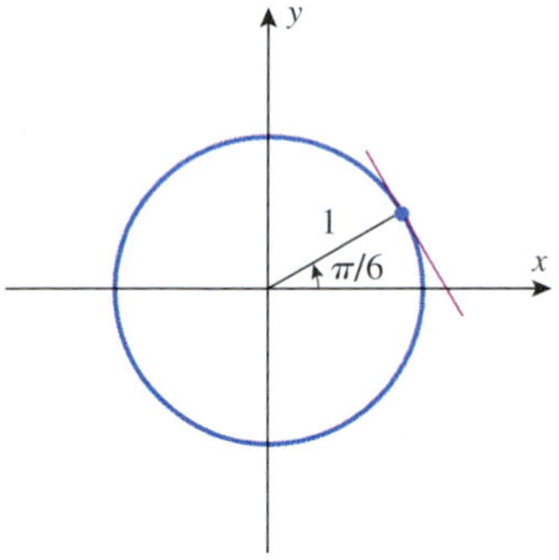

Example 1   Find the slope of the tangent line to the unit circle

$$x = \cos t, \quad y = \sin t \qquad (0 \le t \le 2\pi)$$

at the point where $t = \pi/6$ (Figure 11.2.1).

Solution.   From (1), the slope at a general point on the circle is

$$\frac{dy}{dx} = \frac{dy/dt}{dx/dt} = \frac{\cos t}{-\sin t} = -\cot t \tag{2}$$

Thus, the slope at $t = \pi/6$ is

$$\left.\frac{dy}{dx}\right|_{t=\pi/6} = -\cot\frac{\pi}{6} = -\sqrt{3} \qquad \blacktriangleleft$$

Figure 11.2.1

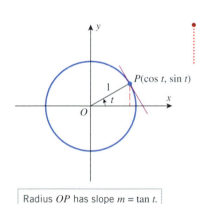

Radius OP has slope $m = \tan t$.

Figure 11.2.2

REMARK.   Note that Formula (2) makes sense geometrically because the radius to the point $P(\cos t, \sin t)$ has slope $m = \tan t$; hence, the tangent line at P, being perpendicular to the radius, has slope $-1/m = -1/\tan t = -\cot t$ (Figure 11.2.2).

It follows from Formula (1) that the tangent line to a parametric curve will be horizontal at those points where $dy/dt = 0$ and $dx/dt \ne 0$, since $dy/dx = 0$ at such points. Two different situations occur when $dx/dt = 0$. At points where $dx/dt = 0$ and $dy/dt \ne 0$, the right side of (1) has a nonzero numerator and a zero denominator; we will agree that the curve has **infinite slope** and a **vertical tangent line** at such points. At points where dx/dt and dy/dt are both zero, the right side of (1) becomes an indeterminate form; we call such points **singular points**. No general statement can be made about the behavior of parametric curves at singular points; they must be analyzed case by case.

Example 2   In a disastrous first flight, an experimental paper airplane follows the trajectory

$$x = t - 3\sin t, \quad y = 4 - 3\cos t \qquad (t \ge 0)$$

but crashes into a wall at time $t = 10$ (Figure 11.2.3).

(a)   At what times was the airplane flying horizontally?

(b)   At what times was it flying vertically?

Solution (a).   The airplane was flying horizontally at those times when $dy/dt = 0$ and $dx/dt \ne 0$. From the given trajectory we have

$$\frac{dy}{dt} = 3\sin t \quad \text{and} \quad \frac{dx}{dt} = 1 - 3\cos t \tag{3}$$

Setting $dy/dt = 0$ yields the equation $3\sin t = 0$, or, more simply, $\sin t = 0$. This equation has four solutions in the time interval $0 \le t \le 10$:

$$t = 0, \quad t = \pi, \quad t = 2\pi, \quad t = 3\pi$$

Since $dx/dt = 1 - 3\cos t \ne 0$ for these values of t (verify), the airplane was flying horizontally at times

$$t = 0, \quad t = \pi \approx 3.14, \quad t = 2\pi \approx 6.28, \quad \text{and} \quad t = 3\pi \approx 9.42$$

which is consistent with Figure 11.2.3.

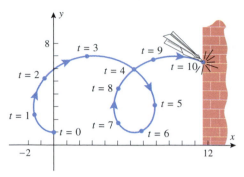

Figure 11.2.3

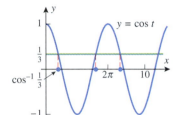

Figure 11.2.4

Solution (b). The airplane was flying vertically at those times when $dx/dt = 0$ and $dy/dt \neq 0$. Setting $dx/dt = 0$ in (3) yields the equation

$$1 - 3\cos t = 0 \quad \text{or} \quad \cos t = \tfrac{1}{3}$$

This equation has three solutions in the time interval $0 \leq t \leq 10$ (Figure 11.2.4):

$$t = \cos^{-1}\tfrac{1}{3}, \quad t = 2\pi - \cos^{-1}\tfrac{1}{3}, \quad t = 2\pi + \cos^{-1}\tfrac{1}{3}$$

Since $dy/dt = 3\sin t$ is not zero at these points (why?), it follows that the airplane was flying vertically at times

$$t = \cos^{-1}\tfrac{1}{3} \approx 1.23, \quad t \approx 2\pi - 1.23 \approx 5.05, \quad t \approx 2\pi + 1.23 \approx 7.51$$

which again is consistent with Figure 11.2.3. ◀

Example 3 The curve represented by the parametric equations

$$x = t^2, \quad y = t^3 \qquad (-\infty < t < +\infty)$$

is called a ***semicubical parabola***. The parameter t can be eliminated by cubing x and squaring y, from which it follows that $y^2 = x^3$. The graph of this equation, shown in Figure 11.2.5, consists of two branches: an upper branch obtained by graphing $y = x^{3/2}$ and a lower branch obtained by graphing $y = -x^{3/2}$. The two branches meet at the origin, which corresponds to $t = 0$ in the parametric equations. This is a singular point because the derivatives $dx/dt = 2t$ and $dy/dt = 3t^2$ are both zero there. ◀

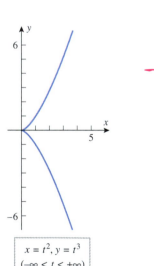

$$x = t^2, \, y = t^3$$
$$(-\infty < t < +\infty)$$

Figure 11.2.5

Example 4 Without eliminating the parameter, find dy/dx and d^2y/dx^2 at the points $(1, 1)$ and $(1, -1)$ on the semicubical parabola given by the parametric equations in Example 3.

Solution. From (1) we have

$$\frac{dy}{dx} = \frac{dy/dt}{dx/dt} = \frac{3t^2}{2t} = \frac{3}{2}t \quad (t \neq 0) \tag{4}$$

and from (1) applied to $y' = dy/dx$ we have

$$\frac{d^2y}{dx^2} = \frac{dy'}{dx} = \frac{dy'/dt}{dx/dt} = \frac{3/2}{2t} = \frac{3}{4t} \tag{5}$$

Since the point $(1, 1)$ on the curve corresponds to $t = 1$ in the parametric equations, it follows from (4) and (5) that

$$\left.\frac{dy}{dx}\right|_{t=1} = \frac{3}{2} \quad \text{and} \quad \left.\frac{d^2y}{dx^2}\right|_{t=1} = \frac{3}{4}$$

Similarly, the point $(1, -1)$ corresponds to $t = -1$ in the parametric equations, so applying (4) and (5) again yields

$$\left.\frac{dy}{dx}\right|_{t=-1} = -\frac{3}{2} \quad \text{and} \quad \left.\frac{d^2y}{dx^2}\right|_{t=-1} = -\frac{3}{4}$$

Note that the values we obtained for the first and second derivatives are consistent with the graph in Figure 11.2.5, since at $(1, 1)$ on the upper branch the tangent line has positive slope and the curve is concave up, and at $(1, -1)$ on the lower branch the tangent line has negative slope and the curve is concave down.

Finally, observe that we were able to apply Formulas (4) and (5) for both $t = 1$ and $t = -1$, even though the points $(1, 1)$ and $(1, -1)$ lie on different branches. In contrast, had we chosen to perform the same computations by eliminating the parameter, we would have had to obtain separate derivative formulas for $y = x^{3/2}$ and $y = -x^{3/2}$. ◀

TANGENT LINES TO POLAR CURVES

Our next objective is to find a method for obtaining slopes of tangent lines to polar curves of the form $r = f(\theta)$ in which r is a differentiable function of θ. We showed in the last section that a curve of this form can be expressed parametrically in terms of the parameter θ by substituting $f(\theta)$ for r in the equations $x = r \cos\theta$ and $y = r \sin\theta$. This yields

$$x = f(\theta)\cos\theta, \quad y = f(\theta)\sin\theta$$

from which we obtain

$$\frac{dx}{d\theta} = -f(\theta)\sin\theta + f'(\theta)\cos\theta = -r\sin\theta + \frac{dr}{d\theta}\cos\theta$$
$$\frac{dy}{d\theta} = f(\theta)\cos\theta + f'(\theta)\sin\theta = r\cos\theta + \frac{dr}{d\theta}\sin\theta \tag{6}$$

Thus, if $dx/d\theta$ and $dy/d\theta$ are continuous and if $dx/d\theta \neq 0$, then y is a differentiable function of x, and Formula (1) with θ in place of t yields

$$\frac{dy}{dx} = \frac{dy/d\theta}{dx/d\theta} = \frac{r\cos\theta + \sin\theta \dfrac{dr}{d\theta}}{-r\sin\theta + \cos\theta \dfrac{dr}{d\theta}} \tag{7}$$

Example 5 Find the slope of the tangent line to the circle $r = 4\cos\theta$ at the point where $\theta = \pi/4$.

Solution. From (7) with $r = 4\cos\theta$ we obtain (verify)

$$\frac{dy}{dx} = \frac{4\cos^2\theta - 4\sin^2\theta}{-8\sin\theta\cos\theta} = \frac{4\cos 2\theta}{-4\sin 2\theta} = -\cot 2\theta$$

Thus, at the point where $\theta = \pi/4$ the slope of the tangent line is

$$m = \frac{dy}{dx}\bigg|_{\theta=\pi/4} = -\cot\frac{\pi}{2} = 0$$

which implies that the circle has a horizontal tangent line at the point where $\theta = \pi/4$ (Figure 11.2.6). ◀

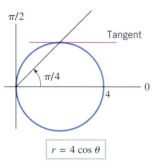

Figure 11.2.6

Example 6 Find the points on the cardioid $r = 1 - \cos\theta$ at which there is a horizontal tangent line, a vertical tangent line, or a singular point.

Solution. A horizontal tangent line will occur where $dy/d\theta = 0$ and $dx/d\theta \neq 0$, a vertical tangent line where $dy/d\theta \neq 0$ and $dx/d\theta = 0$, and a singular point where $dy/d\theta = 0$ and $dx/d\theta = 0$. We could find these derivatives from the formulas in (6). However, an alternative approach is to go back to basic principles and express the cardioid parametrically by substituting $r = 1 - \cos\theta$ in the conversion formulas $x = r\cos\theta$ and $y = r\sin\theta$. This yields

$$x = (1 - \cos\theta)\cos\theta, \quad y = (1 - \cos\theta)\sin\theta \quad (0 \leq \theta \leq 2\pi)$$

Differentiating these equations with respect to θ and then simplifying yields (verify)

$$\frac{dx}{d\theta} = \sin\theta(2\cos\theta - 1), \quad \frac{dy}{d\theta} = (1 - \cos\theta)(1 + 2\cos\theta)$$

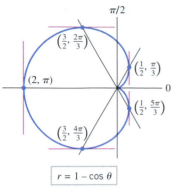

Figure 11.2.7

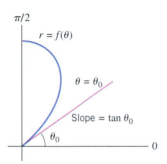

TANGENT LINES TO POLAR CURVES AT THE ORIGIN

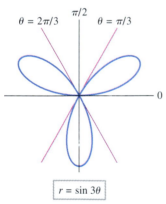

Figure 11.2.8

Thus, $dx/d\theta = 0$ if $\sin\theta = 0$ or $\cos\theta = \frac{1}{2}$, and $dy/d\theta = 0$ if $\cos\theta = 1$ or $\cos\theta = -\frac{1}{2}$. We leave it for you to solve these equations and show that the solutions of $dx/d\theta = 0$ on the interval $0 \le \theta \le 2\pi$ are

$$\frac{dx}{d\theta} = 0: \quad \theta = 0, \ \frac{\pi}{3}, \ \pi, \ \frac{5\pi}{3}, \ 2\pi$$

and the solutions of $dy/d\theta = 0$ on the interval $0 \le \theta \le 2\pi$ are

$$\frac{dy}{d\theta} = 0: \quad \theta = 0, \ \frac{2\pi}{3}, \ \frac{4\pi}{3}, \ 2\pi$$

Thus, horizontal tangent lines occur at $\theta = 2\pi/3$ and $\theta = 4\pi/3$; vertical tangent lines occur at $\theta = \pi/3$, π, and $5\pi/3$; and singular points occur at $\theta = 0$ and $\theta = 2\pi$ (Figure 11.2.7). Note, however, that $r = 0$ at both singular points, so there is really only one singular point on the cardioid—the pole. ◄

Formula (7) reveals some useful information about the behavior of a polar curve $r = f(\theta)$ that passes through the origin. If we assume that $r = 0$ and $dr/d\theta \neq 0$ when $\theta = \theta_0$, then it follows from Formula (7) that the slope of the tangent line to the curve at $\theta = \theta_0$ is

$$\frac{dy}{dx} = \frac{0 + \sin\theta_0 \dfrac{dr}{d\theta}}{0 + \cos\theta_0 \dfrac{dr}{d\theta}} = \frac{\sin\theta_0}{\cos\theta_0} = \tan\theta_0$$

(Figure 11.2.8). However, $\tan\theta_0$ is also the slope of the line $\theta = \theta_0$, so we can conclude that this line is tangent to the curve at the origin. Thus, we have established the following result.

> **11.2.1   THEOREM.**   *If the polar curve $r = f(\theta)$ passes through the origin at $\theta = \theta_0$, and if $dr/d\theta \neq 0$ at $\theta = \theta_0$, then the line $\theta = \theta_0$ is tangent to the curve at the origin.*

This theorem tells us that equations of the tangent lines at the origin to the curve $r = f(\theta)$ can be obtained by solving the equation $f(\theta) = 0$. It is important to keep in mind, however, that $r = f(\theta)$ may be zero for more than one value of θ, so there may be more than one tangent line at the origin. This is illustrated in the next example.

Example 7   The three-petal rose $r = \sin 3\theta$ in Figure 11.2.9 has three tangent lines at the origin, which can be found by solving the equation

$$\sin 3\theta = 0$$

It was shown in Exercise 73 of Section 11.1 that the complete rose is traced once as θ varies over the interval $0 \le \theta < \pi$, so we need only look for solutions in this interval. We leave it for you to confirm that these solutions are

$$\theta = 0, \quad \theta = \frac{\pi}{3}, \quad \text{and} \quad \theta = \frac{2\pi}{3}$$

Since $dr/d\theta = 3\cos 3\theta \neq 0$ for these values of θ, these three lines are tangent to the rose at the origin, which is consistent with the figure. ◄

A formula for the arc length of a polar curve $r = f(\theta)$ can be derived by expressing the curve in parametric form and applying Formula (6) of Section 7.4 for the arc length of a parametric curve. We leave it as an exercise to show the following.

> **11.2.2   ARC LENGTH FORMULA FOR POLAR CURVES.**   If no segment of the polar curve $r = f(\theta)$ is traced more than once as θ increases from α to β, and if $dr/d\theta$ is continuous for $\alpha \le \theta \le \beta$, then the arc length L from $\theta = \alpha$ to $\theta = \beta$ is
>
> $$L = \int_\alpha^\beta \sqrt{r^2 + \left(\frac{dr}{d\theta}\right)^2}\, d\theta \qquad (8)$$

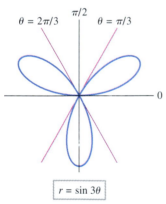

Figure 11.2.9

ARC LENGTH OF A POLAR CURVE

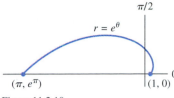

Figure 11.2.10

Example 8 Find the arc length of the spiral $r = e^{\theta}$ in Figure 11.2.10 between $\theta = 0$ and $\theta = \pi$.

Solution.

$$L = \int_{\alpha}^{\beta} \sqrt{r^2 + \left(\frac{dr}{d\theta}\right)^2}\, d\theta = \int_0^{\pi} \sqrt{(e^{\theta})^2 + (e^{\theta})^2}\, d\theta$$

$$= \int_0^{\pi} \sqrt{2}\, e^{\theta}\, d\theta = \sqrt{2}\, e^{\theta}\bigg]_0^{\pi} = \sqrt{2}(e^{\pi} - 1) \approx 31.3 \quad \blacktriangleleft$$

Example 9 Find the total arc length of the cardioid $r = 1 + \cos\theta$.

Solution. The cardioid is traced out once as θ varies from $\theta = 0$ to $\theta = 2\pi$. Thus,

$$L = \int_{\alpha}^{\beta} \sqrt{r^2 + \left(\frac{dr}{d\theta}\right)^2}\, d\theta = \int_0^{2\pi} \sqrt{(1 + \cos\theta)^2 + (-\sin\theta)^2}\, d\theta$$

$$= \sqrt{2} \int_0^{2\pi} \sqrt{1 + \cos\theta}\, d\theta$$

$$= 2 \int_0^{2\pi} \sqrt{\cos^2 \frac{1}{2}\theta}\, d\theta \qquad \boxed{\text{Identity (45) of Appendix E}}$$

$$= 2 \int_0^{2\pi} \left|\cos \frac{1}{2}\theta\right|\, d\theta$$

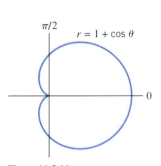

Figure 11.2.11

Since $\cos \frac{1}{2}\theta$ changes sign at π, we must split the last integral into the sum of two integrals: the integral from 0 to π plus the integral from π to 2π. However, the integral from π to 2π is equal to the integral from 0 to π, since the cardioid is symmetric about the polar axis (Figure 11.2.11). Thus,

$$L = 2 \int_0^{2\pi} \left|\cos \frac{1}{2}\theta\right|\, d\theta = 4 \int_0^{\pi} \cos \frac{1}{2}\theta\, d\theta = 8\sin \frac{1}{2}\theta\bigg]_0^{\pi} = 8 \quad \blacktriangleleft$$

EXERCISE SET 11.2 ~ Graphing Utility

1. (a) Find the slope of the tangent line to the parametric curve $x = t^2 + 1$, $y = t/2$ at $t = -1$ and at $t = 1$ without eliminating the parameter.
   (b) Check your answers in part (a) by eliminating the parameter and differentiating an appropriate function of x.

2. (a) Find the slope of the tangent line to the parametric curve $x = 3\cos t$, $y = 4\sin t$ at $t = \pi/4$ and at $t = 7\pi/4$ without eliminating the parameter.
   (b) Check your answers in part (a) by eliminating the parameter and differentiating an appropriate function of x.

3. For the parametric curve in Exercise 1, make a conjecture about the sign of d^2y/dx^2 at $t = -1$ and at $t = 1$, and confirm your conjecture without eliminating the parameter.

4. For the parametric curve in Exercise 2, make a conjecture about the sign of d^2y/dx^2 at $t = \pi/4$ and at $t = 7\pi/4$, and confirm your conjecture without eliminating the parameter.

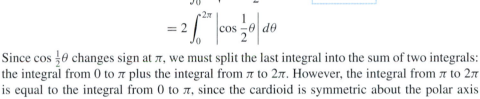

In Exercises 5–10, find dy/dx and d^2y/dx^2 at the given point without eliminating the parameter.

5. $x = \sqrt{t}$, $y = 2t + 4$; $t = 1$

6. $x = \frac{1}{2}t^2$, $y = \frac{1}{3}t^3$; $t = 2$

7. $x = \sec t$, $y = \tan t$; $t = \pi/3$

8. $x = \sinh t$, $y = \cosh t$; $t = 0$

9. $x = 2\theta + \cos\theta$, $y = 1 - \sin\theta$; $\theta = \pi/3$

10. $x = \cos\phi$, $y = 3\sin\phi$; $\phi = 5\pi/6$

11. (a) Find the equation of the tangent line to the curve

$$x = e^t, \quad y = e^{-t}$$

at $t = 1$ without eliminating the parameter.
   (b) Check your answer in part (a) by eliminating the parameter.

12. (a) Find the equation of the tangent line to the curve

$$x = 2t + 4, \quad y = 8t^2 - 2t + 4$$

at $t = 1$ without eliminating the parameter.

(b) Check your answer in part (a) by eliminating the parameter.

In Exercises 13 and 14, find all values of t at which the parametric curve has (a) a horizontal tangent line and (b) a vertical tangent line.

---

13. $x = 2\cos t, \ y = 4\sin t \quad (0 \le t \le 2\pi)$

14. $x = 2t^3 - 15t^2 + 24t + 7, \ y = t^2 + t + 1$

15. As shown in the accompanying figure, the Lissajous curve

$$x = \sin t, \quad y = \sin 2t \quad (0 \le t \le 2\pi)$$

crosses itself at the origin. Find equations for the two tangent lines at the origin.

16. As shown in the accompanying figure, the **prolate cycloid**

$$x = 2 - \pi \cos t, \quad y = 2t - \pi \sin t \quad (-\pi \le t \le \pi)$$

crosses itself at a point on the x-axis. Find equations for the two tangent lines at that point.

 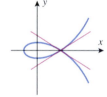

Figure Ex-15   Figure Ex-16

17. Show that the curve $x = t^3 - 4t, \ y = t^2$ intersects itself at the point $(0, 4)$, and find equations for the two tangent lines to the curve at the point of intersection.

18. Show that the curve with parametric equations

$$x = t^2 - 3t + 5, \quad y = t^3 + t^2 - 10t + 9$$

intersects itself at the point $(3, 1)$, and find equations for the two tangent lines to the curve at the point of intersection.

 19. (a) Use a graphing utility to generate the graph of the parametric curve

$$x = \cos^3 t, \quad y = \sin^3 t \quad (0 \le t \le 2\pi)$$

and make a conjecture about the values of t at which singular points occur.

(b) Confirm your conjecture in part (a) by calculating appropriate derivatives.

20. (a) At what values of θ would you expect the cycloid in Figure 1.8.13 to have singular points?

(b) Confirm your answer in part (a) by calculating appropriate derivatives.

In Exercises 21–26, find the slope of the tangent line to the polar curve for the given value of θ.

---

21. $r = 2\cos\theta; \ \theta = \pi/3$ **22.** $r = 1 + \sin\theta; \ \theta = \pi/4$

23. $r = 1/\theta; \ \theta = 2$ **24.** $r = a\sec 2\theta; \ \theta = \pi/6$

25. $r = \cos 3\theta; \ \theta = 3\pi/4$ **26.** $r = 4 - 3\sin\theta; \ \theta = \pi$

In Exercises 27 and 28, calculate the slopes of the tangent lines indicated in the accompanying figures.

---

27. $r = 2 + 2\sin\theta$ **28.** $r = 1 - 2\sin\theta$

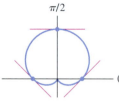 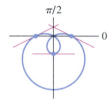

Figure Ex-27   Figure Ex-28

In Exercises 29 and 30, find polar coordinates of all points at which the polar curve has a horizontal or a vertical tangent line.

---

29. $r = a(1 + \cos\theta)$ **30.** $r = a\sin\theta$

In Exercises 31 and 32, use a graphing utility to make a conjecture about the number of points on the polar curve at which there is a horizontal tangent line, and confirm your conjecture by finding appropriate derivatives.

---

 31. $r = \sin\theta\cos^2\theta$ **32.** $r = 1 - 2\sin\theta$

In Exercises 33–38, sketch the polar curve and find polar equations of the tangent lines to the curve at the pole.

---

33. $r = 2\cos 3\theta$ **34.** $r = 4\cos\theta$

35. $r = 4\sqrt{\cos 2\theta}$ **36.** $r = \sin 2\theta$

37. $r = 1 + 2\cos\theta$ **38.** $r = 2\theta$

In Exercises 39–44, use Formula (8) to calculate the arc length of the polar curve.

---

39. The entire circle $r = a$

40. The entire circle $r = 2a\cos\theta$

41. The entire cardioid $r = a(1 - \cos\theta)$

42. $r = \sin^2(\theta/2)$ from $\theta = 0$ to $\theta = \pi$

43. $r = e^{3\theta}$ from $\theta = 0$ to $\theta = 2$

44. $r = \sin^3(\theta/3)$ from $\theta = 0$ to $\theta = \pi/2$

45. (a) What is the slope of the tangent line at time t to the trajectory of the paper airplane in Example 2?

(b) What was the airplane's approximate angle of inclination when it crashed into the wall?

46. Suppose that a bee follows the trajectory

$$x = t - 2\sin t, \quad y = 2 - 2\cos t \quad (t \geq 0)$$

but lands on a wall at time $t = 10$.
(a) At what times was the bee flying horizontally?
(b) At what times was the bee flying vertically?

47. (a) Show that the arc length of one petal of the rose $r = \cos n\theta$ is given by

$$2 \int_0^{\pi/(2n)} \sqrt{1 + (n^2 - 1)\sin^2 n\theta} \, d\theta$$

(b) Use the numerical integration capability of a calculating utility to approximate the arc length of one petal of the four-petal rose $r = \cos 2\theta$.
(c) Use the numerical integration capability of a calculating utility to approximate the arc length of one petal of the n-petal rose $r = \cos n\theta$ for $n = 2, 3, 4, \ldots, 20$; then make a conjecture about the limit of these arc lengths as $n \to +\infty$.

48. (a) Sketch the spiral $r = e^{-\theta/8} \ (0 \leq \theta < +\infty)$.
(b) Find an improper integral for the total arc length of the spiral.
(c) Show that the integral converges and find the total arc length of the spiral.

Exercises 49–54 require the formulas developed in the following discussion: If $f'(t)$ and $g'(t)$ are continuous functions and if no segment of the curve

$$x = f(t), \quad y = g(t) \quad (a \leq t \leq b)$$

is traced more than once, then it can be shown that the area of the surface generated by revolving this curve about the x-axis is

$$S = \int_a^b 2\pi y \sqrt{\left(\frac{dx}{dt}\right)^2 + \left(\frac{dy}{dt}\right)^2} \, dt$$

and the area of the surface generated by revolving the curve about the y-axis is

$$S = \int_a^b 2\pi x \sqrt{\left(\frac{dx}{dt}\right)^2 + \left(\frac{dy}{dt}\right)^2} \, dt$$

[The derivations are similar to those used to obtain Formulas (4) and (5) in Section 7.5.]

---

49. Find the area of the surface generated by revolving $x = t^2$, $y = 2t \ (0 \leq t \leq 4)$ about the x-axis.

50. Find the area of the surface generated by revolving the curve $x = e^t \cos t$, $y = e^t \sin t \ (0 \leq t \leq \pi/2)$ about the x-axis.

51. Find the area of the surface generated by revolving the curve $x = \cos^2 t$, $y = \sin^2 t \ (0 \leq t \leq \pi/2)$ about the y-axis.

52. Find the area of the surface generated by revolving $x = t$, $y = 2t^2 \ (0 \leq t \leq 1)$ about the y-axis.

53. By revolving the semicircle

$$x = r\cos t, \quad y = r\sin t \quad (0 \leq t \leq \pi)$$

about the x-axis, show that the surface area of a sphere of radius r is $4\pi r^2$.

54. The equations

$$x = a\phi - a\sin\phi, \quad y = a - a\cos\phi \quad (0 \leq \phi \leq 2\pi)$$

represent one arch of a cycloid. Show that the surface area generated by revolving this curve about the x-axis is given by $S = 64\pi a^2/3$.

55. As illustrated in the accompanying figure, suppose that a rod with one end fixed at the pole of a polar coordinate system rotates counterclockwise at the constant rate of 1 rad/s. At time $t = 0$ a bug on the rod is 10 mm from the pole and is moving outward along the rod at the constant speed of 2 mm/s.
(a) Find an equation of the form $r = f(\theta)$ for the path of motion of the bug, assuming that $\theta = 0$ when $t = 0$.
(b) Find the distance the bug travels along the path in part (a) during the first 5 seconds. Round your answer to the nearest tenth of a millimeter.

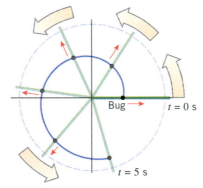

Figure Ex-55

56. Use Formula (6) of Section 7.4 to derive Formula (8).

11.3 AREA IN POLAR COORDINATES

In this section we will show how to find areas of regions that are bounded by polar curves.

AREA IN POLAR COORDINATES

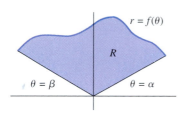

Figure 11.3.1

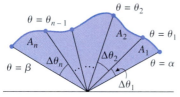

Figure 11.3.2

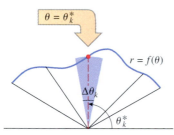

Figure 11.3.3

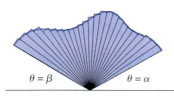

Figure 11.3.4

11.3.1 AREA PROBLEM IN POLAR COORDINATES. Suppose that α and β are angles that satisfy the condition

$$\alpha < \beta \leq \alpha + 2\pi$$

and suppose that $f(\theta)$ is continuous and either nonnegative or nonpositive for $\alpha \leq \theta \leq \beta$. Find the area of the region R enclosed by the polar curve $r = f(\theta)$ and the rays $\theta = \alpha$ and $\theta = \beta$ (Figure 11.3.1).

In rectangular coordinates we solved Area Problem 6.1.1 by dividing the region into an increasing number of vertical strips, approximating the strips by rectangles, and taking a limit. In polar coordinates rectangles are clumsy to work with, and it is better to divide the region into *wedges* by using rays

$$\theta = \theta_1, \ \theta = \theta_2, \ldots, \ \theta = \theta_{n-1}$$

such that

$$\alpha < \theta_1 < \theta_2 < \cdots < \theta_{n-1} < \beta$$

(Figure 11.3.2). As shown in that figure, the rays divide the region R into n wedges with areas $A_1, A_2, \ldots, A_n$ and central angles $\Delta\theta_1, \Delta\theta_2, \ldots, \Delta\theta_n$. The area of the entire region can be written as

$$A = A_1 + A_2 + \cdots + A_n = \sum_{k=1}^{n} A_k \tag{1}$$

If $\Delta\theta_k$ is small, and if we assume for simplicity that $f(\theta)$ is nonnegative, then we can approximate the area A_k of the kth wedge by the area of a sector with central angle $\Delta\theta_k$ and radius $f(\theta_k^*)$, where $\theta = \theta_k^*$ is any ray that lies in the kth wedge (Figure 11.3.3). Thus, from (1) and Formula (5) of Appendix E for the area of a sector, we obtain

$$A = \sum_{k=1}^{n} A_k \approx \sum_{k=1}^{n} \frac{1}{2}[f(\theta_k^*)]^2 \Delta\theta_k \tag{2}$$

If we now increase n in such a way that $\max \Delta\theta_k \to 0$, then the sectors will become better and better approximations of the wedges and it is reasonable to expect that (2) will approach the exact value of the area A (Figure 11.3.4); that is,

$$A = \lim_{\max \Delta\theta_k \to 0} \sum_{k=1}^{n} \frac{1}{2}[f(\theta_k^*)]^2 \Delta\theta_k = \int_{\alpha}^{\beta} \frac{1}{2}[f(\theta)]^2 \, d\theta$$

Thus, we have the following solution of Area Problem 11.3.1.

11.3.2 AREA IN POLAR COORDINATES. If α and β are angles that satisfy the condition

$$\alpha < \beta \leq \alpha + 2\pi$$

and if $f(\theta)$ is continuous and either nonnegative or nonpositive for $\alpha \leq \theta \leq \beta$, then the area A of the region R enclosed by the polar curve $r = f(\theta)$ and the rays $\theta = \alpha$ and $\theta = \beta$ is

$$A = \int_{\alpha}^{\beta} \frac{1}{2}[f(\theta)]^2 \, d\theta = \int_{\alpha}^{\beta} \frac{1}{2}r^2 \, d\theta \tag{3}$$

The hardest part of applying (3) is determining the limits of integration. This can be done as follows:

> **Step 1.**   Sketch the region R whose area is to be determined.
>
> **Step 2.**   Draw an arbitrary "radial line" from the pole to the boundary curve $r = f(\theta)$.
>
> **Step 3.**   Ask, "Over what interval of values must θ vary in order for the radial line to sweep out the region R?"
>
> **Step 4.**   Your answer in Step 3 will determine the lower and upper limits of integration.

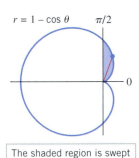

$r = 1 - \cos\theta$ $\pi/2$

The shaded region is swept out by the radial line as θ varies from 0 to $\pi/2$.

Figure 11.3.5

Example 1   Find the area of the region in the first quadrant that is within the cardioid $r = 1 - \cos\theta$.

Solution.   The region and a typical radial line are shown in Figure 11.3.5. For the radial line to sweep out the region, θ must vary from 0 to $\pi/2$. Thus, from (3) with $\alpha = 0$ and $\beta = \pi/2$, we obtain

$$A = \int_0^{\pi/2} \frac{1}{2} r^2 \, d\theta = \frac{1}{2} \int_0^{\pi/2} (1 - \cos\theta)^2 \, d\theta = \frac{1}{2} \int_0^{\pi/2} (1 - 2\cos\theta + \cos^2\theta) \, d\theta$$

With the help of the identity $\cos^2\theta = \frac{1}{2}(1 + \cos 2\theta)$, this can be rewritten as

$$A = \frac{1}{2} \int_0^{\pi/2} \left(\frac{3}{2} - 2\cos\theta + \frac{1}{2}\cos 2\theta \right) d\theta = \frac{1}{2} \left[\frac{3}{2}\theta - 2\sin\theta + \frac{1}{4}\sin 2\theta \right]_0^{\pi/2} = \frac{3}{8}\pi - 1 \quad \blacktriangleleft$$

Example 2   Find the entire area within the cardioid of Example 1.

Solution.   For the radial line to sweep out the entire cardioid, θ must vary from 0 to 2π. Thus, from (3) with $\alpha = 0$ and $\beta = 2\pi$,

$$A = \int_0^{2\pi} \frac{1}{2} r^2 \, d\theta = \frac{1}{2} \int_0^{2\pi} (1 - \cos\theta)^2 \, d\theta$$

If we proceed as in Example 1, this reduces to

$$A = \frac{1}{2} \int_0^{2\pi} \left(\frac{3}{2} - 2\cos\theta + \frac{1}{2}\cos 2\theta \right) d\theta = \frac{3\pi}{2}$$

Alternative Solution.   Since the cardioid is symmetric about the x-axis, we can calculate the portion of the area above the x-axis and double the result. In the portion of the cardioid above the x-axis, θ ranges from 0 to π, so that

$$A = 2 \int_0^{\pi} \frac{1}{2} r^2 \, d\theta = \int_0^{\pi} (1 - \cos\theta)^2 \, d\theta = \frac{3\pi}{2} \quad \blacktriangleleft$$

USING SYMMETRY

Although Formula (3) is applicable if $r = f(\theta)$ is negative, area computations can sometimes be simplified by using symmetry to restrict the limits of integration to intervals where $r \geq 0$. This is illustrated in the next example.

Example 3   Find the area of the region enclosed by the rose curve $r = \cos 2\theta$.

Solution.   Referring to Figure 11.1.10 and using symmetry, the area in the first quadrant that is swept out for $0 \leq \theta \leq \pi/4$ is one-eighth of the total area inside the rose. Thus, from

Formula (3)

$$A = 8 \int_0^{\pi/4} \frac{1}{2} r^2 \, d\theta = 4 \int_0^{\pi/4} \cos^2 2\theta \, d\theta$$

$$= 4 \int_0^{\pi/4} \frac{1}{2} (1 + \cos 4\theta) \, d\theta = 2 \int_0^{\pi/4} (1 + \cos 4\theta) \, d\theta$$

$$= 2\theta + \frac{1}{2} \sin 4\theta \Big]_0^{\pi/4} = \frac{\pi}{2}$$

◀

Sometimes the most natural way to satisfy the restriction $\alpha < \beta \le \alpha + 2\pi$ required by Formula (3) is to use a negative value for α. For example, suppose that we are interested in finding the area of the shaded region in Figure 11.3.6a. The first step would be to determine the intersections of the cardioid $r = 4 + 4\cos\theta$ and the circle $r = 6$, since this information is needed for the limits of integration. To find the points of intersection, we can equate the two expressions for r. This yields

$$4 + 4\cos\theta = 6 \quad \text{or} \quad \cos\theta = \frac{1}{2}$$

which is satisfied by the positive angles

$$\theta = \frac{\pi}{3} \quad \text{and} \quad \theta = \frac{5\pi}{3}$$

However, there is a problem here because the radial lines to the circle and cardioid do not sweep through the shaded region shown in Figure 11.3.6b as θ varies over the interval $\pi/3 \le \theta \le 5\pi/3$. There are two ways to circumvent this problem—one is to take advantage of the symmetry by integrating over the interval $0 \le \theta \le \pi/3$ and doubling the result, and the second is to use a negative lower limit of integration and integrate over the interval $-\pi/3 \le \theta \le \pi/3$ (Figure 11.3.6c). The two methods are illustrated in the next example.

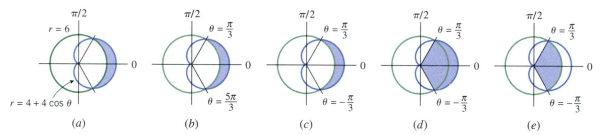

(a) (b) (c) (d) (e)

Figure 11.3.6

Example 4 Find the area of the region that is inside of the cardioid $r = 4 + 4\cos\theta$ and outside of the circle $r = 6$.

Solution Using a Negative Angle. The area of the region can be obtained by subtracting the areas in Figures 11.3.6d and 11.3.6e:

$$A = \int_{-\pi/3}^{\pi/3} \frac{1}{2} (4 + 4\cos\theta)^2 \, d\theta - \int_{-\pi/3}^{\pi/3} \frac{1}{2} (6)^2 \, d\theta \qquad \boxed{\begin{array}{l}\text{Area inside cardioid} \\ \text{minus area inside circle.}\end{array}}$$

$$= \int_{-\pi/3}^{\pi/3} \frac{1}{2} [(4 + 4\cos\theta)^2 - 36] \, d\theta = \int_{-\pi/3}^{\pi/3} (16\cos\theta + 8\cos^2\theta - 10) \, d\theta$$

$$= \left[16\sin\theta + (4\theta + 2\sin 2\theta) - 10\theta \right]_{-\pi/3}^{\pi/3} = 18\sqrt{3} - 4\pi$$

Solution Using Symmetry. Using symmetry, we can calculate the area above the polar axis and double it. This yields (verify)

$$A = 2 \int_0^{\pi/3} \frac{1}{2} [(4 + 4\cos\theta)^2 - 36] \, d\theta = 2(9\sqrt{3} - 2\pi) = 18\sqrt{3} - 4\pi$$

which agrees with the preceding result. ◀

INTERSECTIONS OF POLAR GRAPHS

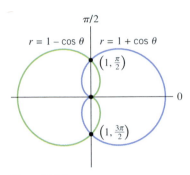

$r = 1 - \cos\theta$ | $r = 1 + \cos\theta$

$\left(1, \frac{\pi}{2}\right)$

$\left(1, \frac{3\pi}{2}\right)$

Figure 11.3.7

The orbits intersect, but the satellites do not collide.

Figure 11.3.8

In the last example we found the intersections of the cardioid and circle by equating their expressions for r and solving for θ. However, because a point can be represented in different ways in polar coordinates, this procedure will not always produce all of the intersections. For example, the cardioids

$$r = 1 - \cos\theta \quad \text{and} \quad r = 1 + \cos\theta \qquad (4)$$

intersect at three points: the pole, the point $(1, \pi/2)$, and the point $(1, 3\pi/2)$ (Figure 11.3.7). Equating the right-hand sides of the equations in (4) yields $1 - \cos\theta = 1 + \cos\theta$ or $\cos\theta = 0$, so

$$\theta = \frac{\pi}{2} + k\pi, \quad k = 0, \pm 1, \pm 2, \ldots$$

Substituting any of these values in (4) yields $r = 1$, so that we have found only two distinct points of intersection, $(1, \pi/2)$ and $(1, 3\pi/2)$; the pole has been missed. This problem occurs because the two cardioids pass through the pole at different values of θ—the cardioid $r = 1 - \cos\theta$ passes through the pole at $\theta = 0$, and the cardioid $r = 1 + \cos\theta$ passes through the pole at $\theta = \pi$.

The situation with the cardioids is analogous to two satellites circling the Earth in intersecting orbits (Figure 11.3.8). The satellites will not collide unless they reach the same point at the same time. In general, when looking for intersections of polar curves, it is a good idea to graph the curves to determine how many intersections there should be.

EXERCISE SET 11.3   ~ Graphing Utility   **C** CAS

1. Write down, but do not evaluate, an integral for the area of each shaded region.

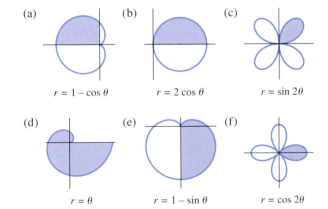

(a)   (b)   (c)

$r = 1 - \cos\theta$ $r = 2\cos\theta$ $r = \sin 2\theta$

(d)   (e)   (f)

$r = \theta$ $r = 1 - \sin\theta$ $r = \cos 2\theta$

2. Evaluate the integrals you obtained in Exercise 1.

3. In each part, find the area of the circle by integration.
   (a) $r = a$ (b) $r = 2a\sin\theta$ (c) $r = 2a\cos\theta$

4. (a) Show that $r = \sin\theta + \cos\theta$ is a circle.
   (b) Find the area of the circle using a geometric formula and then by integration.

In Exercises 5–10, find the area of the region described.

5. The region that is enclosed by the cardioid $r = 2 + 2\cos\theta$.

6. The region in the first quadrant within the cardioid $r = 1 + \sin\theta$.

7. The region enclosed by the rose $r = 4\cos 3\theta$.

8. The region enclosed by the rose $r = 2\sin 2\theta$.

9. The region enclosed by the inner loop of the limaçon $r = 1 + 2\cos\theta$. [*Hint:* $r \leq 0$ over the interval of integration.]

10. The region swept out by a radial line from the pole to the curve $r = 2/\theta$ as θ varies over the interval $1 \leq \theta \leq 3$.

In Exercises 11–14, find the area of the shaded region.

11.   12.

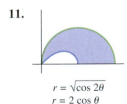

$r = \sqrt{\cos 2\theta}$
$r = 2\cos\theta$

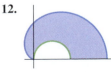

$r = 1 + \cos\theta$
$r = \cos\theta$

13.

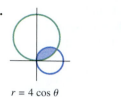

$r = 4 \cos \theta$
$r = 4\sqrt{3} \sin \theta$

14.

$r = 1 + \cos \theta$
$r = 3 \cos \theta$

In Exercises 15–22, find the area of the region described.

15. The region inside the circle $r = 5 \sin \theta$ and outside the limaçon $r = 2 + \sin \theta$.

16. The region outside the cardioid $r = 2 - 2 \cos \theta$ and inside the circle $r = 4$.

17. The region inside the cardioid $r = 2 + 2 \cos \theta$ and outside the circle $r = 3$.

18. The region that is common to the circles $r = 4 \cos \theta$ and $r = 4 \sin \theta$.

19. The region between the loops of the limaçon $r = \frac{1}{2} + \cos \theta$.

20. The region inside the cardioid $r = 2 + 2 \cos \theta$ and to the right of the line $r \cos \theta = \frac{3}{2}$.

21. The region inside the circle $r = 10$ and to the right of the line $r = 6 \sec \theta$.

22. The region inside the rose $r = 2a \cos 2\theta$ and outside the circle $r = a\sqrt{2}$.

23. (a) Find the error: The area that is inside the lemniscate $r^2 = a^2 \cos 2\theta$ is

$$A = \int_0^{2\pi} \frac{1}{2} r^2 \, d\theta = \int_0^{2\pi} \frac{1}{2} a^2 \cos 2\theta \, d\theta$$

$$= \frac{1}{4} a^2 \sin 2\theta \Big]_0^{2\pi} = 0$$

(b) Find the correct area.
(c) Find the area inside the lemniscate $r^2 = 4 \cos 2\theta$ and outside the circle $r = \sqrt{2}$.

24. Find the area inside the curve $r^2 = \sin 2\theta$.

25. A radial line is drawn from the origin to the spiral $r = a\theta$ ($a > 0$ and $\theta \geq 0$). Find the area swept out during the second revolution of the radial line that was not swept out during the first revolution.

26. (a) In the discussion associated with Exercises 49–54 of Section 11.2, formulas were given for the area of the surface of revolution that is generated by revolving a parametric curve about the x-axis or y-axis. Use those formulas to derive the following formulas for the areas of the surfaces of revolution that are generated by revolving the portion of the polar curve $r = f(\theta)$ from $\theta = \alpha$ to $\theta = \beta$ about the polar axis and about the line $\theta = \pi/2$:

$$S = \int_\alpha^\beta 2\pi r \sin \theta \sqrt{r^2 + \left(\frac{dr}{d\theta}\right)^2} \, d\theta \quad \boxed{\text{About } \theta = 0}$$

$$S = \int_\alpha^\beta 2\pi r \cos \theta \sqrt{r^2 + \left(\frac{dr}{d\theta}\right)^2} \, d\theta \quad \boxed{\text{About } \theta = \pi/2}$$

(b) State conditions under which these formulas hold.

In Exercises 27–30, sketch the surface, and use the formulas in Exercise 26 to find the surface area.

27. The surface generated by revolving the circle $r = \cos \theta$ about the line $\theta = \pi/2$.

28. The surface generated by revolving the spiral $r = e^\theta$ ($0 \leq \theta \leq \pi/2$) about the line $\theta = \pi/2$.

29. The "apple" generated by revolving the upper half of the cardioid $r = 1 - \cos \theta$ ($0 \leq \theta \leq \pi$) about the polar axis.

30. The sphere of radius a generated by revolving the semi-circle $r = a$ in the upper half-plane about the polar axis.

C **31. (a)** Show that the Folium of Descartes $x^3 - 3xy + y^3 = 0$ can be expressed in polar coordinates as

$$r = \frac{3 \sin \theta \cos \theta}{\cos^3 \theta + \sin^3 \theta}$$

(b) Use a CAS to show that the area inside of the loop is $\frac{3}{2}$ (Figure 3.6.2).

C **32. (a)** What is the area that is enclosed by one petal of the rose $r = a \cos n\theta$ if n is an even integer?

(b) What is the area that is enclosed by one petal of the rose $r = a \cos n\theta$ if n is an odd integer?

(c) Use a CAS to show that the total area enclosed by the rose $r = a \cos n\theta$ is $\pi a^2/2$ if the number of petals is even. [*Hint:* See Exercise 73 of Section 11.1.]

(d) Use a CAS to show that the total area enclosed by the rose $r = a \cos n\theta$ is $\pi a^2/4$ if the number of petals is odd.

33. One of the most famous problems in Greek antiquity was "squaring the circle"; that is, using a straightedge and compass to construct a square whose area is equal to that of a given circle. It was proved in the nineteenth century that no such construction is possible. However, show that the shaded areas in the accompanying figure are equal, thereby "squaring the crescent."

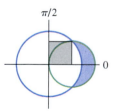

Figure Ex-33

34. Use a graphing utility to generate the polar graph of the equation $r = \cos 3\theta + 2$, and find the area that it encloses.

35. Use a graphing utility to generate the graph of the *bifolium* $r = 2 \cos \theta \sin^2 \theta$, and find the area of the upper loop.

11.4 CONIC SECTIONS IN CALCULUS

In this section we will discuss some of the basic geometric properties of parabolas, ellipses, and hyperbolas. These curves play an important role in calculus and also arise naturally in a broad range of applications in such fields as planetary motion, design of telescopes and antennas, geodetic positioning, and medicine, to name a few.

Some students may already be familiar with the material in this section, in which case it can be treated as a review. Instructors who want to spend some additional time on precalculus review may want to allocate more than one lecture on this material.

CONIC SECTIONS

Circles, ellipses, parabolas, and hyperbolas are called *conic sections* or *conics* because they can be obtained as intersections of a plane with a double-napped circular cone (Figure 11.4.1). If the plane passes through the vertex of the double-napped cone, then the intersection is a point, a pair of intersecting lines, or a single line. These are called *degenerate conic sections*.

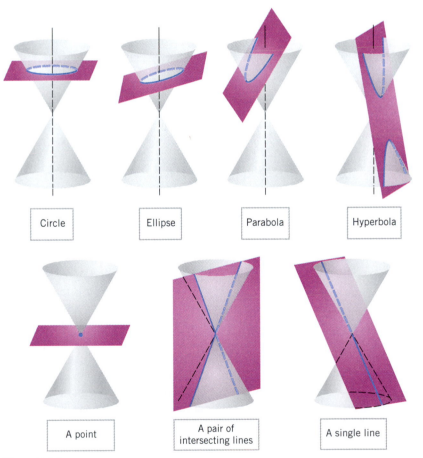

Figure 11.4.1

DEFINITIONS OF THE CONIC SECTIONS

Although we could derive properties of parabolas, ellipses, and hyperbolas by defining them as intersections with a double-napped cone, it will be better suited to calculus if we begin with equivalent definitions that are based on their geometric properties.

11.4.1 DEFINITION. A *parabola* is the set of all points in the plane that are equidistant from a fixed line and a fixed point not on the line.

The line is called the *directrix* of the parabola, and the point is called the *focus* (Figure 11.4.2). A parabola is symmetric about the line that passes through the focus at right angles to the directrix. This line, called the *axis* or the *axis of symmetry* of the parabola, intersects the parabola at a point called the *vertex*.

11.4.2 DEFINITION. An *ellipse* is the set of all points in the plane, the sum of whose distances from two fixed points is a given positive constant that is greater than the distance between the fixed points.

The two fixed points are called the *foci* (plural of "focus") of the ellipse, and the midpoint of the line segment joining the foci is called the *center* (Figure 11.4.3a). To help visualize Definition 11.4.2, imagine that two ends of a string are tacked to the foci and a pencil traces a curve as it is held tight against the string (Figure 11.4.3b). The resulting curve will be an ellipse since the sum of the distances to the foci is a constant, namely the total length of the string. Note that if the foci coincide, the ellipse reduces to a circle. For ellipses other than circles, the line segment through the foci and across the ellipse is called the *major axis* (Figure 11.4.3c), and the line segment across the ellipse, through the center, and perpendicular to the major axis is called the *minor axis*. The endpoints of the major axis are called *vertices*.

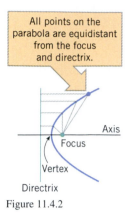

All points on the parabola are equidistant from the focus and directrix.

Axis
Focus
Vertex
Directrix

Figure 11.4.2

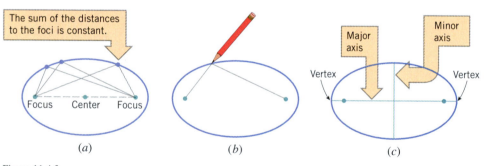

The sum of the distances to the foci is constant.

Focus   Center   Focus

(a)

(b)

Minor axis
Major axis
Vertex   Vertex

(c)

Figure 11.4.3

11.4.3 DEFINITION. A *hyperbola* is the set of all points in the plane, the difference of whose distances from two fixed distinct points is a given positive constant that is less than the distance between the fixed points.

The two fixed points are called the *foci* of the hyperbola, and the term "difference" that is used in the definition is understood to mean the distance to the farther focus minus the distance to the closer focus. As a result, the points on the hyperbola form two *branches*, each "wrapping around" the closer focus (Figure 11.4.4a). The midpoint of the line segment joining the foci is called the *center* of the hyperbola, the line through the foci is called the *focal axis*, and the line through the center that is perpendicular to the focal axis is called the *conjugate axis*. The hyperbola intersects the focal axis at two points called the *vertices*.

Associated with every hyperbola is a pair of lines, called the *asymptotes* of the hyperbola. These lines intersect at the center of the hyperbola and have the property that as a point P moves along the hyperbola away from the center, the vertical distance between P and one of the asymptotes approaches zero (Figure 11.4.4b).

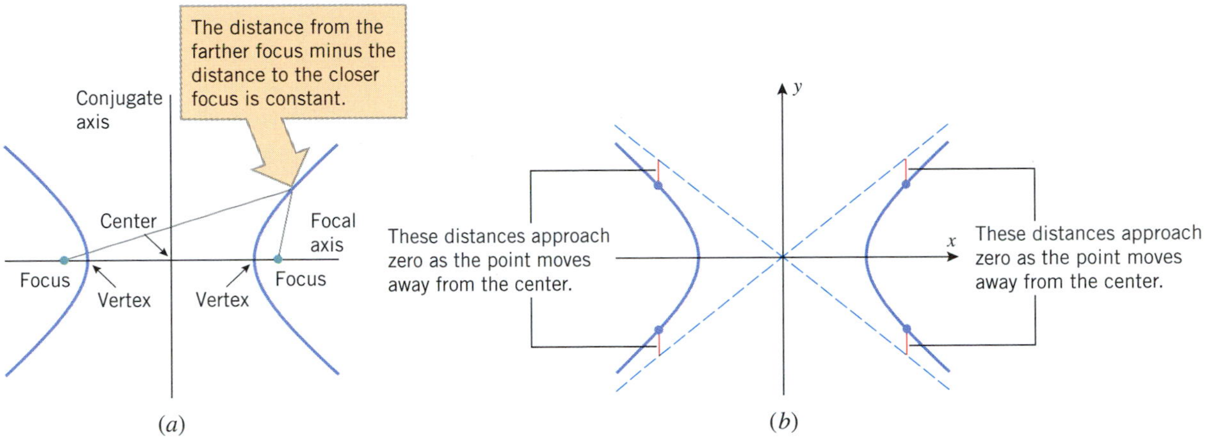

The distance from the farther focus minus the distance to the closer focus is constant.

Conjugate axis

Center

Focal axis

Focus

Vertex Vertex

Focus

(a)

These distances approach zero as the point moves away from the center.

These distances approach zero as the point moves away from the center.

(b)

Figure 11.4.4

EQUATIONS OF PARABOLAS IN STANDARD POSITION

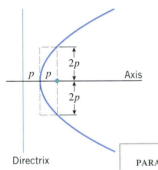

Directrix

Figure 11.4.5

It is traditional in the study of parabolas to denote the distance between the focus and the vertex by p. The vertex is equidistant from the focus and the directrix, so the distance between the vertex and the directrix is also p; consequently, the distance between the focus and the directrix is $2p$ (Figure 11.4.5). As illustrated in that figure, the parabola passes through two of the corners of a box that extends from the vertex to the focus along the axis of symmetry and extends $2p$ units above and $2p$ units below the axis of symmetry.

The equation of a parabola is simplest if the vertex is the origin and the axis of symmetry is along the x-axis or y-axis. The four possible such orientations are shown in Figure 11.4.6. These are called the *standard positions* of a parabola, and the resulting equations are called the *standard equations* of a parabola.

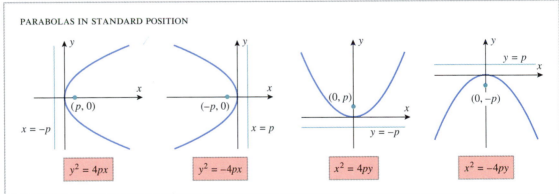

PARABOLAS IN STANDARD POSITION

$y^2 = 4px$

$y^2 = -4px$

$x^2 = 4py$

$x^2 = -4py$

Figure 11.4.6

To illustrate how the equations in Figure 11.4.6 are obtained, we will derive the equation for the parabola with focus $(p, 0)$ and directrix $x = -p$. Let $P(x, y)$ be any point on the parabola. Since P is equidistant from the focus and directrix, the distances PF and PD in Figure 11.4.7 are equal; that is,

$$PF = PD \tag{1}$$

where $D(-p, y)$ is the foot of the perpendicular from P to the directrix. From the distance formula, the distances PF and PD are

$$PF = \sqrt{(x - p)^2 + y^2} \quad \text{and} \quad PD = \sqrt{(x + p)^2} \tag{2}$$

Substituting in (1) and squaring yields

$$(x - p)^2 + y^2 = (x + p)^2 \tag{3}$$

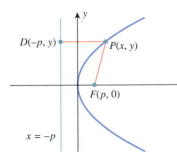

Figure 11.4.7

and after simplifying

$$y^2 = 4px \tag{4}$$

The derivations of the other equations in Figure 11.4.6 are similar.

Parabolas can be sketched from their *standard equations* using four basic steps:

- Determine whether the axis of symmetry is along the x-axis or the y-axis. Referring to Figure 11.4.6, the axis of symmetry is along the x-axis if the equation has a y^2-term, and it is along the y-axis if it has an x^2-term.

- Determine which way the parabola opens. If the axis of symmetry is along the x-axis, then the parabola opens to the right if the coefficient of x is positive, and it opens to the left if the coefficient is negative. If the axis of symmetry is along the y-axis, then the parabola opens up if the coefficient of y is positive, and it opens down if the coefficient is negative.

- Determine the value of p and draw a box extending p units from the origin along the axis of symmetry in the direction in which the parabola opens and extending $2p$ units on each side of the axis of symmetry.

- Using the box as a guide, sketch the parabola so that its vertex is at the origin and it passes through the corners of the box (Figure 11.4.8).

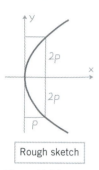

Rough sketch

Figure 11.4.8

Example 1   Sketch the graphs of the parabolas

(a) $x^2 = 12y$ (b) $y^2 + 8x = 0$

and show the focus and directrix of each.

Solution (a). This equation involves x^2, so the axis of symmetry is along the y-axis, and the coefficient of y is positive, so the parabola opens upward. From the coefficient of y, we obtain $4p = 12$ or $p = 3$. Drawing a box extending $p = 3$ units up from the origin and $2p = 6$ units to the left and $2p = 6$ units to the right of the y-axis, then using corners of the box as a guide, yields the graph in Figure 11.4.9.

The focus is $p = 3$ units from the vertex along the axis of symmetry in the direction in which the parabola opens, so its coordinates are $(0, 3)$. The directrix is perpendicular to the axis of symmetry at a distance of $p = 3$ units from the vertex on the opposite side from the focus, so its equation is $y = -3$.

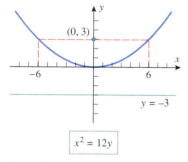

$x^2 = 12y$

Figure 11.4.9

Solution (b). We first rewrite the equation in the standard form

$$y^2 = -8x$$

This equation involves y^2, so the axis of symmetry is along the x-axis, and the coefficient of x is negative, so the parabola opens to the left. From the coefficient of x we obtain $4p = 8$, so $p = 2$. Drawing a box extending $p = 2$ units left from the origin and $2p = 4$ units above and $2p = 4$ units below the x-axis, then using corners of the box as a guide, yields the graph in Figure 11.4.10. ◀

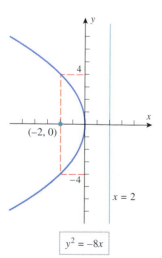

$y^2 = -8x$

Figure 11.4.10

Example 2   Find an equation of the parabola that is symmetric about the y-axis, has its vertex at the origin, and passes through the point $(5, 2)$.

Solution. Since the parabola is symmetric about the y-axis and has its vertex at the origin, the equation is of the form

$$x^2 = 4py \quad \text{or} \quad x^2 = -4py$$

where the sign depends on whether the parabola opens up or down. But the parabola must open up, since it passes through the point $(5, 2)$, which lies in the first quadrant. Thus, the equation is of the form

$$x^2 = 4py \tag{5}$$

Since the parabola passes through (5, 2), we must have $5^2 = 4p \cdot 2$ or $4p = \frac{25}{2}$. Therefore, (5) becomes

$$x^2 = \frac{25}{2}y \qquad \blacktriangleleft$$

EQUATIONS OF ELLIPSES IN STANDARD POSITION

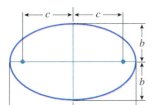

Figure 11.4.11

It is traditional in the study of ellipses to denote the length of the major axis by $2a$, the length of the minor axis by $2b$, and the distance between the foci by $2c$ (Figure 11.4.11). The number a is called the **semimajor axis** and the number b the **semiminor axis** (standard but odd terminology, since a and b are numbers, not geometric axes).

There is a basic relationship between the numbers a, b, and c that can be obtained by examining the sum of the distances to the foci from a point P at the end of the major axis and from a point Q at the end of the minor axis (Figure 11.4.12). From Definition 11.4.2, these sums must be equal, so we obtain

$$2\sqrt{b^2 + c^2} = (a - c) + (a + c)$$

from which it follows that

$$a = \sqrt{b^2 + c^2} \qquad (6)$$

or, equivalently,

$$c = \sqrt{a^2 - b^2} \qquad (7)$$

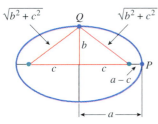

Figure 11.4.12

From (6), the distance from a focus to an end of the minor axis is a (Figure 11.4.13), which implies that for *all* points on the ellipse the sum of the distances to the foci is $2a$.

It also follows from (6) that $a \geq b$ with the equality holding only when $c = 0$. Geometrically, this means that the major axis of an ellipse is at least as large as the minor axis and that the two axes have equal length only when the foci coincide, in which case the ellipse is a circle.

The equation of an ellipse is simplest if the center of the ellipse is at the origin and the foci are on the x-axis or y-axis. The two possible such orientations are shown in Figure 11.4.14. These are called the **standard positions** of an ellipse, and the resulting equations are called the **standard equations** of an ellipse.

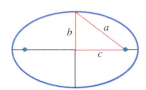

Figure 11.4.13

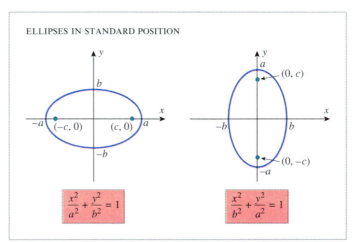

Figure 11.4.14

To illustrate how the equations in Figure 11.4.14 are obtained, we will derive the equation for the ellipse with foci on the x-axis. Let $P(x, y)$ be any point on that ellipse. Since the sum of the distances from P to the foci is $2a$, it follows (Figure 11.4.15) that

$$PF' + PF = 2a$$

so

$$\sqrt{(x + c)^2 + y^2} + \sqrt{(x - c)^2 + y^2} = 2a$$

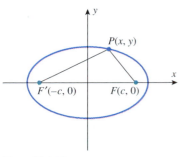

Figure 11.4.15

Transposing the second radical to the right side of the equation and squaring yields

$$(x + c)^2 + y^2 = 4a^2 - 4a\sqrt{(x - c)^2 + y^2} + (x - c)^2 + y^2$$

and, on simplifying,

$$\sqrt{(x - c)^2 + y^2} = a - \frac{c}{a}x \qquad (8)$$

Squaring again and simplifying yields

$$\frac{x^2}{a^2} + \frac{y^2}{a^2 - c^2} = 1$$

which, by virtue of (6), can be written as

$$\frac{x^2}{a^2} + \frac{y^2}{b^2} = 1 \qquad (9)$$

Conversely, it can be shown that any point whose coordinates satisfy (9) has $2a$ as the sum of its distances from the foci, so that such a point is on the ellipse.

....

**A TECHNIQUE FOR SKETCHING
ELLIPSES**

Ellipses can be sketched from their *standard equations* using three basic steps:

- Determine whether the major axis is on the x-axis or the y-axis. This can be ascertained from the sizes of the denominators in the equation. Referring to Figure 11.4.14, and keeping in mind that $a^2 > b^2$ (since $a > b$), the major axis is along the x-axis if x^2 has the larger denominator, and it is along the y-axis if y^2 has the larger denominator. If the denominators are equal, the ellipse is a circle.

- Determine the values of a and b and draw a box extending a units on each side of the center along the major axis and b units on each side of the center along the minor axis.

- Using the box as a guide, sketch the ellipse so that its center is at the origin and it touches the sides of the box where the sides intersect the coordinate axes (Figure 11.4.16).

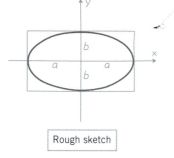

Rough sketch

Figure 11.4.16

Example 3   Sketch the graphs of the ellipses

(a) $\dfrac{x^2}{9} + \dfrac{y^2}{16} = 1$ (b) $x^2 + 2y^2 = 4$

showing the foci of each.

Solution (a). Since y^2 has the larger denominator, the major axis is along the y-axis. Moreover, since $a^2 > b^2$, we must have $a^2 = 16$ and $b^2 = 9$, so

$$a = 4 \quad \text{and} \quad b = 3$$

Drawing a box extending 4 units on each side of the origin along the y-axis and 3 units on each side of the origin along the x-axis as a guide yields the graph in Figure 11.4.17.

The foci lie c units on each side of the center along the major axis, where c is given by (7). From the values of a^2 and b^2 above, we obtain

$$c = \sqrt{a^2 - b^2} = \sqrt{16 - 9} = \sqrt{7} \approx 2.6$$

Thus, the coordinates of the foci are $(0, \sqrt{7})$ and $(0, -\sqrt{7})$, since they lie on the y-axis.

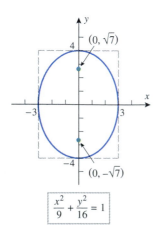

$$\frac{x^2}{9} + \frac{y^2}{16} = 1$$

Figure 11.4.17

Solution (b). We first rewrite the equation in the standard form

$$\frac{x^2}{4} + \frac{y^2}{2} = 1$$

Since x^2 has the larger denominator, the major axis lies along the x-axis, and we have $a^2 = 4$ and $b^2 = 2$. Drawing a box extending $a = 2$ on each side of the origin along the x-axis and extending $b = \sqrt{2} \approx 1.4$ units on each side of the origin along the y-axis as a guide yields the graph in Figure 11.4.18.

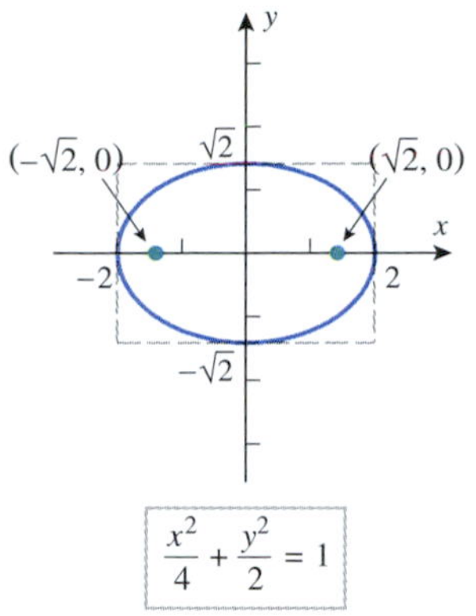

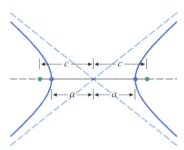

$$\frac{x^2}{4} + \frac{y^2}{2} = 1$$

Figure 11.4.18

From (7), we obtain

$$c = \sqrt{a^2 - b^2} = \sqrt{2} \approx 1.4$$

Thus, the coordinates of the foci are $(\sqrt{2}, 0)$ and $(-\sqrt{2}, 0)$, since they lie on the x-axis. ◄

Example 4 Find an equation for the ellipse with foci $(0, \pm 2)$ and major axis with endpoints $(0, \pm 4)$.

Solution. From Figure 11.4.14, the equation has the form

$$\frac{x^2}{b^2} + \frac{y^2}{a^2} = 1$$

and from the given information, $a = 4$ and $c = 2$. It follows from (6) that

$$b^2 = a^2 - c^2 = 16 - 4 = 12$$

so the equation of the ellipse is

$$\frac{x^2}{12} + \frac{y^2}{16} = 1 \qquad ◄$$

EQUATIONS OF HYPERBOLAS IN STANDARD POSITION

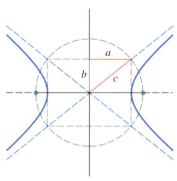

Figure 11.4.19

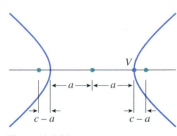

Figure 11.4.20

It is traditional in the study of hyperbolas to denote the distance between the vertices by $2a$, the distance between the foci by $2c$ (Figure 11.4.19), and to define the quantity b as

$$b = \sqrt{c^2 - a^2} \tag{10}$$

This relationship, which can also be expressed as

$$c = \sqrt{a^2 + b^2} \tag{11}$$

is pictured geometrically in Figure 11.4.20. As illustrated in that figure, and as we will show later in this section, the asymptotes pass through the corners of a box extending b units on each side of the center along the conjugate axis and a units on each side of the center along the focal axis. The number a is called the ***semifocal axis*** of the hyperbola and the number b the ***semiconjugate axis***. (As with the semimajor and semiminor axes of an ellipse, these are numbers, not geometric axes.)

If V is one vertex of a hyperbola, then, as illustrated in Figure 11.4.21, the distance from V to the farther focus minus the distance from V to the closer focus is

$$[(c - a) + 2a] - (c - a) = 2a$$

Thus, for *all* points on a hyperbola, the distance to the farther focus minus the distance to the closer focus is $2a$.

The equation of a hyperbola is simplest if the center of the hyperbola is at the origin and the foci are on the x-axis or y-axis. The two possible such orientations are shown in Figure 11.4.22. These are called the ***standard positions*** of a hyperbola, and the resulting equations are called the ***standard equations*** of a hyperbola.

The derivations of these equations are similar to those already given for parabolas and ellipses, so we will leave them as exercises. However, to illustrate how the equations of the asymptotes are derived, we will derive those equations for the hyperbola

$$\frac{x^2}{a^2} - \frac{y^2}{b^2} = 1$$

We can rewrite this equation as

$$y^2 = \frac{b^2}{a^2}(x^2 - a^2)$$

which is equivalent to the pair of equations

$$y = \frac{b}{a}\sqrt{x^2 - a^2} \quad \text{and} \quad y = -\frac{b}{a}\sqrt{x^2 - a^2}$$

Figure 11.4.21

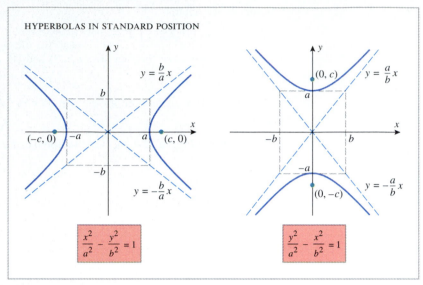

HYPERBOLAS IN STANDARD POSITION

$$\frac{x^2}{a^2} - \frac{y^2}{b^2} = 1$$

$$\frac{y^2}{a^2} - \frac{x^2}{b^2} = 1$$

Figure 11.4.22

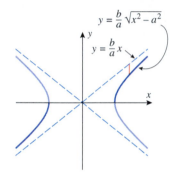

Figure 11.4.23

Thus, in the first quadrant, the vertical distance between the line $y = (b/a)x$ and the hyperbola can be written (Figure 11.4.23) as

$$\frac{b}{a}x - \frac{b}{a}\sqrt{x^2 - a^2}$$

But this distance tends to zero as $x \to +\infty$ since

$$\lim_{x \to +\infty} \left(\frac{b}{a}x - \frac{b}{a}\sqrt{x^2 - a^2} \right) = \lim_{x \to +\infty} \frac{b}{a}(x - \sqrt{x^2 - a^2})$$

$$= \lim_{x \to +\infty} \frac{b}{a} \frac{(x - \sqrt{x^2 - a^2})(x + \sqrt{x^2 - a^2})}{x + \sqrt{x^2 - a^2}}$$

$$= \lim_{x \to +\infty} \frac{ab}{x + \sqrt{x^2 - a^2}} = 0$$

The analysis in the remaining quadrants is similar.

A QUICK WAY TO FIND ASYMPTOTES

There is a trick that can be used to avoid memorizing the equations of the asymptotes of a hyperbola. They can be obtained, when needed, by substituting 0 for the 1 on the right side of the hyperbola equation, and then solving for y in terms of x. For example, for the hyperbola

$$\frac{x^2}{a^2} - \frac{y^2}{b^2} = 1$$

we would write

$$\frac{x^2}{a^2} - \frac{y^2}{b^2} = 0 \quad \text{or} \quad y^2 = \frac{b^2}{a^2}x^2 \quad \text{or} \quad y = \pm \frac{b}{a}x$$

which are the equations for the asymptotes.

A TECHNIQUE FOR SKETCHING HYPERBOLAS

Hyperbolas can be sketched from their *standard equations* using four basic steps:

- Determine whether the focal axis is on the x-axis or the y-axis. This can be ascertained from the location of the minus sign in the equation. Referring to Figure 11.4.22, the focal axis is along the x-axis when the minus sign precedes the y^2-term, and it is along the y-axis when the minus sign precedes the x^2-term.

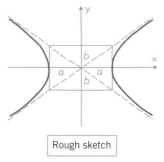

Rough sketch

Figure 11.4.24

- Determine the values of a and b and draw a box extending a units on either side of the center along the focal axis and b units on either side of the center along the conjugate axis. (The squares of a and b can be read directly from the equation.)
- Draw the asymptotes along the diagonals of the box.
- Using the box and the asymptotes as a guide, sketch the graph of the hyperbola (Figure 11.4.24).

Example 5 Sketch the graphs of the hyperbolas

(a) $\dfrac{x^2}{4} - \dfrac{y^2}{9} = 1$ (b) $y^2 - x^2 = 1$

showing their vertices, foci, and asymptotes.

Solution (a). The minus sign precedes the y^2-term, so the focal axis is along the x-axis. From the denominators in the equation we obtain

$$a^2 = 4 \quad \text{and} \quad b^2 = 9$$

Since a and b are positive, we must have $a = 2$ and $b = 3$. Recalling that the vertices lie a units on each side of the center on the focal axis, it follows that their coordinates in this case are $(2, 0)$ and $(-2, 0)$. Drawing a box extending $a = 2$ units along the x-axis on each side of the origin and $b = 3$ units on each side of the origin along the y-axis, then drawing the asymptotes along the diagonals of the box as a guide, yields the graph in Figure 11.4.25.

To obtain equations for the asymptotes, we substitute 0 for 1 in the given equation; this yields

$$\frac{x^2}{4} - \frac{y^2}{9} = 0 \quad \text{or} \quad y = \pm\frac{3}{2}x$$

The foci lie c units on each side of the center along the focal axis, where c is given by (11). From the values of a^2 and b^2 above we obtain

$$c = \sqrt{a^2 + b^2} = \sqrt{4 + 9} = \sqrt{13} \approx 3.6$$

Since the foci lie on the x-axis in this case, their coordinates are $(\sqrt{13}, 0)$ and $(-\sqrt{13}, 0)$.

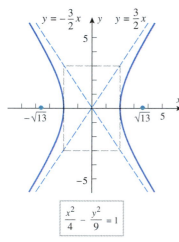

$\dfrac{x^2}{4} - \dfrac{y^2}{9} = 1$

Figure 11.4.25

Solution (b). The minus sign precedes the x^2-term, so the focal axis is along the y-axis. From the denominators in the equation we obtain $a^2 = 1$ and $b^2 = 1$, from which it follows that

$$a = 1 \quad \text{and} \quad b = 1$$

Thus, the vertices are at $(0, -1)$ and $(0, 1)$. Drawing a box extending $a = 1$ unit on either side of the origin along the y-axis and $b = 1$ unit on either side of the origin along the x-axis, then drawing the asymptotes, yields the graph in Figure 11.4.26. Since the box is actually a square, the asymptotes are perpendicular and have equations $y = \pm x$. This can also be seen by substituting 0 for 1 in the given equation, which yields $y^2 - x^2 = 0$ or $y = \pm x$. Also,

$$c = \sqrt{a^2 + b^2} = \sqrt{1 + 1} = \sqrt{2}$$

so the foci, which lie on the y-axis, are $(0, -\sqrt{2})$ and $(0, \sqrt{2})$. ◄

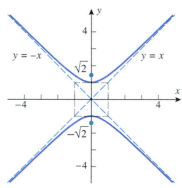

Figure 11.4.26

REMARK.   A hyperbola in which $a = b$, as in part (b) of this example, is called an ***equilateral hyperbola***. Such hyperbolas always have perpendicular asymptotes.

Example 6 Find the equation of the hyperbola with vertices $(0, \pm 8)$ and asymptotes $y = \pm\frac{4}{3}x$.

Solution. Since the vertices are on the y-axis, the equation of the hyperbola has the form $(y^2/a^2) - (x^2/b^2) = 1$ and the asymptotes are

$$y = \pm\frac{a}{b}x$$

From the locations of the vertices we have $a = 8$, so the given equations of the asymptotes yield

$$y = \pm\frac{a}{b}x = \pm\frac{8}{b}x = \pm\frac{4}{3}x$$

from which it follows that $b = 6$. Thus, the hyperbola has the equation

$$\frac{y^2}{64} - \frac{x^2}{36} = 1$$ ◄

TRANSLATED CONICS

Equations of conics that are translated from their standard positions can be obtained by replacing x by $x - h$ and y by $y - k$ in their standard equations. For a parabola, this translates the vertex from the origin to the point (h, k); and for ellipses and hyperbolas, this translates the center from the origin to the point (h, k).

Parabolas with vertex (h, k) and axis parallel to x-axis

$$(y - k)^2 = 4p(x - h) \qquad \text{[Opens right]} \tag{12}$$

$$(y - k)^2 = -4p(x - h) \qquad \text{[Opens left]} \tag{13}$$

Parabolas with vertex (h, k) and axis parallel to y-axis

$$(x - h)^2 = 4p(y - k) \qquad \text{[Opens up]} \tag{14}$$

$$(x - h)^2 = -4p(y - k) \qquad \text{[Opens down]} \tag{15}$$

Ellipse with center (h, k) and major axis parallel to x-axis

$$\frac{(x - h)^2}{a^2} + \frac{(y - k)^2}{b^2} = 1 \quad [b \leq a] \tag{16}$$

Ellipse with center (h, k) and major axis parallel to y-axis

$$\frac{(x - h)^2}{b^2} + \frac{(y - k)^2}{a^2} = 1 \quad [b \leq a] \tag{17}$$

Hyperbola with center (h, k) and focal axis parallel to x-axis

$$\frac{(x - h)^2}{a^2} - \frac{(y - k)^2}{b^2} = 1 \tag{18}$$

Hyperbola with center (h, k) and focal axis parallel to y-axis

$$\frac{(y - k)^2}{a^2} - \frac{(x - h)^2}{b^2} = 1 \tag{19}$$

Example 7 Find an equation for the parabola that has its vertex at $(1, 2)$ and its focus at $(4, 2)$.

Solution. Since the focus and vertex are on a horizontal line, and since the focus is to the right of the vertex, the parabola opens to the right and its equation has the form

$$(y - k)^2 = 4p(x - h)$$

Since the vertex and focus are 3 units apart, we have $p = 3$, and since the vertex is at $(h, k) = (1, 2)$, we obtain

$$(y - 2)^2 = 12(x - 1)$$ ◄

Sometimes the equations of translated conics occur in expanded form, in which case we are faced with the problem of identifying the graph of a quadratic equation in x and y:

$$Ax^2 + Cy^2 + Dx + Ey + F = 0 \tag{20}$$

The basic procedure for determining the nature of such a graph is to complete the squares of the quadratic terms and then try to match up the resulting equation with one of the forms of a translated conic.

Example 8   Describe the graph of the equation

$$y^2 - 8x - 6y - 23 = 0$$

Solution. The equation involves quadratic terms in y but none in x, so we first take all of the y-terms to one side:

$$y^2 - 6y = 8x + 23$$

Next, we complete the square on the y-terms by adding 9 to both sides:

$$(y - 3)^2 = 8x + 32$$

Finally, we factor out the coefficient of the x-term to obtain

$$(y - 3)^2 = 8(x + 4)$$

This equation is of form (12) with $h = -4$, $k = 3$, and $p = 2$, so the graph is a parabola with vertex $(-4, 3)$ opening to the right. Since $p = 2$, the focus is 2 units to the right of the vertex, which places it at the point $(-2, 3)$; and the directrix is 2 units to the left of the vertex, which means that its equation is $x = -6$. The parabola is shown in Figure 11.4.27.   ◄

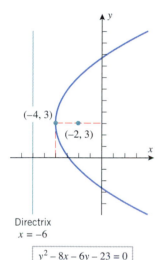

Directrix
x = −6

$y^2 - 8x - 6y - 23 = 0$

Figure 11.4.27

Example 9   Describe the graph of the equation

$$16x^2 + 9y^2 - 64x - 54y + 1 = 0$$

Solution. This equation involves quadratic terms in both x and y, so we will group the x-terms and the y-terms on one side and put the constant on the other:

$$(16x^2 - 64x) + (9y^2 - 54y) = -1$$

Next, factor out the coefficients of x^2 and y^2 and complete the squares:

$$16(x^2 - 4x + 4) + 9(y^2 - 6y + 9) = -1 + 64 + 81$$

or

$$16(x - 2)^2 + 9(y - 3)^2 = 144$$

Finally, divide through by 144 to introduce a 1 on the right side:

$$\frac{(x - 2)^2}{9} + \frac{(y - 3)^2}{16} = 1$$

This is an equation of form (17), with $h = 2$, $k = 3$, $a^2 = 16$, and $b^2 = 9$. Thus, the graph

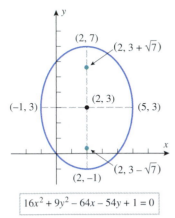

$$16x^2 + 9y^2 - 64x - 54y + 1 = 0$$

Figure 11.4.28

of the equation is an ellipse with center $(2, 3)$ and major axis parallel to the y-axis. Since $a = 4$, the major axis extends 4 units above and 4 units below the center, so its endpoints are $(2, 7)$ and $(2, -1)$ (Figure 11.4.28). Since $b = 3$, the minor axis extends 3 units to the left and 3 units to the right of the center, so its endpoints are $(-1, 3)$ and $(5, 3)$. Since

$$c = \sqrt{a^2 - b^2} = \sqrt{16 - 9} = \sqrt{7}$$

the foci lie $\sqrt{7}$ units above and below the center, placing them at the points $(2, 3 + \sqrt{7})$ and $(2, 3 - \sqrt{7})$.   ◀

Example 10   Describe the graph of the equation

$$x^2 - y^2 - 4x + 8y - 21 = 0$$

Solution.   This equation involves quadratic terms in both x and y, so we will group the x-terms and the y-terms on one side and put the constant on the other:

$$(x^2 - 4x) - (y^2 - 8y) = 21$$

We leave it for you to verify by completing the squares that this equation can be written as

$$\frac{(x - 2)^2}{9} - \frac{(y - 4)^2}{9} = 1 \tag{21}$$

This is an equation of form (18) with $h = 2$, $k = 4$, $a^2 = 9$, and $b^2 = 9$. Thus, the equation represents a hyperbola with center $(2, 4)$ and focal axis parallel to the x-axis. Since $a = 3$, the vertices are located 3 units to the left and 3 units to the right of the center, or at the points $(-1, 4)$ and $(5, 4)$. From (11), $c = \sqrt{a^2 + b^2} = \sqrt{9 + 9} = 3\sqrt{2}$, so the foci are located $3\sqrt{2}$ units to the left and right of the center, or at the points $(2 - 3\sqrt{2}, 4)$ and $(2 + 3\sqrt{2}, 4)$.

The equations of the asymptotes may be found using the trick of substituting 0 for 1 in (21) to obtain

$$\frac{(x - 2)^2}{9} - \frac{(y - 4)^2}{9} = 0$$

This can be written as $y - 4 = \pm(x - 2)$, which yields the asymptotes

$$y = x + 2 \quad \text{and} \quad y = -x + 6$$

With the aid of a box extending $a = 3$ units left and right of the center and $b = 3$ units above and below the center, we obtain the sketch in Figure 11.4.29.   ◀

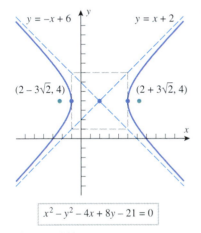

$$x^2 - y^2 - 4x + 8y - 21 = 0$$

Figure 11.4.29

REFLECTION PROPERTIES OF THE CONIC SECTIONS

Parabolas, ellipses, and hyperbolas have certain reflection properties that make them extremely valuable in various applications. In the exercises we will ask you to prove the following results.

11.4.4   THEOREM (*Reflection Property of Parabolas*).   *The tangent line at a point P on a parabola makes equal angles with the line through P parallel to the axis of symmetry and the line through P and the focus (Figure 11.4.30a).*

11.4.5   THEOREM (*Reflection Property of Ellipses*).   *A line tangent to an ellipse at a point P makes equal angles with the lines joining P to the foci (Figure 11.4.30b).*

11.4.6   THEOREM (*Reflection Property of Hyperbolas*).   *A line tangent to a hyperbola at a point P makes equal angles with the lines joining P to the foci (Figure 11.4.30c).*

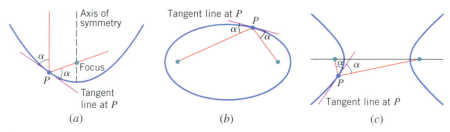

Figure 11.4.30

APPLICATIONS OF THE CONIC SECTIONS

Incoming signals are reflected by the parabolic antenna to the receiver at the focus.

Fermat's principle in optics states that light reflects off of a surface at an angle equal to its angle of incidence. (See Exercise 63 in Section 5.6.) In particular, if a reflecting surface is generated by revolving a parabola about its axis of symmetry, it follows from Theorem 11.4.4 that all light rays entering parallel to the axis will be reflected to the focus (Figure 11.4.31a); conversely, if a light source is located at the focus, then the reflected rays will all be parallel to the axis (Figure 11.4.31b). This principle is used in certain telescopes to reflect the approximately parallel rays of light from the stars and planets off of a parabolic mirror to an eyepiece at the focus; and the parabolic reflectors in flashlights and automobile headlights utilize this principle to form a parallel beam of light rays from a bulb placed at the focus. The same optical principles apply to radar signals and sound waves, which explains the parabolic shape of many antennas.

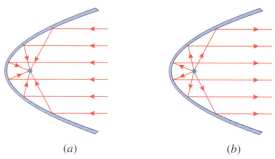

Figure 11.4.31

Figure 11.4.32

Visitors to various rooms in the United States Capitol Building and in St. Paul's Cathedral in Rome are often astonished by the "whispering gallery" effect in which two people at opposite ends of the room can hear one another's whispers very clearly. Such rooms have ceilings with elliptical cross sections and common foci. Thus, when the two people stand at the foci, their whispers are reflected directly to one another off of the elliptical ceiling.

Hyperbolic navigation systems, which were developed in World War II as navigational aids to ships, are based on the definition of a hyperbola. With these systems the ship receives synchronized radio signals from two widely spaced transmitters with known positions. The ship's electronic receiver measures the difference in reception times between the signals and then uses that difference to compute the difference $2a$ in its distance between the two transmitters. This information places the ship somewhere on the hyperbola whose foci are at the transmitters and whose points have $2a$ as the difference in their distances from the foci. By repeating the process with a second set of transmitters, the position of the ship can be approximated as the intersection of two hyperbolas (Figure 11.4.32).

EXERCISE SET 11.4 ~ Graphing Utility [c] CAS

1. In each part, find the equation of the conic.

(a)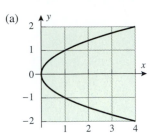

(b)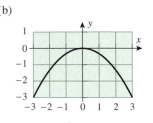

(c)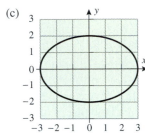

(d)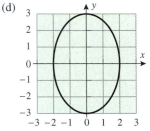

(e)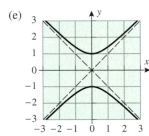

(f)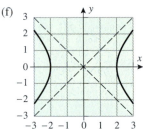

2. (a) Find the focus and directrix for each parabola in Exercise 1.
   (b) Find the foci of the ellipses in Exercise 1.
   (c) Find the foci and the equations of the asymptotes of the hyperbolas in Exercise 1.

In Exercises 3–8, sketch the parabola, and label the focus, vertex, and directrix.

3. (a) $y^2 = 6x$ (b) $x^2 = -9y$

4. (a) $y^2 = -10x$ (b) $x^2 = 4y$

5. (a) $(y-3)^2 = 6(x-2)$ (b) $(x+2)^2 = -(y+2)$

6. (a) $(y+1)^2 = -7(x-4)$ (b) $\left(x - \frac{1}{2}\right)^2 = 2(y-1)$

7. (a) $x^2 - 4x + 2y = 1$ (b) $x = y^2 - 4y + 2$

8. (a) $y^2 - 6y - 2x + 1 = 0$ (b) $y = 4x^2 + 8x + 5$

In Exercises 9–14, sketch the ellipse, and label the foci, the vertices, and the ends of the minor axis.

9. (a) $\dfrac{x^2}{16} + \dfrac{y^2}{9} = 1$ (b) $9x^2 + y^2 = 9$

10. (a) $\dfrac{x^2}{4} + \dfrac{y^2}{25} = 1$ (b) $4x^2 + 9y^2 = 36$

11. (a) $9(x-1)^2 + 16(y-3)^2 = 144$
   (b) $3(x+2)^2 + 4(y+1)^2 = 12$

12. (a) $(x+3)^2 + 4(y-5)^2 = 16$
   (b) $\frac{1}{4}x^2 + \frac{1}{9}(y+2)^2 - 1 = 0$

13. (a) $x^2 + 9y^2 + 2x - 18y + 1 = 0$
   (b) $4x^2 + y^2 + 8x - 10y = -13$

14. (a) $9x^2 + 4y^2 + 18x - 24y + 9 = 0$
   (b) $5x^2 + 9y^2 - 20x + 54y = -56$

In Exercises 15–20, sketch the hyperbola, and label the vertices, foci, and asymptotes.

15. (a) $\dfrac{x^2}{16} - \dfrac{y^2}{4} = 1$ (b) $9y^2 - 4x^2 = 36$

16. (a) $\dfrac{y^2}{9} - \dfrac{x^2}{25} = 1$ (b) $16x^2 - 25y^2 = 400$

17. (a) $\dfrac{(x-2)^2}{9} - \dfrac{(y-4)^2}{4} = 1$
   (b) $(y+3)^2 - 9(x+2)^2 = 36$

18. (a) $\dfrac{(y+4)^2}{3} - \dfrac{(x-2)^2}{5} = 1$
   (b) $16(x+1)^2 - 8(y-3)^2 = 16$

19. (a) $x^2 - 4y^2 + 2x + 8y - 7 = 0$
   (b) $16x^2 - y^2 - 32x - 6y = 57$

20. (a) $4x^2 - 9y^2 + 16x + 54y - 29 = 0$
   (b) $4y^2 - x^2 + 40y - 4x = -60$

In Exercises 21–26, find an equation for the parabola that satisfies the given conditions.

21. (a) Vertex $(0, 0)$; focus $(3, 0)$.
   (b) Vertex $(0, 0)$; directrix $x = 7$.

22. (a) Vertex $(0, 0)$; focus $(0, -4)$.
   (b) Vertex $(0, 0)$; directrix $y = \frac{1}{2}$.

23. (a) Focus $(0, -3)$; directrix $y = 3$.
   (b) Vertex $(1, 1)$; directrix $y = -2$.

24. (a) Focus $(6, 0)$; directrix $x = -6$.
   (b) Focus $(-1, 4)$; directrix $x = 5$.

25. Axis $y = 0$; passes through $(3, 2)$ and $(2, -3)$.

26. Vertex $(5, -3)$; axis parallel to the y-axis; passes through $(9, 5)$.

In Exercises 27–32, find an equation for the ellipse that satisfies the given conditions.

27. (a) Ends of major axis $(\pm 3, 0)$; ends of minor axis $(0, \pm 2)$.
   (b) Length of major axis 26; foci $(\pm 5, 0)$.

28. (a) Ends of major axis $(0, \pm\sqrt{5})$; ends of minor axis $(\pm 1, 0)$.
   (b) Length of minor axis 16; foci $(0, \pm 6)$.

29. (a) Foci $(\pm 1, 0)$; $b = \sqrt{2}$.

(b) $c = 2\sqrt{3}$; $a = 4$; center at the origin; foci on a coordinate axis (two answers).

30. (a) Foci $(\pm 3, 0)$; $a = 4$.

(b) $b = 3$; $c = 4$; center at the origin; foci on a coordinate axis (two answers).

31. (a) Ends of major axis $(\pm 6, 0)$; passes through $(2, 3)$.

(b) Foci $(1, 2)$ and $(1, 4)$; minor axis of length 2.

32. (a) Center at $(0, 0)$; major and minor axes along the coordinate axes; passes through $(3, 2)$ and $(1, 6)$.

(b) Foci $(2, 1)$ and $(2, -3)$; major axis of length 6.

In Exercises 33–38, find an equation for a hyperbola that satisfies the given conditions. (In some cases there may be more than one hyperbola.)

33. (a) Vertices $(\pm 2, 0)$; foci $(\pm 3, 0)$.

(b) Vertices $(\pm 1, 0)$; asymptotes $y = \pm 2x$.

34. (a) Vertices $(0, \pm 3)$; foci $(0, \pm 5)$.

(b) Vertices $(0, \pm 3)$; asymptotes $y = \pm x$.

35. (a) Asymptotes $y = \pm \frac{3}{2} x$; $b = 4$.

(b) Foci $(0, \pm 5)$; asymptotes $y = \pm 2x$.

36. (a) Asymptotes $y = \pm \frac{3}{4} x$; $c = 5$.

(b) Foci $(\pm 3, 0)$; asymptotes $y = \pm 2x$.

37. (a) Vertices $(2, 4)$ and $(10, 4)$; foci 10 units apart.

(b) Asymptotes $y = 2x + 1$ and $y = -2x + 3$; passes through the origin.

38. (a) Foci $(1, 8)$ and $(1, -12)$; vertices 4 units apart.

(b) Vertices $(-3, -1)$ and $(5, -1)$; $b = 4$.

39. (a) As illustrated in the accompanying figure, a parabolic arch spans a road 40 feet wide. How high is the arch if a center section of the road 20 feet wide has a minimum clearance of 12 feet?

(b) How high would the center be if the arch were the upper half of an ellipse?

40. (a) Find an equation for the parabolic arch with base b and height h, shown in the accompanying figure.

(b) Find the area under the arch.

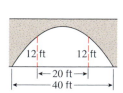

12 ft   12 ft
|← 20 ft →|
|← 40 ft →|

Figure Ex-39

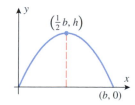

$\left(\frac{1}{2} b, h\right)$

$(b, 0)$

Figure Ex-40

41. Show that the vertex is the closest point on a parabola to the focus. [*Suggestion:* Introduce a convenient coordinate system and use Definition 11.4.1.]

42. As illustrated in the accompanying figure, suppose that a comet moves in a parabolic orbit with the Sun at its focus and that the line from the Sun to the comet makes an angle

of $60°$ with the axis of the parabola when the comet is 40 million miles from the center of the Sun. Use the result in Exercise 41 to determine how close the comet will come to the center of the Sun.

43. For the parabolic reflector in the accompanying figure, how far from the vertex should the light source be placed to produce a beam of parallel rays?

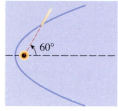

60°

Figure Ex-42

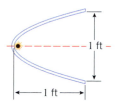

1 ft

1 ft

Figure Ex-43

44. In each part, find the shaded area in the figure.

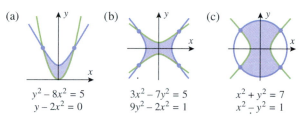

(a) $y^2 - 8x^2 = 5$
$y - 2x^2 = 0$

(b) $3x^2 - 7y^2 = 5$
$9y^2 - 2x^2 = 1$

(c) $x^2 + y^2 = 7$
$x^2 - y^2 = 1$

45. (a) The accompanying figure shows an ellipse with semimajor axis a and semiminor axis b. Express the coordinates of the points P, Q, and R in terms of t.

(b) How does the geometric interpretation of the parameter t differ between a circle

$$x = a \cos t, \quad y = a \sin t$$

and an ellipse

$$x = a \cos t, \quad y = b \sin t?$$

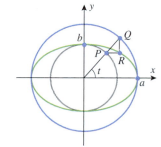

Figure Ex-45

46. (a) Show that the right and left branches of the hyperbola

$$\frac{x^2}{a^2} - \frac{y^2}{b^2} = 1$$

can be represented parametrically as

$$x = a \cosh t, \quad y = b \sinh t \quad (-\infty < t < +\infty)$$
$$x = -a \cosh t, \quad y = b \sinh t \quad (-\infty < t < +\infty)$$

(b) Use a graphing utility to generate both branches of the hyperbola $x^2 - y^2 = 1$ on the same screen.

47. (a) Show that the right and left branches of the hyperbola

    $$\frac{x^2}{a^2} - \frac{y^2}{b^2} = 1$$

    can be represented parametrically as

    $$x = a\sec t, \quad y = b\tan t \quad (-\pi/2 < t < \pi/2)$$
    $$x = -a\sec t, \quad y = b\tan t \quad (-\pi/2 < t < \pi/2)$$

    (b) Use a graphing utility to generate both branches of the hyperbola $x^2 - y^2 = 1$ on the same screen.

48. Find an equation of the parabola traced by a point that moves so that its distance from $(-1, 4)$ is the same as its distance to $y = 1$.

49. Find an equation of the ellipse traced by a point that moves so that the sum of its distances to $(4, 1)$ and $(4, 5)$ is 12.

50. Find the equation of the hyperbola traced by a point that moves so that the difference between its distances to $(0, 0)$ and $(1, 1)$ is 1.

51. Suppose that the base of a solid is elliptical with a major axis of length 9 and a minor axis of length 4. Find the volume of the solid if the cross sections perpendicular to the major axis are squares (see the accompanying figure).

52. Suppose that the base of a solid is elliptical with a major axis of length 9 and a minor axis of length 4. Find the volume of the solid if the cross sections perpendicular to the minor axis are equilateral triangles (see the accompanying figure).

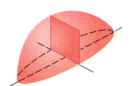

Figure Ex-51

Figure Ex-52

53. Show that an ellipse with semimajor axis a and semiminor axis b has area $A = \pi ab$.

54. (a) Show that the ellipsoid that results when an ellipse with semimajor axis a and semiminor axis b is revolved about the major axis has volume $V = \frac{4}{3}\pi ab^2$.
    (b) Show that the ellipsoid that results when an ellipse with semimajor axis a and semiminor axis b is revolved about the minor axis has volume $V = \frac{4}{3}\pi a^2 b$.

55. Show that the ellipsoid that results when an ellipse with semimajor axis a and semiminor axis b is revolved about the major axis has surface area

    $$S = 2\pi ab\left(\frac{b}{a} + \frac{a}{c}\sin^{-1}\frac{c}{a}\right)$$

    where $c = \sqrt{a^2 - b^2}$.

56. Show that the ellipsoid that results when an ellipse with semimajor axis a and semiminor axis b is revolved about the minor axis has surface area

    $$S = 2\pi ab\left(\frac{a}{b} + \frac{b}{c}\ln\frac{a+c}{b}\right)$$

    where $c = \sqrt{a^2 - b^2}$.

57. Suppose that you want to draw an ellipse that has given values for the lengths of the major and minor axes by using the method shown in Figure 11.4.3b. Assuming that the axes are drawn, explain how a compass can be used to locate the positions for the tacks.

58. The accompanying figure shows Kepler's method for constructing a parabola: a piece of string the length of the left edge of the drafting triangle is tacked to the vertex Q of the triangle and the other end to a fixed point F. A pencil holds the string taut against the base of the triangle as the edge opposite Q slides along a horizontal line L below F. Show that the pencil traces an arc of a parabola with focus F and directrix L.

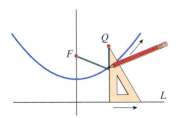

Figure Ex-58

59. The accompanying figure shows a method for constructing a hyperbola: a corner of a ruler is pinned to a fixed point F_1 and the ruler is free to rotate about that point. A piece of string whose length is less than that of the ruler is tacked to a point F_2 and to the free corner Q of the ruler on the same edge as F_1. A pencil holds the string taut against the top edge of the ruler as the ruler rotates about the point F_1. Show that the pencil traces an arc of a hyperbola with foci F_1 and F_2.

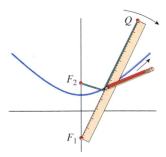

Figure Ex-59

60. Show that if a plane is not parallel to the axis of a right circular cylinder, then the intersection of the plane and cylinder is an ellipse (possibly a circle). [*Hint:* Let θ be the angle shown in Figure Ex-60 (next page), introduce coordinate axes as shown, and express x' and y' in terms of x and y.]

61. As illustrated in the accompanying figure, a carpenter needs to cut an elliptical hole in a sloped roof through which a circular vent pipe of diameter D is to be inserted vertically. The carpenter wants to draw the outline of the hole on the roof using a pencil, two tacks, and a piece of string (as in Figure 11.4.3b). The center point of the ellipse is known,

and common sense suggests that its major axis must be perpendicular to the drip line of the roof. The carpenter needs to determine the length L of the string and the distance T between a tack and the center point. The architect's plans show that the pitch of the roof is p (pitch = rise over run; see the accompanying figure). Find T and L in terms of D and p. [*Note:* This exercise is based on an article by William H. Enos, which appeared in the *Mathematics Teacher*, Feb. 1991, p. 148.]

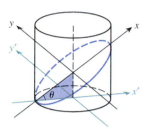

Figure Ex-60

Figure Ex-61

62. Prove: The line tangent to the parabola $x^2 = 4py$ at the point (x_0, y_0) is $x_0 x = 2p(y + y_0)$.

63. Prove: The line tangent to the ellipse

$$\frac{x^2}{a^2} + \frac{y^2}{b^2} = 1$$

at the point (x_0, y_0) has the equation

$$\frac{x x_0}{a^2} + \frac{y y_0}{b^2} = 1$$

64. Prove: The line tangent to the hyperbola

$$\frac{x^2}{a^2} - \frac{y^2}{b^2} = 1$$

at the point (x_0, y_0) has the equation

$$\frac{x x_0}{a^2} - \frac{y y_0}{b^2} = 1$$

65. Use the results in Exercises 63 and 64 to show that if an ellipse and a hyperbola have the same foci, then at each point of intersection their tangent lines are perpendicular.

66. Find two values of k such that the line $x + 2y = k$ is tangent to the ellipse $x^2 + 4y^2 = 8$. Find the points of tangency.

67. Find the coordinates of all points on the hyperbola

$$4x^2 - y^2 = 4$$

where the two lines that pass through the point and the foci are perpendicular.

68. A line tangent to the hyperbola $4x^2 - y^2 = 36$ intersects the y-axis at the point $(0, 4)$. Find the point(s) of tangency.

69. As illustrated in the accompanying figure, suppose that two observers are stationed at the points $F_1(c, 0)$ and $F_2(-c, 0)$ in an xy-coordinate system. Suppose also that the sound of an explosion in the xy-plane is heard by the F_1 observer t seconds before it is heard by the F_2 observer. Assuming that the speed of sound is a constant v, show that the explosion

occurred somewhere on the hyperbola

$$\frac{x^2}{v^2 t^2/4} - \frac{y^2}{c^2 - (v^2 t^2/4)} = 1$$

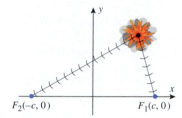

Figure Ex-69

70. As illustrated in the accompanying figure, suppose that two transmitting stations are positioned 100 km apart at points $F_1(50, 0)$ and $F_2(-50, 0)$ on a straight shoreline in an xy-coordinate system. Suppose also that a ship is traveling parallel to the shoreline but 200 km at sea. Find the coordinates of the ship if the stations transmit a pulse simultaneously, but the pulse from station F_1 is received by the ship 100 microseconds sooner than the pulse from station F_2. [*Hint:* Use the formula obtained in Exercise 69, assuming that the pulses travel at the speed of light (299,792,458 m/s).]

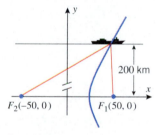

Figure Ex-70

71. As illustrated in the accompanying figure, the tank of an oil truck is 18 feet long and has elliptical cross sections that are 6 feet wide and 4 feet high.

(a) Show that the volume V of oil in the tank (in cubic feet) when it is filled to a depth of h feet is

$$V = 27 \left[4 \sin^{-1} \frac{h - 2}{2} + (h - 2)\sqrt{4h - h^2} + 2\pi \right]$$

(b) Use the numerical root-finding capability of a CAS to determine how many inches from the bottom of a dipstick the calibration marks should be placed to indicate when the tank is $\frac{1}{4}$, $\frac{1}{2}$, and $\frac{3}{4}$ full.

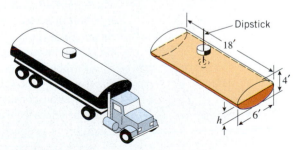

Figure Ex-71

72. Consider the second-degree equation

$$Ax^2 + Cy^2 + Dx + Ey + F = 0$$

where A and C are not both 0. Show by completing the square:

(a) If $AC > 0$, then the equation represents an ellipse, a circle, a point, or has no graph.

(b) If $AC < 0$, then the equation represents a hyperbola or a pair of intersecting lines.

(c) If $AC = 0$, then the equation represents a parabola, a pair of parallel lines, or has no graph.

73. In each part, use the result in Exercise 72 to make a statement about the graph of the equation, and then check your conclusion by completing the square and identifying the graph.

(a) $x^2 - 5y^2 - 2x - 10y - 9 = 0$

(b) $x^2 - 3y^2 - 6y - 3 = 0$

(c) $4x^2 + 8y^2 + 16x + 16y + 20 = 0$

(d) $3x^2 + y^2 + 12x + 2y + 13 = 0$

(e) $x^2 + 8x + 2y + 14 = 0$

(f) $5x^2 + 40x + 2y + 94 = 0$

74. Derive the equation $x^2 = 4py$ in Figure 11.4.6.

75. Derive the equation $(x^2/b^2) + (y^2/a^2) = 1$ given in Figure 11.4.14.

76. Derive the equation $(x^2/a^2) - (y^2/b^2) = 1$ given in Figure 11.4.22.

77. Prove Theorem 11.4.4. [*Hint:* Choose coordinate axes so that the parabola has the equation $x^2 = 4py$. Show that the

tangent line at $P(x_0, y_0)$ intersects the y-axis at $Q(0, -y_0)$ and that the triangle whose three vertices are at P, Q, and the focus is isosceles.]

78. Given two intersecting lines, let L_2 be the line with the larger angle of inclination ϕ_2, and let L_1 be the line with the smaller angle of inclination ϕ_1. We define the **angle θ between L_1 and L_2** by $\theta = \phi_2 - \phi_1$. (See the accompanying figure.)

(a) Prove: If L_1 and L_2 are not perpendicular, then

$$\tan\theta = \frac{m_2 - m_1}{1 + m_1 m_2}$$

where L_1 and L_2 have slopes m_1 and m_2.

(b) Prove Theorem 11.4.5. [*Hint:* Introduce coordinate axes so that the ellipse has the equation $x^2/a^2 + y^2/b^2 = 1$, and use part (a).]

(c) Prove Theorem 11.4.6. [*Hint:* Introduce coordinate axes so the hyperbola has the equation $x^2/a^2 - y^2/b^2 = 1$, and use part (a).]

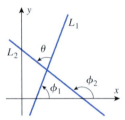

Figure Ex-78

11.5 ROTATION OF AXES; SECOND-DEGREE EQUATIONS

In the preceding section we obtained equations of conic sections with axes parallel to the coordinate axes. In this section we will study the equations of conics that are "tilted" relative to the coordinate axes. This will lead us to investigate rotations of coordinate axes.

QUADRATIC EQUATIONS IN x AND y

We saw in Examples 8–10 of the preceding section that equations of the form

$$Ax^2 + Cy^2 + Dx + Ey + F = 0 \tag{1}$$

can represent conic sections. Equation (1) is a special case of the more general equation

$$Ax^2 + Bxy + Cy^2 + Dx + Ey + F = 0 \tag{2}$$

which, if A, B, and C are not all zero, is called a bisecond-degree equation or **quadratic equation** in x and y. We will show later in this section that the graph of any second-degree equation is a conic section (possibly a degenerate conic section). If $B = 0$, then (2) reduces to (1) and the conic section has its axis or axes parallel to the coordinate axes. However, if $B \neq 0$, then (2) contains a "cross-product" term Bxy, and the graph of the conic section represented by the equation has its axis or axes "tilted" relative to the coordinate axes. As an illustration, consider the ellipse with foci $F_1(1, 2)$ and $F_2(-1, -2)$ and such that the sum of the distances from each point $P(x, y)$ on the ellipse to the foci is 6 units. Expressing this condition as an equation, we obtain (Figure 11.5.1)

$$\sqrt{(x-1)^2 + (y-2)^2} + \sqrt{(x+1)^2 + (y+2)^2} = 6$$

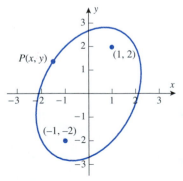

Figure 11.5.1

Squaring both sides, then isolating the remaining radical, then squaring again ultimately yields

$$8x^2 - 4xy + 5y^2 = 36$$

as the equation of the ellipse. This is of form (2) with $A = 8$, $B = -4$, $C = 5$, $D = 0$, $E = 0$, $F = -36$.

ROTATION OF AXES

To study conics that are tilted relative to the coordinate axes it is frequently helpful to rotate the coordinate axes, so that the rotated coordinate axes are parallel to the axes of the conic. Before we can discuss the details, we need to develop some ideas about rotation of coordinate axes.

In Figure 11.5.2a the axes of an xy-coordinate system have been rotated about the origin through an angle θ to produce a new $x'y'$-coordinate system. As shown in the figure, each point P in the plane has coordinates (x', y') as well as coordinates (x, y). To see how the two are related, let r be the distance from the common origin to the point P, and let α be the angle shown in Figure 11.5.2b. It follows that

$$x = r\cos(\theta + \alpha), \quad y = r\sin(\theta + \alpha) \tag{3}$$

and

$$x' = r\cos\alpha, \quad y' = r\sin\alpha \tag{4}$$

Using familiar trigonometric identities, the relationships in (3) can be written as

$$x = r\cos\theta\cos\alpha - r\sin\theta\sin\alpha$$
$$y = r\sin\theta\cos\alpha + r\cos\theta\sin\alpha$$

and on substituting (4) in these equations we obtain the following relationships called the *rotation equations*:

$$\begin{aligned} x &= x'\cos\theta - y'\sin\theta \\ y &= x'\sin\theta + y'\cos\theta \end{aligned} \tag{5}$$

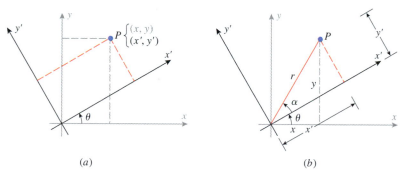

(a)   (b)

Figure 11.5.2

Example 1 Suppose that the axes of an xy-coordinate system are rotated through an angle of $\theta = 45°$ to obtain an $x'y'$-coordinate system. Find the equation of the curve

$$x^2 - xy + y^2 - 6 = 0$$

in $x'y'$-coordinates.

Solution. Substituting $\sin\theta = \sin 45° = 1/\sqrt{2}$ and $\cos\theta = \cos 45° = 1/\sqrt{2}$ in (5) yields the rotation equations

$$x = \frac{x'}{\sqrt{2}} - \frac{y'}{\sqrt{2}} \quad \text{and} \quad y = \frac{x'}{\sqrt{2}} + \frac{y'}{\sqrt{2}}$$

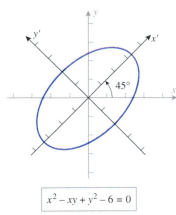

$$x^2 - xy + y^2 - 6 = 0$$

Figure 11.5.3

Substituting these into the given equation yields

$$\left(\frac{x'}{\sqrt{2}} - \frac{y'}{\sqrt{2}}\right)^2 - \left(\frac{x'}{\sqrt{2}} - \frac{y'}{\sqrt{2}}\right)\left(\frac{x'}{\sqrt{2}} + \frac{y'}{\sqrt{2}}\right) + \left(\frac{x'}{\sqrt{2}} + \frac{y'}{\sqrt{2}}\right)^2 - 6 = 0$$

or

$$\frac{x'^2 - 2x'y' + y'^2 - x'^2 + y'^2 + x'^2 + 2x'y' + y'^2}{2} = 6$$

or

$$\frac{x'^2}{12} + \frac{y'^2}{4} = 1$$

which is the equation of an ellipse (Figure 11.5.3). ◀

If the rotation equations (5) are solved for x' and y' in terms of x and y, one obtains (Exercise 14):

$$x' = x \cos\theta + y \sin\theta$$
$$y' = -x \sin\theta + y \cos\theta$$

(6)

Example 2 Find the new coordinates of the point $(2, 4)$ if the coordinate axes are rotated through an angle of $\theta = 30°$.

Solution. Using the rotation equations in (6) with $x = 2$, $y = 4$, $\cos\theta = \cos 30° = \sqrt{3}/2$, and $\sin\theta = \sin 30° = 1/2$, we obtain

$$x' = 2(\sqrt{3}/2) + 4(1/2) = \sqrt{3} + 2$$
$$y' = -2(1/2) + 4(\sqrt{3}/2) = -1 + 2\sqrt{3}$$

Thus, the new coordinates are $(\sqrt{3} + 2, -1 + 2\sqrt{3}\,)$. ◀

ELIMINATING THE CROSS-PRODUCT TERM

In Example 1 we were able to identify the curve $x^2 - xy + y^2 - 6 = 0$ as an ellipse because the rotation of axes eliminated the xy-term, thereby reducing the equation to a familiar form. This occurred because the new $x'y'$-axes were aligned with the axes of the ellipse. The following theorem tells how to determine an appropriate rotation of axes to eliminate the cross-product term of a second-degree equation in x and y.

11.5.1 THEOREM. *If the equation*

$$Ax^2 + Bxy + Cy^2 + Dx + Ey + F = 0$$

(7)

is such that $B \neq 0$, and if an $x'y'$-coordinate system is obtained by rotating the xy-axes through an angle θ satisfying

$$\cot 2\theta = \frac{A - C}{B}$$

(8)

then, in $x'y'$-coordinates, Equation (7) will have the form

$$A'x'^2 + C'y'^2 + D'x' + E'y' + F' = 0$$

Proof. Substituting (5) into (7) and simplifying yields

$$A'x'^2 + B'x'y' + C'y'^2 + D'x' + E'y' + F' = 0$$

where

$$A' = A\cos^2\theta + B\cos\theta\sin\theta + C\sin^2\theta$$
$$B' = B(\cos^2\theta - \sin^2\theta) + 2(C - A)\sin\theta\cos\theta$$
$$C' = A\sin^2\theta - B\sin\theta\cos\theta + C\cos^2\theta$$
$$D' = D\cos\theta + E\sin\theta \tag{9}$$
$$E' = -D\sin\theta + E\cos\theta$$
$$F' = F$$

(Verify.) To complete the proof we must show that $B' = 0$ if

$$\cot 2\theta = \frac{A - C}{B}$$

or equivalently,

$$\frac{\cos 2\theta}{\sin 2\theta} = \frac{A - C}{B} \tag{10}$$

However, by using the trigonometric double-angle formulas, we can rewrite B' in the form

$$B' = B\cos 2\theta - (A - C)\sin 2\theta$$

Thus, $B' = 0$ if θ satisfies (10). ∎

> **REMARK.** It is always possible to satisfy (8) with an angle θ in the range $0 < \theta < \pi/2$. We will always use such a value of θ.

Example 3 Identify and sketch the curve $xy = 1$.

Solution. As a first step, we will rotate the coordinate axes to eliminate the cross-product term. Comparing the given equation to (7), we have

$$A = 0, \quad B = 1, \quad C = 0$$

Thus, the desired angle of rotation must satisfy

$$\cot 2\theta = \frac{A - C}{B} = \frac{0 - 0}{1} = 0$$

This condition can be met by taking $2\theta = \pi/2$ or $\theta = \pi/4 = 45°$. Substituting $\cos\theta = \cos 45° = 1/\sqrt{2}$ and $\sin\theta = \sin 45° = 1/\sqrt{2}$ in (5) yields

$$x = \frac{x'}{\sqrt{2}} - \frac{y'}{\sqrt{2}} \quad \text{and} \quad y = \frac{x'}{\sqrt{2}} + \frac{y'}{\sqrt{2}}$$

Substituting these in the equation $xy = 1$ yields

$$\left(\frac{x'}{\sqrt{2}} - \frac{y'}{\sqrt{2}}\right)\left(\frac{x'}{\sqrt{2}} + \frac{y'}{\sqrt{2}}\right) = 1 \quad \text{or} \quad \frac{x'^2}{2} - \frac{y'^2}{2} = 1$$

which is the equation in the $x'y'$-coordinate system of an equilateral hyperbola with vertices at $(\sqrt{2}, 0)$ and $(-\sqrt{2}, 0)$ in that coordinate system (Figure 11.5.4). ◀

In problems where it is inconvenient to solve

$$\cot 2\theta = \frac{A - C}{B}$$

for θ, the values of $\sin\theta$ and $\cos\theta$ needed for the rotation equations can be obtained by first calculating $\cos 2\theta$ and then computing $\sin\theta$ and $\cos\theta$ from the identities

$$\sin\theta = \sqrt{\frac{1 - \cos 2\theta}{2}} \quad \text{and} \quad \cos\theta = \sqrt{\frac{1 + \cos 2\theta}{2}}$$

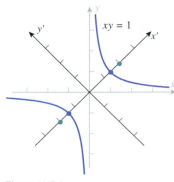

Figure 11.5.4

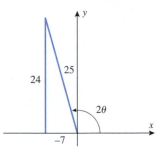

Figure 11.5.5

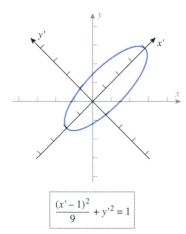

$$\frac{(x'-1)^2}{9} + y'^2 = 1$$

Figure 11.5.6

Example 4   Identify and sketch the curve
$$153x^2 - 192xy + 97y^2 - 30x - 40y - 200 = 0$$

Solution.   We have $A = 153$, $B = -192$, and $C = 97$, so
$$\cot 2\theta = \frac{A - C}{B} = -\frac{56}{192} = -\frac{7}{24}$$

Since θ is to be chosen in the range $0 < \theta < \pi/2$, this relationship is represented by the triangle in Figure 11.5.5. From that triangle we obtain $\cos 2\theta = -\frac{7}{25}$, which implies that

$$\cos \theta = \sqrt{\frac{1 + \cos 2\theta}{2}} = \sqrt{\frac{1 - \frac{7}{25}}{2}} = \frac{3}{5}$$

$$\sin \theta = \sqrt{\frac{1 - \cos 2\theta}{2}} = \sqrt{\frac{1 + \frac{7}{25}}{2}} = \frac{4}{5}$$

Substituting these values in (5) yields the rotation equations
$$x = \tfrac{3}{5}x' - \tfrac{4}{5}y' \quad \text{and} \quad y = \tfrac{4}{5}x' + \tfrac{3}{5}y'$$

and substituting these in turn in the given equation yields
$$\tfrac{153}{25}(3x' - 4y')^2 - \tfrac{192}{25}(3x' - 4y')(4x' + 3y') + \tfrac{97}{25}(4x' + 3y')^2$$
$$- \tfrac{30}{5}(3x' - 4y') - \tfrac{40}{5}(4x' + 3y') - 200 = 0$$

which simplifies to
$$25x'^2 + 225y'^2 - 50x' - 200 = 0$$

or
$$x'^2 + 9y'^2 - 2x' - 8 = 0$$

Completing the square yields
$$\frac{(x'-1)^2}{9} + y'^2 = 1$$

which is the equation in the $x'y'$-coordinate system of an ellipse with center $(1, 0)$ in that coordinate system and semiaxes $a = 3$ and $b = 1$ (Figure 11.5.6).   ◄

THE DISCRIMINANT

It is possible to describe the graph of a second-degree equation without rotating coordinate axes.

> **11.5.2   THEOREM.**   *Consider a second-degree equation*
> $$Ax^2 + Bxy + Cy^2 + Dx + Ey + F = 0 \qquad (11)$$
> *(a)   If $B^2 - 4AC < 0$, the equation represents an ellipse, a circle, a point, or else has no graph.*
> *(b)   If $B^2 - 4AC > 0$, the equation represents a hyperbola or a pair of intersecting lines.*
> *(c)   If $B^2 - 4AC = 0$, the equation represents a parabola, a line, a pair of parallel lines, or else has no graph.*

The quantity $B^2 - 4AC$ in this theorem is called the ***discriminant*** of the quadratic equation. To see why Theorem 11.5.2 is true, we need a fact about the discriminant. It can be shown (Exercise 19) that if the coordinate axes are rotated through any angle θ, and if
$$A'x'^2 + B'x'y' + C'y'^2 + D'x' + E'y' + F' = 0 \qquad (12)$$

is the equation resulting from (11) after rotation, then

$$B^2 - 4AC = B'^2 - 4A'C' \tag{13}$$

In other words, the discriminant of a quadratic equation is not altered by rotating the coordinate axes. For this reason the discriminant is said to be *invariant* under a rotation of coordinate axes. In particular, if we choose the angle of rotation to eliminate the cross-product term, then (12) becomes

$$A'x'^2 + C'y'^2 + D'x' + E'y' + F' = 0 \tag{14}$$

and since $B' = 0$, (13) tells us that

$$B^2 - 4AC = -4A'C' \tag{15}$$

Proof of Theorem 11.5.2(a). If $B^2 - 4AC < 0$, then from (15), $A'C' > 0$, so (14) can be divided through by $A'C'$ and written in the form

$$\frac{1}{C'}\left(x'^2 + \frac{D'}{A'}x'\right) + \frac{1}{A'}\left(y'^2 + \frac{E'}{C'}y'\right) = -\frac{F'}{A'C'}$$

Since $A'C' > 0$, the numbers A' and C' have the same sign. We assume that this sign is positive, since Equation (14) can be multiplied through by -1 to achieve this, if necessary. By completing the squares, we can rewrite the last equation in the form

$$\frac{(x' - h)^2}{(\sqrt{C'})^2} + \frac{(y' - k)^2}{(\sqrt{A'})^2} = K$$

There are three possibilities: $K > 0$, in which case the graph is either a circle or an ellipse, depending on whether or not the denominators are equal; $K < 0$, in which case there is no graph, since the left side is nonnegative for all x' and y'; or $K = 0$, in which case the graph is the single point (h, k), since the equation is satisfied only by $x' = h$ and $y' = k$. The proofs of parts (b) and (c) require a similar kind of analysis. ∎

Example 5 Use the discriminant to identify the graph of

$$8x^2 - 3xy + 5y^2 - 7x + 6 = 0$$

Solution. We have

$$B^2 - 4AC = (-3)^2 - 4(8)(5) = -151$$

Since the discriminant is negative, the equation represents an ellipse, a point, or else has no graph. (Why can't the graph be a circle?) ◀

In cases where a quadratic equation represents a point, a line, a pair of parallel lines, a pair of intersecting lines, or has no graph, we say that equation represents a *degenerate conic section*. Thus, if we allow for possible degeneracy, it follows from Theorem 11.5.2 that *every quadratic equation has a conic section as its graph.*

EXERCISE SET 11.5 C CAS

1. Let an $x'y'$-coordinate system be obtained by rotating an xy-coordinate system through an angle of $\theta = 60°$.
   (a) Find the $x'y'$-coordinates of the point whose xy-coordinates are $(-2, 6)$.
   (b) Find an equation of the curve $\sqrt{3}xy + y^2 = 6$ in $x'y'$-coordinates.
   (c) Sketch the curve in part (b), showing both xy-axes and $x'y'$-axes.

2. Let an $x'y'$-coordinate system be obtained by rotating an xy-coordinate system through an angle of $\theta = 30°$.
   (a) Find the $x'y'$-coordinates of the point whose xy-coordinates are $(1, -\sqrt{3})$.
   (b) Find an equation of the curve $2x^2 + 2\sqrt{3}xy = 3$ in $x'y'$-coordinates.
   (c) Sketch the curve in part (b), showing both xy-axes and $x'y'$-axes.

In Exercises 3–12, rotate the coordinate axes to remove the xy-term. Then name the conic and sketch its graph.

3. $xy = -9$

4. $x^2 - xy + y^2 - 2 = 0$

5. $x^2 + 4xy - 2y^2 - 6 = 0$

6. $31x^2 + 10\sqrt{3}xy + 21y^2 - 144 = 0$

7. $x^2 + 2\sqrt{3}xy + 3y^2 + 2\sqrt{3}x - 2y = 0$

8. $34x^2 - 24xy + 41y^2 - 25 = 0$

9. $9x^2 - 24xy + 16y^2 - 80x - 60y + 100 = 0$

10. $5x^2 - 6xy + 5y^2 - 8\sqrt{2}x + 8\sqrt{2}y = 8$

11. $52x^2 - 72xy + 73y^2 + 40x + 30y - 75 = 0$

12. $6x^2 + 24xy - y^2 - 12x + 26y + 11 = 0$

13. Let an $x'y'$-coordinate system be obtained by rotating an xy-coordinate system through an angle θ. Prove: For every value of θ, the equation $x^2 + y^2 = r^2$ becomes the equation $x'^2 + y'^2 = r^2$. Give a geometric explanation.

14. Derive (6) by solving the rotation equations in (5) for x' and y' in terms of x and y.

15. Let an $x'y'$-coordinate system be obtained by rotating an xy-coordinate system through an angle of $45°$. Use (6) to find an equation of the curve $3x'^2 + y'^2 = 6$ in xy-coordinates.

16. Let an $x'y'$-coordinate system be obtained by rotating an xy-coordinate system through an angle of $30°$. Use (5) to find an equation in $x'y'$-coordinates of the curve $y = x^2$.

17. Show that the graph of the equation

$$\sqrt{x} + \sqrt{y} = 1$$

is a portion of a parabola. [*Hint:* First rationalize the equation and then perform a rotation of axes.]

18. Derive the expression for B' in (9).

19. Use (9) to prove that $B^2 - 4AC = B'^2 - 4A'C'$ for all values of θ.

20. Use (9) to prove that $A + C = A' + C'$ for all values of θ.

21. Prove: If $A = C$ in (7), then the cross-product term can be eliminated by rotating through $45°$.

22. Prove: If $B \neq 0$, then the graph of $x^2 + Bxy + F = 0$ is a hyperbola if $F \neq 0$ and two intersecting lines if $F = 0$.

In Exercises 23–27, use the discriminant to identify the graph of the given equation.

23. $x^2 - xy + y^2 - 2 = 0$

24. $x^2 + 4xy - 2y^2 - 6 = 0$

25. $x^2 + 2\sqrt{3}xy + 3y^2 + 2\sqrt{3}x - 2y = 0$

26. $6x^2 + 24xy - y^2 - 12x + 26y + 11 = 0$

27. $34x^2 - 24xy + 41y^2 - 25 = 0$

28. Each of the following represents a degenerate conic section. Where possible, sketch the graph.
(a) $x^2 - y^2 = 0$
(b) $x^2 + 3y^2 + 7 = 0$
(c) $8x^2 + 7y^2 = 0$
(d) $x^2 - 2xy + y^2 = 0$
(e) $9x^2 + 12xy + 4y^2 - 36 = 0$
(f) $x^2 + y^2 - 2x - 4y = -5$

29. Prove parts (*b*) and (*c*) of Theorem 11.5.2.

C **30.** Consider the conic whose equation is

$$x^2 + xy + 2y^2 - x + 3y + 1 = 0$$

(a) Use the discriminant to identify the conic.
(b) Graph the equation by solving for y in terms of x and graphing both solutions.
(c) Your CAS may be able to graph the equation in the form given. If so, graph the equation in this way.

C **31.** Consider the conic whose equation is

$$2x^2 + 9xy + y^2 - 6x + y - 4 = 0$$

(a) Use the discriminant to identify the conic.
(b) Graph the equation by solving for y in terms of x and graphing both solutions.
(c) Your CAS may be able to graph the equation in the form given. If so, graph the equation in this way.

11.6 CONIC SECTIONS IN POLAR COORDINATES

It will be shown later in the text that if an object moves in a gravitational field that is directed toward a fixed point (such as the center of the Sun), then the path of that object must be a conic section with the fixed point at a focus. For example, planets in our solar system move along elliptical paths with the Sun at a focus, and the comets move along parabolic, elliptical, or hyperbolic paths with the Sun at a focus, depending on the conditions under which they were born. For applications of this type it is usually desirable to express the equations of the conic sections in polar coordinates with the pole at a focus. In this section we will show how to do this.

**THE FOCUS–DIRECTRIX
CHARACTERIZATION OF CONICS**

To obtain polar equations for the conic sections we will need the following theorem.

> **11.6.1** THEOREM (*Focus–Directrix Property of Conics*).   *Suppose that a point P moves in the plane determined by a fixed point (called the **focus**) and a fixed line (called the **directrix**), where the focus does not lie on the directrix. If the point moves in such a way that its distance to the focus divided by its distance to the directrix is some constant e (called the **eccentricity**), then the curve traced by the point is a conic section. Moreover, the conic is a parabola if $e = 1$, an ellipse if $0 < e < 1$, and a hyperbola if $e > 1$.*

REMARK.   It is an unfortunate historical accident that the letter e is used for the base of the natural logarithms and the eccentricity of conic sections. However, the appropriate interpretation will usually be clear from the context in which the letter is used.

We will not give a formal proof of this theorem; rather, we will use the specific cases in Figure 11.6.1 to illustrate the basic ideas. For the parabola, we will take the directrix to be $x = -p$, as usual; and for the ellipse and the hyperbola we will take the directrix to be $x = a^2/c$. We want to show in all three cases that if P is a point on the graph, F is the focus, and D is the directrix, then the ratio PF/PD is some constant e, where $e = 1$ for the parabola, $0 < e < 1$ for the ellipse, and $e > 1$ for the hyperbola. We will give the arguments for the parabola and ellipse and leave the argument for the hyperbola as an exercise.

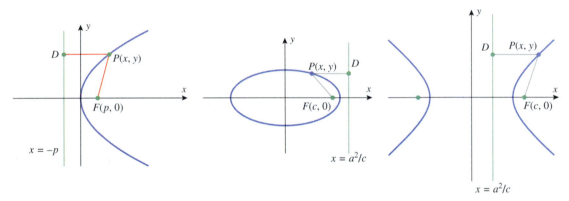

Figure 11.6.1

For the parabola, the distance PF to the focus is equal to the distance PD to the directrix, so that $PF/PD = 1$, which is what we wanted to show. For the ellipse, we rewrite Equation (8) of Section 11.4 as

$$\sqrt{(x-c)^2 + y^2} = a - \frac{c}{a}x = \frac{c}{a}\left(\frac{a^2}{c} - x\right)$$

But the expression on the left side is the distance PF, and the expression in the parentheses on the right side is the distance PD, so we have shown that

$$PF = \frac{c}{a}PD$$

Thus, PF/PD is constant, and the eccentricity is

$$e = \frac{c}{a} \tag{1}$$

If we rule out the degenerate case where $a = 0$ or $c = 0$, then it follows from Formula (7) of Section 11.4 that $0 < c < a$, so $0 < e < 1$, which is what we wanted to show.

We will leave it as an exercise to show that the eccentricity of the hyperbola in Figure 11.6.1 is also given by Formula (1), but in this case it follows from Formula (11) of Section 11.4 that $c > a$, so $e > 1$.

ECCENTRICITY OF AN ELLIPSE AS A MEASURE OF FLATNESS

The eccentricity of an ellipse can be viewed as a measure of its flatness—as e approaches 0 the ellipses become more and more circular, and as e approaches 1 they become more and more flat (Figure 11.6.2). Table 11.6.1 shows the orbital eccentricities of various celestial objects. Note that most of the planets actually have fairly circular orbits.

Table 11.6.1

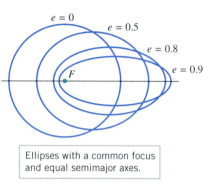

Ellipses with a common focus and equal semimajor axes.

Figure 11.6.2

CELESTIAL BODY	ECCENTRICITY
Mercury	0.206
Venus	0.007
Earth	0.017
Mars	0.093
Jupiter	0.048
Saturn	0.056
Uranus	0.046
Neptune	0.010
Pluto	0.249
Halley's comet	0.970

POLAR EQUATIONS OF CONICS

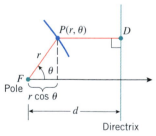

Figure 11.6.3

Our next objective is to derive polar equations for the conic sections from their focus–directrix characterizations. We will assume that the focus is at the pole and the directrix is either parallel or perpendicular to the polar axis. If the directrix is parallel to the polar axis, then it can be above or below the pole; and if the directrix is perpendicular to the polar axis, then it can be to the left or right of the pole. Thus, there are four cases to consider. We will derive the formulas for the case in which the directrix is perpendicular to the polar axis and to the right of the pole.

As illustrated in Figure 11.6.3, let us assume that the directrix is perpendicular to the polar axis and d units to the right of the pole, where the constant d is known. If P is a point on the conic and if the eccentricity of the conic is e, then it follows from Theorem 11.6.1 that $PF/PD = e$ or, equivalently, that

$$PF = ePD \tag{2}$$

However, it is evident from Figure 11.6.3 that $PF = r$ and $PD = d - r\cos\theta$. Thus, (2) can be written as

$$r = e(d - r\cos\theta)$$

which can be solved for r and expressed as

$$r = \frac{ed}{1 + e\cos\theta}$$

(verify). Observe that this single polar equation can represent a parabola, an ellipse, or a hyperbola, depending on the value of e. In contrast, the rectangular equations for these conics all have different forms. The derivations in the other three cases are similar.

> **11.6.2   THEOREM.**   *If a conic section with eccentricity e is positioned in a polar co-*
> *ordinate system so that its focus is at the pole and the corresponding directrix is d units*
> *from the pole and is either parallel or perpendicular to the polar axis, then the equation*
> *of the conic has one of four possible forms, depending on its orientation:*
>
> $$r = \frac{ed}{1 + e\cos\theta} \qquad r = \frac{ed}{1 - e\cos\theta} \qquad (3\text{--}4)$$
> Directrix right of pole   Directrix left of pole
>
> $$r = \frac{ed}{1 + e\sin\theta} \qquad r = \frac{ed}{1 - e\sin\theta} \qquad (5\text{--}6)$$
> Directrix above pole   Directrix below pole

SKETCHING CONICS IN POLAR COORDINATES

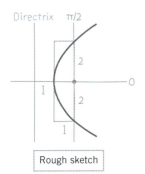

Rough sketch

Figure 11.6.4

Precise graphs of conic sections in polar coordinates can be generated with graphing utilities. However, it is often useful to be able to make quick sketches of these graphs that show their orientations and give some sense of their dimensions. The orientation of a conic relative to the polar axis can be deduced by matching its equation with one of the four forms in Theorem 11.6.2. The key dimensions of a parabola are determined by the constant p (Figure 11.4.5) and those of ellipses and hyperbolas by the constants a, b, and c (Figures 11.4.11 and 11.4.20). Thus, we need to show how these constants can be obtained from the polar equations.

Example 1   Sketch the graph of $r = \dfrac{2}{1 - \cos\theta}$ in polar coordinates.

Solution.   The equation is an exact match to (4) with $d = 2$ and $e = 1$. Thus, the graph is a parabola with the focus at the pole and the directrix 2 units to the left of the pole. This tells us that the parabola opens to the right along the polar axis and $p = 1$. Thus, the parabola looks roughly like that sketched in Figure 11.6.4.   ◄

All of the important geometric information about an ellipse can be obtained from the values of a, b, and c in Figure 11.6.5. One way to find these values from the polar equation of an ellipse is based on finding the distances from the focus to the vertices. As shown in the figure, let r_0 be the distance from the focus to the closest vertex and r_1 the distance to the farthest vertex. Thus,

$$r_0 = a - c \quad\text{and}\quad r_1 = a + c \qquad (7)$$

from which it follows that

$$a = \tfrac{1}{2}(r_1 + r_0) \qquad (8)$$

and

$$c = \tfrac{1}{2}(r_1 - r_0) \qquad (9)$$

Moreover, it also follows from (7) that

$$r_0 r_1 = a^2 - c^2 = b^2$$

Thus,

$$b = \sqrt{r_0 r_1} \qquad (10)$$

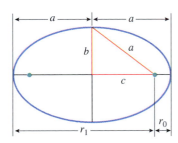

Figure 11.6.5

> REMARK.   In words, Formula (8) states that a is the *arithmetic average* (also called the *arithmetic mean*) of r_0 and r_1, and Formula (10) states that b is the *geometric mean* of r_0 and r_1.

Example 2   Sketch the graph of $r = \dfrac{6}{2 + \cos\theta}$ in polar coordinates.

Solution. This equation does not match any of the forms in Theorem 11.6.2 because they all require a constant term of 1 in the denominator. However, we can put the equation into one of these forms by dividing the numerator and denominator by 2 to obtain

$$r = \frac{3}{1 + \frac{1}{2}\cos\theta}$$

This is an exact match to (3) with $d = 6$ and $e = \frac{1}{2}$, so the graph is an ellipse with the directrix 6 units to the right of the pole. The distance r_0 from the focus to the closest vertex can be obtained by setting $\theta = 0$ in this equation, and the distance r_1 to the farthest vertex can be obtained by setting $\theta = \pi$. This yields

$$r_0 = \frac{3}{1 + \frac{1}{2}\cos 0} = \frac{3}{\frac{3}{2}} = 2, \quad r_1 = \frac{3}{1 + \frac{1}{2}\cos\pi} = \frac{3}{\frac{1}{2}} = 6$$

Thus, from Formulas (8), (10), and (9), respectively, we obtain

$$a = \tfrac{1}{2}(r_1 + r_0) = 4, \quad b = \sqrt{r_0 r_1} = 2\sqrt{3}, \quad c = \tfrac{1}{2}(r_1 - r_0) = 2$$

Thus, the ellipse looks roughly like that sketched in Figure 11.6.6.   ◀

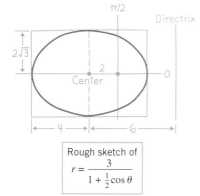

Rough sketch of
$$r = \frac{3}{1 + \frac{1}{2}\cos\theta}$$

Figure 11.6.6

All of the important information about a hyperbola can be obtained from the values of a, b, and c in Figure 11.6.7. As with the ellipse, one way to find these values from the polar equation of a hyperbola is based on finding the distances from the focus to the vertices. As shown in the figure, let r_0 be the distance from the focus to the closest vertex and r_1 the distance to the farthest vertex. Thus,

$$r_0 = c - a \quad \text{and} \quad r_1 = c + a \tag{11}$$

from which it follows that

$$a = \tfrac{1}{2}(r_1 - r_0) \tag{12}$$

and

$$c = \tfrac{1}{2}(r_1 + r_0) \tag{13}$$

Moreover, it also follows from (11) that

$$r_0 r_1 = c^2 - a^2 = b^2$$

from which it follows that

$$b = \sqrt{r_0 r_1} \tag{14}$$

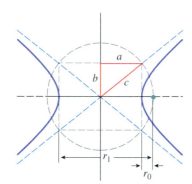

Figure 11.6.7

Example 3   Sketch the graph of $r = \dfrac{2}{1 + 2\sin\theta}$ in polar coordinates.

Solution. This equation is an exact match to (5) with $d = 1$ and $e = 2$. Thus, the graph is a hyperbola with its directrix 1 unit above the pole. However, it is not so straightforward to compute the values of r_0 and r_1, since hyperbolas in polar coordinates are generated in a strange way as θ varies from 0 to 2π. This can be seen from Figure 11.6.8a, which is the graph of the given equation in rectangular coordinates. It follows from this graph that the corresponding polar graph is generated in pieces (see Figure 11.6.8b):

• As θ varies over the interval $0 \le \theta < 7\pi/6$, the value of r is positive and varies from 2 down to 2/3 and then to $+\infty$, which generates part of the lower branch.

- As θ varies over the interval $7\pi/6 < \theta \leq 3\pi/2$, the value of r is negative and varies from $-\infty$ to -2, which generates the right part of the upper branch.
- As θ varies over the interval $3\pi/2 \leq \theta < 11\pi/6$, the value of r is negative and varies from -2 to $-\infty$, which generates the left part of the upper branch.
- As θ varies over the interval $11\pi/6 < \theta \leq 2\pi$, the value of r is positive and varies from $+\infty$ to 2, which fills in the missing piece of the lower right branch.

It is now clear that we can obtain r_0 by setting $\theta = \pi/2$ and r_1 by setting $\theta = 3\pi/2$. Keeping in mind that r_0 and r_1 are positive, this yields

$$r_0 = \frac{2}{1 + 2\sin(\pi/2)} = \frac{2}{3}, \quad r_1 = \left| \frac{2}{1 + 2\sin(3\pi/2)} \right| = \left| \frac{2}{-1} \right| = 2$$

Thus, from Formulas (12), (14), and (13), respectively, we obtain

$$a = \frac{1}{2}(r_1 - r_0) = \frac{2}{3}, \quad b = \sqrt{r_0 r_1} = \frac{2\sqrt{3}}{3}, \quad c = \frac{1}{2}(r_1 + r_0) = \frac{4}{3}$$

Thus, the hyperbola looks roughly like that sketched in Figure 11.6.8c. ◀

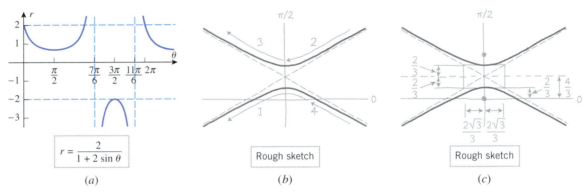

$$r = \frac{2}{1 + 2\sin\theta}$$

(a)

Rough sketch

(b)

Rough sketch

(c)

Figure 11.6.8

APPLICATIONS IN ASTRONOMY

In 1609 Johannes Kepler* published a book known as *Astronomia Nova* (or sometimes *Commentaries on the Motions of Mars*) in which he succeeded in distilling thousands of years of observational astronomy into three beautiful laws of planetary motion (Figure 11.6.9).

---

*JOHANNES KEPLER (1571–1630). German astronomer and physicist, Kepler, whose work provided our contemporary view of planetary motion, led a fascinating but ill-starred life. His alcoholic father made him work in a family-owned tavern as a child, later withdrawing him from elementary school and hiring him out as a field laborer, where the boy contracted smallpox, permanently crippling his hands and impairing his eyesight. In later years, Kepler's first wife and several children died, his mother was accused of witchcraft, and being a Protestant he was often subjected to persecution by Catholic authorities. He was often impoverished, eking out a living as an astrologer and prognosticator. Looking back on his unhappy childhood, Kepler described his father as "criminally inclined" and "quarrelsome" and his mother as "garrulous" and "bad-tempered." However, it was his mother who left an indelible mark on the six-year-old Kepler by showing him the comet of 1577; and in later life he personally prepared her defense against the witchcraft charges. Kepler became acquainted with the work of Copernicus as a student at the University of Tübingen, where he received his master's degree in 1591. He continued on as a theological student, but at the urging of the university officials he abandoned his clerical studies and accepted a position as a mathematician and teacher in Graz, Austria. However, he was expelled from the city when it came under Catholic control, and in 1600 he finally moved on to Prague, where he became an assistant at the observatory of the famous Danish astronomer Tycho Brahe. Brahe was a brilliant and meticulous astronomical observer who amassed the most accurate astronomical data known at that time; and when Brahe died in 1601 Kepler inherited the treasure-trove of data. After eight years of intense labor, Kepler deciphered the underlying principles buried in the data and in 1609 published his monumental work, *Astronomia Nova*, in which he stated his first two laws of planetary motion. Commenting on his discovery of elliptical orbits, Kepler wrote, "I was almost driven to madness in considering and calculating this matter. I could not find out why the planet would rather go on an elliptical orbit (rather than a circle). Oh ridiculous me!" It ultimately remained for Isaac Newton to discover the laws of gravitation that explained the reason for elliptical orbits.

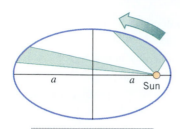

Equal areas are swept out in equal times, and the square of the period T is proportional to a^3.

Figure 11.6.9

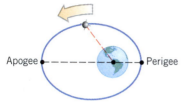

Figure 11.6.10

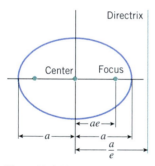

Figure 11.6.11

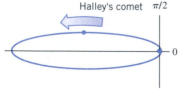

Figure 11.6.12

11.6.3 KEPLER'S LAWS.

- First law (**Law of Orbits**). Each planet moves in an elliptical orbit with the Sun at a focus.
- Second law (**Law of Areas**). The radial line from the center of the Sun to the center of a planet sweeps out equal areas in equal times.
- Third law (**Law of Periods**). The square of a planet's period (the time it takes the planet to complete one orbit about the Sun) is proportional to the cube of the semimajor axis of its orbit.

Kepler's laws, although stated for planetary motion around the Sun, apply to all orbiting celestial bodies that are subjected to a *single* central gravitational force—artificial satellites subjected only to the central force of Earth's gravity and moons subjected only to the central gravitational force of a planet, for example. Later in the text we will derive Kepler's laws from basic principles, but for now we will show how they can be used in basic astronomical computations.

In an elliptical orbit, the closest point to the focus is called the **perigee** and the farthest point the **apogee** (Figure 11.6.10). The distances from the focus to the perigee and apogee are called the **perigee distance** and **apogee distance**, respectively. For orbits around the Sun, it is more common to use the terms **perihelion** and **aphelion**, rather than perigee and apogee, and to measure time in Earth years and distances in astronomical units (AU), where 1 AU is the semimajor axis a of the Earth's orbit (approximately 150×10^6 km or 92.9×10^6 mi). With this choice of units, the constant of proportionality in Kepler's third law is 1, since $a = 1$ AU produces a period of $T = 1$ Earth year. In this case Kepler's third law can be expressed as

$$T = a^{3/2} \tag{15}$$

Shapes of elliptical orbits are often specified by giving the eccentricity e and the semimajor axis a, so it is useful to express the polar equations of an ellipse in terms of these constants. Figure 11.6.11, which can be obtained from the ellipse in Figure 11.6.1 and the relationship $c = ea$, implies that the distance d between the focus and the directrix is

$$d = \frac{a}{e} - c = \frac{a}{e} - ea = \frac{a(1 - e^2)}{e} \tag{16}$$

from which it follows that $ed = a(1 - e^2)$. Thus, depending on the orientation of the ellipse, the formulas in Theorem 11.6.2 can be expressed in terms of a and e as

$$r = \frac{a(1 - e^2)}{1 \pm e \cos \theta} \qquad r = \frac{a(1 - e^2)}{1 \pm e \sin \theta} \tag{17-18}$$

+: Directrix right of pole   +: Directrix above pole
−: Directrix left of pole   −: Directrix below pole

Moreover, it is evident from Figure 11.6.11 that the distances from the focus to the closest and farthest vertices can be expressed in terms of a and e as

$$r_0 = a - ea = a(1 - e) \quad \text{and} \quad r_1 = a + ea = a(1 + e) \tag{19-20}$$

Example 4 Halley's comet (last seen in 1986) has an eccentricity of 0.97 and a semimajor axis of $a = 18.1$ AU.

(a) Find the equation of its orbit in the polar coordinate system shown in Figure 11.6.12.
(b) Find the period of its orbit.
(c) Find its perihelion and aphelion distances.

Halley's comet photographed
April 21, 1910 in Peru

Solution (a). From (17), the polar equation of the orbit has the form

$$r = \frac{a(1 - e^2)}{1 + e \cos \theta}$$

But $a(1 - e^2) = 18.1[1 - (0.97)^2] \approx 1.07$. Thus, the equation of the orbit is

$$r = \frac{1.07}{1 + 0.97 \cos \theta}$$

Solution (b). From (15), with $a = 18.1$, the period of the orbit is

$$T = (18.1)^{3/2} \approx 77 \text{ years}$$

Solution (c). Since the perihelion and aphelion distances are the distances to the closest and farthest vertices, respectively, it follows from (19) and (20) that

$$r_0 = a - ea = a(1 - e) = 18.1(1 - 0.97) \approx 0.543 \text{ AU}$$
$$r_1 = a + ea = a(1 + e) = 18.1(1 + 0.97) \approx 35.7 \text{ AU}$$

or since 1 AU $\approx 150 \times 10^6$ km, the perihelion and aphelion distances in kilometers are

$$r_0 = 18.1(1 - 0.97)(150 \times 10^6) \approx 81,500,000 \text{ km}$$
$$r_1 = 18.1(1 + 0.97)(150 \times 10^6) \approx 5,350,000,000 \text{ km} \qquad \blacktriangleleft$$

FOR THE READER. Use the polar equation of the orbit of Halley's comet to check the values of r_0 and r_1.

Example 5 An Apollo lunar lander orbits the Moon in an elliptic orbit with eccentricity $e = 0.12$ and semimajor axis $a = 2015$ km. Assuming the Moon to be a sphere of radius 1740 km, find the minimum and maximum heights of the lander above the lunar surface (Figure 11.6.13).

Solution. If we let r_0 and r_1 denote the minimum and maximum distances from the center of the Moon, then the minimum and maximum distances from the surface of the Moon will be

$$d_{\min} = r_0 - 1740$$
$$d_{\max} = r_1 - 1740$$

or from Formulas (19) and (20)

$$d_{\min} = r_0 - 1740 = a(1 - e) - 1740 = 2015(0.88) - 1740 = 33.2 \text{ km}$$
$$d_{\max} = r_1 - 1740 = a(1 + e) - 1740 = 2015(1.12) - 1740 = 516.8 \text{ km} \qquad \blacktriangleleft$$

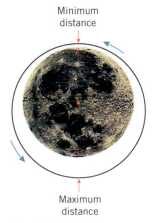

Minimum
distance

Maximum
distance

Figure 11.6.13

EXERCISE SET 11.6 ⌁ Graphing Utility

For the conics in Exercises 1 and 2, find the eccentricity and the distance from the pole to the directrix, and sketch the graph in polar coordinates.

1. (a) $r = \dfrac{3}{2 - 2 \cos \theta}$ (b) $r = \dfrac{3}{2 + \sin \theta}$

(c) $r = \dfrac{4}{2 + 3 \cos \theta}$ (d) $r = \dfrac{5}{3 + 3 \sin \theta}$

2. (a) $r = \dfrac{4}{3 - 2 \cos \theta}$ (b) $r = \dfrac{3}{3 - 4 \sin \theta}$

(c) $r = \dfrac{1}{3 + 3 \sin \theta}$ (d) $r = \dfrac{1}{2 + 6 \sin \theta}$

In Exercises 3 and 4, use Formulas (3)–(6) to name and describe the orientation of the conic, and then check your answer by generating the graph with a graphing utility.

⌁ **3.** (a) $r = \dfrac{8}{1 - \sin \theta}$ (b) $r = \dfrac{16}{4 + 3 \sin \theta}$

(c) $r = \dfrac{4}{2 - 3 \sin \theta}$ (d) $r = \dfrac{12}{4 + \cos \theta}$

⌁ **4.** (a) $r = \dfrac{15}{1 + \cos \theta}$ (b) $r = \dfrac{2}{3 + 3 \cos \theta}$

(c) $r = \dfrac{64}{7 - 12 \sin \theta}$ (d) $r = \dfrac{12}{3 - 2 \cos \theta}$

In Exercises 5–8, find a polar equation for the conic that has its focus at the pole and satisfies the stated conditions. Points are in polar coordinates and directrices in rectangular coordinates for simplicity. (In some cases there may be more than one conic that satisfies the conditions.)

5. (a) Ellipse; $e = \frac{2}{3}$; directrix $x = 1$.
   (b) Parabola; directrix $x = -1$.
   (c) Hyperbola; $e = \frac{3}{2}$; directrix $y = 1$.

6. (a) Ellipse; $e = \frac{2}{3}$; directrix $y = -1$.
   (b) Parabola; directrix $y = 1$.
   (c) Hyperbola; $e = \frac{4}{3}$; directrix $x = -1$.

7. (a) Ellipse; vertices $(6, 0)$ and $(4, \pi)$.
   (b) Parabola; vertex $(1, 3\pi/2)$.
   (c) Hyperbola; vertices $(3, \pi/2)$ and $(-7, 3\pi/2)$.

8. (a) Ellipse; ends of major axis $(1, \pi/2)$ and $(4, 3\pi/2)$.
   (b) Parabola; vertex $(3, \pi)$.
   (c) Hyperbola; equilateral; vertex $(5, 0)$.

In Exercises 9 and 10, find the distances from the pole to the vertices, and then apply Formulas (8)–(10) to find the equation of the ellipse in rectangular coordinates.

9. (a) $r = \dfrac{6}{2 + \sin\theta}$ (b) $r = \dfrac{1}{2 - \cos\theta}$

10. (a) $r = \dfrac{6}{5 + 2\cos\theta}$ (b) $r = \dfrac{8}{4 - 3\sin\theta}$

In Exercises 11 and 12, find the distances from the pole to the vertices, and then apply Formulas (12)–(14) to find the equation of the hyperbola in rectangular coordinates.

11. (a) $r = \dfrac{2}{1 + 3\sin\theta}$ (b) $r = \dfrac{10}{6 - 9\cos\theta}$

12. (a) $r = \dfrac{4}{1 - 2\sin\theta}$ (b) $r = \dfrac{15}{2 + 8\cos\theta}$

In Exercises 13 and 14, find a polar equation for the ellipse that has its focus at the pole and satisfies the stated conditions.

13. (a) Directrix to the right of the pole; $a = 8$; $e = \frac{1}{2}$.
   (b) Directrix below the pole; $a = 4$; $e = \frac{3}{5}$.
   (c) Directrix to the left of the pole; $b = 4$; $e = \frac{3}{5}$.
   (d) Directrix above the pole; $c = 5$; $e = \frac{1}{5}$.

14. (a) Directrix above the pole; $a = 10$; $e = \frac{1}{2}$.
   (b) Directrix to the left of the pole; $a = 6$; $e = \frac{1}{5}$.
   (c) Directrix below the pole; $b = 4$; $e = \frac{3}{4}$.
   (d) Directrix to the right of the pole; $c = 10$; $e = \frac{4}{5}$.

15. (a) Show that the eccentricity of an ellipse can be expressed in terms of r_0 and r_1 as
$$e = \frac{r_1 - r_0}{r_1 + r_0}$$

(b) Show that
$$\frac{r_1}{r_0} = \frac{1 + e}{1 - e}$$

16. (a) Show that the eccentricity of a hyperbola can be expressed in terms of r_0 and r_1 as
$$e = \frac{r_1 + r_0}{r_1 - r_0}$$

(b) Show that
$$\frac{r_1}{r_0} = \frac{e + 1}{e - 1}$$

In Exercises 17–22, use the following values, where needed:

radius of the Earth = 4000 mi = 6440 km
1 year (Earth year) = 365 days (Earth days)
1 AU = 92.9×10^6 mi = 150×10^6 km

17. The planet Pluto has eccentricity $e = 0.249$ and semimajor axis $a = 39.5$ AU.
   (a) Find the period T in years.
   (b) Find the perihelion and aphelion distances.
   (c) Choose a polar coordinate system with the center of the Sun at the pole, and find a polar equation of Pluto's orbit in that coordinate system.
   (d) Make a sketch of the orbit with reasonably accurate proportions.

18. (a) Let a be the semimajor axis of a planet's orbit around the Sun, and let T be its period. Show that if T is measured in days and a is measured in kilometers, then $T = (365 \times 10^{-9})(a/150)^{3/2}$.
   (b) Use the result in part (a) to find the period of the planet Mercury in days, given that its semimajor axis is $a = 57.95 \times 10^6$ km.
   (c) Choose a polar coordinate system with the Sun at the pole, and find an equation for the orbit of Mercury in that coordinate system given that the eccentricity of the orbit is $e = 0.206$.
   (d) Use a graphing utility to generate the orbit of Mercury from the equation obtained in part (c).

19. The Hale–Bopp comet, discovered independently on July 23, 1995 by Alan Hale and Thomas Bopp, has an orbital eccentricity of $e = 0.9951$ and a period of 2380 years.
   (a) Find its semimajor axis in astronomical units (AU).
   (b) Find its perihelion and aphelion distances.
   (c) Choose a polar coordinate system with the center of the Sun at the pole, and find an equation for the Hale–Bopp orbit in that coordinate system.
   (d) Make a sketch of the Hale–Bopp orbit with reasonably accurate proportions.

20. Mars has a perihelion distance of 204,520,000 km and an aphelion distance of 246,280,000 km.
   (a) Use these data to calculate the eccentricity, and compare your answer to the value given in Table 11.6.1.

(b) Find the period of Mars.

(c) Choose a polar coordinate system with the center of the Sun at the pole, and find an equation for the orbit of Mars in that coordinate system.

(d) Use a graphing utility to generate the orbit of Mars from the equation obtained in part (c).

21. *Vanguard 1* was launched in March 1958 into an orbit around the Earth with eccentricity $e = 0.21$ and semimajor axis 8864.5 km. Find the minimum and maximum heights of *Vanguard 1* above the surface of the Earth.

22. The planet Jupiter is believed to have a rocky core of radius 10,000 km surrounded by two layers of hydrogen—a 40,000-km-thick layer of compressed metallic-like hydrogen and a 20,000-km-thick layer of ordinary molecular hydrogen. The visible features, such as the Great Red Spot, are at the outer surface of the molecular hydrogen layer. On November 6, 1997 the spacecraft *Galileo* was placed in a Jovian orbit to study the moon Europa. The orbit had eccentricity 0.814580 and semimajor axis 3,514,918.9 km. Find *Galileo*'s minimum and maximum heights above the molecular hydrogen layer (see the accompanying figure).

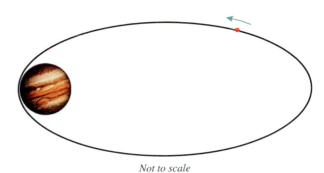

Not to scale

Figure Ex-22

23. What happens to the distance between the directrix and the center of an ellipse if the foci remain fixed and $e \to 0$?

24. (a) Show that the coordinates of the point P on the hyperbola in Figure 11.6.1 satisfy the equation

$$\sqrt{(x-c)^2 + y^2} = \frac{c}{a}x - a$$

(b) Use the result in part (a) to show that $PF/PD = c/a$.

SUPPLEMENTARY EXERCISES

 Graphing Utility   **C** CAS

1. Under what conditions does a parametric curve $x = f(t)$, $y = g(t)$ have a horizontal tangent line? A vertical tangent line? A singular point?

2. Express the point whose *xy*-coordinates are $(-1, 1)$ in polar coordinates with

(a) $r > 0,\ 0 \le \theta < 2\pi$ (b) $r < 0,\ 0 \le \theta < 2\pi$

(c) $r > 0,\ -\pi < \theta \le \pi$ (d) $r < 0,\ -\pi < \theta \le \pi$.

3. In each part, state the name that describes the polar curve most precisely: a rose, a line, a circle, a limaçon, a cardioid, a spiral, a lemniscate, or none of these.

(a) $r = 3\cos\theta$ (b) $r = \cos 3\theta$

(c) $r = \dfrac{3}{\cos\theta}$ (d) $r = 3 - \cos\theta$

(e) $r = 1 - 3\cos\theta$ (f) $r^2 = 3\cos\theta$

(g) $r = (3\cos\theta)^2$ (h) $r = 1 + 3\theta$

4. In each part: (i) Identify the polar graph as a parabola, an ellipse, or a hyperbola; (ii) state whether the directrix is above, below, to the left, or to the right of the pole; and (iii) find the distance from the pole to the directrix.

(a) $r = \dfrac{1}{3 + \cos\theta}$ (b) $r = \dfrac{1}{1 - 3\cos\theta}$

(c) $r = \dfrac{1}{3(1 + \sin\theta)}$ (d) $r = \dfrac{3}{1 - \sin\theta}$

5. The accompanying figure shows the polar graph of the equation $r = f(\theta)$. Sketch the graph of

(a) $r = f(-\theta)$ (b) $r = f\left(\theta - \dfrac{\pi}{2}\right)$

(c) $r = f\left(\theta + \dfrac{\pi}{2}\right)$ (d) $r = -f(\theta)$

(e) $r = f(\theta) + 1$.

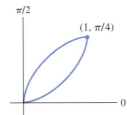

Figure Ex-5

6. Find equations for the two families of circles in the accompanying figure.

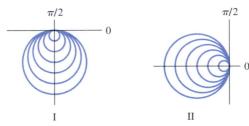

Figure Ex-6

7. In each part, identify the curve by converting the polar equation to rectangular coordinates. Assume that $a > 0$.

(a) $r = a \sec^2 \dfrac{\theta}{2}$

(b) $r^2 \cos 2\theta = a^2$

(c) $r = 4 \csc \left(\theta - \dfrac{\pi}{4}\right)$

(d) $r = 4 \cos \theta + 8 \sin \theta$

8. Use a graphing utility to investigate how the family of polar curves $r = 1 + a \cos n\theta$ is affected by changing the values of a and n, where a is a positive real number and n is a positive integer. Write a brief paragraph to explain your conclusions.

In Exercises 9 and 10, find an equation in xy-coordinates for the conic section that satisfies the given conditions.

---

9. (a) Ellipse with eccentricity $e = \frac{2}{7}$ and ends of the minor axis at the points $(0, \pm 3)$.

(b) Parabola with vertex at the origin, focus on the y-axis, and directrix passing through the point $(7, 4)$.

(c) Hyperbola that has the same foci as the ellipse $3x^2 + 16y^2 = 48$ and asymptotes $y = \pm 2x/3$.

10. (a) Ellipse with center $(-3, 2)$, vertex $(2, 2)$, and eccentricity $e = \frac{4}{5}$.

(b) Parabola with focus $(-2, -2)$ and vertex $(-2, 0)$.

(c) Hyperbola with vertex $(-1, 7)$ and asymptotes $y - 5 = \pm 8(x + 1)$.

11. In each part, sketch the graph of the conic section with reasonably accurate proportions.

(a) $x^2 - 4x + 8y + 36 = 0$

(b) $3x^2 + 4y^2 - 30x - 8y + 67 = 0$

(c) $4x^2 - 5y^2 - 8x - 30y - 21 = 0$

(d) $x^2 + y^2 - 3xy - 3 = 0$

12. If you have a CAS that can graph implicit equations, use it to check your work in Exercise 11.

13. It can be shown that hanging cables form parabolic arcs rather than catenaries if they are subjected to uniformly distributed downward forces along their length. For example, if the weight of the roadway in a suspension bridge is assumed to be uniformly distributed along the supporting cables, then the cables can be modeled by parabolas.

(a) Assuming a parabolic model, find an equation for the cable in the accompanying figure, taking the y-axis to be vertical and the origin at the low point of the cable.

(b) Find the length of the cable between the supports.

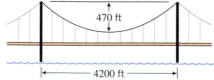

470 ft

4200 ft

Figure Ex-13

14. A parametric curve of the form

$$x = a \cot t + b \cos t, \quad y = a + b \sin t \quad (0 < t < 2\pi)$$

is called a **conchoid of Nicomedes** (see the accompanying figure for the case $0 < a < b$).

(a) Describe how the conchoid

$$x = \cot t + 4 \cos t, \quad y = 1 + 4 \sin t$$

is generated as t varies over the interval $0 < t < 2\pi$.

(b) Find the horizontal asymptote of the conchoid given in part (a).

(c) For what values of t does the conchoid in part (a) have a horizontal tangent line? A vertical tangent line?

(d) Find a polar equation $r = f(\theta)$ for the conchoid in part (a), and then find polar equations for the tangent lines to the conchoid at the pole.

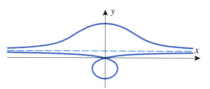

Figure Ex-14

15. Find the area of the region that is common to the circles $r = 1$, $r = 2 \cos \theta$, and $r = 2 \sin \theta$.

16. Find the area of the region that is inside the cardioid $r = a(1 + \sin \theta)$ and outside the circle $r = a \sin \theta$.

17. (a) Find the arc length of the polar curve $r = 1/\theta$ for $\pi/4 \le \theta \le \pi/2$.

(b) What can you say about the arc length of the portion of the curve that lies inside the circle $r = 1$?

18. (a) If a thread is unwound from a fixed circle while being held taut (i.e., tangent to the circle), then the end of the thread traces a curve called an **involute of a circle**. Show that if the circle is centered at the origin, has radius a, and the end of the thread is initially at the point $(a, 0)$, then the involute can be expressed parametrically as

$$x = a(\cos \theta + \theta \sin \theta), \quad y = a(\sin \theta - \theta \cos \theta)$$

where θ is the angle shown in part (a) of Figure Ex-18 (next page).

(b) Assuming that the dog in part (b) of Figure Ex-18 (next page) unwinds its leash while keeping it taut, for what values of θ in the interval $0 \le \theta \le 2\pi$ will the dog be walking North? South? East? West?

(c) Use a graphing utility to generate the curve traced by the dog, and show that it is consistent with your answer in part (b).

19. Let R be the region that is above the x-axis and enclosed between the curve $b^2x^2 - a^2y^2 = a^2b^2$ and the line $x = \sqrt{a^2 + b^2}$.

(a) Sketch the solid generated by revolving R about the x-axis, and find its volume.

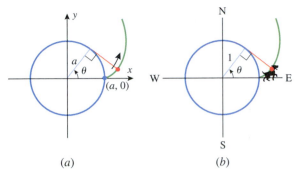

(a)   (b)

Figure Ex-18

(b) Sketch the solid generated by revolving R about the y-axis, and find its volume.

20. (a) Sketch the curves
$$r = \frac{1}{1 + \cos\theta} \quad \text{and} \quad r = \frac{1}{1 - \cos\theta}$$

(b) Find polar coordinates of the intersections of the curves in part (a).

(c) Show that the curves are *orthogonal*, that is, their tangent lines are perpendicular at the points of intersection.

21. How is the shape of a hyperbola affected as its eccentricity approaches 1? As it approaches $+\infty$? Draw some pictures to illustrate your conclusions.

22. Use the formula obtained in part (a) of Exercise 67 of Section 11.1 to find the distance between successive tips of the three-petal rose $r = \sin 3\theta$, and check your answer using trigonometry.

23. (a) Find the minimum and maximum x-coordinates of points on the cardioid $r = 1 + \cos\theta$.

(b) Find the minimum and maximum y-coordinates of points on the cardioid in part (a).

24. (a) Show that the maximum value of the y-coordinate of points on the curve $r = 1/\sqrt{\theta}$ for θ in the interval $(0, \pi]$ occurs when $\tan\theta = 2\theta$.

(b) Use Newton's Method to solve the equation in part (a) for θ to at least four decimal-place accuracy.

(c) Use the result of part (b) to approximate the maximum value of y for $0 < \theta \le \pi$.

25. Define the width of a petal of a rose curve to be the dimension shown in the accompanying figure. Show that the width w of a petal of the four-petal rose $r = \cos 2\theta$ is $w = 2\sqrt{6}/9$. [*Hint:* Express y in terms of θ, and investigate the maximum value of y.]

Petal width

Figure Ex-25

26. A nuclear cooling tower is to have a height of h feet and the shape of the solid that is generated by revolving the

region R enclosed by the right branch of the hyperbola $1521x^2 - 225y^2 = 342{,}225$ and the lines $x = 0$, $y = -h/2$, and $y = h/2$ about the y-axis.

(a) Find the volume of the tower.

(b) Find the lateral surface area of the tower.

27. The amusement park rides illustrated in the accompanying figure consist of two connected rotating arms of length 1—an inner arm that rotates counterclockwise at 1 radian per second and an outer arm that can be programmed to rotate either clockwise at 2 radians per second (the Scrambler ride) or counterclockwise at 2 radians per second (the Calypso ride). The center of the rider cage is at the end of the outer arm.

(a) Show that in the Scrambler ride the center of the cage has parametric equations
$$x = \cos t + \cos 2t, \quad y = \sin t - \sin 2t$$

(b) Find parametric equations for the center of the cage in the Calypso ride, and use a graphing utility to confirm that the center traces the curve shown in the accompanying figure.

(c) Do you think that a rider travels the same distance in one revolution of the Scrambler ride as in one revolution of the Calypso ride? Justify your conclusion.

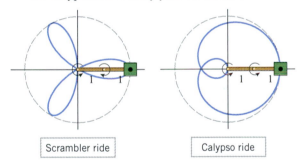

Scrambler ride   Calypso ride

Figure Ex-27

28. Use a graphing utility to explore the effect of changing the rotation rates and the arm lengths in Exercise 27.

29. Use the parametric equations $x = a\cos t$, $y = b\sin t$ to show that the circumference C of an ellipse with semimajor axis a and eccentricity e is
$$C = 4a \int_0^{\pi/2} \sqrt{1 - e^2 \sin^2 u} \, du$$

30. Use Simpson's rule or the numerical integration capability of a graphing utility to approximate the circumference of the ellipse $4x^2 + 9y^2 = 36$ from the integral obtained in Exercise 29.

31. (a) Calculate the eccentricity of the Earth's orbit, given that the ratio of the distance between the center of the Earth and the center of the Sun at perihelion to the distance between the centers at aphelion is $\frac{59}{61}$.

(b) Find the distance between the center of the Earth and the center of the Sun at perihelion, given that the average value of the perihelion and aphelion distances between the centers is 93 million miles.

(c) Use the result in Exercise 29 and Simpson's rule or the numerical integration capability of a graphing utility to approximate the distance that the Earth travels in 1 year (one revolution around the Sun).

32. It will be shown later in this text that if a projectile is launched with speed v_0 at an angle α with the horizontal and at a height y_0 above ground level, then the resulting trajectory relative to the coordinate system in the accompanying figure will have parametric equations

$$x = (v_0 \cos \alpha)t, \quad y = y_0 + (v_0 \sin \alpha)t - \tfrac{1}{2}gt^2$$

where g is the acceleration due to gravity.

(a) Show that the trajectory is a parabola.

(b) Find the coordinates of the vertex.

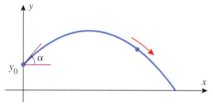

Figure Ex-32

33. Mickey Mantle is recognized as baseball's unofficial king of long home runs. On April 17, 1953 Mantle blasted a pitch by Chuck Stobbs of the hapless Washington Senators out of Griffith Stadium, just clearing the 50-ft wall at the 391-ft marker in left center. Assuming that the ball left the bat at a height of 3 ft above the ground and at an angle of $45°$, use the parametric equations in Exercise 32 with $g = 32$ ft/s² to find

(a) the speed of the ball as it left the bat

(b) the maximum height of the ball

(c) the distance along the ground from home plate to where the ball struck the ground.

c **34.** Recall from Section 6.9 that the Fresnel sine and cosine functions are defined as

$$S(x) = \int_0^x \sin\left(\frac{\pi t^2}{2}\right) dt \quad \text{and} \quad C(x) = \int_0^x \cos\left(\frac{\pi t^2}{2}\right) dt$$

The following parametric curve, which is used to study amplitudes of light waves in optics, is called a **clothoid** or **Cornu spiral** in honor of the French scientist Marie Alfred Cornu (1841–1902):

$$x = C(t) = \int_0^t \cos\left(\frac{\pi u^2}{2}\right) du$$

$$y = S(t) = \int_0^t \sin\left(\frac{\pi u^2}{2}\right) du$$

$$(-\infty < t < +\infty)$$

(a) Use a CAS to graph the Cornu spiral.

(b) Describe the behavior of the spiral as $t \to +\infty$ and as $t \to -\infty$.

(c) Find the arc length of the spiral for $-1 \le t \le 1$.

35. As illustrated in the accompanying figure, let $P(r, \theta)$ be a point on the polar curve $r = f(\theta)$, let ψ be the smallest counterclockwise angle from the extended radius OP to the tangent line at P, and let ϕ be the angle of inclination of the tangent line. Derive the formula

$$\tan \psi = \frac{r}{dr/d\theta}$$

by substituting $\tan \phi$ for dy/dx in Formula (7) of Section 11.2 and applying the trigonometric identity

$$\tan(\phi - \theta) = \frac{\tan \phi - \tan \theta}{1 + \tan \phi \tan \theta}$$

In Exercises 36 and 37, use the formula for ψ obtained in Exercise 35.

36. (a) Use the trigonometric identity

$$\tan \frac{\theta}{2} = \frac{1 - \cos \theta}{\sin \theta}$$

to show that if (r, θ) is a point on the cardioid

$$r = 1 - \cos \theta \quad (0 \le \theta < 2\pi)$$

then $\psi = \theta/2$.

(b) Sketch the cardioid and show the angle ψ at the points where the cardioid crosses the y-axis.

(c) Find the angle ψ at the points where the cardioid crosses the y-axis.

37. Show that for a logarithmic spiral $r = ae^{b\theta}$, the angle from the radial line to the tangent line is constant along the spiral (see the accompanying figure). [*Note:* For this reason, logarithmic spirals are sometimes called **equiangular spirals**.]

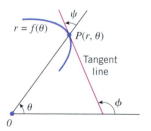

Figure Ex-35

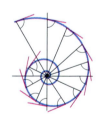

Figure Ex-37

EXPANDING THE CALCULUS HORIZON

Comet Collision

*The Earth lives in a cosmic shooting gallery of comets and asteroids. Although the probability that the Earth will be hit by a comet or asteroid in any given year is small, the consequences of such a collision are so catastrophic that the international community is now beginning to track **near Earth objects** (NEOs). Your job, as part of the international NEO tracking team, is to compute the orbits of incoming comets and asteroids, determine how close they will come to colliding with the Earth, and issue a notification if there is danger of a collision or near miss.*

At the time when the Earth is at its *aphelion* (its farthest point from the Sun), your NEO tracking team receives a notification from the NASA/Caltech Jet Propulsion Laboratory that a previously unknown comet (designation Rogue 2000) is traveling in the plane of Earth's orbit and hurtling in the direction of the Earth. You immediately transmit a request to NASA for the orbital parameters and the current positions of the Earth and Rogue 2000 and receive the following report:

ORBITAL PARAMETERS

EARTH	ROGUE 2000
Eccentricity: $e_1 = 0.017$	Eccentricity: $e_2 = 0.98$
Semimajor axis: $a_1 = 1 \text{ AU} = 1.496 \times 10^8$ km	Semimajor axis: $a_2 = 5 \text{ AU} = 7.48 \times 10^8$ km
Period: $T_1 = 1$ year	Period: $T_2 = 5\sqrt{5}$ years

INITIAL POSITION INFORMATION

The major axes of Earth and Rogue 2000 lie on the same line.

The aphelions of Earth and Rogue 2000 are on the same side of the Sun.

Initial polar angle of Earth: $\theta = 0$ radians.

Initial polar angle of Rogue 2000: $\theta = 0.45$ radian.

The Calculation Strategy

Since the immediate concern is a possible collision at intersection A in Figure 1, your team works out the following plan:

Step 1. Find the polar equations for Earth and Rogue 2000.
Step 2. Find the polar coordinates of intersection A.
Step 3. Determine how long it will take the Earth to reach intersection A.
Step 4. Determine where Rogue 2000 will be when the Earth reaches intersection A.
Step 5. Determine how far Rogue 2000 will be from the Earth when the Earth is at intersection A.

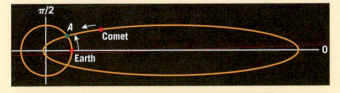

Initial configuration of Earth and Rogue 2000

Figure 1

Polar Equations of the Orbits

Exercise 1   Write polar equations of the form

$$r = \frac{a(1 - e^2)}{1 - e\cos\theta}$$

for the orbits of Earth and Rogue 2000 using AU units for r.

Exercise 2   Use a graphing utility to generate the two orbits on the same screen.

Intersection of the Orbits

The second step in your team's calculation plan is to find the polar coordinates of intersection A in Figure 1.

Exercise 3   For simplicity, let $k_1 = a_1(1 - e_1^2)$ and $k_2 = a_2(1 - e_2^2)$, and use the polar equations obtained in Exercise 1 to show that the angle θ at intersection A satisfies the equation

$$\cos\theta = \frac{k_1 - k_2}{k_1 e_2 - k_2 e_1}$$

Exercise 4   Use the result in Exercise 3 and the inverse cosine capability of a calculating utility to show that the angle θ at intersection A in Figure 1 is $\theta = 0.607$ radian.

Exercise 5   Use the result in Exercise 4 and either polar equation obtained in Exercise 1 to show that if r is in AU units, then the polar coordinates of intersection A are $(r, \theta) = (1.014, 0.607)$.

Time Required for Earth to Reach Intersection A

According to Kepler's second law (see 11.6.3), the radial line from the center of the Sun to the center of an object orbiting around it sweeps out equal areas in equal times. Thus, if t is the time that it takes for the radial line to sweep out an "elliptic sector" from some initial angle θ_I to some final angle θ_F (Figure 2), and if T is the period of the object (the time for one complete revolution), then

$$\frac{t}{T} = \frac{\text{area of the "elliptic sector"}}{\text{area of the entire ellipse}} \tag{1}$$

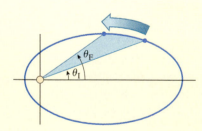

Figure 2

Exercise 6   Use Formula (1) to show that

$$t = \frac{T \displaystyle\int_{\theta_I}^{\theta_F} r^2\, d\theta}{2\pi a^2 \sqrt{1 - e^2}} \tag{2}$$

Exercise 7   Use a calculating utility with a numerical integration capability, Formula (2), and the polar equation for the orbit of the Earth obtained in Exercise 1 to find the time t (in years) required for the Earth to move from its initial position to intersection A.

Position of Rogue 2000 When the Earth Is at Intersection A

The fourth step in your team's calculation strategy is to determine the position of Rogue 2000 when the Earth reaches intersection A.

Exercise 8   During the time that it takes for the Earth to move from its initial position to intersection A, the polar angle of Rogue 2000 will change from its initial value $\theta_I = 0.45$ radian to some final value θ_F that remains to be determined. Apply Formula (2) using the orbital data for Rogue 2000 and the time t obtained in Exercise 7 to show that θ_F satisfies the equation

$$\int_{0.45}^{\theta_F} \left[\frac{a_2(1 - e_2^2)}{1 - e_2 \cos\theta} \right]^2 d\theta = \frac{2t\pi a_2^2 \sqrt{1 - e_2^2}}{5\sqrt{5}} \tag{3}$$

Your team is now faced with the problem of solving Equation (3) for the unknown upper limit θ_F. Some members of the team plan to use a CAS to perform the integration, some plan to use integration tables, and others plan to use hand calculation by making the substitution $u = \tan(\theta/2)$ and applying the formulas in (5) of Section 8.6.

Exercise 9

(a) Evaluate the integral in (3) using a CAS or by hand calculation.

(b) Use the root-finding capability of a calculating utility to find the polar angle of Rogue 2000 when the Earth is at intersection A.

Calculating the Critical Distance

It is the policy of your NEO tracking team to issue a notification to various governmental agencies for any asteroid or comet that will be within 4 million kilometers of the Earth at an orbital intersection. (This distance is roughly 10 times that between the Earth and the Moon.) Accordingly, the final step in your team's plan is to calculate the distance between the Earth and Rogue 2000 when the Earth is at intersection A, and then determine whether a notification should be issued.

Exercise 10   Use the polar equation of Rogue 2000 obtained in Exercise 1 and the result in Exercise 9(b) to find polar coordinates of Rogue 2000 with r in AU units when the Earth is at intersection A.

Exercise 11   Use the distance formula in Exercise 67(a) of Section 11.1 to calculate the distance between the Earth and Rogue 2000 in AU units when the Earth is at intersection A, and then use the conversion factor 1 AU $= 1.496 \times 10^8$ km to determine whether a government notification should be issued.

Note: One of the closest near misses in recent history occurred on October 30, 1937 when the asteroid Hermes passed within 900,000 km of the Earth. More recently, on June 14, 1968 the asteroid Icarus passed within 23,000,000 km of the Earth.

Module by Mary Ann Connors, USMA, West Point, and Howard Anton, Drexel University

APPENDIX A

Real Numbers, Intervals, and Inequalities

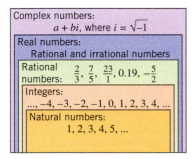

Figure A.1

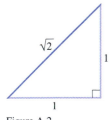

Figure A.2

Figure A.1 describes the various categories of numbers that we will encounter in this text. The simplest numbers are the **natural numbers**

$$1, \ 2, \ 3, \ 4, \ 5, \dots$$

These are a subset of the **integers**

$$\dots, \ -4, \ -3, \ -2, \ -1, \ 0, \ 1, \ 2, \ 3, \ 4, \dots$$

and these in turn are a subset of the **rational numbers**, which are the numbers formed by taking ratios of integers (avoiding division by 0). Some examples are

$$\tfrac{2}{3}, \quad \tfrac{7}{5}, \quad 23 = \tfrac{23}{1}, \quad 0.19 = \tfrac{19}{100}, \quad -\tfrac{5}{2} = \tfrac{-5}{2} = \tfrac{5}{-2}$$

The early Greeks believed that every measurable quantity had to be a rational number. However, this idea was overturned in the fifth century B.C. by Hippasus of Metapontum[*] who demonstrated the existence of **irrational numbers**, that is, numbers that cannot be expressed as the ratio of two integers. Using geometric methods, he showed that the length of the hypotenuse of the triangle in Figure A.2 could not be expressed as a ratio of integers, thereby proving that $\sqrt{2}$ is an irrational number. Some other examples of irrational numbers are

$$\sqrt{3}, \quad \sqrt{5}, \quad 1 + \sqrt{2}, \quad \sqrt[3]{7}, \quad \pi, \quad \cos 19°$$

The rational and irrational numbers together comprise what is called the **real number system**, and both the rational and irrational numbers are called **real numbers**.

Because the square of a real number cannot be negative, the equation

$$x^2 = -1$$

has no solutions in the real number system. In the eighteenth century mathematicians remedied this problem by inventing a new number, which they denoted by

$$i = \sqrt{-1}$$

and which they defined to have the property $i^2 = -1$. This, in turn, led to the development

---

[*] HIPPASUS OF METAPONTUM (circa 500 B.C.). A Greek Pythagorean philosopher. According to legend, Hippasus made his discovery at sea and was thrown overboard by fanatic Pythagoreans because his result contradicted their doctrine. The discovery of Hippasus is one of the most fundamental in the entire history of science.

of the *complex numbers*, which are numbers of the form

$$a + bi$$

where a and b are real numbers. Some examples are

$2 + 3i$	$3 - 4i$	$6i$	$\frac{2}{3}$
$\lvert a = 2, b = 3 \rvert$	$\lvert a = 3, b = -4 \rvert$	$\lvert a = 0, b = 6 \rvert$	$\lvert a = \frac{2}{3}, b = 0 \rvert$

Observe that every real number a is also a complex number because it can be written as

$$a = a + 0i$$

Thus, the real numbers are a subset of the complex numbers. Although we will be concerned primarily with real numbers in this text, complex numbers will arise in the course of solving equations. For example, the solutions of the quadratic equation

$$ax^2 + bx + c = 0$$

which are given by the *quadratic formula*

$$x = \frac{-b \pm \sqrt{b^2 - 4ac}}{2a}$$

are not real if the quantity $b^2 - 4ac$ is negative.

DIVISION BY ZERO

Division by zero is not allowed in numerical computations because it leads to mathematical inconsistencies. For example, if $1/0$ were assigned some numerical value, say p, then it would follow that $0 \cdot p = 1$, which is incorrect.

DECIMAL REPRESENTATION OF REAL NUMBERS

Rational and irrational numbers can be distinguished by their decimal representations. Rational numbers have decimals that are *repeating*, by which we mean that at some point in the decimal some fixed block of numbers begins to repeat indefinitely. For example,

$$\frac{4}{3} = 1.333\ldots, \quad \frac{3}{11} = .272727\ldots, \quad \frac{1}{2} = .50000\ldots, \quad \frac{5}{7} = .714285714285714285\ldots$$

3 repeats   27 repeats   0 repeats   714285 repeats

```
3.1415926535897932384626433832795028884197169
3993751058209749445923078164062862089986280
4825342117067982148086513282306647093844609
5505822317253594081284811174502841027019385
10555964462294895493038196442881097566593344
6128475648233786783165271201909145648566923
6034861045432664821339360726024914127372458
0066063155881748815209209628292540917153643
789259036001133053054882046652138414695194
11609433057270365759591953092186117381932
7931051185480744623799627495673518857527248
122793818301194912983367336244065664308602
9494639522473719070217986094370277053921717
29317675238467481846766940513200056812714526
3560827785771342577896091736371787214684409
0122495343014654958537105079227968925892354
019956112129021960864034418159813629774771
9960518707211349999998372978049951059731732
1609631859502445945534690830264252230825334
685035261931188171010003137838752886587533
83814206171776691473035982534904287554687311
5956286388235378759375195778185778053217122
80661300192787661195909216420198
```

Figure A.3

Decimals in which zero repeats from some point on are called *terminating decimals*. For brevity, it is usual to omit the repetitive zeros in terminating decimals and for other repeating decimals to write the repeating digits only once but with a bar over them to indicate the repetition. For example,

$$\frac{1}{2} = .5, \quad \frac{12}{4} = 3, \quad \frac{8}{25} = .32, \quad \frac{4}{3} = 1.\overline{3}, \quad \frac{3}{11} = .\overline{27}, \quad \frac{5}{7} = .\overline{714285}$$

Irrational numbers have nonrepeating decimals, so we can be certain that the decimals

$$\sqrt{2} = 1.414213562373095\ldots \quad \text{and} \quad \pi = 3.141592653589793\ldots$$

do not repeat from some point on. Moreover, if we stop the decimal expansion of an irrational number at some point, we get only an approximation to the number, never an exact value. For example, even if we compute π to 1000 decimal places, as in Figure A.3, we still have only an approximation.

REMARK.   Beginning mathematics students are sometimes taught to approximate π by $\frac{22}{7}$. Keep in mind, however, that this is only an approximation, since

$$\frac{22}{7} = 3.\overline{142857}$$

is a rational number whose decimal representation begins to differ from π in the third decimal place.

COORDINATE LINES

In 1637 René Descartes[*] published a philosophical work called *Discourse on the Method of Rightly Conducting the Reason*. In the back of that book was an appendix that the British philosopher John Stuart Mill described as "the greatest single step ever made in the progress of the exact sciences." In that appendix René Descartes linked together algebra and geometry, thereby creating a new subject called **analytic geometry**; it gave a way of describing algebraic formulas by geometric curves and, conversely, geometric curves by algebraic formulas.

The key step in analytic geometry is to establish a correspondence between real numbers and points on a line. To do this, choose any point on the line as a reference point, and call it the **origin**; and then arbitrarily choose one of the two directions along the line to be the **positive direction**, and let the other be the **negative direction**. It is usual to mark the positive direction with an arrowhead, as in Figure A.4, and to take the positive direction to the right when the line is horizontal. Next, choose a convenient unit of measure, and represent each positive number r by the point that is r units from the origin in the positive direction, each negative number $-r$ by the point that is r units from the origin in the negative direction from the origin, and 0 by the origin itself (Figure A.5). The number associated with a point P is called the **coordinate** of P, and the line is called a **coordinate line**, a **real number line**, or a **real line**.

| - | Origin | + |

Figure A.4

Figure A.5

INEQUALITY NOTATION

The real numbers can be ordered by size as follows: If $b - a$ is positive, then we write either $a < b$ (read "a is less than b") or $b > a$ (read "b is greater than a"). We write $a \leq b$ to mean $a < b$ or $a = b$, and we write $a < b < c$ to mean that $a < b$ and $b < c$. As one traverses a coordinate line in the positive direction, the real numbers increase in size, so on a horizontal coordinate line the inequality $a < b$ implies that a is to the left of b, and the inequalities $a < b < c$ imply that a is to the left of c, and b lies between a and c. The meanings of such symbols as

$$a \leq b < c, \quad a \leq b \leq c, \quad \text{and} \quad a < b < c < d$$

should be clear. For example, you should be able to confirm that all of the following are true statements:

$$3 < 8, \quad -7 < 1.5, \quad -12 \leq -\pi, \quad 5 \leq 5, \quad 0 \leq 2 \leq 4,$$
$$8 \geq 3, \quad 1.5 > -7, \quad -\pi > -12, \quad 5 \geq 5, \quad 3 > 0 > -1 > -3$$

REVIEW OF SETS

In the following discussion we will be concerned with certain sets of real numbers, so it will be helpful to review the basic ideas about sets. Recall that a **set** is a collection of objects, called **elements** or **members** of the set. In this text we will be concerned primarily with sets whose members are numbers or points that lie on a line, a plane, or in three-dimensional space. We will denote sets by capital letters and elements by lowercase letters. To indicate that a is a member of the set A we will write $a \in A$ (read "a belongs to A"), and to indicate

[*]RENÉ DESCARTES (1596–1650). Descartes, a French aristocrat, was the son of a government official. He graduated from the University of Poitiers with a law degree at age 20. After a brief probe into the pleasures of Paris he became a military engineer, first for the Dutch Prince of Nassau and then for the German Duke of Bavaria. It was during his service as a soldier that Descartes began to pursue mathematics seriously and develop his analytic geometry. After the wars, he returned to Paris where he stalked the city as an eccentric, wearing a sword in his belt and a plumed hat. He lived in leisure, seldom arose before 11 A.M., and dabbled in the study of human physiology, philosophy, glaciers, meteors, and rainbows. He eventually moved to Holland, where he published his *Discourse on the Method*, and finally to Sweden where he died while serving as tutor to Queen Christina. Descartes is regarded as a genius of the first magnitude. In addition to major contributions in mathematics and philosophy, he is considered, along with William Harvey, to be a founder of modern physiology.

that a is not a member of the set A we will write $a \notin A$ (read "a does not belong to A"). For example, if A is the set of positive integers, then $5 \in A$, but $-5 \notin A$. Sometimes sets arise that have no members (e.g., the set of odd integers that are divisible by 2). A set with no members is called an ***empty set*** or a ***null set*** and is denoted by the symbol $\varnothing$.

Some sets can be described by listing their members between braces. The order in which the members are listed does not matter, so, for example, the set A of positive integers that are less than 6 can be expressed as

$$A = \{1, 2, 3, 4, 5\} \quad \text{or} \quad A = \{2, 3, 1, 5, 4\}$$

We can also write A in *set-builder notation* as

$$A = \{x : x \text{ is an integer and } 0 < x < 6\}$$

which is read "A is the set of all x such that x is an integer and $0 < x < 6$." In general, to express a set S in set-builder notation we write $S = \{x : \underline{\quad\quad}\}$ in which the line is replaced by a property that identifies exactly those elements in the set S.

If every member of a set A is also a member of a set B, then we say that A is a ***subset*** of B and write $A \subseteq B$. For example, if A is the set of positive integers and B is the set of all integers, then $A \subseteq B$. If two sets A and B have the same members (i.e., $A \subseteq B$ and $B \subseteq A$), then we say that A and B are ***equal*** and write $A = B$.

INTERVALS

In calculus we will be concerned with sets of real numbers, called ***intervals***, that correspond to line segments on a coordinate line. For example, if $a < b$, then the ***open interval*** from a to b, denoted by (a, b), is the line segment extending from a to b, *excluding* the endpoints; and the ***closed interval*** from a to b, denoted by $[a, b]$, is the line segment extending from a to b, *including* the endpoints (Figure A.6). These sets can be expressed in set-builder notation as

$$(a, b) = \{x : a < x < b\} \qquad \text{The open interval from } a \text{ to } b$$

$$[a, b] = \{x : a \le x \le b\} \qquad \text{The closed interval from } a \text{ to } b$$

The open interval (a, b)

The closed interval $[a, b]$

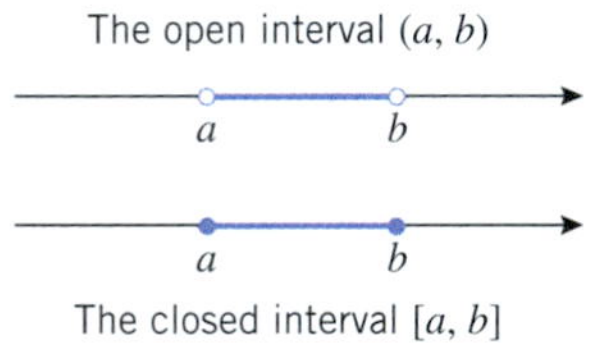

Figure A.6

• REMARK.   Observe that in this notation and in the corresponding Figure A.6, parentheses and open dots mark endpoints that are excluded from the interval, whereas brackets and closed dots mark endpoints that are included in the interval. Observe also that in set-builder notation for the intervals, it is understood that x is a real number, even though it is not stated explicitly.

As shown in Table 1, an interval can include one endpoint and not the other; such intervals are called ***half-open*** (or sometimes ***half-closed***). Moreover, the table also shows that it is possible for an interval to extend indefinitely in one or both directions. To indicate that an interval extends indefinitely in the positive direction we write $+\infty$ (read "positive infinity") in place of a right endpoint, and to indicate that an interval extends indefinitely in the negative direction we write $-\infty$ (read "negative infinity") in place of a left endpoint. Intervals that extend between two real numbers are called ***finite intervals***, whereas intervals that extend indefinitely in one or both directions are called ***infinite intervals***.

• REMARK.   By convention, infinite intervals of the form $[a, +\infty)$ or $(-\infty, b]$ are considered to be closed because they contain their endpoint, and intervals of the form $(a, +\infty)$ and $(-\infty, b)$ are considered to be open because they do not include their endpoint. The interval $(-\infty, +\infty)$, which is the set of all real numbers, has no endpoints and can be regarded as either open or closed, as convenient. This set is often denoted by the special symbol $\mathbb{R}$. To distinguish verbally between the open interval $(0, +\infty) = \{x : x > 0\}$ and the closed interval $[0, +\infty) = \{x : x \ge 0\}$, we will call x ***positive*** if $x > 0$ and ***nonnegative*** if $x \ge 0$. Thus, a positive number must be nonnegative, but a nonnegative number need not be positive, since it might possibly be 0.

Table 1

INTERVAL NOTATION	SET NOTATION	GEOMETRIC PICTURE	CLASSIFICATION
(a, b)	$\{x : a < x < b\}$		Finite; open
$[a, b]$	$\{x : a \leq x \leq b\}$		Finite; closed
$[a, b)$	$\{x : a \leq x < b\}$		Finite; half-open
$(a, b]$	$\{x : a < x \leq b\}$		Finite; half-open
$(-\infty, b]$	$\{x : x \leq b\}$		Infinite; closed
$(-\infty, b)$	$\{x : x < b\}$		Infinite; open
$[a, +\infty)$	$\{x : x \geq a\}$		Infinite; closed
$(a, +\infty)$	$\{x : x > a\}$		Infinite; open
$(-\infty, +\infty)$	$\mathbb{R}$		Infinite; open and closed

UNIONS AND INTERSECTIONS OF INTERVALS

If A and B are sets, then the ***union*** of A and B (denoted by $A \cup B$) is the set whose members belong to A or B (or both), and the ***intersection*** of A and B (denoted by $A \cap B$) is the set whose members belong to both A and B. For example,

$$\{x : 0 < x < 5\} \cup \{x : 1 < x < 7\} = \{x : 0 < x < 7\}$$

$$\{x : x < 1\} \cap \{x : x \geq 0\} = \{x : 0 \leq x < 1\}$$

$$\{x : x < 0\} \cap \{x : x > 0\} = \varnothing$$

or in interval notation,

$$(0, 5) \cup (1, 7) = (0, 7)$$

$$(-\infty, 1) \cap [0, +\infty) = [0, 1)$$

$$(-\infty, 0) \cap (0, +\infty) = \varnothing$$

ALGEBRAIC PROPERTIES OF INEQUALITIES

The following algebraic properties of inequalities will be used frequently in this text. We omit the proofs.

> **A.1** THEOREM (*Properties of Inequalities*). *Let a, b, c, and d be real numbers.*
> (*a*) *If $a < b$ and $b < c$, then $a < c$.*
> (*b*) *If $a < b$, then $a + c < b + c$ and $a - c < b - c$.*
> (*c*) *If $a < b$, then $ac < bc$ when c is positive and $ac > bc$ when c is negative.*
> (*d*) *If $a < b$ and $c < d$, then $a + c < b + d$.*
> (*e*) *If a and b are both positive or both negative and $a < b$, then $1/a > 1/b$.*

If we call the direction of an inequality its *sense*, then these properties can be paraphrased as follows:

(*b*) *The sense of an inequality is unchanged if the same number is added to or subtracted from both sides.*

(*c*) *The sense of an inequality is unchanged if both sides are multiplied by the same positive number, but the sense is reversed if both sides are multiplied by the same negative number.*

(d)   *Inequalities with the same sense can be added.*

(e)   *If both sides of an inequality have the same sign, then the sense of the inequality is reversed by taking the reciprocal of each side.*

REMARK.   These properties remain true if the symbols $<$ and $>$ are replaced by $\leq$ and $\geq$ in Theorem A.1.

Example 1

STARTING INEQUALITY	OPERATION	RESULTING INEQUALITY
$-2 < 6$	Add 7 to both sides.	$5 < 13$
$-2 < 6$	Subtract 8 from both sides.	$-10 < -2$
$-2 < 6$	Multiply both sides by 3.	$-6 < 18$
$-2 < 6$	Multiply both sides by -3.	$6 > -18$
$3 < 7$	Multiply both sides by 4.	$12 < 28$
$3 < 7$	Multiply both sides by -4.	$-12 > -28$
$3 < 7$	Take reciprocals of both sides.	$\frac{1}{3} > \frac{1}{7}$
$-8 < -6$	Take reciprocals of both sides.	$-\frac{1}{8} > -\frac{1}{6}$
$4 < 5, -7 < 8$	Add corresponding sides.	$-3 < 13$

◀

SOLVING INEQUALITIES

A *solution* of an inequality in an unknown x is a value for x that makes the inequality a true statement. For example, $x = 1$ is a solution of the inequality $x < 5$, but $x = 7$ is not. The set of all solutions of an inequality is called its *solution set*. It can be shown that if one does not multiply both sides of an inequality by zero or an expression involving an unknown, then the operations in Theorem A.1 will not change the solution set of the inequality. The process of finding the solution set of an inequality is called *solving* the inequality.

Example 2   Solve $3 + 7x \leq 2x - 9$.

Solution.   We will use the operations of Theorem A.1 to isolate x on one side of the inequality.

$$3 + 7x \leq 2x - 9 \qquad \boxed{\text{Given.}}$$

$$7x \leq 2x - 12 \qquad \boxed{\text{We subtracted 3 from both sides.}}$$

$$5x \leq -12 \qquad \boxed{\text{We subtracted } 2x \text{ from both sides.}}$$

$$x \leq -\tfrac{12}{5} \qquad \boxed{\text{We multiplied both sides by } \tfrac{1}{5}.}$$

Figure A.7

Because we have not multiplied by any expressions involving the unknown x, the last inequality has the same solution set as the first. Thus, the solution set is the interval $\left(-\infty, -\frac{12}{5}\right]$ shown in Figure A.7.   ◀

Example 3   Solve $7 \leq 2 - 5x < 9$.

Solution.   The given inequality is actually a combination of the two inequalities

$$7 \leq 2 - 5x \quad \text{and} \quad 2 - 5x < 9$$

We could solve the two inequalities separately, then determine the values of x that satisfy both by taking the intersection of the two solution sets. However, it is possible to work with the combined inequalities in this problem:

$$7 \leq 2 - 5x < 9 \quad \boxed{\text{Given.}}$$

$$5 \leq -5x < 7 \quad \boxed{\text{We subtracted 2 from each member.}}$$

$$-1 \geq x > -\frac{7}{5} \quad \boxed{\begin{array}{l}\text{We multiplied by } -\frac{1}{5} \text{ and reversed}\\ \text{the sense of the inequalities.}\end{array}}$$

$$-\frac{7}{5} < x \leq -1 \quad \boxed{\begin{array}{l}\text{For clarity, we rewrote the inequalities}\\ \text{with the smaller number on the left.}\end{array}}$$

Thus, the solution set is the interval $\left(-\frac{7}{5}, -1\right]$ shown in Figure A.8. ◄

Figure A.8

(number line image: points at $-\frac{7}{5}$ and -1)

Example 4 Solve $x^2 - 3x > 10$.

Solution. By subtracting 10 from both sides, the inequality can be rewritten as

$$x^2 - 3x - 10 > 0$$

Factoring the left side yields

$$(x + 2)(x - 5) > 0$$

The values of x for which $x + 2 = 0$ or $x - 5 = 0$ are $x = -2$ and $x = 5$. These values divide the coordinate line into three open intervals,

$$(-\infty, -2), \quad (-2, 5), \quad (5, +\infty)$$

on each of which the product $(x + 2)(x - 5)$ has constant sign. To determine those signs we will choose an *arbitrary* number in each interval at which we will determine the sign; these are called ***test values***. As shown in Figure A.9, we will use -3, 0, and 6 as our test values. The results can be organized as follows:

INTERVAL	TEST VALUE	SIGN OF $(x + 2)(x - 5)$ AT THE TEST VALUE
$(-\infty, -2)$	-3	$(-)(-) = +$
$(-2, 5)$	0	$(+)(-) = -$
$(5, +\infty)$	6	$(+)(+) = +$

The pattern of signs in the intervals is shown on the number line in the middle of Figure A.9. We deduce that the solution set is $(-\infty, -2) \cup (5, +\infty)$, which is shown at the bottom of Figure A.9. ◄

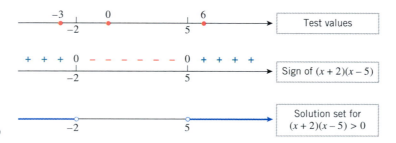

Figure A.9

Example 5 Solve $\dfrac{2x - 5}{x - 2} < 1$.

Solution. We could start by multiplying both sides by $x - 2$ to eliminate the fraction. However, this would require us to consider the cases $x - 2 > 0$ and $x - 2 < 0$ separately because the sense of the inequality would be reversed in the second case, but not the first. The following approach is simpler:

$$\frac{2x-5}{x-2} < 1 \qquad \boxed{\text{Given.}}$$

$$\frac{2x-5}{x-2} - 1 < 0 \qquad \boxed{\begin{array}{l}\text{We subtracted 1 from both sides}\\\text{to obtain a 0 on the right.}\end{array}}$$

$$\frac{(2x-5)-(x-2)}{x-2} < 0 \qquad \boxed{\text{We combined terms.}}$$

$$\frac{x-3}{x-2} < 0 \qquad \boxed{\text{We simplified.}}$$

The quantity $x-3$ is zero if $x=3$, and the quantity $x-2$ is zero if $x=2$. These values divide the coordinate line into three open intervals,

$$(-\infty, 2), \quad (2, 3), \quad (3, +\infty)$$

on each of which the quotient $(x-3)/(x-2)$ has constant sign. Using 0, 2.5, and 4 as test values (Figure A.10), we obtain the following results:

INTERVAL	TEST VALUE	SIGN OF $(x-3)/(x-2)$ AT THE TEST VALUE
$(-\infty, 2)$	0	$(-)/(-) = +$
$(2, 3)$	2.5	$(-)/(+) = -$
$(3, +\infty)$	4	$(+)/(+) = +$

The signs of the quotient are shown in the middle of Figure A.10. From the figure we see that the solution set consists of all real values of x such that $2 < x < 3$. This is the interval $(2, 3)$ shown at the bottom of Figure A.10. ◄

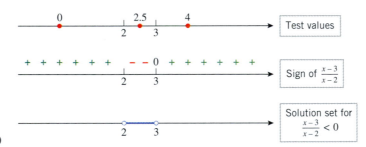

Figure A.10

EXERCISE SET A

1. Among the terms *integer*, *rational*, and *irrational*, which ones apply to the given number?
   (a) $-\frac{3}{4}$ (b) 0   (c) $\frac{24}{8}$
   (d) 0.25   (e) $-\sqrt{16}$ (f) $2^{1/2}$
   (g) 0.020202...   (h) 7.000...

2. Which of the terms *integer*, *rational*, and *irrational* apply to the given number?
   (a) 0.31311311131111...   (b) 0.729999...
   (c) 0.376237623762...   (d) $17\frac{4}{5}$

3. The repeating decimal 0.137137137... can be expressed as a ratio of integers by writing

$$x = 0.137137137...$$
$$1000x = 137.137137137...$$

   and subtracting to obtain $999x = 137$ or $x = \frac{137}{999}$. Use this idea, where needed, to express the following decimals as ratios of integers.
   (a) 0.123123123...   (b) 12.7777...
   (c) 38.07818181...   (d) 0.4296000...

4. Show that the repeating decimal 0.99999...represents the number 1. Since 1.000...is also a decimal representation of 1, this problem shows that a real number can have two different decimal representations. [*Hint:* Use the technique of Exercise 3.]

5. The Rhind Papyrus, which is a fragment of Egyptian mathematical writing from about 1650 B.C., is one of the oldest known examples of written mathematics. It is stated in the papyrus that the area A of a circle is related to its diameter D by

$$A = \left(\tfrac{8}{9}D\right)^2$$

(a) What approximation to π were the Egyptians using?

(b) Use a calculating utility to determine if this approximation is better or worse than the approximation $\frac{22}{7}$.

6. The following are all famous approximations to π:

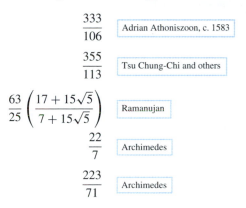

$$\frac{333}{106}$$ Adrian Athoniszoon, c. 1583

$$\frac{355}{113}$$ Tsu Chung-Chi and others

$$\frac{63}{25}\left(\frac{17 + 15\sqrt{5}}{7 + 15\sqrt{5}}\right)$$ Ramanujan

$$\frac{22}{7}$$ Archimedes

$$\frac{223}{71}$$ Archimedes

(a) Use a calculating utility to order these approximations according to size.

(b) Which of these approximations is closest to but larger than π?

(c) Which of these approximations is closest to but smaller than π?

(d) Which of these approximations is most accurate?

7. In each line of the accompanying table, check the blocks, if any, that describe a valid relationship between the real numbers a and b. The first line is already completed as an illustration.

a	b	$a < b$	$a \leq b$	$a > b$	$a \geq b$	$a = b$
1	6	✓	✓			
6	1					
-3	5					
5	-3					
-4	-4					
0.25	$\frac{1}{3}$					
$-\frac{1}{4}$	$-\frac{3}{4}$					

Table Ex-7

8. In each line of the accompanying table, check the blocks, if any, that describe a valid relationship between the real numbers a, b, and c.

a	b	c	$a < b < c$	$a \leq b \leq c$	$a < b \leq c$	$a \leq b < c$
-1	0	2				
2	4	-3				
$\frac{1}{2}$	$\frac{1}{2}$	$\frac{3}{4}$				
-5	-5	-5				
0.75	1.25	1.25				

Table Ex-8

9. Which of the following are always correct if $a \leq b$?

(a) $a - 3 \leq b - 3$ (b) $-a \leq -b$

(c) $3 - a \leq 3 - b$ (d) $6a \leq 6b$

(e) $a^2 \leq ab$ (f) $a^3 \leq a^2 b$

10. Which of the following are always correct if $a \leq b$ and $c \leq d$?

(a) $a + 2c \leq b + 2d$ (b) $a - 2c \leq b - 2d$

(c) $a - 2c \geq b - 2d$

11. For what values of a are the following inequalities valid?

(a) $a \leq a$ (b) $a < a$

12. If $a \leq b$ and $b \leq a$, what can you say about a and b?

13. (a) If $a < b$ is true, does it follow that $a \leq b$ must also be true?

(b) If $a \leq b$ is true, does it follow that $a < b$ must also be true?

14. In each part, list the elements in the set.

(a) $\{x : x^2 - 5x = 0\}$

(b) $\{x : x$ is an integer satisfying $-2 < x < 3\}$

15. In each part, express the set in the notation $\{x : \underline{\hspace{1cm}}\}$.

(a) $\{1, 3, 5, 7, 9, \ldots\}$

(b) the set of even integers

(c) the set of irrational numbers

(d) $\{7, 8, 9, 10\}$

16. Let $A = \{1, 2, 3\}$. Which of the following sets are equal to A?

(a) $\{0, 1, 2, 3\}$ (b) $\{3, 2, 1\}$

(c) $\{x : (x - 3)(x^2 - 3x + 2) = 0\}$

17. In the accompanying figure, let

$S =$ the set of points inside the square

$T =$ the set of points inside the triangle

$C =$ the set of points inside the circle

and let a, b, and c be the points shown. Answer the following as true or false.

(a) $T \subseteq C$

(b) $T \subseteq S$

(c) $a \notin T$

(d) $a \notin S$

(e) $b \in T$ and $b \in C$

(f) $a \in C$ or $a \in T$

(g) $c \in T$ and $c \notin C$

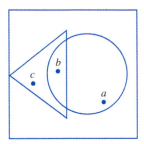

Figure Ex-17

18. List all subsets of

(a) $\{a_1, a_2, a_3\}$

(b) $\varnothing$.

19. In each part, sketch on a coordinate line all values of x that satisfy the stated condition.

(a) $x \le 4$

(b) $x \ge -3$

(c) $-1 \le x \le 7$

(d) $x^2 = 9$

(e) $x^2 \le 9$

(f) $x^2 \ge 9$

20. In parts (a)–(d), sketch on a coordinate line all values of x, if any, that satisfy the stated conditions.

(a) $x > 4$ and $x \le 8$

(b) $x \le 2$ or $x \ge 5$

(c) $x > -2$ and $x \ge 3$

(d) $x \le 5$ and $x > 7$

21. Express in interval notation.

(a) $\{x : x^2 \le 4\}$

(b) $\{x : x^2 > 4\}$

22. In each part, sketch the set on a coordinate line.

(a) $[-3, 2] \cup [1, 4]$

(b) $[4, 6] \cup [8, 11]$

(c) $(-4, 0) \cup (-5, 1)$

(d) $[2, 4) \cup (4, 7)$

(e) $(-2, 4) \cap (0, 5]$

(f) $[1, 2.3) \cup (1.4, \sqrt{2})$

(g) $(-\infty, -1) \cup (-3, +\infty)$

(h) $(-\infty, 5) \cap [0, +\infty)$

In Exercises 23–44, solve the inequality and sketch the solution on a coordinate line.

23. $3x - 2 < 8$

24. $\frac{1}{5}x + 6 \ge 14$

25. $4 + 5x \le 3x - 7$

26. $2x - 1 > 11x + 9$

27. $3 \le 4 - 2x < 7$

28. $-2 \ge 3 - 8x \ge -11$

29. $\frac{x}{x - 3} < 4$

30. $\frac{x}{8 - x} \ge -2$

31. $\frac{3x + 1}{x - 2} < 1$

32. $\frac{\frac{1}{2}x - 3}{4 + x} > 1$

33. $\frac{4}{2 - x} \le 1$

34. $\frac{3}{x - 5} \le 2$

35. $x^2 > 9$

36. $x^2 \le 5$

37. $(x - 4)(x + 2) > 0$

38. $(x - 3)(x + 4) < 0$

39. $x^2 - 9x + 20 \le 0$

40. $2 - 3x + x^2 \ge 0$

41. $\frac{2}{x} < \frac{3}{x - 4}$

42. $\frac{1}{x + 1} \ge \frac{3}{x - 2}$

43. $x^3 - x^2 - x - 2 > 0$

44. $x^3 - 3x + 2 \le 0$

In Exercises 45 and 46, find all values of x for which the given expression yields a real number.

45. $\sqrt{x^2 + x - 6}$

46. $\sqrt{\dfrac{x + 2}{x - 1}}$

47. Fahrenheit and Celsius temperatures are related by the formula $C = \frac{5}{9}(F - 32)$. If the temperature in degrees Celsius ranges over the interval $25 \le C \le 40$ on a certain day, what is the temperature range in degrees Fahrenheit that day?

48. Every integer is either even or odd. The even integers are those that are divisible by 2, so n is even if and only if $n = 2k$ for some integer k. Each odd integer is one unit larger than an even integer, so n is odd if and only if $n = 2k + 1$ for some integer k. Show:

(a) If n is even, then so is n^2

(b) If n is odd, then so is n^2.

49. Prove the following results about sums of rational and irrational numbers:

(a) rational + rational = rational

(b) rational + irrational = irrational.

50. Prove the following results about products of rational and irrational numbers:

(a) rational · rational = rational

(b) rational · irrational = irrational   (provided the rational factor is nonzero).

51. Show that the sum or product of two irrational numbers can be rational or irrational.

52. Classify the following as rational or irrational and justify your conclusion.

(a) $3 + \pi$

(b) $\frac{3}{4}\sqrt{2}$

(c) $\sqrt{8}\sqrt{2}$

(d) $\sqrt{\pi}$

(See Exercises 49 and 50.)

53. Prove: The average of two rational numbers is a rational number, but the average of two irrational numbers can be rational or irrational.

54. Can a rational number satisfy $10^x = 3$?

55. Solve: $8x^3 - 4x^2 - 2x + 1 < 0$.

56. Solve: $12x^3 - 20x^2 \ge -11x + 2$.

57. Prove: If a, b, c, and d are positive numbers such that $a < b$ and $c < d$, then $ac < bd$. (This result gives conditions under which inequalities can be "multiplied together.")

58. Is the number represented by the decimal

$0.101001000100001000001\ldots$

rational or irrational? Explain your reasoning.

ABSOLUTE VALUE

> **B.1** DEFINITION. The *absolute value* or *magnitude* of a real number a is denoted by $|a|$ and is defined by
>
> $$|a| = \begin{cases} a & \text{if} \quad a \geq 0 \\ -a & \text{if} \quad a < 0 \end{cases}$$

Example 1

$$|5| = 5 \qquad \left|-\tfrac{4}{7}\right| = -\left(-\tfrac{4}{7}\right) = \tfrac{4}{7} \qquad |0| = 0 \qquad \blacktriangleleft$$

| Since $5 > 0$ | Since $-\tfrac{4}{7} < 0$ | Since $0 \geq 0$ |

Note that the effect of taking the absolute value of a number is to strip away the minus sign if the number is negative and to leave the number unchanged if it is nonnegative.

Example 2   Solve $|x - 3| = 4$.

Solution. Depending on whether $x - 3$ is positive or negative, the equation $|x - 3| = 4$ can be written as

$$x - 3 = 4 \quad \text{or} \quad x - 3 = -4$$

Solving these two equations gives $x = 7$ and $x = -1$. $\blacktriangleleft$

Example 3   Solve $|3x - 2| = |5x + 4|$.

Solution. Because two numbers with the same absolute value are either equal or differ in sign, the given equation will be satisfied if either

$$3x - 2 = 5x + 4 \quad \text{or} \quad 3x - 2 = -(5x + 4)$$

Solving the first equation yields $x = -3$ and solving the second yields $x = -\tfrac{1}{4}$; thus, the given equation has the solutions $x = -3$ and $x = -\tfrac{1}{4}$. $\blacktriangleleft$

RELATIONSHIP BETWEEN SQUARE ROOTS AND ABSOLUTE VALUES

Recall from algebra that a number is called a *square root* of a if its square is a. Recall also that every positive real number has two square roots, one positive and one negative; the positive square root is denoted by $\sqrt{a}$ and the negative square root by $-\sqrt{a}$. For example, the positive square root of 9 is $\sqrt{9} = 3$, and the negative square root of 9 is $-\sqrt{9} = -3$.

REMARK. Readers who may have been taught to write $\sqrt{9}$ as ± 3 should stop doing so, since it is incorrect.

It is a common error to replace $\sqrt{a^2}$ by a. Although this is correct when a is nonnegative, it is false for negative a. For example, if $a = -4$, then

$$\sqrt{a^2} = \sqrt{(-4)^2} = \sqrt{16} = 4 \neq a$$

A result that is correct for all a is given in the following theorem.

B.2 THEOREM. *For any real number a,*

$$\sqrt{a^2} = |a|$$

Proof. Since $a^2 = (+a)^2 = (-a)^2$, the numbers $+a$ and $-a$ are square roots of a^2. If $a \geq 0$, then $+a$ is the nonnegative square root of a^2, and if $a < 0$, then $-a$ is the nonnegative square root of a^2. Since $\sqrt{a^2}$ denotes the nonnegative square root of a^2, it follows that

$$\sqrt{a^2} = +a \quad \text{if} \quad a \geq 0$$
$$\sqrt{a^2} = -a \quad \text{if} \quad a < 0$$

That is, $\sqrt{a^2} = |a|$. ∎

PROPERTIES OF ABSOLUTE VALUE

B.3 THEOREM. *If a and b are real numbers, then*

(a) $|-a| = |a|$ A number and its negative have the same absolute value.

(b) $|ab| = |a||b|$ The absolute value of a product is the product of the absolute values.

(c) $|a/b| = |a|/|b|$ The absolute value of a ratio is the ratio of the absolute values.

We will prove parts (a) and (b) only.

Proof (a). From Theorem B.2,

$$|-a| = \sqrt{(-a)^2} = \sqrt{a^2} = |a|$$

Proof (b). From Theorem B.2 and a basic property of square roots,

$$|ab| = \sqrt{(ab)^2} = \sqrt{a^2 b^2} = \sqrt{a^2}\sqrt{b^2} = |a||b| \qquad ∎$$

REMARK. In part (c) of Theorem B.3 we did not explicitly state that $b \neq 0$, but this must be so since division by zero is not allowed. Whenever divisions occur in this text, it will be assumed that the denominator is not zero, even if we do not mention it explicitly.

The result in part (b) of Theorem B.3 can be extended to three or more factors. More precisely, for any n real numbers, $a_1, a_2, \ldots, a_n$, it follows that

$$|a_1 a_2 \cdots a_n| = |a_1||a_2| \cdots |a_n| \tag{1}$$

In the special case where $a_1, a_2, \ldots, a_n$ have the same value, a, it follows from (1) that

$$|a^n| = |a|^n \tag{2}$$

GEOMETRIC INTERPRETATION OF ABSOLUTE VALUE

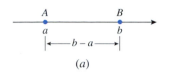

(a)

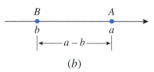

(b)

Figure B.1

The notion of absolute value arises naturally in distance problems. For example, suppose that A and B are points on a coordinate line that have coordinates a and b, respectively. Depending on the relative positions of the points, the distance d between them will be $b - a$ or $a - b$ (Figure B.1). In either case, the distance can be written as $d = |b - a|$, so we have the following result.

> **B.4**   THEOREM (*Distance Formula*).   *If A and B are points on a coordinate line with coordinates a and b, respectively, then the distance d between A and B is $d = |b - a|$.*

This theorem provides useful geometric interpretations of some common mathematical expressions:

EXPRESSION	GEOMETRIC INTERPRETATION ON A COORDINATE LINE
$|x - a|$	The distance between x and a
$|x + a|$	The distance between x and $-a$ (since $|x + a| = |x - (-a)|$)
$|x|$	The distance between x and the origin (since $|x| = |x - 0|$)

INEQUALITIES WITH ABSOLUTE VALUES

Inequalities of the form $|x - a| < k$ and $|x - a| > k$ arise so often that we have summarized the key facts about them in Table 1.

Table 1

INEQUALITY ($k > 0$)	GEOMETRIC INTERPRETATION	FIGURE	ALTERNATIVE FORMS OF THE INEQUALITY
$|x - a| < k$	x is within k units of a.	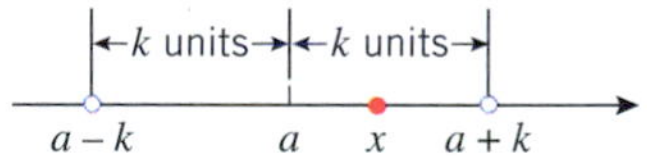 ←k units→←k units→ $a-k$ a x $a+k$	$-k < x - a < k$ $a - k < x < a + k$
$|x - a| > k$	x is more than k units away from a.	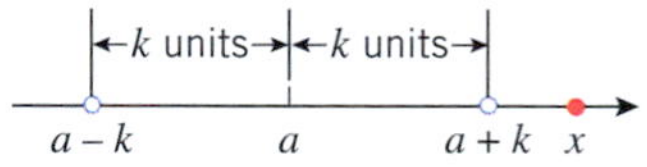 ←k units→←k units→ $a-k$ a $a+k$ x	$x - a < -k$ or $x - a > k$ $x < a - k$ or $x > a + k$

REMARK.   The statements in this table remain true if $<$ is replaced by $\leq$ and $>$ by $\geq$, and if the open dots are replaced by closed dots in the illustrations.

Example 4   Solve

(a) $|x - 3| < 4$ (b) $|x + 4| \geq 2$ (c) $\dfrac{1}{|2x - 3|} > 5$

Solution (a). The inequality $|x - 3| < 4$ can be rewritten as

$$-4 < x - 3 < 4$$

Adding 3 throughout yields

$$-1 < x < 7$$

which can be written in interval notation as $(-1, 7)$. Observe that this solution set consists of all x that are within 4 units of 3 on a number line (Figure B.2), which is consistent with Table 1.

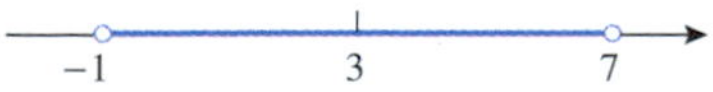

−1   3   7

Figure B.2

Solution (b). The inequality $|x + 4| \geq 2$ will be satisfied if

$$x + 4 \leq -2 \quad \text{or} \quad x + 4 \geq 2$$

Solving for x in the two cases yields

$$x \leq -6 \quad \text{or} \quad x \geq -2$$

which can be expressed in interval notation as

$$(-\infty, -6] \cup [-2, +\infty)$$

Observe that the solution set consists of all x that are at least 2 units away from -4 on a number line (Figure B.3), which is consistent with Table 1 and the remark that follows it.

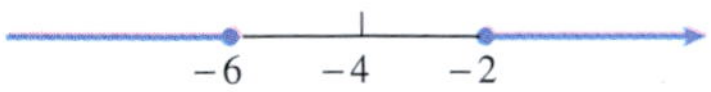

Figure B.3

Solution (c). Observe first that $x = \frac{3}{2}$ results in a division by zero, so this value of x cannot be in the solution set. Putting this aside for the moment, we will begin by taking reciprocals on both sides and reversing the sense of the inequality in accordance with Theorem A.1(e) of Appendix A; then we will use Theorem B.3 to rewrite the inequality $1/|2x - 3| > 5$ in a more familiar form:

$$|2x - 3| < \tfrac{1}{5}$$

$$|2||x - \tfrac{3}{2}| < \tfrac{1}{5} \qquad \boxed{\text{Theorem B.3}(b)}$$

$$|x - \tfrac{3}{2}| < \tfrac{1}{10} \qquad \boxed{\text{We multiplied both sides by } 1/|2| = 1/2.}$$

$$-\tfrac{1}{10} < x - \tfrac{3}{2} < \tfrac{1}{10} \qquad \boxed{\text{Table 1}}$$

$$\tfrac{7}{5} < x < \tfrac{8}{5} \qquad \boxed{\text{We added 3/2 throughout.}}$$

As noted earlier, we must eliminate $x = \frac{3}{2}$ to avoid a division by zero, so the solution set is

$$\tfrac{7}{5} < x < \tfrac{3}{2} \quad \text{or} \quad \tfrac{3}{2} < x < \tfrac{8}{5}$$

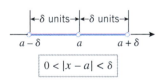

Figure B.4

which can be expressed in interval notation as $\left(\frac{7}{5}, \frac{3}{2}\right) \cup \left(\frac{3}{2}, \frac{8}{5}\right)$. (See Figure B.4.) ◄

AN INEQUALITY FROM CALCULUS

One of the most important inequalities in calculus is

$$0 < |x - a| < \delta \tag{3}$$

where δ (Greek "delta") is a positive real number. This is equivalent to the two inequalities

$$0 < |x - a| \quad \text{and} \quad |x - a| < \delta$$

the first of which is satisfied by all x except $x = a$, and the second of which is satisfied by all x that are within δ units of a on a coordinate line. Combining these two restrictions, we conclude that the solution set of (3) consists of all x in the interval $(a - \delta, a + \delta)$ except $x = a$ (Figure B.5). Stated another way, the solution set of (3) is

$$(a - \delta, a) \cup (a, a + \delta) \tag{4}$$

Figure B.5

THE TRIANGLE INEQUALITY

It is *not* generally true that $|a + b| = |a| + |b|$. For example, if $a = 1$ and $b = -1$, then $|a + b| = 0$, whereas $|a| + |b| = 2$. It is true, however, that *the absolute value of a sum is always less than or equal to the sum of the absolute values*. This is the content of the following useful theorem, called the ***triangle inequality***.

> **B.5**   THEOREM (*Triangle Inequality*).   *If a and b are any real numbers, then*
>
> $$|a + b| \leq |a| + |b| \tag{5}$$

Proof. Observe first that a satisfies the inequality

$$-|a| \leq a \leq |a|$$

because either $a = |a|$ or $a = -|a|$, depending on the sign of a. The corresponding inequality for b is

$$-|b| \le b \le |b|$$

Adding the two inequalities we obtain

$$-(|a| + |b|) \le a + b \le (|a| + |b|) \tag{6}$$

Let us now consider the cases $a + b \ge 0$ and $a + b < 0$ separately. In the first case, $a + b = |a + b|$, so the right-hand inequality in (6) yields the triangle inequality (5). In the second case, $a + b = -|a + b|$, so the left-hand inequality in (6) can be written as

$$-(|a| + |b|) \le -|a + b|$$

which yields the triangle inequality (5) on multiplying by -1. ∎

> **REMARK.**   The name "triangle inequality" arises from a geometric interpretation of the inequality that can be made when a and b are complex numbers. A more detailed explanation is outside the scope of this text.

EXERCISE SET B

1. Compute $|x|$ if
   (a) $x = 7$
   (b) $x = -\sqrt{2}$
   (c) $x = k^2$
   (d) $x = -k^2$.

2. Rewrite $\sqrt{(x - 6)^2}$ without using a square root or absolute value sign.

> In Exercises 3–10, find all values of x for which the given statement is true.

3. $|x - 3| = 3 - x$
4. $|x + 2| = x + 2$
5. $|x^2 + 9| = x^2 + 9$
6. $|x^2 + 5x| = x^2 + 5x$
7. $|3x^2 + 2x| = x|3x + 2|$
8. $|6 - 2x| = 2|x - 3|$
9. $\sqrt{(x + 5)^2} = x + 5$
10. $\sqrt{(3x - 2)^2} = 2 - 3x$
11. Verify $\sqrt{a^2} = |a|$ for $a = 7$ and $a = -7$.
12. Verify the inequalities $-|a| \le a \le |a|$ for $a = 2$ and for $a = -5$.
13. Let A and B be points with coordinates a and b. In each part find the distance between A and B.
   (a) $a = 9, \ b = 7$
   (b) $a = 2, \ b = 3$
   (c) $a = -8, \ b = 6$
   (d) $a = \sqrt{2}, \ b = -3$
   (e) $a = -11, \ b = -4$
   (f) $a = 0, \ b = -5$
14. Is the equality $\sqrt{a^4} = a^2$ valid for all values of a? Explain.
15. Let A and B be points with coordinates a and b. In each part, use the given information to find b.
   (a) $a = -3$, B is to the left of A, and $|b - a| = 6$.
   (b) $a = -2$, B is to the right of A, and $|b - a| = 9$.
   (c) $a = 5$, $|b - a| = 7$, and $b > 0$.
16. Let E and F be points with coordinates e and f. In each part, determine whether E is to the left or to the right of F on a coordinate line.
   (a) $f - e = 4$
   (b) $e - f = 4$
   (c) $f - e = -6$
   (d) $e - f = -7$

> In Exercises 17–24, solve for x.

17. $|6x - 2| = 7$
18. $|3 + 2x| = 11$
19. $|6x - 7| = |3 + 2x|$
20. $|4x + 5| = |8x - 3|$
21. $|9x| - 11 = x$
22. $2x - 7 = |x + 1|$
23. $\left| \dfrac{x + 5}{2 - x} \right| = 6$
24. $\left| \dfrac{x - 3}{x + 4} \right| = 5$

> In Exercises 25–36, solve for x and express the solution in terms of intervals.

25. $|x + 6| < 3$
26. $|7 - x| \le 5$
27. $|2x - 3| \le 6$
28. $|3x + 1| < 4$
29. $|x + 2| > 1$
30. $|\frac{1}{2}x - 1| \ge 2$
31. $|5 - 2x| \ge 4$
32. $|7x + 1| > 3$
33. $\dfrac{1}{|x - 1|} < 2$
34. $\dfrac{1}{|3x + 1|} \ge 5$
35. $\dfrac{3}{|2x - 1|} \ge 4$
36. $\dfrac{2}{|x + 3|} < 1$

37. For which values of x is $\sqrt{\left(x^2 - 5x + 6\right)^2} = x^2 - 5x + 6$?
38. Solve $3 \le |x - 2| \le 7$ for x.
39. Solve $|x - 3|^2 - 4|x - 3| = 12$ for x. [*Hint:* Begin by letting $u = |x - 3|$.]
40. Verify the triangle inequality $|a + b| \le |a| + |b|$ (Theorem B.5) for
   (a) $a = 3, \ b = 4$
   (b) $a = -2, \ b = 6$
   (c) $a = -7, \ b = -8$
   (d) $a = -4, \ b = 4$.
41. Prove: $|a - b| \le |a| + |b|$.
42. Prove: $|a| - |b| \le |a - b|$.
43. Prove: $\big| \, |a| - |b| \, \big| \le |a - b|$. [*Hint:* Use Exercise 42.]

APPENDIX C

Coordinate Planes and Lines

Just as points on a coordinate line can be associated with real numbers, so points in a plane can be associated with pairs of real numbers by introducing a *rectangular coordinate system* (also called a *Cartesian coordinate system*). A rectangular coordinate system consists of two perpendicular coordinate lines, called *coordinate axes*, that intersect at their origins. Usually, but not always, one axis is horizontal with its positive direction to the right, and the other is vertical with its positive direction up. The intersection of the axes is called the *origin* of the coordinate system.

It is common to call the horizontal axis the *x-axis* and the vertical axis the *y-axis*, in which case the plane and the axes together are referred to as the *xy-plane* (Figure C.1). Although labeling the axes with the letters x and y is common, other letters may be more appropriate in specific applications. Figure C.2 shows a uv-plane and a ts-plane—the first letter in the name of the plane always refers to the horizontal axis and the second to the vertical axis.

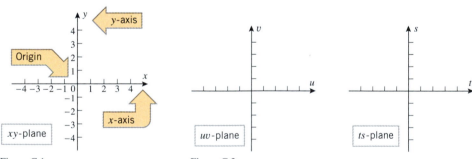

Figure C.1   Figure C.2

Every point P in a coordinate plane can be associated with a unique ordered pair of real numbers by drawing two lines through P, one perpendicular to the x-axis and the other perpendicular to the y-axis (Figure C.3). If the first line intersects the x-axis at the point with coordinate a and the second line intersects the y-axis at the point with coordinate b, then we associate the ordered pair of real numbers (a, b) with the point P. The number a is called the *x-coordinate* or *abscissa* of P and the number b is called the *y-coordinate* or *ordinate* of P. We will say that P has *coordinates* (a, b) and write $P(a, b)$ when we want to emphasize that the coordinates of P are (a, b). We can also reverse the above procedure and find the point P associated with the coordinates (a, b) by locating the intersection of the dashed

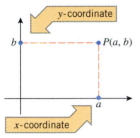

Figure C.3

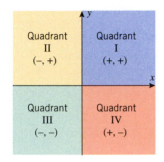

Figure C.4

lines in Figure C.3. Because of this one-to-one correspondence between coordinates and points, we will sometimes blur the distinction between points and ordered pairs of numbers by talking about the *point* (a, b).

REMARK.   Recall that the symbol (a, b) also denotes the open interval between a and b; the appropriate interpretation will usually be clear from the context.

In a rectangular coordinate system the coordinate axes divide the rest of the plane into four regions called **quadrants**. These are numbered counterclockwise with roman numerals as shown in Figure C.4. As indicated in that figure, it is easy to determine the quadrant in which a given point lies from the signs of its coordinates: a point with two positive coordinates $(+, +)$ lies in Quadrant I, a point with a negative x-coordinate and a positive y-coordinate $(-, +)$ lies in Quadrant II, and so forth. Points with a zero x-coordinate lie on the y-axis and points with a zero y-coordinate lie on the x-axis.

To **plot** a point $P(a, b)$ means to locate the point with coordinates (a, b) in a coordinate plane. For example, in Figure C.5 we have plotted the points

$$P(2, 5), \quad Q(-4, 3), \quad R(-5, -2), \quad \text{and} \quad S(4, -3)$$

Observe how the signs of the coordinates identify the quadrants in which the points lie.

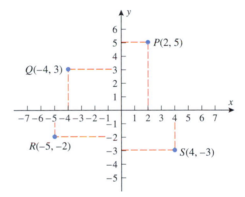

Figure C.5

GRAPHS

The correspondence between points in a plane and ordered pairs of real numbers makes it possible to visualize algebraic equations as geometric curves, and, conversely, to represent geometric curves by algebraic equations. To understand how this is done, suppose that we have an xy-coordinate system and an equation involving two variables x and y, say

$$6x - 4y = 10, \quad y = \sqrt{x}, \quad x = y^3 + 1, \quad \text{or} \quad x^2 + y^2 = 1$$

We define a **solution** of such an equation to be any ordered pair of real numbers (a, b) whose coordinates satisfy the equation when we substitute $x = a$ and $y = b$. For example, the ordered pair $(3, 2)$ is a solution of the equation $6x - 4y = 10$, since the equation is satisfied by $x = 3$ and $y = 2$ (verify). However, the ordered pair $(2, 0)$ is not a solution of this equation, since the equation is not satisfied by $x = 2$ and $y = 0$ (verify).

The following definition makes the association between equations in x and y and curves in the xy-plane.

C.1   DEFINITION.   The set of all solutions of an equation in x and y is called the **solution set** of the equation, and the set of all points in the xy-plane whose coordinates are members of the solution set is called the **graph** of the equation.

One of the main themes in calculus is to identify the exact shape of a graph. Point plotting is one approach to obtaining a graph, but this method has limitations, as discussed in the following example.

Example 1   Sketch the graph of $y = x^2$.

Solution.   The solution set of the equation has infinitely many members, since we can substitute an arbitrary value for x into the right side of $y = x^2$ and compute the associated y to obtain a point (x, y) in the solution set. The fact that the solution set has infinitely many members means that we cannot obtain the *entire* graph of $y = x^2$ by point plotting. However, we can obtain an *approximation* to the graph by plotting some sample members of the solution set and connecting them with a smooth curve, as in Figure C.6. The problem with this method is that we cannot be sure how the graph behaves *between* the plotted points. For example, the curves in Figure C.7 also pass through the plotted points and hence are legitimate candidates for the graph in the absence of additional information. Moreover, even if we use a graphing calculator or a computer program to generate the graph, as in Figure C.8, we have the same problem because graphing technology uses point-plotting algorithms to generate graphs. Indeed, in Section 1.3 of the text we see examples where graphing technology can be fooled into producing grossly inaccurate graphs.   ◄

x	$y = x^2$	(x, y)
0	0	$(0, 0)$
1	1	$(1, 1)$
2	4	$(2, 4)$
3	9	$(3, 9)$
−1	1	$(−1, 1)$
−2	4	$(−2, 4)$
−3	9	$(−3, 9)$

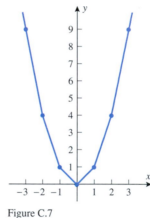

Figure C.6

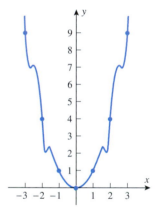

Figure C.7

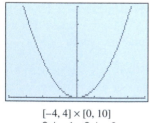

$[-4, 4] \times [0, 10]$
xScl $= 1$, yScl $= 2$

$y = x^2$

Figure C.8

In spite of its limitations, point plotting by hand or with the help of graphing technology can be useful, so here are two more examples.

Example 2   Sketch the graph of $y = \sqrt{x}$.

Solution.   If $x < 0$, then $\sqrt{x}$ is an imaginary number. Thus, we can only plot points for which $x \geq 0$, since points in the xy-plane have real coordinates. Figure C.9 shows the graph obtained by point plotting and a graph obtained with a graphing calculator.   ◄

Example 3   Sketch the graph of $y^2 - 2y - x = 0$.

Solution.   To calculate coordinates of points on the graph of an equation in x and y, it is desirable to have y expressed in terms of x or x in terms of y. In this case it is easier to express x in terms of y, so we rewrite the equation as

$$x = y^2 - 2y$$

Members of the solution set can be obtained from this equation by substituting arbitrary values for y in the right side and computing the associated values of x (Figure C.10).   ◄

x	$y = \sqrt{x}$	(x, y)
0	0	$(0, 0)$
1	1	$(1, 1)$
2	$\sqrt{2}$	$(2, \sqrt{2}) \approx (2, 1.4)$
3	$\sqrt{3}$	$(3, \sqrt{3}) \approx (3, 1.7)$
4	2	$(4, 2)$

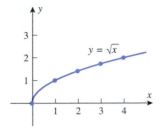

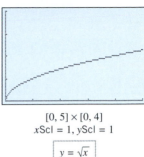

$[0, 5] \times [0, 4]$
xScl $= 1$, yScl $= 1$

$y = \sqrt{x}$

Figure C.9

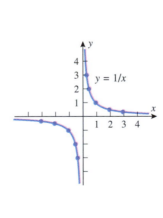

y	$x = y^2 - 2y$	(x, y)
-2	8	$(8, -2)$
-1	3	$(3, -1)$
0	0	$(0, 0)$
1	-1	$(-1, 1)$
2	0	$(0, 2)$
3	3	$(3, 3)$
4	8	$(8, 4)$

Figure C.10

REMARK.   Most graphing calculators and computer graphing programs require that y be expressed in terms of x to generate a graph in the xy-plane. In Section 1.8 we discuss a method for circumventing this restriction.

Example 4   Sketch the graph of $y = 1/x$.

Solution.   Because $1/x$ is undefined at $x = 0$, we can only plot points for which $x \neq 0$. This forces a break, called a *discontinuity*, in the graph at $x = 0$ (Figure C.11).   ◄

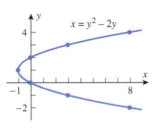

x	$y = 1/x$	(x, y)
$\frac{1}{3}$	3	$\left(\frac{1}{3}, 3\right)$
$\frac{1}{2}$	2	$\left(\frac{1}{2}, 2\right)$
1	1	$(1, 1)$
2	$\frac{1}{2}$	$\left(2, \frac{1}{2}\right)$
3	$\frac{1}{3}$	$\left(3, \frac{1}{3}\right)$
$-\frac{1}{3}$	-3	$\left(-\frac{1}{3}, -3\right)$
$-\frac{1}{2}$	-2	$\left(-\frac{1}{2}, -2\right)$
-1	-1	$(-1, -1)$
-2	$-\frac{1}{2}$	$\left(-2, -\frac{1}{2}\right)$
-3	$-\frac{1}{3}$	$\left(-3, -\frac{1}{3}\right)$

Figure C.11

INTERCEPTS

Points where a graph intersects the coordinate axes are of special interest in many problems. As illustrated in Figure C.12, intersections of a graph with the x-axis have the form $(a, 0)$ and intersections with the y-axis have the form $(0, b)$. The number a is called an ***x-intercept*** of the graph and the number b a ***y-intercept***.

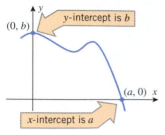

Figure C.12

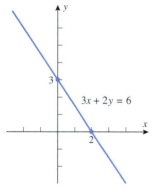

Figure C.13

Example 5   Find all intercepts of

(a) $3x + 2y = 6$ (b) $x = y^2 - 2y$ (c) $y = 1/x$

Solution (a). To find the x-intercepts we set $y = 0$ and solve for x:

$$3x = 6 \quad \text{or} \quad x = 2$$

To find the y-intercepts we set $x = 0$ and solve for y:

$$2y = 6 \quad \text{or} \quad y = 3$$

As we will see later, the graph of $3x + 2y = 6$ is the line shown in Figure C.13.

Solution (b). To find the x-intercepts, set $y = 0$ and solve for x:

$$x = 0$$

Thus, $x = 0$ is the only x-intercept. To find the y-intercepts, set $x = 0$ and solve for y:

$$y^2 - 2y = 0$$

$$y(y - 2) = 0$$

So the y-intercepts are $y = 0$ and $y = 2$. The graph is shown in Figure C.10.

Solution (c). To find the x-intercepts, set $y = 0$:

$$\frac{1}{x} = 0$$

This equation has no solutions (why?), so there are no x-intercepts. To find y-intercepts we would set $x = 0$ and solve for y. But, substituting $x = 0$ leads to a division by zero, which is not allowed, so there are no y-intercepts either. The graph of the equation is shown in Figure C.11.   ◀

SLOPE

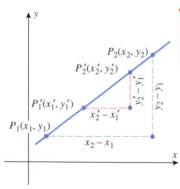

Figure C.14

To obtain equations of lines we will first need to discuss the concept of *slope*, which is a numerical measure of the "steepness" of a line.

Consider a particle moving left to right along a *nonvertical* line from a point $P_1(x_1, y_1)$ to a point $P_2(x_2, y_2)$. As shown in Figure C.14, the particle moves $y_2 - y_1$ units in the y-direction as it travels $x_2 - x_1$ units in the positive x-direction. The vertical change $y_2 - y_1$ is called the **rise**, and the horizontal change $x_2 - x_1$ the **run**. The ratio of the rise over the run can be used to measure the steepness of the line, which leads us to the following definition.

> **C.2   DEFINITION.**   If $P_1(x_1, y_1)$ and $P_2(x_2, y_2)$ are points on a nonvertical line, then the **slope** m of the line is defined by
> $$m = \frac{\text{rise}}{\text{run}} = \frac{y_2 - y_1}{x_2 - x_1} \qquad (1)$$

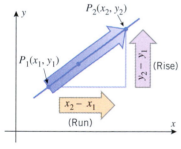

Figure C.15

REMARK.   Observe that this definition does not apply to vertical lines. For such lines we have $x_2 = x_1$ (a zero run), which means that the formula for m involves a division by zero. For this reason, the slope of a vertical line is **undefined**, which is sometimes described informally by stating that a vertical line has **infinite slope**.

When calculating the slope of a nonvertical line from Formula (1), it does not matter which two points on the line you use for the calculation, as long as they are distinct. This can be proved using Figure C.15 and similar triangles to show that

$$m = \frac{y_2 - y_1}{x_2 - x_1} = \frac{y_2' - y_1'}{x_2' - x_1'}$$

Moreover, once you choose two points to use for the calculation, it does not matter which one you call P_1 and which one you call P_2 because reversing the points reverses the sign of both the numerator and denominator of (1) and hence has no effect on the ratio.

Example 6   In each part find the slope of the line through

(a)   the points $(6, 2)$ and $(9, 8)$

(b)   the points $(2, 9)$ and $(4, 3)$

(c)   the points $(-2, 7)$ and $(5, 7)$.

Solution.

(a) $m = \dfrac{8 - 2}{9 - 6} = \dfrac{6}{3} = 2$ (b) $m = \dfrac{3 - 9}{4 - 2} = \dfrac{-6}{2} = -3$ (c) $m = \dfrac{7 - 7}{5 - (-2)} = 0$ ◄

Example 7   Figure C.16 shows the three lines determined by the points in Example 6 and explains the significance of their slopes.   ◄

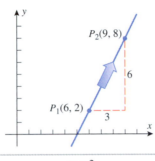

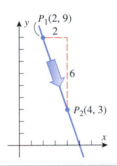

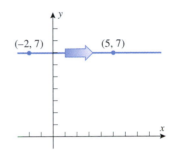

$m = 2$
Traveling left to right, a point on the line rises two units for each unit it moves in the positive x-direction.

$m = -3$
Traveling left to right, a point on the line falls three units for each unit it moves in the positive x-direction.

$m = 0$
Traveling left to right, a point on the line neither rises nor falls.

Figure C.16

As illustrated in this example, the slope of a line can be positive, negative, or zero. A positive slope means that the line is inclined upward to the right, a negative slope means that the line is inclined downward to the right, and a zero slope means that the line is horizontal. An undefined slope means that the line is vertical. Figure C.17 shows various lines through the origin with their slopes.

The following theorem shows how slopes can be used to tell whether two lines are parallel or perpendicular.

PARALLEL AND PERPENDICULAR LINES

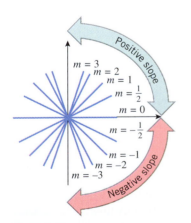

Figure C.17

C.3   THEOREM.

(a)   *Two nonvertical lines with slopes m_1 and m_2 are parallel if and only if they have the same slope, that is,*

$$m_1 = m_2$$

(b)   *Two nonvertical lines with slopes m_1 and m_2 are perpendicular if and only if the product of their slopes is -1, that is,*

$$m_1 m_2 = -1$$

This relationship can also be expressed as $m_1 = -1/m_2$ or $m_2 = -1/m_1$, which states that nonvertical lines are perpendicular if and only if their slopes are negative reciprocals of one another.

A complete proof of this theorem is a little tedious, but it is not hard to motivate the results informally. Let us start with part (a).

Suppose that L_1 and L_2 are nonvertical parallel lines with slopes m_1 and m_2, respectively. If the lines are parallel to the x-axis, then $m_1 = m_2 = 0$, and we are done. If they are not parallel to the x-axis, then both lines intersect the x-axis; and for simplicity assume that they are oriented as in Figure C.18a. On each line choose the point whose run relative to the point of intersection with the x-axis is 1. On line L_1 the corresponding rise will be m_1 and on L_2 it will be m_2. However, because the lines are parallel, the shaded triangles in the figure must be congruent (verify), so $m_1 = m_2$. Conversely, the condition $m_1 = m_2$ can be used to show that the shaded triangles are congruent, from which it follows that the lines make the same angle with the x-axis and hence are parallel (verify).

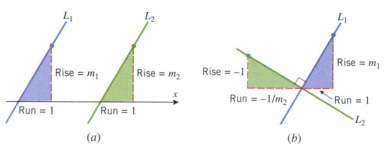

(a)   (b)

Figure C.18

Now suppose that L_1 and L_2 are nonvertical perpendicular lines with slopes m_1 and m_2, respectively; and for simplicity assume that they are oriented as in Figure C.18b. On line L_1 choose the point whose run relative to the point of intersection of the lines is 1, in which case the corresponding rise will be m_1; and on line L_2 choose the point whose rise relative to the point of intersection is -1, in which case the corresponding run will be $-1/m_2$. Because the lines are perpendicular, the shaded triangles in the figure must be congruent (verify), and hence the ratios of corresponding sides of the triangles must be equal. Taking into account that for line L_2 the vertical side of the triangle has length 1 and the horizontal side has length $-1/m_2$ (since m_2 is negative), the congruence of the triangles implies that $m_1/1 = (-1/m_2)/1$ or $m_1 m_2 = -1$. Conversely, the condition $m_1 = -1/m_2$ can be used to show that the shaded triangles are congruent, from which it can be deduced that the lines are perpendicular (verify).

Example 8   Use slopes to show that the points $A(1, 3)$, $B(3, 7)$, and $C(7, 5)$ are vertices of a right triangle.

Solution.   We will show that the line through A and B is perpendicular to the line through B and C. The slopes of these lines are

$$m_1 = \frac{7-3}{3-1} = 2 \quad \text{and} \quad m_2 = \frac{5-7}{7-3} = -\frac{1}{2}$$

| Slope of the line through A and B | | Slope of the line through B and C |

Since $m_1 m_2 = -1$, the line through A and B is perpendicular to the line through B and C; thus, ABC is a right triangle (Figure C.19).   ◄

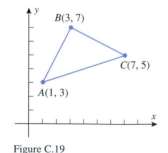

Figure C.19

LINES PARALLEL TO THE COORDINATE AXES

We now turn to the problem of finding equations of lines that satisfy specified conditions. The simplest cases are lines parallel to the coordinate axes. A line parallel to the y-axis intersects the x-axis at some point $(a, 0)$. This line consists precisely of those points whose x-coordinates equal a (Figure C.20). Similarly, a line parallel to the x-axis intersects the y-axis at some point $(0, b)$. This line consists precisely of those points whose y-coordinates equal b (Figure C.20). Thus, we have the following theorem.

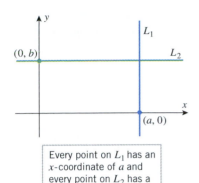

Every point on L_1 has an x-coordinate of a and every point on L_2 has a y-coordinate of b.

Figure C.20

C.4   THEOREM.   *The vertical line through $(a, 0)$ and the horizontal line through $(0, b)$ are represented, respectively, by the equations*

$$x = a \qquad and \qquad y = b$$

Example 9   The graph of $x = -5$ is the vertical line through $(-5, 0)$, and the graph of $y = 7$ is the horizontal line through $(0, 7)$ (Figure C.21).   ◀

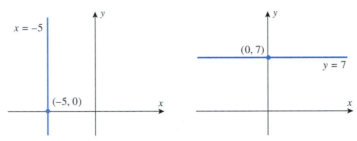

Figure C.21

LINES DETERMINED BY POINT AND SLOPE

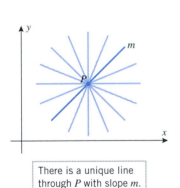

There is a unique line through P with slope m.

Figure C.22

There are infinitely many lines that pass through any given point in the plane. However, if we specify the slope of the line in addition to a point on it, then the point and the slope together determine a unique line (Figure C.22).

Let us now consider how to find an equation of a nonvertical line L that passes through a point $P_1(x_1, y_1)$ and has slope m. If $P(x, y)$ is any point on L, different from P_1, then the slope m can be obtained from the points $P(x, y)$ and $P_1(x_1, y_1)$; this gives

$$m = \frac{y - y_1}{x - x_1}$$

which can be rewritten as

$$y - y_1 = m(x - x_1) \tag{2}$$

With the possible exception of (x_1, y_1), we have shown that every point on L satisfies (2). But $x = x_1$, $y = y_1$ satisfies (2), so that all points on L satisfy (2). We leave it as an exercise to show that every point satisfying (2) lies on L.

In summary, we have the following theorem.

C.5   THEOREM.   *The line passing through $P_1(x_1, y_1)$ and having slope m is given by the equation*

$$y - y_1 = m(x - x_1) \tag{3}$$

*This is called the **point-slope form** of the line.*

Example 10   Find the point-slope form of the line through $(4, -3)$ with slope 5.

Solution.   Substituting the values $x_1 = 4$, $y_1 = -3$, and $m = 5$ in (3) yields the point-slope form $y + 3 = 5(x - 4)$.   ◀

LINES DETERMINED BY SLOPE AND y-INTERCEPT

A nonvertical line crosses the y-axis at some point $(0, b)$. If we use this point in the point-slope form of its equation, we obtain

$$y - b = m(x - 0)$$

which we can rewrite as $y = mx + b$. To summarize:

> **C.6**   THEOREM.   *The line with y-intercept b and slope m is given by the equation*
>
> $$y = mx + b \qquad\qquad\qquad\qquad\qquad (4)$$
>
> *This is called the **slope-intercept form** of the line.*

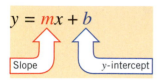

Figure C.23

REMARK.   Note that y is alone on one side of Equation (4). When the equation of a line is written in this way the slope of the line and its y-intercept can be determined by inspection of the equation—the slope is the coefficient of x and the y-intercept is the constant term (Figure C.23).

Example 11

EQUATION	SLOPE	y-INTERCEPT
$y = 3x + 7$	$m = 3$	$b = 7$
$y = -x + \frac{1}{2}$	$m = -1$	$b = \frac{1}{2}$
$y = x$	$m = 1$	$b = 0$
$y = \sqrt{2}x - 8$	$m = \sqrt{2}$	$b = -8$
$y = 2$	$m = 0$	$b = 2$

◀

Example 12   Find the slope-intercept form of the equation of the line that satisfies the stated conditions:

(a)   slope is -9; crosses the y-axis at $(0, -4)$

(b)   slope is 1; passes through the origin

(c)   passes through $(5, -1)$; perpendicular to $y = 3x + 4$

(d)   passes through $(3, 4)$ and $(2, -5)$.

Solution (a). From the given conditions we have $m = -9$ and $b = -4$, so (4) yields $y = -9x - 4$.

Solution (b). From the given conditions $m = 1$ and the line passes through $(0, 0)$, so $b = 0$. Thus, it follows from (4) that $y = x + 0$ or $y = x$.

Solution (c). The given line has slope 3, so the line to be determined will have slope $m = -\frac{1}{3}$. Substituting this slope and the given point in the point-slope form (3) and then simplifying yields

$$y - (-1) = -\tfrac{1}{3}(x - 5)$$

$$y = -\tfrac{1}{3}x + \tfrac{2}{3}$$

Solution (d). We will first find the point-slope form, then solve for y in terms of x to obtain the slope-intercept form. From the given points the slope of the line is

$$m = \frac{-5 - 4}{2 - 3} = 9$$

We can use either of the given points for (x_1, y_1) in (3). We will use $(3, 4)$. This yields the point-slope form

$$y - 4 = 9(x - 3)$$

Solving for y in terms of x yields the slope-intercept form

$$y = 9x - 23$$

We leave it for the reader to show that the same equation results if $(2, -5)$ rather than $(3, 4)$ is used for (x_1, y_1) in (3). ◀

THE GENERAL EQUATION OF A LINE

An equation that is expressible in the form

$$Ax + By + C = 0 \qquad (5)$$

where A, B, and C are constants and A and B are not both zero, is called a ***first-degree equation*** in x and y. For example,

$$4x + 6y - 5 = 0$$

is a first-degree equation in x and y since it has form (5) with

$$A = 4, \quad B = 6, \quad C = -5$$

In fact, all the equations of lines studied in this section are first-degree equations in x and y.

The following theorem states that the first-degree equations in x and y are precisely the equations whose graphs in the xy-plane are straight lines.

C.7 THEOREM. *Every first-degree equation in x and y has a straight line as its graph and, conversely, every straight line can be represented by a first-degree equation in x and y.*

Because of this theorem, (5) is sometimes called the ***general equation*** of a line or a ***linear equation*** in x and y.

Example 13 Graph the equation $3x - 4y + 12 = 0$.

Solution. Since this is a linear equation in x and y, its graph is a straight line. Thus, to sketch the graph we need only plot any two points on the graph and draw the line through them. It is particularly convenient to plot the points where the line crosses the coordinate axes. These points are $(0, 3)$ and $(-4, 0)$ (verify), so the graph is the line in Figure C.24. ◀

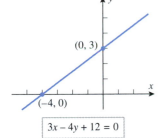

$$3x - 4y + 12 = 0$$

Figure C.24

Example 14 Find the slope of the line in Example 13.

Solution. Solving the equation for y yields

$$y = \tfrac{3}{4}x + 3$$

which is the slope-intercept form of the line. Thus, the slope is $m = \tfrac{3}{4}$. ◀

EXERCISE SET C

1. Draw the rectangle, three of whose vertices are $(6, 1)$, $(-4, 1)$, and $(6, 7)$, and find the coordinates of the fourth vertex.

2. Draw the triangle whose vertices are $(-3, 2)$, $(5, 2)$, and $(4, 3)$, and find its area.

In Exercises 3 and 4, draw a rectangular coordinate system and sketch the set of points whose coordinates (x, y) satisfy the given conditions.

3. (a) $x = 2$ (b) $y = -3$ (c) $x \geq 0$
   (d) $y = x$ (e) $y \geq x$ (f) $|x| \geq 1$

4. (a) $x = 0$ (b) $y = 0$
   (c) $y < 0$ (d) $x \geq 1$ and $y \leq 2$
   (e) $x = 3$ (f) $|x| = 5$

In Exercises 5–12, sketch the graph of the equation. (A calculating utility will be helpful in some of these problems.)

5. $y = 4 - x^2$

6. $y = 1 + x^2$

7. $y = \sqrt{x} - 4$

8. $y = -\sqrt{x+1}$

9. $x^2 - x + y = 0$

10. $x = y^3 - y^2$

11. $x^2 y = 2$

12. $xy = -1$

13. Find the slope of the line through
(a) $(-1, 2)$ and $(3, 4)$ (b) $(5, 3)$ and $(7, 1)$
(c) $(4, \sqrt{2})$ and $(-3, \sqrt{2})$ (d) $(-2, -6)$ and $(-2, 12)$.

14. Find the slopes of the sides of the triangle with vertices $(-1, 2)$, $(6, 5)$, and $(2, 7)$.

15. Use slopes to determine whether the given points lie on the same line.
(a) $(1, 1)$, $(-2, -5)$, and $(0, -1)$
(b) $(-2, 4)$, $(0, 2)$, and $(1, 5)$

16. Draw the line through $(4, 2)$ with slope
(a) $m = 3$ (b) $m = -2$ (c) $m = -\frac{3}{4}$.

17. Draw the line through $(-1, -2)$ with slope
(a) $m = \frac{3}{5}$ (b) $m = -1$ (c) $m = \sqrt{2}$.

18. An equilateral triangle has one vertex at the origin, another on the x-axis, and the third in the first quadrant. Find the slopes of its sides.

19. List the lines in the accompanying figure in the order of increasing slope.

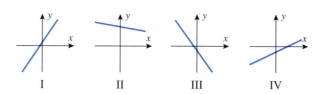

I   II   III   IV

20. List the lines in the accompanying figure in the order of increasing slope.

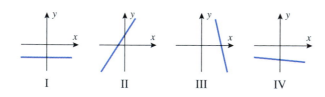

I   II   III   IV

21. A particle, initially at $(1, 2)$, moves along a line of slope $m = 3$ to a new position (x, y).
(a) Find y if $x = 5$.   (b) Find x if $y = -2$.

22. A particle, initially at $(7, 5)$, moves along a line of slope $m = -2$ to a new position (x, y).
(a) Find y if $x = 9$.   (b) Find x if $y = 12$.

23. Let the point $(3, k)$ lie on the line of slope $m = 5$ through $(-2, 4)$; find k.

24. Given that the point $(k, 4)$ is on the line through $(1, 5)$ and $(2, -3)$, find k.

25. Find x if the slope of the line through $(1, 2)$ and $(x, 0)$ is the negative of the slope of the line through $(4, 5)$ and $(x, 0)$.

26. Find x and y if the line through $(0, 0)$ and (x, y) has slope $\frac{1}{2}$, and the line through (x, y) and $(7, 5)$ has slope 2.

27. Use slopes to show that $(3, -1)$, $(6, 4)$, $(-3, 2)$, and $(-6, -3)$ are vertices of a parallelogram.

28. Use slopes to show that $(3, 1)$, $(6, 3)$, and $(2, 9)$ are vertices of a right triangle.

29. Graph the equations
(a) $2x + 5y = 15$ (b) $x = 3$
(c) $y = -2$ (d) $y = 2x - 7$.

30. Graph the equations
(a) $\dfrac{x}{3} - \dfrac{y}{4} = 1$ (b) $x = -8$
(c) $y = 0$ (d) $x = 3y + 2$.

31. Graph the equations
(a) $y = 2x - 1$ (b) $y = 3$
(c) $y = -2x$.

32. Graph the equations
(a) $y = 2 - 3x$ (b) $y = \frac{1}{4}x$
(c) $y = -\sqrt{3}$.

33. Find the slope and y-intercept of
(a) $y = 3x + 2$ (b) $y = 3 - \frac{1}{4}x$
(c) $3x + 5y = 8$ (d) $y = 1$
(e) $\dfrac{x}{a} + \dfrac{y}{b} = 1$.

34. Find the slope and y-intercept of
(a) $y = -4x + 2$ (b) $x = 3y + 2$
(c) $\dfrac{x}{2} + \dfrac{y}{3} = 1$ (d) $y - 3 = 0$
(e) $a_0 x + a_1 y = 0$ $(a_1 \neq 0)$.

In Exercises 35 and 36, use the graph to find the equation of the line in slope-intercept form.

35.

(a)   *(b)*

Figure Ex-35

36.

(a)   *(b)*

Figure Ex-36

In Exercises 37–48, find the slope-intercept form of the line satisfying the given conditions.

37. Slope $= -2$, y-intercept $= 4$.

38. $m = 5$, $b = -3$.

39. The line is parallel to $y = 4x - 2$ and its y-intercept is 7.

40. The line is parallel to $3x + 2y = 5$ and passes through $(-1, 2)$.

41. The line is perpendicular to $y = 5x + 9$ and its y-intercept is 6.

42. The line is perpendicular to $x - 4y = 7$ and passes through $(3, -4)$.

43. The line passes through $(2, 4)$ and $(1, -7)$.

44. The line passes through $(-3, 6)$ and $(-2, 1)$.

45. The y-intercept is 2 and the x-intercept is -4.

46. The y-intercept is b and the x-intercept is a.

47. The line is perpendicular to the y-axis and passes through $(-4, 1)$.

48. The line is parallel to $y = -5$ and passes through $(-1, -8)$.

49. In each part, classify the lines as parallel, perpendicular, or neither.
(a) $y = 4x - 7$ and $y = 4x + 9$
(b) $y = 2x - 3$ and $y = 7 - \frac{1}{2}x$
(c) $5x - 3y + 6 = 0$ and $10x - 6y + 7 = 0$
(d) $Ax + By + C = 0$ and $Bx - Ay + D = 0$
(e) $y - 2 = 4(x - 3)$ and $y - 7 = \frac{1}{4}(x - 3)$

50. In each part, classify the lines as parallel, perpendicular, or neither.
(a) $y = -5x + 1$ and $y = 3 - 5x$
(b) $y - 1 = 2(x - 3)$ and $y - 4 = -\frac{1}{2}(x + 7)$
(c) $4x + 5y + 7 = 0$ and $5x - 4y + 9 = 0$
(d) $Ax + By + C = 0$ and $Ax + By + D = 0$
(e) $y = \frac{1}{2}x$ and $x = \frac{1}{2}y$

51. For what value of k will the line $3x + ky = 4$
(a) have slope 2
(b) have y-intercept 5
(c) pass through the point $(-2, 4)$
(d) be parallel to the line $2x - 5y = 1$
(e) be perpendicular to the line $4x + 3y = 2$?

52. Sketch the graph of $y^2 = 3x$ and explain how this graph is related to the graphs of $y = \sqrt{3x}$ and $y = -\sqrt{3x}$.

53. Sketch the graph of $(x - y)(x + y) = 0$ and explain how it is related to the graphs of $x - y = 0$ and $x + y = 0$.

54. Graph $F = \frac{9}{5}C + 32$ in a CF-coordinate system.

55. Graph $u = 3v^2$ in a uv-coordinate system.

56. Graph $Y = 4X + 5$ in a YX-coordinate system.

57. A point moves in the xy-plane in such a way that at any time t its coordinates are given by $x = 5t + 2$ and $y = t - 3$. By expressing y in terms of x, show that the point moves along a straight line.

58. A point moves in the xy-plane in such a way that at any time t its coordinates are given by $x = 1 + 3t^2$ and $y = 2 - t^2$. By expressing y in terms of x, show that the point moves along a straight-line path and specify the values of x for which the equation is valid.

59. Find the area of the triangle formed by the coordinate axes and the line through $(1, 4)$ and $(2, 1)$.

60. Draw the graph of $4x^2 - 9y^2 = 0$.

61. In each part, name an appropriate coordinate system for graphing the equation [e.g., an $\alpha\beta$-coordinate system in part (a)], and state whether the graph of the equation is a line in that coordinate system.
(a) $3\alpha - 2\beta = 5$
(b) $A = 2000(1 + 0.06t)$
(c) $A = \pi r^2$
(d) $E = mc^2$ (c constant)
(e) $V = C(1 - rt)$ (r and C constant)
(f) $V = \frac{1}{3}\pi r^2 h$ (r constant)
(g) $V = \frac{1}{3}\pi r^2 h$ (h constant)

APPENDIX D

Distance, Circles, and Quadratic Equations

DISTANCE BETWEEN TWO POINTS IN THE PLANE

Suppose that we are interested in finding the distance d between two points $P_1(x_1, y_1)$ and $P_2(x_2, y_2)$ in the xy-plane. If, as in Figure D.1, we form a right triangle with P_1 and P_2 as vertices, then it follows from Theorem B.4 in Appendix B that the sides of that triangle have lengths $|x_2 - x_1|$ and $|y_2 - y_1|$. Thus, it follows from the Theorem of Pythagoras that

$$d = \sqrt{|x_2 - x_1|^2 + |y_2 - y_1|^2} = \sqrt{(x_2 - x_1)^2 + (y_2 - y_1)^2}$$

and hence we have the following result.

D.1 THEOREM. *The distance d between two points $P_1(x_1, y_1)$ and $P_2(x_2, y_2)$ in a coordinate plane is given by*

$$d = \sqrt{(x_2 - x_1)^2 + (y_2 - y_1)^2} \tag{1}$$

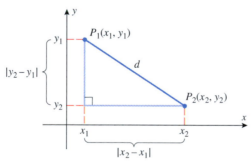

Figure D.1

REMARK. To apply Formula (1) the scales on the coordinate axes must be the same; otherwise, we would not have been able to use the Theorem of Pythagoras in the derivation. Moreover, when using Formula (1) it does not matter which point is labeled P_1 and which one is labeled P_2, since reversing the points changes the signs of $x_2 - x_1$ and $y_2 - y_1$; this has no effect on the value of d because these quantities are squared in the formula. When it is important to emphasize the points, the distance between P_1 and P_2 is denoted by $d(P_1, P_2)$ or $d(P_2, P_1)$.

Example 1 Find the distance between the points $(-2, 3)$ and $(1, 7)$.

Solution. If we let (x_1, y_1) be $(-2, 3)$ and let (x_2, y_2) be $(1, 7)$, then (1) yields

$$d = \sqrt{[1 - (-2)]^2 + [7 - 3]^2} = \sqrt{3^2 + 4^2} = \sqrt{25} = 5$$ ◀

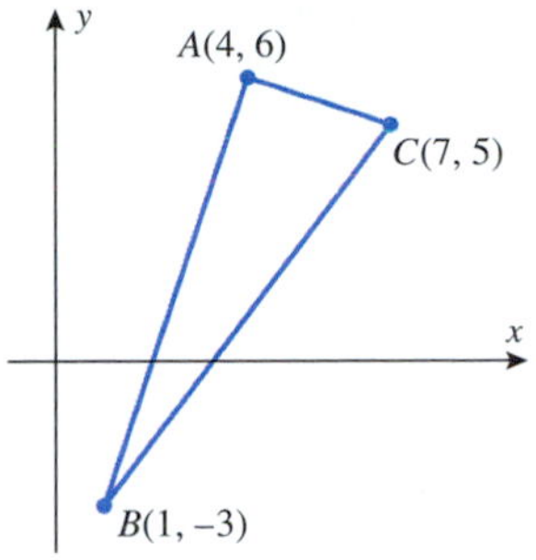

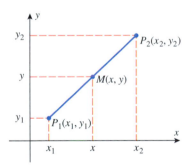

Figure D.2

Example 2 It can be shown that the converse of the Theorem of Pythagoras is true; that is, if the sides of a triangle satisfy the relationship $a^2 + b^2 = c^2$, then the triangle must be a right triangle. Use this result to show that the points $A(4, 6)$, $B(1, -3)$, and $C(7, 5)$ are vertices of a right triangle.

Solution. The points and the triangle are shown in Figure D.2. From (1), the lengths of the sides of the triangles are

$$d(A, B) = \sqrt{(1-4)^2 + (-3-6)^2} = \sqrt{9 + 81} = \sqrt{90}$$

$$d(A, C) = \sqrt{(7-4)^2 + (5-6)^2} = \sqrt{9 + 1} = \sqrt{10}$$

$$d(B, C) = \sqrt{(7-1)^2 + [5-(-3)]^2} = \sqrt{36 + 64} = \sqrt{100} = 10$$

Since

$$[d(A, B)]^2 + [d(A, C)]^2 = [d(B, C)]^2$$

it follows that $\triangle ABC$ is a right triangle with hypotenuse BC. ◀

THE MIDPOINT FORMULA

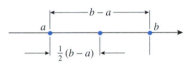

Figure D.3

It is often necessary to find the coordinates of the midpoint of a line segment joining two points in the plane. To derive the midpoint formula, we will start with two points on a coordinate line. If we assume that the points have coordinates a and b and that $a \le b$, then, as shown in Figure D.3, the distance between a and b is $b - a$, and the coordinate of the midpoint between a and b is

$$a + \tfrac{1}{2}(b - a) = \tfrac{1}{2}a + \tfrac{1}{2}b = \tfrac{1}{2}(a + b)$$

which is the arithmetic average of a and b. Had the points been labeled with $b \le a$, the same formula would have resulted (verify). Therefore, *the midpoint of two points on a coordinate line is the arithmetic average of their coordinates, regardless of their relative positions.*

If we now let $P_1(x_1, y_1)$ and $P_2(x_2, y_2)$ be any two points in the plane and $M(x, y)$ the midpoint of the line segment joining them (Figure D.4), then it can be shown using similar triangles that x is the midpoint of x_1 and x_2 on the x-axis and y is the midpoint of y_1 and y_2 on the y-axis, so

$$x = \tfrac{1}{2}(x_1 + x_2) \quad \text{and} \quad y = \tfrac{1}{2}(y_1 + y_2)$$

Thus, we have the following result.

Figure D.4

D.2 THEOREM (*The Midpoint Formula*). *The midpoint of the line segment joining two points (x_1, y_1) and (x_2, y_2) in a coordinate plane is*

$$\left(\tfrac{1}{2}(x_1 + x_2), \tfrac{1}{2}(y_1 + y_2)\right) \tag{2}$$

Example 3 Find the midpoint of the line segment joining $(3, -4)$ and $(7, 2)$.

Solution. From (2) the midpoint is

$$\left(\tfrac{1}{2}(3 + 7), \tfrac{1}{2}(-4 + 2)\right) = (5, -1) \quad ◀$$

CIRCLES

If (x_0, y_0) is a fixed point in the plane, then the circle of radius r centered at (x_0, y_0) is the set of all points in the plane whose distance from (x_0, y_0) is r (Figure D.5). Thus, a point (x, y) will lie on this circle if and only if

$$\sqrt{(x - x_0)^2 + (y - y_0)^2} = r$$

or equivalently,

$$(x - x_0)^2 + (y - y_0)^2 = r^2 \tag{3}$$

This is called the *standard form of the equation of a circle.*

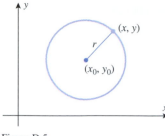

Figure D.5

Example 4 Find an equation for the circle of radius 4 centered at $(-5, 3)$.

Solution. From (3) with $x_0 = -5$, $y_0 = 3$, and $r = 4$ we obtain

$$(x + 5)^2 + (y - 3)^2 = 16$$

If desired, this equation can be written in an expanded form by squaring the terms and then simplifying:

$$(x^2 + 10x + 25) + (y^2 - 6y + 9) - 16 = 0$$
$$x^2 + y^2 + 10x - 6y + 18 = 0 \qquad \blacktriangleleft$$

Example 5 Find an equation for the circle with center $(1, -2)$ that passes through $(4, 2)$.

Solution. The radius r of the circle is the distance between $(4, 2)$ and $(1, -2)$, so

$$r = \sqrt{(1 - 4)^2 + (-2 - 2)^2} = 5$$

We now know the center and radius, so we can use (3) to obtain the equation

$$(x - 1)^2 + (y + 2)^2 = 25 \quad \text{or} \quad x^2 + y^2 - 2x + 4y - 20 = 0 \qquad \blacktriangleleft$$

FINDING THE CENTER AND RADIUS OF A CIRCLE

When you encounter an equation of form (3), you will know immediately that its graph is a circle; its center and radius can then be found from the constants that appear in the equation:

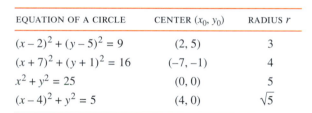

Example 6

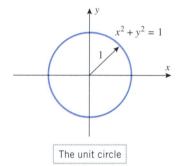

The unit circle

Figure D.6

EQUATION OF A CIRCLE	CENTER (x_0, y_0)	RADIUS r
$(x - 2)^2 + (y - 5)^2 = 9$	$(2, 5)$	3
$(x + 7)^2 + (y + 1)^2 = 16$	$(-7, -1)$	4
$x^2 + y^2 = 25$	$(0, 0)$	5
$(x - 4)^2 + y^2 = 5$	$(4, 0)$	$\sqrt{5}$

$\blacktriangleleft$

The circle $x^2 + y^2 = 1$, which is centered at the origin and has radius 1, is of special importance; it is called the **unit circle** (Figure D.6).

OTHER FORMS FOR THE EQUATION OF A CIRCLE

An alternative version of Equation (3) can be obtained by squaring the terms and simplifying. This yields an equation of the form

$$x^2 + y^2 + dx + ey + f = 0 \qquad (4)$$

where d, e, and f are constants. (See the final equations in Examples 4 and 5.)

Still another version of the equation of a circle can be obtained by multiplying both sides of (4) by a nonzero constant A. This yields an equation of the form

$$Ax^2 + Ay^2 + Dx + Ey + F = 0 \qquad (5)$$

where A, D, E, and F are constants and $A \neq 0$.

If the equation of a circle is given by (4) or (5), then the center and radius can be found by first rewriting the equation in standard form, then reading off the center and radius from that equation. The following example shows how to do this using the technique of **completing the square**. In preparation for the example, recall that completing the square is a method for rewriting an expression of the form

$$x^2 + bx$$

as a difference of two squares. The procedure is to take half the coefficient of x, square it, and then add and subtract that result from the original expression to obtain

$$x^2 + bx = x^2 + bx + (b/2)^2 - (b/2)^2 = [x + (b/2)]^2 - (b/2)^2$$

Example 7   Find the center and radius of the circle with equation

(a) $x^2 + y^2 - 8x + 2y + 8 = 0$ (b) $2x^2 + 2y^2 + 24x - 81 = 0$

Solution (a). First, group the x-terms, group the y-terms, and take the constant to the right side:

$$(x^2 - 8x) + (y^2 + 2y) = -8$$

Next we want to add the appropriate constant within each set of parentheses to complete the square, and subtract the same constant outside the parentheses to maintain equality. The appropriate constant is obtained by taking half the coefficient of the first-degree term and squaring it. This yields

$$(x^2 - 8x + 16) - 16 + (y^2 + 2y + 1) - 1 = -8$$

from which we obtain

$$(x - 4)^2 + (y + 1)^2 = -8 + 16 + 1 \quad \text{or} \quad (x - 4)^2 + (y + 1)^2 = 9$$

Thus from (3), the circle has center $(4, -1)$ and radius 3.

Solution (b). The given equation is of form (5) with $A = 2$. We will first divide through by 2 (the coefficient of the squared terms) to reduce the equation to form (4). Then we will proceed as in part (a) of this example. The computations are as follows:

$$x^2 + y^2 + 12x - \tfrac{81}{2} = 0 \quad \boxed{\text{We divided through by 2.}}$$

$$(x^2 + 12x) + y^2 = \tfrac{81}{2}$$

$$(x^2 + 12x + 36) + y^2 = \tfrac{81}{2} + 36 \quad \boxed{\text{We completed the square.}}$$

$$(x + 6)^2 + y^2 = \tfrac{153}{2}$$

From (3), the circle has center $(-6, 0)$ and radius $\sqrt{\tfrac{153}{2}}$.   ◀

DEGENERATE CASES OF A CIRCLE

There is no guarantee that an equation of form (5) represents a circle. For example, suppose that we divide both sides of (5) by A, then complete the squares to obtain

$$(x - x_0)^2 + (y - y_0)^2 = k$$

Depending on the value of k, the following situations occur:

- $(k > 0)$ The graph is a circle with center (x_0, y_0) and radius $\sqrt{k}$.
- $(k = 0)$ The only solution of the equation is $x = x_0$, $y = y_0$, so the graph is the single point (x_0, y_0).
- $(k < 0)$ The equation has no real solutions and consequently no graph.

Example 8   Describe the graphs of

(a) $(x - 1)^2 + (y + 4)^2 = -9$ (b) $(x - 1)^2 + (y + 4)^2 = 0$

Solution (a). There are no real values of x and y that will make the left side of the equation negative. Thus, the solution set of the equation is empty, and the equation has no graph.

Solution (b). The only values of x and y that will make the left side of the equation 0 are $x = 1$, $y = -4$. Thus, the graph of the equation is the single point $(1, -4)$.   ◀

The following theorem summarizes our observations.

D.3 THEOREM. *An equation of the form*

$$Ax^2 + Ay^2 + Dx + Ey + F = 0 \qquad (6)$$

where $A \neq 0$, represents a circle, or a point, or else has no graph.

REMARK. The last two cases in Theorem D.3 are called *degenerate cases*. In spite of the fact that these degenerate cases can occur, (6) is often called the *general equation of a circle*.

THE GRAPH of $y = ax^2 + bx + c$

An equation of the form

$$y = ax^2 + bx + c \quad (a \neq 0) \qquad (7)$$

is called a *quadratic equation in x*. Depending on whether a is positive or negative, the graph, which is called a *parabola*, has one of the two forms shown in Figure D.7. In both cases the parabola is symmetric about a vertical line parallel to the y-axis. This line of symmetry cuts the parabola at a point called the *vertex*. The vertex is the low point on the curve if $a > 0$ and the high point if $a < 0$.

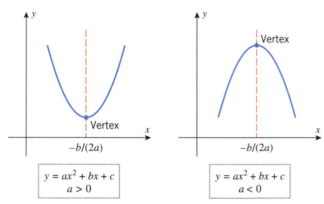

Figure D.7

x	$y = x^2 - 2x - 2$
-1	1
0	-2
1	-3
2	-2
3	1

In the exercises (Exercise 78) we will help the reader show that the x-coordinate of the vertex is given by the formula

$$x = -\frac{b}{2a} \qquad (8)$$

With the aid of this formula, a reasonably accurate graph of a quadratic equation in x can be obtained by plotting the vertex and two points on each side of it.

Example 9 Sketch the graph of

(a) $y = x^2 - 2x - 2$ (b) $y = -x^2 + 4x - 5$

Solution (a). The equation is of form (7) with $a = 1$, $b = -2$, and $c = -2$, so by (8) the x-coordinate of the vertex is

$$x = -\frac{b}{2a} = 1$$

Using this value and two additional values on each side, we obtain Figure D.8.

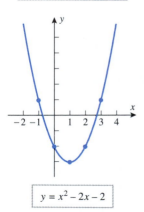

$y = x^2 - 2x - 2$

Figure D.8

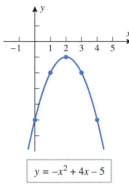

x	$y = -x^2 + 4x - 5$
0	-5
1	-2
2	-1
3	-2
4	-5

$y = -x^2 + 4x - 5$

Figure D.9

Solution (b). The equation is of form (7) with $a = -1$, $b = 4$, and $c = -5$, so by (8) the x-coordinate of the vertex is

$$x = -\frac{b}{2a} = 2$$

Using this value and two additional values on each side, we obtain the table and graph in Figure D.9. ◄

Quite often the intercepts of a parabola $y = ax^2 + bx + c$ are important to know. The y-intercept, $y = c$, results immediately by setting $x = 0$. However, in order to obtain the x-intercepts, if any, we must set $y = 0$ and then solve the resulting quadratic equation $ax^2 + bx + c = 0$.

Example 10   Solve the inequality

$$x^2 - 2x - 2 > 0$$

Solution.   Because the left side of the inequality does not have readily discernible factors, the test-value method illustrated in Example 4 of Appendix A is not convenient to use. Instead, we will give a graphical solution. The given inequality is satisfied for those values of x where the graph of $y = x^2 - 2x - 2$ is above the x-axis. From Figure D.8 those are the values of x to the left of the smaller intercept or to the right of the larger intercept. To find these intercepts we set $y = 0$ to obtain

$$x^2 - 2x - 2 = 0$$

Solving by the quadratic formula gives

$$x = \frac{-b \pm \sqrt{b^2 - 4ac}}{2a} = \frac{2 \pm \sqrt{12}}{2} = 1 \pm \sqrt{3}$$

Thus, the x-intercepts are

$$x = 1 + \sqrt{3} \approx 2.7 \quad \text{and} \quad x = 1 - \sqrt{3} \approx -0.7$$

and the solution set of the inequality is

$$(-\infty, 1 - \sqrt{3}) \cup (1 + \sqrt{3}, +\infty) \qquad ◄$$

REMARK.   Note that the decimal approximations of the intercepts calculated in the preceding example agree with the graph in Figure D.8. Observe, however, that we used the exact values of the intercepts to express the solution. The choice of exact versus approximate values is often a matter of judgment that depends on the purpose for which the values are to be used. Numerical approximations often provide a sense of size that exact values do not, but they can introduce severe errors if not used with care.

Example 11   From Figure D.9 we see that the parabola $y = -x^2 + 4x - 5$ has no x-intercepts. This can also be seen algebraically by solving for the x-intercepts. Setting $y = 0$ and solving the resulting equation

$$-x^2 + 4x - 5 = 0$$

by the quadratic formula yields

$$y = \frac{-4 \pm \sqrt{16 - 20}}{-2} = 2 \pm i$$

Because the solutions are not real numbers, there are no x-intercepts. ◄

Example 12   A ball is thrown straight up from the surface of the Earth at time $t = 0$ s with an initial velocity of 24.5 m/s. If air resistance is ignored, it can be shown that the

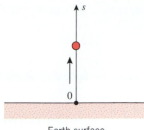

Earth surface

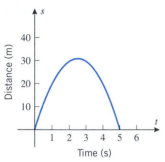

Time (s)

Figure D.10

distance s (in meters) of the ball above the ground after t seconds is given by

$$s = 24.5t - 4.9t^2 \tag{9}$$

(a)   Graph s versus t, making the t-axis horizontal and the s-axis vertical.

(b)   How high does the ball rise above the ground?

Solution (a). Equation (9) is of form (7) with $a = -4.9$, $b = 24.5$, and $c = 0$, so by (8) the t-coordinate of the vertex is

$$t = -\frac{b}{2a} = -\frac{24.5}{2(-4.9)} = 2.5 \text{ s}$$

and consequently the s-coordinate of the vertex is

$$s = 24.5(2.5) - 4.9(2.5)^2 = 30.625 \text{ m}$$

The factored form of (9) is

$$s = 4.9t(5 - t)$$

so the graph has t-intercepts $t = 0$ and $t = 5$. From the vertex and the intercepts we obtain the graph shown in Figure D.10.

Solution (b). From the s-coordinate of the vertex we deduce that the ball rises 30.625 m above the ground.   ◀

THE GRAPH of $x = ay^2 + by + c$

If x and y are interchanged in (7), the resulting equation,

$$x = ay^2 + by + c$$

is called a ***quadratic equation in y***. The graph of such an equation is a parabola with its line of symmetry parallel to the x-axis and its vertex at the point with y-coordinate $y = -b/(2a)$ (Figure D.11). Some problems relating to such equations appear in the exercises.

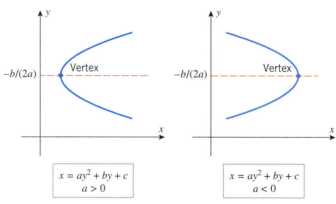

Figure D.11

EXERCISE SET D

1. Where in this section did we use the fact that the same scale was used on both coordinate axes?

    In Exercises 2–5, find
    (a) the distance between A and B
    (b) the midpoint of the line segment joining A and B.

2. $A(2, 5)$, $B(-1, 1)$ 3. $A(7, 1)$, $B(1, 9)$

4. $A(2, 0)$, $B(-3, 6)$ 5. $A(-2, -6)$, $B(-7, -4)$

    In Exercises 6–10, use the distance formula to solve the given problem.

6. Prove that $(1, 1)$, $(-2, -8)$, and $(4, 10)$ lie on a straight line.

7. Prove that the triangle with vertices $(5, -2)$, $(6, 5)$, $(2, 2)$ is isosceles.

8. Prove that $(1, 3)$, $(4, 2)$, and $(-2, -6)$ are vertices of a right triangle and then specify the vertex at which the right angle occurs.

9. Prove that $(0, -2)$, $(-4, 8)$, and $(3, 1)$ lie on a circle with center $(-2, 3)$.

10. Prove that for all values of t the point $(t, 2t - 6)$ is equidistant from $(0, 4)$ and $(8, 0)$.

11. Find k, given that $(2, k)$ is equidistant from $(3, 7)$ and $(9, 1)$.

12. Find x and y if $(4, -5)$ is the midpoint of the line segment joining $(-3, 2)$ and (x, y).

In Exercises 13 and 14, find an equation of the given line.

13. The line is the perpendicular bisector of the line segment joining $(2, 8)$ and $(-4, 6)$.

14. The line is the perpendicular bisector of the line segment joining $(5, -1)$ and $(4, 8)$.

15. Find the point on the line $4x - 2y + 3 = 0$ that is equidistant from $(3, 3)$ and $(7, -3)$. [*Hint:* First find an equation of the line that is the perpendicular bisector of the line segment joining $(3, 3)$ and $(7, -3)$.]

16. Find the distance from the point $(3, -2)$ to the line
    (a) $y = 4$ (b) $x = -1$.

17. Find the distance from $(2, 1)$ to the line $4x - 3y + 10 = 0$. [*Hint:* Find the foot of the perpendicular dropped from the point to the line.]

18. Find the distance from $(8, 4)$ to the line $5x + 12y - 36 = 0$. [*Hint:* See the hint in Exercise 17.]

19. Use the method described in Exercise 17 to prove that the distance d from (x_0, y_0) to the line $Ax + By + C = 0$ is

$$d = \frac{|Ax_0 + By_0 + C|}{\sqrt{A^2 + B^2}}$$

20. Use the formula in Exercise 19 to solve Exercise 17.

21. Use the formula in Exercise 19 to solve Exercise 18.

22. Prove: For any triangle, the perpendicular bisectors of the sides meet at a point. [*Hint:* Position the triangle with one vertex on the y-axis and the opposite side on the x-axis, so that the vertices are $(0, a)$, $(b, 0)$, and $(c, 0)$.]

In Exercises 23 and 24, find the center and radius of each circle.

23. (a) $x^2 + y^2 = 25$
    (b) $(x - 1)^2 + (y - 4)^2 = 16$
    (c) $(x + 1)^2 + (y + 3)^2 = 5$
    (d) $x^2 + (y + 2)^2 = 1$

24. (a) $x^2 + y^2 = 9$
    (b) $(x - 3)^2 + (y - 5)^2 = 36$
    (c) $(x + 4)^2 + (y + 1)^2 = 8$
    (d) $(x + 1)^2 + y^2 = 1$

In Exercises 25–32, find the standard equation of the circle satisfying the given conditions.

25. Center $(3, -2)$; radius $= 4$.

26. Center $(1, 0)$; diameter $= \sqrt{8}$.

27. Center $(-4, 8)$; circle is tangent to the x-axis.

28. Center $(5, 8)$; circle is tangent to the y-axis.

29. Center $(-3, -4)$; circle passes through the origin.

30. Center $(4, -5)$; circle passes through $(1, 3)$.

31. A diameter has endpoints $(2, 0)$ and $(0, 2)$.

32. A diameter has endpoints $(6, 1)$ and $(-2, 3)$.

In Exercises 33–44, determine whether the equation represents a circle, a point, or no graph. If the equation represents a circle, find the center and radius.

33. $x^2 + y^2 - 2x - 4y - 11 = 0$

34. $x^2 + y^2 + 8x + 8 = 0$

35. $2x^2 + 2y^2 + 4x - 4y = 0$

36. $6x^2 + 6y^2 - 6x + 6y = 3$

37. $x^2 + y^2 + 2x + 2y + 2 = 0$

38. $x^2 + y^2 - 4x - 6y + 13 = 0$

39. $9x^2 + 9y^2 = 1$

40. $(x^2/4) + (y^2/4) = 1$

41. $x^2 + y^2 + 10y + 26 = 0$

42. $x^2 + y^2 - 10x - 2y + 29 = 0$

43. $16x^2 + 16y^2 + 40x + 16y - 7 = 0$

44. $4x^2 + 4y^2 - 16x - 24y = 9$

45. Find an equation of
    (a) the bottom half of the circle $x^2 + y^2 = 16$
    (b) the top half of the circle $x^2 + y^2 + 2x - 4y + 1 = 0$.

46. Find an equation of
    (a) the right half of the circle $x^2 + y^2 = 9$
    (b) the left half of the circle $x^2 + y^2 - 4x + 3 = 0$.

47. Graph
    (a) $y = \sqrt{25 - x^2}$ (b) $y = \sqrt{5 + 4x - x^2}$.

48. Graph
    (a) $x = -\sqrt{4 - y^2}$ (b) $x = 3 + \sqrt{4 - y^2}$.

49. Find an equation of the line that is tangent to the circle

$$x^2 + y^2 = 25$$

at the point $(3, 4)$ on the circle.

50. Find an equation of the line that is tangent to the circle at the point P on the circle
    (a) $x^2 + y^2 + 2x = 9$; $P(2, -1)$
    (b) $x^2 + y^2 - 6x + 4y = 13$; $P(4, 3)$.

51. For the circle $x^2 + y^2 = 20$ and the point $P(-1, 2)$:
    (a) Is P inside, outside, or on the circle?

(b) Find the largest and smallest distances between P and points on the circle.

52. Follow the directions of Exercise 51 for the circle

$$x^2 + y^2 - 2y - 4 = 0$$

and the point $P\left(3, \frac{5}{2}\right)$.

53. Referring to the accompanying figure, find the coordinates of the points T and T', where the lines L and L' are tangent to the circle of radius 1 with center at the origin.

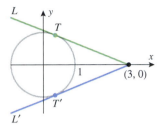

Figure Ex-53

54. A point (x, y) moves so that its distance to $(2, 0)$ is $\sqrt{2}$ times its distance to $(0, 1)$.
(a) Show that the point moves along a circle.
(b) Find the center and radius.

55. A point (x, y) moves so that the sum of the squares of its distances from $(4, 1)$ and $(2, -5)$ is 45.
(a) Show that the point moves along a circle.
(b) Find the center and radius.

56. Find all values of c for which the system of equations

$$\begin{cases} x^2 - y^2 = 0 \\ (x - c)^2 + y^2 = 1 \end{cases}$$

has 0, 1, 2, 3, or 4 solutions. [*Hint:* Sketch a graph.]

In Exercises 57–70, graph the parabola and label the coordinates of the vertex and the intersections with the coordinate axes.

57. $y = x^2 + 2$

58. $y = x^2 - 3$

59. $y = x^2 + 2x - 3$

60. $y = x^2 - 3x - 4$

61. $y = -x^2 + 4x + 5$

62. $y = -x^2 + x$

63. $y = (x - 2)^2$

64. $y = (3 + x)^2$

65. $x^2 - 2x + y = 0$

66. $x^2 + 8x + 8y = 0$

67. $y = 3x^2 - 2x + 1$

68. $y = x^2 + x + 2$

69. $x = -y^2 + 2y + 2$

70. $x = y^2 - 4y + 5$

71. Find an equation of
(a) the right half of the parabola $y = 3 - x^2$
(b) the left half of the parabola $y = x^2 - 2x$.

72. Find an equation of
(a) the upper half of the parabola $x = y^2 - 5$
(b) the lower half of the parabola $x = y^2 - y - 2$.

73. Graph
(a) $y = \sqrt{x + 5}$ (b) $x = -\sqrt{4 - y}$.

74. Graph
(a) $y = 1 + \sqrt{4 - x}$ (b) $x = 3 + \sqrt{y}$.

75. If a ball is thrown straight up with an initial velocity of 32 ft/s, then after t seconds the distance s above its starting height, in feet, is given by $s = 32t - 16t^2$.
(a) Graph this equation in a ts-coordinate system (t-axis horizontal).
(b) At what time t will the ball be at its highest point, and how high will it rise?

76. A rectangular field is to be enclosed with 500 ft of fencing along three sides and by a straight stream on the fourth side. Let x be the length of each side perpendicular to the stream, and let y be the length of the side parallel to the stream.
(a) Express y in terms of x.
(b) Express the area A of the field in terms of x.
(c) What is the largest area that can be enclosed?

77. A rectangular plot of land is to be enclosed using two kinds of fencing. Two opposite sides will have heavy-duty fencing costing $3/ft, and the other two sides will have standard fencing costing $2/ft. A total of $600 is available for the fencing. Let x be the length of each side with the heavy-duty fencing, and let y be the length of each side with the standard fencing.
(a) Express y in terms of x.
(b) Find a formula for the area A of the rectangular plot in terms of x.
(c) What is the largest area that can be enclosed?

78. (a) By completing the square, show that the quadratic equation $y = ax^2 + bx + c$ can be rewritten as

$$y = a\left(x + \frac{b}{2a}\right)^2 + \left(c - \frac{b^2}{4a}\right)$$

if $a \neq 0$.
(b) Use the result in part (a) to show that the graph of the quadratic equation $y = ax^2 + bx + c$ has its high point at $x = -b/(2a)$ if $a < 0$ and its low point there if $a > 0$.

In Exercises 79 and 80, solve the given inequality.

79. (a) $2x^2 + 5x - 1 < 0$ (b) $x^2 - 2x + 3 > 0$

80. (a) $x^2 + x - 1 > 0$ (b) $x^2 - 4x + 6 < 0$

81. At time $t = 0$ a ball is thrown straight up from a height of 5 ft above the ground. After t seconds its distance s, in feet, above the ground is given by $s = 5 + 40t - 16t^2$.
(a) Find the maximum height of the ball above the ground.
(b) Find, to the nearest tenth of a second, the time when the ball strikes the ground.
(c) Find, to the nearest tenth of a second, how long the ball will be more than 12 ft above the ground.

82. Find all values of x at which points on the parabola $y = x^2$ lie below the line $y = x + 3$.

APPENDIX E
Trigonometry Review

TRIGONOMETRIC FUNCTIONS AND IDENTITIES

ANGLES

Angles in the plane can be generated by rotating a ray about its endpoint. The starting position of the ray is called the *initial side* of the angle, the final position is called the *terminal side* of the angle, and the point at which the initial and terminal sides meet is called the *vertex* of the angle. We allow for the possibility that the ray may make more than one complete revolution. Angles are considered to be *positive* if generated counterclockwise and *negative* if generated clockwise (Figure E.1).

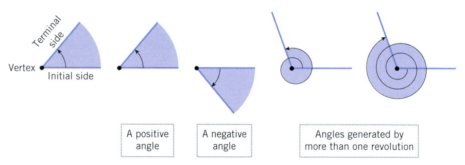

Figure E.1

There are two standard measurement systems for describing the size of an angle: *degree measure* and *radian measure*. In degree measure, one degree (written $1°$) is the measure of an angle generated by $1/360$ of one revolution. Thus, there are $360°$ in an angle of one revolution, $180°$ in an angle of one-half revolution, $90°$ in an angle of one-quarter revolution (a *right angle*), and so forth. Degrees are divided into sixty equal parts, called *minutes*, and minutes are divided into sixty equal parts, called *seconds*. Thus, one minute (written $1'$) is $1/60$ of a degree, and one second (written $1''$) is $1/60$ of a minute. Smaller subdivisions of a degree are expressed as fractions of a second.

In radian measure, angles are measured by the length of the arc that the angle subtends on a circle of radius 1 when the vertex is at the center. One unit of arc on a circle of radius 1 is called one *radian* (written 1 radian or 1 rad) (Figure E.2), and hence the entire circumference of a circle of radius 1 is 2π radians. It follows that an angle of $360°$ subtends an arc of 2π radians, an angle of $180°$ subtends an arc of π radians, an angle of $90°$ subtends an arc of $\pi/2$ radians, and so forth. Figure E.3 and Table 1 show the relationship between degree measure and radian measure for some important positive angles.

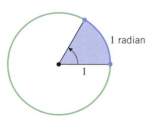

Figure E.2

REMARK. Observe that in Table 1, angles in degrees are designated by the degree symbol, but angles in radians have no units specified. This is standard practice—when no units are specified for an angle, it is understood that the units are radians.

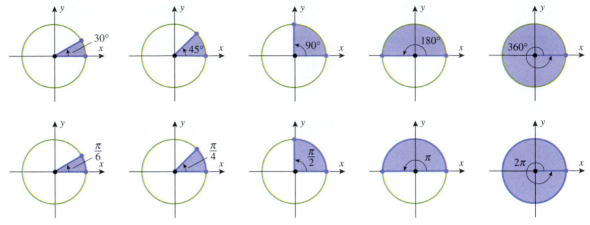

Figure E.3

Table 1

DEGREES	30°	45°	60°	90°	120°	135°	150°	180°	270°	360°
RADIANS	$\dfrac{\pi}{6}$	$\dfrac{\pi}{4}$	$\dfrac{\pi}{3}$	$\dfrac{\pi}{2}$	$\dfrac{2\pi}{3}$	$\dfrac{3\pi}{4}$	$\dfrac{5\pi}{6}$	π	$\dfrac{3\pi}{2}$	2π

From the fact that π radians corresponds to $180°$, we obtain the following formulas, which are useful for converting from degrees to radians and conversely.

$$1° = \frac{\pi}{180}\text{rad} \approx 0.01745 \text{ rad} \tag{1}$$

$$1 \text{ rad} = \left(\frac{180}{\pi}\right)^{\circ} \approx 57° \, 17' \, 44.8'' \tag{2}$$

Example 1
(a) Express $146°$ in radians. (b) Express 3 radians in degrees.

Solution (a). From (1), degrees can be converted to radians by multiplying by a conversion factor of $\pi/180$. Thus,

$$146° = \left(\frac{\pi}{180} \cdot 146\right) \text{rad} = \frac{73\pi}{90} \text{ rad} \approx 2.5482 \text{ rad}$$

Solution (b). From (2), radians can be converted to degrees by multiplying by a conversion factor of $180/\pi$. Thus,

$$3 \text{ rad} = \left(3 \cdot \frac{180}{\pi}\right)^{\circ} = \left(\frac{540}{\pi}\right)^{\circ} \approx 171.9° \qquad \blacktriangleleft$$

RELATIONSHIPS BETWEEN ARC LENGTH, ANGLE, RADIUS, AND AREA

There is a theorem from plane geometry which states that for two concentric circles, the ratio of the arc lengths subtended by a central angle is equal to the ratio of the corresponding radii (Figure E.4). In particular, if s is the arc length subtended on a circle of radius r by a central angle of θ radians, then by comparison with the arc length subtended by that angle on a circle of radius 1 we obtain

$$\frac{s}{\theta} = \frac{r}{1}$$

from which we obtain the following relationships between the central angle θ, the radius r, and the subtended arc length s when θ is in radians (Figure E.5):

$$\theta = s/r \qquad \text{and} \qquad s = r\theta \qquad\qquad (3\text{--}4)$$

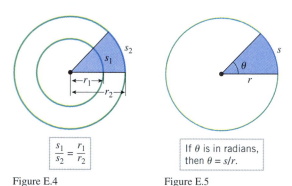

| $\dfrac{s_1}{s_2} = \dfrac{r_1}{r_2}$ | If θ is in radians, then $\theta = s/r$. |

Figure E.4   Figure E.5

The shaded region in Figure E.5 is called a **sector**. It is a theorem from plane geometry that the ratio of the area A of this sector to the area of the entire circle is the same as the ratio of the central angle of the sector to the central angle of the entire circle; thus, if the angles are in radians, we have

$$\frac{A}{\pi r^2} = \frac{\theta}{2\pi}$$

Solving for A yields the following formula for the area of a sector in terms of the radius r and the angle θ in radians:

$$A = \tfrac{1}{2} r^2 \theta \qquad\qquad (5)$$

TRIGONOMETRIC FUNCTIONS FOR RIGHT TRIANGLES

The **sine**, **cosine**, **tangent**, **cosecant**, **secant**, and **cotangent** of a positive acute angle θ can be defined as ratios of the sides of a right triangle. Using the notation from Figure E.6, these definitions take the following form:

$$\sin\theta = \frac{\text{side opposite } \theta}{\text{hypotenuse}} = \frac{y}{r}, \qquad \csc\theta = \frac{\text{hypotenuse}}{\text{side opposite } \theta} = \frac{r}{y}$$

$$\cos\theta = \frac{\text{side adjacent to } \theta}{\text{hypotenuse}} = \frac{x}{r}, \qquad \sec\theta = \frac{\text{hypotenuse}}{\text{side adjacent to } \theta} = \frac{r}{x} \qquad (6)$$

$$\tan\theta = \frac{\text{side opposite } \theta}{\text{side adjacent to } \theta} = \frac{y}{x}, \qquad \cot\theta = \frac{\text{side adjacent to } \theta}{\text{side opposite } \theta} = \frac{x}{y}$$

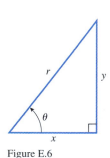

Figure E.6

We will call sin, cos, tan, csc, sec, and cot the **trigonometric functions**. Because similar triangles have proportional sides, the values of the trigonometric functions depend only on the size of θ and not on the particular right triangle used to compute the ratios. Moreover, in these definitions it does not matter whether θ is measured in degrees or radians.

Example 2   Recall from geometry that the two legs of a $45°\text{--}45°\text{--}90°$ triangle are of equal size and that the hypotenuse of a $30°\text{--}60°\text{--}90°$ triangle is twice the shorter leg, where the shorter leg is opposite the $30°$ angle. These facts and the Theorem of Pythagoras yield Figure E.7. From that figure we obtain the results in Table 2.

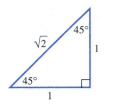 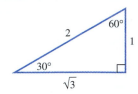

Figure E.7

Table 2

$\sin 45° = 1/\sqrt{2}$,	$\cos 45° = 1/\sqrt{2}$,	$\tan 45° = 1$
$\csc 45° = \sqrt{2}$,	$\sec 45° = \sqrt{2}$,	$\cot 45° = 1$
$\sin 30° = 1/2$,	$\cos 30° = \sqrt{3}/2$,	$\tan 30° = 1/\sqrt{3}$
$\csc 30° = 2$,	$\sec 30° = 2/\sqrt{3}$,	$\cot 30° = \sqrt{3}$
$\sin 60° = \sqrt{3}/2$,	$\cos 60° = 1/2$,	$\tan 60° = \sqrt{3}$
$\csc 60° = 2/\sqrt{3}$,	$\sec 60° = 2$,	$\cot 60° = 1/\sqrt{3}$

◄

**ANGLES IN RECTANGULAR
COORDINATE SYSTEMS**

Because the angles of a right triangle are between $0°$ and $90°$, the formulas in (6) are not directly applicable to negative angles or to angles greater than $90°$. To extend the trigonometric functions to include these cases, it will be convenient to consider angles in rectangular coordinate systems. An angle is said to be in ***standard position*** in an xy-coordinate system if its vertex is at the origin and its initial side is on the positive x-axis (Figure E.8).

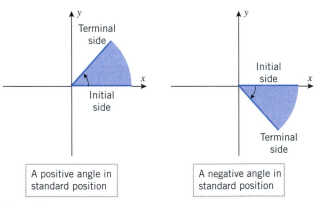

Figure E.8

To define the trigonometric functions of an angle θ in standard position, construct a circle of radius r, centered at the origin, and let $P(x, y)$ be the intersection of the terminal side of θ with this circle (Figure E.9). We make the following definition.

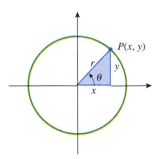

Figure E.9

E.1   DEFINITION.

$$\sin\theta = \frac{y}{r}, \quad \cos\theta = \frac{x}{r}, \quad \tan\theta = \frac{y}{x}$$

$$\csc\theta = \frac{r}{y}, \quad \sec\theta = \frac{r}{x}, \quad \cot\theta = \frac{x}{y}$$

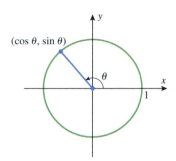

Figure E.10

Note that the formulas in this definition agree with those in (6), so there is no conflict with the earlier definition of the trigonometric functions for triangles. However, this definition applies to all angles (except for cases where a zero denominator occurs).

In the special case where $r = 1$, we have $\sin\theta = y$ and $\cos\theta = x$, so the terminal side of the angle θ intersects the unit circle at the point $(\cos\theta, \sin\theta)$ (Figure E.10). It follows from Definition E.1 that the remaining trigonometric functions of θ are expressible as (verify)

$$\tan\theta = \frac{\sin\theta}{\cos\theta}, \quad \cot\theta = \frac{\cos\theta}{\sin\theta} = \frac{1}{\tan\theta}, \quad \sec\theta = \frac{1}{\cos\theta}, \quad \csc\theta = \frac{1}{\sin\theta} \quad (7\text{--}10)$$

These observations suggest the following procedure for evaluating the trigonometric functions of common angles:

- Construct the angle θ in standard position in an xy-coordinate system.
- Find the coordinates of the intersection of the terminal side of the angle and the unit circle; the x- and y-coordinates of this intersection are the values of $\cos\theta$ and $\sin\theta$, respectively.
- Use Formulas (7) through (10) to find the values of the remaining trigonometric functions from the values of $\cos\theta$ and $\sin\theta$.

Example 3 Evaluate the trigonometric functions of $\theta = 150°$.

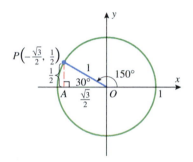

Figure E.11

Solution. Construct a unit circle and place the angle $\theta = 150°$ in standard position (Figure E.11). Since $\angle AOP$ is $30°$ and $\triangle OAP$ is a $30°$–$60°$–$90°$ triangle, the leg AP has length $\frac{1}{2}$ (half the hypotenuse) and the leg OA has length $\sqrt{3}/2$ by the Theorem of Pythagoras. Thus, the coordinates of P are $(-\sqrt{3}/2, 1/2)$, from which we obtain

$$\sin 150° = \frac{1}{2}, \quad \cos 150° = -\frac{\sqrt{3}}{2}, \quad \tan 150° = \frac{\sin 150°}{\cos 150°} = \frac{1/2}{-\sqrt{3}/2} = -\frac{1}{\sqrt{3}}$$

$$\csc 150° = \frac{1}{\sin 150°} = 2, \quad \sec 150° = \frac{1}{\cos 150°} = -\frac{2}{\sqrt{3}}$$

$$\cot 150° = \frac{1}{\tan 150°} = -\sqrt{3} \qquad \blacktriangleleft$$

Example 4 Evaluate the trigonometric functions of $\theta = 5\pi/6$.

Solution. Since $5\pi/6 = 150°$, this problem is equivalent to that of Example 3. From that example we obtain

$$\sin\frac{5\pi}{6} = \frac{1}{2}, \quad \cos\frac{5\pi}{6} = -\frac{\sqrt{3}}{2}, \quad \tan\frac{5\pi}{6} = -\frac{1}{\sqrt{3}}$$

$$\csc\frac{5\pi}{6} = 2, \quad \sec\frac{5\pi}{6} = -\frac{2}{\sqrt{3}}, \quad \cot\frac{5\pi}{6} = -\sqrt{3} \qquad \blacktriangleleft$$

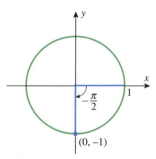

Figure E.12

Example 5 Evaluate the trigonometric functions of $\theta = -\pi/2$.

Solution. As shown in Figure E.12, the terminal side of $\theta = -\pi/2$ intersects the unit circle at the point $(0, -1)$, so

$$\sin(-\pi/2) = -1, \quad \cos(-\pi/2) = 0$$

and from Formulas (7) through (10),

$$\tan(-\pi/2) = \frac{\sin(-\pi/2)}{\cos(-\pi/2)} = \frac{-1}{0} \quad \text{(undefined)}$$

$$\cot(-\pi/2) = \frac{\cos(-\pi/2)}{\sin(-\pi/2)} = \frac{0}{-1} = 0$$

$$\sec(-\pi/2) = \frac{1}{\cos(-\pi/2)} = \frac{1}{0} \quad \text{(undefined)}$$

$$\csc(-\pi/2) = \frac{1}{\sin(-\pi/2)} = \frac{1}{-1} = -1$$

◀

The reader should be able to obtain all of the results in Table 3 by the methods illustrated in the last three examples. The dashes indicate quantities that are undefined.

Table 3

	$\theta = 0$ (0°)	$\pi/6$ (30°)	$\pi/4$ (45°)	$\pi/3$ (60°)	$\pi/2$ (90°)	$2\pi/3$ (120°)	$3\pi/4$ (135°)	$5\pi/6$ (150°)	π (180°)	$3\pi/2$ (270°)	2π (360°)
$\sin\theta$	0	$1/2$	$1/\sqrt{2}$	$\sqrt{3}/2$	1	$\sqrt{3}/2$	$1/\sqrt{2}$	$1/2$	0	-1	0
$\cos\theta$	1	$\sqrt{3}/2$	$1/\sqrt{2}$	$1/2$	0	$-1/2$	$-1/\sqrt{2}$	$-\sqrt{3}/2$	-1	0	1
$\tan\theta$	0	$1/\sqrt{3}$	1	$\sqrt{3}$	—	$-\sqrt{3}$	-1	$-1/\sqrt{3}$	0	—	0
$\csc\theta$	—	2	$\sqrt{2}$	$2/\sqrt{3}$	1	$2/\sqrt{3}$	$\sqrt{2}$	2	—	-1	—
$\sec\theta$	1	$2/\sqrt{3}$	$\sqrt{2}$	2	—	-2	$-\sqrt{2}$	$-2/\sqrt{3}$	-1	—	1
$\cot\theta$	—	$\sqrt{3}$	1	$1/\sqrt{3}$	0	$-1/\sqrt{3}$	-1	$-\sqrt{3}$	—	0	—

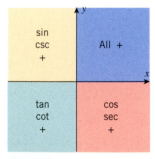

Figure E.13

REMARK. It is only in special cases that exact values for trigonometric functions can be obtained; usually, a calculating utility or a computer program will be required.

The signs of the trigonometric functions of an angle are determined by the quadrant in which the terminal side of the angle falls. For example, if the terminal side falls in the first quadrant, then x and y are positive in Definition E.1, so all of the trigonometric functions have positive values. If the terminal side falls in the second quadrant, then x is negative and y is positive, so sin and csc are positive, but all other trigonometric functions are negative. The diagram in Figure E.13 shows which trigonometric functions are positive in the various quadrants. The reader will find it instructive to check that the results in Table 3 are consistent with Figure E.13.

TRIGONOMETRIC IDENTITIES

A *trigonometric identity* is an equation involving trigonometric functions that is true for all angles for which both sides of the equation are defined. One of the most important identities in trigonometry can be derived by applying the Theorem of Pythagoras to the triangle in Figure E.9 to obtain

$$x^2 + y^2 = r^2$$

Dividing both sides by r^2 and using the definitions of $\sin\theta$ and $\cos\theta$ (Definition E.1), we obtain the following fundamental result:

$$\sin^2\theta + \cos^2\theta = 1 \tag{11}$$

The following identities can be obtained from (11) by dividing through by $\cos^2\theta$ and $\sin^2\theta$, respectively, then applying Formulas (7) through (10):

$$\tan^2\theta + 1 = \sec^2\theta \tag{12}$$

$$1 + \cot^2\theta = \csc^2\theta \tag{13}$$

If (x, y) is a point on the unit circle, then the points $(-x, y)$, $(-x, -y)$, and $(x, -y)$ also lie on the unit circle (why?), and the four points form corners of a rectangle with sides parallel to the coordinate axes (Figure E.14a). The x- and y-coordinates of each corner represent the cosine and sine of an angle in standard position whose terminal side passes through the corner; hence we obtain the identities in parts (b), (c), and (d) of Figure E.14 for sine and cosine. Dividing those identities leads to identities for the tangent. In summary:

$$\sin(\pi - \theta) = \sin\theta, \qquad \sin(\pi + \theta) = -\sin\theta, \qquad \sin(-\theta) = -\sin\theta \tag{14--16}$$
$$\cos(\pi - \theta) = -\cos\theta, \qquad \cos(\pi + \theta) = -\cos\theta, \qquad \cos(-\theta) = \cos\theta \tag{17--19}$$
$$\tan(\pi - \theta) = -\tan\theta, \qquad \tan(\pi + \theta) = \tan\theta, \qquad \tan(-\theta) = -\tan\theta \tag{20--22}$$

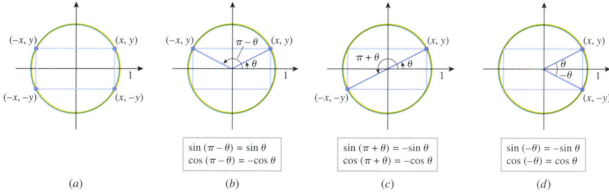

(a) (b) (c) (d)

Figure E.14

Two angles in standard position that have the same terminal side must have the same values for their trigonometric functions since their terminal sides intersect the unit circle at the same point. In particular, two angles whose radian measures differ by a multiple of 2π have the same terminal side and hence have the same values for their trigonometric functions. This yields the identities

$$\sin\theta = \sin(\theta + 2\pi) = \sin(\theta - 2\pi) \tag{23}$$
$$\cos\theta = \cos(\theta + 2\pi) = \cos(\theta - 2\pi) \tag{24}$$

and more generally,

$$\sin\theta = \sin(\theta \pm 2n\pi), \quad n = 0, 1, 2, \ldots \tag{25}$$
$$\cos\theta = \cos(\theta \pm 2n\pi), \quad n = 0, 1, 2, \ldots \tag{26}$$

Identity (21) implies that

$$\tan\theta = \tan(\theta + \pi) \qquad \text{and} \qquad \tan\theta = \tan(\theta - \pi) \tag{27--28}$$

Identity (27) is just (21) with the terms in the sum reversed, and identity (28) follows from (21) by substituting $\theta - \pi$ for θ. These two identities state that adding or subtracting π from an angle does not affect the value of the tangent of the angle. It follows that the same is true

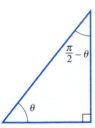

Figure E.15

for any multiple of π; thus,

$$\tan\theta = \tan(\theta \pm n\pi), \quad n = 0, 1, 2, \ldots \tag{29}$$

Figure E.15 shows complementary angles θ and $(\pi/2) - \theta$ of a right triangle. It follows from (6) that

$$\sin\theta = \frac{\text{side opposite }\theta}{\text{hypotenuse}} = \frac{\text{side adjacent to }(\pi/2) - \theta}{\text{hypotenuse}} = \cos\left(\frac{\pi}{2} - \theta\right)$$

$$\cos\theta = \frac{\text{side adjacent to }\theta}{\text{hypotenuse}} = \frac{\text{side opposite }(\pi/2) - \theta}{\text{hypotenuse}} = \sin\left(\frac{\pi}{2} - \theta\right)$$

which yields the identities

$$\sin\left(\frac{\pi}{2} - \theta\right) = \cos\theta, \quad \cos\left(\frac{\pi}{2} - \theta\right) = \sin\theta, \quad \tan\left(\frac{\pi}{2} - \theta\right) = \cot\theta \tag{30–32}$$

where the third identity results from dividing the first two. These identities are also valid for angles that are not acute and for negative angles as well.

THE LAW OF COSINES

The next theorem, called the ***law of cosines***, generalizes the Theorem of Pythagoras. This result is important in its own right and is also the starting point for some important trigonometric identities.

E.2   THEOREM (***Law of Cosines***).   *If the sides of a triangle have lengths a, b, and c, and if θ is the angle between the sides with lengths a and b, then*

$$c^2 = a^2 + b^2 - 2ab\cos\theta$$

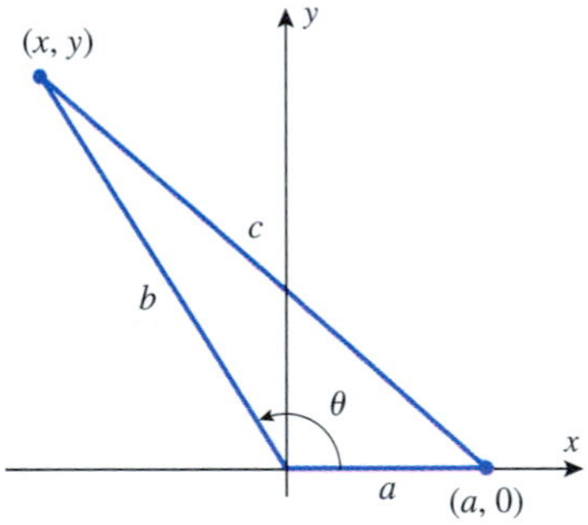

Figure E.16

Proof.   Introduce a coordinate system so that θ is in standard position and the side of length a falls along the positive x-axis. As shown in Figure E.16, the side of length a extends from the origin to $(a, 0)$ and the side of length b extends from the origin to some point (x, y). From the definition of $\sin\theta$ and $\cos\theta$ we have $\sin\theta = y/b$ and $\cos\theta = x/b$, so

$$y = b\sin\theta, \quad x = b\cos\theta \tag{33}$$

From the distance formula in Theorem D.1 of Appendix D, we obtain

$$c^2 = (x - a)^2 + (y - 0)^2$$

so that, from (33),

$$c^2 = (b\cos\theta - a)^2 + b^2\sin^2\theta$$

$$= a^2 + b^2(\cos^2\theta + \sin^2\theta) - 2ab\cos\theta$$

$$= a^2 + b^2 - 2ab\cos\theta$$

which completes the proof.   ∎

We will now show how the law of cosines can be used to obtain the following identities, called the ***addition formulas*** for sine and cosine:

$$\sin(\alpha + \beta) = \sin\alpha\cos\beta + \cos\alpha\sin\beta \tag{34}$$

$$\cos(\alpha + \beta) = \cos\alpha\cos\beta - \sin\alpha\sin\beta \tag{35}$$

$$\sin(\alpha - \beta) = \sin\alpha\cos\beta - \cos\alpha\sin\beta \tag{36}$$

$$\cos(\alpha - \beta) = \cos\alpha\cos\beta + \sin\alpha\sin\beta \tag{37}$$

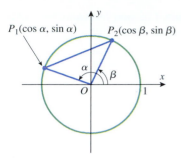

Figure E.17

We will derive (37) first. In our derivation we will assume that $0 \le \beta < \alpha < 2\pi$ (Figure E.17). As shown in the figure, the terminal sides of α and β intersect the unit circle at the points $P_1(\cos\alpha, \sin\alpha)$ and $P_2(\cos\beta, \sin\beta)$. If we denote the lengths of the sides of triangle OP_1P_2 by OP_1, P_1P_2, and OP_2, then $OP_1 = OP_2 = 1$ and, from the distance formula in Theorem D.1 of Appendix D,

$$
\begin{aligned}
(P_1P_2)^2 &= (\cos\beta - \cos\alpha)^2 + (\sin\beta - \sin\alpha)^2 \\
&= (\sin^2\alpha + \cos^2\alpha) + (\sin^2\beta + \cos^2\beta) - 2(\cos\alpha\cos\beta + \sin\alpha\sin\beta) \\
&= 2 - 2(\cos\alpha\cos\beta + \sin\alpha\sin\beta)
\end{aligned}
$$

But angle $P_2OP_1 = \alpha - \beta$, so that the law of cosines yields

$$
\begin{aligned}
(P_1P_2)^2 &= (OP_1)^2 + (OP_2)^2 - 2(OP_1)(OP_2)\cos(\alpha - \beta) \\
&= 2 - 2\cos(\alpha - \beta)
\end{aligned}
$$

Equating the two expressions for $(P_1P_2)^2$ and simplifying, we obtain

$$
\cos(\alpha - \beta) = \cos\alpha\cos\beta + \sin\alpha\sin\beta
$$

which completes the derivation of (37).

We can use (31) and (37) to derive (36) as follows:

$$
\begin{aligned}
\sin(\alpha - \beta) &= \cos\left[\frac{\pi}{2} - (\alpha - \beta)\right] = \cos\left[\left(\frac{\pi}{2} - \alpha\right) - (-\beta)\right] \\
&= \cos\left(\frac{\pi}{2} - \alpha\right)\cos(-\beta) + \sin\left(\frac{\pi}{2} - \alpha\right)\sin(-\beta) \\
&= \cos\left(\frac{\pi}{2} - \alpha\right)\cos\beta - \sin\left(\frac{\pi}{2} - \alpha\right)\sin\beta \\
&= \sin\alpha\cos\beta - \cos\alpha\sin\beta
\end{aligned}
$$

Identities (34) and (35) can be obtained from (36) and (37) by substituting $-\beta$ for β and using the identities

$$
\sin(-\beta) = -\sin\beta, \quad \cos(-\beta) = \cos\beta
$$

We leave it for the reader to derive the identities

$$
\tan(\alpha + \beta) = \frac{\tan\alpha + \tan\beta}{1 - \tan\alpha\tan\beta} \qquad \tan(\alpha - \beta) = \frac{\tan\alpha - \tan\beta}{1 + \tan\alpha\tan\beta} \qquad \text{(38–39)}
$$

Identity (38) can be obtained by dividing (34) by (35) and then simplifying. Identity (39) can be obtained from (38) by substituting $-\beta$ for β and simplifying.

In the special case where $\alpha = \beta$, identities (34), (35), and (38) yield the **double-angle formulas**

$$
\sin 2\alpha = 2\sin\alpha\cos\alpha \tag{40}
$$

$$
\cos 2\alpha = \cos^2\alpha - \sin^2\alpha \tag{41}
$$

$$
\tan 2\alpha = \frac{2\tan\alpha}{1 - \tan^2\alpha} \tag{42}
$$

By using the identity $\sin^2\alpha + \cos^2\alpha = 1$, (41) can be rewritten in the alternative forms

$$
\cos 2\alpha = 2\cos^2\alpha - 1 \qquad \text{and} \qquad \cos 2\alpha = 1 - 2\sin^2\alpha \qquad \text{(43–44)}
$$

If we replace α by $\alpha/2$ in (43) and (44) and use some algebra, we obtain the **half-angle formulas**

$$
\cos^2\frac{\alpha}{2} = \frac{1 + \cos\alpha}{2} \qquad \text{and} \qquad \sin^2\frac{\alpha}{2} = \frac{1 - \cos\alpha}{2} \qquad \text{(45–46)}
$$

We leave it for the exercises to derive the following ***product-to-sum formulas*** from (34) through (37):

$$\sin\alpha\cos\beta = \frac{1}{2}[\sin(\alpha - \beta) + \sin(\alpha + \beta)] \tag{47}$$

$$\sin\alpha\sin\beta = \frac{1}{2}[\cos(\alpha - \beta) - \cos(\alpha + \beta)] \tag{48}$$

$$\cos\alpha\cos\beta = \frac{1}{2}[\cos(\alpha - \beta) + \cos(\alpha + \beta)] \tag{49}$$

We also leave it for the exercises to derive the following ***sum-to-product formulas***:

$$\sin\alpha + \sin\beta = 2\sin\frac{\alpha + \beta}{2}\cos\frac{\alpha - \beta}{2} \tag{50}$$

$$\sin\alpha - \sin\beta = 2\cos\frac{\alpha + \beta}{2}\sin\frac{\alpha - \beta}{2} \tag{51}$$

$$\cos\alpha + \cos\beta = 2\cos\frac{\alpha + \beta}{2}\cos\frac{\alpha - \beta}{2} \tag{52}$$

$$\cos\alpha - \cos\beta = -2\sin\frac{\alpha + \beta}{2}\sin\frac{\alpha - \beta}{2} \tag{53}$$

FINDING AN ANGLE FROM THE VALUE OF ITS TRIGONOMETRIC FUNCTIONS

There are numerous situations in which it is necessary to find an unknown angle from a known value of one of its trigonometric functions. The following example illustrates a method for doing this.

Example 6   Find θ if $\sin\theta = \frac{1}{2}$.

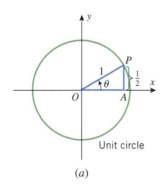

Unit circle

(a)

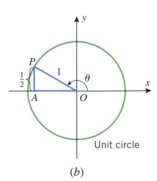

Unit circle

(b)

Figure E.18

Solution.   We begin by looking for positive angles that satisfy the equation. Because $\sin\theta$ is positive, the angle θ must terminate in the first or second quadrant. If it terminates in the first quadrant, then the hypotenuse of $\triangle OAP$ in Figure E.18a is double the leg AP, so

$$\theta = 30° = \frac{\pi}{6}\text{ radians}$$

If θ terminates in the second quadrant (Figure E.18b), then the hypotenuse of $\triangle OAP$ is double the leg AP, so $\angle AOP = 30°$, which implies that

$$\theta = 180° - 30° = 150° = \frac{5\pi}{6}\text{ radians}$$

Now that we have found these two solutions, all other solutions are obtained by adding or subtracting multiples of $360°$ (2π radians) to or from them. Thus, the entire set of solutions is given by the formulas

$$\theta = 30° \pm n\cdot 360°, \quad n = 0, 1, 2, \ldots$$

and

$$\theta = 150° \pm n\cdot 360°, \quad n = 0, 1, 2, \ldots$$

or in radian measure,

$$\theta = \frac{\pi}{6} \pm n\cdot 2\pi, \quad n = 0, 1, 2, \ldots$$

and

$$\theta = \frac{5\pi}{6} \pm n\cdot 2\pi, \quad n = 0, 1, 2, \ldots \qquad \blacktriangleleft$$

EXERCISE SET E

In Exercises 1 and 2, express the angles in radians.

1. (a) 75°   (b) 390°   (c) 20°   (d) 138°

2. (a) 420°   (b) 15°   (c) 225°   (d) 165°

In Exercises 3 and 4, express the angles in degrees.

3. (a) $\pi/15$ (b) 1.5   (c) $8\pi/5$ (d) 3π

4. (a) $\pi/10$ (b) 2   (c) $2\pi/5$ (d) $7\pi/6$

In Exercises 5 and 6, find the exact values of all six trigonometric functions of θ.

5. (a)   (b)   (c)

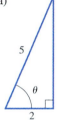
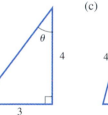

6. (a)   (b)   (c)

In Exercises 7–12, the angle θ is an acute angle of a right triangle. Solve the problems by drawing an appropriate right triangle. Do *not* use a calculator.

7. Find $\sin\theta$ and $\cos\theta$ given that $\tan\theta = 3$.

8. Find $\sin\theta$ and $\tan\theta$ given that $\cos\theta = \frac{2}{3}$.

9. Find $\tan\theta$ and $\csc\theta$ given that $\sec\theta = \frac{5}{2}$.

10. Find $\cot\theta$ and $\sec\theta$ given that $\csc\theta = 4$.

11. Find the length of the side adjacent to θ given that the hypotenuse has length 6 and $\cos\theta = 0.3$.

12. Find the length of the hypotenuse given that the side opposite θ has length 2.4 and $\sin\theta = 0.8$.

In Exercises 13 and 14, the value of an angle θ is given. Find the values of all six trigonometric functions of θ without using a calculator.

13. (a) 225°   (b) −210°   (c) $5\pi/3$ (d) $-3\pi/2$

14. (a) 330°   (b) −120°   (c) $9\pi/4$ (d) -3π

In Exercises 15 and 16, use the information to find the exact values of the remaining five trigonometric functions of θ.

15. (a) $\cos\theta = \frac{3}{5},\ 0 < \theta < \pi/2$

(b) $\cos\theta = \frac{3}{5},\ -\pi/2 < \theta < 0$

(c) $\tan\theta = -1/\sqrt{3},\ \pi/2 < \theta < \pi$

(d) $\tan\theta = -1/\sqrt{3},\ -\pi/2 < \theta < 0$

(e) $\csc\theta = \sqrt{2},\ 0 < \theta < \pi/2$

(f) $\csc\theta = \sqrt{2},\ \pi/2 < \theta < \pi$

16. (a) $\sin\theta = \frac{1}{4},\ 0 < \theta < \pi/2$

(b) $\sin\theta = \frac{1}{4},\ \pi/2 < \theta < \pi$

(c) $\cot\theta = \frac{1}{3},\ 0 < \theta < \pi/2$

(d) $\cot\theta = \frac{1}{3},\ \pi < \theta < 3\pi/2$

(e) $\sec\theta = -\frac{5}{2},\ \pi/2 < \theta < \pi$

(f) $\sec\theta = -\frac{5}{2},\ \pi < \theta < 3\pi/2$

In Exercises 17 and 18, use a calculating utility to find x to four decimal places.

17. (a)   (b)

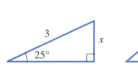

18. (a)   (b)

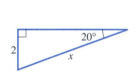

19. In each part, let θ be an acute angle of a right triangle. Express the remaining five trigonometric functions in terms of a.

(a) $\sin\theta = a/3$ (b) $\tan\theta = a/5$ (c) $\sec\theta = a$

In Exercises 20–27, find all values of θ (in radians) that satisfy the given equation. Do not use a calculator.

20. (a) $\cos\theta = -1/\sqrt{2}$ (b) $\sin\theta = -1/\sqrt{2}$

21. (a) $\tan\theta = -1$ (b) $\cos\theta = \frac{1}{2}$

22. (a) $\sin\theta = -\frac{1}{2}$ (b) $\tan\theta = \sqrt{3}$

23. (a) $\tan\theta = 1/\sqrt{3}$ (b) $\sin\theta = -\sqrt{3}/2$

24. (a) $\sin\theta = -1$ (b) $\cos\theta = -1$

25. (a) $\cot\theta = -1$ (b) $\cot\theta = \sqrt{3}$

26. (a) $\sec\theta = -2$ (b) $\csc\theta = -2$

27. (a) $\csc\theta = 2/\sqrt{3}$ (b) $\sec\theta = 2/\sqrt{3}$

> In Exercises 28 and 29, find the values of all six trigonometric functions of θ.

28.

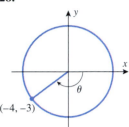

$(-4, -3)$

29.

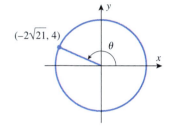

$(-2\sqrt{21}, 4)$

30. Find all values of θ (in radians) such that
   (a) $\sin\theta = 1$ (b) $\cos\theta = 1$ (c) $\tan\theta = 1$
   (d) $\csc\theta = 1$ (e) $\sec\theta = 1$ (f) $\cot\theta = 1$.

31. Find all values of θ (in radians) such that
   (a) $\sin\theta = 0$ (b) $\cos\theta = 0$ (c) $\tan\theta = 0$
   (d) $\csc\theta$ is undefined   (e) $\sec\theta$ is undefined
   (f) $\cot\theta$ is undefined.

32. How could you use a ruler and protractor to approximate $\sin 17°$ and $\cos 17°$?

33. Find the length of the circular arc on a circle of radius 4 cm subtended by an angle of
   (a) $\pi/6$ (b) $150°$.

34. Find the radius of a circular sector that has an angle of $\pi/3$ and a circular arc length of 7 units.

35. A point P moving counterclockwise on a circle of radius 5 cm traverses an arc length of 2 cm. What is the angle swept out by a radius from the center to P?

36. Find a formula for the area A of a circular sector in terms of its radius r and arc length s.

37. As shown in the accompanying figure, a right circular cone is made from a circular piece of paper of radius R by cutting out a sector of angle θ radians and gluing the cut edges of the remaining piece together. Find
   (a) the radius r of the base of the cone in terms of R and θ
   (b) the height h of the cone in terms of R and θ.

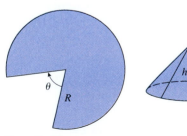

Figure Ex-37

38. As shown in the accompanying figure, let r and L be the radius of the base and the slant height of a right circular cone. Show that the lateral surface area, S, of the cone is $S = \pi r L$. [*Hint:* As shown in the figure in Exercise 37, the lateral surface of the cone becomes a circular sector when cut along a line from the vertex to the base and flattened.]

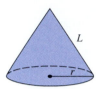

Figure Ex-38

39. Two sides of a triangle have lengths of 3 cm and 7 cm and meet at an angle of $60°$. Find the area of the triangle.

40. Let ABC be a triangle whose angles at A and B are $30°$ and $45°$. If the side opposite the angle B has length 9, find the lengths of the remaining sides and the size of the angle C.

41. A 10-foot ladder leans against a house and makes an angle of $67°$ with level ground. How far is the top of the ladder above the ground? Express your answer to the nearest tenth of a foot.

42. From a point 120 feet on level ground from a building, the angle of elevation to the top of the building is $76°$. Find the height of the building. Express your answer to the nearest foot.

43. An observer on level ground is at a distance d from a building. The angles of elevation to the bottoms of the windows on the second and third floors are α and β, respectively. Find the distance h between the bottoms of the windows in terms of α, β, and d.

44. From a point on level ground, the angle of elevation to the top of a tower is α. From a point that is d units closer to the tower, the angle of elevation is β. Find the height h of the tower in terms of α, β, and d.

> In Exercises 45 and 46, do *not* use a calculator.

45. If $\cos\theta = \frac{2}{3}$ and $0 < \theta < \pi/2$, find
   (a) $\sin 2\theta$ (b) $\cos 2\theta$.

46. If $\tan\alpha = \frac{3}{4}$ and $\tan\beta = 2$, where $0 < \alpha < \pi/2$ and $0 < \beta < \pi/2$, find
   (a) $\sin(\alpha - \beta)$ (b) $\cos(\alpha + \beta)$.

47. Express $\sin 3\theta$ and $\cos 3\theta$ in terms of $\sin\theta$ and $\cos\theta$.

> In Exercises 48–58, derive the given identities.

48. $\dfrac{\cos\theta \sec\theta}{1 + \tan^2\theta} = \cos^2\theta$

49. $\dfrac{\cos\theta \tan\theta + \sin\theta}{\tan\theta} = 2\cos\theta$

50. $2\csc 2\theta = \sec\theta \csc\theta$ **51.** $\tan\theta + \cot\theta = 2\csc 2\theta$

52. $\dfrac{\sin 2\theta}{\sin \theta} - \dfrac{\cos 2\theta}{\cos \theta} = \sec \theta$

53. $\dfrac{\sin \theta + \cos 2\theta - 1}{\cos \theta - \sin 2\theta} = \tan \theta$

54. $\sin 3\theta + \sin \theta = 2 \sin 2\theta \cos \theta$

55. $\sin 3\theta - \sin \theta = 2 \cos 2\theta \sin \theta$

56. $\tan \dfrac{\theta}{2} = \dfrac{1 - \cos \theta}{\sin \theta}$ **57.** $\tan \dfrac{\theta}{2} = \dfrac{\sin \theta}{1 + \cos \theta}$

58. $\cos \left(\dfrac{\pi}{3} + \theta \right) + \cos \left(\dfrac{\pi}{3} - \theta \right) = \cos \theta$

Exercises 59 and 60 refer to an arbitrary triangle ABC in which the side of length a is opposite angle A, the side of length b is opposite angle B, and the side of length c is opposite angle C.

---

59. Prove: The area of a triangle ABC can be written as

$$\text{area} = \tfrac{1}{2} bc \sin A$$

Find two other similar formulas for the area.

60. Prove the *law of sines*: In any triangle, the ratios of the sides to the sines of the opposite angles are equal; that is,

$$\frac{a}{\sin A} = \frac{b}{\sin B} = \frac{c}{\sin C}$$

61. Use identities (34) through (37) to express each of the following in terms of $\sin \theta$ or $\cos \theta$.

(a) $\sin \left(\dfrac{\pi}{2} + \theta \right)$ (b) $\cos \left(\dfrac{\pi}{2} + \theta \right)$

(c) $\sin \left(\dfrac{3\pi}{2} - \theta \right)$ (d) $\cos \left(\dfrac{3\pi}{2} + \theta \right)$

62. Derive identities (38) and (39).

63. Derive identity
(a) (47)   (b) (48)   (c) (49).

64. If $A = \alpha + \beta$ and $B = \alpha - \beta$, then $\alpha = \tfrac{1}{2}(A + B)$ and $\beta = \tfrac{1}{2}(A - B)$ (verify). Use this result and identities (47) through (49) to derive identity
(a) (50)   (b) (52)   (c) (53).

65. Substitute $-\beta$ for β in identity (50) to derive identity (51).

66. (a) Express $3 \sin \alpha + 5 \cos \alpha$ in the form

$$C \sin(\alpha + \phi)$$

(b) Show that a sum of the form

$$A \sin \alpha + B \cos \alpha$$

can be rewritten in the form $C \sin(\alpha + \phi)$.

67. Show that the length of the diagonal of the parallelogram in the accompanying figure is

$$d = \sqrt{a^2 + b^2 + 2ab \cos \theta}$$

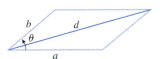

Figure Ex-67

APPENDIX F

Solving Polynomial Equations

We will assume in this appendix that you know how to divide polynomials using long division and synthetic division. If you need to review those techniques, refer to an algebra book.

A BRIEF REVIEW OF POLYNOMIALS

Recall that if n is a nonnegative integer, then a **polynomial of degree n** is a function that can be written in the following forms, depending on whether you want the powers of x in ascending or descending order:

$$c_0 + c_1x + c_2x^2 + \cdots + c_nx^n \quad (c_n \neq 0)$$
$$c_nx^n + c_{n-1}x^{n-1} + \cdots + c_1x + c_0 \quad (c_n \neq 0)$$

The numbers $c_0, c_1, \ldots, c_n$ are called the **coefficients** of the polynomial. The coefficient c_n (which multiplies the highest power of x) is called the **leading coefficient**, the term c_nx^n is called the **leading term**, and the coefficient c_0 is called the **constant term**. Polynomials of degree $1, 2, 3, 4,$ and 5 are called **linear**, **quadratic**, **cubic**, **quartic**, and **quintic**, respectively. For simplicity, general polynomials of low degree are often written without subscripts on the coefficients:

$$p(x) = a \qquad \text{Constant polynomial}$$
$$p(x) = ax + b \quad (a \neq 0) \qquad \text{Linear polynomial}$$
$$p(x) = ax^2 + bx + c \quad (a \neq 0) \qquad \text{Quadratic polynomial}$$
$$p(x) = ax^3 + bx^2 + cx + d \quad (a \neq 0) \qquad \text{Cubic polynomial}$$

When you attempt to factor a polynomial completely, one of three things can happen:

- You may be able to decompose the polynomial into distinct linear factors using only real numbers; for example,

$$x^3 + x^2 - 2x = x(x^2 + x - 2) = x(x - 1)(x + 2)$$

- You may be able to decompose the polynomial into linear factors using only real numbers, but some of the factors may be repeated; for example,

$$x^6 - 3x^4 + 2x^3 = x^3(x^3 - 3x + 2) = x^3(x - 1)^2(x + 2) \tag{1}$$

- You may be able to decompose the polynomial into linear and quadratic factors using only real numbers, but you may not be able to decompose the quadratic factors into linear factors using only real numbers (such quadratic factors are said to be **irreducible** over the real numbers); for example,

$$x^4 - 1 = (x^2 - 1)(x^2 + 1) = (x - 1)(x + 1)(x^2 + 1)$$
$$= (x - 1)(x + 1)(x - i)(x + i)$$

Here, the factor $x^2 + 1$ is irreducible over the real numbers.

In general, if $p(x)$ is a polynomial of degree n with leading coefficient a, and if complex numbers are allowed, then $p(x)$ can be factored as

$$p(x) = a(x - r_1)(x - r_2) \cdots (x - r_n) \tag{2}$$

where $r_1, r_2, \ldots, r_n$ are called the **zeros** of $p(x)$ or the **roots** of the equation $p(x) = 0$, and (2) is called the **complete linear factorization** of $p(x)$. If some of the factors in (2) are repeated, then they can be combined; for example, if the first k factors are distinct and the rest are repetitions of the first k, then (2) can be expressed in the form

$$p(x) = a(x - r_1)^{m_1}(x - r_2)^{m_2} \cdots (x - r_k)^{m_k} \tag{3}$$

where $r_1, r_2, \ldots, r_k$ are the *distinct* roots of $p(x) = 0$. The exponents $m_1, m_2, \ldots, m_k$ tell us how many times the various factors occur in the complete linear factorization; for example, in (3) the factor $(x - r_1)$ occurs m_1 times, the factor $(x - r_2)$ occurs m_2 times, and so forth. Some techniques for factoring polynomials are discussed later in this appendix. In general, if a factor $(x - r)$ occurs m times in the complete linear factorization of a polynomial, then we say that r is a root or zero of **multiplicity m**, and if $(x - r)$ has no repetitions (i.e., r has multiplicity 1), then we say that r is a **simple** root or zero. For example, it follows from (1) that the equation $x^6 - 3x^4 + 2x^3 = 0$ can be expressed as

$$x^3(x - 1)^2(x + 2) = 0 \tag{4}$$

so this equation has three distinct roots—a root $x = 0$ of multiplicity 3, a root $x = 1$ of multiplicity 2, and a simple root $x = -2$.

Note that in (3) the multiplicities of the roots must add up to n, since $p(x)$ has degree n; that is,

$$m_1 + m_2 + \cdots + m_k = n$$

For example, in (4) the multiplicities add up to 6, which is the same as the degree of the polynomial.

It follows from (2) that a polynomial of degree n can have at most n distinct roots; if all of the roots are simple, then there will be *exactly* n, but if some are repeated, then there will be fewer than n. However, when counting the roots of a polynomial, it is standard practice to count multiplicities, since that convention allows us to say that a polynomial of degree n has n roots. For example, from (1) the six roots of the polynomial $p(x) = x^6 - 3x^4 + 2x^3$ are

$$r = 0, \ 0, \ 0, \ 1, \ 1, \ -2$$

In summary, we have the following important theorem.

F.1 THEOREM. *If complex roots are allowed, and if roots are counted according to their multiplicities, then a polynomial of degree n has exactly n roots.*

THE REMAINDER THEOREM

When two positive integers are divided, the numerator can be expressed as the quotient plus the remainder over the divisor, where the remainder is less than the divisor. For example,

$$\tfrac{17}{5} = 3 + \tfrac{2}{5}$$

If we multiply this equation through by 5, we obtain

$$17 = 5 \cdot 3 + 2$$

which states that the *numerator is the divisor times the quotient plus the remainder.*

The following theorem, which we state without proof, is an analogous result for division of polynomials.

> **F.2**   THEOREM.   *If $p(x)$ and $s(x)$ are polynomials, and if $s(x)$ is not the zero polynomial, then $p(x)$ can be expressed as*
>
> $$p(x) = s(x)q(x) + r(x)$$
>
> *where $q(x)$ and $r(x)$ are the quotient and remainder that result when $p(x)$ is divided by $s(x)$, and either $r(x)$ is the zero polynomial or the degree of $r(x)$ is less than the degree of $s(x)$.*

In the special case where $p(x)$ is divided by a first-degree polynomial of the form $x - c$, the remainder must be some constant r, since it is either zero or has degree less than 1. Thus, Theorem F.2 implies that

$$p(x) = (x - c)q(x) + r$$

and this in turn implies that $p(c) = r$. In summary, we have the following theorem.

> **F.3**   THEOREM (*Remainder Theorem*).   *If a polynomial $p(x)$ is divided by $x - c$, then the remainder is $p(c)$.*

Example 1   According to the Remainder Theorem, the remainder on dividing

$$p(x) = 2x^3 + 3x^2 - 4x - 3$$

by $x + 4$ should be

$$p(-4) = 2(-4)^3 + 3(-4)^2 - 4(-4) - 3 = -67$$

Show that this is so.

Solution.   By long division

$$
\begin{array}{r}
2x^2 - 5x + 16 \\
x + 4 \overline{\smash{\big)}\, 2x^3 + 3x^2 - 4x - 3} \\
\underline{2x^3 + 8x^2} \phantom{{}-4x-3} \\
-5x^2 - 4x \\
\underline{-5x^2 - 20x} \\
16x - 3 \\
\underline{16x + 64} \\
-67
\end{array}
$$

which shows that the remainder is -67.

Alternative Solution.   Because we are dividing by an expression of the form $x - c$ (where $c = -4$), we can use synthetic division rather than long division. The computations are

$$
\begin{array}{r|rrrr}
-4 & 2 & 3 & -4 & -3 \\
   &   & -8 & 20 & -64 \\
\hline
   & 2 & -5 & 16 & -67
\end{array}
$$

which again shows that the remainder is -67.   ◄

THE FACTOR THEOREM

To *factor* a polynomial $p(x)$ is to write it as a product of lower-degree polynomials, called *factors* of $p(x)$. For $s(x)$ to be a factor of $p(x)$ there must be no remainder when $p(x)$ is divided by $s(x)$. For example, if $p(x)$ can be factored as

$$p(x) = s(x)q(x) \tag{5}$$

then

$$\frac{p(x)}{s(x)} = q(x) \tag{6}$$

so dividing $p(x)$ by $s(x)$ produces a quotient $q(x)$ with no remainder. Conversely, (6) implies (5), so $s(x)$ is a factor of $p(x)$ if there is no remainder when $p(x)$ is divided by $s(x)$.

In the special case where $x - c$ is a factor of $p(x)$, the polynomial $p(x)$ can be expressed as

$$p(x) = (x - c)q(x)$$

which implies that $p(c) = 0$. Conversely, if $p(c) = 0$, then the Remainder Theorem implies that $x - c$ is a factor of $p(x)$, since the remainder is 0 when $p(x)$ is divided by $x - c$. These results are summarized in the following theorem.

F.4   THEOREM (*Factor Theorem*).   *A polynomial $p(x)$ has a factor $x - c$ if and only if $p(c) = 0$.*

It follows from this theorem that the statements below say the same thing in different ways:

- $x - c$ is a factor of $p(x)$.
- $p(c) = 0$.
- c is a zero of $p(x)$.
- c is a root of the equation $p(x) = 0$.
- c is a solution of the equation $p(x) = 0$.
- c is an x-intercept of $y = p(x)$.

Example 2   Confirm that $x - 1$ is a factor of

$$p(x) = x^3 - 3x^2 - 13x + 15$$

by dividing $x - 1$ into $p(x)$ and checking that the remainder is zero.

Solution.   By long division

$$
\begin{array}{r}
x^2 - 2x - 15 \\
x - 1\overline{\smash{)}\,x^3 - 3x^2 - 13x + 15} \\
\underline{x^3 - x^2} \\
-2x^2 - 13x \\
\underline{-2x^2 + 2x} \\
- 15x + 15 \\
\underline{- 15x + 15} \\
0
\end{array}
$$

which shows that the remainder is zero.

Alternative Solution.   Because we are dividing by an expression of the form $x - c$, we can use synthetic division rather than long division. The computations are

$$
\begin{array}{r|rrrr}
\underline{1|} & 1 & -3 & -13 & 15 \\
 & & 1 & -2 & -15 \\
\hline
 & 1 & -2 & -15 & 0
\end{array}
$$

which again confirms that the remainder is zero.   ◀

USING ONE FACTOR TO FIND OTHER FACTORS

If $x - c$ is a factor of $p(x)$, and if $q(x) = p(x)/(x - c)$, then

$$p(x) = (x - c)q(x) \tag{7}$$

so that additional linear factors of $p(x)$ can be obtained by factoring the quotient $q(x)$.

Example 3 Factor

$$p(x) = x^3 - 3x^2 - 13x + 15 \tag{8}$$

completely into linear factors.

Solution. We showed in Example 2 that $x - 1$ is a factor of $p(x)$ and we also showed that $p(x)/(x - 1) = x^2 - 2x - 15$. Thus,

$$x^3 - 3x^2 - 13x + 15 = (x - 1)(x^2 - 2x - 15)$$

Factoring $x^2 - 2x - 15$ by inspection yields

$$x^3 - 3x^2 - 13x + 15 = (x - 1)(x - 5)(x + 3)$$

which is the complete linear factorization of $p(x)$. ◀

METHODS FOR FINDING ROOTS

A general quadratic equation $ax^2 + bx + c = 0$ can be solved by using the quadratic formula to express the solutions of the equation in terms of the coefficients. Versions of this formula were known since Babylonian times, and by the seventeenth century formulas had been obtained for solving general cubic and quartic equations. However, attempts to find formulas for the solutions of general fifth-degree equations and higher proved fruitless. The reason for this became clear in 1829 when the French mathematician Evariste Galois (1811–1832) proved that it is impossible to express the solutions of a general fifth-degree equation or higher in terms of its coefficients using algebraic operations.

Today, we have powerful computer programs for finding the zeros of specific polynomials. For example, it takes only seconds for a computer algebra system, such as *Mathematica*, *Maple*, or *Derive*, to show that the zeros of the polynomial

$$p(x) = 10x^4 - 23x^3 - 10x^2 + 29x + 6 \tag{9}$$

are

$$x = -1, \quad x = -\tfrac{1}{5}, \quad x = \tfrac{3}{2}, \quad \text{and} \quad x = 2 \tag{10}$$

The algorithms that these programs use to find the integer and rational zeros of a polynomial, if any, are based on the following theorem, which is proved in advanced algebra courses.

F.5   THEOREM.   *Suppose that*

$$p(x) = c_n x^n + c_{n-1} x^{n-1} + \cdots + c_1 x + c_0$$

is a polynomial with integer coefficients.

(a)   *If r is an integer zero of $p(x)$, then r must be a divisor of the constant term c_0.*

(b)   *If $r = a/b$ is a rational zero of $p(x)$ in which all common factors of a and b have been canceled, then a must be a divisor of the constant term c_0, and b must be a divisor of the leading coefficient c_n.*

For example, in (9) the constant term is 6 (which has divisors $\pm 1, \pm 2, \pm 3$, and ± 6) and the leading coefficient is 10 (which has divisors $\pm 1, \pm 2, \pm 5$, and ± 10). Thus, the only possible integer zeros of $p(x)$ are

$$\pm 1, \quad \pm 2, \quad \pm 3, \quad \pm 6$$

and the only possible noninteger rational zeros are

$$\pm\tfrac{1}{2}, \quad \pm\tfrac{1}{5}, \quad \pm\tfrac{1}{10}, \quad \pm\tfrac{2}{5}, \quad \pm\tfrac{3}{2}, \quad \pm\tfrac{3}{5}, \quad \pm\tfrac{3}{10}, \quad \pm\tfrac{6}{5}$$

Using a computer, it is a simple matter to evaluate $p(x)$ at each of the numbers in these lists to show that its only rational zeros are the numbers in (10).

Example 4 Solve the equation $x^3 + 3x^2 - 7x - 21 = 0$.

Solution. The solutions of the equation are the zeros of the polynomial

$$p(x) = x^3 + 3x^2 - 7x - 21$$

We will look for integer zeros first. All such zeros must divide the constant term, so the only possibilities are ± 1, ± 3, ± 7, and ± 21. Substituting these values into $p(x)$ (or using the method of Exercise 6) shows that $x = -3$ is an integer zero. This tells us that $x + 3$ is a factor of $p(x)$ and that $p(x)$ can be written as

$$x^3 + 3x^2 - 7x - 21 = (x + 3)q(x)$$

where $q(x)$ is the quotient that results when $x^3 + 3x^2 - 7x - 21$ is divided by $x + 3$. We leave it for you to perform the division and show that $q(x) = x^2 - 7$; hence,

$$x^3 + 3x^2 - 7x - 21 = (x + 3)(x^2 - 7) = (x + 3)(x + \sqrt{7})(x - \sqrt{7})$$

which tells us that the solutions of the given equation are $x = 3$, $x = \sqrt{7} \approx 2.65$, and $x = -\sqrt{7} \approx -2.65$. ◀

EXERCISE SET F [c] CAS

In Exercises 1 and 2, find the quotient $q(x)$ and the remainder $r(x)$ that result when $p(x)$ is divided by $s(x)$.

1. (a) $p(x) = x^4 + 3x^3 - 5x + 10$; $s(x) = x^2 - x + 2$
(b) $p(x) = 6x^4 + 10x^2 + 5$; $s(x) = 3x^2 - 1$
(c) $p(x) = x^5 + x^3 + 1$; $s(x) = x^2 + x$

2. (a) $p(x) = 2x^4 - 3x^3 + 5x^2 + 2x + 7$; $s(x) = x^2 - x + 1$
(b) $p(x) = 2x^5 + 5x^4 - 4x^3 + 8x^2 + 1$; $s(x) = 2x^2 - x + 1$
(c) $p(x) = 5x^6 + 4x^2 + 5$; $s(x) = x^3 + 1$

In Exercises 3 and 4, use synthetic division to find the quotient $q(x)$ and the remainder r that result when $p(x)$ is divided by $s(x)$.

3. (a) $p(x) = 3x^3 - 4x - 1$; $s(x) = x - 2$
(b) $p(x) = x^4 - 5x^2 + 4$; $s(x) = x + 5$
(c) $p(x) = x^5 - 1$; $s(x) = x - 1$

4. (a) $p(x) = 2x^3 - x^2 - 2x + 1$; $s(x) = x - 1$
(b) $p(x) = 2x^4 + 3x^3 - 17x^2 - 27x - 9$; $s(x) = x + 4$
(c) $p(x) = x^7 + 1$; $s(x) = x - 1$

5. Let $p(x) = 2x^4 + x^3 - 3x^2 + x - 4$. Use synthetic division and the Remainder Theorem to find $p(0)$, $p(1)$, $p(-3)$, and $p(7)$.

6. Let $p(x)$ be the polynomial in Example 4. Use synthetic division and the Remainder Theorem to evaluate $p(x)$ at $x = \pm 1$, ± 3, ± 7, and ± 21.

7. Let $p(x) = x^3 + 4x^2 + x - 6$. Find a polynomial $q(x)$ and a constant r such that
(a) $p(x) = (x - 2)q(x) + r$
(b) $p(x) = (x + 1)q(x) + r$.

8. Let $p(x) = x^5 - 1$. Find a polynomial $q(x)$ and a constant r such that

(a) $p(x) = (x + 1)q(x) + r$
(b) $p(x) = (x - 1)q(x) + r$.

9. In each part, make a list of all possible candidates for the rational zeros of $p(x)$.
(a) $p(x) = x^7 + 3x^3 - x + 24$
(b) $p(x) = 3x^4 - 2x^2 + 7x - 10$
(c) $p(x) = x^{35} - 17$

10. Find all integer zeros of

$$p(x) = x^6 + 5x^5 - 16x^4 - 15x^3 - 12x^2 - 38x - 21$$

In Exercises 11–15, factor the polynomials completely.

11. $p(x) = x^3 - 2x^2 - x + 2$

12. $p(x) = 3x^3 + x^2 - 12x - 4$

13. $p(x) = x^4 + 10x^3 + 36x^2 + 54x + 27$

14. $p(x) = 2x^4 + x^3 + 3x^2 + 3x - 9$

15. $p(x) = x^5 + 4x^4 - 4x^3 - 34x^2 - 45x - 18$

[c] **16.** For each of the factorizations that you obtained in Exercises 11–15, check your answer using a CAS.

In Exercises 17–21, find all real solutions of the equations.

17. $x^3 + 3x^2 + 4x + 12 = 0$

18. $2x^3 - 5x^2 - 10x + 3 = 0$

19. $3x^4 + 14x^3 + 14x^2 - 8x - 8 = 0$

20. $2x^4 - x^3 - 14x^2 - 5x + 6 = 0$

21. $x^5 - 2x^4 - 6x^3 + 5x^2 + 8x + 12 = 0$

[c] **22.** For each of the equations you solved in Exercises 17–21, check your answer using a CAS.

23. Find all values of k for which $x - 1$ is a factor of the polynomial $p(x) = k^2x^3 - 7kx + 10$.

24. Is $x + 3$ a factor of $x^7 + 2187$? Justify your answer.

c **25.** A 3-cm-thick slice is cut from a cube, leaving a volume of 196 cm^3. Use a CAS to find the length of a side of the original cube.

26. (a) Show that there is no positive rational number that exceeds its cube by 1.

(b) Does there exist a real number that exceeds its cube by 1? Justify your answer.

27. Use the Factor Theorem to show each of the following.

(a) $x - y$ is a factor of $x^n - y^n$ for all positive integer values of n.

(b) $x + y$ is a factor of $x^n - y^n$ for all positive even integer values of n.

(c) $x + y$ is a factor of $x^n + y^n$ for all positive odd integer values of n.

APPENDIX G

Selected Proofs

PROOFS OF BASIC LIMIT
THEOREMS

An extensive excursion into proofs of limit theorems would be too time consuming to undertake, so we have selected a few proofs of results from Section 2.2 that illustrate some of the basic ideas.

G.1 THEOREM. *Let a be any real number, let k be a constant, and suppose that* $\lim_{x \to a} f(x) = L_1$ *and that* $\lim_{x \to a} g(x) = L_2$. *Then*

(a) $\lim_{x \to a} k = k$

(b) $\lim_{x \to a} [f(x) + g(x)] = \lim_{x \to a} f(x) + \lim_{x \to a} g(x) = L_1 + L_2$

(c) $\lim_{x \to a} [f(x)g(x)] = \left(\lim_{x \to a} f(x) \right) \left(\lim_{x \to a} g(x) \right) = L_1 L_2$

Proof (a). We will apply Definition 2.4.1 with $f(x) = k$ and $L = k$. Thus, given $\epsilon > 0$, we must find a number $\delta > 0$ such that

$$|k - k| < \epsilon \quad \text{if} \quad 0 < |x - a| < \delta$$

or equivalently,

$$0 < \epsilon \quad \text{if} \quad 0 < |x - a| < \delta$$

But the condition on the left side of this statement is *always* true, no matter how δ is chosen. Thus, any positive value for δ will suffice.

Proof (b). We must show that given $\epsilon > 0$ we can find a number $\delta > 0$ such that

$$|(f(x) + g(x)) - (L_1 + L_2)| < \epsilon \quad \text{if} \quad 0 < |x - a| < \delta \tag{1}$$

However, from the limits of f and g in the hypothesis of the theorem we can find numbers δ_1 and δ_2 such that

$$|f(x) - L_1| < \epsilon/2 \quad \text{if} \quad 0 < |x - a| < \delta_1$$

$$|g(x) - L_2| < \epsilon/2 \quad \text{if} \quad 0 < |x - a| < \delta_2$$

Moreover, the inequalities on the left sides of these statements *both* hold if we replace δ_1 and δ_2 by any positive number δ that is less than both δ_1 and δ_2. Thus, for any such δ it follows that

$$|f(x) - L_1| + |g(x) - L_2| < \epsilon \quad \text{if} \quad 0 < |x - a| < \delta \tag{2}$$

However, it follows from the triangle inequality [Theorem B.5] that

$$|(f(x) + g(x)) - (L_1 + L_2)| = |(f(x) - L_1) + (g(x) - L_2)|$$
$$\leq |f(x) - L_1| + |g(x) - L_2|$$

so that (1) follows from (2).

Proof (c).   We must show that given $\epsilon > 0$ we can find a number $\delta > 0$ such that

$$|f(x)g(x) - L_1 L_2| < \epsilon \quad \text{if} \quad 0 < |x - a| < \delta \tag{3}$$

To find δ it will be helpful to express (3) in a different form. If we rewrite $f(x)$ and $g(x)$ as

$$f(x) = L_1 + (f(x) - L_1) \quad \text{and} \quad g(x) = L_2 + (g(x) - L_2)$$

then the inequality on the left side of (3) can be expressed as (verify)

$$|L_1(g(x) - L_2) + L_2(f(x) - L_1) + (f(x) - L_1)(g(x) - L_2)| < \epsilon \tag{4}$$

Since

$$\lim_{x \to a} f(x) = L_1 \quad \text{and} \quad \lim_{x \to a} g(x) = L_2$$

we can find positive numbers $\delta_1, \delta_2, \delta_3,$ and δ_4 such that

$$
\begin{array}{ll}
|f(x) - L_1| < \sqrt{\epsilon/3} & \text{if} \quad 0 < |x - a| < \delta_1 \\[2mm]
|f(x) - L_1| < \dfrac{\epsilon}{3(1 + |L_2|)} & \text{if} \quad 0 < |x - a| < \delta_2 \\[4mm]
|g(x) - L_2| < \sqrt{\epsilon/3} & \text{if} \quad 0 < |x - a| < \delta_3 \\[2mm]
|g(x) - L_2| < \dfrac{\epsilon}{3(1 + |L_1|)} & \text{if} \quad 0 < |x - a| < \delta_4
\end{array}
\tag{5}
$$

Moreover, the inequalities on the left sides of these four statements *all* hold if we replace $\delta_1, \delta_2, \delta_3,$ and δ_4 by any positive number δ that is smaller than $\delta_1, \delta_2, \delta_3,$ and δ_4. Thus, for any such δ it follows with the help of the triangle inequality that

$$|L_1(g(x) - L_2) + L_2(f(x) - L_1) + (f(x) - L_1)(g(x) - L_2)|$$

$$\leq |L_1(g(x) - L_2)| + |L_2(f(x) - L_1)| + |(f(x) - L_1)(g(x) - L_2)|$$

$$= |L_1||g(x) - L_2| + |L_2||f(x) - L_1| + |f(x) - L_1||g(x) - L_2|$$

$$< |L_1|\frac{\epsilon}{3(1 + |L_1|)} + |L_2|\frac{\epsilon}{3(1 + |L_2|)} + \sqrt{\epsilon/3}\sqrt{\epsilon/3} \qquad \boxed{\text{From (5)}}$$

$$= \frac{\epsilon}{3}\frac{|L_1|}{1 + |L_1|} + \frac{\epsilon}{3}\frac{|L_2|}{1 + |L_2|} + \frac{\epsilon}{3}$$

$$< \frac{\epsilon}{3} + \frac{\epsilon}{3} + \frac{\epsilon}{3} = \epsilon \qquad \boxed{\text{Since } \dfrac{|L_1|}{1 + |L_1|} < 1 \text{ and } \dfrac{|L_2|}{1 + |L_2|} < 1}$$

which shows that (4) holds for the δ selected. ∎

REMARK.   Do not be alarmed if the proof of part (*c*) seems difficult; it takes some experience with proofs of this type to develop a feel for choosing a valid δ. Your initial goal should be to understand the ideas and the computations.

PROOF OF A BASIC CONTINUITY PROPERTY

Next we will prove Theorem 2.5.5 for two-sided limits.

G.2   THEOREM (*Theorem 2.5.5*).   *If $\lim_{x \to c} g(x) = L$ and if the function f is continuous at L, then $\lim_{x \to c} f(g(x)) = f(L)$. That is,*

$$\lim_{x \to c} f(g(x)) = f\left(\lim_{x \to c} g(x) \right)$$

Proof.   We must show that given $\epsilon > 0$, we can find a number $\delta > 0$ such that

$$|f(g(x)) - f(L)| < \epsilon \quad \text{if} \quad 0 < |x - c| < \delta \tag{6}$$

Since f is continuous at L, we have

$$\lim_{u \to L} f(u) = f(L)$$

and hence we can find a number $\delta_1 > 0$ such that

$$|f(u) - f(L)| < \epsilon \quad \text{if} \quad |u - L| < \delta_1$$

In particular, if $u = g(x)$, then

$$|f(g(x)) - f(L)| < \epsilon \quad \text{if} \quad |g(x) - L| < \delta_1 \tag{7}$$

But $\lim_{x \to c} g(x) = L$, and hence there is a number $\delta > 0$ such that

$$|g(x) - L| < \delta_1 \quad \text{if} \quad 0 < |x - c| < \delta \tag{8}$$

Thus, if x satisfies the condition on the right side of statement (8), then it follows that $g(x)$ satisfies the condition on the right side of statement (7), and this implies that the condition on the left side of statement (6) is satisfied, completing the proof. ▋

....

PROOF OF THE CHAIN RULE

Next we will prove the chain rule (Theorem 3.5.2), but first we need a preliminary result.

G.3   THEOREM.   *Suppose that a function f is defined on an open interval I and that f is differentiable at a value x in I. Then, for w in I,*

$$f(w) - f(x) = (f'(x) + \epsilon)(w - x)$$

where $\epsilon \to 0$ as $w \to x$ and $\epsilon = 0$ if $w = x$.

Proof. For w in I define

$$\epsilon = \begin{cases} \dfrac{f(w) - f(x)}{w - x} - f'(x), & w \neq x \\[2mm] 0, & w = x \end{cases} \tag{9}$$

Note that

$$\lim_{w \to x} \epsilon = \lim_{w \to x} \left(\frac{f(w) - f(x)}{w - x} - f'(x) \right) = f'(x) - f'(x) = 0$$

If $w \neq x$, it follows from (9) that

$$\epsilon \cdot (w - x) = [f(w) - f(x)] - f'(x)(w - x)$$

or

$$f(w) - f(x) = (f'(x) + \epsilon)(w - x) \tag{10}$$

Note that (10) also holds if $w = x$ (why?), so (10) is valid for all values of w in I. ▋

We are now ready to prove the chain rule.

G.4   THEOREM (*Theorem 3.5.2*).   *If g is differentiable at x and f is differentiable at $g(x)$, then the composition $f \circ g$ is differentiable at x. Moreover,*

$$(f \circ g)'(x) = f'(g(x))g'(x)$$

Proof. Since g is differentiable at x, it follows from Theorem G.3 that

$$g(w) - g(x) = (g'(x) + \epsilon_1)(w - x) \tag{11}$$

where $\epsilon_1 \to 0$ as $w \to x$ and $\epsilon_1 = 0$ if $w = x$. Since f is differentiable at $g(x)$, it follows from Theorem G.3 that

$$f(v) - f(g(x)) = [f'(g(x)) + \epsilon_2](v - g(x)) \tag{12}$$

where $\epsilon_2 \to 0$ as $v \to g(x)$ and $\epsilon_2 = 0$ if $v = g(x)$.

Since $g(w) \to g(x)$ as $w \to x$ (why?), we can replace v by $g(w)$ and $v \to g(x)$ by $w \to x$ in (12):

$$f(g(w)) - f(g(x)) = [f'(g(x)) + \epsilon_2](g(w) - g(x)) \tag{13}$$

where $\epsilon_2 \to 0$ as $w \to x$ and $\epsilon_2 = 0$ if $w = x$. Substituting (11) into (13) yields

$$f(g(w)) - f(g(x)) = [f'(g(x)) + \epsilon_2](g'(x) + \epsilon_1)(w - x) \tag{14}$$

where $\epsilon_1 \to 0$ and $\epsilon_2 \to 0$ as $w \to x$. Thus, from (14)

$$
\begin{aligned}
(f \circ g)'(x) &= \lim_{w \to x} \frac{f(g(w)) - f(g(x))}{w - x} \\
&= \lim_{w \to x} ([f'(g(x)) + \epsilon_2](g'(x) + \epsilon_1)) \\
&= [f'(g(x)) + 0](g'(x) + 0) = f'(g(x))g'(x) \quad \blacksquare
\end{aligned}
$$

PROOF OF THE REMAINDER ESTIMATION THEOREM

G.5   THEOREM (*Theorem 10.1.4*).   *If the function f can be differentiated $n + 1$ times on an interval I containing the number x_0, and if M is an upper bound for $|f^{(n+1)}(x)|$ on I, that is, $|f^{(n+1)}(x)| \le M$ for all x in I, then*

$$|R_n(x)| \le \frac{M}{(n+1)!}|x - x_0|^{n+1}$$

for all x in I.

Proof.   We are assuming that f can be differentiated $n + 1$ times on an interval I containing the number x_0 and that

$$|f^{(n+1)}(x)| \le M \tag{15}$$

for all x in I. We want to show that

$$|R_n(x)| \le \frac{M}{(n+1)!}|x - x_0|^{n+1} \tag{16}$$

for all x in I, where

$$R_n(x) = f(x) - \sum_{k=0}^{n} \frac{f^{(k)}(x_0)}{k!}(x - x_0)^k \tag{17}$$

In our proof we will need the following two properties of $R_n(x)$:

$$R_n(x_0) = R_n'(x_0) = \cdots = R_n^{(n)}(x_0) = 0 \tag{18}$$

$$R_n^{(n+1)}(x) = f^{(n+1)}(x) \quad \text{for all } x \text{ in } I \tag{19}$$

These properties can be obtained by analyzing what happens if the expression for $R_n(x)$ in Formula (17) is differentiated j times and x_0 is then substituted in that derivative. If $j < n$, then the jth derivative of the summation in Formula (17) consists of a constant term $f^{(j)}(x_0)$ plus terms involving powers of $x - x_0$ (verify). Thus, $R_n^{(j)}(x_0) = 0$ for $j < n$, which proves all but the last equation in (18). For the last equation, observe that the nth derivative of the summation in (17) is the constant $f^{(n)}(x_0)$, so $R_n^{(n)}(x_0) = 0$. Formula (19) follows from the observation that the $(n + 1)$-st derivative of the summation in (17) is zero (why?).

Now to the main part of the proof. For simplicity we will give the proof for the case where $x \ge x_0$ and leave the case where $x < x_0$ for the reader. It follows from (15) and (19) that $|R_n^{(n+1)}(x)| \le M$, and hence

$$-M \le R_n^{(n+1)}(x) \le M$$

Thus,

$$\int_{x_0}^{x} -M\,dt \le \int_{x_0}^{x} R_n^{(n+1)}(t)\,dt \le \int_{x_0}^{x} M\,dt \tag{20}$$

However, it follows from (18) that $R_n^{(n)}(x_0) = 0$, so

$$\int_{x_0}^{x} R_n^{(n+1)}(t)\,dt = R_n^{(n)}(t)\Big]_{x_0}^{x} = R_n^{(n)}(x)$$

Thus, performing the integrations in (20) we obtain the inequalities

$$-M(x - x_0) \le R_n^{(n)}(x) \le M(x - x_0)$$

Now we will integrate again. Replacing x by t in these inequalities, integrating from x_0 to x, and using $R_n^{(n-1)}(x_0) = 0$ yields

$$-\frac{M}{2}(x - x_0)^2 \le R_n^{(n-1)}(x) \le \frac{M}{2}(x - x_0)^2$$

If we keep repeating this process, then after $n + 1$ integrations we will obtain

$$-\frac{M}{(n + 1)!}(x - x_0)^{n+1} \le R_n(x) \le \frac{M}{(n + 1)!}(x - x_0)^{n+1}$$

which we can rewrite as

$$|R_n(x)| \le \frac{M}{(n + 1)!}(x - x_0)^{n+1}$$

This completes the proof of (16), since the absolute value signs can be omitted in that formula when $x \ge x_0$ (which is the case we are considering). ∎

PROOF OF THE LIMIT COMPARISON TEST

> **G.6** THEOREM (*Theorem 10.6.4*). *Let $\sum a_k$ and $\sum b_k$ be series with positive terms and suppose that*
>
> $$\rho = \lim_{k \to +\infty} \frac{a_k}{b_k}$$
>
> *If ρ is finite and $\rho > 0$, then the series both converge or both diverge.*

Proof. We need only show that $\sum b_k$ converges when $\sum a_k$ converges and that $\sum b_k$ diverges when $\sum a_k$ diverges, since the remaining cases are logical implications of these (why?). The idea of the proof is to apply the comparison test to $\sum a_k$ and suitable multiples of $\sum b_k$. For this purpose let ϵ be any positive number. Since

$$\rho = \lim_{k \to +\infty} \frac{a_k}{b_k}$$

it follows that eventually the terms in the sequence $\{a_k/b_k\}$ must be within ϵ units of ρ; that is, there is a positive integer K such that for $k \ge K$ we have

$$\rho - \epsilon < \frac{a_k}{b_k} < \rho + \epsilon$$

In particular, if we take $\epsilon = \rho/2$, then for $k \ge K$ we have

$$\frac{1}{2}\rho < \frac{a_k}{b_k} < \frac{3}{2}\rho \quad \text{or} \quad \frac{1}{2}\rho b_k < a_k < \frac{3}{2}\rho b_k$$

Thus, by the comparison test we can conclude that

$$\sum_{k=K}^{\infty} \frac{1}{2}\rho b_k \quad \text{converges if} \quad \sum_{k=K}^{\infty} a_k \quad \text{converges} \tag{21}$$

$$\sum_{k=K}^{\infty} \frac{3}{2}\rho b_k \quad \text{diverges if} \quad \sum_{k=K}^{\infty} a_k \quad \text{diverges} \tag{22}$$

But the convergence or divergence of a series is not affected by deleting finitely many terms or by multiplying the general term by a nonzero constant, so (21) and (22) imply that

$$\sum_{k=1}^{\infty} b_k \quad \text{converges if} \quad \sum_{k=1}^{\infty} a_k \quad \text{converges}$$

$$\sum_{k=1}^{\infty} b_k \quad \text{diverges if} \quad \sum_{k=1}^{\infty} a_k \quad \text{diverges}$$

∎

G.7 **THEOREM** (*Theorem 10.6.5*). *Let $\sum u_k$ be a series with positive terms and suppose that*

$$\rho = \lim_{k \to +\infty} \frac{u_{k+1}}{u_k}$$

(a) *If $\rho < 1$, the series converges.*
(b) *If $\rho > 1$ or $\rho = +\infty$, the series diverges.*
(c) *If $\rho = 1$, the series may converge or diverge, so that another test must be tried.*

Proof (a). The number ρ must be nonnegative since it is the limit of u_{k+1}/u_k, which is positive for all k. In this part of the proof we assume that $\rho < 1$, so that $0 \le \rho < 1$.

We will prove convergence by showing that the terms of the given series are eventually less than the terms of a convergent geometric series. For this purpose, choose any real number r such that $0 < \rho < r < 1$. Since the limit of u_{k+1}/u_k is ρ, and $\rho < r$, the terms of the sequence $\{u_{k+1}/u_k\}$ must eventually be less than r. Thus, there is a positive integer K such that for $k \ge K$ we have

$$\frac{u_{k+1}}{u_k} < r \quad \text{or} \quad u_{k+1} < r u_k$$

This yields the inequalities

$$
\begin{aligned}
u_{K+1} &< r u_K \\
u_{K+2} &< r u_{K+1} < r^2 u_K \\
u_{K+3} &< r u_{K+2} < r^3 u_K \\
u_{K+4} &< r u_{K+3} < r^4 u_K \\
&\quad \vdots
\end{aligned}
\tag{23}
$$

But $0 < r < 1$, so

$$r u_K + r^2 u_K + r^3 u_K + \cdots$$

is a convergent geometric series. From the inequalities in (23) and the comparison test it follows that

$$u_{K+1} + u_{K+2} + u_{K+3} + \cdots$$

must also be a convergent series. Thus, $u_1 + u_2 + u_3 + \cdots + u_k + \cdots$ converges by Theorem 10.5.3(c).

Proof (b). In this part we will prove divergence by showing that the limit of the general term is not zero. Since the limit of u_{k+1}/u_k is ρ and $\rho > 1$, the terms in the sequence $\{u_{k+1}/u_k\}$ must eventually be greater than 1. Thus, there is a positive integer K such that for $k \ge K$ we have

$$\frac{u_{k+1}}{u_k} > 1 \quad \text{or} \quad u_{k+1} > u_k$$

This yields the inequalities

$$
\begin{aligned}
u_{K+1} &> u_K \\
u_{K+2} &> u_{K+1} > u_K \\
u_{K+3} &> u_{K+2} > u_K \\
u_{K+4} &> u_{K+3} > u_K \\
&\quad \vdots
\end{aligned}
\tag{24}
$$

Since $u_K > 0$, it follows from the inequalities in (24) that $\lim_{k \to +\infty} u_k \neq 0$, and thus the series $u_1 + u_2 + \cdots + u_k + \cdots$ diverges by part (a) of Theorem 10.5.1. The proof in the case where $\rho = +\infty$ is omitted.

Proof (c). The divergent harmonic series and the convergent p-series with $p = 2$ both have $\rho = 1$ (verify), so the ratio test does not distinguish between convergence and divergence when $\rho = 1$. ∎

ANSWERS TO ODD-NUMBERED EXERCISES

▶ **Exercise Set 1.1 (Page 14)**

1. **(a)** 1943 **(b)** 1960; 4200 **(c)** no, you need the year's population **(d)** war, marketing **(e)** news of health risk, social pressure, antismoking campaigns, increased taxation

3. **(a)** $-2.9, -2.0, 2.35, 2.9$ **(b)** none **(c)** 0 **(d)** $-1.75 \leq x \leq 2.15$ **(e)** $y_{max} = 2.8$ at $x = -2.6$; $y_{min} = -2.2$ at $x = 1.2$

5. **(a)** 2, 4 **(b)** none **(c)** $x \leq 2; 4 \leq x$ **(d)** $y_{min} = -1$; no maximum **7. (a)** no; war, pestilence, flood, earthquakes **(b)** decreases for 8 hours, takes a jump upward, and repeats

9. **(a)** $L = x + 2000/x$ **(c)** 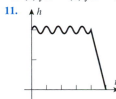 **(b)** $x > 0$; x must be smaller than the width of the building, which was not given. **(d)** 89.44

11. **(a)** $r \approx 3.4, h \approx 13.8$ **(b)** taller **(c)** $r \approx 3.1$ cm, $h \approx 16.0$ cm, $C \approx 4.76$ cents

▶ **Exercise Set 1.2 (Page 25)**

1. **(a)** $-2; 10; 10; 25; 4; 27t^2 - 2$ **(b)** $0; 4; -4; 6; 2\sqrt{2}$; $f(3t) = 1/3t$ for $t > 1$ and $f(3t) = 6t$ for $t \leq 1$

3. **(a)** $x \neq 3$ **(b)** $x \leq -\sqrt{3}, x \geq \sqrt{3}$ **(c)** $(-\infty, +\infty)$ **(d)** $x \neq 0$ **(e)** $x \neq \left(2n + \frac{1}{2}\right)\pi, n = 0, \pm 1, \pm 2, \ldots$

5. **(a)** $x \leq 3$ **(b)** $-2 \leq x \leq 2$ **(c)** $x \geq 0$ **(d)** all x **(e)** all x

7. **(a)** yes **(b)** yes **(c)** no **(d)** no **9.** $h = L(1 - \cos\theta)$

11.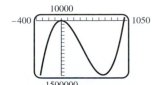
13. **(a)** $f(x) = \begin{cases} 2x + 1, & x < 0 \\ 4x + 1, & x \geq 0 \end{cases}$

(b) $g(x) = \begin{cases} 1 - 2x, & x < 0 \\ 1, & 0 \leq x < 1 \\ 2x - 1, & x \geq 1 \end{cases}$

15. **(a)** $V = (8 - 2x)(15 - 2x)x$ **(b)** $-\infty < x < \infty, -\infty < V < \infty$ **(c)** $0 < x < 4$ **17. (i)** $x = 1, -2$ **(ii)** $g(x) = x + 1$, all x

19. **(a)** $25°$F **(b)** $2°$F **(c)** $-15°$F

21. $5°$F **23.** $D(t) = 1000 - 20t$ ft

▶ **Exercise Set 1.3 (Page 36)**

1. **(e)** **3.** (b), (c) **5.** $[-3, 3] \times [0, 5]$
9. $[-5, 14] \times [-60, 40]$ **11.** $[-0.1, 0.1] \times [-3, 3]$
13. $[-400, 1050] \times [-1500000, 10000]$ **15.** $[-2, 2] \times [-20, 20]$

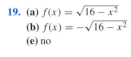

19. **(a)** $f(x) = \sqrt{16 - x^2}$ **(b)** $f(x) = -\sqrt{16 - x^2}$ **(e)** no

▶ **Exercise Set 1.4 (Page 48)**

1. **(a)**  **(b)**
(c) 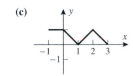 **(d)**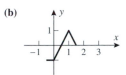

3. **(a)** 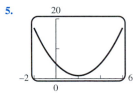 **(b)**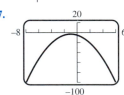
(c) 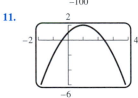 **(d)**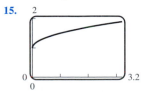

5. 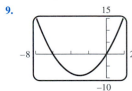 7.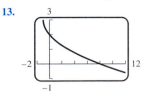

9. 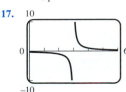 11.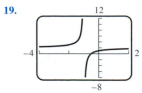

13. 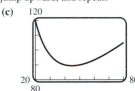 15.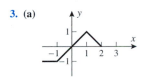

17. 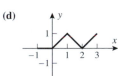 19.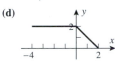

21.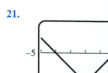

23.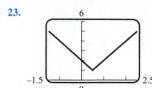

25.

27.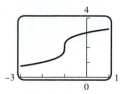

29. (a)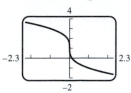

(b) $y = \begin{cases} 0, & x \le 0 \\ 2x, & x > 0 \end{cases}$

31. $x^2 + 2x + 1$, all x; $2x - x^2 - 1$, all x; $2x^3 + 2x$, all x; $2x/(x^2 + 1)$, all x

33. $3\sqrt{x-1}$, $x \ge 1$; $\sqrt{x-1}$, $x \ge 1$; $2x - 2$, $x \ge 1$; 2, $x > 1$

35. (a) 3 **(b)** 9 **(c)** 2 **(d)** 2

37. (a) $t^4 + 1$ **(b)** $t^2 + 4t + 5$ **(c)** $x^2 + 4x + 5$ **(d)** $\dfrac{1}{x^2} + 1$

(e) $x^2 + 2xh + h^2 + 1$ **(f)** $x^2 + 1$ **(g)** $x + 1$ **(h)** $9x^2 + 1$

39. $2x^2 - 2x + 1$, all x; $4x^2 + 2x$, all x

41. $1 - x$, $x \le 1$; $\sqrt{1 - x^2}$, $|x| \le 1$

43. $\dfrac{1}{1 - 2x}$, $x \ne \dfrac{1}{2}$, 1; $-\dfrac{1}{2x} - \dfrac{1}{2}$, $x \ne 0, 1$ **45.** $x^{-6} + 1$

47. (a) $g(x) = \sqrt{x}$, $h(x) = x + 2$ **(b)** $g(x) = |x|$, $h(x) = x^2 - 3x + 5$

49. (a) $g(x) = x^2$, $h(x) = \sin x$ **(b)** $g(x) = 3/x$, $h(x) = 5 + \cos x$

51. (a) $f(x) = x^3$, $g(x) = 1 + \sin x$, $h(x) = x^2$

(b) $f(x) = \sqrt{x}$, $g(x) = 1 - x$, $h(x) = \sqrt[3]{x}$

53.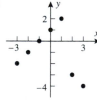

55.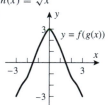

57. ± 2 **59.** $6x + 3h$, $3w + 3x$ **61.** $-\dfrac{1}{x(x + h)}$, $-\dfrac{1}{xw}$

63. (a) origin
(b) x-axis
(c) y-axis
(d) none

65. (a)

x	-3	-2	-1	0	1	2	3
$f(x)$	1	-5	-1	0	-1	-5	1

(b)

x	-3	-2	-1	0	1	2	3
$f(x)$	1	5	-1	0	1	-5	-1

67. (a) even **(b)** odd
(c) odd **(d)** neither

69. (a) even **(b)** odd **(c)** even
(d) neither **(e)** odd **(f)** even

71. (a) y-axis
(b) origin
(c) x-axis, y-axis, origin

73.

77. (a) 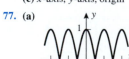 **(b)**

79. yes; $f(x) = x^k$, $g(x) = x^n$

▶ **Exercise Set 1.5 (Page 59)**
1. (a) $-\frac{3}{2}, -\frac{1}{18}, \frac{2}{3}$ **(b)** yes **3.** III $<$ II $<$ IV $<$ I
5. (a) The slopes are equal; the points lie on the same line.
(b) The slopes -1, 3, $\frac{1}{3}$ are not equal; the points do not lie on a line.
7. (a) 14 **(b)** $-\frac{1}{3}$ **9.** $\frac{13}{7}$
11. (a) $153°$ **(b)** $45°$ **(c)** $117°$ **(d)** $89°$
13. (a) $60°$ **(b)** $117°$ **15.** $y = -2x + 4$
17. $y = 4x + 7$ **19.** $y = -\frac{1}{5}x + 6$ **21.** $y = 11x - 18$
23. (a) parallel **(b)** perpendicular **(c)** parallel **(d)** perpendicular
(e) neither **25. (a)** $y = \frac{3}{2}x - 3$ **(b)** $y = -\frac{3}{4}x$
27. (a) $\frac{9}{10}$ ft/s **(b)** -4 **(c)** -2.2 **(d)** $\frac{80}{9}$ s
29. (a) $-\frac{4}{3}$ ft/s^2 **(b)** $v = -\frac{4}{3}t + \frac{13}{3}$ **(c)** $v = \frac{13}{3}$ ft/s
31. (b) $-\frac{9}{10}$ cm/s **(c)** $\frac{9}{10}$ cm/s
33. (a) 0 mi/h **(b)** 48 mi/h **(c)** 240 mi
35. (a)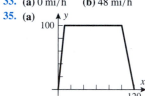

(b) $v = \begin{cases} 10t & \text{if} & 0 \le t \le 10 \\ 100 & \text{if} & 10 \le t \le 100 \\ 600 - 5t & \text{if} & 100 \le t \le 120 \end{cases}$

37. (a) $y = x/9$ **(b)**   **(c)** 26.11 in **(d)** 135 lb

39. $y = 1.2x + 2$ **41. (a)** $T_C = \frac{5}{9}(T_F - 32)$ **(b)** $\frac{5}{9}$
(c) $-40°$ (F or C) **(d)** $37°$C **43. (a)** $p = 0.098h + 1$ **(b)** 10.20 m
45. (a) $r = -0.0125t + 0.8$ **47. (a)** $C_1 = 2x$, $C_2 = 25 + (x/4)$
(b) 64 days

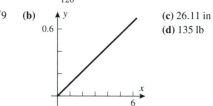

(b) $x = 15$

▶ **Exercise Set 1.6 (Page 75)**
1. (a) $y = 3x + b$ **(c)**
(b) $y = 3x + 6$

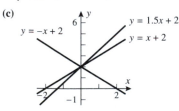

3. (a) $y = mx + 2$ **(c)**
(b) $y = -x + 2$

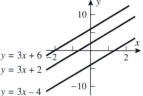

5. **(a)** slope: -1

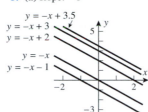

$y = -x + 3.5$
$y = -x + 3$
$y = -x + 2$
$y = -x$
$y = -x - 1$

(b) y-intercept: $y = -1$

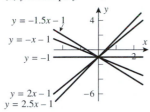

$y = -1.5x - 1$
$y = -x - 1$
$y = -1$
$y = 2x - 1$
$y = 2.5x - 1$

(c) pass through $(-4, 2)$

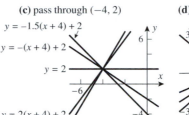

$y = -1.5(x + 4) + 2$
$y = -(x + 4) + 2$
$y = 2$
$y = 2(x + 4) + 2$
$y = 2.5(x + 4) + 2$

(d) x-intercept: $x = 1$

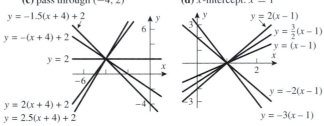

$y = 2(x - 1)$
$y = \frac{3}{2}(x - 1)$
$y = (x - 1)$
$y = -2(x - 1)$
$y = -3(x - 1)$

7. $y = \pm \dfrac{9 - x_0 x}{\sqrt{9 - x_0^2}}$

9.

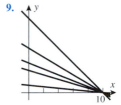

11. **(a)** VI
(b) IV
(c) III
(d) V
(e) I
(f) II

13. **(a)**

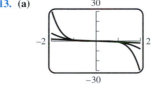

(b)

(c)

15. **(a)**

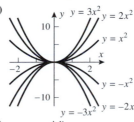

$y = 3x^2$ $y = 2x^2$
$y = x^2$
$y = -x^2$
$y = -3x^2$ $y = -2x^2$

(b)

(c)

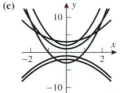

17. **(a)**

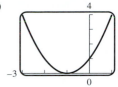

(b)

(c)

(d)

19. **(a)**

(b)

(c)

(d)

21.

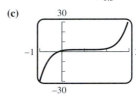

23. **(a)** newton-meters (N·m)   **(b)** 20 N·m

(c)

V (L)	0.25	0.5	1.0	1.5	2.0
P (N/m²)	80×10^3	40×10^3	20×10^3	13.3×10^3	10×10^3

(d)

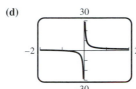

25. **(a)** $k = 0.000045$ N·m²
(b) 0.000005 N

(c)

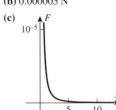

(d) The force becomes infinite; the force tends to zero.

27. **(a)** II; $y = 1, x = -1, 2$
(b) I; $y = 0, x = -2, 3$
(c) IV; $y = 2$
(d) III; $y = 0, x = -2$

29. Order the six trigonometric functions as sin, cos, tan, cot, sec, csc:
(a) pos, pos, pos, pos, pos, pos
(b) neg, zero, undefined, zero, undefined, neg
(c) pos, neg, neg, neg, neg, pos   **(d)** neg, pos, neg, neg, pos, neg
(e) neg, neg, pos, pos, neg, neg   **(f)** neg, pos, neg, neg, pos, neg

31. (a) $\sin(\pi - x) = \sin x$; 0.588
   (b) $\cos(-x) = \cos x$; 0.924
   (c) $\sin(2\pi + x) = \sin x$; 0.588
   (d) $\cos(\pi - x) = -\cos x$; -0.924
   (e) $\cos^2 x = 1 - \sin^2 x$; 0.655
   (f) $\sin^2 2x = 4\sin^2 x(1 - \sin^2 x)$; 0.905
33. (a) $-a$ (b) b (c) $-c$ (d) $\pm\sqrt{1 - a^2}$ (e) $-b$ (f) $-a$
   (g) $\pm 2b\sqrt{1 - b^2}$ (h) $2b^2 - 1$ (i) $1/b$ (j) $-1/a$ (k) $1/c$
   (l) $(1 - b)/2$ **35.** 87,458 km
37. The second quarter revolves twice ($720°$) about its own center.
39. (a) $y = 3\sin(x/2)$ (b) $y = 4\cos 2x$ (c) $y = -5\sin 4x$
41. (a) $y = \sin[x + (\pi/2)]$ (b) $y = 3 + 3\sin(2x/9)$
   (c) $y = 1 + 2\sin\left[2\left(x - \dfrac{\pi}{4}\right)\right]$
43. (a) $3, \pi/2, 0$ (b) $2, 2, 0$

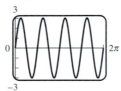

(c) $1, 4\pi, 0$

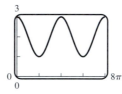

45. (b) $A = \sqrt{A_1^2 + A_2^2}, \theta = \tan^{-1}(A_2/A_1)$
   (c) $x = \dfrac{5\sqrt{13}}{2}\sin\left(2\pi t + \tan^{-1}\dfrac{1}{2\sqrt{3}}\right)$

▶ **Exercise Set 1.7 (Page 84)**

1. II   **3.** $1.5388t - 2842.9, 0.8341$

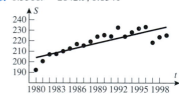

5. (a) $p = 0.0146T + 3.98, 0.9999$ (b) 3.25 atm   (c) $\approx -272°$C
7. (a) $R = 0.00723T + 1.55$ (b) $\approx -214°$C
9. (a) $S = 0.50179w - 0.00643$ (b) ≈ 16.0 lb
11. (a) $y = 0.00630h - 0.266, 0.313$ **13.** (a) 181.8 km/s/Mly
   (b)   (b) 1.492×10^{10} years

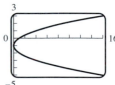
   (c) increase

15. (a) $p = 0.322t^2 + 0.0671t + 0.00837$ (b) ≈ 1.43 m
17. $T = 849.5 + 143.5\sin\left[\dfrac{\pi}{183}t - \dfrac{\pi}{2}\right]$ **19.** $t = 0.445\sqrt{d}$

▶ **Exercise Set 1.8 (Page 95)**

1. (a)

(c)

t	0	1	2	3	4	5
x	-1	0	1	2	3	4
y	1	2	3	4	5	6

3.

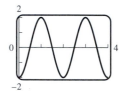

5.

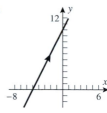

7.

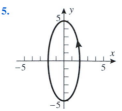

9.

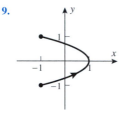

11.

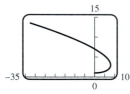

13. $x = 5\cos t, \ y = -5\sin t, \ 0 \le t \le 2\pi$
15. $x = 2, \ y = t$
17. $x = t^2, \ y = t, \ -1 \le t \le 1$

19. (a) IV   (b) II   (c) V   (d) VI   (e) III   (f) I
21. (a)

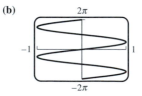

(b)

t	0	1	2	3	4	5
x	0	5.5	8	4.5	-8	-32.5
y	1	1.5	3	5.5	9	13.5

(c) $t = 0, 2\sqrt{3}$
(d) $0 < t < 2\sqrt{2}$
(e) 2

23. (a)   (b)

25. (a) $\dfrac{x - x_0}{x_1 - x_0} = \dfrac{y - y_0}{y_1 - y_0}$ (c) $x = 1 + t, \ y = -2 + 6t$
   (d) $x = 2 - t, \ y = 4 - 6t$ **27.** (b) $\frac{1}{2}$ (c) $\frac{3}{4}$

31. (b)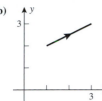

33.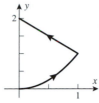

33. (a) 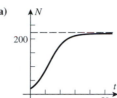 **(b)** about 10 years **(c)** 220 sheep

35. (a) $x = 4\cos t$, $y = 3\sin t$ **(b)** $x = -1 + 4\cos t$, $y = 2 + 3\sin t$

37. (a) $x = 400\sqrt{2}t$, $y = 400\sqrt{2}t - 4.9t^2$
(b) 16,326.53 m **(c)** 65,306.12 m

39. (a) ellipses with fixed center, varying axes of symmetry
(b) (assume $a \neq 0, b \neq 0$) ellipses with varying center, fixed axes of symmetry
(c) circles of radius 1 with centers on line $y = x - 1$

41. (a)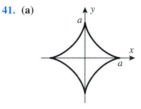

35. (a) 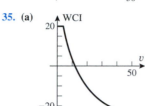 **(b)** $3°\text{F}, -11°\text{F}, -18°\text{F}, -22°\text{F}$
(c) $v = 35, 19, 12, 7$ mi/h

37. $-0.7245 \le x \le 1.2207$; $-1.0551 \le y \le 1.4902$

39. (a) 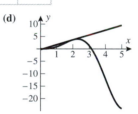 **(c)** For large t the velocity approaches c.
(d) No, but it comes arbitrarily close.
(e) 3.013 s

▶ **Chapter 1 Supplementary Exercises (Page 98)**

1. 1940–1945 **3.** 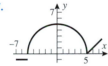 **5.** $C = 5x^2 + (64/x)$

41. (a)

1.90	1.92	1.94	1.96	1.98	2.00
3.4161	3.4639	3.5100	3.5543	3.5967	3.6372

2.02	2.04	2.06	2.08	2.10	
3.6756	3.7119	3.7459	3.7775	3.8068	

7. 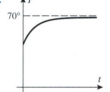 **9. (a)** $V = (6 - 2x)(5 - x)x$ ft³
(b) $0 < x < 3$
(c) 3.57 ft × 3.79 ft × 1.21 ft
11. $0, -2$
13. $1/(2 - x^2)$

(b) $y = 1.9590x - 0.29101$ **(d)**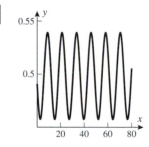
(c) $y = 1.9590x - 0.2808$

15.

x	-4	-3	-2	-1	0	1	2	3	4
$f(x)$	0	-1	2	1	3	-2	-3	4	-4
$g(x)$	3	2	1	-3	-1	-4	4	-2	0
$(f \circ g)(x)$	4	-3	-2	-1	1	0	-4	2	3
$(g \circ f)(x)$	-1	-3	4	-4	-2	1	2	0	3

17. (a) odd **(b)** even **(c)** neither **(d)** even **19. (b)** 295.72 ft
21. $A: \left(-\frac{2}{3}\pi, 1 - \sqrt{3}\right)$; $B: \left(\frac{1}{3}\pi, 1 + \sqrt{3}\right)$;
$C: \left(\frac{2}{3}\pi, 1 + \sqrt{3}\right)$; $D: \left(\frac{5}{3}\pi, 1 - \sqrt{3}\right)$
23. (a) circles of radius 1 centered on the parabola $y = x^2$
(b) parabolas that open up with vertices on the line $y = x/2$

25. 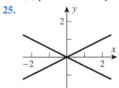 **27.**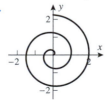

29. $d = \sqrt{(x - 1)^2 + (\sqrt{x} - 2)^2}$; $x = 1.358094$ **31.** 0.48 ft

43. $T = 0.537 + 0.495 \sin\left[\frac{\pi}{6}(t - 6.5)\right]$

▶ **Exercise Set 2.1 (Page 118)**
1. (a) -1 **(b)** 3 **(c)** does not exist **(d)** 1 **(e)** -1 **(f)** 3
3. (a) 1 **(b)** 1 **(c)** 1 **(d)** 1 **(e)** $-\infty$ **(f)** $+\infty$
5. (a) 0 **(b)** 0 **(c)** 0 **(d)** 3 **(e)** $+\infty$ **(f)** $+\infty$
7. (a) $-\infty$ **(b)** $+\infty$ **(c)** does not exist **(d)** undefined **(e)** 2 **(f)** 0
9. (a) $-\infty$ **(b)** $-\infty$ **(c)** $-\infty$ **(d)** 1 **(e)** 1 **(f)** 2
11. (a) 0 **(b)** 0 **(c)** 0 **(d)** 0 **(e)** does not exist **(f)** does not exist
13. for all $x_0 \neq -4$

15.

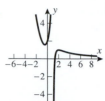

17.

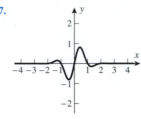

19. (a) $\frac{1}{3}$ **(b)** $+\infty$ **(c)** $-\infty$ **21. (a)** 3 **(b)** does not exist **23.** 21 ft/s

25. (a) $y = 2$ **(b)** $y = 20.086$ **(c)** no horizontal asymptote

27. $\lim\limits_{t \to +\infty} n(t) = +\infty$; $\lim\limits_{t \to +\infty} e(t) = c$

29. (a) (i) 1 **(ii)** $\lim\limits_{t \to 0^+} \dfrac{\sin t}{t}$ **(b) (i)** -1 **(ii)** $\lim\limits_{t \to 0^+} \dfrac{t-1}{t+1}$

(c) (i) e^2 **(ii)** $\lim\limits_{t \to 0^-} (1 + 2t)^{1/t}$

35. (a) catastrophic subtraction when the x-interval is small (the size depending on the calculating utility) **(c)** no

▶ **Exercise Set 2.2 (Page 129)**

1. (a) 7 **(b)** π **(c)** -6 **(d)** 36

3. (a) -6 **(b)** 13 **(c)** -8 **(d)** 16 **(e)** 2 **(f)** $-\frac{1}{2}$

(g) The denominator tends to zero but the numerator does not.

(h) The denominator tends to zero but the numerator does not.

5. 0 **7.** 8 **9.** 4 **11.** $-\frac{4}{5}$ **13.** $\frac{3}{2}$ **15.** $+\infty$

17. does not exist **19.** $-\infty$ **21.** $+\infty$ **23.** does not exist

25. $+\infty$ **27.** $+\infty$ **29.** 6 **33. (a)** 2 **(b)** 2 **(c)** 2

37. (a) Theorem 2.2.2(a) does not apply.

(b) $\lim\limits_{x \to 0^+} \left(\dfrac{1}{x} - \dfrac{1}{x^2} \right) = \lim\limits_{x \to 0^+} \left(\dfrac{x-1}{x^2} \right) = -\infty$ **39.** $\frac{1}{4}$

41. The left and/or right limits could be $\pm\infty$; or the limit could exist and equal any preassigned real number.

▶ **Exercise Set 2.3 (Page 136)**

1. (a) -3 **(b)** $-\infty$ **3. (a)** -12 **(b)** 21 **(c)** -15 **(d)** 25

(e) 2 **(f)** $-\frac{3}{5}$ **(g)** 0 **(h)** does not exist

5. $+\infty$ **7.** $-\infty$ **9.** $+\infty$ **11.** $\frac{3}{2}$ **13.** 0 **15.** 0

17. $\dfrac{-\sqrt[3]{5}}{2}$ **19.** $-\sqrt{5}$ **21.** $1/\sqrt{6}$ **23.** $\sqrt{3}$ **25.** $-\infty$

27. $-\frac{1}{7}$ **29. (a)** $+\infty$ **(b)** -5 **31.** 0 **33.** $a/2$

35. $\lim\limits_{x \to +\infty} p(x) = (-1)^n \infty$ and $\lim\limits_{x \to -\infty} p(x) = +\infty$

37. For $m > n$, the limits are both zero; for $m = n$, the limits are equal to the leading coefficient of p; for $n > m$, the limits are $\pm\infty$.

39. $\lim\limits_{x \to -\infty} = \begin{cases} 0 & \text{if } m > n \\ -3 & \text{if } m = n \\ +\infty & \text{if } m < n \text{ and } n - m \text{ odd} \\ -\infty & \text{if } m < n \text{ and } n - m \text{ even} \end{cases}$

41. $x + 2$ **43.** $1 - x^2$ **45.** $\sin x$

▶ **Exercise Set 2.4 (Page 145)**

1. (a) $|x| < 0.1$ **(b)** $|x - 3| < 0.0025$ **(c)** $|x - 4| < 0.000125$

3. (a) $x_1 = 3.8025$, $x_2 = 4.2025$ **(b)** $\delta = 0.1975$

5. $\delta = 0.0442$

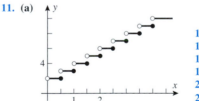

7. $\delta = 0.13$ **9.** $\delta = 0.05$

11. $\delta = \frac{1}{700}$ **13.** $\delta = 0.05$

15. $\delta = \frac{1}{9000}$ **17.** $\delta = 1$

19. $\delta = \frac{1}{3}\epsilon$ **21.** $\delta = \frac{1}{2}\epsilon$

23. $\delta = \epsilon$

25. $\delta = \min\left(1, \frac{1}{6}\epsilon\right)$

27. $\delta = \min\left(\dfrac{1}{6}, \dfrac{\epsilon}{18}\right)$

29. $\delta = 2\epsilon$

31. $\delta = \epsilon$ **33. (a)** $\sqrt{10}$ **(b)** 99 **(c)** -10 **(d)** -101

35. (a) $-\sqrt{\dfrac{1-\epsilon}{\epsilon}}; \sqrt{\dfrac{1-\epsilon}{\epsilon}}$ **(b)** $\sqrt{\dfrac{1-\epsilon}{\epsilon}}$ **(c)** $-\sqrt{\dfrac{1-\epsilon}{\epsilon}}$

37. 10 **39.** 999 **41.** -202 **43.** -57.5 **45.** $\dfrac{1}{\sqrt{\epsilon}}$

47. $-2 - \dfrac{1}{\epsilon}$ **49.** $\dfrac{1}{\epsilon} - 1$ **51.** $-\dfrac{5}{2} - \dfrac{11}{2\epsilon}$

53. (a) $|x| < \frac{1}{10}$ **(b)** $|x - 1| < \frac{1}{1000}$ **(c)** $|x - 3| < \frac{1}{10\sqrt{10}}$

(d) $|x| < \frac{1}{10}$ **55.** $\delta = 1/\sqrt{M}$ **57.** $\delta = 1/M$

59. $\delta = 1/(-M)^{1/4}$ **61.** $\delta = \epsilon$ **63.** $\delta = \epsilon^2$ **65.** $\delta = \epsilon$

67. (a) $\delta = -1/M$ **(b)** $\delta = 1/M$

69. (a) $N = M - 1$ **(b)** $N = M - 1$ **71.** $\delta = \min\left(2, \frac{1}{8}\epsilon\right)$

▶ **Exercise Set 2.5 (Page 156)**

1. (a) not continuous, $x = 2$ **(b)** not continuous, $x = 2$

(c) not continuous, $x = 2$ **(d)** continuous **(e)** continuous

(f) continuous

3. (a) not continuous, $x = 1, 3$ **(b)** continuous

(c) not continuous, $x = 1$ **(d)** continuous

(e) not continuous, $x = 3$ **(f)** continuous

5. (a) $x = 3$, no limits **(b)** $x = -2$, limit $\neq F(-2)$

(c) $c = 3$, no limits **7. (a)** 3 **(b)** 3

9. (a)

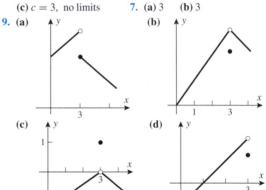

11. (a)

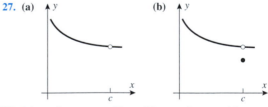

(b) One second could cost you one dollar.

13. none

15. none

17. f is not defined at $x = \pm 4$.

19. f is not defined at $x = \pm 3$.

21. none **23.** none

25. (a) $k = 5$ **(b)** $k = \frac{4}{3}$

27. (a)

(b)

29. (a) $x = 0$, not removable **(b)** $x = -3$, removable

(c) $x = 2$, removable; $x = -2$, not removable

31. (a) $x = \frac{1}{2}$, not removable; at $x = -3$, removable

(b) $(2x - 1)(x + 3)$

35. (a) $f(x) = k$ for $x \neq c$, $f(c) = 0$; $g(x) = \ell$ for $x \neq c$, $g(c) = 0$.

If $k = -\ell$, then $f + g$ is continuous; otherwise it is not.

(b) $f(x) = k$ for $x \neq c$, $f(c) = 1$; $g(x) = \ell$ for $x \neq c$, $g(c) = 1$.

If $k\ell = 1$, then fg is continuous; otherwise it is not.

39. $f(x) = 1$ for $0 \le x < 1$, $f(x) = -1$ for $1 \le x \le 2$
45. $x = -1.25$, $x = 0.75$ **47.** $x = -1.605$, $x = 1.375$
49. $x = 2.24$ **51.** $x = 4.847$ cm

▶ **Exercise Set 2.6 (Page 163)**

1. none   **3.** $x = n\pi$, $n = 0, \pm 1, \pm 2, \ldots$
5. $x = n\pi$, $n = 0, \pm 1, \pm 2, \ldots$ **7.** none
9. $2n\pi + (\pi/6)$, $2n\pi + (5\pi/6)$, $n = 0, \pm 1, \pm 2, \ldots$
11. (a) $\sin x$, $x^3 + 7x + 1$ (b) $|x|$, $\sin x$ (c) x^3, $\cos x$, $x + 1$
    (d) $\sqrt{x}$, $3 + x$, $\sin x$, $2x$ (e) $\sin x$, $\sin x$ (f) $x^5 - 2x^3 + 1$, $\cos x$
13. 1   **15.** $-\sqrt{3}/2$ **17.** 3   **19.** -1 **21.** 0   **23.** $\frac{7}{3}$
25. 1   **27.** 2   **29.** 0   **31.** $-\frac{25}{49}$ **33.** does not exist
35. 3   **37.** $\frac{1}{2}$ **39.** $\frac{1}{3}$ **41.** $k = \frac{1}{2}$ **43.** (a) 1   (b) 0   (c) 1
45. $-\pi$ **47.** $-|x| \le x\cos(50\pi/x) \le |x|$
49. $\lim\limits_{x \to 0} f(x) = 1$ by the Squeezing Theorem.

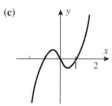

$y = \cos x$
$y = f(x)$
$y = 1 - x^2$

51. $g(x) = -\dfrac{1}{x}$, $h(x) = \dfrac{1}{x}$; $\lim\limits_{x \to +\infty} \dfrac{\sin x}{x} = 0$ by the Squeezing Theorem.
55. (a) 0.17365   (b) 0.17453
57. (a) 0.08749   (b) 0.08727
61. (a) symmetry about the equatorial plane

▶ **Chapter 2 Supplementary Exercises (Page 165)**

1. (a) 1   (b) does not exist   (c) does not exist   (d) 1   (e) 3
    (f) 0   (g) 0   (h) 2   (i) $\frac{1}{2}$ **5.** (a) $\frac{1}{4}$ (b) 4
7. (a) 0.4051   **17.** (a) 1.449   (b) $x = 0, \pm 1.896$
19. (a) $\sqrt{5}$, does not exist, $\sqrt{10}$, $\sqrt{10}$, does not exist, $+\infty$, does not exist
    (b) 5, 10, 0, 0, 10, $-\infty$, $+\infty$ **21.** a/b **23.** does not exist
25. 0   **27.** $3 - k$ **31.** 2.71828   **33.** 0.54030   **35.** 0.49996
37. 0.07747   **39.** (b)

(d) 1, 1.26, 1.31, 1.322, 1.324, 1.3246, 1.3247

41. $x = \sqrt[5]{x + 2}$; 1.267168

▶ **Exercise Set 3.1 (Page 175)**

1. (a) 4 m/s   **3.** (a) t_0 (b) 0   (c) speeding up   (d) slowing down
5. straight line with slope equal to the velocity
7. (a) $\frac{7}{2}$ (d)
    (b) 3
    (c) x_0

Secant
Tangent

9. (a) $-\frac{1}{6}$ (d)
    (b) $-\frac{1}{4}$
    (c) $-1/x_0^2$

Tangent
Secant

11. (a) $2x_0$ (b) 4
13. (a) $1/(2\sqrt{x_0})$ (b) $\frac{1}{2}$
15. (a) $72°$ F at about 4:30 P.M.
    (b) $4°$ F/h
    (c) $-7°$ F/h at about 9 P.M.
17. (a) first year   (b) 6 cm/year   (c) 10 cm/year at about age 14

(d)

Growth rate (cm/year)
t (yr)

19. (a) 320,000 ft
    (b) 8000 ft/s
    (c) 45 ft/s
    (d) 24,000 ft/s
21. (a) 720 ft/min
    (b) 192 ft/min

▶ **Exercise Set 3.2 (Page 188)**

1. 2, 0, -2, -1 **5.**
3. (b) 3
    (c) 3

7. $y = 5x - 16$
9. $6x$; $y = 18x - 27$
11. $3x^2$; $y = 0$
13. $\dfrac{1}{2\sqrt{x + 1}}$; $y = \frac{1}{6}x + \frac{5}{3}$
15. $-1/x^2$
17. $2ax$
19. $-1/2x^{3/2}$

21. $8t + 1$ **25.** (a)
23. (a) D
    (b) F
    (c) B
    (d) C
    (e) A
    (f) E

(b)

(c)

27. (a) x^2, 3
    (b) $\sqrt{x}$, 1
29. 8

31. $y = -2x + 1$

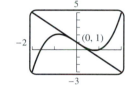

$(0, 1)$

33. (b)

h	0.5	0.1	0.01	0.001	0.0001	0.00001
$[f(1 + h) - f(1)]/h$	1.6569	1.4355	1.3911	1.3868	1.3863	1.3863

35. (a) dollars per foot   (b) the price per additional foot
    (c) positive   (d) $1000
37. (a) $F \approx 200$ lb, $dF/d\theta \approx 50$ lb/rad   (b) $\mu \approx 0.25$
39. (a) $T \approx 115°$F, $dT/dt \approx -3.35°$F/min   (b) $k = -0.084$
47. $f(1) = 0$, $f'(1) = 5$

▶ **Exercise Set 3.3 (Page 198)**

1. $28x^6$ **3.** $24x^7 + 2$ **5.** 0   **7.** $-\frac{1}{3}(7x^6 + 2)$
9. $3ax^2 + 2bx + c$ **11.** $24x^{-9} + (1/\sqrt{x})$ **13.** $-3x^{-4} - 7x^{-8}$
15. $18x^2 - \frac{3}{2}x + 12$ **17.** $-15x^{-2} - 14x^{-3} + 48x^{-4} + 32x^{-5}$
19. $12x(3x^2 + 1)$ **21.** $-\frac{5}{4}$ **23.** $\dfrac{3}{(2t + 1)^2}$ **25.** $\frac{7}{16}$
27. -29 **31.** 0   **33.** $32t$ **35.** $3\pi r^2$
37. (a) $4\pi r^2$ (b) 100π **39.** (a) $-\frac{37}{4}$ (b) $-\frac{23}{16}$
41. (a) 10   (b) 19   (c) 9   (d) -1 **43.** $y = 5x + 17$
45. (a) $42x - 10$ (b) 24   (c) $2/x^3$ (d) $700x^3 - 96x$
47. (a) $-210x^{-8} + 60x^2$ (b) $-6x^{-4}$ (c) $6a$
49. (a) 0   (b) 112   (c) 360   **53.** $F''(x) = xf''(x) + 2f'(x)$
55. $(1, \frac{5}{6})$ $(2, \frac{2}{3})$

57. $y = 3x^2 - x - 2$
59. $x = \frac{1}{2}$
61. $2 \pm \sqrt{3}$
63. $-2x_0$
67. $-\dfrac{2GmM}{r^3}$
69. $f'(x) > 0$ for all $x \ne 0$

73. (a) $2(1 + x^{-1})(x^{-3} + 7) + (2x + 1)(-x^{-2})(x^{-3} + 7) +$
$(2x + 1)(1 + x^{-1})(-3x^{-4})$ **(b)** $3(7x^6 + 2)(x^7 + 2x - 3)^2$
75. differentiable at $x = 1$ **77.** not differentiable at $x = 1$
79. (a) $x = \frac{2}{3}$ **(b)** $x = \pm 2$
83. (a) $n(n-1)(n-2)\cdots 1$ **(b)** 0 **(c)** $a_n n(n-1)(n-2)\cdots 1$
85. (b) f and all its derivatives up to $f^{(n-1)}(x)$ are continuous on (a, b).

▶ **Exercise Set 3.4 (Page 203)**
1. $-2\sin x - 3\cos x$ **3.** $\dfrac{x\cos x - \sin x}{x^2}$
5. $x^3 \cos x + (3x^2 + 5)\sin x$ **7.** $\sec x \tan x - \sqrt{2}\sec^2 x$
9. $\sec^3 x + \sec x \tan^2 x$ **11.** $-\csc^3 x - \csc x \cot^2 x$
13. $-\dfrac{\csc x}{1 + \csc x}$ **15.** 0 **17.** $\dfrac{1}{(1 + x\tan x)^2}$
19. $-x\cos x - 2\sin x$ **21.** $-x\sin x + 5\cos x$ **23.** $-4\sin x \cos x$
25. (a) $y = x$ **(b)** $y = 2x - (\pi/2) + 1$ **(c)** $y = 2x + (\pi/2) - 1$
29. (a) $x = \pm\pi/2, \pm 3\pi/2$ **(b)** $x = -3\pi/2, \pi/2$ **(c)** no horizontal
tangent line   **(d)** $x = \pm 2\pi, \pm\pi, 0$ **31.** 0.087 ft/degree
33. 1.75 m/degree   **35. (a)** $-\cos x$ **(b)** $\cos x$
37. (a) all x **(b)** all x **(c)** $x \ne (\pi/2) + n\pi, n = 0, \pm 1, \pm 2, \ldots$
   (d) $x \ne n\pi, n = 0, \pm 1, \pm 2, \ldots$
   (e) $x \ne (\pi/2) + n\pi, n = 0, \pm 1, \pm 2, \ldots$
   (f) $x \ne n\pi, n = 0, \pm 1, \pm 2, \ldots$
   (g) $x \ne (2n+1)\pi, n = 0, \pm 1, \pm 2, \ldots$
   (h) $x \ne n\pi/2, n = 0, \pm 1, \pm 2, \ldots$ **(i)** all x
39. $3, 7, 11, \ldots$ **41.** $\sec^2 y$

▶ **Exercise Set 3.5 (Page 208)**
1. 6 **3. (a)** $(2x - 3)^5, 10(2x - 3)^4$ **(b)** $2x^5 - 3, 10x^4$
5. (a) -7 **(b)** -8 **7.** $37(x^3 + 2x)^{36}(3x^2 + 2)$
9. $-2\left(x^3 - \dfrac{7}{x}\right)^{-3}\left(3x^2 + \dfrac{7}{x^2}\right)$ **11.** $\dfrac{24(1 - 3x)}{(3x^2 - 2x + 1)^4}$
13. $\dfrac{3}{4\sqrt{3x}\sqrt{4 + \sqrt{3x}}}$ **15.** $3x^2 \cos(x^3)$ **17.** $-20\cos^4 x \sin x$
19. $-\dfrac{2}{x^3}\cos\left(\dfrac{1}{x^2}\right)$ **21.** $28x^6 \sec^2(x^7)\tan(x^7)$ **23.** $-\dfrac{5\sin(5x)}{2\sqrt{\cos(5x)}}$
25. $-3\left[x + \csc(x^3 + 3)\right]^{-4}\left[1 - 3x^2 \csc(x^3 + 3)\cot(x^3 + 3)\right]$
27. $10x^3 \sin 5x \cos 5x + 3x^2 \sin^2 5x$
29. $-x^3 \sec\left(\dfrac{1}{x}\right)\tan\left(\dfrac{1}{x}\right) + 5x^4 \sec\left(\dfrac{1}{x}\right)$ **31.** $\sin(\cos x)\sin x$
33. $-6\cos^2(\sin 2x)\sin(\sin 2x)\cos 2x$
35. $12(5x + 8)^{13}(x^3 + 7x)^{11}(3x^2 + 7) + 65(x^3 + 7x)^{12}(5x + 8)^{12}$
37. $\dfrac{33(x - 5)^2}{(2x + 1)^4}$ **39.** $-\dfrac{2(2x + 3)^2(52x^2 + 96x + 3)}{(4x^2 - 1)^9}$
41. $5\left[x\sin 2x + \tan^4(x^7)\right]^4\left[2x\cos 2x + \sin 2x + 28x^6 \tan^3(x^7)\sec^2(x^7)\right]$
43. $y = -x$ **45.** $y = -1$ **47.** $y = 8\sqrt{\pi}x - 8\pi$
49. $y = \frac{7}{2}x - \frac{3}{2}$ **51.** $-25x\cos(5x) - 10\sin(5x) - 2\cos(2x)$
53. $4(1 - x)^{-3}$ **55.** $3\cot^2\theta \csc^2\theta$ **57.** $\pi(b - a)\sin 2\pi\omega$
59. (a)   **(c)** $\dfrac{4 - 2x^2}{\sqrt{4 - x^2}}$

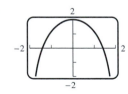

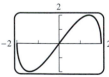

(d) $y - \sqrt{3} = \dfrac{2}{\sqrt{3}}(x - 1)$

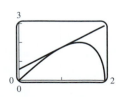

61. (c) $f = 1/T$ **(d)** amplitude $= 0.6$ cm, $T = 2\pi/15$ seconds
per oscillation, $f = 15/(2\pi)$ oscillations per second
63. (a) 10 lb/in^2, -2 lb/in^2/mi   **(b)** -0.6 lb/in^2/s
65. $\begin{cases} \cos x, & 0 < x < \pi \\ -\cos x, & -\pi < x < 0 \end{cases}$ **67. (c)** $-\dfrac{1}{x}\cos\dfrac{1}{x} + \sin\dfrac{1}{x}$
   (d) limit as x goes to 0 does not exist   **69. (a)** 21 **(b)** -36
71. $1/2x$ **73.** $\frac{2}{3}x$ **77.** $f'(g(h(x)))g'(h(x))h'(x)$

▶ **Exercise Set 3.6 (Page 217)**
1. $\frac{2}{3}(2x - 5)^{-2/3}$ **3.** $\dfrac{9}{2(x + 2)^2}\left[\dfrac{x - 1}{x + 2}\right]^{1/2}$
5. $\frac{1}{3}x^2(5x^2 + 1)^{-5/3}(25x^2 + 9)$ **7.** $-\dfrac{15[\sin(3/x)]^{3/2}\cos(3/x)}{2x^2}$
9. (a) $\dfrac{2 - 3x^2 - y}{x}$ **(b)** $-\dfrac{1}{x^2} - 2x$ **11.** $-\dfrac{x}{y}$ **13.** $\dfrac{1 - 2xy - 3y^3}{x^2 + 9xy^2}$
15. $-\dfrac{y^2}{x^2}$ **17.** $\dfrac{1 - 2xy^2 \cos(x^2 y^2)}{2x^2 y \cos(x^2 y^2)}$
19. $\dfrac{1 - 3y^2 \tan^2(xy^2 + y)\sec^2(xy^2 + y)}{3(2xy + 1)\tan^2(xy^2 + y)\sec^2(xy^2 + y)}$ **21.** $-\dfrac{21}{16y^3}$
23. $\dfrac{2y}{x^2}$ **25.** $\dfrac{\sin y}{(1 + \cos y)^3}$ **27.** $-1, +1$ **29.** -0.1312
31. $-\frac{9}{13}$ **35. (a)**   **(b)** ± 1.1547
   (c) $x = \pm\dfrac{2}{\sqrt{3}}$

37. $\dfrac{2t^3 + 3a^2}{2a^3 - 6at}$ **39.** $-\dfrac{b^2\lambda}{a^2\omega}$ **41.** $a = \frac{1}{4}, b = \frac{5}{4}$
43. $y = (\sqrt{3}/3)x, y = -(\sqrt{3}/3)x$ **45.** $-\dfrac{2y^3 + 3t^2 y}{(6ty^2 + t^3)\cos t}$
47. $\dfrac{dy}{dt} = \dfrac{3\cos 3x - y^2}{2xy}\dfrac{dx}{dt}$ **49.** $-1, \frac{2}{3}$

▶ **Exercise Set 3.7 (Page 223)**
1. (a) 6 **(b)** $-\frac{1}{3}$ **3. (a)** $-\dfrac{1}{\sqrt{3}}$ **(b)** 2
5. (b) $A = x^2$ **(c)** $\dfrac{dA}{dt} = 2x\dfrac{dx}{dt}$ **(d)** 12 ft^2/min
7. (a) $\dfrac{dV}{dt} = \pi\left(r^2\dfrac{dh}{dt} + 2rh\dfrac{dr}{dt}\right)$ **(b)** -20π in^3/s; decreasing
9. (a) $\dfrac{d\theta}{dt} = \dfrac{\cos^2\theta}{x^2}\left(x\dfrac{dy}{dt} - y\dfrac{dx}{dt}\right)$ **(b)** $-\frac{5}{16}$ rad/s; decreasing
11. $\dfrac{4\pi}{15}$ in^2/min   **13.** $\dfrac{1}{\sqrt{\pi}}$ mi/h   **15.** 4860π cm^3/min
17. $\frac{5}{6}$ ft/s   **19.** $\dfrac{125}{\sqrt{61}}$ ft/s   **21.** 704 ft/s
23. (a) 500 mi, 1716 mi   **(b)** 1354 mi; 27.7 mi/min
25. $\dfrac{9}{20\pi}$ ft/min   **27.** 125π ft^3/min   **29.** 250 mi/h
31. $\dfrac{36\sqrt{69}}{25}$ ft/min   **33.** $\dfrac{8\pi}{5}$ km/s   **35.** $600\sqrt{7}$ mi/h
37. (a) $-\frac{60}{7}$ units per second   **(b)** falling   **39.** -4 units per second
41. $x = \pm\sqrt{\dfrac{-5 + \sqrt{33}}{2}}$ **43.** 4.5 cm/s; away   **47.** $\dfrac{20}{9\pi}$ cm/s

▶ **Exercise Set 3.8 (Page 232)**
1. (a) $f(x) \approx 1 + 3(x - 1)$ **(b)** $f(1 + \Delta x) \approx 1 + 3\,\Delta x$ **(c)** 1.06

3. **(a)** $1 + \frac{1}{2}x$, 0.95, 1.05

(b)

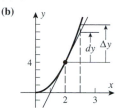

13. $|x| < 1.692$

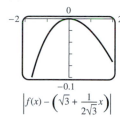

$$\left| f(x) - \left(\sqrt{3} + \frac{1}{2\sqrt{3}} x \right) \right|$$

15. 0.31583   **17.** **(a)** 0.0174533   **(b)** $x_0 = 45°$ **(c)** 0.694765

19. 83.16   **21.** 8.0625   **23.** 8.9944   **25.** 0.1   **27.** 0.8573

29. **(a)** 4, 5   **31.** **(a)** 0.5, 1

(b)   **(b)**

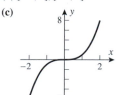

33. $3x^2\,dx$, $3x^2\,\Delta x + 3x(\Delta x)^2 + (\Delta x)^3$

35. $(2x - 2)\,dx$, $2x\,\Delta x + (\Delta x)^2 - 2\,\Delta x$

37. **(a)** $(12x^2 - 14x)\,dx$ **(b)** $(-x\sin x + \cos x)\,dx$

39. **(a)** $\dfrac{2 - 3x}{2\sqrt{1 - x}}\,dx$ **(b)** $-17(1 + x)^{-18}\,dx$ **41.** 0.0225

43. 0.0048   **45.** **(a)** $\pm 2\ \text{ft}^2$ **(b)** side: $\pm 1\%$; area: $\pm 2\%$

47. **(a)** opposite: ± 0.151 in; adjacent: ± 0.087 in   **(b)** opposite: $\pm 3.0\%$;

adjacent: $\pm 1.0\%$ **49.** $\pm 10\%$ **51.** $\pm 0.017\ \text{cm}^2$

53. $\pm 6\%$ **55.** $\pm 0.5\%$ **57.** $0.236\ \text{cm}^3$

59. **(a)** $\alpha = 1.5 \times 10^{-5}/°\text{C}$ **(b)** 180.1 cm long

▶ **Chapter 3 Supplementary Exercises** (Page 234)

5. $x = -\frac{7}{2}, 2, -\frac{1}{2}$ **7.** $x = 1, -\frac{1}{15}$

9. **(a)** $x = -2, -1, 1, 3$ **(b)** $(-\infty, -2), (-1, 1), (3, +\infty)$

     (c) $(-2, -1), (1, 3)$ **(d)** 4   **11.** $y = -16x$, $y = -145x/4$

13. $x = n\pi \pm (\pi/4), n = 0, \pm 1, \pm 2, \ldots$

17. **(a)** $-0.5, 1, 0.5$ **(b)** $\pi/4, 1, \pi/2$ **(c)** $3, -1, 0$

19. **(a)** 2000 gal/min   **(b)** 2500 gal/min

21. **(a)** between 139.48 m and 144.55 m   **(b)** $|d\phi| \leq 0.98°$

23. **(a)** 3.6   **(b)** -0.777778 **25.** $4\ln 2$

27. 58.75 ft/s   **29.** ± 0.535428 **39.** $y = 1$

▶ **Exercise Set 4.1** (Page 250)

1. **(a)** yes   **(b)** no   **(c)** yes   **(d)** no   **3.** **(a)** yes   **(b)** no

5. **(a)** yes   **(b)** yes   **(c)** no   **(d)** yes   **(e)** no   **(f)** no

7. **(a)** no   **(b)** no   **(c)** yes

9. **(b)** $[-2, 2]$, $[-8, 8]$

(c)

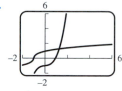

11. **(a)** no   **(b)** yes   **(c)** yes

13. $x^{1/5}$

15. $\frac{1}{7}(x + 6)$

17. $\sqrt[3]{(x + 5)/3}$

19. $(x^3 + 1)/2$

21. $-\sqrt{3/x}$

23. $\begin{cases} (5/2) - x, & x > 1/2 \\ 1/x, & 0 < x \leq 1/2 \end{cases}$

25. $x^{1/4} - 2$ for $x \geq 16$

27. $\frac{1}{2}(3 - x^2)$ for $x \leq 0$ **29.** $\frac{1}{10}(1 + \sqrt{1 - 20x})$ for $x \leq -4$

31. **(a)** $y = (6.214 \times 10^{-4})x$ **(b)** $x = \dfrac{10^4}{6.214}y$

     (c) how many meters in y miles

33. **(b)**

(c) No, because $f(g(x)) = x$ for $x > 1$ but the domain of g is $x \geq 0$.

35. **(b)** symmetric about the line $y = x$ **37.** **(b)** $1 - (\sqrt{3}/3)$

39. 10

41.   **43.**

45. $\dfrac{1}{15y^2 + 1}$ **51.**   **53.** 25

47. $\dfrac{1}{10y^4 + 3y^2}$

▶ **Exercise Set 4.2** (Page 260)

1. **(a)** -4 **(b)** 4   **(c)** $\frac{1}{4}$ **3.** **(a)** 2.9690   **(b)** 0.0341

5. **(a)** 4   **(b)** -5 **(c)** 1   **(d)** $\frac{1}{2}$ **7.** **(a)** 1.3655   **(b)** -0.3011

9. **(a)** $2r + \dfrac{s}{2} + \dfrac{t}{2}$ **(b)** $s - 3r - t$ **11.** **(a)** $1 + \log x + \frac{1}{2}\log(x - 3)$

     (b) $2\ln|x| + 3\ln\sin x - \frac{1}{2}\ln(x^2 + 1)$ **13.** $\log\frac{256}{3}$

15. $\ln\dfrac{\sqrt[3]{x}(x + 1)^2}{\cos x}$ **17.** 0.01   **19.** e^2 **21.** 4   **23.** 10^5

25. $\sqrt{3/2}$ **27.** $-\dfrac{\ln 3}{2\ln 5}$ **29.** $\frac{1}{3}\ln\frac{7}{2}$ **31.** -2 **33.** $0, -\ln 2$

35. **(a)**   **(b)**

37. $2.8777, -0.3174$ **39.**   **41.** $x = 3.6541$

                                           $y = 1.2958$

43. **(a)** no   **(b)** $y = 2^{x/4}$ **45.** $\log\frac{1}{2} < 0$, so $3\log\frac{1}{2} < 2\log\frac{1}{2}$.

     (c) $y = 2^{-x}$ **(d)** $y = (\sqrt{5})^x$ **47.** 201 days

49. **(a)** 7.4, basic

           (b) 4.2, acidic

           (c) 6.4, acidic

           (d) 5.9, acidic

51. **(a)** 140 dB, damage

           (b) 120 dB, damage

           (c) 80 dB, no damage

           (d) 75 dB, no damage

53. ≈ 200 **55.** **(a)** $\approx 5 \times 10^{16}$ J   **(b)** ≈ 0.67 **57.** e^{-2}

▶ **Exercise Set 4.3** (Page 267)

1. $\dfrac{1}{x}$ **3.** $\dfrac{2\ln x}{x}$ **5.** $\dfrac{\sec^2 x}{\tan x}$ **7.** $\dfrac{1 - x^2}{x(1 + x^2)}$

9. $\dfrac{3x^2 - 14x}{x^3 - 7x^2 - 3}$ **11.** $\dfrac{1}{2x\sqrt{\ln x}}$ **13.** $-\dfrac{1}{x}\sin(\ln x)$

15. $3x^2\log_2(3 - 2x) - \dfrac{2x^3}{(\ln 2)(3 - 2x)}$ **17.** $\dfrac{2x(1 + \log x) - x/(\ln 10)}{(1 + \log x)^2}$

19. $7e^{7x}$ **21.** $x^2 e^x(x + 3)$ **23.** $\dfrac{4}{(e^x + e^{-x})^2}$

25. $(x \sec^2 x + \tan x)e^{x \tan x}$ **27.** $(1 - 3e^{3x})e^{x - e^{3x}}$ **29.** $\dfrac{x - 1}{e^x - x}$

31. $-\dfrac{y}{x(y + 1)}$ **33.** $-\tan x + \dfrac{3x}{4 - 3x^2}$

35. $x\sqrt[3]{1 + x^2}\left[\dfrac{1}{x} + \dfrac{2x}{3(1 + x^2)}\right]$

37. $\dfrac{(x^2 - 8)^{1/3}\sqrt{x^3 + 1}}{x^6 - 7x + 5}\left[\dfrac{2x}{3(x^2 - 8)} + \dfrac{3x^2}{2(x^3 + 1)} - \dfrac{6x^5 - 7}{x^6 - 7x + 5}\right]$

39. $2^x \ln 2$ **41.** $\pi^{\sin x}(\ln \pi)\cos x$

43. $(x^3 - 2x)^{\ln x}\left[\dfrac{3x^2 - 2}{x^3 - 2x}\ln x + \dfrac{1}{x}\ln(x^3 - 2x)\right]$

45. $(\ln x)^{\tan x}\left[\dfrac{\tan x}{x \ln x} + (\sec^2 x)\ln(\ln x)\right]$ **47.** exe^{-1}

49. (a) $-\dfrac{1}{x(\ln x)^2}$ **(b)** $-\dfrac{\ln 2}{x(\ln x)^2}$ **51. (a)** $k^n e^{kx}$ **(b)** $(-1)^n k^n e^{-kx}$

53. $-\dfrac{1}{\sqrt{2\pi}\sigma^3}(x - \mu)\exp\left[-\dfrac{1}{2}\left(\dfrac{x - \mu}{\sigma}\right)^2\right]$ **57. (a)** 1   **(b)** $\ln 10$

▶ **Exercise Set 4.4 (Page 273)**

1. (a) $-\pi/2$ **(b)** π **(c)** $-\pi/4$ **(d)** 0
3. $1/2, -\sqrt{3}, -1/\sqrt{3}, 2, -2/\sqrt{3}$ **5.** $\frac{4}{5}, \frac{3}{5}, \frac{3}{4}, \frac{5}{3}, \frac{5}{4}$
7. (a) $\pi/7$ **(b)** 0   **(c)** $2\pi/7$ **(d)** $201\pi - 630$ **9. (a)** $0 \le x \le \pi$
   (b) $-1 \le x \le 1$ **(c)** $-\pi/2 < x < \pi/2$ **(d)** $-\infty < x < +\infty$
11. $\frac{24}{25}$ **13. (a)** $\dfrac{1}{\sqrt{1 + x^2}}$ **(b)** $\dfrac{\sqrt{1 - x^2}}{x}$ **(c)** $\dfrac{\sqrt{x^2 - 1}}{x}$ **(d)** $\dfrac{1}{\sqrt{x^2 - 1}}$

15. (a)

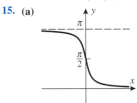

(b) domain of $\cot^{-1} x$ is $(-\infty, +\infty)$, range is $(0, \pi)$; domain of $\csc^{-1} x$ is $(-\infty, -1] \cup [1, +\infty)$, range is $[-\pi/2, 0) \cup (0, \pi/2]$.

17. (a) $55.0°$ **(b)** $33.6°$ **(c)** $25.8°$
21. (a) $x = 3.6964$ rad   **(b)** $\theta = -76.7°$
23. (a) $\dfrac{1}{\sqrt{9 - x^2}}$ **(b)** $-\dfrac{2}{\sqrt{1 - (2x + 1)^2}}$
25. (a) $\dfrac{7}{|x|\sqrt{x^{14} - 1}}$ **(b)** $-\dfrac{1}{\sqrt{e^{2x} - 1}}$
27. (a) $-\dfrac{1}{|x|\sqrt{x^2 - 1}}$ **(b)** $\begin{cases} 1, & \sin x > 0 \\ -1, & \sin x < 0 \end{cases}$
29. (a) $\dfrac{e^x}{|x|\sqrt{x^2 - 1}} + e^x \sec^{-1} x$ **(b)** $\dfrac{3x^2(\sin^{-1} x)^2}{\sqrt{1 - x^2}} + 2x(\sin^{-1} x)^3$
31. $\dfrac{(3x^2 + \tan^{-1} y)(1 + y^2)}{(1 + y^2)e^y - x}$

33. (a)

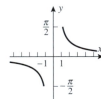

(b)

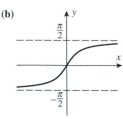

35. $23°$ **37.** $32°$ or $58°$; $32°$ **39.** $29°$

▶ **Exercise Set 4.5 (Page 283)**

1. (a) $\frac{2}{3}$ **(b)** $\frac{2}{3}$ **3.** 1   **5.** 1   **7.** 1   **9.** -1 **11.** 0
13. $-\infty$ **15.** 0   **17.** 2   **19.** 0   **21.** π **23.** $-\frac{5}{3}$ **25.** e^{-3}

27. e^2 **29.** $e^{2/\pi}$ **31.** 0   **33.** $\frac{1}{2}$ **35.** $+\infty$ **39. (b)** 2

41. 0   **43.** e^3

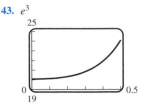

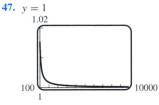

45. no horizontal asymptote   **47.** $y = 1$

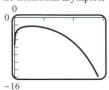

49. (a) 0   **(b)** $+\infty$ **(c)** 0   **(d)** $-\infty$ **(e)** $+\infty$ **(f)** $-\infty$
51. 1   **53.** does not exist   **55.** Vt/L
59. $k = -1, \ell = \pm 2\sqrt{2}$ **61.** does not exist

▶ **Chapter 4 Supplementary Exercises (Page 285)**

3. (a) $\frac{1}{2}(x + 1)^{1/3}$ **(b)** none   **(c)** $\frac{1}{2}\ln(x - 1)$ **(d)** $\dfrac{x + 2}{x - 1}$
5. $15x + 2$ **7. (a)** $-\dfrac{3}{x^2}$ **(b)** $\dfrac{2}{x}$
9. (a)

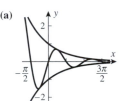

(b) $x = -\dfrac{\pi}{4}, \dfrac{3\pi}{4}$ **11. (b)** $x = 3.654$

15. (a) $3x^2$ **(b)** $\dfrac{abe^{-x}}{(1 + be^{-x})^2}$ **(c)** $\dfrac{5x + 3}{6x(x + 1)} - \cot x - \tan x$

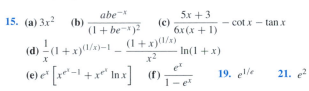

(d) $\dfrac{1}{x}(1 + x)^{(1/x) - 1} - \dfrac{(1 + x)^{(1/x)}}{x^2}\ln(1 + x)$
(e) $e^x\left[x e^{x - 1} + x e^x \ln x\right]$ **(f)** $\dfrac{e^x}{1 - e^x}$ **19.** $e^{1/e}$ **21.** e^2

23. $-\dfrac{k_0 q}{2T^2}\exp\left(-\dfrac{q(T - T_0)}{2T_0 T}\right)$

25. (a)

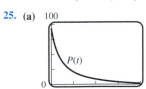

(b) The population tends to 19.
(c) The *rate* tends to zero.

27. (b)

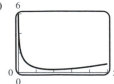

(d) must take the value zero in between
(e) $x = 2$

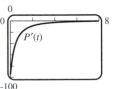

► **Exercise Set 5.1 (Page 297)**

1. **(a)** $f' > 0$, $f'' > 0$ **(b)** $f' > 0$, $f'' < 0$ **(c)** $f' < 0$, $f'' > 0$

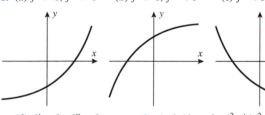

(d) $f' < 0$, $f'' < 0$

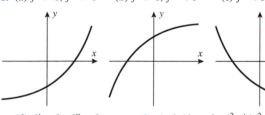

3. A: $dy/dx < 0$, $d^2y/dx^2 > 0$
   B: $dy/dx > 0$, $d^2y/dx^2 < 0$
   C: $dy/dx < 0$, $d^2y/dx^2 < 0$

5. $x = -1, 0, 1, 2$

7. **(a)** $[4, 6]$ **(b)** $[1, 4]$, $[6, 7]$
   (c) $(1, 2)$, $(3, 5)$ **(d)** $(2, 3)$, $(5, 7)$
   (e) $x = 2, 3, 5$

9. **(a)** $[1, 3]$ **(b)** $(-\infty, 1]$, $[3, +\infty)$ **(c)** $(-\infty, 2)$, $(4, +\infty)$
   (d) $(2, 4)$ **(e)** $x = 2, 4$

11. **(a)** $\left[\frac{5}{2}, +\infty\right)$ **(b)** $\left(-\infty, \frac{5}{2}\right]$ **(c)** $(-\infty, +\infty)$ **(d)** none   **(e)** none

13. **(a)** $(-\infty, +\infty)$ **(b)** none   **(c)** $(-2, +\infty)$ **(d)** $(-\infty, -2)$ **(e)** -2

15. **(a)** $[1, +\infty)$ **(b)** $(-\infty, 1]$ **(c)** $(-\infty, 0)$, $\left(\frac{2}{3}, +\infty\right)$
    (d) $\left(0, \frac{2}{3}\right)$ **(e)** $0, \frac{2}{3}$

17. **(a)** $[0, +\infty)$ **(b)** $(-\infty, 0]$ **(c)** $(-\sqrt{2/3}, \sqrt{2/3})$
    (d) $(-\infty, -\sqrt{2/3})$, $(\sqrt{2/3}, +\infty)$ **(e)** $-\sqrt{2/3}, \sqrt{2/3}$

19. **(a)** $(-\infty, +\infty)$ **(b)** none   **(c)** $(-\infty, -2)$ **(d)** $(-2, +\infty)$ **(e)** -2

21. **(a)** $[-1, +\infty)$ **(b)** $(-\infty, -1]$ **(c)** $(-\infty, 0)$, $(2, +\infty)$
    (d) $(0, 2)$ **(e)** $0, 2$

23. **(a)** $(-\infty, 0]$ **(b)** $[0, +\infty)$ **(c)** $(-\infty, -1)$, $(1, +\infty)$
    (d) $(-1, 1)$ **(e)** $-1, 1$

25. **(a)** $[0, +\infty)$ **(b)** $(-\infty, 0]$ **(c)** $(-1, 1)$ **(d)** $(-\infty, -1)$, $(1, +\infty)$
    (e) $-1, 1$

27. **(a)** $[\pi, 2\pi]$
    (b) $[0, \pi]$
    (c) $(\pi/2, 3\pi/2)$
    (d) $(0, \pi/2)$, $(3\pi/2, 2\pi)$
    (e) $\pi/2, 3\pi/2$

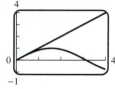

29. **(a)** $(-\pi/2, \pi/2)$
    (b) none
    (c) $(0, \pi/2)$
    (d) $(-\pi/2, 0)$
    (e) 0

30. (graph)

31. **(a)** $[0, \pi/4]$, $[3\pi/4, \pi]$
    (b) $[\pi/4, 3\pi/4]$
    (c) $(\pi/2, \pi)$
    (d) $(0, \pi/2)$
    (e) $\pi/2$

33. **(a)**   **(b)**   **(c)**

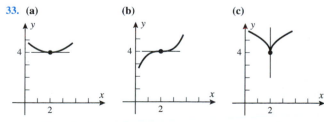

35. **(a)** $(a, 0)$
    (b) none

37. 2.5

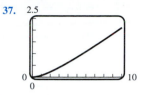

39. $x \geq \sin x$ 4

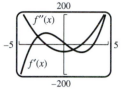

43. points of inflection at $x = -2, 2$;
    concave up on $[-5, -2]$, $[2, 5]$;
    concave down on $[-2, 2]$;
    increasing on $[-3.5829, 0.2513]$
    and $[3.3316, 5]$;
    decreasing on $[-5, -3.5829]$,
    $[0.2513, 3.3316]$

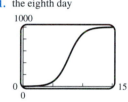

45. $-2.464202, 0.662597, 2.701605$

49. **(a)** true   **(b)** false   51. **(c)** 1

57.

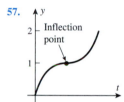

Inflection point

59. **(a)** $\dfrac{LAk}{(1 + A)^2}$ **(c)** $\dfrac{1}{k}\ln A$

61. the eighth day

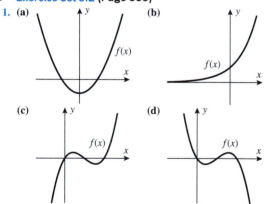

63. **(a)** 0   **(b)** 1   **(c)** $\displaystyle\lim_{x \to +\infty} g'(x) = 0$

► **Exercise Set 5.2 (Page 305)**

1. **(a)**   **(b)**
   (c)   **(d)**

$f(x)$

5. **(b)** nothing   **(c)** f has a relative minimum at $x = 1$, g has no relative extremum at $x = 1$.

7. **(a)** $x = -3, 1$ (stationary points)   **(b)** $x = 0, \pm\sqrt{3}$ (stationary points)

9. **(a)** $x = \pm\sqrt{2}$ (stationary points)   **(b)** no stationary points

11. **(a)** $x = -1$ (stationary point)   **(b)** $n\pi/3$, $n = 0, \pm1, \pm2, \ldots$
    (stationary points)   13. **(a)** none   **(b)** $x = 1$ **(c)** none

15. **(a)** 2   **(b)** 0   **(c)** 1, 3

17. **(a)** $x = 0$, relative max; $x = \pm\sqrt{5}$, relative min
    (b) $x = -1$, relative max; $x = 1$, relative min

19. **(a)** $x = 0$, relative min   **(b)** $x = \ln 2$, relative min

21. relative max of 5 at $x = -2$

23. relative min of 0 at $x = \pi$, relative max of 1 at $x = \pi/2, 3\pi/2$

25. no relative extrema

27. relative min of 0 at $x = 1$, relative max of $\frac{4}{27}$ at $x = \frac{1}{3}$

29. relative min of 0 at $x = 0$, relative max of 1 at $x = 1, -1$

31. relative min of 0 at $x = 0$ **33.** relative min of 0 at $x = 0$

35. relative min of 0 at $x = 0$

37. relative min of 0 at $x = 2, -2$, relative max of 4 at $x = 0$

39. relative min of 0 at $x = \pi/2, \pi, 3\pi/2$;
relative max of 1 at $x = \pi/4$,
$3\pi/4, 5\pi/4, 7\pi/4$

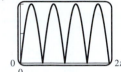

41. relative min of 0 at $x = \pi/2, 3\pi/2$;
relative max of 1 at $x = \pi$

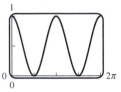

43. relative min of $-1/e$ at $x = 1/e$

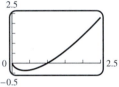

45. relative min of 0 at $x = 0$;
relative max of $1/e^2$ at $x = 1$

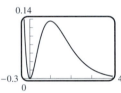

47. relative minima at $x = -3.58, 3.33$;
relative max at $x = 0.25$

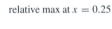

49. relative min at $x = -1.20$;   **51.** relative extrema at $x = 0, 0.59$
relative max at $x = 1.80$

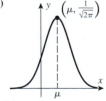

53. (a) 54   **(b)** 9   **55. (a)** $(-2.2, 4), (2, 1.2), (4.2, 3)$
   (b) critical numbers at $x = -5.1, -2, 0.2, 2$; local min at $x = -5.1, 2$;
   local max at $x = -2$; no extrema at $x = 0.2$; $f''(1) \approx -1.2$

57. $a = -2, b = 3, c = d = 0$

59. (b)

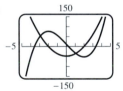

61. (a)   **(b)**

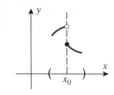
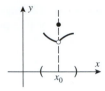

$f(x_0)$ is not an extreme value $f(x_0)$ is a relative maximum

(c)

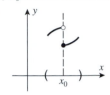

$f(x_0)$ is a relative minimum

▶ **Exercise Set 5.3** (Page 319)

1.   **3.**

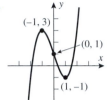

5.   **7.** $\left(-\frac{1}{\sqrt{2}}, \frac{7\sqrt{2}}{8}\right)$

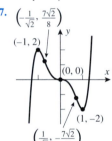

9.   **11.**

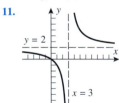

13.   **15.**

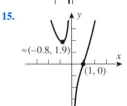

17.   **19.**

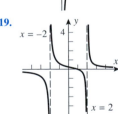

21.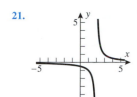

23.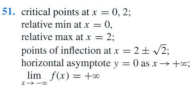

25.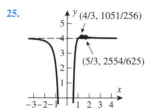

27. (a) VI
(b) I
(c) III
(d) V
(e) IV
(f) II

29.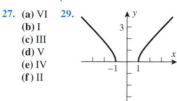

31.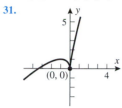

33.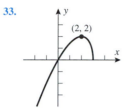

35.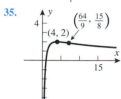

37.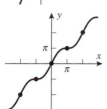

39.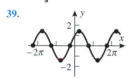

41.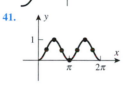

43. (a) $+\infty$, 0
(b)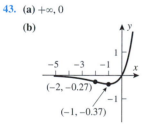

45. (a) 0, $+\infty$
(b)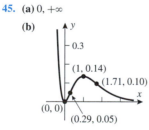

47. (a) $+\infty$, $-\infty$
(b)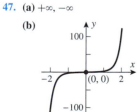

49. $\lim\limits_{x \to +\infty} f(x) = +\infty$;
critical point at $x = 1$;
relative min at $x = 1$;
no points of inflection;
vertical asymptote $x = 0$;
horizontal asymptote $y = 0$ for $x \to -\infty$
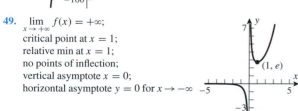

51. critical points at $x = 0, 2$;
relative min at $x = 0$,
relative max at $x = 2$;
points of inflection at $x = 2 \pm \sqrt{2}$;
horizontal asymptote $y = 0$ as $x \to +\infty$;
$\lim\limits_{x \to -\infty} f(x) = +\infty$
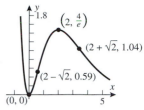

53. (a) 0; $+\infty$
(b)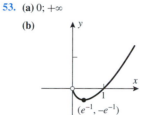

55. (a) $-\infty$; 0
(b)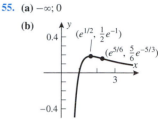

57. (a) $+\infty$, not defined   (b)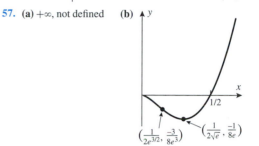

59. (a) 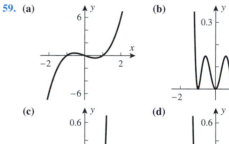 (b)

(c) (d)

61. (a) 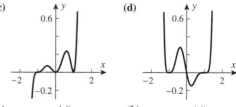 (b)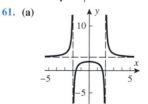

(c) 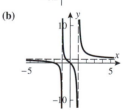 (d)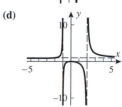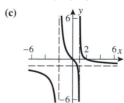

63. (a)

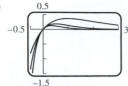

65. (a) The limit does not exist.

(b)

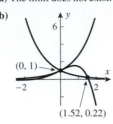

(0, 1)

(1.52, 0.22)

13. (a) $v(t) = -(3\pi/2)\sin(\pi t/2)$, $a(t) = -(3\pi^2/4)\cos(\pi t/2)$
(b) $s(1) = 0$ ft, $v(1) = -3\pi/2$ ft/s, $|v(1)| = 3\pi/2$ ft/s, $a(1) = 0$ ft/s^2
(c) 0, 2, 4   **(d)** speeding up for $0 < t < 1$, $2 < t < 3$, and $4 < t < 5$;
slowing down for $1 < t < 2$ and $3 < t < 4$ **(e)** 15 ft

15. (a) $\sqrt{5}$ **(b)** $\sqrt{5}/10$

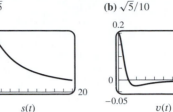

$s(t)$ $v(t)$

(c) speeding up for $\sqrt{5} < t < \sqrt{15}$,
slowing down for $0 < t < \sqrt{5}$ and $\sqrt{15} < t$

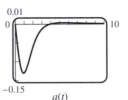

$a(t)$

(c)

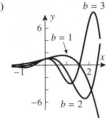

$b = 3$, $b = 1$, $b = 2$

$a = 3$, $a = 2$, $a = 1$

17. Constant speed

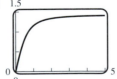

$t = 0$ s 2

67.

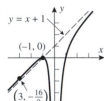

$y = x$

69.

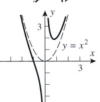

(2, 0)

(−4, −13.5)

$y = x − 6$

19.

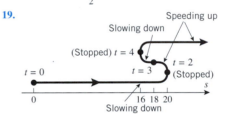

Speeding up

Slowing down

(Stopped) $t = 4$ $t = 2$

$t = 0$ $t = 3$ (Stopped) s

16 18 20

Slowing down

71.

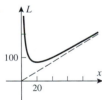

$y = x + 1$

(−1, 0)

$\left(3, -\frac{16}{9}\right)$

73.

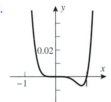

$y = x^2$

21.

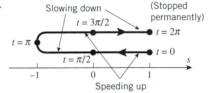

Slowing down   (Stopped permanently)

$t = 3\pi/2$ $t = 2\pi$

$t = \pi$ $t = 0$

$t = \pi/2$ s

−1   0   1

Speeding up

75.

L 100   20

77.

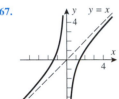

0.02   −1

23. (a) 12 ft/s   **(b)** $t = 2.2$, $s = -24.2$
25. (a) 11.025 m   **(b)** -14.7 m/s   **(c)** 1.2245 s   **(d)** $t = 4.5175$ s
27. (a) 6.12 s   **(b)** 183.67 m   **(c)** 6.12 s   **(d)** 60 m/s
29. 113.42 ft/s   **31.** 29.39 m   **33. (b)** 113.42 ft/s
35. (a)   1.5

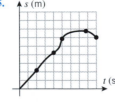

(b) $\sqrt{2}$
37. (b) $\frac{2}{3}$ unit
(c) $0 \le t < 1$ and $t > 2$

▶ **Exercise Set 5.4 (Page 329)**

1. (a) positive, negative, slowing down
(b) positive, positive, speeding up
(c) negative, positive, slowing down
3. (a) left
(b) negative
(c) speeding up
(d) slowing down

5.

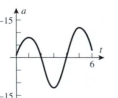

s (m)

t (s)

7.

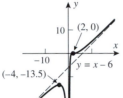

s 25   6

a −15 t 6   −15

9. (a) 6.7 ft/s^2
(b) $t = 0$ s

▶ **Exercise Set 5.5 (Page 339)**

1. relative maxima at $x = 2, 6$; absolute max at $x = 6$; relative and
absolute min at $x = 4$

3. (a)

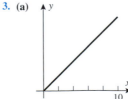

y x 10

(b)

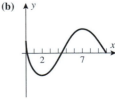

y x 2   7

11. (a) $v(t) = 3t^2 - 12t$, $a(t) = 6t - 12$ **(b)** $s(1) = -5$ ft, $v(1) = -9$
ft/s, $|v(1)| = 9$ ft/s, $a(1) = -6$ ft/s^2 **(c)** 0, 4   **(d)** speeding up for
$0 < t < 2$ and $4 < t$, slowing down for $2 < t < 4$ **(e)** 39 ft

(c)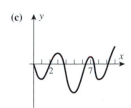

5. maximum value 1 at $x = 0, 1$; minimum value 0 at $x = \frac{1}{2}$

7. maximum value 27 at $x = 4$, minimum value -1 at $x = 0$

9. maximum value $3/\sqrt{5}$ at $x = 1$, minimum value $-3/\sqrt{5}$ at $x = -1$

11. maximum value $1 - (\pi/4)$ at $x = -\pi/4$, minimum value $(\pi/4) - 1$ at $x = \pi/4$ **13.** maximum value 17 at $x = -5$, minimum value 1 at $x = -3$ **15.** minimum value $f\left(\frac{3}{2}\right) = -\frac{13}{4}$, no maximum

17. maximum value $f(1) = 1$, no minimum   **19.** no maximum or minimum   **21.** maximum value $f(-2) = -4$, no minimum

23. minimum value 0 for $x = \pm 1$, no maximum   **25.** maximum value 48 at $x = 8$, minimum value 0 at $x = 0, 20$

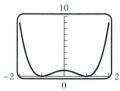   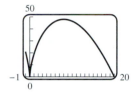

27. no maximum or minimum   **29.** maximum value 2 at $x = 0$, minimum value $\sqrt{3}$ at $x = \pi/6$

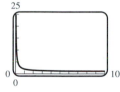   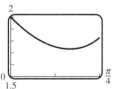

31. maximum value $\frac{27}{8}e^{-3}$ at $x = \frac{3}{2}$, minimum value $64/e^8$ at $x = 4$

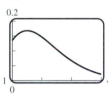

33. maximum value $f(-2 - \sqrt{3}) \approx 0.84$, minimum value $f(-2 + \sqrt{3}) \approx -0.06$

35. maximum value $\sin(1) \approx 0.84147$, minimum value $-\sin(1) \approx -0.84147$ **37.** maximum value 2 minimum value $-\frac{1}{4}$

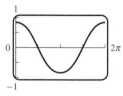

39. maximum value $3\sqrt{3}/2$ at $x = (\pi/6) + n\pi$, minimum value $-3\sqrt{3}/2$ at $x = (5\pi/6) + n\pi, n = 0, \pm 1, \pm 2, \dots$

43. $f'(1) = 2$ **47.** $\left(\frac{1}{2}, -\frac{1}{4}\right)$ is closest, $(-1, -1)$ is farthest

49. maximum $y = 4$ at $t = \pi, 3\pi$; minimum $y = 0$ at $t = 0, 2\pi$

▶ **Exercise Set 5.6 (Page 350)**

1. 5, 5   **3. (a)** 1   **(b)** $\frac{1}{2}$ **5.** 500 ft × 750 ft   **7.** 5 in × $\frac{12}{5}$ in
9. $10\sqrt{2} \times 10\sqrt{2}$ **11.** 80 ft ($1 fencing), 40 ft ($2 fencing)
15. (a) largest number $\approx 161{,}788$, smallest number $= 125{,}000$
    (b) $t = 40$ **17.** 2 in square   **19.** $\frac{200}{27}$ ft^3
21. base 10 cm square, height 20 cm
23. ends $\sqrt[3]{3V/4}$ units square, height $\frac{4}{3}\sqrt[3]{3V/4}$
25. height $= 2\sqrt{(5 - \sqrt{5})/10}\, R$, radius $= \sqrt{(5 + \sqrt{5})/10}\, R$
29. height $=$ radius $= \sqrt[3]{500/\pi}$ cm   **31.** $L/12$ by $L/12$ by $L/12$

33. height $= L/\sqrt{3}$, radius $= \sqrt{2/3}\, L$
35. radius $= \sqrt[6]{450/\pi^2}$ cm, height $= \dfrac{30}{\pi}\sqrt[3]{\pi^2/450}$ cm
37. height $= 4R$, radius $= \sqrt{2}R$ **39.** $\pi/3$ **41.** $5\sqrt{5}$ ft
43. (a) 7000 units   **(b)** yes   **(c)** $15   **45.** 13,722 lb   **47.** $1/\sqrt{5}$
51. $(-\sqrt{2}, 1), (\sqrt{2}, 1)$ **53.** $\left(\sqrt{2}, \frac{1}{2}\right)$ **55.** $\left(-1/\sqrt{3}, \frac{3}{4}\right)$
57. $4(1 + 2^{2/3})^{3/2}$ ft   **59.** 30 cm from the weaker source
61. $x = 1 + 2\sqrt{2}$ **65. (c)** $\frac{1}{4}$ mile downstream from the house

▶ **Exercise Set 5.7 (Page 358)**
1. 1.414213562   **3.** 1.817120593
5. -1.671699882 **7.** 1.224439550
9. -1.452626879 **11.** 1.895494267

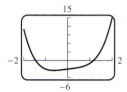   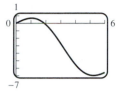

13. 4.493409458   **15.** -1.165373043

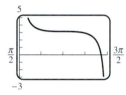   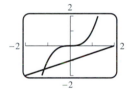

17. $-0.474626618, 1.395336994$ **19.** $x = 0, x \approx 1.292695719$
21. (b) 3.162277660
23. -4.098859132
25. $x \approx -0.18, x = 1$
27. $(0.589754512, 0.347810385)$
29. (b) $171°$
31. $-1.220744085, 0.724491959$
33. $i = 0.053362$

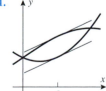

▶ **Exercise Set 5.8 (Page 363)**
1. $[0, 4], c = 3$ **3.** $c = 3$ **5.** $c = \pi$ **7.** $c = 1$
9. $c = 1.54$
11. 1
13. $\frac{5}{4}$
15. $-\sqrt{5}$
17. (a) $[-2, 1]$
    (b) $c = -1.29$
    (c) -1.2885843

19. (b) $\tan x$ is not continuous on $[0, \pi]$.
29. $f(x) = x^3 - 4x + 5$
41.   **45.** $a = 6, b = -3$

▶ **Chapter 5 Supplementary Exercises (Page 365)**
1. (a) $f(x_1) < f(x_2); \ f(x_1) > f(x_2); \ f(x_1) = f(x_2)$
    (b) $f' > 0; \ f' < 0; \ f' = 0$
7. (a) relative max at $x = 1$, relative min at $x = 7$, neither at $x = 0$
    (b) relative max at $x = \pi/2, 3\pi/2$; relative min at $x = 7\pi/6, 11\pi/6$
    (c) relative max at $x = 5$

9. $\lim\limits_{x \to -\infty} f(x) = +\infty$, $\lim\limits_{x \to +\infty} f(x) = +\infty$;
relative min at $x = 0$;
points of inflection at $x = \frac{1}{2}, 1$;
no asymptotes

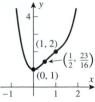

11. $\lim\limits_{x \to \pm\infty} f(x)$ does not exist;
critical point at $x = 0$;
relative min at $x = 0$;
point of inflection when
$1 + 4x^2 \tan(x^2 + 1) = 0$;
vertical asymptotes at
$x = \pm\sqrt{\pi(n + \frac{1}{2}) - 1}, n = 0, 1, 2, \ldots$

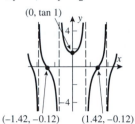

13. critical points at $x = -5, 0$;
relative max at $x = -5$,
relative min at $x = 0$;
points of inflection at
$x = -7.26, -1.44, 1.20$;
horizontal asymptote $y = 1$
for $x \to \pm\infty$

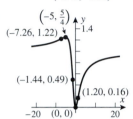

15. $\lim\limits_{x \to -\infty} f(x) = +\infty$,
$\lim\limits_{x \to +\infty} f(x) = -\infty$;
critical point at $x = 0$;
no extrema;
inflection point at $x = 0$
(f changes concavity);
no asymptotes

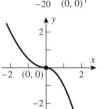

17. (a)

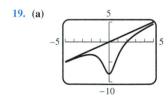

(b) relative max at $x = -\frac{1}{20}$,
relative min at $x = \frac{1}{20}$

(c) The finer details can be seen
when graphing over
a much smaller x-window.

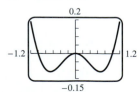

19. (a)

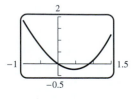

21. $y = 2x$, $y = 3$

23. $f(x) = \dfrac{x^2 + x - 7}{3x^2 + x - 1}$, $x \neq \frac{1}{2}$

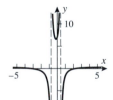

25. $-\dfrac{b}{2a} \leq 0$

27. (a) relative min -0.232466
at $x = 0.450184$

(b) relative max 0 at $x = 0$;
relative min -0.107587
at $x = \pm 0.674841$

(c) relative max 0.876839
at $x = 0.886352$;
relative min -0.355977
at $x = -1.244155$

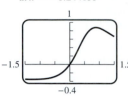

29. (a) yes **(b)** yes

31. (a) 0.3501 s, 0.0820 s

33. (a) $v = -2\dfrac{t(t^4 + 2t^2 - 1)}{(t^4 + 1)^2}$,
$a = 2\dfrac{3t^8 + 10t^6 - 12t^4 - 6t^2 + 1}{(t^4 + 1)^3}$

(c) $t = 0.64$, $s = 1.2$

(d) $0 \leq t \leq 0.64$ s

(e) speeding up when $0 \leq t < 0.36$
and $0.64 < t < 1.1$, otherwise
slowing down

(f) maximum speed $= 1.05$ m/s
when $t = 1.10$ s

37. (a) true **(b)** false

39. (a) $M = -\frac{1}{2}$ at $x = -2$; $m = -1$ at $x = -1$

(b) $M = \frac{27}{256}$ at $x = \frac{3}{4}$; $m = -2$ at $x = -1$

(c) $m \approx -1.9356$ at $x = \frac{12}{7}$; $M = 9$ at $x = 3$

(d) $m = e^2/4$ at $x = 2$

43. 2.3561945 **45. (a)** yes, $c = 0$ **(b)** no **(c)** yes, $c = \sqrt{\pi/2}$

47. $r = 2P/(8 + 3\pi)$ ft

49.

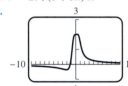

(b) minimum:
$(-2.111985, -0.355116)$;
maximum:
$(0.372591, 2.012931)$

51. 149.988×10^6 km

▶ **Exercise Set 6.1 (Page 377)**

1.

n	2	5	10	50	100
A_n	0.853553	0.749739	0.710509	0.676095	0.671463

3.

n	2	5	10	50	100
A_n	1.57080	1.93376	1.98352	1.99935	1.99984

5.

n	2	5	10	50	100
A_n	0.583333	0.645635	0.668771	0.688172	0.690653

7.

n	2	5	10	50	100
A_n	0.433013	0.659262	0.726130	0.774567	0.780106

9. $3(x - 1)$ **11.** $x(x + 2)$ **13.** $(x + 3)(x - 1)$

▶ **Exercise Set 6.2 (Page 385)**

1. (a) $\displaystyle\int \dfrac{x}{\sqrt{1 + x^2}}\, dx = \sqrt{1 + x^2} + C$

(b) $\displaystyle\int (x + 1)e^x\, dx = xe^x + C$

3. $\dfrac{d}{dx}\left[\sqrt{x^3 + 5}\right] = \dfrac{3x^2}{2\sqrt{x^3 + 5}}$, so $\displaystyle\int \dfrac{3x^2}{2\sqrt{x^3 + 5}}\, dx = \sqrt{x^3 + 5} + C$.

5. $\dfrac{d}{dx}\left[\sin(2\sqrt{x})\right] = \dfrac{\cos(2\sqrt{x})}{\sqrt{x}}$, so $\displaystyle\int \dfrac{\cos(2\sqrt{x})}{\sqrt{x}}\, dx = \sin(2\sqrt{x}) + C$.

7. (a) $(x^9/9) + C$ **(b)** $\frac{7}{12}x^{12/7} + C$ **(c)** $\frac{2}{9}x^{9/2} + C$

9. (a) $-\frac{1}{4}x^{-2} + C$ **(b)** $(u^4/4) - u^2 + 7u + C$

11. $-\frac{1}{2}x^{-2} + \frac{2}{3}x^{3/2} - \frac{12}{5}x^{5/4} + \frac{1}{3}x^3 + C$

13. $(x^2/2) + (x^5/5) + C$ **15.** $3x^{4/3} - \frac{12}{7}x^{7/3} + \frac{3}{10}x^{10/3} + C$

17. $\dfrac{x^2}{2} - \dfrac{2}{x} + \dfrac{1}{3x^3} + C$ **19.** $2\ln|x| + 3e^x + C$

21. $-4\cos x + 2\sin x + C$

23. $\tan x + \sec x + C$ **25.** $\tan\theta + C$ **27.** $\sec x + C$

29. $\theta - \cos\theta + C$ **31.** $\frac{1}{2}\sin^{-1} x - 3\tan^{-1} x + C$

33. $\tan x - \sec x + C$

35. (a)   **(b)**   **(c)**

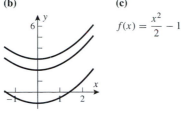   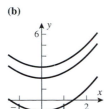

$f(x) = \dfrac{x^2}{2} - 1$

37.

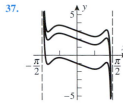

39. $f(x) = \cos x + 1$

41. (a) $y(x) = \frac{3}{4}x^{4/3} + \frac{5}{4}$

(b) $y = -\cos t + t + 1 - \pi/3$

(c) $y(x) = \frac{2}{3}x^{3/2} + 2x^{1/2} - \frac{8}{3}$

43. (a) $y = 4e^x - 3$

(b) $y = \ln|t| + 5$

45. $f(x) = \frac{4}{15}x^{5/2} + C_1 x + C_2$

47. $y = x^2 + x - 6$

49. $y = x^3 - 6x + 7$

51. (b) $F(0) - G(0) = \frac{8}{3}$ **53.** $\tan x - x + C$

55. (a) $\frac{1}{2}(x - \sin x) + C$ **(b)** $\frac{1}{2}(x + \sin x) + C$

57. $v = \dfrac{1087}{\sqrt{273}}T^{1/2}$ ft/s

▶ **Exercise Set 6.3 (Page 392)**

1. (a) $\dfrac{(x^2+1)^{24}}{24} + C$ **(b)** $-\dfrac{\cos^4 x}{4} + C$ **(c)** $-2\cos\sqrt{x} + C$

(d) $\frac{3}{4}\sqrt{4x^2+5}+C$ **3. (a)** $-\frac{1}{2}\cot^2 x + C$ **(b)** $\frac{1}{10}(1+\sin t)^{10}+C$

(c) $\frac{1}{2}\sin 2x + C$ **(d)** $\frac{1}{2}\tan(x^2) + C$

5. (a) $\ln|\ln x| + C$ **(b)** $-\frac{1}{5}e^{-5x} + C$ **(c)** $-\frac{1}{3}\ln(1+\cos 3\theta) + C$

(d) $\ln(1+e^x) + C$ **7.** $-\dfrac{(2-x^2)^4}{8} + C$ **9.** $\frac{1}{8}\sin 8x + C$

11. $\frac{1}{4}\sec 4x + C$ **13.** $\frac{1}{2}e^{2x} + C$ **15.** $\frac{1}{2}\sin^{-1}(2x) + C$

17. $\frac{1}{21}(7t^2+12)^{3/2} + C$ **19.** $\frac{2}{3}\sqrt{x^3+1} + C$

21. $-\frac{1}{16}(4x^2+1)^{-2} + C$ **23.** $e^{\sin x} + C$ **25.** $-\frac{1}{6}e^{-2x^3} + C$

27. $\tan^{-1} e^x + C$ **29.** $\frac{1}{5}\cos(5/x) + C$ **31.** $\frac{1}{3}\tan(x^3) + C$

33. $\frac{1}{18}\sin^6 3t + C$ **35.** $-\frac{1}{6}(2-\sin 4\theta)^{3/2}+C$ **37.** $\sin^{-1}(\tan x)+C$

39. $\frac{1}{6}\sec^3 2x + C$ **41.** $-e^{-x} + C$ **43.** $2e^{\sqrt{y+1}} + C$

45. $\frac{2}{5}(x-3)^{5/2} + 2(x-3)^{3/2} + C$ **47.** $-\frac{1}{2}\cos 2\theta + \frac{1}{6}\cos^3 2\theta + C$

49. $t + \ln|t| + C$ **51.** $\displaystyle\int [\ln(e^x) + \ln(e^{-x})]\,dx = C$

53. (a) $\sin^{-1}(\frac{1}{3}x) + C$ **(b)** $\dfrac{1}{\sqrt{5}}\tan^{-1}\left(\dfrac{x}{\sqrt{5}}\right) + C$

(c) $\dfrac{1}{\sqrt{\pi}}\sec^{-1}\left(\dfrac{x}{\sqrt{\pi}}\right) + C$ **55.** $\dfrac{1}{b}\dfrac{(a+bx)^{n+1}}{n+1} + C$

57. $\dfrac{1}{b(n+1)}\sin^{n+1}(a+bx) + C$

59. (a) $\frac{1}{2}\sin^2 x + C_1;\ -\frac{1}{2}\cos^2 x + C_2$ **(b)** They differ by a constant.

61. $y(x) = \frac{2}{9}(3x+1)^{3/2} + \frac{29}{9}$ **63.** $-2e^{-t} + 3$

65.

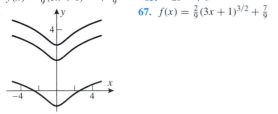

67. $f(x) = \frac{2}{9}(3x+1)^{3/2} + \frac{7}{9}$

▶ **Exercise Set 6.4 (Page 404)**

1. (a) 36   **(b)** 55   **(c)** 40   **(d)** 6   **(e)** 11   **(f)** 0   **3.** $\displaystyle\sum_{k=1}^{10} k$

5. $\displaystyle\sum_{k=1}^{10} 2k$ **7.** $\displaystyle\sum_{k=1}^{6}(-1)^{k+1}(2k-1)$ **9. (a)** $\displaystyle\sum_{k=1}^{50} 2k$ **(b)** $\displaystyle\sum_{k=1}^{50}(2k-1)$

11. 5050   **13.** 2870   **15.** 214,365   **17.** $\frac{3}{2}(n+1)$

19. $\frac{1}{4}(n-1)^2$ **23.** $\dfrac{n+1}{2n};\frac{1}{2}$ **25.** $\dfrac{5(n+1)}{2n};\frac{5}{2}$

27. (a) $\displaystyle\sum_{j=0}^{5} 2^j$ **(b)** $\displaystyle\sum_{j=1}^{6} 2^{j-1}$ **(c)** $\displaystyle\sum_{j=2}^{7} 2^{j-2}$

29. (a) 46   **(b)** 52   **(c)** 58   **31. (a)** $\dfrac{\pi}{4}$ **(b)** 0   **(c)** $-\dfrac{\pi}{4}$

33. (a) 0.7188, 0.7058, 0.6982   **(b)** 0.6688, 0.6808, 0.6882

(c) 0.6928, 0.6931, 0.6931

35. (a) 4.8841, 5.1156, 5.2488   **(b)** 5.6841, 5.5156, 5.4088

(c) 5.3471, 5.3384, 5.3346

37. $\frac{15}{4}$ **39.** 18   **41.** 320   **43.** $\frac{15}{4}$ **45.** 18   **47.** $\frac{1}{3}$

49. 0   **51.** $\frac{2}{3}$ **53.** $\frac{1}{2}m(b^2-a^2)$ **55. (b)** $\frac{1}{4}(b^4-a^4)$

57. $\dfrac{n^2+2n}{4}$ if n is even; $\dfrac{(n+1)^2}{4}$ if n is odd   **59. (a)** yes   **(b)** yes

▶ **Exercise Set 6.5 (Page 413)**

1. (a) $\frac{71}{6}$ **(b)** 2   **3. (a)** $-\frac{117}{16}$ **(b)** 3   **5.** $\displaystyle\int_{-1}^{2} x^2\,dx$

7. (a) $\displaystyle\int_{-3}^{3} 4x(1-3x)\,dx$ **9. (a)** $\displaystyle\lim_{\max \Delta x_k \to 0}\sum_{k=1}^{n} 2x_k^*\Delta x_k; a=1, b=2$

(b) $\displaystyle\lim_{\max \Delta x_k \to 0}\sum_{k=1}^{n} \dfrac{x_k^*}{x_k^*+1}\Delta x_k; a=0, b=1$

11. (a) $A = \frac{9}{2}$ **(b)** $-A = -\frac{3}{2}$

    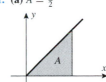

(c) $-A_1 + A_2 = \frac{15}{2}$ **(d)** $-A_1 + A_2 = 0$

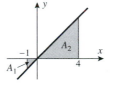   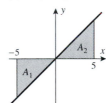

13. (a) $A = 10$ **(b)** $A_1 - A_2 = 0$ by symmetry

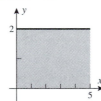   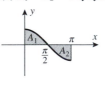

(c) $A_1 + A_2 = \frac{13}{2}$ **(d)** $\pi/2$ **15. (a)** 0.8

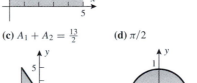   **(b)** -2.6

(c) -1.8

(d) -0.3

17. -1

19. 3

21. (a) $(1+\pi)/2$

(b) -4

23. (a) negative

(b) positive

25. $\frac{25}{2}\pi$ **27.** $\frac{5}{2}$ **29. (a)** yes   **(b)** yes   **(c)** no   **(d)** yes   **33.** $\frac{16}{3}$

▶ **Exercise Set 6.6 (Page 424)**

1. (a) $\displaystyle\int_{0}^{2}(2-x)\,dx = 2$ **(b)** $\displaystyle\int_{-1}^{1} 2\,dx = 4$ **(c)** $\displaystyle\int_{1}^{3}(x+1)\,dx = 6$

3. $\frac{65}{4}$ **5.** $\frac{52}{3}$ **7.** $e^3 - e$ **9.** 48 **11.** $\frac{2}{3}$ **13.** $\frac{844}{5}$

15. 0 **17.** $\sqrt{2}$ **19.** $5e^3 - 10$ **21.** $\pi/4$ **23.** $\pi/12$

25. $-\frac{55}{3}$ **27.** $\frac{\pi^2}{9} + 2\sqrt{3}$ **29.** (a) $\frac{5}{2}$ (b) $2 - \frac{\sqrt{2}}{2}$

31. (a) $e + (1/e) - 2$ (b) 1

33. (a) $\frac{17}{6}$ (b) $F(x) = \begin{cases} \dfrac{x^2}{2}, & x \le 1 \\ \dfrac{x^3}{3} + \dfrac{1}{6}, & x > 1 \end{cases}$ **35.** $0.6659; \frac{2}{3}$

37. 3.1060; 3.1148 **39.** 12 **41.** $\frac{9}{2}$ **43.** (a) $\sin^{-1} 0.8$ (b) 0.93

45. (a) The integral is zero. (c) $\displaystyle\int_{-a}^{a} f(x)\,dx = 2\int_0^a f(x)\,dx$

47. (a) $x^3 + 1$ **49.** (a) $\sin\sqrt{x}$ (b) e^{x^2}

51. $-\dfrac{x}{\cos x}$ **53.** (a) 0 (b) $\sqrt{13}$ (c) $6/\sqrt{13}$

55. (a) $x = 3$ (b) increasing on $[3, +\infty)$, decreasing on $(-\infty, 3]$

(c) concave up on $(-1, 7)$, concave down on $(-\infty, -1)$ and

$(7, +\infty)$ **57.** (a) $(0, +\infty)$ (b) $x = 1$

59. (a) 4 (b) $-\dfrac{1 - \sqrt{13}}{3}$ **61.** $3\sqrt{2} \le \displaystyle\int_0^3 \sqrt{x^3 + 2}\,dx \le 3\sqrt{29}$ **65.** 1

▶ **Exercise Set 6.7 (Page 436)**

1. (a) the increase in height in inches, during the first 10 years

(b) the change in the radius in cm, during the time interval $t = 1$ to

$t = 2$ seconds

(c) the change in the speed of sound in ft/s, during an increase in

temperature from $t = 32°\text{F}$ to $t = 100°\text{F}$

(d) the displacement of the particle in cm, during the time interval

$t = t_1$ to $t = t_2$ seconds

3. (a) displacement $= -\frac{1}{2}$; distance $= \frac{3}{2}$

(b) displacement $= \frac{3}{2}$; distance $= 2$

5. (a) 35.3 m/s (b) 51.4 m/s **7.** (a) $\frac{1}{4}t^4 - \frac{2}{3}t^3 + t + 1$

(b) $-\cos 2t - t - 2$ **9.** (a) $t^2 - 3t + 7$ (b) $-\cos t + t - (\pi/2)$

11. (a) displacement $= 1$ m; distance $= 1$ m

(b) displacement $= -1$ m; distance $= 3$ m

13. (a) displacement $= \frac{9}{4}$ m; distance $= \frac{11}{4}$ m

(b) displacement $= 2\sqrt{3} - 6$ m; distance $= 6 - 2\sqrt{3}$ m

15. displacement $= -6$ m; distance $= \frac{13}{2}$ m

17. displacement $= \frac{616}{75}$ m; distance $= \frac{616}{75}$ m

19. (a) $s = 2/\pi, v = 1, |v| = 1, a = 0$ **21.** $\frac{22}{3}$

(b) $s = \frac{1}{2}, v = -\frac{3}{2}, |v| = \frac{3}{2}, a = -3$ **23.** $A = \frac{4}{3}(\sqrt{2} - 1)$

25. $e + (1/e) - 2$ **27.** $t \approx 1.27$ s

29. (a)

$s(t)$

(b)

$v(t)$

(c)

$a(t)$

(b) $\frac{5}{2} - \sin 5 + 5\cos 5$

31. (a) positive then negative

then positive

$v(t)$

33. (a) The displacement is always positive. (b) $\dfrac{t}{2} + (t + 1)e^{-t}$

35. (a) $a(t) = \begin{cases} 0, & t < 4 \\ -10, & t > 4 \end{cases}$ (b) $v(t) = \begin{cases} 25, & t < 4 \\ 65 - 10t, & t > 4 \end{cases}$

(c) $x(t) = \begin{cases} 25t, & t < 4 \\ 65t - 5t^2 - 80, & t > 4 \end{cases}$ so $x(8) = 120, x(12) = -20$.

(d) $x(6.5) = 131.25$

37. (a) $-\frac{22}{15}$ ft/s^2 (b) $\frac{1}{7200}$ km/s^2 **39.** (a) $-\frac{121}{5}$ ft/s^2

(b) $\frac{70}{33}$ s (c) $\frac{60}{11}$ s **41.** 280 m **43.** 100 s; 10,000 ft

45. (a) -48 ft/s (b) 196 ft (c) 112 ft/s **47.** (a) 1 s (b) $\frac{1}{2}$ s

49. (a) $(5 + 5\sqrt{33})/8$ s (b) $20\sqrt{33}$ ft/s **51.** (a) 5 s (b) 272.5 m

(c) 10 s (d) -49 m/s (e) 12.46 s (f) 73.1 m/s

53. 4.04 m/s **55.** 6 **57.** $2/\pi$ **59.** $\dfrac{1}{e - 1}$

61. (a) $\frac{4}{3}$ (c)

(b) $2/\sqrt{3}$

63. (a) $\frac{263}{4}$

(b) 31

65. 1404π lb

67. (a) 120 gal

(b) 420 gal

(c) 2076.36 gal

69. (b) no

▶ **Exercise Set 6.8 (Page 444)**

1. (a) $\displaystyle\int_1^3 u^7\,du$ (b) $-\dfrac{1}{2}\int_7^4 u^{1/2}\,du$ (c) $\dfrac{1}{\pi}\int_{-\pi}^{\pi} \sin u\,du$

(d) $\displaystyle\int_{-3}^0 (u + 5)u^{20}\,du$

3. (a) $\dfrac{1}{2}\displaystyle\int_{-1}^1 e^u\,du$ (b) $\displaystyle\int_1^2 u\,du$ **5.** $\frac{121}{5}$ **7.** 10

9. $\frac{1192}{15}$ **11.** $8 - (4\sqrt{2})$ **13.** $-\frac{1}{48}$ **15.** $\ln\frac{21}{13}$ **17.** $\pi/6$

19. $\frac{25}{12}\pi$ **21.** $\pi/8$ **23.** $2/\pi$ **25.** $\frac{1}{24}$ **27.** $\pi/18$ **29.** $\frac{1}{21}$

31. $\frac{1}{8}(1 - e^{-8})$ **33.** $\frac{2}{3}$ **35.** $\frac{2}{3}(\sqrt{10} - 2\sqrt{2})$ **37.** $2(\sqrt{7} - \sqrt{3})$

39. 0 **41.** 0 **43.** $(\sqrt{3} - 1)/3$ **45.** $\frac{106}{405}$ **47.** $\ln 2$

49. $\dfrac{\pi}{6\sqrt{3}}$ **51.** $\frac{1}{18}\pi$ **53.** $\frac{23}{4480}$ **55.** (a) $\frac{5}{3}$ (b) $\frac{5}{3}$ (c) $-\frac{1}{2}$

59. $\approx 48,233,500,000$ **61.** (a) ≈ 328.69 ft (b) yes

63. $k \approx 5.08$ **65.** (b) 169.7 V **67.** (b) $\frac{3}{2}$ (c) $\pi/4$

▶ **Exercise Set 6.9 (Page 455)**

1. (a)

(b)

(c)

3. (a) 7

(b) -5

(c) -3

(d) 6

5. 1.603210678;

magnitude of error is < 0.0063

7. (a) $x^{-1}, x > 0$ (b) $x^2, x \ne 0$ (c) $-x^2, -\infty < x < +\infty$

(d) $-x, -\infty < x < +\infty$ (e) $x^3, x > 0$ (f) $\ln x + x, x > 0$

(g) $x - \sqrt[3]{x}, -\infty < x < +\infty$ (h) $\dfrac{e^x}{x}, x > 0$

9. (a) $e^{\pi \ln 3}$ (b) $e^{\sqrt{2}\ln 2}$ **11.** (a) e^2 (b) e^2 **13.** $x^2 - x$

15. (a) $3/x$ (b) 1 **17.** (a) 0 (b) $\frac{1}{3}$ (c) 0

19. (a) $2x^3\sqrt{1+x^2}$ (b) $-\frac{2}{3}(x^2+1)^{3/2} + \frac{2}{5}(x^2+1)^{5/2} - \frac{4\sqrt{2}}{15}$

21. (a) $-\sin x^2$ (b) $-\tan^2 x$ **23.** $-3\frac{3x-1}{9x^2+1} + 2x\frac{x^2-1}{x^4+1}$

25. (a) $3x^2\sin^2(x^3) - 2x\sin^2(x^2)$ (b) $\frac{2}{1-x^2}$

27. (a) $F(0) = 0$, $F(3) = 0$, $F(5) = 6$, $F(7) = 6$, $F(10) = 3$

(b) increasing on $\left[\frac{3}{2}, 6\right]$ and $\left[\frac{37}{4}, 10\right]$, decreasing on $\left[0, \frac{3}{2}\right]$ and $\left[6, \frac{37}{4}\right]$

(c) maximum $\frac{15}{2}$ at $x = 6$, minimum $-\frac{9}{4}$ at $x = \frac{3}{2}$

(d)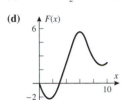

29. $F(x) = \begin{cases} (1-x^2)/2, & x < 0 \\ (1+x^2)/2, & x \geq 0 \end{cases}$

31. $y(x) = \frac{5}{4} + \frac{3}{4}x^{4/3}$

33. $y(x) = \tan x + \cos x - (\sqrt{2}/2)$

35. $P(x) = P_0 + \int_0^x r(t)\, dt$ individuals

37. I is the derivative of II.

39. (a) $t = 3$ (f)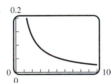

(b) $t = 1, 5$

(c) $t = 5$

(d) $t = 3$

(e) F is concave up on $\left(0, \frac{1}{2}\right)$ and $(2, 4)$, concave down on $\left(\frac{1}{2}, 2\right)$ and $(4, 5)$.

41. (a) relative maxima at $x = \pm\sqrt{4k+1}$, $k = 0, 1, \ldots$; relative minima at $x = \pm\sqrt{4k-1}$, $k = 1, 2, \ldots$

(b) $x = \pm\sqrt{2k}$, $k = 1, 2, \ldots$, and at $x = 0$

43. $f(x) = 3e^{3x}$, $a = \frac{1}{3}\ln 2$ **45.** 0.06

Chapter 6 Supplementary Exercises (Page 458)

7. (a) $\frac{3}{4}$ (b) $-\frac{3}{2}$ (c) $-\frac{35}{4}$ (d) -2 (e) not enough information   (f) not enough information

9. (a) $2 + (\pi/2)$ (b) $\frac{1}{3}(10^{3/2} - 1) - \frac{9\pi}{4}$ (c) $\pi/8$ **11.** $\frac{35\pi}{128}$

15. (d) $n \geq 1000$ **21.** $3^{17} - 3^4$ **23.** $-\frac{399}{400}$ **25.** (b) $\frac{1}{2}$

29. (a) $\frac{3}{2}(3^{20} - 1)$ (b) $2^{31} - 2^5$ (c) $-\frac{2}{3}\left(1 + \frac{1}{2^{101}}\right)$

33. $(x^{2/3} + 1)^{3/2} + C$ **35.** $A \approx 8$ **37.** $e - 1$

39. $0.35122, 0.42054, 0.38650$ **41.** $\int_0^1 e^x\, dx = e - 1$

43. $\int_A^x \frac{1}{1+e^t}\, dt$ with (a) $A = 1$, (b) $A = -\ln(e^2 + e - 1)$

47. (a) $F(x)$ is 0 if $x = 1$, positive if $x > 1$, and negative if $x < 1$.

(b) $F(x)$ is 0 if $x = -1$, positive if $-1 < x \leq 2$, and negative if $-2 \leq x < -1$.   **49.** (a) no   (b) $25 < t < 40$ (c) 3.54 ft/s

(d) 141.5 ft   (e) no   (f) no   **51.** $\frac{1}{3}\sqrt{5 + 2\sin 3x} + C$

53. $-\frac{1}{3a}\frac{1}{ax^3+b} + C$ **55.** $\frac{389}{192}$ **57.** $\ln 2$ **59.** $\frac{3}{8} + \frac{1}{2}(\sin 1 - \sin\frac{1}{4})$

61. 1.007514 **63.** (a) $k = 2.073948$ (b) $k = 1.837992$

65. (a) (b) 0.7651976866

(c) $x = 2.404826$

Exercise Set 7.1 (Page 473)

1. $9/2$ **3.** 1 **5.** (a) $32/3$ (b) $32/3$ **7.** $49/192$ **9.** $1/2$

11. $\sqrt{2}$ **13.** $\frac{1}{2}$ **15.** $\pi - 1$ **17.** 24 **19.** $37/12$

21. $4\sqrt{2}$ **23.** $\frac{1}{2}$ **25.** $\ln 2 - \frac{1}{2}$ **27.** $k \approx 0.9973$

29. $9152/105$ **31.** $9/\sqrt[3]{4}$ **33.** (a) $4/3$ (b) $m = 2 - \sqrt[3]{4}$

37. 1.180898334 **39.** $0.4814, 2.3639, 1.1897$

41. (a) 1800 ft   (b) $\frac{3}{2}T^2 - \frac{1}{60}T^3$ ft   **43.** $a^2/6$

Exercise Set 7.2 (Page 480)

1. 8π **3.** $13\pi/6$ **5.** $32\pi/5$ **7.** $(1 - \sqrt{2}/2)\pi$ **9.** $256\pi/3$

11. $2048\pi/15$ **13.** 4π **15.** $\pi^2/4$ **17.** $3\pi/5$ **19.** 8π

21. 2π **23.** $72\pi/5$ **25.** $\frac{\pi}{2}(e^2 - 1)$ **27.** $4\pi ab^2/3$ **29.** π

31. $648\pi/5$ **33.** $\pi/2$ **35.** $40,000\pi$ ft^3 **37.** $1/30$

39. (a) $2\pi/3$ (b) $16/3$ (c) $4\sqrt{3}/3$ **41.** 0.710172176 **43.** π

47. (b) left ≈ 11.157; right ≈ 11.771; $V \approx$ average $= 11.464$ cm^3

49. $V = \begin{cases} 3\pi h^2, & 0 \leq h < 2 \\ \frac{1}{3}\pi(12h^2 - h^3 - 4), & 2 \leq h \leq 4 \end{cases}$ **51.** $\frac{2}{3}r^3\tan\theta$ **53.** $16r^3/3$

Exercise Set 7.3 (Page 485)

1. $15\pi/2$ **3.** $\pi/3$ **5.** $2\pi/5$ **7.** 4π **9.** $20\pi/3$ **11.** $\pi\ln 2$

13. $\pi/2$ **15.** $\pi/5$ **17.** $2\pi^2$ **19.** (a) $7\pi/30$ (b) easier

21. $9\pi/14$ **23.** $\pi r^2 h/3$

25. $V = \frac{4\pi}{3}[r^3 - (r^2 - a^2)^{3/2}]$ **27.** $b = 1$

Exercise Set 7.4 (Page 490)

1. $L = \sqrt{5}$ **3.** $(85\sqrt{85} - 8)/243$ **5.** $\frac{1}{27}(80\sqrt{10} - 13\sqrt{13})$

7. $\frac{17}{6}$ **9.** $(2\sqrt{2} - 1)/3$ **11.** π **13.** $L = \sqrt{2}(e^{\pi/2} - 1)$

15. $L = \ln(1 + \sqrt{2})$

19. (a) 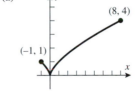   (b) dy/dx does not exist at $x = 0$.

$(8, 4)$ (c) $L = (13\sqrt{13} + 80\sqrt{10} - 16)/27$

$(-1, 1)$

21. 4.645975301 **23.** 3.820197788

27. (b) 9.69 (c) 5.16 cm   **29.** $k = 1.83$

Exercise Set 7.5 (Page 494)

1. $35\pi\sqrt{2}$ **3.** 8π **5.** $40\pi\sqrt{82}$ **7.** 24π **9.** $16\pi/9$

11. $16,911\pi/1024$ **13.** $2\pi(\sqrt{2} + \ln(\sqrt{2} + 1))$ **15.** $S \approx 22.94$

21. $\frac{8}{3}\pi(17\sqrt{17} - 1)$ **23.** $\frac{\pi}{24}(17\sqrt{17} - 1)$

Exercise Set 7.6 (Page 502)

1. (a) 210 ft·lb   (b) $5/6$ ft·lb   **3.** 100 ft·lb   **5.** 160 J

7. 20 lb/ft   **9.** $900\pi\rho$ ft·lb   **11.** $261,600$ J

13. (a) $926,640$ ft·lb   (b) hp of motor $= 0.468$

15. $75,000$ ft·lb   **17.** $120,000$ ft·tons

19. (a) $2,400,000,000/x^2$ lb   (b) $(9.6 \times 10^{10})/(x + 4000)^2$ lb

(c) 2.5344×10^{10} ft·lb   **21.** $v_f = 100$ m/s

23. (a) decrease of 4.5×10^{14} J   (b) ≈ 0.107 (c) ≈ 8.24 bombs

Exercise Set 7.7 (Page 507)

1. (a) $F = 31,200$ lb; $P = 312$ lb/ft^2

(b) $F = 2,452,500$ N; $P = 98.1$ kPa

3. 499.2 lb   **5.** 8.175×10^5 N   **7.** $1,098,720$ N   **9.** yes

11. $\rho a^3/\sqrt{2}$ lb   **13.** $14,976\sqrt{17}$ lb   **15.** 9.81×10^9 N

Exercise Set 7.8 (Page 517)

1. (a) ≈ 10.0179 (b) ≈ 3.7622 (c) $\approx 15/17 \approx 0.8824$

(d) ≈ -1.4436 (e) ≈ 1.7627 (f) ≈ 0.9730

3. (a) $\frac{4}{3}$ (b) $\frac{5}{4}$ (c) $\frac{312}{313}$ (d) $-\frac{63}{16}$

5.

	$\sinh x_0$	$\cosh x_0$	$\tanh x_0$	$\coth x_0$	$\operatorname{sech} x_0$	$\operatorname{csch} x_0$
(a)	2	$\sqrt{5}$	$2/\sqrt{5}$	$\sqrt{5}/2$	$1/\sqrt{5}$	$1/2$
(b)	3/4	5/4	3/5	5/3	4/5	4/3
(c)	4/3	5/3	4/5	5/4	3/5	3/4

9. $4\cosh(4x - 8)$ **11.** $-\dfrac{1}{x}\operatorname{csch}^2(\ln x)$

13. $\dfrac{1}{x^2}\operatorname{csch}\left(\dfrac{1}{x}\right)\coth\left(\dfrac{1}{x}\right)$ **15.** $\dfrac{2 + 5\cosh(5x)\sinh(5x)}{\sqrt{4x + \cosh^2(5x)}}$

17. $x^{5/2}\tanh(\sqrt{x})\operatorname{sech}^2(\sqrt{x}) + 3x^2\tanh^2(\sqrt{x})$ **19.** $\dfrac{1}{\sqrt{9 + x^2}}$

21. $\dfrac{1}{(\cosh^{-1} x)\sqrt{x^2 - 1}}$ **23.** $-\dfrac{(\tanh^{-1} x)^{-2}}{1 - x^2}$

25. $\dfrac{\sinh x}{|\sinh x|} = \begin{cases} 1, & x > 0 \\ -1, & x < 0 \end{cases}$ **27.** $-\dfrac{e^x}{2x\sqrt{1 - x}} + e^x\operatorname{sech}^{-1} x$

31. $\frac{1}{7}\sinh^7 x + C$ **33.** $\frac{2}{3}(\tanh x)^{3/2} + C$ **35.** $\ln(\cosh x) + C$

37. $37/375$ **39.** $\frac{1}{3}\sinh^{-1} 3x + C$ **41.** $-\operatorname{sech}^{-1}(e^x) + C$

43. $-\operatorname{csch}^{-1}|2x| + C$ **45.** $\frac{1}{2}\ln 3$ **49.** $16/9$ **51.** 5π

53. $\frac{3}{4}$ **61.** $|u| < 1: \tanh^{-1} u + C; |u| > 1: \tanh^{-1}(1/u) + C$

63. (a) $+\infty$ (b) $-\infty$ (c) 1   (d) -1 (e) $+\infty$ (f) $+\infty$ **71.** 405.9 ft

▶ **Chapter 7 Supplementary Exercises (Page 519)**

7. (a) $\displaystyle\int_a^b (f(x) - g(x))\,dx + \int_b^c (g(x) - f(x))\,dx +$
$\displaystyle\int_c^d (f(x) - g(x))\,dx$ (b) $\frac{11}{4}$ **9.** $9a/8$

13. (a)

(b) 1.42 in
(c) The length of the centerline is 192.026 in.

15. $S = \dfrac{2\sqrt{2}}{5}\pi(2e^\pi + 1)$

17. (a) $W = \frac{1}{16}$ J   (b) 5 m

19. (a) $F = \displaystyle\int_0^1 \rho x 3\,dx$ N   (b) $F = \displaystyle\int_1^4 \rho(1 + x)2x\,dx$ lb/ft^2

(c) $\displaystyle\int_{-10}^0 9810|y|2\sqrt{\tfrac{125}{8}(y + 10)}\,dy$ N

23. Set $a = 68.7672$, $b = 0.0100333$, $c = 693.8597$, $d = 299.2239$.

(a)

(b) 1480.2798 ft
(c) 283.6249 ft
(d) 82°

25. $k \approx 0.724611$

▶ **Exercise Set 8.1 (Page 525)**

1. $-\frac{1}{8}(3 - 2x)^4 + C$ **3.** $\frac{1}{2}\tan(x^2) + C$ **5.** $-\frac{1}{3}\ln(2 + \cos 3x) + C$

7. $\cosh(e^x) + C$ **9.** $-e^{\cot x} + C$ **11.** $-\frac{1}{42}\cos^6 7x + C$

13. $\ln(e^x + \sqrt{e^{2x} + 4}) + C$ **15.** $2e^{\sqrt{x-2}} + C$ **17.** $2\sinh\sqrt{x} + C$

19. $-\dfrac{2}{\ln 3}3^{-\sqrt{x}} + C$ **21.** $\frac{1}{2}\coth\dfrac{2}{x} + C$ **23.** $-\frac{1}{4}\ln\left|\dfrac{2 + e^{-x}}{2 - e^{-x}}\right| + C$

25. $\sin^{-1}(e^x) + C$ **27.** $\frac{1}{2}\sin(x^2) + C$ **29.** $-\dfrac{1}{\ln 16}4^{-x^2} + C$

▶ **Exercise Set 8.2 (Page 533)**

1. $-xe^{-x} - e^{-x} + C$ **3.** $x^2e^x - 2xe^x + 2e^x + C$

5. $-\frac{1}{2}x\cos 2x + \frac{1}{4}\sin 2x + C$ **7.** $x^2\sin x + 2x\cos x - 2\sin x + C$

9. $\frac{2}{3}x^{3/2}\ln x - \frac{4}{9}x^{3/2} + C$ **11.** $x(\ln x)^2 - 2x\ln x + 2x + C$

13. $x\ln(2x + 3) - x + \frac{3}{2}\ln(2x + 3) + C$ **15.** $x\sin^{-1} x + \sqrt{1 - x^2} + C$

17. $x\tan^{-1}(2x) - \frac{1}{4}\ln(1 + 4x^2) + C$ **19.** $\frac{1}{2}e^x(\sin x - \cos x) + C$

21. $\dfrac{e^{ax}}{a^2 + b^2}(a\sin bx - b\cos bx) + C$

23. $(x/2)[\sin(\ln x) - \cos(\ln x)] + C$ **25.** $x\tan x + \ln|\cos x| + C$

27. $\frac{1}{2}x^2 e^{x^2} - \frac{1}{2}e^{x^2} + C$ **29.** $(1 - 6e^{-5})/25$ **31.** $(2e^3 + 1)/9$

33. $5\ln 5 - 4$ **35.** $\dfrac{5\pi}{6} - \sqrt{3} + 1$ **37.** $-\pi/8$

39. $\frac{1}{3}\left(2\sqrt{3}\pi - \dfrac{\pi}{2} - 2 + \ln 2\right)$ **41.** (a) $2(\sqrt{x} - 1)e^{\sqrt{x}} + C$
(b) $2\sqrt{x}\sin\sqrt{x} + 2\cos\sqrt{x} + C$ **43.** $-(3x^2 + 5x + 7)e^{-x} + C$

45. $(4x^4 - 12x^2 + 6)\sin(2x) + (8x^3 - 12x)\cos(2x) + C$

47. (a) $A = 1$ (b) $V = \pi(e - 2)$ **49.** $V = 2\pi^2$

51. distance $= -37e^{-5} + 2$

53. (a) $-\frac{1}{3}\sin^2 x \cos x - \frac{2}{3}\cos x + C$ (b) $\dfrac{3\pi}{32} - \dfrac{1}{4}$

57. (a) $\frac{1}{3}\tan^3 x - \tan x + x + C$ (b) $\frac{1}{3}\sec^2 x \tan x + \frac{2}{3}\tan x + C$
(c) $x^3 e^x - 3x^2 e^x + 6xe^x - 6e^x + C$

63. $(x + 1)\ln(x + 1) - x + C$ **65.** $\frac{1}{2}(x^2 + 1)\tan^{-1} x - \frac{1}{2}x + C$

▶ **Exercise Set 8.3 (Page 541)**

1. $-\frac{1}{6}\cos^6 x + C$ **3.** $\dfrac{1}{2a}\sin^2 ax + C$, $a \neq 0$

5. $\frac{1}{2}\theta - \frac{1}{20}\sin 10\theta + C$ **7.** $\sin\theta - \frac{2}{3}\sin^3\theta + \frac{1}{5}\sin^5\theta + C$

9. $\frac{1}{6}\sin^3 2t - \frac{1}{10}\sin^5 2t + C$ **11.** $\frac{1}{8}x - \frac{1}{32}\sin 4x + C$

13. $-\frac{1}{6}\cos 3x + \frac{1}{2}\cos x + C$ **15.** $-\frac{1}{3}\cos(3x/2) - \cos(x/2) + C$

17. $(5\sqrt{2})/12$ **19.** 0   **21.** $\frac{1}{24}$ **23.** $\frac{1}{3}\tan(3x + 1) + C$

25. $\frac{1}{2}\ln|\cos(e^{-2x})| + C$ **27.** $\frac{1}{2}\ln|\sec 2x + \tan 2x| + C$

29. $\frac{1}{3}\tan^3 x + C$ **31.** $\frac{1}{16}\tan^4 4x + \frac{1}{24}\tan^6 4x + C$

33. $\frac{1}{7}\sec^7 x - \frac{1}{5}\sec^5 x + C$

35. $\frac{1}{4}\sec^3 x \tan x - \frac{5}{8}\sec x \tan x + \frac{3}{8}\ln|\sec x + \tan x| + C$

37. $\frac{1}{6}\sec^3 2t + C$ **39.** $\tan x + \frac{1}{3}\tan^3 x + C$

41. $\frac{1}{3}\tan^3 x - \tan x + x + C$ **43.** $\frac{2}{3}\tan^{3/2} x + \frac{2}{7}\tan^{7/2} x + C$

45. $\dfrac{\sqrt{3}}{2} - \dfrac{\pi}{6}$ **47.** $-\frac{1}{2} + \ln 2$ **49.** $-\frac{1}{5}\csc^5 x + \frac{1}{3}\csc^3 x + C$

51. $-\frac{1}{2}\csc^2 x - \ln|\sin x| + C$ **55.** $L = \ln(\sqrt{2} + 1)$ **57.** $V = \pi/2$

63. $-\dfrac{1}{\sqrt{a^2 + b^2}}\ln\left[\dfrac{\sqrt{a^2 + b^2} + a\cos x - b\sin x}{a\sin x + b\cos x}\right] + C$

65. (a) $\frac{2}{3}$ (b) $3\pi/16$ (c) $\frac{8}{15}$ (d) $5\pi/32$

▶ **Exercise Set 8.4 (Page 547)**

1. $2\sin^{-1}(x/2) + \frac{1}{2}x\sqrt{4 - x^2} + C$ **3.** $\frac{9}{2}\sin^{-1}(x/3) - \frac{1}{2}x\sqrt{9 - x^2} + C$

5. $\frac{1}{16}\tan^{-1}(x/2) + \dfrac{x}{8(4 + x^2)} + C$ **7.** $\sqrt{x^2 - 9} - 3\sec^{-1}(x/3) + C$

9. $-2\sqrt{2 - x^2} + \frac{1}{3}(2 - x^2)^{3/2} + C$ **11.** $\dfrac{\sqrt{4x^2 - 9}}{9x} + C$

13. $\dfrac{x}{\sqrt{1 - x^2}} + C$ **15.** $\ln|x + \sqrt{x^2 - 1}| + C$ **17.** $-(x/\sqrt{9x^2 - 1}) + C$

19. $\frac{1}{2}\sin^{-1}(e^x) + \frac{1}{2}e^x\sqrt{1 - e^{2x}} + C$ **21.** $\frac{2048}{15}$ **23.** $(\sqrt{3} - \sqrt{2})/2$

25. $\dfrac{10\sqrt{3} + 18}{243}$ **27.** $\frac{1}{2}\ln(x^2 + 4) + C$

29. $L = \sqrt{5} - \sqrt{2} + \ln\dfrac{2 + 2\sqrt{2}}{1 + \sqrt{5}}$ **31.** $S = \dfrac{\pi}{32}[18\sqrt{5} - \ln(2 + \sqrt{5})]$

33. (a) $\sinh^{-1}(x/3) + C$ (b) $\ln\left(\dfrac{\sqrt{x^2 + 9}}{3} + \dfrac{x}{3}\right) + C$

(c) $\frac{1}{2}x\sqrt{x^2 - 1} - \frac{1}{2}\cosh^{-1} x + C$ **35.** $\frac{1}{3}\tan^{-1}\left(\dfrac{x - 2}{3}\right) + C$

37. $\sin^{-1}\left(\dfrac{x - 1}{3}\right) + C$ **39.** $\ln(x - 3 + \sqrt{(x - 3)^2 + 1}) + C$

41. $2\sin^{-1}\left(\dfrac{x + 1}{2}\right) + \frac{1}{2}(x + 1)\sqrt{3 - 2x - x^2} + C$

43. $\dfrac{1}{\sqrt{10}}\tan^{-1}\sqrt{\tfrac{2}{5}}(x + 1) + C$ **45.** $\pi/6$

47. $u = \sin^2 x$, $\dfrac{1}{2}\displaystyle\int\sqrt{1 - u^2}\,du = \dfrac{1}{4}[\sin^2 x\sqrt{1 - \sin^4 x}$
$+ \sin^{-1}(\sin^2 x)] + C$

▶ **Exercise Set 8.5** (Page 554)

1. $\dfrac{A}{(x-2)} + \dfrac{B}{(x+5)}$ 3. $\dfrac{A}{x} + \dfrac{B}{x^2} + \dfrac{C}{x-1}$

5. $\dfrac{A}{x} + \dfrac{B}{x^2} + \dfrac{C}{x^3} + \dfrac{Dx+E}{x^2+1}$ 7. $\dfrac{Ax+B}{x^2+5} + \dfrac{Cx+D}{(x^2+5)^2}$

9. $\dfrac{1}{5}\ln\left|\dfrac{x-1}{x+4}\right| + C$ 11. $\frac{5}{2}\ln|2x-1| + 3\ln|x+4| + C$

13. $\ln\left|\dfrac{x(x+3)^2}{x-3}\right| + C$ 15. $\frac{1}{2}x^2 - 2x + 6\ln|x+2| + C$

17. $3x + 12\ln|x-2| - \dfrac{2}{x-2} + C$ 19. $\frac{1}{3}x^3 + x + \ln\left|\dfrac{(x+1)(x-1)^2}{x}\right| + C$

21. $3\ln|x| - \ln|x-1| - \dfrac{5}{x-1} + C$ 23. $\ln\dfrac{(x-3)^2}{|x+1|} + \dfrac{1}{x-3} + C$

25. $\ln|x+2| + \dfrac{4}{x+2} - \dfrac{2}{(x+2)^2} + C$

27. $-\frac{7}{34}\ln|4x-1| + \frac{6}{17}\ln(x^2+1) + \frac{3}{17}\tan^{-1}x + C$

29. $3\tan^{-1}x + \frac{1}{2}\ln(x^2+3) + C$ 31. $\frac{1}{2}x^2 - 3x + \frac{1}{2}\ln(x^2+1) + C$

33. $\dfrac{1}{6}\ln\left(\dfrac{1-\sin\theta}{5+\sin\theta}\right) + C$ 35. $V = \pi\left(\frac{19}{5} - \frac{9}{4}\ln 5\right)$

37. $\dfrac{1}{\sqrt2}\tan^{-1}\left(\dfrac{x+1}{\sqrt2}\right) + \dfrac{1}{x^2+2x+3} + C$

39. $\frac{1}{8}\ln|x-1| - \frac{1}{5}\ln|x-2| + \frac{1}{12}\ln|x-3| - \frac{1}{120}\ln|x+3| + C$

▶ **Exercise Set 8.6** (Page 563)

1. Formula (60): $\frac{3}{16}[4x + \ln|-1 + 4x|] + C$

3. Formula (65): $\dfrac{1}{5}\ln\left|\dfrac{x}{5+2x}\right| + C$

5. Formula (102): $\frac{1}{5}(x+1)(-3+2x)^{3/2} + C$

7. Formula (108): $\dfrac{1}{2}\ln\left|\dfrac{\sqrt{4-3x}-2}{\sqrt{4-3x}+2}\right| + C$

9. Formula (69): $\dfrac{1}{2\sqrt5}\ln\left|\dfrac{x+\sqrt5}{x-\sqrt5}\right| + C$

11. Formula (73): $\dfrac{x}{2}\sqrt{x^2-3} - \dfrac{3}{2}\ln|x+\sqrt{x^2-3}| + C$

13. Formula (95): $\dfrac{x}{2}\sqrt{x^2+4} - 2\ln(x+\sqrt{x^2+4}) + C$

15. Formula (74): $\dfrac{x}{2}\sqrt{9-x^2} + \dfrac{9}{2}\sin^{-1}\dfrac{x}{3} + C$

17. Formula (79): $\sqrt{3-x^2} - \sqrt3\ln\left|\dfrac{\sqrt3+\sqrt{3-x^2}}{x}\right| + C$

19. Formula (38): $-\frac{1}{10}\sin(5x) + \frac{1}{2}\sin x + C$

21. Formula (50): $\dfrac{x^4}{16}[4\ln x - 1] + C$

23. Formula (42): $\dfrac{e^{-2x}}{13}[-2\sin(3x) - 3\cos(3x)] + C$

25. Formula (62): $\dfrac{1}{2}\displaystyle\int\dfrac{u\,du}{(4-3u)^2} = \dfrac{1}{18}\left[\dfrac{4}{4-3e^{2x}} + \ln\left|4-3e^{2x}\right|\right] + C$

27. Formula (68): $\dfrac{2}{3}\displaystyle\int\dfrac{du}{u^2+4} = \dfrac{1}{3}\tan^{-1}\dfrac{3\sqrt x}{2} + C$

29. Formula (76): $\dfrac{1}{3}\displaystyle\int\dfrac{du}{\sqrt{u^2-4}} = \dfrac{1}{3}\ln|3x+\sqrt{9x^2-4}| + C$

31. Formula (81): $\dfrac{1}{54}\displaystyle\int\dfrac{u^2\,du}{\sqrt{5-u^2}} = -\dfrac{x^2}{36}\sqrt{5-9x^4} + \dfrac{5}{108}\sin^{-1}\dfrac{3x^2}{\sqrt5} + C$

33. Formula (26): $\displaystyle\int\sin^2 u\,du = \dfrac{1}{2}\ln x + \dfrac{1}{4}\sin(2\ln x) + C$

35. Formula (51): $\dfrac{1}{4}\displaystyle\int ue^u\,du = \dfrac{1}{4}(-2x-1)e^{-2x} + C$

37. $u = \cos 3x$, Formula (67):

$-\displaystyle\int\dfrac{du}{u(u+1)^2} = -\dfrac{1}{3}\left[\dfrac{1}{1+\cos 3x} + \ln\left|\dfrac{\cos 3x}{1+\cos 3x}\right|\right] + C$

39. $u = 4x^2$, Formula (70): $\dfrac{1}{8}\displaystyle\int\dfrac{du}{u^2-1} = \dfrac{1}{16}\ln\left|\dfrac{4x^2-1}{4x^2+1}\right| + C$

41. $u = 2e^x$, Formula (74):

$\dfrac{1}{2}\displaystyle\int\sqrt{3-u^2}\,du = \dfrac{1}{2}e^x\sqrt{3-4e^{2x}} + \dfrac{3}{4}\sin^{-1}\left(\dfrac{2e^x}{\sqrt3}\right) + C$

43. $u = 3x$, Formula (112): $\dfrac{1}{3}\displaystyle\int\sqrt{\dfrac{5}{3}u - u^2}\,du = \dfrac{18x-5}{36}\sqrt{5x-9x^2} + \dfrac{25}{216}\sin^{-1}\left(\dfrac{18x-5}{5}\right) + C$

45. $u = 3x$, Formula (44): $\dfrac{1}{9}\displaystyle\int u\sin u\,du = \dfrac{1}{9}(\sin 3x - 3x\cos 3x) + C$

47. $u = -\sqrt x$, Formula (51): $2\displaystyle\int ue^u\,du = -2(\sqrt x + 1)e^{-\sqrt x} + C$

49. $x^2 + 4x - 5 = (x+2)^2 - 9$; $u = x+2$, Formula (70):

$\displaystyle\int\dfrac{du}{u^2-9} = \dfrac{1}{6}\ln\left|\dfrac{x-1}{x+5}\right| + C$

51. $x^2 - 4x - 5 = (x-2)^2 - 9$, $u = x-2$, Formula (77):

$\displaystyle\int\dfrac{u+2}{\sqrt{9-u^2}}\,du = -\sqrt{5+4x-x^2} + 2\sin^{-1}\left(\dfrac{x-2}{3}\right) + C$

53. $u = \sqrt{x-2}$, $\frac{2}{5}(x-2)^{5/2} + \frac{4}{3}(x-2)^{3/2} + C$

55. $u = \sqrt{x^3+1}$, $\dfrac{2}{3}\displaystyle\int u^2(u^2-1)\,du = \dfrac{2}{15}(x^3+1)^{5/2} - \dfrac{2}{9}(x^3+1)^{3/2} + C$

57. $u = x^{1/6}$, $\displaystyle\int\dfrac{6u^5}{u^3+u^2}\,du = 2x^{1/2} - 3x^{1/3} + 6x^{1/6} - 6\ln(x^{1/6}+1) + C$

59. $u = x^{1/4}$, $4\displaystyle\int\dfrac{1}{u(1-u)}\,du = 4\ln\dfrac{x^{1/4}}{|1-x^{1/4}|} + C$

61. $u = x^{1/6}$, $6\displaystyle\int\dfrac{u^3}{u-1}\,du = 2x^{1/2} + 3x^{1/3} + 6x^{1/6} + 6\ln|x^{1/6}-1| + C$

63. $u = \sqrt{1+x^2}$, $\displaystyle\int(u^2-1)\,du = \dfrac{1}{3}(1+x^2)^{3/2} - (1+x^2)^{1/2} + C$

65. $u = \sqrt x$, $2\displaystyle\int u\sin u\,du = 2\sin\sqrt x - 2\sqrt x\cos\sqrt x + C$

67. $\displaystyle\int\dfrac{1}{1 + \dfrac{2u}{1+u^2} + \dfrac{1-u^2}{1+u^2}}\cdot\dfrac{2}{1+u^2}\,du = \displaystyle\int\dfrac{1}{u+1}\,du = $

$\ln|\tan(x/2) + 1| + C$

69. $\displaystyle\int\dfrac{d\theta}{1-\cos\theta} = \displaystyle\int\dfrac{1}{u^2}\,du = -\cot(\theta/2) + C$

71. $2\displaystyle\int\dfrac{1-u^2}{(3u^2+1)(u^2+1)}\,du$, $\dfrac{4}{\sqrt3}\tan^{-1}[\sqrt3\tan(x/2)] - x + C$

73. $x = \dfrac{4e^2}{1+e^2}$

75. $A = 6 + \frac{25}{2}\sin^{-1}\frac{4}{5}$

77. $A = \frac{1}{40}\ln 9$

79. $V = \pi(\pi - 2)$

81. $V = 2\pi(1 - 4e^{-3})$

83. $L = \sqrt{65} + \frac{1}{8}\ln(8 + \sqrt{65})$

85. $S = 2\pi[\sqrt2 + \ln(1 + \sqrt2)]$

87.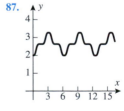

▶ **Exercise Set 8.7** (Page 576)

1. exact value $= 14/3 \approx 4.666666667$
   (a) 4.667600663, $|E_M| \approx 0.000933996$
   (b) 4.664795679, $|E_T| \approx 0.001870988$
   (c) 4.666651630, $|E_S| \approx 0.000015037$

3. exact value $= 2$
   (a) 2.008248408, $|E_M| \approx 0.008248408$
   (b) 1.983523538, $|E_T| \approx 0.016476462$
   (c) 2.000109517, $|E_S| \approx 0.000109517$

5. exact value $= e^{-1} - e^{-3} \approx 0.318092373$
   (a) 0.317562837, $|E_M| \approx 0.000529536$
   (b) 0.319151975, $|E_T| \approx 0.001059602$
   (c) 0.318095187, $|E_S| \approx 0.000002814$

7. (a) $|E_M| \leq \dfrac{27}{2400}(1/4) = 0.002812500$

(b) $|E_T| \leq \dfrac{27}{1200}(1/4) = 0.005625000$

(c) $|E_S| \leq \dfrac{243}{180 \times 10^4}(15/16) \approx 0.000126563$

9. (a) $|E_M| \leq \dfrac{\pi^3}{2400}(1) \approx 0.012919282$

(b) $|E_T| \leq \dfrac{\pi^3}{1200}(1) \approx 0.025838564$

(c) $|E_S| \leq \dfrac{\pi^5}{180 \times 10^4}(1) \approx 0.000170011$

11. (a) $|E_M| \leq \dfrac{8}{2400}(e^{-1}) \approx 0.001226265$

(b) $|E_T| \leq \dfrac{8}{1200}(e^{-1}) \approx 0.002452530$

(c) $|E_S| \leq \dfrac{32}{180 \times 10^4}(e^{-1}) \approx 0.000006540$

13. (a) $n = 24$ **(b)** $n = 34$ **(c)** $2n = 8$

15. (a) $n = 36$ **(b)** $n = 51$ **(c)** $2n = 8$

17. (a) $n = 351$ **(b)** $n = 496$ **(c)** $2n = 16$

19. $g(x) = \frac{1}{24}x^2 - \frac{3}{8}x + \frac{13}{12}$ **21.** $0.746824948, 0.746824133$

23. $2.129861595, 2.129861293$ **25.** $0.805376152, 0.804776489$

27. (a) $3.142425985, |E_M| \approx 0.000833331$

(b) $3.139925989, |E_T| \approx 0.001666665$

(c) $3.141592614, |E_S| \approx 0.000000040$

29. $S_{14} = 0.693147984, |E_S| \approx 0.000000803 = 8.03 \times 10^{-7}$

31. $n = 116$ **35.** ≈ 3.82019 **37.** 1604 ft   **39.** 37.9 mi

41. 9.3 L   **45. (a)** $\max|f''(x)| \approx 3.844880$ **(b)** $n = 18$ **(c)** 0.904741

47. (a) $\max|f^{(4)}(x)| \approx 42.551816$ **(b)** $2n = 8$ **(c)** 0.904524

▶ **Exercise Set 8.8 (Page 585)**

1. (a) improper; infinite discontinuity at $x = 3$ **(b)** not improper

(c) improper; infinite discontinuity at $x = 0$ **(d)** improper;

infinite interval of integration   **(e)** improper; infinite interval of

integration and infinite discontinuity at $x = 1$ **(f)** not improper

3. 1 **5.** $\ln \frac{5}{3}$ **7.** $\frac{1}{2}$ **9.** $-\frac{1}{4}$ **11.** $\frac{1}{3}$ **13.** divergent

15. 0 **17.** divergent   **19.** divergent   **21.** $\pi/2$ **23.** 1

25. divergent   **27.** $\frac{9}{2}$ **29.** divergent   **31.** 2 **33.** 2 **35.** $\frac{1}{2}$

37. (a) 2.726585 **(b)** 2.804364 **(c)** 0.219384 **(d)** 0.504067 **39.** 12

41. -1 **43.** $\frac{1}{3}$ **45. (a)** $V = \pi/2$ **(b)** $S = \pi[\sqrt{2} + \ln(1 + \sqrt{2})]$

47. (b) $1/e$ **(c)** It is convergent.   **51.** $\dfrac{8\sqrt{2}}{5}$

53. $\dfrac{2\pi N I}{kr}\left(1 - \dfrac{a}{\sqrt{r^2 + a^2}}\right)$ **55. (b)** 2.4×10^7 mi·lb

57. (a) $\dfrac{1}{s^2}$ **(b)** $\dfrac{2}{s^3}$ **(c)** $\dfrac{e^{-3s}}{s}$ **61. (a)** 1.047 **65.** 1.809

▶ **Chapter 8 Supplementary Exercises (Page 588)**

1. (a) parts   **(b)** substitution   **(c)** reduction formula

(d) substitution   **(e)** substitution   **(f)** substitution   **(g)** parts

(h) substitution   **(i)** substitution

5. (a) 40 **(b)** 57 **(c)** 113 **(d)** 108 **(e)** 52 **(f)** 71

7. (a) $-\frac{1}{8}\sin^3(2x)\cos 2x - \frac{3}{16}\cos 2x \sin 2x + \frac{3}{8}x + C$

(b) $\frac{1}{10}\cos^4(x^2)\sin(x^2) + \frac{2}{15}\cos^2(x^2)\sin(x^2) + \frac{4}{15}\sin(x^2) + C$

9. (a) $2\sin^{-1}(\sqrt{x/2}) + C; -2\sin^{-1}(\sqrt{2-x}/\sqrt{2}) + C; \sin^{-1}(x - 1) + C$

11.

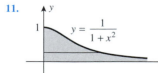

$y = \dfrac{1}{1 + x^2}$

13. $V = 2\pi$

15. $-\frac{2}{3}\cos^{3/2}\theta + C$

17. $\frac{1}{6}\tan^3(x^2) + C$

19. $\dfrac{x}{3\sqrt{3 + x^2}} + C$

21. $\sqrt{x^2 + 2x + 2} + 2\ln(\sqrt{x^2 + 2x + 2} + x + 1) + C$

23. $-\frac{1}{6}\ln|x - 1| + \frac{1}{15}\ln|x + 2| + \frac{1}{10}\ln|x - 3| + C$ **25.** $4 - \pi$

27. $\ln\dfrac{\sqrt{e^x + 1} - 1}{\sqrt{e^x + 1} + 1} + C$ **29.** $\dfrac{1}{2(a^2 + 1)}$ **31.** $\frac{1}{4}\sin^{-1}(x^4) + C$

33. $\dfrac{\sqrt{2}}{3}[(x + 2)^{3/2} - (x - 2)^{3/2}] + C$

35. (a) $(x + 4)(x - 5)(x^2 + 1)^2; \dfrac{A}{x + 4} + \dfrac{B}{x - 5} + \dfrac{Cx + D}{x^2 + 1} + \dfrac{Ex + F}{(x^2 + 1)^2}$

(b) $-\dfrac{3}{x + 4} + \dfrac{2}{x - 5} - \dfrac{x - 2}{x^2 + 1} - \dfrac{3}{(x^2 + 1)^2}$ **(c)** $-3\ln|x + 4| +$

$2\ln|x - 5| + 2\tan^{-1}x - \dfrac{1}{2}\ln(x^2 + 1) - \dfrac{3}{2}\left(\dfrac{x}{x^2 + 1} + \tan^{-1}x\right) + C$

▶ **Exercise Set 9.1 (Page 605)**

3. (a) first order   **(b)** second order

7. (a) $y = Ce^{-3x}$ **(b)** $y = Ce^{2t}$ **9.** $y = e^{-2x} + Ce^{-3x}$

11. $y = e^{-x}\sin(e^x) + Ce^{-x}$ **13.** $y = \dfrac{C}{\sqrt{x^2 + 1}}$ **15.** $y = Cx$

17. $y = Ce^{-\sqrt{1+x^2}} - 1, C \neq 0$ **19.** $\ln|y| + y^2/2 = e^x + C$ and $y = 0$

21. $y = \ln(\sec x + C)$ **23.** $y = \dfrac{1}{1 - C(\csc x - \cot x)}, C \neq 0$ and $y = 0$

25. (a) $y = \dfrac{x}{2} + \dfrac{3}{2x}$ **(b)** $y = \dfrac{x}{2} - \dfrac{5}{2x}$ **27.** $y = -1 + 4e^{x^2/2}$

29. $3y^2 + 6\sin y = 8x^3 + 3\pi^2 - 8$ **31.** $y^2 - 2y = t^2 + t + 3$

33. (a)

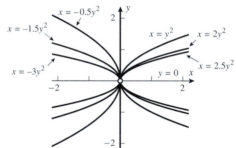

(b) $y^2 = x/4$

35. $y = \dfrac{C}{\sqrt{x^2 + 4}}$ **37.** $x^3 + y^3 - 3y = C$

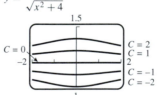

41. $x^2 + 2e^{-y} = 6$ **43. (a)** $200 - 175e^{-t/25}$ oz   **(b)** 136 oz

45. 25 lb   **49. (a)** $I(t) = \frac{6}{5}(1 - e^{-5t/2})$ A   **(b)** It tends to $\frac{6}{5}$ A.

51. (a) $v = c\ln\dfrac{m_0}{m_0 - kt} - gt$ **(b)** 3044 m/s

53. (a) $h \approx (2 - 0.003979t)^2$ **(b)** 8.4 min

55. $v = \dfrac{50}{2t + 1}$ cm/s, $x = 25\ln(2t + 1)$ cm

57. $\dfrac{dy}{dx} = -\sin x + e^{-x^2}, y(0) = 1$

▶ **Exercise Set 9.2 (Page 613)**

1.

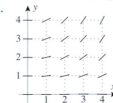

3.

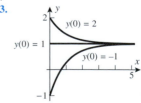

5.

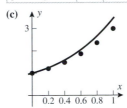

9. (a) IV
(b) VI
(c) V
(d) II
(e) I
(f) III

11. (a)

n	0	1	2	3	4	5
x_n	0	0.2	0.4	0.6	0.8	1.0
y_n	1	1.20	1.48	1.86	2.35	2.98

(b) $y = -(x+1) + 2e^x$

x_n	0	0.2	0.4	0.6	0.8	1.0
$y(x_n)$	1	1.24	1.58	2.04	2.65	3.44
absolute error	0	0.04	0.10	0.19	0.30	0.46
percentage error	0	3	6	9	11	13

(c)

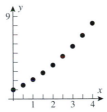

13.

n	0	1	2	3	4	5	6	7	8
x_n	0	0.5	1	1.5	2	2.5	3	3.5	4
y_n	1	1.50	2.11	2.84	3.68	4.64	5.72	6.91	8.23

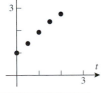

15.

n	0	1	2	3	4
t_n	0	0.5	1	1.5	2
y_n	1	1.42	1.92	2.39	2.73

17. $y(1) \approx 1.00$

n	0	1	2	3	4	5
t_n	0	0.2	0.4	0.6	0.8	1.0
y_n	1.00	1.20	1.26	1.10	0.94	1.00

19. (b) $y(1/2) = \sqrt{3}/2$

▶ **Exercise Set 9.3 (Page 622)**

1. (a) $\dfrac{dy}{dt} = ky^2, y(0) = y_0 (k > 0)$ **(b)** $\dfrac{dy}{dt} = -ky^2, y(0) = y_0 (k > 0)$

3. (a) $\dfrac{ds}{dt} = \dfrac{1}{2}s$ **(b)** $\dfrac{d^2 s}{dt^2} = 2\dfrac{ds}{dt}$ **5. (a)** $\dfrac{dy}{dt} = 0.01y, y_0 = 10,000$
(b) $y = 10,000e^{t/100}$ **(c)** 69.31 h **(d)** 150.41 h

7. (a) $\dfrac{dy}{dt} = -ky, k \approx 0.1810$ **(b)** $y = 5.0 \times 10^7 e^{-0.181t}$
(c) $\approx 219,000$ atoms **(d)** 12.72 days **9.** 196 days

11. 3.30 days **13. (a)** $y \approx 2e^{0.1386t}$ **(b)** $y = 5e^{0.015t}$
(c) $y \approx 0.5995e^{0.5117t}$ **(d)** $y \approx 0.8706e^{0.1386t}$

17. (b) 70 years **(c)** 20 years **(d)** 7%

21. $y_0 \approx 2, L \approx 8, k \approx 0.5493$ **23. (a)** $y_0 = 5$ **(b)** $L = 12$
(c) $k = 1$ **(d)** $t = 0.3365$ **(e)** $\dfrac{dy}{dt} = \dfrac{1}{12}y(12 - y), y(0) = 5$
25. (a) $L = 10$ **(b)** $k = 10$ **(c)** $y = 5$
27. Assume that $y(t)$ students have had the flu t days after semester break.
Then $y(0) = 20$, $y(5) = 35$. **(a)** $\dfrac{dy}{dt} = ky(1000 - y), y_0 = 20$

(b) $y = \dfrac{1000}{1 + 49e^{-0.115t}}$; $k = 0.115$

(c)

t	0	1	2	3	4	5	6	7	8	9	10	11	12	13	14
$y(t)$	20	22	25	28	31	35	39	44	49	54	61	67	75	83	93

(d)

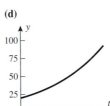

29. (a) $\dfrac{dT}{dt} = -k(T - 21), T(0) = 95$;

$T = 21 + 74e^{-kt}$

(b) 6.22 min
33. (c) $y = 4e^{t \ln 2}$ **(d)** $y = 4e^{-t \ln 2}$

▶ **Exercise Set 9.4 (Page 631)**

3. $y = c_1 e^x + c_2 e^{-4x}$ **5.** $y = c_1 e^x + c_2 x e^x$
7. $y = c_1 \cos \sqrt{5}x + c_2 \sin \sqrt{5}x$ **9.** $y = c_1 + c_2 e^x$
11. $y = c_1 e^{-2t} + c_2 t e^{-2t}$ **13.** $y = e^{2x}(c_1 \cos 3x + c_2 \sin 3x)$
15. $y = c_1 e^{-x/4} + c_2 e^{x/2}$ **17.** $y = 2e^x - e^{-3x}$
19. $y = (2 - 5x)e^{3x}$ **21.** $y = -e^{-2x}(3 \cos x + 6 \sin x)$
23. (a) $y'' - 3y' - 10y = 0$ **(b)** $y'' - 8y' + 16y = 0$
(c) $y'' + 2y' + 17y = 0$ **25. (a)** $k < 0$ or $k > 4$ **(b)** $0, 4$
27. (a) $y = (1/x)[c_1 \cos(\ln x) + c_2 \sin(\ln x)]$
(b) $y = c_1 x^{1+\sqrt{3}} + c_2 x^{1-\sqrt{3}}$

33. (a) $y = 0.3 \cos(t/2)$ **35. (a)** $y = -0.12 \cos 14t$
(b) $T = 4\pi$ s, $f = 1/(4\pi)$ Hz **(b)** $T = \pi/7$ s, $f = 7/\pi$ Hz
(c) **(c)**

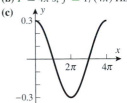 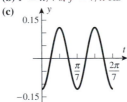

(d) $t = \pi$ s **(d)** $t = \pi/28$ s
(e) $t = 2\pi$ s **(e)** $t = \pi/14$ s
37. (a) Maximum speed occurs when $y = 0$.
(b) Minimum speed occurs when $y = \pm y_0$.
39. $Mx''(t) + kx(t) = 0, x(0) = x_0, x'(0) = 0$
41. (a) $y = e^{-1.2t} + 3.2t e^{-1.2t}$ **(b)** 1.427364 cm
43. (a) $y = e^{-t/2} \cos(\sqrt{19}\,t/2) - \frac{6}{19}\sqrt{19}e^{-t/2}\sin(\sqrt{19}t/2)$
(c) -3.210357 cm/s
(d) $3.210357/s^2$

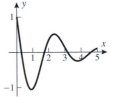

45. (a) $y = (4 + 2v_0)e^{-3t/2} - (3 + 2v_0)e^{-2t}$
(b) $8e^{-3t/2} - 7e^{-2t}, 2e^{-3t/2} - e^{-2t}$,
$-4e^{-3t/2} + 5e^{-2t}$

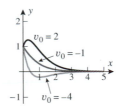

▶ **Chapter 9 Supplementary Exercises (Page 634)**

5. (a) linear **(b)** both **(c)** separable **(d)** neither **7.** $y = L/2$

9. $r = 4 - t$ m **11. (a)** $P = 4(1 - e^{-t/12,000})\%$ **(b)** 36.05 min

13. $y^{-4} + 4\ln(x/y) = 1$ **15.** $y = \dfrac{1}{3 - 2\tan 2x}$

17. (a) $y = \left(-\frac{3}{10}x - \frac{3}{50}\right)\cos 3x + \left(-\frac{1}{10}x + \frac{2}{25}\right)\sin 3x + \frac{53}{50}e^x$

(c)

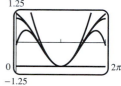

19. (a) no **(b)** same, $r\%$

21. (a) $y = C_1 e^x + C_2 e^{2x}$

(b) $y = C_1 e^{x/2} + C_2 x e^{x/2}$

(c) $y = e^{-x/2}\left[c_1 \cos \dfrac{\sqrt{7}x}{2} + C \sin \dfrac{\sqrt{7}x}{2}\right]$

23. (a) 7.77 years

(b) $\dfrac{dy}{dt} = k\left(1 - \dfrac{y}{95}\right)y$, $y(0) = 19$

27. (a) \$1491.82 **(b)** \$4493.29 **(c)** 8.7 years

▶ **Exercise Set 10.1 (Page 646)**

1. (a) $1 - x + \frac{1}{2}x^2$, $1 - x$ **(b)** $1 - \frac{1}{2}x^2$, 1 **(c)** $1 - \frac{1}{2}(x - \pi/2)^2$, 1
(d) $1 + \frac{1}{2}(x - 1) - \frac{1}{8}(x - 1)^2$, $1 + \frac{1}{2}(x - 1)$

3. (a) $1 + \frac{1}{2}(x - 1) - \frac{1}{8}(x - 1)^2$ **(b)** 1.04875 **5.** 1.80397443

7. $p_0(x) = 1$, $p_1(x) = 1 - x$, $p_2(x) = 1 - x + \frac{1}{2}x^2$,

$p_3(x) = 1 - x + \dfrac{1}{2}x^2 - \dfrac{1}{3!}x^3$,

$p_4(x) = 1 - x + \dfrac{1}{2}x^2 - \dfrac{1}{3!}x^3 + \dfrac{1}{4!}x^4$; $\displaystyle\sum_{k=0}^{n} \dfrac{(-1)^k}{k!}x^k$

9. $p_0(x) = 1$, $p_1(x) = 1$, $p_2(x) = 1 - \dfrac{\pi^2}{2!}x^2$;

$p_3(x) = 1 - \dfrac{\pi^2}{2!}x^2$, $p_4(x) = 1 - \dfrac{\pi^2}{2!}x^2 + \dfrac{\pi^4}{4!}x^4$; $\displaystyle\sum_{k=0}^{[n/2]} \dfrac{(-1)^k \pi^{2k}}{(2k)!}x^{2k}$

11. $p_0(x) = 0$, $p_1(x) = x$, $p_2(x) = x - \dfrac{1}{2}x^2$, $p_3(x) = x - \dfrac{1}{2}x^2 + \dfrac{1}{3}x^3$,

$p_4(x) = x - \dfrac{1}{2}x^2 + \dfrac{1}{3}x^3 - \dfrac{1}{4}x^4$; $\displaystyle\sum_{k=1}^{n} \dfrac{(-1)^{k+1}}{k}x^k$

13. $p_0(x) = 1$, $p_1(x) = 1$, $p_2(x) = 1 + \dfrac{x^2}{2}$, $p_3(x) = 1 + \dfrac{x^2}{2}$,

$p_4(x) = 1 + \dfrac{x^2}{2} + \dfrac{x^4}{4!}$; $\displaystyle\sum_{k=0}^{[n/2]} \dfrac{1}{(2k)!}x^{2k}$

15. $p_0(x) = 0$, $p_1(x) = 0$, $p_2(x) = x^2$, $p_3(x) = x^2$, $p_4(x) = x^2 - \dfrac{1}{6}x^4$;

$\displaystyle\sum_{k=0}^{[n/2]-1} \dfrac{(-1)^k}{(2k+1)!}x^{2k+2}$

17. $p_0(x) = e$, $p_1(x) = e + e(x - 1)$, $p_2(x) = e + e(x - 1) + \dfrac{e}{2}(x - 1)^2$,

$p_3(x) = e + e(x - 1) + \dfrac{e}{2}(x - 1)^2 + \dfrac{e}{3!}(x - 1)^3$,

$p_4(x) = e + e(x - 1) + \dfrac{e}{2}(x - 1)^2 + \dfrac{e}{3!}(x - 1)^3 + \dfrac{e}{4!}(x - 1)^4$;

$\displaystyle\sum_{k=0}^{n} \dfrac{e}{k!}(x - 1)^k$

19. $p_0(x) = -1$, $p_1(x) = -1 - (x+1)$, $p_2(x) = -1 - (x+1) - (x+1)^2$,
$p_3(x) = -1 - (x + 1) - (x + 1)^2 - (x + 1)^3$,

$p_4(x) = -1 - (x+1) - (x+1)^2 - (x+1)^3 - (x+1)^4$; $\displaystyle\sum_{k=0}^{n}(-1)(x+1)^k$

21. $p_0(x) = p_1(x) = 1$, $p_2(x) = p_3(x) = 1 - \dfrac{\pi^2}{2}\left(x - \dfrac{1}{2}\right)^2$,

$p_4(x) = 1 - \dfrac{\pi^2}{2}\left(x - \dfrac{1}{2}\right)^2 + \dfrac{\pi^4}{4!}\left(x - \dfrac{1}{2}\right)^4$; $\displaystyle\sum_{k=0}^{[n/2]} \dfrac{(-1)^k \pi^{2k}}{(2k)!}\left(x - \dfrac{1}{2}\right)^{2k}$

23. $p_0(x) = 0$, $p_1(x) = (x - 1)$, $p_2(x) = (x - 1) - \dfrac{1}{2}(x - 1)^2$,

$p_3(x) = (x - 1) - \dfrac{1}{2}(x - 1)^2 + \dfrac{1}{3}(x - 1)^3$,

$p_4(x) = (x - 1) - \dfrac{1}{2}(x - 1)^2 + \dfrac{1}{3}(x - 1)^3 - \dfrac{1}{4}(x - 1)^4$;

$\displaystyle\sum_{k=1}^{n} \dfrac{(-1)^{k-1}}{k}(x - 1)^k$

25. (a) $1 + 2x - x^2 + x^3$ **(b)** $1 + 2(x - 1) - (x - 1)^2 + (x - 1)^3$

27. $p_0(x) = 1$, $p_1(x) = 1 - 2x$,
$p_2(x) = 1 - 2x + 2x^2$,
$p_3(x) = 1 - 2x + 2x^2 - \frac{4}{3}x^3$

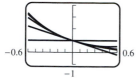

29. $p_0(x) = -1$, $p_2(x) = -1 + \frac{1}{2}(x - \pi)^2$,

$p_4(x) = -1 + \frac{1}{2}(x - \pi)^2 - \frac{1}{24}(x - \pi)^4$,

$p_6(x) = -1 + \frac{1}{2}(x - \pi)^2 - \frac{1}{24}(x - \pi)^4 + \frac{1}{720}(x - \pi)^6$

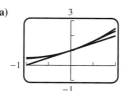

31. 1.64870 **33.** IV **37. (a)**

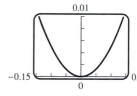

(b)

x	−1.000	−0.750	−0.500	−0.250	0.000	0.250	0.500	0.750	1.000
$f(x)$	0.431	0.506	0.619	0.781	1.000	1.281	1.615	1.977	2.320
$p_1(x)$	0.000	0.250	0.500	0.750	1.000	1.250	1.500	1.750	2.000
$p_2(x)$	0.500	0.531	0.625	0.781	1.000	1.281	1.625	2.031	2.500

(c) $|e^{\sin x} - (1 + x)| < 0.01$
for $-0.14 < x < 0.14$

(d) $\left|e^{\sin x} - \left(1 + x + \dfrac{x^2}{2}\right)\right| < 0.01$
for $-0.50 < x < 0.50$

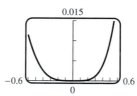

39. (a) $(-0.569, 0.569)$

▶ **Exercise Set 10.2 (Page 656)**

1. (a) $\dfrac{1}{3^{n-1}}$ **(b)** $\dfrac{(-1)^{n-1}}{3^{n-1}}$ **(c)** $\dfrac{2n-1}{2n}$ **(d)** $\dfrac{n^2}{\pi^{1/(n+1)}}$

3. (a) $2, 0, 2, 0$ **(b)** $1, -1, 1, -1$ **(c)** $2(1 + (-1)^n)$; $2 + 2\cos n\pi$

5. $\frac{1}{3}, \frac{2}{4}, \frac{3}{5}, \frac{4}{6}, \frac{5}{7}$; converges, $\displaystyle\lim_{n\to+\infty} \dfrac{n}{n+2} = 1$

7. $2, 2, 2, 2, 2$; converges, $\displaystyle\lim_{n\to+\infty} 2 = 2$

9. $\dfrac{\ln 1}{1}, \dfrac{\ln 2}{2}, \dfrac{\ln 3}{3}, \dfrac{\ln 4}{4}, \dfrac{\ln 5}{5}$; converges, $\displaystyle\lim_{n\to+\infty} \dfrac{\ln n}{n} = 0$

11. $0, 2, 0, 2, 0$; diverges   **13.** $-1, \frac{16}{9}, -\frac{54}{28}, \frac{128}{65}, -\frac{250}{126}$; diverges

15. $\frac{6}{2}, \frac{12}{8}, \frac{20}{18}, \frac{30}{32}, \frac{42}{50}$; converges, $\lim\limits_{n \to +\infty} \frac{1}{2}\left(1 + \frac{1}{n}\right)\left(1 + \frac{2}{n}\right) = \frac{1}{2}$

17. $\cos 3, \cos \frac{3}{2}, \cos 1, \cos \frac{3}{4}, \cos \frac{3}{5}$; converges, $\lim\limits_{n \to +\infty} \cos(3/n) = 1$

19. $e^{-1}, 4e^{-2}, 9e^{-3}, 16e^{-4}, 25e^{-5}$; converges, $\lim\limits_{n \to +\infty} n^2 e^{-n} = 0$

21. $2, \left(\frac{5}{3}\right)^2, \left(\frac{6}{4}\right)^3, \left(\frac{7}{5}\right)^4, \left(\frac{8}{6}\right)^5$; converges, $\lim\limits_{n \to +\infty} \left[\frac{n+3}{n+1}\right]^n = e^2$

23. $\left\{\frac{2n-1}{2n}\right\}_{n=1}^{+\infty}$; converges, $\lim\limits_{n \to +\infty} \frac{2n-1}{2n} = 1$

25. $\left\{\frac{1}{3^n}\right\}_{n=1}^{+\infty}$; converges, $\lim\limits_{n \to +\infty} \frac{1}{3^n} = 0$

27. $\left\{\frac{1}{n} - \frac{1}{n+1}\right\}_{n=1}^{+\infty}$; converges, $\lim\limits_{n \to +\infty} \left(\frac{1}{n} - \frac{1}{n+1}\right) = 0$

29. $\{\sqrt{n+1} - \sqrt{n+2}\}_{n=1}^{+\infty}$; converges, $\lim\limits_{n \to +\infty} (\sqrt{n+1} - \sqrt{n+2}) = 0$

31. **(a)** $1, 2, 1, 4, 1, 6$ **(b)** $a_n = \begin{cases} n, & n \text{ odd} \\ 1/2^n, & n \text{ even} \end{cases}$

(c) $a_n = \begin{cases} 1/n, & n \text{ odd} \\ 1/(n+1), & n \text{ even} \end{cases}$ **(d)** (a) diverges; (b) diverges;

(c) $\lim\limits_{n \to +\infty} a_n = 0$ **33.** $\lim\limits_{n \to +a} \sqrt[n]{n^3} = 1$

37. **(a)** $1, \frac{3}{4}, \frac{2}{3}, \frac{5}{8}$ **(c)** $\lim\limits_{n \to +\infty} a_n = \frac{1}{2}$

41. **(a)** $(0.5)^{2^n}$ **(c)** $\lim\limits_{n \to +\infty} a_n = 0$ **(d)** $-1 \le a_0 \le 1$

43. **(a)** 30

(b) $\lim\limits_{n \to +\infty} (2^n + 3^n)^{1/n} = 3$

45. converges to 0

47. **(a)** $N = 3$
(b) $N = 11$
(c) $N = 1001$

▶ **Exercise Set 10.3 (Page 663)**

1. strictly decreasing   **3.** strictly increasing   **5.** strictly decreasing
7. strictly increasing   **9.** strictly decreasing   **11.** strictly increasing
13. strictly increasing   **15.** strictly decreasing   **17.** strictly decreasing
19. eventually strictly increasing   **21.** eventually strictly decreasing
23. eventually strictly increasing
25. **(a)** Yes; the limit lies in the interval $[1, 2]$.
    (b) No, but if so, then the limit is ≤ 2.
27. $\sqrt{2}, \sqrt{2 + \sqrt{2}}, \sqrt{2 + \sqrt{2 + \sqrt{2}}}$ **(e)** $L = 2$

▶ **Exercise Set 10.4 (Page 670)**

1. **(a)** $2, \frac{12}{5}, \frac{62}{25}, \frac{312}{125}, \frac{5}{2}\left(1 - \left(\frac{1}{5}\right)^n\right)$, $\lim\limits_{n \to +\infty} s_n = \frac{5}{2}$, converges

(b) $\frac{1}{4}, \frac{3}{4}, \frac{7}{4}, \frac{15}{4}, -\frac{1}{4}(1 - 2^n)$, $\lim\limits_{n \to +\infty} s_n = +\infty$, diverges

(c) $\frac{1}{6}, \frac{1}{4}, \frac{3}{10}, \frac{1}{3}, \frac{1}{2} - \frac{1}{n+2}$, $\lim\limits_{n \to +\infty} s_n = \frac{1}{2}$, converges

3. $\frac{4}{7}$ **5.** 6   **7.** $\frac{1}{3}$ **9.** $\frac{1}{6}$ **11.** diverges   **13.** $\frac{448}{3}$

15. $\frac{4}{9}$ **17.** $\frac{532}{99}$ **19.** $\frac{79}{101}$ **21.** 70 m

23. **(a)** $s_n = -\ln(n+1)$, $\lim\limits_{n \to +\infty} s_n = -\infty$, diverges

(b) $s_n = \sum\limits_{k=2}^{n+1} \left[\ln \frac{k-1}{k} - \ln \frac{k}{k+1}\right]$, $\lim\limits_{n \to +\infty} s_n = -\ln 2$

25. **(a)** converges for $|x| < 1$; $S = \dfrac{x}{1 + x^2}$

(b) converges for $|x| > 2$; $S = \dfrac{1}{x^2 - 2x}$

(c) converges for $x > 0$; $S = \dfrac{1}{e^x - 1}$

31. $a_n = \dfrac{1}{2^{n-1}} a_1 + \dfrac{1}{2^{n-1}} + \dfrac{1}{2^{n-2}} + \cdots + \dfrac{1}{2}$, $\lim\limits_{n \to +\infty} a_n = 1$

33. The series converges only if $-1 < x < 1$.

37. **(b)** $A = 1, B = -2$ **(c)** $s_n = 2 - \dfrac{2^{n+1}}{3^{n+1} - 2^{n+1}}$,

$\lim\limits_{n \to +\infty} s_n = \lim\limits_{n \to +\infty} \left[2 - \dfrac{(2/3)^{n+1}}{1 - (2/3)^{n+1}}\right] = 2$

▶ **Exercise Set 10.5 (Page 677)**

1. **(a)** $\frac{4}{3}$ **(b)** $-\frac{3}{4}$ **3.** **(a)** $p = 3$, converges   **(b)** $p = \frac{1}{2}$, diverges
(c) $p = 1$, diverges   **(d)** $p = \frac{2}{3}$, diverges
5. **(a)** diverges   **(b)** diverges   **(c)** diverges   **(d)** no information
7. **(a)** diverges   **(b)** converges   **9.** diverges   **11.** diverges
13. diverges   **15.** diverges   **17.** diverges   **19.** converges
21. diverges   **23.** converges   **25.** converges for $p > 1$
27. **(a)** $(\pi^2/2) - (\pi^4/90)$ **(b)** $(\pi^2/6) - (5/4)$ **(c)** $\pi^4/90$
29. **(a)** diverges   **(b)** diverges   **(c)** converges
31. **(d)** $\frac{1}{11} < \frac{1}{6}\pi^2 - s_{10} < \frac{1}{10}$

33. **(a)** $1/46, 1/42$ **(b)** $\frac{\pi}{2} - \tan^{-1}(11), \frac{\pi}{2} - \tan^{-1}(10)$ **(c)** $\dfrac{12}{e^{11}}, \dfrac{11}{e^{10}}$
35. $S \approx 1.08$ **37.** converges for $a > e$
39. **(c)**   **(d)** $S \approx 0.6865$

n	10	20	30	40	50
s_n	0.681980	0.685314	685966	0.686199	0.686307
n	60	70	80	90	100
s_n	0.686367	0.686403	686426	0.686442	0.686454

▶ **Exercise Set 10.6 (Page 684)**

1. **(a)** converges   **(b)** diverges   **5.** converges   **7.** converges
9. diverges   **11.** converges   **13.** inconclusive   **15.** diverges
17. diverges   **19.** converges   **21.** converges   **23.** converges
25. converges   **27.** converges   **29.** diverges   **31.** converges
33. diverges   **35.** converges   **37.** converges   **39.** diverges
41. converges   **43.** converges
45. $u_k = \dfrac{k!}{1 \cdot 3 \cdot 5 \cdots (2k-1)}$, $\rho = \lim\limits_{k \to +\infty} \dfrac{k+1}{2k+1} = \dfrac{1}{2}$; converges
47. converges   **49.** diverges   **51.** **(a)** converges   **(b)** diverges

▶ **Exercise Set 10.7 (Page 692)**

3. diverges   **5.** converges   **7.** converges absolutely   **9.** diverges
11. converges absolutely   **13.** conditionally convergent   **15.** divergent
17. conditionally convergent   **19.** conditionally convergent
21. divergent   **23.** conditionally convergent
25. absolutely convergent   **27.** conditionally convergent
29. absolutely convergent   **31.** $|\text{error}| < 0.125$ **33.** $|\text{error}| < 0.1$
35. $n = 9999$ **37.** $n = 39,999$
39. $|\text{error}| < 0.00074$; $s_{10} \approx 0.4995$; $S = 0.5$ **41.** 0.84   **43.** 0.41
45. **(c)** $n = 50$ **51.** **(a)** $124.58 < d < 124.77$ **(b)** $1243 < s < 1424$

▶ **Exercise Set 10.8 (Page 700)**

1. $\sum\limits_{k=0}^{\infty} \dfrac{(-1)^k}{k!} x^k$ **3.** $\sum\limits_{k=0}^{\infty} \dfrac{(-1)^k \pi^{2k}}{(2k)!} x^{2k}$ **5.** $\sum\limits_{k=1}^{\infty} \dfrac{(-1)^{k+1}}{k} x^k$

7. $\sum\limits_{k=0}^{\infty} \dfrac{1}{(2k)!} x^{2k}$ **9.** $\sum\limits_{k=0}^{\infty} \dfrac{(-1)^k}{(2k+1)!} x^{2k+2}$ **11.** $\sum\limits_{k=0}^{\infty} \dfrac{e}{k!}(x-1)^k$

13. $\sum\limits_{k=0}^{\infty} (-1)(x+1)^k$ **15.** $\sum\limits_{k=0}^{\infty} \dfrac{(-1)^k \pi^{2k}}{(2k)!}\left(x - \dfrac{1}{2}\right)^{2k}$

17. $\sum\limits_{k=1}^{\infty} \dfrac{(-1)^{k-1}}{k}(x-1)^k$ **19.** $-1 < x < 1, \dfrac{1}{1+x}$

21. $1 < x < 3, \dfrac{1}{3-x}$ **23.** **(a)** $-2 < x < 2$ **(b)** $f(0) = 1$; $f(1) = \frac{2}{3}$

25. $R = 1, [-1, 1)$ **27.** $R = +\infty, (-\infty, +\infty)$ **29.** $R = \frac{1}{5}, \left[-\frac{1}{5}, \frac{1}{5}\right]$

31. $R = 1, [-1, 1]$　　**33.** $R = 1, (-1, 1]$　　**35.** $R = +\infty, (-\infty, +\infty)$
37. $R = +\infty, (-\infty, +\infty)$　　**39.** $R = 1, [-1, 1]$
41. $R = 1, (-2, 0]$　　**43.** $R = \frac{4}{3}, \left(-\frac{19}{3}, -\frac{11}{3}\right)$　　**45.** $R = 1, [-2, 0]$
47. $R = +\infty, (-\infty, +\infty)$　　**49.** $(-\infty, +\infty)$
51.

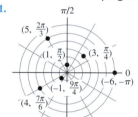

55. (a) radius $= R$
　　(b) radius $= R$
　　(c) radius $\geq \min(R_1, R_2)$

▶ **Exercise Set 10.9 (Page 709)**

1. 0.069756　**3.** 0.99500　**5.** 0.99619　**7.** 0.5208
11. (a) $\displaystyle\sum_{k=1}^{\infty} 2 \frac{(1/9)^{2k-1}}{2k-1}$　(b) 0.223
13. (a) 0.4635, 0.3218　(b) 3.1412　(c) no
17. (a) $\displaystyle\sum_{k=0}^{\infty}(-1)^k x^k$　(b) $1 + \frac{x}{3} + \displaystyle\sum_{k=2}^{\infty}(-1)^{k-1}\frac{2 \cdot 5 \cdots (3k-4)}{3^k k!}x^k$
　(c) $\displaystyle\sum_{k=0}^{\infty}(-1)^k\frac{(k+2)(k+1)}{2}x^k$
23. (a) 0.785398163397448309615660 8
　(b)

n	s_n
0	0.3183098 78 ...
1	0.3183098 861837906 067 ...
2	0.3183098 861837906 7153776 695 ...
3	0.3183098 861837906 7153776 752674502 34 ...
$1/\pi$	0.3183098 861837906 7153776 752674502 87 ...

▶ **Exercise Set 10.10 (Page 717)**

1. (a) $1 - x + x^2 - \cdots + (-1)^k x^k + \cdots; \ R = 1$
　(b) $1 + x^2 + x^4 + \cdots + x^{2k} + \cdots; \ R = 1$
　(c) $1 + 2x + 4x^2 + \cdots + 2^k x^k + \cdots; \ R = \frac{1}{2}$
　(d) $\frac{1}{2} + \frac{1}{2^2}x + \frac{1}{2^3}x^2 + \cdots + \frac{1}{2^{k+1}}x^k + \cdots; \ R = 2$
3. (a) $(2+x)^{-1/2} = \frac{1}{2^{1/2}} - \frac{1}{2^{5/2}}x + \frac{1 \cdot 3}{2^{9/2} \cdot 2!}x^2 - \frac{1 \cdot 3 \cdot 5}{2^{13/2} \cdot 3!}x^3 + \cdots$
　(b) $(1-x^2)^{-2} = 1 + 2x^2 + 3x^4 + 4x^6 + \cdots$
5. (a) $2x - \frac{2^3}{3!}x^3 + \frac{2^5}{5!}x^5 - \frac{2^7}{7!}x^7 + \cdots; \ R = +\infty$
　(b) $1 - 2x + 2x^2 - \frac{4}{3}x^3 + \cdots; \ R = +\infty$
　(c) $1 + x^2 + \frac{1}{2!}x^4 + \frac{1}{3!}x^6 + \cdots; \ R = +\infty$
　(d) $x^2 - \frac{\pi^2}{2}x^4 + \frac{\pi^4}{4!}x^6 - \frac{\pi^6}{6!}x^8 + \cdots; \ R = +\infty$
7. (a) $x^2 - 3x^3 + 9x^4 - 27x^5 + \cdots; \ R = \frac{1}{3}$
　(b) $2x^2 + \frac{2^3}{3!}x^4 + \frac{2^5}{5!}x^6 + \frac{2^7}{7!}x^8 + \cdots; \ R = +\infty$
　(c) $x - \frac{3}{2}x^3 + \frac{3}{8}x^5 + \frac{1}{16}x^7 + \cdots; \ R = 1$
9. (a) $x^2 - \frac{2^3}{4!}x^4 + \frac{2^5}{6!}x^6 - \frac{2^7}{8!}x^8 + \cdots$　(b) $12x^3 - 6x^6 + 4x^9 - 3x^{12} + \cdots$
11. (a) $1 - (x-1) + (x-1)^2 - \cdots + (-1)^k(x-1)^k + \cdots$　(b) $(0, 2)$
13. (a) $x + x^2 + \frac{x^3}{3} - \frac{x^5}{30} + \cdots$　(b) $x - \frac{x^3}{24} + \frac{x^4}{24} - \frac{71}{1920}x^5 + \cdots$
15. (a) $1 + \frac{1}{2}x^2 + \frac{5}{24}x^4 + \frac{61}{720}x^6 + \cdots$　(b) $x - x^2 + \frac{1}{3}x^3 - \frac{1}{30}x^5 + \cdots$
19. $2 - 4x + 2x^2 - 4x^3 + 2x^4 + \cdots$
25. (a) $\displaystyle\sum_{k=0}^{\infty} x^{2k+1}$　(b) $f^{(5)}(0) = 5!, \ f^{(6)}(0) = 0$
　(c) $f^{(n)}(0) = n!c_n = \begin{cases} n! & \text{if } n \text{ odd} \\ 0 & \text{if } n \text{ even} \end{cases}$　**27.** (a) 1　(b) $-\frac{1}{3}$

29. 0.3103　　**31.** 0.200　　**35.** (a) $x - \frac{1}{6}x^3 + \frac{3}{40}x^5 - \frac{5}{112}x^7 + \cdots$
　(b) $x + \displaystyle\sum_{k=1}^{\infty}(-1)^k\frac{1 \cdot 3 \cdot 5 \cdots (2k-1)}{2^k k!(2k+1)}x^{2k+1}$　(c) $R = 1$
37. (a) $y(t) = y_0 \displaystyle\sum_{k=0}^{\infty}\frac{(-1)^k(0.000121)^k t^k}{k!}$　(c) $0.9998790073 y_0$
39. (a) $T \approx 2.00709$　(b) $T \approx 2.008044621$　(c) 2.008045644
41. (a) $F = mg\left(1 - \frac{2h}{R} + \frac{3h^2}{R^2} - \frac{4h^3}{R^3} + \cdots\right)$　(d) about 0.27% less

▶ **Chapter 10 Supplementary Exercises (Page 720)**

9. (a) true　(b) sometimes false　(c) sometimes false　(d) true
　(e) sometimes false　(f) sometimes false　(g) false
　(h) sometimes false　(i) true　(j) true　(k) sometimes false
　(l) sometimes false
11. (a) converges　(b) converges　(c) diverges
13. (a) converges　(b) diverges　(c) converges　**15.** $\frac{1}{4 \cdot 5^{99}}$
17. (a) $p_0(x) = 1, \ p_1(x) = 1 - 7x, \ p_2(x) = 1 - 7x + 5x^2,$
　$p_3(x) = 1 - 7x + 5x^2 + 4x^3, \ p_4(x) = 1 - 7x + 5x^2 + 4x^3$
21. (a) converges　(b) diverges　**23.** (a) $u_{100} = \frac{1}{9900}$　(b) 0　(c) 2
25. (a) $e^2 - 1$　(b) 0　(c) $\cos e$　(d) $\frac{1}{3}$　**27.** 23.406%
31. (a) $x + \frac{1}{2}x^2 + \frac{3}{14}x^3 + \frac{3}{35}x^4; \ R = 3$
　(b) $-x^3 + \frac{2}{3}x^5 - \frac{2}{5}x^7 + \frac{8}{35}x^9; \ R = \sqrt{2}$

▶ **Exercise Set 11.1 (Page 734)**

1.

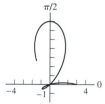

3. (a) $(3\sqrt{3}, 3)$
　(b) $(-7/2, 7\sqrt{3}/2)$
　(c) $(3\sqrt{3}, 3)$
　(d) $(0, 0)$
　(e) $(-7\sqrt{3}/2, 7/2)$
　(f) $(-5, 0)$

5. (a) both $(5, \pi)$　(b) $(4, 11\pi/6), (4, -\pi/6)$　(c) $(2, 3\pi/2), (2, -\pi/2)$
　(d) $(8\sqrt{2}, 5\pi/4), (8\sqrt{2}, -3\pi/4)$　(e) both $(6, 2\pi/3)$
　(f) both $(\sqrt{2}, \pi/4)$　**7.** (a) $(5, 0.6435)$　(b) $(\sqrt{29}, 5.0929)$
　(c) $(1.2716, 0.6658)$　**9.** (a) circle　(b) line　(c) circle　(d) line
11. (a) $r\cos\theta = 7$　(b) $r = 3$　(c) $r = 6\sin\theta$　(d) $r^2\sin 2\theta = 9/2$
13.

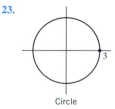

15.

17. (a) $r = 5$　(b) $r = 6\cos\theta$　(c) $r = 1 - \cos\theta$
19. (a) $r = 3\sin 2\theta$　(b) $r = 3 + 2\sin\theta$　(c) $r^2 = 9\cos 2\theta$
21.

Line

23.

Circle

25.

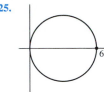

Circle

27.

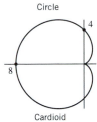

Circle

29.

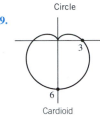

Cardioid

31.

Cardioid

33.

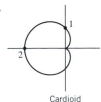

Cardioid

35.

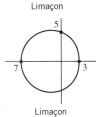

Limaçon

37.

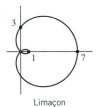

Limaçon

39.

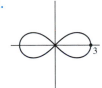

Limaçon

41.

Lemniscate

43.

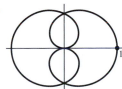

Lemniscate

45.

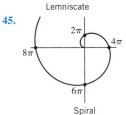

Spiral

47.

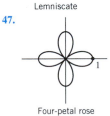

Four-petal rose

49.

Eight-petal rose

51.

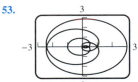

53.

55.

57. $-4\pi < \theta < 4\pi$

61. (a) $r = 1 + \dfrac{\sqrt{2}}{2}(\cos\theta + \sin\theta)$
  (b) $r = 1 + \sin\theta$
  (c) $r = 1 - \cos\theta$
  (d) $r = 1 - \dfrac{\sqrt{2}}{2}(\cos\theta + \sin\theta)$

63.

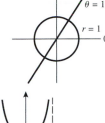

65. $(3/2, \pi/3)$

67. (c) $\sqrt{13 - 6\sqrt{3}} \approx 1.615$
  (d) $A = 1$

71.

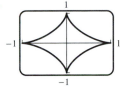

▶ **Exercise Set 11.2** (Page 741)

1. (a) $-1/4,\ 1/4$ **3.** positive when $t = -1$, negative when $t = 1$

5. $4, 4$ **7.** $2/\sqrt{3}, -1/(3\sqrt{3})$ **9.** $\dfrac{-1}{4 - \sqrt{3}},\ \dfrac{8(1 - \sqrt{3})}{(4 - \sqrt{3})^3}$

11. (a) $y = -e^{-2}x + 2e^{-1}$ **19.**

13. (a) $t = \pi/2 + n\pi$ for $n = 0, \pm 1, \dots$
  (b) $t = n\pi$ for $n = 0, \pm 1, \dots$

15. $y = -2x,\ y = 2x$

17. $y = \frac{1}{2}x + 4,\ y = -\frac{1}{2}x + 4$

21. $1/\sqrt{3}$ **23.** $\dfrac{\tan 2 - 2}{2\tan 2 + 1}$ **25.** -2 **27.** $1, 0, -1$

29. horizontal: $(3a/2, \pi/3),\ (0, \pi),\ (3a/2, 5\pi/3)$;
  vertical: $(2a, 0),\ (a/2, 2\pi/3),\ (a/2, 4\pi/3)$

31. $(0, 0),\ (\sqrt{2}/4, \pi/4),\ (\sqrt{2}/4, 3\pi/4)$

33.   **35.**   **37.**

$\theta_0 = \pi/6, \pi/2, 5\pi/6$ $\theta_0 = \pm\pi/4$ $\theta_0 = 2\pi/3, 4\pi/3$

39. $L = 2\pi a$ **41.** $L = 8a$ **43.** $L = \sqrt{10}(e^6 - 1)/3$

45. (a) $\dfrac{dy}{dx} = \dfrac{3\sin t}{1 - 3\cos t}$ (b) $\theta = -0.4345$

47. (b) ≈ 2.42

(c)

n	2	3	4	5	6	7
L	2.42211	2.22748	2.14461	2.10100	2.07501	2.05816

n	8	9	10	11	12	13	14
L	2.04656	2.03821	2.03199	2.02721	2.02346	2.02046	2.01802

n	15	16	17	18	19	20
L	2.01600	2.01431	2.01288	2.01167	2.01062	2.00971

49. $S = \dfrac{8\pi}{3}(17\sqrt{17} - 1)$ **51.** $S = \sqrt{2}\pi$

55. (a) $r = 2\theta + 10$ (b) 75.7 mm

▶ **Exercise Set 11.3** (Page 747)

1. (a) $\displaystyle\int_{\pi/2}^{\pi} \frac{1}{2}(1 - \cos\theta)^2\, d\theta$ (b) $\displaystyle\int_{0}^{\pi/2} \frac{1}{2}4\cos^2\theta\, d\theta$

(c) $\displaystyle\int_{0}^{\pi/2} \frac{1}{2}\sin^2 2\theta\, d\theta$ (d) $\displaystyle\int_{0}^{2\pi} \frac{1}{2}\theta^2\, d\theta$

(e) $\displaystyle\int_{-\pi/2}^{\pi/2} \frac{1}{2}(1 - \sin\theta)^2\, d\theta$ (f) $2\displaystyle\int_{0}^{\pi/4} \frac{1}{2}\cos^2 2\theta\, d\theta$

3. (a) πa^2 (b) πa^2 (c) πa^2 **5.** 6π **7.** 4π **9.** $\pi - 3\sqrt{3}/2$

11. $\pi/2 - \frac{1}{4}$ **13.** $10\pi/3 - 4\sqrt{3}$ **15.** $8\pi/3 + \sqrt{3}$

17. $9\sqrt{3}/2 - \pi$ **27.** π^2

19. $(\pi + 3\sqrt{3})/4$

21. $100\cos^{-1}(3/5) - 48$

23. (b) a^2 (c) $2\sqrt{3} - \dfrac{2\pi}{3}$

25. $8\pi^3 a^2$

29. $32\pi/5$ **35.** $\pi/16$

▶ **Exercise Set 11.4 (Page 762)**

1. (a) $x = y^2$ (b) $-3y = x^2$ (c) $\dfrac{x^2}{9} + \dfrac{y^2}{4} = 1$

(d) $\dfrac{x^2}{4} + \dfrac{y^2}{9} = 1$ (e) $y^2 - x^2 = 1$ (f) $\dfrac{x^2}{4} - \dfrac{y^2}{4} = 1$

3. (a)   (b)

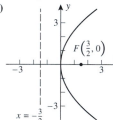

5. (a)   (b)

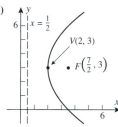

7. (a)   (b)

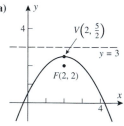

9. (a)   (b)

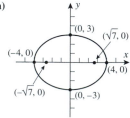

11. (a)   (b)

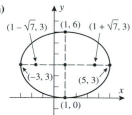
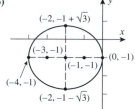

13. (a)   (b)

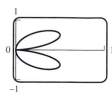
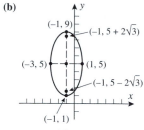

15. (a)   (b)

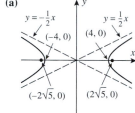
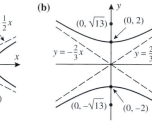

17. (a)   (b)

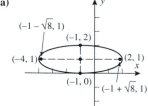

19. (a)

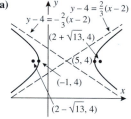
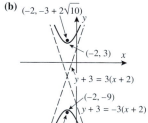

(b)

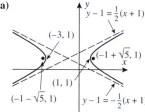
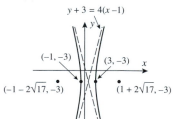

21. (a) $y^2 = 12x$
(b) $y^2 = -28x$

23. (a) $x^2 = -12y$
(b) $(x - 1)^2 = 12(y - 1)$

25. $y^2 = -5\left(x - \frac{19}{5}\right)$

27. (a) $\frac{1}{9}x^2 + \frac{1}{4}y^2 = 1$
(b) $\frac{1}{169}x^2 + \frac{1}{144}y^2 = 1$

29. (a) $\frac{1}{3}x^2 + \frac{1}{2}y^2 = 1$
(b) $\frac{1}{4}x^2 + \frac{1}{16}y^2 = 1$

31. (a) $\frac{1}{36}x^2 + \frac{8}{81}y^2 = 1$ (b) $(x - 1)^2 + \frac{1}{2}(y - 3)^2 = 1$

33. (a) $\frac{1}{4}x^2 - \frac{1}{5}y^2 = 1$ (b) $x^2 - \frac{1}{4}y^2 = 1$

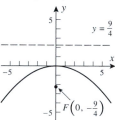
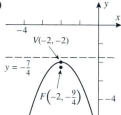
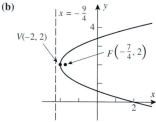
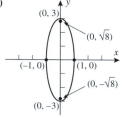

35. (a) $\frac{9}{64}x^2 - \frac{1}{16}y^2 = 1$, $\frac{1}{36}y^2 - \frac{1}{16}x^2 = 1$ **(b)** $\frac{1}{20}y^2 - \frac{1}{5}x^2 = 1$

37. (a) $\frac{1}{16}(x-6)^2 - \frac{1}{9}(y-4)^2 = 1$ **(b)** $\frac{1}{3}(y-2)^2 - \frac{4}{3}\left(x - \frac{1}{2}\right)^2 = 1$

39. (a) 16 ft **(b)** $8\sqrt{3}$ ft

43. $\frac{1}{16}$ ft

45. (a) $P : (b\cos t, b\sin t)$;
  $Q : (a\cos t, a\sin t)$;
  $R : (a\cos t, b\sin t)$

49. $\frac{1}{32}(x-4)^2 + \frac{1}{36}(y-3)^2 = 1$

51. 96

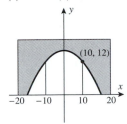

61. $L = D\sqrt{1+p^2}$, $T = \frac{1}{2}pD$

67. $\left(\pm\frac{3}{\sqrt{5}}, \frac{4}{\sqrt{5}}\right), \left(\pm\frac{3}{\sqrt{5}}, -\frac{4}{\sqrt{5}}\right)$

71. (b) 14.30465, 24, 33.69535 in

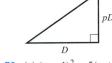

73. (a) $(x-1)^2 - 5(y+1)^2 = 5$, hyperbola
  (b) $x^2 - 3(y+1)^2 = 0$, $x = \pm\sqrt{3}(y+1)$, two lines
  (c) $4(x+2)^2 + 8(y+1)^2 = 4$, ellipse
  (d) $3(x+2)^2 + (y+1)^2 = 0$, the point $(-2, -1)$ (degenerate case)
  (e) $(x+4)^2 + 2y = 2$, parabola
  (f) $5(x+4)^2 + 2y = -14$, parabola

▶ **Exercise Set 11.5 (Page 771)**
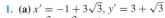

1. (a) $x' = -1 + 3\sqrt{3}$, $y' = 3 + \sqrt{3}$ **3.** $y'^2 + y'^2 = 18$, hyperbola
  (b) $3x'^2 - y'^2 = 12$
  (c)
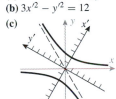

5. $\frac{1}{3}x'^2 - \frac{1}{2}y'^2 = 1$, hyperbola **7.** $y' = x'^2$, parabola

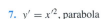

9. $y'^2 = 4(x' - 1)$, parabola **11.** $\frac{1}{4}(x' + 1)^2 + y'^2 = 1$, ellipse

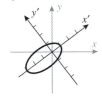

15. $x^2 + xy + y^2 = 3$ **25.** parabola **27.** ellipse

31. (a) hyperbola **(b)** $y = -\frac{9}{2}x - \frac{1}{2} + \frac{1}{2}\sqrt{73x^2 + 42x + 17}$
  (c)
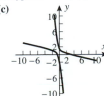

▶ **Exercise Set 11.6 (Page 779)**

1. (a) $e = 1$, $d = \frac{3}{2}$ **(b)** $e = \frac{1}{2}$, $d = 3$
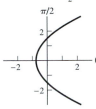

(c) $e = \frac{3}{2}$, $d = \frac{4}{3}$ **(d)** $e = 1$, $d = \frac{5}{3}$
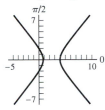

3. (a) parabola, opens up **(b)** ellipse, directrix above the pole
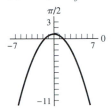

(c) hyperbola, **(d)** ellipse,
directrix below the pole directrix to the right of the pole
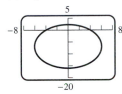

5. (a) $r = \dfrac{2}{3 + 2\cos\theta}$ **(b)** $r = \dfrac{1}{1 - \cos\theta}$ **(c)** $r = \dfrac{3}{2 + 3\sin\theta}$

7. (a) $r = \dfrac{24}{5 - 5\cos\theta}$ **(b)** $r = \dfrac{2}{1 - \sin\theta}$ **(c)** $r = \dfrac{21}{2 + 5\sin\theta}$

9. (a) $d = 6$, $\frac{1}{12}x^2 + \frac{1}{16}(y+2)^2 = 1$ **(b)** $d = 1$, $\frac{9}{4}\left(x - \frac{1}{3}\right)^2 + 3y^2 = 1$

11. (a) $d = \frac{2}{3}$, $-2x^2 + 16\left(y - \frac{3}{4}\right)^2 = 1$
  (b) $d = \frac{10}{9}$, $\frac{9}{16}(x+2)^2 - \frac{9}{20}y^2 = 1$

13. (a) $r = \dfrac{12}{2 + \cos\theta}$ **(b)** $r = \dfrac{64}{25 - 15\sin\theta}$
  (c) $r = \dfrac{16}{5 - 3\cos\theta}$ **(d)** $r = \dfrac{120}{5 + \sin\theta}$

17. (a) $T \approx 248$ yr **(d)**
  (b) $r_0 \approx 4{,}449{,}675{,}000$ km
  $r_1 \approx 7{,}400{,}325{,}000$ km
  (c) $r \approx \dfrac{37.05}{1 + 0.249\cos\theta}$ AU
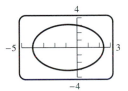

19. (a) $a \approx 178.26$ AU **(d)**
  (b) $r_0 \approx 0.8735$ AU,
  $r_1 \approx 355.64$ AU
  (c) $r \approx \dfrac{1.74}{1 + 0.9951\cos\theta}$ AU
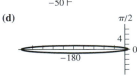

21. 563 km, 4286 km

▶ **Chapter 11 Supplementary Exercises (Page 781)**

3. (a) circle **(b)** rose **(c)** line **(d)** limaçon **(e)** limaçon
(f) none **(g)** none **(h)** spiral

5. (a) $\pi/2$ **(b)**

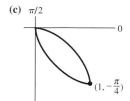

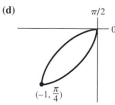

$(1, \frac{3\pi}{4})$

(c) $\pi/2$ **(d)**

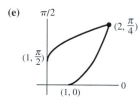

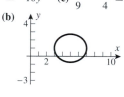

$(1, -\frac{\pi}{4})$ $(-1, \frac{\pi}{4})$

(e) $\pi/2$

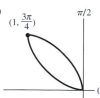

$(2, \frac{\pi}{4})$
$(1, \frac{\pi}{2})$
$(1, 0)$

7. (a) parabola
(b) hyperbola
(c) line
(d) circle

9. (a) $\frac{5}{49}x^2 + \frac{1}{9}y^2 = 1$ **(b)** $x^2 = -16y$ **(c)** $\frac{x^2}{9} - \frac{y^2}{4} = 1$

11. (a) **(b)** **(c)** **(c)**

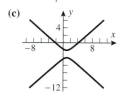

13. (a) $y = \dfrac{470}{2100^2}x^2$ **(b)** $L \approx 4336.3$ ft **15.** $A = \dfrac{5\pi}{12} - \dfrac{\sqrt{3}}{2}$

17. (a) $L = \displaystyle\int_{\pi/4}^{\pi/2} \frac{1}{\theta^2}\sqrt{1 + \theta^2}\, d\theta \approx 0.9457$
(b) The arc length is infinite.

19. (a) $V = \dfrac{\pi b^2}{3a^2}(b^2 - 2a^2)\sqrt{a^2 + b^2} + \dfrac{2}{3}ab^2\pi$ **(b)** $V = \dfrac{2b^4}{3a}\pi$

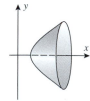

23. (a) $-1/4, 2$ **(b)** $-3\sqrt{3}/4, 3\sqrt{3}/4$
27. (b) $x = \cos t + \cos 2t$, $y = \sin t + \sin 2t$ **(c)** yes
31. (a) $1/60$ **(b)** 91,450,000 mi **(c)** 584,295,652.5 mi
33. (a) 119.3 ft/s **(b)** 114.1 ft **(c)** 447.4 ft

▶ **Exercise Set A (Page A8)**

1. (a) rational **(b)** integer, rational **(c)** integer, rational **(d)** rational
(e) integer, rational **(f)** irrational **(g)** rational
(h) integer, rational **3. (a)** $\frac{41}{333}$ **(b)** $\frac{115}{9}$ **(c)** $\frac{20943}{550}$ **(d)** $\frac{537}{1250}$
5. (a) $\frac{256}{81}$ **(b)** worse **7.**

Line	2	3	4	5	6	7
Blocks	3, 4	1, 2	3, 4	2, 4, 5	1, 2	3, 4

9. (a), (d), (f) **11. (a)** all values **(b)** none **13. (a)** yes **(b)** no
15. (a) $\{x : x$ is a positive odd integer$\}$ **(b)** $\{x : x$ is an even integer$\}$
(c) $\{x : x$ is irrational$\}$ **(d)** $\{x : x$ is an integer and $7 \le x \le 10\}$
17. (a) false **(b)** true **(c)** true **(d)** false **(e)** true **(f)** true **(g)** true
19. (a) **(b)**
4 -3
(c) **(d)**
-1 7 -3 3
(e) **(f)**
-3 3 -3 3

21. (a) $[-2, 2]$ **(b)** $(-\infty, -2) \cup (2, +\infty)$
23. $\left(-\infty, \frac{10}{3}\right)$ **25.** $\left(-\infty, -\frac{11}{2}\right]$
$\frac{10}{3}$ $-\frac{11}{2}$

27. $\left(-\frac{3}{2}, \frac{1}{2}\right]$ **29.** $(-\infty, 3) \cup (4, +\infty)$
$-\frac{3}{2}$ $\frac{1}{2}$ 3 4

31. $\left(-\frac{3}{2}, 2\right)$ **33.** $(-\infty, -2] \cup (2, +\infty)$
$-\frac{3}{2}$ 2 -2 2

35. $(-\infty, -3) \cup (3, +\infty)$ **37.** $(-\infty, -2) \cup (4, +\infty)$
-3 3 -2 4

39. $[4, 5]$ **41.** $(-8, 0) \cup (4, +\infty)$
4 5 -8 0 4

43. $(2, +\infty)$ **45.** $(-\infty, -3) \cup [2, +\infty)$
2

47. $77 \le F \le 104$ **55.** $\left(-\infty, -\frac{1}{2}\right)$

▶ **Exercise Set B (Page A15)**

1. (a) 7 **(b)** $\sqrt{2}$ **(c)** k^2 **(d)** k^2 **3.** $x \le 3$ **5.** all real x
7. $x \ge 0$ or $x = -\frac{2}{3}$ **9.** $x \ge -5$ **13. (a)** 2 **(b)** 1 **(c)** 14
(d) $3 + \sqrt{2}$ **(e)** 7 **(f)** 5 **15. (a)** -9 **(b)** 7 **(c)** 12
17. $-\frac{5}{6}, \frac{3}{2}$ **19.** $\frac{1}{2}, \frac{5}{2}$ **21.** $-\frac{11}{10}, \frac{11}{8}$ **23.** $1, \frac{17}{5}$
25. $(-9, -3)$ **27.** $\left[-\frac{3}{2}, \frac{9}{2}\right]$ **29.** $(-\infty, -3) \cup (-1, +\infty)$
31. $\left(-\infty, \frac{1}{2}\right] \cup \left[\frac{9}{2}, +\infty\right)$ **33.** $\left(-\infty, \frac{1}{2}\right) \cup \left(\frac{3}{2}, +\infty\right)$
35. $\left[\frac{1}{8}, \frac{1}{2}\right) \cup \left(\frac{1}{2}, \frac{7}{8}\right]$ **37.** $x \in (-\infty, 2] \cup [3, +\infty)$ **39.** $-3, 9$

▶ **Exercise Set C (Page A25)**

1. $(-4, 7)$ **3. (a)**

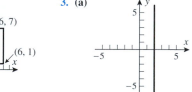

$(-4, 7)$ $(6, 7)$
$(-4, 1)$ $(6, 1)$

(b) **(c)**

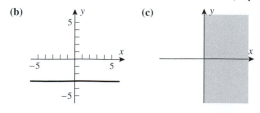

(d)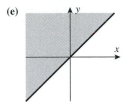

(e)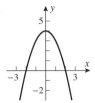

(f)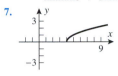

5.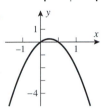

7.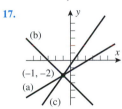

9.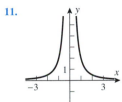

11.

13. (a) $\frac{1}{2}$

(b) -1

(c) 0

(d) not defined

15. (a) yes

(b) no

17.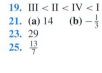

19. $\text{III} < \text{II} < \text{IV} < \text{I}$

21. (a) 14 **(b)** $-\frac{1}{3}$

23. 29

25. $\frac{13}{7}$

29. (a)   **(b)**

(c) **(d)**

31. (a) 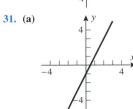 **(b)**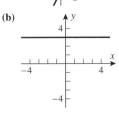

(c)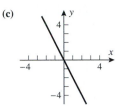

33.

	(a)	(b)	(c)	(d)	(e)
Slope	3	$-1/4$	$-3/5$	0	$-b/a$
y-intercept	2	3	$8/5$	1	b

35. (a) $y = \frac{3}{2}x - 3$ **(b)** $y = -\frac{3}{4}x$ **37.** $y = -2x + 4$

39. $y = 4x + 7$ **41.** $y = -\frac{1}{5}x + 6$ **43.** $y = 11x - 18$

45. $y = \frac{1}{2}x + 2$ **47.** $y = 1$ **49. (a)** parallel

(b) perpendicular **(c)** parallel **(d)** perpendicular **(e)** neither

51. (a) $-\frac{3}{2}$ **53.** the union of the graphs of $x - y = 0$ and $x + y = 0$

(b) $\frac{4}{5}$

(c) $\frac{5}{2}$

(d) $-\frac{15}{2}$

(e) -4

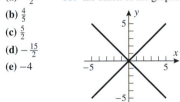

55. 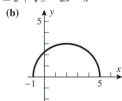 **59.** $\frac{49}{6}$ **61. (a)** yes

(b) yes

(c) no

(d) yes

(e) yes

(f) yes

(g) no

▶ **Exercise Set D (Page A34)**

1. in the proof of Theorem D.1 **3. (a)** 10 **(b)** (4, 5)

5. (a) $\sqrt{29}$ **(b)** $\left(-\frac{9}{2}, -5\right)$ **11.** 0 **13.** $y = -3x + 4$

15. $\left(-\frac{29}{8}, -\frac{23}{4}\right)$ **17.** 3 **21.** 4 **23. (a)** (0, 0); 5 **(b)** (1, 4); 4

(c) $(-1, -3)$; $\sqrt{5}$ **(d)** (0, −2); 1 **25.** $(x - 3)^2 + (y + 2)^2 = 16$

27. $(x + 4)^2 + (y - 8)^2 = 64$ **29.** $(x + 3)^2 + (y + 4)^2 = 25$

31. $(x - 1)^2 + (y - 1)^2 = 2$ **33.** circle; center (1, 2), radius 4

35. circle; center $(-1, 1)$, radius $\sqrt{2}$ **37.** the point $(-1, -1)$

39. circle; center (0, 0), radius $\frac{1}{3}$ **41.** no graph

43. circle; center $\left(-\frac{5}{4}, -\frac{1}{2}\right)$, radius $\frac{3}{2}$

45. (a) $y = -\sqrt{16 - x^2}$ **(b)** $y = 2 + \sqrt{3 - 2x - x^2}$

47. (a) **(b)**

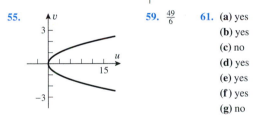

49. $y = -\frac{3}{4}x + \frac{25}{4}$ **51. (a)** inside **(b)** largest $3\sqrt{5}$, smallest $\sqrt{5}$

53. $(1/3, \pm\sqrt{8}/3)$ **55. (a)** equation: $2x^2 + 2y^2 - 12x + 8y + 1 = 0$

(b) center $(3, -2)$, radius $5/\sqrt{2}$

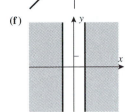

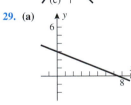

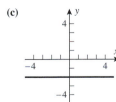

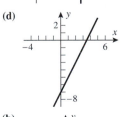

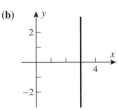

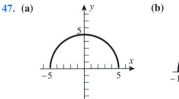

57.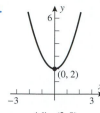

59.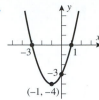

61.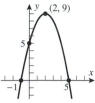

63.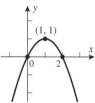

65.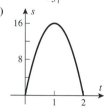

67.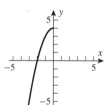

69.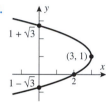

71. (a) $x = \sqrt{3-y}$
(b) $x = 1 - \sqrt{y+1}$

73. (a)  (b)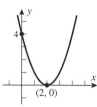

75. (a)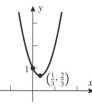

77. (a) $y = 150 - \frac{3}{2}x$
(b) $A = 150x - \frac{3}{2}x^2$
(c) 3750 ft^2

79. (a) $(-5 - \sqrt{33})/4 < x < (-5 + \sqrt{33})/4$ (b) $-\infty < x < +\infty$
81. (a) 30 ft   (b) 2.6 s   (c) 2.1 s

▶ **Exercise Set E (Page A47)**
1. (a) $\frac{5}{12}\pi$ (b) $\frac{13}{6}\pi$ (c) $\frac{1}{9}\pi$ (d) $\frac{23}{30}\pi$
3. (a) $12°$ (b) $(270/\pi)°$ (c) $288°$ (d) $540°$
5.

	$\sin\theta$	$\cos\theta$	$\tan\theta$	$\csc\theta$	$\sec\theta$	$\cot\theta$
(a)	$\sqrt{21}/5$	$2/5$	$\sqrt{21}/2$	$5/\sqrt{21}$	$5/2$	$2/\sqrt{21}$
(b)	$3/4$	$\sqrt{7}/4$	$3/\sqrt{7}$	$4/3$	$4/\sqrt{7}$	$\sqrt{7}/3$
(c)	$3/\sqrt{10}$	$1/\sqrt{10}$	3	$\sqrt{10}/3$	$\sqrt{10}$	$1/3$

7. $\sin\theta = 3/\sqrt{10}, \cos\theta = 1/\sqrt{10}$

9. $\tan\theta = \sqrt{21}/2, \csc\theta = 5/\sqrt{21}$ **11.** 1.8
13.

	θ	$\sin\theta$	$\cos\theta$	$\tan\theta$	$\csc\theta$	$\sec\theta$	$\cot\theta$
(a)	$225°$	$-1/\sqrt{2}$	$-1/\sqrt{2}$	1	$-\sqrt{2}$	$-\sqrt{2}$	1
(b)	$-210°$	$1/2$	$-\sqrt{3}/2$	$-1/\sqrt{3}$	2	$-2/\sqrt{3}$	$-\sqrt{3}$
(c)	$5\pi/3$	$-\sqrt{3}/2$	$1/2$	$-\sqrt{3}$	$-2/\sqrt{3}$	2	$-1/\sqrt{3}$
(d)	$-3\pi/2$	1	0	$-$	1	$-$	0

15.

	$\sin\theta$	$\cos\theta$	$\tan\theta$	$\csc\theta$	$\sec\theta$	$\cot\theta$
(a)	$4/5$	$3/5$	$4/3$	$5/4$	$5/3$	$3/4$
(b)	$-4/5$	$3/5$	$-4/3$	$-5/4$	$5/3$	$-3/4$
(c)	$1/2$	$-\sqrt{3}/2$	$-1/\sqrt{3}$	2	$-2/\sqrt{3}$	$-\sqrt{3}$
(d)	$-1/2$	$\sqrt{3}/2$	$-1/\sqrt{3}$	-2	$2/\sqrt{3}$	$-\sqrt{3}$
(e)	$1/\sqrt{2}$	$1/\sqrt{2}$	1	$\sqrt{2}$	$\sqrt{2}$	1
(f)	$1/\sqrt{2}$	$-1/\sqrt{2}$	-1	$\sqrt{2}$	$-\sqrt{2}$	-1

17. (a) 1.2679   (b) 3.5753
19.

	$\sin\theta$	$\cos\theta$	$\tan\theta$	$\csc\theta$	$\sec\theta$	$\cot\theta$
(a)	$a/3$	$\sqrt{9-a^2}/3$	$a/\sqrt{9-a^2}$	$3/a$	$3/\sqrt{9-a^2}$	$\sqrt{9-a^2}/a$
(b)	$a/\sqrt{a^2+25}$	$5/\sqrt{a^2+25}$	$a/5$	$\sqrt{a^2+25}/a$	$\sqrt{a^2+25}/5$	$5/a$
(c)	$\sqrt{a^2-1}/a$	$1/a$	$\sqrt{a^2-1}$	$a/\sqrt{a^2-1}$	a	$1/\sqrt{a^2-1}$

21. (a) $3\pi/4 \pm n\pi, n = 0, 1, 2, \ldots$
(b) $\pi/3 \pm 2n\pi$ and $5\pi/3 \pm 2n\pi, n = 0, 1, 2, \ldots$
23. (a) $\pi/6 \pm n\pi, n = 0, 1, 2, \ldots$
(b) $4\pi/3 \pm 2n\pi$ and $5\pi/3 \pm 2n\pi, n = 0, 1, 2, \ldots$
25. (a) $3\pi/4 \pm n\pi, n = 0, 1, 2, \ldots$ (b) $\pi/6 \pm n\pi, n = 0, 1, 2, \ldots$
27. (a) $\pi/3 \pm 2n\pi$ and $2\pi/3 \pm 2n\pi, n = 0, 1, 2, \ldots$
(b) $\pi/6 \pm 2n\pi$ and $11\pi/6 \pm 2n\pi, n = 0, 1, 2, \ldots$
29. $\sin\theta = 2/5, \cos\theta = -\sqrt{21}/5, \tan\theta = -2/\sqrt{21}, \csc\theta = 5/2,$
$\sec\theta = -5/\sqrt{21}, \cot\theta = -\sqrt{21}/2$
31. (a) $\theta = \pm n\pi, n = 0, 1, 2, \ldots$ (b) $\theta = \pi/2 \pm n\pi, n = 0, 1, 2, \ldots$
(c) $\theta = \pm n\pi, n = 0, 1, 2, \ldots$ (d) $\theta = \pm n\pi, n = 0, 1, 2, \ldots$
(e) $\theta = \pi/2 \pm n\pi, n = 0, 1, 2, \ldots$ (f) $\theta = \pm n\pi, n = 0, 1, 2, \ldots$
33. (a) $2\pi/3$ cm   (b) $10\pi/3$ cm   **35.** $\frac{2}{5}$
37. (a) $\dfrac{2\pi - \theta}{2\pi}R$ (b) $\dfrac{\sqrt{4\pi\theta - \theta^2}}{2\pi}R$ **39.** $\frac{21}{4}\sqrt{3}$ **41.** 9.2 ft
43. $h = d(\tan\beta - \tan\alpha)$ **45.** (a) $4\sqrt{5}/9$ (b) $-\frac{1}{9}$
47. $\sin 3\theta = 3\sin\theta\cos^2\theta - \sin^3\theta, \cos 3\theta = \cos^3\theta - 3\sin^2\theta\cos\theta$
61. (a) $\cos\theta$ (b) $-\sin\theta$ (c) $-\cos\theta$ (d) $\sin\theta$

▶ **Exercise Set F (Page A55)**
1. (a) $x^2 + 4x + 2, -11x + 6$ (b) $2x^2 + 4, 9$
(c) $x^3 - x^2 + 2x - 2, 2x + 1$
3. (a) $3x^2 + 6x + 8, 15$ (b) $x^3 - 5x^2 + 20x - 100, 504$
(c) $x^4 + x^3 + x^2 + x + 1, 0$
5.

x	0	1	-3	7
$p(x)$	-4	-3	101	5001

7. (a) $x^2 + 6x + 13, 20$
(b) $x^2 + 3x - 2, -4$

9. (a) $\pm 1, \pm 2, \pm 3, \pm 4, \pm 6, \pm 8, \pm 12, \pm 24$
(b) $\pm 1, \pm 2, \pm 5, \pm 10, \pm\frac{1}{3}, \pm\frac{2}{3}, \pm\frac{5}{3}, \pm\frac{10}{3}$
(c) $\pm 1, \pm 17$ **11.** $(x+1)(x-1)(x-2)$ **13.** $(x+3)^3(x+1)$
15. $(x+3)(x+2)(x+1)^2(x-3)$ **17.** -3 **19.** $-2, -\frac{2}{3}$
21. $-2, 2, 3$ **23.** 2, 5   **25.** 7 cm

PHOTO CREDITS

INDEX

DATE DUE

RATIONAL FUNCTIONS CONTAINING POWERS OF $a + bu$ IN THE DENOMINATOR

60. $\displaystyle\int \frac{u\,du}{a+bu} = \frac{1}{b^2}[bu - a\ln|a+bu|] + C$

61. $\displaystyle\int \frac{u^2\,du}{a+bu} = \frac{1}{b^3}\left[\frac{1}{2}(a+bu)^2 - 2a(a+bu) + a^2\ln|a+bu|\right] + C$

62. $\displaystyle\int \frac{u\,du}{(a+bu)^2} = \frac{1}{b^2}\left[\frac{a}{a+bu} + \ln|a+bu|\right] + C$

63. $\displaystyle\int \frac{u^2\,du}{(a+bu)^2} = \frac{1}{b^3}\left[bu - \frac{a^2}{a+bu} - 2a\ln|a+bu|\right] + C$

64. $\displaystyle\int \frac{u\,du}{(a+bu)^3} = \frac{1}{b^2}\left[\frac{a}{2(a+bu)^2} - \frac{1}{a+bu}\right] + C$

65. $\displaystyle\int \frac{du}{u(a+bu)} = \frac{1}{a}\ln\left|\frac{u}{a+bu}\right| + C$

66. $\displaystyle\int \frac{du}{u^2(a+bu)} = -\frac{1}{au} + \frac{b}{a^2}\ln\left|\frac{a+bu}{u}\right| + C$

67. $\displaystyle\int \frac{du}{u(a+bu)^2} = \frac{1}{a(a+bu)} + \frac{1}{a^2}\ln\left|\frac{u}{a+bu}\right| + C$

RATIONAL FUNCTIONS CONTAINING $a^2 \pm u^2$ IN THE DENOMINATOR ($a > 0$)

68. $\displaystyle\int \frac{du}{a^2+u^2} = \frac{1}{a}\tan^{-1}\frac{u}{a} + C$

69. $\displaystyle\int \frac{du}{a^2-u^2} = \frac{1}{2a}\ln\left|\frac{u+a}{u-a}\right| + C$

70. $\displaystyle\int \frac{du}{u^2-a^2} = \frac{1}{2a}\ln\left|\frac{u-a}{u+a}\right| + C$

71. $\displaystyle\int \frac{bu+c}{a^2+u^2}\,du = \frac{b}{2}\ln(a^2+u^2) + \frac{c}{a}\tan^{-1}\frac{u}{a} + C$

INTEGRALS OF $\sqrt{a^2+u^2}$, $\sqrt{a^2-u^2}$, $\sqrt{u^2-a^2}$ AND THEIR RECIPROCALS ($a > 0$)

72. $\displaystyle\int \sqrt{u^2+a^2}\,du = \frac{u}{2}\sqrt{u^2+a^2} + \frac{a^2}{2}\ln(u+\sqrt{u^2+a^2}) + C$

73. $\displaystyle\int \sqrt{u^2-a^2}\,du = \frac{u}{2}\sqrt{u^2-a^2} - \frac{a^2}{2}\ln|u+\sqrt{u^2-a^2}| + C$

74. $\displaystyle\int \sqrt{a^2-u^2}\,du = \frac{u}{2}\sqrt{a^2-u^2} + \frac{a^2}{2}\sin^{-1}\frac{u}{a} + C$

75. $\displaystyle\int \frac{du}{\sqrt{u^2+a^2}} = \ln(u+\sqrt{u^2+a^2}) + C$

76. $\displaystyle\int \frac{du}{\sqrt{u^2-a^2}} = \ln|u+\sqrt{u^2-a^2}| + C$

77. $\displaystyle\int \frac{du}{\sqrt{a^2-u^2}} = \sin^{-1}\frac{u}{a} + C$

POWERS OF u MULTIPLYING OR DIVIDING $\sqrt{a^2-u^2}$ OR ITS RECIPROCAL

78. $\displaystyle\int u^2\sqrt{a^2-u^2}\,du = \frac{u}{8}(2u^2-a^2)\sqrt{a^2-u^2} + \frac{a^4}{8}\sin^{-1}\frac{u}{a} + C$

79. $\displaystyle\int \frac{\sqrt{a^2-u^2}\,du}{u} = \sqrt{a^2-u^2} - a\ln\left|\frac{a+\sqrt{a^2-u^2}}{u}\right| + C$

80. $\displaystyle\int \frac{\sqrt{a^2-u^2}\,du}{u^2} = -\frac{\sqrt{a^2-u^2}}{u} - \sin^{-1}\frac{u}{a} + C$

81. $\displaystyle\int \frac{u^2\,du}{\sqrt{a^2-u^2}} = -\frac{u}{2}\sqrt{a^2-u^2} + \frac{a^2}{2}\sin^{-1}\frac{u}{a} + C$

82. $\displaystyle\int \frac{du}{u\sqrt{a^2-u^2}} = -\frac{1}{a}\ln\left|\frac{a+\sqrt{a^2-u^2}}{u}\right| + C$

83. $\displaystyle\int \frac{du}{u^2\sqrt{a^2-u^2}} = -\frac{\sqrt{a^2-u^2}}{a^2u} + C$

POWERS OF u MULTIPLYING OR DIVIDING $\sqrt{u^2 \pm a^2}$ OR THEIR RECIPROCALS

84. $\displaystyle\int u\sqrt{u^2+a^2}\,du = \frac{1}{3}(u^2+a^2)^{3/2} + C$

85. $\displaystyle\int u\sqrt{u^2-a^2}\,du = \frac{1}{3}(u^2-a^2)^{3/2} + C$

86. $\displaystyle\int \frac{du}{u\sqrt{u^2+a^2}} = -\frac{1}{a}\ln\left|\frac{a+\sqrt{u^2+a^2}}{u}\right| + C$

87. $\displaystyle\int \frac{du}{u\sqrt{u^2-a^2}} = \frac{1}{a}\sec^{-1}\left|\frac{u}{a}\right| + C$

88. $\displaystyle\int \frac{\sqrt{u^2-a^2}\,du}{u} = \sqrt{u^2-a^2} - a\sec^{-1}\left|\frac{u}{a}\right| + C$

89. $\displaystyle\int \frac{\sqrt{u^2+a^2}\,du}{u} = \sqrt{u^2+a^2} - a\ln\left|\frac{a+\sqrt{u^2+a^2}}{u}\right| + C$

90. $\displaystyle\int \frac{du}{u^2\sqrt{u^2\pm a^2}} = \mp\frac{\sqrt{u^2\pm a^2}}{a^2u} + C$

91. $\displaystyle\int u^2\sqrt{u^2+a^2}\,du = \frac{u}{8}(2u^2+a^2)\sqrt{u^2+a^2} - \frac{a^4}{8}\ln(u+\sqrt{u^2+a^2}) + C$

92. $\displaystyle\int u^2\sqrt{u^2-a^2}\,du = \frac{u}{8}(2u^2-a^2)\sqrt{u^2-a^2} - \frac{a^4}{8}\ln|u+\sqrt{u^2-a^2}| + C$

93. $\displaystyle\int \frac{\sqrt{u^2+a^2}}{u^2}\,du = -\frac{\sqrt{u^2+a^2}}{u} + \ln(u+\sqrt{u^2+a^2}) + C$

94. $\displaystyle\int \frac{\sqrt{u^2-a^2}}{u^2}\,du = -\frac{\sqrt{u^2-a^2}}{u} + \ln|u+\sqrt{u^2-a^2}| + C$

95. $\displaystyle\int \frac{u^2}{\sqrt{u^2+a^2}}\,du = \frac{u}{2}\sqrt{u^2+a^2} - \frac{a^2}{2}\ln(u+\sqrt{u^2+a^2}) + C$

96. $\displaystyle\int \frac{u^2}{\sqrt{u^2-a^2}}\,du = \frac{u}{2}\sqrt{u^2-a^2} + \frac{a^2}{2}\ln|u+\sqrt{u^2-a^2}| + C$

INTEGRALS CONTAINING $(a^2+u^2)^{3/2}$, $(a^2-u^2)^{3/2}$, $(u^2-a^2)^{3/2}$ ($a > 0$)

97. $\displaystyle\int \frac{du}{(a^2-u^2)^{3/2}} = \frac{u}{a^2\sqrt{a^2-u^2}} + C$

98. $\displaystyle\int \frac{du}{(u^2\pm a^2)^{3/2}} = \pm\frac{u}{a^2\sqrt{u^2\pm a^2}} + C$

99. $\displaystyle\int (a^2-u^2)^{3/2}\,du = -\frac{u}{8}(2u^2-5a^2)\sqrt{a^2-u^2} + \frac{3a^4}{8}\sin^{-1}\frac{u}{a} + C$

100. $\displaystyle\int (u^2+a^2)^{3/2}\,du = \frac{u}{8}(2u^2+5a^2)\sqrt{u^2+a^2} + \frac{3a^4}{8}\ln(u+\sqrt{u^2+a^2}) + C$

101. $\displaystyle\int (u^2-a^2)^{3/2}\,du = \frac{u}{8}(2u^2-5a^2)\sqrt{u^2-a^2} + \frac{3a^4}{8}\ln|u+\sqrt{u^2-a^2}| + C$